CMS/CAIMS Books in Mathematics

Volume 13

CMS/CAIMS Books in Mathematics is a collection of monographs and graduate-level textbooks published in cooperation jointly with the Canadian Mathematical Society- Societé mathématique du Canada and the Canadian Applied and Industrial Mathematics Society-Societé Canadienne de Mathématiques Appliquées et Industrielles. This series offers authors the joint advantage of publishing with two major mathematical societies and with a leading academic publishing company. The series is edited by Karl Dilcher, Frithjof Lutscher, Nilima Nigam, and Keith Taylor. The series publishes high-impact works across the breadth of mathematics and its applications. Books in this series will appeal to all mathematicians, students and established researchers. The series replaces the CMS Books in Mathematics series that successfully published over 45 volumes in 20 years.

CMS
SMC

CAIMS
SCMAI

Kenneth R. Davidson

Functional Analysis and Operator Algebras

Springer

Kenneth R. Davidson
Department of Pure Mathematics
University of Waterloo
Waterloo, ON, Canada

ISSN 2730-650X ISSN 2730-6518 (electronic)
CMS/CAIMS Books in Mathematics
ISBN 978-3-031-63664-6 ISBN 978-3-031-63665-3 (eBook)
https://doi.org/10.1007/978-3-031-63665-3

Mathematics Subject Classification: 46L10, 47L25, 47L30, 47B35, 46A03, 46L05, 46-01

This Springer imprint is published by the registered company Springer Nature Switzerland AG
The registered company address is: Gewerbestrasse 11, 6330 Cham, Switzerland

Dedication

I dedicate this book to my many former doctoral and master's students and post-docs. It has been a pleasure to interact with so many bright young minds, and to see them develop and enhance their skills in research. It has provided me with many opportunities to teach interesting material, and it has made this book possible as I began with course notes covering the majority of the material covered here. They have all gone on to a variety of challenging careers, and have made me proud to have been a part of their development.

PREFACE

This book began with material for a two-semester course in functional analysis for graduate students. This material has been expanded in various ways during the writing of the book. This is followed by additional material on operator theory and operator algebras. A course in analysis on metric spaces and some knowledge of Lebesgue measure is assumed. More general measure theory is an asset, and certainly will be needed later in the book.

Part I is based on a one-semester course in functional analysis that I taught many times. Chapter 1 covers some necessary background in point-set topology. This is not a comprehensive treatment, but we will need to use nets and weak topologies. Normally I do not cover everything in this chapter until it is needed. Chapter 2 treats Banach spaces and establishes basic results such as the Banach–Steinhaus theorem, the Open Mapping theorem and the Hahn–Banach theorem. In the Chap. 3, topological vector spaces are introduced. Geometric versions of the Hahn–Banach theorem are established as well as the Krein–Milman and Krein–Smulian theorems. I do not usually cover all of the section on fixed point theorems nor weak compactness. They are included as a resource for some later material. The next two chapters deal with bounded linear operators between Banach spaces. In particular, the structure of compact operators and basic Fredholm theory is covered. Specializing to Hilbert spaces, the spectral theorem for compact normal operators is established.

Part II is based on our second-semester course, an introduction to Banach and C*-algebras. Chapter 6 covers the basic material on spectrum and parallels the treatment already given for operators. It then develops the Riesz–Dunford functional calculus, which is needed for operators as well. Chapter 7 deals with Gelfand theory for commutative Banach algebras, with extensive analysis of the examples of $L^1(G)$ and commutative C*-algebras. Chapter 8 deals with the representation theory of general Banach algebras. I usually don't have time for the Cohen factorization theorem. Chapter 9 is an introduction to the basics of C*-algebras, culminating in the Gelfand–Naimark theorem that every abstract C*-algebra has a faithful representation on Hilbert space. It is followed by a number of additional topics including the spectral theorem and group C*-algebras. I usually finish my course with the spectral theorem. Chapter 10 covers additional material on von Neumann algebras. This is an important subject, but there is no time for this in our syllabus.

Part III of the notes covers some specialized topics that I have covered in further courses. This is a range of topics in operator theory, dilation theory, nonself-adjoint operator algebras and noncommutative convexity. The treatment of Hardy spaces is chosen to provide tools needed to do operator theory on H^2, and to elucidate some of the nice Banach algebra properties of the disc algebra and H^∞. The chapter on Toeplitz operators was in large part inspired by Douglas's book [98].

The chapter on nest algebras takes some material from my out-of-print text [69], and contains a shorter, more direct proof of the Similarity theorem as well as Orr's interpolation theorem. Dilation theory is an important tool in operator theory and operator algebras. The subject of completely positive maps and its role in operator algebras is central to the modern theory. Our treatment was strongly influenced by Paulsen's book [226] and the treatise of Sz.-Nagy–Foiaş [297]. It has many applications. We provide a chapter with introductions to semicrossed products and multivariable operator theory, both commutative and non-commutative. The final chapter is an introduction to noncommutative convexity and non-commutative Choquet theory.

The bibliography goes back about a century. We have not referenced original papers earlier than that. This affects the Part I on functional analysis, as the later material begins with around 1929 with the spectral theorem for normal operators due independently to Stone and von Neumann. The notes and references in Dunford and Schwarz [103] provide additional references going back further.

I would like to thank many people who have provided useful feedback on this book. In particular, I warmly thank Martin Argerami, Adam Dor-On, Michael Hartz, Evgenios Kakariadis, Matthew Kennedy, Orr Shalit and Dilian Yang. I also thank Albrecht Böttcher, Heydar Radjavi and Nico Spronk for extended discussions on several issues in the book. I also thank the referees for helpful comments.

Kenneth R. Davidson
Orleans, ON, Canada
April 2024

Contents

Part III. Operator Theory

Part I

Functional Analysis

CHAPTER 1

Set Theory and Topology

The set theory covered here is minimal—basically a treatment of the Axiom of Choice and some of its common variants. This is an essential tool in functional analysis. In particular, we do not discuss the standard axioms of set theory, but a basic understanding of how to work with sets is assumed.

Then we have a brief treatment of point set topology as used by analysts. This includes a treatment of nets, since in many important topologies that arise, sequences do not suffice. This includes various weak topologies that play an important role. We expect that most readers will have some familiarity with continuity and compactness in metric spaces. We will extend these notions to general topological spaces. Our particular interest is in (locally) compact Hausdorff spaces.

Not every topology that arises in analysis is Hausdorff, but these will not arise often, and do not play a role in the material in this book.

The instructor can pick and choose the results needed here, and leave the remainder to the interested reader. Early on, one will need to understand how to use Zorn's lemma and nets. So I recommend that you read Sections 1.1, 1.2, 1.4 and 1.5. You should also know how to work with continuity and compactness. The main ideas are the same in topological spaces as in metric spaces except that sequences are inadequate. Weak topologies (Section 1.8) can be introduced when they are studied later. The topology of compact and locally compact spaces is about important properties of continuous functions that won't arise much in Part I.

1.1. Orders on Sets

First we define some important properties of orders which we require.

1.1.1. DEFINITION. A *partial order* on a set X is a relation $\leq$ satisfying

(1) $x \leq x$ for $x \in X$ (reflexive).

(2) $x \leq y$ and $y \leq x$ implies that $x = y$ (antisymmetric).

(3) $x \leq y$ and $y \leq z$ implies that $x \leq z$ (transitive).

There need not be a relation between two points $x, y \in X$. We call $(X, \leq)$ a *poset*.

A poset $(X, \leq)$ is a *total order* if given $x, y \in X$, then either $x \leq y$ or $x \geq y$. A totally ordered set $(X, \leq)$ is *well-ordered* if every non-empty subset of X has a

K. R. Davidson, *Functional Analysis and Operator Algebras*, CMS/CAIMS Books in Mathematics 13,
https://doi.org/10.1007/978-3-031-63665-3_1

smallest element. $Y \subset X$ is an *initial segment* of a well-ordered set X if $y < x$ for all $y \in Y$ and $x \in X \setminus Y$.

A poset is *upward directed* if for every $x_1, x_2 \in X$, there is a $y \in X$ so that $x_1 \leq y$ and $x_2 \leq y$. A totally ordered subset C of a poset X is called a *chain* . A poset is *inductive* if for every chain $C \subset X$, there is an *upper bound* $y \in X$ for C, i.e. $x \leq y$ for every $x \in C$.

1.1.2. EXAMPLES.

(1) Let X be a set, and let $\mathcal{P}(X)$ denote the collection of all subsets of X. Put a relation on $\mathcal{P}(X)$ by setting $A \leq B$ if $A \subseteq B$. It is easy to see that this is a partial order. Moreover it is upward directed because given $A, B \in \mathcal{P}(X)$, $A \cup B$ is an upper bound. Indeed this set is inductive because if $\mathcal{C} \subset \mathcal{P}(X)$ is a chain, then $B = \bigcup_{A \in \mathcal{C}} A$ is an upper bound for $\mathcal{C}$.

(2) The real line $\mathbb{R}$ is totally ordered but not well-ordered. For example, $(0, 1)$ has no least element.

(3) The set of natural numbers $\mathbb{N}$ is well-ordered.

(4) If $(X, \leq)$ is well-ordered and infinite, then X has a least element x_1. Also $X \setminus \{x_1\}$ has a least element x_2. Recursively, we define x_{n+1} to be the least element of $X \setminus \{x_1, \dots, x_n\}$. Then $N = \{x_n : n \in \mathbb{N}\}$ is an initial segment of X which is order isomorphic to $\mathbb{N}$. This can be continued if $N \neq X$. Set x_ω to be the minimal element of $X \setminus N$, and set $x_{\omega+1}$ to be the minimal element of $X \setminus \{N \cup \{x_\omega\}\}$, etc.

1.2. The Axiom of Choice

The Axiom of Choice (AC) seems very natural, but it has some surprising consequences. It is known to be independent of the usual axioms of set theory (which we do not study), meaning that one can safely assume that it is valid, but can equally well assume other axioms instead which contradict AC. Most mathematicians other than set theorists and logicians assume the Axiom of Choice because it has many important positive consequences.

1.2.1. DEFINITION.

Axiom of Choice: for each non-empty set X, there is a *choice function*

$c : \mathcal{P}(X) \setminus \{\varnothing\} \to X$ so that $c(A) \in A$ for $\varnothing \neq A \subset X$.

Well-ordering Principle: every set X has a well-order.

Zorn's Lemma: every inductive partial order $(P, \leq)$ has a maximal element.

All of these properties are useful. Zorn's lemma may seem to be the least intuitive. But in fact, it is probably the most useful of the three. The following result shows that they are all equivalent.

1.2.2. THEOREM. *The following are equivalent:*

(1) *Axiom of Choice.*

(2) *Well-Ordering Principle.*

(3) *Zorn's Lemma.*

PROOF. (2) implies (1). Put a well-order $\leq$ on X. For each non-empty set A, define $c(A)$ to be the least element in A in this order.

(3) implies (2). Let X be a set. Let

$$\mathcal{P} = \{(F, <_F) : F \subset X, \ <_F \text{ is a well-order on } F\}.$$

Say that $(F, <_F) \leq (G, <_G)$ if $F \subseteq G$, $<_F = <_G|_{F\times F}$ *and* F is an initial segment of G. It is easy to check that this is a partial order.

We will show that $(\mathcal{P}, \leq)$ is inductive: suppose that $\mathcal{C} = \{(F_\lambda, <_{F_\lambda}) : \lambda \in \Lambda\}$ is a chain in $\mathcal{P}$. Define $G = \bigcup_{\lambda\in\Lambda} F_\lambda$ and $<_G = \bigcup_{\lambda\in\Lambda} <_{F_\lambda}$. By the latter, I mean that if $x, y \in G$, then there are $\lambda, \mu \in \Lambda$ so that $x \in F_\lambda$ and $y \in F_\mu$. Since $\mathcal{C}$ is totally ordered, either $F_\lambda \subset F_\mu$ or $F_\mu \subset F_\lambda$; say $x, y \in F_\lambda$. Then since F_λ is a well-order, either $x <_{F_\lambda} y$, $x = y$ or $y <_{F_\lambda} x$. Moreover, if $F_\lambda \subset F_\gamma$, then since the orders $<_{F_\lambda}$ and $<_{F_\gamma}$ agree on F_λ, the order on x and y is independent of the choice of λ. So $(G, <_G)$ is a total order. Finally if A is a non-empty subset of G, then there is some λ so that $A \cap F_\lambda \neq \varnothing$. Let a be the least element of $A \cap F_\lambda$ in $<_{F_\lambda}$. Then because F_λ is an initial segment of G, the element a satisfies $a <_G b$ for all $b \in A \setminus F_\lambda$. Therefore a is the least element of A in $(G, <_G)$. So G is well-ordered. It is now clear that $(G, <_G)$ is an upper bound for $\mathcal{C}$. Therefore $\mathcal{P}$ is inductive.

By Zorn's lemma, $\mathcal{P}$ has a maximal element $(F, <_F)$. If $F \neq X$, let $x \in X \setminus F$. Define a well-order on $G = F \cup \{x\}$ by $a <_G b$ if $b = x$ or if $a, b \in F$ and $a <_F b$. Then G is well-ordered and F is an initial segment of G. Thus $(F, <_F) \lneqq (G, <_G)$, contradicting the maximality of $(F, <_F)$. Therefore $F = X$ and we have a well-order on X.

(1) implies (3). Let $(\mathcal{P}, \leq)$ be an inductive partial order. Suppose that $\mathcal{P}$ has no maximal elements. Then for each $x \in \mathcal{P}$, the set $U_x = \{y \in \mathcal{P} : x < y\}$ is non-empty. By the Axiom of Choice, there is a choice function c on non-empty subsets of $\mathcal{P}$. Define $f(x) = c(U_x)$.

For each chain $\mathcal{C} \subset \mathcal{P}$, let $U_\mathcal{C}$ denote the non-empty set of all upper bounds for $\mathcal{C}$. Define a function $g(\mathcal{C}) = c(U_\mathcal{C})$. Then $h(\mathcal{C}) = f(g(\mathcal{C}))$ is a *strict* upper bound for $\mathcal{C}$. If $x \in \mathcal{C}$, set $I(\mathcal{C}, x) = \{y \in \mathcal{C} : y < x\}$.

We will say (for this proof only) that $A \subset \mathcal{P}$ is *conforming* if $(A, \leq)$ is well-ordered and for every $x \in A$, $h(I(A, x)) = x$. In particular, if a_1 is the least element of A, then $I(A, a_1) = \varnothing$ and $a_1 = h(\varnothing)$. Also if a_2 is the least element of $A \setminus \{a_1\}$, then $I(A, a_2) = \{a_1\}$ and $a_2 = h(a_1)$.

We claim that if A, B are conforming subsets of $\mathcal{P}$, then either A is an initial segment of B or B is an initial segment of A. To this end, let $\mathcal{X}$ consist of all initial segments which are common to A and B. Then $C = \bigcup \mathcal{X}$ is the largest initial segment common to A and B. If $C \in \{A, B\}$, then we are done. Otherwise let a be the least element of $A \setminus C$ and let b be the least element of $B \setminus C$. Then $I(A, a) = C = I(B, b)$. Since A and B are conforming, $a = h(C) = b$. Therefore $C \cup \{a\}$ is a larger initial segment of both A and B, a contradiction. So either $C = A$ or $C = B$.

Now $D = \bigcup\{A : A \text{ is conforming}\}$ must be the largest conforming subset of $\mathcal{P}$. However $D \cup h(D)$ is a strictly larger conforming set! This contradiction shows that the assumption that $\mathcal{P}$ has no maximal elements must be false. So (3) holds. ■

1.3. Ordinals

We will not have a lot of use for ordinals in this book, so this section can be skipped as it is primarily for interest. However, ordinals have an important place in set theory, and some familiarity with them may be useful. Some of these results will be left to the exercises.

The well-ordering principle allows for an extended version of induction called *transfinite induction* .

1.3.1. Transfinite Induction. *Suppose that Λ is a well-ordered set and that $P(\lambda)$ is a statement for each $\lambda \in \Lambda$. Suppose that for each λ, if $P(\mu)$ is true for every $\mu < \lambda$, then $P(\lambda)$ is true. Then $P(\lambda)$ is true for all $\lambda \in \Lambda$.*

Proof. Suppose that $S = \{\lambda \in \Lambda : P(\lambda) \text{ is false}\}$ is non-empty. By the well-ordering property, there is a smallest element $\lambda_0 \in S$. So $P(\mu)$ is true for all $\mu < \lambda_0$ but $P(\lambda_0)$ is false, contrary to the hypotheses. This contradicts the fact that S is non-empty. Therefore $P(\lambda)$ must be true for all $\lambda \in \Lambda$. ■

The proof of Theorem 1.2.2 points to a structure of well-ordered sets. We say that two posets $(X, <)$ and $(Y, \prec)$ are *isomorphic* ($X \simeq Y$) if there is a bijection $f : X \to Y$ such that both f and f^{-1} are order preserving. In particular, f and f^{-1} are strictly increasing. If $a \in X$, let $X(a) = \{x \in X : x < a\}$ be an *initial segment* of X. Every proper initial segment J has this form with a equal to the minimal element of $X \setminus J$.

1.3.2. Lemma. *If $(X, <)$ is well-ordered and $f : X \to X$ is strictly increasing, then $f(x) \geq x$ for $x \in X$.*

Proof. If $A = \{x \in X : f(x) < x\}$ is non-empty, let a be the least element. So $f(a) = b < a$. Hence $f(b) < f(a) = b$, contradicting the minimality of a. Therefore A is empty and $f(x) \geq x$. ■

This lemma has immediate consequences. See Exercise **1.3**. This includes

1.3.3. COROLLARY. *No well-order is isomorphic to a proper initial segment of itself.*

1.3.4. THEOREM. *If X and Y are two well-orders, then exactly one of the following holds:*

(1) *X is isomorphic to an initial segment of Y.*

(2) *X and Y are isomorphic.*

(3) *Y is isomorphic to an initial segment of X.*

PROOF. Define a relation on $X \times Y$ by

$$R = \{(x, y) \in X \times Y : X(x) \simeq Y(y)\}.$$

Let $X_1 = \pi_1(R)$ and $Y_1 = \pi_2(R)$, where π_i are the coordinate projections. For each $x \in X_1$, there is a unique $y \in Y$ such that $(x, y) \in R$ by Corollary 1.3.3. So there is a function $f : X_1 \to Y_1$ such that $R = \{(x, f(x)) : x \in X_1\}$. That same result shows that f is one to one. If $x \in X_1$ and $x' < x$, then the isomorphism of $X(x)$ onto $Y(f(x))$ must carry the initial segment $X(x')$ onto an initial segment of $Y(f(x))$. Hence X_1 is an initial segment of X and f is strictly increasing. Similarly Y_1 is an initial segment of Y, and f^{-1} is strictly increasing. Therefore f is an isomorphism of X_1 onto Y_1.

If $X_1 \subsetneq X$ and $Y_1 = Y$, then (1) holds. If $X_1 = X$ and $Y_1 = Y$, then (2) holds. And if $X_1 = X$ and $Y_1 \subsetneq Y$, then (3) holds. If $X_1 \subsetneq X$ and $Y_1 \subsetneq Y$, let x_0 be the least element of $X \setminus X_1$ and let y_0 be the least element of $Y \setminus Y_1$. Then $X_1 = X(x_0)$ and $Y_1 = Y(y_0)$ are isomorphic. Now $X_0 = X_1 \cup \{x_0\}$ and $Y_0 = Y_1 \cup \{y_0\}$ are isomorphic initial segments, so $(x_0, y_0) \in R$. This contradicts the definition of X_1 and Y_1, and hence this case does not occur. ■

This theorem tells us that well-ordered sets are very prescribed. This led to the notion of *ordinals* . Von Neumann found a canonical definition of ordinals. It begins with the empty set $\varnothing =: 0$. Then one forms the single element set $1 := \{\varnothing\} = \{0\}$. This follows with

$$2 := \{0, 1\} = \{\varnothing, \{\varnothing\}\}, \quad 3 := \{0, 1, 2\} = \{\varnothing, \{\varnothing\}, \{\varnothing, \{\varnothing\}\}\}, \quad \text{etc.}$$

Beyond that, there is $\omega = \{0, 1, 2, \dots\} = \mathbb{N}_0$. An ordinal is defined to be a set α with the properties

(1) every element of α is a subset of α and

(2) the elements of α are well-ordered by containment.

1.3.5. LEMMA.

(1) *If α is an ordinal and $\beta \in \alpha$, then β is an ordinal.*

(2) *If $\alpha \neq \beta$ are ordinals, then either $\alpha \subset \beta$ or $\beta \subset \alpha$.*

(3) *If A is a set of ordinals, then $\bigcap_{\alpha \in A} \alpha$ is the least element.*

(4) *If A is a set of ordinals, then $\bigcup_{\alpha \in A} \alpha$ is an ordinal.*

We refer to Jech [160] for the straightforward proof of the following result.

1.3.6. THEOREM. *Every well-ordered set is isomorphic to a unique ordinal.*

Every ordinal α has a *successor* $\alpha + 1 := \alpha \cup \{\alpha\}$. An ordinal which is not a successor is called a *limit ordinal* . Starting with $0, 1, 2,$ we define the natural numbers $\mathbb{N}_0$ by taking the sequence of successors. Then we obtain the first limit ordinal $\omega = \mathbb{N}_0 = \{0, 1, 2, \dots\}$. This is followed by $\omega + 1, \omega + 2, \dots, \omega + \omega =: \omega 2$. There are ordinals ωn, ω^2, ω^n, ω^ω, etc.

A transfinite induction argument often has two cases, depending on whether the ordinal is a successor or a limit ordinal.

We can well-order the reals $\mathbb{R}$ by the well-ordering principle, say by $\prec$. Since $\mathbb{R}$ is uncountable, the set of elements r such that the initial segment $\mathbb{R}(r)$ is uncountable is either empty or has a minimal element. In either case, we obtain an uncountable well-ordered set such that every initial segment is countable. This means that there is a first uncountable ordinal, usually called ω_1 or Ω.

1.4. Topological spaces

Point set topology begins with a very abstract notion of the properties that the collection of all open sets of a topological space must possess. Examples show that some topological spaces have little useful structure, so further properties are generally added to get interesting continuous functions. This section deals with the basics. The reader is expected to have a solid understanding of the topology of metric spaces. This is a useful guide for understanding, but the reader should pay attention when metric space notions break down.

1.4.1. DEFINITION. A *topology* τ on a set X is a collection of subsets of X such that

(1) $\varnothing,\ X \in \tau$.

(2) If $\{U_\lambda : \lambda \in \Lambda\} \subset \tau$, then $\bigcup_{\lambda \in \Lambda} U_\lambda \in \tau$.

(3) If $U_1, \dots, U_n \in \tau$, then $\bigcap_{i=1}^n U_i \in \tau$.

The elements $U \in \tau$ are called *open sets*. Say (X, τ) is a *topological space* .

1.4.2. EXAMPLES.

(1) If (X, d) is a metric space, then U is open if for every $x \in U$, there is an $r > 0$ so that the open ball $b_r(x) = \{y \in X : d(x, y) < r\} \subset U$.

(2) If X is any set, the *discrete topology* has $\tau_d = \mathcal{P}(X)$, the collection of all subsets of X.

(3) If X is any set, the *trivial topology* has $\tau = \{\varnothing, X\}$.

(4) If $(X, \leq)$ is a totally ordered set, the *order topology* is generated by the open order intervals $(a, b) = \{x \in X : a < x < b\}$, $(-\infty, b) = \{x \in X : x < b\}$ and $(a, \infty) = \{x \in X : x > a\}$. The topology consists of arbitrary unions of such intervals.

(5) If (X, τ) is a topology and $Y \subset X$, the *induced topology* on Y is given by $\tau|_Y = \{U \cap Y : U \in \tau\}$.

1.4.3. DEFINITION. A set $F \subset X$ is *closed* if F^c is open. If $A \subset X$, the *closure* of A is $\overline{A} = \bigcap\{F : A \subset F,\ F \text{ closed}\}$. A point in $\overline{A}$ is called a *limit point* of A. A point $a \in A$ is a *cluster point* of A if $a \in \overline{A \setminus \{a\}}$.

If $A \subset X$, then $a \in A$ is an *interior point* of A if there exists $U \in \tau$ with $a \in U \subset A$. If $A \subset X$, the *interior* of A is A^o or $\operatorname{int} A = \bigcup\{U \in \tau : U \subset A\}$.

If $x \in X$, a *neighbourhood* of x is a set N such that $x \in N^o$.

Caveat: a neighbourhood does not need to be an open set.

1.4.4. PROPOSITION.

(1) *Finite unions and arbitrary intersections of closed sets are closed.*

(2) $\overline{A}$ *is the smallest closed set containing* A.

(3) $x \in \overline{A}$ *if and only if every* $U \in \tau$ *with* $x \in U$ *has* $A \cap U \neq \varnothing$.

(4) $\overline{A} = A^{coc}$ *is the complement of the interior of* A^c.

PROOF. Since open sets are closed under arbitrary unions and finite intersections, the collection of closed sets is closed under arbitrary intersections and finite unions. Hence, the intersection of all closed sets $F \supset A$ is closed, and is thus the smallest closed set containing A. Now $x \in \overline{A}$ if and only if $x \in F$ for every closed $F \supset A$ if and only if $x \notin U$ if U is open and disjoint from A. Finally

$$X \setminus \overline{A} = \bigcup\{U \in \tau : U \cap A = \varnothing\} = \bigcup\{U \in \tau : U \subset A^c\} = A^{co}. \qquad \blacksquare$$

1.4.5. DEFINITION. If σ and τ are two topologies on X, we say that σ is a *weaker topology* than τ, and τ is a *stronger topology* than σ, if $\sigma \subset \tau$.

1.4.6. Proposition. *If $S \subset \mathcal{P}(X)$, then there is a weakest topology τ containing S. It consists of arbitrary unions of sets which are intersections of finitely many elements of S.*

PROOF. Clearly if $\tau \supset S$ is a topology, then it contains all intersections of finitely many elements of S, and arbitrary unions of these sets. The intersection of no sets is X by convention, and $\varnothing$ is the union of no sets, so they both belong to τ. This collection is clearly closed under arbitrary unions. To check that it is stable under intersection, observe that if $A_{\alpha,i}$ and $B_{\beta,j}$ are in S, then

$$\bigcup_{\alpha\in A} A_{\alpha,1} \cap \cdots \cap A_{\alpha,n_\alpha} \cap \bigcup_{\beta\in B} B_{\beta,1} \cap \cdots \cap B_{\beta,m_\beta}$$
$$= \bigcup_{\alpha\in A,\, \beta\in B} A_{\alpha,1} \cap \cdots \cap A_{\alpha,n_\alpha} \cap B_{\beta,1} \cap \cdots \cap B_{\beta,m_\beta}.$$

Hence this collection is a topology. By construction, this is the weakest topology containing S. ■

1.4.7. Definition. Say that $S \subset \mathcal{P}(X)$ is a *base* for a topology τ if every open set $U \in \tau$ is the union of elements of S. Also S is a *subbase* for a topology τ if the collection of finite intersections of elements of S is a base for τ.

1.4.8. Examples.

(1) If (X, d) is a metric space, then $\{b_{1/n}(x) : x \in X,\ n \geq 1\}$ is a base for the topology.

(2) $\{(r, s) : r < s \in \mathbb{Q}\}$ is a base for the topology of $\mathbb{R}$.

(3) Let $C[0, 1]$ denote the space of continuous functions on $[0, 1]$. For each $x \in [0, 1]$, $a \in \mathbb{C}$ and $r > 0$, let $U(x, a, r) = \{f \in C[0, 1] : f(x) \in b_r(a)\}$. Let τ be the topology generated by these sets. This is the topology of pointwise convergence. An open neighbourhood of f must contain a set of the form

$$\{g \in C[0, 1] : |g(x_i) - f(x_i)| < r \text{ for } 1 \leq i \leq n\}$$

for $x_1, \dots, x_n \in [0, 1]$ and $r > 0$.

1.4.9. Definition. A set A is *dense* in X if $X = \overline{A}$. X is *separable* if it has a countable dense subset. X is *first countable* if for each $x \in X$, there is a countable family $\{U_i\} \subset \tau$ with $x \in U_i$ which forms a *countable base of neighbourhoods* of x, i.e. if $x \in V$ is open, then there is some i so that $U_i \subset V$. X is *second countable* if there is a countable family of open sets which is a base for τ.

1.4.10. Examples.

(1) If (X, d) is a metric space and $x \in X$, then $\{b_{1/n}(x) : n \geq 1\}$ is a countable base of neighbourhoods of x. If X is separable, and $\{x_i : i \geq 1\}$ is dense in X, then $\big\{b_{1/n}(x_i) : i \geq 1,\ n \geq 1\big\}$ is a base for τ. Indeed, suppose that $x \in U$ is open. Pick $r > 0$ so that $b_r(x) \subset U$ and x_i so that $d(x, x_i) < \frac{1}{n} < \frac{r}{2}$. Then $x \in b_{1/n}(x_i) \subset U$. So X is second countable. In particular, compact metric spaces are separable and so second countable.

(2) Consider the discrete topology τ_d on a set X. Since the topology is generated by $\big\{\{x\} : x \in X\big\}$, X is always first countable. However it is second countable if and only if X is countable if and only if X is separable.

1.4.11. Definition. A topological space is T_0 if for every pair $x \neq y \in X$, there is an open set containing one of these points, but not the other. A topological space is T_1 if points are closed. A topological space is *Hausdorff* or T_2 if for all $x \neq y \in X$, there are open sets $U \ni x$ and $V \ni y$ such that $U \cap V = \varnothing$.

1.4.12. Examples.
(1) Let $X = \{0, 1\}$ and $\tau = \big\{\varnothing, \{0\}, X\big\}$. Then $\{1\}$ is closed, but $\{0\}$ is not, and $\overline{\{0\}} = X$. Also $\{0\}$ is an open neighbourhood of 0 disjoint from 1. This space is T_0 but not T_1.

(2) Let $X = [0, 1) \cup \{a, b\}$. Let the open sets in τ be $U \subset [0, 1)$ which are open in the usual metric on $[0, 1)$ together with sets $U \cup (r, 1) \cup \{a\}$, $U \cup (r, 1) \cup \{b\}$ and $U \cup (r, 1) \cup \{a, b\}$ for $r < 1$. Here the points $\{a\}$ and $\{b\}$ are closed because the complement is open. However if $a \in U$ and $b \in V$ are open sets, then $U \cap V \supset (r, 1)$ for some $r < 1$. That means that you cannot separate a and b from one another by open sets. Therefore X is T_1, but is not Hausdorff.

(3) Metric spaces are Hausdorff because if $x \neq y \in (X, d)$, then $d(x, y) = r > 0$. So $b_{r/2}(x)$ and $b_{r/2}(y)$ are disjoint open sets.

1.5. Nets

Sequences are not sufficient for dealing with convergence in general topological spaces, including many that arise in normal contexts. The replacement is the notion of a net, which you can think of as a very wide and very long generalized sequence.

1.5.1. Definition. A *net* in X is an upward directed poset Λ with a function $j : \Lambda \to X$, say $x_\lambda = j(\lambda)$. We usually write the net as $(x_\lambda)_\Lambda$. A net $(x_\lambda)_\Lambda$ *converges to* x in (X, τ) if for every open set $U \ni x$, there is $\lambda_0 \in \Lambda$ so that $x_\lambda \in U$ for every $\lambda \geq \lambda_0$. We write $\lim_\Lambda x_\lambda = x$.

A *subnet* $(y_\gamma)_\Gamma$ of $(x_\lambda)_\Lambda$ is given by a cofinal function $\varphi : \Gamma \to \Lambda$ so that $y_\gamma = x_{\varphi(\gamma)}$, where we say that φ is *cofinal* if for all $\lambda \in \Lambda$, there is a $\gamma_0 \in \Gamma$ so that

$\varphi(\gamma) \geq \lambda$ for all $\gamma \geq \gamma_0$. It is convenient if φ is *monotone*, meaning that $\gamma_1 \leq \gamma_2$ implies that $\varphi(\gamma_1) \leq \varphi(\gamma_2)$. But this is not necessary.

We present a detailed example to explain why nets are needed, and how to use them.

1.5.2. EXAMPLE. Let $X = \mathbb{N}_0 \times \mathbb{N}_0$. Define $\pi_1 : X \to \mathbb{N}_0$ by $\pi_1(m, n) = m$. Declare that $U \subset X$ is open if $(0,0) \notin U$, and that a set $U \ni (0,0)$ is open if $\{m : \pi_1^{-1}(m) \cap U$ is cofinite in $\mathbb{N}_0\}$ is cofinite in $\mathbb{N}_0$. It is easy to verify that this defines a topology. The reader should check this!

(a) X is Hausdorff because $\{(m,n)\}$ is open if $m+n \geq 1$ and $\{(m,n)\}^c$ is an open neighbourhood of $(0,0)$.

(b) $(0,0) \in \overline{X \setminus \{(0,0)\}}$ because every open set U containing $(0,0)$ intersects $X \setminus \{(0,0)\}$.

(c) However no sequence $x_k = (m_k, n_k)$ in $X \setminus \{(0,0)\}$ converges to $(0,0)$. There are two cases. If $\{m_k : k \geq 1\}$ is bounded, pick m_0 so that $m_k = m_0$ infinitely often. The set $U = \{(m,n) : m \neq m_0\} \cup \{(0,0)\}$ is an open neighbourhood of $(0,0)$, and the sequence is not eventually in U. Otherwise, there is a sequence $k_i \to \infty$ so that $m_{k_i} < m_{k_{i+1}}$ for $i \geq 1$. Then $U = X \setminus \{x_{k_i} : i \geq 1\}$ is an open neighbourhood of $(0,0)$, and the sequence is not eventually in U.

(d) There is a net in $X \setminus \{(0,0)\}$ converging to $(0,0)$. Let $\Lambda = \{U \in \tau : (0,0) \in U\}$ where $U \leq V$ if $U \supset V$. (We say that Λ is *ordered by containment* .) This is directed because $U, V \leq U \cap V$. Order $X \setminus \{(0,0)\}$ by

$$(0,1), (1,0), (0,2), (1,1), (2,0), (0,3), (1,2), (2,1), (3,0), \ldots.$$

Define x_U to be the least element in this list which belongs to U. (This avoids any issues with the Axiom of Choice.) Then $(x_U)_\Lambda$ converges to $(0,0)$ because given an open neighbourhood $U \ni (0,0)$, we have $x_V \in V \subset U$ whenever $U \leq V$.

(e) The sequence $(0,1), (1,0), (0,2), (1,1), (2,0), (0,3), (1,2), (2,1), \ldots$ has a subnet converging to $(0,0)$. Let Λ be the net just constructed. Define $\varphi(U) = x_U$ considered as an element in this sequence. To see that this map is cofinal, let (m_0, n_0) be in this sequence, and set $N_0 = m_0 + n_0$. Let

$$U_0 = X \setminus \{(m,n) : 1 \leq m+n \leq N_0\}.$$

Then if $U_0 \leq U$, it follows that $x_U = (m,n)$ with $m+n > N_0$ and thus $\varphi(U)$ follows (m_0, n_0) in the sequence. Therefore, this is a subnet of the sequence which converges to $(0,0)$.

1.5.3. PROPOSITION. *Let $A \subset X$. Then $x \in \overline{A}$ if and only if there is a net $(a_\lambda)_\Lambda$ in A such that* $\lim_\Lambda a_\lambda = x$.

PROOF. Suppose that $x \in \overline{A}$. By Proposition 1.4.4(3), every open neighbourhood U of x intersects A. Let $\mathcal{O}(x)$ be the open neighbourhoods of x ordered by containment. By the Axiom of Choice, we can pick a point $a_U \in A \cap U$ for each $U \in \mathcal{O}(x)$. The net $(a_U)_{\mathcal{O}(x)}$ converges to x by construction.

Conversely, suppose that $(a_\lambda)_\Lambda$ is a net in A such that $\lim_\Lambda a_\lambda = x$. Then for any neighbourhood U of x, there is a λ so that $a_\lambda \in U$. In particular, $A \cap U \neq \varnothing$. Hence by Proposition 1.4.4(3), $x \in \overline{A}$. ■

1.6. Continuity

The notion of a continuous function is the beginning of doing analysis on topological spaces. The definition is a familiar non-sequential definition that already arises in metric spaces.

1.6.1. Definition. A function $f : (X, \tau) \to (Y, \sigma)$ between topological spaces is *continuous* if for all $V \subset Y$ open, the set $f^{-1}(V)$ is open in X. Say that f is a *homeomorphism* if f is a bijection such that both f and f^{-1} are continuous.

1.6.2. Examples.

(1) The identity maps $(X, \text{discrete}) \xrightarrow{\iota_1} (X, \tau) \xrightarrow{\iota_2} (X, \text{trivial})$ are continuous bijections, however in both cases ι_j^{-1} will be discontinuous provided that τ satisfies $\{\varnothing, X\} \subsetneq \tau \subsetneq \mathcal{P}(X)$.

(2) A function $f : (X, \text{trivial}) \to \mathbb{R}$ is continuous only if it is constant, while every function $f : (X, \text{discrete}) \to \mathbb{R}$ is continuous. On the other hand, a function $f : \mathbb{R} \to (X, \text{discrete})$ is continuous only if it is constant, while every function $f : \mathbb{R} \to (X, \text{trivial})$ is continuous.

(3) $f : (-1, 1) \to \mathbb{R}$ given by $f(x) = \tan \frac{\pi x}{2}$ is a homeomorphism.

(4) Consider Example 1.4.12(1): $X = \{0, 1\}$ and $\tau = \{\varnothing, X, \{0\}\}$. If $f : X \to \mathbb{R}$ is continuous, then $f^{-1}(b_r(f(1)))$ is open and contains 1, so $f^{-1}(b_r(f(1)) = X$. Therefore f is constant.

(5) Consider Example 1.4.12(2): $X = [0, 1) \cup \{a, b\}$. If $a \in U$ and $b \in V$ are open sets, then $U \cap V \supset (r, 1)$ for some $r < 1$. Hence if $f : X \to \mathbb{R}$ is continuous and $\varepsilon > 0$, then

$$f^{-1}(b_\varepsilon(f(a))) \cap f^{-1}(b_\varepsilon(f(b))) \supset (r, 1) \quad \text{for some } r < 1.$$

Thus $|f(a) - f(b)| < 2\varepsilon$. Therefore $f(a) = f(b)$.

The following easy result shows that composition of continuous functions is continuous. The proof is left as an exercise.

1.6.3. PROPOSITION. *Suppose that* $f : (X, \tau) \to (Y, \sigma)$ *and* $g : (Y, \sigma) \to (Z, \rho)$ *are continuous functions. Then* $g \circ f : (X, \tau) \to (Z, \rho)$ *is continuous.*

1.6.4. THEOREM. *Let* $f : (X, \tau) \to (Y, \sigma)$. *Then* f *is continuous if and only if whenever* $(x_\lambda)_\Lambda$ *is a net in* X *converging to* x, *it follows that* $f(x) = \lim_\Lambda f(x_\lambda)$.

PROOF. Suppose that f is continuous, and let $(x_\lambda)_\Lambda$ be a net in X converging to x. Let V be an open neighbourhood of $f(x)$. Then $U = f^{-1}(V)$ is an open neighbourhood of x. By convergence, there is a $\lambda_0 \in \Lambda$ so that $x_\lambda \in U$ for all $\lambda \geq \lambda_0$. Hence $f(x_\lambda) \in f(U) \subset V$ for all $\lambda \geq \lambda_0$. That means that $f(x) = \lim_\Lambda f(x_\lambda)$.

Conversely, suppose f is not continuous. Thus there is an open set $V \subset Y$ such that $U = f^{-1}(V)$ is not open. Then $\overline{U^c} \cap U$ contains a point x. By Proposition 1.5.3, there is a net $(x_\lambda)_\Lambda$ in U^c with limit x. Therefore $f(x_\lambda) \in f(U^c) \subset V^c$. Since V^c is closed, any limit point of this net must remain in V^c by Proposition 1.5.3. Therefore it cannot converge to $f(x)$ which lies in V. So $f(x) \neq \lim_\Lambda f(x_\lambda)$. ■

1.6.5. DEFINITION. Let $C^b(X)$ and $C^b_{\mathbb{R}}(X)$ or $C^b(X, \mathbb{R})$ denote the vector space of bounded continuous functions from X into $\mathbb{C}$ and $\mathbb{R}$, respectively, and set $\|f\|_\infty = \sup_X |f(x)|$. Similarly, $C(X)$ and $C_{\mathbb{R}}(X)$ or $C(X, \mathbb{R})$ denote the vector space of continuous functions from X into $\mathbb{C}$ and $\mathbb{R}$, respectively.

1.6.6. REMARK. The function $\|\cdot\|_\infty$ on $C^b(X)$ is a norm (see Definition 2.1.1). This yields a metric on $C^b(X)$ by $d_\infty(f, g) = \|f - g\|_\infty$. Recall that $f_n \in C^b(X)$ *converge uniformly* to a function f if $\|f - f_n\|_\infty \to 0$. That is, convergence in this metric is uniform convergence on X, a notion that should be familiar even if general norms are not. Norms will be studied systematically in the next chapter.

1.6.7. PROPOSITION. *If* $C^b(X)$ *separates points of* X, *i.e. for* $x \neq y$ *in* X, *there is a continuous function* $f \in C^b(X)$ *so that* $f(x) \neq f(y)$, *then* X *is Hausdorff.*

PROOF. If $f(x) = \alpha$ and $f(y) = \beta$ and $r = |\alpha - \beta|/2 > 0$, then $x \in U = f^{-1}(b_r(\alpha))$ and $y \in V = f^{-1}(b_r(\beta))$ and $U \cap V = \varnothing$. ■

Consider Examples 1.6.2 (4) and (5) again in light of this proposition.

The following standard result for metric spaces extends easily.

1.6.8. PROPOSITION. *The uniform limit* f *of a sequence* $f_n \in C^b(X)$ *is continuous.*

PROOF. Let U be open in $\mathbb{C}$ and let $x \in f^{-1}(U)$. Then there is an $r > 0$ so that $b_r(f(x)) \subset U$. Choose n so large that $\|f - f_n\|_\infty < r/3$. Then $x \in V =$

$f_n^{-1}(b_{r/3}(f_n(x)))$ is open. If $y \in V$, then $|f_n(y) - f_n(x)| < r/3$, so

$$\begin{aligned}|f(y) - f(x)| &\le |f(y) - f_n(y)| + |f_n(y) - f_n(x)| + |f_n(x) - f(x)| \\ &< \|f - f_n\|_\infty + \frac{r}{3} + \|f - f_n\|_\infty < r.\end{aligned}$$

Hence $f(y) \in b_r(f(x)) \subset U$. Thus $x \in V \subset f^{-1}(U)$. So $f^{-1}(U)$ is open, and therefore f is continuous. ■

The norm $\|f\|_\infty$ makes $C^b(X)$ into a normed vector space. In particular, it is a metric space with distance function $\rho(f, g) = \|f - g\|_\infty$. In view of Proposition 1.6.7, the following is most interesting when X is Hausdorff.

1.6.9. THEOREM. *For any topological space, $C^b(X)$ is complete.*

PROOF. Let $(f_n)_{n\ge1}$ be a Cauchy sequence in $C^b(X)$. If $\varepsilon > 0$, there is an N so that if $N \le m < n$, then $\|f_n - f_m\|_\infty < \varepsilon$. In particular, for $x \in X$, the sequence $\big(f_n(x)\big)_{n\ge1}$ is Cauchy in $\mathbb{C}$. So we may define $f(x) = \lim_{n\to\infty} f_n(x)$ pointwise. However for $m \ge N$,

$$|f(x) - f_m(x)| = \lim_{n\to\infty} |f_n(x) - f_m(x)| \le \varepsilon.$$

Hence $\|f - f_m\|_\infty \le \varepsilon$. So convergence is uniform. By Proposition 1.6.8, f is continuous. Also $\|f\|_\infty = \lim_{n\to\infty} \|f_n\|_\infty < \infty$, and so f lies in $C^b(X)$. Therefore $C^b(X)$ is complete. ■

1.7. Compactness

The notion of compactness should be familiar from metric space theory. It is extremely useful in topology, as it turns out that every net in a compact set has a convergent subnet, analogous to the sequential result for metrics. The sequential definition of compactness is not appropriate, but the open cover version works.

1.7.1. DEFINITION. An *open cover* of a set $A \subset X$ is a collection of open sets $\{U_\lambda : \lambda \in \Lambda\}$ such that $A \subset \bigcup_\Lambda U_\lambda$. A set A is *compact* if every open cover has a finite subcover, i.e. a finite subset $U_{\lambda_1}, \dots, U_{\lambda_n}$ such that $A \subset \bigcup_{i=1}^n U_{\lambda_i}$.

A set is *sequentially compact* if every sequence has a convergent subsequence.

1.7.2. EXAMPLES.

(1) In any metric space, compact sets are the same as sequentially compact sets.

(2) In Example 1.4.12(1), the point $\{0\}$ is compact but not closed.

(3) A compact set need not be sequentially compact (Exercise **1.22**) and conversely a sequentially compact set need not be compact (Exercise **1.25**).

1.7.3. PROPOSITION. *If X is compact and $A \subset X$ is closed, then A is compact.*

If X is Hausdorff and $A \subset X$ is compact, then A is closed. Moreover, if $x \notin A$, there are disjoint open sets $U \supset A$ and $V \ni x$.

PROOF. If $\mathcal{U} = \{U_\lambda : \lambda \in \Lambda\}$ is an open cover of A, then $\mathcal{U} \cup \{A^c\}$ is an open cover of X. By compactness, it has a finite subcover $U_{\lambda_1}, \dots, U_{\lambda_n}, A^c$. Hence $U_{\lambda_1}, \dots, U_{\lambda_n}$ covers A, whence A is compact.

Suppose that X is Hausdorff and $A \subset X$ is compact, and let $x \in A^c$. For each $a \in A$, there are open sets $a \in U_a$ and $x \in V_a$ so that $U_a \cap V_a = \varnothing$. Clearly $\{U_a : a \in A\}$ is an open cover of A. By compactness, there is a finite subcover $U_{a_1}, \dots, U_{a_n}$. Let $V = \bigcap_{i=1}^n V_{a_i}$. Then $x \in V$ is open, and

$$V \cap A \subset \bigcup_{i=1}^{n} V \cap U_{a_i} = \varnothing.$$

Hence $x \notin \overline{A}$. So A is closed. Moreover $A \subset U = \bigcup_{i=1}^n U_{a_i}$, and $U \cap V = \varnothing$. ■

1.7.4. DEFINITION. A family $\{A_\lambda : \lambda \in \Lambda\}$ of subsets of X has the *finite intersection property* (FIP) if whenever $\lambda_1, \dots, \lambda_n$ are finitely many elements of Λ, then $\bigcap_{i=1}^n A_{\lambda_i} \neq \varnothing$.

1.7.5. PROPOSITION. *Let X be a topological space. The following are equivalent:*

(1) *X is compact.*

(2) *every family $\mathcal{F} = \{A_\lambda : \lambda \in \Lambda\}$ of closed subsets of X with FIP has non-empty intersection $\bigcap \mathcal{F} := \bigcap_\Lambda A_\lambda \neq \varnothing$.*

(3) *every net in X has a convergent subnet.*

PROOF. Suppose that X is compact and $\mathcal{F}$ has FIP. Define open sets $U_\lambda = A_\lambda^c$. If $\bigcap \mathcal{F} = \varnothing$, then $\bigcup_\Lambda U_\lambda = \left(\bigcap \mathcal{F}\right)^c = X$. So $\mathcal{U} = \{U_\lambda : \lambda \in \Lambda\}$ is an open cover of X. By compactness, there is a finite subcover $U_{\lambda_1}, \dots, U_{\lambda_n}$. Hence $\bigcap_{i=1}^n A_{\lambda_i} = \left(\bigcup_{i=1}^n U_{\lambda_i}\right)^c = \varnothing$, contradicting FIP. Therefore $\bigcap \mathcal{F} \neq \varnothing$.

Conversely, suppose that $\mathcal{U} = \{U_\lambda : \lambda \in \Lambda\}$ is an open cover of X. Define closed sets $A_\lambda = U_\lambda^c$. If there is no finite subcover, then $\bigcap_{i=1}^n A_{\lambda_i} = \left(\bigcup_{i=1}^n U_{\lambda_i}\right)^c \neq \varnothing$, and thus $\mathcal{F} = \{A_\lambda : \lambda \in \Lambda\}$ has FIP. But then $\bigcap \mathcal{F} \neq \varnothing$. Therefore $\bigcup_\Lambda U_\lambda = \left(\bigcap \mathcal{F}\right)^c \neq X$, contradicting the fact that $\mathcal{U}$ is an open cover. Hence $\mathcal{F}$ does not have FIP, so there is a finite set $A_{\lambda_1}, \dots, A_{\lambda_n}$ such that $\bigcap_{i=1}^n A_{\lambda_i} = \varnothing$. Hence $\bigcup_{i=1}^n U_{\lambda_i} = \left(\bigcap_{i=1}^n A_{\lambda_i}\right)^c = X$. Therefore X is compact.

Suppose that (3) holds. Consider a collection $\mathcal{F} = \{A_\delta : \delta \in \Delta\}$ of closed sets with the FIP. Let $\Lambda = \{F \subset \Delta : F \text{ is finite, non-empty}\}$ ordered by inclusion, i.e. $F \leq G$ if $F \subset G$. This is an upward directed poset: $F_1, F_2 \leq F_1 \cup F_2$. For each $F \in \Lambda$, use the Axiom of Choice to select a point $x_F \in \bigcap_{\delta \in F} A_\delta$. This is possible since the finite intersection is non-empty. Then $(x_F)_\Lambda$ is a net in X. By (3), there is a subnet $(y_\gamma)_\Gamma$ with limit x, where $\varphi : \Gamma \to \Lambda$ and $y_\gamma = x_{\varphi(\gamma)}$. For any $\delta \in \Delta$, there is a $\gamma_\delta \in \Gamma$ so that $\gamma \geq \gamma_\delta$ implies that $\varphi(\gamma) \geq \{\delta\}$. Hence $y_\gamma \in A_\delta$ for all $\gamma \geq \gamma_\delta$. Since A_δ is closed, the limit point x lies in A_δ. This holds for all $\delta \in \Delta$. Therefore x belongs to $\bigcap \mathcal{F}$. Thus (2) holds.

Conversely suppose that (2) holds. Let $(x_\lambda)_\Lambda$ be a net in X. For each $\lambda \in \Lambda$, define $A_\lambda = \overline{\{x_\mu : \mu \geq \lambda\}}$. Then $\mathcal{F} = \{A_\lambda : \lambda \in \Lambda\}$ is a collection of non-empty closed sets. It has FIP because if $\lambda_1, \dots, \lambda_n \in \Lambda$, the upward directed property ensures that there is some $\lambda_0 \in \Lambda$ so that $\lambda_i \leq \lambda_0$ for $1 \leq i \leq n$. Hence $\bigcap_{i=1}^n A_{\lambda_i} \supset A_{\lambda_0} \neq \varnothing$. Therefore, there is a point $x \in \bigcap_\Lambda A_\lambda$.

Now we build a subnet with limit x. Let $\mathcal{O}(x)$ be the set of all open neighbourhoods of x. Let $\Gamma = \Lambda \times \mathcal{O}(x)$ with order $(\lambda, U) \leq (\mu, V)$ if $\lambda \leq \mu$ and $U \supset V$. Let $S_{\lambda,U} = \{\mu \in \Lambda : \mu \geq \lambda \text{ and } x_\mu \in U\}$. This set is non-empty because $x \in A_\lambda \cap U = \overline{\{x_\mu : \mu \geq \lambda\}} \cap U$, and thus by Proposition 1.5.3, $x_\mu \in U$ for some $\mu \geq \lambda$. Use the Axiom of Choice to select $\mu = \varphi(\lambda, U) \in S_{\lambda,U}$ for each $(\lambda, U) \in \Gamma$. The map $\varphi : \Gamma \to \Lambda$ is cofinal because if $\lambda_0 \in \Lambda$, then every $(\lambda, U) \geq (\lambda_0, X)$ will have $\varphi(\lambda, U) = \mu \geq \lambda \geq \lambda_0$. So $y_{\lambda,U} = x_{\varphi(\lambda,U)}$ defines a subnet $(y_{\lambda,U})_\Gamma$ of $(x_\lambda)_\Lambda$. Finally, we claim that $\lim_\Gamma y_{\lambda,U} = x$. Indeed, let $U \in \mathcal{O}(x)$ be any open neighbourhood of x. Fix some $\lambda_0 \in \Lambda$. Whenever $(\lambda, V) \geq (\lambda_0, U)$, we have $y_{\lambda,V} \in V \subset U$. Thus this net converges to x. ∎

1.7.6. Proposition. *If $f : (X, \tau) \to (Y, \sigma)$ is continuous and $A \subset X$ is compact, then $f(A)$ is compact.*

Proof. Let $\{V_\lambda : \lambda \in \Lambda\}$ be an open cover of $f(A)$ in Y. Define $U_\lambda = f^{-1}(V_\lambda)$. These are open sets by continuity, and they cover A. Thus there is a finite subcover $U_{\lambda_1}, \dots, U_{\lambda_n}$. Then since $V_\lambda \supset f(U_\lambda)$, it follows that $V_{\lambda_1}, \dots, V_{\lambda_n}$ covers $f(A)$. Hence $f(A)$ is compact. ∎

The following important consequence follows directly.

1.7.7. Extreme Value Theorem. *If (X, τ) is compact and $f \in C(X)$, then $|f|$ attains its maximum. In particular, $\|f\|_\infty < \infty$.*

1.7.8. Corollary. *If X is a compact topological space, then $\|\cdot\|_\infty$ is a norm on $C(X)$, and $C(X)$ is complete.*

1.8. Weak Topologies

A weak topology on a set X is the smallest (weakest) topology that makes a family of functions continuous. This arises frequently in functional analysis, where the set of functions is often a vector space of linear scalar-valued functions. Different choices yield different topologies.

This will allow us to define the product topology on an infinite product space. When this is a product of compact spaces, the product turns out to be compact. This result requires the Axiom of Choice.

1.8.1. DEFINITION. Given a set X, let $\mathcal{F} = \{f_\alpha : X \to (Y_\alpha, \tau_\alpha) \mid \alpha \in A\}$ be a family of functions. There is a weakest topology $\tau_{\mathcal{F}}$ for which each f_α is continuous called the *weak topology* determined by $\mathcal{F}$. This is the topology generated by $\{f_\alpha^{-1}(U) : U \in \tau_\alpha,\ \alpha \in \Lambda\}$. A base for this topology is given by

$$\{f_{\alpha_1}^{-1}(U_1) \cap \cdots \cap f_{\alpha_n}^{-1}(U_n) : \alpha_i \in A,\ U_i \in \tau_{\alpha_i},\ n \geq 1\}.$$

1.8.2. EXAMPLE. Product topology. If (X_α, τ_α) are topological spaces for $\alpha \in A$, we define the *product space* to be $X = \prod_A X_\alpha = \{x = (x_\alpha) : x_\alpha \in X_\alpha\}$ with the *product topology* τ, the weakest topology which makes the coordinate projections $\pi_\alpha : X \to X_\alpha$ given by $\pi_\alpha(x) = x_\alpha$ continuous.

The sets $\pi_\alpha^{-1}(U) = \prod_{\beta \in A \setminus \{\alpha\}} X_\beta \times U$ for $U \in \tau_\alpha$ are open and form a subbase for the topology. The product topology τ consists of arbitrary unions of finite intersections of the subbase. So if $\alpha_1, \ldots, \alpha_n \in A$ and $U_i \in \tau_{\alpha_i}$, then the sets of the form

$$U_1 \times \cdots \times U_n \times \prod_{\beta \in A \setminus \{\alpha_i, 1 \leq i \leq n\}} X_\beta$$

form a base for the topology.

If A is finite, this is a familiar construction in the metric space case. Indeed, if (X_i, d_i) are metric spaces for $1 \leq i \leq n$, then

$$D\big((x_1, \ldots, x_n), (y_1, \ldots, y_n)\big) = \max\big\{d_i(x_i, y_i) : 1 \leq i \leq n\big\}$$

is a metric on the product, and the metric topology coincides with the product topology.

When A is infinite, it often requires the Axiom of Choice to be able to say that X is non-empty.

1.8.3. PROPOSITION. *Let $\mathcal{F}$ be a family of functions $f_\alpha : X \to (Y_\alpha, \tau_\alpha)$ for $\alpha \in A$. Then a net $(x_\lambda)_\Lambda$ converges to x in $(X, \tau_{\mathcal{F}})$ if and only if $f_\alpha(x_\lambda) \to f_\alpha(x)$ for all $\alpha \in A$.*

PROOF. If $(x_\lambda)_\Lambda \to x$ in $(X, \tau_{\mathcal{F}})$, then since each f_α is continuous by definition, $f_\alpha(x_\lambda) \to f_\alpha(x)$ by Theorem 1.6.4.

Conversely, let U be an open neighbourhood of x. Then x has a basic open neighbourhood $x \in V = f_{\alpha_1}^{-1}(U_1) \cap \cdots \cap f_{\alpha_n}^{-1}(U_n) \subset U$ for some $\alpha_i \in A$ and $U_i \in \tau_{\alpha_i}$. For each $1 \leq i \leq n$, there is some $\lambda_i \in \Lambda$ so that $f_{\alpha_i}(x_\lambda) \in U_i$ for all $\lambda \geq \lambda_i$. Since Λ is upward directed, there is some $\lambda_0 \geq \lambda_i$ for $1 \leq i \leq n$. When $\lambda \geq \lambda_0$, $f_{\alpha_i}(x_\lambda) \in U_i$ for $1 \leq i \leq n$. Therefore $x_\lambda \in V \subset U$. It follows that $(x_\lambda)_\Lambda$ converges to x. ■

1.8.4. COROLLARY. *Let $\mathcal{F}$ be a family of functions $f_\alpha : X \to (Y_\alpha, \tau_\alpha)$ for $\alpha \in A$. Then $g : (Y, \tau) \to (X, \tau_{\mathcal{F}})$ is continuous if and only if $f_\alpha \circ g : Y \to Y_\alpha$ are continuous for all $\alpha \in A$.*

PROOF. The composition of continuous functions is continuous by Proposition 1.6.3. So if g is continuous, so are $f_\alpha \circ g$ for all $\alpha \in A$. Conversely, suppose that a net $(y_\lambda)_\Lambda$ converges to $y \in Y$. By continuity, $f_\alpha(g(y_\lambda))$ converges to $f_\alpha(g(y))$ in Y_α for all $\alpha \in A$. By Proposition 1.8.3, $g(y_\lambda)$ converges to $g(y)$ in Y. Therefore g is continuous by Theorem 1.6.4. ■

1.8.5. PROPOSITION. *Let $\mathcal{F}$ be a family of functions $f_\alpha : X \to (Y_\alpha, \tau_\alpha)$ for $\alpha \in A$. If the Y_α are all Hausdorff and $\mathcal{F}$ separates points, then $(X, \tau_{\mathcal{F}})$ is Hausdorff.*

PROOF. Let $x \neq y \in X$. Since $\mathcal{F}$ separates points, there is some α so that $f_\alpha(x) \neq f_\alpha(y)$. Since Y_α is Hausdorff, there are disjoint open neighbourhoods $U \ni f_\alpha(x)$ and $V \ni f_\alpha(y)$. Thus $f_\alpha^{-1}(U) \ni x$ and $f_\alpha^{-1}(V) \ni y$ are disjoint open sets in X. Therefore X is Hausdorff. ■

1.8.6. COROLLARY. *If (X_α, τ_α) are Hausdorff spaces for $\alpha \in A$, then $X = \prod_{\alpha \in A} X_\alpha$ is Hausdorff in the product topology.*

1.8.7. TYCHONOFF THEOREM. *The product of compact spaces is compact.*

PROOF. Let $X = \prod_{\alpha \in A} X_\alpha$ where each X_α is compact. Suppose that there is an open cover with no finite subcover. Order

$$\mathcal{P} = \{\mathcal{U} : \text{open covers with no finite subcover}\}$$

by inclusion. If $\mathcal{C} = \{\mathcal{U}_\beta : \beta \in B\}$ is a chain in $\mathcal{P}$, let $\mathcal{U} = \bigcup_{\beta \in B} \mathcal{U}_\beta$. This is an open cover of X. If it has a finite subcover $U_1, \ldots, U_n$, the chain property shows that there is some β so that $U_i \in \mathcal{U}_\beta$ for $1 \leq i \leq n$. Hence $\mathcal{U}_\beta$ contains a finite subcover, contrary to fact. Therefore $\mathcal{U} \in \mathcal{P}$ is an upper bound for $\mathcal{C}$. Hence $\mathcal{P}$ is inductive. By Zorn's lemma, there is a maximal open cover $\mathcal{U}_0$ with no finite subcover.

We claim the $\mathcal{U}_0$ has the following properties:

(1) If $U \in \mathcal{U}_0$ and $V \subset U$ is open, then $V \in \mathcal{U}_0$. This is because V can't help cover anything that U doesn't cover.

(2) If $U_1, U_2 \in \mathcal{U}_0$, then $U_1 \cup U_2 \in \mathcal{U}_0$. This is because $U_1 \cup U_2$ can't cover anything not covered by U_1 and U_2 together.

(3) If V_1, V_2 are open and $V_1 \cap V_2 \in \mathcal{U}_0$, then one of $V_i \in \mathcal{U}_0$. Indeed, if both $V_i \notin \mathcal{U}_0$, then $\{V_i, \mathcal{U}_0\}$ has a finite subcover, necessarily of the form:

$$V_1 \cup U_1 \cup \cdots \cup U_m = X = V_2 \cup U_{m+1} \cup \ldots U_{m+n}.$$

However then $X = (V_1 \cap V_2) \cup U_1 \cup \cdots \cup U_{m+n}$ is a finite subcover from $\mathcal{U}_0$, a contradiction.

For each $\alpha \in A$, let $\mathcal{Y}_\alpha = \{U \in \tau_\alpha : \pi_\alpha^{-1}(U) \in \mathcal{U}_0\}$ and let $Y_\alpha = \bigcup \mathcal{Y}_\alpha$. Then $Y_\alpha \neq X_\alpha$. For otherwise, $\mathcal{Y}_\alpha$ would be an open cover of X_α, and thus would have a finite subcover $U_1, \ldots, U_n$. Then $\pi_\alpha^{-1}(U_1), \ldots, \pi_\alpha^{-1}(U_n)$ would be a finite subcover of X from $\mathcal{U}_0$. Hence $X_\alpha \setminus Y_\alpha$ is non-empty for each $\alpha \in A$. By the Axiom of Choice, there is a point $x = (x_\alpha) \in X$ so that $x_\alpha \in X_\alpha \setminus Y_\alpha$ for each $\alpha \in A$. There must be a set $U \in \mathcal{U}_0$ with $x \in U$. Hence there is a basic open neighbourhood $x \in V = \pi_{\alpha_1}^{-1}(U_1) \cap \cdots \cap \pi_{\alpha_n}^{-1}(U_n) \subset U$. Therefore, $\bigcap_{i=1}^n \pi_{\alpha_i}^{-1}(U_i) \in \mathcal{U}_0$ by property (1). Repeated application of property (3) now shows that there is some i so that $\pi_{\alpha_i}^{-1}(U_i) \in \mathcal{U}_0$. But this means that $x_{\alpha_i} \in Y_{\alpha_i}$, contrary to our construction of x. This contradiction shows that $\mathcal{P}$ is empty, and thus every open cover has a finite subcover. ∎

We finish off this section by showing that the Tychonoff theorem implies the Axiom of Choice. So it is in fact equivalent to Choice, since Choice was used to prove Tychonoff's theorem.

1.8.8. THEOREM. *The Tychonoff theorem is equivalent to the Axiom of Choice.*

PROOF. Beyond basic rules of set theory, we used the Axiom of Choice and the equivalent Zorn's lemma to establish the Tychonoff theorem.

Conversely, suppose that X_α for $\alpha \in A$ are non-empty sets. Define topological spaces $Y_\alpha = X_\alpha \dot\cup \{p_\alpha\}$, where we have added a distinguished point p_α with the topology $\tau_\alpha = \{\varnothing, \{p_\alpha\}, X_\alpha, Y_\alpha\}$. Clearly each Y_α is compact because every cover is finite. By the Tychonoff theorem, $Y = \prod_{\alpha \in A} Y_\alpha$ is compact. Since X_α is closed, $C_\alpha = X_\alpha \times \prod_{\beta \neq \alpha} Y_\beta$ is closed in Y. A closed subset of a compact set is compact, so C_α is compact. The collection $\mathcal{F} = \{C_\alpha : \alpha \in A\}$ has FIP because if $\alpha_1, \ldots, \alpha_n$ are given, we can choose $x_i \in X_{\alpha_i}$ for $1 \leq i \leq n$. Then $\bigcap_{i=1}^n C_{\alpha_i} \ni x = (x_1, \ldots, x_n, p_\beta) : \beta \in A \setminus \{\alpha_1, \ldots, \alpha_n\}$. By Proposition 1.7.5, the intersection $C = \bigcap_{\alpha \in A} C_\alpha$ is non-empty. Let $x = (x_\alpha)$ be a point in the intersection. Note that $x_\alpha \in X_\alpha$ for all $\alpha \in A$. Thus $c(\alpha) = x_\alpha$ is a choice function. Hence AC holds. ∎

1.9. Compact Hausdorff Spaces

There is a whole hierarchy of *separation properties* to classify how nice a topological space is. T_0, T_1 and T_2 have already been defined. We now add now stronger properties which affect the behaviour of continuous functions. In particular, we shall see that compact Hausdorff spaces are normal, the strongest of these new properties.

1.9.1. DEFINITION. A topological space is T_3 if it is T_1 and *regular*: given a closed set A and a point $x \notin A$, there are disjoint open sets $U \supset A$ and $V \ni x$.
A topological space is $T_{3.5}$ or *Tychonoff* if it is T_1 and *completely regular*: given a closed set A and a point $x \notin A$, there is a continuous function $f : X \to [0,1]$ so that $f(x) = 1$ and $f|A = 0$.
A topological space is T_4 if it is T_1 and *normal*: given disjoint closed sets A, B, there are disjoint open sets $U \supset A$ and $V \supset B$.

1.9.2. REMARKS.
(1) The T_0 property is very weak. Example 1.6.2(4) is T_0 but not T_1.

(2) T_1 is also a very weak property. It implies T_0 since if $x \neq y$, the set $\{x\}^c$ is an open set containing y but not x. To be T_1 it is enough to find open sets $U \ni x$ with $y \notin U$ and $V \ni y$ with $x \notin V$ (Exercise **1.11**). Example 1.6.2(5) is T_1 but not Hausdorff. Points are closed in Hausdorff spaces, so T_2 implies T_1.

(3) The trivial topology is regular because there are no points which are disjoint from a non-empty closed set. But throwing in the T_1 condition makes a T_3 space Hausdorff because you can take your closed set to be $\{y\}$.

(4) Completely regular spaces are regular, because if $x \notin A$ and A is closed, let $f : X \to [0,1]$ be a continuous function with $f(x) = 1$ and $f|_A = 0$. Then $U = \{y : f(y) > 1/2\}$ and $V = \{y : f(x) < 1/2\}$ are disjoint open sets with $x \in U$ and $A \subset V$. Again we need to add the T_1 property to exclude examples like the trivial topology.

(5) The T_1 property ensures that T_4 spaces are Hausdorff. A metric space (X, d) is normal. If A and B are disjoint closed sets, let $U = \{x : d(x, A) < d(x, B)\}$ and $V = \{x : d(x, A) > d(x, B)\}$. It is easy to check that these are disjoint open sets.

1.9.3. PROPOSITION. *Compact Hausdorff spaces are normal, whence T_4.*

PROOF. Let A, B be disjoint closed subsets of a compact Hausdorff space X. Then A and B are compact by Proposition 1.7.3. Moreover since X is Hausdorff, that same Proposition shows that for each point $x \in B$, there are disjoint open sets $U_x \supset A$ and $V_x \ni x$. The collection $\{V_x : x \in B\}$ is an open cover of B.

Let $V_{x_1}, \dots, V_{x_n}$ be a finite subcover. Set $V = \bigcup_{i=1}^n V_{x_i}$ and $U = \bigcap_{i=1}^n U_{x_i}$. These are disjoint open sets with $A \subset U$ and $B \subset V$. ■

Now we prove that normal Hausdorff spaces have lots of continuous functions.

1.9.4. URYSOHN LEMMA. *Let X be a normal Hausdorff space, and let A and B be disjoint closed sets. Then there is a continuous function $f : X \to [0, 1]$ such that $f|_A = 0$ and $f|_B = 1$.*

PROOF. Normality implies the following property: if A is closed and W is open and $A \subset W$, then there is an open set U such that $A \subset U \subset \overline{U} \subset W$. To see this, take $B = W^c$. Use normality to find disjoint open sets $U \supset A$ and $V \supset B$. Then $\overline{U} \subset V^c \subset W$.

Start with $U_1 = B^c$. Find an open $U_{1/2}$ so that $A \subset U_{1/2} \subset \overline{U_{1/2}} \subset U_1$. Repeating this procedure recursively, we find open sets $U_{k/2^n}$ for $1 \le k \le 2^n$ and $n \ge 1$ so that

$$A \subset U_{k/2^n} \subset \overline{U_{k/2^n}} \subset U_{(k+1)/2^n} \quad \text{for} \quad 1 \le k < 2^n.$$

Let $D = \{k/2^n : 1 \le k \le 2^n,\ n \ge 1\}$. Define $f(x) = \inf\{r \in D : x \in U_r\}$ if $x \in U_1$ and $f|_B = 1$. Clearly $0 \le f \le 1$ and $f|_A = 0$.

Claim: f is continuous. Note that

$$f^{-1}\big([0, t)\big) = \bigcup_{r<t,\, r\in D} U_r \quad \text{is open for } t \in [0, 1].$$

Also for $0 \le t < 1$, since $t < r < s$ for $r, s \in D$ implies that $\overline{U_r} \subset U_s$,

$$f^{-1}\big([0, t]\big) = \bigcap_{r>t,\, r\in D} f^{-1}\big([0, r)\big) = \bigcap_{r>t,\, r\in D} U_r = \bigcap_{r>t,\, r\in D} \overline{U_r}.$$

This is closed, and therefore $f^{-1}\big((t, 1]\big) = \big(\bigcap_{r>t,\, r\in D} \overline{U_r}\big)^c$ is open. Hence, $f^{-1}\big((s, t)\big)$ is open for $s < t$, and so f is continuous. ■

1.9.5. REMARK. If (X, d) is a metric space and A and B are disjoint closed sets, define

$$f(x) = \frac{d(x, A)}{d(x, A) + d(x, B)}.$$

This satisfies the conclusion of Urysohn's lemma.

1.9.6. COROLLARY. *If X is a compact Hausdorff space, $C(X)$ separates points.*

The Urysohn lemma implies the following significant strengthening.

1.9.7. TIETZE EXTENSION THEOREM. *Let X be a normal Hausdorff space, and let $A \subset X$ be a closed set. If $f : A \to [a, b]$ is continuous, there is a continuous function $g : X \to [a, b]$ such that $g|_A = f$.*

PROOF. After scaling, we may assume that the range is $[-1, 1]$. Let $A_1 = f^{-1}([-1, -\frac{1}{3}])$ and $B_1 = f^{-1}([\frac{1}{3}, 1])$. By Urysohn's lemma, there is a function $g_1 : X \to [-\frac{1}{3}, \frac{1}{3}]$ so that $g_1|_{A_1} = -\frac{1}{3}$ and $g_1|_{B_1} = \frac{1}{3}$. Then $f_1 = f - g_1|_A$ has range in $[-\frac{2}{3}, \frac{2}{3}]$. Repeat the process, setting $A_2 = f_1^{-1}([-\frac{2}{3}, -\frac{2}{9}])$ and $B_2 = f_1^{-1}([\frac{2}{9}, \frac{2}{3}])$, and finding $g_2 : X \to [-\frac{2}{9}, \frac{2}{9}]$ with $g_2|_{A_2} = -\frac{2}{9}$ and $g_2|_{B_2} = \frac{2}{9}$.

Then $f_2 = f_1 - g_2|_A$ has range in $[-(\frac{2}{3})^2, (\frac{2}{3})^2]$. Recursively we obtain functions $g_n : X \to [-2 \cdot 3^{-n}, 2 \cdot 3^{-n}]$ so that $f_n = f - \sum_{i=1}^{n} g_n|_A$ has range in $[-(\frac{2}{3})^n, (\frac{2}{3})^n]$. Let $g = \sum_{n \geq 1} g_n$. Then $g|_A = f$ and

$$\|g\|_\infty \leq \sum_{n \geq 1} \|g_n\|_\infty = \sum_{n \geq 1} 2 \cdot 3^{-n} = 1.$$

Thus this series converges uniformly, whence $g \in C(X)$. ■

1.10. Locally compact spaces

Locally compact Hausdorff spaces includes the familiar spaces $\mathbb{R}^n$ and $\mathbb{C}^n$, and many topological groups. They arise in both functional analysis and measure theory. In this section, we provide a substitute for the Urysohn lemma and the Tietze extension theorem which are valid in this greater generality.

1.10.1. DEFINITION. A Hausdorff space X is *locally compact* (LCH) if every point has a compact neighbourhood, i.e. for $x \in X$, there is a compact set K with $x \in \operatorname{int} K$.

1.10.2. PROPOSITION. *Locally compact Hausdorff spaces are regular.*

PROOF. Let A be a closed set in a LCH space X, and $x \notin A$. Let K be a compact neighbourhood of x. Then $A \cap K$ is compact. So by Proposition 1.7.3, there are disjoint open sets $U \ni x$ and $V \supset A \cap K$. Then $W = U \cap \operatorname{int} K \subset K \setminus V$ is an open neighbourhood of x and $\overline{W} \subset K \setminus V \subset A^c$. Hence $\overline{W}^c \supset A$ is open and disjoint from W. Therefore X is regular. ■

1.10.3. DEFINITION. If $f : X \to \mathbb{C}$, the *support* of f is

$$\operatorname{supp}(f) = \overline{\{x : f(x) \neq 0\}}.$$

If X is LCH, $C_c(X)$ denotes the space of continuous $\mathbb{C}$-valued functions with compact support. Let $C_0(X)$ be the closure of $C_c(X)$ in $(C^b(X), \|\cdot\|_\infty)$.

There is a weaker version of Urysohn's lemma valid for LCH spaces.

1.10.4. PROPOSITION. *Locally compact Hausdorff spaces are completely regular. Moreover, if X is LCH, $A \subset X$ is closed, $B \subset X$ is compact and $A \cap B = \varnothing$, then there is $f \in C_c(X)$ with $0 \le f \le 1$, $f|_A = 0$ and $f|_B = 1$.*

PROOF. Arguing as in the previous proof, for each $x \in B$, there is a compact neighbourhood $\overline{W_x} \ni x$ which is disjoint from A. We can take $W_x = \operatorname{int}\overline{W_x}$. Then $\{W_x : x \in B\}$ is an open cover of B. Let $W_{x_1}, \dots, W_{x_n}$ be a finite subcover. Then $W = \bigcup_{i=1}^n W_{x_i}$ is open containing B and $\overline{W} = \bigcup_{i=1}^n \overline{W_{x_i}}$ is compact and disjoint from A.

Let $C = \overline{W} \setminus W$. Then B and C are disjoint closed sets in the compact Hausdorff space $\overline{W}$. By Urysohn's lemma, there is a continuous function $f : \overline{W} \to [0, 1]$ with $f|_B = 1$ and $f|_C = 0$. Extend f to a function on X by setting $f|_{\overline{W}^c} = 0$. Now $A \subset W^c$, and thus $f|_A = 0$. To see that f is continuous, note that if $a \ge 0$, $f^{-1}(a, b)$ is open in W, and hence open in X. If $a < 0 < b$, then $D = f^{-1}([b, \infty)$ is closed in $\overline{W}$, and thus $f^{-1}(a, b) = D^c$ is open in X. Since $\operatorname{supp}(f) \subset \overline{W}$, we have $f \in C_c(X)$. ■

There is also a weaker version of the Tzietze theorem valid for LCH spaces.

1.10.5. COROLLARY. *Suppose that X is LCH, $A \subset X$ is closed, $B \subset X$ is compact, $A \cap B = \varnothing$ and $g \in C(B)$. Then there is an $f \in C_c(X)$ with $f|_A = 0$ and $f|_B = g$, and $\|f\|_\infty = \|g\|_\infty$.*

PROOF. The setup is the same as the previous proof. Define $f|B = g$ and $f|_C = 0$, and extend this to a continuous function on $\overline{W}$ using the Tietze extension theorem. Then extend f to all of X by setting $f|_{\overline{W}^c} = 0$. This is continuous with compact support as in the previous proof. ■

1.10.6. DEFINITION. If U is open, say $f \prec U$ for a continuous function f if $0 \le f \le 1$ and $\operatorname{supp} f \subset U$. Suppose that $\mathcal{U} = \{U_\lambda : \lambda \in \Lambda\}$ is an open cover of $K \subset X$. Then a *partition of unity* for K *relative to* $\mathcal{U}$ is a collection of functions $g_\lambda \prec U_\lambda$ which is *locally finite*, i.e. for each $x \in X$, $\{\lambda : g_\lambda(x) \ne 0\}$ is finite, and $\sum_\Lambda g_\lambda(x) = 1$ for all $x \in K$.

1.10.7. PROPOSITION. *Let X be LCH, and let $K \subset X$ be compact. Suppose that $\mathcal{U} = \{U_1, \dots, U_n\}$ is an open cover of K. Then there are $g_i \prec U_i$ in $C_c(X)$ which is a partition of unity for K relative to $\mathcal{U}$.*

PROOF. For each $x \in K$, there is some $U_i \ni x$. Pick a compact neighbourhood with $x \in C_x \subset U_i$. Now $\{\operatorname{int} C_x : x \in K\}$ is an open cover with finite subcover $C_{x_1}, \ldots, C_{x_m}$. Let $K_i = \bigcup\{C_{x_j} : C_{x_j} \subset U_i\}$. Then $K_i \subset U_i$ and $K \subset \bigcup_{i=1}^n K_i =: C$ are compact. Let W be an open neighbourhood of C with $\overline{W}$ compact. By Urysohn's lemma, there are functions h_i so that $\chi_{K_i} \le h_i \prec U_i \cap W$. Let $g_i = h_i / \sum_{i=1}^n h_i$. ■

Exercises for Chapter 1

1.1 Show that every vector space has a basis. (Requires AC.)

1.2 (a) Assuming AC, show (‡) if $f : X \to Y$ is surjective, there is a function $g : Y \to X$ such that $f \circ g = id_Y$.
(b) Show that property (‡) implies AC.

1.3 Deduce the following from Lemma 1.3.2.
(a) Show that the only automorphism of a well-ordered set is the identity.
(b) If two well-orders are isomorphic, the isomorphism is unique.

1.4 Prove Lemma 1.3.5 about properties of ordinals.

1.5 Show that the open sets defined in Example 1.5.2 form a topology.

1.6 Let $h : X \to Y$ be a continuous bijection of X onto Y. Suppose that X is compact and Y is Hausdorff. Prove that h is a homeomorphism.

1.7 (*Kuratowski closure axioms*). A closure operation $A \to \overline{A}$ on all subsets of a set X has the following properties:
(1) $A \subset \overline{A}$.
(2) $\overline{\overline{A}} = \overline{A}$.
(3) $\overline{A \cup B} = \overline{A} \cup \overline{B}$.
(4) $\overline{\varnothing} = \varnothing$.
Say that a set A is closed if $A = \overline{A}$, and U is open if U^c is closed. Prove that the collection of open sets forms a topology.

1.8 In a topological space X, say that a net $(x_\lambda)_{\lambda \in \Lambda}$ has x as a *cluster point* if for each $\lambda_0 \in \Lambda$ and each open neighbourhood $U \ni x$, there is some $\lambda \ge \lambda_0$ so that $x_\lambda \in U$. Prove that x is a cluster point of this net if and only if there is a subnet with limit x. Be explicit about use of the Axiom of Choice.

1.9 Let V be the vector space of complex-valued functions on $\mathbb{R}$. Put a topology τ on V with a subbase given by $U_{t,a,r} = \{f \in V : |f(t) - a| < r\}$ for $t \in \mathbb{R}$, $a \in \mathbb{C}$, $r > 0$.

(a) Show that a net $(f_\lambda)_{\lambda\in\Lambda}$ in V converges if and only if $\lim f_\lambda(t)$ exists for each $t \in \mathbb{R}$, i.e. pointwise convergence. (See Example 1.4.8.) In particular, $\varepsilon_t(f) := f(t)$ is continuous on V for each $t \in \mathbb{R}$.

(b) Let $E = \{f \in V : f(t) = 2 \text{ except for finitely many } t \in \mathbb{R}\}$. Construct a net in E with limit 0.

(c) Show that no sequence of points in E can converge to 0.

1.10 Consider $\mathbb{R}$ with its usual metric topology τ, which is also the order topology on $\mathbb{R}$. Let Y be a subset of $\mathbb{R}$. Consider the topology ρ on Y induced by the topology on $\mathbb{R}$, and let σ be the order topology on Y given by the induced order.

(a) What relationship always holds between ρ and σ?

(b) Find a subset Y where these two topologies are different.

(c) Find reasonable conditions on Y which ensure that these two topologies agree. **Remark.** Try to find necessary and sufficient conditions. Reasonable means that your conditions should handle the cases where Y is open, closed, $\mathbb{Q}$ and your example in part (b).

1.11 Show that (X, τ) is T_1 if for $x \neq y \in X$, there are open sets $U \ni x$ with $y \notin U$ and $V \ni y$ with $x \notin V$.

1.12 Prove that the composition of continuous functions is continuous (Proposition 1.6.3).

1.13 Let τ be the topology on $\mathbb{R}$ generated by $\{[x, y) : x < y\}$. Prove that $(\mathbb{R}, \tau)$ is normal.

1.14 Let X be a second countable LCH space, and let $W \subset X$ be open. Prove that W is σ-compact, i.e. that W is the union of countably many compact sets.

1.15 Let (X, τ) be a topological space, and let A and B be closed subsets such that $A \cup B = X$. Suppose that $f : A \to Y$ and $g : B \to Y$ are continuous functions into (Y, σ) such that $f|_{A\cap B} = g|_{A\cap B}$. Define $h(x) = f(x)$ for $x \in A$ and $h(x) = g(x)$ for $x \in B$. Prove that h is continuous on X.

1.16 Let $\overline{\mathbb{D}}$ be the closed unit disc in $\mathbb{C}$ and let $\mathbb{T}$ be the unit circle. The identity function $\iota : \mathbb{T} \to \mathbb{T}$ by $\iota(z) = z$ is continuous. Show that this extends to a continuous function of $\overline{\mathbb{D}}$ into $\mathbb{C}$, but not to a continuous function of $\overline{\mathbb{D}}$ into $\mathbb{T}$.
HINT: If f were such an extension, look at the winding number of $f(r\mathbb{T})$ for $0 \le r \le 1$.

1.17 Suppose that $(X_\lambda, \tau_\lambda)$ are non-empty LCH spaces. Show that $\prod_\Lambda X_\lambda$ with the product topology is LCH if and only if all but finitely many of the X_λ are compact.

1.18 Let (X, τ) be LCH. Show that if $U \subset X$ is open and $C \subset X$ is closed, then $U \cap C$ is LCH in the induced topology.

1.19 Let (X, τ) be LCH, and let C be a compact G_δ set. Prove that there is a function $f \in C_0(X)$ with $0 \le f \le 1$ such that $f^{-1}(1) = C$.
HINT: Write $C = \bigcap_{n \ge 1} U_n$ where U_n are open, and apply Proposition 1.10.4 to C and U_n^c for each n. Define f as a linear combination of these functions.

1.20 The *one-point compactification* of X. Let (X, τ) be LCH, but not compact. Form the space $X^* = X \cup \{p\}$. Define a topology τ^* by specifying that $\{p\}$ is closed, $\tau^* \cap X = \tau$, and open neighbourhoods of p have the form $\{p\} \cup (X \setminus K)$ for K compact in X. Prove that X^* is compact and Hausdorff.

1.21 The *Stone–Čech compactification* of X. Let (X, τ) be LCH, but not compact. For each $f \in C^b_{\mathbb{R}}(X)$, let $I_f = [-\|f\|_\infty, \|f\|_\infty]$. Then the product space $Y = \prod I_f$ is compact and Hausdorff. Define $\varepsilon : X \to Y$ by $\varepsilon(x)(f) = f(x)$.
(a) Show that ε is continuous.
(b) Define $\beta X = \overline{\varepsilon(X)}$. Observe that βX is compact Hausdorff and X is dense in βX.
(c) Prove that if K is compact and Hausdorff and $g : X \to K$ is continuous, then there is a (unique) continuous function $\tilde{g} : \beta X \to K$ such that $\tilde{g} \circ \varepsilon = g$.

1.22 (a) Let X be the closed unit ball of $B[0, 1]$, the space of bounded functions on $[0, 1]$, with the topology of pointwise convergence. Prove that X is compact.
(b) Show that X is not sequentially compact.
HINT: For $x \in [0, 1]$, consider its binary expansion, choosing a finite expansion if possible. Let $f_n(x)$ be the nth digit in this binary expansion. Prove that there is no convergent subsequence.

1.23 (*Baire Theorem*). A topological space is a *Baire space* if every countable family $\{U_n : n \ge 1\}$ of dense open sets has a dense intersection.
(a) Let X be a compact Hausdorff space. Show that X is Baire.
HINT: Take $x_0 \in X$ and a neighbourhood U. Recursively select open sets V_i so that $\overline{V_i} \subset U \cap V_{i-1} \cap \bigcap_{j<i} U_j$.
(b) Show that every locally compact Hausdorff space X is Baire.
(c) Show that a G_δ subset Y of a compact Hausdorff space is Baire.
HINT: Assume $Y = \bigcap_{n \ge 1} W_n$ is dense in X. Modify the argument in (a) by alternating the use of U_i and W_i.
(d) Show that a (non-empty) Baire space is not a countable union of closed nowhere dense subsets. (A is *nowhere dense* if $\operatorname{int}(\overline{A}) = \varnothing$.)

1.24 (*Arzela–Ascoli Theorem*). Let X be compact Hausdorff. A subset $\mathcal{F} \subset C(X)$ is *equicontinuous* at $x \in X$ if for each $\varepsilon > 0$, there is an open neighbourhood $U_\varepsilon \ni x$ such that $U_\varepsilon \subset f^{-1}\big(b_\varepsilon(f(x))\big)$ for all $f \in \mathcal{F}$. It is equicontinuous if it is equicontinuous at every point $x \in X$. Prove that $\mathcal{F}$ is a compact subset of $C(X)$ if and only if it is closed, bounded and equicontinuous. Use the following outline.
(a) First show that if $\mathcal{F}$ is compact, then it is closed, bounded and equicontinuous.

(b) $C(X)$ is a metric space, so use sequential compactness. Fix a sequence (f_n) in $\mathcal{F}$. Let $\{x_j : j \geq 1\}$ be a countable dense subset of X. Use boundedness and the diagonal technique to extract a subsequence $(f_{n_i})_{i\geq 1}$ such that $\lim_i f_{n_i}(x_j) = a_j$ exists for all $j \geq 1$.

(c) Fix $x \in X$. For $\varepsilon = \frac{1}{3n}$, let U_n be the open set from equicontinuity at x. Choose $x_{j_n} \in U_n$ and find I_n so that for $i \geq I_n$, $|f_{n_i}(x_{j_n}) - a_{j_n}| < \frac{1}{3n}$. Show that for any $y \in U_n$ that $|f_{n_i}(y) - a_{j_n}| < \frac{1}{n}$ for $i \geq I_n$.

(d) Deduce that $\big(f_{n_i}(x)\big)_{i\geq 1}$ is Cauchy. Call the limit $f(x)$.

(e) Deduce that $f^{-1}\big(b_{1/n}(f(x))\big)$ contains U_n, whence f is continuous.

1.25 (a) Prove that there is an uncountable well-ordered set Ω such that the initial segments $\Omega(a) = \{x \in \Omega : x < a\}$ are countable for every $a \in \Omega$. (This is the first uncountable ordinal.) Endow Ω with the order topology.
HINT: Well-order $\mathbb{R}$ and select an appropriate initial segment.

(b) Show that every increasing sequence in Ω converges.

(c) Show that every sequence in Ω has a convergent subsequence (i.e. Ω is *sequentially compact*).

(d) Show that Ω is not compact.

(e) Show that every continuous function $f : \Omega \to \mathbb{R}$ is eventually constant.
HINT: Show there is an $a_k \in \Omega$ so that if $b > a_k$, then $|f(b) - f(a_k)| < \frac{1}{k}$.

1.26 The *Moore plane* is $\Gamma = \{(x, y) : x \in \mathbb{R},\ y \geq 0\}$ where for each (x, y) with $y > 0$, the usual open balls $B_r((x, y))$ of radius $r \leq y$ are open; and for $(x, 0) \in R = \{(x, 0) : x \in \mathbb{R}\}$, the sets $\{(x, 0)\} \cup B_r((x, r))$ are open.

(a) Show that this determines a Hausdorff topology.

(b) Show that Γ is separable, but that R with the induced topology is not.

(c) Show that Γ is first countable, but not second countable.

(d) Show that $Q = \{(x, 0) : x \in \mathbb{Q}\}$ and $I = \{(x, 0) : x \notin \mathbb{Q}\}$ are disjoint closed sets, but if $Q \subset U$ and $I \subset V$, where U and V are open, then $U \cap V \neq \varnothing$, i.e. Γ is not normal.

(e) Show that if $C \subset \Gamma$ is closed and $p \notin C$, there is a continuous function $f : \Gamma \to [0, 1]$ so that $f|_C = 0$ and $f(p) = 1$. (Γ is completely regular.)

1.27 A filter on $\mathbb{N}$ is a non-empty collection $\mathcal{U}$ of non-empty subsets of $\mathbb{N}$ with the property that if $A, B \in \mathcal{U}$, then $A \cap B \in \mathcal{U}$ and if $C \supset A$, then $C \in \mathcal{U}$.

(a) Use Zorn's lemma to show that there exist a maximal filter $\mathcal{U}$ of $\mathbb{N}$ containing all cofinite sets. (This is known as a *free ultrafilter*.)

(b) Show that for each $A \subset \mathbb{N}$, either $A \in \mathcal{U}$ or $A^c \in \mathcal{U}$.

(c) Show that there is no subsequence of $\mathbb{N}$ such that $\lim_{i\to\infty} x_{n_i}$ exists for all bounded sequences $x = (x_n)$.

(d) For $A \in \mathcal{U}$, let $n_A = \min\{k : k \in A\}$. Prove that n_A is a (cofinal) subnet of the sequence $1, 2, 3, \ldots$ with the property that $\lim_{A\in\mathcal{U}} x_{n_A}$ does exist for all bounded sequences $x = (x_n)$.

Notes on Chapter 1

The beginnings of set theory are due to Cantor. He developed the notion of ordinals in order to classify countable subsets of the real numbers. He is well known for showing that the reals are uncountable. He thought that the cardinality of $\mathbb{R}$ should be the smallest uncountable cardinal number. The Continuum Hypothesis (CH) states that if a set A is uncountable, then its cardinality is at least as great as that of $\mathbb{R}$, namely, $2^{\aleph_0}$.

The standard axioms for set theory were first described by Zermelo and modified by Fraenkel, and are now known as Zermelo–Fraenkel (ZF) set theory. Gödel [128] proved that the Axiom of Choice and the Continuum Hypothesis are consistent with ZF in 1938. In 1963, Cohen [60] proved that ZF was also consistent with the failure of AC and the failure of CH, and that ZFC (ZF+AC) is consistent with the failure of CH. These works of Gödel and Cohen are two of the major results in set theory. See Jech [160] for more information. For a nice introduction to the basics, see Halmos [136].

Before the introduction of abstract topology, Fréchet developed a large part of metric space theory. In 1914, Hausdorff published a book called *Grundzüge der Mengenlehre* (Basics of Set Theory) which was hugely influential. He is credited with the transition to the modern axiomatic style, and to the view that all areas of mathematics are interconnected. Abstract point set topology was a part of Hausdorff's book which was completely new. His axioms included the use of neighbourhood systems and the T_2 property. Product topologies were introduced by Tychonoff [307]. Urysohn was a leader of the Russian school of topology, and, after his lemma [309], is best known for conditions that determine when a topological space is metrizable. The Tietze extension theorem was in [304].

The material covered here is standard and is contained in various books on topology. Willard [325] is an excellent source for point set topology. Kelley [176] is a classic textbook on the subject. Manheim [197] is an interesting book on the history.

CHAPTER 2

Banach Spaces

Completeness of various normed vector spaces is the foundation of functional analysis. We shall see that this property is critical at many points of the theory. For example, completeness is critical to the results which rely on the Baire category theorem. The study of linear functionals is basic, and gives functional analysis its name. The Hahn–Banach theorem is the cornerstone result.

2.1. Examples

This short section defines Banach spaces and provides a few examples. This book is full of many others.

2.1.1. DEFINITION. A *norm* on a vector space V over a field $\mathbb{F} \in \{\mathbb{R}, \mathbb{C}\}$ is a function $\|\cdot\| : V \to [0, \infty)$ such that

(1) $\|x\| = 0 \iff x = 0$ (definite).

(2) $\|\lambda x\| = |\lambda|\,\|x\|$ for all $\lambda \in \mathbb{F}$ and $x \in V$ (homogeneous).

(3) $\|x + y\| \leq \|x\| + \|y\|$ (triangle inequality).

A normed space is a metric space with distance function $d(x, y) := \|x - y\|$. A *Banach space* is a complete normed vector space (i.e. complete is this metric).

2.1.2. EXAMPLE. $C(X)$ is the space of continuous complex-valued functions on a compact Hausdorff space X with $\|f\|_\infty = \sup_{x\in X} |f(x)|$. A sequence is Cauchy in the sup norm precisely when it converges uniformly. The uniform limit of continuous functions is continuous by Proposition 1.6.8. This leads to Proposition 1.6.9 that $C(X)$ is complete. Variants include $C_{\mathbb{R}}(X)$ the real vector space of real-valued continuous functions.

Also if X is a locally compact Hausdorff space, then $C_0(X)$ consists of continuous functions on X such that $K_\varepsilon = \{x : |f(x)| \geq \varepsilon\}$ is compact for all $\varepsilon > 0$. This is also a Banach space with the sup norm. Note that the space $C_c(X)$ of continuous functions on X with compact support, where $\operatorname{supp}(f) = \overline{\{x : f(x) \neq 0\}}$, is a normed vector space with the sup norm, but it is not complete unless X is compact. The completion of $C_c(X)$ in the sup norm is $C_0(X)$.

K. R. Davidson, *Functional Analysis and Operator Algebras*, CMS/CAIMS Books in Mathematics 13,
https://doi.org/10.1007/978-3-031-63665-3_2

Two very important examples in this family of spaces are

$$c_0 = \{(a_i)_{i\ge1} : \lim_{i\to\infty} a_i = 0\} \quad \text{and} \quad c = \{(a_i)_{i\ge1} : \lim_{i\to\infty} a_i \text{ exists}\}$$

with $\|(a_i)\|_\infty = \sup_{i\ge1} |a_i|$.

2.1.3. EXAMPLE. For $1 \le p < \infty$,

$$l_p = \{(a_i)_{i\ge1} : \|(a_i)\|_p = \Big(\sum_{i\ge1} |a_i|^p\Big)^{1/p} < \infty\}$$

and $l_\infty = \{(a_i)_{i\ge1} : \|(a_i)\|_\infty = \sup|a_i| < \infty\}$. Likewise if $I \subset \mathbb{R}$ is an interval (finite or infinite), one can put the p-norm on $C_c(I)$ by $\|f\|_p = \Big(\int_I |f(x)|^p\,dx\Big)^{1/p}$. The completion is called $L^p(I)$ for $1 \le p < \infty$. Here the measure involved is Lebesgue measure on I.

In greater generality, if μ is a measure on a measure space $(X, \mathcal{B})$, then "$L^p(\mu)$" consists of all measurable functions on X such that $\|f\|_p = \big(\int_I |f|^p\,d\mu\big)^{1/p} < \infty$. Then we define $L^p(\mu) =$ "$L^p(\mu)$"$/\mathcal{N}$ where $\mathcal{N} = \{f : f = 0 \text{ a.e.}(\mu)\}$. That is, elements of $L^p(\mu)$ are equivalence classes of functions. This is necessary so that the norm is definite. The space $L^\infty(\mu)$ consists of all bounded measurable functions modulo $\mathcal{N}$ with

$$\|f\|_\infty = \text{ess sup}|f| = \sup\{r \ge 0 : \mu(|f| > r) > 0\}.$$

In all of these cases, the triangle inequality is called Minkowski's inequality (see Exercise **2.1**). Completeness is a theorem of Riesz usually established in measure theory courses. (See Exercise **2.2** for l_p.)

When $1 \le p < \infty$, define q so that $\frac{1}{p} + \frac{1}{q} = 1$ ($q = \infty$ for $p = 1$). Hölder's inequality (see Exercise **2.3**) states that if $f \in L^p(\mu)$ and $g \in L^q(\mu)$, then

$$\Big|\int fg\,d\mu\Big| \le \|fg\|_1 \le \|f\|_p\,\|g\|_q.$$

2.1.4. EXAMPLE. $C^{(n)}[a,b] = \{f(x) : f \text{ has } n \text{ continuous derivatives on } [a,b]\}$. Define $\|f\|_{C^{(n)}} = \sum_{k=0}^n \|f^{(k)}\|_\infty$. In this norm, a sequence $f_n \to f$ if and only if $f_n^{(k)}$ converges uniformly to $f^{(k)}$ for $0 \le k \le n$. The completeness of $C^{(n)}[a,b]$ follows from the fact that the uniform limit of continuous functions is continuous, and the fact that the Riemann integral of a uniform limit is the uniform limit of the integral, combined with the Fundamental Theorem of Calculus. See Exercise **2.4**.

2.1.5. EXAMPLE. Hilbert spaces. An inner product space is a vector space V over $\mathbb{F} \in \{\mathbb{R}, \mathbb{C}\}$ with a positive definite inner product. That is, $\langle\cdot,\cdot\rangle : V \times V \to \mathbb{F}$ satisfies

(1) $\langle x, x\rangle \ge 0$ for $x \in V$ with equality only when $x = 0$ (positive definite).

(2) $\langle \lambda x + \mu y, z\rangle = \lambda\langle x, z\rangle + \mu\langle y, z\rangle$ for $x, y, z \in V$ and $\lambda, \mu \in \mathbb{F}$ (linear in first variable).

(3) $\langle y, x\rangle = \overline{\langle x, y\rangle}$ for $x, y \in V$.

From (2) and (3), one sees that inner products are conjugate linear in second variable

$$\langle x, \lambda y + \mu z\rangle = \overline{\lambda}\langle x, y\rangle + \overline{\mu}\langle x, z\rangle \text{ for } x, y, z \in V \text{ and } \lambda, \mu \in \mathbb{F}.$$

This is called a *sesquilinear form* because it is linear in the first variable and conjugate linear in the second. A norm is defined by $\|x\| = \langle x, x\rangle^{1/2}$. A complete inner product space is called a *Hilbert space* .

The Cauchy–Schwarz inequality is $|\langle x, y\rangle| \le \|x\|\,\|y\|$, which is valid in all inner product spaces. To see this, note that

$$0 \le \langle x - \lambda y, x - \lambda y\rangle = \|x\|^2 - 2\,\mathrm{Re}(\overline{\lambda}\langle x, y\rangle) + |\lambda|^2\|y\|^2.$$

Substitute $\lambda = \langle x, y\rangle/\|y\|^2$ and simplify to get $|\langle x, y\rangle| \le \|x\|\,\|y\|$. The triangle inequality follows from

$$\begin{aligned}\|x + y\|^2 &= \langle x + y, x + y\rangle = \|x\|^2 + 2\,\mathrm{Re}\langle x, y\rangle + \|y\|^2\\ &\le \|x\|^2 + 2\|x\|\,\|y\| + \|y\|^2 = (\|x\| + \|y\|)^2.\end{aligned}$$

Examples of Hilbert spaces include l_2, $L^2(I)$ and for any measure, one obtains $L^2(\mu)$ with $\langle f, g\rangle = \displaystyle\int f\bar{g}\,d\mu$ and $\|f\|^2 = \displaystyle\int |f|^2\,d\mu$.

A less obvious example is the following. Let Ω be a bounded open subset of $\mathbb{C}$. Define $L^2_a(\Omega)$ to be the vector space of all analytic functions on Ω such that

$$\|f\|_2 = \Big(\int_\Omega |f(z)|^2\,dA\Big)^{1/2} < \infty$$

where $dA = dx\,dy$ is planar measure on Ω. Define $\langle f, g\rangle = \displaystyle\int_\Omega f(z)\overline{g(z)}\,dA$. The non-obvious fact is completeness.

Let $z_0 \in \Omega$. There is some $\rho > 0$ so that $\overline{b_\rho(z_0)} \subset \Omega$. Then using polar coordinates ($dA = r\,dr\,d\theta$)

$$\int_{b_\rho(z_0)} f(z)\,dA = \int_0^\rho\int_0^{2\pi} f(z_0 + re^{i\theta})d\theta\,r\,dr = \int_0^\rho 2\pi f(z_0)d\theta\,r\,dr = \pi\rho^2 f(z_0).$$

Therefore by Cauchy–Schwarz,

$$\begin{aligned}|f(z_0)| &= \Big|\frac{1}{\pi\rho^2}\int_{b_\rho(z_0)} f(z)\,dA\Big|\\ &\le \frac{1}{\pi\rho^2}\Big(\int_{b_\rho(z_0)} |f(z)|^2\,dA\Big)^{1/2}\Big(\int_{b_\rho(z_0)} 1\,dA\Big)^{1/2}\\ &\le \frac{1}{\pi\rho^2}\|f\|_2\sqrt{\pi\rho^2} = \frac{\|f\|_2}{\sqrt{\pi}\rho}.\end{aligned}$$

This says that evaluation of $f \in L^2_a(\Omega)$ at z_0 is continuous. (This is false for $L^2(I)$ where elements are only defined a.e.)

Now suppose that f_n is a Cauchy sequence in $L^2_a(\Omega)$. So for $\varepsilon > 0$, there is an N so that for $m, n \geq N$, $\|f_m - f_n\|_2 < \varepsilon$. Hence for $\rho = \frac{1}{k}$,

$$z \in \Omega_k = \{z \in \Omega : \operatorname{dist}(z, \Omega^c) \geq \tfrac{1}{k}\},$$

we have $|f_m(z) - f_n(z)| \leq k\varepsilon/\sqrt{\pi}$. This means that $f_n(z)$ is uniformly Cauchy on Ω_n, and thus converges uniformly there to a function $f(z)$. That is, $f_n \to f$ uniformly on compact subsets of Ω. Therefore the limit $f(z)$ is analytic. It is routine to check that $\|f\|_2 = \lim_{n\to\infty} \|f_n\|_2 < \infty$. So $f \in L^2_a(\Omega)$ and $f_n \to f$ in the $L^2_a(\Omega)$ norm.

2.2. Constructions of Banach Spaces

In this section, we will look at several constructions for building new Banach spaces from known examples. We look at subspaces, quotient spaces and direct sums.

The following easy proposition is basic to functional analysis, providing a simple characterization of continuous linear maps.

2.2.1. PROPOSITION. *Let X and Y be normed vector spaces, and let $T : X \to Y$ be a linear map. The following are equivalent:*

(1) $\|T\| := \sup_{\|x\|\leq 1} \|Tx\| < \infty$. *I.e. T is* bounded.

(2) *T is Lipschitz, and hence uniformly continuous.*

(3) *T is continuous.*

(4) *T is continuous at $x = 0$.*

PROOF. (1) $\Rightarrow$ (2). $\|Tx_1 - Tx_2\| = \|T(x_1 - x_2)\| \leq \|T\| \, \|x_1 - x_2\|$. So T is Lipschitz with constant $\|T\|$. In particular, given $\varepsilon > 0$, if $\|x_1 - x_2\| < \varepsilon/\|T\|$, then $\|Tx_1 - Tx_2\| < \varepsilon$. So T is uniformly continuous.

(2) $\Rightarrow$ (3) $\Rightarrow$ (4) is trivial.

(4) $\Rightarrow$ (1) is equivalent to "not (1)" $\Rightarrow$ "not (4)". So suppose that $\|T\| = \infty$. Then there are $x_n \in X$ with $\|x_n\| \leq 1$ and $\|Tx_n\| > n^2$. Hence $\frac{1}{n}x_n \to 0$ but $\|T(\frac{1}{n}x_n)\| \to \infty$. So T is not continuous at $x = 0$. ∎

2.2.2. DEFINITION. Let X and Y be normed vector spaces. Then the space of *bounded linear operators* from X to Y is

$$\mathcal{B}(X, Y) = \{T : X \to Y \text{ continuous linear maps}\} \text{ with } \|T\| = \sup_{\|x\|\leq 1} \|Tx\|.$$

When $Y = X$, we write $\mathcal{B}(X)$ for $\mathcal{B}(X, X)$. When $Y = \mathbb{F}$ is the scalar field for X, we write X^* for $\mathcal{B}(X, \mathbb{F})$. This is called the *dual space* of X. Elements of X^* are called (continuous linear) *functionals* .

It is easy to check that $\|\cdot\|$ is a norm on $\mathcal{B}(X, Y)$. This is left to the reader.

2.2.3. PROPOSITION. *Let X be a normed vector space and let Y be a Banach space. Then $\mathcal{B}(X, Y)$ is a Banach space.*

PROOF. We will verify completeness. Let $(T_n)_{n\geq 1}$ be a Cauchy sequence in $\mathcal{B}(X, Y)$. For $\varepsilon > 0$, there is an N_ε so that if $N_\varepsilon \leq n < m$, then $\|T_m - T_n\| < \varepsilon$. Hence for each $x \in X$,

$$\|T_m x - T_n x\| \leq \|T_m - T_n\| \, \|x\| < \varepsilon\|x\|.$$

This means that $(T_n x)$ is a Cauchy sequence in Y for every $x \in X$. Since Y is complete, we may define $Tx := \lim_{n\to\infty} T_n x$. It is easy to check that T is linear. Moreover if $n \geq N_\varepsilon$ and $\|x\| \leq 1$,

$$\|Tx - T_n x\| = \lim_{m\to\infty} \|T_m x - T_n x\| \leq \varepsilon.$$

It follows that $\|T\| \leq \|T_{N_\varepsilon}\| + \varepsilon < \infty$ and that $\|T - T_n\| \to 0$ as $n \to \infty$. Thus $\mathcal{B}(X, Y)$ is complete. ∎

2.2.4. THEOREM. *Let $1 < p < \infty$ and let q satisfy $\frac{1}{p} + \frac{1}{q} = 1$. Then the map taking $y = (y_j) \in l_q$ to the linear functional $\varphi_y(x) = \sum_{j\geq 1} x_j y_j$ is an isometric map of l_q onto l_p^*.*

PROOF. Let e_j denote the element of l_p with a 1 in the jth entry and 0s elsewhere. If $y = (y_j) \in l_q$ and $x = \sum_{j=1}^N x_j e_j$, then $\varphi_y(x) = \sum_{j=1}^N x_j y_j$. By Hölder's inequality (see Exercise **2.3**),

$$|\varphi_y(x)| \leq \sum_{j=1}^{N} |x_j| \, |y_j| \leq \|x\|_p \, \|y\|_q.$$

This shows that $\|\varphi_y\| \leq \|y\|_q$ on the span$\{e_j : j \geq 1\}$, and so it extends by continuity to all of l_p. Moreover, $\varphi_y(x) = \sum_{j=1}^{\infty} x_j y_j$ converges absolutely for all $x \in l_p$. This shows that $\|\varphi_y\| \leq \|y\|_q$. Taking $x_j = \overline{\text{sign}(y_j)}|y_j|^{q/p}$, we obtain

$$\|x\|_p^p = \sum_{j\geq 1} |y_j|^q = \|y\|_q^q < \infty;$$

so that $x \in l_p$ with $\|x\|_p = \|y\|_q^{q/p}$. Evaluating φ_y at $x/\|x\|_p$ shows

$$\|\varphi_y\| \geq \frac{|\varphi_y(x)|}{\|x\|_p} = \frac{1}{\|y\|_q^{q/p}} \sum_{j\geq 1} |y_j|^{q/p} |y_j|^{q/q}$$
$$= \frac{1}{\|y\|_q^{q/p}} \sum_{j\geq 1} |y_j|^q = \|y\|_q^{q(1-1/p)} = \|y\|_q.$$

Therefore the map $y \to \varphi_y$ is an isometric map of l_q into l_p^*.

If $\varphi \in l_p^*$, let $y_j = \varphi(e_j)$. Again we define $x_j = \overline{\operatorname{sign}(y_j)}|y_j|^{q/p}$. For any finite N, we have $u_N = \sum_{j=1}^N x_j e_j \in l_p$ and $v_N = \sum_{j=1}^N y_j f_j \in l_q$, where f_j is the sequence in l_q with a 1 in the jth entry and 0s elsewhere. By the previous paragraph, $\|u_N\|_p = \|v_N\|_q^{q/p}$ and

$$\|\varphi\| \geq \frac{|\varphi(u_N)|}{\|u_N\|_p} = \|v_N\|_q.$$

Hence $y = (y_j)$ satisfies

$$\|y\|_q = \lim_{N\to\infty} \|v_N\|_q \leq \|\varphi\| < \infty.$$

So $y \in l_q$. Moreover for any $x = \sum_{j=1}^N x_j e_j \in \operatorname{span}\{e_j : j \geq 1\}$, we have

$$\varphi(x) = \sum_{j=1}^N x_j \varphi(e_j) = \sum_{j=1}^N x_j y_j = \varphi_y(x).$$

The two continuous functionals φ and φ_y agree on a dense subset of l_p, and thus they are equal. This shows that the isometry from l_q into l_p^* is surjective. ■

2.2.5. REMARKS. Theorem 2.2.4 is a very special case of a theorem of Riesz which is established in a course on measure theory. Namely that if $1 < p < \infty$, then $L^p(\mu)^* = L^q(\mu)$ for arbitrary measures. This is also true for $p = 1$ (with $q = \infty$) provided that μ is σ-finite.

The other big theorem on dual spaces from measure theory is the Riesz representation theorem. If X is a locally compact Hausdorff space, then $C_0(X)^* = M(X)$, the space of complex Radon measures on X. When X is compact, these are the regular complex Borel measures.

2.2.6. EXAMPLE. $c_0^* = l_1$. Again let e_j denote the sequence with a 1 in the jth coordinate and 0s elsewhere, and let f_j be the corresponding sequence in l_1. If $\varphi \in c_0^*$, let $a_j = \varphi(e_j)$. Then

$$\|\varphi\| \geq \varphi\left(\tfrac{\bar{a}_1}{|a_1|}, \dots, \tfrac{\bar{a}_j}{|a_j|}, \dots, \tfrac{\bar{a}_n}{|a_n|}, 0, 0 \dots, \right) = \sum_{j=1}^n |a_j|.$$

Letting $n \to \infty$, we deduce that $y = (a_j)$ belongs to l_1 and $\|y\|_1 \le \|\varphi\|$. Moreover $\varphi(x) = \varphi_y(x) = \sum_{j\ge1} x_j a_j$. Conversely, each $y \in l_1$ determines a linear functional with $\|\varphi_y\| = \|y\|_1$.

Now consider $\mathcal{B}(c_0)$. If $T \in \mathcal{B}(c_0)$, define an infinite matrix $[t_{ij}]$ by $Te_j = \sum_{i\ge1} t_{ij} e_i$. The columns Te_j belong to c_0, so $\lim_{i\to\infty} t_{ij} = 0$. Also the row $(t_{i1}, t_{i2}, t_{i3}, \dots)$ acts on a vector $x \in c_0$ by $\varphi_i(x) = \sum_{j\ge1} t_{ij} x_j = \varphi_{f_i}(Tx)$. This determines the ith coordinate of $Tx = \sum_{i\ge1} \varphi_i(x) e_i$. So each $\varphi_i = \varphi_{f_i} \circ T$ is a continuous linear functional on c_0, so there is a vector $y_i \in l_1$ such that $\varphi_i = \varphi_{y_i}$. Moreover $\|\varphi_i\| = \|y_i\|_1 \le \|T\|$. In fact,

$$\mathcal{B}(c_0) = \{T = [t_{ij}] : (t_{ij})_{i\ge1} \in c_0 \text{ for } j \ge 1 \text{ and } \sup_{i\ge1} \sum_{j\ge1} |t_{ij}| < \infty\}.$$

To see this, note that the boundedness of the rows in the l_1 norm ensures that T is a bounded linear map from c_0 into l_∞. The condition on the columns ensures that $Te_j \in c_0$ for $g \ge 1$. So $T \operatorname{span}\{e_j : j \ge 1\} \subset c_0$. The subspace $\operatorname{span}\{e_j : j \ge 1\}$ is dense in c_0 and T is continuous. Thus, since c_0 is closed in l_∞, the range of T is contained in c_0. Therefore $T \in \mathcal{B}(c_0)$.

This is a rather special property of c_0. There are very few Banach spaces where the space of bounded linear maps can be simply characterized.

An easy way to obtain a new Banach space from another is to take a *closed subspace* , which as the term suggests is a subspace which is norm closed. Since a closed subset of a complete metric space is complete, a closed subspace of a Banach space is a Banach space.

2.2.7. EXAMPLE. Disc algebra. Let $\mathbb{T} = \{e^{i\theta} : \theta \in [-\pi, \pi]\}$ be the unit circle in $\mathbb{C}$. The (complex) Fourier coefficients of $f \in C(\mathbb{T})$ are given by

$$\hat{f}(n) = \frac{1}{2\pi} \int_{-\pi}^{\pi} f(e^{i\theta}) e^{-in\theta}\, d\theta \quad \text{for} \quad n \in \mathbb{Z}.$$

Define the *disc algebra* as $A(\mathbb{D}) = \{f \in C(\mathbb{T}) : \hat{f}(n) = 0 \text{ for } n < 0\}$. The linear functionals $\varphi_n(f) = \hat{f}(n)$ are continuous since $|\hat{f}(n)| \le \|f\|_\infty$, so $\|\varphi_n\| \le 1$. They are exactly norm 1 since $\|e^{in\theta}\|_\infty = 1$ and $\varphi_n(e^{in\theta}) = 1$. Hence the kernel, $\ker \varphi_n$, is a closed subspace. It follows that $A(\mathbb{D}) = \bigcap_{n<0} \ker \varphi_n$ is a closed subspace of $C(\mathbb{T})$.

Every continuous function $f \in C(\mathbb{T})$ extends uniquely to a harmonic function on the open disc by

$$\begin{aligned}\tilde{f}(re^{i\theta}) &= \frac{1}{2\pi} \int_{-\pi}^{\pi} f(e^{it}) P_r(\theta - t)\, dt \\ &= \frac{1}{2\pi} \int_{-\pi}^{\pi} f(e^{it}) \frac{1 - r^2}{1 - 2r\cos(\theta - t) + r^2}\, dt = \sum_{n=-\infty}^{\infty} \hat{f}(n) r^{|n|} e^{in\theta}.\end{aligned}$$

Poisson's theorem shows that $\widetilde{f}(re^{i\theta})$ converges uniformly to $f(e^{i\theta})$ as $r \to 1$. (See Theorem 11.1.10 (1).) Hence $\widetilde{f}$ extends to be continuous on the closed disc, and coincides with f on the boundary circle. For $f \in A(\mathbb{D})$, all negative Fourier coefficients vanish. Thus, writing $z = re^{i\theta}$, we get that $\widetilde{f}(z) = \sum_{n\geq 0} \hat{f}(n)z^n$. This power series converges on the open unit disc $\mathbb{D}$, and converges uniformly when $|z| \leq r < 1$. So for $f \in A(\mathbb{D})$, $\widetilde{f}(z)$ is analytic on $\mathbb{D}$ and continuous on $\overline{\mathbb{D}}$. Conversely every function which has these properties has a power series which converges on the open disc. By Cauchy's theorem, $f(z)$ for $z \in \mathbb{D}$ is determined by the boundary values. Indeed, analytic functions are harmonic, so this is the Poisson extension of the boundary values. It is clear that each $f_r(e^{i\theta}) := \widetilde{f}(re^{i\theta})$ lies in $A(\mathbb{D})$ since the Fourier series converges uniformly. Thus (f_r) converges uniformly to f as $r \to 1^-$ by continuity on $\overline{\mathbb{D}}$, and so $f \in A(\mathbb{D})$.

The next construction in this section is the notion of a quotient space.

2.2.8. Definition. Let X be a Banach space and let M be a closed subspace. Then X/M is a vector space. Define a norm on X/M by

$$\|x + M\| = \inf\{\|x - m\| : m \in M\}.$$

2.2.9. Proposition. *Let X be a Banach space and let M be a closed subspace. Then X/M is a Banach space.*

Proof. First check that this is a bona fide norm. Note that $\|x + M\| = 0$ only if there are $m_n \in M$ with $\|x - m_n\| \to 0$. That means that x lies in $\overline{M} = M$. So $x + M = \dot{0}$ is the zero element. Positive homogeneity and the triangle inequality are straightforward.

For completeness, let $\dot{x}_n = x_n + M$ for $n \geq 1$ be a Cauchy sequence in X/M. Find a subsequence $\dot{x}_{n_i}$ so that $\|\dot{x}_{n_i} - \dot{x}_{n_{i+1}}\| < 2^{-i}$. Recursively choose $m_i \in M$ so that $x'_{n_i} = x_{n_i} - m_i$ satisfy $\|x'_{n_i} - x'_{n_{i+1}}\| < 2^{-i}$. Then $(x'_{n_i})_{i\geq 1}$ is Cauchy in X. Let x be the limit. Then it is clear that $\dot{x}_{n_i}$ converge to $\dot{x}$. A standard metric space argument now shows that the whole sequence converges to $\dot{x}$. Hence X/M is complete, and thus is a Banach space. ■

2.2.10. Proposition. *Let X be a Banach space and let M be a proper closed subspace. Define a linear map $Q : X \to X/M$ by $Qx = \dot{x}$. Then $Q \in \mathcal{B}(X, X/M)$ is surjective, $\|Q\| = 1$ and* $\ker Q = M$.

Proof. It is clear that Q is a linear surjection with $\ker Q = M$. Moreover, since $\|\dot{x}\| \leq \|x\|$, we have $\|Q\| \leq 1$. Since M is proper, pick $x \in X \setminus M$; so that $\dot{x} \neq \dot{0}$. Replace x by $x/\|\dot{x}\|$ so that $1 = \|\dot{x}\| = \inf\{\|x - m\| : m \in M\}$. Choose

$m_n \in M$ so that $\|x - m_n\| \to 1$. Then $\|Q\| \geq \sup_{n\geq 1} \frac{\|\dot{x}\|}{\|x-m_n\|} = 1$. Therefore $\|Q\| = 1$. ■

The map $Q : X \to X/M$ is called the *quotient map* .

2.2.11. EXAMPLE. Let X be a compact Hausdorff space, and let E be a closed subset of X. Then $I(E) = \{f \in C(X) : f|_E = 0\}$ is a closed subspace (and, in fact, an ideal). Consider $C(X)/I(E)$.

Let $h \in C(X)$. Then

$$\|\dot{h}\| = \inf\{\|h - f\|_\infty : f|_E = 0\} \geq \sup_{x\in E} |h(x)| = \|h|_E\|_\infty.$$

On the other hand, suppose that $\|h|_E\|_\infty = 1$. Define a continuous function $g : \mathbb{C} \to \overline{\mathbb{D}}$ by $g(z) = \begin{cases} z & \text{if } |z| \leq 1 \\ \frac{z}{|z|} & \text{if } |z| > 1 \end{cases}$. Then $g(h(x))$ has norm 1, and $g \circ h|_E = h|_E$. Then $f = h - g \circ h$ vanishes on E, so $f \in I(E)$. Moreover $h - f = g$, so that $\|\dot{h}\| \leq \|g\|_\infty = 1$. We deduce that $\|\dot{h}\| = \|h|_E\|_\infty$.

Let $Q : C(X) \to C(X)/I(E)$ be the quotient map. Define a linear map $T : C(X)/I(E) \to C(E)$ by $T\dot{h} = h|_E$. This makes sense since $h - f|_E = h|_E$ for all $f \in I(E)$. Moreover, we have just shown that $\|T\dot{h}\| = \|\dot{h}\|$ for all $h \in C(X)/I(E)$. That is, T is an *isometry* . Observe that $TQ = R$, where R is the restriction map $Rh = h|_E$ from $C(X)$ onto $C(E)$. Also T maps onto $C(E)$ by Tietze's extension theorem 1.9.7. So $C(X)/I(E)$ is *isometrically isomorphic* to $C(E)$.

The final construction in this section is the notion of a direct sum.

2.2.12. DEFINITION. Let X and Y be Banach spaces. Consider the *direct sum* $X \oplus Y = \{(x, y) : x \in X, y \in Y\}$. For $1 \leq p \leq \infty$, let $X \oplus_p Y$ denote the direct sum with the norm $\|(x, y)\|_p = \left(\|x\|^p + \|y\|^p\right)^{1/p}$ if $p < \infty$ and $\|(x, y)\|_\infty = \max\{\|x\|, \|y\|\}$.

The following result is straightforward, and the proof is left to the reader.

2.2.13. PROPOSITION. *Let X and Y be Banach spaces. Then $X \oplus_p Y$ is a Banach space for $1 \leq p \leq \infty$. Moreover*

$$\|(x, y)\|_\infty \leq \|(x, y)\|_p \leq \|(x, y)\|_1 \leq 2\|(x, y)\|_\infty.$$

2.2.14. DEFINITION. We say that two norms on X are *equivalent* if there are constants $0 < C_1 \leq C_2 < \infty$ so that $C_1\|x\|_2 \leq \|x\|_1 \leq C_2\|x\|_2$ for all $x \in X$.

2.2.15. REMARK. When two norms on X are equivalent, the identity map *id* : $(X, \|\cdot\|_1) \to (X, \|\cdot\|_2)$ is continuous with $\|\, id\,\| \le C_1^{-1}$ and id^{-1} is continuous with $\|\, id^{-1}\,\| \le C_2$. Conversely, if *id* and id^{-1} are continuous, then the two norms are equivalent. The topologies arising from the two norms are the same. Moreover, they have the same Cauchy sequences.

2.2.16. PROPOSITION. *Let X be a finite dimensional vector space. Then any two norms on X are equivalent. In particular, X is complete in any norm. So every finite dimensional subspace of a Banach space is closed.*

PROOF. Let $\mathbb{F}$ be the field of scalars for X. Fix a basis $x_1, \dots, x_n$ for X, and consider a norm $\|\cdot\|$ on X. Define a linear isomorphism $T : (\mathbb{F}^n, \|\cdot\|_2) \to X$ by $T\sum_{i=1}^n a_i e_i = \sum_{i=1}^n a_i x_i$. Then for $u = \sum_{i=1}^n a_i e_i \in \mathbb{F}^n$,

$$\|Tu\| = \Big\|\sum_{i=1}^n a_i x_i\Big\| \le \sum_{i=1}^n |a_i|\,\|x_i\| \le \|u\|_2 \Big(\sum_{i=1}^n \|x_i\|^2\Big)^{1/2} = C\|u\|_2$$

by the Cauchy–Schwarz inequality. Therefore T is a continuous linear bijection.

Let $S = \{u \in \mathbb{F}^n : \|u\|_2 = 1\}$ be the unit sphere. This is compact, and therefore TS is compact in X. As $0 \notin TS$, the continuous function $f(x) = \|x\|$ is never 0 on TS. By compactness, it attains a minimum value on TS, say $\min_{u\in S}\|Tu\| = r > 0$. It follows that $T^{-1}b_r(X) \subset b_1(\mathbb{F}^n)$; for if $\|T^{-1}x\| = t \ge 1$, then $x/t \in b_r(X)$ and $T^{-1}(x/t)$ would belong to S. So $\|T^{-1}\| \le \frac{1}{r}$. Therefore T is a continuous linear isomorphism, and hence is a biLipschitz homeomorphism. It follows that $(X, \|\cdot\|)$ is complete, and this norm is equivalent to the Euclidean norm $\|x\|_2 := \|T^{-1}x\|_2$.

If $|\!|\!|\cdot|\!|\!|$ is another norm on X, then it is also equivalent to $\|\cdot\|_2$, and thus is equivalent to $\|\cdot\|$. If X is a finite dimensional subspace of a Banach space Y, then X is complete in this norm, and hence it is closed in Y. ∎

2.2.17. PROPOSITION. *The closed unit ball of a Banach space is compact if and only if it is finite dimensional.*

PROOF. If X has finite dimension n, then the norm is equivalent to the Euclidean norm on $\mathbb{F}^n$. Thus every closed bounded set is compact.

If X has infinite dimension, we can recursively choose a sequence of unit vectors $(x_n)_{n\ge1}$ so that $\operatorname{dist}(x_{n+1}, \operatorname{span}\{x_1, \dots, x_n\}) > \frac{1}{2}$. Suppose that $x_1, \dots, x_n$ are chosen, and let $M_n = \operatorname{span}\{x_1, \dots, x_n\}$. Then M_n is closed, and X/M_n is an infinite dimensional Banach space. Let $\dot{x} = x + M_n \in X/M_n$ with

$$\tfrac{1}{2} < \|\dot{x}\| = \inf_{m\in M_n} \|x + m\| < 1.$$

Choose $x + m$ so that $\|x + m\| < 1$, and define $x_{n+1} = \frac{x+m}{\|x+m\|}$. Then there is no finite $\frac{1}{4}$-net in the unit ball since a point can be within $\frac{1}{4}$ of only a single x_n. Therefore $\overline{b_1(X)}$ is not compact. ∎

2.3. Hilbert spaces

In this section, we develop some of the special structures enjoyed by Hilbert spaces (Example 2.1.5). This makes them the centre of attention for much of this book. Some of these results are likely familiar from linear algebra. Many arguments are the same, but one has to take their infinite dimensional nature into account.

2.3.1. Definition. In an inner product space, $x \perp y$ means $\langle x, y\rangle = 0$ and the vectors are *orthogonal* . A set $\{e_\alpha : \alpha \in A\}$ is *orthonormal* if $\langle e_\alpha, e_\beta\rangle = \delta_{\alpha\beta}$. If $S \subset H$, then $S^\perp = \{x \in H : x \perp y \text{ for all } y \in S\}$.

If $x \perp y$, then $\|x + y\|^2 = \|x\|^2 + \|y\|^2$, the Pythagorean law. Therefore if $e_1, \dots, e_n$ are orthonormal, then $\big\| \sum_{i=1}^n a_i e_i \big\|^2 = \sum_{i=1}^n |a_i|^2$.

2.3.2. Definition. A series $\sum_{\alpha\in A} a_\alpha e_\alpha$ *converges unconditionally* in a Hilbert space if for the upward directed poset Λ of finite subsets $F \Subset A$ (ordered by inclusion), the net $x_F = \sum_{\alpha\in F} a_\alpha e_\alpha$ converges.

2.3.3. Theorem. *Let H be a Hilbert space, let $\{e_\alpha : \alpha \in A\}$ be a non-empty orthonormal set in H, and let $M = \overline{\text{span}}\{e_\alpha : \alpha \in A\}$. Then*

(1) *The set $\{e_\alpha : \alpha \in A\}$ is linearly independent, and*

$$\mathit{dist}\big(e_\alpha, \overline{\text{span}\{e_\beta : \beta \neq \alpha\}}\big) = 1.$$

(2) *(Bessel's inequality) If $x \in H$ and $a_\alpha = \langle x, e_\alpha\rangle$, then $\sum_{\alpha\in A} |a_\alpha|^2 \le \|x\|^2$.*

(3) *If $a_\alpha \in \mathbb{C}$ and $\sum_{\alpha\in A} |a_\alpha|^2 = L^2 < \infty$, then $\sum_{\alpha\in A} a_\alpha e_\alpha$ converges unconditionally to a vector x with $\|x\| = L$. Moreover $\langle x, e_\alpha\rangle = a_\alpha$.*

(4) *If $x \in H$, let $a_\alpha = \langle x, e_\alpha\rangle$ and define $Px = \sum_{\alpha\in A} a_\alpha e_\alpha$. Then P is in $\mathcal{B}(H)$ with $\|P\| = 1$, $P^2 = P$ and $P(H) = M$.*

(5) ker $P = M^\perp$; *$Px = x$ if and only if $x \in M$, and $M + M^\perp = H$.*

(6) *(Parseval's identity): If $y \in M$, then $\|y\| = \big(\sum_{\alpha\in A} |\langle y, e_\alpha\rangle|^2\big)^{1/2}$.*

The operator P constructed here is the *orthogonal projection* of H onto M.

PROOF. (1) Elements of span$\{e_\alpha : \alpha \in A\}$ (before closure) have the form $\sum_{\beta\in F} a_\beta e_\beta$ for $F \Subset A$ finite. By the Pythagorean law, if $F \subset A \setminus \{\alpha\}$ is finite,

$$\Big\| e_\alpha - \sum_{\beta\in F} a_\beta e_\beta \Big\|^2 = 1 + \sum_{\beta\in F} |a_\beta|^2.$$

Thus the minimum value, 1, is obtained when $a_\beta = 0$ for all $\beta \neq \alpha$. This extends to the closure by continuity of the norm. In particular, $\{e_\alpha : \alpha \in A\}$ are linearly independent.

(2) For finite $F \Subset A$, let $x_F = \sum_{\alpha\in F} a_\alpha e_\alpha$. Then

$$\begin{aligned} 0 \le \|x - x_F\|^2 &= \|x\|^2 - 2\,\mathrm{Re}\langle x, x_F\rangle + \|x_F\|^2 \\ &= \|x\|^2 - 2\sum_{\alpha\in F} |a_\alpha|^2 + \sum_{\alpha\in F} |a_\alpha|^2 = \|x\|^2 - \sum_{\alpha\in F} |a_\alpha|^2. \end{aligned}$$

Taking the sup over $F \Subset A$ yields $\sum_{\alpha\in A} |a_\alpha|^2 \le \|x\|^2$.

(3) Again define $x_F = \sum_{\alpha\in F} a_\alpha e_\alpha$. Then $\|x_F\|^2 = \sum_{\alpha\in F} |a_\alpha|^2 \le L^2$. Note that $L^2 = \sup_{F\Subset A} \|x_F\|^2$. Given $\varepsilon > 0$, choose finite $F_0 \Subset A$ so that $\|x_{F_0}\|^2 > L^2 - \varepsilon$. Then for $F, G \ge F_0$,

$$\|x_F - x_G\|^2 = \sum_{\alpha\in F\triangle G} |a_\alpha|^2 \le \sum_{\alpha\in A\setminus F_0} |a_\alpha|^2 < L^2 - (L^2 - \varepsilon) = \varepsilon.$$

Therefore $\{x_F\}_\Lambda$ is a Cauchy net. Since H is complete, it has a limit $x \in H$. Since each $x_F \in M$ and M is closed, $x \in M$. Hence $\|x\|^2 = \lim_\Lambda \|x_F\|^2 = L^2$. Compute $\langle x, e_\alpha\rangle = \lim_\Lambda \langle x_F, e_\alpha\rangle = a_\alpha$ since $\langle x_F, e_\alpha\rangle = a_\alpha$ when $F \ge \{\alpha\}$.

(4) By (2) and (3), for each $x \in H$ with $a_\alpha = \langle x, e_\alpha\rangle$, the sum $Px = \sum_{\alpha\in A} a_\alpha e_\alpha$ is well defined and lies in M. The map P is evidently linear. Moreover $\|Px\|^2 = \sum_{\alpha\in A} |a_\alpha|^2 \le \|x\|^2$ by (3) and Bessel's inequality. Therefore $\|P\| \le 1$. One can check that $Pe_\alpha = e_\alpha$, and thus $\|P\| = 1$. (3) also shows that $\langle Px, e_\alpha\rangle = a_\alpha$. Therefore $P^2x = Px$. The range of P contains the unclosed span$\{e_\alpha : \alpha \in A\}$, and $P\sum_{\alpha\in F} a_\alpha e_\alpha = \sum_{\alpha\in F} a_\alpha e_\alpha$. That is, P is the identity on this subspace. By continuity, $Py = y$ for all $y \in M$.

(5) $Px = 0$ iff $\langle x, e_\alpha\rangle = 0$ for all $\alpha \in A$ iff $\langle x, \sum_{\alpha\in F} a_\alpha e_\alpha\rangle = 0$ for all vectors in span$\{e_\alpha : \alpha \in A\}$ iff $\langle x, y\rangle = 0$ for all $y \in M$ by continuity iff $x \in M^\perp$. We showed that $Py = y$ for $y \in M$, and since $Px \in M$, we have $Px = x$ only if $x \in M$. If $x \in H$, then $x = Px + (I - P)x$, and $Px \in M$ and $P(I - P)x = (P - P^2)x = 0$, so $(I - P)x \in M^\perp$. Hence $M + M^\perp = H$.

(6) This is immediate from (3) and (5). ∎

2.3.4. DEFINITION. An *orthonormal basis* for a Hilbert space H is an orthonormal set $\{e_\alpha : \alpha \in A\}$ such that $\overline{\mathrm{span}\{e_\alpha : \alpha \in A\}} = H$.

2.3.5. PROPOSITION. *Every Hilbert space has an orthonormal basis.*

PROOF. Order the collection $\mathcal{O}$ of all orthonormal sets by inclusion. Then $\mathcal{O}$ is inductive because if $\mathcal{C}$ is a chain of orthonormal sets, then the union is an orthonormal set containing every element of $\mathcal{C}$. By Zorn's lemma, there is a maximal orthonormal set, say $\{e_\alpha : \alpha \in A\}$. Let $M = \overline{\operatorname{span}\{e_\alpha : \alpha \in A\}}$. If $M = H$, we are done. Otherwise $M^\perp \neq \{0\}$. Let f be a unit vector in $M^\perp$. Then $\{f, e_\alpha : \alpha \in A\}$ is a larger orthonormal set, contradicting the maximality of $\{e_\alpha : \alpha \in A\}$. Hence we must have $M = H$. ∎

2.3.6. THEOREM. *Let H be a Hilbert space. For each $f \in H^*$, there is a unique vector $y \in H$ so that $f(x) = \langle x, y\rangle$. The map $\Upsilon : H^* \to H$ given by $\Upsilon(f) = y$ is a conjugate linear isometry of H^* onto H. In particular, H^* is a Hilbert space with the inner product $\langle f, g\rangle = \overline{\langle \Upsilon f, \Upsilon g\rangle}$. If $\{e_\alpha : \alpha \in A\}$ is an orthonormal basis for H, the set $\{\Upsilon^{-1} e_\alpha : \alpha \in A\}$ forms an orthonormal basis for H^*.*

PROOF. Let $\{e_\alpha : \alpha \in A\}$ be an orthonormal basis for H. Define $a_\alpha = f(e_\alpha)$. For every finite set $F \Subset A$, let $y_F = \sum_{\alpha\in F} \overline{a_\alpha} e_\alpha$. Then

$$|f(y_F)| = \sum_{\alpha\in F} |a_\alpha|^2 \le \|f\|\,\|y_F\| = \|f\|\Big(\sum_{\alpha\in F} |a_\alpha|^2\Big)^{1/2}.$$

Hence $\sum_{\alpha\in F} |a_\alpha|^2 \le \|f\|^2$. Taking the supremum over all finite subsets F yields $\sum_{\alpha\in A} |a_\alpha|^2 \le \|f\|^2$. Therefore $y = \sum_{\alpha\in A} \overline{a_\alpha} e_\alpha$ belongs to H by Theorem 2.3.3, and $\|y\| \le \|f\|$.

If $x = \sum_{\alpha\in F} b_\alpha e_\alpha$, then

$$f(x) = \sum_{\alpha\in F} b_\alpha a_\alpha = \langle x, y\rangle.$$

Therefore f agrees with the linear functional $g(x) = \langle x, y\rangle$ on the algebraic span of the basis, which is dense in H. Both are continuous, so they are equal. Note that $|f(x)| = |\langle x, y\rangle| \le \|x\|\,\|y\|$. Take the supremum over all $x \in H$ with $\|x\| \le 1$ to get $\|f\| \le \|y\|$. Hence $\|y\| = \|f\|$. The uniqueness of y follows from the fact that any vector determining the functional f must satisfy

$$\langle y, e_\alpha\rangle = \overline{\langle e_\alpha, y\rangle} = \overline{f(e_\alpha)} = \overline{a_\alpha}.$$

By Theorem 2.3.3, $y = \sum_{\alpha\in A} \overline{a_\alpha} e_\alpha$.

Therefore there is a map $\Upsilon : H^* \to H$ given by $\Upsilon f = y$. It is isometric because $\|f\| = \|y\|$. It is surjective because for each $y \in H$, $f(x) = \langle x, y\rangle$ belongs to H^* and $\Upsilon f = y$. However this is not a linear map. If $g \in H^*$ with $\Upsilon g = z$ and $\lambda, \mu \in \mathbb{C}$, then

$$(\lambda f + \mu g)(x) = \lambda\langle x, y\rangle + \mu\langle x, z\rangle = \langle x, \bar\lambda y + \bar\mu z\rangle.$$

Thus

$$\Upsilon(\lambda f + \mu g) = \bar\lambda y + \bar\mu z = \bar\lambda \Upsilon f + \bar\mu \Upsilon g.$$

This shows that Υ is a conjugate linear map.

Now define an inner product on H^* by $\langle f, g\rangle = \overline{\langle \Upsilon f, \Upsilon g\rangle}$. It is easy to check that this satisfies the properties of an inner product. We will verify linearity in the first variable.

$$\begin{aligned}\langle \lambda f + \mu g, h\rangle &= \overline{\langle \Upsilon(\lambda f + \mu g), \Upsilon h\rangle} = \overline{\langle \overline{\lambda}\Upsilon f + \overline{\mu}\Upsilon g, \Upsilon h\rangle} \\ &= \overline{\overline{\lambda}\langle \Upsilon f, \Upsilon h\rangle} + \overline{\overline{\mu}\langle \Upsilon g, \Upsilon h\rangle} = \lambda\langle f, h\rangle + \mu\langle g, h\rangle.\end{aligned}$$

It is also clear that $\{\Upsilon^{-1}e_\alpha : \alpha \in A\}$ is orthonormal. Moreover since Υ is isometric,

$$\overline{\operatorname{span}\{\Upsilon^{-1}e_\alpha : \alpha \in A\}} = \Upsilon^{-1}\overline{\operatorname{span}\{e_\alpha : \alpha \in A\}} = \Upsilon^{-1}H = H^*.$$

So $\{\Upsilon^{-1}e_\alpha : \alpha \in A\}$ is an orthonormal basis. ■

2.3.7. DEFINITION. The *dimension* of a Hilbert space H is the cardinality of an orthonormal basis.

2.3.8. PROPOSITION. *The dimension of a Hilbert space is well defined.*

PROOF. We need to show that any two orthonormal bases have the same cardinality. If H is finite dimensional, this is just the cardinality of a basis, which is shown to be well defined in linear algebra.

If H is infinite dimensional, let $\{e_\alpha : \alpha \in A\}$ and $\{f_\beta : \beta \in B\}$ be two orthonormal bases. For each $\alpha \in A$, let $B_\alpha = \{\beta \in B : \langle e_\alpha, f_\beta\rangle \neq 0\}$. Then because

$$1 = \|e_\alpha\|^2 = \sum_{\beta \in B_\alpha} |\langle e_\alpha, f_\beta\rangle|^2,$$

it follows that B_α is non-empty and countable or finite. Indeed, there can be at most n βs such that $|\langle e_\alpha, f_\beta\rangle|^2 \geq \frac{1}{n}$. Thus there are at most countably many elements in B_α. Also $B = \bigcup_{\alpha\in A} B_\alpha$ because similarly every f_β has non-zero inner product with some e_α. Therefore $|B| \leq |A|\aleph_0 = |A|$. Similarly $|A| \leq |B|$. So by the Schroeder–Bernstein theorem, $|A| = |B|$. ■

The following easy result is left as an exercise.

2.3.9. PROPOSITION. *Two Hilbert spaces are (isometrically) isomorphic if and only if they have the same dimension.*

2.3.10. REMARK. If H and K are Hilbert spaces, then $H \oplus_2 K$ is also a Hilbert space. We will write this as $H \oplus K$. Then H and K are orthogonal complements in the direct sum.

2.3.11. DEFINITION. A *unitary operator* $U \in \mathcal{B}(H, K)$ from one Hilbert space H to another, K, is a surjective isometry.

Note that U is a linear isomorphism which preserves norm. That means that it also preserves the inner product. This follows from the *polarization identity* :

$$\langle x,y\rangle = \frac{1}{4}\sum_{\varepsilon\in\{\pm1,\pm i\}} \varepsilon\|x+\varepsilon y\|^2 \quad\text{or}\quad \langle x,y\rangle = \frac{1}{2}\big(\|x+y\|^2 - \|x-y\|^2\big)$$

in the complex and real cases, respectively. Since U preserves norms, it preserves the inner product as well.

2.3.12. EXAMPLE. Let $H = L^2(\mathbb{T})$, the space of square integrable functions (modulo equality almost everywhere) with respect to Lebesgue measure $\frac{1}{2\pi}\,d\theta$. For $n\in\mathbb{Z}$, let $e_n(z) = z^n = e^{in\theta}$. Note that

$$\langle e_n, e_m\rangle = \frac{1}{2\pi}\int_{-\pi}^{\pi} e^{in\theta}\overline{e^{im\theta}}\,d\theta = \frac{1}{2\pi}\int_{-\pi}^{\pi} e^{i(n-m)\theta}\,d\theta = \begin{cases} 1 & \text{if } n=m \\ 0 & \text{if } n\neq m \end{cases}.$$

So $\{e_n : n\in\mathbb{Z}\}$ is an orthonormal set.

It is a fact that the trigonometric polynomials of the form $\sum_{n=-N}^{N} a_n e^{in\theta}$ for $a_n\in\mathbb{C}$ and $N\in\mathbb{N}$ are uniformly dense in $C(\mathbb{T})$. This is a classical theorem of Weierstrass, which also follows from Féjer's theorem or the Stone–Weierstrass theorem. It is also a fact from measure theory that $C(\mathbb{T})$ is dense in $L^2(\mathbb{T})$ in the 2-norm. This follows from Lusin's theorem. Hence $\{e_n : n\in\mathbb{Z}\}$ has dense linear span. Therefore $\{e_n : n\in\mathbb{Z}\}$ is an orthonormal basis.

Consider the linear map $U : L^2(\mathbb{T}) \to l_2(\mathbb{Z})$ given by $Uf = \big(\hat{f}(n)\big)_{n\in\mathbb{Z}}$ that takes a function to its Fourier series:

$$\hat{f}(n) = \frac{1}{2\pi}\int_{-\pi}^{\pi} f(e^{i\theta})e^{-in\theta}\,d\theta = \langle f, e_n\rangle.$$

This is the Fourier operator. This map takes the orthonormal basis $\{e_n : n\in\mathbb{Z}\}$ onto the standard basis $\{\delta_n : n\in\mathbb{Z}\}$ for $l_2(\mathbb{Z})$. The Parseval identity shows that U is an isometry. The range is therefore closed, and is dense, so is surjective. Thus U is unitary.

2.4. Category Theorems

In this section, we obtain some striking results which follow from completeness. Our main tool is the Baire category theorem. In spite of its relatively easy proof, it is surprisingly powerful. Most likely you have already seen this in a course on metric spaces, but we reprove it for completeness. Compare with Exercise **1.23**.

2.4.1. DEFINITION. A subset $A \subset X$ of a topological space X is *nowhere dense* if $\overline{A}$ has no interior. A subset of a complete metric space is *first category* if it is the countable union of nowhere dense sets.

2.4.2. BAIRE CATEGORY THEOREM. *A non-empty complete metric space X is not a countable union of nowhere dense sets. Indeed, if U_n are dense open subsets of a complete metric space X, then $\bigcap_{n\geq 1} U_n$ is dense in X.*

PROOF. Note that $U_n = \overline{A_n}^c$ is a dense open set, and $\big(\bigcup_{n\geq 1} \overline{A_n}\big)^c = \bigcap_{n\geq 1} U_n$. Let $x \in X$ and $r > 0$. We will find a point in $b_r(x) \cap \bigcap_{n\geq 1} U_n$.

Since $\overline{A_1}$ has no interior, $V_1 := b_r(x) \cap U_1$ is non-empty and open. Thus there is a point $x_1 \in V_1$ and $0 < r_1 < r/2$ so that $\overline{b_{r_1}(x_1)} \subset V_1$. Proceed recursively. At stage n, we will have $x_1, \dots, x_n$ and $r_1, \dots, r_n$ so that $r_i < r/2^i$ and

$$\overline{b_{r_i}(x_i)} \subset V_i = b_{r_{i-1}}(x_{i-1}) \cap U_i \quad \text{for} \quad 1 \leq i \leq n.$$

Set $V_{n+1} = b_{r_n}(x_n) \cap U_{n+1}$. Since A_{n+1} is nowhere dense, this is a non-empty open set. So we may find a point x_{n+1} and an $r_{n+1} < r_n/2$ so that $\overline{b_{r_{n+1}}(x_{n+1})} \subset V_{n+1}$. This completes the inductive step.

The balls $\overline{b_{r_n}(x_n)}$ form a decreasing nested sequence of closed sets. We claim that the sequence $(x_n)_{n\geq 1}$ is Cauchy. Indeed, if $N \leq m < n$, then x_n, x_m lie in $\overline{b_{r_N}(x_N)}$ and hence $d(x_n, x_m) \leq 2r_N < 2^{1-N} r$. Since X is complete, this sequence has a limit, say $x_0 = \lim\limits_{n\to\infty} x_n$. Hence x_0 belongs to $\bigcap_{n\geq 1} \overline{b_{r_n}(x_n)}$. Since $\overline{b_{r_n}(x_n)}$ is disjoint from $\overline{A_n}$, we have $x_0 \in \bigcap_{n\geq 1} U_n$. Moreover, $x_0 \in V_1 \subset b_r(x)$, so that $d(x, x_0) < r$. Therefore $\bigcap_{n\geq 1} U_n$ is dense in X. ■

The following important result is sometimes called the Uniform Boundedness Principle.

2.4.3. BANACH–STEINHAUS THEOREM. *Let X be a Banach space, let Y be a normed vector space, and let $\mathcal{A} \subset \mathcal{B}(X, Y)$. Suppose that for each $x \in X$, $\sup_{A\in\mathcal{A}} \|Ax\| = K_x < \infty$. Then $\sup_{A\in\mathcal{A}} \|A\| < \infty$.*

PROOF. Let $X_n = \{x \in X : K_x \leq n\}$. This is a closed set since if $x_k \in X_n$ and $x_k \to x$ and $A \in \mathcal{A}$, then $\|Ax\| = \lim_{k\to\infty} \|Ax_k\| \leq n$. By assumption, $X = \bigcup_{n\geq 1} X_n$. By the Baire category theorem, there is an n_0 so that X_{n_0} has interior, say $b_r(x_0) \subset A_{n_0}$. If $\|x\| < 1$, then $x_0 + rx \in A_{n_0}$. Hence

$$\|Ax\| = \frac{1}{r}\|A(x_0 + rx) - Ax_0\| \leq \frac{2n_0}{r}.$$

Therefore $\sup_{A\in\mathcal{A}} \|A\| \leq \frac{2n_0}{r}$. ■

2.4.4. Corollary. *Let X be a Banach space, let Y be a normed vector space, and let $\mathcal{A} \subset \mathcal{B}(X,Y)$ such that $\sup_{A\in\mathcal{A}} \|A\| = \infty$. Then*

$$B = \{x \in X : \mathcal{A}x \text{ is bounded }\}$$

is first category.

Proof. $B = \bigcup_{n\geq 1} B_n$ where $B_n = \{x : \sup_{A\in\mathcal{A}} \|Ax\| \leq n\}$. As in the previous proof, B_n is closed. If some B_{n_0} has interior, then as in the previous proof, $\mathcal{A}$ is bounded. Since it isn't bounded, each B_n is nowhere dense. Hence B is first category. ■

The next corollary shows that $\mathcal{B}(X,Y)$ is closed under pointwise limits.

2.4.5. Corollary. *If X and Y are Banach spaces, $T_n \in \mathcal{B}(X,Y)$ and $\lim\limits_{n\to\infty} T_n x$ exists for every $x \in X$, then $Tx := \lim\limits_{n\to\infty} T_n x$ belongs to $\mathcal{B}(X,Y)$ and $\|T\| \leq \sup \|T_n\| < \infty$.*

Proof. Apply the uniform boundedness principle to $\{T_n : n \geq 1\}$. Since convergent sequences are bounded, $\{T_n x : n \geq 1\}$ is bounded for each $x \in X$. Therefore $\sup \|T_n\| = M < \infty$. The map T is clearly linear and $\|T\| \leq M$, so $T \in \mathcal{B}(X,Y)$. ■

Now we establish a second important result using the Baire category theorem.

2.4.6. Open Mapping Theorem. *Let X and Y be Banach spaces, and let $T \in \mathcal{B}(X,Y)$. If $TX = Y$, then T is open, i.e. if $U \subset X$ is open, then TU is open.*

Proof. Observe that $Y = TX = \bigcup_{n\geq 1} Tb_n(X) = \bigcup_{n\geq 1} \overline{Tb_n(X)}$, where $b_r(X) = b_r(0_X)$. Since Y is complete, the Baire category theorem shows that there is some n so that $\overline{Tb_n(X)} = n\overline{Tb_1(X)}$ has interior. Thus $\overline{Tb_1(X)}$ has interior, so it contains $b_r(y_0)$ for some $y_0 \in Y$ and $r > 0$. Therefore

$$\overline{Tb_1(X)} = \tfrac{1}{2}\overline{Tb_1(X)} - \tfrac{1}{2}\overline{Tb_1(X)} \supset \tfrac{1}{2}b_r(y_0) - \tfrac{1}{2}b_r(y_0) = b_r(0_Y).$$

We claim that for $\varepsilon > 0$, $\overline{Tb_1(X)} \subset Tb_{1+\varepsilon}(X)$. Let $y \in \overline{Tb_1(X)}$. Pick $x_1 \in b_1(X)$ with $\|y - Tx_1\| < \frac{\varepsilon r}{2}$. Then $y_2 := y - Tx_1 \in \overline{Tb_{\varepsilon/2}(X)}$. So there is an $x_2 \in X$ with $\|x_2\| < \frac{\varepsilon}{2}$ so that $\|y - Tx_1 - Tx_2\| < \frac{\varepsilon r}{4}$. Recursively we find $x_n \in X$ with $\|x_n\| < 2^{1-n}\varepsilon$ so that $\left\|y - \sum_{k=1}^{n} Tx_k\right\| < 2^{-n}\varepsilon r$. Let $x = \sum_{k\geq 1} x_k$. Then

$$\|x\| < 1 + \sum_{k\geq 1} 2^{1-k}\varepsilon = 1 + \varepsilon.$$

By continuity, $Tx = y$. So $b_r(Y) \subset \overline{Tb_1(X)} \subset Tb_{1+\varepsilon}(X)$. If $\|y\| = (1-\varepsilon)r$, then $y \in Tb_{(1+\varepsilon)(1-\varepsilon)}(X) \subset Tb_1(X)$. Hence $b_r(Y) \subset Tb_1(X)$. By translation, $b_r(Tx) \subset Tb_1(x)$ for any $x \in X$.

Now if U is open in X and $x \in U$, then $b_\rho(x) \subset U$ for some $\rho > 0$. Hence $TU \supset Tb_\rho(x) \supset b_{r\rho}(Tx)$. Therefore TU is open. ■

One of the most cited consequences of the Open Mapping theorem (OMT) is the following.

2.4.7. Banach Isomorphism Theorem. *Let X and Y be Banach spaces, and let $T \in \mathcal{B}(X,Y)$. If T is a bijection, then T is invertible.*

Proof. Clearly T^{-1} is a well-defined linear map. We need to show that it is continuous. By the OMT, T is open. So $Tb_1(X) \supset b_r(Y)$ for some $r > 0$. Therefore $T^{-1}b_1(Y) \subset b_{1/r}(X)$. That is, $\|T^{-1}\| \leq \frac{1}{r}$. ■

There are many other consequences of the Open Mapping theorem, and we collect a few here.

2.4.8. Corollary. *Let X and Y be Banach spaces, and let $T \in \mathcal{B}(X,Y)$. If T maps X onto Y, then $Y \simeq X/\ker T$.*

Proof. Since T is continuous, $\ker T$ is a closed subspace of X, and thus $X/\ker T$ is a Banach space. Define $S : X/\ker T \to Y$ by $S(x + \ker T) = Tx$. This is well defined, linear and bijective. Moreover if $\dot{x} = x + \ker T$,

$$\begin{aligned} \|S\dot{x}\| &= \inf\{\|Tv\| : v \in x + \ker T\} \\ &\leq \inf\{\|T\|\,\|v\| : v \in x + \ker T\} = \|T\|\,\|\dot{x}\|. \end{aligned}$$

So $\|S\| \leq \|T\| < \infty$. Thus S is continuous, and so is an isomorphism by the Banach isomorphism theorem. ■

2.4.9. Corollary. *Let X be a Banach space with two complete norms $\|\cdot\|_1$ and $\|\cdot\|_2$. If there is a constant C so that $\|x\|_1 \leq C\|x\|_2$ for all $x \in X$, then there is a constant C' so that $\|x\|_2 \leq C'\|x\|_1$, i.e. the two norms are equivalent.*

Proof. Consider the identity map *id* : $(X, \|\cdot\|_2) \to (X, \|\cdot\|_1)$. By hypothesis this is a continuous bijection. Hence the inverse map is continuous by the Banach isomorphism theorem. ■

2.4.10. Definition. A linear map $T : D \subset X \to Y$, where D is a not necessarily closed subspace of X, is a *closed linear map* if the graph

$$\mathcal{G}(T) = \{(x, Tx) : x \in D\}$$

is closed in $X \oplus_\infty Y$.

2.4.11. Closed Graph Theorem. *Let X and Y be Banach spaces, and let $T : X \to Y$ be a linear map defined on all of X. If T is closed, then T is continuous.*

PROOF. Since $\mathcal{G}(T)$ is a closed subspace of $X \oplus_\infty Y$, it is a Banach space. The coordinate projections $\pi_1 : \mathcal{G}(T) \to X$ and $\pi_2 : \mathcal{G}(T) \to Y$ are continuous. Moreover π_1 is a bijection. By the Banach isomorphism theorem, π_1^{-1} is continuous. Therefore $T = \pi_2 \pi_1^{-1}$ is continuous. ∎

2.4.12. Corollary. *Let X and Y be Banach spaces, and let $T : X \to Y$ be a linear map defined on all of X. Then T is continuous if*

$$(\ddagger) \qquad x_n \in X,\ x_n \to 0 \text{ and } Tx_n \to y \quad \Longrightarrow \quad y = 0.$$

PROOF. We show that condition (‡) implies that T is closed. Suppose that $(x_n, Tx_n) \in \mathcal{G}(T)$ converge to $(x_0, y_0) \in X \oplus_\infty Y$. Then $x_n - x_0 \to 0$ and $T(x_n - x_0) \to y_0 - Tx_0$. By (‡), $Tx_0 = y_0$, so that $(x_0, y_0) \in \mathcal{G}(T)$. Thus $\mathcal{G}(T)$ is closed. Hence T is continuous by the Closed Graph theorem. ∎

2.4.13. Example. Hellinger–Toeplitz. Let T be a linear operator on a Hilbert space H such that $\langle Tx, y\rangle = \langle x, Ty\rangle$ for all $x, y \in H$. Suppose that $x_n \in H$, $x_n \to 0$ and $Tx_n \to y_0$. Then

$$\|y_0\|^2 = \lim_{n\to\infty} \langle Tx_n, y_0\rangle = \lim_{n\to\infty} \langle x_n, Ty_0\rangle = 0.$$

Thus $y_0 = 0$. By the Closed Graph theorem, T is continuous.

2.5. Fourier series

Fourier analysis is a central part of analysis. In this section, we quickly obtain a few important results.

If $f \in L^1(\mathbb{T})$, the Fourier coefficients are

$$\hat{f}(n) = \frac{1}{2\pi} \int_{-\pi}^{\pi} f(e^{i\theta}) e^{-in\theta}\, d\theta \quad \text{for} \quad n \in \mathbb{Z}.$$

It is easy to see that $|\hat{f}(n)| \le \|f\|_1$. The Riemann–Lebesgue lemma states that $\lim_{|n|\to\infty} |\hat{f}(n)| = 0$ (Exercise **2.15**). This follows immediately from the density of the trigonometric polynomials in $L^1(\mathbb{T})$. Hence $\Gamma f = \big(\hat{f}(n)\big)_{n\in\mathbb{Z}}$ belongs to $c_0(\mathbb{Z})$.

Define $s_n(f)(e^{it}) = \sum_{k=-n}^{n} \hat{f}(k)e^{ikt}$ to be the partial sums of the Fourier series. Compute

$$\begin{aligned}
s_n(f)(e^{it}) &= \sum_{k=-n}^{n} \frac{1}{2\pi}\int_{-\pi}^{\pi} f(e^{i\theta})e^{-ik\theta}\,d\theta e^{ikt} \\
&= \frac{1}{2\pi}\int_{-\pi}^{\pi} f(e^{i\theta})\sum_{k=-n}^{n} e^{ik(t-\theta)}\,d\theta = \frac{1}{2\pi}\int_{-\pi}^{\pi} f(e^{i\theta})D_n(t-\theta)\,d\theta
\end{aligned}$$

and we sum the geometric series to get a formula for the *Dirichlet kernel*

$$\begin{aligned}
D_n(t) &= \sum_{k=-n}^{n} e^{ikt} = \frac{e^{i(n+1)t} - e^{-int}}{e^{it} - 1} = \frac{e^{i(n+\frac{1}{2})t} - e^{-i(n+\frac{1}{2})t}}{e^{it/2} - e^{-it/2}} \\
&= \frac{e^{i(n+\frac{1}{2})t} - e^{-i(n+\frac{1}{2})t}}{2i} \cdot \frac{2i}{e^{it/2} - e^{-it/2}} = \frac{\sin(n+\frac{1}{2})t}{\sin\frac{1}{2}t}.
\end{aligned}$$

Now we apply this when $f \in C(\mathbb{T})$. It is a classical result that there are continuous functions such that the Fourier series does not converge at some points. In fact, this behaviour is typical.

2.5.1. THEOREM. *For $t_0 \in [-\pi, \pi]$, the set*

$$\{f \in C(\mathbb{T}) : s_n(f)(e^{it_0}) \to f(e^{it_0})\}$$

is first category.

PROOF. The map $\varphi_n(f) = s_n(f)(e^{it_0}) = \frac{1}{2\pi}\int_{-\pi}^{\pi} f(e^{i\theta})D_n(t_0-\theta)\,d\theta$ is a continuous linear functional on $C(\mathbb{T})$. We will show that $\|\varphi_n\| = \|D_n\|_1 \to \infty$. Clearly if $\|f\|_\infty \le 1$, then

$$|\varphi_n(f)| \le \frac{1}{2\pi}\int_{-\pi}^{\pi} |D_n(t_0-\theta)|\,d\theta \le \|D_n\|_1.$$

For $\varepsilon > 0$, define

$$f_\varepsilon(t) = \begin{cases} 1 & \text{if } D_n(t_0-\theta) \ge \varepsilon \\ -1 & \text{if } D_n(t_0-\theta) \le -\varepsilon\ . \\ \text{p.w.linear} & \text{in between} \end{cases}$$

Then $\|f_\varepsilon\|_\infty = 1$ and for ε sufficiently small, $\varphi_n(f_\varepsilon) \approx \|D_n\|_1$.

Now we estimate this norm.

$$\begin{aligned}
\|D_n\|_1 &= \frac{1}{2\pi}\int_{-\pi}^{\pi} |D_n(t)|\,dt = \frac{1}{\pi}\int_0^{\pi} \frac{|\sin(n+\frac{1}{2})t|}{\sin\frac{1}{2}t} \\
&\geq \frac{1}{\pi}\int_0^{\pi} \frac{|\sin(n+\frac{1}{2})t|}{t/2} = \frac{2}{\pi}\int_0^{(n+\frac{1}{2})\pi} |\sin x|\,\frac{dx}{x} \\
&> \frac{2}{\pi}\int_0^{\pi/2} \sin x\,\frac{dx}{x} + \sum_{k=1}^{2n} \frac{1}{k\pi/2}\,\frac{2}{\pi}\int_{k\pi/2}^{(k+1)\pi/2} |\sin x|\,dx \\
&\geq \frac{4}{\pi^2}\sum_{k=1}^{2n}\frac{1}{k} \approx \frac{4}{\pi^2}\ln 2n \to \infty.
\end{aligned}$$

Therefore $\mathcal{A} = \{\varphi_n : n \geq 1\}$ is unbounded. By Corollary 2.4.4,

$$\{f \in C(\mathbb{T}) : s_n(f)(e^{it_0}) \text{ is bounded}\}$$

is first category. This contains any f whose Fourier series converges at e^{it_0}. ■

2.5.2. THEOREM. *The map* $\Gamma : L^1(\mathbb{T}) \to c_0(\mathbb{Z})$ *is continuous and one to one, but is not surjective.*

PROOF. Clearly Γ is linear, and $\|\Gamma f\|_\infty = \sup|\hat{f}(n)| \leq \|f\|_1$, so Γ is norm 1. The Riemann–Lebesgue lemma shows that the range of Γ lies in $c_0(\mathbb{Z})$.

For $f \in L^1(\mathbb{T})$, define $f_r(e^{i\theta}) = \sum_{n\in\mathbb{Z}} \hat{f}(n) r^{|n|} e^{in\theta}$. This series converges absolutely and uniformly to a continuous function for $0 \leq r < 1$. To see that Γ is injective, we quote a result from Fourier analysis: $f_r \to f$ in $L^1(\mathbb{T})$ as $r \to 1^-$. (See Theorem 11.1.10.) This shows that f can be recovered from its Fourier series, and thus Γ is one to one.

If Γ were onto, then the Banach isomorphism theorem would show that Γ^{-1} is continuous, i.e. there would be a constant C so that $\|f\|_1 \leq C\|\Gamma f\|_\infty$. However, we saw in the previous proof that $\|D_n\|_1 \to \infty$ while $\hat{D}_n(k) = \begin{cases} 1 & \text{if } |k| \leq n \\ 0 & \text{if } |k| > n \end{cases}$. So $\|\Gamma \hat{D}_n\|_\infty = \|\hat{D}_n\|_\infty = 1$. Thus Γ^{-1} is not bounded, whence Γ is not surjective. ■

Since the Fourier series does not converge even for many continuous functions, an alternative method was introduced. The *Cesàro means* are defined as

$$\Sigma_n(f)(e^{it}) = \frac{1}{n+1}\sum_{k=0}^{n} s_n(f)(e^{it}) = \sum_{k=-n}^{n} \left(1 - \tfrac{k}{n+1}\right)\hat{f}(k)e^{ikt}.$$

2.5.3. LEMMA. *If $f \in L^1(\mathbb{T})$, then $\Sigma_n(f)(e^{it}) = \frac{1}{2\pi}\int_{-\pi}^{\pi} f(e^{i\theta})k_n(t-\theta)\,d\theta$ where*

$$k_n(t) = \sum_{k=-n}^{n} \left(1 - \tfrac{|k|}{n+1}\right)e^{ikt} = \frac{1}{n+1}\Big(\frac{\sin(n+1)t/2}{\sin t/2}\Big)^2.$$

The function k_n is even, continuous and 2π-periodic. Moreover $k_n(0) = n+1$, $k_n \geq 0$ and $\|k_n\|_1 = 1$. The sequence k_n converges to 0 uniformly for $|t| \in [\delta, \pi]$ if $\delta > 0$.

PROOF. It is clear that

$$\Sigma_n(f)(e^{it}) = \frac{1}{2\pi}\int_{-\pi}^{\pi} f(e^{i\theta})\frac{1}{n+1}\sum_{k=0}^{n} D_k(t-\theta)\,d\theta,$$

so that

$$k_n(t) = \frac{1}{n+1}\sum_{k=0}^{n}\sum_{j=-k}^{k} e^{ikt} = \sum_{k=-n}^{n}\left(1 - \tfrac{|k|}{n+1}\right)e^{ikt}.$$

Now $\sin^2\frac{t}{2} = \frac{1}{4}(2 - e^{it} - e^{-it})$, and thus

$$\begin{aligned}
4\sin^2\tfrac{t}{2}\,k_n(t) &= (2 - e^{it} - e^{-it})\sum_{k=-n}^{n}\left(1 - \tfrac{|k|}{n+1}\right)e^{ikt} \\
&= -\tfrac{e^{-i(n+1)t}}{n+1} + \sum_{k=-n}^{n}\Big(2\big(1-\tfrac{|k|}{n+1}\big) - \big(1-\tfrac{|k-1|}{n+1}\big) - \big(1-\tfrac{|k+1|}{n+1}\big)\Big)e^{ikt} - \tfrac{e^{i(n+1)t}}{n+1} \\
&= \frac{-e^{-i(n+1)t} + 2 - e^{i(n+1)t}}{n+1} = \frac{4}{n+1}\sin^2\big((n+1)t/2\big).
\end{aligned}$$

Hence $k_n(t) = \dfrac{1}{n+1}\Big(\dfrac{\sin(n+1)t/2}{\sin t/2}\Big)^2$.

It is now evident that $k_n(t)$ is continuous since $\lim_{t\to 0} k_n(t) = n+1 = k_n(0)$. It is even, positive and 2π-periodic. Moreover

$$\sup_{\delta\leq|t|\leq\pi} k_n(t) \leq \frac{1}{n+1}\Big(\frac{1}{\sin\delta/2}\Big)^2.$$

Hence $k_n(t)$ converges uniformly to 0 on $[-\pi,-\delta]\cup[\delta,\pi]$. Finally, compute

$$\|k_n\|_1 = \frac{1}{2\pi}\int_{-\pi}^{\pi} k_n(t)\,dt = \sum_{k=-n}^{n}\left(1 - \tfrac{|k|}{n+1}\right)\frac{1}{2\pi}\int_{-\pi}^{\pi} e^{ikt}\,dt = 1. \qquad \blacksquare$$

These properties are all that is needed for uniform convergence at every continuous function. The sequence of functions $\{k_n\}$ is known as the *Fejér kernel* .

2.5.4. FEJÉR THEOREM. *If $f \in C(\mathbb{T})$, then $\Sigma_n(f)$ converges to f uniformly.*

PROOF. Let $\varepsilon > 0$. By uniform continuity, there is a $\delta > 0$ so that if $|s-t| \le \delta$, then $|f(e^{is}) - f(e^{it})| \le \varepsilon$. Also there is an N so that $\sup_{\delta\le|t|\le\pi} k_n(t) \le \varepsilon$ for $n \ge N$. We compute for $n \ge N$,

$$\begin{aligned}|f(e^{i\theta}) - \Sigma_n(f)(e^{i\theta})| &= \left|f(e^{i\theta})\frac{1}{2\pi}\int_{-\pi}^{\pi} k_n(t)\,dt - \frac{1}{2\pi}\int_{-\pi}^{\pi} f(e^{i(\theta-t)})k_n(t)\,dt\right| \\ &\le \frac{1}{2\pi}\int_{-\delta}^{\delta} |f(e^{i\theta}) - f(e^{i(\theta-t)})|k_n(t)\,dt \\ &\quad + \frac{1}{2\pi}\int_{\delta\le|t|\le\pi} |f(e^{i\theta}) - f(e^{i(\theta-t)})|k_n(t)\,dt \\ &\le \frac{\varepsilon}{2\pi}\int_{-\delta}^{\delta} k_n(t)\,dt + \frac{2\|f\|_\infty}{2\pi}\int_{\delta\le|t|\le\pi} k_n(t)\,dt \\ &< \varepsilon\|k_n\|_1 + 2\|f\|_\infty\varepsilon = (1+2\|f\|_\infty)\varepsilon.\end{aligned}$$

Hence $\Sigma_n(f)$ converges uniformly to f. ■

The functions $\Sigma_n(f)$ are trigonometric polynomials, i.e. they are finite linear combinations of $\{e^{ikt} : k \in \mathbb{Z}\}$. We get the following well-known approximation result as an immediate consequence.

2.5.5. COROLLARY. (Weierstrass). *The trigonometric polynomials are dense in $C(\mathbb{T})$.*

Similar results hold for $L^1(\mathbb{T})$. See Exercise **2.16**.

2.6. The Hahn–Banach Theorems

In this section, we establish one of the most fundamental results in functional analysis. Its use is ubiquitous in the subject. We will see a number of interesting corollaries here. Later in Section 3.3, we will derive various geometric variants.

2.6.1. DEFINITION. Let X be a vector space over $\mathbb{F} \in \{\mathbb{R}, \mathbb{C}\}$, and let $A \subset X$. Then A is *convex* if whenever $a, b \in A$ and $0 \le t \le 1$, then $ta + (1-t)b \in A$. The *convex hull* of A, $\operatorname{conv}(A)$, consists of all points $x = \sum_{i=1}^n t_i x_i$ for $n \ge 1$, $x_i \in X$, and $t_i \ge 0$ such that $\sum_{i=1}^n t_i = 1$. Say that A is *balanced* if $a \in A$ and $\lambda \in \mathbb{F}$ with $|\lambda| \le 1$ implies that $\lambda a \in A$. Then A is *absolutely convex* if it is convex and balanced. The *absolute convex hull* of A, $\operatorname{aconv}(A)$, consists of all points $x = \sum_{i=1}^n \lambda_i x_i$ for $n \ge 1$, $x_i \in X$, and $\lambda_i \in \mathbb{F}$ such that $\sum_{i=1}^n |\lambda_i| = 1$. If X has a topology, we let $\overline{\operatorname{conv}}(A)$ denote the closure of $\operatorname{conv}(A)$ and let $\overline{\operatorname{aconv}}(A)$ denote the closure of $\operatorname{aconv}(A)$.

2.6.2. EXAMPLES.
(1) It is routine to check that $\text{conv}(A)$ is convex and $\text{aconv}(A)$ is absolutely convex.
(2) The balls $b_r(x)$ and $\overline{b_r(x)}$ in a normed vector space are convex.
(3) Subspaces are convex. However, if f is a discontinuous linear functional defined everywhere on a Banach space X, then $\ker f$ is a proper convex set with closure X. So if $f(x_0) \neq 0$, the two convex sets $\ker f$ and $x_0 + \ker f$ are disjoint, but contain each other in their closures. This non-intuitive aspect is quite different from what happens in finite dimensions.
(4) The intersection of convex sets is convex.

An important device for the construction of linear functionals is the following.

2.6.3. DEFINITION. Let X be a vector space. A function $p : X \to \mathbb{R}$ is *sublinear* if

(1) $p(tx) = tp(x)$ for $t \geq 0$ and $x \in X$, and

(2) $p(x + y) \leq p(x) + p(y)$ for $x, y \in X$.

2.6.4. EXAMPLES.
(1) On any normed space, $p(x) = \|x\|$ is sublinear.
(2) If $0 \in U \subset X$ is open and convex, then $X = \bigcup_{n\geq 1} nU$. The *Minkowski functional* is

$$p_U(x) = \inf\{t > 0 : x \in tU\}.$$

Note that if $x \in sU$ and $y \in tU$, then $x + y \in (s + t)U$ by convexity of U. That is, $\frac{x}{s}$ and $\frac{y}{t}$ are in U, so $\frac{s}{s+t}\frac{x}{s} + \frac{t}{s+t}\frac{y}{t} = \frac{x+y}{s+t} \in U$. Positive homogeneity is clear. So p_U is sublinear.

We start with a general result showing that functionals dominated by a sublinear functional on a subspace can be extended to one on the whole space. It starts with a "soft" part using Zorn's lemma, followed by a "hard" part where an explicit extension is constructed increasing the space by one dimension.

2.6.5. THEOREM (Hahn–Banach). *Let X be a $\mathbb{R}$ vector space, and let M_0 be a subspace (no topology). Let p be a sublinear map on X. Suppose that $f_0 : M \to \mathbb{R}$ is linear and $f_0(x) \leq p(x)$ for $x \in M_0$. Then there is a linear functional $f{:}X \to \mathbb{R}$ so that $f|_{M_0} = f_0$ and $f(x) \leq p(x)$ for all $x \in X$.*

PROOF. Consider pairs (f, M) where $M_0 \subset M \subset X$ is a subspace and $f : M \to \mathbb{R}$ is a linear functional such that $f|_{M_0} = f_0$ and $f(x) \leq p(x)$ for all $x \in M$. Say $(f_1, M_1) \leq (f_2, M_2)$ is $M_1 \subset M_2$ and $f_2|_{M_1} = f_1$. This is a partial order on the collection of all such pairs.

We will show that it is inductive. Suppose that $\{(f_\alpha, M_\alpha) : \alpha \in A\}$ is a chain. Let $M = \bigcup_{\alpha \in A} M_\alpha$ and set $f = \bigcup_{\alpha \in A} f_\alpha$. Then it is easy to see that $(f_\alpha, M_\alpha) \leq (f, M)$ for all $\alpha \in A$. Hence by Zorn's lemma, there exists a maximal element, say (f, M).

Suppose that $M \neq X$. Pick $x \in X \setminus M$. We will try to extend f to a functional f' on $M' = M + \mathbb{R}x$ by setting $f'(x) = a$. We obtain necessary and sufficient conditions on a:

$$f'(m + tx) = f(m) + ta \leq p(m + tx) \quad \text{for all} \quad m \in M \text{ and } t \in \mathbb{R}.$$

This is true for $t = 0$. When $t > 0$, we set $m' = m/t$ and obtain

$$a \leq \frac{p(m + tx) - f(m)}{t} = p(m' + x) - f(m') \quad \text{for all} \quad m' \in M.$$

When $t < 0$, we set $m' = m/|t|$ and obtain

$$\frac{f(m) - p(m + tx)}{|t|} = f(m') - p(m' - x) \leq a \quad \text{for all} \quad m' \in M.$$

Therefore

$$\sup_{m \in M} f(m) - p(m - x) \leq a \leq \inf_{m \in M} p(m + x) - f(m).$$

Such a value a exists if and only if the LHS $\leq$ RHS.

If LHS $>$ RHS, there are vectors $m_1, m_2 \in M$ so that

$$f(m_1) - p(m_1 - x) > p(m_2 + x) - f(m_2).$$

Rearranging yields

$$f(m_1 + m_2) > p(m_1 - x) + p(m_2 + x) \geq p(m_1 + m_2).$$

This contradicts the condition on f. So we must have LHS $\leq$ RHS. Choosing any value of a satisfying the inequality yields an extension to (f', M'). This contradicts the maximality of (f, M). So we have $M = X$, which completes the proof. ■

2.6.6. Hahn–Banach Theorem. *Let X be a normed space and let M be a (not necessarily closed) subspace of X. Suppose that f_0 is a continuous linear functional on M. Then there is an $f \in X^*$ with $f|_M = f_0$ and $\|f\| = \|f_0\|$.*

PROOF. Real case. Let $p(x) = \|f_0\| \, \|x\|$. This is a sublinear functional, and $f_0(m) \leq \|f_0\| \, \|m\| = p(m)$ for all $m \in M$. By Theorem 2.6.5, there is an extension $f : X \to \mathbb{R}$ so that $f|_M = f_0$ and

$$\pm f(x) = f(\pm x) \leq p(\pm x) = \|f_0\| \, \|x\|.$$

Thus $|f(x)| \leq \|f_0\| \, \|x\|$. So $\|f\| = \|f_0\|$.

Complex case. Consider X as a real Banach space. Let $g_0(x) = \text{Re } f_0(x)$ for $x \in M$. Extend g_0 to a real linear functional g on X with $\|g\| = \|g_0\| \leq \|f_0\|$. Define $f(x) = g(x) + ig(-ix)$. Then $f : X \to \mathbb{C}$ is real linear, and

$$f(ix) = g(ix) - ig(x) = i\big(g(x) + ig(-ix)\big) = if(x).$$

Therefore f is complex linear. Now if $x \in M$,

$$\begin{aligned} f(x) &= g(x) + ig(-ix) = \text{Re } f_0(x) + i\,\text{Re } f_0(-ix) \\ &= \text{Re } f_0(x) + i\,\text{Re} -if_0(x) = \text{Re } f_0(x) + i\,\text{Im } f_0(x) = f_0(x). \end{aligned}$$

Thus f extends f_0. Finally, if $f(x) = e^{i\theta}|f(x)|$, then

$$|f(x)| = f(e^{-i\theta}x) = g(e^{-i\theta}x) \leq \|g_0\|\,\|e^{-i\theta}x\| \leq \|f_0\|\,\|x\|.$$

Hence $\|f\| \leq \|f_0\|$. ∎

2.6.7. Corollary. *Let X be a normed vector space. If $x_0 \in X$, then there is an $f \in X^*$ with $\|f\| = 1$ and $f(x_0) = \|x_0\|$. Hence*

$$\sup_{\substack{\|f\|=1 \\ f\in X^*}} |f(x)| = \|x\| \qquad \textit{for all } x \in X.$$

Proof. Define f_0 on $M = \mathbb{C}x_0$ by $f_0(\lambda x_0) = \lambda\|x_0\|$. Then $\|f_0\| = 1$ and $f_0(x_0) = \|x_0\|$. By Hahn–Banach, extend f_0 to some $f \in X^*$ with $\|f\| = \|f_0\|$. This proves the non-trivial part of the equality in the supremum. ∎

2.6.8. Corollary. *If X is a normed vector space, then X^* separates points of X.*

Proof. If $x \neq y$, there is some $f \in X^*$ so that $f(x - y) = \|x - y\| > 0$. So $f(x) \neq f(y)$. ∎

We state the last corollary for normed vector spaces, but it is usually applied only when X is a Banach space. Of course, X^{**} is a dual space, so it is always a Banach space.

2.6.9. Corollary. *If X is a normed vector space, then X imbeds isometrically into X^{**} by $\iota : x \to \hat{x}$ where $\hat{x}(f) := f(x)$.*

Proof. Clearly ι is linear. Moreover

$$\|\hat{x}\| = \sup_{\substack{\|f\|=1 \\ f\in X^*}} |\hat{x}(f)| = \sup_{\substack{\|f\|=1 \\ f\in X^*}} |f(x)| = \|x\|.$$

Hence this map is an isometry. ∎

2.6.10. DEFINITION. A Banach space X is *reflexive* if $\iota X = X^{**}$.

2.6.11. REMARK. Hilbert spaces are reflexive by two applications of Theorem 2.3.6. The Riesz theorem that $L^p(\mu)^* = L^q(\mu)$ for $1 < p < \infty$ and $\frac{1}{p} + \frac{1}{q} = 1$ shows that $L^p(\mu)$ is reflexive. Finite dimensional Banach spaces are reflexive because if V is n-dimensional, then so are V^* and V^{**}. So ιX is an n-dimensional subspace of X^{**}, and therefore they are equal.

The space c_0 is not reflexive since $c_0^* = l_1$ and $l_1^* = l_\infty$. If $L^1(\mu)$ is infinite dimensional, then it is not reflexive either. Similarly $C(X)$ is not reflexive if it is infinite dimensional.

We will often drop the ι and consider X as a canonical subspace of X^{**}.

2.6.12. COROLLARY. *If X be a Banach space, M is a closed subspace of X and $x_0 \in X \setminus M$, then there is a functional $f \in X^*$ with $\|f\| = 1$ such that $f|_M = 0$ and $f(x_0) = dist(x_0, M)$.*

PROOF. Let $Y = X/M$ and $\dot{x}_0 = x_0 + M \in Y$. By definition of the quotient norm, $\|\dot{x}_0\| = \text{dist}(x_0, M)$. By Corollary 2.6.7, there is a functional $g \in Y^*$ with $\|g\| = 1$ and $g(\dot{x}_0) = \|\dot{x}_0\|$. Define $f = gQ$ where $Q : X \to Y$ is the quotient map. Then $\|f\| \le \|g\| \, \|Q\| = 1$, $\ker f \supset \ker Q = M$ and

$$f(x_0) = g(\dot{x}_0) = \|\dot{x}_0\| = \text{dist}(x_0, M). \qquad \blacksquare$$

2.6.13. DEFINITION. If X is a Banach space and $Y \subset X$, the *annihilator* of Y in X^* is $Y^\perp = \{f \in X^* : f|_Y = 0\}$. If $Z \subset X^*$, the *preannihilator* of Z in X is $Z_\perp = \{x \in X : \hat{x}|_Z = 0\} = Z^\perp \cap X$.

A net $\{f_\lambda\}_\Lambda$ in X^* is said to converge to $f \in X^*$ in the *weak-$*$ topology* if $\lim_\Lambda f_\lambda(x) = f(x)$ for all $x \in X$. A subspace $Y \subset X^*$ is *weak-$*$ closed* if it contains all weak-* limits of nets in Y.

2.6.14. REMARK. The weak-$*$ topology is the weakest topology on X^* that makes evaluation at points of X continuous. Various weak topologies, including the weak-$*$ topology on a dual space, will be studied systematically in Section 3.1; or you can refer back to Section 1.8. At the moment, we will not verify that there is a topology that conforms to this notion of convergence.

2.6.15. LEMMA. *If $Y \subset X$, then $Y^\perp$ is a weak-$*$ closed subspace of X^*. If $Z \subset X^*$, then $Z_\perp$ is a norm-closed subspace of X. Moreover*

$$(Y^\perp)_\perp = \overline{\text{span}\, Y} \quad \textit{and} \quad (Z_\perp)^\perp = \overline{\text{span}\, Z}^{w^*}.$$

PROOF. The first two statements are straightforward. Suppose that $M = \overline{\text{span}\, Y}$. Then $Y^\perp = M^\perp$. Suppose that $x \notin M$. Then by Corollary 2.6.12, there is an

$f \in M^\perp$ so that $f(x) \neq 0$. Hence $x \notin (M^\perp)_\perp$. That is $(M^\perp)_\perp = M$. The last claim is similar. ■

2.6.16. PROPOSITION. *Let M be a closed subspace of a Banach space X. Then there are canonical isometric isomorphisms*

$$M^* \simeq X^*/M^\perp \quad \textit{and} \quad (X/M)^* \simeq M^\perp.$$

PROOF. Define $R : X^* \to M^*$ to be the restriction map $Rf = f|_M$. Then $\ker R = M^\perp$. Thus there is a well-defined map $\widetilde{R} : X^*/M^\perp \to M^*$ given by $\widetilde{R}(f+M^\perp) = Rf$. The Hahn–Banach theorem shows that if $h \in M^*$, then there is a functional $f \in X^*$ with $\|f\| = \|h\|$ and $f|_M = h$. Thus $h = Rf = \widetilde{R}(f + M^\perp)$. Moreover $\|h\| \leq \|f + M^\perp\| \leq \|f\| = \|h\|$. Therefore, the map $\widetilde{R}$ is a surjective isometry, and hence an isometric isomorphism.

Let $Q : X \to X/M$ be the quotient map. Define a map $Q^* : (X/M)^* \to M^\perp$ by $Q^*f = f \circ Q$. Indeed, Q^*f is a linear functional on X that annihilates M, and thus belongs to $M^\perp$. On the other hand, if $g \in M^\perp$, define $f(x + M) = g(x)$. This is easily seen to be well defined, and $Q^*f = g$. So Q^* is surjective. Now $\|Q^*f\| = \|fQ\| \leq \|f\|\,\|Q\| = \|f\|$. On the other hand, suppose that $Q^*f = g$ and $1 > \|x + M\| = \inf_{m\in M} \|x + m\|$. Then there is some $y = x + m$ with $\|y\| < 1$. Thus $|f(x + M)| = |g(y)| < \|g\|$. Taking the supremum over all $x + M$ in the open unit ball yields $\|f\| \leq \|g\|$. Therefore Q^* is a surjective isometry, and thus an isometric isomorphism. ■

2.6.17. COROLLARY. *If N is a weak-$*$ closed subspace of a dual space X^*, then N is a dual space.*

PROOF. By Lemma 2.6.15, $N = (N_\perp)^\perp$. Therefore by Proposition 2.6.16, $(X/N_\perp)^* \simeq N$. ■

2.6.18. PROPOSITION. *X is reflexive if and only if X^* is reflexive.*

PROOF. If X is reflexive, then $X^{***} = (X^{**})^* = X^*$. So X^* is reflexive.

If $X^{**} \neq \iota X$, pick $z \in X^{**}\backslash\iota X$. By the previous corollary, there is a functional $f \in X^{***}$ such that $f|_X = 0$ and $f(z) \neq 0$. So $f \in X^{***} \setminus \iota X^*$. ■

2.6.19. PROPOSITION. *If X is reflexive, then all closed subspaces and quotients of X are reflexive.*

PROOF. Let $Y \subset X$ be a closed subspace. Then $Y^* \simeq X^*/Y^\perp$ and $Y^{**} \simeq Y^{\perp\perp} \subset X^{**} = \iota X$. Let $\Phi \in Y^{**}$ identified as an element of X^{**}. There is some

$x \in X$ so that $\iota x = \Phi \in Y^{\perp\perp}$. Therefore $x \in (Y^\perp)_\perp = Y$. Hence $\iota Y = Y^{**}$, so Y is reflexive.

Now $(X/Y)^* = Y^\perp \subset X^*$. By the first part, $Y^\perp$ is reflexive. Therefore X/Y is reflexive by Proposition 2.6.18. ■

Here is one more result, which combines the Hahn–Banach theorem with the Uniform Boundedness Principle.

2.6.20. PROPOSITION. *Let X be a Banach space. A subset $A \subset X$ is bounded if and only if $f(A)$ is bounded for each $f \in X^*$.*

PROOF. Clearly if A is bounded, then $f(A)$ is bounded for $f \in X^*$. Conversely, we identify A with $\iota A \subset X^{**}$, and consider it as a subset of $\mathcal{B}(X^*, \mathbb{F})$. The hypothesis is that it is bounded on each $f \in X^*$. By the Banach–Steinhaus theorem, ιA is bounded. Since ι is isometric by Corollary 2.6.9, A is bounded. ■

2.6.21. EXAMPLE. Adjoint of an operator. If X, Y are Banach spaces and $T \in \mathcal{B}(X, Y)$, we define the *adjoint* $T^* \in \mathcal{B}(Y^*, X^*)$ by $T^*g = g \circ T$. From this, it is clear that $\|T^*\| \le \|T\|$, so that T^* is continuous. In fact,

$$\begin{aligned} \|T^*\| &= \sup_{\substack{\|g\|\le 1\\ g\in Y^*}} \|T^*g\| = \sup_{\substack{\|g\|\le 1\\ g\in Y^*}} \sup_{\substack{\|x\|\le 1\\ x\in X}} |T^*g(x)| \\ &= \sup_{\substack{\|x\|\le 1\\ x\in X}} \sup_{\substack{\|g\|\le 1\\ g\in Y^*}} |g(Tx)| = \sup_{\substack{\|x\|\le 1\\ x\in X}} \|Tx\| = \|T\|. \end{aligned}$$

The Hahn–Banach theorem is required for the second last equality. This will be studied systematically in Section 4.1.

2.6.22. EXAMPLE. Banach Limits. There is a linear functional $L \in l_\infty^{\mathbb{R}\,*}$ such that

(1) $\liminf x_n \le L((x_n)) \le \limsup x_n$ and

(2) If $y_{n+k} = x_n$ for $n \ge N$ and some $k \in \mathbb{Z}$, then $L((y_n)) = L((x_n))$.

This yields a generalized limit since $L((x_n)) = \lim x_n$ when the limit exists, and it is translation invariant: $L((x_{n+1})) = L((x_n))$.

PROOF. Let S be the shift map: if $x = (x_n)$, then $(Sx)_n = x_{n+1}$, i.e. $Sx = (x_{n+1})$. Let $M = \{x - Sx : x \in l_\infty^{\mathbb{R}}\}$. This is a subspace because S is linear.

Claim: $\operatorname{dist}(\mathbf{1}, M) = 1$, where $\mathbf{1} = (1, 1, 1, \dots)$. Indeed, $\|\mathbf{1} - 0\| = 1$. Suppose that $\|\mathbf{1} - (x - Sx)\| \le r < 1$. Then $1 - (x_n - x_{n+1}) \le r$, so that $x_{n+1} \le x_n - (1-r)$. Recursively we get that $x_{n+1} \le x_1 - n(1-r)$. Hence (x_n) is not bounded, a contradiction. By Corollary 2.6.12, there is a linear functional $L \in l_\infty^{\mathbb{R}}(\mathbb{Z})^*$ with $1 = \|L\| = L(\mathbf{1})$ and $L|_M = 0$.

Claim: $c_0 \subset \overline{M} \subset \ker L$. If $x \in c_0$, then

$$x - S^n x = (x - Sx) + (Sx - S^2 x) + \cdots + (S^{n-1}x - S^n x) \in M \quad \text{for} \quad n \geq 1.$$

Moreover $\lim_{n\to\infty} S^n x = 0$. Therefore $x \in \overline{M}$.

Thus $L(S^k x) = L(x)$ and the initial terms of the sequence are irrelevant to $L(x)$, so (2) holds. Let $x \in l_\infty^{\mathbb{R}}$ and let $\alpha = \liminf x_n$ and $\beta = \limsup x_n$. If $\alpha = \beta$, i.e. if $\lim_{n\to\infty} x_n = \alpha$, then $x - \alpha\mathbf{1} \in c_0$, so that $L(x) = \alpha$. If $\alpha < \beta$, let $y = \frac{2}{\beta-\alpha}(x - \frac{\alpha+\beta}{2}\mathbf{1})$, so that $\liminf y_n = -1$ and $\limsup y_n = 1$. Then there is a $z \in c_0$ so that $\|y - z\|_\infty = 1$. Thus $L(y) = L(y - z) \in [-1, 1]$. Rescaling, we get that $\alpha \leq L(x) \leq \beta$. ■

If we want a functional on complex l_∞, we just complexify: $\widetilde{L}(x + iy) := L(x) + iL(y)$. You can check that $\|\widetilde{L}\| = 1$ and that it is translation invariant, kills c_0, and yields the limit value when it exists.

2.6.23. EXAMPLE. Runge Theorem. *Let $K \subset \mathbb{C}$ be a compact set such that $\mathbb{C} \setminus K$ is connected. Then every function $f(z)$ which is analytic on an open neighbourhood $U \supset K$ is a uniform limit of polynomials on K.*

PROOF. Let $P(K)$ be the closure of the polynomials in $C(K)$ and let $R(K)$ be the closure of the rational functions with poles in K^c. There is a union $\mathcal{C}$ of finitely many closed curves in $U \setminus K$ so that the winding number of $\mathcal{C}$ satisfies $\operatorname{ind}_{\mathcal{C}}(z) = \begin{cases} 1 & \text{if } z \in K \\ 0 & \text{if } z \in U^c \end{cases}$. Therefore by Cauchy's theorem, for $z \in K$,

$$f(z) = \frac{1}{2\pi i}\int_{\mathcal{C}} \frac{f(w)}{w - z}\,dw \approx \frac{1}{2\pi i}\sum \frac{f(w_k)}{w_k - z}(w_k - w_{k-1})$$

where the last term is a Riemann sum that uniformly approximates the integral for $z \in K$. It follows that $f|_K$ lies in the closure, $\overline{\operatorname{span}\{\frac{1}{a-z}|_K : a \in K^c\}} \subset R(K)$.

Suppose that $|a| > M := \sup\{|z| : z \in K\}$. Then

$$\frac{1}{a - z} = \frac{1}{a}\frac{1}{1 - z/a} = \frac{1}{a}\sum_{n\geq 0}\left(\frac{z}{a}\right)^n.$$

This series converges uniformly on K, and thus $\frac{1}{a-z}|_K \in P(K)$.

Now let $\varphi \in C(K)^*$ such that $\ker\varphi \supset P(K)$. Define $h(a) = \varphi(\frac{1}{w-z})$ for $a \in \mathbb{C} \setminus K$. Claim: h is analytic on $\mathbb{C} \setminus K$. Suppose that $b_{2r}(a) \subset \mathbb{C} \setminus K$. Then for $|w| < r$,

$$\frac{1}{(a + w) - z} = \frac{1}{a - z}\frac{1}{1 - \frac{w}{z-a}} = \sum_{n\geq 0}\frac{-1}{(z - a)^{n+1}}w^n.$$

This converges uniformly for $z \in K$ because $|w| \le r < 2r \le |z-a|$. Hence we can apply φ to get

$$h(a+w) = \sum_{n\ge 0} \varphi\Big(\frac{-1}{(z-a)^{n+1}}\Big) w^n.$$

Since this is valid for $|w| < r$, h is analytic on a neighbourhood of a. Now h vanishes on $\{a : |a| > M\}$, so $h = 0$ on the connected component containing this annulus, which is $\mathbb{C} \setminus K$ by hypothesis. Hence $\ker\varphi \supset \operatorname{span}\{\frac{1}{a-z}|_K : a \in K^c\}$.

We deduce that $\varphi(f) = 0$. However, if $f \notin P(K)$, there would be a functional φ annihilating $P(K)$ but non-zero on f. Therefore $f \in P(K)$. In particular, this applies to any rational function with poles off of K. So $R(K) = P(K)$. ■

2.7. Complemented Subspaces

In a Hilbert space, every closed subspace has a closed complementary subspace. However, in general Banach spaces, this is a special property.

2.7.1. Definition. Let X be a Banach space. A closed subspace $Y \subset X$ is *complemented* if there is a closed subspace $Z \subset X$ so that $Y \cap Z = \{0\}$ and $Y + Z = X$. This is called an *internal direct sum* , and we write $X = Y \dot{+} Z$.

Define $T : Y \oplus_1 Z \to X$ by $T(y, z) = y + z$. By assumption, T is a bijection. Indeed, if $T(y, z) = 0$, then $y = -z \in Y \cap Z = \{0\}$, so $(y, z) = (0, 0)$. Therefore T is one to one. It is onto by the second condition. Also

$$\|y + z\| \le \|y\| + \|z\| = \|(y, z)\|.$$

So $\|T\| \le 1$. By Banach's isomorphism theorem, T is an isomorphism, i.e. T^{-1} is continuous. Thus $X \simeq Y \oplus_1 Z$. In particular, $Z \simeq X/Y$.

Define $P : Y \oplus_1 Z \to Y \oplus_1 0$ by $P(y, z) = (y, 0)$. This is *idempotent* : $P^2 = P$ and $\|P\| = 1$. Then TPT^{-1} is a continuous projection of X onto Y with kernel Z. It is easy to see that this is the unique idempotent with this property.

Conversely suppose that $Q \in \mathcal{B}(X)$ is a continuous idempotent with range Y, and let $Z = \ker Q = \operatorname{Ran}(I - Q)$. Then $X \ni x = Qx + (I - Q)x \in Y + Z$, and if $x \in Y \cap Z$, then $x = (I - Q)x' = Qx = Q(I - Q)x' = 0$. So Y is complemented.

2.7.2. Example. X^* is always complemented in X^{***}. Let $\iota : X^* \to X^{***}$ be the canonical injection, and let $R : X^{***} \to X^*$ be the restriction map $(R\varphi) = \varphi|_X$. Note that $R\iota = I_{X^*}$. Then $P = \iota R$ is a continuous projection of X^{***} onto X^* because $P^2 = \iota(R\iota)R = P$.

Now we will show that a familiar subspace of a Banach space is not complemented.

2.7.3. LEMMA. *There are uncountably many infinite subsets $\{A_r \subset \mathbb{N} : r \in \mathbb{R}\}$ such that $A_r \cap A_s$ is a finite set if $r \neq s$.*

PROOF. List $\mathbb{Q} = \{q_n : n \geq 1\}$. For each $r \in \mathbb{R}$, pick a sequence $(q_{n(r,i)})_{i\geq 1}$ of rational numbers with r as its limit. Let $A_r = \{n(r,i) : i \geq 1\}$. Sequences converging to distinct points can only have finitely many terms in common. ■

2.7.4. THEOREM. *c_0 is not complemented in l_∞.*

PROOF. Suppose that $l_\infty \simeq c_0 \oplus Z$ where $Z \simeq l_\infty/c_0$. Observe that if $z = (z_n) + c_0 \in l_\infty/c_0$, then $\|z\| = \limsup |z_n|$. Let A_r be the sets that were constructed in Lemma 2.7.3, and let $x_r = \chi_{A_r} + c_0 \in l_\infty/c_0$. Claim: if $r_1, \ldots, r_n$ are distinct, $\|\sum_{i=1}^n a_i x_{r_i}\| = \max\{|a_i| : 1 \leq i \leq n\}$. Let $B_i = A_{r_i} \setminus \bigcup_{j\neq i} A_{r_j}$; so $A_{r_i} \setminus B_i$ is finite and $B_1, \ldots, B_n$ are disjoint and infinite. Thus $x_{r_i} = \chi_{B_i} + c_0$. Hence

$$\Big\|\sum_{i=1}^n a_i x_{r_i}\Big\| \leq \Big\|\sum_{i=1}^n a_i \chi_{B_i} + c_0\Big\| = \max\{|a_i| : 1 \leq i \leq n\}.$$

Now suppose that $T : l_\infty/c_0 \to l_\infty$ is a continuous one-to-one map into l_∞. Let $y_r = Tx_r \neq 0$; so there is an integer n_r so that $y_r(n_r) \neq 0$. Now

$$\mathbb{R} = \bigcup_{n,k\in\mathbb{N}} \{r : |y_r(n)| > 1/k\}.$$

Since this is a countable union, there is some n, k so that $\{r : |y_r(n)| > 1/k\}$ is uncountable. Let r_j for $j \geq 1$ be distinct reals in this set. Write $y_{r_j}(n) = e^{i\theta_j}|y_{r_j}(n)|$. Then $\|\sum_{j=1}^N e^{-i\theta_j} x_{r_i}\| = \max\{|e^{-i\theta_j}|\} = 1$, but

$$\Big\|T\sum_{j=1}^N e^{-i\theta_j} x_{r_i}\Big\| \geq \Big|\sum_{j=1}^N e^{-i\theta_j} y_{r_i}(n)\Big| \geq \frac{N}{k}.$$

This tends to ∞ as $N \to \infty$, and thus T is not bounded. This contradiction shows that c_0 is not complemented in l_∞. ■

We can parlay this result into a more general fact.

2.7.5. THEOREM. *c_0 is not isomorphic to a complemented subspace of any dual space X^*.*

PROOF. Suppose that $X^* \simeq c_0 \oplus_1 Z$ for some complementary Banach space Z. Then $X^{***} \simeq l_\infty \oplus_1 Z^{**}$. Let $J : c_0 \to X^*$ be the continuous injection onto the isomorphic copy in X. Then J^{**} denotes the corresponding injection of l_∞ into

X^{***}. Let $R : X^{***} \to X^*$ be the restriction map to X. One can readily check that $RJ^{**}\iota = J$. Let Q be a projection of X^* onto Jc_0 with kernel Z. Define

$$P = \iota J^{-1}QRJ^{**} : l_\infty \xrightarrow{J^{**}} X^{***} \xrightarrow{R} X^* \xrightarrow{J^{-1}Q} c_0 \xrightarrow{\iota} l_\infty.$$

Then if $x \in c_0$,

$$Px = \iota J^{-1}Q(RJ^{**}\iota)x = \iota J^{-1}QJx = \iota J^{-1}Jx = \iota x$$

because $Jc_0 \subset X^*$ and $R|_{X^*} = I_{X^*}$ and $Jx = QJx$. Therefore $P^2 = P$, and this is a continuous projection of l_∞ onto c_0, which is impossible by Theorem 2.7.4. ■

2.7.6. COROLLARY. *If (K, d) is an infinite compact metric space, then $C(K)$ is not a dual space.*

PROOF. Pick a sequence of distinct points in K, say $(x_i)_{i\geq 0}$, with $\lim_{i\to\infty} x_i = x_0$. Define $r_i = \frac{1}{2}\min\{d(x_i, x_j) : j \neq i\}$. Then $r_i > 0$. Define functions $g_i(x) = \frac{1}{r_i}\max\{0, r_i - d(x, x_i)\}$ in $C(K)$. Then these are positive functions of norm 1 with $g_i g_j = 0$ if $i \neq j$. Define $J : c_0 \to C(K)$ by $J((a_i)) = \sum_{i\geq 1} a_i g_i$. Because $a_i \to 0$, this sum converges uniformly, and thus belongs to $C(K)$ and $Ja(x_i) = a_i$ and $Ja(x_0) = 0$. Moreover $\|Ja\|_\infty = \|a\|_\infty$, i.e. J is an isometry. In particular it has closed range, say Y.

Define $T : C(K) \to I(x_0) = \{f : f(x_0) = 0\}$ by $TF = f - f(x_0)$. Then define $S : I(x_0) \to c_0$ by $Sf = (f(x_i))_{i\geq 1}$. Note that $STJa = SJa = a$ for $a \in c_0$. Let $P = JST$ and note that $P^2 = J(STJ)ST = JST = P$ is a continuous projection onto Y.

Since c_0 is isomorphic to a complemented subspace of $C(K)$, Theorem 2.7.5 shows that $C(K)$ is not a dual space. ■

2.7.7. DEFINITION. A subspace M of a Banach space X is an *M-ideal* if $M^\perp$ has a complement N in X^* so that $X^* = M^\perp \oplus_1 N$, i.e. if $\varphi \in M^\perp$ and $\psi \in N$, then $\|\varphi + \psi\| = \|\varphi\| + \|\psi\|$. The subspace $M^\perp$ is an *L-summand* of X^*.

2.7.8. EXAMPLE. Consider c_0 as a subspace of l_∞. Let $\{e_n : n \geq 1\}$ be the standard basis for c_0. Then $c_0^\perp = \{\psi \in l_\infty^* : \psi(e_n) = 0 \text{ for } n \geq 1\}$. Note that $l_1 = c_0^*$ and $l_1^* = l_\infty$, so that there is a canonical embedding of l_1 into $l_\infty^* = l_1^{**}$. We claim that $l_\infty^* = c_0^\perp \oplus_1 l_1$.

Indeed, let $\chi \in l_\infty^*$. Define $\varphi = \chi|_{c_0} = \sum_{i\geq 1} a_i\delta_i$ considered as an element of $l_1 \subset l_\infty^*$. So $\|\varphi\|_1 = \sum_{i\geq 1}|a_i|$. Then $\psi := \chi - \varphi \in c_0^\perp$. Since $\chi = \varphi + \psi$, we have $\|\chi\| \leq \|\varphi\| + \|\psi\|$. Let $\varepsilon > 0$. Let P_n denote the projection of l_∞ given by

$$P_n(f)(i) = \begin{cases} 0 & \text{if } i \leq n \\ f(i) & \text{if } i > n. \end{cases}$$

Pick a large n so that $\|\varphi P_n\|_1 = \sum_{i>n} |a_i| < \varepsilon$. Then select a unit vector f in $\operatorname{span}\{e_1, \dots, e_n\}$ so that $\varphi(f) > \|\varphi\| - \varepsilon$. Now choose another unit vector $g \in l_\infty$ so that $\psi(g) > \|\psi\| - \varepsilon$. Since $\psi \in c_0^\perp$, we can replace g with $P_n g$ without affecting $\psi(g)$ and possibly reducing the norm of g. This assures that f and g have disjoint supports, thus $\|f + g\| = \max\{\|f\|_\infty, \|g\|_\infty\} = 1$. Then

$$\begin{aligned}\chi(f+g) &= \varphi(f) + \varphi(g) + \psi(g)\\ &> \|\varphi\| - \varepsilon - \|\varphi P_n\| + \|\psi\| - \varepsilon > \|\varphi\| + \|\psi\| - 3\varepsilon.\end{aligned}$$

Thus $\|\chi\| = \|\varphi\| + \|\psi\|$. Therefore c_0 is an M-ideal of l_∞.

2.7.9. Proposition. *Let M be an M-ideal in X and let $X^* = M^\perp \oplus_1 N$. Then the restriction map $\rho : N \to M^*$ by $\rho(\psi) = \psi|_M$ is an isometric isomorphism of N onto M^*. For $\chi \in M^*$, the inverse $\rho^{-1}(\chi)$ is the unique Hahn–Banach extension of χ to a functional on X.*

Proof. If $\psi \in N$, then $\chi = \psi|_M \in M^*$ and $\|\chi\| \le \|\psi\|$.

Conversely, if $\chi \in M^*$, the Hahn–Banach theorem provides a functional $\widetilde{\chi} \in X^*$ with $\|\widetilde{\chi}\| = \|\chi\|$ and $\widetilde{\chi}|_M = \chi$. Write $\widetilde{\chi} = \varphi \oplus_1 \psi$ where $\varphi \in M^\perp$ and $\psi \in N$. Then $\|\widetilde{\chi}\| = \|\varphi\| + \|\psi\| \ge \|\psi\|$. Any two extensions of χ differ by an element of $M^\perp$, and, in particular, $\rho(\psi) = \chi$. Moreover,

$$\|\chi\| \le \|\psi\| \le \|\varphi\| + \|\psi\| = \|\widetilde{\chi}\| = \|\chi\|.$$

Hence $\varphi = 0$ and $\widetilde{\chi} = \psi$. So the Hahn–Banach extension is unique and $\|\rho(\psi)\| = \|\psi\|$. Therefore ρ is a surjective isometric map. ∎

2.7.10. Lemma. *Let M be an M-ideal in X. Suppose that $x \in X$ such that $d = dist(x, M) < \|x\| = d + r$. Given $\varepsilon > 0$, there is an $m \in M$ with $\|m\| \le r$ and $\|x - m\| < d + \varepsilon$.*

Proof. If this is false, then the open ball $B = b_{d+\varepsilon}(0)$ and the closed convex set $A = \{x - m : m \in M,\ \|m\| \le r\}$ are disjoint. By the Hahn–Banach theorem, there is a non-zero functional $\chi \in X^*$ so that

$$\sup_{b\in B} \operatorname{Re} \chi(b) \le \inf_{a\in A} \operatorname{Re} \chi(a).$$

We write $\chi = \varphi \oplus_1 \psi \in M^\perp \oplus_1 N$. Note that $|\varphi(x)| \le \|\varphi\| \operatorname{dist}(x, M)$. Evaluating the inequality above, we obtain

$$\begin{aligned}(d+\varepsilon)\|\chi\| &\le \operatorname{Re} \chi(x) - r\|\chi|_M\| = \operatorname{Re} \varphi(x) + \psi(x) - r\|\psi\|\\ &\le d\|\varphi\| + (d+r)\|\psi\| - r\|\psi\| = d\|\chi\|.\end{aligned}$$

This contradiction shows that $A \cap B$ is non-empty as claimed. ∎

The next result is an important property of M ideals. Quotient maps automatically map the open ball onto the open ball, but this is generally false for closed balls.

2.7.11. THEOREM. *If M is an M-ideal in X, then the quotient $Q : X \to X/M$ maps the closed unit ball of X onto the closed unit ball of X/M.*

PROOF. Let $y = Qx$ have $1 = \|y\| = \text{dist}(x, M)$. Let $\varepsilon > 0$. By Lemma 2.7.10, if $\|x\| = 1 + r$, choose $m_0 \in M$ with $\|m_0\| \le r$ so that $\|x - m_0\| < 1 + \varepsilon$. Recursively, we choose $m_{k+1} \in M$ for $k \ge 0$ so that $\|m_k - m_{k+1}\| < 2^{-k}\varepsilon$ so that $\|x - m_{k+1}\| < 1 + 2^{-k-1}\varepsilon$. Then $\{m_k\}$ is Cauchy with limit $m \in M$ such that $\|x - m\| = 1$. Thus $Q(x - m) = y$, so Q maps the closed unit ball of X onto the closed unit ball of X/M. ∎

The following consequence is immediate.

2.7.12. COROLLARY. *If M is an M-ideal in X and $x \in X$, there is some $m \in M$ such that $\|x - m\| = \text{dist}(x, M)$.*

Exercises for Chapter 2

2.1 (*Minkowski*) Prove that the norm on $L^p(\mu)$ for $1 < p < \infty$ satisfies the triangle inequality as follows: (you can take μ to be Lebesgue measure on $(0, 1)$)

(a) Let $\|f\|_p = A$ and $\|g\|_p = B$ with $AB > 0$ and let $f_0 = f/A$ and $g_0 = g/B$. Show that

$$\frac{1}{A+B}|f(x) + g(x)| \le \frac{A}{A+B}|f_0(x)| + \frac{B}{A+B}|g_0(x)|.$$

(b) Show that $\varphi(x) = x^p$ is strictly convex on $[0, \infty)$. Hence deduce that

$$\Big(\frac{A}{A+B}|f_0(x)| + \frac{B}{A+B}|g_0(x)|\Big)^p \le \frac{A}{A+B}|f_0(x)|^p + \frac{B}{A+B}|g_0(x)|^p.$$

(c) Integrate. When is the inequality an equality?

2.2 Prove that l_p is complete for $1 \le p \le \infty$.

2.3 For $1 < p < \infty$, define q so that $\frac{1}{p} + \frac{1}{q} = 1$. Prove Hölder's inequality: if $f \in L^p(\mu)$ and $g \in L^q(\mu)$, then $\|fg\|_1 \le \|f\|_p \|g\|_q$ as follows: (you can take μ to be Lebesgue measure on $(0, 1)$)

(a) If $a, b \in (0, \infty)$ and $0 \le t \le 1$, then $a^t b^{1-t} \le ta + (1 - t)b$ with equality only when $a = b$ or $t = 0$ or 1.

(b) Let $\|f\|_p = A$ and $\|g\|_p = B$ with $AB > 0$ and let $f_0 = f/A$ and $g_0 = g/B$. Show that
$$\frac{|f(x)g(x)|}{AB} \le \frac{1}{pA^p}|f(x)|^p + \frac{1}{qB^q}|g(x)|^q.$$
(c) Integrate. When is the inequality an equality?

2.4 (a) Prove that $C^{(1)}[a,b]$ is complete.
HINT: If (f_n) is Cauchy, show that $\lim_n f'_n = g$ exists uniformly. Let $c = \lim_n f_n(a)$. Show that $f(x) = c + \int_a^x g(t)\,dt$ is the limit.
(b) Show that $C^{(n)}[a,b]$ is complete.

2.5 Let $(V, \|\cdot\|)$ be a normed vector space. Prove that V is complete if and only if for every sequence (v_n) in V with $\sum_n \|v_n\| < \infty$, the series $\sum_n v_n$ *converges in* V.
HINT: For a Cauchy sequence (x_n), drop to a subsequence so $\sum_i \|x_{n_i} - x_{n_{i+1}}\| < \infty$.

2.6 (a) Prove that no infinite dimensional Banach space has a countable basis *as a vector space* (i.e. a collection $\{e_n : n \ge 1\}$ so that every vector in X is a *finite* linear combination of $\{e_n : n \ge 1\}$).
(b) Let X be an infinite dimensional Banach space. Recursively construct a sequence of unit vectors x_n so that $\operatorname{dist}(x_n, \operatorname{span}\{x_i : i < n\}) > 1 - 2^{-n}$. Hence deduce that the unit ball of X is not compact.

2.7 (a) Prove that $l_1^* = l_\infty$.
(b) Describe all infinite matrices $T = \big[t_{ij}\big]_{i,j=1}^{\infty}$ which act as bounded operators from l_1 to itself and find a formula for $\|T\|$.

2.8 (a) If H is a Hilbert space and $\xi, \eta \in H$, prove the *parallelogram law*
$$\|\xi + \eta\|^2 + \|\xi - \eta\|^2 = 2\|\xi\|^2 + 2\|\eta\|^2.$$
(b) If a Banach space X satisfies the parallelogram law, show X is a Hilbert space.
HINT: define
$$\langle \xi, \eta \rangle = \frac{1}{4}\big(\|\xi + \eta\|^2 - \|\xi - \eta\|^2 + i\|i\xi + \eta\|^2 - i\|i\xi - \eta\|^2\big).$$
(c) Note that this proves the polarization identity.
(d) Show that if $2\|\xi\|^2 + 2\|\eta\|^2 \le \|\xi + \eta\|^2 + \|\xi - \eta\|^2$ for all $\xi, \eta \in X$, then the parallelogram law holds.

2.9 Prove Proposition 2.3.9 that two Hilbert spaces are isomorphic if and only if they have the same dimension.

2.10 Recall that $L^2_a(\mathbb{D})$ is the Hilbert space of analytic functions on the open unit disc which are square integrable with respect to area measure. Let $\zeta_n(z) = z^n$ for $n \ge 0$. Find constants α_n so that $\{\alpha_n \zeta_n : n \ge 0\}$ is an orthonormal basis for $L^2_a(\mathbb{D})$. Use this basis to find vectors $k_w \in L^2_a(\mathbb{D})$ so that $\langle h, k_w \rangle = h(w)$ for each $z \in \mathbb{D}$ and $h \in L^2_a(\mathbb{D})$. Be sure to sum the series and express k_w as a closed-form function.

2.11 Let H be a Hilbert space which is a vector space of functions on a set X endowed with an inner product with the property that the linear functionals $\varepsilon_x(f) = f(x)$ are continuous for each $x \in X$.

(a) Show that for each $x \in X$, there is an element $k_x \in H$ so that $\langle f, k_x \rangle = f(x)$ for $x \in H$. Prove that the closed linear span of $\{k_x : x \in X\}$ is H.

(b) A multiplier of H is a function h on X so that $hf \in H$ for every $f \in H$. Prove that the linear map $M_h f = hf$ is continuous.

2.12 (*Riesz lemma*). Let Y be a proper closed subspace of a Banach space X. Given $\varepsilon > 0$, show that there is a unit vector $x \in X$ such that $\text{dist}(x, Y) > 1 - \varepsilon$. If X is finite dimensional, find a unit vector $x \in X$ such that $\text{dist}(x, Y) = 1$.

2.13 Let $X = C[0, 1]$. Define

$$Y = \{g \in X : g(0) = 0\} \quad \text{and} \quad Z = \Big\{g \in Y : \int_0^1 g(x)\, dx = 0\Big\}.$$

(a) Show that every $f \in X \setminus Y$ has infinitely many closest points $g \in Y$.

(b) Show that $f(x) = x$ has no closest point in Z.

(c) Deduce that the quotient map of X onto X/Z does not take the closed unit ball of X onto the closed unit ball of X/Z.

2.14 Let X be a complete, countable metric space.

(a) Prove that X has isolated points.

HINT: Baire category theorem.

(b) The *derived set* Y' of a topological space Y is the set of cluster points. Set $X_0 = X$. For each ordinal α, define $X_{\alpha+1} = X'_\alpha$. For a limit ordinal β, set $X_\beta = \bigcap_{\alpha<\beta} X_\alpha$. Prove that there is a countable ordinal such that X_α is empty.

2.15 Prove the Riemann–Lebesgue lemma: for $f \in L^1(\mathbb{T})$, $\lim\limits_{|n|\to\infty} |\hat{f}(n)| = 0$.

HINT: Use Example 2.3.12 and the fact that $L^2(\mathbb{T})$ is dense in $L^1(\mathbb{T})$.

2.16 (a) Show that $\{f \in L^1(\mathbb{T}) : s_n(f) \to f \text{ in } L^1(\mathbb{T})\}$ is first category.

(b) Prove that $\Sigma_n(f)$ converges to f in the $L^1(\mathbb{T})$ norm for $f \in L^1(\mathbb{T})$.

HINT: Use Fubini's theorem and the fact that $s \mapsto f_s(e^{it}) = f(e^{i(t-s)})$ is $L^1(\mathbb{T})$-norm continuous.

2.17 A *Schauder basis* for a Banach space X is a sequence $\{e_n : n \geq 1\}$ such that for each $x \in X$, there are unique scalars $\{c_n\}$ such that $x = \lim_{n\to\infty} \sum_{i=1}^n c_i e_i$. For convenience, normalize so that $\|e_n\| = 1$ for $n \geq 1$.

(a) Show that $\varphi_n(x) = c_n$ is a linear functional.

(b) Define $S_n x = \sum_{i=1}^n c_i e_i$, and set $|||x||| = \sup_{n\geq 1} \|S_n x\|$. Prove $|||\cdot|||$ is a norm.

(c) Show that $(X, |||\cdot|||)$ is complete.

(d) Prove that the identity map T from $(X, |||\cdot|||)$ to $(X\|\cdot\|)$ is an isomorphism. Hence deduce that $\sup_{n\geq 1} \|S_n\| = \|T^{-1}\|$ and that each φ_n is continuous.

2.18 The *Haar basis* for $L^p(0,1)$, $1 \le p < \infty$, is given by $h_0 = 1$ and for $n = 2^j + k$, where $j \ge 0$ and $0 \le k < 2^j$,

$$h_n = \chi_{[k2^{-j},(2k+1)2^{-j-1})} - \chi_{[2k+1)2^{-j-1},(k+1)2^{-j})}.$$

(a) Find a formula for each φ_n, $n \ge 0$ and describe $S_n f$ as a step function with values equal to the average over each interval.
(b) Prove that $\left\| \sum_{i=0}^{n} c_i h_i \right\|_p \le \left\| \sum_{i=0}^{n+1} c_i h_i \right\|_p$ for any c_i.
(c) Prove that this is a basis.

2.19 Consider $C[0,1]$. Let $x_1 = 0$, $x_2 = 1$ and $\{x_n\}_{n \ge 3}$ be a dense set of distinct points in $(0,1)$. Let $S_1 f = f(x_1)$ and for $n \ge 2$, let $S_n f$ denote the piecewise linear continuous function with $S_n f(x_i) = f(x_i)$ for $1 \le i \le n$.
(a) Prove that S_n are the partial sums for a basis of $C[0,1]$.
(b) Prove that $\|S_n\| = 1$ for $n \ge 1$ and $\|S_n f\|_\infty \le \|S_{n+1} f\|_\infty$ for $n \ge 1$.

2.20 Let $e_n(z) = z^n$ for $z \in \mathbb{T}$. Prove that $\{e_0, e_1, e_{-1}, e_2, e_{-2}, \dots\}$ is not a basis for $C(\mathbb{T})$ or for $L^1(\mathbb{T})$.
HINT: Use Fourier series.

2.21 Let X be a Banach space, and suppose that Y and Z are two closed subspaces such that $Y + Z = X$. Prove that there is a constant C so that every $x \in X$ can be written as $x = y + z$ with $\|y\| + \|z\| \le C\|x\|$.

2.22 Let X, Y, Z be Banach spaces. Let $B : X \times Y \to Z$ be a bilinear map (i.e. $B(x, y_0)$ is linear in x for fixed y_0 and $B(x_0, y)$ is linear in y for fixed x_0). Suppose that for every $f \in Z^*$, f^*B is separately continuous in x and y. Prove that there is a constant C so that $\|B(x,y)\| \le C\|x\|\,\|y\|$.

2.23 Let X, Y be Banach spaces and let $T \in \mathcal{B}(X,Y)$ map onto Y. Show that there is a constant C so that whenever (y_n) is a sequence in Y with $\lim y_n = y_0$, there is a sequence (x_n) in X such that $Tx_n = y_n$ and $\lim x_n = x_0$ exists.

2.24 Let $\varphi(f) = \displaystyle\int_0^1 f(t)\,dt$ for $f \in C_{\mathbb{R}}[0,1]$. Let Φ be any Hahn–Banach extension of φ to the Banach space $B_{\mathbb{R}}[0,1]$ of all bounded real-valued functions on $[0,1]$ with the sup norm.
(a) What are the possible values for $\Phi(\chi_{[0,.5]})$?
(b) What are the possible values for $\Phi(\chi_{\mathbb{Q}\cap[0,1]})$?

2.25 Let A be a subset of a vector space X.
(a) Prove that conv(A) is convex, and is the smallest convex set containing A.
(b) Prove that aconv(A) is absolutely convex, and is the smallest absolutely convex set containing A.

2.26 Let X be a Banach space and let Y be a closed subspace. Show that if Y and X/Y are reflexive, then X is reflexive.

2.27 Let X be a *separable* Banach space.

(a) Show that X is isometrically isomorphic to a subspace of l_∞.
HINT: Find a countable set of linear functionals $\{\varphi_n\}$ of norm one so that $\sup|\varphi_n(x)| = \|x\|$ for all $x \in X$.

(b) Show that X is isometrically isomorphic to a quotient of l_1.
HINT: Define a norm one linear map T of l_1 *onto* X so that the image of the unit ball of l_1 is dense in the ball of X. Prove that the quotient norm on $l_1/\ker T$ coincides with the norm on X.

2.28 (a) Let X be a Banach space. If Y is a weak-$*$ closed subspace of X^*, let $Y_\perp = \{x \in X : f(x) = 0 \text{ for all } f \in Y\}$. Show that $Y = (Y_\perp)^\perp$.

(b) Hence show that Y is a dual space, and that the weak-$*$ topology on Y as this dual space coincides with the weak-$*$ topology of X^* restricted to Y.

2.29 A Banach space X is *uniformly convex* if for all $\varepsilon > 0$, there is a $\delta > 0$ so that $\|x\| \le 1$, $\|y\| \le 1$ and $\|x - y\| > \varepsilon$ implies that $\|(x+y)/2\| < 1 - \delta$. If X is uniformly convex, C is a closed convex set in X and $x_0 \notin C$, prove that there exists a unique closest point to x_0 in C.
HINT: Say $\operatorname{dist}(x_0, C) = 1$. Given $\varepsilon > 0$, consider the diameter of $C \cap \overline{b_{1+\delta}(x_0)}$.

2.30 Let H be a Hilbert space, and let M and N be closed subspaces.

(a) Show that $M + N$ is closed if N is finite dimensional.

(b) Provide an example for which $M + N$ is not closed.

(c) Show that $(M+N)^\perp = M^\perp \cap N^\perp$ and $(M \cap N)^\perp = \overline{M^\perp + N^\perp}$.

2.31 Let c be the subspace of l_∞ consisting of sequences which have a limit.

(a) Show that c^* is isometrically isomorphic to l_1.
HINT: Riesz representation theorem.

(b) Show that c is isomorphic to c_0, but if $T : c \to c_0$ is an isomorphism, then $\|T\|\,\|T^{-1}\| \ge 2$. So they are not isometrically isomorphic.
HINT: Normalize so that $\|T^{-1}\| = 1$. Let $\mathbf{1}$ be the sequence of all 1s, and e_n the standard basis for c_0. Pick k_n so that $|Te_n(k_n)| \ge 1$. Prove that $\limsup |T(\mathbf{1} - 2e_n)(k_n)| \ge 2$. (Actually $\|T\|\,\|T^{-1}\| \ge 3$.)

(c) Show that if $P : c \to c_0$ is a projection onto c_0, then $\|P\| \ge 2$.

2.32 (*Auerbach bases*). Let X be an n-dimensional Banach space over $\mathbb{F} \in \{\mathbb{R}, \mathbb{C}\}$. Prove that there are vectors $x_1, \dots, x_n \in X$ and $f_1, \dots, f_n \in X^*$ such that $\|x_i\| = \|f_j\| = 1$ and $f_j(x_i) = \delta_{ij}$ for $1 \le i, j \le n$.
HINT: Choose $x_1, \dots, x_n$ in $\overline{b_1(X)}$ which satisfy

$$|\det(x_1, \dots, x_n)| = \max_{y_i \in \overline{b_1(X)}} |\det(y_1, \dots, y_n)|,$$

and multiply x_1 by a scalar of modulus 1 so that $\det(x_1, \dots, x_n) > 0$. Define

$$f_j(x) = \frac{\det(x_1, \dots, x_{j-1}, x, x_{j+1}, \dots, x_n)}{\det(x_1, \dots, x_n)}.$$

Notes on Chapter 2

The work of Volterra and Fredholm considering integral equations as linear operators led to work of Hilbert, Riesz and others. Riesz established the completeness and duality for the L^p spaces after Lebesgue developed his measure theory. The uniform boundedness principle evolved from work of Lebesgue, Hahn, Steinhaus and eventually in the final form by Banach and Steinhaus [24]. The open mapping theorem is due to Banach [22]. The Hahn–Banach theorem was a result of Hahn [134] and extended by Banach [22]. All of this developed from about 1905 to 1930. This was all treated in Banach's book [23]. The trick to extend the real case to the complex case came a decade later [35]. The notion of an M-ideal is more modern, due to Alfsen and Effros [7].

A very comprehensive book on functional analysis is Dunford and Schwarz [104], where we cite the first of three volumes. It contains a comprehensive survey of the history and has an extensive bibliography. More modern treatments of functional analysis include Rudin [269] and Conway [61]. Specialized topics in Banach space theory are covered in Diestel [90] and Albiac–Kalton [5]. The study of Fourier series is treated in the classic book by Katznelson [175]. See [140] for a monograph on M-ideals.

CHAPTER 3

LCTVSs and Weak Topologies

Banach spaces can be endowed with other, weaker, topologies such as the weak topology on X and the weak-$*$ topology on X^*. The advantage of this is that convergence is more common, and sometimes one can obtain compactness of closed balls. More generally, there is a whole family of infinite dimensional vector spaces with tractable topologies which contain the ones above as special cases. In this chapter, we will investigate general properties of these spaces. Highlights include a geometric version of the Hahn–Banach theorem, and the existence of "enough" extreme points for compact convex sets.

3.1. Locally convex topological vector spaces

3.1.1. DEFINITION. A *seminorm* on a vector space V is a function $p : V \to [0, \infty)$ which is homogeneous (i.e. $p(tx) = |t|p(x)$) and satisfies the triangle inequality, but non-zero vectors of norm 0 are allowed.

3.1.2. DEFINITION. A *locally convex topological vector space* (LCTVS) is a vector space X over $\mathbb{F} \in \{\mathbb{R}, \mathbb{C}\}$ together with a family $\mathcal{P}$ of seminorms on X such that $p(x) = 0$ for all $p \in \mathcal{P}$ implies that $x = 0$. Put a topology $\tau_{\mathcal{P}}$ on X generated by the convex sets

$$U(x_0, p, r) = \{x \in X : p(x - x_0) < r\} \quad \text{for} \quad x_0 \in X, p \in \mathcal{P} \text{ and } r > 0.$$

3.1.3. THEOREM. *Let X be a LCTVS.*

(1) *A neighbourhood base at* 0 *is provided by the convex open sets*

$$U_{F,r} = \{x \in X : p(x) < r,\ p \in F\} \text{ for finite } F \Subset \mathcal{P} \text{ and } r > 0.$$

The neighbourhood base at x_0 has the form $x_0 + U_{F,r}$.

(2) *X is Hausdorff.*

(3) *$X \times X \ni (x, y) \to x + y \in X$ is continuous.*

(4) *$\mathbb{F} \times X \ni (\alpha, x) \to \alpha x \in X$ is continuous.*

(5) *A net $(x_\lambda)_\Lambda$ in X converges to x if and only if $p(x - x_\lambda) \to 0$ for all $p \in \mathcal{P}$.*

K. R. Davidson, *Functional Analysis and Operator Algebras*, CMS/CAIMS Books in Mathematics 13,
https://doi.org/10.1007/978-3-031-63665-3_3

PROOF. Observe that $\tau_{\mathcal{P}}$ has a base obtained from finite intersections of sets $U(x_i, p_i, r_i)$ for $1 \le i \le n$. Suppose that $0 \in U = \bigcap_{i=1}^{n} U(x_i, p_i, r_i)$. Let $r = \min_{1 \le i \le n} r_i - p_i(x_i - 0)$ and $F = \{p_1, \dots, p_n\}$. Then $0 \in U_{F,r} \subset U$ because if $x \in U_{F,r}$, then

$$p_i(x_i - x) \le p_i(x_i) + p_i(x) < p_i(x_i) + r_i - p(x_i) = r_i;$$

whence $x \in U$. So the sets $U_{F,r}$ provide a neighbourhood base around 0. The topology is translation invariant because $U(x_0, p, r) = x_0 + U(0, p, r)$.

If $x \neq y$, then $x - y \neq 0$ so there is some $p \in \mathcal{P}$ so that $p(x - y) = r > 0$. Hence $U(x, p, r/2)$ and $U(y, p, r/2)$ are disjoint open sets separating x and y.

Suppose that $x_0, y_0 \in X$ and $x_0 + y_0 \in U$ is open. Let $A(x, y) = x + y$ be the addition map. Choose $F \subset \mathcal{P}$ and $r > 0$ so that $x_0 + y_0 + U_{F,r} \subset U$. If $x \in x_0 + U_{F,r/2}$ and $y \in y_0 + U_{F,r/2}$, then for $p \in F$,

$$p(x_0 + y_0 - x + y) \le p(x_0 - x) + p(y_0 - y) < \frac{r}{2} + \frac{r}{2} = r.$$

Therefore $(x_0 + U_{F,r/2}) \times (y_0 + U_{F,r/2}) \subset A^{-1}(x_0 + y_0 + U_{F,r})$. Hence A is continuous. The proof that scalar multiplication is continuous is similar.

Observe that $\lim_\Lambda x_\lambda = x$ means that for all finite $F \Subset \mathcal{P}$ and $r > 0$, there is a $\lambda_0 \in \Lambda$ so that $x_\lambda \in x + U_{F,r}$ for all $\lambda \ge \lambda_0$. Taking $F = \{p\}$ shows that $\lim_\Lambda p(x - x_\lambda) = 0$ for all $p \in \mathcal{P}$. Conversely, when this holds, given $F = \{p_1, \dots, p_n\}$ and r, we can choose λ_i so that $p_i(x - x_\lambda) < r$ for $\lambda \ge \lambda_i$. Select λ_0 so that $\lambda_0 \ge \lambda_i$. Then for $\lambda \ge \lambda_0$, $x_\lambda \in x + U_{F,r}$. ■

3.1.4. EXAMPLES.

(1) Any normed space $(X, \|\cdot\|)$ is a LCTVS.

(2) Let X be a Banach space, and let $Y \subset X^*$ be a (not necessarily closed) subspace which separates points of X. Define seminorms $p_f(x) = |f(x)|$ for $f \in Y$. These seminorms determine the topology (X, τ_Y). Note that since Y is closed under scalar multiplication, $U_{F,r} = U_{\frac{1}{r}F,1}$. So the open sets $U_F := U_{F,1}$ form a base of neighbourhoods of 0.

If Z is a topological space, a map $T : Z \to (X, \tau_Y)$ is continuous if and only if $f \circ T$ is continuous for all $f \in Y$. This follows from Corollary 1.8.4.

Important special cases are (X, τ_{X^*}), the *weak topology* on X, and (X^*, τ_X), the *weak-∗ topology* on X^*.

(3) Two other important examples are topologies on $\mathcal{B}(X, Y)$ where X, Y are Banach spaces. The *weak operator topology* is $(\mathcal{B}(X, Y), \text{WOT}) = (\mathcal{B}(X, Y), \tau_Z)$ where Z is the set of linear functions

$$f_{x,\varphi}(T) = \varphi(Tx) \quad \text{for} \quad x \in X \text{ and } \varphi \in Y^*.$$

Then $T_\lambda \xrightarrow{\text{WOT}} T$ if $\varphi(T_\lambda x) \to \varphi(Tx)$ for all $x \in X$ and $\varphi \in Y^*$.

The *strong operator topology* on $\mathcal{B}(X, Y)$ is $(\mathcal{B}(X, Y), \text{SOT})$ given by the seminorms $p_x(T) = \|Tx\|$ for $x \in X$. Then $T_\lambda \xrightarrow{\text{SOT}} T$ if $T_\lambda x \to Tx$ for all $x \in X$. This is the topology of pointwise convergence.

3.1.5. EXAMPLE. Fréchet spaces. Let X be a LCTVS given by a *countable* family of seminorms $\{p_i : i \geq 1\}$. Then there is a translation invariant metric on X which yields the same topology. Define

$$d(x, y) = \sum_{i \geq 1} 2^{-i} \frac{p_i(x-y)}{1+p(x-y)} \quad \text{for} \quad x, y \in X.$$

Since this depends only on $x-y$, it is translation invariant. The function $f(t) = \frac{t}{1+t}$ is increasing. Therefore

$$\frac{p(x+y)}{1+p(x+y)} \leq \frac{p(x)+p(y)}{1+p(x)+p(y)} \leq \frac{p(x)}{1+p(x)} + \frac{p(y)}{1+p(y)}.$$

This is valid for each p_i, and so d satisfies the triangle inequality. Since $d(x, y) = 0$ only if $p_i(x-y) = 0$ for all $i \geq 1$, this implies that $x = y$. Hence d is a metric. Now $x_\lambda \to x$ if and only if $p_i(x - x_\lambda) \to 0$ for $i \geq 1$ if and only if $\frac{p_i(x-x_\lambda)}{1+p(x-x_\lambda)} \to 0$ for $i \geq 1$. If $d(x, x_\lambda) \to 0$, then clearly $\frac{p_i(x-x_\lambda)}{1+p(x-x_\lambda)} \to 0$ for $i \geq 1$. Conversely, if $\frac{p_i(x-x_\lambda)}{1+p(x-x_\lambda)} \to 0$ for $i \geq 1$ and $\varepsilon > 0$, choose N so that $2^{-N} < \varepsilon/2$. Then choose $\lambda_0 \in \Lambda$ so that $\frac{p_i(x-x_\lambda)}{1+p(x-x_\lambda)} < \frac{\varepsilon}{2}$ for $1 \leq i \leq N$ and $\lambda \geq \lambda_0$. Then

$$d(x_\lambda, x) < \sum_{i=1}^{N} \frac{\varepsilon}{2} 2^{-i} + \sum_{i>N} 2^{-i} < \frac{\varepsilon}{2} + 2^{-N} < \varepsilon.$$

Therefore, the topology on X is equivalent to the metric topology given by d.

We say X is a *Fréchet space* if it is complete in this metric. It is an easy exercise to show that a sequence is Cauchy in this metric if and only if it is Cauchy in each seminorm p_i for $i \geq 1$. Thus it is complete in the metric if and only if every sequence which is Cauchy in each seminorm converges in X. See Exercise **3.15**.

(a) Let $\Omega \subset \mathbb{C}$ be an open subset of the complex plane. Let $H(\Omega)$ be the space of analytic functions on Ω. For each compact $K \subset \Omega$, define $p_K(f) = \sup_{z \in K} |f(z)|$. Then $\mathcal{P} = \{p_K : K \subset \Omega, \text{ compact}\}$ is a family of seminorms that makes $H(\Omega)$ a LCTVS. Observe that $f_\lambda \to f$ in $H(\Omega)$ if and only if f_λ converges to f uniformly on compact subsets of Ω, often called u.c.c. convergence. It is straightforward to show that the u.c.c. limit of analytic functions is analytic. It is not necessary to use all compact sets. Define $K_n = \{z \in \Omega : |z| \leq n, \ \text{dist}(z, \Omega^c) \geq \frac{1}{n}\}$. Then each K_n is compact, $K_n \subset K_{n+1}$, and $\bigcup_{n \geq 1} \operatorname{int} K_n = \Omega$. Thus for any compact set $K \subset \Omega$, there is some n so that $K \subset K_n$. It follows that the seminorms p_{K_n} for $n \geq 1$ determine the topology. Since $H(\Omega)$ is complete in the topology of u.c.c. convergence, it is a Fréchet space.

(b) Let Ω be an open subset of $\mathbb{R}^n$, and let $C^\infty(\Omega)$ be the space of all C^∞ functions on Ω. For $\alpha = (\alpha_1, \dots, \alpha_n) \in \mathbb{N}_0^n$, let $|\alpha| = \sum_{i=1}^n \alpha_i$. Define $D^\alpha f = \dfrac{\partial^{|\alpha|} f}{\partial x^{\alpha_1} \cdots \partial x_n^{\alpha_n}}$. For each compact $K \subset \Omega$ and $k \in \mathbb{N}_0$, define

$$p_{K,k}(f) = \sup_{x \in K,\, \alpha \in \mathbb{N}_0^n,\, |\alpha| \le k} |D^\alpha f(x)|.$$

Then $\mathcal{P} = \{p_{K,k} : k \ge 0,\ K \subset \Omega, \text{ compact}\}$ is a family of seminorms that makes $C^\infty(\Omega)$ a LCTVS. It is possible to select a sequence of compact sets K_j with $\Omega = \bigcup_{j \ge 0} K_j$ and $K_j \subset \operatorname{int} K_{j+1}$. Then $p_j := p_{K_j,j}$ for $j \ge 0$ is an increasing sequence of seminorms which eventually dominates all $p_{K,k}$.

Suppose that (f_n) is a sequence in $C^\infty(\Omega)$ which is Cauchy with respect to every p_j. Then each sequence of derivatives $\left(D^\alpha f_n\right)_{n \ge 1}$ converges uniformly on every compact subset of Ω. It is a standard result from calculus that if the sequence of derivatives of a function converge uniformly and the sequence of functions converge uniformly, then the derivatives converge to the derivative of the limit. Therefore, the limit function $f = \lim_n f_n$ is C^∞ and $D^\alpha f = \lim_n D^\alpha f_n$ uniformly on compact subsets of Ω. That is, $C^\infty(\Omega)$ is complete for this family of seminorms, i.e. it is a Fréchet space.

If $K \subset \mathbb{R}^n$ is compact, let $\mathcal{D}_K$ denote the space of all $f \in C^\infty(\mathbb{R}^n)$ with support in K, i.e. $\overline{\{x : f(x) \ne 0\}} \subset K$. If $K \subset \Omega$, then $\mathcal{D}_K \subset C^\infty(\Omega)$. Moreover, if (f_n) is a Cauchy sequence in $\mathcal{D}_K$, then it is clear that its limit in $C^\infty(\Omega)$ still has support in K. Hence $\mathcal{D}_K$ is closed, and thus is also a Fréchet space. Then one defines $C_c^\infty(\Omega)$ or $\mathcal{D}(\Omega)$ as $\bigcup_{K \subset \Omega, \text{ compact}} \mathcal{D}_K$. This is the space of all C^∞ functions with compact support in Ω, known as the *test functions* on Ω. This is not closed, and in fact is dense in $C^\infty(\Omega)$. To see this, suppose that $f \in C^\infty(\Omega)$. For each $j \ge 1$, there is a C^∞ function $h_j \in \mathcal{D}_{K_{j+1}}$ such that $h|_{K_j} = 1$. Then $fh_j \in \mathcal{D}_{K_{j+1}}$ and $f|_{K_j} = fh|_{K_j}$, so that $p_j(f - fh_j) = 0$. Hence fh_j converges to f in $C^\infty(\Omega)$.

3.1.6. PROPOSITION. *Let X be an LCTVS, and let $f : X \to \mathbb{F}$ be a linear map. The following are equivalent*

(1) *f is continuous.*

(2) *f is continuous at* 0.

(3) ker *f is closed.*

(4) *There is a finite subset $F = \{p_1, \dots, p_n\} \subset \mathcal{P}$ and $t_i \in \mathbb{R}^+$ so that $|f(x)| \le \sum_{i=1}^n t_i p_i(x)$ for all $x \in X$.*

PROOF. Clearly (1) implies (2). If (2) holds and $x_\lambda \in \ker f$ such that $x_\lambda \to x$, then $x - x_\lambda \to 0$. Thus $f(x) - f(x_\lambda) \to 0$ by continuity at 0. Since $f(x_\lambda) = 0$, we have $f(x) = 0$, so ker f is closed.

Suppose that (3) holds. We may assume that $f \neq 0$. Pick an $x_0 \in X$ so that $f(x_0) = 1$. Since $\ker f$ is closed, there is a neighbourhood $x_0 + U_{F,r}$ disjoint from $\ker f$. Since $x_0 + U_{F,r}$ is convex, so is $f(x_0 + U_{F,r}) = 1 + f(U_{F,r})$. Therefore $-1 \notin f(U_{F,r})$. However $U_{F,r}$ is *balanced*, meaning that if $x \in U_{F,r}$ and $|\lambda| \leq 1$ for $\lambda \in \mathbb{F}$, then $\lambda x \in U_{F,r}$. This is because $p_i(\lambda x) = |\lambda| p_i(x) \leq p_i(x) < r$ for $p_i \in F$. Therefore $f(U_{F,r}) \subset \mathbb{D} = \{z \in \mathbb{F} : |z| < 1\}$. It follows that if $\sum_{i=1}^n p_i(x) < r$, then $p_i(x) < r$ for $1 \leq i \leq n$, and so $|f(x)| < 1$. Scaling shows that $|f(x)| \leq \sum_{i=1}^n \frac{1}{r} p_i(x)$. So (4) holds.

Suppose that (4) holds: $|f(x)| \leq \sum_{i=1}^n t_i p_i(x)$ for all $x \in X$. Then if $x_\lambda \to x$, then $p_i(x - x_\lambda) \to 0$ and so $|f(x) - f(x_\lambda)| \leq \sum_{i=1}^n t_i p_i(x - x_\lambda) \to 0$. Therefore $f(x_\lambda) \to f(x)$, and f is continuous. ■

We will write X^* for the vector space of continuous linear functionals on a LCTVS X. The natural topology on X^* is the τ_X topology, which makes X^* a LCTVS. However, in the case of a Banach space, there was also a natural norm topology on X^* which is stronger than the τ_X topology. In general, there are no such norms.

One might wonder, though, what happens when X is a Fréchet space which is not a Banach space. We may suppose that the topology on X is given by a sequence of seminorms $\{p_n : n \geq 1\}$ such that $p_n \leq p_{n+1}$. If necessary, we just take the max or sum of the first n seminorms. Then (X, p_n) or more precisely $(X/\ker p_n, p_n)$ are normed spaces and have a Banach space dual $Y_n := (X, p_n)^*$. Proposition 3.1.6 shows that $X^* = \bigcup_{n \geq 1} Y_n$. Since $p_n \leq p_{n+1}$, any functional satisfying $|f(x)| \leq Cp_n(x)$ satisfies $|f(x)| \leq Cp_{n+1}(x)$, so that $Y_n \subset Y_{n+1}$. Each Y_n has a norm p_n^* induced by the norm of the functional on (X, p_n), but this norm does not extend to a norm on X^*. So it raises the question of whether there is some mysterious complete metric on X^* making it into a Fréchet space. It would be important that the metric topology be stronger than the τ_X topology, so that the maps $f \mapsto f(x)$ are continuous for each $x \in X$. It turns out that there is no such metric unless X is already a Banach space. See Exercise **3.14**.

We now discuss weak topologies on X in more detail. First a useful lemma.

3.1.7. LEMMA. *Let $f_1, \ldots, f_n$ be linear functionals on a vector space X. Suppose that f is a linear functional on X such that* $\ker f \supset \bigcap_{i=1}^n \ker f_i$. *Then* $f \in \operatorname{span}\{f_1, \ldots, f_n\}$.

PROOF. We may assume that $f_1, \ldots, f_n$ are linearly independent because if say $f_n = \sum_{i=1}^{n-1} a_i f_i$, then $\ker f_n \supset \bigcap_{i=1}^{n-1} \ker f_i$. Thus there is no loss in dropping f_n from the list.

Define $T : X \to \mathbb{F}^n$ by $Tx = (f_1(x), \ldots, f_n(x))$. By linear independence, this map is surjective. Note that $\ker T = \bigcap_{i=1}^n \ker f_i$. So $X/\ker T \simeq \mathbb{F}^n$. Since $\ker f \supset \bigcap_{i=1}^n \ker f_i$, there is a well-defined functional $\widetilde{f}$ on $X/\ker T \simeq \mathbb{F}^n$ given by $\widetilde{f}(x + \ker T) = f(x)$. Now every linear functional on $\mathbb{F}^n$ has the form $\varphi(y) =$

$\sum_{i=1}^n a_i y_i$. It follows that there are scalars a_i so that $f(x) = \sum_{i=1}^n a_i f_i(x)$. Hence $f \in \text{span}\{f_1, \dots, f_n\}$. ■

3.1.8. COROLLARY. *A linear functional $f : (X, \tau_Y) \to \mathbb{F}$ is continuous if and only if $f \in Y$.*

PROOF. By definition, if $f \in Y$, then it is continuous. On the other hand, if f is continuous, then by Proposition 3.1.6, there are $f_1, \dots, f_n \in Y$ so that $|f(x)| \le \sum_{i=1}^n t_i |f_i(x)|$. This forces $\ker f \supset \bigcap_{i=1}^n \ker f_i$. Hence by Lemma 3.1.7, f lies in $\text{span}\{f_1, \dots, f_n\} \subset Y$. ■

Under certain circumstances, convergent *sequences* must be bounded.

3.1.9. LEMMA. *Let X be a Banach space, and let Y be a closed subspace of X^* which norms X, i.e.* $\sup\{|f(x)| : f \in Y,\ \|f\| \le 1\} = \|x\|$. *Then if a sequence $x_n \xrightarrow{\tau_Y} x$, the set $\{x_n : n \ge 1\}$ is bounded.*

PROOF. Consider $\hat{x}_n$ as linear functionals on Y by $\hat{x}_n(f) = f(x_n)$. The norming condition ensures that $\|\hat{x}_n\| = \|x_n\|$. Now $x_n \xrightarrow{\tau_Y} x$ means that $\hat{x}_n(f) \to \hat{x}(f)$ for all $f \in Y$. By the Banach–Steinhaus theorem, $\{\hat{x}_n : n \ge 1\}$ is bounded. Therefore $\{x_n : n \ge 1\}$ is bounded. ■

3.1.10. REMARK. This lemma applies in particular to the weak topology τ_{X^*} on X and to the weak-$*$ topology (X^*, τ_X).

If Y is not closed in X^*, then the space of continuous functionals on (X, τ_Y) is not complete. An example of this is the weak operator topology on $\mathcal{B}(X, Y)$ (Example 3.1.4) when X, Y are infinite dimensional. So Lemma 3.1.9 does not apply directly. However, in this case, the conclusion is still valid. It is obtained using two applications of the Banach–Steinhaus theorem using the completeness of X and Y separately. See Exercise **3.3**(b).

3.1.11. EXAMPLE. In Example 2.6.21, we defined the adjoint of $T \in \mathcal{B}(X, Y)$, where X, Y are Banach spaces. We showed that $T^* \in \mathcal{B}(Y^*, X^*)$. Here we will show that T^* is continuous as a map from (Y^*, τ_Y) to (X^*, τ_X), i.e. T^* is weak-$*$–weak-$*$ continuous. Let $g_0 \in Y^*$ and let $F \Subset X$ be a finite set. Then $T^* g_0 + U_{F,r}$ is a basic open neighbourhood of $T^* g_0$. Observe that

$$\begin{aligned}(T^*)^{-1}(T^* g_0 + U_{F,r}) &= \{g \in Y^* : T^*(g - g_0) \in U_{F,r}\} \\ &= \{g \in Y^* : |T^*(g - g_0)(x_i)| < r,\ x_i \in F\} \\ &= \{g \in Y^* : |(g - g_0)(T x_i)| < r,\ x_i \in F\} = g_0 + V_{G,r}\end{aligned}$$

where $G = \{T x_i : x_i \in F\}$. This is τ_Y open in Y^*, whence T^* is weak-$*$–weak-$*$ continuous.

3.1.12. EXAMPLE. Since $l_1 = c_0^*$, it has an associated weak-$*$ topology τ_{c_0}. Caveat: l_1 has many different preduals, and hence many different weak-$*$ topologies. Claim: a bounded net $(x_\lambda)_\Lambda$ converges τ_{c_0} to x_0 if and only if each coordinate $x_{\lambda,i} \to x_{0,i}$ for $i \geq 1$. One direction is clear since the element $\varepsilon_i \in c_0$ with a 1 in the ith coordinate and 0s elsewhere is in c_0; whence $x_{\lambda,i} = \varepsilon_i(x_\lambda) \to \varepsilon_i(x_0) = x_{0,i}$.

Conversely, suppose that $\sup_\Lambda \|x_\lambda\|_1 = M$ and that $x_{\lambda,i} \to x_{0,i}$ for $i \geq 1$. Let $y = (y_i) \in c_0$ and $\varepsilon > 0$. Pick K so that $\sup_{i>K} |y_i| < \varepsilon$. Then pick λ_0 so that for $\lambda \geq \lambda_0$, $\sum_{i=1}^K |x_{\lambda,i} - x_{0,i}|\,|y_i| < \varepsilon$, which is possible since the first K terms $|x_{\lambda,i} - x_{0,i}|$ each converge to 0. Then for $\lambda \geq \lambda_0$, since $\|x_\lambda - x_0\| \leq M + \|x_0\|$,

$$\begin{aligned}|\langle x_\lambda - x_0, y\rangle| &\leq \sum_{i=1}^{K} |x_{\lambda,i} - x_{0,i}|\,|y_i| + \sum_{i>K} |x_{\lambda,i} - x_{0,i}|\,|y_i| \\ &\leq \varepsilon + (M + \|x\|)\varepsilon.\end{aligned}$$

Since $\varepsilon > 0$ is arbitrary, this shows that the net $(x_\lambda)_\Lambda$ converges in τ_{c_0} to x_0.

Next we construct an unbounded net that converges to 0 in τ_{c_0}. Consider the net $\Lambda = \{F \Subset c_0 : |F| < \infty\}$ of finite subsets of c_0 ordered by inclusion. Use the Axiom of Choice to pick $x_F \in \bigcap_{y \in F} \ker y$ with $\|x_F\|_1 = |F|$. Since each $\ker y$ has codimension 1, this intersection has finite codimension, and hence is non-zero. This is clearly an unbounded net. However for any $y \in c_0$ and $F \supset \{y\}$, $\langle x_F, y\rangle = 0$. Hence $(x_F)_\Lambda \xrightarrow{\tau_{c_0}} 0$.

3.2. Schwartz Space and Distributions

This section provides a quick look at an important Fréchet space and its dual which has many applications. The following development can be carried out in $\mathbb{R}^n$, but we will just work with domain $\mathbb{R}$.

3.2.1. DEFINITION. The *Schwartz space* $\mathcal{S}(\mathbb{R})$, or space of *rapidly decreasing functions*, is the set of all C^∞ functions f on $\mathbb{R}$ such that

$$\sup_{x\in\mathbb{R}} \left|x^n f^{(m)}(x)\right| < \infty \quad \text{for all} \quad m, n \geq 0.$$

Define seminorms on $\mathcal{S}(\mathbb{R})$ by $p_k(f) = \sup_{x\in\mathbb{R},\, 0\leq m,n\leq k} \left|x^n f^{(m)}(x)\right|$ for $k \geq 0$. So $\mathcal{S}(\mathbb{R})$ is a LCTVS with the topology determined by $\{p_k : k \geq 0\}$.

3.2.2. REMARK. It is easy to check that $\mathcal{S}(\mathbb{R})$ is a vector space. A C^∞ function f belongs to $\mathcal{S}(\mathbb{R})$ if and only if for each $m, n \geq 0$, there is a constant $C_{m,n}$ so that

$$\left|f^{(m)}(x)\right| \leq \frac{C_{m,n}}{1 + |x|^n} \quad \text{for} \quad x \in \mathbb{R}.$$

Indeed, if $f \in \mathcal{S}(\mathbb{R})$, then for $k = \max\{m, n\}$,

$$\left|f^{(m)}(x)\right| \leq \begin{cases} p_k(f) \leq \frac{2p_k(f)}{1+|x|^n} & \text{if } |x| \leq 1 \\ \frac{p_k(f)}{|x|^n} \leq \frac{2p_k(f)}{1+|x|^n} & \text{if } |x| \geq 1. \end{cases}$$

Since $\mathcal{S}(\mathbb{R})$ is a subspace of $C_0(\mathbb{R})$, it follows that p_0 is a norm, and thus p_k is a norm for all $k \geq 0$.

3.2.3. EXAMPLES.
(1) If $f \in C_c^\infty(\mathbb{R})$ is a C^∞ function on $\mathbb{R}$ with compact support, then $f \in \mathcal{S}(\mathbb{R})$ because $x^n f^{(m)}(x)$ is also a C^∞ function with compact support, and thus it is bounded.

(2) $f(x) = e^{-x^2} \in \mathcal{S}(\mathbb{R})$ because

$$e^{-x^2} = \frac{1}{e^{x^2}} \leq \frac{1}{1 + x^2 + \frac{1}{2!}x^4 + \cdots + \frac{1}{n!}x^{2n}} \leq \frac{n!}{1 + x^{2n}} \qquad \text{for } n \geq 1.$$

Since $f^{(m)}(x) = e^{-x^2} q_m(x)$ for some polynomial q_m, similar estimates hold for $f^{(m)}(x)$.

(3) If $f \in \mathcal{S}(\mathbb{R})$ and p is a polynomial, then $pf \in \mathcal{S}(\mathbb{R})$. Indeed, if p has degree d, then $|p(x)| \leq C(1 + |x|^d)$ for some constant C. Thus

$$\left|x^n p(x) f(x)\right| \leq \frac{C_{0,n+d} C(1 + |x|^d)}{1 + |x|^{n+d}} \leq \frac{C'}{1 + |x|^n}.$$

Similarly, $(pf)^{(m)}$ is a finite sum of functions of the same type, so one obtains similar bounds.

3.2.4. PROPOSITION. *$\mathcal{S}(\mathbb{R})$ is a Fréchet space.*

PROOF. The topology on $\mathcal{S}(\mathbb{R})$ is given by a countable family of seminorms, so it suffices to establish completeness. Let (f_i) be a sequence in $\mathcal{S}(\mathbb{R})$ which is Cauchy in each p_k norm for $k \geq 0$. Then in particular, for each $m, n \geq 0$, the sequence $x^n f_i^{(m)}$ is Cauchy in the sup norm. Thus it converges uniformly to a function $g_{m,n}$. It is a well-known result from calculus that if a sequence h_i of differentiable functions converges uniformly to h and h_i' converges uniformly to k, then h is differentiable and $h' = k$. We deduce that $g_{m+1,0} = g'_{m,0}$, and thus $g = g_{0,0}$ is C^∞ and $g_{m,0} = g^{(m)}$. Moreover, it also follows easily that $g_{m,n} = x^n g_{m,0} = x^n g^{(m)}$. Therefore $g \in \mathcal{S}(\mathbb{R})$ and $g = \lim f_i$ in each p_k norm, i.e. $\mathcal{S}(\mathbb{R})$ is complete. ■

3.2.5. REMARK. The space $C_c^\infty(\mathbb{R})$ is dense in $\mathcal{S}(\mathbb{R})$. The proof is similar to the proof for $C^\infty(\mathbb{R})$ in Example 3.1.5(b).

3.2.6. DEFINITION. A *tempered distribution* is a continuous linear functional on $\mathcal{S}(\mathbb{R})$. The space of all tempered distributions is the dual space $\mathcal{S}(\mathbb{R})^*$.

3.2.7. EXAMPLES.

(1) Suppose that μ is a finite regular Borel measure on $\mathbb{R}$ and h is Borel and satisfies $|h(x)| \leq C(1+|x|^d)$ for some $d \in \mathbb{N}$. (We say that h has *polynomially bounded growth.*) Consider the measure $h\mu$ as a functional on $\mathcal{S}(\mathbb{R})$ given by

$$h\mu(f) = \int hf\,d\mu \quad \text{for} \quad f \in \mathcal{S}(\mathbb{R}).$$

Note that if $f \in \mathcal{S}(\mathbb{R})$, then $|f(x)| \leq \frac{2p_d(f)}{1+|x|^d}$ by Remark 3.2.2. Hence

$$|h\mu(f)| \leq \|\mu\|\,\|hf\|_\infty \leq 2C\|\mu\|\,p_d(f).$$

Therefore, this functional is continuous by Proposition 3.1.6.

More generally, suppose that ν is a Borel measure on $\mathbb{R}$ with polynomially bounded growth, meaning that there is a constant C and $n \in \mathbb{N}$ such that

$$\nu([-x,x]) \leq C(1+x^n).$$

If h has polynomially bounded growth as above, then again $h\nu$ belongs to $\mathcal{S}(\mathbb{R})^*$. In particular, if m is Lebesgue measure on $\mathbb{R}$, then $hm \in \mathcal{S}(\mathbb{R})^*$. We will not write m, and consider h itself as the functional on $\mathcal{S}(\mathbb{R})$ by $h(f) = \int_{\mathbb{R}} h(x)f(x)\,dx$.

(2) The delta function $\varepsilon_x(f) = f(x)$ is given by integration against the unit point mass δ_x. Likewise $\varepsilon_x^{(m)}(f) := (-1)^m f^{(m)}(x)$ is a continuous linear functional. We will explain soon why these may be considered as derivatives of ε_x. For $m \geq 1$, these functionals are not obtained by integration against some measure as in (1). See Exercise **3.17**.

(3) We note that the canonical injection of $\mathcal{S}$ into $L^p(\mathbb{R})$ for $1 \leq p \leq \infty$ is continuous. Indeed, $p_0(f) = \|f\|_\infty$, so the injection of $\mathcal{S}$ into $L^\infty(\mathbb{R})$ is continuous. Observe that

$$\begin{aligned}\|f\|_1 &= \int_{\mathbb{R}} \frac{1}{1+x^2}(1+x^2)|f|\,dx \leq \Big\|\frac{1}{1+x^2}\Big\|_1 \Big\|(1+x^2)f\Big\|_\infty \\ &\leq \pi(\|f\|_\infty + \|x^2 f\|_\infty) \leq 2\pi p_2(f).\end{aligned}$$

If $1 < p < \infty$, define q so that $\frac{1}{p} + \frac{1}{q} = 1$. Then by Holder's inequality,

$$\|f\|_p^p = \int_{\mathbb{R}} |f||f|^{p/q}\,dx \leq \|f\|_1 \|f\|_\infty^{p/q}.$$

Hence $\|f\|_p \leq 2\pi p_2(f)$. Therefore the imbedding of $\mathcal{S}$ into $L^p(\mathbb{R})$ is continuous. It follows that if $g \in L^q(\mathbb{R})$, then $\psi(f) = \int_{\mathbb{R}} f(x)g(x)\,dx$ is continuous on $\mathcal{S}$ with $|\psi(f)| \leq 2\pi\|g\|_q p_2(f)$.

(4) The principle value of integration against $\frac{1}{x}$ is continuous on $\mathcal{S}(\mathbb{R})$. That is,

$$\mathcal{P}(\tfrac{1}{x})(f) = \lim_{\varepsilon\to 0^+} \int_{|x|\geq\varepsilon} \frac{f(x)}{x}\,dx = \lim_{\varepsilon\to 0^+} \int_{\varepsilon}^{\infty} \frac{f(x)-f(-x)}{x}\,dx.$$

Observe that $g(x) = \begin{cases} \frac{f(x)-f(-x)}{x} & \text{if } x>0 \\ 2f'(0) & \text{if } x=0 \end{cases}$ is continuous on $[0,\infty)$. Now

$$\left|\frac{f(x)-f(-x)}{x}\right| \leq \frac{1}{x}\int_{-x}^{x} |f'(t)|\,dt \leq 2\|f'\|_\infty \leq 2p_1(f).$$

Also $|xf(x)| \leq p_1(f)$. Thus

$$\begin{aligned}\mathcal{P}(\tfrac{1}{x})(f) &\leq \int_0^1 \frac{f(x)-f(-x)}{x}\,dx + \int_1^\infty |xf(x)| + |-xf(-x)|\,\frac{dx}{x^2} \\ &\leq \int_0^1 2p_1(f)\,dx + \int_1^\infty 2p_1(f)\,\frac{dx}{x^2} = 4p_1(f).\end{aligned}$$

Hence $P(\frac{1}{x})$ is continuous.

Integration by parts yields the formula for $f,g\in\mathcal{S}(\mathbb{R})$:

$$\int_{\mathbb{R}} f'(x)g(x)\,dx = -\int_{\mathbb{R}} f(x)g'(x)\,dx$$

because the boundary terms vanish. This motivates the following definition.

3.2.8. DEFINITION. If ψ is a tempered distribution, the *derivative* of ψ is given by

$$\psi'(f) = -\psi(f') \quad \text{for} \quad f\in\mathcal{S}(\mathbb{R}).$$

3.2.9. LEMMA. *If $\psi\in\mathcal{S}(\mathbb{R})^*$, then ψ' is a tempered distribution. Tempered distributions have derivatives of all orders given by $\psi^{(d)}(f) = (-1)^d\psi(f^{(d)})$.*

PROOF. We first observe that the derivative map D on $\mathcal{S}(\mathbb{R})$ given by $Df = f'$ is continuous. Indeed, if $f_i\in\mathcal{S}(\mathbb{R})$ converges to f, then for all $k\geq 0$, we have $p_{k+1}(f_i - f)\to 0$. That is, $\|x^n(f_i^{(m)} - f^{(m)})\|_\infty \to 0$ for $0\leq m,n\leq k+1$. In particular, $\|x^n(f_i'^{(m)} - f'^{(m)})\|_\infty \to 0$ for $0\leq m,n\leq k$. Therefore it follows that $p_k(f_i' - f')\to 0$ for all $k\geq 0$. Hence D is sequentially continuous. As $\mathcal{S}(\mathbb{R})$ is metrizable, D is continuous.

Therefore $\psi' = -\psi D$ is continuous and linear, and so lies in $\mathcal{S}(\mathbb{R})^*$. It follows by an easy induction that $\psi^{(d)}(f) = (-1)^d\psi(f^{(d)})$. ∎

3.2.10. EXAMPLE. Let $g(x) = \begin{cases} x & \text{if } x \geq 0 \\ 0 & \text{if } x < 0 \end{cases}$. Then g has polynomial growth, so determines a functional on $\mathcal{S}(\mathbb{R})$ as in Example 3.2.7(1). Compute the derivative by integration by parts:

$$g'(f) = -g(f') = -\int_0^\infty x f'(x)\,dx = -xf\Big|_0^\infty + \int_0^\infty f(x)\,dx = H(f)$$

where H is the Heavyside function $H(x) = \begin{cases} 1 & \text{if } x \geq 0 \\ 0 & \text{if } x < 0 \end{cases}$. Note that g is continuous on $\mathbb{R}$, but is not differentiable at $x = 0$.

Now we compute the derivative of H as a distribution, even though H is not even continuous on $\mathbb{R}$.

$$H'(f) = -\int_0^\infty f'(x)\,dx = -f\Big|_0^\infty = f(0) = \varepsilon_0(f).$$

Thus $H' = \varepsilon_0$ corresponds to the point mass at 0.

3.3. Geometric Hahn–Banach Theorems

By a geometric Hahn–Banach theorem, we mean a result which takes two disjoint convex sets and separates them by a continuous linear functional, i.e. the functional takes disjoint values on the two sets. Some topological hypothesis is necessary to make this separation possible. (See Exercise **3.5**.)

A subspace $M \subset X$ has *finite codimension* if there are finitely many vectors $x_1, \dots, x_n$ such that $X = \operatorname{span}\{M, x_1, \dots, x_n\}$, and the codimension is the minimal number n which works. If $f \neq 0$ is a linear functional on X (not necessarily continuous), then pick some x_0 with $f(x_0) = 1$. Then we can write $x = (x - f(x)x_0) + f(x)x_0 \in \ker f + \mathbb{F}x_0$ for every $x \in X$. So $\ker f$ has codimension 1. By Proposition 3.1.6, $\ker f$ is closed precisely when it is continuous. So if f is discontinuous, $\overline{\ker f} = X$. It is easy to check that the intersection of finitely many subspaces of finite codimension is still finite codimension.

3.3.1. DEFINITION. Two subsets A, B in a vector space X are *separated* if there is a linear functional f such that $f(A)$ and $f(B)$ are disjoint.

We will generally try to separate convex sets. In this case, $f(A)$ and $f(B)$ are convex. So there is a line in the complex plane separating the two set. We can rotate this picture by replacing f by some $e^{i\theta} f$ so that the line has the form $\operatorname{Re} z = \alpha$. In particular, $A \subset \{x \in X : \operatorname{Re} f(x) \leq \alpha\}$ and $B \subset \{x \in X : \operatorname{Re} f(x) \geq \alpha\}$. The set $H = \{x \in X : \operatorname{Re} f(x) = \alpha\}$ is an *affine* set (i.e. $x, y \in H$ implies that $tx + (1-t)y \in H$ for $0 \leq t \leq 1$) which divides X into two convex halves

$H^+ = \{x : \operatorname{Re} f(x) > a\}$ and $H^- = \{x : \operatorname{Re} f(x) < a\}$. In the real case, H is a translate of a codimension 1 subspace (a *hyperplane*), and is often called an *affine hyperplane*. In the complex case, it is the translate of a *real codimension 1* subspace, $\ker \operatorname{Re} f$. In either case, H is closed if and only if f is continuous.

3.3.2. Lemma. *Let X be a LCTVS, $0 \in U$ an open convex set, and let p_U be the Minkowski functional $p_U(x) = \inf\{r > 0 : x \in rU\}$. Then p_U is continuous and $\{x : p_U(x) < 1\} = U$.*

Proof. Since $0 \in U$ and U is convex, $rU \subset sU$ if $0 < r < s$. Hence if $p_U(x) < 1$, then there is $r < 1$ so that $x \in rU \subset U$. Conversely, if $x \in U$, then since $(t, x) \to tx$ is continuous, there is some $\varepsilon > 0$ so that $(1+\varepsilon)x \in U$. Hence $p_U(x) \le \frac{1}{1+\varepsilon} < 1$. Therefore $U = \{x : p_U(x) < 1\}$.

Suppose that $V \subset \mathbb{R}$ is open and $p_U(x_0) = t_0 \subset (t_0 - r, t_0 + r) \subset V$. If $x \in x_0 + r(U \cap -U)$, then $p_U(x - x_0) < r$ and $p_U(x_0 - x) < r$. By sublinearity,

$$p(x) \le p(x - x_0) + p(x_0) < p(x_0) + r$$

and

$$p(x_0) \le p(x_0 - x) + p(x) < p(x) + r.$$

Combining, we get $|p_U(x) - p_U(x_0)| < r$. Hence $x_0 + r(U \cap -U) \subset p_U^{-1}(V)$. Therefore p_U is continuous. ■

3.3.3. Hyperplane Theorem. *Let X be a LCTVS and let $U \subset X$ be an open convex set such that $0 \notin U$. Then there is a continuous functional $f \in X^*$ so that* $\operatorname{Re} f(x) > 0$ *for all $x \in U$, i.e. $H = \{x : \operatorname{Re} f(x) = 0\}$ is a closed real subspace of real codimension 1 which is disjoint from U.*

Proof. Case: $\mathbb{F} = \mathbb{R}$. Fix $x_0 \in U$. Then $V = x_0 - U$ is an open convex neighbourhood of 0. By Lemma 3.3.2, p_V is continuous. Since $x_0 \notin V$, we have that $p_V(x_0) \ge 1$. Define $f_0(tx_0) = t$ for $t \in \mathbb{R}$. Then

$$f_0(tx_0) = \begin{cases} t \le tp_V(x_0) = p_V(tx_0) & \text{if } t \ge 0 \\ t \le 0 \le p_V(tx_0) & \text{if } t < 0. \end{cases}$$

So $f_0 \le p_V$ on $\mathbb{R}x_0$. By the Hahn–Banach theorem 2.6.5, there is a linear functional f on X so that $f(x) \le p_V(x)$ for all $x \in X$ and $f(x_0) = 1$. Note that

$$\begin{aligned} f^{-1}(b_1(0)) &= \{x : |f(x)| < 1\} = \{x : f(x) < 1\} \cap \{x : f(-x) < 1\} \\ &\supseteq \{x : p_V(x) < 1\} \cap \{x : p_V(-x) < 1\} = V \cap -V. \end{aligned}$$

Thus $f^{-1}(b_1(0))$ contains an open neighbourhood of 0. Therefore f is continuous. Finally if $x \in U$, then $x_0 - x \in V$, so that $f(x_0 - x) < p_V(x_0 - x) < 1$. Hence $f(x) > f(x_0) - 1 = 0$. Then $H = \ker f$ is a closed hyperplane disjoint from U.

Case: $\mathbb{F} = \mathbb{C}$. Treat X as a real vector space and find a continuous real functional f using the real case. Set $g(x) = f(x) + if(-ix)$. As in the proof of the Hahn–Banach theorem, g is a continuous, complex linear functional such that $\operatorname{Re} g = f$. Thus if $x \in U$, we have $\operatorname{Re} g(x) = f(x) > 0$. ■

The following two special cases are immediate.

3.3.4. COROLLARY. *If X is a normed vector space, $U \subset X$ is convex and open, and $0 \notin U$, then there is some $f \in X^*$ so that* $\operatorname{Re} f(x) > 0$ *for all $x \in U$.*

3.3.5. COROLLARY. *If X is a normed vector space, $Y \subset X^*$ separates points, $U \subset X$ is convex and τ_Y-open, and $0 \notin U$, then there is some $f \in Y$ so that* $\operatorname{Re} f(x) > 0$ *for all $x \in U$.*

Next we consider the separation of two disjoint convex sets. Again a topological condition is required.

3.3.6. SEPARATION THEOREM. *Let X be a LCTVS and let A, B be disjoint convex subsets of X. If A is open, then there is a continuous linear functional $f \in X^*$ and $\alpha \in \mathbb{R}$ so that*

$$\operatorname{Re} f(b) \leq \alpha < \operatorname{Re} f(a) \quad \textit{for all} \quad a \in A,\ b \in B.$$

PROOF. Let $U = A - B = \{a - b : a \in A,\ b \in B\} = \bigcup_{b \in B} A - b$. This is open, convex and $0 \notin U$. By the Hyperplane theorem, there is an $f \in X^*$ so that $\operatorname{Re} f(x) > 0$ for all $x \in U$. Thus for $a \in A$ and $b \in B$,

$$0 < \operatorname{Re} f(a - b) = \operatorname{Re} f(a) - \operatorname{Re} f(b).$$

Hence

$$\sup_{b \in B} \operatorname{Re} f(b) := \alpha \leq \inf_{a \in A} \operatorname{Re} f(a).$$

Now $f(A)$ is open. Indeed, suppose that $a \in U_{F,r}(a) \subset A$. If $U_{F,r}(0) \subset \ker f$, then $\ker f \supset \bigcup_{n \geq 1} U(F, nr)(0) = X$ and so $f = 0$. This isn't possible, and thus $f(U_{F,r}(0))$ contains some $z \neq 0$. Since $U_{F,r}(0)$ is balanced and convex, so is $f(U_{F,r}(0))$, and so $f(U_{F,r}(0))$ contains an open disc about 0. It follows that $f(A)$ is open, and thus the infimum $\inf_{a \in A} \operatorname{Re} f(a)$ is not attained. Therefore $\alpha < f(a)$ for all $a \in A$. ■

The following is immediate.

3.3.7. COROLLARY. *Let X be a LCTVS and let A, B be disjoint open convex subsets of X. Then there is a continuous linear functional $f \in X^*$ and $\alpha \in \mathbb{R}$ so that*

$$\operatorname{Re} f(b) < \alpha < \operatorname{Re} f(a) \quad \textit{for all} \quad a \in A,\ b \in B.$$

3.3.8. LEMMA. *Let X be a LCTVS. Suppose that K is a compact subset of X, V is open in X and $K \subset V$. Then there is a convex open neighbourhood $U \ni 0$ such that $K + U \subset V$.*

PROOF. For each $x \in K$, there is a basic neighbourhood $x + U(F_x, r_x) \subset V$. The sets $\{x + U(F_x, \frac{r_x}{2}) : x \in K\}$ cover K, so there is a finite subcover $x_i + U(F_i, \frac{r_i}{2})$ for $1 \le i \le n$. Let $F = \bigcup_{i=1}^n F_i$ and $r = \min\{r_i : 1 \le i \le n\}$, and set $U = U(F, \frac{r}{2})$. If $x \in K$, then $x \in x_i + U(F_i, \frac{r_i}{2})$ for some i. Thus

$$\begin{aligned} x + U &\subset x_i + U(F_i, \tfrac{r_i}{2}) + U(F, \tfrac{r}{2}) \\ &\subset x_i + U(F_i, \tfrac{r_i}{2}) + U(F_i, \tfrac{r_i}{2}) = x_i + U(F_i, r_i) \subset V. \end{aligned}$$

Hence $K + U \subset V$. ■

3.3.9. COROLLARY. *Let X be a LCTVS and let A, B be disjoint closed convex subsets of X. If B is compact, then there is an $f \in X^*$ and $\beta < \alpha \in \mathbb{R}$ so that*

$$\sup_{b \in B} \operatorname{Re} f(b) = \beta < \alpha = \inf_{a \in A} \operatorname{Re} f(a).$$

PROOF. By Lemma 3.3.8, there is a convex open set $U \ni 0$ so that $B + U \subset A^c$. Then $B + U$ is convex and open, and disjoint from A. Hence there is $f \in X^*$ and $\alpha \in \mathbb{R}$ so that

$$\operatorname{Re} f(b) < \alpha = \inf_{a \in A} \operatorname{Re} f(a) \quad \text{for all} \quad b \in B + U.$$

Since B is compact and f is continuous, the value $\beta = \sup_{b \in B} \operatorname{Re} f(b)$ is attained, and thus $\beta < \alpha$. ■

3.3.10. EXAMPLE. Consider (l_1, τ_{c_0}). Let $\{\delta_i : i \ge 1\}$ be the standard basis for l_1. Define

$$A = \{x = (x_n) \in l_1 : \sum_{n \ge 1} x_n = 0\} \quad \text{and} \quad B = \{\delta_1\}.$$

Clearly A and B are disjoint convex sets, and B is compact. However, no functional $f \in c_0 = (l_1, \tau_{c_0})^*$ separates A from B. Indeed, suppose that $0 \ne f = (a_n) \in c_0$, say $a_m \ne 0$. Then $\delta_m - \delta_n \in A$ for $m < n$, and $f(\delta_m - \delta_n) = a_m - a_n \to a_m \ne 0$ as $n \to \infty$. So the subspace A is not contained in ker f, whence $f(A) = \mathbb{C}$. Therefore $f(\delta_1)$ cannot be disjoint from $f(A)$.

What went wrong is that A is not closed in the τ_{c_0} topology. Indeed, $A = \ker g$ where $g = (1, 1, 1, \dots) \in l_\infty = l_1^*$. By Proposition 3.1.6, A is not τ_{c_0} closed.

On the other hand, A is norm closed. The functional g separates A and B since $\sup_{a \in A} \operatorname{Re} g(a) = 0 < 1 = g(\delta_1)$.

Taking $A = \{x\}$ and $B = \{y\}$ yields

3.3.11. COROLLARY. *Let X be a LCTVS. Then X^* separates points of X.*

3.3.12. COROLLARY. *Let X be a LCTVS and let $A \subset X$. Then the closed convex hull of A is the intersection of all closed half-spaces containing A:*

$$\overline{\operatorname{conv}}(A) = \bigcap \Big\{ \{x : \operatorname{Re} f(x) \le \alpha\} : f \in X^*, \ \sup_{a \in A} \operatorname{Re} f(x) = \alpha < \infty \Big\}.$$

Hence if A is closed and convex, then $x \in A$ if and only if

$$\operatorname{Re} f(x) \le \sup_{a \in A} \operatorname{Re} f(x) \quad \textit{for all} \quad f \in X^*.$$

PROOF. The intersection of closed convex sets is closed and convex, and contains A by design. If $x \notin \overline{\operatorname{conv}}(A)$, then by Corollary 3.3.9, there is $f \in X^*$ so that $\operatorname{Re} f(x) > \alpha = \sup_{a \in A} \operatorname{Re} f(x)$. Thus the intersection is exactly $\overline{\operatorname{conv}}(A)$. ∎

3.3.13. PROPOSITION.

(1) *Every norm-closed convex set in a Banach space X is weakly closed.*

(2) *Every norm-closed ball in X^* is weak-$*$ closed.*

PROOF. A closed half space

$$\{x : \operatorname{Re} f(x) \le \alpha\} = f^{-1}(\{z \in \mathbb{C} : \operatorname{Re} z \le \alpha\})$$

for $f \in X^*$ is weakly closed. Thus a closed convex set is weakly closed by Corollary 3.3.12. The ball

$$\overline{b_r(x_0^*)} = \{x^* \in X^* : \|x^* - x_0^*\| \le r\} = \bigcap_{x \in \overline{b_1(X)}} \{x^* \in X^* : |\hat{x}(x^* - x_0^*)| \le r\}$$

is the intersection of weak-$*$ closed sets. ∎

3.3.14. GOLDSTINE THEOREM. *Let X be a Banach space. Then $b_1(X)$ is weak-$*$ dense in $b_1(X^{**})$. In particular, X is weak-$*$ dense in X^{**}.*

PROOF. Let $A = \overline{\iota b_1(X)}^{\tau_{X^*}}$. This set is τ_{X^*}-closed and convex. It is contained in $\overline{b_1(X^{**})}$ because the closed ball is also weak-$*$ closed by Proposition 3.3.13. If A is strictly smaller, let $x^{**} \in \overline{b_1(X^{**})} \setminus A$. By Corollary 3.3.9, there is some $f \in X^*$ so that

$$\sup_{a \in A} \operatorname{Re} f(a) = \alpha < f(x^{**}) \le \|f\|.$$

However

$$\alpha = \sup_{x \in b_1(X)} \operatorname{Re} f(x) = \|f\|.$$

This contradiction establishes the claim. ∎

3.4. Compactness in Weak Topologies

The unit ball of any infinite dimensional Banach space is not compact. See Exercise **2.6**(b). A useful substitute is that the unit ball of a dual space is weak-$*$ compact. This sometimes allows us to use compactness in the Banach space setting.

3.4.1. BANACH–ALAOGLU THEOREM. *Let X be a Banach space. Then the closed unit ball of X^* is weak-$*$ compact.*

PROOF. For each $x \in X$, let $\mathbb{D}_x = \{z \in \mathbb{F} : |z| \le \|x\|\}$. These are compact sets, and so by Tychonoff's theorem 1.8.7, $\mathcal{D} = \prod_{x\in X} \mathbb{D}_x$ is compact. Define

$$\Phi : (\overline{b_1(X^*)}, \tau_X) \to \mathcal{D} \quad \text{by} \quad \Phi(f)(x) = f(x) \text{ for } x \in X.$$

Claim 1: Φ is continuous. A basic open set in $\mathcal{D}$ is given by $x_1, \dots, x_n \in X$ and $U_i \subset \mathbb{D}_{x_i}$ open,

$$U = \{d \in \mathcal{D} : d(x_i) \in U_i,\ 1 \le i \le n\}.$$

Then

$$\Phi^{-1}(U) = \{f \in \overline{b_1(X^*)} : f(x_i) \in U_i\} = \overline{b_1(X^*)} \cap \bigcap_{i=1}^{n} \hat{x}_i^{-1}(U_i).$$

This is the intersection of finitely many weak-$*$ open sets with $\overline{b_1(X^*)}$, so it is weak-$*$ open in the relative topology.

Claim 2: Φ is one to one. This is because X separates points of X^*.

Claim 3: $\Phi\big(\overline{b_1(X^*)}\big)$ is closed in $\mathcal{D}$. Suppose that $f_\lambda \in \overline{b_1(X^*)}$ is a net such that $\Phi(f_\lambda) = d_\lambda$ converges in $\mathcal{D}$ to $d \in \mathcal{D}$. Then $d : X \to \mathbb{F}$ by evaluation at each coordinate, and $d(x) \in \mathbb{D}_x$, so $|d(x)| \le \|x\|$. Check to see that d is linear: if $x, y \in X$ and $\alpha, \beta \in \mathbb{F}$, then

$$d(\alpha x + \beta y) = \lim_{\Lambda} f_\lambda(\alpha x + \beta y) = \lim_{\Lambda} \alpha f_\lambda(x) + \beta f_\lambda(y) = \alpha d(x) + \beta d(y).$$

Hence $d \in X^*$ and $\|d\| \le 1$. Therefore $\Phi(\overline{b_1(X^*)})$ is a closed subset of a compact set, and thus it is compact.

Claim 4: Φ^{-1} is continuous on $\Phi(\overline{b_1(X^*)})$, so that Φ is a homeomorphism onto its image. A basic weak-$*$ open set in X^* has the form:

$$V = \{f \in X^* : f(x_i) \in U_i, 1 \le i \le n\} \quad \text{for} \quad U_i \text{ open in } \mathbb{F}.$$

Observe that

$$\Phi(V \cap \overline{b_1(X^*)}) = \{d \in \mathcal{D} : d(x_i) \in U_i, 1 \le i \le n\} \cap \Phi(\overline{b_1(X^*)}).$$

This is relatively open in $\Phi(\overline{b_1(X^*)})$. Therefore $\overline{b_1(X^*)}$ is the continuous image of a compact set, and hence it is compact. ∎

3.4.2. COROLLARY. *If X is a reflexive Banach space, then every norm-closed bounded convex set is weakly compact.*

PROOF. A closed convex set X is weakly closed by Proposition 3.3.13(1). Since it is bounded, it is contained in a large closed ball, which is weak-$*$ compact by the Banach–Alaoglu theorem. Since X is reflexive, the weak and weak-$*$ topologies are the same. So C is a weakly closed subset of a weakly compact set, and thus is weakly compact. ■

3.4.3. COROLLARY. $(\overline{b_1(X)}, \tau_{X^*})$ *is compact if and only if X is reflexive.*

PROOF. We just proved the only if direction, so suppose that $\overline{b_1(X)}$ is weakly compact. By Goldstine's theorem, $\iota(\overline{b_1(X)}, \tau_{X^*})$ is dense in $(\overline{b_1(X^{**})}, \tau_{X^*})$. Since $\overline{b_1(X)}$ is weakly compact and ι is continuous, the image is compact and therefore closed. Hence $\overline{b_1(X^{**})} = \overline{b_1(X)}$, and $X^{**} = X$. ■

3.5. Extreme points

Closed bounded convex sets have a nice boundary, and some points on the boundary are special in that they are not convex combinations of other points in the set. In finite dimensions, it is a well-known result that a convex set can be recovered from these extreme points. In this section, we extend this to compact convex sets in LCTVSs. In particular, by the previous section, this applies to closed bounded convex sets in dual Banach spaces, because we have compactness in the weak-$*$ topology.

3.5.1. DEFINITION. Let K be a convex set in a vector space V. Then $F \subset K$ is a *face* of K if $x, y \in K$, $0 < t < 1$ and $tx + (1-t)y \in F$ implies that $x, y \in F$. A point $x \in K$ is an *extreme point* of K if $\{x\}$ is a face. Let ext(K) denote the set of all extreme points of K.

3.5.2. EXAMPLES.
(1) Let $K = \{(x, y) \in \mathbb{R}^2 : y \geq 0,\ x^2 + y^2 \leq 1\}$ be a semicircle. Then $F = [-1, 1] \times \{0\}$ is a face. The extreme points are

$$\operatorname{ext} K = \{(x, y) \in \mathbb{R}^2 : y > 0,\ x^2 + y^2 = 1\} \cup \{(\pm 1, 0)\}.$$

Note that $\operatorname{int} K$ is convex and has no extreme points. Also $\mathbb{R}$ is convex and has no extreme points. So we look for extreme points in closed bounded (generally compact) convex sets.

(2) Let $C = \{(x, y, 0) \in \mathbb{R}^3 : x^2 + y^2 = 1\}$. Let $K = \mathrm{conv}(C \cup \{(0, 1, \pm 1)\})$. Then K is compact. $F = \{(0, 1, z) : -1 \leq z \leq 1\}$ is a face. The extreme points are $\{(0, 1, \pm 1)\} \cup C \setminus \{(0, 1, 0)\}$. Thus $\operatorname{ext} K$ is not closed.

(3) A Banach space $(X, \|\cdot\|)$ is *strictly convex* if $\|x\| = \|y\| = \left\|\frac{x+y}{2}\right\|$ implies that $x = y$. Examples include Hilbert spaces and $L^p(\mu)$ for $1 < p < \infty$. If $1 = \|f\|_p = \|g\|_p$ and $2 = \|f+g\|_p \leq \|f\|_p + \|g\|_p = 2$, then equality in the triangle inequality shows that $f = g$. When X is strictly convex, $\operatorname{ext} \overline{b_1(X)} = \{x \in X : \|x\| = 1\}$.

(4) Consider $K = \overline{b_1(L^1(0, 1))}$. Let $f \in L^1(0, 1)$ with $\|f\|_1 = 1$. Then there is some $t_0 \in (0, 1)$ so that $\int_0^{t_0} |f| \, dx = \frac{1}{2}$. Let $f_1 = 2f\chi_{[0,t_0)}$ and $f_2 = 2f\chi_{[t_0,1)}$. Then $\|f_1\|_1 = \|f_2\|_1 = 1$ and $f = \frac{1}{2}(f_1 + f_2)$. Therefore K has no extreme points.

(5) Let $K \subset \mathbb{R}^n$ be a compact convex set. Then $\operatorname{ext} K \neq \varnothing$ because $f(x) = \|x\|_2^2$ is continuous and takes its maximum on K at some point x_0. This is extreme in the unit ball of radius $\|x\|$, and so is also extreme in K.

It is a theorem of Carathéodory that if $E \subset \mathbb{R}^n$, every point in $\mathrm{conv}(E)$ is a convex combination of at most $n + 1$ points of E. This can be used to prove a theorem of Minkowski: in finite dimensions, a compact convex set $K = \mathrm{conv}(\operatorname{ext} K)$ without requiring closure. (See Exercise **3.31**.)

3.5.3. Krein–Milman Theorem. *Let X be a LCTVS, and let $K \subset X$ be a compact convex set. Then $\overline{\mathrm{conv}}(\operatorname{ext} K) = K$.*

Proof. Let $\mathcal{F}$ be the set of closed faces of K ordered by containment, i.e. $F \leq G$ if $F \supset G$. This is a poset. Let's show that it is inductive. Suppose that $\mathcal{C} = \{F_\alpha : \alpha \in A\}$ is a chain of faces. Then $\mathcal{C}$ has the finite intersection property because the intersection of finitely many faces in $\mathcal{C}$ equals the smallest one. By compactness of K, $F = \bigcap_{\alpha \in A} F_\alpha$ is non-empty. Suppose that $x, y \in K, 0 < t < 1$ and $tx + (1-t)y \in F$. Then $tx + (1-t)y \in F_\alpha$ for $\alpha \in A$. As each F_α is a face, $x, y \in F_\alpha$, and thus $x, y \in F$. Hence F is a face, and is an upper bound for $\mathcal{C}$. By Zorn's lemma, $\mathcal{F}$ contains a maximal element, i.e. a minimal face, say F_0.

We claim that F_0 is a single point. If not, then since X^* separates points of X, there is a continuous linear functional f on X which is not constant on F_0. By scaling, we may suppose that $\operatorname{Re} f$ is not constant on F_0. Let $\alpha = \sup_{x \in F_0} \operatorname{Re} f(x)$ and $G = \{x \in F_0 : \operatorname{Re} f(x) = \alpha\}$. Now G is a face for if $tx + (1-t)y \in G \subset F$ for $x, y \in K$ and $0 < t < 1$, then $x, y \in F$ as it is a face. But

$$\alpha = \operatorname{Re}(tx + (1-t)y) = t \operatorname{Re} f(x) + (1-t) \operatorname{Re} f(y) \leq t\alpha + (1-t)\alpha = \alpha.$$

Thus $\operatorname{Re} f(x) = \operatorname{Re} f(y) = \alpha$, so that $x, y \in G$. Hence G is a proper face of F_0, contradicting minimality. So $F_0 = \{x_0\}$ is an extreme point. Therefore $\operatorname{ext} K \neq \varnothing$.

Let $L = \overline{\text{conv}}(\text{ext}\, K)$. This is a closed convex subset of K, and thus is compact. If $L \neq K$, pick $x \in K \setminus L$. By the Separation theorem, there is an $f \in X^*$ so that

$$\sup_{x \in L} \text{Re}\, f(x) = \alpha < \text{Re}\, f(x) \leq \sup_{x \in K} \text{Re}\, f(x) = \beta.$$

By compactness, the supremum is attained, so $F = \{x \in K : \text{Re}\, f(x) = \beta\}$ is non-empty. It is easy to see that F is a compact face of K. Hence F has an extreme point x_0. A face of a face is a face, so $x_0 \in \text{ext}\, K$, contradicting the definition of L. Therefore $K = \overline{\text{conv}}(\text{ext}\, K)$. ∎

3.5.4. Corollary (Krein–Milman). *Let X be a Banach space. The closed unit ball of X^* is the weak-$*$ closed convex hull of its extreme points.*

Proof. By the Banach–Alaoglu theorem, $\overline{b_1(X^*)}$ is weak-$*$ compact and convex. Thus the Krein–Milman theorem applies in (X^*, τ_X). ∎

3.5.5. Corollary. *c_0 and $L^1(0, 1)$ are not dual spaces.*

Proof. The unit balls have no extreme points. ∎

3.5.6. Example. $C[0, 1]^* = M[0, 1]$ is the space of complex regular Borel measures on $[0, 1]$. The point masses δ_x for $x \in [0, 1]$ are extreme in $K = \overline{b_1(M[0, 1])}$, as are $e^{i\theta}\delta_x$. If μ is a complex measure with $\|\mu\| = 1$ which is not supported on a single point, then there is a Borel set $A \subset [0, 1]$ so that $0 < |\mu|(A) < 1$. Set $\mu_1(B) = \frac{1}{|\mu|(A)}\mu(B \cap A)$ and $\mu_2(B) = \frac{1}{|\mu|(A^c)}\mu(B \cap A^c)$. Then

$$\mu = |\mu|(A)\mu_1 + (1 - |\mu|(A))\mu_2,$$

so it is not extreme. Thus $\text{ext}\, K = \{e^{i\theta}\delta_x : \theta \in \mathbb{R}, x \in [0, 1]\}$. Now $\text{conv}(\text{ext}\, K)$ is the set of atomic measures of norm at most 1, not the whole ball. This is the intersection of K with the norm-closed subspace of atomic measures. The weak-$*$ closure is needed to obtain the other points in K.

The same works for $M(X) = C_0(X)^*$ where X is any locally compact Hausdorff space.

3.5.7. Lemma. *If $K_1, \ldots, K_n$ are compact convex sets in a LCTVS X, then $\text{conv}(K_1 \cup \cdots \cup K_n)$ is compact.*

Proof. It is enough to establish the $n = 2$ case, since the general case then follows from repeated application.

Consider a point x in $\text{conv}(K_1 \cup K_2)$ given by $x_j \in K_1 \cup K_2$, $0 < t_j < 1$ for $1 \leq j \leq m$, $\sum_{j=1}^m t_j = 1$ and $x = \sum_{j=1}^m t_j x_j$. Let $A_1 = \{j : x_j \in K_1\}$ and $A_2 = \{1, \ldots, m\} \setminus A_1$. Let $s_1 = \sum_{j \in A_1} t_j$ and $s_2 = 1 - s_1 = \sum_{j \in A_2} t_j$. Then

$y_i = \sum_{j\in A_i} \frac{t_j}{s_i} x_j$ is a convex combination of $\{x_j : j \in A_i\}$, and hence it belongs to K_i. Finally $x = s_1 y_1 + s_2 y_2$.

It follows from this analysis that the map $\Phi : K_1 \times K_2 \times [0,1] \to X$ given by $\Phi(x_1, x_2, s) = s x_1 + (1-s) x_2$ maps onto $\mathrm{conv}(K_1 \cup K_2)$. It is clear that Φ is a continuous map from a compact set into X. Therefore the image is compact. ■

There is a partial converse to the Krein–Milman theorem.

3.5.8. Milman Theorem. *Let X be a LCTVS, and let $K \subset X$ be a compact convex set. Suppose that $E \subset K$ is closed and $\overline{\mathrm{conv}}(E) = K$. Then* $\mathrm{ext}\, K \subset E$.

Proof. Suppose that $x_0 \in \mathrm{ext}\, K \setminus E$. Since E is closed, there is a neighbourhood $x_0 + U(F,r)$ which is disjoint from E. Now $\{x + U(F, r/2) : x \in E\}$ is an open cover of E, which is compact, so there is a finite subcover $x_i + U(F, r/2)$ for $1 \le i \le n$. Let $K_i = \overline{\mathrm{conv}}(E \cap \overline{x_i + U(F, r/2)})$. This is contained in the convex set $\overline{x_i + U(F, r/2)}$, and thus $x_0 \notin K_i$. Then

$$\mathrm{conv}(E) \subset \mathrm{conv}(K_1 \cup \cdots \cup K_n) \subset \overline{\mathrm{conv}}(E) = K.$$

By Lemma 3.5.7, the middle term is compact, so that $\mathrm{conv}(K_1 \cup \cdots \cup K_n) = K$. In particular, x_0 is a convex combination of points in $K_1 \cup \cdots \cup K_n$, which contradicts the fact that x_0 is extreme and not in this union. Therefore $\mathrm{ext}\, K \subset E$. ■

As an application of the new ideas in this course, we provide a proof of the Stone–Weierstrass theorem due to de Bruijn.

3.5.9. Stone–Weierstrass Theorem. *Let X be a compact Hausdorff space. Suppose that A is a closed subalgebra of $C_{\mathbb{R}}(X)$ which separates points and doesn't vanish at any point in X. Then $A = C_{\mathbb{R}}(X)$.*

Proof. Suppose that $A \ne C_{\mathbb{R}}(X)$. Then by the Hahn–Banach theorem, $A^\perp$ is a non-zero weak-$*$ closed subspace of $C_{\mathbb{R}}(X)^*$. Let $K = \overline{b_1(C_{\mathbb{R}}(X)^*)} \cap A^\perp$. By the Banach–Alaoglu theorem and the weak-$*$ closure of $A^\perp$, K is weak-$*$ compact and convex. By the Krein–Milman theorem, K has an extreme point, say ψ. By the Riesz representation theorem, there is a finite regular signed Borel measure μ on X with $\|\mu\| = 1$ so that $\psi(f) = \int f\, d\mu$.

Claim: the support of μ is a single point. If not, let x, y be distinct points in $\mathrm{supp}(\mu)$. Since A separates points, there is some $f \in A$ so that $f(x) \ne f(y)$. Scaling and adding a constant produces a function $g \in A + \mathbb{R}\mathbf{1}$ so that $0 \le g \le 1$ and $g(x) \ne g(y)$. Then $g\mu$ is not a scalar multiple of μ. Note that $g\mu \in A^\perp$ since for $h \in A$, $\int hg\, d\mu = 0$ because $hg \in A$. Let $\mu_1 = g\mu / \|g\|_1$ and $\mu_2 = (1-g)\mu / \|1-g\|_1$ where $\|g\|_1 = \int g\, d\mu$ and $\|g\|_1 + \|1-g\|_1 = 1$. Then μ_1 and μ_2 lie in $A^\perp$ and have norm 1, so belong to K. Then $\mu = \|g\|_1 \mu_1 + \|1-g\|_1 \mu_2$ is not extreme. This contradiction shows that μ is supported on a single point, say x_0. That is, $\mu = e^{i\theta}\delta_{x_0}$.

Since A does not vanish at x_0, there is some $f \in A$ so that

$$\psi(f) = \int f\,d\mu = e^{i\theta} f(x_0) \neq 0.$$

This contradicts the fact that $\psi \in A^\perp$. Therefore, the original assumption was incorrect, and $A = C_{\mathbb{R}}(X)$. ∎

Recall that a subset A of $C(X)$ is *self-adjoint* if $f \in A$ implies that $\bar{f} \in A$. The following well-known corollary is left as an exercise.

3.5.10. COROLLARY (Stone–Weierstrass). *Let X be a compact Hausdorff space. Suppose that A is a self-adjoint, closed subalgebra of $C(X)$ which separates points and doesn't vanish at any point in X. Then $A = C(X)$.*

3.6. The Krein–Smulian Theorem

In this section, we establish a useful result for checking weak-$*$ closure of unbounded convex sets, including subspaces.

3.6.1. DEFINITION. For $A \subset X$, define the *polar* of A to be

$$A^\circ = \{f \in X^* : \operatorname{Re} f(x) \le 1 \text{ for } x \in A\}.$$

Note that A° is always a weak-$*$ closed convex set. In particular, observe that $\overline{b_r(X)}^\circ = \overline{b_{1/r}(X^*)}$. If A is a subspace, then $A^\circ = A^\perp$.

3.6.2. KREIN–SMULIAN THEOREM. *Let X be a Banach space, and let $K \subset X^*$ be a convex set. Then K is weak-$*$ closed if and only of $K \cap \overline{b_r(X^*)}$ is weak-$*$ closed for all $r > 0$.*

PROOF. The only if direction is trivial.

Suppose that $K \cap \overline{b_r(X^*)}$ is weak-$*$ closed for all $r > 0$. By the Banach–Alaoglu theorem, they are weak-$*$ compact. Then K is norm closed because norm convergent sequences are bounded, and so lie in the closed set $K \cap \overline{b_r(X^*)}$ for some large r. Let $f \in K^c$ and choose an $\varepsilon > 0$ so that $\overline{b_\varepsilon(f)} \cap K = \varnothing$. After translating and scaling, we may suppose that $f = 0$ and $\varepsilon = 1$, so $\overline{b_1(0)} \cap K = \varnothing$. We need to show that $0 \notin \overline{K}^{w*}$.

We will choose finite sets $\{A_n : n \geq 2\}$ in X so that

$$A_n \subset \overline{b_{\frac{1}{n-1}}(X)} \quad \text{and} \quad \overline{b_n(X^*)} \cap K \cap (A_2 \cup \cdots \cup A_n)^\circ = \varnothing.$$

We can begin with $A_1 = \varnothing$ and note that $\overline{b_1(X^*)} \cap K = \varnothing$. Suppose that $A_2, \ldots, A_n$ have been chosen with these properties. Let $D = \overline{b_{n+1}(X^*)} \cap K \cap (A_2 \cup \cdots \cup A_n)^\circ$. This is weak-$*$ compact and convex, and by assumption, $D \cap \overline{b_n(X^*)} = \varnothing$. Now $\overline{b_n(X^*)} = \overline{b_{1/n}(X)}^\circ$, so

$$\varnothing = D \cap \overline{b_n(X^*)} = D \cap \bigcap_{\substack{x \in X \\ \|x\| \leq 1/n}} \{x\}^\circ.$$

By compactness of D and the FIP, there is a finite subset $A_{n+1} \subset \overline{b_{1/n}(X)}$ so that $D \cap A_{n+1}^\circ = \varnothing$. This completes the induction.

List $\bigcup_{n \geq 2} A_n = \{x_1, x_2, \ldots\}$ as a countable set with $\lim_{i \to \infty} x_i = 0$. Define a linear map $T : X^* \to c_0$ by $(Tf) = \big(f(x_i)\big)_{i \geq 1}$. We have $\|T\| = \sup_{i \geq 1} \|x_i\| \leq 1$.

If $f \in K$, then $\|f\| \leq n$ for n large enough, and thus $f \notin (A_2 \cup \cdots \cup A_n)^\circ$. Hence, there is some $x_i \in A_2 \cup \cdots \cup A_n$ so that $\operatorname{Re} f(x_i) > 1$; so that $\|Tf\| > 1$. Therefore TK is convex and $TK \cap b_1(c_0) = \varnothing$. Since $b_1(c_0)$ is open, the Separation theorem yields $0 \neq \varphi = (a_1, a_2, \ldots) \in c_0^* = l_1$ so that

$$\sup_{z \in b_1(c_0)} \operatorname{Re} \varphi(z) = \|\varphi\|_1 \leq \inf_{f \in K} \operatorname{Re} \varphi(Tf) = \inf_{f \in K} \operatorname{Re} \sum_{i \geq 1} a_i f(x_i) = \inf_{f \in K} \operatorname{Re} f(x_0),$$

where $x_0 = \sum_{i \geq 1} a_i x_i$, which converges since the $\{x_i, i \geq 1\}$ is bounded and $\sum |a_i| = \|\varphi\|_1 < \infty$. Therefore $x_0 \in X$ and $\operatorname{Re} f(x_0) \geq \|\varphi\|_1 > 0$ for all $f \in K$. Thus $\operatorname{Re} f(x_0) \geq \|\varphi\|_1$ for all $f \in \overline{K}^{w*}$, whence $0 \notin \overline{K}^{w*}$. Hence K is weak-$*$ closed. ■

When K is a subspace, $K \cap \overline{b_r(X^*)} = r(K \cap \overline{b_1(X^*)})$. So the following is immediate.

3.6.3. Corollary. *If Y is a subspace of a dual space X^*, then Y is weak-$*$ closed if and only if $Y \cap \overline{b_1(X^*)}$ is weak-$*$ closed.*

3.6.4. Corollary. *For $\varphi \in X^{**}$, the following are equivalent:*

(1) *$\varphi \in \iota X$.*

(2) *φ is continuous on (X^*, τ_X), i.e. φ is weak-$*$ continuous.*

(3) *φ is weak-$*$ continuous on $\overline{b_1(X^*)}$.*

Proof. The equivalence of (1) and (2) is Lemma 3.1.8. Clearly (2) implies (3). To show that (3) implies (2), Proposition 3.1.6 shows that it suffices to show that $\ker \varphi$ is weak-$*$ closed. Now (3) implies that $\ker \varphi \cap \overline{b_1(X^*)}$ is weak-$*$ closed. So Corollary 3.6.3 completes the proof. ■

3.6.5. DEFINITION. A subset $K \subset X$ of a LCTVS X is *sequentially closed* if whenever a sequence x_n in K has a limit x, then $x \in K$. So a subset K is *weak-∗ sequentially closed* in a Banach dual space X^* if it is sequentially closed in (X^*, τ_X).

3.6.6. LEMMA. *Let X be a Banach space. Suppose that $Y \subset X^*$ is a separable subspace which separates points of X. Then $(\overline{b_1(X)}, \tau_Y)$ is metrizable.*

PROOF. Let $\{g_n : n \geq 1\}$ be a dense subset of $b_1(Y)$. Define a metric on $\overline{b_1(X)}$ by

$$d(x, y) = \sum_{n \geq 1} 2^{-n} |g_n(x - y)| \quad \text{for} \quad x, y \in \overline{b_1(X)}.$$

This is definite since Y separates points, and the triangle inequality follows since each term satisfies the triangle inequality. Let $\beta_r(x)$ denote the d-ball about x of radius r.

Suppose that $x \in \overline{b_1(X)}$, U is relatively open in $(\overline{b_1(X)}, \tau_Y)$ and $x \in U$. Find a finite $F = \{f_1, \ldots, f_m\} \Subset b_1(Y)$ and $r > 0$ so that $x + U(F, r) \cap \overline{b_1(X)} \subset U$. For each i, select n_i so that $\|f_i - g_{n_i}\| < \frac{r}{4}$ and let $N = \max\{n_i : 1 \leq i \leq m\}$. Suppose that $y \in \overline{b_1(X)}$ and $d(x, y) < \varepsilon = 2^{-1-N} r$. Then

$$\begin{aligned} |f_i(y) - f_i(x)| &\leq |(f_i - g_{n_i})(y)| + |g_{n_i}(y - x)| + |(f_i - g_{n_i})(x)| \\ &< \frac{r}{4} + 2^{n_i} d(y, x) + \frac{r}{4} \leq \frac{r}{2} + 2^N 2^{-1-N} r = r. \end{aligned}$$

Therefore $\beta_\varepsilon(x) \subset x + U(F, r) \subset U$.

Conversely, suppose that V is d-open and $x \in V$. Find an $\varepsilon > 0$ such that $\beta_\varepsilon(x) \subset V$. Find N so that $2^{1-N} < \frac{\varepsilon}{2}$, and let $F = \{g_1, \ldots, g_N\}$. Suppose that $y \in x + U(F, \varepsilon/2)$. Then

$$\begin{aligned} d(y, x) &= \sum_{n \geq 1} 2^{-n} |f_n(y - x)| \\ &< \sum_{n=1}^{N} 2^{-n} \frac{\varepsilon}{2} + \sum_{n > N} 2^{-n} 2 \|f_n\| < \frac{\varepsilon}{2} + 2^{1-N} < \varepsilon. \end{aligned}$$

Thus $x + U(F, \varepsilon/2) \subset \beta_\varepsilon(x)$.

This shows that $(\overline{b_1(X)}, \tau_Y)$ and $(\overline{b_1(X)}, d)$ have the same open sets, and hence they determine the same topology. ■

3.6.7. COROLLARY. *Let X be a separable Banach space. Suppose that a convex set $K \subset X^*$ is weak-∗ sequentially closed. Then K is weak-∗ closed.*

PROOF. By the Krein–Smulian theorem, K is weak-∗ closed if and only if $K \cap \overline{b_n(X^*)}$ is weak-∗ closed for $n \geq 1$. But $(\overline{b_n(X^*)}, \tau_X)$ is metrizable by

Lemma 3.6.6. Now the result follows because in a metric space, a set is closed if and only if it is sequentially closed. ■

3.6.8. PREDUALS. A *predual* of a Banach space X is a Banach space Y such that Y^* is isometrically isomorphic to X. We have seen that some Banach spaces are not dual spaces. When a Banach space is a dual space, it may have a unique predual, or it may have many non-isomorphic preduals. Since every predual of X has a canonical imbedding into X^*, we can try to identify subspaces of X^* which are preduals of X.

3.6.9. EXAMPLE. Consider l_1. We know that $c_0^* = l_1$, but it has a plethora of preduals. We only mention a special class. Consider $C(K)$ where K is a countable locally compact metric space. Then the Riesz representation theorem states that $C(K)^* = M(K)$, the space of complex regular Borel measures on K. Since K is countable, every measure has the form $\mu = \sum_{x\in K} a_x \delta_x$ with $\|\mu\| = \sum_{x\in K} |a_x|$. It is clear that $M(K)$ is isometrically isomorphic to l_1. These spaces yield a variety of different Banach spaces. In particular, the space

$$c = \{(a_1, a_2, \dots) : \lim_{n\to\infty} a_n = a_0 \text{ exists}\} \simeq C(K) \quad \text{for} \quad K = \{0, \tfrac{1}{n} : n \ge 1\}$$

is not isometrically isomorphic to c_0. (See Exercise **2.31**.)

3.6.10. PROPOSITION. *Let X be a Banach space. Then a closed subspace Y of X^* satisfies $Y^* = X$ if and only if*

(1) *Y norms X and*

(2) *$\overline{b_1(X)}$ is τ_Y-compact.*

PROOF. If $X = Y^*$, then (1) is immediate and (2) follows from the Banach–Alaoglu theorem.

Assume that (1) and (2) hold. Let $J : X \to Y^*$ by $Jx(y) = y(x)$. By (1), J is isometric: $\|Jx\| = \sup_{\|y\|\le 1,\ y\in Y} |y(x)| = \|x\|$. Also J is continuous from (X, τ_Y) to (Y^*, τ_Y). Indeed, if E is a finite subset of Y, then $U_{E,r}$ is a basic open neighbourhood of 0. Then

$$J^{-1}U(E, r) = \{x \in X : |y(x)| < r,\ y \in E\}$$

is τ_Y-open in X. By (2), $\overline{b_1(X)}$ is τ_Y-compact. Therefore $J\overline{b_1(X)}$ is τ_Y-compact in Y^*, and hence it is weak-$*$ closed. The isometric property ensures that

$$J\overline{b_1(X)} = \overline{b_1(JX)} = JX \cap \overline{b_1(Y^*)}.$$

By the Krein–Smulian theorem, JX is weak-$*$ closed in Y^*. Now

$$(JX)_\perp = \{y \in Y : 0 = JX(y) = y(X)\} = \{0\}.$$

Finally, by the Hahn–Banach theorem, $JX = ((JX)_\perp)^\perp = \{0\}^\perp = Y^*$. So X is isometric to Y^*. ■

3.6.11. Example. We will show that l_∞ has a unique predual, namely, l_1. Let ε_n be the sequence in l_∞ with a 1 in the nth coordinate and zeros elsewhere, and let δ_n be the corresponding sequence in l_1. If $l_\infty = Y^*$, then $\overline{b_1(l_\infty)}$ is τ_Y-compact. Therefore $\overline{b_1(\varepsilon_n)} \cap \overline{b_1(-\varepsilon_n)} = \overline{b_1(\ker \delta_n)}$ is τ_Y-compact. By the Krein–Smulian theorem, $\ker \delta_n$ is τ_Y-closed. By Proposition 3.1.6 and Lemma 3.1.8, $\delta_n \in Y$. Therefore $l_1 = \overline{\operatorname{span}\{\delta_n : n \geq 1\}} \subseteq Y$. If $Y \neq l_1$, there is some $0 \neq \varphi \in l_\infty = Y^*$ such that $\varphi|_{l_1} = 0$. This is impossible, and therefore $Y = l_1$.

3.7. Weakly Compact Sets

In this section, we first establish a theorem of Eberlein and Smulian characterizing weakly compact sets in terms of countable structures. We also include a characterization by Grothendieck in terms of double limits. Then we use this to prove a result of Krein and Smulian that the norm-closed convex hull of a weakly compact set is weakly compact.

Recall that the weak topology on a Banach space X is the τ_{X^*} topology. We write $\overline{A}^w$ to denote the weak closure. Likewise the weak-$*$ topology on X^* is the τ_X topology, and we write $\overline{B}^{w^*}$ for the weak-$*$ closure of $B \subset X^*$.

3.7.1. Definition. Let X be a Banach space and let $A \subset X$ and $B \subset X^*$ be bounded sets. Say that A and B have the *interchangeable double limit property* (IDLP) if for every pair of sequences $(x_m) \subset A$ and $(f_n) \subset B$ such that both

$$\lim_{m\to\infty} \lim_{n\to\infty} f_n(x_m) \quad \text{and} \quad \lim_{n\to\infty} \lim_{m\to\infty} f_n(x_m)$$

exist, then the limits are equal.

3.7.2. Lemma. *Let $A \subset X$ and $B \subset X^*$ be bounded subsets with IDLP.*

(1) *If (x_m) is a sequence in A and $x_0 \in X$ satisfies* $\lim_{m\to\infty} f(x_m) = f(x_0)$ *for $f \in B$, then this also holds for all $f \in \overline{B}^{w^*}$.*

(2) *If x_0 lies in the τ_B-closure of A, then there is a sequence (x_m) in A such that* $\lim_{m\to\infty} f(x_m) = f(x_0)$ *for $f \in \overline{B}^{w^*}$.*

(3) *Suppose $\overline{B}^{w^*} \supset \overline{b_1(X^*)}$ and $\widetilde{A}$ is the weak-$*$ (i.e. τ_{X^*}) closure of ιA in X^{**}. For $\xi \in \widetilde{A}$, there is a sequence (x_m) in A converging to ξ weakly (i.e. $\tau_{X^{***}}$). Moreover $\widetilde{A} = \overline{A}^w$ is contained in X and is weakly compact.*

(4) *Let $\overline{\mathbb{D}}B = \{\lambda f : f \in B,\ \lambda \in \overline{\mathbb{D}}\}$. Then A and $\overline{\mathbb{D}}B$ have IDLP.*

PROOF. For (1), fix $f \in \overline{B}^{w^*}$. For each $n \geq 1$, select $f_n \in B$ such that $|f_n(x_m) - f(x_m)| < \frac{1}{n}$ for $0 \leq m \leq n$. Then $\lim_{n\to\infty} f_n(x_m) = f(x_m)$ for $m \geq 0$. Hence

$$\lim_{n\to\infty}\lim_{m\to\infty} f_n(x_m) = \lim_{n\to\infty} f_n(x_0) = f(x_0)$$

and

$$\lim_{m\to\infty}\lim_{n\to\infty} f_n(x_m) = \lim_{m\to\infty} f(x_m).$$

The sequence $\big(f(x_m)\big)$ is bounded in $\mathbb{C}$, and so if L is any cluster point, there is a subsequence $\big(f(x_{m_j})\big)$ so that $\lim_{j\to\infty} f(x_{m_j}) = L$. By IDLP for (x_{m_j}) and (f_n), $L = f(x_0)$, whence there is a unique cluster point. Therefore $\lim_{m\to\infty} f(x_m) = f(x_0)$.

For (2), fix x_0. Recursively construct points x_m in A and sequences $(f_{m,k})_{k\geq 1}$ in B as follows. For $m \geq 1$, set $F_m = \operatorname{span}\{x_0, \ldots, x_{m-1}\}$. Then $B|_{F_m} \subset F_m^*$ is separable, so we can select $(f_{m,k})_{k\geq 1}$ in B so that $\{f_{m,k}|_{F_m} : k \geq 1\}$ is dense in $B|_{F_m}$. As x_0 lies in the τ_B-closure of A, we can select $x_m \in A$ so that

$$|f_{j,k}(x_0) - f_{j,k}(x_m)| < \frac{1}{m} \quad \text{for} \quad 1 \leq j, k \leq m.$$

Define $F = \overline{\operatorname{span}}\{x_m : m \geq 0\}$. Set $A_0 = A \cap F$ and $B_0 = \{f_{j,k}|_F : j, k \geq 1\}$. Observe that A_0 and B_0 have IDLP since $B_0 \subset B|_F$. Also by construction, B_0 approximates B arbitrarily well on any finite subset of $\bigcup_{m\geq 1} F_m$. Thus B_0 is weak-$*$ (i.e. τ_F) dense in $B|_F$. Finally, the construction shows that

$$\lim_{m\to\infty} f_{j,k}(x_m) = f_{j,k}(x_0) \quad \text{for all} \quad f_{j,k} \in B_0.$$

By (1), this also holds for $f \in \overline{B_0}^{w^*} = \overline{B|_F}^{w^*} \supseteq \overline{B}^{w^*}|_F$.

For (3), consider ιA as a subset of X^{**} and $\overline{\iota B}^{w^*} \supset \overline{b_1(X^*)}$ as a subset of X^{***} via the canonical injections. Fix $\xi \in \widetilde{A} = \overline{\iota A}^{\tau_{X^*}}$. By (2), since ξ lies in the $\tau_{\overline{B}^{w^*}} = \tau_{X^*}$ closure of A, there is a sequence (x_m) in A such that $\lim_{m\to\infty} f(x_m) = f(\xi)$ for all $f \in \overline{b_1(X^*)}^{\tau_{X^{**}}}$, as $\tau_{X^{**}}$ is the weak-$*$ topology on X^{***}. By Goldstine's theorem, $\overline{b_1(X^*)}^{\tau_{X^{**}}} = \overline{b_1(X^{***})}$. Thus (x_m) converges to ξ weakly (i.e. $\tau_{X^{***}}$). Subspaces are weakly closed, and thus $\widetilde{A}$ must be contained in ιX. By the Banach–Alaoglu theorem, $\widetilde{A}$ is weak-$*$ compact in X^{**}, and therefore is weakly compact as a subset of X.

For (4), suppose that $(x_m) \subset A$ and $(\lambda_n f_n)$ for $x_m \in A$, $\lambda_n \in \mathbb{D}_n$ and $f_n \in B$ satisfy the double limit property

$$\lim_{m\to\infty}\lim_{n\to\infty} \lambda_n f_n(x_m) = L \quad \text{and} \quad \lim_{n\to\infty}\lim_{m\to\infty} \lambda_n f_n(x_m) = M.$$

Drop to a subsequence of $(\lambda_n f_n)$ so that $\lim_n \lambda_n = \lambda$ exists. If $\lambda = 0$, then clearly $L = 0 = M$. Otherwise,

$$\lim_{m\to\infty}\lim_{n\to\infty} f_n(x_m) = \lambda^{-1} L \quad \text{and} \quad \lim_{n\to\infty}\lim_{m\to\infty} f_n(x_m) = \lambda^{-1} M.$$

By IDLP for A and B, these limits are equal, and thus $L = M$. So A and $\mathbb{D}B$ have IDLP. ■

In the following, the results of Eberlein and Smulian, commonly known as the Eberlein–Smulian theorem, are the equivalence of the following excluding (4) and (4′), which are due to Grothendieck. A topological space is *countably compact* if every sequence has a convergent cofinal subnet.

3.7.3. EBERLEIN–SMULIAN–GROTHENDIECK THEOREM. *Let A be a subset of a Banach space X. The following are equivalent:*

(1) $\overline{A}^w$ *is weakly compact.*

(2) $\overline{A}^w$ *is weakly sequentially compact.*

(2′) *Every sequence in A has a subsequence weakly convergent in X.*

(3) $\overline{A}^w$ *is weakly countably compact*

(3′) *Every countably infinite subset of A has a weak cluster point in X.*

(4) *A is bounded, and A and $\overline{b_1(X^*)}$ have the IDLP.*

(4′) *A is bounded, and A and B have the IDLP for $B \subset X^*$ satisfying $\overline{B}^{w^*} \supset \{f \in X^* : \|f\| = 1\}$.*

PROOF. Notice first that each of these conditions imply that A is bounded.

Clearly (1) implies (3), which implies (3′). Also (2) implies (2′) and (3), and (2′) implies (3′), and (4) implies (4′). We will show that (3′) implies (4) and (4′) implies (1) and (2).

Suppose that (3′) holds, and let $(x_m) \subset A$ and $(f_n) \subset \overline{b_1(X^*)}$ be sequences such that both

$$\lim_{m\to\infty}\lim_{n\to\infty} f_n(x_m) \quad \text{and} \quad \lim_{n\to\infty}\lim_{m\to\infty} f_n(x_m)$$

exist. By (3′), the sequence (x_m) has a cluster point $x \in X$. Also $\overline{b_1(X^*)}$ is weak-$*$ compact by the Banach–Alaoglu theorem, so (f_n) has a weak-$*$ cluster point $f \in X^*$. Since $\lim_{n\to\infty} f_n(x_m)$ exists for fixed $m \geq 1$, this limit must be $f(x_m)$, and likewise $\lim_{m\to\infty} f_n(x_m) = f_n(x)$ for each $n \geq 1$. Thus

$$\lim_{m\to\infty}\lim_{n\to\infty} f_n(x_m) = \lim_{m\to\infty} f(x_m) = f(x)$$

again because the limit exists and x is a cluster point of (x_m), and similarly

$$\lim_{n\to\infty}\lim_{m\to\infty} f_n(x_m) = \lim_{n\to\infty} f_n(x) = f(x).$$

Therefore the two limits agree. So (4) holds, and thus (4′) holds as well.

Suppose that (4′) holds for A and B. By Lemma 3.7.2(4), we may suppose that $B = \mathbb{D}B$. Hence $\overline{B}^{w^*} \supset \overline{b_1(X^*)}$. By Lemma 3.7.2(3), $\overline{A}^w$ is weakly compact. So (1) holds. Finally, let (x_m) be a sequence in A. By weak compactness, there is a weak cluster point $x_0 \in X$. Let $F = \overline{\operatorname{span}}\{x_m : m \geq 0\}$. This is separable, so as in the proof of Lemma 3.7.2(2), there is a countable weak-$*$ dense subset B_0 of $B|_F$. Write $B_0 = \{f_n : n \geq 1\}$. Note that $A_0 = A \cap F$ and B_0 have the IDLP. Construct a subsequence $(x_{m_j})_{j\geq 1}$ such that $|f_k(x_0) - f_k(x_{m_j})| < \frac{1}{j}$ for $1 \leq k \leq j$. Then $\lim_{j\to\infty} f(x_{m_j}) = f(x_0)$ for all $f \in B_0$. By Lemma 3.7.2(1), this also holds for $f \in \overline{B_0}^{w^*} = \overline{b_1(F^*)}$. That is, x_{m_j} converges weakly to x_0 in F, and hence in X. Therefore $\overline{A}^w$ is weakly sequentially compact, i.e. (2) holds. ∎

It is easy to show that the closed convex hull of a compact set in a Banach space is compact (Exercise **3.23**). It is also easy to show that the weak-$*$ closed convex hull of a weak-$*$ compact set is weak-$*$ compact (Exercise **3.33**). It turns out to also be true for weakly compact sets in a Banach space, but the proof is much more subtle. This result was established by Krein and Smulian in the same paper as the Krein–Smulian theorem 3.6.2.

First we need a lemma.

3.7.4. LEMMA. *Let X be a separable Banach space and let $\varphi \in X^{**}$. Suppose that for every sequence $(f_n)_{n\geq 1}$ in X^* which converges weak-$*$ to f, we have $\varphi(f) = \lim_{n\to\infty} \varphi(f_n)$. Then $\varphi \in \iota X$.*

PROOF. If $\varphi \notin \iota X$, then $d = \operatorname{dist}(\varphi, \iota X) > 0$ and there is a $\Psi \in X^{***}$ such that $\|\Psi\| = 1$, $\Psi|_{\iota X} = 0$ and $\Psi(\varphi) = d$. By Goldstine's theorem, the unit ball of X^* is $\tau_{X^{**}}$-dense in the ball of X^{***}. Let $\{x_i : i \geq 1\}$ be a dense subset of X. Then the $\tau_{X^{**}}$-open set

$$U_n = \{\Phi \in X^{***} : |\Phi(x_i)| < \tfrac{1}{n} \text{ for } 1 \leq i \leq n \text{ and } |\Phi(\varphi)| > d/2\}$$

is a neighbourhood of Ψ, and so contains some $f_n \in X^*$ with $\|f_n\| \leq 1$. By construction, f_n converges to 0 weak-$*$ but $|\varphi(f_n)| > d/2$, so that $0 = \varphi(0) \neq \lim_{n\to\infty} \varphi(f_n)$. ∎

3.7.5. THEOREM (Krein–Smulian). *Let K be a weakly compact set in a Banach space X. Then the norm-closed convex hulls $\overline{\operatorname{conv}}(K)$ and $\overline{\operatorname{aconv}(K)}$ are weakly compact.*

PROOF. By the Eberlein–Smulian theorem, it suffices to show that $\overline{\operatorname{aconv}(K)}$ is weakly sequentially compact. Given a sequence $\{x_i\}_{i\geq 1}$ in $\overline{\operatorname{aconv}(K)}$, set $Y = \overline{\operatorname{span}}(\{x_i\})$, and restrict our attention to $K \cap Y$. This is weakly closed since Y is weakly closed, and thus weakly compact. Thus we may assume that X is separable.

Considering K with its weak topology, the space $C(K)$ of (weakly) continuous functions on K is a Banach space. The restriction map $R : X^* \to C(K)$ by $(Rf)(x) = f(x)$ for $x \in K$ is continuous. Thus $R^* : C(K)^* \to X^{**}$ is weak-$*$ continuous by Example 3.1.11. By the Riesz representation theorem, $C(K)^* = M(K)$ is the space of regular complex Borel measures on K.

Fix $\mu \in M(K)$. Consider a weak-$*$ convergent sequence $(f_n)_{n\geq 1}$ in X^* with limit f. Then this sequence is bounded and converges pointwise on K to f. By the Lebesgue dominated convergence theorem,

$$R^*\mu(f) = \int_K Rf(x)\,d\mu(x) = \lim_{n\to\infty} \int_K Rf_n(x)\,d\mu(x) = \lim_{n\to\infty} R^*\mu(f_n).$$

By Lemma 3.7.4, $R^*\mu \in \iota X$. Hence $R^*(M(K)) \subset \iota X$. In particular, for $x \in K$, the point mass δ_x is mapped to ιx because

$$R^*\delta_x(f) = f(x) = \iota x(f) \quad \text{for} \quad f \in X^*.$$

The unit ball $B = \overline{b_1(M(K))}$ is weak-$*$ compact, and R^* is weak-$*$ continuous. Thus R^*B is a τ_{X^*} compact balanced convex set in X^{**}. Since it lies in ιX, we have that $\widetilde{B} = \iota^{-1}R^*B$ is a τ_{X^*}-compact absolutely convex set in X. Now $\widetilde{B}$ contains K and therefore contains the norm-closed absolutely convex hull of K. Since closed convex sets are weakly closed, $\overline{\text{aconv}}(K)$ and $\overline{\text{conv}}(K)$ are weakly compact. ∎

3.8. Fixed Point Theorems

In this section, we establish theorems that guarantee a fixed point for a family of mappings. In particular, we prove the Schauder–Tychonoff fixed point theorem, which is used to prove a theorem of Lomonosov about invariant subspaces for linear operators in Section 5.5. We will also establish the Kakutani fixed point theorem, and use it to prove the existence of Haar measure for compact groups. The Ryll-Nardzewski fixed point theorem will be used in Section 10.12 to construct traces on von Neumann algebras and in Section 7.7 to construct a mean on the space of weakly almost periodic functions on a topological group.

Perhaps the most basic fixed point theorem is the *Contraction mapping principle: let (X, d) be a complete metric space, and let $T : X \to X$ be a mapping with Lipschitz constant $c < 1$, i.e. $d(Tx, Ty) < cd(x, y)$ for $x, y \in X$. Then T has a unique fixed point x_*.* Indeed, the proof shows that for any $x_0 \in X$, the sequence $x_n := T^n x_0$ converges to x_*. This is generally taught in a course on metric space theory, and will not be repeated here.

The basis for the Schauder–Tychonoff theorem is the famous Brouwer fixed point theorem for a continuous map of the closed unit ball B_n of $\mathbb{R}^n$ into itself. For $n = 1$, this is just the Intermediate Value theorem, and for $n = 2$, it has a relatively easy proof using winding number. But for $n \geq 3$, this is a result in algebraic topology or homotopy theory. There are a variety of proofs (see the Notes), and we present a curious but elementary argument due to Milnor [206].

3.8.1. LEMMA. *Let A be a compact subset of $\mathbb{R}^n$, let $U \supset A$ be open, and let $v : U \to \mathbb{R}^n$ be C^1. Define $f_t(x) = x + tv(x)$. Then for t small enough, the volume $V_n(f_t(A))$ is a polynomial in t.*

PROOF. Since v is C^1, the mean value theorem shows that it is locally Lipschitz. Thus there is an open set V with $A \subset V \subset \overline{V} \subset U$ on which v is Lipschitz with constant $c < \infty$. If $|t| < c^{-1}$, then f_t is one to one because

$$\|f_t(x) - f_t(y)\| \geq \|x - y\| - |t|\|v(x) - v(y)\| \geq (1 - c|t|)\|x - y\| > 0.$$

The Jacobian of f_t is just $J_t(x) = I_n + t\big[\frac{\partial v_i}{\partial x_j}\big]$. Thus $\det J_t(x)$ is a polynomial in t of degree n with $\det J_0(x) = 1$. By continuity and the compactness of A, this determinant remains positive on A for all $|t| \leq \delta$ for some $0 < \delta < c^{-1}$. Therefore

$$V_n(f_t(A)) = \int_A |\det J_t(x)|\, dV_n(x) = \int_A \det J_t(x)\, dV_n(x)$$

is a polynomial in t for $|t| \leq \delta$. ∎

3.8.2. LEMMA. *Let $S^{n-1} = \{x \in \mathbb{R}^n : \|x\| = 1\}$ be the unit sphere in $\mathbb{R}^n$, and let $A = \{x \in \mathbb{R}^n : \frac{1}{2} \leq \|x\| \leq \frac{3}{2}\}$. Suppose that $v : S^{n-1} \to S^{n-1}$ is a C^1 function taking each $u \in S^{n-1}$ to a unit tangent vector, i.e. $\langle u, v(u)\rangle = 0$. Extend v to $\mathbb{R}^n \setminus \{0\}$ by $v(ru) = rv(u)$ for $u \in S^{n-1}$ and $r > 0$. Then for t small enough, $f_t(x) = x + tv(x)$ maps A onto $(1 + t^2)^{1/2}A$.*

PROOF. Let c be a Lipschitz constant for v on A. Choose $\delta < \min\{\frac{1}{3}, \frac{1}{c}\}$. For $|t| \leq \delta$, note that since $\|v(x)\| = \|x\|$ and $\langle x, v(x)\rangle = 0$, we have

$$\|f_t(x)\|^2 = \|x\|^2 + t^2\|v(x)\|^2 = (1 + t^2)\|x\|^2.$$

Thus $f_t(rS^{n-1}) \subset (1 + t^2)^{1/2} rS^{n-1}$.

Now fix $u_0 \in S^{n-1}$, and consider $g_t(x) = u_0 - tv(x)$ defined on A. Note that

$$\big|\|g_t(x)\| - 1\big| \leq |t|\|v(x)\| \leq \delta\|x\| < \tfrac{1}{3}\tfrac{3}{2} = \tfrac{1}{2}.$$

Thus $g_t(A) \subset A$. Also

$$\|g_t(x) - g_t(y)\| = |t|\|v(x) - v(y)\| \leq \delta c\|x - y\|.$$

So g_t is a contractive mapping. By the contraction mapping principle, g_t has a unique fixed point $g_t(x_0) = x_0 = u_0 - tv(x_0)$, and hence $f_t(x_0) = u_0$. By the first paragraph, we see that f_t maps $(1 + t^2)^{-1/2}S^{n-1}$ onto S^{n-1}. However $f_t(ru) = rf_t(u)$ for $u \in S^{n-1}$ and $r > 0$, so $f_t(A) = (1 + t^2)^{1/2}A$. ∎

The following well-known result from topology is called the "hairy ball theorem" because it implies that you cannot smoothly comb the hairs on a hairy 2-sphere. You can do this on an odd sphere S^{2n-1} by defining

$$v(x_1, \dots, x_{2n}) = (x_2, -x_1, x_4, -x_3, \dots, x_{2n}, -x_{2n-1}) \quad \text{for} \quad x \in S^{2n-1}.$$

3.8.3. Hairy Ball Theorem. *For $n \geq 1$, S^{2n} does not have a continuous field of non-zero tangent vectors.*

Proof. First suppose that $u : S^{2n} \to \mathbb{R}^{2n+1} \setminus \{0\}$ is C^1 and $\langle x, u(x)\rangle = 0$ for $x \in S^{2n}$. Define $v(x) = u(x)\langle u(x), u(x)\rangle^{-1/2}$, which is also C^1, and extend it to $\mathbb{R}^{2n+1} \setminus \{0\}$ by $v(rx) = rv(x)$ for $x \in S^{2n}$ and $r > 0$. By Lemma 3.8.2, $f_t(x) = x + tv(x)$ maps $A = \{x \in \mathbb{R}^n : \frac{1}{2} \leq \|x\| \leq \frac{3}{2}\}$ onto $(1+t^2)^{1/2}A$ for $|t| \leq \delta$. Therefore $V_{2n+1}(f_t(A)) = (1+t^2)^{n+1/2}V_{2n+1}(A)$. This is not a polynomial, and so contradicts Lemma 3.8.1. We conclude that there is no non-vanishing C^1 field of tangent vectors.

Now suppose that $u : S^{2n} \to \mathbb{R}^{2n+1} \setminus \{0\}$ is a continuous function such that $\langle x, u(x)\rangle = 0$ for $x \in S^{2n}$. By compactness, there are $R, \varepsilon > 0$ so that $\varepsilon \leq \|u(x)\| \leq R$ for $x \in S^{2n}$. By the Stone–Weierstrass theorem, the polynomials in $x_1, \dots, x_{2n+1}$ are dense in $C_{\mathbb{R}}(A)$. Thus there is a polynomial vector field p on S^{2n} with $\|u - p\| < \frac{\varepsilon^2}{2R}$, so p is non-zero. Define a tangential polynomial vector field by

$$v(x) = p(x) - \langle p(x), x\rangle x \quad \text{for} \quad x \in S^{2n}.$$

This is still non-zero because

$$\begin{aligned}\langle v(x), u(x)\rangle &= \langle p(x), u(x)\rangle - \langle p(x), x\rangle\langle x, u(x)\rangle = \langle p(x), u(x)\rangle \\ &= \|u(x)\|^2 + \langle p(x) - u(x), u(x)\rangle \geq \varepsilon^2 - \frac{\varepsilon^2}{2R}R = \frac{\varepsilon^2}{2}.\end{aligned}$$

This is a contradiction, and the result follows. ∎

We can now establish Brouwer's theorem.

3.8.4. Brouwer Fixed Point Theorem. *For $n \geq 1$, every continuous map of the closed unit ball B_n of $\mathbb{R}^n$ into itself has a fixed point.*

Proof. Let $f : B_n \to B_n$ be a continuous map with no fixed point. Define $v(x) = x - f(x)$, so v does not vanish. Define a function on

$$S^{n+1} = \left\{(x,t) : x \in B_n,\ t = \pm\sqrt{1 - \|x\|^2}\right\}$$

by $w(x,t) = \big(-tv(x), \langle x, v(x)\rangle\big)$. Then w is continuous and non-zero since $t = 0$ implies that $\|x\| = 1$ and $\langle x, v(x)\rangle = 1 - \langle x, f(x)\rangle > 0$. Compute

$$\big\langle (x,t), w(x,t)\big\rangle = -t\langle x, v(x)\rangle + t\langle x, v(x)\rangle = 0.$$

Thus w is a non-vanishing tangent field on S^{n+1}. If n is odd, this contradicts the hairy ball theorem, and thus f must have a fixed point.

If n is even, consider $B_{n+1} = \{(x,t) \in \mathbb{R}^n \times \mathbb{R} : x \in B_n \text{ and } t^2 \leq 1 - \|x\|^2\}$ and define a map g on B_{n+1} by $g(x,t) = (f(x), 0)$. By the odd case, g has a fixed point $(x_0, 0)$ and thus $f(x_0) = x_0$ is a fixed point for f. ∎

Every compact convex subset of $\mathbb{R}^n$ is homeomorphic to a ball in the smallest affine subset containing it. However rather than prove that we use a different approach in the following corollary.

3.8.5. COROLLARY. *Let K be a compact convex subset of $\mathbb{R}^n$. Then every continuous map $f : K \to K$ has a fixed point.*

PROOF. Let B be a compact ball containing K. For each $x \in \mathbb{R}^n$, the strict convexity of the 2-norm implies that there is a unique closest point $p(x) \in K$ to x. Moreover p is continuous (see Exercise **3.22**). Therefore $g = f \circ p$ is a continuous map of B into K. By Brouwer's fixed point theorem, there is a point $x \in B$ with $x = g(x) = f(p(x))$. Hence $x \in K$ and $p(x) = x$, so $x = f(x)$. ■

Now we wish to extend this result to compact convex sets in a LCTVS. In the following version, there is no topological constraint on K except convexity. If our space X is complete in some sense, then we could show that the closed convex hull of a compact set is compact (see Exercise **3.23**).

3.8.6. SCHAUDER–TYCHONOFF THEOREM. *Let X be a LCTVS, let K be a convex subset of X and let $K_0 \subset K$ be compact. Suppose that $f : K \to K_0$ is continuous. Then f has a fixed point.*

PROOF. The set $\Lambda = \{U_{F,r} : F \Subset \mathcal{P},\ r > 0\}$ is the set of all basic convex open neighbourhoods of 0. Then Λ is a directed set if we define $U_{F,r} \prec U_{G,s}$ for $F \subseteq G$ and $r \geq s$.

For $U \in \Lambda$, the compactness of K_0 provides a finite subset $\xi_1, \ldots, \xi_n \in K_0$ so that

$$K_0 \subset \bigcup_{i=1}^{n} \xi_i + U.$$

By Proposition 1.10.7, there is a partition of unity $g_1, \ldots, g_n \in C(K_0)$ relative to this open cover. Define

$$f_U(\xi) = \sum_{i=1}^{n} g_i\big(f(\xi)\big)\xi_i \quad \text{for} \quad \xi \in K.$$

Then f_U is continuous and has range in $K_U := \operatorname{conv}\big(\{\xi_1, \ldots, \xi_n\}\big)$, which lies in a finite dimensional subspace and is compact by Exercise **3.31**. Also $K_U \subset K$ by convexity, so f is defined on K_U. By Corollary 3.8.5, there is a point $\xi_U \in K_U$ such that $f_U(\xi_U) = \xi_U$. Observe that

$$f(\xi_U) - \xi_U = f(\xi_U) - f_U(\xi_U) = \sum_{i=1}^{n} g_i\big(f(\xi_U)\big)\big(f(\xi_U) - \xi_i\big).$$

Now $g_i(f(\xi_U)) = 0$ except when $f(\xi_U) \in \xi_i + U$, so the RHS is a convex combination of points in U. Since U is convex, $f(\xi_U) - \xi_U \in U$.

The net $(f(\xi_U))_\Lambda$ in the compact set K_0 has a cofinal convergent subnet Γ with limit η. Fix $V \in \Lambda$, and choose $V_0 \in \Lambda$ such that $V_0 - V_0 \subset V$. Since f is continuous, there is $W \in \Lambda$ such that $V_0 \prec W$ and $\xi - \eta \in W - W$ implies that $f(\xi) - f(\eta) \in V_0$. There is a $U \in \Lambda$ such that $W \prec U$ and $f(\xi_U) - \eta \in W$. Therefore

$$\xi_U - \eta = \big(f(\xi_U) - \eta\big) - \big(f(\xi_U) - \xi_U\big) \in W - U \subset W - W.$$

Hence $f(\xi_U) - f(\eta) \in V_0$. It follows that

$$f(\eta) - \eta = \big(f(\xi_U) - \eta\big) - \big(f(\xi_U) - f(\eta)\big) \in W - V_0 \subset V_0 - V_0 \subset V.$$

Since V was arbitrary, we see that $f(\eta) = \eta$. ■

When $K_0 = K$, we obtain the following important special case.

3.8.7. Corollary. *Let X be a LCTVS, and let K be a compact convex subset of X. Then every continuous map $f : K \to K$ has a fixed point.*

Recall that a mapping A on a vector space V is *affine* if

$$A(t\xi_1 + (1-t)\xi_2) = tA\xi_1 + (1-t)A\xi_2 \quad \text{for} \quad \xi_1, \xi_2 \in V.$$

If A is a continuous affine map and $AK \subset K$, then $A\overline{\text{conv}}(K) \subset \overline{\text{conv}}(K)$.

We will derive this next result as a corollary.

3.8.8. Markov–Kakutani Theorem. *Let X be a LCTVS, and let K be a non-empty compact convex subset of X. Suppose that $\mathcal{A}$ is a family of commuting continuous affine maps of K into itself. Then there is a common fixed point $\xi \in K$, i.e. $A\xi = \xi$ for all $A \in \mathcal{A}$.*

Proof. If $A \in \mathcal{A}$, define $A^{(n)} = \frac{1}{n}\sum_{i=1}^n A^i$. Then $A^{(n)}$ is continuous, affine, and maps K into itself. If $B \in \mathcal{A}$, then $A^{(n)}$ and $B^{(m)}$ commute. It follows that

$$A^{(n)}B^{(m)}K = B^{(m)}A^{(n)}K \subset A^{(n)}K \cap B^{(m)}K.$$

Hence $\mathcal{K} = \{A_1^{(n_1)} \dots A_k^{(n_k)}K : A_i \in \mathcal{A}, n_i \geq 1, k \geq 1\}$ has the finite intersection property. These are all compact sets, and therefore $C = \bigcap \mathcal{K}$ is a non-empty compact subset of K.

Let $\eta \in C$. For $A \in \mathcal{A}$, $A^{(n)}K \supset C$, so there is $\xi \in K$ with $A^{(n)}\xi = \eta$. Compute

$$A\eta - \eta = \frac{1}{n}\sum_{i=1}^n A^{i+1}\xi - \frac{1}{n}\sum_{i=1}^n A^i\xi = \frac{1}{n}(A^{n+1}\xi - A\xi) \in \frac{1}{n}(K - K).$$

If U is a basic neighbourhood of 0, then there is some n so that $K - K \subset nU$, from which we see that $A\eta - \eta \in U$. This holds for all U, whence $A\eta = \eta$, i.e. η is a fixed point for every $A \in \mathcal{A}$. ∎

Our last fixed point results work for non-commuting families which are non-contracting or distal.

3.8.9. DEFINITION. Let X be a LCTVS, and let $K \subset X$. A semigroup $\mathcal{S}$ of maps from K to itself is *distal* if for any two points $\xi \neq \eta$ in K,

$$0 \notin \overline{\{A\xi - A\eta : A \in \mathcal{S}\}}.$$

Note that $0 \notin \overline{\{A\xi - A\eta : A \in \mathcal{S}\}}$ precisely when there is a basic open neighbourhood $U_{F,r} = \{\zeta \in X : p_i(\zeta) < r,\ p_i \in F\}$ of 0 for some finite set of continuous seminorms $F \Subset \mathcal{P}$ on X which is disjoint from $\{A\xi - A\eta : A \in \mathcal{S}\}$. Setting $p = \sum_F p_i$, we have $U_{p,r} \subset U_{F,r}$. Thus distal is equivalent to the existence of a continuous seminorm p such that $\inf_{A \in \mathcal{S}} p(A\xi - A\eta) \geq r > 0$.

One common example of a distal action is a semigroup of affine isometries on a Banach space.

We first prove an easier result than our main objective.

3.8.10. HAHN FIXED POINT THEOREM. *Let X be a LCTVS, and let K be a non-empty compact convex subset of X. Suppose that $\mathcal{S}$ is a semigroup of continuous affine maps of K into itself which is distal. Then there is a common fixed point $\xi \in K$, i.e. $A\xi = \xi$ for all $A \in \mathcal{S}$.*

PROOF. Consider the compact convex subsets $C \subset K$ which are $\mathcal{S}$-invariant, i.e. $AC \subset C$ for all $A \in \mathcal{S}$. A decreasing chain of such sets has non-empty intersection by compactness, and this intersection is also compact, convex and $\mathcal{S}$-invariant. Hence by Zorn's lemma, there is a minimal compact, convex, $\mathcal{S}$-invariant set. By restricting our attention to this set, we may suppose that K is minimal. If K is a single point, we are done.

Otherwise, let ξ, η be distinct points in K. Set $\zeta = \frac{1}{2}(\xi + \eta)$. Define $L = \overline{\mathcal{S}\zeta}$. This is an $\mathcal{S}$-invariant compact subset of K. Then $\overline{\operatorname{conv}}(L)$ is a closed, convex $\mathcal{S}$-invariant subset of K. By minimality, $\overline{\operatorname{conv}}(L) = K$. By the Krein–Milman theorem, K has an extreme point γ. By Milman's theorem 3.5.8, $\gamma \in L$.

Thus there is a net $(A_\lambda)_\Lambda$ in $\mathcal{S}$ such that

$$\gamma = \lim_\Lambda A_\lambda \zeta = \lim_\Lambda \frac{1}{2}(A_\lambda \xi + A_\lambda \eta).$$

By compactness, we may drop to a cofinal subnet and ensure that $\lim_\Lambda A_\lambda \xi = \xi'$ and $\lim_\Lambda A_\lambda \eta = \eta'$ exist. Observe that $\gamma = \frac{1}{2}(\xi' + \eta')$. Since γ is an extreme point, $\xi' = \gamma = \eta'$. Hence

$$\lim_\Lambda A_\lambda \xi - A_\lambda \eta = \gamma - \gamma = 0.$$

This implies that $\mathcal{S}$ is not distal, contrary to our hypothesis. Therefore K is a singleton which is a fixed point for $\mathcal{S}$. ■

3.8.11. Definition. Let K be a subset of a LCTVS X. A set $\mathcal{S}$ of maps from K to itself is *equicontinuous* if for every open neighbourhood $V \ni 0$, there is an open neighbourhood $W \ni 0$ such that $\xi, \eta \in K$ and $\xi - \eta \in W$ implies that $A\xi - A\eta \in V$ for all $A \in \mathcal{S}$.

We can derive another well-known fixed point result.

3.8.12. Kakutani Fixed Point Theorem *Let K be a compact convex subset of a LCTVS X. Suppose that $\mathcal{G}$ is a group of affine maps of K into itself which is equicontinuous. Then $\mathcal{G}$ has a fixed point.*

Proof. It suffices to show that $\mathcal{G}$ is distal on K. Let $\xi, \eta \in K$ be distinct points. Choose an open set $V \ni 0$ so that $\eta \notin \xi + V$. By equicontinuity, find an open set $W \ni 0$ such that $\xi', \eta' \in K$ and $\xi' - \eta' \in W$ implies that $A\xi' - A\eta' \in V$ for all $A \in \mathcal{G}$. If there were an $A \in \mathcal{G}$ such that $A\xi - A\eta \in W$, then

$$\xi - \eta = A^{-1}(A\xi) - A^{-1}(A\eta) \in V,$$

which is a contradiction. Thus $\mathcal{G}$ is distal, and hence has a fixed point by the Hahn fixed point theorem. ■

Assuming compactness of K in the norm topology is a very strong condition. The following result weakens this by assuming that K is weakly compact. We will require that $\mathcal{S}$ is weakly continuous, but the distal property is with respect to the stronger topology.

Let X be a LCTVS with topology τ. The weak topology $\tau_w = \tau_{X^*}$ has the same closed convex sets by the geometric Hahn–Banach theorem. The other ingredient of the proof is Baire's theorem (Exercise **1.23**), which is the Baire category theorem in a topological setting. We will only require this result for compact Hausdorff spaces.

3.8.13. Ryll-Nardzewski Theorem. *Let X be a LCTVS with topology τ, and let $\tau_w = \tau_{X^*}$ be the weak topology on X. Suppose that K is a non-empty weakly compact convex subset of X. Suppose that $\mathcal{S}$ is a semigroup of weakly continuous affine maps of K into itself which is τ-distal. Then there is a common fixed point $\xi \in K$, i.e. $A\xi = \xi$ for all $A \in \mathcal{S}$.*

Proof. Suppose that the result is known for finitely generated semigroups. For each $F \Subset \mathcal{S}$, let K_F be the set of common fixed points for F. This is a compact subset, and clearly $\{K_F : F \Subset \mathcal{S}\}$ would have the finite intersection property. Thus any point $\xi \in \bigcap_{F \Subset \mathcal{S}} K_F$, which is non-empty by compactness, is a common fixed point. So we assume that $\mathcal{S}$ is finitely generated, and, in particular, countable.

Arguing as in the proof of Hahn's theorem, we may suppose that K is a minimal weakly compact, convex, $\mathcal{S}$-invariant set. Thus for any $\xi \in K$, the τ-closed set $\overline{\operatorname{conv}}^{\tau}(\{A\xi : A \in \mathcal{S}\})$ is an $\mathcal{S}$-invariant convex subset of K. By the geometric Hahn–Banach theorem, it is also weakly closed, and hence is equal to K by minimality. Therefore K is τ-separable. Let B be a countable, τ-dense subset.

A similar argument shows that there is a minimal, weakly compact, $\mathcal{S}$-invariant set $L \subset K$. If L is a singleton, we are done; so we assume that L contains distinct points ξ, η. The τ-distal property implies that there is a convex τ-open neighbourhood $U \ni 0$ such that $A\xi - A\eta \notin U$ for $A \in \mathcal{S}$. Select a convex τ-closed neighbourhood V of 0 with $V - V \subset U$. Then V is also weakly closed. Since B is τ-dense in K, if $\zeta \in K$, $(\zeta - V) \cap B \neq \varnothing$; so

$$L \subset K \subset \bigcup_{\beta \in B} \beta + V \,, \quad \text{whence} \quad L = \bigcup_{\beta \in B} (\beta + V) \cap L.$$

The sets $(\beta + V) \cap L$ for $\beta \in B$ are also weakly closed. By Baire's theorem (Exercise **1.23**), L is not a countable union of weakly closed nowhere dense sets. Therefore, there is a τ_w-open set W and $\beta_0 \in B$ such that

$$\varnothing \neq L \cap W \subset (\beta_0 + V) \cap L \subset \beta_0 + V.$$

Since $\overline{\operatorname{conv}}^{\tau}(L)$ is a τ-closed convex $\mathcal{S}$-invariant subset of K, it equals K as above. Let γ be any extreme point of K, which exist by the Krein–Milman theorem. By Milman's theorem 3.5.8, $\gamma \in L$. The minimality of L guarantees that $\overline{\mathcal{S}\gamma}^{\tau_w} = L$. Thus we can pick $A_0 \in \mathcal{S}$ such that $A_0\gamma \in L \cap W$.

Set $\zeta = \frac{1}{2}(\xi + \eta)$. By minimality of K and the fact that it is weakly closed,

$$K = \overline{\operatorname{conv}}^{\tau}(\mathcal{S}\zeta) = \overline{\operatorname{conv}}^{\tau}(\overline{\mathcal{S}\zeta}^{\tau_w}).$$

Using Milman's theorem again, we see that $\gamma \in \overline{\mathcal{S}\zeta}^{\tau_w}$. Thus there is a net $(A_\lambda)_\Lambda$ in $\mathcal{S}$ such that

$$\gamma = \tau_w\text{–}\lim_{\Lambda} A_\lambda \zeta = \tau_w\text{–}\lim_{\Lambda} \frac{1}{2}(A_\lambda \xi + A_\lambda \eta).$$

By weak compactness of L, we may drop to a cofinal subnet and ensure that $\tau_w\text{–}\lim_\Lambda A_\lambda \xi = \xi'$ and $\tau_w\text{–}\lim_\Lambda A_\lambda \eta = \eta'$ exist. Observe that $\gamma = \frac{1}{2}(\xi' + \eta')$. Since γ is an extreme point, $\xi' = \gamma = \eta'$. Hence, since $A \in \mathcal{S}$ are τ_w-continuous, there is some $\lambda_0 \in \Lambda$ so that

$$A_0 A_{\lambda_0} \xi \in L \cap W \subset \beta_0 + V \quad \text{and} \quad A_0 A_{\lambda_0} \eta \in L \cap W \subset \beta_0 + V.$$

Therefore $A_0 A_{\lambda_0} \xi - A_0 A_{\lambda_0} \eta \in V - V \subset U$. This contradicts the distal assumption. Consequently L is a singleton, and $\mathcal{S}$ has a fixed point. ∎

A common application of this result is the following.

3.8.14. COROLLARY. *Let X be a Banach space, let K be a non-empty weakly compact convex subset of X and let $\mathcal{S}$ be a semigroup of linear isometries on X which leave K invariant. Then $\mathcal{S}$ has a fixed point $\xi \in K$.*

PROOF. As noted earlier, the distal property with respect to the norm is immediate. Bounded linear maps on a Banach space are weakly continuous, since if $(\xi_\lambda)_\Lambda$ is a net converging weakly to ξ and $T \in \mathcal{B}(X)$, then for $\varphi \in X^*$,

$$\lim_\Lambda \varphi(T\xi_\lambda) = \lim_\Lambda T^*\varphi(\xi_\lambda) = T^*\varphi(\xi) = \varphi(T\xi).$$

Hence $(T\xi_\lambda)_\Lambda$ converges weakly to $T\xi$. Therefore, the Ryll-Nardzewski theorem applies to yield a fixed point. ■

3.9. Haar Measure on Compact Topological Groups

In this short section, we will use some of the preceding results to establish the existence of a translation invariant probability measure on compact groups. It is not our intention to make a serious study of topological groups.

3.9.1. DEFINITION. A topological group G is a group with a Hausdorff topology such that multiplication is jointly continuous, i.e. the map $(x, y) \to xy$ from $G \times G$ to G is continuous, and the inverse operation $x \to x^{-1}$ is continuous on G.

We have stipulated that topological groups are Hausdorff. This is not universally done, but all commonly studied groups are Hausdorff, and frequently also locally compact. Some basic properties are left to Exercise **3.48**, including the fact that a T_0 group is Hausdorff.

3.9.2. EXAMPLES.
(1) Any group G is a topological group in the discrete topology.

(2) The groups $\mathbb{T}^n$, $\mathbb{R}^n$ and $\mathbb{C}^n$ are locally compact topological groups. A finite product of LCH topological groups is a LCH topological group. The group $\mathbb{R}^{\aleph_0}$ is a topological group in its product topology, but it is not LCH.

(3) The groups $\mathrm{GL}_n(\mathbb{C})$ and $\mathrm{GL}_n(\mathbb{R})$ of invertible matrices with the Euclidean topology inherited from $M_n(\mathbb{C})$ and $M_n(\mathbb{R})$ are LCH topological groups. The unitary and orthogonal groups, $\mathcal{U}_n$ and $\mathcal{O}_n$, are compact topological groups.

(4) In general, a subgroup of a topological group is a topological group. If H is a normal subgroup of G, then G/H has the quotient topology. This topology will be Hausdorff only when H is closed. If G is LCH and H is closed, then G/H is also LCH.

Let G be a LCH topological group. Let $C_c(G)$ denote the space of compactly supported continuous functions on G with the supremum norm. Its completion is $C_0(G)$. Recall that a Radon measure on G is a Borel measure which is outer regular, inner regular on open sets, and finite on compact sets. The Riesz–Markov theorem states that every positive linear functional on $C_c(G)$ is obtained by integration against a unique Radon measure, and conversely every Radon measure yields a positive linear functional. The left translation operators $L_s f(t) = f(st)$ for $s, t \in G$ are isometries on $C_0(G)$. It is a well-known, non-trivial result from measure theory that there is a positive Radon measure μ on G which is left invariant, i.e. $\mu(L_s f) = \mu(f)$ for every $s \in G$ and $f \in C_c(G)$. Moreover, this measure is unique up to multiplication by a positive scalar. It is called the *Haar measure* for G. When G is compact, it is standard to normalize it so that $\mu(G) = 1$. We will show that when G is compact, we can derive the existence of Haar measure using a fixed point theorem.

3.9.3. THEOREM. *Let G be a compact group. Then G has a unique left invariant probability measure μ. This measure is also the unique right invariant probability measure, and*

$$\int f(s^{-1})\, d\mu(s) = \int f(s)\, d\mu(s).$$

PROOF. We have $C(G)^* = M(G)$, the space of all complex regular (finite) Borel measures on X via the Riesz representation theorem. Let K denote the weak-$*$ compact convex set of probability measures on G. The operators L_s^* are weak-$*$ continuous by Example 3.1.11, and are isometries which map K onto itself. Let $\mathcal{G} = \{L_s^* : s \in G\}$. This is a group isomorphic to G since for $\mu \in K$, $f \in C(G)$ and $s, t, u \in G$,

$$L_t L_s f(u) = (L_s f)(tu) = f(stu) = L_{st} f(u)$$

and hence

$$\langle L_s^* L_t^* \mu, f \rangle = \langle \mu, L_t L_s f \rangle = \langle \mu, L_{st} f \rangle = \langle L_{st}^* \mu, f \rangle.$$

Let us show that $\mathcal{G}$ is equicontinuous on K. A basic weak-$*$ open neighbourhood of 0 is given by

$$U_F = \{\mu \in M(G) : |\mu(f)| < 1 \text{ for } f \in F\}$$

where $F = \{f_1, \dots, f_n\} \Subset C(G)$ is finite. The map $\Phi : G \to C(G)^F$ by $\Phi(s) = (L_s f_1, \dots, L_s f_n)$ is continuous, and hence has a norm-compact image. Thus there is a finite set $S \Subset G$ such that $\Phi(S)$ is a $\frac{1}{3}$-net in $\Phi(G)$. Define a weak-$*$ open neighbourhood of 0 in $M(G)$ by

$$W = \{\mu \in M(G) : |\mu(L_s f_i)| < \tfrac{1}{3},\ s \in S,\ f_i \in F\}.$$

Suppose that $\mu, \nu \in K$ and $\mu - \nu \in W$. Then for $t \in G$, there is some $s \in S$ such that $\|L_t f_i - L_s f_i\| < \frac{1}{3}$ for $f_i \in F$. Hence for $f_i \in F$,

$$\begin{aligned}|L_t^*(\mu - \nu)(f_i)| &= |(\mu - \nu)(L_t f_i)| \le |(\mu - \nu)(L_s f_i)| + |(\mu - \nu)(L_t f_i - L_s f_i)| \\ &< \frac{1}{3} + \|\mu - \nu\| \|L_t f_i - L_s f_i\| < \frac{1}{3} + \frac{2}{3} = 1.\end{aligned}$$

Thus $L_t^*\mu - L_t^*\nu \in U_F$, showing that $\mathcal{G}$ is equicontinuous.

By Kakutani's fixed point theorem, there is some $\mu \in K$ such that $L_s^*\mu = \mu$ for all $s \in G$. This is a left invariant measure.

The same argument shows that there is a right invariant probability measure ν. For $f \in C(G)$, by Fubini's theorem,

$$\begin{aligned}\int f(s)\, d\nu(s) &= \int d\mu(t) \int f(st)\, d\nu(s) \\ &= \int d\nu(s) \int f(st)\, d\mu(t) = \int f(t)\, d\mu(t).\end{aligned}$$

Hence $\mu = \nu$. In particular, this shows that μ is unique, and it is also right invariant.

The last statement is part of Exercise **3.49**. ∎

Exercises for Chapter 3

3.1 Let $(X, \mathcal{P})$ and $(Y, \mathcal{Q})$ be LCTVSs, and let $T : X \to Y$ be a linear map. Prove that the following are equivalent:

(1) T is continuous.
(2) T is continuous at 0.
(3) For each $q \in \mathcal{Q}$, there is a finite subset $F = \{p_1, \dots, p_n\} \subset \mathcal{P}$ and $t_i \in \mathbb{R}^+$ so that $q(Tx) \le \sum_{i=1}^n t_i p_i(x)$ for all $x \in X$.

3.2 If $(x_\lambda)_\Lambda$ is a net in a Banach space X which converges weakly to x, find a sequence of points in the convex hull of the net converging to x in norm.
HINT: Show that the convex hull of the net intersects $b_{1/n}(x)$.

3.3 (a) Let X be a Banach space, and let Y be a closed subspace of X^* which norms X. Show that a *sequence* in X which converges in the τ_Y topology must be bounded.
(b) Let H be a Hilbert space. Show that if a *sequence* $T_n \in \mathcal{B}(H)$ converges to T in the weak operator topology, then $\{T_n\}$ is bounded. Note: The set of functionals determining this topology is not closed, so part (a) doesn't apply.

3.4 (a) Suppose that φ is a continuous functional on $(\mathcal{B}(H), \text{SOT})$. Show that there are vectors $x_i, y_i \in H$ for $1 \le i \le n$ so that $\varphi(T) = \sum_{i=1}^n \langle Tx_i, y_i \rangle$.
HINT: Use that $\varphi^{-1}(\mathbb{D})$ is SOT-open to show that there are vectors x_i for $1 \le i \le n$ so that φ factors through the map $\Phi : \mathcal{B}(H) \to H^{(n)}$ given by $\Phi(T) = (Tx_1, Tx_2, \dots, Tx_n)$.

(b) Conclude that $(\mathcal{B}(H), \text{SOT})$ and $(\mathcal{B}(H), \text{WOT})$ have the same continuous linear functionals. Deduce that these two topologies have the same *closed convex* sets.

(c) Consider the operator M_z on $L^2(\mathbb{T})$ given by $M_z f = zf$, (where z is the function $z(e^{i\theta}) = e^{i\theta}$). Show that M_z^n converges to 0 in the WOT topology, but not in the SOT topology. So these two locally convex topologies are different.

3.5 Let X be an infinite dimensional Banach space.

(a) Show that there is a discontinuous linear functional φ defined on all of X. HINT: Use an algebraic basis for X.

(b) Show that $A_r = \{x \in X : \operatorname{Re} \varphi(x) = r\}$ are disjoint convex sets for $r \in \mathbb{R}$, and each one is dense in X. So they cannot be separated by a continuous linear functional.

3.6 Let X be a Banach space, and let $Y \subset X^*$ be a subspace which norms X. If there is a non-empty τ_Y-open bounded subset of X, show that X is finite dimensional.

3.7 Let $A = \{e_m + me_n : 1 \leq m < n\}$, where $\{e_n : n \geq 1\}$ is an orthonormal basis for l_2, Show that 0 lies in the weak closure of A, but is not the weak limit of any sequence in A.

3.8 Show that if A is a *bounded* subset of l_2 and x lies in the weak closure of A, then there is a sequence in A converging weakly to x.

3.9 Show that if X is a Banach space such that $(\overline{b_1(X^*)}, \tau_X)$ is metrizable, then X is separable.
HINT: find τ_X open sets $U_n = \{\varphi \in X^* : |\varphi(x)| < 1,\ x \in F_n\}$ for finite $F_n \Subset X$ such that $\overline{b_1(X^*)} \cap \bigcap_{n\geq 1} U_n = \{0\}$. Use the Hahn–Banach theorem to show that $X = \overline{\operatorname{span}}(\bigcup_{n\geq 1} F_n)$.

3.10 Show that if X is a Banach space such that $(\overline{b_1(X)}, \tau_{X^*})$ is metrizable, then X^* is separable.
HINT: Proceed as in the previous example. After using the Hahn–Banach theorem to get $\psi \in X^{**}$, use Goldstine's theorem.

3.11 Let X be a LCTVS given by a family of seminorms $\mathcal{P}$. A subset $E \subset X$ is *bounded* if for every open neighbourhood $U \ni 0$, there is a constant $c < \infty$ such that $E \subset cU$.

(a) Prove that E is bounded if and only if each $p \in \mathcal{P}$ is bounded on E.

(b) Show that every compact subset of X is bounded.

(c) Show that if X has a bounded neighbourhood of 0, then there is a norm on X generating its topology.
HINT: Find a convex balanced bounded neighbourhood U of 0. If V is a neighbourhood of 0, find $\varepsilon > 0$ so that $\varepsilon U \subset V$. Show p_U is the desired norm.

3.12 Show that finite dimensional subspaces of an LCTVS are closed.

3.13 Show that if there is a compact neighbourhood of 0 in a LCTVS, then it is finite dimensional.
HINT: Let U be an open neighbourhood of 0 with $\overline{U}$ compact. Cover $\overline{U}$ with sets $x_i + \frac{1}{2}U$ for $x_i \in \overline{U}$, $1 \le i \le m$. Set $Y = \operatorname{span}\{x_1, \dots, x_m\}$. So $\overline{U} \subset Y + \frac{1}{2}U$. Show by induction that $\overline{U} \subset Y + 2^{-k}U$.

3.14 Prove that if X is a Fréchet space which is not a Banach space, then there is no complete metric on X^* whose topology is stronger than the τ_X topology. (See the discussion after Proposition 3.1.6.)
HINT: Let $\{p_n : n \ge 1\}$ be an increasing family of seminorms determining the topology on X, and let $Y_n = (X, p_n)^*$. Now $X^* = \bigcup_{n \ge 1} \overline{b_n(Y_n)}$. Apply the Baire category theorem. Hence show that $X^* = Y_n$ for some n.

3.15 Let X be a LCTVS with a countable family $\{p_i : i \ge 1\}$ of seminorms determining the topology. Let d be the metric defined in Example 3.1.5.

(a) Prove that the topology on X is equivalent to the metric topology.
(b) Show that X is complete in the metric if and only if every sequence which is Cauchy in every seminorm converges in X.

3.16 Let Ω be an open domain in $\mathbb{R}^n$.

(a) Show that the derivative maps $D^i f = \frac{\partial}{\partial x_i} f$ on $C^\infty(\Omega)$ are continuous.
(b) Show that a closed bounded subset E of $C^\infty(\Omega)$ is compact.
HINT: For $K \Subset \Omega$ compact, use the boundedness of derivatives of order $k+1$ to show that the derivatives of order k of E are equicontinuous on K. Apply the Arzela–Ascoli theorem.
(c) Deduce that there are no bounded neighbourhoods of 0. Hence $C^\infty(\Omega)$ is not normable.
HINT: Exercise **3.13**.

3.17 Let ψ be a tempered distribution.

(a) Show that there is a positive integer k so that ψ extends to a continuous linear functional on $C_0(\mathbb{R})^{(k+1)}$, the space of $k+1$-tuples of functions in $C_0(\mathbb{R})$.
HINT: Show that ψ depends only on the first k derivatives, $(f, f', \dots, f^{(k)})$. So ψ can be considered as a functional on a subspace of $C_0(\mathbb{R})^{(k+1)}$.
(b) Hence show that there are finite regular Borel measures $\mu_0, \dots, \mu_k$ on $\mathbb{R}$ so that $\psi(f) = \sum_{i=0}^k \int_{\mathbb{R}} f^{(i)}(x)\, d\mu_i(x)$.
(c) Show that $\varepsilon_x^{(m)}(f)$ is not given by integration of f against a measure for $m \ge 1$.

3.18 Let s be the space of all complex sequences $x = (x_n)_{n \ge 1}$. Define seminorms $p_n(x) = |x_n|$ for $n \ge 1$, making s into a Fréchet space. Let s_0 be the subspace of s consisting of sequences which are zero except finitely often. An element $x = (x_n)$ of s determines a linear functional on s_0 by $\langle (y_n), (x_n) \rangle = \sum_i y_n x_n$. Put the τ_{s} topology on s_0. Show that $\mathrm{s}^* = \mathrm{s}_0$ and $\mathrm{s}_0^* = \mathrm{s}$.

3.19 Let $0 < p < 1$ and let $X = L^p(0,1)$ be the set of Lebesgue integrable functions (modulo functions vanishing a.e.) such that $d_p(f) = \int |f|^p \, dm < \infty$.
(a) Show that X is a metrizable vector space with $d(f,g) := d_p(f-g)$.
(b) Let $B = \{f \in X : d_p(f) < 1\}$. Show that $\operatorname{conv} B = X$.
(c) Show that X is not a LCTVS.
(d) Show that there are no non-zero continuous linear functionals on X.

3.20 Let X be a LCTVS and let $K \subset X$ be a closed convex set with non-empty interior. Show that if $x_0 \in \partial K = K \setminus \operatorname{int}(K)$, there is a non-zero continuous linear functional so that $\sup_{x \in K} \operatorname{Re} \varphi(x) = \varphi(x_0)$.

3.21 Let X be a LCTVS and let $A \subset X$. Show that $\overline{\operatorname{span}(A)}$ equals the intersection of all closed hyperplanes containing A.

3.22 Let K be a compact convex subset of a Hilbert space H.
(a) Prove that for each $\xi \in H$, there is a unique closest point $p(\xi)$ in K.
HINT: If η and ζ are two closest points, consider their midpoint.
(b) Show that $\langle \xi - p(\xi), \eta - p(\xi) \rangle \le 0$ for all $\eta \in K$, and conversely, $p(\xi)$ is determined by this property.
HINT: $\|\xi - p(\xi)\|^2 < \|\xi - (t\eta + (1-t)p(\xi))\|^2$.
(c) Show that if $\xi, \eta \in H$, then $\|p(\xi) - p(\eta)\| \le \|\xi - \eta\|$.
HINT: Show that $\big\langle (\xi - \eta) + (p(\xi) - p(\eta)), p(\xi) - p(\eta) \big\rangle \le 0$.

3.23 (a) Let H be a Hilbert space with an orthonormal set $\{e_n : n \ge 1\}$. Show that $K = \{0, \frac{1}{n} e_n : n \ge 1\}$ is a compact set such that $\operatorname{conv}(K)$ is not closed.
(b) Let K be a compact subset of a Fréchet space X. Show $\overline{\operatorname{conv}(K)}$ is compact.
HINT: Show that $\operatorname{conv}(K)$ is totally bounded.

3.24 Let $K = \{x = (x_n) : x_n \ge 0,\ x \in l_p\} \subset l_p$ for $1 \le p < \infty$.
(a) Show that K is weak-$*$ closed and convex with no interior.
(b) Show that if $a = (a_n) \in K$, there is a non-zero linear functional φ such that $\sup_{x \in K} \operatorname{Re} \varphi(x) = \varphi(a)$ if and only if $a_n = 0$ for some n.

3.25 Let H be a Hilbert space. Prove that the closed unit ball of $\mathcal{B}(H)$ is compact in the weak operator topology.
HINT: Modify the proof of the Banach–Alaoglu theorem.

3.26 (a) Find all extreme points of $\overline{b_1(l_p)}$ for $1 \le p < \infty$.
(b) Compute the norm-closed convex hull of $\operatorname{ext}(\overline{b_1(l_p)})$.
HINT: Equality in Minkowski's inequality.

3.27 Describe all extreme points of the closed unit ball of $C(K)$ where K is the Cantor set. What is the norm-closed convex hull of $\operatorname{ext}(\overline{b_1(C(K))})$?

3.28 Show that if the set of extreme points of the unit ball of an infinite dimensional Banach space X is finite, then X is not a dual space. Apply this to the space $c_{\mathbb{R}}$ of all real convergent sequences with the sup norm.

3.29 Show that $c \oplus_\infty c$ is not isometrically isomorphic to c, but $(c \oplus_\infty c)^* \simeq l_1$.

3.30 Give an example of a closed convex set K in $\mathbb{R}^2$ and a point $p \in \mathbb{R}^2$ such that $\operatorname{conv}(K \cup \{p\})$ is not closed.

3.31 The following deals with convex sets in finite dimensional spaces.
(a) Show that if $S \subset \mathbb{R}^n$, then every element of $\operatorname{conv}(S)$ is a convex combination of $n+1$ elements of S.
HINT: If $s_1, \ldots, s_m \in S$ with $m \geq n+2$, use the linear dependence of $s_i - s_m$ for $1 \leq i \leq m-1$.
(b) Hence show that the (non-closed) convex hull of a compact set is compact.
(c) Show that if $s_1, \ldots, s_m$ are points in $\mathbb{R}^n$ with $m \geq n+2$, then you can split $\{1, \ldots, m\}$ into two disjoint sets I and J so that $\operatorname{conv}(I) \cap \operatorname{conv}(J) \neq \varnothing$.
(d) (*Helly's theorem*) Let C_k be compact convex subsets of $\mathbb{R}^n$ for $k \geq 1$. Suppose that any $n+1$ of these sets have non-empty intersection. Show that $\cap_{k \geq 1} C_k$ is non-empty.
HINT: Use part (c) and induction to show that any m sets intersect.

3.32 Derive Corollary 3.5.10 from the Stone–Weierstrass theorem 3.5.9.

3.33 Let K be a weak-$*$ compact set in a Banach space X^*. Show that $\overline{\operatorname{conv}(K)}^{w^*}$ is weak-$*$ compact.

3.34 (a) Show that a *sequence* $\mathbf{x}_n = (x_{n,i})_{i \geq 1}$ in l_p, $1 < p < \infty$, converges weakly to 0 if and only if $\{\mathbf{x}_n\}_{n \geq 1}$ is bounded and $\lim_{n \to \infty} x_{n,i} = 0$ for $i \geq 1$.
(b) Show that a *sequence* $\mathbf{x}_n = (x_{n,i})_{i \geq 1}$ in l_1 converges weakly if and only if it converges in norm. This is known as the *Schur property*.
(c) Find a *bounded net* in l_1 which converges weakly to 0 but not in norm.

3.35 Let $A \subset X^*$ for a Banach space X. Suppose that for every *bounded* net in A converging weak-$*$ to $\varphi \in X^*$, one has $\varphi \in A$. Show that A is weak-$*$ closed.

3.36 Let X and Y be Banach spaces, and let $T \in \mathcal{B}(X^*, Y^*)$. Suppose that for every bounded net $(x_\lambda)_\Lambda$ in X^* converging weak-$*$ to x, the net $(Tx_\lambda)_\Lambda$ converges weak-$*$ in Y^* to Tx. Show that T is weak-$*$–weak-$*$ continuous.

3.37 (a) A Banach space is *uniformly convex* if for every $\varepsilon > 0$, there is a $\delta > 0$ so that if $\|x\| \leq 1$, $\|y\| \leq 1$ and $\|x - y\| > \varepsilon$, then $\|(x+y)/2\| < 1 - \delta$. Show that Hilbert space is uniformly convex, but c_0 is not.
(b) Show that every uniformly convex Banach space is reflexive.
HINT: Let $z \in X^{**}$ with $\|z\| = 1$. Let $\varepsilon > 0$, get $\delta > 0$ from the uniformly

convexity. Pick $f \in b_1(X^*)$ so that $|f(z) - 1| < \delta$. Let

$$C = \{x \in b_1(X) : |f(x) - 1| < \delta\}.$$

Show that z is in the weak* closure of C in X^{**}. Prove that $\|x - y\| < \varepsilon$ for all $x, y \in C$. Hence deduce that there is an $x \in C$ so that $\|z - x\| \le \varepsilon$.

3.38 Let X be a LCTVS with dual Y, and let $A \subset X$.
(a) Show that A° is a τ_X-closed convex subset of Y containing 0.
(b) Show that if $0 \in A$ is a τ_Y-closed convex subset of X, then $(A^\circ)^\circ = A$.
(c) A *convex cone* is a convex set C such that $x \in C$ and $t \ge 0$ implies that $tx \in C$. Show that $C^\circ = C^\dagger := \{y \in Y : \operatorname{Re}\langle x, y\rangle \le 0 \text{ for all } x \in C\}$ is a τ_X-closed convex cone.
(d) Show that if C is a τ_Y-closed convex cone, then $C = (C^\dagger)^\dagger$.

3.39 Suppose that $A \subset X$ and $B \subset X^*$ have IDLP. If $\overline{\mathbb{D}B}^{w^*}$ has interior, show that $\overline{A}^w$ is weakly compact.

3.40 Let A be a bounded subset of $C(X)$ for a compact Hausdorff space X. Show that if X_0 is a dense subset of X, and A and $\varepsilon(X_0)$ (point evaluations) have IDLP, then $\overline{A}^w$ is weakly compact.
HINT: If A is bounded by R, consider A as a subset of $\overline{b_R(X)}^X$ as in the proof of Banach–Alaoglu. Show that points in $\overline{A}$ correspond to continuous functions.

3.41 Prove that l_1 contains no infinite dimensional reflexive subspace.
HINT: Let X be a reflexive subspace. Show that $\overline{b_1(X)}$ is weakly sequentially compact. Apply Exercise **3.34** to show that $\overline{b_1(X)}$ is norm sequentially compact.

3.42 Let X be a separable Banach space, and let $K \subset X$ be weakly compact.
(a) Find a countable subset $\{f_n\}$ of $\overline{b_1(X^*)}$ which norms X.
(b) Define a metric on K by $d(x, y) = \sum_{n\ge1} 2^{-n}|f_n(x - y)|$.
(c) Prove that the identity map from (K, τ_w) to (K, d) is a homeomorphism. Hence deduce that K is metrizable in its weak topology.

3.43 A *retract* of a set B to its boundary ∂B is a continuous function $r : B \to \partial B$ such that $r|_{\partial B} = id$. Prove that there is no retract of B_n to S^{n-1}.
HINT: Use a retract to construct a continuous $f : B_n \to B_n$ with no fixed points.

3.44 Suppose that $f : \overline{\mathbb{D}} \to \overline{\mathbb{D}}$ is a continuous map with no fixed point. Define $g(z)$ to be the unique point in $\{f(z) + t(x - f(x)) : t \ge 0\} \cap \mathbb{T}$. This is a continuous retract of $\overline{\mathbb{D}}$ onto $\mathbb{T}$. Compute the winding number of $g(r\mathbb{T})$ for $0 \le r \le 1$ and reach a contradiction.

3.45 (*Kakutani*). Let B be the closed unit ball of l_2. For $\varepsilon > 0$, define $f_\varepsilon : B \to B$ by $f_\varepsilon(x) = (\varepsilon(1 - \|x\|), x_1, x_2, x_3, \dots)$. Show that f_ε is uniformly continuous, even Lipschitz with Lipschitz constant $\sqrt{1 + \varepsilon^2}$, but has no fixed points.

3.46 Let f_1 be taken from the previous example. Define

$$g(\xi) = \begin{cases} \frac{1}{2} f_1(2\xi) & \text{if } \|\xi\| \le \frac{1}{2} \\ (1 - \|\xi\|) f_1(\xi/\|\xi\|) & \text{if } \frac{1}{2} \le \|\xi\| \le 1 \end{cases} \quad \text{and} \quad h(\xi) = \frac{\xi - g(\xi)}{\|\xi - g(\xi)\|}.$$

Show that h is a retract of B onto ∂B.

3.47 (*Alspach*) Let $K = \{f \in L^1(0,1) : 0 \le f \le 2 \text{ and } \int f = 1\}$. Define

$$Af(t) = \begin{cases} 2f(2t) \wedge 2 & \text{if } 0 \le t \le \frac{1}{2} \\ (2f(2t-1) - 1) - 2) \vee 0 & \text{if } \frac{1}{2} < t \le 1. \end{cases}$$

(a) Prove that K is weakly compact.
HINT: First show that $\{f \in L^1(0,1) : 0 \le f \le 2\}$ is weakly compact.
(b) Prove that A is an isometry of K into itself.
(c) Show that $Af = f$ implies $f = 2\chi_B$ for measurable $B \subset [0,1]$ with $m(B) = \frac{1}{2}$.
HINT: $\{t : f(t) = 2\} = \{t/2 : 1 \le f(t) \le 2\} \cup \{(1+t)/2 : f(t) = 2\}$.
(d) Show that B must intersect each diadic interval $\left[\frac{k}{2^n}, \frac{k+1}{2^n}\right]$ in a set of measure 2^{-n-1}. Hence conclude that A has no fixed points.

3.48 Let G be a topological group, but not necessarily Hausdorff.
(a) Show that the continuity conditions on G are equivalent to the assumption that $(x, y) \to x^{-1}y$ is continuous on $G \times G$.
(b) Show that for $x \in G$, the maps $L_x y = xy$, $R_x y = yx$ and the inverse map $Jy = y^{-1}$ are homeomorphisms of G.
(c) Show that if $1 \in U$ is open, then xU and Ux are open neighbourhoods of x and U^{-1} is a neighbourhood of 1.
(d) Show that if $1 \in U$ is open, then there is an open neighbourhood $V \ni 1$ such that $V = V^{-1}$ and $V^2 \subset U$.
(e) If $1 \in U$ is open, find an open neighbourhood $V \ni 1$ such that $\overline{V} \subset U$.
HINT: Show that $\overline{V} \subset V^2$.
(f) Show that if G is T_0, then it is Hausdorff.
(g) Show that open subgroups are closed.

3.49 Let G be a compact group with Haar measure μ.
(a) Verify that the measure ν defined by $\int f(x)\,d\nu(x) = \int f(x^{-1})\,d\mu(x)$ is right invariant, and thus equals μ.
(b) Show that every non-empty open set U has $\mu(U) > 0$.

3.50 Let G be a compact group and let $K = M_1^+(G)$ with the weak-$*$ topology.
(a) For $\mu \in M(G)$, show that the map $T_\mu(s) = L_s^*\mu$ is continuous.
(b) Show that $\mathcal{G} = \{L_s^* : s \in G\}$ is weak-$*$ distal on K.
HINT: Separate 0 from the range of $T_{\mu-\nu}$.
(c) Use Hahn's theorem to find Haar measure.

Notes on Chapter 3

Early notions of topological spaces go back to Fréchet, and the notion of weak convergence was used by Hilbert and Riesz. However, the definition of a topological vector space is due to Kolmogoroff [181] in 1934, and the notion of a locally convex space is due to von Neumann [315] in 1935. The spaces of C^∞ functions and their duals, distributions of various types, were introduced by Schwartz [281].

The first geometric separation theorem is due to Mazur [201] in 1933, with refinements by Dieudonné [89]. Goldstine [127] proved his theorem in 1938, but it was reformulated close to the modern version by Kakutani [169]. Banach had a sequentially compact version of the weak-$*$ compactness of the ball in X^* in his book [23]. Alaoglu, a Canadian mathematician, proved the modern version in 1940. The Krein–Milman [184] and Krein–Smulian [185] theorems date from 1940 as well. The Eberlein–Smulian theorem was proved in parts by Smulian [288] and Eberlein [109]. The IDLP is due to Grothendieck [129]. The result on the closed convex hull of a weakly compact set is in Krein–Smulian [185]. The proofs provided of these last two results are due to Kremp [186] and Whitley [323].

Brouwer established his fixed point theorem in 1911. Various proofs can be found in [36], [148] , [149] and Dunford and Schwarz [104]. Milnor's novel proof is taken from [206]. Schauder established his fixed point for compact convex subsets of a Banach space. Tychonoff generalized it to LCTVSs in [308]. Markov [198] established the fixed point theorem for commuting affine maps in 1936. Kakutani [168] provided a different approach and also establishes the theorem for equicontinuous groups of affine maps. Ryll-Nardzewski's fixed point theorem was announced in 1962, with a detailed proof appearing later in [271]. F. Hahn's fixed point theorem [133] actually came a bit later. The proofs of these last few theorems given here are taken from Namioka's survey [215].

Haar measure was constructed by Haar [131] for compact groups and Weil [320] for locally compact groups. Treatments can be found in Folland [114] and Hewitt and Ross [147]. A simpler proof for amenable groups can be found in Izzo [156].

A more detailed history with complete references can be found in Dunford and Schwarz [104]. This book contains most of the results in this chapter and other related material. Rudin [269] and Conway [61], and most other books on functional analysis, cover this material as well. Rudin [269] contains a nice treatment of distributions; see also Reed–Simon [256].

CHAPTER 4

Linear Operators

In large part, the study of Banach and Hilbert spaces concentrates on the structure of linear maps between them. In particular, the notion of spectrum extends the notion of eigenvalues for matrices. In this chapter, we develop some of the basic ideas of linear operators and their spectra.

4.1. Adjoint Operators

4.1.1. DEFINITION. Let X, Y be Banach spaces. If $T \in \mathcal{B}(X, Y)$, define the *adjoint operator* $T^* \in \mathcal{B}(Y^*, X^*)$ by $T^*g = g \circ T$; i.e.

$$(T^*g)(x) = g(Tx) \quad \text{for } g \in Y^* \text{ and } x \in X.$$

Clearly $\|T^*\| = \sup_{\|g\|\le 1} \|gT\| \le \sup_{\|g\|\le 1} \|g\|\,\|T\| = \|T\|$, so T^* is continuous. In Example 2.6.21, we showed that $\|T^*\| = \|T\|$. In Example 3.1.11, we showed that T^* is weak-$*$–weak-$*$ continuous. Different proofs are given below.

We sometimes write the pairing between a Banach space and its adjoint as a bilinear form: $\langle x, f\rangle = f(x)$ for $x \in X$ and $f \in X^*$. Unlike an inner product, this form is linear in each variable separately:

$$\langle a_1x_1 + a_2x_2, f\rangle = a_1\langle x_1, f\rangle + a_2\langle x_2, f\rangle \quad \text{for } x_1, x_2 \in X,\ a_1, a_2 \in \mathbb{F},\ f \in X^*$$

and

$$\langle x, b_1f_1 + b_2, f_2\rangle = b_1\langle x, f_1\rangle + b_2\langle x, f_2\rangle \quad \text{for } x \in X,\ b_1, b_2 \in \mathbb{F},\ f_1, f_2 \in X^*.$$

Using this notation, we get

$$\langle x, T^*g\rangle = \langle Tx, g\rangle \quad \text{for } g \in Y^* \text{ and } x \in X.$$

4.1.2. THEOREM. *Let X, Y, Z be Banach spaces, and let $T \in \mathcal{B}(X, Y)$ and $S \in \mathcal{B}(Y, Z)$.*

(1) *The map $T \to T^*$ is linear.*

(2) $I_X^* = I_{X^*}$.

(3) $(ST)^* = T^*S^*$.

(4) $T^{**}|_X = T$.

K. R. Davidson, *Functional Analysis and Operator Algebras*, CMS/CAIMS Books in Mathematics 13,
https://doi.org/10.1007/978-3-031-63665-3_4

(5) $\|T^*\| = \|T\|$.

(6) T^* *is weak-*$*$*–weak-*$*$ *continuous; i.e.* T^* *is continuous from* (Y^*, τ_Y) *to* (X^*, τ_X).

PROOF. (1) For $T_1, T_2 \in \mathcal{B}(X, Y)$, $a_1, a_2 \in \mathbb{F}$ and $g \in Y^*$,

$$(a_1T_1 + a_2T_2)^*g = g(a_1T_1 + a_2T_2) = a_1gT_1 + a_2gT_2 = a_1T_1^*g + a_2T_2^*g.$$

Therefore $(a_1T_1 + a_2T_2)^* = a_1T_1^* + a_2T_2^*$.

(2) $I^*f = fI = f$ for $f \in X^*$. Thus $I_X^* = I_{X^*}$.

(3) For $h \in Z^*$, $(ST)^*h = hST = (hS)T = (S^*h)T = T^*(S^*h)$.

(4) For $x \in X$ and $g \in Y^*$,

$$\langle T^{**}\iota x, g\rangle = \langle \iota x, T^*g\rangle = \langle x, T^*g\rangle = \langle Tx, g\rangle.$$

Therefore $T^{**}|_X = T$.

(5) We observed that $\|T^*\| \leq \|T\|$. By (4), $\|T\| \leq \|T^{**}\| \leq \|T^*\|$. Hence $\|T^*\| = \|T\|$.

(6) Let $(g_\lambda)_\Lambda$ be a net in Y^* converging weak-$*$ to g. Then for $x \in X$,

$$\langle x, T^*g_\lambda\rangle = \langle Tx, g_\lambda\rangle \to \langle Tx, g\rangle = \langle x, T^*g\rangle.$$

Therefore $T^*g_\lambda \xrightarrow{w*} T^*g$. So T^* is weak-$*$–weak-$*$ continuous. ∎

4.1.3. EXAMPLE. Let $T : \mathbb{C}^m \to \mathbb{C}^n$. Fix a basis $e_1, \dots, e_m$ for $\mathbb{C}^m$ and $f_1, \dots, f_n$ for $\mathbb{C}^n$. Then T has an $n \times m$ matrix $T = [t_{ij}]$ so that $Te_j = \sum_{i=1}^n t_{ij}f_i$. Now $\mathbb{C}^{m*}$ has a *dual basis* $\delta_1, \dots, \delta_m$ given by $\delta_j(e_i) = \delta_{ij}$. Similarly $\mathbb{C}^{n*}$ has dual basis $\varepsilon_1, \dots, \varepsilon_n$. Then we calculate:

$$(T^*\varepsilon_j)(e_i) = \varepsilon_j(Te_i) = \varepsilon_j \sum_{k=1}^n t_{ki}f_k = t_{ji}.$$

Thus $T^*\varepsilon_j = \sum_{i=1}^m t_{ji}\delta_i$. Hence the matrix for T^* with respect to the dual bases is the $m \times n$ matrix $[t_{ji}] = [t_{ij}]^t$, which is the transpose of the matrix for T.

4.1.4. PROPOSITION. *Let* X, Y *be Banach spaces and let* $T : X \to Y$ *be a linear map. Then* T *is bounded (norm continuous) if and only if it is weak–weak continuous (i.e. from* (X, τ_{X^*}) *to* (Y, τ_{Y^*})*).*

PROOF. Suppose that T is norm continuous. Let $x_0 \in X$. If $G \subset Y^*$ is a finite set and $r > 0$, then $Tx_0 + U(G,r)$ is a basic τ_{Y^*}-open neighbourhood of Tx_0. Now

$$\begin{aligned} T^{-1}(Tx_0 + U(G,r)) &= x_0 + T^{-1}(\{y \in Y : |g(y)| < r,\ g \in G\}) \\ &= x_0 + \{x \in X : |g(Tx)| = |T^* g(x)| < r,\ g \in G\} \\ &= x_0 + U(T^*G, r) \end{aligned}$$

is τ_{X^*} open. Hence T is weak–weak continuous.

Conversely, suppose that T is weak–weak continuous. This is equivalent to saying that $g \circ T$ is weakly continuous on X for all $g \in Y^*$; i.e. $g \circ T \in X^*$. Therefore $\sup_{\|x\| \le 1} \|g(Tx)\| = \|g \circ T\| < \infty$ for all $g \in Y^*$. By the Banach–Steinhaus theorem, $\|T\| = \sup_{\|x\| \le 1} \|Tx\| < \infty$. So T is bounded. ■

4.1.5. PROPOSITION. *Let X, Y be Banach spaces and let $T : Y^* \to X^*$ be a linear map. Then $T = S^*$ for some $S \in \mathcal{B}(X,Y)$ if and only if it is weak-$*$–weak-$*$ continuous (i.e. from (Y^*, τ_Y) to (X^*, τ_X)).*

PROOF. The direct direction is Theorem 4.1.2(6).

Suppose that T is weak-$*$–weak-$*$ continuous. Let $\hat{x} = \iota x \in X^{**}$. Hence $\hat{x} \circ T$ is weak-$*$ continuous on Y^* for every $x \in X$, so that $\hat{x} \circ T \in Y$. Therefore

$$\sup_{\|g\| \le 1} |(Tg)(x)| = \|\hat{x} \circ T\| < \infty \quad \text{for all} \quad x \in X.$$

By the Banach–Steinhaus theorem, $\|T\| = \sup_{\|g\| \le 1} \|Tg\| < \infty$. So T is bounded.

Therefore $T^* : X^{**} \to Y^{**}$. We will show that $T^*X \subset Y$. Let $x \in X$ and $g \in Y^*$; and $T^*\hat{x} = y \in Y^{**}$. Then

$$y(g) = (T^*\hat{x})(g) = \hat{x}(Tg) = Tg(x).$$

If we show that y is a τ_Y-continuous functional on Y^*, then it belongs to Y. Let $(g_\lambda)_\Lambda$ be a net converging weak-$*$ to g in Y^*. Then since T is weak-$*$–weak-$*$ continuous, $(Tg_\lambda)_\Lambda$ converges weak-$*$ to Tg. Thus

$$y(g_\lambda) = Tg_\lambda(x) \to Tg(x) = y(g).$$

This shows that y is a τ_Y continuous linear functional on Y^* and hence $y \in Y$; i.e. $T^*X \subset Y$.

Let $S = T^*|_X \in \mathcal{B}(X,Y)$. Then for $x \in X$ and $g \in Y^*$,

$$(S^*g)(x) = g(Sx) = g(T^*x) = (T^*\hat{x})(g) = \hat{x}(Tg) = (Tg)(x).$$

This is true for all $x \in X$ and $g \in Y^*$, and therefore $S^* = T$. ■

4.1.6. EXAMPLES.

(1) One might suspect that $\overline{\operatorname{Ran} T^*}$ is always weak-$*$ closed, but this is not true. Let $T \in \mathcal{B}(l_1)$ given by $T(x_n) = (\frac{1}{n}x_n)$. Then it is easy to see that $T^* \in \mathcal{B}(l_\infty)$ also is given by $T^*(y_n) = (\frac{1}{n}y_n)$. Then $\operatorname{Ran} T^* \subset c_0$, and it is easy to check that $\overline{\operatorname{Ran} T^*} = c_0$. This is weak-$*$ dense in l_∞; and in particular is not weak-$*$ closed.

(2) Let X be a Banach space. Let $\iota_X : X \to X^{**}$ and $\iota_{X^*} : X^* \to X^{***}$ be the canonical injections. Then $\iota_X^* : X^{***} \to X^*$ is the restriction map to X, i.e. $\iota_X^*\Phi(x) = \Phi(\iota_X x) = \Phi|_X(x)$. Hence it follows that $\iota_X^*\iota_{X^*} = I_{X^*}$. Let $P = \iota_{X^*}\iota_X^* \in \mathcal{B}(X^{***})$. Then

$$P^2 = \iota_{X^*}(\iota_X^*\iota_{X^*})\iota_X^* = \iota_{X^*}I_{X^*}\iota_X^* = P.$$

Moreover $\operatorname{Ran} P$ is contained in $\operatorname{Ran} \iota_{X^*}$ and if $\Phi = \iota_{X^*}f$, then

$$P\Phi = \iota_{X^*}(\iota_X^*\iota_{X^*})f = \iota_{X^*}f = \Phi.$$

Thus P is the canonical projection of X^{***} onto X^*. Also $\|P\| \leq \|\iota_{X^*}\|\,\|\iota_X^*\| = 1$, so that $\|P\| = 1$.

4.1.7. LEMMA. *Let $T \in \mathcal{B}(X, Y)$. Then*

$$\ker T^* = (\operatorname{Ran} T)^\perp \quad \textit{and} \quad \ker T = (\operatorname{Ran} T^*)_\perp.$$

PROOF. Let $g \in Y^*$. Then

$$g \perp \operatorname{Ran} T \iff 0 = \langle Tx, g\rangle = \langle x, T^*g\rangle \text{ for all } x \in X \iff T^*g = 0.$$

Similarly, for $x \in X$,

$$x \perp \operatorname{Ran} T^* \iff 0 = \langle x, T^*g\rangle = \langle Tx, g\rangle \text{ for all } g \in Y^* \iff Tx = 0.$$

The last equivalence is a consequence of the Hahn–Banach theorem. ■

4.2. The Hilbert Space Adjoint

When studying operators on a Hilbert space, there is a different operation which is also called the adjoint. This can lead to some confusion at first. The notion is based on Theorem 2.3.6 that the dual of a Hilbert space is conjugate linearly isomorphic to the original space.

4.2.1. DEFINITION. Let H, K be Hilbert spaces. A *bounded sesquilinear form* on $H \times K$ is a map $[\cdot,\cdot] : H \times K \to \mathbb{F}$ which is linear in the first variable and conjugate linear in the second variable which satisfies $|[x, y]| \leq C\|x\|\,\|y\|$ for some constant C and all $x \in H$ and $y \in K$.

4.2.2. PROPOSITION. *Let H, K be Hilbert spaces. Suppose that there is a sesquilinear form $[\cdot,\cdot]$ on $H \times K$ bounded by C. Then there is a unique operator B in $\mathcal{B}(K, H)$ so that $[x, y] = \langle x, By\rangle$ for $x \in H$ and $y \in K$. Moreover $\|B\| \leq C$.*

PROOF. Fix $y \in K$, and define a linear functional $\varphi_y(x) = [x, y]$ on H. Note that

$$\|\varphi_y\| = \sup_{\|x\|\le 1} \big|[x, y]\big| \le \sup_{\|x\|\le 1} C\|x\|\,\|y\| = C\|y\|.$$

By Theorem 2.3.6, there is a unique vector $B(y) \in H$ so that $[x, y] = \langle x, B(y)\rangle$; and $\|B(y)\| = \|\varphi_y\| \le C\|y\|$. At this point, B is just a function from K to H. However, it is clear that B is uniquely defined.

We verify that B is linear: if $y_1, y_2 \in K$ and $\lambda_1, \lambda_2 \in \mathbb{F}$, then

$$\begin{aligned}\langle x, B(\lambda_1 y_1 + \lambda_2 y_2)\rangle &= [x, \lambda_1 y_1 + \lambda_2 y_2] = \overline{\lambda_1}[x, y_1] + \overline{\lambda_2}[x, y_2] \\ &= \overline{\lambda_1}\langle x, B(y_1)\rangle + \overline{\lambda_2}\langle x, B(y_2)\rangle = \langle x, \lambda_1 B(y_1) + \lambda_2 B(y_2)\rangle.\end{aligned}$$

This holds for all $x \in H$, and thus $B(\lambda_1 y_1 + \lambda_2 y_2) = \lambda_1 B(y_1) + \lambda_2 B(y_2)$. That is, B is a linear map. We already have shown that $\|B\| \le C$. ■

If $T \in \mathcal{B}(H, K)$, there is a sesquilinear form on $H \times K$ given by $[x, y] = \langle Tx, y\rangle$; and $[x, y]| \le \|T\|\,\|x\|\,\|y\|$. An immediate application of this proposition shows that there is a unique operator $B \in \mathcal{B}(K, H)$ so that $\langle Tx, y\rangle = \langle x, By\rangle$.

4.2.3. DEFINITION. Let H, K be Hilbert spaces. If $T \in \mathcal{B}(H, K)$, the *Hilbert space adjoint* or just *adjoint* of T is the unique operator $T^* \in \mathcal{B}(K, H)$ such that $\langle Tx, y\rangle = \langle x, T^*y\rangle$ for all $x \in H$ and $y \in K$.

The proof of the following proposition is very similar to the proof of Theorem 4.1.2. The details are left to the reader.

4.2.4. PROPOSITION. *Let H, K, L be Hilbert spaces. If $S, T \in \mathcal{B}(H, K)$ and $R \in \mathcal{B}(K, L)$, then*

(1) $(\lambda S + \mu T)^* = \bar{\lambda} S^* + \bar{\mu} T^*$ *for $\lambda, \mu \in \mathbb{F}$. So the adjoint operation on $\mathcal{B}(H, K)$ is a conjugate linear map.*

(2) $I_H^* = I_H$.

(3) $(RT)^* = T^*R^*$.

(4) $T^{**} = T$.

(5) $\|T^*\| = \|T\|$.

4.2.5. EXAMPLE. Let $T \in \mathcal{B}(H)$ where $\dim H = n < \infty$. Fix an orthonormal basis $e_1, \dots, e_n$. Then T has a matrix $\big[t_{ij}\big]$ with respect to this basis. Then

$$\langle T^*e_j, e_i\rangle = \langle e_j, Te_i\rangle = \overline{t_{ji}} \quad \text{for} \quad 1 \le i, j \le n.$$

Therefore T^* has the matrix $\big[\,\overline{t_{ji}}\,\big]$, which is the conjugate transpose of $\big[t_{ij}\big]$.

4.3. Invertible Operators and the Spectrum

In this section, we define the spectrum of a bounded linear operator. Basic properties and examples are examined.

4.3.1. DEFINITION. An operator $T \in \mathcal{B}(X,Y)$ is *bounded below* if there is some $\varepsilon > 0$ so that $\|Tx\| \geq \varepsilon\|x\|$ for $x \in X$.

4.3.2. EXAMPLES.
(1) Let $K \in \mathcal{B}(l_p)$ for $1 \leq p < \infty$ given by $K(x_n) = (\frac{1}{n}x_n)$. Then K is one-to-one, and the range of K is dense since it contains the standard basis $\{e_n : n \geq 1\}$. It is not bounded below since $\|Ke_n\| = \frac{1}{n} \to 0$. Also K is not surjective because $\operatorname{Ran} K$ does not contain $x = (n^{-1-1/p})$. You can easily check that $x \in l_p$ but $(n^{-1/p})$ does not. So K is not invertible.

(2) Let $S \in \mathcal{B}(l_p)$ for $1 \leq p < \infty$ given by $S(x_n) = (0, x_1, x_2, \dots)$, known as a *unilateral shift*. Then S is isometric: $\|Sx\|_p = \|x\|_p$. In particular, S is bounded below, and hence one-to-one. The range of S is $\{x = (x_n) : x_1 = 0\}$, which is a proper closed subspace. Thus S is not surjective and hence is not invertible.

(3) Let $T \in \mathcal{B}(l_p)$ be given by $T(x_n) = (x_2, x_3, x_4, \dots)$. This is the *backward shift*. Then $\ker T = \mathbb{C}e_1$, and thus T is not invertible. However, T is surjective. Moreover, $TS = I$. So S is left invertible and T is right invertible, but neither is invertible. Now $ST(x_1, x_2, x_3, \dots) = (0, x_2, x_3, \dots)$; so ST has kernel and proper closed range.

(4) Let $T \in M_n = \mathcal{B}(\mathbb{C}^n)$, the space of $n \times n$ matrices. Then T is invertible if and only if it is one-to-one if and only if it is surjective. Thus if T is either left or right invertible, then T is invertible. The previous examples show that these results for invertible matrices all fail for operators on infinite dimensional spaces.

4.3.3. PROPOSITION. *Let $T \in \mathcal{B}(X,Y)$. The following are equivalent:*

(1) *T is invertible.*

(2) *T is one-to-one and onto.*

(3) *T is bounded below and has dense range.*

(4) *T and T^* are bounded below.*

(5) *T^* is invertible.*

PROOF. (1) implies (2) is clear; and (2) implies (1) is the Banach Isomorphism theorem.

(1) implies (3) because $\operatorname{Ran} T = Y$ and $\|x\| = \|T^{-1}(Tx)\| \le \|T^{-1}\|\,\|Tx\|$. Thus $\|Tx\| \ge \frac{1}{\|T^{-1}\|}\|x\|$ is bounded below.

(3) implies (2). Clearly bounded below implies one-to-one. We have that $\|Tx\| \ge \varepsilon_0\|x\|$ for some $\varepsilon_0 > 0$. Let $y \in Y$. Since T has dense range, there is a sequence $(x_n) \in X$ so that $Tx_n = y_n \to y$. Since y_n converges, it is a Cauchy sequence. So given $\varepsilon > 0$, there is an N so that $N \le m < n$ implies that $\|T(x_m - x_n)\| = \|y_m - y_n\| < \varepsilon\varepsilon_0$. Therefore $\|x_m - x_n\| < \varepsilon$. So (x_n) is also Cauchy. Since X is complete, $x = \lim x_n$ exists. Therefore $Tx = \lim Tx_n = y$. Hence T is onto.

(1) implies (5). We have that $T^{-1}T = I_X$ and $TT^{-1} = I_Y$. Taking adjoints yields $T^*T^{-1*} = I_{X^*}$ and $T^{-1*}T^* = I_{Y^*}$. Hence $(T^*)^{-1} = (T^{-1})^*$; whence T^* is invertible.

(5) implies (4). Since T^* is invertible, it is bounded below. Also T^{**} is invertible, and hence bounded below. Since $T = T^{**}|_X$, we see that T is bounded below.

(4) implies (3). We have T is bounded below; and $(\operatorname{Ran} T)^\perp = \ker T^* = \{0\}$ by Lemma 4.1.7 and the fact that T^* is bounded below. It follows from the Hahn–Banach theorem that $\operatorname{Ran} T$ is dense because if $M = \overline{\operatorname{Ran} T}$ is a proper subspace, then there is a non-zero linear functional $g \in Y^*$ in $M^\perp$, and hence in $\ker T^*$. ■

When we discuss the spectrum of a linear operator, we will always work in complex vector spaces. In the study of matrices, the eigenvalues of a real matrix may be complex. Things are more convenient when we assume that $\mathbb{C}$ is our field.

4.3.4. Definition. Let $T \in \mathcal{B}(X)$ where X is a complex Banach space.

(1) The *spectrum* of T is $\sigma(T) = \{\lambda \in \mathbb{C} : \lambda I - T \text{ is not invertible}\}$.

(2) The *resolvent* of T is $\rho(T) = \mathbb{C} \setminus \sigma(T)$.

(3) The *resolvent function* of T on $\rho(T)$ is $R(T, \lambda) = (\lambda I - T)^{-1}$.

(4) The *point spectrum* of T is $\sigma_p(T) = \{\lambda \in \mathbb{C} : \ker(\lambda I - T) \ne \{0\}\}$, the *eigenvalues* of T.

(5) The *approximate point spectrum* of T is

$$\pi(T) = \{\lambda \in \mathbb{C} : \lambda I - T \text{ is not bounded below}\}.$$

(6) The *compression spectrum* of T is

$$\gamma(T) = \{\lambda \in \mathbb{C} : \operatorname{Ran}(\lambda I - T) \text{ is not dense in } X\}.$$

By Proposition 4.3.3, $\sigma(T) = \pi(T) \cup \gamma(T)$.

4.3.5. Proposition. *Let X be a Banach space. Then $\mathcal{B}(X)^{-1}$ is open and contains $b_1(I)$.*

PROOF. If $\|A\| < 1$, then $(I - A)^{-1} = \sum_{n \ge 0} A^n$. This series converges in $\mathcal{B}(X)$ because

$$\sum_{n\ge 0} \|A^n\| \le \sum_{n \ge 0} \|A\|^n = \frac{1}{1 - \|A\|} < \infty.$$

Then one sees that

$$(I - A)\sum_{n\ge 0} A^n = \lim_{N\to\infty} (I - A)\sum_{n=0}^{N} A^n = \lim_{N\to\infty} I - A^{N+1} = I.$$

Similarly, $\sum_{n\ge 0} A^n(I - A) = I$. So $b_1(I) \subset \mathcal{B}(X)^{-1}$.

Now suppose that T is invertible and $\|A\| < 1/\|T^{-1}\|$. Then

$$T - A = T(I - T^{-1}A).$$

Since $\|T^{-1}A\| \le \|T^{-1}\|\,\|A\| < 1$, it follows that $(I - T^{-1}A)$ is invertible. So $(T-A)^{-1} = (I-T^{-1}A)^{-1}T^{-1}$. Thus $b_{1/\|T^{-1}\|}(T) \subset \mathcal{B}(X)^{-1}$. Therefore $\mathcal{B}(X)^{-1}$ is open. ■

4.3.6. PROPOSITION. *Let $T \in \mathcal{B}(X)$. Then $\sigma(T)$ is a compact subset of $\overline{b_{\|T\|}(0)}$. Thus $\rho(T)$ is open.*

PROOF. Let $R : \mathbb{C} \to \mathcal{B}(X)$ by $R(\lambda) = \lambda I - T$. Clearly R is continuous, and thus $\rho(T) = R^{-1}(\mathcal{B}(X)^{-1})$ is open. Therefore $\sigma(T)$ is closed. If $|\lambda| > \|T\|$, then

$$\lambda I - T = \lambda(I - \lambda^{-1}T) \quad \text{and} \quad \|\lambda^{-1}T\| = \frac{\|T\|}{|\lambda|} < 1.$$

Therefore

$$(\lambda I - T)^{-1} = \lambda^{-1}(I - \lambda^{-1}T)^{-1} = \lambda^{-1}\sum_{n\ge 0}(\lambda^{-1}T)^n = \sum_{n\ge 0}\lambda^{-n-1}T^n.$$

Thus $\sigma(T) \subset \overline{b_{\|T\|}(0)}$ and is closed; whence it is compact. ■

4.3.7. PROPOSITION. *The map $T \to T^{-1}$ is continuous on $\mathcal{B}(X)^{-1}$.*

PROOF. If $T_k \to I$, write $T_k = I - A_k$ where $\|A_k\| \to 0$. Once $\|A_k\| < 1$, we have that

$$\|I - (I - A_k)^{-1}\| = \Big\|\sum_{n\ge 1} A_k^n\Big\| \le \sum_{n\ge 1} \|A_k\|^n = \frac{\|A_k\|}{1 - \|A_k\|} \to 0.$$

Hence the inverse map is continuous at I. Now suppose that T is invertible and $S_k \to T$. Then $S_kT^{-1} \to I$ since multiplication is continuous. Then $TS_k^{-1} \to I$ by continuity of the inverse at I. Left multiplying by T^{-1} yields $S_k^{-1} \to T^{-1}$. ■

4.3.8. Example. Multiplication operators. Take $L^p(0,1)$ for $1 \le p < \infty$, and let $f \in L^\infty(0,1)$. Define the *multiplication operator* $M_f h = fh$ on $L^p(0,1)$. Recall that

$$\|f\|_\infty = \operatorname{ess\,sup}|f| = \sup\{r : m(\{x : |f(x)| > r\}) > 0\}.$$

Note that $|f(x)| \le \|f\|_\infty$ a.e. Claim: $\|M_f\| = \|f\|_\infty$. Indeed, if $\|h\|_p \le 1$, then

$$\|M_f h\|_p^p = \int |f|^p |h|^p \, dm \le \|f\|_\infty^p \int |h|^p \, dm = \big(\|f\|_\infty \|h\|_p\big)^p.$$

So $\|M_f\| \le \|f\|_\infty$. Conversely, if $r < \|f\|_\infty$, then $A_r = \{x : |f(x)| > r\}$ has positive measure. Let $h = \dfrac{\overline{\operatorname{sign}(f)}}{m(A_r)^{1/p}} \chi_{A_r}$. Then $\|h\|_p = 1$ and

$$\|M_f h\|_p = \Big\| \frac{|f|}{\sqrt{m(A_r)}} \chi_{A_r} \Big\|_p > r\|h\|_p = r.$$

Therefore $\|M_f\| \ge \|f\|_\infty$, and hence we have equality.

Next note that if $g \in L^\infty$, then $M_f M_g = M_{fg} = M_g M_f$. So the map from $L^\infty(0,1)$ into $\mathcal{B}(L^p(0,1))$ is an isometric algebra isomorphism. Define

$$\operatorname{ess\,ran}(f) = \{\lambda \in \mathbb{C} : m(f^{-1}(b_\varepsilon(\lambda))) > 0 \text{ for all } \varepsilon > 0\}.$$

If $\lambda \notin \operatorname{ess\,ran}(f)$, then there is some $\varepsilon > 0$ so that $f^{-1}(b_\varepsilon(\lambda))$ has measure 0. Thus $g = (\lambda - f)^{-1}$ is essentially bounded by ε^{-1}, and thus lies in $L^\infty(0,1)$. Therefore $(\lambda I - M_f)M_g = M_g(\lambda I - M_f) = M_1 = I$. Consequently, $\sigma(M_f) \subset \operatorname{ess\,ran}(f)$. On the other hand, if $\lambda \in \operatorname{ess\,ran}(f)$, then $A_\varepsilon = f^{-1}(b_\varepsilon(\lambda))$ has positive measure for all $\varepsilon > 0$. Observe that

$$\|(\lambda I - M_f)\chi_{A_\varepsilon}\| = \|(\lambda - f)\chi_{A_\varepsilon}\| < \varepsilon \|\chi_{A_\varepsilon}\|.$$

Therefore $\lambda I - M_f$ is not bounded below, and hence it is not invertible. Thus $\sigma(M_f) = \operatorname{ess\,ran}(f)$. Note that the entire spectrum is approximate point spectrum.

The operator M_x has no point spectrum because if $xh = \lambda h$ a.e., we need $h = 0$ a.e.. Also $\operatorname{Ran}(\lambda I - M_x)$ is always dense since it is onto unless $\lambda \in [0,1]$, and in that case, $\operatorname{Ran}(\lambda I - M_x)$ contains $L^p(0, \lambda - \varepsilon) + L^p(\lambda + \varepsilon, 1)$ for all $\varepsilon > 0$. The union of these subspaces is dense in $L^p(0,1)$. So M_x has no compression spectrum. Therefore $\sigma(M_x) = \pi(M_x) = [0,1]$ and $\sigma_p(M_x) = \varnothing = \gamma(M_x)$.

Consider $M_f^* \in \mathcal{B}(L^q(0,1))$ where $\frac{1}{p} + \frac{1}{q} = 1$. If $h \in L^p(0,1)$ and $k \in L^q(0,1)$, then

$$\langle M_f h, k\rangle = \int (fh)k \, dm = \int h(fk) \, dm = \langle h, M_f^* k\rangle.$$

Thus $M_f^* = M_f \in \mathcal{B}(L^q(0,1))$.

In the case of $p = 2$, we are in the Hilbert space situation where the adjoint is defined differently. For $h, k \in L^2(0,1)$

$$\langle M_f h, k\rangle = \int (fh)\overline{k} \, dm = \int h(\overline{\overline{f}k}) \, dm = \langle h, \overline{f}k\rangle = \langle h, M_{\overline{f}} k\rangle.$$

Thus $M_f^* = M_{\bar{f}}$ in this situation. Since $M_f M_f^* = M_{|f|^2} = M_f^* M_f$, we say that M_f is a *normal* operator.

4.3.9. EXAMPLE. A similar but easier analysis can be carried out for the diagonal operators on l_p for $1 \le p < \infty$. Given a bounded sequence $(d_n) \in l_\infty$, define $D \in \mathcal{B}(l_p)$ by $D(a_n) = (d_n a_n)$. It is left to the reader (see Exercise **4.9**) to show that $\sigma_p(D) = \{d_n : n \ge 1\}$ and $\sigma(D) = \overline{\{d_n : n \ge 1\}}$. We will write $D = \operatorname{diag}(d_1, d_2, d_3, \dots)$ since the "matrix" of D is diagonal.

4.3.10. EXAMPLE. Shift operators. Let S and T be the forward and backward shifts on l_p for $1 \le p < \infty$ defined in Example 4.3.2(2,3). We saw that S is an isometry with proper closed range, and so is not invertible. Also $TS = I$ but $\ker ST = \mathbb{C}e_1$. We claim that $\sigma(S) = \sigma(T) = \overline{\mathbb{D}}$. Since $\|S\| = \|T\| = 1$, their spectra are contained in the closed disc. If $|\lambda| < 1$, let $x_\lambda = (1, \lambda, \lambda^2, \dots)$. Then

$$\|x_\lambda\|_p^p = \sum_{n \ge 0} |\lambda|^{np} = \frac{1}{1 - |\lambda|^p} < \infty.$$

Observe that $Tx_\lambda = (\lambda, \lambda^2, \lambda^3, \dots) = \lambda x_\lambda$. Hence $\sigma_p(T) \supset \mathbb{D}$, and hence $\sigma(T) = \overline{\mathbb{D}}$. Even though l_p is separable, T has uncountably many eigenvalues. However S does not have any eigenvalues at all: if $\lambda(a_n) = S(a_n) = (0, a_1, a_2, \dots)$, then $\lambda = 0$; but S is an isometry, so has no kernel.

Next we identify the adjoint. We will write S_p and T_p for the shifts on l_p. We calculate the adjoints of S_p and T_p. Suppose that $h = (a_n) \in l_p$ and $k = (b_n) \in l_q$, where $\frac{1}{p} + \frac{1}{q} = 1$. Then

$$\langle S_p h, k\rangle = \sum_{n=2}^{\infty} a_n b_{n+1} = \langle (a_n), (b_2, b_3, \dots)\rangle = \langle h, T_q k\rangle.$$

Therefore $S_p^* = T_q$. The calculation to show that $T_p^* = S_q$ is similar. Now compute for $\lambda \in \mathbb{D}$ and $h \in l_p$,

$$\langle (\lambda I - S_p)h, x_\lambda\rangle = \langle h, (I - T_q)x_\lambda\rangle = 0.$$

(You can check that $T_\infty x_\lambda = \lambda x_\lambda$ in the $p = 1$ case.) Therefore $\operatorname{Ran}(\lambda I - S_p)$ annihilates the non-zero vector $x_\lambda \in l_q$, and thus the closure of this range is proper. Hence $\lambda I - S_p$ is not invertible. So $\gamma(S_p) \supset \mathbb{D}$ and $\sigma(S_p) = \overline{\mathbb{D}}$.

4.4. Analyticity and the Resolvent

The theory of analytic functions plays an important role in functional analysis. In this section, we show that the resolvent function is analytic. Later on, in Section 6.4, we develop a holomorphic functional calculus for any bounded operator.

There are at least two natural definitions of what analytic should mean for functions with values in a Banach space.

4.4.1. DEFINITION. Let $\Omega \subset \mathbb{C}$ be open, and let X be a Banach space. A function $f : \Omega \to X$ is *weakly analytic* if $\varphi \circ f : \Omega \to \mathbb{C}$ is analytic for all $\varphi \in X^*$. The function f is *strongly analytic* if for every $z_0 \in \Omega$, there is a sequence $(x_n)_{n\geq 0} \subset X$ so that $f(z) = \sum_{n\geq 0} x_n(z-z_0)^n$ on a ball $b_r(z_0)$ for $r > 0$.

Clearly a strongly analytic function is weakly analytic. The converse is also true. (See Exercise **4.25**.) We don't need this for our study of the spectrum because we can prove directly that the resolvent function is strongly analytic.

4.4.2. THEOREM. *Let X be a Banach space, and let $T \in \mathcal{B}(X)$. Then*

(1) $\dfrac{R(T,\lambda) - R(T,\mu)}{\lambda - \mu} = -R(T,\lambda)R(T,\mu)$ *for* $\lambda, \mu \in \rho(T)$.

(2) $\lambda \to R(T,\lambda)$ *is strongly analytic.*

(3) $\lim_{|\lambda|\to\infty} R(T,\lambda) = 0$.

PROOF. Note that $\lambda I - T$ and $(\mu I - T)^{-1}$ commute for $\lambda \in \mathbb{C}$ and $\mu \in \rho(T)$. Thus (1) follows from rearranging the identity

$$(R(T,\lambda) - R(T,\mu))(\lambda I - T)(\mu I - T) = (\mu I - T) - (\lambda I - T) = \mu - \lambda.$$

For (2), let $z_0 \in \rho(T)$ and $|w| < \|R(T,z_0)\|^{-1}$ and compute

$$\begin{aligned}\big((z_0+w)I - T\big)^{-1} &= \big((z_0 I - T) - wI\big)^{-1} = \big((z_0 I - T)(I - wR(T,z_0))\big)^{-1} \\ &= R(T,z_0)\big(I - wR(T,z_0)\big)^{-1} \\ &= R(T,z_0)\sum_{n\geq 0} R(T,z_0)^n w^n = \sum_{n\geq 0} R(T,z_0)^{n+1} w^n.\end{aligned}$$

(3) For $|\lambda| > \|T\|$, we have

$$\|R(T,\lambda)\| = \Big\|\sum_{n\geq 0} \lambda^{-n-1} T^n\Big\| \leq \frac{1}{|\lambda| - \|T\|}.$$

This converges to 0 as $|\lambda| \to \infty$. ∎

4.4.3. THEOREM. *Let X be a Banach space. If $T \in \mathcal{B}(X)$, then $\sigma(T)$ is nonempty.*

PROOF. If $\sigma(T)$ is empty, then $R(T,\lambda)$ is an entire function (an analytic function defined on all of $\mathbb{C}$). Since $\lim_{|\lambda|\to\infty} \|R(T,\lambda)\| = 0$, we see that $R(T,\lambda)$ is bounded. Therefore for each $\varphi \in X^*$, $\varphi \circ R(T,\lambda)$ is a bounded entire function.

Thus this function is constant by Liouville's theorem. By the Hahn–Banach theorem, $R(T, \lambda)$ is constant. This is absurd, and so the spectrum is non-empty. ∎

4.4.4. Proposition. *Let $T \in \mathcal{B}(X)$. If $\lambda \in \rho(T)$, then*

$$\|R(T, \lambda)\| \geq \frac{1}{\operatorname{dist}(\lambda, \sigma(T))}.$$

The boundary of the spectrum, $\partial\sigma(T)$, is in the approximate point spectrum $\pi(T)$.

Proof. Pick $\lambda_0 \in \sigma(T)$ so that $|\lambda - \lambda_0| = \operatorname{dist}(\lambda, \sigma(T))$. Then

$$(\lambda_0 I - T)(\lambda I - T)^{-1} = (\lambda_0 - \lambda)(\lambda I - T)^{-1} + I$$

is not invertible. It follows that $|\lambda_0 - \lambda| \, \|(\lambda I - T)^{-1}\| \geq 1$. Therefore

$$\|(\lambda I - T)^{-1}\| \geq \frac{1}{|\lambda_0 - \lambda|} = \frac{1}{\operatorname{dist}(\lambda, \sigma(T))}.$$

Fix $\lambda_0 \in \partial\sigma(T)$. For $\varepsilon > 0$ there is a point $\lambda \in \rho(T)$ with $|\lambda - \lambda_0| < \varepsilon$. Hence $\|R(T, \lambda)\| > 1/\varepsilon$. So there is some vector $x \in X$ with $\|x\| = 1$ and $\|y\| = \|R(T, \lambda)x\| > 1/\varepsilon$. Hence

$$\frac{\|(\lambda I - T)y\|}{\|y\|} = \frac{\|x\|}{\|y\|} < \varepsilon.$$

Thus

$$\frac{\|(\lambda_0 I - T)y\|}{\|y\|} \leq \frac{|\lambda_0 - \lambda| \|y\| + \|x\|}{\|y\|} < |\lambda_0 - \lambda| + \varepsilon < 2\varepsilon.$$

Therefore $\lambda_0 I - T$ is not bounded below. ∎

4.4.5. Spectral Mapping for Rational Functions. *Let $f(z) = \frac{p(z)}{q(z)}$ be a rational function with no poles in $\sigma(T)$. Then $\sigma(f(T)) = f(\sigma(T))$.*

Proof. We may suppose that p and q have no common factors. Write $q(z) = \prod_{i=1}^{n}(z - \alpha_i)$ where $\alpha_i \in \rho(T)$. Therefore $q(T)$ is invertible, and so we see that $f(T) = p(T)q(T)^{-1}$ is defined. For $\lambda \in \mathbb{C}$,

$$\lambda - f(z) = \frac{p_\lambda(z)}{q(z)} = \frac{(z - \beta_1)\dots(z - \beta_m)}{(z - \alpha_1)\dots(z - \alpha_n)}.$$

Thus $\lambda I - T = \prod_{j=1}^{m}(T - \beta_j I)\, q(T)^{-1}$. Therefore $\lambda \in \sigma(f(T))$ if and only if some $\beta_j \in \sigma(T)$ if and only if there is $\beta \in \sigma(T)$ so that $0 = \dfrac{p_\lambda(\beta)}{q(\beta)} = \lambda - f(\beta)$ if and only if $\lambda \in f(\sigma(T))$. ∎

4.4.6. Definition. The *spectral radius* of T is

$$\operatorname{spr}(T) = \sup\{|\lambda| : \lambda \in \sigma(T)\}.$$

The following formula is due to Gelfand.

4.4.7. Theorem. *Let $T \in \mathcal{B}(X)$. Then* $\operatorname{spr}(T) = \lim_{n\to\infty} \|T^n\|^{1/n}$.

Proof. By the Spectral Mapping theorem, $\sigma(T^n) = \sigma(T)^n$, and hence $\operatorname{spr}(T^n) = \operatorname{spr}(T)^n$. Therefore

$$\operatorname{spr}(T) = \operatorname{spr}(T^n)^{1/n} \le \|T^n\|^{1/n}.$$

Hence $\operatorname{spr}(T) \le \inf_{n\ge1} \|T^n\|^{1/n}$.

Now $R(T,\lambda)$ is analytic on $\rho(T) \supset \{\lambda : |\lambda| > \operatorname{spr}(T)\}$. The Laurent expansion about ∞ is the series $R(T,\lambda) = \sum_{n\ge0} T^n \lambda^{-n-1}$. We know that this is valid for $|\lambda| > \|T\|$. However the function is analytic on the larger annulus $|\lambda| > \operatorname{spr}(T)$. So by analogy with the scalar case, we expect that this should converge absolutely and uniformly if $|\lambda| \ge r > \operatorname{spr}(T)$.

To see that this is indeed correct, note that for $\varphi \in X^*$,

$$\varphi(R(T,\lambda)) = \sum_{n\ge0} \varphi(T^n)\lambda^{-n-1}$$

is analytic on $\{\lambda : |\lambda| > \operatorname{spr}(T)\}$, and therefore this series converges absolutely for $|\lambda| > \operatorname{spr}(T)$. In particular, $|\varphi(T^n)\lambda^{-n-1}| \to 0$; whence $|\varphi(T^n)\lambda^{-n-1}| \le C$ for $n \ge 0$. Therefore $|\varphi(T^n)| \le Cr^{n+1}$ for any $r > \operatorname{spr}(T)$. Rearranging, we get that $\sup_{n\ge0} \left|\varphi\left(\frac{T^n}{r^n}\right)\right| \le Cr < \infty$. By the Banach–Steinhaus theorem, we get $\sup_{n\ge0} \left\|\frac{T^n}{r^n}\right\| = C' < \infty$. Therefore $\|T^n\| \le C'r^n$; whence $\limsup \|T^n\|^{1/n} \le r$ for any $r > \operatorname{spr}(T)$. This means that

$$\limsup \|T^n\|^{1/n} \le \operatorname{spr}(T) \le \inf \|T^n\|^{1/n} \le \liminf \|T^n\|^{1/n}.$$

Therefore $\operatorname{spr}(T) = \lim_{n\to\infty} \|T^n\|^{1/n}$. ■

We prove a lemma which will prove useful later.

4.4.8. Lemma. *Suppose that $X = Y \dotplus Z$, and $T \in \mathcal{B}(X)$ satisfies $TY \subset Y$ and $TZ \subset Z$. Then $\sigma(T) = \sigma(T|_Y) \cup \sigma(T|_Z)$.*

Proof. If $\lambda \in \rho(T)$, then $(\lambda I - T)Y \subset Y$ and $(\lambda I - T)Z \subset Z$. But $\lambda I - T$ is surjective, and hence $(\lambda I - T)Y = Y$ and $(\lambda I - T)Z = Z$. Also $\ker(\lambda I - T) = \{0\}$, so that $(\lambda I - T)|_Y$ is a continuous bijection on Y and $(\lambda I - T)|_Z$ is a continuous bijection on Z. So both are invertible by Banach's Isomorphism theorem. Hence $\rho(T) \subset \rho(T|_Y) \cap \rho(T|_Z)$.

Conversely, suppose that $\lambda \in \rho(T|_Y) \cap \rho(T|_Z)$. Let $A = (\lambda I - T)|_Y^{-1} \in \mathcal{B}(Y)$ and $B = (\lambda I - T)|_Z^{-1} \in \mathcal{B}(Z)$. Define $S \in \mathcal{B}(X)$ by $S(y+z) = Ay + Bz$ for $x = y + z \in X$. Then for $y \in Y$ and $z \in Z$,

$$S(\lambda I - T)(y+z) = A(\lambda I - T)|_Y y + B(\lambda I - T)|_Z z = y + z$$

and

$$(\lambda I - T)S(y + z) = (\lambda I - T)|_Y Ay + (\lambda I - T)|_Z Bz = y + z.$$

Thus $S = (\lambda I - T)^{-1}$. Therefore $\rho(T|_Y) \cap \rho(T|_Z) \subset \rho(T)$. ∎

Exercises for Chapter 4

4.1 Let $T \in \mathcal{B}(X, Y)$ for Banach spaces X, Y. Show that T has dense range if and only if T^* is injective, and T^* has dense range if and only if T is injective.

4.2 Let $T \in \mathcal{B}(X, Y)$ for Banach spaces X, Y. Show that T has closed range if and only if T^* has closed range.

4.3 Let $T \in \mathcal{B}(H, K)$ where H, K are Hilbert spaces.
(a) Show that T is injective if and only if T^*T is injective.
(b) Show that T is surjective if and only if TT^* is surjective.

4.4 If $T \in \mathcal{B}(X)$ is invertible, show that $\|T^{-1}\| \geq \|T\|^{-1}$.

4.5 Suppose that $T_n \in \mathcal{B}(X)$ are invertible and $\lim_n T_n = T$. Show that T is invertible if and only if $\sup\{\|T_n^{-1}\| : n \geq 1\} < \infty$.

4.6 Prove Proposition 4.2.4 regarding the Hilbert space adjoint.

4.7 (a) Let $T \in \mathcal{B}(X)$ for a Banach space X. Show that $\sigma(T^*) = \sigma(T)$ and $R(T^*, \lambda) = R(T, \lambda)^*$ for $\lambda \in \rho(T)$.
(b) Let $T \in \mathcal{B}(H)$ for a Hilbert space H. Show that $\sigma(T^*) = \sigma(T)^*$, where $C^* = \{\overline{\lambda} : \lambda \in C\}$, and $R(T^*, \overline{\lambda}) = R(T, \lambda)^*$ for $\lambda \in \rho(T)$.

4.8 Two operators $A, B \in \mathcal{B}(X)$ are *similar* ($A \sim B$) if there is an invertible operator $S \in \mathcal{B}(X)$ such that $B = SAS^{-1}$. If X is a Hilbert space and S is unitary, we say that A and B are *unitarily equivalent* ($A \simeq B$).
(a) Show that if A and B are similar, then $\sigma(A) = \sigma(B)$.
(b) Show that if A and B are unitarily equivalent, then $\|A\| = \|B\|$.
(c) Give an example of similar operators A and B such that $\|A\| \neq \|B\|$.

4.9 Let D be a diagonal operator on l_p for $1 \leq p < \infty$, as in Example 4.3.9. Compute $\sigma(D)$, $\sigma_p(D)$, $\pi(D)$ and $\gamma(D)$.

4.10 Let $T \in M_n$, and let p_T be its minimal polynomial.
(a) Show that if $q \in \mathbb{C}[x]$, then $q(T)$ is invertible if and only if $\gcd(q, p_T) = 1$.
(b) Show that every rational function of T (with poles in $\rho(T)$) is equal to a polynomial in T.
HINT: Euclidean algorithm for polynomials.

4.11 Let $T = (I_4 + J_4) \oplus (2I_2 + J_2) \oplus 3I_2 \in M_8$ where J_k is the nilpotent in M_k defined on a basis $e_1, \dots, e_k$ by $J_k e_i = \begin{cases} 0 & \text{if } i = 1 \\ e_{i-1} & \text{if } i > 1. \end{cases}$

(a) Find a polynomial q_1 such that $q_1(T)$ is the projection onto $\ker(T - I)^4$ with kernel $\ker(T - 2I)^2(T - 3I)$.

(b) Describe the algebra $A(T) = \{q(T) : q \in \mathbb{C}[x]\}$ explicitly.

4.12 Let $T \in \mathcal{B}(X)$, for a Banach space X.

(a) Show that $\sigma_p(T^*) = \gamma(T)$ and $\gamma(T^*) = \sigma_p(T)$.

(b) Show that $\sigma(T^*) = \pi(T) \cup \pi(T^*)$.

(c) What are the corresponding results for Hilbert space operators?

4.13 Let $T \in \mathcal{B}(X)$, for a Banach space X, and let f be a rational function with no poles in $\sigma(T)$.

(a) Show that $\sigma_p(f(T)) = f(\sigma_p(T))$.

(b) Show that $\pi(f(T)) = f(\pi(T))$.

(c) Show that $\gamma(f(T)) = f(\gamma(T))$.

4.14 (*upper semicontinuity of the spectrum*).

(a) Let X be a Banach space and $T \in \mathcal{B}(X)$. If $\sigma(T) \subset U$ for an open subset U, show that there is some $\delta > 0$ so that $\|S - T\| < \delta$ implies that $\sigma(S) \subset U$.
HINT: show that $\sup_{\lambda \in U^c} \|(\lambda I - T)^{-1}\| < \infty$.

(b) Show that the spectral radius is upper semicontinuous; i.e. if $\varepsilon > 0$, there exists $\delta > 0$ so that $\|S - T\| < \delta$ implies that $\operatorname{spr}(S) < \operatorname{spr}(T) + \varepsilon$.

4.15 (a) A weighted shift W on a Hilbert space H acts on an orthonormal basis $\{e_n : n \geq 1\}$ by $We_n = w_n e_{n+1}$ for $n \geq 1$, where $\{w_n\}$ is a bounded sequence in $\mathbb{C}$. Compute $\|W\|$.

(b) Show that there is a diagonal unitary U_θ so that $U_\theta W U_\theta^* = e^{i\theta} W$ for each $\theta \in (0, 2\pi)$. Deduce that $\sigma(W)$ has circular symmetry.

(c) The *Kakutani shift* W_K has weights $w_n = 1/\gcd(n, 2^n)$ for $n \geq 1$. Compute $\operatorname{spr}(W_K)$.

(d) Let W_k be the weighted shift with weights $w_n = 1/\gcd(n, 2^n)$ if 2^k doesn't divide n, and $w_{2^k m} = 0$ for $m \geq 1$. Compute the spectral radius of W_k.

(e) Show that $\lim_{k \to \infty} W_k = W_K$. Deduce that spectral radius is not continuous on $\mathcal{B}(H)$.

4.16 Let $S \in \mathcal{B}(H, K)$ for Hilbert spaces H, K.

(a) Show that S is an isometry if and only if $S^*S = I$.

(b) Show that S is unitary if and only if $S^*S = I = SS^*$.

4.17 Suppose that $P, Q \in \mathcal{B}(X)$ are idempotents such that $PQ = QP$.

(a) Show that $\operatorname{Ran} P$ is closed.

(b) Compute $\sigma(P)$.

(c) Show that PQ is idempotent with $\operatorname{Ran} PQ = \operatorname{Ran} P \cap \operatorname{Ran} Q$.

(d) Show that $P+Q-PQ$ is idempotent with range Ran $P+$Ran Q. In particular, Ran $P+$ Ran Q is closed.

4.18 Let H be a Hilbert space and let $E \in \mathcal{B}(H)$ be idempotent. Prove that the following are equivalent:
(1) $\ker E$ is orthogonal to Ran E.
(2) $E = E^*$.
(3) $\|E\| = 1$.

4.19 If $A, B, C, D \in \mathcal{B}(H)$, then $T = \begin{bmatrix} A & B \\ C & D \end{bmatrix}$ (and $S = \begin{bmatrix} W & X \\ Y & Z \end{bmatrix}$) are bounded operators on the Hilbert space $H \oplus H$ given by $\begin{bmatrix} A & B \\ C & D \end{bmatrix} \begin{bmatrix} x \\ y \end{bmatrix} = \begin{bmatrix} Ax + By \\ Cx + Dy \end{bmatrix}$.
(a) Show that every operator T in $\mathcal{B}(H \oplus H)$ has this form, and show that TS is computed by matrix multiplication.
(b) Find T^*. (This is the Hilbert space adjoint.)
(c) Find all operators $E \in \mathcal{B}(H \oplus H)$ such that $E = E^2$ and Ran $E = H \oplus \{0\}$.
(d) If A is invertible, show that T is invertible if and only if $D - CA^{-1}B$ is also invertible.
HINT: find X and Y to factor $T = \begin{bmatrix} A & 0 \\ C & I \end{bmatrix} \begin{bmatrix} I & 0 \\ 0 & X \end{bmatrix} \begin{bmatrix} I & Y \\ 0 & I \end{bmatrix}$.

4.20 Suppose that $A_n \in \mathcal{B}(X)$ are finite rank, and $A = \lim_{n \to \infty} A_n$. Prove that rank $A \le \liminf_n$ rank A_n.

4.21 Let H be a Hilbert space and let $P \in \mathcal{B}(H)$ be an orthogonal projection. Show that for $T \in \mathcal{B}(H)$ that

$$\|TP - PT\| = \max\left\{\|P^\perp TP\|, \|PTP^\perp\|\right\}.$$

HINT: $P^\perp TP$ and $PTP^\perp$ have orthogonal ranges and domains.

4.22 (a) Suppose that $A, B \in \mathcal{B}(X)$ commute, i.e. $AB = BA$. Prove that $\operatorname{spr}(AB) \le \operatorname{spr}(A) \operatorname{spr}(B)$.
(b) Show by example that this fails if A and B do not commute, even in M_2.

4.23 (*Numerical range*). Let $T \in \mathcal{B}(H)$ for a Hilbert space H. The *numerical range* of T is $W(T) = \{\langle T\xi, \xi\rangle : \|\xi\| = 1\}$. The *numerical radius* of T is $w(T) = \sup\{|\lambda| : \lambda \in W(T)\}$.
(a) Show that if $T \in M_2$ is a 2×2 matrix, then $W(T)$ is an ellipse with foci at the eigenvalues of T.
HINT: put T in upper triangular form.
(b) (*Hausdorff–Toeplitz*) Prove that $W(T)$ is convex. .
HINT: reduce to the 2×2 case.
(c) Show that $\sigma(T) \subset \overline{W(T)}$. .
HINT: show that if $\operatorname{Re}\lambda \ge \varepsilon > 0$ for $\lambda \in W(T)$, then T is invertible.
(d) Prove that $\frac{1}{2}\|T\| \le w(T) \le \|T\|$.

4.24 The sequence $\{\|T^n\| : n \geq 1\}$ is an example of a *submultiplicative sequence* $\{a_n : n \geq 1\}$ such that $a_n \geq 0$ and $a_{m+n} \leq a_m a_n$ for all $m, n \geq 1$.

(a) Show that $\limsup a_n^{1/n} = \inf a_n^{1/n} =: L$ for any submultiplicative sequence.
HINT: if $\varepsilon > 0$ and $a_n^{1/n} < L + \varepsilon$, estimate $a_m^{1/m}$ for large $m = qn + r$.

(b) Formulate and prove a result about *subadditive sequences.*

4.25 Prove that a weakly analytic function f on $\Omega \subset \mathbb{C}$ is strongly analytic as follows.

(a) For $z_0 \in \overline{b_r(z_0)} \subset \Omega$, show that $\left\{\varphi\left(\frac{f(z_0+w)-f(z_0)}{w}\right) : 0 < |w| \leq r\right\}$ is a bounded set for each $\varphi \in X^*$.

(b) Show that f is continuous on $\overline{b_r(z_0)}$.

(c) Define $x_n = \frac{1}{2\pi r^n} \int_0^{2\pi} f(z_0 + re^{i\theta})e^{-in\theta}\, d\theta$ as a Riemann integral.

(d) Show that $f(z_0 + w) = \sum_{n\geq 0} x_n w^n$ for $|w| < r$.

4.26 Recall that $T \in \mathcal{B}(l_1)$ has a matrix $[t_{ij}]$ such that the columns are uniformly bounded (Chapter 2, Exercise **2.7**). Show that T is the adjoint of an operator $S \in B(c_0)$ if and only if the rows of the matrix are in c_0.

Notes on Chapter 4

Volterra and Fredholm were the first to treat integral operators as linear maps on a vector space of functions in the late 19th century. Hilbert studied self-adjoint operators on Hilbert space and proved a version of the spectral theorem. Banach [22] introduced the notion of adjoint for general operators on Banach spaces. The notion of the spectrum comes from linear algebra and also integral equations. Unifying these ideas for operators was due to Hilbert and Moore, and extended by Riesz. A. Taylor [302] established that the spectrum was non-empty in 1938. The spectral radius formula is due to Gelfand [119] in 1941 in the context of Banach algebras.

In addition to the books on functional analysis by Dunford and Schwarz [104], Rudin [269] and Conway [61], books on Banach algebras such as Bonsall and Duncan [37] treat the general properties of the spectrum.

CHAPTER 5

Compact Operators

The notion of a compact operator is an extension of the class of finite rank operators. They behave more like matrices than general operators. They have a structure theory which is suggested by results in linear algebra, although there are differences. We also introduce the notion of Fredholm operators, which are invertible modulo compact operators. The spectral theorem for compact normal operators on Hilbert space is also established. Finally we show that compact operators have many invariant subspaces, leading to a triangular form and a theorem of Ringrose.

In this chapter, X, Y, Z will denote Banach spaces; and H, K will be Hilbert spaces.

5.1. Compact Operators

5.1.1. DEFINITION. Let X, Y be Banach spaces. An operator $T \in \mathcal{B}(X, Y)$ is a *compact operator* if $\overline{Tb_1(X)}$ is compact in Y. We write $\mathcal{K}(X, Y)$ for the set of all compact operators in $\mathcal{B}(X, Y)$, and we write $\mathcal{K}(X)$ for $\mathcal{K}(X, X)$.

5.1.2. THEOREM. *$\mathcal{K}(X, Y)$ is a norm closed $\mathcal{B}(Y)$–$\mathcal{B}(X)$ bimodule. In particular, $\mathcal{K}(X)$ is a closed 2-sided ideal of $\mathcal{B}(X)$.*

PROOF. Let $K_1, K_2 \in \mathcal{K}(X, Y)$ be compact operators, and let $C_i = \overline{K_i b_1(X)}$. Then $C_1 \times C_2$ is compact. If $\lambda_1, \lambda_2 \in \mathbb{C}$, then $\lambda_1 C_1 + \lambda_2 C_2$ is compact because it is the continuous image of $C_1 \times C_2$ under the map $(x, y) \to \lambda_1 x + \lambda_2 y$. However

$$\overline{(\lambda_1 K_1 + \lambda_2 K_2) b_1(X)} \subset \{\lambda_1 x + \lambda_2 y : x \in C_1,\ y \in C_2\} = \lambda_1 C_1 + \lambda_2 C_2.$$

A closed subset of a compact set is compact, and therefore $\lambda_1 K_1 + \lambda_2 K_2$ is compact. Hence $\mathcal{K}(X, Y)$ is a subspace of $\mathcal{B}(X, Y)$.

If $S \in \mathcal{B}(Y)$, $T \in \mathcal{B}(X)$ and $K \in \mathcal{K}(X, Y)$, then

$$\overline{SKTb_1(X)} \subset \overline{SKb_{\|T\|}(X)} = S\|T\|\overline{Kb_1(X)}$$

is the continuous image of a compact set, and hence is compact. Thus $\mathcal{K}(X, Y)$ is a $\mathcal{B}(Y)$–$\mathcal{B}(X)$ bimodule.

Finally we show that $\mathcal{K}(X, Y)$ is norm closed. So suppose that $K_n \in \mathcal{K}(X, Y)$ and $K_n \to K \in \mathcal{B}(X, Y)$. Let $\varepsilon > 0$. Pick N so that $\|K - K_N\| < \varepsilon/3$. Since

K. R. Davidson, *Functional Analysis and Operator Algebras*, CMS/CAIMS Books in Mathematics 13,
https://doi.org/10.1007/978-3-031-63665-3_5

$K_N b_1(X)$ has compact closure, it is totally bounded. Thus we can pick $y_i = K_N x_i$ with $\|x_i\| < 1$ for $1 \le i \le m$ to be a finite $\varepsilon/3$-net for $K_N b_1(X)$. Let $y_i' = Kx_i$ in $Kb_1(X)$. For any $x \in b_1(X)$, there is some i so that $\|K_N x - K_N x_i\| < \varepsilon/3$. Hence

$$\begin{aligned}\|Kx - y_i'\| &\le \|(K - K_N)x\| + \|K_N x - K_N x_i\| + \|(K_N - K)x_i\| \\ &< \frac{\varepsilon}{3} + \frac{\varepsilon}{3} + \frac{\varepsilon}{3} = \varepsilon.\end{aligned}$$

Thus y_i' for $1 \le i \le m$ is an ε-net for $Kb_1(X)$; and hence $\overline{Kb_1(X)}$ is compact. Therefore K is a compact operator. ■

5.1.3. Example. Finite rank operators are compact. Suppose $F \in \mathcal{B}(X, Y)$ has a finite rank. Then $\operatorname{Ran} F = M$ is an n-dimensional subspace of Y, and therefore it is closed. Hence $\overline{Fb_1(X)}$ is a closed bounded subset of $M \simeq \mathbb{C}^n$. Thus it is compact. So F is compact.

We can deduce that any limit of finite rank operators is also compact. Consider a diagonal operator $D = \operatorname{diag}(d_1, d_2, \dots)$ on l_p for $1 \le p \le \infty$ is given by $D(a_n) = (d_n a_n)$. We have seen that D is bounded if and only if $\sup |d_n| < \infty$. We claim that it is compact if and only of $\lim_{n\to\infty} d_n = 0$. If the limit exists and is 0, then given $\varepsilon > 0$, there is an N so that $|d_n| < \varepsilon$ for all $n > N$. The diagonal operator $D_N = \operatorname{diag}(d_1, \dots, d_N, 0, 0 \dots)$ is finite rank, and thus compact. Also $\|D - D_N\| = \sup_{n>N} |d_n| < \varepsilon$. Hence $D = \lim_{n\to\infty} D_n$ is compact.

On the other hand, if 0 is not the limit, then there is some $\delta > 0$ and a sequence $n_i \to \infty$ so that $|d_{n_i}| \ge \delta$ for $i \ge 1$. The vectors $De_{n_i} = d_{n_i} e_{n_i}$ all belong to $\overline{Db_1(X)}$ and $\|De_{n_i} - De_{n_j}\| = \|d_{n_i} e_{n_i} - d_{n_j} e_{n_j}\| \ge 2^{1/p}\delta$. Therefore $\overline{Db_1(X)}$ is not totally bounded, because there is no finite δ net. So D is not compact.

5.1.4. Example. Integral Operators. Let $k(x, y) \in L^2((0, 1)^2)$, and define an operator $K \in \mathcal{B}(L^2(0, 1))$ by

$$Kh(x) = \int_0^1 h(y) k(x, y)\, dy.$$

This is a *Hilbert–Schmidt integral operator* . Compute using Cauchy–Schwarz

$$\begin{aligned}\|Kh\|_2^2 &= \int_0^1 \Big| \int_0^1 h(y) k(x, y)\, dy \Big|^2 dx \le \int_0^1 \big(\|h\|_2 \, \|k(x, \cdot)\|_2\big)^2 dx \\ &= \|h\|_2^2 \int_0^1 \int_0^1 |k(x, y)|^2 \, dy\, dx = \|k\|_2^2 \, \|h\|_2^2.\end{aligned}$$

Therefore $\|K\| \le \|k\|_2$.

Let $\{e_i(x) : i \ge 1\}$ be an orthonormal basis for $L^2(0, 1)$. Then you can check (Exercise **5.4**) that $\{e_i(x) e_j(y) : i, j \ge 1\}$ is an orthonormal basis for $L^2((0, 1)^2)$. Write $k = \sum_{i,j=1}^\infty a_{ij} e_i(x) e_j(y)$, where $\|k\|_2^2 = \sum_{i,j=1}^\infty |a_{ij}|^2$. Let

$k_N(x,y) = \sum_{i,j=1}^{N} a_{ij} e_i(x) e_j(y)$. Define $K_N \in \mathcal{B}(X,Y)$ by

$$K_N h(x) = \int_0^1 h(y) k_N(x,y)\,dy = \sum_{i,j=1}^{N} a_{ij} \int_0^1 h(y) e_i(x) e_j(y)\,dy$$
$$= \sum_{i,j=1}^{N} a_{ij} \langle h, e_j \rangle e_i(x).$$

This operator is finite rank, with $\operatorname{Ran} K_N \subset \operatorname{span}\{e_i : 1 \le i \le N\}$. By the first paragraph, $\|K - K_N\| \le \|k - k_N\|_2 = \sum_{i>N \text{ or } j>N} |a_{ij}|^2 \to 0$ as $N \to \infty$. Therefore K is a norm limit of finite rank operators and thus is compact.

5.1.5. Example. Volterra Operator. Define an operator $V \in \mathcal{B}(L^2(0,1))$ by

$$Vh(x) = \int_0^x h(y)\,dy.$$

This is a Hilbert–Schmidt integral operator with kernel $k(x,y) = \begin{cases} 1 & \text{if } y \le x \\ 0 & \text{if } y > x \end{cases}$.

Therefore $\|V\| \le \|k\|_2 = \dfrac{1}{\sqrt{2}}$.

First we show that V has no eigenvalues. Suppose that

$$\lambda h(x) = Vh(x) = \int_0^x h(y)\,dy.$$

If $\lambda = 0$, then $h = 0$a.e. If $\lambda \ne 0$, then $h(x) = \lambda^{-1} \displaystyle\int_0^x h(y)\,dy$. The RHS is a continuous function, and hence $h \in C[0,1]$ with $h(0) = 0$. Then by the Fundamental Theorem of Calculus, the RHS is differentiable. Therefore $h'(x) = \lambda^{-1} h(x)$. This implies that $h(x) = c e^{x/\lambda}$. Since $h(0) = 0 = c$, we have $h = 0$. Therefore $\sigma_p(V) = \varnothing$.

Now we compute the powers of V.

$$V^2 h(x) = \int_0^x \int_0^y h(t)\,dt\,dy = \int_0^x \int_t^x dy\,h(t)\,dt = \int_0^x (x-t) h(t)\,dt.$$
$$V^3 h(x) = \int_0^x \int_0^y (y-t) h(t)\,dt\,dy = \int_0^x \int_t^x (y-t)\,dy\,h(t)\,dt = \int_0^x \frac{(x-t)^2}{2} h(t)\,dt.$$

We will show by induction that $V^n h(x) = \displaystyle\int_0^x \frac{(x-t)^{n-1}}{(n-1)!} h(t)\,dt$. This holds for

$n = 1, 2, 3$ and assuming the formula for V^n,

$$V^{n+1}h(x) = \int_0^x \int_0^y \frac{(y-t)^{n-1}}{(n-1)!} h(t)\,dt\,dy$$
$$= \int_0^x \int_t^x \frac{(y-t)^{n-1}}{(n-1)!}\,dy\,h(t)\,dt = \int_0^x \frac{(y-t)^n}{n!} h(t)\,dt.$$

Now from Example 5.1.4, we know that

$$\|V^n\| \le \left\| \frac{(x-t)^{n-1}}{(n-1)!} \chi_{t\le x} \right\|_2 \le \frac{1}{(n-1)!}.$$

Therefore $\operatorname{spr}(V) \le \lim_{n\to\infty} \left(\frac{1}{(n-1)!}\right)^{1/n} = 0$. Hence $\sigma(V) = \{0\}$. An operator with spectrum $\{0\}$ is called *quasinilpotent* .

5.1.6. PROPOSITION. *Let H be a Hilbert space. Then every $K \in \mathcal{K}(H)$ is a limit of finite rank operators.*

PROOF. Let $\varepsilon = \frac{1}{n}$. Choose an ε-net for $Kb_1(H)$, say $y_1, \dots, y_m$. Let P_n be the orthogonal projection onto $\operatorname{span}\{y_1, \dots, y_m\}$. Then P_n is finite rank, and thus P_nK is also finite rank. Compute:

$$\|K - P_nK\| = \sup_{\|x\|\le 1} \|P_n^\perp Kx\| \le \sup_{\|x\|\le 1} \min_{i\le m} \|Kx - y_i\| \le \frac{1}{n}.$$

Therefore P_nK converge to K in norm. ∎

5.1.7. REMARK. This same argument will work in any Banach space X in which there is a sequence (or net) of finite rank projections (P_n) which are uniformly bounded and converge strongly to the identity, i.e. $P_nx \to x$ for every $x \in X$. This includes l_p for $1 \le p < \infty$. Here the projection P_n onto the span of the first n standard basis vectors does the job.

The existence of Banach spaces in which there are compact operators that are not limits of finite rank operators is a difficult result of Enflo [110].

5.1.8. SCHAUDER THEOREM. *Let X, Y be Banach spaces. Then $T \in \mathcal{B}(X, Y)$ is compact if and only if $T^* \in \mathcal{B}(Y^*, X^*)$ is compact.*

PROOF. Assume that T is compact, and let $K = \overline{Tb_1(X)}$. This is a compact metric space. Define $R : \overline{b_1(Y^*)} \to C(K)$ be the restriction map $\psi \to \psi|_K$. If $\psi \in \overline{b_1(Y^*)}$ and $y_i \in K$, $|\psi(y_1) - \psi(y_2)| \le \|y_1 - y_2\|$. So the functions $R\psi$ are bounded by $\|T\|$ and equicontinuous—all have Lipschitz constant 1. By the Arzela–Ascoli theorem, $\overline{Rb_1(Y^*)}$ is compact in $C(K)$.

Therefore if $\{\psi_n : n \ge 1\}$ is any sequence in $\overline{b_1(Y^*)}$, there is a subsequence ψ_{n_i} such that $R\psi_{n_i}$ converge uniformly. That is, $\psi_{n_i}|_K$ converge uniformly to a function $\Psi \in C(K)$. This implies $\psi_{n_i}(Tx) = T^*\psi_{n_i}(x)$ converges uniformly to $\Psi(Tx) = \Psi \circ T(x)$ for $x \in b_1(X)$. Hence $\varphi = \Psi \circ T \in X^*$ and $T^*\psi_{n_i} \to \varphi$ uniformly over the unit ball; i.e. converges in norm. This shows that $T^*b_1(Y^*)$ is precompact; whence T^* is a compact operator.

The converse is now easy. If T^* is compact, then T^{**} is compact. Then considering X as a subspace of X^{**} and Y as a subspace of Y^{**}, we see that $\overline{Tb_1(X)} \subset \overline{T^{**}b_1(X^{**})}$. This is a closed subset of a compact set and hence is compact. Therefore T is a compact operator. ∎

5.1.9. REMARK. For the Hilbert space adjoint, there is an easier proof based on Proposition 5.1.6. If T is compact in $\mathcal{B}(H)$, then $T = \lim_{n\to\infty} F_n$ where F_n are finite rank. Then $T^* = \lim_{n\to\infty} F_n^*$ is also a limit of finite rank operators, and thus is compact.

5.2. Structure of Compact Operators

In this section, we show that the spectrum of a compact operator can be used to provide a structural picture of the operator. In particular, non-zero points in the spectrum are eigenvalues and are isolated points that correspond to finite dimensional summands. The point 0 in the spectrum can hide some pathology.

5.2.1. LEMMA. *Let X be a Banach space, and let V be a closed subspace such that either V or $\dim X/V$ is finite dimensional. Then V is complemented.*

PROOF. If $\dim V = n < \infty$, let $v_1, \dots, v_n$ be a basis for V. There is a dual basis $\varphi_1, \dots, \varphi_n$ for V^* given by $\varphi_i(v_j) = \delta_{ij}$. Using the Hahn–Banach theorem, we can extend each φ_j to some $\widetilde{\varphi}_j \in X^*$. Then $P = \sum_{i=1}^n v_i\widetilde{\varphi}_i \in \mathcal{B}(X)$. Observe that $Px = \sum_{i=1}^n v_i\widetilde{\varphi}_i(x) \in V$ and $Pv_j = v_j$. Therefore P is a continuous idempotent with range V. Hence $X = V \dotplus \ker P$.

If $\dim X/V = n < \infty$, let $\dot{x}_1, \dots, \dot{x}_n$ be a basis for X/V. Choose $x_i \in X$ so that $\dot{x}_i = x_i + V$. Let $W = \operatorname{span}\{x_1, \dots, x_n\}$. Then $V + W = X$ because if $x \in X$, then $\dot{x} := x + V = \sum_{i=1}^n a_i\dot{x}_i$. Therefore $x - \sum_{i=1}^n a_ix_i \in V$. Also if $x \in V \cap W$, write $x = \sum_{i=1}^n a_ix_i$ then $\dot{x} = \sum_{i=1}^n a_i\dot{x}_i = 0$; so that all $a_i = 0$ and thus $x = 0$. Therefore $X = V \dotplus W$. ∎

5.2.2. KEY LEMMA. *Let $K \in \mathcal{K}(X)$. Suppose that there are closed subspaces $V_0 \subsetneq V_1 \subsetneq V_2 \subsetneq \dots$ and scalars $\alpha_i \in \mathbb{C}$ so that $(\alpha_i I - K)V_i \subset V_{i-1}$ for $i \geq 1$. Then $\lim_{i\to\infty} \alpha_i = 0$.*

PROOF. Choose unit vectors $x_i \in V_i$ with $\operatorname{dist}(x_i, V_{i-1}) \geq \frac{1}{2}$ for $i \geq 1$. Then $Kx_i = \alpha_ix_i + y_i$ where $y_i \in V_{i-1}$. Suppose that there is a subsequence with

$|\alpha_{i_n}| \geq \delta > 0$ for $i_1 < i_2 < \dots$. Then if $m < n$, since $y_{i_n} - Kx_{i_m} \in V_{i_n - 1}$,

$$\|Kx_{i_n} - Kx_{i_m}\| = \|\alpha_{i_n} x_{i_n} + (y_{i_n} - Kx_{i_m})\| \geq \frac{1}{2}|\alpha_{i_n}| \geq \frac{\delta}{2}.$$

This shows that $\overline{Kb_1(X)}$ is not compact, a contradiction. ■

Write $\operatorname{null}(T) = \dim \ker T$ for the *nullity* of $T \in \mathcal{B}(X)$.

5.2.3. THEOREM. *Let $K \in \mathcal{K}(X)$. Then for $\lambda \neq 0$, $\ker(\lambda I - K)$ is finite dimensional and $\operatorname{Ran}(\lambda I - K)$ is closed and finite codimension.*

PROOF. By replacing $\lambda I - K$ by $I - \lambda^{-1}K$ and K by $\lambda^{-1}K$, we may suppose that $\lambda = 1$. Let $B = \overline{b_1(X)} \cap \ker(I - K)$. Then the unit ball of $N = \ker(I - K)$ is $B = KB \subset \overline{Kb_1(X)}$ which is compact. Therefore $\operatorname{null}(I - K) < \infty$ by Proposition 2.2.17.

Choose a complement for N so that $X = N \dotplus V$. Then $(I - K)|_V$ is injective and $\operatorname{Ran}(I - K) = (I - K)V$. We claim that $(I - K)|_V$ is bounded below. If not, there are unit vectors $v_n \in V$ so that $\|(I - K)v_n\| \to 0$. By compactness, there is a subsequence v_{n_i} so that $Kv_{n_i} \to y$. Since $(I - K)v_{n_i} \to 0$, we get that $v_{n_i} \to y$. Therefore

$$\|y\| = \lim_{i \to \infty} \|v_{n_i}\| = 1 \quad \text{and} \quad (I - K)y = \lim_{i \to \infty} (I - K)v_{n_i} = 0.$$

So $y \in V \cap N$, which is impossible. Hence $I - K$ is bounded below on V, and thus $R = (I - K)V = \operatorname{Ran}(I - K)$ is closed.

If X/R is infinite dimensional, let $Q : X \to X/R$ be the quotient map. Choose vectors $x_i \in X$ so that Qx_i are linearly independent for $i \geq 1$. Define $R_n = \operatorname{span}\{R, x_1, \dots, x_n\}$. Then R_n/R is finite dimensional, and thus closed in X/R; so that $R_n = Q^{-1}(R_n/R)$ is closed. Clearly

$$R \subsetneq R_1 \subsetneq R_n \subsetneq R_{n+1} \subsetneq \dots$$

However $(I - K)R_n \subset \operatorname{Ran}(I - K) = R$. By the Key lemma, we have that $\lim_{n \to \infty} 1 = 0$, which is absurd. Therefore R has finite codimension. ■

5.2.4. DEFINITION. If $\mathcal{S} \subset \mathcal{B}(X)$, the *commutant* of $\mathcal{S}$ is

$$\mathcal{S}' = \{T \in \mathcal{B}(X) : TS = ST \text{ for all } S \in \mathcal{S}\}.$$

We write $\mathcal{S}''$ for $(\mathcal{S}')'$.

Note that $\mathcal{S}'$ is an algebra (a vector space closed under multiplication) containing the identity. Moreover it is closed in the weak operator topology. (Exercise **5.15**.)

5.2.5. Theorem. *Let $K \in \mathcal{K}(X)$. Then for $\lambda \neq 0$, there is a least integer n_0 so that*

$$N(\lambda) := \ker(\lambda I - K)^{n_0} = \ker(\lambda I - K)^{n_0+k}$$

and

$$R(\lambda) := \operatorname{Ran}(\lambda I - K)^{n_0} = \operatorname{Ran}(\lambda I - K)^{n_0+k}$$

for $k \geq 1$. Moreover $X = N(\lambda) \dotplus R(\lambda)$. The subspaces $N_n = \ker(\lambda I - K)^n$ and $R_n = \operatorname{Ran}(\lambda I - K)^n$ are invariant for $\{K\}'$ and

$$\operatorname{null}(\lambda I - K)^n = \dim X / \operatorname{Ran}(\lambda I - K)^n = \operatorname{null}(\lambda I - K)^{*n} \quad \text{for} \quad n \geq 1.$$

Proof. Let $N_n = \ker(\lambda I - K)^n$ for $n \geq 0$. Since

$$(\lambda I - K)^n = \lambda^n I - K_n \quad \text{where} \quad K_n = \sum_{i=1}^{n} \binom{n}{i} (-1)^{i-1} K^i$$

and K_n is compact, these are finite dimensional subspaces. Note that $N_{n-1} \subset N_n$ and $(\lambda I - K)N_n \subset N_{n-1}$ for $n \geq 1$. If $N_n \supsetneq N_{n-1}$ for all $n \geq 1$, the Key lemma would imply that $\lim_{n\to\infty} \lambda = 0$, which is absurd. Therefore for some smallest $n_0 \geq 0$, $N_{n_0} = N_{n_0+1}$. We show by induction that $N_{n_0} = N_{n_0+k}$ for $k \geq 1$. It is valid for $k = 1$. Suppose that it holds for k. If $x \in N_{n_0+k+1}$, then $(\lambda I - K)x \in N_{n_0+k} = N_{n_0}$. Therefore $x \in N_{n_0+1} = N_{n_0}$. So $N_{n_0+k+1} = N_{n_0}$.

Now let $R_n = \operatorname{Ran}(\lambda I - K)^n$ for $n \geq 1$. Since $(\lambda I - K)^n = \lambda^n I - K_n$ where K_n is compact, these are closed subspaces of finite codimension. Note that $R_{n-1} \supset R_n$ and $(\lambda I - K)R_{n-1} = R_n$ for $n \geq 1$. Therefore by Lemma 4.1.7, $N_n^* := R_n^\perp = \ker(\lambda I - K^*)^n$ are finite dimensional subspaces of X^* such that $N_{n-1}^* \subset N_n^*$ and $(\lambda I - K^*)N_n^* \subset N_{n-1}^*$. By Schauder's theorem, K^* is compact. So arguing as in the previous paragraph, there is a least integer m_0 so that $N_{m_0+k}^* = N_{m_0}^*$ and thus $R_{m_0+k} = R_{m_0}$ for all $k \geq 1$.

Note that if $T \in \{K\}'$ and $x \in N_k$, then $0 = T(\lambda I - K)^k x = (\lambda I - K)^k Tx$; whence $TN_k \subset N_k$. Also if $y = (\lambda I - K)^k x \in R_k$, then

$$Ty = T(\lambda I - K)^k x = (\lambda I - K)^k Tx;$$

whence $TR_k \subset R_k$. So N_k and R_k are $\{K\}'$ invariant subspaces for each $k \geq 1$.

Let $p = \max\{m_0, n_0\}$ (they will turn out to be equal). Let $x \in X$. Since $(\lambda I - K)^p x \in R_p = R_{2p}$, there is $y \in X$ so that $(\lambda I - K)^p x = (\lambda I - K)^{2p} y$. Let $z = x - (\lambda I - K)^p y$. Observe that

$$(\lambda I - K)^p z = (\lambda I - K)^p x - (\lambda I - K)^{2p} y = 0;$$

whence $z \in N_p$. Therefore $x = z + (\lambda I - K)^p y \in N_p + R_p$. On the other hand, if $x \in N_p \cap R_p$, there is $y \in X$ so that $(\lambda I - K)^p y = x$. Therefore $(\lambda I - K)^{2p} y = (\lambda I - K)^p x = 0$, so that $y \in N_{2p} = N_p$. Hence $x = 0$. So $N_p \cap R_p = \{0\}$. Therefore $X = N_p \dotplus R_p$.

Next observe that

$$R_p = (\lambda I - K)^p N_p + (\lambda I - K)^p R_p = (\lambda I - K)^p R_p.$$

Therefore $(\lambda I - K)^p$ maps R_p one to one and onto itself. That is, $(\lambda I - K)^p|_{R_p}$ is invertible. Since $\lambda I - K$ maps R_p into itself, it is also bijective on R_p. Therefore

$$\lambda I - K = (\lambda I - K)|_{N_p} \dotplus (\lambda I - K)|_{R_p},$$

and $F_\lambda = (\lambda I - K)|_{N_p}$ acts on a finite dimensional space, and is nilpotent of order at most p. It follows that $\sigma(F_\lambda) = \{0\}$ and so $\sigma(K|_{N_p}) = \{\lambda\}$.

Since $(\lambda I - K)|_{R_p}$ is invertible, we have that

$$\ker(\lambda I - K)^n = \ker(\lambda I - K)^n|_{N_p}$$

and

$$X/\operatorname{Ran}(\lambda I - K)^n = N_p/\operatorname{Ran}(\lambda I - K)^n|_{N_p}.$$

For a linear map A on a finite dimensional space N, we have

$$\operatorname{null}(A) + \dim \operatorname{Ran} A = \dim N.$$

Therefore $\operatorname{null}(A) = \dim N/\operatorname{Ran} A = \operatorname{null}(A^*)$. In particular

$$\operatorname{null}(\lambda I - K)^n = \dim X/\operatorname{Ran}(\lambda I - K)^n = \operatorname{null}(\lambda I - K)^{*n} \quad \text{for} \quad n \geq 1.$$

In particular, we have $m_0 = n_0$. ■

5.2.6. Remark. Since $(\lambda I - K)|_{N(\lambda)}$ is nilpotent on a finite dimensional space, we have $\sigma(K|_{N(\lambda)}) = \{\lambda\}$. By a familiar result in linear algebra, this restriction is similar to a sum of Jordan blocks for the eigenvalue λ.

This theorem was the main technical result. We put everything together in the following.

5.2.7. Structure of Compact Operators. *Let X be an infinite dimensional Banach space, and let $K \in \mathcal{K}(X)$. Then*

(1) $0 \in \sigma(K)$. *If* $0 \neq \lambda \in \sigma(K)$, *then* $\lambda \in \sigma_p(K)$.

(2) *The spectrum is finite or is a countable set* $\{0, \lambda_n : n \geq 1\}$ *such that* $\lim_{n\to\infty} \lambda_n = 0$.

(3) *For each* $\lambda \in \sigma(K) \setminus \{0\}$, *there is a least integer* n_λ *so that*

$$N(\lambda) = \ker(\lambda I - K)^{n_\lambda} = \ker(\lambda I - K)^{n_\lambda + k} \quad \textit{for} \quad k \geq 1,$$
$$R(\lambda) = \operatorname{Ran}(\lambda I - K)^{n_\lambda} = \operatorname{Ran}(\lambda I - K)^{n_\lambda + k} \quad \textit{for} \quad k \geq 1,$$

$N(\lambda)$ *is finite dimensional,* $R(\lambda)$ *is closed and* $X = N(\lambda) \dotplus R(\lambda)$.

(4) $\sigma(K|_{N(\lambda)}) = \{\lambda\}$ *and* $\sigma(K|_{R(\lambda)}) = \sigma(K) \setminus \{\lambda\}$.

(5) *There is a unique finite rank idempotent* E_λ *in* $\{K\}''$ *with range* $N(\lambda)$ *and kernel* $R(\lambda)$, *both invariant for* $\{K\}'$.

(6) *If* $\lambda, \mu \in \sigma(K) \setminus \{0\}$ *are distinct, then* $E_\lambda E_\mu = 0$.

PROOF. Compact operators on an infinite dimensional Banach space can never be surjective because the Open Mapping theorem would then show that $Kb_1(X)$ contains a ball $b_r(X)$. So $0 \in \sigma(K)$. Theorem 5.2.5 shows that if $\lambda \neq 0$, then either $\ker(\lambda I - K) \neq \{0\}$ or $\lambda I - K$ is invertible. Thus $\sigma(K) = \sigma_p(K) \cup \{0\}$. If $\lambda \in \sigma(K) \setminus \{0\}$, then $X = N(\lambda) \dotplus R(\lambda)$ where both $N(\lambda)$ and $R(\lambda)$ are $\{K\}'$-invariant. By Lemma 4.4.8,

$$\sigma(K) = \sigma(K|_{N(\lambda)}) \cup \sigma(K|_{R(\lambda)}) = \{\lambda\} \cup \sigma(K|_{R(\lambda)}).$$

Since $(\lambda I - K)|_{R(\lambda)}$ is invertible, $\sigma(K|_{R(\lambda)}) = \sigma(K) \setminus \{\lambda\}$. Therefore λ is an isolated point of the spectrum, and hence the spectrum is at most countable. If $\sigma(K) \setminus \{0\} = \{\lambda_n : n \geq 1\}$ is countable, pick a unit vector $x_n \in \ker(\lambda_n I - K)$. Define $V_n = \operatorname{span}\{x_1, \dots, x_n\}$. Observe that $V_{n-1} \subsetneq V_n$ and $(\lambda_n - K)V_n \subset V_{n-1}$ for $n \geq 1$. By the Key lemma, $\lim_{n\to\infty} \lambda_n = 0$.

For (5), there is a unique idempotent with range $N(\lambda)$ and kernel $R(\lambda)$ by Section 2.7. If $T \in \{K\}'$, then T leaves the range and kernel of E_λ invariant. That means that

$$TE_\lambda = E_\lambda T E_\lambda \quad \text{and} \quad T(I - E_\lambda) = (I - E_\lambda)T(I - E_\lambda).$$

This implies that $TE_\lambda = E_\lambda T E_\lambda = E_\lambda T$. Thus $E_\lambda \in \{K\}''$.

Suppose that μ is another non-zero point in $\sigma(K)$. Then $E_\mu \in \{K\}'' \subset \{K\}'$, and hence $E_\mu E_\lambda = E_\lambda E_\mu$. Therefore $\operatorname{Ran}(E_\mu E_\lambda) \subset N(\lambda) \cap N(\mu) = \{0\}$ because $(\lambda I - K)^{n(\lambda)}$ annihilates $N(\lambda)$ while $(\mu I - K)^{n(\mu)}$ is invertible on $N(\lambda)$. Therefore $E_\lambda E_\mu = 0$. ■

5.3. Fredholm Operators

This section deals with operators which are almost invertible, up to a finite dimensional obstruction. One of our main results is that these are the operators which have an inverse *modulo compacts*. An integer-valued invariant called the Fredholm index is defined and its basic properties are developed.

5.3.1. DEFINITION. An operator $T \in \mathcal{B}(X, Y)$ is *Fredholm* if $\operatorname{null}(T) < \infty$ and $\operatorname{Ran} T$ has finite codimension. The *Fredholm index* of T is

$$\operatorname{ind} T = \operatorname{null}(T) - \dim(Y/TX) = \operatorname{null}(T) - \operatorname{null}(T^*).$$

The following consequence is so important that it is often included in the definition of Fredholm operator.

5.3.2. PROPOSITION. *If $T \in \mathcal{B}(X, Y)$ is Fredholm, then* Ran T *is closed.*

PROOF. Suppose that $Y = \text{span}\{TX, y_1, \dots, y_m\}$ where $\dim(Y/TX) = m$. Define

$$S : X/\ker T \oplus \mathbb{C}^m \to Y \quad \text{by} \quad S(\dot{x}, a) = Tx + \sum_{i=1}^{m} a_i y_i.$$

This is a continuous bijection, and hence an isomorphism. Therefore $TX = S(X/\ker T)$ is closed. ∎

5.3.3. EXAMPLES.
(1) Every invertible operator is Fredholm of index 0.

(2) The unilateral shift $S \in \mathcal{B}(l_p)$ for $1 \le p < \infty$ is an isometry, and thus has no kernel and has a closed range. Since $l_p/Sl_p \simeq \mathbb{C}$, we see that S is Fredholm and $\text{ind}\, S = -1$. Similarly if T is the backward shift, then $Tl_p = l_p$ and $\ker T = \mathbb{C}e_1$. So T is Fredholm and $\text{ind}\, T = 1$. Recall that $TS = I$ and $ST = I - e_1\delta_1$ both lie in $I + \mathcal{K}(l_p)$. (Here $\delta_i(\sum_n a_n e_n) = a_i$ and $(e_1\delta_1)(x) = \delta_1(x)e_1$.)

5.3.4. PROPOSITION. *If $\lambda \neq 0$ and $K \in \mathcal{K}(X)$, then $\lambda I - K$ is Fredholm and* $\text{ind}(\lambda I - K) = 0$.

PROOF. By Theorem 5.2.5, $\lambda I - K$ has closed range of finite codimension, and $\text{null}(\lambda I - K) = \dim\big(X/\text{Ran}(\lambda I - K)\big) = \text{null}(\lambda I - K)^*$. Hence

$$\text{ind}(\lambda I - K) = \text{null}(\lambda I - K) - \text{null}(\lambda I - K)^* = 0. \qquad ∎$$

5.3.5. THEOREM. *The set $\mathcal{F}(X, Y)$ of Fredholm operators in $\mathcal{B}(X, Y)$ is open. Index is a continuous integer-valued function, and hence is locally constant.*

PROOF. Let T be Fredholm. Set $N = \ker T$ and choose a closed complement V; so that $X = N \dotplus V$. Then $TV = TX$. Since TX is closed and has finite codimension, choose a finite dimensional subspace W so that $Y = TX \dotplus W$. Define $\widetilde{T} : V \oplus_1 W \to Y$ by $\widetilde{T}(v, w) = Tv + w$. Note that

$$\|\widetilde{T}(v, w)\| \le \|Tv\| + \|w\| \le \max\{\|T\|, 1\}(\|v\| + \|w\|).$$

So $\widetilde{T}$ is continuous. It is easy to check that $\widetilde{T}$ is a bijection, and hence it is invertible by the Banach Isomorphism theorem.

Now suppose that $S \in \mathcal{B}(X, Y)$ such that $\|S - T\| < 1/\|\widetilde{T}^{-1}\|$. Define $\widetilde{S} : V \oplus W \to Y$ by $\widetilde{S}(v, w) = Sv + w$. Then $\widetilde{S}$ is continuous and

$$\|\widetilde{S} - \widetilde{T}\| = \|(S - T)|_V\| < 1/\|\widetilde{T}^{-1}\|.$$

Therefore $\widetilde{S}$ is invertible. It follows that $SV = \widetilde{S}V$ is closed and

$$Y = \widetilde{S}(V \oplus W) = SV \dotplus W.$$

Therefore $\dim Y/SX \leq \dim Y/SV = \dim W < \infty$. Also $\ker S \cap V = \{0\}$, and so the quotient $Q : X \to X/V$ is injective on $\ker S$; so that

$$\operatorname{null} S \leq \dim X/V = \dim N < \infty.$$

Therefore S is Fredholm.

The subspace $V \dotplus \ker S = Q^{-1}((\ker S + V)/V)$ is closed and finite codimension. Choose a finite dimensional complement Z so that $X = V \dotplus \ker S \dotplus Z = \ker S \dotplus (V \dotplus Z)$. Then $SX = SV \dotplus SZ$ because S is injective on $V \dotplus Z$ and Z is finite dimensional. Let $P : Y = SV \dotplus W \to W$ be the projection onto W with kernel SV. Then $PSX = PSZ \simeq Z$ because S is injective on Z and P is injective on SZ. Therefore

$$Y/SX = (Y/SV)/(SX/SV) \simeq W/PSZ.$$

We compute

$$\begin{aligned}
\operatorname{ind} S &= \operatorname{null} S - \dim Y/SX \\
&= \operatorname{null} S - (\dim W - \dim Z) \\
&= \dim(\ker S \dotplus Z) - \dim W \\
&= \operatorname{null} T - \dim Y/TX = \operatorname{ind} T.
\end{aligned}$$

Thus index is constant on an open ball around T, so that it is locally constant, whence continuous. ■

5.3.6. COROLLARY. *Fredholm index is constant on connected components of $\mathcal{F}(X, Y)$.*

We also get the following result which is a consequence of our proof. We showed that if $\|S - T\| < 1/\|\widetilde{T}^{-1}\|$, then $\operatorname{null} S \leq \operatorname{null} T$.

5.3.7. COROLLARY. *If $T \in \mathcal{B}(X, Y)$ is Fredholm, then*

$$\limsup_{S \to T} \operatorname{null} S \leq \operatorname{null} T.$$

Let $\pi : \mathcal{B}(X) \to \mathcal{B}(X)/\mathcal{K}(X)$ be the quotient map. The quotient $\mathcal{B}(X)/\mathcal{K}(X)$ is a Banach space, and since $\mathcal{K}(X)$ is an ideal, it is also a ring. Multiplication is continuous: if $a, b \in \mathcal{B}(X)/\mathcal{K}(X)$ then we can choose $A, B \in \mathcal{B}(X)$ so that $\pi(A) = a$ and $\|A\| < (1 + \varepsilon)\|a\|$ and $\pi(B) = b$ and $\|B\| < (1 + \varepsilon)\|b\|$. Then

$$\|ab\| = \|\pi(AB)\| \leq \|A\|\,\|B\| < (1 + \varepsilon)^2 \|a\|\,\|b\|.$$

Thus $\|ab\| \leq \|a\|\,\|b\|$. Therefore $\mathcal{B}(X)/\mathcal{K}(X)$ is a *Banach algebra* , namely a Banach space which is also a ring in which multiplication is continuous. In particular, we can talk about invertible elements in $\mathcal{B}(X)/\mathcal{K}(X)$. See Chapter 6.

5.3.8. Atkinson Theorem. *$T \in \mathcal{B}(X)$ is Fredholm if and only if $\pi(T)$ is invertible in $\mathcal{B}(X)/\mathcal{K}(X)$.*

PROOF. Suppose that T is Fredholm. Let $N = \ker T$ and let V be a complement for N, so that $X = N \dot{+} V$. Also $TX = TV$ is closed with finite codimension, and we pick a (finite dimensional) complement W so that $X = TV \dot{+} W$. If we consider $T|_V$ as an element of $\mathcal{B}(V, TV)$, then this is a continuous bijection. Hence by Banach's Isomorphism theorem, it has a continuous inverse $S \in \mathcal{B}(TV, V)$. Define $\widetilde{S} \in \mathcal{B}(X)$ by $\widetilde{S}(Tv + w) = v$ for $v \in V$ and $w \in W$. Note that $\widetilde{S}T(u + v) = \widetilde{S}Tv = v$ for $u \in N$ and $v \in V$. That is, $\widetilde{S}T = P_V$ is the projection of X onto V with kernel N. Now $I - P_V = P_N$ is finite rank. Therefore $\pi(\widetilde{S})\pi(T) = \pi(I - P_N) = \pi(I)$. Similarly, $T\widetilde{S}(Tv + w) = Tv = P_{TV}(Tv + w)$ is the projection onto TV with kernel W. Again $I - P_{TV} = P_W$ is finite rank, so that $\pi(T)\pi(\widetilde{S}) = \pi(I)$. Therefore $\pi(T)$ is invertible.

Conversely, suppose that $\pi(T)$ has an inverse $\pi(S)$ in $\mathcal{B}(X)/\mathcal{K}(X)$. Then there are compact operators $K, L \in \mathcal{K}(X)$ so that $ST = I + K$ and $TS = I + L$. Then $\ker T \subseteq \ker I + K$ is finite dimensional and $\operatorname{Ran} T \supseteq \operatorname{Ran}(I + L)$, so it is closed and has finite codimension. Therefore T is Fredholm. ■

5.3.9. Corollary. *If $T \in \mathcal{B}(X)$ is Fredholm and $K \in \mathcal{K}(X)$, then $T + K$ is Fredholm and* $\operatorname{ind}(T + K) = \operatorname{ind}(T)$.

PROOF. By Atkinson's theorem, $\pi(T + K) = \pi(T)$ is invertible and hence $T + K$ is Fredholm. Thus $T + tK$ for $0 \le t \le 1$ is a continuous path of Fredholm operators. Since index is locally constant, we have $\operatorname{ind}(T + K) = \operatorname{ind}(T)$. ■

It follows that the following definition is well defined.

5.3.10. Definition. If $a \in \mathcal{B}(X)/\mathcal{K}(X)$ is invertible, define $\operatorname{ind}(a) = \operatorname{ind}(T)$ for any T with $\pi(T) = a$.

Next we show that the index behaves nicely for products. Hence the index will be a homomorphism of $(\mathcal{B}(X)/\mathcal{K}(X))^{-1}$ into $\mathbb{Z}$.

5.3.11. Theorem. *If $T \in \mathcal{B}(X, Y)$ and $S \in \mathcal{B}(Y, Z)$ are both Fredholm, then ST is Fredholm and* $\operatorname{ind}(ST) = \operatorname{ind}(S) + \operatorname{ind}(T)$.

PROOF. As before, let $N_S = \ker S$ and choose a complement $Y_0 \subset Y$ so that $Y = N_S \dot{+} Y_0$. Also choose a finite dimensional complement Z_0 for SY so that $Z = SY \dot{+} Z_0$. Choose a complement X_0 for $N_T = \ker T$ so that $X = N_T \dot{+} X_0$. Finally we require a complement W for TX in Y, but we do this carefully in two steps. First choose a complement $W_0 \subset N_S$ for $N_S \cap TX$ in N_S, which is possible since N_S is finite dimensional, so that $N_S = (N_S \cap TX) \dotplus W_0$. Note that

$$W_0 \cap TX \subset W_0 \cap (N_S \cap TX) = \{0\},$$

so that $TX + W_0 = TX \dotplus W_0 = TX + N_S$. Now choose a complement W_1 for $TX + N_S$, so that $Y = TX \dotplus W_0 \dotplus W_1$.

Recall that $TX = TV$ and $T|_V$ maps V bijectively onto TX. We compute

$$\ker ST = \ker T + \{v \in V : Tv \in N_S\} = N_T \dotplus (T|_V)^{-1}(N_S \cap TX).$$

Therefore $\operatorname{null}(ST) = \operatorname{null}(T) + \dim(N_S \cap TX)$. Next,

$$SY = S(TX \dotplus W_0 \dotplus W_1) = STX \dotplus SW_1.$$

The latter sum is a direct sum because if $w_1 \in W_1$ and $Sw_1 = STx$, then $w_1 - Tx$ lies in N_S, and hence $w_1 \in W_1 \cap (N_S + TX) = \{0\}$. Moreover, W_1 does not intersect N_S, so that S is injective on W_1. Therefore

$$\dim Z/SY = \dim Z/(STX \dotplus SW_1) = \dim Z/STX - \dim W_1.$$

Finally we can calculate

$$\begin{aligned}
\operatorname{ind} ST &= \operatorname{null} ST - \dim Z/STX \\
&= \operatorname{null} T + \dim(N_S \cap TX) - \big(\dim Z/SY + \dim W_1\big) \\
&= \operatorname{null} T + \big(\dim(N_S \cap TX) + \dim W_0\big) - \dim(W_0 \dotplus W_1) - \dim Z/SY \\
&= \operatorname{null} T + \operatorname{null} S - \dim Y/TX - \dim Z/SY = \operatorname{ind} S + \operatorname{ind} T.
\end{aligned}$$

■

The invertible elements of any ring form a group under multiplication. When we restate Theorem 5.3.11 as a result about $\mathcal{B}(X)/\mathcal{K}(X)$, we obtain:

5.3.12. COROLLARY. *The Fredholm index is a group homomorphism from* $(\mathcal{B}(X)/\mathcal{K}(X))^{-1}$ *into* $(\mathbb{Z}, +)$.

5.4. Normal Operators

In finite dimensions, a normal matrix on the inner product space $l_2^{(n)}$ is unitarily equivalent to a diagonal matrix. It turns out that this extends to compact normal operators on Hilbert space. Later in Section 9.9, a truly infinite dimensional version will be established for arbitrary normal operators.

5.4.1. DEFINITION. Let H be a Hilbert space. An operator $T \in \mathcal{B}(H)$ is *self-adjoint* if $T^* = T$. It is *positive* if $T^* = T$ and $\langle Tx, x\rangle \geq 0$ for $x \in H$. It is *unitary* if $T^* = T^{-1}$. And it is *normal* if $T^*T = TT^*$.

5.4.2. REMARKS.

(1) If $T : H \to H$ is everywhere defined and formally self-adjoint, meaning that

$\langle Tx, y\rangle = \langle x, Ty\rangle$ for all $x, y \in H$, then T is automatically bounded. See Example 2.4.13. This is known as the Hellinger–Toeplitz theorem.

(2) If $T = T^*$, then $\sigma(T) \subset \mathbb{R}$. This is because if $\lambda = r + is$ with $s \neq 0$, then

$$|\langle(\lambda I - T)x, x\rangle| \geq |\operatorname{Im}\langle(\lambda I - T)x, x\rangle| = |s|\|x\|^2.$$

Therefore $\lambda I - T$ is bounded below, as is $(\lambda I - T)^* = \bar{\lambda} I - T$. Hence $\lambda I - T$ is invertible.

If T is positive, and $r > 0$, $\langle(T + rI)x, x\rangle \geq r\|x\|^2$. So $T + rI$ is bounded below. Since this is self-adjoint, $(T + rI)^* = T + rI$, the adjoint is also bounded below, and thus it is invertible. So $\sigma(T) \subset [0, \|T\|]$.

(3) If H is a complex Hilbert space and $T \in \mathcal{B}(H)$ satisfies $\langle Tx, x\rangle \geq 0$ for $x \in H$, then the polarization identity shows that

$$\langle Tx, y\rangle = \frac{1}{4}\sum_{k=0}^{3} i^k \langle T(x + i^k y), x + i^k y\rangle$$

and therefore

$$\begin{aligned}\langle T^*x, y\rangle = \langle x, Ty\rangle = \overline{\langle Ty, x\rangle} &= \overline{\frac{1}{4}\sum_{k=0}^{3} i^k \langle T(y + i^k x), y + i^k x\rangle} \\ &= \frac{1}{4}\sum_{k=0}^{3} (-i)^k \langle T(y + i^k x), y + i^k x\rangle \\ &= \frac{1}{4}\sum_{k=0}^{3} (-i)^k \langle T(x + (-i)^k y), x + (-i)^k y\rangle = \langle Tx, y\rangle\end{aligned}$$

Hence $T^* = T$ is automatically self-adjoint.

If H is a real Hilbert space, then $T = \left[\begin{smallmatrix} 0 & -1 \\ 1 & 0\end{smallmatrix}\right] \in M_2(\mathbb{R})$ satisfies

$$\left\langle \begin{bmatrix} 0 & -1 \\ 1 & 0\end{bmatrix}\begin{bmatrix} x \\ y\end{bmatrix}, \begin{bmatrix} x \\ y\end{bmatrix}\right\rangle = -yx + xy = 0$$

for all $(x, y) \in \mathbb{R}^2$. However, this matrix represents a 90 degree rotation and is clearly not self-adjoint because $T^* = \left[\begin{smallmatrix} 0 & 1 \\ -1 & 0\end{smallmatrix}\right] = -T$, even though $\langle Tx, x\rangle \geq 0$ for all $x \in \mathbb{R}^2$.

(4) If U is unitary, then $\langle Ux, Uy\rangle = \langle U^*Ux, y\rangle = \langle x, y\rangle$. Thus unitary maps are invertible isometries and they preserve the inner product. Since $\|U\| = 1$, we have $\sigma(U) \subset \overline{\mathbb{D}}$. But $U^* = U^{-1}$ shows that $\sigma(U^*) = \{\lambda^{-1} : \lambda \in \sigma(U)\}$ is also contained in $\overline{\mathbb{D}}$. Therefore $\sigma(U) \subset \mathbb{T}$.

5.4.3. Examples.

(1) In Example 4.3.8, we showed that the multiplication operator M_f on $L^2(0, 1)$

is bounded when $f \in L^\infty(0,1)$ and that $M_f^* = M_{\bar{f}}$. It is clear that this is a normal operator. It is self-adjoint if and only if $f = \bar{f}$ a.e., i.e. when f is real valued. It is positive if and only if $f \geq 0$ a.e.. And it is unitary if and only if $|f|^2 = 1$a.e.

A similar analysis applies to $D = \operatorname{diag}(d_1, d_2, d_3, \dots) \in \mathcal{B}(l_2)$.

(2) If $A \in \mathcal{B}(H)$, then A^*A is positive. Indeed, $(A^*A)^* = A^*A^{**} = A^*A$ and $\langle A^*Ax, x\rangle = \langle Ax, Ax\rangle = \|Ax\|^2 \geq 0$. Moreover $\|A^*A\| \leq \|A^*\|\,\|A\| = \|A\|^2$ and

$$\|A^*A\| = \sup_{\|x\|\leq 1} \|A^*Ax\| \geq \sup_{\|x\|\leq 1} \langle A^*Ax, x\rangle = \sup_{\|x\|\leq 1} \|Ax\|^2 = \|A\|^2.$$

Hence $\|A^*A\| = \|A\|^2$.

We collect a few basic properties of normal operators.

5.4.4. Proposition. *Let $N \in \mathcal{B}(H)$ be a normal operator. Then*

(1) $\|Nx\| = \|N^*x\|$ *for all $x \in H$.*

(2) $\|N\| = \operatorname{spr}(N)$.

(3) $\ker(\lambda I - N) = \ker(\lambda I - N)^n = \ker(\lambda I - N)^*$ *for all $\lambda \in \mathbb{C}$ and $n \geq 1$.*

(4) $\ker(\lambda I - N)^\perp = \overline{\operatorname{Ran}(\lambda I - N)}$ *for all $\lambda \in \mathbb{C}$.*

(5) $\ker(\lambda I - N) \perp \ker(\mu I - N)$ *if $\lambda \neq \mu \in \mathbb{C}$.*

(6) *If $p \in \mathbb{C}[z]$ is a polynomial, then* $\|p(N)\| = \sup\{|p(\lambda)| : \lambda \in \sigma(N)\}$.

Proof.

$$\begin{aligned}\|Nx\|^2 &= \langle Nx, Nx\rangle = \langle N^*Nx, x\rangle \\ &= \langle NN^*x, x\rangle = \langle N^*x, N^*x\rangle = \|N^*x\|^2.\end{aligned}$$

So (1) holds. For (2), $\|N^2x\| = \|N^*(Nx)\| \geq \langle N^*Nx, x\rangle = \|Nx\|^2$. Taking the supremum over $\|x\| \leq 1$ yields $\|N^2\| \geq \|N\|^2$, while $\|N^2\| \leq \|N\|^2$ is true for all operators. Thus $\|N^2\| = \|N\|^2$. Iterating this yields $\|N^{2^k}\| = \|N\|^{2^k}$ for $k \geq 1$. Therefore

$$\operatorname{spr}(N) = \lim_{k\to\infty} \|N^{2^k}\|^{1/2^k} = \|N\|.$$

Since $\lambda I - N$ is normal, we have by (1) that $\|(\lambda I - N)x\| = \|(\lambda I - N)^*x\|$. In particular, $\ker(\lambda I - N) = \ker(\lambda I - N)^*$. Also by the proof of (2),

$$\|(\lambda I - N)^{2^k}x\| \geq \|(\lambda I - N)x\|^{2^k}.$$

Hence $x \in \ker(\lambda I - N)^{2^k}$ implies that $x \in \ker(\lambda I - N)$. The other inclusion is trivial. So $\ker(\lambda I - N)^{2^k} = \ker(\lambda I - N)$. Hence $\ker(\lambda I - N)^n = \ker(\lambda I - N)$ for all $n \geq 1$. So (3) holds.

(4) $\overline{\operatorname{Ran}(\lambda I - N)} = \big(\ker(\lambda I - N)^*\big)^\perp = \big(\ker(\lambda I - N)\big)^\perp$.

(5) If $x \in \ker(\lambda I - N)$ and $y \in \ker(\mu I - N)$, then $Nx = \lambda x$ and $Ny = \mu y$, whence $N^*y = \overline{\mu}y$ since $y \in \ker(\mu I - N) = \ker(\bar{\mu}I - N^*)$. Hence

$$\lambda\langle x, y\rangle = \langle Nx, y\rangle = \langle x, N^*y\rangle = \langle x, \overline{\mu}y\rangle = \mu\langle x, y\rangle.$$

Since $\lambda \neq \mu$, this forces $\langle x, y\rangle = 0$. Therefore $\ker(\lambda I - N) \perp \ker(\mu I - N)$.

(6) It is easy to check that $p(N)$ is normal. By the Spectral Mapping theorem, $\sigma(p(N)) = p(\sigma(N))$. Therefore

$$\|p(N)\| = \operatorname{spr}(p(N)) = \sup\{|\mu| : \mu \in \sigma(p(N))\} = \sup\{|p(\lambda)| : \lambda \in \sigma(N)\}. \quad ■$$

We obtain an immediate corollary from (3).

5.4.5. COROLLARY. *A normal Fredholm operator has index* 0.

PROOF. If N is normal and Fredholm, then $\operatorname{ind} N = \operatorname{null} N - \operatorname{null} N^* = 0$. ■

Now we can establish the main result about compact normal operators. Like the finite dimensional case, compact normal operators are diagonalizable with respect to an orthonormal basis.

5.4.6. SPECTRAL THEOREM FOR COMPACT NORMAL OPERATORS. *Suppose that $K \in \mathcal{B}(H)$ is compact and normal. Let $\sigma(K) = \{\lambda_n : n \geq 0\}$ where $\lambda_0 = 0 = \lim_{n\to\infty} \lambda_n$. Let $M_n = \ker(\lambda_n I - K)$ for $n \geq 0$. These are finite dimensional for $n \geq 1$ and $M_m \perp M_n$ for $m \neq n$. Moreover $H = \sum_{n\geq 0}^{\oplus} M_n$ and $K = \sum_{n\geq 0} \lambda_n P_{M_n}$ is diagonalizable by an orthonormal basis.*

PROOF. By the Structure Theorem for Compact Operators 5.2.7, $\sigma(K)$ contains 0, and is either finite or countable, $\{\lambda_n : n \geq 0\}$, where $\lambda_0 = 0 = \lim_{n\to\infty} \lambda_n$. For each $0 \neq \lambda_n \in \sigma(K) \setminus \{0\}$, the space

$$M_n = \ker(\lambda_n I - K) = \ker(\lambda_n I - K)^2$$

is finite dimensional. By Proposition 5.4.4(5), $M_m \perp M_n$ for $1 \leq m < n$.

Let $N = \sum_{n\geq 1}^{\oplus} M_n$ be the orthogonal direct sum of M_n for $n \geq 1$. Then this subspace is invariant for K, and by Proposition 5.4.4(3), it is also invariant for K^*. By Proposition 5.4.4(4), $N^\perp = \bigcap_{n\geq 1} \overline{\operatorname{Ran}(\lambda_n I - K)}$, which is also invariant for K and K^*. Write $H = N \oplus N^\perp$. Both N and $N^\perp$ are invariant for K and K^*. Thus

$$K \simeq \begin{bmatrix} K|_N & 0 \\ 0 & K|_{N^\perp} \end{bmatrix} \quad \text{and} \quad K^* \simeq \begin{bmatrix} K^*|_N & 0 \\ 0 & K^*|_{N^\perp} \end{bmatrix}.$$

Computation of $0 = K^*K - KK^*$ shows that $K|_N$ and $K|_{N^\perp}$ are both normal. All of the non-zero eigenvectors of K lie in N, so that $\sigma(K|_{N^\perp}) = \{0\}$. By Proposition 5.4.4(2), $\|K|_{N^\perp}\| = 0$. So $K|_{N^\perp} = 0$, which shows that $N^\perp = \ker K = M_0$ and thus $H = \sum_{n\geq 0}^{\oplus} M_n$.

Let P_{M_n} be the orthogonal projection onto M_n. Then $L = \sum_{n\geq 0} \lambda_n P_{M_n}$ is a normal operator with $\ker(\lambda_n I - L) = M_n$ for $n \geq 0$. These spaces sum to the whole space, so L and K agree on each M_n, and therefore on all of H. If we choose an orthonormal basis for each M_n, we obtain an orthonormal basis for H consisting of eigenvalues for K. So K is diagonalized by this basis. ∎

Let $K \in \mathcal{K}(H)$ where H is a separable Hilbert space. Then K^*K is a positive compact operator. Hence it is diagonalizable. Thus there is an orthonormal basis, say $\{e_n : n \geq 1\}$ so that $K^*Ke_n = s_n^2 e_n$, where $s_n \geq 0$ and $\lim_{n\to\infty} s_n = 0$. Define a positive compact operator by $|K|e_n = s_n e_n$ for $n \geq 1$. Then $|K|^2 = K^*K$. For $x \in H$,

$$\begin{aligned}\| \, |K|x\|^2 &= \langle |K|x, |K|x\rangle = \langle |K|^2x, x\rangle \\ &= \langle K^*Kx, x\rangle = \langle Kx, Kx\rangle = \|Kx\|^2.\end{aligned}$$

Define $U_0 : \operatorname{Ran}|K| \to \operatorname{Ran} K$ by $U_0(|K|x) = Kx$. This map is isometric, and so it extends to an isometry from $\overline{\operatorname{Ran}|K|}$ onto $\overline{\operatorname{Ran} K}$. Now

$$(\operatorname{Ran}|K|)^\perp = \ker|K| = \ker K;$$

so $H = \overline{\operatorname{Ran}|K|} \oplus \ker K$. Define $U \in \mathcal{B}(H)$ by

$$U(x \oplus y) = U_0 x \quad \text{for} \quad x \in \overline{\operatorname{Ran}|K|} \text{ and } y \in \ker K.$$

Then by construction, $K = U|K|$, and U has the property that $\ker U = \ker K$ and $\operatorname{Ran} U = \overline{\operatorname{Ran} K}$. Algebraically, we have U^*U is the orthogonal projection onto $\overline{\operatorname{Ran}|K|} = (\ker K)^\perp$ and UU^* is the orthogonal projection onto $\overline{\operatorname{Ran} K}$. An operator like U, which is an isometry on the orthogonal complement of its kernel, is called a *partial isometry* .

We think of the factorization $K = U|K|$ as the *polar decomposition* by analogy with factoring a complex number as $z = e^{i\theta}|z|$. Let $S = \{n : s_n > 0\}$. Then

$$\ker K = \ker|K| = \overline{\operatorname{span}\{e_n : n \notin S\}}$$

and

$$(\ker K)^\perp = \overline{\operatorname{Ran}|K|} = \overline{\operatorname{span}\{e_n : n \in S\}}.$$

Note that $f_n = Ue_n$ for $n \in S$ is an orthonormal set because U_0 is an isometry on $(\ker K)^\perp$. We can write $|K| = \sum_{n\in S} s_n e_n e_n^*$. Therefore

$$K = U|K| = U \sum_{n\in S} s_n e_n e_n^* = \sum_{n\in S} s_n f_n e_n^*.$$

The values $(s_n)_{n\in S}$, reordered so that they form a decreasing sequence, are the *singular values* of K. If K is finite rank, then we set $s_k = 0$ for $k > \operatorname{rank} K$.

When H is not separable, a compact operator is always supported on a separable subspace $M = \overline{\operatorname{Ran} K + (\ker K)^\perp}$. Thus K also has a polar decomposition, and the non-zero singular values form a finite or countable set with limit 0.

5.4.7. DEFINITION. The *Schatten p-class* on H is the set

$$\mathfrak{S}_p = \{K \in \mathcal{K}(H) : (s_n) \in l_p\}$$

with norm $\|K\|_p := \|(s_n)\|_p$. The *trace class operators* are $\mathfrak{S}_1$.

It can be shown (Exercise **5.35**) that each $\mathfrak{S}_p$ is an ideal of $\mathcal{B}(H)$, and

$$\|SKT\|_p \le \|S\|\,\|K\|_p\,\|T\| \quad \text{for} \quad K \in \mathfrak{S}_p \text{ and } S, T \in \mathcal{B}(H).$$

It contains all finite rank operators, and thus is operator norm dense in $\mathcal{K}(H)$; and it is complete in the $\|\cdot\|_p$ norm. For $1 < p < \infty$, $\mathfrak{S}_p^* = \mathfrak{S}_q$ where $\frac{1}{p} + \frac{1}{q} = 1$. And it is an important theorem that $\mathcal{K}(H)^* = \mathfrak{S}_1$ and $\mathfrak{S}_1^* = \mathcal{B}(H)$. In particular, $\mathcal{B}(H)$ is a dual space, and so has a weak-$*$ topology. See Section 10.1.

5.5. Invariant Subspaces

In finite dimensions, every complex matrix is unitarily equivalent to a matrix in upper triangular form. That is, given $T \in M_n(\mathbb{C})$, there is an orthonormal basis $e_1, \dots, e_n$ such that $Te_i \in \operatorname{span}\{e_1, \dots, e_i\} =: M_i$. The subspaces M_i are invariant for T, in that $TM_i \subset M_i$.

Examples of Enflo [111] and Read [252, 253] show that bounded operators on a Banach space may not have a proper closed invariant subspace. In fact, Read constructs such an operator on l_1. It remains a famous open problem, the *invariant subspace problem*, whether every bounded operator on Hilbert space has an invariant subspace. In fact, there is no counterexample known on any reflexive Banach space.

However compact operators always have lots of invariant subspaces, as we shall see in this section.

5.5.1. Definition. If $\mathcal{A} \subset \mathcal{B}(X)$, a closed subspace $M \subset X$ is an *invariant subspace* for $\mathcal{A}$ if $AM \subset M$ for all $A \in \mathcal{A}$. It is a *proper invariant subspace* if $\{0\} \ne M \ne X$. We write $\operatorname{Lat}\mathcal{A} = \{M : \text{ invariant subspaces for } \mathcal{A}\}$ for the *lattice of invariant subspaces*. A subspace M which is invariant for the commutant $\mathcal{A}' = \{T \in \mathcal{B}(X) : TA = AT \text{ for } A \in \mathcal{A}\}$ is called a *hyperinvariant subspace* for $\mathcal{A}$. A non-zero algebra $\mathcal{A}$ of operators is *transitive* if $\operatorname{Lat}\mathcal{A} = \{\{0\}, H\}$.

$\operatorname{Lat}\mathcal{A}$ is a *complete lattice*. The operations are intersection, $M \wedge N = M \cap N$, and closed span, $M \vee N = \overline{M + N}$. It is easy to check that these are both invariant subspaces. Completeness indicates that we can take the supremum and infimum over arbitrary subsets of $\operatorname{Lat}\mathcal{A}$. We have

$$\bigwedge_{\alpha\in A} M_\alpha = \bigcap_{\alpha\in A} M_\alpha \quad \text{and} \quad \bigvee_{\alpha\in A} M_\alpha = \overline{\sum_{\alpha\in A} M_\alpha}$$

where in the sum, we allow only finitely many non-zero vectors. Again it is easy to check that these subspaces are invariant.

If $\mathcal{A}$ is transitive, $0 \neq \xi \in H$, and $\overline{\mathcal{A}\xi} = \{0\}$, then $\mathbb{C}\xi$ is invariant. But then H is 1 dimensional and $\mathcal{A} \neq \{0\}$, so this can't happen. Therefore $\overline{\mathcal{A}\xi} = H$; i.e. every non-zero vector is cyclic.

The following result uses the Schauder–Tychonoff fixed point theorem.

5.5.2. Lomonosov Theorem. *If $T \in \mathcal{B}(H) \setminus \mathbb{C}I$ and $\{T\}'$ contains a non-zero compact operator, then T has a proper hyperinvariant subspace.*

Proof. Suppose that $\mathcal{A} = \{T\}'$ is transitive and contains a non-zero compact operator K. We may normalize so that $\|K\| = 1$ and pick $\zeta \in H$ with $\|K\zeta\| > 1$. Let $S = \overline{b_1(\zeta)}$ and $D = \overline{KS}$. Since S is bounded and convex, D is convex and compact. The choice of ζ guarantees that $\operatorname{dist}(0, D) \geq \|K\zeta\| - 1 > 0$.

For each $\xi \in D$, the transitivity of $\mathcal{A}$ means that there is some $A \in \mathcal{A}$ such that $\|A\xi - \zeta\| < 1$. So there is an open ball $U \ni \xi$ such that $AU \subset S$. By compactness, there is a finite open cover $U_1, \dots, U_n$ of D and $A_i \in \mathcal{A}$ satisfying $A_i U_i \subset S$. Let $g_1, \dots, g_n$ be a partition of unity (Proposition 1.10.7) in $C(D)$ with respect to this cover. Define a map $f : D \to D$ by

$$f(\xi) = \sum_{i=1}^{n} g_i(\xi) K A_i \xi.$$

For $\xi \in D$, this is a convex combination of the points $KA_i\xi$, and $g_i(\xi) \neq 0$ only when $\xi \in U_i$, so that $KA_i\xi \in KS \subset D$. Hence f is a continuous map of the compact convex set D into itself. By the Schauder–Tychonoff fixed point theorem 3.8.6, f has a fixed point ξ_0. Set $A = \sum_{i=1}^{n} g_i(\xi_0) A_i$. Then $KA\xi_0 = f(\xi_0) = \xi_0$; so that 1 is an eigenvalue of KA.

Hence $M = \ker(KA - I)$ is finite dimensional, and since T commutes with $KA - I$,

$$(KA - I)TM = T(KA - I)M = \{0\};$$

whence $TM \subset M$. Since M is finite dimensional, $T|_M$ has an eigenvalue λ_0. Therefore $N = \ker(T - \lambda_0 I)$ is non-zero. Since T is not a scalar, N is proper. Arguing as above, N is invariant for $\mathcal{A}$. ■

5.5.3. Remark. This proof does *not* show that if T commutes with a non-zero compact operator, then it has an eigenvector. This consequence was only obtained using the (false) assumption that $\mathcal{A}$ is transitive. Indeed, the Volterra operator, which is compact, has no eigenvalues.

The following is an immediate consequence.

5.5.4. Corollary. *Every non-zero compact operator has a hyperinvariant subspace.*

5.5.5. EXAMPLE. Consider the Volterra operator $V \in \mathcal{B}(L^2(0,1))$ from Example 5.1.5, $Vh(x) = \displaystyle\int_0^x h(t)\,dt$. This is compact and has no eigenvalues. It has the invariant subspace $N_t = \{h \in L^2(0,1) : h|_{(0,t)} = 0 \text{a.e.}\}$ for $0 \le t \le 1$. In fact, these are the only invariant subspaces for V.

5.5.6. COROLLARY. *If $K \in \mathcal{K}(X)$, then there is a maximal chain of subspaces (with respect to containment) which are invariant for $\{K\}'$.*

PROOF. By Zorn's lemma, there is a maximal chain $\mathcal{N}$ of subspaces in $\mathrm{Lat}\{K\}'$. If $M_\alpha \in \mathcal{N}$ for $\alpha \in A$, then $M = \bigwedge M_\alpha \in \mathrm{Lat}\{K\}'$. Any other subspace $N \in \mathcal{N}$ is either larger than some M_α, and hence $N \supset M$, or is contained in all of them, and hence $N \subset M$. In either case, it shows that $\mathcal{N} \cup \{M\}$ is a chain of subspaces; and thus $M \in \mathcal{N}$ by maximality. Similarly, $\bigvee M_\alpha \in \mathcal{N}$. Thus the chain $\mathcal{N}$ is complete.

If $N \in \mathcal{N}$, let $N_- = \bigvee\{M \in \mathcal{N} : M \subsetneq N\}$. If $N_- \subsetneq N$, I claim that $\dim N/N_- = 1$. Otherwise, we define an operator

$$\widetilde{K} \in \mathcal{B}(N/N_-) \quad \text{by} \quad \widetilde{K}(x + N_-) = Kx + N_-.$$

Note that $\overline{\widetilde{K}b_1(N/N_-)} = Q(\overline{Kb_1(N)})$, where $Q : N \to N/N_-$ is the quotient map. The continuous image of a compact set is compact, and therefore $\widetilde{K}$ is a compact operator. The algebra $\{\widetilde{K}\}'$ has a proper invariant subspace, say M. Then $Q^{-1}M =: L$ is invariant for $\{K\}'$ and $N_- \subsetneq L \subsetneq N$. The reason is that for $T \in \{K\}'$, the operator $\widetilde{T}(x + N_-) = Tx + N_-$ in $\mathcal{B}(N/N_-)$ lies in $\{\widetilde{K}\}'$ and so leaves M invariant. Pulling this back to X shows that L is invariant for T. However this contradicts the maximality of $\mathcal{N}$. Hence the gap N/N_- is either 0 or 1 dimensional.

This chain $\mathcal{N}$ must be a maximal chain in the lattice of all subspaces of X. For if $\mathcal{N} \dot\cup \{M\}$ were a larger chain of subspaces, let $N_- = \bigvee\{N \in \mathcal{N} : N \subset M\}$ and $N_+ = \bigwedge\{N \in \mathcal{N} : M \subset N\}$. Then by completeness, $N_\pm \in \mathcal{N}$ and $N_- \subsetneq M \subsetneq N_+$. Moreover $N_- = (N_+)_-$ and the gap N_+/N_- is at least 2-dimensional. This is false, and hence $\mathcal{N}$ is maximal. ∎

For the rest of this section, we will restrict our attention to operators on Hilbert space. This is only for the convenience of presentation.

5.5.7. DEFINITION. A *nest of subspaces* of H is a complete chain of subspaces containing $\{0\}$ and H. The *nest algebra* of a nest $\mathcal{N}$ is defined as

$$\mathcal{T}(\mathcal{N}) = \{T \in \mathcal{B}(H) : TN \subset N \ \text{ for all } \ N \in \mathcal{N}\}.$$

Again we define $N_- = \bigvee\{M \in \mathcal{N} : M \subsetneq N\}$. The *atoms* of $\mathcal{N}$ are the subspaces $A = N \ominus N_- = \{x \in N : x \perp N_-\}$ when $N_- \subsetneq N$; and let $\mathbb{A}$ denote the set of all atoms of $\mathcal{N}$. Define $\Phi_A(T) = P_A T|_A$ be the compression of $T \in \mathcal{T}(\mathcal{N})$ to the subspace A. When $\mathcal{N}$ is a maximal nest, the atoms are 1 dimensional and so $\Phi_A(T)$ are scalars.

Note that $\mathcal{T}(\mathcal{N})$ is a vector space closed under multiplication and contains I. If $T_\alpha \in \mathcal{T}(\mathcal{N})$ converge in the weak operator topology to T, i.e. $\langle T_\alpha x, y\rangle \to \langle Tx, y\rangle$ for each $x, y \in H$, then for every $x \in N$ and $y \in N^\perp$, we have $0 = \langle T_\alpha x, y\rangle$, and thus $0 = \langle Tx, y\rangle$. Therefore $Tx \in (N^\perp)^\perp = N$; so $T \in \mathcal{T}(\mathcal{N})$. Therefore $\mathcal{T}(\mathcal{N})$ is a weak operator topology closed algebra of operators. The fact that each $N \in \mathcal{N}$ is invariant can be seen as an upper triangular form for elements of $\mathcal{T}(\mathcal{N})$.

Ringrose's theorem shows that the non-zero diagonal entries in this triangular form are precisely the non-zero eigenvalues including multiplicity. We will use the notation of Theorem 5.2.7. The proof will be given in a series of lemmas.

5.5.8. Ringrose Theorem. *Suppose that $K \in \mathcal{K}(H)$ and that $\mathcal{N}$ is a maximal nest in* Lat K. *Let $\mathbb{A}$ be the atoms of $\mathcal{N}$. Then*

$$\sigma(K) = \{0\} \cup \{\Phi_A(K) : A \in \mathbb{A}\}.$$

Moreover each non-zero eigenvalue λ is repeated $n_\lambda = \dim N(\lambda)$ times.

The proof is broken down into the following four lemmas.

5.5.9. Lemma. $\sigma(K) = \{0\} \cup \{\Phi_A(K) : A \in \mathbb{A}\}$.

Proof. Let $A \in \mathbb{A}$ such that $0 \neq \lambda = \Phi_A(K)$. Write $A = N \ominus N_-$ for some $N \in \mathcal{N}$. Since A is 1 dimensional, pick a unit vector $x \in A$. Then

$$\lambda = \Phi_A(K) = \langle Kx, x\rangle.$$

Since $N = N_- \oplus \mathbb{C}x$, we see that $Kx = \lambda x + y$ for some $y \in N_-$. Hence $(\lambda I - K)x = -y \in N_-$. This shows that $(\lambda I - K)|_N$ is not surjective, and so $\lambda \in \sigma(K|_N)$. By Theorem 5.2.7, $\lambda \in \sigma_p(K|_N)$. Thus

$$\ker(\lambda I - K) \supset \ker(\lambda I - K)|_N \neq \{0\}.$$

Hence $\lambda \in \sigma(K)$.

Conversely suppose that $\lambda \in \sigma(K)\setminus\{0\}$. Then $E = \ker(\lambda I - K)$ is a non-zero finite dimensional subspace. Thus the sphere $S = \{x \in E : \|x\| = 1\}$ is compact. Let

$$N = \bigwedge\{M \in \mathcal{N} : E \cap M \neq \{0\}\}.$$

If $E \cap M \neq \{0\}$, then $S \cap M$ is non-zero and compact. This collection has the finite intersection property because

$$(S \cap M_1) \cap (S \cap M_2) = S \cap (M_1 \cap M_2),$$

and $M_1 \cap M_2$ is the smaller of M_1 and M_2. By compactness, the intersection $S \cap N$ is non-empty. Thus $E \cap N \neq \{0\}$. Therefore $\lambda I - K|_N$ has a non-trivial kernel, and thus $\operatorname{Ran}(\lambda I - K|_N)$ is a proper closed subspace of N.

Now if $M \in \mathcal{N}$ and $M < N$, then $E \cap M = \{0\}$. Hence $\lambda I - K|_M$ is injective, and thus by Theorem 5.2.7, $\operatorname{Ran}(\lambda I - K|_M) = M$. Since N_- is the closed span of $M \in \mathcal{N}$ with $M < N$, $\operatorname{Ran}(\lambda I - K|_{N_-}) \supset \bigcup_{M<N} M$. This is dense in N_- and the range is closed, so equals N_-. This shows that $N_- \subsetneq N$. It follows that $\operatorname{Ran}(\lambda I - K|_N) \supset N_-$ and is a proper subspace of N. Since $A = N \ominus N_-$ is 1 dimensional, the range is N_-. Thus $0 = \Phi_A(\lambda I - K) = \lambda - \Phi_A(K)$. Therefore $\Phi_A(K) = \lambda$. ∎

Recall that E_λ is the projection onto $N(\lambda)$ with kernel $R(\lambda)$, and $E_\lambda \in \{K\}''$.

5.5.10. LEMMA. *Let $\lambda \in \sigma(K) \setminus \{0\}$, and let $M \in \operatorname{Lat} K$. Then M is invariant for E_λ.*

PROOF. The restriction $K|_M$ is compact. Let

$$N = \ker(\lambda I - K|_M)^{n_\lambda} = N(\lambda) \cap M \quad \text{and} \quad R = \operatorname{Ran}(\lambda I - K|_M)^{n_\lambda} \subset R(\lambda) \cap M.$$

Then by Theorem 5.2.7, $M = N \dotplus R$. Note that $E_\lambda|_N = I|_N$ and $E_\lambda|_R = 0$. Therefore $E_\lambda M = N \subset M$. ∎

5.5.11. LEMMA. *Let $\lambda \in \sigma(K) \setminus \{0\}$, and let $A \in \mathbb{A}$. Then*

$$\Phi_A(E_\lambda) = \begin{cases} 1 & \text{if } \Phi_A(K) = \lambda \\ 0 & \text{otherwise.} \end{cases}$$

PROOF. We have $H = N(\lambda) \dotplus R(\lambda)$ is a direct sum of subspaces in $\operatorname{Lat} K$. Then $K_\lambda := K|_{N(\lambda)}$ has the form $\lambda I_\lambda + J$, where I_λ is the identity on $N(\lambda)$ and $J \in \mathcal{B}(N(\lambda))$ is nilpotent because $N(\lambda) = \ker(\lambda I - K)^{n_\lambda}$. Let $L = K|_{R(\lambda)}$. This is compact and $\lambda \notin \sigma(L)$. With respect to the decomposition $H = N(\lambda) \dotplus R(\lambda)$, K, E_λ and $K + zE_\lambda$ have matrices

$$K = \begin{bmatrix} \lambda I_\lambda + J & 0 \\ 0 & L \end{bmatrix}, \quad E = \begin{bmatrix} I_\lambda & 0 \\ 0 & 0 \end{bmatrix} \quad \text{and} \quad K + zE_\lambda = \begin{bmatrix} (\lambda + z) I_\lambda + J & 0 \\ 0 & L \end{bmatrix}.$$

By Lemma 4.4.8,

$$\sigma(K + zE_\lambda) = \sigma((\lambda + z)I_\lambda + J) \cup \sigma(L) = \{\lambda + z\} \cup \sigma(K) \setminus \{\lambda\}.$$

For any $A \in \mathbb{A}$ and $z \in \mathbb{C}$, we have that

$$\Phi_A(K + zE_\lambda) = \Phi_A(K) + z\Phi_A(E_\lambda) \in \{\lambda + z\} \cup \sigma(K) \setminus \{\lambda\}.$$

If $\Phi_A(K) = \lambda$, then $\Phi_A(K + zE_\lambda)$ will be near to λ for small z, and therefore $\Phi_A(K + zE_\lambda) = \lambda + z$. Thus $\Phi_A(E_\lambda) = 1$. On the other hand, if $\Phi_A(K) = \mu \neq \lambda$, then $\mu + z\Phi_A(E_\lambda) \in \sigma(K) \setminus \{\lambda\}$ for very large z, and hence $\Phi_A(E_\lambda) = 0$. ∎

The following lemma completes the proof of Ringrose's theorem.

5.5.12. LEMMA. *The set* $\mathbb{A}(\lambda) = \{A \in \mathbb{A} : \Phi_A(K) = \lambda\}$ *has cardinality* $n_\lambda = \operatorname{rank} E_\lambda$.

PROOF. Let $\mathbb{A}(\lambda) = \{A_1, \dots, A_n\}$ be a finite or countable set, and define $P = \sum_{A_i \in \mathbb{A}(\lambda)} A_i$. This is an orthogonal projection of rank $|\mathbb{A}(\lambda)|$. Each $A_i = N_i \ominus N_{i-} = \mathbb{C}e_i$ for a unit vector e_i. Then $E_\lambda e_i = e_i + y_i$ where $y_i \in N_{i-}$. In particular, if we order the N_i by containment, then we see that the compression $PE_\lambda|_{PH}$ is upper triangular with 1's on the diagonal. Therefore $n \leq \operatorname{rank} E_\lambda$.

The projection P also lies in $\mathcal{T}(\mathcal{N})$ since either $e_i \in N$ or $e_i \in N^\perp$, and so

$$PN = \operatorname{span}\{e_i : e_i \in N\} = N \cap \operatorname{Ran} P.$$

By choice of P, we have $\Phi_A(E_\lambda - P) = 0$ for $A \in \mathbb{A}(\lambda)$. Also by construction, $\Phi_A(P) = 0 = \Phi_A(E_\lambda)$ for $A \in \mathbb{A} \setminus \mathbb{A}(\lambda)$. By Lemma 5.5.9 applied to the finite rank operator $E_\lambda - P$, $\sigma(E_\lambda - P) = \{0\}$. If $\operatorname{rank} P < \operatorname{rank} E_\lambda$, there would be a unit vector x in $N(\lambda) \cap P^\perp H$, and $(E_\lambda - P)x = x$. Hence

$$|\mathbb{A}(\lambda)| = \operatorname{rank} P = \operatorname{rank} E_\lambda = n_\lambda. \qquad \blacksquare$$

For vectors $x, y \in H$, we use the notation xy^* for the rank one operator $(xy^*)(z) = \langle z, y\rangle x$. This is a natural notation because

$$xy^* = \begin{bmatrix} x_1 \\ x_2 \\ x_3 \\ \vdots \end{bmatrix} \begin{bmatrix} \overline{y_1} & \overline{y_2} & \overline{y_3} & \dots \end{bmatrix} = \begin{bmatrix} x_1\overline{y_1} & x_1\overline{y_2} & x_1\overline{y_3} & \cdots \\ x_2\overline{y_1} & x_2\overline{y_2} & x_2\overline{y_3} & \cdots \\ x_3\overline{y_1} & x_3\overline{y_2} & x_3\overline{y_3} & \cdots \\ \vdots & \vdots & \vdots & \ddots \end{bmatrix}.$$

Also we have the identity $A(xy^*)B = (Ax)(B^*y)^*$ for any $A, B \in \mathcal{B}(H)$.

5.5.13. LEMMA. *Let* $\mathcal{A} \subset \mathcal{B}(H)$. *Then* $\operatorname{Lat} \mathcal{A}^* = \{M^\perp : M \in \operatorname{Lat} \mathcal{A}\}$.

PROOF. If $M \in \operatorname{Lat} \mathcal{A}$, $x \in M$, $y \in M^\perp$ and $A \in \mathcal{A}$, then $\langle A^*y, x\rangle = \langle y, Ax\rangle = 0$. This shows that $\mathcal{A}^*M^\perp$ is orthogonal to M, whence $\mathcal{A}^*M^\perp \subset M^\perp$. Since $\mathcal{A} = (\mathcal{A}^*)^*$, this argument is reversible. $\blacksquare$

We finish this section with a useful finite dimensional result.

5.5.14. BURNSIDE THEOREM. *If* $\mathcal{A} \subset M_n$ *is a transitive subalgebra of the* $n \times n$ *complex matrices, i.e.* $\mathcal{A}x = \mathbb{C}^n$ *for all* $x \neq 0$, *then* $\mathcal{A} = M_n$.

PROOF. Proceed by induction on n. For $n = 1$, the only proper subalgebra is $\{0\}$, which doesn't satisfy the hypothesis. So the result is true. Assume the result is true for all $k < n$.

First we show that there is some $F \in \mathcal{A}$ with $0 < \operatorname{rank} F < n$. $\mathcal{A}$ cannot consist solely of scalar multiples of I since $n \geq 2$, so pick a non-scalar $A \in \mathcal{A}$. If A is not invertible, then $0 < \operatorname{rank} A < n$ as desired. Otherwise pick $\lambda \in \sigma(A)$, so $A - \lambda I$ is non-zero and not invertible. The rank is unchanged if we multiply by the invertible A, so $F = A(A - \lambda) = A^2 - \lambda A \in \mathcal{A}$ works.

Let $M = \operatorname{Ran} F$. The set $\mathcal{B} = F\mathcal{A}|_M$ is an algebra since

$$(FA|_M)(FB|_M) = F(AFB)|_M \in \mathcal{B} \quad \text{for} \quad A, B \in \mathcal{A}.$$

If $0 \neq x \in M$, then $\mathcal{B}x = F\mathcal{A}x = F\mathbb{C}^n = M$. By the induction hypothesis, $\mathcal{B} = \mathcal{L}(M)$. In particular, $\mathcal{B}$ contains a rank one operator $R = FA_0|_M$. Therefore $FA_0F \in \mathcal{A}$ has rank 1, say $FA_0F = x_0y_0^*$.

Let $x, y \in \mathbb{C}^n$. Pick $A \in \mathcal{A}$ so that $Ax_0 = x$. Now $\mathcal{A}^*y_0$ is invariant for $\mathcal{A}^*$, so by Lemma 5.5.13, it is $\{0\}$ or $\mathbb{C}^n$. But if $\mathcal{A}^*y_0 = 0$, then

$$0 = \langle \mathcal{A}^*y_0, x_0 \rangle = \langle y_0, \mathcal{A}x_0 \rangle$$

which contradicts that $y_0 \in \mathcal{A}x_0$. Pick $B^* \in \mathcal{A}^*$ so that $B^*y_0 = y$. Then $\mathcal{A}$ contains $A(x_0y_0^*)B = (Ax_0)(B^*y_0)^* = xy^*$. Every matrix is a sum of rank one matrices, so $\mathcal{A} = M_n$. ∎

Exercises for Chapter 5

5.1 If $E = E^2 \in \mathcal{B}(X)$ is compact, show that E is finite rank.

5.2 Show that if $K \in \mathcal{B}(X)$ is compact and $M \subset X$ is a closed invariant subspace, then $K|_M$ is compact.

5.3 Show that if $f \in C[0, 1]$, then the multiplication operator M_f on $L^2(0, 1)$ is not compact unless $f = 0$.

5.4 Let $\{e_i(x) : i \geq 1\}$ be an orthonormal basis for $L^2(0, 1)$. Show that $\{e_i(x)e_j(y) : i, j \geq 1\}$ is an orthonormal basis for $L^2((0, 1)^2)$.

5.5 Let F be a finite rank operator on $L^2(0, 1)$. Show that F is a Hilbert–Schmidt integral operator as in Example 5.1.4.

5.6 Find a compact operator on $L^2(0, 1)$ which is not a Hilbert–Schmidt integral operator.
HINT: HS operators have an infinite matrix with square summable coefficients.

5.7 (a) Let X, Y be Banach spaces. Prove that if $K \in \mathcal{K}(X, Y)$ is compact, then whenever a *sequence* $x_n \in X$ converges weakly to x_0, the sequence Kx_n converges to Kx_0 in norm.

(b) Let X be a *separable reflexive* Banach space. Suppose $T \in \mathcal{B}(X, Y)$ and whenever a sequence $x_n \in X$ converges weakly to x_0, the sequence Tx_n converges to Tx_0 in the norm. Prove that T is compact.
HINT: use the Banach–Alaoglu theorem.

(c) Find a *non-compact* operator $T \in \mathcal{B}(l_2)$ so that $\lim_{n\to\infty} \|Te_n\| = 0$ for the usual orthonormal basis $\{e_n\}$.

5.8 Let $\{A_\lambda\}_\Lambda$ be a bounded net in $\mathcal{B}(X)$ which converges to I in the strong operator topology. Show that if $K \in \mathcal{K}(X)$, then $A_\lambda K$ converges to K in norm.

5.9 Let H be a Hilbert space with orthonormal basis $\{e_n : n \geq 1\}$. Let $A = [a_{ij}]$ be an infinite matrix such that $\sum_{i,j} |a_{ij}| < \infty$. Prove that A represents a compact operator.

5.10 (a) Show that every compact operator $K \in \mathcal{B}(l_p)$ for $1 \leq p < \infty$ is a norm limit of compact operators.
(b) Show that if X, Y are Banach spaces and Y has a Schauder basis (Exercise **2.17**), then every compact $K \in \mathcal{K}(X, Y)$ is a norm limit of finite rank operators.
HINT: consider $S_n x_i$ for an ε-net in $\overline{Kb_1(X)}$.

5.11 Let $K \in \mathcal{B}(X)$ be a compact operator, and let $\mu \in \sigma(K) \setminus \{0\}$. Define n_μ as in Theorem 5.2.7. Show that $R(T, \lambda)$ has a pole at μ of order n_μ.

5.12 Let H_n be separable Hilbert spaces, and let $H = \bigoplus_{n\geq 1} H_n$ be the Hilbert space of all vectors $\xi = (\xi_n)_{n\geq 1} = (\xi_1, \xi_2, \dots)$ such that $\xi_n \in H_n$ and $\|\xi\|^2 = \sum_{n\geq 1} \|\xi_n\|^2 < \infty$. Suppose that $T_n \in \mathcal{B}(H_n)$ and define $T = \bigoplus_{n\geq 1} T_n$ by $T\xi = (T_n \xi_n)_{n\geq 1}$.
(a) Show that T is bounded if and only if $\sup_n \|T_n\| < \infty$.
(b) When is T compact? Find necessary and sufficient conditions.

5.13 Let $T \in \mathcal{B}(X, Y)$. Show that if $T|_M$ is compact for every separable subspace $M \subset X$, then T is compact.

5.14 Let X be a reflexive Banach space.
(a) Let $T \in \mathcal{B}(X, l_1)$. Prove that T is compact.
HINT: First assume that X is separable, so $(\overline{b_1(X)}, \tau_{X^*})$ is compact and metrizable. Use Exercise **3.34**(2) to show that if $x_n \in b_1(X)$ converge weakly to 0, then Tx_n converges to 0 in norm.
(b) Show that if $S \in \mathcal{B}(c_0, X)$, then S is compact.

5.15 If $\mathcal{S} \subset \mathcal{B}(X)$, show that $\mathcal{S}'$ is a unital algebra (subspace closed under multiplication containing I) which is closed in the weak operator topology.

5.16 (a) Suppose that $T \in \mathcal{B}(X, Y)$ is Fredholm. Prove that T^* is Fredholm, and find the relationship between $\operatorname{ind} T$ and $\operatorname{ind} T^*$.
(b) If X, Y are Hilbert spaces and T^* is the Hilbert space adjoint, what happens?

5.17 (*Fredholm alternative*) Let $K \in \mathcal{B}(X)$ be a compact operator. Show that $K\xi = \lambda\xi$ has a unique solution ξ in X if and only if $(K - \lambda I)\xi = \eta$ has a solution for every $\eta \in X$.

5.18 Let S be the unilateral shift on l_2. Define $T = S \oplus (S^* + \frac{1}{2}I)$ in $\mathcal{B}(l_2 \oplus l_2)$. Compute $\operatorname{ind}(T - \lambda I)$ for $\lambda \in \{-\frac{3}{4}, 0, \frac{5}{4}\}$.

5.19 The *Donaghue operator* acts on l_2 by $Ae_0 = 0$ and $Ae_n = 2^{-n}e_{n-1}$ for $n \geq 1$.

(a) Show that A is compact, and compute $\|A\|$.

(b) Show that the proper closed invariant subspaces for A are precisely the subspaces $\mathcal{M}_n = \operatorname{span}\{e_k : 0 \leq k \leq n\}$ for $n \geq 0$.
HINT: If an invariant subspace $\mathcal{M}$ contains a vector $x = \sum_{n\geq 0} x_n e_n$ with $x_n \neq 0$ infinitely often, show that $e_0 \in \mathcal{M}$. Pick $n_i \to \infty$ such that $|x_{n_i}| \geq |x_n|$ for all $n \geq n_i$. Consider appropriate multiples of $A^{n_i}x$.

5.20 Let V be the Volterra operator on $L^2(0,1)$, $Vf(x) = \int_0^x f(t)\,dt$.

(a) Express V^* as an integral operator.

(b) Show that $V + V^*$ is a rank 1 projection.

(c) Suppose that f is an eigenvector of VV^*. Show that f is C^∞ and satisfies a second order ODE with boundary conditions.

(d) Hence diagonalize VV^* and compute $\|V\|$.

5.21 Let $N \in M_n(\mathbb{C})$ be a normal matrix. Assume that $e_1, \ldots, e_n$ is an orthonormal basis such that the matrix for N in this basis is upper triangular. Prove that N is diagonal in this basis.

5.22 Let $N \in \mathcal{B}(H)$ be compact and normal. Show the following are equivalent:

(1) N has a *cyclic vector* ξ, i.e. $\{p(N)\xi : p \in \mathbb{C}[x]\}$ is dense in H.

(2) $\operatorname{null} \lambda I - N \leq 1$ for all $\lambda \in \sigma(N)$.

(3) $\{N\}'$ is abelian.

5.23 (a) If N is a compact normal operator on H, show that the closure of the numerical range, $\overline{W(N)}$, is the convex hull of $\sigma(N)$. (See Exercise **4.23**.)

(b) If M_f is a multiplication operator on $L^2(0,1)$ (Example 4.3.8), show that the closure of the numerical range, $\overline{W(M_f)}$, is the convex hull of $\sigma(M_f)$.

5.24 Let $N_1, N_2 \in \mathcal{B}(H)$ be compact normal operators. Show that they are unitarily equivalent if and only if $\operatorname{null} \lambda I - N_1 = \operatorname{null} \lambda I - N_2$ for all $\lambda \in \mathbb{C}$.

5.25 Let H be a Hilbert space. An operator $R \in \mathcal{B}(H)$ is a *symmetry* if $R = R^*$ and $R^2 = I$. Prove that the following are equivalent:

(1) R is a symmetry.

(2) $\frac{1}{2}(R + I)$ is an orthogonal projection.

(3) $R^2 = I$ and R is normal.

(4) There is a closed subspace M such that $(I + R)\xi \in M$ and $(I - R)\xi \in M^\perp$ for all $\xi \in H$.

5.26 (*Schur test*) Let H be a Hilbert space with orthonormal basis $\{e_i : i \geq 1\}$, and let A be an infinite matrix $A = \big[a_{ij}\big]$. Suppose that there are constants b, c,

$p_i > 0$ and $q_j > 0$ for $i, j \geq 1$ satisfying

$$\sum_{i \geq 1} |a_{ij}| p_i \leq b q_j \quad \text{and} \quad \sum_{j \geq 1} |a_{ij}| q_j \leq c p_i \quad \text{for} \quad i, j \geq 1.$$

Prove that A is bounded with $\|A\| \leq \sqrt{bc}$.
HINT: for a vector $\xi = (x_1, x_2, \dots)$, use

$$\sum_{i \geq 1} \Big| \sum_{j \geq 1} a_{ij} x_j \Big|^2 \leq \sum_{i \geq 1} \Big| \sum_{j \geq 1} \big(|a_{ij}|^{1/2} q_j^{1/2} \big) \big(|a_{ij}|^{1/2} q_j^{-1/2} |x_j| \big) \Big|^2 .$$

Apply Cauchy–Schwarz and then interchange the order of summation.

5.27 (*Hilbert matrix*). In the previous exercise, define $a_{ij} = \frac{1}{i+j-1}$. $A = \left[\frac{1}{i+j-1} \right]$ is called the Hilbert matrix. Prove that $\|A\| \leq \pi$.
HINT: use $p_i = q_i = \left(i - \frac{1}{2} \right)^{-1/2}$ and the integral test.

5.28 An operator $T \in \mathcal{B}(H)$, for a Hilbert space H, is *left Fredholm* if null $T < \infty$ and Ran T is closed. Prove that the following are equivalent:
(1) T is left Fredholm.
(2) There is $S \in \mathcal{B}(H)$ such that ST is the orthogonal projection onto $(\ker T)^\perp$.
(3) There is $S \in \mathcal{B}(H)$ such that $ST - I$ is finite rank.
(4) There is $S \in \mathcal{B}(H)$ such that $ST - I$ is compact.

5.29 (*Hilden*). Show that if K is a non-zero compact operator, then K has a hyperinvariant subspace as follows.
(a) If $\sigma(K) \neq \{0\}$, find a hyperinvariant subspace.
(b) Let S, D, the cover U_i and operators $A_i \in \mathcal{A} = \{K\}'$ be as in the proof of Lomonosov's theorem. Show that there is a sequence $i_1, i_2, \dots$ with $1 \leq i_j \leq n$ such that $A_{i_k} K A_{i_{k-1}} K \dots A_{i_1} K \zeta \in S$ for all $k \geq 1$. Show that if K is quasinilpotent, this leads to a contradiction.
HINT: estimate the norm of $A_{i_k} K A_{i_{k-1}} K \dots A_{i_1} K \zeta$.

5.30 Show that the unilateral shift on l_2 does not commute with any non-zero compact operator.

5.31 Show that if $T \in \mathcal{B}(X)$ and $TK = KT$ for every compact operator K, then $T = \lambda I$ for some $\lambda \in \mathbb{C}$.

5.32 Let $T \in \mathcal{B}(X)$ where X is a non-separable Banach space. Prove that T has a proper invariant subspace.
HINT: what is the smallest invariant subspace containing a vector ξ?

5.33 Show that $\mathcal{K}(H)$ is not a dual space for any infinite dimensional Hilbert space.
HINT: polar decomposition and extreme points.

5.34 (*generalized Schur complement*).

(a) (*Schur complement*). Suppose that $T = T^* \in \mathcal{B}(H)$ and P is a projection such that $A = PT|_{PH}$ is positive and invertible in $\mathcal{B}(PH)$. Write $T = \left[\begin{smallmatrix} A & B \\ B^* & D \end{smallmatrix}\right]$ with respect to $H = PH \oplus P^\perp H$. Show that $T \geq 0$ if and only if $D \geq B^* A^{-1} B$.
HINT: Look at $\langle T(-A^{-1}B\xi, \xi), (-A^{-1}B\xi, \xi)\rangle$.
(b) (*Douglas lemma*). Let $A, B \in \mathcal{B}(H)$. Show that if $B^*B \leq A^*A$, then there is $L \in \mathcal{B}(H)$ with $\|L\| \leq 1$ such that $B = LA$. If $L|_{\ker A} = 0$, then it is unique.
HINT: define $L(A\xi) = B\xi$ on $\operatorname{Ran} A$ and $L = 0$ on $\ker A$.
(c) Suppose that $T \in \mathcal{B}(H)$ is positive and P is a projection. Write $T = \left[\begin{smallmatrix} A & B \\ B^* & D \end{smallmatrix}\right]$ with respect to $H = PH \oplus P^\perp H$. Show that there is $X \in \mathcal{B}(P^\perp H, PH)$ such that $B = A^{1/2}X$ and $\operatorname{Ran} X \subset \overline{\operatorname{Ran} A}$.
HINT: replace D by $\|D\|I$ to get T' and look at $(\|D\|P - B^*)T'(\|D\|P - B)$.
(d) (*Lance*). With T, P and X as in (c), show that $D \geq X^*X$; and conversely if $B = A^{1/2}X$ as in (c) and $D \geq X^*X$, then $T \geq 0$.

5.35 (*Schatten ideals* .) Let $K \in \mathcal{K}(H)$ with singular values $s_n(K)$.
(a) Show that $s_n(K) = \inf\{\|K - F\| : \operatorname{rank} F \leq n - 1\}$.
(b) If $T \in \mathcal{B}(H)$, show that $s_n(TK) \leq \|T\| s_n(K)$ and $s_n(KT) \leq \|T\| s_n(K)$.
(c) Show that if $K \in \mathfrak{S}_p$ for $1 \leq p < \infty$ and $S, T \in \mathcal{B}(H)$, then $SKT \in \mathfrak{S}_p$ and $\|SKT\|_p \leq \|S\| \, \|K\|_p \|T\|$.

5.36 (a) Let $T \in \mathcal{B}(X)$. Show that if $\ker T$ is a proper subspace, then it is hyperinvariant. Likewise show the same for $\overline{\operatorname{Ran} T}$.
(b) Suppose that $S \in \mathcal{B}(X)$ and both $\ker S$ and $(\operatorname{Ran} S)^\perp$ are non-zero and finite dimensional. If $TS - ST = F$ has rank 1, show that either T or T^* has an eigenvalue; and hence T has a hyperinvariant subspace.
HINT: Choose $0 \neq \xi \in \ker S$ and let $M = \overline{\operatorname{span}\{T^n\xi : n \geq 0\}}$. Either $\dim M < \infty$ (so T has an eigenvalue) or not. If not, $M \not\subset \ker S$. Choose n_0 smallest so that $ST^{n_0}\xi \neq 0$. Show $0 \neq FT^{n_0-1}\xi \in \operatorname{Ran} S$ (and therefore $\operatorname{Ran} F \subset \operatorname{Ran} S$). Deduce that $T\,\overline{\operatorname{Ran} S} \subset \overline{\operatorname{Ran} S}$ (so T^* has an eigenvalue).

5.37 An operator $T \in \mathcal{B}(X, Y)$ between Banach spaces is *weakly compact* if $\overline{Tb_1(X)}$ is weakly compact in Y.
(a) Show that T is weakly compact if and only if $T^{**}b_1(X^{**})$ is contained in ιY, where $\iota : Y \to Y^{**}$ is the canonical imbedding. Show that if X or Y is reflexive, then every $T \in \mathcal{B}(X, Y)$ is weakly compact.
(b) Show that the set of weakly compact operators from X to Y is norm closed.
(c) If $A \in \mathcal{B}(W, X)$, $B \in \mathcal{B}(Y, Z)$ and T is weakly compact, then so is ATB.

5.38 (a) Show that $T \in \mathcal{B}(X, Y)$ is weakly compact if and only if T^* is continuous from (Y^*, τ_Y) to $(X^*, \tau_{X^{**}})$.
(b) Show that $T \in \mathcal{B}(X, Y)$ is weakly compact if and only if T^* is weakly compact.

Notes on Chapter 5

Compact operators were defined by Hilbert, calling them *completely continuous*. He established a spectral theorem for self-adjoint operators in 1906 in the context of integral equations. Reisz established many of the structural results about compact operators. Schauder proved Theorem 5.1.8 in 1930. Stone [290] and von Neumann [314] independently established the spectral theorem for arbitrary normal operators in 1929-30. Fredholm operators were defined by Atkinson [21] in 1951, proving most of section 5.3, generalizing results of Fredholm on integral operators.

In 1954, Aronszajn and Smith [12] established that every compact operator has an invariant subspace. In 1973, Lomonosov [193] showed that every operator T which commutes with a non-zero compact operator has a hyperinvariant subspace. Ringrose [260] established Theorem 5.5.8. The proof of Burnside's theorem given here is due to Lomonosov and P. Rosenthal [194].

Conway [61] and Douglas [99] treat compact operators and Fredholm theory. See Radjavi–Rosenthal [250] and Chalendar–Partington [46] for material on invariant subspaces. Davidson [69] and Radjavi–Rosenthal [251] have treatments of Ringrose's theorem.

Part II

Banach and C*-algebras

CHAPTER 6

Banach Algebras

In this book, we will study many complex algebras which are simultaneously Banach spaces and rings with continuous operations. Some arise as algebras of linear operators, but others occur in other settings or even have an abstract definition. Many of the ideas that apply to bounded operators apply more generally. In this chapter, we look at a few examples and explore the spectrum and an analytic functional calculus.

6.1. Banach algebra basics

Recall that a $\mathbb{C}$-algebra A is an associative ring which is also a vector space over $\mathbb{C}$ such that the two operations are compatible: namely A is a $\mathbb{C}$-module. We will call it a *$\mathbb{C}$-algebra* or just an *algebra*, since we will always use $\mathbb{C}$ as our scalar field.

6.1.1. DEFINITION. A *Banach algebra* is a Banach space $\mathfrak{A}$ over $\mathbb{C}$ which is a $\mathbb{C}$-algebra such that

$$\|xy\| \le \|x\|\,\|y\| \quad \text{for all} \quad x, y \in \mathfrak{A}.$$

If $\mathfrak{A}$ has a (multiplicative) unit e, we insist that $\|e\| = 1$. In this case, we will write 1 for the unit e and λ for λe.

6.1.2. REMARK. Multiplication is jointly continuous because

$$\begin{aligned}\|x_1y_1 - x_2y_2\| &= \|x_1y_1 - x_2y_1 + x_2y_1 - x_2y_2\| \\ &\le \|x_1y_1 - x_2y_1\| + \|x_2y_1 - x_2y_2\| \\ &\le \|x_1 - x_2\|\,\|y_1\| + \|x_2\|\,\|y_1 - y_2\|.\end{aligned}$$

Thus if $x_n \to x$ and $y_n \to y$, both sequences are bounded, say by M, and hence

$$\|x_ny_n - xy\| \le M\|x_n - x\| + M\|y_n - y\| \to 0.$$

Therefore $x_ny_n \to xy$.

K. R. Davidson, *Functional Analysis and Operator Algebras*, CMS/CAIMS Books in Mathematics 13,
https://doi.org/10.1007/978-3-031-63665-3_6

6.1.3. EXAMPLE. If X is a Banach space, then $\mathcal{B}(X)$ is a Banach algebra with the operator norm $\|T\| = \sup_{\|x\|\leq 1} \|Tx\|$. If $S, T \in \mathcal{B}(X)$, then

$$\|ST\| = \sup_{\|x\|\leq 1} \|STx\| \leq \sup_{\|x\|\leq 1} \|S\| \, \|Tx\| = \|S\| \, \|T\|.$$

6.1.4. EXAMPLE. If X is a compact Hausdorff space, then $C(X)$ is a commutative Banach algebra with supremum norm $\|f\|_\infty = \sup_{x\in X} |f(x)|$. Multiplication is the usual pointwise product. Clearly $\|fg\|_\infty \leq \|f\|_\infty \|g\|_\infty$. If X is locally compact, then we define $C^b(X)$ to be the algebra of all bounded continuous functions on X with the supremum norm. Since a uniform limit of bounded continuous functions is bounded and continuous, $C^b(X)$ is complete and thus is a Banach algebra. We let $C_0(X)$ be the closure in $C^b(X)$ of the non-closed algebra $C_c(X)$ of continuous functions f on X with compact support; i.e. $\overline{f^{-1}(\mathbb{C} \setminus \{0\})}$ is compact.

6.1.5. EXAMPLE. Let $C^{(n)}[a, b]$ be the vector space of functions on $[a, b]$ with n continuous derivatives. This becomes a Banach algebra with the norm

$$\|f\|_{C^{(n)}} = \sum_{i=0}^{n} \frac{\|f^{(i)}\|_\infty}{i!}.$$

In Exercise **6.1**, you are asked to verify that this satisfies the norm inequality for products.

6.1.6. EXAMPLE. We say that G is a *locally compact abelian group* (LCA group) if it is an abelian group with a locally compact topology such that the map $(x, y) \to x + y$ is continuous from $G \times G$ into G, and the map $x \to -x$ is continuous on G. The usual examples are $(\mathbb{Z}^n, +)$, $(\mathbb{R}^n, +)$ and $(\mathbb{T}^n, \cdot)$. These groups are given their usual metric topologies.

It is a standard result from measure theory that a LCA group G has a unique translation invariant Radon measure $m = m_G$ up to a scalar multiple, called the Haar measure. That is, $m(E) = m(E + s)$ for any Borel set $E \subset G$ and $s \in G$; and that $m(K) < \infty$ for K compact. For example, see [114, §11.1]. In Section 3.9, we handle the compact case. The uniqueness of the Haar measure implies that $m(-E) = m(E)$ as well. So the map sending $f(s)$ to $f(-s)$ is isometric on $L^1(G, m)$. It is normal practice to normalize m so that if G is discrete, then points have mass 1; and if G is compact and infinite, to make $m(G) = 1$; and if $G = \mathbb{R}^n$, to take m to be Lebesgue measure. It is also normal practice to write dt instead of $dm(t)$.

Then $L^1(G) = L^1(G, m)$ is the Banach space of m-integrable functions. A product is defined on $L^1(G)$ by *convolution*.

$$f * g(t) = \int_G f(s)g(t - s)\, ds \quad \text{for} \quad f, g \in L^1(G) \text{ and } t \in G.$$

If $g \in L^1(G)$, let $g_s(t) = g(t-s)$ be a translation of g. By translation invariance, $\|g_s\|_1 = \|g\|_1$ for any $s \in G$. Then for $f, g \in L^1(G)$, using Tonelli's theorem,

$$\begin{aligned}\|f * g\|_1 &= \int_G \Big| \int_G f(s)g(t-s)\,ds \Big|\,dt \\ &\le \int_G \int_G |f(s)|\,|g(t-s)|\,ds\,dt \\ &= \int_G |f(s)| \int_G |g_s(t)|\,dt\,ds \\ &= \int_G |f(s)|\,\|g_s\|_1\,ds = \int_G |f(s)|\,\|g\|_1\,ds = \|f\|_1\|g\|_1.\end{aligned}$$

This establishes the Banach algebra norm condition. It is easy to check that multiplication is commutative. It is another exercise using Fubini's theorem to show that convolution is associative. Thus $L^1(G)$ is a commutative Banach algebra.

If G is discrete, then for each $s \in G$, the function δ_s belongs to $L^1(G)$. In this case, δ_0 is the unit. Indeed, if $f = \sum_{s\in G} a_s\delta_s$, then

$$(f * \delta_0)(t) = \sum_{s\in G} a_s\delta_0(t-s) = a_t.$$

6.1.7. Examples. A closed subalgebra of a Banach algebra is evidently both a Banach space and an algebra satisfying the Banach algebra identity; and thus is a Banach algebra. Here are some examples.

(1) The disc algebra of Example 2.2.7 consists of continuous functions on the closed unit disc which are analytic on the interior:

$$A(\mathbb{D}) = \{f \in C(\overline{\mathbb{D}}) : f|_{\mathbb{D}} \text{ is analytic}\}.$$

It is clearly a subalgebra. It is closed because a uniform limit of analytic functions on the open set $\mathbb{D}$ is analytic.

(2) If $T \in \mathcal{B}(X)$, then the unital algebra that it generates can be closed to obtain a Banach algebra.

$$\mathrm{Alg}(T) = \overline{\{p(T) : p \in \mathbb{C}[z]\}}.$$

(3) If H is a Hilbert space, then the C*-algebra of T is $C^*(T)$, the smallest closed and self-adjoint Banach subalgebra of $\mathcal{B}(H)$ containing T and I.

6.1.8. Example. If $(X, \mathcal{B}, \mu)$ is a measure space, the Banach space $L^\infty(\mu)$ of essentially bounded μ-measurable functions (modulo identification of functions which agree a.e. μ) is closed under multiplication. As in Example 4.3.8, where we did this on an interval using Lebesgue measure, for any σ-finite measure μ, we can define a *multiplication operator* M_f in $\mathcal{B}(L^2(\mu))$ for any $f \in L^\infty(\mu)$ by $M_f h = fh$. As in that example, one can verify that $\|M_f\| = \|f\|_\infty$ and

$M_f M_g = M_{fg}$ for $f, g \in L^\infty(\mu)$. Thus $L^\infty(\mu)$ is isometrically isomorphic to the subalgebra of $\mathcal{B}(L^2(\mu))$ consisting of all bounded multiplication operators. We also note that $M_f^* = M_{\bar{f}}$ is also a multiplication operator. So this algebra is self-adjoint, making it a C*-algebra.

6.1.9. REMARK. Every Banach algebra $\mathfrak{A}$ without a unit is contained in a unital Banach algebra $\widetilde{\mathfrak{A}} = \mathfrak{A} \oplus \mathbb{C}$ with multiplication

$$(a + \lambda)(b + \mu) = (ab + \lambda b + \mu a) + \lambda\mu.$$

The norm is not unique, but a convenient choice is $\|a + \lambda\| = \|a\| + |\lambda|$. Then

$$\begin{aligned}\|(a + \lambda)(b + \mu)\| &= \|ab + \lambda b + \mu a\| + |\lambda\mu| \\ &\le \|a\|\,\|b\| + |\lambda|\|b\| + |\mu|\|a\| + |\lambda|\,|\mu| \\ &= (\|a\| + |\lambda|)(\|b\| + |\mu|) = \|a + \lambda\|\,\|b + \mu\|\end{aligned}$$

Moreover $\mathfrak{A}$ is a (closed) maximal ideal in $\widetilde{\mathfrak{A}}$.

The same construction works even if $\mathfrak{A}$ has a unit, and then the unit of $\widetilde{\mathfrak{A}}$ is distinct from the unit of $\mathfrak{A}$. So while this is possible, it is not very useful.

6.1.10. PROPOSITION. *Every Banach algebra is isometrically isomorphic to a (closed) subalgebra of $\mathcal{B}(X)$ for some Banach space X.*

PROOF. If $\mathfrak{A}$ is unital, let $X = \mathfrak{A}$ and if it isn't, let $X = \widetilde{\mathfrak{A}}$. Define $L : \mathfrak{A} \to \mathcal{B}(X)$ by $a \to L_a$ where $L_a x = ax$. It is easy to check that this is linear and multiplicative.

$$\|L_a\| = \sup_{\|x\| \le 1} \|ax\| \le \sup_{\|x\| \le 1} \|a\|\,\|x\| \le \|a\|.$$

On the other hand, $\|L_a 1\| = \|a\| = \|a\|\,\|1\|$. Therefore $\|L_a\| = \|a\|$; whence L is an isometric isomorphism onto its image, a closed subalgebra of $\mathcal{B}(X)$. ∎

6.1.11. REMARK. A simple example shows why the unitization is required in this proof when $\mathfrak{A}$ is not unital. Consider $\mathfrak{A} = \mathbb{C}a$ with $a^2 = 0$ and $\|a\| = 1$. Then the map taking a to multiplication by a on $\mathfrak{A}$ is the zero map; so it is not isometric.

6.1.12. REMARK. Suppose that $\mathfrak{A}$ is a Banach algebra with a unit e which is not norm 1. Then the map $L : \mathfrak{A} \to \mathcal{B}(\mathfrak{A})$ by $L(a) = L_a$ is still linear, multiplicative and norm at most 1. Since $e = e^2$, we have $0 \ne \|e\| \le \|e\|^2$; whence $\|e\| \ge 1$. Therefore

$$\|a\| \ge \|L_a\| \ge \frac{\|L_a e\|}{\|e\|} = \frac{\|a\|}{\|e\|}.$$

Hence $\|L_a\|$ is an equivalent norm on $\mathfrak{A}$ in which $\|L_e\| = \|I\| = 1$. So every Banach algebra with a unit has a Banach algebra norm with $\|e\| = 1$.

6.2. Spectrum

The results of this section parallel the results of sections 4.3 and 4.4. The proofs are the same, so we treat this as a review and will not repeat the arguments. There are a few places where things differ, and we will try to clarify those points.

6.2.1. DEFINITION. Let $\mathfrak{A}$ be a unital Banach algebra with unit 1. For $a \in \mathfrak{A}$, the *spectrum* of a *relative to* $\mathfrak{A}$ is

$$\sigma(a) = \sigma_{\mathfrak{A}}(a) = \{\lambda \in \mathbb{C} : \lambda - a \text{ is not invertible in } \mathfrak{A}\}.$$

The *resolvent* of a is $\rho(a) = \mathbb{C} \setminus \sigma(a)$. The *resolvent function* defined for $\lambda \in \rho(a)$ is given by $R(a, \lambda) = (\lambda - a)^{-1}$.

In the case of operators on a Banach space, we could talk about parts of the spectrum: point spectrum or eigenvalues, approximate point spectrum, and compression spectrum. These definitions require a spatial representation of $\mathfrak{A}$ and are not defined for abstract Banach algebras.

6.2.2. PROPOSITION. *Let* $\mathfrak{A}$ *be an unital Banach algebra. Then*

$$\mathfrak{A}^{-1} = \{a \in \mathfrak{A} : a \text{ is invertible }\}$$

is open, and is a group under multiplication.

PROOF. The proof is the same as for operators. First we show that if $\|a\| < 1$, then

$$(1-a)^{-1} = \sum_{n \geq 0} a^n$$

where $a^0 = 1$. This series converges absolutely, and you check that

$$(1-a)\sum_{n\geq 0} a^n = 1 = \sum_{n\geq 0} a^n(1-a).$$

If $b \in \mathfrak{A}^{-1}$ and $\|a\| < 1/\|b^{-1}\|$, we may factor $b - a = b(1 - b^{-1}a)$. Since $\|b^{-1}a\| \leq \|b^{-1}\|\,\|a\| < 1$, $1 - b^{-1}a$ is invertible; whence $b - a$ is invertible. Thus the open ball $B_{1/\|b^{-1}\|}(b) \subset \mathfrak{A}^{-1}$. Therefore, $\mathfrak{A}^{-1}$ is open. The set of invertible elements is a group under multiplication, as in any ring. ■

6.2.3. COROLLARY. *For* $a \in \mathfrak{A}$*, a unital Banach algebra,* $\sigma(a)$ *is a compact subset of* $\{\lambda : |\lambda| \leq \|a\|\}$.

PROOF. Since $\mathfrak{A}^{-1}$ is open, so is $\rho(a)$. If $|\lambda| > \|a\|$, then $\lambda - a = \lambda(1 - \lambda^{-1}a)$. Since $\|\lambda^{-1}a\| < 1$, this is invertible. Thus $\sigma(a)$ is a closed subset of the compact set $\{\lambda : |\lambda| \leq \|a\|\}$; and hence it is compact. ■

6.2.4. COROLLARY. *Let $\mathfrak{A}$ be an unital Banach algebra. Then the map taking $a \in \mathfrak{A}^{-1}$ to a^{-1} is a homeomorphism and an anti-automorphism.*

PROOF. As for operators, use the power series for $(1-a)^{-1}$ to show that the inverse map is continuous at 1. Then translate the result to other points.

As for any ring, we have $(ab)^{-1} = b^{-1}a^{-1}$. So this map reverses the order of multiplication; whence it is an anti-automorphism. ■

6.2.5. COROLLARY. *Let $\mathfrak{A}$ be a unital Banach algebra. If $z \in \rho(a)$, then*

$$\operatorname{dist}(z, \sigma(a)) \geq \frac{1}{\|(z-a)^{-1}\|}.$$

PROOF. Suppose that $\lambda \in \sigma(a)$. Then since $B_{1/\|(z-a)^{-1}\|}(z-a) \subset \mathfrak{A}^{-1}$, it does not include $\lambda - a$. Therefore

$$|z-\lambda| = \|(z-a)-(\lambda-a)\| \geq \frac{1}{\|(z-a)^{-1}\|}.$$

Take the infimum over $\sigma(a)$. ■

6.2.6. EXAMPLE. Let X be a compact Hausdorff space. An element $f \in C(X)$ is invertible if and only if $f(x) \neq 0$ for all $x \in X$, for then $g(x) = 1/f(x)$ is also continuous. Therefore

$$\sigma(f) = \operatorname{Ran}(f) = \{f(x) : x \in X\}.$$

6.2.7. EXAMPLE. One difference with the operator situation is that the spectrum of an element can depend on the ambient algebra. Consider the disc algebra $A(\mathbb{D})$. For an element $f \in A(\mathbb{D})$ to have an inverse, we need some $g \in A(\mathbb{D})$ so that $fg = 1$. Clearly such a function g exists in $C(\overline{\mathbb{D}})$ if and only if f does not vanish. Let $g(z) = \dfrac{1}{f(z)}$. However since $f|_{\mathbb{D}}$ is never 0, $\dfrac{1}{f(z)}$ is also analytic on $\mathbb{D}$. Thus f is invertible in $A(\mathbb{D})$ as well. Therefore

$$\sigma_{A(\mathbb{D})}(f) = \sigma_{C(\overline{\mathbb{D}})}(f) = \operatorname{Ran}(f) = \{f(z) : |z| \leq 1\}.$$

Now consider the map from $A(\mathbb{D})$ into $C(\mathbb{T})$ by restriction, $f \to f|_{\mathbb{T}}$, where $\mathbb{T}$ is the unit circle bounding $\mathbb{D}$. Clearly this map is contractive. However it is actually isometric because the maximum modulus principle shows that $|f|$ attains its maximum on the boundary. Thus this is an isometric imbedding, and we may consider $A(\mathbb{D})$ as a subalgebra of $C(\mathbb{T})$. The element $\zeta(z) = z$ exhibits the difference in two spectra. We have

$$\sigma_{A(\mathbb{D})}(\zeta) = \zeta(\overline{\mathbb{D}}) = \overline{\mathbb{D}} \quad \text{and} \quad \sigma_{C(\mathbb{T})}(\zeta) = \zeta(\mathbb{T}) = \mathbb{T}.$$

Since the restriction to the boundary of $f \in A(\mathbb{D})$ determines $f(z)$ inside the disc, one might ask how to determine the spectrum from the boundary values. This is accomplished by the argument principle. If $f(z)$ vanishes on $\mathbb{T}$, then it isn't invertible. If it is non-zero on $\mathbb{T}$, then we can compute the number of zeros as

$$\frac{1}{2\pi i}\int_{\mathbb{T}} \frac{f'(z)}{f(z)}\,dz.$$

Here the circle $\mathbb{T}$ is considered in the counterclockwise orientation. Geometrically, this computes the winding number $\mathrm{ind}_{f(\mathbb{T})}(0)$ of the curve $f(\mathbb{T})$ around 0. Translating this to other points, we obtain for $f \in A(\mathbb{D})$,

$$\sigma_{A(\mathbb{D})}(f) = f(\mathbb{T}) \cup \{\lambda : \mathrm{ind}_{f(\mathbb{T})}(\lambda) \neq 0\}.$$

We let ∂X denote the boundary of a set X. So for $X \subset \mathbb{C}$,

$$\partial X = \overline{X} \cap \overline{\mathbb{C}\setminus X}.$$

6.2.8. PROPOSITION. *Let $\mathfrak{A}$ be a unital Banach algebra contained in a larger Banach algebra $\mathfrak{B}$ with the same unit. Then for $a \in \mathfrak{A}$,*

$$\partial\sigma_{\mathfrak{A}}(a) \subseteq \sigma_{\mathfrak{B}}(a) \subseteq \sigma_{\mathfrak{A}}(a).$$

PROOF. If a is invertible in $\mathfrak{A}$, then this same element is the inverse in $\mathfrak{B}$. Therefore $\sigma_{\mathfrak{B}}(a) \subseteq \sigma_{\mathfrak{A}}(a)$.

Now suppose $\lambda \in \partial\sigma_{\mathfrak{A}}(a)$. We can select a sequence $z_n \in \rho(a)$ converging to λ. By Corollary 6.2.5,

$$\|(z_n - a)^{-1}\| \geq \frac{1}{\mathrm{dist}(z_n, \sigma(a))} \geq \frac{1}{|z_n - \lambda|} \to +\infty.$$

If there is an element $b \in B$ so that $b(\lambda - a) = e = (\lambda - a)b$, then by Corollary 6.2.4, $b = \lim_{n\to\infty}(z_n - a)^{-1}$. This contradicts the fact that the norms of $(z_n - a)^{-1}$ are unbounded. So $\lambda \in \sigma_{\mathfrak{B}}(a)$. ∎

Now recall Definition 4.4.1 that a function $f : \Omega \to X$ from an open domain $\Omega \subset \mathbb{C}$ into a Banach space X is weakly analytic if $\varphi \circ f$ is analytic for all $\varphi \in X^*$ and is strongly analytic if for any $z_0 \in \Omega$, there are elements $x_n \in X$ and $r > 0$ so that $f(z) = \sum_{n\geq 0} x_n(z - z_0)^n$ for $|z - z_0| < r$.

The following has an identical proof to Theorem 4.4.2.

6.2.9. THEOREM. *Let $\mathfrak{A}$ be a unital Banach algebra, and let $a \in \mathfrak{A}$. Then*

(1) $\dfrac{R(a,\lambda) - R(a,\mu)}{\lambda - \mu} = -R(a,\lambda)R(a,\mu)$ *for* $\lambda, \mu \in \rho(a)$.

(2) $\lambda \to R(a,\lambda)$ *is strongly analytic.*

(3) $\lim_{|\lambda|\to\infty} R(a,\lambda) = 0$.

As for operators, this has the following important consequence.

6.2.10. THEOREM. *Let $\mathfrak{A}$ be an unital Banach algebra, and let $a \in \mathfrak{A}$. Then $\sigma(a)$ is non-empty.*

PROOF. Again the proof is the same as for operators. If $\sigma(a)$ were empty, the function $R(a, \lambda)$ would be an entire strongly analytic function. Moreover, $R(a, \lambda)$ tends to 0 as $|\lambda| \to \infty$, and therefore $R(a, \lambda)$ is bounded. For any continuous linear functional $\varphi \in \mathfrak{A}^*$, $\varphi(R(a, \lambda))$ is a bounded entire function and thus is constant by Liouville's theorem. So by the behaviour at infinity, $\varphi(R(a, \lambda)) = 0$. By the Hahn–Banach theorem, we have $R(a, \lambda) = 0$ for all $\lambda \in \mathbb{C}$, which is absurd. Hence $\sigma(a)$ is non-empty. ■

6.2.11. MAZUR THEOREM. *If $\mathfrak{A}$ is a Banach algebra and a division ring, then it is isomorphic to $\mathbb{C}$.*

PROOF. Division rings have an identity, so $\mathfrak{A}$ has a unit. If $a \in \mathfrak{A}$, then $\sigma(a)$ contains some $\lambda \in \mathbb{C}$. Thus $a - \lambda$ is not invertible. Hence $a - \lambda = 0$. Consequently, the map $\lambda \to \lambda e$ is a surjective homomorphism of $\mathbb{C}$ onto $\mathfrak{A}$. Therefore $\mathfrak{A}$ is isomorphic to $\mathbb{C}$. ■

6.2.12. DEFINITION. Let $\mathfrak{A}$ be a unital Banach algebra, and let $a \in \mathfrak{A}$. The *spectral radius* of a is

$$\operatorname{spr}(a) = \sup\{|\lambda| : \lambda \in \sigma(a)\}.$$

The analyticity of the resolvent leads to the spectral radius formula. Again the proof is identical to the operator case. Indeed, since the spectral radius only depends on the boundary of the spectrum, we could use the fact that $\mathfrak{A}$ embeds into $\mathcal{B}(X)$ for some Banach space X and apply the formula for operators.

6.2.13. THEOREM. *Let $\mathfrak{A}$ be an unital Banach algebra, and let $a \in \mathfrak{A}$. Then*

$$\operatorname{spr}(a) = \lim_{n \to \infty} \|a^n\|^{1/n}.$$

As for operators, there is a natural *rational functional calculus*. If $p(z) = \sum_{k=0}^{n} c_k z^k$ is a polynomial, then for $a \in \mathfrak{A}$, we evaluate $p(a) = \sum_{k=0}^{n} c_k a^k$. It is easy to verify that this map is a homomorphism from $\mathbb{C}[z]$ into $\mathfrak{A}$. If $q(z) = \prod_{k=1}^{m}(z - \lambda_k)$, then $q(a) = \prod_{k=1}^{m}(a - \lambda_k)$. Thus this element is invertible if and only if no λ_k belongs to $\sigma(a)$. Therefore we can define

$$\left(\tfrac{p}{q}\right)(a) = p(a)q(a)^{-1}.$$

This is defined on

$$\operatorname{Rat}(\sigma(a)) = \left\{\frac{p}{q} : p, q \in \mathbb{C}[z],\ \text{no poles in } \sigma(a)\right\}.$$

It is easy to check that this map is a homomorphism.

As for operators, we obtain a *spectral mapping theorem for rational functions*. The proof is the same as for the operator case, Theorem 4.4.5. In any case, we are going to generalize this in Section 6.4.

6.2.14. THEOREM. *Let $\mathfrak{A}$ be an unital Banach algebra, and let $a \in \mathfrak{A}$. Let $f(z) = \frac{p(z)}{q(z)} \in \operatorname{Rat}(\sigma(a))$ be a rational function with no poles in $\sigma(a)$. Then*

$$\sigma(f(T)) = f(\sigma(T)).$$

6.2.15. REMARK. (Non-unital Banach algebras) When $\mathfrak{A}$ has no unit, there are two ways to proceed. One way is to adjoin a unit as in Remark 6.1.9 to imbed $\mathfrak{A}$ as a codimension 1 ideal in an unital Banach algebra $\widetilde{\mathfrak{A}} = \mathfrak{A} \oplus \mathbb{C}$. The other way is to define a *quasi-product* $a \circ b = a + b - ab$. We say that a has the right quasi-inverse b if $a \circ b = 0$. If $a \circ b = 0 = b \circ a$, we say that a is *quasi-invertible*. Then $1 - a$ is invertible in $\widetilde{\mathfrak{A}}$ with inverse $1 - b$ if and only if b is the quasi-inverse of a. See Exercise **6.30**. We will adjoin a unit in this book.

6.3. Riemann integration in Banach space

We need to extend the Riemann integral to continuous, Banach space-valued functions. We assume the reader recalls how a Riemann integral is defined in the scalar context.

Let X be a Banach space, and let C be an oriented rectifiable curve in $\mathbb{C}$. Suppose that $f : C \to X$ is a continuous function. Parameterize C by arclength, say $\gamma : [0, L] \to \mathbb{C}$ so that the length of $\gamma([s_1, s_2])$ is $s_2 - s_1$ for $0 \le s_1 \le s_2 \le L$. Equivalently $\frac{d|\gamma|}{ds} = 1$. A partition is a finite ordered set

$$\mathcal{P} = \{0 = s_0 < s_1 < \cdots < s_n = L\}.$$

The *mesh* of $\mathcal{P}$ is $\operatorname{mesh}(\mathcal{P}) = \max\{s_i - s_{i-1} : 1 \le i \le n\}$. Let $\xi_i \in \gamma([s_{i-1}, s_i])$ for $1 \le i \le n$ be arbitrary points, and let $\Xi = \{\xi_i : 1 \le i \le n\}$ be the sample set. The corresponding Riemann sum is

$$R(\mathcal{P}, \Xi) = \sum_{i=1}^{n} f(\xi_i)\big(\gamma(s_i) - \gamma(s_{i-1})\big).$$

The *Riemann integral*

$$\int_C f(z)\,dz = x$$

is said to exist provided that for any $\varepsilon > 0$, there is a $\delta > 0$ so that for any partition $\mathcal{P}$ with $\operatorname{mesh}(\mathcal{P}) < \delta$ and any choice of sample set Ξ,

$$\|R(\mathcal{P}, \Xi) - x\| < \varepsilon.$$

If $\mathcal{P}$ and $\mathcal{Q}$ are two partitions, their common refinement $\mathcal{P} \vee \mathcal{Q}$ is the ordered union of $\mathcal{P} \cup \mathcal{Q}$, removing duplications.

6.3.1. THEOREM. *Let X be a Banach space, let $\mathcal{C}$ be an oriented rectifiable curve in $\mathbb{C}$, and let $f : \mathcal{C} \to X$ be a continuous function. Then $\displaystyle\int_{\mathcal{C}} f(z)\,dz$ exists as a Riemann integral.*

PROOF. Use the uniform continuity of f to select $\delta > 0$ so that $|s - t| < \delta$ implies $\|f(\gamma(s)) - f(\gamma(t))\| < \frac{\varepsilon}{2L}$. Let $(\mathcal{P}_j, \Xi_j)$ for $j = 1, 2$ be two partitions and sample sets with $\operatorname{mesh}(\mathcal{P}_i) < \delta$. Form the partition $\mathcal{P} = \mathcal{P}_1 \vee \mathcal{P}_2$. Then choose a sample set $\Omega = \{\omega_i\}$ for $\mathcal{P}$. We will compare each Riemann sum $R(\mathcal{P}_j, \Xi_j)$ to $R(\mathcal{P}, \Omega)$.

Write $\mathcal{P} = \{0 = s_0 < s_1 < \cdots < s_n = L\}$. Then

$$\mathcal{P}_1 = \{0 = s_0 < s_{i_1} < \cdots < s_{i_m} = s_n = L\}$$

for some sequence $0 = i_0 < i_1 < \cdots < i_m = n$. For each $i_{k-1} < i \le i_k$, the sample points ω_i and ξ_k all belong to $\gamma([s_{i_{k-1}}, s_{i_k}])$ and thus $\|f(\omega_i) - f(\xi_k)\| < \frac{\varepsilon}{2L}$. Therefore

$$\begin{aligned}
&\|R(\mathcal{P}, \Omega) - R(\mathcal{P}_1, \Xi_1)\| \\
&\quad= \Big\| \sum_{i=1}^{n} f(\omega_i)\big(\gamma(s_i) - \gamma(s_{i-1})\big) - \sum_{k=1}^{m} f(\xi_k)\big(\gamma(s_{i_k}) - \gamma(s_{i_{k-1}})\big) \Big\| \\
&\quad= \Big\| \sum_{i=1}^{n} f(\omega_i)\big(\gamma(s_i) - \gamma(s_{i-1})\big) - \sum_{k=1}^{m} f(\xi_k) \sum_{i=i_{k-1}+1}^{i_k} \big(\gamma(s_i) - \gamma(s_{i-1})\big) \Big\| \\
&\quad\le \sum_{i=1}^{n} \|f(\omega_i) - f(\xi_k)\| \big|\gamma(s_i) - \gamma(s_{i-1})\big| \\
&\quad< \frac{\varepsilon}{2L} \sum_{i=1}^{n} \big|\gamma(s_i) - \gamma(s_{i-1})\big| \le \frac{\varepsilon}{2}.
\end{aligned}$$

Similarly, $\|R(\mathcal{P}, \Omega) - R(\mathcal{P}_2, \Xi_2)\| < \frac{\varepsilon}{2}$. Hence

$$\|R(\mathcal{P}_2, \Xi_2) - R(\mathcal{P}_1, \Xi_1)\| < \varepsilon.$$

This means that the net $R(\mathcal{P}, \Xi)$ is Cauchy. Indeed, we can make a choice $\mathcal{Q}_n$ with $t_i = Li/n$ and sample set $\Omega_n = \{\omega_i = \gamma(t_i) : 1 \le i \le n\}$. With this choice, $x_n = R(\mathcal{Q}_n, \Omega_n)$ is a Cauchy sequence in X. Indeed, given $\varepsilon > 0$, we choose δ as in the previous paragraph. Taking $N = \lceil L\delta^{-1} \rceil$, we have $\operatorname{mesh}(\mathcal{Q}_n) < \delta$ if $n \ge N$. Thus $\|x_n - x_m\| < \varepsilon$ for $N \le n < m$. Let $x = \lim_{n\to\infty} x_n$ in X. Again the previous paragraph shows that given $\varepsilon > 0$,

$$\|R(P_j, \Xi_j) - x\| = \lim_{n\to\infty} \|R(P_j, \Xi_j) - R(\mathcal{Q}_n, \Omega_n)\| \le \varepsilon.$$

Therefore f is Riemann integrable with integral x. ■

6.3.2. COROLLARY. *Let X be a Banach space, let $\mathcal{C}$ be an oriented rectifiable curve in $\mathbb{C}$, and let $f : \mathcal{C} \to X$ be a continuous function. Let $\varphi \in X^*$. Then*

$$\varphi\Big(\int_{\mathcal{C}} f(z)\, dz \Big) = \int_{\mathcal{C}} \varphi(f(z))\, dz.$$

PROOF. The LHS is the limit of the terms

$$\varphi(R(\mathcal{P}, \Xi)) = \sum_{i=1}^{n} \varphi(f(\xi_i))\big(\gamma(s_i) - \gamma(s_{i-1})\big).$$

However, these terms are the Riemann sums for the scalar-valued function $\varphi(f(z))$, and these converge to the integral on the RHS. ■

6.4. Riesz–Dunford functional calculus

The purpose of this section is to find a method for evaluating $f(a)$ for an analytic function $f(z)$ defined on a neighbourhood of $\sigma(a)$. It turns out that this has an elegant solution using Cauchy's theorem. A strong functional calculus is a powerful tool. We will see some applications that are not available just using rational functions.

We recall some facts from complex analysis.

If $U \subset \mathbb{C}$ is open, let $\mathcal{O}(U)$ denote the space of holomorphic (analytic) functions on U. Suppose that $f(z) \in \mathcal{O}(U)$. If $\mathcal{C}$ is a finite union of oriented piecewise smooth closed curves in U, the *winding number* of $\mathcal{C}$ around a point $w \in \mathbb{C} \setminus \mathcal{C}$ is given by

$$\operatorname{ind}_{\mathcal{C}}(w) = \frac{1}{2\pi i} \int_{\mathcal{C}} \frac{dz}{z - w}.$$

This measures the change in argument of $z - w$ as you traverse $\mathcal{C}$, divided by 2π; and thus it is always an integer.

Suppose that $\operatorname{ind}_{\mathcal{C}}(w) = 0$ for $w \notin U$. Then Cauchy's theorem states that for $w \in U \setminus \mathcal{C}$,

$$\frac{1}{2\pi i} \int_{\mathcal{C}} \frac{f(z)}{z - w}\, dz = f(w) \operatorname{ind}_{\mathcal{C}}(w).$$

Our goal in this section is to provide a method for defining $f(a)$ if $a \in \mathfrak{A}$ and f is analytic on a neighbourhood of $\sigma(a)$. This map should be a homomorphism extending the rational functional calculus.

We will define $f(a)$ using Cauchy's theorem. Suppose that $\sigma(a) \subset U$ is open. A finite union of piecewise smooth closed curves, $\mathcal{C}$, in $U \setminus \sigma(a)$ is called *compatible* with $\sigma(a)$ if $\operatorname{ind}_C(w) = \begin{cases} 1 & \text{if } w \in \sigma(a) \\ 0 & \text{if } w \notin U \end{cases}$. Given such a curve, we define

$$f(a) = \frac{1}{2\pi i} \int_C f(z) R(a, z)\, dz.$$

These curves always exist. We will need to verify that this integral does not depend on the choice of curve C in order to consider this a proper definition.

6.4.1. THEOREM. *Let $\mathfrak{A}$ be an unital Banach algebra, and let $a \in \mathfrak{A}$. Let $f, g \in \mathcal{O}(U)$ for an open set $U \supset \sigma(a)$.*

(1) *$f(a)$ is well defined, independent of the choice of compatible curve C.*

(2) *$(f + g)(a) = f(a) + g(a)$.*

(3) *$(\lambda f)(a) = \lambda f(a)$ for $\lambda \in \mathbb{C}$.*

(4) *$(fg)(a) = f(a)g(a)$.*

(5) *If $f(z) = \sum_{n \geq 0} c_n (z - z_0)^n$ is analytic on $U = \{z : |z - z_0| < r\}$ containing $\sigma(a)$, then $f(a) = \sum_{n \geq 0} c_n (a - z_0)^n$.*

(6) *If $f = \frac{p}{q} \in \operatorname{Rat}(\sigma(a))$, then $f(a) = p(a)q(a)^{-1}$.*

PROOF. Let C_1 and C_2 be two compatible curves in $U \setminus \sigma(a)$. Let $C = C_1 - C_2$, which is the union of C_1 and C_2, but the orientation is reversed on C_2. Then C is a finite union of smooth curves with winding number 0 around every point in $\sigma(a) \cup U^c$; i.e. C is homologous to zero in $U \setminus \sigma(a)$. Take an arbitrary functional $\varphi \in \mathfrak{A}^*$ and consider

$$\varphi\Big(\frac{1}{2\pi i} \int_C f(z) R(a, z)\, dz\Big) = \frac{1}{2\pi i} \int_C f(z) \varphi(R(a, z))\, dz = 0.$$

The function $f(z)\varphi(R(a, z))$ is a scalar-valued analytic function on $U \setminus \sigma(a)$. By the version of Cauchy's theorem stated above, this integral is 0. By the Hahn–Banach theorem, the element $\frac{1}{2\pi i} \int_C f(z)(z-a)^{-1}\, dz = 0$. Hence the two integrals agree.

(2) and (3) follow from the linearity of the integral.

(4) Consider two functions f and g holomorphic on a common neighbourhood U of $\sigma(a)$. Choose a compatible curve C_1. Then let $V = \{z \in U \setminus C_1 : \operatorname{ind}_{C_1}(z) = 1\}$. This is a union of components of $U \setminus C_1$, and thus it is open and contains $\sigma(a)$. Choose a second compatible curve C_2 in $V \setminus \sigma(a)$. So in particular, $\operatorname{ind}_{C_2}(z) = 0$ for $z \in C_1$. Then

$$\begin{aligned} f(a)g(a) &= \frac{1}{2\pi i} \int_{C_1} f(z) R(a, z)\, dz \, \frac{1}{2\pi i} \int_{C_2} g(w) R(a, w)\, dw \\ &= \frac{1}{(2\pi i)^2} \int_{C_1} \int_{C_2} f(z) g(w) R(a, z) R(a, w)\, dw\, dz \end{aligned}$$

$$
\begin{aligned}
&= \frac{1}{(2\pi i)^2} \int_{\mathcal{C}_1} \int_{\mathcal{C}_2} f(z) g(w) \frac{R(a,z) - R(a,w)}{z-w}\, dw\, dz \\
&= \frac{1}{2\pi i} \int_{\mathcal{C}_1} f(z) R(a,z) \frac{1}{2\pi i} \int_{\mathcal{C}_2} \frac{g(w)}{z-w}\, dw\, dz \\
&\quad + \frac{1}{2\pi i} \int_{\mathcal{C}_2} g(w) R(a,w) \frac{1}{2\pi i} \int_{\mathcal{C}_1} \frac{f(z)}{w-z}\, dz\, dw \\
&= -\frac{1}{2\pi i} \int_{\mathcal{C}_1} f(z) R(a,z) g(z) \operatorname{ind}_{\mathcal{C}_2}(z)\, dz \\
&\quad + \frac{1}{2\pi i} \int_{\mathcal{C}_2} g(w) R(a,w) f(w) \operatorname{ind}_{\mathcal{C}_1}(w)\, dw \\
&= +\frac{1}{2\pi i} \int_{\mathcal{C}_2} f(w) g(w) R(a,w)\, dw = (fg)(a).
\end{aligned}
$$

(5) Suppose that $f(z) = \sum_{n \ge 0} c_n (z - z_0)^n$ is analytic on $U = \{z : |z - z_0| < r\}$, a disc containing $\sigma(a)$. Then we can take $\mathcal{C} = \{\rho e^{i\theta} : 0 \le \theta \le 2\pi\}$ with the counterclockwise orientation for $\operatorname{spr}(a) < \rho < r$. Now

$$
R(a, z_0 + \rho e^{i\theta}) = (\rho e^{i\theta} - (a - z_0))^{-1} = \sum_{k \ge 0} \frac{(a - z_0)^k}{(\rho e^{i\theta})^{k+1}}.
$$

Therefore, since everything is converging absolutely and uniformly,

$$
\begin{aligned}
f(a) &= \frac{1}{2\pi i} \int_{\mathcal{C}} \sum_{n \ge 0} c_n (z - z_0)^n R(a,z)\, dz \\
&= \sum_{n \ge 0} c_n \frac{1}{2\pi} \int_0^{2\pi} (\rho e^{i\theta})^n \sum_{k \ge 0} \frac{(a - z_0 e)^k}{(\rho e^{i\theta})^{k+1}} \rho e^{i\theta}\, d\theta \\
&= \sum_{n \ge 0} c_n \sum_{k \ge 0} (a - z_0)^k \frac{1}{2\pi} \int_0^{2\pi} (\rho e^{i\theta})^{n-k}\, d\theta \\
&= \sum_{n \ge 0} c_n (a - z_0)^n.
\end{aligned}
$$

In particular, this yields the correct answer for polynomials. Items (2–4) show that this map is a ring homomorphism. Hence $\left(\frac{p}{q}\right)(a) = p(a) q(a)^{-1}$. Thus the Riesz–Dunford functional calculus extends the rational functional calculus. ■

6.4.2. EXAMPLE. Let K be a compact subset of $\mathbb{C}$ and let $\{\alpha_n : n \ge 1\}$ be a countable dense subset of K. Define $D = \operatorname{diag}(\alpha_1, \alpha_2, \dots)$ as an operator on H with respect to an orthonormal basis $\{e_n : n \ge 1\}$. Then $\sigma_p(D) = \{\alpha_n : n \ge 1\}$ and $\sigma(D) = K$. (Think of D as a multiplication operator.) If $z \notin K$, then

$$
R(D, z) = \operatorname{diag}\left((z - \alpha_1)^{-1}, (z - \alpha_2)^{-1}, (z - \alpha_3)^{-1}, \dots\right).
$$

Let $U \supset K$ be open, and let $\mathcal{C}$ be an compatible curve in $U \setminus K$ for K. If $f \in \mathcal{O}(U)$, then by Cauchy's theorem,

$$f(D)e_n = \frac{1}{2\pi i}\int_{\mathcal{C}} f(z)R(D,z)e_n\,dz = \frac{1}{2\pi i}\int_{\mathcal{C}} \frac{f(z)}{z-\alpha_n}\,dz\,e_n = f(\alpha_n)e_n.$$

Therefore $f(D) = \operatorname{diag}\big(f(\alpha_1), f(\alpha_2), f(\alpha_3), \dots\big)$.

6.4.3. Example. Let $\lambda_0 \in \mathbb{C}$ and $n \geq 2$. Consider the $n \times n$ Jordan block

$$J = \begin{bmatrix} \lambda_0 & 1 & 0 & \dots & 0 & 0 \\ 0 & \lambda_0 & 1 & \ddots & \ddots & 0 \\ 0 & \ddots & \ddots & \ddots & \ddots & \vdots \\ \vdots & \ddots & \ddots & \ddots & \ddots & 0 \\ 0 & \ddots & \ddots & \ddots & \lambda_0 & 1 \\ 0 & 0 & \dots & 0 & 0 & \lambda_0 \end{bmatrix}$$

Then $\sigma(J) = \{\lambda_0\}$. If U is a neighbourhood of λ_0 and $f \in \mathcal{O}(U)$, then f has a Taylor series $f(z) = \sum_{k\geq 0} c_k(z-\lambda_0)^k$ valid on an open ball centred at λ_0 and contained in U. Recall that $c_k = \dfrac{f^{(k)}(\lambda_0)}{k!}$ for $k \geq 0$. Compute

$$f(J) = \sum_{k\geq 0} c_k(J-\lambda_0 I)^k = \begin{bmatrix} c_0 & c_1 & c_2 & \dots & c_{n-2} & c_{n-1} \\ 0 & c_0 & c_1 & \ddots & \ddots & c_{n-2} \\ 0 & \ddots & \ddots & \ddots & \ddots & \vdots \\ \vdots & \ddots & \ddots & \ddots & \ddots & \vdots \\ 0 & \ddots & \ddots & \ddots & c_0 & c_1 \\ 0 & 0 & \dots & 0 & 0 & c_0 \end{bmatrix}.$$

Next we strengthen our spectral mapping theorem from rational functions to analytic functions.

6.4.4. Spectral Mapping Theorem. *Let $\mathfrak{A}$ be an unital Banach algebra, and let $a \in \mathfrak{A}$. Let $f \in \mathcal{O}(U)$ for an open set $U \supset \sigma(a)$. Then*

$$\sigma(f(a)) = f(\sigma(a)).$$

Proof. For $\lambda \notin f(\sigma(a))$, let $g(z) = \dfrac{1}{\lambda - f(z)}$. Then $g \in \mathcal{O}(U \setminus \{f^{-1}(\lambda)\})$ and $U \setminus \{f^{-1}(\lambda)\} \supset \sigma(a)$. Observe that

$$g(a)(\lambda - f(a)) = \big(g(z)(\lambda - f(z))\big)(a) = 1 = (\lambda - f(a))g(a).$$

Thus $\lambda \notin \sigma(f(a))$.

Conversely, if $\lambda \in f(\sigma(a))$, pick $z_0 \in \sigma(a)$ so that $f(z_0) = \lambda$. Then $f(z) - \lambda$ has a zero at $z = z_0$. Thus

$$h(z) = \begin{cases} \frac{f(z)-\lambda}{z-z_0} & \text{if } z \in U \setminus \{z_0\} \\ f'(z_0) & \text{if } z = z_0 \end{cases}$$

is analytic on U. Observe that $(z - z_0)h(z) = f(z) - \lambda$. Therefore

$$f(a) - \lambda = (a - z_0)h(a).$$

The RHS is not invertible because $z_0 \in \sigma(a)$. Thus the LHS is not invertible. Therefore $\lambda \in \sigma(f(a))$. Together this shows that $\sigma(f(a)) = f(\sigma(a))$. ■

An important application of these ideas shows that a linear operator with disconnected spectrum can be split into pieces based on the spectrum.

6.4.5. THEOREM. *Let $T \in \mathcal{B}(X)$ for a Banach space X. Suppose that $\sigma(T) = \sigma_1 \dot{\cup} \sigma_2$ is a disjoint union, where each σ_i is both closed and open in $\sigma(T)$. Then there are idempotents E_i such that*

(1) $E_1 + E_2 = I$.

(2) *$X_i = E_i X$ is invariant for all $g(T)$, $g \in \mathcal{O}(V)$ for $V \supset \sigma(T)$, and in particular X_i is invariant for T.*

(3) $\sigma(T|_{X_i}) = \sigma_i$.

(4) *For $f \in \mathcal{O}(\sigma(T))$, $f(T|_{X_i}) = f(T)|_{X_i}$.*

PROOF. Since σ_1 and σ_2 are disjoint closed sets in $\mathbb{C}$, there are disjoint open sets $U_i \supset \sigma_i$; for example, if $\delta = \inf\{|z_i - z_2| : z_i \in \sigma_i\}$, then take

$$U_i = \{z \in \mathbb{C} : \text{dist}(z, \sigma_i) < \delta/2\}.$$

Select curves $\mathcal{C}_i \in U_i \setminus \sigma_i$ which are compatible for σ_i. Let $U = U_1 \cup U_2$, and let $\mathcal{C} = \mathcal{C}_1 \cup \mathcal{C}_2$. Define

$$f_1(z) = \begin{cases} 1 & \text{if } z \in U_1 \\ 0 & \text{if } z \in U_2 \end{cases}.$$

Then $f_1 \in \mathcal{O}(U)$, so that $E_1 = f_1(T)$ is defined. Since $f_1 = f_1^2$, we have that $E_1 = E_1^2$ is an idempotent. If we set $f_2(z) = 1 - f_1(z)$ and $E_2 = f_2(T) = I - E_1$, then E_2 is also an idempotent and $E_1 + E_2 = I$.

If $g \in \mathcal{O}(V)$ where $V \supset \sigma(T)$, then f_i and g are both holomorphic on $U \cap V$. Thus $g(T)E_1 = (gf_1)(T) = (f_1 g)(T) = E_1 g(T)$. Thus

$$g(T)X_1 = g(T)E_1X_1 = E_1g(T)X_1 \subset X_1.$$

Let $T_i = T|_{X_i} \in \mathcal{B}(X_i)$. Then $X = X_1 \dot{+} X_2$ is a sum of two complementary subspaces, and $T = T_1 \dot{+} T_2$, meaning that if $x = x_1 + x_2$ for $x_i \in X_i$, then $Tx = T_1x_1 \dot{+} T_2x_2$.

If $\lambda \in \rho(T)$, then $R(T,\lambda) = R_1 \dot{+} R_2$ where $R_i = R(T,\lambda)|_{X_i}$. So

$$I = R(T,\lambda)(\lambda I - T) = (\lambda I - T)R(T,\lambda)$$

becomes

$$\begin{aligned} I = E_1 \dot{+} E_2 &= (R_1 \dot{+} R_2)((\lambda E_1 - T_1) \dot{+} (\lambda E_2 - T_2)) \\ &= (R_1(\lambda E_1 - T_1)) \dot{+} (R_2(\lambda E_2 - T_2)). \end{aligned}$$

Thus $R_1(\lambda E_1 - T_1) = E_1$, and similarly $(\lambda E_1 - T_1)R_1 = E_1$. So that

$$(\lambda I_{X_1} - T_1)^{-1} = R_1 = R(T,\lambda)|_{X_1}.$$

Next, if $\lambda \in \sigma_2$, then let

$$f(z) = \begin{cases} (\lambda - z)^{-1} & \text{if } z \in U_1 \\ 0 & \text{if } z \in U_2 \end{cases}.$$

This is a holomorphic function on U. Observe that $T_1 \dot{+} 0_2 = (zf_1)(T)$. Let $S = f(T)$. Then $S = S_1 \dot{+} 0_2$. Observe that

$$S((\lambda E_1 - T_1) \dot{+} 0_2) = f(T)\big((\lambda - z)f_1\big)(T) = f_1(T) = E_1.$$

This shows that $S_1(\lambda I_{X_1} - T_1) = I_{X_1}$. Similarly $(\lambda I_{X_1} - T_1)S_1 = I_{X_1}$. Thus $\sigma(T_1) \subset \sigma_1$.

A similar analysis shows that $\sigma(T_2) \subset \sigma_2$. Finally we will get equality, $\sigma(T_i) = \sigma_i$, if we can show that

$$\sigma(T) = \sigma(T_1) \cup \sigma(T_2).$$

Indeed, if $\lambda \in \rho(T_1) \cap \rho(T_2)$, let $R_i \in \mathcal{B}(X_i)$ satisfy

$$R_i(\lambda I_{X_i} - T_i) = I_{X_i} = (\lambda I_{X_i} - T_i)R_i.$$

Let $R = R_1 \dot{+} R_2$. Then

$$RT = (R_1 \dot{+} R_2)(T_1 \dot{+} T_2) = I_{X_1} \dot{+} I_{X_2} = I = TR.$$

So $\lambda \in \rho(T)$. The result follows. ∎

6.4.6. REMARK. Suppose that $\sigma_1 = \sigma(T)$ and $\sigma_2 = \varnothing$. Then since $\sigma(T|X_2) = \varnothing$, we see that $X_2 = \{0\}$. Therefore $E_1 = I$. Thus if $\mathcal{C}$ is a compatible curve for $\sigma(T)$, we have

$$\frac{1}{2\pi i}\int_{\mathcal{C}} R(T,z)\,dz = I.$$

6.4.7. EXAMPLE. Let $X = X_1 \oplus_2 X_2$. Let I_i be the identity operator on X_i. An operator in $\mathcal{B}(X)$ can be written as a 2×2 matrix $T = \begin{bmatrix} T_{11} & T_{12} \\ T_{21} & T_{22} \end{bmatrix}$ where $T_{ij} \in B(X_j, X_i)$. Consider

$$T = \begin{bmatrix} A & B \\ 0 & D \end{bmatrix}$$

where $\sigma(A) \cap \sigma(D) = \varnothing$. Observe that if $\lambda \in \rho(A) \cap \rho(D)$, then

$$\begin{bmatrix} \lambda I_1 - A & -B \\ 0 & \lambda I_2 - D \end{bmatrix} \begin{bmatrix} (\lambda I_1 - A)^{-1} & (\lambda I_1 - A)^{-1} B (\lambda I_2 - D)^{-1} \\ 0 & (\lambda I_2 - D)^{-1} \end{bmatrix} = \begin{bmatrix} I_1 & 0 \\ 0 & I_2 \end{bmatrix}.$$

Similarly, $\lambda I - T$ has a left inverse, and hence

$$R(T, \lambda) = \begin{bmatrix} R(A, \lambda) & R(A, \lambda) B R(D, \lambda) \\ 0 & R(D, \lambda) \end{bmatrix}.$$

Therefore $\sigma(T) \subset \sigma(A) \cup \sigma(B)$,

Let U_i be disjoint open sets with $\sigma(A) \subset U_1$ and $\sigma(D) \subset U_2$. Let C_1 be a curve in $U_1 \setminus \sigma(A)$ compatible with $\sigma(A)$; and similarly let C_2 be a curve in $U_2 \setminus \sigma(D)$ compatible with $\sigma(D)$. Let $f_1(z) = 1$ on U_1 and $f_1(z) = 0$ on U_2. We compute

$$\begin{aligned} E_1 = f_1(T) &= \frac{1}{2\pi i} \int_C f_1(z) R(T, z)\, dz = \frac{1}{2\pi i} \int_{C_1} R(T, z)\, dz \\ &= \begin{bmatrix} \frac{1}{2\pi i} \int_{C_1} R(A, z)\, dz & \frac{1}{2\pi i} \int_{C_1} R(A, z) B R(D, z)\, dz \\ 0 & \frac{1}{2\pi i} \int_{C_1} R(D, z)\, dz \end{bmatrix} = \begin{bmatrix} I_1 & V \\ 0 & 0 \end{bmatrix}. \end{aligned}$$

For the $1, 1$ entry, since $\sigma(A)$ is compatible with C_1, Remark 6.4.6 shows that the integral is I_1. Similarly, since $R(D, z)$ is analytic on all of U_1, the $2, 2$ entry is 0. The $1, 2$ entry is

$$V = \frac{1}{2\pi i} \int_{C_1} R(A, z) B R(D, z)\, dz.$$

Thus

$$E_1 = \begin{bmatrix} I_1 & V \\ 0 & 0 \end{bmatrix} \quad \text{and} \quad E_2 = \begin{bmatrix} 0 & -V \\ 0 & I_2 \end{bmatrix}.$$

Next observe that

$$\begin{aligned} &A\, R(A, z) B R(D, z) - R(A, z) B R(D, z)\, D \\ &\qquad = (A - z I_1) R(A, z) B R(D, z) - R(A, z) B R(D, z) (D - z I_2) \\ &\qquad = -B R(D, z) + R(A, z) B. \end{aligned}$$

Therefore

$$\begin{aligned} AV - VD &= \frac{1}{2\pi i} \int_{C_1} R(A, z) B - B R(D, z)\, dz \\ &= \frac{1}{2\pi i} \int_{C_1} R(A, z)\, dz\, B - B \frac{1}{2\pi i} \int_{C_1} R(D, z)\, dz \\ &= I_1 B - B(0) = B. \end{aligned}$$

Now $S := \begin{bmatrix} I_1 & V \\ 0 & -I_2 \end{bmatrix}$ is its own inverse, and

$$TS = \begin{bmatrix} A & B \\ 0 & D \end{bmatrix} \begin{bmatrix} I_1 & V \\ 0 & -I_2 \end{bmatrix} = \begin{bmatrix} A & AV - B \\ 0 & -D \end{bmatrix} = \begin{bmatrix} I_1 & V \\ 0 & -I_2 \end{bmatrix} \begin{bmatrix} A & 0 \\ 0 & D \end{bmatrix}.$$

Therefore

$$S^{-1}TS = \begin{bmatrix} A & 0 \\ 0 & D \end{bmatrix}.$$

This shows that $\sigma(T) = \sigma(A \oplus D) = \sigma(A) \cup \sigma(D)$. To see the last identity, note that if $\lambda \in \sigma(A)$, then $\lambda I - A \oplus D = (\lambda I_1 - A) \oplus (\lambda I_2 - D)$. Now $\lambda I_1 - A$ is not invertible. If it has kernel or fails to be bounded below, then $\lambda I - A \oplus D$ has kernel or is not bounded below, so is not invertible. If $\lambda I_1 - A$ is not surjective, then likewise $\lambda I - A \oplus D$ does not contain X_1 in its range; so it is not invertible. Thus $\sigma(A) \subset \sigma(A \oplus D) = \sigma(T)$. Similarly $\sigma(D) \subset \sigma(T)$. Combining this with the reverse inequality establishes the identity.

We can also show that the operator V satisfying $AV - VD = B$ has a unique solution. Indeed, for $\lambda \notin \sigma(A) \cup \sigma(D)$, $V(\lambda I_2 - D) - (\lambda I_1 - A)V = B$. Therefore

$$R(A, \lambda)V - VR(D, \lambda) = R(A, \lambda)BR(D, \lambda).$$

Integrate over $\mathcal{C}_1$ and divide by $2\pi i$ to get

$$V = \frac{1}{2\pi i} \int_{\mathcal{C}_1} R(A, z)BR(D, z)\, dz.$$

6.4.8. REMARK. Note that the equality of $\sigma(T) = \sigma(A) \cup \sigma(D)$ only holds in general if $\sigma(A)$ and $\sigma(D)$ are disjoint. See Exercise **6.20**.

The argument shows that if $A \in \mathcal{B}(X)$ and $D \in \mathcal{B}(Y)$ with $\sigma(A) \cap \sigma(D) = \varnothing$, the map $\rho_{A,D}(V) = AV - VD$ on $\mathcal{B}(Y, X)$ is invertible. This is known as the *Rosenblum operator*. A simpler proof is contained in Exercise **7.18**.

It is straightforward to check that if $T \in \mathcal{B}(X)$ and f is holomorphic on a neighbourhood of $\sigma(T)$, then $f(T^*) = f(T)^*$. See Exercise **6.15**. When we are dealing with operators on Hilbert space, we wish to relate the functional calculus of T^* to that of T. Recall that for Hilbert space operators,

$$\sigma(T^*) = \sigma(T)^* = \{\bar{\lambda} : \lambda \in \sigma(T)\}.$$

6.4.9. PROPOSITION. *Let H be a Hilbert space, and let $T \in \mathcal{B}(H)$. Let U be an open neighbourhood of $\sigma(T)$, and let $f \in \mathcal{O}(U)$. Set $U^* = \{\bar{\lambda} : \lambda \in U\}$. Define $f^* \in \mathcal{O}(U^*)$ by $f^*(z) = \overline{f(\bar{z})}$. Then $f^*(T^*) = f(T)^*$.*

First we see how to take the complex conjugate of a Riemann integral.

6.4.10. LEMMA. *Let $T \in \mathcal{B}(H)$ for a Hilbert space H. Then*

$$\Big(\frac{1}{2\pi i} \int_{\mathcal{C}} f(z)R(T, z)\, dz\Big)^* = \frac{1}{2\pi i} \int_{-\overline{\mathcal{C}}} f^*(w)R(T^*, w)\, dw.$$

PROOF. Suppose that $\mathcal{C}$ is a single closed curve given by $\gamma : [a,b] \to \mathcal{C}$. Then $\overline{\gamma}(t) = \overline{\gamma(t)}$ is also a closed curve, but the orientation is reversed.

The LHS is approximated by Riemann sums

$$R(\mathcal{P}, \Xi) = \frac{1}{2\pi i} \sum_{i=1}^{n} f(\xi_i)(\xi_i I - T)^{-1}\big(\big(\gamma(s_i) - \gamma(s_{i-1})\big)$$

as mesh $\mathcal{P} \to 0$. Compute

$$\begin{aligned} R(\mathcal{P}, \Xi)^* &= \frac{-1}{2\pi i} \sum_{i=1}^{n} \overline{f(\xi_i)}(\overline{\xi_i} I - T^*)^{-1}\big(\overline{\gamma(s_i)} - \overline{\gamma(s_{i-1})}\big) \\ &= \frac{-1}{2\pi i} \sum_{i=1}^{n} f^*(\overline{\xi_i}) R(T^*, \overline{\xi_i})\big(\overline{\gamma}(s_i) - \overline{\gamma}(s_{i-1})\big). \end{aligned}$$

As the Riemann sums converge to the LHS, the adjoints converge to

$$\frac{-1}{2\pi i} \int_{\overline{\mathcal{C}}} f^*(w) R(T^*, w)\, dw = \frac{1}{2\pi i} \int_{-\overline{\mathcal{C}}} f^*(w) R(T^*, w)\, dw. \qquad \blacksquare$$

PROOF OF PROPOSITION 6.4.9. Let $\mathcal{C}$ be a curve in $U \setminus \sigma(T)$ which is compatible with $\sigma(T)$. Then $\overline{\mathcal{C}}$ is a curve in $U^* \setminus \sigma(T^*)$ and

$$\operatorname{ind}_{\overline{\mathcal{C}}}(z) = \begin{cases} 0 & \text{if } z \notin U^* \\ -1 & \text{if } z \in \sigma(T^*) \end{cases}.$$

Therefore $-\overline{\mathcal{C}}$ is a compatible curve in $U^* \setminus \sigma(T^*)$ relative to $\sigma(T^*)$. Compute using the lemma

$$f(T)^* = \Big(\tfrac{1}{2\pi i}\int_{\mathcal{C}} f(z) R(T,z)\, dz\Big)^* = \tfrac{1}{2\pi i}\int_{-\overline{\mathcal{C}}} f^*(w) R(T^*, w)\, dw = f^*(T^*).$$

$\blacksquare$

Exercises for Chapter 6

6.1 Verify that $C^{(n)}[a,b]$ is a Banach algebra with the norm

$$\|f\|_{C^{(n)}} = \sum_{i=0}^{n} \frac{\|f^{(i)}\|_\infty}{i!}.$$

HINT: to establish the norm of a product, use the Leibnitz rule for higher derivatives of a product.

6.2 Prove that convolution in $L^1(G)$ for an LCA group G is associative.

6.3 Let $w : \mathbb{N}_0 \to (0, \infty)$ such that $w(m+n) \le w(m)w(n)$ for $m, n \in \mathbb{N}_0$. Let $l_1(w)$ be the space of sequences with $\|f\|_w = \sum_{n \ge 0} |f(n)|w(n) < \infty$. For $f, g \in l_1(w)$, define $f * g(n) = \sum_{k=0}^{n} f(k)g(n-k)$. Prove that $\|f * g\|_w \le \|f\|_w \|g\|_w$, making $l_1(w)$ into a commutative Banach algebra.

6.4 Suppose that $\mathfrak{A}_\lambda$ are Banach algebras for $\lambda \in \Lambda$. Form $\mathfrak{A} = \prod_\Lambda \mathfrak{A}_\lambda$ consisting of $\{a = (a_\lambda)_\Lambda : a_\lambda \in A_\lambda, \ \|a\| := \sup_\Lambda \|a_\lambda\| < \infty\}$. Show that $\mathfrak{A}$ is a Banach algebra.

6.5 Let X be an infinite dimensional Banach space. Show that every non-zero ideal of $\mathcal{B}(X)$ contains the (non-closed) ideal of all finite rank operators.

6.6 Let $\mathfrak{A}$ be a Banach algebra and let $\mathfrak{J}$ be a closed (2-sided) ideal. Show that $\mathfrak{A}/\mathfrak{J}$ is a Banach algebra.

6.7 Let $\mathfrak{A}$ be a finite dimensional Banach algebra. Prove that $\mathfrak{A}$ is isomorphic to an algebra of matrices.
HINT: Proposition 6.1.10.

6.8 (a) Let $\mathfrak{A}$ be an unital Banach algebra. Prove that $\sigma(ab) \cup \{0\} = \sigma(ba) \cup \{0\}$ for all $a, b \in \mathfrak{A}$.
HINT: if $0 \ne \lambda \in \rho(ab)$, compute $(\lambda - ba)b(\lambda - ab)^{-1}$.
(b) Find operators $A, B \in \mathcal{B}(H)$ with $\sigma(AB) \ne \sigma(BA)$.

6.9 Let $\mathfrak{A}$ be an unital Banach algebra, and let $a, b \in \mathfrak{A}$. Show that if both ab and ba are invertible, then a and b are invertible.

6.10 Let $\mathfrak{A}$ be a non-commutative unital Banach algebra, and let $a \in \mathfrak{A}$.
(a) Show that there is a maximal commutative subalgebra $\mathfrak{B}$ of $\mathfrak{A}$ containing a. Note that $1 \in \mathfrak{B}$.
(b) Prove that $\sigma_{\mathfrak{B}}(a) = \sigma_{\mathfrak{A}}(a)$.

6.11 (*Upper semicontinuity of the spectrum.*) Let $\mathfrak{A}$ be an unital Banach algebra; and let $a \in \mathfrak{A}$. Recall that Chapter 4, Exercise **4.15** shows that the spectrum is not a continuous function on a Banach algebra.
(a) Given $\varepsilon > 0$, find $\delta > 0$ so that $\|a - b\| < \delta$ implies that $\sigma(b)$ is contained in $\sigma(a)_\varepsilon := \{z : \text{dist}(z, \sigma(a)) < \varepsilon\}$.
HINT: Enclose $\sigma(a)$ with a curve $\mathcal{C} \subset \sigma(a)_\varepsilon$. Then $R(a, z)|_{\mathcal{C}}$ is bounded.
(b) Suppose $\sigma(a) = \sigma_1 \cup \sigma_2$, where σ_i are disjoint non-empty clopen subsets of $\sigma(a)$. Enclose σ_1 inside curves $\mathcal{C}_i$ so that $\overline{\Omega_i} = \mathcal{C}_i \cup \{z : \text{ind}_{\mathcal{C}_i}(z) \ne 0\}$ are disjoint (how?). Find $\delta > 0$ so that $\|a - b\| < \delta$ implies that $\sigma(b) \cap \Omega_i \ne \varnothing$.
HINT: Use part (a) to get δ. Let $b_t = (1-t)a + tb$ for $0 \le t \le 1$. Consider the Riesz projection e_t for the operator b_t and its spectrum in Ω_1.
(c) If $\text{spr}(a) = 0$, then spr is continuous at a.

6.12 Let $\mathfrak{A}$ be an unital Banach algebra. For $a \in \mathfrak{A}$, let $L_a \in \mathcal{B}(\mathfrak{A})$ by $L_a b = ab$. Prove that L_a is invertible in $\mathcal{B}(\mathfrak{A})$ if and only if a is invertible in $\mathfrak{A}$.
HINT: if $T = L_a^{-1}$, let $b = T1$. Show that $T = L_b$.

6.13 Show that if $\operatorname{spr}(a) < 1$, then $1 - a$ is invertible.
HINT: show that the proof in Proposition 6.2.2 still works.

6.14 Let a be an element of an unital Banach algebra $\mathfrak{A}$. Suppose that $f(z)$ is analytic in a neighbourhood of $\sigma(a)$ and that $g(w)$ is analytic in a neighbourhood of $f(\sigma(a))$. Show that $(g \circ f)(a) = g(f(a))$.

6.15 Let X be a Banach space and let $T \in \mathcal{B}(X)$. Suppose that f is analytic on a neighbourhood of $\sigma(T)$. Prove that $f(T^*) = f(T)^*$.

6.16 Let S be the unilateral shift on l_2 with respect to the basis $\{e_n : n \geq 0\}$. Let $f(z) = \sum_{n\geq 0} a_n z^n$ be analytic on $r\mathbb{D}$ for some $r > 1$. Find the matrix form for $f(S)$ with respect to this basis.

6.17 Let $\mathfrak{A}$ be a non-unital Banach algebra. Let $a \in \mathfrak{A}$ and let f be analytic on a neighbourhood of $\sigma(a)$. Then $f(a) \in \widetilde{\mathfrak{A}}$. When does $f(a)$ belong to $\mathfrak{A}$?

6.18 Let $h \in L^\infty(0, 1)$, and let M_h be the corresponding multiplication operator on $L^2(0, 1)$. Let $f(z)$ be holomorphic on a neighbourhood of $\sigma(M_h)$. Describe $f(M_h)$ as a multiplication operator.

6.19 Let $a \in \mathfrak{A}$, where $\mathfrak{A}$ is an unital Banach algebra. Define $e^a = \sum_{n\geq 0} \frac{1}{n!} a^n$.
(a) If $ab = ba$, show that $e^{a+b} = e^a e^b$. Thus e^a is invertible.
(b) Give an example with $ab \neq ba$ such that $e^{a+b} \neq e^a e^b$.
HINT: look in M_2.
(c) Let $a \in \mathfrak{A}^{-1}$ such that 0 is in the unbounded component of the resolvent. Prove that there is an element $b \in \mathfrak{A}$ such that $a = e^b$. (This includes special case $\|1 - a\| < 1$.)
HINT: define a logarithm and use Exercise **6.14**.

6.20 Let S be the unilateral shift on l_2 with respect to the basis $\{e_n : n \geq 0\}$. Let $P = e_0 e_0^*$ be the rank 1 projection onto $\mathbb{C}e_0$. Let $T = \begin{bmatrix} S^* & 0 \\ 0 & S \end{bmatrix}$ and $U = \begin{bmatrix} S^* & 0 \\ P & S \end{bmatrix}$. Find $\sigma(T)$ and $\sigma(U)$. Compare this with Example 6.4.7.

6.21 An element $a \in \mathfrak{A}$ is a *left topological divisor of zero* if there are $b_n \in \mathfrak{A}$ with $\|b_n\| = 1$ such that $\lim_n \|ab_n\| = 0$; and a *joint topological divisor of zero* if there are $b_n \in \mathfrak{A}$ with $\|b_n\| = 1$ such that $\lim_n \|ab_n\| = 0 = \lim_n \|b_n a\|$.
(a) Show that if a lies in the boundary of $\mathfrak{A}^{-1}$, then a is a joint topological divisor of zero.
HINT: if a_n are invertible with limit a which is not invertible, show that $\lim_n \|a_n^{-1}\| = \infty$.
(b) Show that if a is a left topological divisor of zero in $\mathfrak{A}$, then it is not invertible in any Banach algebra containing $\mathfrak{A}$.

6.22 Let $A = \{z \in \mathbb{C} : \frac{1}{2} \leq |z| \leq 2\}$. Let $\zeta(z) = z$ be the identity function. Let $\mathfrak{A}$ be the closed subalgebra of $C(A)$ generated by 1 and ζ; and let $\mathfrak{B}$ be the closed subalgebra generated by $1, \zeta$ and ζ^{-1}. Find $\sigma_{\mathfrak{A}}(\zeta)$ and $\sigma_{\mathfrak{B}}(\zeta)$.

6.23 (a) Let $\mathfrak{A}_0^{-1}$ denote the set of all invertible elements of a Banach algebra $\mathfrak{A}$ which may be written as a finite product of exponentials e^{b_i}. Show that this is a normal subgroup of the group $\mathfrak{A}^{-1}$ of invertible elements of $\mathfrak{A}$.

(b) Show that $\mathfrak{A}_0^{-1}$ consists of all invertible elements which can be connected to the identity by a continuous path of invertible elements. Hence conclude that it is an open subgroup (and thus also closed).
HINT: Exercise **6.19**(c).

(c) Let $\mathfrak{A} = C(\mathbb{T})$. Describe $C(\mathbb{T})_0^{-1}$ and $C(\mathbb{T})^{-1}/C(\mathbb{T})_0^{-1}$.

6.24 Let $\mathfrak{A}$ be a Banach algebra. A linear map D from $\mathfrak{A}$ to itself is a *derivation* if $D(ab) = aD(b) + D(a)b$ for $a, b \in \mathfrak{A}$.

(a) Prove the Leibnitz rule: for $n \geq 1$ and $a, b \in \mathfrak{A}$,

$$D^n(ab) = \sum_{i=0}^{n} \binom{n}{i} D^i(a)\, D^{n-i}(b).$$

(b) Suppose that D is bounded. Show that $\alpha(a) = e^D(a) = \sum_{n\geq 0} \frac{1}{n!} D^n(a)$ is a continuous automorphism of $\mathfrak{A}$.

6.25 Let $\mathfrak{A}$ be a Banach algebra. A *left approximate identity* is a net $(e_\lambda)_\Lambda$ in $\mathfrak{A}$ such that $\lim_\Lambda e_\lambda a = a$ for all $a \in \mathfrak{A}$. It is *bounded* if the net is bounded. Similarly define a *right approximate identity*. Show that if $(e_\lambda)_\Lambda$ is a bounded left approximate identity and $(f_\mu)_M$ is a bounded right approximate identity, then $a_{\lambda,\mu} = e_\lambda + f_\mu - f_\mu e_\lambda$ is a bounded two-sided approximate identity.

6.26 Show that $L^1(\mathbb{R})$ has the bounded approximate identity $e_n = \frac{n}{2}\chi_{(-1/n,1/n)}$.

6.27 Let $\mathfrak{A}$ be a Banach algebra. Say $\mathfrak{A}$ has *left approximate units* if for every $a \in \mathfrak{A}$ and $\varepsilon > 0$, there is a $u \in \mathfrak{A}$ so that $\|ua - a\| < \varepsilon$. Also $\mathfrak{A}$ has *bounded left approximate units* if there is $M < \infty$ so that one can take $\|u\| \leq M$ as above. Prove that the following are equivalent:

(1) $\mathfrak{A}$ has left approximate units bounded by M.

(2) For $\varepsilon > 0$ and $a_1, \ldots, a_n \in \mathfrak{A}$, there is a $u \in \mathfrak{A}$ with $\|u\| \leq M$ so that $\|ua_i - a_i\| < \varepsilon$ for $1 \leq i \leq n$.

(3) $\mathfrak{A}$ has a left approximate identity bounded by M.

HINT: for (1) $\Rightarrow$ (2), successively choose $u_i \in \mathfrak{A}$ with $\|u_i\| \leq M$ so that $\|(1-u_j)\cdots(1-u_1)a_j\| < \frac{\varepsilon}{2(1+M)^{n+1-i}}$. Set $1 - v = (1-u_n)\cdots(1-u_1)$ and choose u so that $\|u\| \leq M$ and $\|uv - v\| < \varepsilon\big(2\max_i\{\|a_i\|\}\big)^{-1}$. Note that $1 \in \widetilde{\mathfrak{A}}$ but $v \in \mathfrak{A}$. Use $x_i - ux_i = (x_i - vx_i) + (v - uv)x_i + u(vx_i - x_i)$.

6.28 A *normed algebra* $\mathfrak{A}$ is a normed $\mathbb{C}$-vector space which is also a ring satisfying $\|ab\| \leq \|a\|\,\|b\|$ for $a, b \in \mathfrak{A}$. Show that $\mathfrak{A}$ is a dense subalgebra of a Banach algebra.
HINT: let X be the Banach space completion of $\mathfrak{A}$. Show that the multiplication on $\mathfrak{A}$ extends to a multiplication on X.

6.29 Show that the algebra $\mathbb{C}[x]$ of polynomials can be a normed algebra, but it cannot be complete in such a norm.
HINT: Baire category theorem.

6.30 Let $\mathfrak{A}$ be a non-unital Banach algebra. Define a quasi-inverse as in Remark 6.2.15. Define $\sigma(a) = \{0\} \cup \{\lambda \in \mathbb{C}^* : \lambda^{-1}a \text{ is not quasi-invertible}\}$.
(a) Prove that this coincides with the spectrum of a in $\widetilde{\mathfrak{A}}$.
(b) Show that the set of quasi-invertible elements is open.

Notes on Chapter 6

Von Neumann began the study of *rings of operators* in [314]. Murray and von Neumann wrote a definitive series of papers [211–213, 317] about what are now called von Neumann algebras. These were concrete subalgebras of $\mathcal{B}(H)$. Nagumo [214] initiated a study of abstract Banach algebras under the name *linear metric rings*. However it was Gelfand [119] who began the systematic study of complete normed algebras. Gelfand established the spectral radius formula, non-emptiness of the spectrum and other basic properties.

Riesz developed the functional calculus in his study of compact operators, recognizing that the resolvent function was analytic. Dunford [103] expanded the functional calculus as described here.

See Dunford and Schwarz [104, 105] for the material in this chapter and extended historical remarks. Bonsall and Duncan [37], Conway [61] and Kaniuth [172] cover this material as well.

CHAPTER 7

Commutative Banach Algebras

7.1. Multiplicative linear functionals

In this chapter, we develop the special features of commutative Banach algebras. This first definition is valid in the non-commutative case, but its importance is in relation to commutative algebras.

7.1.1. DEFINITION. If $\mathfrak{A}$ is a Banach algebra, a *non-zero* linear map $\varphi : \mathfrak{A} \to \mathbb{C}$ is a *multiplicative linear functional* if $\varphi(xy) = \varphi(x)\varphi(y)$ for all $x, y \in \mathfrak{A}$.

7.1.2. PROPOSITION. *Let $\mathfrak{A}$ be a Banach algebra, and let φ be a multiplicative linear functional. Then φ is continuous with $\|\varphi\| \leq 1$.*

If $\mathfrak{A}$ has a unit, then $\varphi(1) = 1 = \|\varphi\|$.

PROOF. If $\|\varphi\| > 1$, pick $x \in \mathfrak{A}$ with $\|x\| < |\varphi(x)|$. Set $y = x/\varphi(x)$ and $z = \sum_{n\geq 1} y^n$, which converges since $\|y\| < 1$. Note that $z = y + yz$. Applying φ, we get

$$\varphi(z) = \varphi(y) + \varphi(y)\varphi(z) = 1 + \varphi(z).$$

This yields a contradiction. Thus $\|\varphi\| \leq 1$.

If $\mathfrak{A}$ has a unit, then take some $x \in \mathfrak{A}$ with $\varphi(x) \neq 0$. Then

$$\varphi(x) = \varphi(x1) = \varphi(x)\varphi(1)$$

forces $\varphi(1) = 1$. Hence $\|\varphi\| = 1$. ■

7.1.3. THEOREM. *Let $\mathfrak{A}$ be a commutative Banach algebra with unit. There is a bijective correspondence between the set of multiplicative linear functionals of $\mathfrak{A}$ and the set of maximal ideals of $\mathfrak{A}$ by $\varphi \to \ker \varphi$.*

PROOF. If φ is a multiplicative functional, then $M = \ker \varphi$ is a closed ideal of codimension 1, and thus is maximal.

Conversely, suppose that M is a maximal ideal of $\mathfrak{A}$. Since M is a proper ideal, it contains no invertible elements, and thus it does not intersect $b_1(1)$. Therefore, the closure $\overline{M}$ does not intersect $b_1(1)$. It is readily checked that $\overline{M}$ is a proper closed ideal containing M. Since M is maximal, it is closed. Therefore,

K. R. Davidson, *Functional Analysis and Operator Algebras*, CMS/CAIMS Books in Mathematics 13,
https://doi.org/10.1007/978-3-031-63665-3_7

$\mathfrak{B} = \mathfrak{A}/M$ is a unital commutative Banach algebra. Since M is maximal, $\mathfrak{B}$ has no proper ideals. For any $a \in \mathfrak{A} \setminus M$, $\dot{a} = a + M$ generates the non-zero ideal $\mathfrak{B}\dot{a}$ in $\mathfrak{B}$. This must equal $\mathfrak{B}$, and hence $\dot{a}$ is invertible. Thus $\mathfrak{B}$ is a field. By Mazur's theorem 6.2.11, $\mathfrak{B} = \mathbb{C}$. Hence, the quotient map $\varphi : \mathfrak{A} \to \mathfrak{A}/M \simeq \mathbb{C}$ is a multiplicative linear functional with $\ker \varphi = M$. Moreover, φ is uniquely determined by M since $\varphi(1) = 1$ and $\mathfrak{B} = \mathbb{C}\dot{1}$. ■

7.1.4. DEFINITION. If $\mathfrak{A}$ is a commutative Banach algebra with unit, the *maximal ideal space* of $\mathfrak{A}$ is the set $\mathcal{M}(\mathfrak{A})$ of all multiplicative linear functionals on $\mathfrak{A}$ endowed with the $\tau_{\mathfrak{A}}$ (weak-$*$) topology. The *Gelfand transform* is the map $\Gamma : \mathfrak{A} \to C(\mathcal{M}(\mathfrak{A}))$ given by

$$\Gamma(a)(\varphi) = \varphi(a) \quad \text{for} \quad a \in \mathfrak{A} \text{ and } \varphi \in \mathcal{M}(\mathfrak{A}).$$

We now establish the main structure theorem for unital commutative Banach algebras.

7.1.5. GELFAND THEOREM. *Let $\mathfrak{A}$ be a commutative Banach algebra with unit. Then $\mathcal{M}(\mathfrak{A})$ is a non-empty compact Hausdorff space. Γ is a norm decreasing algebra homomorphism. The subalgebra $\Gamma(\mathfrak{A})$ separates points of $C(\mathcal{M}(\mathfrak{A}))$.*

PROOF. The collection of proper ideals of $\mathfrak{A}$ is non-empty since $\{0\}$ is an ideal. Order the set of ideals by inclusion. If $\mathcal{C} = \{J_\alpha\}$ is a (nested) chain of proper ideals, let $J = \bigcup J_\alpha$. This is an ideal containing every $J_\alpha \in \mathcal{C}$. Moreover, it is proper because each J_α is disjoint from $b_1(\mathfrak{A})$, and thus so is the union. Therefore, J is an upper bound for $\mathcal{C}$. By Zorn's lemma, there are maximal ideals. So $\mathcal{M}(\mathfrak{A})$ is non-empty.

Next we show that $\mathcal{M}(\mathfrak{A})$ is weak-$*$ closed. Suppose that a net of multiplicative linear functionals $(\varphi_\lambda)_\Lambda$ converges weak-$*$ to $\varphi \in \mathfrak{A}^*$. Then for $a, b \in \mathfrak{A}$,

$$\varphi(ab) = \lim_\Lambda \varphi_\lambda(ab) = \lim_\Lambda \varphi_\lambda(a)\varphi_\lambda(b) = \varphi(a)\varphi(b).$$

That is, φ is multiplicative. The unit ball of $\mathfrak{A}^*$ is weak-$*$ compact by the Banach–Alaoglu theorem. Since $\mathcal{M}(\mathfrak{A})$ is a weak-$*$ closed subset of the unit ball, it is compact and Hausdorff.

By definition of the weak-$*$ topology, each $\hat{a} = \Gamma(a)$ is a continuous function. The map Γ is contractive since each $\varphi \in \mathcal{M}(\mathfrak{A})$ has norm 1. As each φ is a homomorphism into $\mathbb{C}$, Γ is also a homomorphism:

$$\widehat{ab}(\varphi) = \varphi(ab) = \varphi(a)\varphi(b) = \hat{a}(\varphi)\hat{b}(\varphi) \quad \text{for} \quad a, b \in \mathfrak{A} \text{ and } \varphi \in \mathcal{M}(\mathfrak{A}).$$

Finally, $\Gamma(\mathfrak{A})$ separates points of $\mathcal{M}(\mathfrak{A})$ because $\mathfrak{A}$ separates points of $\mathfrak{A}^*$. ■

Next we show that the Gelfand transform preserves the spectrum.

7.1.6. THEOREM. *Let $\mathfrak{A}$ be a commutative Banach algebra with unit, and let $a \in \mathfrak{A}$.*

(1) *a is invertible in $\mathfrak{A}$ if and only if $\hat{a}$ is invertible in $C(\mathcal{M}(\mathfrak{A}))$ if and only if $\hat{a}$ does not vanish on $\mathcal{M}(\mathfrak{A})$.*

(2) $\sigma(a) = \sigma(\hat{a}) = \operatorname{Ran}(\hat{a})$.

(3) $\|\hat{a}\| = \operatorname{spr}(a)$.

PROOF. If a^{-1} exists in $\mathfrak{A}$, then $\hat{a}^{-1} = \Gamma(a^{-1})$. Hence, $\hat{a}$ does not vanish on $\mathcal{M}(\mathfrak{A})$. Conversely, if a is not invertible, then $J = \mathfrak{A}a$ is a proper ideal of $\mathfrak{A}$, and thus is contained in a maximal ideal M. Let φ be the multiplicative linear functional with $\ker\varphi = M$. Then $\varphi(a) = 0$, so that $\hat{a}$ has a zero in $\mathcal{M}(\mathfrak{A})$, and thus is not invertible in $C(\mathcal{M}(\mathfrak{A}))$.

(2) and (3) are immediate consequences of (1). ∎

This first corollary is immediate.

7.1.7. COROLLARY. *Let $\mathfrak{A}$ be a commutative Banach algebra with unit. Then*

$$\ker\Gamma = \{a \in \mathfrak{A} : \sigma(a) = \{0\}\} = \{a \in \mathfrak{A} : \lim_{n\to\infty} \|a^n\|^{1/n} = 0\}.$$

7.1.8. COROLLARY. *Let $\mathfrak{A}$ be a commutative Banach algebra with unit. Then Γ is isometric if and only if $\|a^2\| = \|a\|^2$ for all $a \in \mathfrak{A}$.*

PROOF. If $\|\hat{a}\| = \operatorname{spr}(a) = \|a\|$, then

$$\|\hat{a}\| = \operatorname{spr}(a) = \lim_{n\to\infty} \|a^{2^n}\|^{1/2^n} = \|a\|.$$

Since the quantities $\|a^{2^n}\|^{1/2^n}$ are decreasing, we must have $\|a^2\|^{1/2} = \|a\|$. Conversely, if this holds for all $a \in \mathfrak{A}$, then $\|a^{2^n}\| = \|a\|^{2^n}$ and the spectral radius formula shows that $\|\hat{a}\| = \operatorname{spr}(a) = \|a\|$. ∎

7.1.9. DEFINITION. A commutative Banach algebra is *semisimple* if Γ is injective. The kernel of Γ is called the *radical* of $\mathfrak{A}$, $\operatorname{rad}(\mathfrak{A}) = \ker\Gamma$.

7.1.10. EXAMPLE. Let $\mathfrak{A} = C(X)$ for a compact Hausdorff space X. For each $x \in X$, the point evaluation $\varepsilon_x(f) = f(x)$ is clearly multiplicative. Conversely, if $M \lhd C(X)$ is a maximal ideal, then M is closed. Let

$$\ker M = \{x \in X : f(x) = 0 \text{ for all } f \in M\}.$$

This cannot be more than one point, for then it would be properly contained in $\ker\varepsilon_x$ for any $x \in \ker M$. If it is a singleton $\{x\}$, then $M = \{f : f(x) = 0\}$ and the corresponding functional is ε_x. Finally if $\ker M$ is empty, then for each $x \in X$, select $f_x \in M$ with $f_x(x) = 1$. Then $\{U_x = f_x^{-1}(\{z : |z| > 1/2\}) : x \in X\}$ is an open cover of X. Let $U_{x_1}, \dots, U_{x_n}$ be a finite subcover. Then M contains

$$g = \sum_{i=1}^{n} f_{x_i} \overline{f_{x_i}} \geq \frac{1}{4}.$$

Thus g is invertible, which forces $M = C(X)$, a contradiction. Hence $\mathcal{M}(C(X))$ is identified with X as a set. The map $\varepsilon : X \to \mathcal{M}(C(X))$ by $\varepsilon(x) = \varepsilon_x$ is continuous from X to $\mathcal{M}(C(X))$ since if $x_\alpha \to x$ in X, then

$$\varepsilon_{x_\alpha}(f) = f(x_\alpha) \to f(x) = \varepsilon_x(f).$$

So $\varepsilon_{x_\alpha} \to \varepsilon_x$ weak-$*$. Thus ε is a continuous bijection. Since both spaces are compact and Hausdorff, ε is a homeomorphism. Therefore $\mathcal{M}(C(X)) = X$. Thus the Gelfand map is the identity map. So $C(X)$ is semisimple.

7.1.11. Example. Let $\mathfrak{A}$ be the set of $2n \times 2n$ matrices of the form $\left[\begin{smallmatrix} \lambda I_n & A \\ 0_n & \lambda I_n \end{smallmatrix}\right]$ where $\lambda \in \mathbb{C}$ and $A \in M_n(\mathbb{C})$. It is easy to check that this is commutative. Note that if $\varphi \in \mathcal{M}(\mathfrak{A})$, then

$$\varphi\Big(\left[\begin{smallmatrix} 0_n & A \\ 0_n & 0_n \end{smallmatrix}\right]\Big)^2 = \varphi\Big(\left[\begin{smallmatrix} 0_n & A \\ 0_n & 0_n \end{smallmatrix}\right]^2\Big) = \varphi(0) = 0.$$

So

$$\ker\varphi \supseteq M_0 := \Big\{ \left[\begin{smallmatrix} 0_n & A \\ 0_n & 0_n \end{smallmatrix}\right] : A \in M_n(\mathbb{C}) \Big\}.$$

Since M_0 is a maximal ideal, there is a unique multiplicative linear functional $\varphi_0\Big(\left[\begin{smallmatrix} \lambda I_n & A \\ 0_n & \lambda I_n \end{smallmatrix}\right]\Big) = \lambda$. This algebra has a large radical, $\operatorname{rad}\mathfrak{A} = M_0$.

7.1.12. Example. Recall Example 6.1.6. If we take our group to be $\mathbb{Z}$ with the discrete topology, we can make $l_1(\mathbb{Z})$ into a Banach algebra by using convolution as the multiplication. In this setting, if $f, g \in l_1(\mathbb{Z})$, let

$$f * g(n) = \sum_{k=-\infty}^{\infty} f(k)g(n-k) \quad \text{for} \quad n \in \mathbb{Z}.$$

Let δ_k be the standard basis for $l_1(\mathbb{Z})$ given by $\delta_k(n) = \delta_{kn}$. We showed that δ_0 is a unit for $l_1(\mathbb{Z})$. Also it is easy to check that $\delta_k * \delta_l = \delta_{k+l}$ for $k, l \in \mathbb{Z}$. So $l_1(\mathbb{Z})$ is a unital commutative Banach algebra.

Let φ be a multiplicative linear functional on $l_1(\mathbb{Z})$. Set $\lambda = \varphi(\delta_1)$. Then $|\lambda| \leq \|\delta_1\|_1 = 1$. Since $\delta_k = \delta_1^k$ for $k \in \mathbb{Z}$, we have $\varphi(\delta_k) = \lambda^k$. In particular, $\varphi(\delta_{-1}) = \lambda^{-1}$, so that $|\lambda^{-1}| \leq \|\delta_{-1}\|_1 = 1$. Hence $\lambda \in \mathbb{T} = \{\lambda \in \mathbb{C} : |\lambda| = 1\}$. Therefore

$$\varphi(f) = \varphi_\lambda(f) = \sum_{k \in \mathbb{Z}} f(k)\lambda^k.$$

This converges absolutely for any $\lambda \in \mathbb{T}$, and $\|\varphi_\lambda\| = 1$. It is clear that φ_λ is multiplicative on the algebraic linear span of $\{\delta_k : k \in \mathbb{Z}\}$. Since φ_λ is a continuous linear functional, it extends to be multiplicative on $l_1(\mathbb{Z})$. One readily checks that if $\lambda_i \to \lambda$ in $\mathbb{T}$, then $\varphi_{\lambda_i}(f) \to \varphi_\lambda(f)$ for $f \in l_1(\mathbb{Z})$. Arguing as in the previous example, the topology on $\mathcal{M}(l_1) = \mathbb{T}$ is the usual one.

The map $\Gamma(f) = \hat{f}$ is a continuous function on $\mathbb{T}$ with Fourier series $(f(n))$. Thus $\hat{f}(\lambda) = \sum_{k\in\mathbb{Z}} f(k)\lambda^k$ is a function with absolutely convergent Fourier series, and conversely every continuous function with absolutely convergent Fourier series is the image of some $f \in l_1(\mathbb{Z})$. This map is injective because we can recover the Fourier coefficients by integration:

$$f(n) = \frac{1}{2\pi i}\int_0^{2\pi} \hat{f}(e^{i\theta})e^{-in\theta}\,d\theta \quad \text{for} \quad n \in \mathbb{Z}.$$

Hence $l_1(\mathbb{Z})$ is semisimple.

Let $A(\mathbb{T})$ denote the algebra of continuous functions on $\mathbb{T}$ with absolutely convergent Fourier series with the norm

$$\|h\|_{A(\mathbb{T})} = \sum_{n\in\mathbb{Z}} |a_n| \quad \text{where } h \sim \sum_{n\in\mathbb{Z}} a_n e^{in\theta}$$

is its Fourier series. Then considering Γ as a map from $l_1(\mathbb{Z})$ onto $A(\mathbb{T})$, we have an isometric isomorphism.

An element $f \in l_1(\mathbb{Z})$ is invertible if and only if $\hat{f}(\lambda) \neq 0$ for all $\lambda \in \mathbb{T}$. Translating this to $A(\mathbb{T})$, we obtain the following famous result of Wiener.

7.1.13. WIENER THEOREM. *If $f \in C(\mathbb{T})$ has an absolutely convergent Fourier series and does not vanish on $\mathbb{T}$, then $1/f$ also has an absolutely convergent Fourier series.*

This result was very difficult to prove before the theory of commutative Banach algebras was developed. This elementary proof by Gelfand of this formerly difficult result was a convincing argument for the importance of Gelfand's new theory.

7.1.14. EXAMPLE. The disc algebra $A(\mathbb{D})$ is a commutative unital Banach algebra. For any $\lambda \in \overline{\mathbb{D}}$, the functional $\varepsilon_\lambda(f) = f(\lambda)$ is multiplicative. Conversely, suppose that φ is a multiplicative linear functional and let $\lambda = \varphi(z)$, where $z(\zeta) = \zeta$ is the identity function on $\overline{\mathbb{D}}$. Then $|\lambda| \leq \|z\| = 1$. Moreover $\ker\varphi$ contains $M = \overline{(z-\lambda)A(\mathbb{D})}$. For $\lambda \in \mathbb{D}$, any $g \in A(\mathbb{D})$ such that $g(\lambda) = 0$ factors as $g(z) = (z-\lambda)h(z)$ for some $h \in A(\mathbb{D})$. Thus $M = \ker\varepsilon_\lambda$ and so $\varphi = \varepsilon_\lambda$.

If $\lambda \in \mathbb{T}$, the closure is needed. Suppose $f \in A(\mathbb{D})$ such that $f(\lambda) = 0$. For $r \in (0,1)$ sufficiently close to 1, $f_r(z) = f(rz)$ satisfies $\|f - f_r\| < \varepsilon/2$. Let $g(z) = f_r(z) - f_r(\lambda)$ for $z \in \frac{1}{r}\overline{\mathbb{D}}$. Then, in particular, $g \in A(\mathbb{D})$, and

$$\|f - g\| \leq \|f - f_r\| + |f_r(\lambda)| < \varepsilon.$$

Since $g(\lambda) = 0$ and g is holomorphic on a neighbourhood of $\overline{\mathbb{D}}$, the argument of the previous paragraph shows that $g(z) = (z - \lambda)h(z)$ for some $h \in A(\mathbb{D})$. Thus $g \in M$. As $\varepsilon > 0$ was arbitrary, we have $f \in M$. Thus as before, $M = \ker \varepsilon_\lambda$ and so $\varphi = \varepsilon_\lambda$. Therefore $\mathcal{M}(A(\mathbb{D})) = \overline{\mathbb{D}}$. The Gelfand map is the identity map. So $A(\mathbb{D})$ is semisimple.

A similar analysis works for $l_1(\mathbb{Z}^+)$. This is a closed subalgebra of $l_1(\mathbb{Z})$. Arguing as for the disc algebra, we find that $\mathcal{M}(l_1(\mathbb{Z}^+)) = \overline{\mathbb{D}}$. The Gelfand map Γ maps $l_1(\mathbb{Z}^+)$ onto $\{h \in A(\mathbb{D}) : h|_\mathbb{T} \in A(\mathbb{T})\}$.

7.2. The Non-unital case

In this section, we extend the results of the previous section to the non-unital case. In this section, $\mathfrak{A}$ will be non-unital and commutative. We could adjoin a unit and apply Gelfand's theorem (which we will do), but first we take an intrinsic approach.

7.2.1. DEFINITION. An ideal $I \lhd \mathfrak{A}$ is *modular* if $\mathfrak{A}/I$ has a unit. That is, there is a $u \in \mathfrak{A}$ so that $a - au \in I$ for all $a \in \mathfrak{A}$. We say that u is a unit mod I. The ideal I is *maximal modular* if it is modular, and maximal among all modular ideals.

7.2.2. REMARKS. The definition of modular ideal makes sense for any Banach algebra, unital or not, and commutative or not; though we will generally talk about modular *left* ideals in the non-commutative case.

If $\mathfrak{A}$ is unital, then every proper ideal is modular.

If I is a modular ideal and u is a unit mod I, then u is also a unit mod J for any proper ideal $J \supset I$. Therefore, every ideal containing I is modular. Thus, every maximal ideal containing I is maximal modular. In particular, the maximal modular ideals are precisely the maximal ideals which are modular.

In the unital case, we showed that maximal ideals exist by using the fact that all proper ideals are disjoint from $b_1(e)$. That argument is not available in the non-unital case, but a simple trick makes it work.

7.2.3. PROPOSITION. *Let $\mathfrak{A}$ be a commutative Banach algebra, and let $I \lhd \mathfrak{A}$ be a modular ideal. Then I is contained in a maximal modular ideal.*

PROOF. Let u be a unit modulo I. Therefore $\mathfrak{A}/I$ is a Banach algebra with unit $e = u + I$. It may not be norm 1, but by Remark 6.1.12 there is an equivalent norm $|||\cdot|||$ on $\mathfrak{A}/I$ with $|||e||| = 1$. By Gelfand's theorem, $\mathfrak{A}/I$ contains a maximal ideal N. Thus the inverse image $M = q^{-1}(N)$, where q is the quotient map, is a proper maximal ideal in $\mathfrak{A}$ containing I. Hence M is modular, so is a maximal modular ideal containing I. ∎

7.2.4. Theorem. *Let $\mathfrak{A}$ be a non-unital commutative Banach algebra. The map $\varphi \to \ker\varphi$ yields a natural bijection between multiplicative linear functionals and maximal modular ideals of $\mathfrak{A}$.*

Proof. Let φ be a (non-zero) multiplicative linear functional on $\mathfrak{A}$, and let $M = \ker\varphi$. Since φ is continuous by Proposition 7.1.2, M is a closed ideal of codimension 1. In particular it is maximal. Since $\mathfrak{A}/M \simeq \mathbb{C}$ has a unit, M is modular.

Conversely, suppose that M is a maximal modular ideal with modular unit u. Then $\mathfrak{A}/M$ is a Banach algebra with unit $e = u + M$. Let $\varphi : \mathfrak{A} \to \mathfrak{A}/M$ be the quotient map. If $a \notin M$ but $\varphi(a)$ is not invertible, then $J = \varphi(a)\mathfrak{A}/M$ is a proper ideal containing $\varphi(a) \neq 0$. Thus $\varphi^{-1}(J)$ is a proper ideal of $\mathfrak{A}$ containing M. This contradicts the maximality of M, and hence $\varphi(a)$ must be invertible. That is, $\mathfrak{A}/M$ is a Banach field. By Mazur's theorem 6.2.11, $\mathfrak{A}/M \simeq \mathbb{C}$. After making this identification, the quotient map φ is a multiplicative linear functional with kernel M. Since $\varphi(u) = 1$ is necessary, φ is uniquely determined by M. So this correspondence is bijective. ∎

Now we can extend Gelfand's theorem to the non-unital case. As before, we let $\mathcal{M}(\mathfrak{A})$ denote the set of multiplicative linear functionals on $\mathfrak{A}$ endowed with the $\tau_{\mathfrak{A}}$ (weak-$*$) topology. Recall Exercise **1.20** about the one-point compactification of an LCH space.

7.2.5. Theorem. *Let $\mathfrak{A}$ be a non-unital commutative Banach algebra, and let $\widetilde{\mathfrak{A}}$ be its unitization. Let $\widetilde{\varphi}_\infty : \widetilde{\mathfrak{A}} \to \widetilde{\mathfrak{A}}/\mathfrak{A} \simeq \mathbb{C}$. Then the restriction map from $\widetilde{\mathfrak{A}}$ to $\mathfrak{A}$ yields a homeomorphism of $\mathcal{M}(\widetilde{\mathfrak{A}}) \setminus \{\widetilde{\varphi}_\infty\}$ onto $\mathcal{M}(\mathfrak{A})$. Moreover $\mathcal{M}(\mathfrak{A})$ is a locally compact Hausdorff space, and $\mathcal{M}(\widetilde{\mathfrak{A}})$ is its one-point compactification.*

The Gelfand transform Γ carries $\mathfrak{A}$ into $C_0(\mathcal{M}(\mathfrak{A}))$, and the image separates points of $\mathcal{M}(\mathfrak{A})$.

Proof. If $\widetilde{\varphi} \in \mathcal{M}(\widetilde{\mathfrak{A}})$, then $\varphi = \widetilde{\varphi}|_{\mathfrak{A}}$ is multiplicative. Only $\widetilde{\varphi}_\infty$ restricts to the zero map, and hence is not in $\mathcal{M}(\mathfrak{A})$.

Conversely, suppose that $\varphi \in \mathcal{M}(\mathfrak{A})$, and let $M = \ker\varphi$. Define

$$\widetilde{\varphi}(a + \lambda) = \varphi(a) + \lambda.$$

It is easy to check that this is a homomorphism and hence lies in $\mathcal{M}(\widetilde{\mathfrak{A}})$. Clearly $\widetilde{\varphi}|_{\mathfrak{A}} = \varphi$. Moreover $\widetilde{\varphi}$ is unique since $\widetilde{\varphi}(1) = 1$ is necessary. Therefore, we have a bijection between $\mathcal{M}(\widetilde{\mathfrak{A}}) \setminus \{\widetilde{\varphi}_\infty\}$ and $\mathcal{M}(\mathfrak{A})$ via the restriction map.

In $\mathcal{M}(\mathfrak{A})$, a basic open neighbourhood of a point φ_0 has the form

$$U(\varphi_0, F) = \{\varphi \in \mathcal{M}(\mathfrak{A}) : |\varphi(a) - \varphi_0(a)| < 1 \text{ for } a \in F\}$$

for F a finite subset of $\mathfrak{A}$. Similarly, a basic open neighbourhood of $\widetilde{\varphi}_0$ is

$$V(\widetilde{\varphi}_0, G) = \{\widetilde{\varphi} \in \mathcal{M}(\widetilde{\mathfrak{A}}) : |\widetilde{\varphi}(b) - \widetilde{\varphi}_0(b)| < 1 \text{ for } b \in G\}$$

for G a finite subset of $\widetilde{\mathfrak{A}}$. If the elements of G are $b_i = a_i + \lambda_i$ for $a_i \in \mathfrak{A}$ and $\lambda_i \in \mathbb{C}$, for $1 \leq i \leq n$, then observe that

$$|\widetilde{\varphi}(b_i) - \widetilde{\varphi}_0(b_i)| = |\widetilde{\varphi}(a_i + \lambda_i) - \widetilde{\varphi}_0(a_i + \lambda_i)| = |\widetilde{\varphi}(a_i) - \widetilde{\varphi}_0(a_i)|.$$

Thus setting $F = \{a_i : 1 \leq i \leq n\}$, we have $V(\widetilde{\varphi}_0, G) = V(\widetilde{\varphi}_0, F)$. Moreover,

$$V(\widetilde{\varphi}_0, F) = \begin{cases} \{\widetilde{\varphi} : \varphi \in U(\varphi_0, F)\} & \text{if } \exists a \in F \text{ s.t. } |\varphi_0(a)| \geq 1 \\ \{\widetilde{\varphi} : \varphi \in U(\varphi_0, F)\} \cup \{\widetilde{\varphi}_\infty\} & \text{if } |\varphi_0(a)| < 1 \text{ for all } a \in F. \end{cases}$$

Thus if we identify $\mathcal{M}(\mathfrak{A})$ with $\mathcal{M}(\widetilde{\mathfrak{A}}) \setminus \{\widetilde{\varphi}_\infty\}$, we have that

$$V(\widetilde{\varphi}_0, F) \cap \mathcal{M}(\mathfrak{A}) = U(\varphi_0, F).$$

This shows that the topology on $\mathcal{M}(\mathfrak{A})$ is induced from the topology of $\mathcal{M}(\widetilde{\mathfrak{A}})$ by removing the point $\widetilde{\varphi}_\infty$. If $\varphi_0(u) = 1$, then $\overline{V(\widetilde{\varphi}_0, \{2u\})}$ is a compact subset of $\mathcal{M}(\widetilde{\mathfrak{A}})$ disjoint from $\{\widetilde{\varphi}_\infty\}$. Therefore φ_0 has a compact neighbourhood in $\mathcal{M}(\mathfrak{A})$. Hence, $\mathcal{M}(\mathfrak{A})$ is locally compact and $\mathcal{M}(\widetilde{\mathfrak{A}})$ is its one-point compactification.

The Gelfand transform of $a \in \mathfrak{A}$ is the restriction of the Gelfand transform on $\widetilde{\mathfrak{A}}$ to $\mathcal{M}(\mathfrak{A})$. If $a \in \mathfrak{A}$, then $\hat{a}(\widetilde{\varphi}_\infty) = 0$. Hence $\hat{a}|_{\mathcal{M}(\mathfrak{A})}$ vanishes at infinity, so lies in $C_0(\mathcal{M}(\mathfrak{A}))$. Distinct functionals are separated on $\mathfrak{A}$ by definition. ∎

7.3. Semisimplicity and automatic continuity

Automatic continuity refers to a phenomenon in which algebraic conditions on a map between Banach algebras force the map to be continuous. Results of this nature point to a strong link between the algebra and the analysis of Banach algebras. One of the simplest results of this type is the following.

7.3.1. THEOREM. *Suppose that $\mathfrak{A}$ and $\mathfrak{B}$ are Banach algebras, and that $\mathfrak{B}$ is commutative and semisimple. Then every homomorphism $\theta : \mathfrak{A} \to \mathfrak{B}$ is continuous.*

PROOF. For every $\varphi \in \mathcal{M}(\mathfrak{B})$, $\varphi \circ \theta$ is a multiplicative linear functional on $\mathfrak{A}$. Hence it is continuous. Now we apply the Closed Graph theorem. Suppose that $a_n \in \mathfrak{A}$ such that $a_n \to 0$ and $\theta(a_n) \to b$. Then for each $\varphi \in \mathcal{M}(\mathfrak{B})$,

$$\varphi(b) = \lim_{n\to\infty} \varphi(\theta(a_n)) = \lim_{n\to\infty} \varphi \circ \theta(a_n) = 0.$$

Therefore $b \in \bigcap_{\varphi \in \mathcal{M}(\mathfrak{B})} \ker \varphi = \operatorname{rad} \mathfrak{B} = \{0\}$. Since $b = 0$, the Closed Graph theorem implies that θ is continuous. ■

We provide two interesting applications of this result. The first is the *uniqueness of norm* for commutative semisimple Banach algebras.

7.3.2. COROLLARY. *Let $\mathfrak{A}$ be a commutative semisimple Banach algebra. Then any two Banach algebra norms on $\mathfrak{A}$ are equivalent. Also every automorphism of $\mathfrak{A}$ is continuous.*

PROOF. Suppose that $|\!|\!|\cdot|\!|\!|$ is another Banach algebra norm on $\mathfrak{A}$. Then the identity map $id : (\mathfrak{A}, |\!|\!|\cdot|\!|\!|) \to (\mathfrak{A}, \|\cdot\|)$ is an algebra isomorphism. By Theorem 7.3.1, id is continuous. Since it is a bijection, Banach's Isomorphism theorem shows that the inverse map is also continuous. Hence the two norms are equivalent.

For an automorphism, this is an immediate application of Theorem 7.3.1. ■

7.3.3. COROLLARY. *There is no Banach algebra norm on $C^\infty[0, 1]$.*

PROOF. Suppose that $\|\cdot\|$ is a Banach algebra norm on $C^\infty[0, 1]$. The natural injection J of $C^\infty[0, 1]$ into $C[0, 1]$ is a monomorphism. Since $C[0, 1]$ is semisimple (see Example 7.1.10), J is continuous. Thus with $C = \|J\|$, we have $\|f\|_\infty \leq C\|f\|$ for all $f \in C^\infty[0, 1]$.

Next we show that the derivation operator $Df = f'$ is continuous on $C^\infty[0, 1]$. Indeed, D is everywhere defined. So suppose that $f_n \to 0$ and $f_n' \to g$ in $(C^\infty[0, 1], \|\cdot\|)$. By the previous paragraph, $\|f_n\|_\infty \to 0$ and $\|f_n' - g\|_\infty \to 0$. Now suppose that $0 \leq x < y \leq 1$. Compute

$$\begin{aligned}
\Big|\int_x^y g(t)\,dt\Big| &= \Big|\int_x^y f_n'(t) + (g(t) - f_n'(t))\,dt\Big| \\
&\leq \Big|\int_x^y f_n'(t)\,dt\Big| + \int_x^y \|f_n' - g\|_\infty\,dt \\
&= |f_n(y) - f_n(x)| + (y - x)\|f_n' - g\|_\infty \\
&\leq 2\|f_n\|_\infty + \|f_n' - g\|_\infty \to 0.
\end{aligned}$$

Therefore $\displaystyle\int_x^y g(t)\,dt = 0$ for all $0 \leq x < y \leq 1$. Since g is continuous, the Fundamental Theorem of Calculus implies that $g = 0$. By the Closed Graph theorem, D is continuous.

Let $f(x) = e^{2\|D\|x}$. Then $Df = f' = 2\|D\|f$. Hence

$$2\|D\|\|f\| = \|f'\| \leq \|D\|\,\|f\|.$$

Since $f \neq 0$ and $D \neq 0$, this yields a contradiction. Therefore no such Banach algebra norm exists. ■

7.4. $\mathbf{L}^1(\mathbf{G})$

In the next two sections, we explore the important algebras $L^1(G)$ for LCA groups. In particular, we show that the maximal ideal space is another LCA group. Recall from Example 6.1.6 that if G is a locally compact abelian group, then $L^1(G)$ is a commutative Banach algebra.

7.4.1. DEFINITION. If G is an LCA group, a *character* is a continuous group homomorphism $\gamma : G \to \mathbb{T}$. If γ and δ are characters, then $(\gamma\delta)(x) = \gamma(x)\delta(x)$ and $\gamma^{-1}(x) = \gamma(x)^{-1}$ are also characters. The set $\widehat{G}$ consisting of all characters on G is called the *dual group* of G.

Note that there is always the trivial character $\gamma_0(s) = 1$ for all $s \in G$. Since $\widehat{G}$ is closed under products and inverses, it is an abelian group. Soon we will put a topology on it to make it a locally compact group.

Recall that the translation of an element $f \in L^1(G)$ by $s \in G$ is defined as $f_s(t) = f(t - s)$.

7.4.2. LEMMA. *If G is an LCA group and $f \in L^1(G)$, then $s \to f_s$ is continuous.*

PROOF. Recall that $C_c(G)$ is dense in $L^1(G)$. Thus given $\varepsilon > 0$, we can choose $g \in C_c(G)$ with $\|f - g\| < \varepsilon/3$. Now g is a continuous function with compact support K, and hence it is uniformly continuous. Thus there is a neighbourhood W of 0 in G so that $\|g_s - g\|_\infty < \varepsilon/6m(K)$ for $s \in W$. Therefore

$$\|g_s - g\|_1 \le \|g_s - g\|_\infty m(K \cup (K + s)) < \frac{\varepsilon}{6m(K)}(2m(K)) = \frac{\varepsilon}{3}.$$

Hence for $s \in W$,

$$\|f_s - f\|_1 \le \|(f - g)_s\|_1 + \|g_s - g\|_1 + \|g - f\|_1 < \frac{\varepsilon}{3} + \frac{\varepsilon}{3} + \frac{\varepsilon}{3} = \varepsilon.$$

This shows that translation is continuous at $s = 0$. Hence it is continuous on G. ∎

7.4.3. THEOREM. *Let G be a locally compact abelian group. Then every character $\gamma \in \widehat{G}$ determines a multiplicative linear functional on $L^1(G)$ by*

$$\varphi_\gamma(f) = \int f(s)\overline{\gamma(s)}\, ds.$$

The map taking γ to φ_γ is a bijection of $\widehat{G}$ onto $\mathcal{M}(L^1(G))$.

PROOF. Since γ is continuous and bounded, it lies in $L^\infty(G)$ and $\|\varphi_\gamma\| = \|\gamma\|_\infty = 1$. For $f, g \in L^1(G)$, compute

$$\begin{aligned}\varphi_\gamma(f * g) &= \int_G \overline{\gamma(s)} f * g(s)\, ds \\ &= \int_G \overline{\gamma(s)} \int_G f(t) g(s-t)\, dt\, ds\end{aligned}$$

Now $\overline{\gamma(s)} f(t) g(s-t)$ belongs to $L^1(G \times G, m^2)$. Thus by Fubini's theorem,

$$= \int_G f(t)\overline{\gamma(t)} \int_G g(s-t)\overline{\gamma(s-t)}\, ds\, dt$$

and hence by translation invariance,

$$= \int_G f(t)\overline{\gamma(t)} \int_G g(u)\overline{\gamma(u)}\, du\, dt = \varphi_\gamma(f)\varphi_\gamma(g).$$

Therefore φ_γ is multiplicative. If γ and δ are distinct characters, then $\|\gamma - \delta\|_\infty > 0$ and hence $\varphi_\gamma \neq \varphi_\delta$. So $\gamma \to \varphi_\gamma$ is an injective map of $\widehat{G}$ into $\mathcal{M}(L^1(G))$.

Conversely, let $\varphi \in \mathcal{M}(L^1(G))$. Since $\|\varphi\| \leq 1$, there is a function $\chi \in L^\infty(G)$ with $\|\chi\|_\infty \leq 1$ with $\varphi(f) = \int_G f(s)\chi(s)\, ds$. Fix $g \in L^1(G)$ such that $\varphi(g) = 1$. Then for $f \in L^1(G)$, using Fubini's theorem again,

$$\begin{aligned}\int_G f(s)\chi(s)\, ds &= \varphi(f) = \varphi(f)\varphi(g) = \varphi(f * g) \\ &= \int_G \chi(s)(f * g)(s)\, ds = \int_G \chi(s) \int_G f(t) g(s-t)\, dt\, ds \\ &= \int_G f(t) \int_G g_t(s)\chi(s)\, ds\, dt = \int_G f(t)\, \varphi(g_t)\, dt.\end{aligned}$$

Since this identity holds for all $f \in L^1(G)$, we have $\chi(s) = \varphi(g_s)$ a.e.(m). By Lemma 7.4.2, $\varphi(g_s)$ is continuous. Let $\gamma(s) = \overline{\varphi(g_s)}$.

Next we need an identity: substitute $v = x + s$ and use translation invariance

$$\begin{aligned}(g * g_{s+t})(u) &= \int_G g(x) g_{s+t}(u-x)\, dx = \int_G g(x) g(u-x-s-t)\, dx \\ &= \int_G g(v-s) g(u-v-t)\, dv \\ &= \int_G g_s(v) g_t(u-v)\, dv = (g_s * g_t)(u).\end{aligned}$$

Therefore

$$\begin{aligned}\gamma(s+t) &= \overline{\varphi(g_{s+t})} = \overline{\varphi(g)\varphi(g_{s+t})} = \overline{\varphi(g * g_{s+t})} \\ &= \overline{\varphi(g_s * g_t)} = \overline{\varphi(g_s)\varphi(g_t)} = \gamma(s)\gamma(t).\end{aligned}$$

So γ is multiplicative on G. In particular, $\gamma(-s) = \gamma(s)^{-1}$. Moreover,

$$|\gamma(s)| = |\varphi(g_s)| \leq \|g_s\|_1 = \|g\|_1$$

is bounded. Therefore for any $s \in G$,

$$|\gamma(s)|^n = |\gamma(s^n)| \leq \|g\|_1 \quad \text{for all} \quad n \in \mathbb{Z}.$$

It follows that $|\gamma(s)| = 1$. Thus γ is a character. We have

$$\varphi(f) = \int f(s)\overline{\gamma(s)}\, ds = \varphi_\gamma(f) \quad \text{for} \quad f \in L^1(G).$$

Therefore $\varphi = \varphi_\gamma$. ■

We have identified $\mathcal{M}(L^1(G))$ with a group $\widehat{G}$. If we put the topology of $\mathcal{M}(L^1(G))$ on $\widehat{G}$, it becomes a locally compact space. It will be compact if $L^1(G)$ is unital, which we know includes (but actually equals) the case in which G is discrete. We now want to identify this topology more precisely, and show that $\widehat{G}$ is a locally compact group.

7.4.4. DEFINITION. The Gelfand map $\Gamma : L^1(G) \to C_0(\widehat{G})$ is called the *Fourier transform*

$$\Gamma f(\gamma) = \hat{f}(\gamma) = \int_G f(s)\overline{\gamma(s)}\, ds.$$

Let $A(\widehat{G}) = \Gamma(L^1(G))$ be the *Fourier algebra* of $\widehat{G}$.

We collect some easy but useful properties of the Fourier transform and $A(\widehat{G})$.

7.4.5. THEOREM.

(1) *If $f, g \in L^1(G)$, then $\widehat{f * g}(\gamma) = \hat{f}(\gamma)\hat{g}(\gamma)$.*

(2) *If $f \in L^1(G)$, then $\|\hat{f}\|_\infty \leq \|f\|_1$.*

(3) *If $f \in L^1(G)$ and $t \in G$, then $\hat{f_t}(\gamma) = \overline{\gamma(t)}\hat{f}(\gamma)$.*

(4) *If $f \in L^1(G)$ and $\gamma_1 \in \widehat{G}$, then $(\widehat{\gamma_1 f})(\gamma) = (\hat{f})_{\gamma_1}(\gamma)$.*

(5) *$A(\widehat{G})$ is invariant under translations by $\widehat{G}$, and thus the topology of $\widehat{G}$ is translation invariant.*

(6) *If $f \in L^1(G)$, let $f^*(s) = \overline{f(-s)}$ for $s \in G$. Then $\widehat{f^*}(\gamma) = \overline{\hat{f}(\gamma)}$.*

(7) *$A(\widehat{G})$ is a dense self-adjoint subalgebra of $C_0(\widehat{G})$.*

PROOF. (1) follows since the Gelfand map is a homomorphism. Next note that

$$\|\hat{f}\|_\infty = \sup_{\gamma \in \widehat{G}} |\varphi_\gamma(\hat{f})| \leq \|f\|_1$$

because each φ_γ has norm 1. So (2) holds. For (3), compute using translation invariance of Haar measure

$$\begin{aligned}\hat{f}_t(\gamma) &= \int_G f(s-t)\overline{\gamma(s)}\,ds = \int_G f(u)\overline{\gamma(u+t)}\,du \\ &= \overline{\gamma(t)} \int_G f(u)\overline{\gamma(u)}\,du = \overline{\gamma(t)}\hat{f}(\gamma).\end{aligned}$$

For (4), compute

$$\begin{aligned}\widehat{(\gamma_1 f)}(\gamma) &= \int_G \gamma_1(s)f(s)\overline{\gamma(s)}\,ds = \int_G f(s)\overline{\gamma(s)\gamma_1^{-1}(s)}\,ds \\ &= \hat{f}(\gamma\gamma_1^{-1}) = (\hat{f})_{\gamma_1}(\gamma).\end{aligned}$$

Therefore, $A(\widehat{G})$ is invariant under translation by $\widehat{G}$. The topology on $\widehat{G}$ is the weak-$*$ topology, which is the weakest topology that makes every $\hat{f} \in A(\widehat{G})$ continuous. Thus this topology is also translation invariant, proving (5). For (6), we again compute

$$\begin{aligned}\widehat{f^*}(\gamma) &= \int_G f^*(s)\overline{\gamma(s)}\,ds = \int_G \overline{f(-s)\gamma(s)}\,ds \\ &= \overline{\int_G f(-s)\overline{\gamma(-s)}\,ds} = \overline{\hat{f}(\gamma)}.\end{aligned}$$

Here we use the fact that $m(-E) = m(E)$ so that the change of variables from s to $-s$ has Jacobian 1. This shows that $A(\widehat{G})$ is closed under complex conjugation. Since it is a self-adjoint subalgebra of $C_0(\widehat{G})$ which separates points, the Stone–Weierstrass theorem shows that it is dense in $C_0(\widehat{G})$. ∎

7.4.6. EXAMPLE. The case of $G = \mathbb{Z}$ was examined in Example 7.1.12. Here $\widehat{\mathbb{Z}} = \mathbb{T}$ with its usual topology. To see this, we know that $\hat{\delta}_1 = z$ is continuous, so that the topology on $\mathbb{T}$ contains the usual metric topology. But then the identity map from $\widehat{\mathbb{Z}}$ onto $\mathbb{T}$ is a continuous bijection of compact Hausdorff spaces. Therefore, it is a homeomorphism. The algebra $A(\mathbb{T})$ is the algebra of continuous functions with absolutely convergent Fourier series.

Consider $G = \mathbb{Z}^n$ with generators g_i for $1 \le i \le n$. A character γ on $\mathbb{Z}^n$ is then determined by $\lambda_i = \gamma(g_i) \in \mathbb{T}$. So $\widehat{\mathbb{Z}^n}$ is identified with $\mathbb{T}^n$. A similar argument shows that it is endowed with the usual metric topology.

7.4.7. EXAMPLE. Let $G = \mathbb{T}$. Observe that $\gamma_k(z) = z^k$ is a character on $\mathbb{T}$ for any $k \in \mathbb{Z}$. The corresponding multiplicative linear functional is

$$\hat{f}(k) = \frac{1}{2\pi}\int_0^{2\pi} f(e^{it})e^{-ikt}\,dt \quad \text{for} \quad f \in L^1(\mathbb{T}),$$

which computes the usual Fourier coefficients.

Let $f_n(z) = z^n \in C(\mathbb{T}) \subset L^1(\mathbb{T})$ for $n \in \mathbb{Z}$. Compute

$$f_m * f_n(z) = \frac{1}{2\pi}\int_0^{2\pi} f_m(e^{it}) f_n(ze^{-it})\,dt = \frac{1}{2\pi}\int_0^{2\pi} e^{int} z^n e^{-int}\,dt$$
$$= \frac{f_n(z)}{2\pi}\int_0^{2\pi} e^{i(m-n)t}\,dt = \delta_{mn} f_n(z).$$

Therefore, each f_n is an idempotent in $L^1(\mathbb{T})$ and they annihilate each other.

Suppose that γ is a character on $\mathbb{T}$ for a multiplicative linear functional φ_γ. Then $\varphi_\gamma(f_n) = \varphi_\gamma(f_n)^2$ and $\varphi_\gamma(f_m)\varphi_\gamma(f_n) = 0$ if $m \neq n$. Therefore $\varphi_\gamma(f_n)$ is 0 or 1 and at most one is non-zero. They cannot all be 0 because $\operatorname{span}\{f_n : n \in \mathbb{Z}\}$ is dense in $L^1(\mathbb{T})$, and this would make $\varphi_\gamma = 0$. Thus, there is a unique integer k so that $\varphi_\gamma(f_k) = 1$ and $\varphi_\gamma(f_n) = 0$ for $n \neq k$. Thus, φ_γ acts by computing the kth Fourier coefficient, and so $\gamma = \gamma_k$. This shows that $\widehat{\mathbb{T}} = \mathbb{Z}$.

7.4.8. Example. Consider $G = \mathbb{R}$. We will show that $\widehat{\mathbb{R}} = \mathbb{R}$. First observe that if $s \in \mathbb{R}$, then $\gamma_s(t) = e^{ist}$ is a character on $\mathbb{R}$. Conversely, suppose that $\gamma : \mathbb{R} \to \mathbb{T}$ is continuous and multiplicative. Then $\gamma(0) = 1$. Fix some $\delta > 0$ so that $c = \displaystyle\int_0^\delta \gamma(s)\,ds \neq 0$. Then

$$c\gamma(t) = \gamma(t)\int_0^\delta \gamma(s)\,ds = \int_0^\delta \gamma(t+s)\,ds = \int_t^{t+\delta} \gamma(s)\,ds.$$

Since γ is continuous, the RHS is differentiable by the Fundamental Theorem of Calculus. Hence γ is differentiable, and

$$c\gamma'(t) = \gamma(t+\delta) - \gamma(t) = \gamma(t)\big(\gamma(\delta) - 1\big).$$

Therefore $\gamma(t) = Ce^{ist}$ for $s = \dfrac{\gamma(\delta)-1}{ic}$. Since $\gamma(0) = 1$, we have $C = 1$ and since the range of γ is contained in $\mathbb{T}$, $s \in \mathbb{R}$. This shows that $\widehat{\mathbb{R}} = \mathbb{R}$ as a set.

If $s_i \to s$ in $\mathbb{R}$, then e^{is_it} converges weak-$*$ to e^{ist} in $L^\infty(\mathbb{R}) = L^1(\mathbb{R})^*$. Thus $\hat{f}(s_i) \to \hat{f}(s)$ for all $f \in L^1(\mathbb{R})$. Hence φ_{s_i} converges to φ_s in $\mathcal{M}(L^1(\mathbb{R}))$.

The Fourier transform of $f \in L^1(\mathbb{R})$ is the function

$$\hat{f}(s) = \varphi_s(f) = \int_{\mathbb{R}} f(t)e^{-ist}\,dt \quad \text{for} \quad s \in \mathbb{R}.$$

The previous paragraph shows that $\hat{f}(s)$ is continuous on $\mathbb{R}$ in its usual topology. Recall the *Riemann–Lebesgue lemma*: *if* $f \in L^1(\mathbb{R})$, *then* $\displaystyle\lim_{s\to\pm\infty} \hat{f}(s) = 0$.

One way to prove this is to first approximate f by a function $g \in C_c(\mathbb{R})$ in the $L^1(\mathbb{R})$ norm, say within $\varepsilon/2$. Then approximate g by a $C^1_c(\mathbb{R})$ function h in the uniform norm, say within $\varepsilon/2k$ where $K = [-k, k]$ is a compact set on which g and h are supported. Then $\|g - h\|_1 < \varepsilon/2$; so that $\|f - h\|_1 < \varepsilon$. Now compute using integration by parts

$$\hat{h}(s) = \int_{-k}^{k} h(t)e^{-ist}\,dt = \frac{h(t)e^{-ist}}{-is}\Big|_{-k}^{k} + \frac{1}{is}\int_{-k}^{k} h'(t)e^{-ist}\,dt = \frac{1}{is}\widehat{h'}(s).$$

Thus

$$|\hat{h}(s)| \le \frac{|\widehat{h'}(s)|}{|s|} \le \frac{\|h'\|_1}{|s|} \to 0 \quad \text{as } |s| \to \infty.$$

If $|\hat{h}(s)| < \varepsilon$ for $|s| > N$, then

$$|\hat{f}(s)| \le |\hat{h}(s)| + |(\widehat{f-h})(s)| < \varepsilon + \|f-h\|_1 < 2\varepsilon \quad \text{for} \quad |s| > N.$$

Therefore $\lim_{s\to\pm\infty} \hat{f}(s) = 0$. Hence $\hat{f} \in C_0(\mathbb{R})$.

Conversely, if φ_{s_i} converges to φ_s pointwise on $\mathcal{M}(L^1(\mathbb{R}))$, then e^{is_it} converges weak-$*$ to e^{ist} in $L^\infty(\mathbb{R})$. The Riemann–Lebesgue lemma shows that if $|s_i| \to \infty$, then e^{is_it} converges weak-$*$ to 0. If s_i had more than one limit point in $\mathbb{R}$, then we would have subsequences converging weak-$*$ to distinct points, which is impossible. Thus s_i must converge to s in $\mathbb{R}$. Therefore, the two topologies coincide.

7.4.9. Definition. The *compact-open topology* defined on $\widehat{G}$ is given by a neighbourhood base of open sets

$$V(K,U) = \{\gamma \in \widehat{G} : \gamma(K) \subset U\} \quad \text{for} \quad K \subset G \text{ compact and } U \subset \mathbb{T} \text{ open}.$$

In particular, a neighbourhood base of the identity element γ_0 is given by

$$V_0(K,r) = \{\gamma \in \widehat{G} : |\gamma(s) - 1| < r \text{ for } s \in K\} \quad \text{for} \quad K \subset G \text{ compact}, \ r > 0.$$

We now show that the Gelfand topology coincides with the compact-open topology, and makes $\widehat{G}$ an LCA group.

7.4.10. Theorem.

(1) *The map $(s,\gamma) \to \gamma(s)$ is continuous on $G \times \widehat{G}$.*

(2) *If $K \subset G$ and $C \subset \widehat{G}$ are compact and $r > 0$, then $V(K,r)$ is open in $\widehat{G}$ and $U(C,r) = \{s \in G : |\gamma(s) - 1| < r$ for all $\gamma \in C\}$ is open in G.*

(3) *The sets $\gamma_1 V(K,r)$ for $K \subset G$, $\gamma_1 \in \widehat{G}$ and $r > 0$ form a base for the topology of $\widehat{G}$, i.e. the Gelfand topology is the compact-open topology.*

(4) *$\widehat{G}$ is an LCA group.*

Proof. Fix $\gamma_1 \in \widehat{G}$ and pick $f \in L^1(G)$ so that $\hat{f}(\gamma_1) = 1$. Now $s \to f_s$ is continuous by Lemma 7.4.2, and the Gelfand map is continuous into $C_0(\widehat{G})$.

Therefore $(s, \gamma) \to \widehat{f_s}(\gamma)$ is continuous on $G \times \widehat{G}$. By Theorem 7.4.5(3), $\widehat{f_s}(\gamma) = \overline{\gamma(s)}\hat{f}(\gamma)$. Therefore

$$\overline{\gamma(s)} = \frac{\widehat{f_s}(\gamma)}{\hat{f}(\gamma)}$$

is continuous in a neighbourhood of γ_1 on which $\hat{f}(\gamma) \neq 0$ and arbitrary $s \in G$.

Fix a compact $K \subset G$ and $r > 0$. Let $\gamma_1 \in V(K, r)$. Thus $|\gamma_1(s) - 1| < r$ for all $s \in K$. By (1), for each $s \in K$, there is a neighbourhood $U_s \ni s$ and $W_s \ni \gamma_1$ so that $|\gamma(t) - 1| < r$ for $(t, \gamma) \in U_s \times W_s$. Let $U_{s_1}, \dots, U_{s_n}$ be a finite subcover of K. Set $W = \bigcap_{i=1}^n W_{s_i}$. Then W is an open neighbourhood of γ_1. For $\gamma \in W$ and $t \in K$, $(t, \gamma) \in U_{s_i} \times W$ for some i, whence $|\gamma(t) - 1| < r$. That shows that $\gamma_1 \in W \subset V(K, r)$. Thus $V(K, r)$ is open in $\widehat{G}$. Similarly, $U(C, r)$ is open in G.

Now suppose that W is an open neighbourhood of γ_1 in $\widehat{G}$. Since the topology on $\widehat{G}$ is translation invariant, we may assume that $\gamma_1 = \gamma_0$ is the identity of $\widehat{G}$. We want to find a set $V(K, r)$ contained in W. Since W is open, there are elements $f_1, \dots, f_n \in L^1(G)$ and $\varepsilon > 0$ so that

$$\gamma_0 \in \bigcap_{i=1}^{n} \{\gamma : |\hat{f}_i(\gamma) - \hat{f}_i(\gamma_0)| < \varepsilon\} \subset W.$$

Since $C_c(G)$ is dense in $L^1(G)$, it is a routine approximation argument to replace f_i by continuous functions of compact support, say K, with a possible change of ε. Define $r = \varepsilon\big(\max\{\|f_i\|_1 : 1 \leq i \leq n\}\big)^{-1}$. Then if $\gamma \in V(K, r)$, then

$$|\hat{f}_i(\gamma) - \hat{f}_i(\gamma_0)| = \int_K |f_i(s)|\, |\overline{\gamma(s) - 1}|\, ds < r\|f_i\|_1 \leq \varepsilon.$$

Therefore $V(K, r) \subset W$. This shows that translations of sets of the form $V(K, r)$ form a base for the topology of $\widehat{G}$. That is, the Gelfand topology on $\widehat{G}$ coincides with the compact-open topology.

Note that $V(K, \frac{r}{2})V(K, \frac{r}{2}) \subset V(K, r)$ because if $\gamma_i \in V(K, \frac{r}{2})$, then for $s \in K$,

$$|\gamma_1\gamma_2(s) - 1| \leq |\gamma_1(s)|\, |\gamma_2(s) - 1| + |\gamma_1(s) - 1| < \tfrac{r}{2} + \tfrac{r}{2} = \varepsilon.$$

Hence, multiplication is continuous at (γ_0, γ_0). Also since $\gamma^{-1} = \overline{\gamma}$, the neighbourhood $V(K, r)$ is symmetric. So inverse is continuous at γ_0. The topology is translation invariant, and thus multiplication and inverses are continuous. So $\widehat{G}$ is an LCA group. ■

7.4.11. REMARK. Note that we did not show that the sets $U(C, r)$ form a neighbourhood base at 0 is G. This takes more work, and eventually one can show that $\widehat{\widehat{G}} = G$. This is known as *Pontryagin duality*. A proof can be found in [268, § 1.7].

7.5. More about $\mathbf{L}^1(\mathbf{G})$

In this section, we develop further properties of $L^1(G)$. In particular, we show that it is semisimple.

7.5.1. THEOREM. *Let G be an LCA group. If G is discrete, then $\widehat{G}$ is compact. If G is compact, then $\widehat{G}$ is discrete.*

PROOF. If G is discrete, then $L^1(G)$ is unital, and thus the Gelfand space $\widehat{G}$ is compact.

If G is compact, then the constant function $1 = \gamma_0$ belongs to L^1. If $\gamma \in \widehat{G}$ and $\gamma \neq \gamma_0$, then by translation invariance,

$$\int_G \gamma(t)\,dt = \gamma(s)\int_G \gamma(t-s)\,dt = \gamma(s)\int_G \gamma(t)\,dt.$$

Since $\gamma \neq 1$, there is some $s \in G$ so that $\gamma(s) \neq 1$, and hence $\displaystyle\int_G \gamma(t)\,dt = 0$. Consider the open set

$$\Big\{\gamma \in \widehat{G} : \Big|\int_G \gamma(t)\,dt - 1\Big| < 1\Big\} = \{\gamma_0\}.$$

This shows that $\{\gamma_0\}$ is open. The topology is translation invariant, and so $\widehat{G}$ has the discrete topology. ∎

When a Banach algebra does not have a unit, there is often a useful substitute.

7.5.2. DEFINITION. If $\mathfrak{A}$ is a Banach algebra, an *approximate identity* is a net $\{e_\lambda : \lambda \in \Lambda\}$ in $\mathfrak{A}$ such that

$$\lim_\Lambda e_\lambda a = a = \lim_\Lambda a e_\lambda \quad \text{for all} \quad a \in \mathfrak{A}.$$

It is a *bounded approximate identity* (b.a.i.) if $\sup \|e_\lambda\| = M < \infty$ and a *contractive approximate identity* (c.a.i.) if $M = 1$.

Similarly we define a *(bounded, contractive) left/right approximate identity* if we only have the limit for multiplication on one side.

7.5.3. REMARKS. If $\{e_\lambda : \lambda \in \Lambda\}$ is a bounded left approximate identity for $\mathfrak{A}$ and $\{f_\mu : \mu \in \Lambda'\}$ is a bounded right approximate identity for $\mathfrak{A}$, then $\{f_\mu + e_\lambda - f_\mu e_\lambda : (\lambda, \mu) \in \Lambda \times \Lambda'\}$ is a bounded approximate identity. Indeed, if $a \in \mathfrak{A}$, then

$$\|(f_\mu + e_\lambda - f_\mu e_\lambda)a - a\| \le \|e_\lambda a - a\| + \|f_\mu(a - e_\lambda a)\| \le \big(1 + \|f_\mu\|\big)\|e_\lambda a - a\|.$$

A similar inequality holds on the right.

If $\mathfrak{A}$ is separable and $\{e_\lambda : \lambda \in \Lambda\}$ is a bounded (contractive) approximate identity for $\mathfrak{A}$, then there is a sequential bounded (contractive) approximate identity for $\mathfrak{A}$. Let $\{a_i : i \geq 1\}$ be a dense subset of $b_1(\mathfrak{A})$, and let $K = \sup_\Lambda \|e_\lambda\|$. For each $n \geq 1$, we can choose $f_n = e_{\lambda_n}$ so that $\|f_n a_i - a_i\| < \frac{1}{n}$ and $\|a_i f_n - a_i\| < \frac{1}{n}$ for $1 \leq i \leq n$. If $a \in \mathfrak{A}$ has norm at most 1 and $\varepsilon > 0$, pick a_j so that $\|a - a_j\| < \frac{\varepsilon}{2K+2}$. For $n \geq \max\{j, 2\varepsilon^{-1}\}$, we have

$$\begin{aligned}\|f_n a - a\| &\leq \|f_n(a - a_j)\| + \|f_n a_j - a_j\| + \|a_j - a\| \\ &\leq (K+1)\frac{\varepsilon}{2K+2} + \frac{1}{n} < \frac{\varepsilon}{2} + \frac{\varepsilon}{2} = \varepsilon.\end{aligned}$$

A similar argument works for right multiplication.

7.5.4. Proposition. *$L^1(G)$ has a contractive approximate identity.*

Proof. Consider the set Λ of compact neighbourhoods of 0 in G with the partial order $K \geq L$ if $K \subset L$. This is a directed set because $\bigcap_{i=1}^n K_i \geq K_i$ for $1 \leq i \leq n$. For each compact neighbourhood K of 0, using Urysohn's lemma for LCH spaces (Proposition 1.10.4), there is a continuous function f in $C_c(G)$ so that $0 \leq f(x) \leq 1 = f(0)$ so that $f(s) = 0$ for $s \notin K$. Let $e_K = f/\|f\|_1$, so that $e_K \geq 0$, $\operatorname{supp}(e_K) \subset K$ and $\|e_K\|_1 = 1$. We will show that $\{e_K\}$ is a c.a.i.

Let $f \in L^1(G)$ and let $\varepsilon > 0$. By Lemma 7.4.2, the function $s \to f_s$ is continuous. Choose a neighbourhood U of 0 so that $\|f_s - f\|_1 < \varepsilon$ for $s \in U$. Then select a compact neighbourhood K of 0 such that $K \subset U$. If $L \geq K$,

$$\begin{aligned}\|e_L * f - f\|_1 &= \int_G |(e_L * f - f)(s)|\, ds \\ &= \int_G \left| \int_L e_L(t) f(s-t)\, dt - \int_L e_L(t)\, dt\, f(s) \right| ds \\ &\leq \int_G \int_L e_L(t) |f_t(s) - f(s)|\, dt\, ds \\ &= \int_L e_L(t) \int_G |f_t(s) - f(s)|\, ds\, dt \\ &= \int_L e_L(t) \|f_t - f\|_1\, dt < \varepsilon \|e_L\|_1 = \varepsilon.\end{aligned}$$

Therefore $\{e_K\}$ is contractive approximate identity. ■

Next we want to highlight the significance of Theorem 7.4.5 (6).

7.5.5. Definition. A *Banach $*$-algebra* is a Banach algebra with a continuous involution $a \to a^*$ which is an anti-automorphism, i.e.

(1) $(a^*)^* = a$ for all $a \in \mathfrak{A}$.

(2) $(\lambda a)^* = \bar{\lambda} a^*$ for $\lambda \in \mathbb{C}$ and $a \in \mathfrak{A}$.

(3) $(a+b)^* = a^* + b^*$ for $a, b \in \mathfrak{A}$.

(4) $(ab)^* = b^* a^*$ for $a, b \in \mathfrak{A}$.

A homomorphism $\theta : \mathfrak{A} \to \mathfrak{B}$ between Banach $*$-algebras is a $*$-homomorphism if $\theta(a^*) = \theta(a)^*$ for $a \in \mathfrak{A}$.

7.5.6. EXAMPLES.

(1) $C_0(X)$ for an LCH space X. The $*$ operation is conjugation: $f^*(x) = \overline{f(x)}$.

(2) $\mathcal{B}(H)$ for a Hilbert space H. The $*$ operator is the Hilbert space adjoint.

(3) $L^1(G)$ with $f^*(s) = \overline{f(-s)}$ is a Banach $*$-algebra. All of the properties are easy to verify. Theorem 7.4.5 (6) says that the Fourier transform is a $*$-homomorphism of $L^1(G)$ into $C_0(\widehat{G})$. If we define a c.a.i. as above using only compact neighbourhoods of 0 of the form $K = -K$, then we can replace e_K with $(e_K + e_K^*)/2$, so that $e_K = e_K^*$.

7.5.7. LEMMA. *If $f \in L^1(G)$ and $g \in C_0(G)$, then $f * g \in C_0(G)$.*

PROOF. Note that g is uniformly continuous. Thus given $\varepsilon > 0$, there is a neighbourhood $U = -U \ni 0$ so that $\|g - g_s\|_\infty < \varepsilon$ for $s \in U$. For $s, t \in G$, we have

$$\begin{aligned}(f*g)(s) - (f*g)(t) &= \int_G f(u)\big(g(s-u) - g(t-u)\big)\,du \\ &\le \|f\|_1 \|g_{-s} - g_{-t}\|_\infty = \|f\|_1 \|g_{t-s} - g\|_\infty < \|f\|_1 \varepsilon\end{aligned}$$

provided $t - s \in U$. Therefore $f * g$ is continuous. Also $\|f * g\|_\infty \le \|f\|_1 \|g\|_\infty$.

Now choose $f_1, g_1 \in C_c(G)$ so that $\|f - f_1\|_1 < \varepsilon$ and $\|g - g_1\|_\infty < \varepsilon$, and suppose that the supports of f_1 and g_1 are contained in a compact set K. Then $\operatorname{supp}(f_1 * g_1) \subset K + K$ is compact. A routine estimate shows that

$$|(f*g)(s) - (f_1*g_1)(s)| \le \|f - f_1\|_1 \|g\|_\infty + \|f_1\|_1 \|g - g_1\|_\infty < (\|f\|_1 + \|g\|_\infty + \varepsilon)\varepsilon.$$

This shows that $f * g(s)$ is small off of the compact set $K + K$. Therefore $f * g$ lies in $C_0(G)$. ∎

7.5.8. THEOREM. *Let G be a LCA group. Then $L^1(G)$ is semisimple.*

PROOF. Suppose that $f \ne 0$ belongs to $\operatorname{rad} L^1(G) = \ker \Gamma$, i.e. $\operatorname{spr}(f) = 0$. This is a closed ideal. Let $\{e_K\}$ be a contractive approximate identity for $L^1(G)$ with $e_K \in C_c(G)$. Then $f * e_K$ belong to $\operatorname{rad} L^1(G)$ and this net converges to f. Thus eventually they are non-zero, and they lie in $C_0(G)$ by Lemma 7.5.7. So we may suppose that $f \in L^1(G) \cap C_0(G)$. Note that it also belongs to $L^2(G)$ because

$\|f\|_2^2 \le \|f\|_1 \|f\|_\infty$. Observe that $f * f^*$ also belongs to $L^1(G) \cap C_0(G)$ and to $\operatorname{rad} L^1(G)$. Moreover

$$(f * f^*)(0) = \int_G f(s) f^*(-s)\, ds = \int_G |f(s)|^2\, ds = \|f\|_2^2 > 0.$$

Thus $f * f^*$ is non-zero on a neighbourhood of 0, and hence is also a non-zero element of $\operatorname{rad} L^1(G)$.

Define a linear functional on $L^1(G)$ by $F(g) = (f * f^* * g)(0)$. This is continuous since $|F(g)| \le \|f * f^*\|_\infty \|g\|_1$. Next define a sesquilinear form on $L^1(G)$ by $\langle g, h\rangle = F(g * h^*)$. You should check that this is linear in g and conjugate linear in h and jointly continuous. Moreover

$$\langle g, g\rangle = (f * f^* * g * g^*)(0) = (f * g) * (f * g)^*(0) = \|f * g\|_2^2 \ge 0.$$

So this inner product is positive semidefinite. Also

$$\begin{aligned}\langle h, g\rangle &= \gamma_0(f * f^* * h * g^*) = \overline{\gamma_0((f * f^* * h * g^*)^*)} \\ &= \overline{\gamma_0(f * f^* * h^* * g)} = \overline{\langle g, h\rangle}\end{aligned}$$

because Γ is a $*$-homomorphism. Therefore the inner product satisfies the Cauchy–Schwarz inequality

$$|\langle g, h\rangle| \le \langle g, g\rangle^{1/2} \langle h, h\rangle^{1/2}.$$

Let $\{e_K\}$ be a c.a.i. with $e_K = e_K^*$. For $g \in L^1(G)$, we have

$$\begin{aligned}|F(g)| &= \lim_K |F(g * e_K)| = \lim_K |\langle g, e_K\rangle| \\ &\le \lim_K \langle g, g\rangle^{1/2} \langle e_K, e_K\rangle^{1/2} \\ &= \langle g, g\rangle^{1/2} \lim_K \big((f * f^* * e_K * e_K)(0)\big)^{1/2} \\ &= F(g * g^*)^{1/2} \big((f * f^*)(0)\big)^{1/2} = F(g * g^*)^{1/2} \|f\|_2.\end{aligned}$$

Applying this to $(g * g^*)^{2^n}$ for $n \ge 0$, we get

$$F\big((g * g^*)^{2^n}\big) \le F\big((g * g^*)^{2^{n+1}}\big)^{1/2} \|f\|_2.$$

Therefore

$$\begin{aligned}|F(g)| &\le F\big((g * g^*)^2\big)^{1/4} \|f\|_2^{1+\frac12} \le F\big((g * g^*)^{2^2}\big)^{1/8} \|f\|_2^{1+\frac12+\frac14} \\ &\le F\big((g * g^*)^{2^n}\big)^{1/2^{n+1}} \|f\|_2^{1+2^{-1}+\dots+2^{-n}} \\ &\le \|F\|^{1/2^{n+1}} \|(g * g^*)^{2^n}\|^{1/2^{n+1}} \|f\|_2^{2-2^{-n}} \to \operatorname{spr}(g * g^*)^{1/2} \|f\|_2^2.\end{aligned}$$

Now take $g = f * f^*$. This lies in $\operatorname{rad} L^1(G)$, so that $\operatorname{spr}(g * g^*) = 0$. Therefore

$$0 < \|f * f^*\|_2^2 \le 0.$$

This contradiction shows that $\operatorname{rad} L^1(G) = \{0\}$, i.e. $L^1(G)$ is semisimple. ■

7.5.9. COROLLARY. *If $f \in L^1(G)$ and $\hat{f} = 0$, then $f = 0$.*

7.6. Commutative C*-algebras

Later we will have a couple of chapters on C*-algebras. Here we will deal with the commutative case, which is an application of the Gelfand theory. This leads to a (weak) version of the spectral theorem, and a continuous functional calculus for normal operators.

7.6.1. DEFINITION. A *C*-algebra* is a Banach $*$-algebra such that

$$\|a^*a\| = \|a\|^2 \quad \text{for all} \quad a \in \mathfrak{A}.$$

An element a of $\mathfrak{A}$ is *self-adjoint* if $a = a^*$, n is *normal* if $nn^* = n^*n$, and u is *unitary* if $u^* = u^{-1}$. Say a is *positive* if $a = a^*$ and $\sigma(a) \subset [0, \infty)$.

Note that $\|a\|^2 = \|a^*a\| \le \|a^*\| \, \|a\|$ implies that $\|a\| \le \|a^*\| \le \|a^{**}\| = \|a\|$. Thus $\|a^*\| = \|a\|$ and $\|aa^*\| = \|a^*a\|$.

We establish one general result valid in the non-commutative case.

7.6.2. PROPOSITION. *If $\mathfrak{A}$ is a non-unital C*-algebra, then it imbeds into a unital C*-algebra $\widetilde{\mathfrak{A}}$ as a maximal ideal of codimension* 1.

PROOF. As usual, $\widetilde{\mathfrak{A}} = \mathfrak{A} + \mathbb{C}1$. The issue is how to define the norm. We set

$$\|a + \lambda\| := \|L_a + \lambda I\|_{\mathcal{B}(\mathfrak{A})} = \sup\bigl\{\|(a + \lambda)b\| : b \in \mathfrak{A}, \ \|b\| \le 1\bigr\}.$$

Then for $a \in \mathfrak{A}$, we have

$$\|L_a\| \le \|a\| = \frac{\|aa^*\|}{\|a^*\|} = \|L_a \tfrac{a^*}{\|a^*\|}\| \le \|L_a\|.$$

Thus $\|L_a\| = \|a\|$, so we are not changing the norm on $\mathfrak{A}$. To check the C*-norm condition on $\widetilde{\mathfrak{A}}$, we compute

$$\begin{aligned}
\|a + \lambda\|^2 &= \sup_{\|b\| \le 1} \|ab + \lambda b\|^2 = \sup_{\|b\| \le 1} \|(ab + \lambda b)^*(ab + \lambda b)\| \\
&= \sup_{\|b\| \le 1} \|b^*a^*ab + \lambda b^*a^*b + \overline{\lambda} b^*ab + |\lambda|^2 b^*b\| \\
&= \sup_{\|b\| \le 1} \|b^*(a^*a + \lambda a^* + \overline{\lambda} a + |\lambda|^2)b\| \\
&\le \|a^*a + \lambda a^* + \overline{\lambda} a + |\lambda|^2\| \\
&= \|(a + \lambda)^*(a + \lambda)\| \le \|(a + \lambda)^*\| \, \|a + \lambda\|.
\end{aligned}$$

Hence $\|a + \lambda\| \le \|(a + \lambda)^*\|$. Interchanging a and a^* yields $\|a + \lambda\| = \|(a + \lambda)^*\|$. Thus

$$\|a + \lambda\|^2 \le \|(a + \lambda)^*(a + \lambda)\| \le \|a + \lambda\|^2,$$

forcing equality. Therefore $\widetilde{\mathfrak{A}}$ is a C*-algebra. ■

7.6.3. THEOREM. *Let $\mathfrak{A}$ be an abelian C*-algebra. Then the Gelfand map $\Gamma : \mathfrak{A} \to C_0(\mathcal{M}(\mathfrak{A}))$ is an isometric $*$-isomorphism.*

PROOF. First assume that $\mathfrak{A}$ is unital. Let $a = a^* \in \mathfrak{A}$ and $\varphi \in \mathcal{M}(\mathfrak{A})$. We claim that $\varphi(a)$ is real. To this end, we define

$$u_t = e^{ita} = \sum_{n\geq 0} \frac{(ita)^n}{n!} \quad \text{for} \quad t \in \mathbb{R}.$$

Then $u_t^{-1} = u_{-t}$ (see Exercise **6.19**), and thus

$$u_t^* = \sum_{n\geq 0} \frac{(ita)^{n*}}{n!} = \sum_{n\geq 0} \frac{(-ita)^n}{n!} = u_{-t} = u_t^{-1}.$$

Hence u_t is unitary. Thus $\|u_t\|^2 = \|u_t^* u_t\| = \|1\| = 1$. We obtain

$$1 \geq |\varphi(u_t)| = \Big| \sum_{n\geq 0} \frac{\varphi(ita)^n}{n!} \Big| = \big|e^{i\varphi(a)t}\big| \quad \text{for} \quad t \in \mathbb{R}.$$

Therefore $\varphi(a) \in \mathbb{R}$.

For $a \in \mathfrak{A}$, set $x = \operatorname{Re} a = \frac{1}{2}(a + a^*)$ and $y = \operatorname{Im} a = \frac{1}{2i}(a - a^*)$. Then $x = x^*$, $y = y^*$ and $a = x + iy$ is written as a sum of its real and imaginary parts. It follows that

$$\varphi(a^*) = \varphi(x - iy) = \varphi(x) - i\varphi(y) = \overline{\varphi(x) + i\varphi(y)} = \overline{\varphi(a)}.$$

Therefore $\Gamma(a^*) = \Gamma(a)^*$, so Γ is a $*$-homomorphism.

If $a = a^*$, then $\|a\|^2 = \|a^*a\| = \|a^2\|$. Repeated application of this fact yields that $\|a\|^{2^n} = \|a^{2^n}\|$. Thus

$$\|\hat{a}\|_\infty = \operatorname{spr}(a) = \lim_{n\to\infty} \|a^{2^n}\|^{2^{-n}} = \|a\|.$$

For general $a \in \mathfrak{A}$,

$$\|a\|^2 = \|a^*a\| = \|\widehat{a^*a}\|_\infty = \|\hat{a}\|_\infty^2.$$

Therefore Γ is isometric.

In particular, $\Gamma(\mathfrak{A})$ is a norm-closed, unital, self-adjoint subalgebra of $C(\mathcal{M}(\mathfrak{A}))$ which separates points. By the Stone–Weierstrass theorem, $\Gamma(\mathfrak{A}) = C(\mathcal{M}(\mathfrak{A}))$.

Now consider the non-unital case. Let $\widetilde{\mathfrak{A}}$ be the unital C*-algebra containing $\mathfrak{A}$ as a maximal ideal of codimension 1. Then $\Gamma(\widetilde{\mathfrak{A}}) = C(\mathcal{M}(\widetilde{\mathfrak{A}}))$, and $\mathcal{M}(\mathfrak{A}) = \mathcal{M}(\widetilde{\mathfrak{A}}) \setminus \{\varphi_\infty\}$, where $\varphi_\infty(a + \lambda) = \lambda$. The restriction of Γ to $\mathfrak{A}$ is the Gelfand map of $\mathfrak{A}$, and it is mapped into $\ker \varphi_\infty = C_0(\mathcal{M}(\mathfrak{A}))$. It is an isometric $*$-isomorphism which separates points, and does not vanish except at φ_∞. Thus it is surjective. ■

This has immediate application to the structure of normal operators on Hilbert space, and more generally to normal elements of C*-algebras. By definition, they generate an abelian C*-algebra.

7.6.4. COROLLARY. *Let n be a normal element of a C*-algebra $\mathfrak{A}$. Then $C^*(n)$ is $*$-isomorphic to $C(\sigma(n))$ via $\Gamma(n) = z$, where $z(t) = t$. The (not necessarily unital) subalgebra $C^*(n, n^*)$ is isomorphic to $C_0(\sigma(n) \setminus \{0\})$.*

PROOF. The non-closed $*$-algebra generated by $1, n$ and n^* consists of polynomials $p(n, n^*)$ for $p \in \mathbb{C}[x, y]$ and $C^*(n)$ is the norm closure. This is an abelian unital C*-algebra, and thus it is $*$-isomorphic to $C(\mathcal{M}(C^*(n)))$. A functional $\varphi \in \mathcal{M}(C^*(n))$ is determined by $\lambda = \varphi(n)$ because $\varphi(n^*) = \bar{\lambda}$, and thus $\varphi(p(n, n^*)) = p(\lambda, \bar{\lambda})$. By continuity, this determines φ on $C^*(n)$. Now the range of $\hat{n}$ is $\sigma(n)$, and hence we have a bijection between $\mathcal{M}(C^*(n))$ and $\sigma(n)$ via $\varphi \to \varphi(n)$. So $C^*(n) \simeq C(\sigma(n))$. Under this identification, $\hat{n}(\lambda) = \varphi_\lambda(n) = \lambda$.

The subalgebra generated by n and n^* is isomorphic to the subalgebra of $C(\sigma(n))$ generated by z and $\bar{z}$. This is a self-adjoint subalgebra which separates points, and either vanishes at 0 if $0 \in \sigma(n)$ or does not vanish otherwise. In the first case, the Stone–Weierstrass theorem shows that this is the ideal of function as vanishing at 0. In the second case, it is all of $C(\sigma(n))$. ∎

The inverse of the Gelfand map yields a very important corollary for normal operators, allowing us to evaluate continuous functions of a normal element.

7.6.5. CONTINUOUS FUNCTIONAL CALCULUS. *Let n be a normal element of a unital C*-algebra $\mathfrak{A}$. Then*

(1) *there is an isometric $*$-isomorphism from $C(\sigma(n))$ onto $C^*(n)$ such that $z(n) = n$, where $z(t) = t$ is the identity function.*

(2) *spectral mapping: $\sigma(f(n)) = f(\sigma(n))$ for $f \in C(\sigma(n))$.*

(3) *composition: if g is continuous on $f(\sigma(n))$, then $g(f(n)) = (g \circ f)(n)$.*

PROOF. There isn't much to do. By Corollary 7.6.4, $\mathfrak{B} = C^*(n)$ is a unital commutative C*-algebra, and the Gelfand map is an isomorphism onto $C(\sigma(n))$ with $\Gamma(n) = z$. The homomorphism for (1) is Γ^{-1}. Thus we define $f(n) := \Gamma^{-1}(f)$. Since Γ preserves spectrum, so does Γ^{-1}, and (2) holds.

For (3), note that $f(n)$ is normal, so $g(f(n))$ makes sense by the continuous functional calculus for $f(n)$. Also $g \circ f \in C(\sigma(n))$, so that $(g \circ f)(n)$ is defined. If $g(z) = z^j \bar{z}^k$, then $g \circ f(z) = f(z)^j \overline{f(z)}^k$. Thus

$$g(f(n)) = f(n)^j f(n)^{*k} = (g \circ f)(n).$$

By linearity, this extends to all polynomials $p(z, \bar{z})$. By the Stone–Weierstrass theorem, these polynomials are uniformly dense in $C(\sigma(f(n)))$. Therefore, by continuity of both functional calculi, this extends to all continuous functions $g \in C(\sigma(f(n)))$. ∎

In the non-unital case, the functional calculus holds for functions vanishing at 0.

7.6.6. COROLLARY. *Let n be a normal element of a non-unital C*-algebra $\mathfrak{A}$. Then $0 \in \sigma(n)$, and there is an isometric $*$-isomorphism from $C_0(\sigma(n) \setminus \{0\})$ onto $C^*(n, n^*)$ such that $z(n) = n$, where $z(t) = t$ is the identity function.*

PROOF. In this case, the C*-algebra $\mathfrak{B}$ generated by n is abelian, but not unital. In particular, n is not invertible; so $0 \in \sigma(n)$. So the Gelfand map Γ carries $\mathfrak{B}$ onto $C_0(\sigma(n) \setminus \{0\})$. The functional calculus is given by Γ^{-1}. ■

It is an important fact that the functional calculus commutes with homomorphisms.

7.6.7. COROLLARY. *Let $\pi : \mathfrak{A} \to \mathfrak{B}$ be a continuous $*$-homomorphism of a C*-algebra $\mathfrak{A}$ into a C*-algebra $\mathfrak{B}$. Let n be a normal element of $\mathfrak{A}$, and let $f \in C(\sigma(n))$ if $\mathfrak{A}$ is unital, or $f \in C_0(\sigma(n) \setminus \{0\})$ in the non-unital case. Then $\pi(f(n)) = f(\pi(n))$.*

PROOF. Let $m = \pi(n)$. Then $m^*m = \pi(n^*n) = \pi(nn^*) = mm^*$, so m is normal. For any polynomial $p(z, \bar{z}) \in \mathbb{C}[z, \bar{z}]$, we have $\pi(p(n, n^*)) = p(m, m^*)$. In the non-unital case, we only use polynomials with constant term 0. By the Stone–Weierstrass theorem, these polynomials are dense in $C(\sigma(n))$ or $C_0(\sigma(n) \setminus \{0\})$ in the non-unital case. By continuity, this map extends to $\pi(f(n)) = f(m)$. ■

7.7. Weakly Almost Periodic Functions

In this section, we will look at some commutative C*-algebras that arise from topological groups. This also provides a nice application of results on weak compactness and fixed point theorems.

Let G be a topological group, and let $C^b(G)$ denote the Banach space of bounded continuous functions on G. The left and right translation operators on $C^b(G)$ are given by

$$L_s f(t) = (f \cdot s)(t) = f(st) \text{ and } R_s f(t) = (s \cdot f)(t) = f(ts) \text{ for } t \in G.$$

Note that

$$(s \cdot (s' \cdot f))(t) = (s' \cdot f)(ts) = f(tss') = (ss' \cdot f)(t)$$

and

$$((f \cdot s) \cdot s')(t) = (f \cdot s)(s't) = f(ss't) = (f \cdot ss')(t).$$

7.7.1. DEFINITION. A function $f \in C^b(G)$ is *almost periodic* (AP) if $R_G f = \{s \cdot f : s \in G\}$ has norm-compact closure. A function $f \in C^b(G)$ is *weakly almost periodic* (WAP) if $R_G f$ has weakly compact weak closure. Let AP(G) and WAP(G) denote the set of all (weakly) almost periodic functions.

7.7.2. Examples.

(1) Let X be a *finite dimensional* Banach space, and let $U : G \to \mathcal{B}(X)^{-1}$ be a bounded, norm-continuous representation of G. That is, $U_s U_t = U_{st}$ for $s, t \in G$, $U_e = I$, $\|U\| := \sup_G \|U_s\| < \infty$ and $s \to U_s$ is continuous from G into $\mathcal{B}(X)$. Let $\xi \in X$ and $\varphi \in X^*$, and define

$$f(s) = \varphi(U_s \xi) \quad \text{for} \quad s \in G.$$

We claim that f is almost periodic. Note that $\mathcal{O}(\xi) = \overline{\{U_s \xi : s \in G\}}$ is a closed bounded set. Hence $\mathcal{O}(\xi)$ is compact since $\dim X < \infty$. Define $V : X \to C^b(G)$ by $V\xi(s) = \varphi(U_s \xi)$. This is linear and $\|V\| \le \|\varphi\| \, \|U\|$ is bounded. Observe that

$$V(U_s \xi)(t) = \varphi(U_t U_s \xi) = \varphi(U_{ts} \xi) = V\xi(ts) = (s \cdot V\xi)(t).$$

Thus $VU_s = R_s V$, whence $V\xi = f$ and $VU_s \xi = s \cdot f$. So $V(\{U_s \xi : s \in G\}) = R_G f$. This is contained in $V\mathcal{O}(\xi)$, which is compact, and hence $\overline{R_G f}$ is compact. Therefore f is almost periodic.

(2) Let X be a *reflexive* Banach space, and let $U : G \to \mathcal{B}(X)^{-1}$ be a bounded, strongly continuous representation of G. That is, $U_s U_t = U_{st}$ for $s, t \in G$ is a homomorphism, $U_e = I$, $\|U\| := \sup_G \|U_s\| < \infty$ and $s \to U_s \xi$ is continuous from G into X for every $\xi \in X$. Let $\xi \in X$ and $\varphi \in X^*$, and define $f(s) = \varphi(U_s \xi)$ for $s \in G$. We claim that f is weakly almost periodic. Here we define $\overline{\mathcal{O}(\xi)}^w = \overline{\{U_s \xi : s \in G\}}^w$. This is a weakly closed bounded set. Since X is reflexive, the weak-$*$ and weak topologies on X coincide. Hence $\overline{\mathcal{O}(\xi)}^w$ is weakly compact. Again we define $V : X \to C^b(G)$ by $V\xi(s) = \varphi(U_s \xi)$. Then V is a bounded linear operator, and hence is continuous from the weak topology on X to the weak topology on $C^b(G)$. Moreover $VU_s = R_s V$, so that $V\overline{\mathcal{O}(\xi)}^w \supset \overline{R_G f}^w$ is weakly compact. Therefore f is weakly almost periodic.

(3) Suppose that G is locally compact (a common assumption). Then $C_0(G)$ is the closure in $C^b(G)$ of $C_c(G)$, the space of continuous functions with compact support. It is difficult at this stage to describe the continuous linear functionals on $C^b(G)$, but the linear functionals on $C_0(G)$ are given by integration against (finite) complex Radon measures on G. We will show that $C_0(G) \subset \mathrm{WAP}(G)$.

By the Eberlein–Smulian theorem, it suffices to show that if $f \in C_c(G)$ and $(s_i \cdot f)$ is a sequence, then there is a subsequence which converges weakly. There are two cases: either $K = \overline{\{s_i : i \ge 1\}}$ is compact or not.

Suppose that K is compact. Since f has compact support, it is uniformly continuous, i.e. there is an open neighbourhood $U = U^{-1} \ni e$ such that $st^{-1} \in U$ implies that $|f(s) - f(t)| < \varepsilon$. This implies that $\|s \cdot f - t \cdot f\|_\infty < \varepsilon$, whence the map $s \to s \cdot f$ is norm continuous. Therefore $\{s \cdot f : s \in K\}$ is norm compact. Thus there is a subsequence $(s_{i_j} \cdot f)_{j \ge 1}$ which converges in norm, and hence also weakly.

Otherwise K is not compact. Let C be the compact support of f. Observe that $s \cdot f$ is supported on Cs^{-1}. Now $Cs^{-1} \cap C \neq \varnothing$ if and only if $s \in CC^{-1}$. Since the inverse is continuous and multiplication is jointly continuous, CC^{-1} is compact. Thus there is some i_1 such that $s_{i_1}C \cap C = \varnothing$. A repetition of this argument yields a subsequence $(s_{i_j} \cdot f)_{j \geq 1}$ of functions with pairwise disjoint supports. If μ is a continuous linear functional on $C^b(G)$, the restriction to $C_0(G)$ is given by integration against a finite measure $\bar{\mu}$. Since $\sum_{j \geq 1} |\bar{\mu}|(Cs_{i_j}^{-1}) \leq \|\bar{\mu}\| \leq \|\mu\|$, it follows that $\lim_j \mu(s_{i_j} \cdot f) = 0$. Therefore $(s_{i_j} \cdot f)_{j \geq 1}$ converges weakly to 0.

We deduce that $\overline{R_G f}^w$ is weakly compact. So $C_c(G) \subset \mathrm{WAP}(G)$. Below we show that $\mathrm{WAP}(G)$ is norm closed, and hence $C_0(G) \subset \mathrm{WAP}(G)$.

Proofs are generally more difficult in the weak topology. They usually also apply in the norm case, but there may be much simpler proofs. By the convexity Theorem 3.7.5 of Krein and Smulian, the (weak) compactness of $\overline{R_G f}$ (or $\overline{R_G f}^w$) is equivalent to saying that $\overline{\mathrm{conv}}(R_G f)$ or even $\overline{\mathrm{aconv}}(R_G f)$ is norm (weakly) compact, respectively.

The reader might suppose that this property should be called *right* (weakly) almost periodic. However, the left and right definitions yield the same family of functions. So no such adjective is required.

7.7.3. THEOREM. *Let G be a topological group, and let $f \in C^b(G)$. If $f \in \mathrm{AP}(G)$, then $\overline{L_G R_G f}$ is compact.*

PROOF. Let $\varepsilon > 0$. Since $\overline{R_G f}$ is compact, $R_G f$ is totally bounded. Select a finite set of points $S = \{s_i : 1 \leq i \leq m\} \subset G$ so that $\{s_i \cdot f : 1 \leq i \leq m\}$ is an ε-net for $R_G f$. Fix $t \in G$. Then for each i, there is some $j(i)$ such that $\|ts_i \cdot f - s_{j(i)} \cdot f\| < \varepsilon$. Conversely, suppose that $j : \{1, \dots, m\} \to \{1, \dots, m\}$ is a function for which there is some such t. Choose one, say t_j, and let $T = \{t_j\}$, which is a finite set of size $n \leq m^m$. Then for each $t \in G$, there is some $t_j \in T$ so that

$$\|ts_i \cdot f - t_j s_i \cdot f\| \leq \|ts_i \cdot f - s_{j(i)} \cdot f\| + \|s_{j(i)} \cdot f - t_j s_i \cdot f\| < 2\varepsilon \quad \text{for } 1 \leq i \leq m.$$

Now if $x \in G$, pick $s_i \in S$ with $\|x \cdot f - s_i \cdot f\| < \varepsilon$. Then for $t \in G$, using t_j as above,

$$\begin{aligned} \|tx \cdot f - t_j x \cdot f\| &\leq \|tx \cdot f - ts_i \cdot f\| + \|ts_i \cdot f - t_j s_i \cdot f\| + \|t_j s_i \cdot f - t_j x \cdot f\| \\ &= \|x \cdot f - s_i \cdot f\| + \|ts_i \cdot f - t_j s_i \cdot f\| + \|s_i \cdot f - x \cdot f\| \\ &< \varepsilon + 2\varepsilon + \varepsilon = 4\varepsilon. \end{aligned}$$

Evaluating this at $y \in G$ shows that $|f(ytx) - f(yt_j x)| < 4\varepsilon$ for all $x, y \in G$. Setting $x = e$ shows that $\|t \cdot f - t_j \cdot f\| < 4\varepsilon$; and setting $y = e$ shows that if

$t' \in G$ is matched to $t_{j'}$, that $\|f \cdot t' - f \cdot t_{j'}\| < 4\varepsilon$. Therefore

$$\begin{aligned}\|t \cdot f \cdot t' - t_j \cdot f \cdot t_{j'}\| &\le \|t \cdot f \cdot t' - t_j \cdot f \cdot t'\| + \|t_j \cdot f \cdot t' - t_j \cdot f \cdot t_{j'}\| \\ &= \|t \cdot f - t_j \cdot f\| + \|f \cdot t' - f \cdot t_{j'}\| < 4\varepsilon + 4\varepsilon = 8\varepsilon.\end{aligned}$$

Hence $\{t_i \cdot f \cdot t_j : 1 \le i, j \le n\}$ is an 8ε-net for $L_G R_G f = \{t \cdot f \cdot t' : t, t' \in G\}$. So $L_G R_G f$ is totally bounded, whence $\overline{L_G R_G F}$ is compact. ■

7.7.4. Corollary. *Let G be a topological group, and let $f \in C^b(G)$. Then $\overline{L_G f}$ is compact if and only if $\overline{R_G f}$ is compact.*

Proof. Theorem 7.7.3 shows that $\overline{R_G f}$ compact implies that $\overline{L_G R_G f}$ is compact, and thus $\overline{L_G f}$ is compact. By symmetry, one can run the proof the other direction beginning with compactness of $\overline{L_G f}$. ■

For the corresponding result for WAP(G), we need the Eberlein–Smulian–Grothendieck theorem 3.7.3.

7.7.5. Theorem. *Let G be a topological group, and let $f \in C^b(G)$. Then $\overline{L_G f}^w$ is weakly compact if and only if $\overline{R_G f}^w$ is weakly compact.*

Proof. By symmetry, it is enough to show that if f is weakly almost periodic, then $\overline{L_G f}^w$ is weakly compact. Consider $A = \operatorname{aconv} L_G f$ (no closure). Let $B = \operatorname{aconv}\{\varepsilon_s : s \in G\}$, where ε_s is the linear functional of evaluation at s. We claim that A and B satisfy IDLP.

Let (μ_m) be a sequence in A and (ν_n) be a sequence in B such that all the limits in the iterated double limits exist:

$$\lim_{m\to\infty} \lim_{n\to\infty} \nu_n(\mu_m) \quad \text{and} \quad \lim_{n\to\infty} \lim_{m\to\infty} \nu_n(\mu_m).$$

We can write $\mu_m = \sum_i \alpha_{m,i} f \cdot s_{m,i}$ and $\nu_n = \sum_j \beta_{n,j} \varepsilon_{t_{n,j}}$ where $s_{m,i}$ and $t_{n,j}$ are in G, $\alpha_{m,i}$ and $\beta_{n,j}$ are in $\mathbb{C}$ and $\sum_i |\alpha_{m,i}| = 1 = \sum_j |\beta_{n,j}|$. Then

$$\begin{aligned}\nu_n(\mu_m) &= \sum_{i,j} \alpha_{m,i}\beta_{n,j}\varepsilon_{t_{n,j}}(f \cdot s_{m,i}) = \sum_{i,j} \alpha_{m,i}\beta_{n,j} f(s_{m,i}t_{n,j}) \\ &= \sum_{i,j} \alpha_{m,i}\beta_{n,j}\varepsilon_{s_{m,i}}(t_{n,k} \cdot f) = \widetilde{\mu}_m(\widetilde{\nu}_n)\end{aligned}$$

where $\widetilde{\mu}_m = \sum_i \alpha_{m,i}\varepsilon_{s_{m,i}} \in B$ and $\widetilde{\nu} = \sum_j \beta_{n,j}(t_{n,k} \cdot f) \in \operatorname{aconv} R_G f =: A'$. Since $\overline{A'}^w$ is weakly compact, the Eberlein–Smulian–Grothendieck theorem shows that A' and B satisfy IDLP. Now

$$\lim_m \lim_n \widetilde{\mu}_m(\widetilde{\nu}_n) = \lim_m \lim_n \nu_n(\mu_m) \quad \text{and} \quad \lim_n \lim_m \widetilde{\mu}_m(\widetilde{\nu}_n) = \lim_n \lim_m \nu_n(\mu_m)$$

both exist by assumption, and thus these two limits are equal. This means that A and B also satisfy IDLP.

Next we show that $\overline{B}^{w^*} = \overline{b_1(X^*)}$. Indeed, $\overline{B}^{w^*}$ is a balanced convex weak-$*$ closed set. If $\mu \notin \overline{B}^{w^*}$, then by the separation theorem, there is an element $f \in C^b(G)$ such that $\sup_{\nu\in B} \operatorname{Re} \nu(f) = 1 < \mu(f)$. As B is balanced, the LHS equals $\sup_{\nu\in B} |\nu(f)|$. Since $\varepsilon_s \in B$, we see that $\|f\|_\infty = \sup_G |f(s)| \le 1$. Since $\mu(f) > 1$, we have $\|\mu\| > 1$. Therefore $\overline{b_1(X^*)} \subset \overline{B}^{w^*}$. The reverse inclusion is clear. It now follows from the Eberlein–Smulian–Grothendieck theorem that $\overline{\operatorname{aconv}}(L_G f)$ is weakly compact. ∎

The next thing we wish to establish is that AP(G) and WAP(G) are norm-closed subalgebras of $C^b(G)$ with additional properties. In particular, they are commutative C*-algebras.

7.7.6. THEOREM. *Let G be a topological group. Then* AP(G) *and* WAP(G) *are norm-closed, left and right translation invariant, self-adjoint subspaces of* $C^b(G)$. *Also if* $f \in$ AP(G) *or* WAP(G), *then* $f \circ J \in$ AP(G) *or* WAP(G), *where* $Jx = x^{-1}$. *Moreover* AP(G) *and* WAP(G) *are algebras, and so are commutative C*-algebras.*

PROOF. Let $f, g \in$ WAP(G). For $\lambda \in \mathbb{C}$, the map $\sigma : \overline{R_G f}^w \times \overline{R_G g}^w \to C^b(G)$ by $\sigma(f', g') = \lambda f' + g'$ is weakly continuous. Therefore, the range is weakly compact, and contains $R_G(\lambda f + g)$. Thus $\lambda f + g$ is weakly almost periodic, whence WAP(G) is a vector space. Since $R_G(s \cdot f) = R_G f$, WAP(G) is invariant under right translations. By the previous theorem, it is also left translation invariant. Also $\bar{f}$ satisfies $R_G \bar{f} = \overline{R_G f}$, where the latter means complex conjugate. So WAP(G) is self-adjoint. Note that $f \circ J$ is left weakly almost periodic, and hence is weakly almost periodic by the previous theorem. We say that WAP(G) is *inversion invariant*. Identical arguments work for AP(G).

For norm closure, there is an easy argument for AP(G). Compactness of $\overline{R_G f}$ is equivalent to total boundedness of $R_G f$. If (f_n) is a Cauchy sequence in AP(G) with limit f and $\varepsilon > 0$, $\|f - f_n\| < \varepsilon/2$ and F is a finite $\varepsilon/2$-net for $R_G f_n$, then it is an ε-net for $R_G f$. Hence $R_G f$ is totally bounded.

It is also easy to see that AP(G) is an algebra because multiplication is jointly continuous on $C^b(G)$. Thus if $f, g \in$ AP(G), the map $m : \overline{R_G f} \times \overline{R_G g} \to C^b(G)$ by $m(f', g') = f'g'$ is continuous, so has compact image. This set contains $\overline{R_G fg}$, and hence $fg \in$ AP(G).

Now consider WAP(G), where we need the Eberlein–Smulian theorem. Suppose that $f_n \in$ WAP(G) and $\sum_n \|f_n\| < \infty$, and let $f = \sum_n f_n$. It suffices to show that f is weakly almost periodic (see Exercise **2.5**). As in the first paragraph, the sets $X_n = \sum_{i=1}^n \overline{R_G f_i}^w$ are weakly compact, and $\overline{R_G f}^w$ is contained in $\overline{X}^w$ where $X = \bigcup_n X_n$. To show that $\overline{X}^w$ is weakly compact, it is enough to show that every sequence in X has a weakly convergent subsequence by the Eberlein–Smulian theorem. Let (g_n) be a sequence in X, and write $g_n = \sum_i g_{n,i}$ as a finite sum with $g_{n,i} \in \overline{R_G f_i}^w$. Since each $\overline{R_G f_i}^w$ is weakly sequentially compact by the

Eberlein–Smulian theorem, a standard diagonal argument produces a subsequence (which for convenience we denote still as (g_n)) such that w–$\lim_n g_{n,i} = g_i$ exists in $\overline{R_G f_i}^w$ for each $i \geq 1$. Clearly $\|g_i\| \leq \|f_i\|$, and thus $g = \sum_i g_i$ converges in $C^b(G)$. Let $\mu \in C^b(G)^*$. Then

$$|\mu(g - g_n)| \leq \sum_{i=1}^{k} |\mu(g_i - g_{n,i})| + \sum_{i>k} \|\mu\| \, \|g_i - g_{n_i}\|$$
$$< \sum_{i=1}^{k} |\mu(g_i - g_{n,i})| + \sum_{i>k} 2\|\mu\|\|f_i\|.$$

Given $\varepsilon > 0$, choosing k sufficiently large makes the second term at most $\varepsilon/2$. Then for n large enough, the first sum is at most $\varepsilon/2$. Hence (g_n) converges weakly to g. Again by the Eberlein–Smulian theorem, $\overline{X}^w$ is weakly compact. This contains $\overline{R_G f}^w$, and so f is weakly almost periodic.

Now we will show that WAP(G) is also closed under multiplication. By Theorem 7.6.3, the maximal ideal space Y of $C^b(G)$ is a compact Hausdorff space such that $C^b(G) \simeq C(Y)$. By the Riesz representation theorem, linear functionals on $C^b(G)$ are given by regular complex Borel measures on Y, i.e. if $\mu \in C^b(G)^*$, there is a measure $\bar{\mu}$ on Y so that $\mu(f) = \int_Y \hat{f}\, d\bar{\mu}$. Point evaluations $\varepsilon_s(f) = f(s)$ for $s \in G$ are multiplicative linear functionals. By continuity of f, if (s_λ) is a net in G converging to s, then $(\varepsilon_{s_\lambda})$ converges pointwise on $C_b(G)$ to ε_s. In other words, it converges weak-$*$ in $C^b(G)^*$ and thus in Y.

To show that WAP(G) is closed under multiplication, we show that multiplication is jointly weakly sequentially continuous. If (f_n) and (g_n) are sequences in WAP(G) converging weakly to f and g, respectively, then these sequences are bounded and the extensions $(\hat{f}_n)$ and $(\hat{g}_n)$ in $C(Y)$ converge pointwise on Y to $\hat{f}$ and $\hat{g}$. (Caveat: convergence on G is a weaker notion.) Therefore $(\widehat{f_n g_n})$ converges pointwise to $\widehat{fg}$. Let $\mu \in C(G)^*$ and let $\bar{\mu}$ be the associated measure on X. By the Lebesgue dominated convergence theorem,

$$\lim_n \mu(f_n g_n) = \lim_n \int_Y \widehat{f_n g_n}\, d\bar{\mu} = \int_Y \widehat{fg}\, d\bar{\mu} = \mu(fg).$$

That is, the sequence $(f_n g_n)$ converges weakly to fg.

Now if $f, g \in$ WAP(G), the sets $\overline{R_G f}^w$ and $\overline{R_G g}^w$ are weakly sequentially compact by the Eberlein–Smulian theorem 3.7.3. The map $m : \overline{R_G f}^w \times \overline{R_G g}^w \to C^b(G)$ by $m(f', g') = f'g'$ is weakly sequentially continuous. Therefore, the image is weakly sequentially compact, and hence weakly compact, again by the Eberlein–Smulian theorem. This image contains $R_G fg$, and therefore $fg \in$ WAP(G).

So WAP(G) is a unital C*-subalgebra of $C^b(G)$. We can again apply Gelfand's theorem to see that WAP$(G) \simeq C(Z)$, where Z is the maximal ideal space of WAP(G). As for $C^b(G)$, there is a map $\varepsilon : G \to Z$ given by $\varepsilon(s) = \varepsilon_s$. This map is continuous from G into Z for the same reason as above.

If G is locally compact, then $C_0(G) \subset \mathrm{WAP}(G)$ by Example 7.7.2(3). This shows that $\mathrm{WAP}(G)$ separates points. Therefore, the map from G onto $\varepsilon(G) \subset Z$ taking $s \to \varepsilon_s$ is a continuous injection. A function in $f \in \mathrm{WAP}(G)$ which vanishes on ε_G must be 0, so $\varepsilon(G)$ is dense in Z. ∎

We won't prove it here, but the multiplication on G extends to a multiplication on Z which is separately continuous, making Z into a compact topological semigroup known as the WAP-compactification of G.

7.7.7. DEFINITION. A *mean* on a unital self-adjoint space $\mathcal{S} \subset C^b(G)$ is a linear functional μ such that $\mu(1) = 1 = \|\mu\|$. A mean is *left (right) invariant* if $\mu(L_s f) = \mu(f)$ (respectively $\mu(R_s f) = \mu(f)$) for all $f \in \mathcal{S}$ and $s \in G$.

We will see in Chapter 9 that this forces μ to be positive (taking positive functions to positive scalars), and thus if $a \le f \le b$ for $a, b \in \mathbb{R}$, we get $a \le \mu(f) \le b$. So in some sense, $\mu(f)$ is an average value of f. For example, if G is a compact group and μ is normalized Haar measure, then $\mu(f) = \int_G f\,d\mu$ is a left and right translation invariant mean. A very important notion in relation to this is amenability, which is studied for discrete groups in Section 9.13. Here we wish to show that there is a unique right invariant mean on $\mathrm{AP}(G)$ and $\mathrm{WAP}(G)$, and that it is also left invariant.

Not all continuous linear functionals on $C^b(G)$ are given by integration against a measure on G. However, we have shown that there is a compact Hausdorff space Z containing G as a dense subset such that continuous linear functionals on $\mathrm{WAP}(G)$ are given by integration against measures on Z. We will compute the invariant mean for each function separately.

7.7.8. THEOREM. *There is a unique right invariant mean μ on* $\mathrm{AP}(G)$ *and* $\mathrm{WAP}(G)$*, and it is also left invariant. Moreover, if J is the inverse map on G and $J^*\mu(f) := \mu(f \circ J)$, then $\mu = J^*\mu$.*

PROOF. Let $K = \overline{\mathrm{conv}}(R_G f)$. If $f \in \mathrm{AP}(G)$, since $\overline{R_G f}$ is compact, so is K by Exercise **3.23**. If $f \in \mathrm{WAP}(G)$, then K is weakly compact by Theorem 3.7.5. Clearly $R_G f$ is right invariant, and since R_s is linear and continuous, K is also R_G invariant. The maps R_G are isometries and hence is a distal family. Therefore, there is a common fixed point $g \in K$ by Hahn's theorem for $\mathrm{AP}(G)$, or by Corollary 3.8.14 of the Ryll-Nardzewski theorem in the weakly almost periodic case. The fact that $R_s g = g$ for all $s \in G$ forces g to be a constant γ. Moreover, it is clear that $\gamma \in \overline{\mathrm{conv}}(f(G))$, i.e. it is some sort of average value of f.

Since f is also left (weakly) almost periodic, we can similarly find a function $h \in \overline{\mathrm{conv}}(L_G f)$ which is invariant for L_G, and hence is a constant η. If we can show that $\gamma = \eta$, this will also show that the fixed point must be unique. Suppose that $|\gamma - \eta| = \delta > 0$. In both cases, $g \in \overline{\mathrm{conv}}(R_G f)$ and $h \in \overline{\mathrm{conv}}(L_G f)$. A dense subset of the convex hull is obtained from averages of sums of translates of f. Hence there are $s_1, \dots, s_m$ and $t_1, \dots, t_n$ in G such that

$$\|g - \frac{1}{m}\sum_{i=1}^{m} s_i \cdot f\| < \frac{\delta}{2} \quad \text{and} \quad \|h - \frac{1}{n}\sum_{j=1}^{n} f \cdot t_j\| < \frac{\delta}{2}.$$

Since g and h are constant, and thus left and right invariant,

$$\|g - \frac{1}{mn}\sum_{i=1}^{m}\sum_{j=1}^{n} s_i \cdot f \cdot t_j\| < \frac{\delta}{2}$$

and similarly

$$\|h - \frac{1}{mn}\sum_{i=1}^{m}\sum_{j=1}^{n} s_i \cdot f \cdot t_j\| < \frac{\delta}{2}.$$

Thus $|\gamma - \eta| = \|g - h\| < \delta$, a contradiction. Therefore, $\gamma = \eta$ and there is a unique constant function in K.

Define $\mu(f) = \gamma$. This is well defined, $|\mu(f)| \le \|f\|$, $\mu(\lambda f) = \lambda\mu(f)$ for $\lambda \in \mathbb{C}$, $\mu(1) = 1$ and $\mu(R_s f) = \gamma = \mu(L_t f)$. We need to verify that μ is additive.

Suppose that $f_1, f_2 \in \mathrm{WAP}(G)$ and $\gamma_i = \mu(f_i)$. Given $\delta > 0$, find $s_1, \dots, s_m$ in G such that $\|\gamma_1 - \frac{1}{m}\sum_{i=1}^{m} s_i \cdot f_1\| < \delta$. Let $f_3 = \frac{1}{m}\sum_{i=1}^{m} s_i \cdot f_2$ and note that $R_G f_3 \subset \mathrm{conv}(R_G f_2)$. It follows that $\mu(f_3)$ equals the unique constant function in $\overline{\mathrm{conv}}(R_G f_2)$, namely, γ_2. Find $t_1, \dots, t_n$ in G such that

$$\delta > \|\gamma_2 - \frac{1}{n}\sum_{j=1}^{n} t_j \cdot f_3\| = \|\gamma_2 - \frac{1}{mn}\sum_{i=1}^{m}\sum_{j=1}^{n} t_j \cdot s_i \cdot f_2\|.$$

However as constants are right invariant, we also have

$$\delta > \|\gamma_1 - \frac{1}{mn}\sum_{i=1}^{m}\sum_{j=1}^{n} t_j \cdot s_i \cdot f_1\|.$$

It follows that

$$\|(\gamma_1 + \gamma_2) - \frac{1}{mn}\sum_{i=1}^{m}\sum_{j=1}^{n} t_j \cdot s_i \cdot (f_1 + f_2)\| < 2\delta.$$

But δ is arbitrary, and hence $\mu(f_1 + f_2) = \gamma_1 + \gamma_2$. This shows that μ is a continuous linear functional which is both left and right invariant.

Finally, we observe that $L_G(f \circ J) = (R_G f) \circ J$, from which it follows that $\mu(f)$ is the unique constant in $\overline{\mathrm{conv}}(L_G(f \circ J))$. Therefore $\mu(f \circ J) = \mu(f)$. ∎

7.8. The Shilov Boundary

The notion of a boundary in this section makes sense for all commutative Banach algebras. But it is most useful for *uniform algebras*, which are norm-closed subalgebras of $C(X)$ which separate points.

7.8.1. DEFINITION. Let $\mathfrak{A}$ be a commutative Banach algebra. A *boundary* for $\mathfrak{A}$ is a subset $X \subset \mathcal{M}_{\mathfrak{A}}$ such that each $\hat{f} \in \Gamma(\mathfrak{A})$ attains its maximum modulus on X.

It turns out that there is a smallest *closed* boundary $\partial_{\mathfrak{A}}$, known as the *Shilov boundary*. This is the main goal of this section.

7.8.2. LEMMA. *Let* $a_1, \dots, a_n \in \mathfrak{A}$ *and let*

$$V = \{\varphi \in \mathcal{M}_{\mathfrak{A}} : |\varphi(a_i)| < 1 \textit{ for } 1 \le i \le n\}.$$

Then either

(1) *whenever* E *is a closed boundary, then* $E \setminus V$ *is a closed boundary or*

(2) *there is an* $a \in \mathfrak{A}$ *so that* $\{\varphi \in \mathcal{M}_{\mathfrak{A}} : |\varphi(a)| = \|\hat{a}\|_\infty\} \subset V$.

PROOF. Suppose that (1) fails. Find a closed boundary E such that $E \setminus V$ is not a boundary. Then there is some $a \in \mathfrak{A}$ so that $\|\hat{a}\|_\infty = 1$ and $\sup_{E \setminus V} |\varphi(a)| = 1 - \delta < 1$. Then there is an integer k so that

$$\sup_{E \setminus V} |\varphi(a_i a^k)| \le \|\hat{a}_i\|_\infty (1-\delta)^k < 1 \quad \text{for} \quad 1 \le i \le n.$$

Also if $\varphi \in V$, then $|\varphi(a_i a^k)| \le |\varphi(a_i)| < 1$. So $|\varphi(a_i a^k)| < 1$ for all $\varphi \in E$. Because E is a boundary, $\|\widehat{a_i a^k}\|_\infty < 1$. For every $\varphi \notin V$, there is some i so that $|\varphi(a_i)| \ge 1$, which forces $|\varphi(a)| < 1$. Therefore $\{\varphi \in \mathcal{M}_{\mathfrak{A}} : |\varphi(a)| = 1\} \subset V$. So (2) holds. ∎

7.8.3. THEOREM. *The intersection of all closed boundaries for a commutative Banach algebra* $\mathfrak{A}$ *is a closed boundary.*

PROOF. Let $\partial_{\mathfrak{A}}$ denote the intersection of all closed boundaries. If this is not a boundary, then we can find $a \in \mathfrak{A}$ so that $\|\hat{a}\|_\infty = 1$ and $\sup_{\partial_{\mathfrak{A}}} |\varphi(a)| = 1 - \delta < 1$. Let $F = \{\varphi \in \mathcal{M}_{\mathfrak{A}} : |\varphi(a)| = 1\}$. This is a compact subset of $\mathcal{M}_{\mathfrak{A}}$ which is disjoint from $\partial_{\mathfrak{A}}$.

For each $\varphi \in F$, there is a closed boundary E_φ so $\varphi \notin E_\varphi$. Thus, there is a basic open neighbourhood $\varphi \in V_\varphi = \{\psi : |\psi(a_{\varphi,i})| < 1 \text{ for } 1 \le i \le n_\varphi\}$ such that $V_\varphi \cap E_\varphi = \varnothing$. Note that since V_φ is disjoint from the boundary E_φ, it does not have property (2) of the previous lemma. Hence it has property (1), and can be removed from any boundary and leave a boundary behind.

The collection $\{V_\varphi : \varphi \in F\}$ is an open cover. Let $V_1, \dots, V_m$ be a finite subcover of F. Then since $\mathcal{M}_\mathfrak{A}$ is a boundary, it follows that $\mathcal{M}_\mathfrak{A} \setminus (V_1 \cup \dots V_m)$ is a boundary. However this set is disjoint from F, and thus $\hat{a}$ does not achieve its maximum modulus on this set. This contradiction shows that $\partial_\mathfrak{A}$ is indeed a closed boundary for $\mathfrak{A}$. ■

7.8.4. EXAMPLES. The prototypical example is the disc algebra $A(\mathbb{D})$. The maximal ideal space is $\overline{\mathbb{D}}$. However by the maximum modulus principle, every $f \in A(\mathbb{D})$ attains its maximum modulus on $\mathbb{T}$. Indeed, $\partial_{A(\mathbb{D})} = \mathbb{T}$ because if $e^{i\theta} \in \mathbb{T}$, the function $f(z) = z + e^{i\theta}$ has norm 2 and attains that value only at $e^{i\theta}$. So $\mathbb{T}$ is the minimal boundary.

A related but more interesting example is $A(\mathbb{D}^2)$, the algebra of continuous functions on $\overline{\mathbb{D}} \times \overline{\mathbb{D}}$ which are analytic (in 2 variables) on $\mathbb{D}^2$. One can obtain this as the closure in the sup norm of the polynomials in two complex variables w, z. If $f(w, z) \in A(\mathbb{D}^2)$, then for each $w \in \overline{\mathbb{D}}$, the function $f_w(z) = f(w, z) \in A(\mathbb{D})$. This attains its maximum modulus on $\{w\} \times \mathbb{T}$. Therefore $\overline{\mathbb{D}} \times \mathbb{T}$ is a closed boundary. Similarly $\mathbb{T} \times \overline{\mathbb{D}}$ is a closed boundary. The intersection is $\mathbb{T} \times \mathbb{T}$, and this is the Shilov boundary. For example, the function $(w + e^{i\omega})(z + e^{i\theta})$ takes its maximum modulus 4 only at $(e^{i\omega}, e^{i\theta})$. One important observation is that $\mathbb{T}^2$ is strictly smaller than the topological boundary of $\overline{\mathbb{D}}^2$.

If $\mathfrak{A}$ is a commutative C*-algebra $C(X)$ or $C_0(X)$, then $\partial_{C_0(X)} = X$ because for any proper closed set Y and a point $x \notin Y$, there is an $f \in C_0(X)$ such that $f|_Y = 0$ and $f(x) = 1$. So Y is not a boundary.

7.8.5. DEFINITION. A *uniform algebra* $\mathcal{A}$ is a closed subalgebra of $C(X)$ for some compact Hausdorff space X which contains the constant functions and separates points of X. If $\varphi \in \mathcal{A}^*$ satisfies $1 = \varphi(1) = \|\varphi\|$, a regular Borel probability measure μ on X is a *representing measure* for φ if

$$f(x) = \int_Y f \, d\mu \quad \text{for all} \quad f \in \mathcal{A}.$$

Note that by the Hahn–Banach theorem, φ extends to a linear functional $\widetilde{\varphi}$ on $C(X)$ with $\|\widetilde{\varphi}\| = 1$. Then by the Riesz representation theorem, there is a regular Borel measure of norm 1 on X representing $\widetilde{\varphi}$. Since $\widetilde{\varphi}(1) = 1 = \|\mu\|$, it follows that μ is positive and so a probability measure. This establishes the existence of representing measures.

In the following, we use the fact that a uniform algebra $\mathcal{A}$ on X is also a uniform algebra on $Y = \partial_\mathcal{A}$.

7.8.6. PROPOSITION. *Let X be a compact Hausdorff space and let $\mathcal{A} \subset C(X)$ be a uniform algebra. Let $Y = \partial_\mathcal{A}$ be its Shilov boundary. For each $x \in X$, there* *is a Borel probability measure μ on Y representing evaluation at x:*

$$f(x) = \int_Y f\, d\mu \quad \textit{for all} \quad f \in \mathcal{A}.$$

Exercises for Chapter 7

7.1 Let X be a locally compact Hausdorff space.

(a) Let J be a closed ideal of $C_0(X)$ and let

$$Y = \ker J = \{x \in X : f(x) = 0 \text{ for all } f \in J\}.$$

Prove that $J = I(Y) = \{f \in C_0(X) : f(x) = 0 \text{ for all } x \in Y\}$.

(b) Show that every closed ideal of $C_0(X)$ is the intersection of all maximal ideals which contain it.

(c) Let J be an ideal of $C_0(X)$, not necessarily closed, and let $E = \ker J$. Let $I_0(E)$ be the ideal of functions that vanish on an open neighbourhood of E. Prove that $I_0(E) \subset J \subset I(E)$.

7.2 (a) Find all maximal ideals of $C^{(n)}[a, b]$ for $n \geq 1$.

(b) Find a closed ideal which is not the intersection of maximal ideals.

(c) Find all closed ideals of $C^{(n)}[a, b]$.

7.3 Let $\mathfrak{A} = \mathbb{C}[z]/(z^n)$ with norm of $p(z) = \sum_{i\geq 0} a_i z_i$ given by $\|p\| = \sum_{i=0}^{n-1} |a_i|$. Show that this commutative Banach algebra has a unique maximal ideal.

7.4 (*Wiener–Levy*) Let $A(\mathbb{T})$ be the Banach algebra of functions on $\mathbb{T}$ with absolutely convergent Fourier series. Show that if $f \in A(\mathbb{T})$ and $g(z)$ is analytic on a neighbourhood of $f(\mathbb{T})$, then $g(f(z)) \in A(\mathbb{T})$.

7.5 Suppose that $\mathfrak{A}$ is a unital Banach algebra generated by a single element a. Prove that $\hat{a}$ is a homeomorphism of $\mathcal{M}_{\mathfrak{A}}$ onto $\sigma(a)$.

7.6 Let $V \in \mathcal{B}(L^2(0, 1))$ be the Volterra operator (Example 5.1.5). Find all multiplicative functionals on $A(V)$, the unital Banach algebra generated by V.

7.7 Let $\mathfrak{A}$ be a unital Banach algebra, and let $a, b \in \mathfrak{A}$ commute. Prove that $\operatorname{spr}(a + b) \leq \operatorname{spr}(a) + \operatorname{spr}(b)$.
HINT: Work in the commutative Banach algebra generated by $a, b, 1$.

7.8 Let $l_1(w)$ be the Banach algebra from Exercise **6.3**. Let $\lim_{n\to\infty} w_n^{1/n} = L$ (see Exercise **4.24**). Show that $l_1(w)$ is semisimple if $L > 0$ and radical if $L = 0$.

7.9 Let $\mathfrak{A}$ and $\mathfrak{B}$ be unital commutative Banach algebras. Suppose that $\theta : \mathfrak{A} \to \mathfrak{B}$ is an injective homomorphism, and $\mathfrak{B}$ is semisimple. Prove that $\mathfrak{A}$ is semisimple.
HINT: If $a \in \operatorname{rad}(\mathfrak{A})$, what can you say about $\sigma(\theta(a))$?

7.10 Let l_p, for $1 \le p < \infty$, be a Banach algebra with pointwise multiplication.
(a) Find its maximal modular ideals.
(b) Prove that there are maximal ideals which are not modular.
HINT: Any subspace containing $l_p^2 = \{xy : x, y \in l_p\}$ is an ideal.
(c) Show that l_p has an unbounded approximate identity, but no bounded one.

7.11 Let $\mathfrak{A} = C^{(1)}[0,1]$ with $\|f\| = \|f\|_\infty + \|f'\|_\infty$. Let

$$J = \{f \in \mathfrak{A} : f(0) = f'(0) = 0\}.$$

Show that $\mathfrak{A}$ is semisimple but $\mathfrak{A}/J$ is not.

7.12 Show that the spectral radius is continuous on commutative Banach algebras.

7.13 Let a, b belong to a unital Banach algebra $\mathfrak{A}$ such that $ab = ba$. Show that

$$\sigma(a+b) \subset \sigma(a) + \sigma(b) := \{\alpha + \beta : \alpha \in \sigma(a),\ \beta \in \sigma(b)\}.$$

HINT: Work in the commutative Banach subalgebra generated by a, b and 1.

7.14 Let $\mathfrak{A}$ be a unital commutative Banach algebra. If $a_1, \dots, a_n \in \mathfrak{A}$, define the *joint spectrum* to be

$$\sigma(a_1, \dots, a_n) = \{\lambda = (\varphi(a_1), \dots, \varphi(a_n)) \in \mathbb{C}^n : \varphi \in \mathcal{M}(\mathfrak{A})\}.$$

(a) Show that $\lambda = (\lambda_1, \dots, \lambda_n) \notin \sigma(a_1, \dots, a_n)$ if and only if there are $b_1, \dots, b_n$ in $\mathfrak{A}$ such that $\sum_{i=1}^n (a_i - \lambda_i) b_i = 1$.
(b) We say $a_1, \dots, a_n$ generate $\mathfrak{A}$ if the set of polynomials in $a_1, \dots, a_n$ is dense in $\mathfrak{A}$. If this occurs, show that the map $\Phi : \mathcal{M}(\mathfrak{A}) \to \sigma(a_1, \dots, a_n)$ given by $\Phi(\varphi) = (\varphi(a_1), \dots, \varphi(a_n))$ is a homeomorphism.
(c) If $a_1, \dots, a_n$ generate $\mathfrak{A}$, show that $\sigma(a_1, \dots, a_n)$ is *polynomially convex*, i.e. if $\lambda \in \mathbb{C}^n$ and $|p(\lambda)| \le \sup\{|p(z)| : z \in \sigma(a_1, \dots, a_n)\}$ for all $p \in \mathbb{C}[z]$, then $\lambda \in \sigma(a_1, \dots, a_n)$.
HINT: If $\lambda \notin \sigma(a_1, \dots, a_n)$, use (a) and approximate $1 - \sum_{i=1}^n (a_i - \lambda_i) b_i = 0$ by a polynomial.

7.15 Let G be a compact abelian group. Show that distinct characters are orthogonal in $L^2(G)$.
HINT: If $\gamma \in \widehat{G} \setminus \{1\}$, show $\int_G \gamma(s)\, ds = \gamma(s_0) \int_G \gamma(s)\, ds$ for any $s_0 \in G$.

7.16 Show that $L^1(\mathbb{Z})$ is not a C*-algebra.
HINT: Show that the Gelfand map is not isometric.

7.17 Let G be a LCA group. Here is another proof of semisimplicity of $L^1(G)$.
(a) Show that for $f \in L^1(G)$, the operator $\lambda_f(g) = f * g$ is bounded on $L^2(G)$ with $\|\lambda_f\| \le \|f\|_1$.
(b) Prove that $\Lambda : L^1(G) \to \mathcal{B}(L^2(G))$ by $\Lambda(f) = \lambda_f$ is a homomorphism.
(c) Show that Λ is injective with range in a commutative C*-algebra.
HINT: $L^2(G)$ contains an approximate identity for $L^1(G)$.
(d) Apply Exercise **7.9** to conclude that $L^1(G)$ is semisimple.

7.18 Let $A \in \mathcal{B}(X)$ and $B \in \mathcal{B}(Y)$ be operators on Banach spaces X and Y such that $\sigma(A) \cap \sigma(B) = \varnothing$. Define linear maps in $\mathcal{B}(\mathcal{B}(Y, X))$ by $L_A T = AT$ and $R_B T = TB$. Prove that

$$\sigma(L_A - R_B) \subset \sigma(A) - \sigma(B) = \{\lambda - \mu : \lambda \in \sigma(A) \text{ and } \mu \in \sigma(B)\}.$$

In particular, the map $L_A - L_B$ is invertible.
HINT: The Banach algebra generated by L_A and L_B is commutative.

7.19 Show that if S and T are two normal operators, then there is a $*$-isomorphism between $C^*(S)$ and $C^*(T)$ taking S to T if and only if $\sigma(S) = \sigma(T)$.

7.20 Let a be a positive element of a C*-algebra $\mathfrak{A}$. Prove that a has a unique positive square root.
HINT: For uniqueness, if b is the square root obtained from the functional calculus and c is another, then both b, c belong to the abelian algebra $C^*(c)$.

7.21 Put an involution on $A(\mathbb{D})$ by $f^*(z) = \overline{f(\bar{z})}$. Show that this is an isometric involution which fails to satisfy the C*-algebra identity.

7.22 Let G be a topological group, and let $f \in \mathrm{AP}(G)$. Show that $s \to s \cdot f$ is norm continuous on G.
HINT: This is pointwise continuous (τ_p) on $K = \overline{R_G f}$. Show that the map from K to (K, τ_p) is a homeomorphism.

7.23 Let θ be a homomorphism of $C(X)$ into $C(Y)$, where X and Y are compact Hausdorff spaces. Prove that there is a continuous map $h : Y \to X$ so that $\theta(f) = f \circ h$.

7.24 Let $\beta\mathbb{N}$ be the maximal ideal space of l_∞, and let δ_n be evaluation at n for $n \in \mathbb{N}$. Consider $\mathbb{N}$ as a subset of $\beta\mathbb{N}$ by identifying it with $\{\delta_n : n \in \mathbb{N}\}$.

(a) Show that $\mathbb{N}$ is dense in $\beta\mathbb{N}$.
(b) Show that the Gelfand map carries l_∞ onto $\mathrm{C}(\beta\mathbb{N})$ isometrically.
(c) Show that the closure $\overline{S}$ in $\beta\mathbb{N}$ of each subset $S \subset \mathbb{N}$ is open, and that these clopen sets form a basis for the topology of $\beta\mathbb{N}$.
HINT: simple functions are dense in l_∞.
(d) For $x \in \beta\mathbb{N}$, let $\mathcal{U}_x = \{S \subset \mathbb{N} : x \in \overline{S}\}$. Show that this is an ultrafilter. Conversely, show that for every ultrafilter $\mathcal{U}$ on $\mathbb{N}$, there is a unique point in $\bigcap\{\overline{S} : S \in \mathcal{U}\}$.
HINT: Look up what an ultrafilter is.

7.25 Let X be a Hausdorff space and let $\mathfrak{A} = C^b(X)$, the space of all bounded continuous functions on X with the sup norm. Let βX denote the maximal ideal space of X.

(a) Show that βX is a compact Hausdorff space and there is a natural continuous injection of X onto a dense subset of βX. (Compare Exercise **1.21**.)

(b) Let K be a compact Hausdorff space, and suppose that $f : X \to K$ is continuous. Prove that f extends uniquely to a continuous function $\widetilde{f} : \beta X \to K$.
HINT: Consider the map sending $h \in C(K)$ to $h \circ f$ in $C^b(X)$.
(c) Show that X is homeomorphic to its image in βX if and only if X is completely regular.
HINT: Compact Hausdorff spaces are completely regular.

7.26 Let $\mathfrak{A}$ be a unital commutative Banach algebra. An element $\varphi \in \mathcal{M}_{\mathfrak{A}}$ is a *peak point* if there is some $a \in \mathfrak{A}$ such that $|\hat{a}(\varphi)| > |\hat{a}(\psi)|$ for $\psi \in \mathcal{M}_{\mathfrak{A}} \setminus \{\varphi\}$.
(a) Show that the Shilov boundary contains all peak points.
(b) Use this to find the Shilov boundary of $C(X)$ for X compact metric.

7.27 Let $\mathfrak{A}$ be a unital commutative Banach algebra, and let $a \in \mathfrak{A}$. Show that the boundary of $\sigma(a)$ is contained in $\hat{a}(\partial_{\mathfrak{A}})$.

7.28 Let $\mathfrak{A}$ be a unital commutative Banach algebra with infinite maximal ideal space. Prove that there are elements of $\mathfrak{A}$ with infinite spectrum.
HINT: Let $\{\varphi_n\}_{n\geq 1}$ be distinct points in $\mathcal{M}(\mathfrak{A})$. Consider $\bigcup_{m\neq n} \ker(\varphi_m - \varphi_n)$.

7.29 Let $\mathcal{E}$ denote the space of all entire functions on $\mathbb{C}$.
(a) Show $\mathcal{E}$ is a unital commutative normed algebra with $\|f\|_1 = \sup_{|z|\leq 1} |f(z)|$, but it is not complete.
(b) Show that $\mathcal{E}$ is a Fréchet algebra for the family of norms $\|f\|_n = \sup_{|z|\leq n} |f(z)|$.
(c) Show that every multiplicative linear functional has the form $\varepsilon_z(f) = f(z)$ for $z \in \mathbb{C}$.
(d) Compute the spectrum of $id(z) = z$.
(e) Let $J = \{f \in \mathcal{E} : \exists N \in \mathbb{N},\ f(n) = 0 \text{ for } n \in \mathbb{N},\ n \geq N\}$. Show that J is contained in a maximal ideal I such that $\mathcal{E}/I$ is infinite dimensional.
HINT: Let $I = \{f \in \mathcal{E} : Z(f) \cap \mathbb{N} \in \mathcal{U}\}$ where $\mathcal{U}$ is a non-principal ultrafilter on $\mathbb{N}$.

Notes on Chapter 7

The formal study of abstract Banach algebras is due to Gelfand [119]. He developed the commutative theory in several papers from 1939 to 1944. Many details were explained in the 1946 book, available in translation [122]. It was perhaps his new simple proof of Wiener's theorem on absolutely convergent Fourier series which demonstrated the power of this theory.

Theorem 7.3.1 and the uniqueness of norm is due to Rickart [257] in 1950. The algebras $L^1(G)$, especially for $\mathbb{R}$ and $\mathbb{T}$, were widely studied using Fourier analysis for a long time. The semisimplicity of $L^1(G)$ can be deduced from the inversion formula for the Fourier transform, which shows that the Fourier transform is injective. See Rudin [268] or Hewitt and Ross [147].

Commutative C*-algebras were introduced by Gelfand and Naimark [120] in 1943. They establish Theorem 7.6.3 and the continuous functional calculus for normal operators. Shilov proved the existence of a minimal closed boundary (see [122]). He also developed a functional calculus of several complex variables and used it to prove the Shilov idempotent theorem. This states that if K is a clopen subset of $\mathcal{M}_{\mathfrak{A}}$, then there is an $a \in \mathfrak{A}$ with $\hat{a} = \chi_K$.

Almost periodic functions are studied in Loomis [195] and Dixmier [95]. Weakly almost periodic functions and the compactification of groups and semigroups is covered in Burckel [44] and Berglund–Junghenn–Milnes [27]. The application of the Ryll-Nardzewski theorem to obtain an invariant mean on WAP(G) is in [271].

Scc Kaniuth [172] or Bonsall and Duncan [37] for more on this material.

CHAPTER 8

Noncommutative Banach algebras

The representation theory of non-commutative Banach algebras parallels ideas from ring theory. We begin with a study of irreducible modules and representations. We define the radical and characterize it. The Jacobson density theorem shows that irreducible modules are highly transitive. Then we cover some automatic continuity results of B. Johnson. Finally we establish Cohen's factorization theorem, which shows that a Banach algebra with a bounded approximate identity satisfies $\mathfrak{A}^2 = \mathfrak{A}$.

8.1. Representation Theory

In parallel with ideas used in the commutative case, we make a definition.

8.1.1. DEFINITION. A proper left ideal I of $\mathfrak{A}$ is a *modular left ideal* if it has a *right* modular unit $e \in \mathfrak{A}$, meaning that $a - ae \in I$ for all $a \in \mathfrak{A}$. One similarly defines right modular ideals.

8.1.2. REMARK. If $\mathfrak{A}$ is unital, then all left (right) ideals are modular. A left ideal I is modular if $\mathfrak{A}(1-e) \subset I$, where 1 is understood to belong to $\widetilde{\mathfrak{A}}$ if $\mathfrak{A}$ is not unital. If I is a modular left ideal with right modular unit e, then every left ideal $J \supset I$ is modular with right modular unit e because $\mathfrak{A}(1-e) \subset I \subset J$.

If $I \lhd \mathfrak{A}$ is a 2-sided ideal, and it is both left and right modular, then the same e works on both sides. Indeed, if $\mathfrak{A}(1-e_1) \subset I$ and $(1-e_2)\mathfrak{A} \subset I$, then $e_2 - e_2e_1$ and $e_1 - e_2e_1$ belong to I, and thus $e_1 - e_2 \in I$. Therefore

$$(1-e_1)\mathfrak{A} \subset (1-e_2)\mathfrak{A} + (e_1-e_2)\mathfrak{A} \subset I + I = I.$$

There is a close relationship between modular left ideals of $\mathfrak{A}$ and ideals of $\widetilde{\mathfrak{A}}$.

8.1.3. THEOREM. *Let $\mathfrak{A}$ be a Banach algebra without a unit. If J is a proper left ideal of $\widetilde{\mathfrak{A}}$ not contained in $\mathfrak{A}$, then $I = J \cap \mathfrak{A}$ is a proper left modular ideal of $\mathfrak{A}$. Conversely if I is a proper left modular ideal of $\mathfrak{A}$ with right modular unit e, then $J = I + \mathbb{C}(1-e)$ is a proper left ideal of $\widetilde{\mathfrak{A}}$ not contained in $\mathfrak{A}$ and $I = J \cap \mathfrak{A}$.*

K. R. Davidson, *Functional Analysis and Operator Algebras*, CMS/CAIMS Books in Mathematics 13,
https://doi.org/10.1007/978-3-031-63665-3_8

PROOF. Let J be a proper left ideal of $\widetilde{\mathfrak{A}}$ not contained in $\mathfrak{A}$. Evidently $I = J \cap \mathfrak{A}$ is a left ideal of $\mathfrak{A}$. It is proper since $I = \mathfrak{A}$ implies that $J = \mathfrak{A}$ or $J = \widetilde{\mathfrak{A}}$, which are both excluded. Since $\dim \widetilde{\mathfrak{A}}/\mathfrak{A} = 1$, we have $\dim J/I \leq 1$, and it equals 1 since $J \neq I$. Pick $u \in J \setminus I$ and write $u = a + \lambda$ for $a \in \mathfrak{A}$ and $\lambda \in \mathbb{C}$. Thus $1 + \lambda^{-1}a \in J$. Set $e = -\lambda^{-1}a \in \mathfrak{A}$. Then $1 - e \in J$, so that $\mathfrak{A}(1-e) \subset \mathfrak{A} \cap J = I$. Hence e is the right modular unit for I.

Conversely, suppose that I is a proper left modular ideal with the right modular unit e. Define $J = I + \mathbb{C}(1-e)$. Then

$$\widetilde{\mathfrak{A}}J \subset \mathfrak{A}I + I + \mathfrak{A}(1-e) + \mathbb{C}(1-e) \subset I + \mathbb{C}(1-e) = J.$$

This is a left ideal not contained in $\mathfrak{A}$ with $J \cap \mathfrak{A} = I$. Since I is proper, so is J. ■

8.1.4. COROLLARY. *If I is a proper left modular ideal of $\mathfrak{A}$ with right modular unit e, then $\|a - e\| \geq 1$ for all $a \in I$.*

PROOF. If $\|a - e\| < 1$ for $a \in I$, then $1 - e + a$ is invertible in $\widetilde{\mathfrak{A}}$ and belongs to $J = I + \mathbb{C}(1-e)$. This would mean that $J = \widetilde{\mathfrak{A}}$ is not proper. ■

This allows us to establish the existence of maximal modular ideals in the non-unital case.

8.1.5. PROPOSITION. *If I is a modular left ideal of $\mathfrak{A}$, then I is contained in a maximal modular left ideal. Maximal modular left ideals are closed and are maximal left ideals.*

PROOF. Consider the set of proper left ideals containing I. They are all modular with right modular unit e, and thus are disjoint from $b_1(e)$ by Corollary 8.1.4. The union of any nested family of such ideals is an ideal which is disjoint from $b_1(e)$. So by Zorn's lemma, maximal ideals containing I exists, and they are modular. The closure of this ideal is still a left ideal and is still disjoint from $b_1(e)$, and thus is a possibly larger proper left ideal. Therefore maximal modular ideals are closed. ■

8.1.6. DEFINITION. If X is a vector space and $\mathcal{L}(X)$ is the space of linear maps on X, a *representation* of $\mathfrak{A}$ is an algebra homomorphism $\pi : \mathfrak{A} \to \mathcal{L}(X)$. This makes X (actually (X, π)) into a *left $\mathfrak{A}$-module* by $ax = \pi(a)x$ for $a \in \mathfrak{A}$ and $x \in X$. It is a *left Banach $\mathfrak{A}$-module* if X is a Banach space and π is a continuous algebra homomorphism into $\mathcal{B}(X)$. Thus there is a constant $K = \|\pi\|$ such that

$$\|ax\| \leq K\|a\|\|x\| \quad \text{for} \quad a \in \mathfrak{A} \text{ and } x \in X.$$

A module X is *non-trivial* if $\pi(\mathfrak{A})X \neq \{0\}$. We say that X is *irreducible* if X is non-trivial and has no proper submodules; i.e. no proper subspaces which are invariant under the action of $\mathfrak{A}$.

We first obtain some basic properties of irreducible modules.

8.1.7. PROPOSITION. *Let $\mathfrak{A}$ be a Banach algebra, and let X be an irreducible left $\mathfrak{A}$ module. Fix a vector $0 \neq x_0 \in X$. Then*

(1) $\mathfrak{A}x_0 = X$.

(2) *Let $I_{x_0} = \{a \in \mathfrak{A} : ax_0 = 0\}$. If $e \in \mathfrak{A}$ such that $ex_0 = x_0$, then e is a right modular unit for I_{x_0}.*

(3) *I_{x_0} is a maximal modular left ideal.*

(4) *If π is the representation associated to X, then*

$$\ker \pi = \bigcap_{x \in X} I_x = I_{x_0} : \mathfrak{A} := \{a \in \mathfrak{A} : a\mathfrak{A} \subset I_{x_0}\}$$

is a closed 2-sided ideal.

PROOF. First, $\mathfrak{A}x_0$ is a submodule, so either $\mathfrak{A}x_0 = X$ or $\mathfrak{A}x_0 = \{0\}$. If the latter, then $\mathbb{C}x_0$ is a submodule. As it is non-zero, we have $X = \mathbb{C}x_0$ and $\pi = 0$. This means that X is trivial, a contradiction. Therefore $\mathfrak{A}x_0 = X$.

Clearly I_{x_0} is a left ideal, which is proper by (1). Since $\mathfrak{A}x_0 = X$, there is some $e \in \mathfrak{A}$ so that $ex_0 = x_0$. For any $a \in \mathfrak{A}$, $(a - ae)x_0 = ax_0 - ax_0 = 0$. Hence $a - ae \in I_{x_0}$. Thus e is a right modular unit for I_{x_0}.

Suppose that J is a left ideal properly containing I_{x_0}. Then $Jx_0 \neq \{0\}$ is a submodule of X, and thus $Jx_0 = X$. Thus there is some $e \in J$ so that $ex_0 = x_0$. If $a \in \mathfrak{A}$, then $a = (a - ae) + ae \in I_{x_0} + J = J$. Therefore $J = \mathfrak{A}$; whence I_{x_0} is maximal.

Clearly $\ker \pi = \bigcap_{x \in X \setminus \{0\}} I_x$. By maximality, each I_x is closed, so $\ker \pi$ is closed. Since π is a homomorphism, $\ker \pi$ is a 2-sided ideal. Finally,

$$\begin{aligned} \ker \pi &= \{a \in \mathfrak{A} : ax = 0 \text{ for all } x \in X\} \\ &= \{a \in \mathfrak{A} : a(bx_0) = 0 \text{ for all } b \in \mathfrak{A}\} \\ &= \{a \in \mathfrak{A} : a\mathfrak{A} \subset I_{x_0}\} = I_{x_0} : \mathfrak{A}. \end{aligned}$$ ∎

Next we show that there is an irreducible Banach module with the same properties.

8.1.8. PROPOSITION. *Let I be a maximal modular left ideal. Then there is an irreducible Banach module (X, π) and element $x_0 \in X$ such that $I_{x_0} = I$.*

PROOF. Take $X = \mathfrak{A}/I$ with action $a(b+I) = ab+I$. Let e be a right modular unit for I, and set $x_0 = e + I$. Then for $x = b + I \in X$,

$$\|ax\| = \|ab + I\| = \inf_{i \in I} \|ab + i\| \leq \inf_{i \in I} \|ab + ai\| \leq \inf_{i \in I} \|a\| \, \|b + i\| = \|a\| \, \|x\|.$$

So X is a Banach $\mathfrak{A}$-module.

$$I_{x_0} = \{a \in \mathfrak{A} : ax_0 = 0\} = \{a \in \mathfrak{A} : ae \in I\}.$$

But $a = (a - ae) + ae \in ae + I$; so $ae \in I$ if and only if $a \in I$. Hence $I_{x_0} = I$. ∎

Now we show that the ideal I_{x_0} determines the module.

8.1.9. PROPOSITION. *Let (X, π) be an irreducible module. Fix $0 \neq x_0 \in X$ and set $I = I_{x_0}$. Then the map $\theta : \mathfrak{A}/I \to X$ by $\theta(a + I) = ax_0$ is a module isomorphism. The norm $|\!|\!|x|\!|\!| = \|\theta^{-1}x\|$ makes X into a Banach module. If X is already a Banach module with norm $\|\cdot\|$, then θ is a homeomorphism.*

PROOF. Clearly θ is a well-defined linear bijection of $\mathfrak{A}/I$ onto $\mathfrak{A}x_0 = X$. Also

$$\theta(a(b + I)) = \theta(ab + I) = abx_0 = a(bx_0) = a\theta(b + I).$$

Hence θ is a module homomorphism. The norm $|\!|\!|x|\!|\!|$ makes X isometrically module isomorphic to $\mathfrak{A}/I$; whence it is a Banach module.

If X is a Banach module already, then

$$\begin{aligned} \|\theta(a + I)\| = \|ax_0\| &= \inf_{i \in I} \|(a + i)x_0\| \\ &\leq \|\pi\| \inf_{i \in I} \|a + i\| \, \|x_0\| = \big(\|\pi\| \, \|x_0\|\big) \|a + I\|. \end{aligned}$$

Therefore θ is a continuous linear bijection, and thus is invertible by the Banach isomorphism theorem. ∎

8.1.10. DEFINITION. A 2-sided ideal of a Banach algebra $\mathfrak{A}$ is *primitive* if it is the kernel of an irreducible representation. The *radical* $\operatorname{rad}\mathfrak{A}$ is the intersection of all primitive ideals:

$$\operatorname{rad}\mathfrak{A} = \{a \in \mathfrak{A} : \pi(a) = 0 \text{ for all } \pi \text{ irreducible}\}.$$

If $\operatorname{rad}\mathfrak{A} = \{0\}$, then $\mathfrak{A}$ is *semisimple*. If $\mathfrak{A}$ has no irreducible representations, then $\mathfrak{A}$ is a *radical Banach algebra*.

The previous results make the following immediate.

8.1.11. COROLLARY. *The primitive ideals of $\mathfrak{A}$ are the 2-sided ideals of the form $I : \mathfrak{A} = \{a \in \mathfrak{A} : a\mathfrak{A} \subset I\}$ for the maximal modular left ideals I of $\mathfrak{A}$.*

8.1.12. REMARK. If $\mathfrak{A}$ is commutative and (X, π) is an irreducible module with $I = I_{x_0}$, then I is a maximal (2-sided) modular ideal. Thus π is a multiplicative linear functional and $X \simeq \mathbb{C}$. Therefore the primitive ideals are precisely the maximal modular ideals, and we see that the definition of radical reduces to our earlier definition in the commutative case.

8.1.13. EXAMPLE. Let $\mathfrak{A} = \mathcal{T}_n$, the set of $n \times n$ upper triangular matrices. This algebra is unital, so all ideals are modular.

First let's identify all 2-sided ideals J. If $T = [t_{ij}] \in J$ and $a_{i_0j_0} \neq 0$, then

$$t_{i_0j_0}^{-1} E_{i_0i_0} T E_{j_0j_0} = E_{i_0j_0} \in J.$$

Moreover if $i \leq i_0$ and $j \geq j_0$, then

$$E_{ii_0} E_{i_0j_0} E_{j_0j} = E_{ij} \in J.$$

Hence a 2-sided ideal J is determined by the monotone increasing function

$$f(i) = \min\{j : E_{ij} \in J\},$$

where the minimum of the empty set is ∞. Then

$$J = \operatorname{span}\{E_{ij} : j \geq f(i), 1 \leq i, j \leq n\}.$$

This makes it easy to see that the maximal 2-sided ideals are

$$J_i = \{T \in \mathcal{T}_n : t_{ii} = 0\} \quad \text{for} \quad 1 \leq i \leq n.$$

Clearly these are also maximal left ideals. We will show that they are exactly the maximal left ideals. To this end, suppose that I is a left idea which is not contained in any J_i. Then for each $1 \leq i \leq n$, there is some $T_i \in I$ with non-zero ii entry t_{ii}. So $T = \sum_{i=1}^n t_{ii}^{-1} E_{ii} T_i$ is an element of I of the form $T = I_n + A$ where A is strictly upper triangular. Thus $\sigma(T) = \{1\}$, and in particular T is invertible. Indeed, $T^{-1} = \sum\limits_{i=0}^{n-1} (-A)^i$ because $A^n = 0$. So $I = \mathcal{T}_n$.

Since a maximal left ideal J_i is actually 2-sided, $J_i : \mathcal{T}_n = J_i$. The quotient is $\mathcal{T}_n / J_i \simeq \mathbb{C}$. This determines the irreducible representation $\pi_i(T) = t_{ii}$; and $\ker \pi_i = J_i$ is a primitive ideal. Therefore

$$\operatorname{rad} \mathcal{T}_n = \bigcap_{i=1}^n J_i = \mathcal{T}_n^0 = \{T \in \mathcal{T}_n : t_{ii} = 0 \text{ for } 1 \leq i \leq n\}$$

is the ideal of strictly upper triangular matrices.

The quotient $\mathcal{T}_n / \operatorname{rad} \mathcal{T}_n$ is isomorphic to the algebra $\mathcal{D}_n$ of $n \times n$ diagonal matrices. This is a commutative algebra which is isomorphic to $C(X_n)$ where $X_n = \{1, 2, \ldots, n\}$ is an n-point space with the discrete topology. This has been shown to be semisimple.

8.1.14. EXAMPLE. Suppose that $\mathcal{T}_n^0$ has a modular left ideal I with the right modular unit e. Then $\mathcal{T}_n^0 (I_n - e) \subset I$. In particular, if $1 \leq i < j \leq n$, then I contains

$$E_{ij} - E_{ij} e = E_{ij} + \sum_{k > j} a_{ijk} E_{ik}.$$

Taking $j = n$, it follows that I contains all E_{in} for $i < n$. By stages, one sees that $E_{ij} \in I$ for all $1 \leq i < j \leq n$. Hence $I = \mathcal{T}_n^0$, which is not proper. Therefore $\mathcal{T}_n^0$

has no maximal modular left ideals, and so has no irreducible modules. Thus the strictly upper triangular matrices $\mathcal{T}_n^0$ is a radical Banach algebra.

8.1.15. EXAMPLE. Let $\mathfrak{A} = M_n$, the set of all $n \times n$ matrices. This algebra is unital, so all ideals are modular. The identity representation, $id(A) = A$, is irreducible since $M_n x = \mathbb{C}^n$ for all non-zero vectors x. It is easier to describe the maximal right ideals, but one can check that the transpose map takes right ideals to left ideals and vice versa.

Suppose that $I = IM_n$ is a maximal right ideal. Let

$$V = \operatorname{Ran} I = \{x = Ay : A \in I,\ y \in \mathbb{C}^n\}.$$

Since I is a linear space, so is V. Let P_V be the orthogonal projection onto V.

We will show first that V is a proper subspace. Otherwise, for an orthonormal basis $e_1, \dots, e_n$, there are elements $A_i \in I$ and $y_i \in \mathbb{C}^n$ so that $A_i y_i = e_i$. Thus I contains the rank one matrix $A_i(y_i e_i^*) = (A_i y_i)e_i^* = e_i e_i^*$. Summing over i yields the identity $I_n \in I$. So $I = M_n$ is not proper, a contradiction. This also shows that if W is a subspace of codimension 1, then $P_W M_n$ is a maximal right ideal.

Hence V is a proper subspace. So there is a subspace W of codimension 1 with $V \subset W$. Then $I \subset P_V M_n \subset P_W M_n$. Since I is maximal and $P_W M_n$ is a maximal right ideal, $V = W$ and $I = P_V M_n$. Taking transposes, the maximal left ideals have the form $I_V = M_n P_V$ where V is a subspace of codimension 1.

Let v be a unit vector in $V^\perp$. Then $I_V = \{A \in M_n : Av = 0\}$. The module

$$X_V = M_n / I_V \simeq M_n P_V^\perp = \{xv^* : x \in \mathbb{C}^n\}$$

is n-dimensional. Let's take $x_0 = I_n + I_V = vv^* + I_V$. Then

$$I_{x_0} = \{A \in M_n : Avv^* + I_v = I_v\} = \{A \in M_n : Av = 0\} = I_V.$$

The corresponding representation acts by

$$\pi(A)(xv^* + I) = (Ax)v^* + I.$$

The map $\theta : \mathbb{C}^n \to X_V$ by $\theta(x) = xv^* + I_V$ is readily seen to be a module isomorphism. In other words, π is equivalent to the identity representation on $\mathbb{C}^n$. So M_n has a unique irreducible representation up to equivalence. It is faithful, so $\operatorname{rad} M_n = \ker id = \{0\}$. Thus M_n is semisimple.

8.2. The radical

The goal of this section is to characterize the radical in terms of invertibility. We need an easy algebraic lemma.

8.2.1. Lemma. *Let $\mathfrak{A}$ be a unital Banach algebra, and let $a, b \in \mathfrak{A}$. If $0 \neq \lambda \in \mathbb{C}$, then $ab - \lambda$ is left (right) invertible if and only if $ba - \lambda$ is left (right) invertible. In particular, $\sigma(ab) \cup \{0\} = \sigma(ba) \cup \{0\}$.*

Proof. If $u(ab - \lambda) = 1$, then

$$-\lambda^{-1}(1 - bua)(ba - \lambda) = -\lambda^{-1}\big(ba - \lambda - bu(ab - \lambda)a\big) = 1.$$

A similar identity works on the right. In particular, if $\lambda \neq 0$, then $\lambda \in \sigma(ab)$ if and only if $\lambda \in \sigma(ba)$. ■

We now present several characterizations of the radical. In particular, this shows that the choice of left or right modules/ideals does not matter.

8.2.2. Theorem. *Let $\mathfrak{A}$ be a Banach algebra. If it is not unital, then we will work with invertibility in $\widetilde{\mathfrak{A}}$. The following sets are equal:*

(1) $J_1 = \operatorname{rad} \mathfrak{A}$

(2l) J_{2l}, *the intersection of all maximal modular left ideals.*

(2r) J_{2r}, *the intersection of all maximal modular right ideals.*

(3) $J_3 = \{a \in \mathfrak{A} : \sigma(ab) = \{0\}$ *for all* $b \in \mathfrak{A}\}$.

(3′) $J_{3'} = \{a \in \mathfrak{A} : \sigma(ba) = \{0\}$ *for all* $b \in \mathfrak{A}\}$.

(4l) $J_{4l} = \{a \in \mathfrak{A} : ab - \lambda$ *is left invertible for all* $b \in \mathfrak{A}$ *and* $0 \neq \lambda \in \mathbb{C}\}$.

(4l′) $J_{4l'} = \{a \in \mathfrak{A} : ba - \lambda$ *is left invertible for all* $b \in \mathfrak{A}$ *and* $0 \neq \lambda \in \mathbb{C}\}$.

(4r) $J_{4r} = \{a \in \mathfrak{A} : ab - \lambda$ *is right invertible for all* $b \in \mathfrak{A}$ *and* $0 \neq \lambda \in \mathbb{C}\}$.

(4r′) $J_{4r'} = \{a \in \mathfrak{A} : ba - \lambda$ *is right invertible for all* $b \in \mathfrak{A}$ *and* $0 \neq \lambda \in \mathbb{C}\}$.

Proof. Proposition 8.1.7(4) shows that $J_1 = J_{2l}$. Lemma 8.2.1 shows that $J_3 = J_{3'}$, $J_{4l} = J_{4l'}$ and $J_{4r} = J_{4r'}$. Also $J_3 = J_{4l} \cap J_{4r}$.

$J_1 \subset J_{4r}$. If $ab - \lambda$ is not left invertible for some $b \in \mathfrak{A}$ and $0 \neq \lambda \in \mathbb{C}$, then $J = \mathfrak{A}(ab - \lambda) = \mathfrak{A}(1 - (\lambda^{-1}ab))$ is a proper left ideal with right modular unit $\lambda^{-1}ab$. Let I be a maximal modular left ideal containing J. Consider the irreducible module $\mathfrak{A}/I$ corresponding to a representation π. Then

$$\pi(a)b + I = ab + I = \lambda + I \neq 0.$$

So $a \notin \ker \pi$, and hence $a \notin \operatorname{rad} \mathfrak{A}$. Similarly, $J_1 \subset J_{4l}$. Hence $J_1 \subset J_{4l} \cap J_{4r} = J_3$.

$J_{4l'} \subset J_{2l}$. If I is a maximal modular left ideal with right modular unit e, and $a \notin I$, then $a + I \neq 0$ and the corresponding representation on $\mathfrak{A}/I$ is irreducible by Proposition 8.1.8. Hence there is a $b \in \mathfrak{A}$ so that $ba + I = e + I$. Thus $ba - 1 \in J = I + \mathbb{C}(1 - e)$, a proper left ideal of $\widetilde{\mathfrak{A}}$. Therefore $ba - 1$ is not left invertible, so $J_{4l'} \subset J_{2l}$. Similarly $J_{4r} \subset J_{2r}$.

Combining these observations, we have

$$J_1 \subset J_3 = J_{3'} \subset J_{4l} = J_{4l'} \subset J_{2l} = J_1.$$

Thus $J_1 = J_3 = J_{3'} = J_{4l} = J_{4l'} = J_{2l}$. Similarly, since the ideal J_3 is left-right blind, we obtain $J_3 = J_{2r} = J_{4r} = J_{4r'}$. ■

8.2.3. COROLLARY. *Let $\mathfrak{A}$ be a Banach algebra. Then* rad $\mathfrak{A}$ *is a radical Banach algebra and* $\mathfrak{A}/\operatorname{rad}\mathfrak{A}$ *is semisimple.*

PROOF. If $a \subset \operatorname{rad}\mathfrak{A}$, then $\sigma_{\widetilde{\mathfrak{A}}}(ba) = 0$ for all $b \in \mathfrak{A}$. This is equivalent to $\lim_{n\to\infty} \|(ba)^n\|^{1/n} = 0$. If $b \in \operatorname{rad}\mathfrak{A}$, then this still holds. So if $\mathfrak{B} = \widetilde{\operatorname{rad}\mathfrak{A}}$, then $\sigma_{\mathfrak{B}}(ba) = \{0\}$. Thus $a \in \operatorname{rad}(\operatorname{rad}\mathfrak{A})$. So $\operatorname{rad}(\operatorname{rad}\mathfrak{A}) = \operatorname{rad}\mathfrak{A}$.

Suppose that J is a maximal modular left ideal in $\mathfrak{A}/\operatorname{rad}\mathfrak{A}$ with the right modular unit e. Let $q : \mathfrak{A} \to \mathfrak{A}/\operatorname{rad}\mathfrak{A}$ be the quotient map, and set $I = q^{-1}(J)$. Then I is a maximal left ideal of $\mathfrak{A}$. Moreover if $q(f) = e$, then

$$q(\mathfrak{A}(1 - f)) = \mathfrak{A}/\operatorname{rad}\mathfrak{A}(\dot{1} - e) \subset J;$$

so that $\mathfrak{A}(1-f) \subset I$ and hence I is modular. Conversely, if I is a maximal modular ideal in $\mathfrak{A}$ with right modular unit f, then since $I \supset \operatorname{rad}\mathfrak{A}$, $J = I/\operatorname{rad}\mathfrak{A}$ is a maximal modular left ideal of $\mathfrak{A}/\operatorname{rad}\mathfrak{A}$ with right modular unit $e = q(f)$. Thus the intersection of the left modular ideals of $\mathfrak{A}/\operatorname{rad}\mathfrak{A}$ coincides with $q(\operatorname{rad}\mathfrak{A}) = \{0\}$. Therefore $\mathfrak{A}/\operatorname{rad}\mathfrak{A}$ is semisimple. ■

The reader might think that a good alternative to irreducibility is the topological version. Indeed this is an important concept. However, it turns out to be a distinct notion.

8.2.4. DEFINITION. A Banach module X is *topologically irreducible* if X is non-trivial and has no proper closed submodules.

8.2.5. EXAMPLE. If $T \in \mathcal{B}(X)$, let $\mathcal{A}(T)$ be the unital closed subalgebra of $\mathcal{B}(X)$ that it generates. Consider the identity representation *id* : $\mathcal{A}(T) \to \mathcal{B}(X)$ by *id*$(A) = A$. The closed submodules of X are the invariant subspace of T. There are examples known of operators on infinite dimensional spaces like l_1 with no proper closed invariant subspaces. Since $\mathcal{A}(T)$ is abelian, the only possible irreducible representations arise from multiplicative linear functionals, which are 1-dimensional irreducible representations. So *id* is topologically irreducible but not irreducible.

One can improve on this as follows. Charles Read [254] has constructed a quasinilpotent operator on l_1 with no proper invariant subspaces. Let $\mathcal{A}_0(T)$ be the norm closed subalgebra of $\mathcal{B}(l_1)$ generated by all positive powers of T. If φ is a multiplicative linear functional on $\mathcal{A}_0(T)$, then $\varphi(T) \in \sigma(T) = \{0\}$. Thus $\varphi(p(T)) = 0$ for all polynomials with 0 constant coefficient. These operators are dense in $\mathcal{A}_0(T)$, and hence $\varphi = 0$. Thus $\mathcal{A}_0(T)$ is a radical Banach algebra with a faithful topologically irreducible representation.

8.2.6. Example. Since the previous example is difficult to describe in detail, we present a more elementary example of a topologically irreducible representation which is not irreducible.

Let $\mathbb{F}_2^+$ denote the free semigroup on two letters. This consists of all finite words in an alphabet $\{x, y\}$ including the empty word $\varnothing$. Multiplication is just concatenation; if u and v are words in $\mathbb{F}_2^+$, then uv is their product. Let

$$l_1(\mathbb{F}_2^+) = \{a = \sum_{u \in \mathbb{F}_2^+} \alpha_u u : \|a\|_1 = \sum_{u \in \mathbb{F}_2^+} |\alpha_u| < \infty\}.$$

The product is given by $(\alpha u)(\beta v) = (\alpha\beta)uv$ for $\alpha, \beta \in \mathbb{C}$ and the distributive law.

Let H be a Hilbert space with orthonormal basis $e_0, e_1, e_2, \dots$. Let S be the unilateral shift on this basis. Define a representation $\pi : l_1(\mathbb{F}_2^+) \to \mathcal{B}(H)$ by $\pi(x) = S$ and $\pi(y) = S^*$. Thus for $k_i, l_i \geq 0$,

$$\pi(x^{k_1} y^{l_1} x^{k_2} y^{l_2} \dots x^{k_m} y^{l_m}) = S^{k_1} S^{*l_1} S^{k_2} S^{*l_2} \dots S^{k_m} S^{*l_m}.$$

Since $\|\pi(u)\| \leq 1$ for all words u, it follows readily that $\|\pi\| = 1$. The norm-closure of $\pi(l_1(\mathbb{F}_2^+))$ in $\mathcal{B}(H)$ is the closed algebra generated by S and S^*, so this is $C^*(S)$. Now $S^*S - SS^* = I - SS^* = e_0 e_0^*$ is a rank 1 projection. Observe that

$$S^i(S^*S - SS^*)S^{*j} = (S^i e_0)(S^j e_0)^* = e_i e_j^* = E_{ij}$$

is the matrix unit sending e_j to e_i for any $i, j \geq 0$. The closed linear span of these rank one operators is the set of all compact operators. This algebra is transitive: there is a compact operator (even a rank one operator) sending any non-zero vector to any other vector. Thus $C^*(S)$ is algebraically irreducible. Since $\pi(l_1(\mathbb{F}_2^+))$ is dense in $C^*(S)$, it is topologically irreducible.

To see that $\pi(l_1(\mathbb{F}_2^+))$ is not irreducible, consider $\pi(l_1(\mathbb{F}_2^+))e_0$. Now $\pi(u)e_0$ is either a basis vector e_k or it is 0. Indeed, counting from the right in a word u, if at any stage, the number of y's is greater than the number of x's, then the result is 0. Otherwise, if k is the difference between the number of x's and the number of y's, the image is e_k. Let W_k denote the set of words such that $\pi(u)e_0 = e_k$ for $k \geq 0$. For any $a \in l_1(\mathbb{F}_2^+)$,

$$\pi(a)e_0 = \sum_{k \geq 0} \Big(\sum_{u \in W_k} \alpha_u \Big) e_k.$$

Hence

$$\sum_{k \geq 0} \Big| \sum_{u \in W_k} \alpha_u \Big| \leq \sum_{u \in l_1(\mathbb{F}_2^+)} |\alpha_u| = \|a\|.$$

This shows that $\pi(l_1(\mathbb{F}_2^+))e_0 = \{\sum_{k \geq 0} \beta_k e_k : (\beta_k) \in l_1\}$, which is a proper (non-closed but dense) subspace of H. Therefore π is not irreducible.

8.3. Jacobson Density Theorem

Jacobson's theorem shows that irreducible representations are more than just transitive. This result has a variant in ring theory, but we stick to the Banach algebra case, which is easier.

We begin with an important observation. Let $\mathcal{L}(X)$ denote the space of linear maps from X into itself, with no continuity condition.

8.3.1. SCHUR LEMMA. *Let X be an irreducible $\mathfrak{A}$ module. Define*

$$\mathcal{D} = \{T \in \mathcal{L}(X) : aT = Ta \textit{ for all } a \in \mathfrak{A}\}.$$

Then $\mathcal{D} = \mathbb{C}I_X$ consists only of scalars.

PROOF. Clearly $\mathcal{D} \supset \mathbb{C}I_X$ and $\mathcal{D}$ is a $\mathbb{C}$-algebra. Take a non-zero $T \in \mathcal{D}$. Then

$$aTX = TaX \subset TX \quad \text{for} \quad a \in \mathfrak{A};$$

so TX is an $\mathfrak{A}$-submodule. As $T \neq 0$, $TX \neq \{0\}$ and hence $TX = X$. If $x \in \ker T$, then $Tax = aTx = 0$; so $\ker T$ is also an $\mathfrak{A}$-submodule. Since $T \neq 0$, $\ker T \neq X$; so we must have $\ker T = \{0\}$. Thus T is invertible. Moreover

$$aT^{-1} = T^{-1}TaT^{-1} = T^{-1}aTT^{-1} = T^{-1}a \quad \text{for all} \quad a \in \mathfrak{A}.$$

Hence $T^{-1} \in \mathcal{D}$. Therefore $\mathcal{D}$ is a division algebra.

Fix $0 \neq x_0 \in X$. Without loss of generality, using Propositions 8.1.9 and 8.1.8, we can consider that $X = \mathfrak{A}/I_{x_0}$ is a Banach module. If $x \in X$, we write it as $(a+i)x_0$ for some $a \in \mathfrak{A}$ and arbitrary $i \in I_{x_0}$. Then

$$\|Tx\| = \|T(a+i)x_0\| = \|(a+i)Tx_0\| \leq \inf_{i \in I_{x_0}} \|a+i\| \, \|Tx_0\| = \|Tx_0\| \|x\|.$$

Therefore T is bounded. Next we show that $\mathcal{D}$ is closed: if $T_n \in \mathcal{D}$ and $T_n \to T$, then

$$aT = \lim_{n\to\infty} aT_n = \lim_{n\to\infty} T_n a = Ta \quad \text{for} \quad a \in \mathfrak{A}.$$

Thus $\mathcal{D}$ is a Banach division ring. By Mazur's theorem 6.2.11, $\mathcal{D} \simeq \mathbb{C}$. Therefore $\mathcal{D} = \mathbb{C}I_X$. ■

8.3.2. DEFINITION. Let X be an $\mathfrak{A}$-module. Say that $\mathfrak{A}$ is *strictly transitive* on X if for all $0 \neq x \in X$, $\mathfrak{A}x = X$. Say that $\mathfrak{A}$ is k*-transitive* if for any k linearly independent vectors $x_1, \ldots, x_k$ in X and any k vectors $y_1, \ldots, y_k$, there is an $a \in \mathfrak{A}$ so that $ax_i = y_i$ for $1 \leq i \leq k$. Finally $\mathfrak{A}$ is *strictly dense* on X if it is k-transitive for all $k \geq 1$.

Strictly transitive is the same as 1-transitive. Every irreducible module is strictly transitive, as we have seen.

8.3.3. Jacobson Density Theorem.
Let $\mathfrak{A}$ be a Banach algebra, and let X be an irreducible $\mathfrak{A}$ module. Then $\mathfrak{A}$ is strictly dense on X.

We first establish two lemmas.

8.3.4. Lemma. *Let X be an irreducible $\mathfrak{A}$-module. If x_1, x_2 are linearly independent in X, then there is an $a \in \mathfrak{A}$ so that $ax_1 = 0 \neq ax_2$.*

Proof. If the lemma is false, then for $a \in \mathfrak{A}$, $ax_1 = 0$ implies that $ax_2 = 0$. Define $T \in \mathcal{L}(X)$ by $T(ax_1) = ax_2$. This is well defined because if $ax_1 = bx_1$, then $(a-b)x_1 = 0$. Thus $(a-b)x_2 = 0$, so that $ax_2 = bx_2$, showing that the value of $T(ax_1)$ is independent of the choice of a. Clearly T is linear.

Suppose that $b \in \mathfrak{A}$ and $y \in X$. Since X is irreducible, there is some $a \in \mathfrak{A}$ so that $y = ax_1$. Then

$$T(by) = T(bax_1) = bax_2 = bT(ax_1) = bTy.$$

Thus T commutes with b for all $b \in \mathfrak{A}$. By Schur's lemma, T is scalar and $x_2 = \lambda x_1$. This contradicts the linear independence of x_1 and x_2. Therefore the lemma must be correct. ∎

8.3.5. Lemma. *Let X be an irreducible $\mathfrak{A}$-module. If $x_1, \dots, x_n$ for $n \geq 2$ are linearly independent in X, then there is an $a \in \mathfrak{A}$ so that*

$$ax_1 = \cdots = ax_{n-1} = 0 \neq ax_n.$$

Proof. We proceed by induction on n. Lemma 8.3.4 establishes the lemma for $n = 2$. Assume the result for $n-1$ for all Banach algebras $\mathfrak{B}$ and all irreducible $\mathfrak{B}$-modules. Let $M = \operatorname{span}\{x_1, \dots, x_{n-2}\}$. Define

$$\mathfrak{J}_n = \bigcap_{i=1}^{n-2} I_{x_i} = \{a \in \mathfrak{A} : ax_1 = \cdots = ax_{n-2} = 0\}.$$

Now $\mathfrak{J}_n$ is a left ideal of $\mathfrak{A}$, and we consider it as a Banach algebra. Let $Y = X/M$. This is a left $\mathfrak{J}_n$-module since $\mathfrak{J}_n M = 0$. Moreover if $y_1 = z_1 + M \neq 0$ and $y_2 = z_2 + M$, then $x_1, \dots, x_{n-2}, z_1$ are linearly independent. The induction hypothesis shows that there is an $a \in \mathfrak{A}$ so that $ax_1 = \cdots = ax_{n-2} = 0 \neq az_1$. Then there is a $b \in \mathfrak{A}$ so that $baz_1 = z_2$. In particular, $a \in \mathfrak{J}_n$ and so $ba \in \mathfrak{J}_n$. We have $bay_1 = y_2$; whence $\mathfrak{J}_n y_1 = Y$. It follows that Y is an irreducible $\mathfrak{J}_n$-module.

Since $x_1, \dots, x_n$ are linearly independent in X, $x_{n-1} + M$ and $x_n + M$ are linearly independent in Y. By Lemma 8.3.4, there is $c \in \mathfrak{J}_n$ so that $cx_{n-1} + M = 0 \neq cx_n + M$. Choose another element $d \in \mathfrak{J}_n$ so that $dcx_n + M \neq 0$; whence $dcx_n \neq 0$. Since $cx_{n-1} \in M$, we get $dcx_{n-1} = 0$. Clearly $dcx_1 = \cdots = dcx_{n-2}$. This verifies case n. ∎

The proof of Jacobson's density theorem is now routine.

PROOF OF THEOREM 8.3.3. Let X be an irreducible $\mathfrak{A}$-module, and suppose that $x_1, \dots, x_n$ are linearly independent in X. By Lemma 8.3.5, there are elements $a_i \in \mathfrak{A}$ so that $a_i x_j = 0$ if $i \neq j$ and $a_i x_i = z_i \neq 0$ for $1 \leq i, j \leq n$. Since X is irreducible, there are elements $b_i \in \mathfrak{A}$ so that $b_i z_i = y_i$. Thus $a = \sum_{i=1}^n b_i a_i$ satisfies $ax_i = y_i$ for $1 \leq i \leq n$. Hence $\mathfrak{A}$ is n-transitive on X for all $n \geq 1$. That is, $\mathfrak{A}$ is strictly dense on X. ■

As an immediate consequence, we obtain a result of Wedderburn which was the inspiration for Jacobson's theorem.

8.3.6. COROLLARY. *Let $\mathfrak{A}$ be a Banach algebra, and let (X, π) be a finite dimensional irreducible $\mathfrak{A}$ module. Then $\pi(\mathfrak{A}) = \mathcal{L}(X) \simeq M_n$, where $n = \dim X$.*

8.4. Automatic Continuity

In this section, we extend the results from Section 7.3 to the non-commutative setting. These results, due to B. Johnson, are much deeper than the commutative case. It leads, for example, to the uniqueness of the norm for semisimple Banach algebras.

8.4.1. THEOREM. *Let X be a Banach space and let $\mathfrak{A}$ be a Banach algebra. Then every irreducible representation $\pi : \mathfrak{A} \to \mathcal{B}(X)$ is continuous.*

PROOF. $J = \ker \pi$ is a primitive ideal, and thus it is closed and $\mathfrak{B} = \mathfrak{A}/J$ is a Banach algebra. Let $q : \mathfrak{A} \to \mathfrak{B}$ be the quotient map. There is a unique representation $\dot{\pi} : \mathfrak{B} \to \mathcal{B}(X)$ by $\dot{\pi}(b) = \pi(a)$ for any $a \in q^{-1}(a)$. Moreover $\dot{\pi}$ is injective and irreducible. If $\dot{\pi}$ is continuous, then $\pi = \dot{\pi} q$ is also continuous. Thus we may suppose that $\ker \pi = \{0\}$.

If X is finite dimensional, then $\mathfrak{A}$ is finite dimensional, and every linear map from $\mathfrak{A}$ into $\mathcal{B}(X)$ is continuous. So we may also assume that X is infinite dimensional.

For each $x \in X$, define a linear map $T_x : \mathfrak{A} \to X$ by $T_x a = ax$. Let

$$Y = \{x \in X : T_x \text{ is continuous}\}.$$

Then Y is a (not necessarily closed) subspace of X. For $y \in Y$ and $b \in \mathfrak{A}$,

$$\|T_{by} a\| = \|aby\| = \|T_y ab\| \leq \big(\|T_y\| \|b\|\big) \|a\|.$$

Thus $\mathfrak{A} Y \subset Y$; i.e. Y is a submodule of X. Since X is irreducible, we have $Y = X$ or $Y = \{0\}$.

If $Y = X$, then for each $x \in X$,

$$\sup_{\|a\|\le 1} \|\pi(a)x\| = \sup_{\|a\|\le 1} \|ax\| = \|T_x\| < \infty.$$

By the Uniform Boundedness Principle applied to $\{\pi(a) : \|a\| \le 1\}$, we have

$$\|\pi\| = \sup\{\|\pi(a)\| : \|a\| \le 1\} < \infty.$$

Thus π is continuous.

Now consider the case $Y = \{0\}$. Since X is infinite dimensional, choose a linear independent sequence of unit vectors in X, $x_1, x_2, \ldots$. Let $L_n = \bigcap_{i=1}^{n-1} I_{x_i}$. These are closed left ideals. By the Jacobson density theorem, there are elements $a_n \in \mathfrak{A}$ so that $a_n x_i = 0$ for $1 \le i \le n-1$ and $a_n x_n \ne 0$. Thus $a_n \in L_n \setminus I_{x_n}$. Recursively, we use the fact that $T_{a_n x_n}$ is unbounded to choose $b_n \in \mathfrak{A}$ with

$$\|b_n\| < \frac{2^{-n}}{\|a_n\|} \quad \text{and} \quad \|b_n a_n x_n\| > n + \Big\|\Big(\sum_{i=1}^{n-1} b_i a_i\Big) x_n\Big\|.$$

Then $b_n a_n \in L_n$ and $\|b_n a_n\| < 2^{-n}$. Define a convergent sum in $\mathfrak{A}$

$$b = \sum_{i\ge 1} b_i a_i = \sum_{i=1}^{n} b_i a_i + \sum_{i\ge n+1} b_i a_i.$$

Then since $b_i a_i \in I_{x_n}$ for all $i > n$ and this left ideal is closed, the tail of the series belongs to I_{x_n}; so $\big(\sum_{i\ge n+1} b_i a_i\big) x_n = 0$. Therefore

$$\|\pi(b)x_n\| = \|bx_n\| \ge \|b_n a_n x_n\| - \Big\|\Big(\sum_{i=1}^{n-1} b_i a_i\Big) x_n\Big\| > n.$$

This shows that $\pi(b)$ is not bounded, contrary to our hypothesis.

Hence $Y = X$ and π is continuous. ∎

8.4.2. DEFINITION. Let T be a linear map from a Banach space X into a Banach space Y. The *separating space* of T is

$$\mathcal{S}(T) = \{y \in Y : \exists x_n \in X,\ x_n \to 0 \text{ and } Tx_n \to y\}.$$

This definition is motivated by the Closed Graph theorem, which says that if $\mathcal{S}(T) = \{0\}$, then T is continuous.

8.4.3. THEOREM. *Let $\mathfrak{A}$ and $\mathfrak{B}$ be Banach algebras and let $\theta : \mathfrak{A} \to \mathfrak{B}$ be a surjective algebra homomorphism. Then $\mathcal{S}(\theta)$ is contained in* rad $\mathfrak{B}$.

PROOF. Let $b \in \mathcal{S}(\theta)$, say $a_n \in \mathfrak{A}$, $a_n \to 0$ and $\theta(a_n) \to b$. Let (X, π) be an irreducible left Banach $\mathfrak{B}$-module. Because θ is surjective, $\pi\theta : \mathfrak{A} \to \mathcal{B}(X)$ is an irreducible representation. Therefore it is continuous. Thus

$$\pi(b) = \lim_{n\to\infty} \pi\theta(a_n) = 0.$$

Hence $b \in \ker\pi$, a primitive ideal. Every irreducible representation is equivalent to a Banach module. Thus $\mathcal{S}(\theta) \subset \bigcap_{J \lhd \mathfrak{A} \text{ primitive}} J = \operatorname{rad}\mathfrak{B}$. ∎

The following corollary is immediate.

8.4.4. COROLLARY. *Every homomorphism of a Banach algebra onto a semisimple Banach algebra is continuous. In particular, every automorphism of a semisimple Banach algebra is continuous.*

We now can obtain the uniqueness of norm result. This fails for Banach algebras which are not semisimple. See Exercise **8.16**.

8.4.5. JOHNSON THEOREM. *A semisimple Banach algebra has a unique Banach algebra norm up to equivalence.*

PROOF. If $\|\cdot\|$ and $|\!|\!|\cdot|\!|\!|$ are two Banach algebra norms on $\mathfrak{A}$, and $(\mathfrak{A}, \|\cdot\|)$ is semisimple, consider the identity map *id* : $(\mathfrak{A}, |\!|\!|\cdot|\!|\!|) \to (\mathfrak{A}, \|\cdot\|)$. This is a surjective homomorphism, and thus it is a continuous bijection by Corollary 8.4.4. By Banach's Isomorphism theorem, it is invertible. Thus the two norms are equivalent. ∎

8.5. Cohen Factorization Theorem

The main result of this section has a very technical proof, but it is frequently useful. It shows that if $\mathfrak{A}$ has a bounded approximate identity, every $a \in \mathfrak{A}$ factors as $a = bc$ for $b, c \in \mathfrak{A}$; i.e. $\mathfrak{A}^2 = \mathfrak{A}$.

8.5.1. DEFINITION. If X is a Banach $\mathfrak{A}$-module, then a *bounded approximate identity for* X is a bounded left approximate identity $\{e_\lambda : \lambda \in \Lambda\}$ for $\mathfrak{A}$ such that

$$\lim_{\Lambda} \|x - e_\lambda x\| = 0 \quad \text{for all} \quad x \in X.$$

A Banach $\mathfrak{A}$-module X is *non-degenerate* if $\overline{\mathfrak{A}X} = X$.

8.5.2. PROPOSITION. *Let $\mathfrak{A}$ be a Banach algebra with bounded approximate identity $\{e_\lambda : \lambda \in \Lambda\}$, and let X be a Banach $\mathfrak{A}$-module. Then $\{e_\lambda\}$ is a b.a.i. for X if and only if X is non-degenerate.*

PROOF. If $\{e_\lambda\}$ is a b.a.i. for X and $x \in X$, then $x = \lim_{\Lambda} e_\lambda x \in \overline{\mathfrak{A}X}$. So X is non-degenerate.

Conversely, if (X,π) is non-degenerate, let $L = \|\pi\|$. If $x = ay$, then

$$\lim_\Lambda e_\lambda x = \lim_\Lambda (e_\lambda a)y = ay = x.$$

In general, for $\varepsilon > 0$, there is some ay so that $\|x - ay\| < \varepsilon$. There is some λ_0 so that $\|ay - e_\lambda ay\| < \varepsilon$ for $\lambda \geq \lambda_0$. Hence if $\sup \|e_\lambda\| = K$, then for $\lambda \geq \lambda_0$,

$$\|x - e_\lambda x\| \leq \|x - ay\| + \|ay - e_\lambda ay\| + L\|e_\lambda\|\|ay - x\| < (2 + KL)\varepsilon.$$

Therefore $\{e_\lambda\}$ is a b.a.i. for X. ∎

8.5.3. LEMMA. *Let $\mathfrak{A}$ be an unital Banach algebra, and let $e \in \mathfrak{A}$. Given a constant $0 < c < \dfrac{1}{2(\|e\|+1)}$, then $a = 1 - c + ce$ has $\|a^{-1}\| < 2$. If $x \in X$, then $\|x - a^{-1}x\| \leq \|x - ex\|$.*

PROOF. Since $a = 1 - c(1-e)$ and $\|c(1-e)\| \leq c(\|e\|+1) < \frac{1}{2}$, we have $a^{-1} = \sum_{n\geq 0} \big(c(1-e)\big)^n$. Thus

$$\|a^{-1}\| \leq \sum_{n\geq 0} c^n \|1-e\|^n < \sum_{n\geq 0} 2^{-n} = 2.$$

Moreover, since $c < \frac{1}{2}$,

$$\begin{aligned} \|x - a^{-1}x\| &= \Big\| x - \sum_{n\geq 0} \big(c(1-e)\big)^n x \Big\| \\ &\leq \sum_{n\geq 1} c^n \|1-e\|^{n-1} \|x - ex\| \\ &\leq 2c\|x - ex\| < \|x - ex\|. \end{aligned}$$ ∎

8.5.4. COHEN FACTORIZATION THEOREM. *Let (X,π) be a Banach $\mathfrak{A}$-module with bounded left approximate identity $\{e_\lambda\}_\Lambda$. Then $\mathfrak{A}X = X$.*

More precisely, if $x \in X$ and $\varepsilon > 0$, there is an $a \in \mathfrak{A}$ and $y \in X$ with $\|a\| \leq K+2$ and $\|x - y\| < \varepsilon$ so that $ay = x$, where $K = \sup\{\|\pi\|, \|e_\lambda\| : \lambda \in \Lambda\}$.

PROOF. Fix $x \in X$ and $\varepsilon > 0$. We may suppose that $\|x\| = 1$. Let K be a bound for the b.a.i. $\{e_\lambda\}_\Lambda$ and for $\|\pi\|$, and let $0 < c < \dfrac{\varepsilon}{2(K+1)}$. We will construct a sequence $e_i = e_{\lambda_i}$ for $i \geq 1$ so that with $a_0 = 1$, for $n \geq 1$

$$(\ddagger) \quad a_n = 1 + c\sum_{i=1}^{n} (1-c)^{i-1}(e_i - 1) \in \widetilde{\mathfrak{A}}^{-1} \quad \text{and} \quad \|a_n^{-1}x - a_{n-1}^{-1}x\| < \frac{\varepsilon}{2^n}.$$

Once this is accomplished, we observe that $a_n^{-1}x$ is a Cauchy sequence, so that $y = \lim_{n\to\infty} a_n^{-1}x$ exists; and

$$\|y-x\| = \Big\| \sum_{n\ge1} a_n^{-1}x - a_{n-1}^{-1}x \Big\| \le \sum_{n\ge1} \|a_n^{-1}x - a_{n-1}^{-1}x\| < \sum_{n\ge1} \frac{\varepsilon}{2^n} = \varepsilon.$$

Also

$$a = \lim_{n\to\infty} a_n = 1 + c\sum_{i=1}^{\infty}(1-c)^{i-1}(e_i - 1)$$

is absolutely convergent with

$$\|a\| \le 1 + c(K+1)\sum_{i=1}^{\infty}(1-c)^{i-1} = 1 + \frac{c(K+1)(1-c)}{1-(1-c)} < K+2.$$

Finally $ay = \lim_{n\to\infty} a_n(a_n^{-1}x) = x$.

We now establish (‡) recursively. Suppose that it holds for all $1 \le k \le n$, which is vacuous if $n = 0$ and $a_0 = 1$. Define $\varepsilon_{n+1} = \dfrac{\varepsilon}{2^{n+4}K\|a_n^{-1}\|^2}$. Use the b.a.i. property to select $e_{n+1} = e_{\lambda_{n+1}}$ so that

$$\|e_{n+1}e_i - e_i\| < \varepsilon_{n+1} \text{ for } 1 \le i \le n \quad \text{and} \quad \|e_{n+1}x - x\| < \varepsilon_{n+1}.$$

Define $a_{n+1} = a_n + c(1-c)^n(e_{n+1} - 1)$ as in (‡). Observe that

$$c\sum_{i=1}^{n}(1-c)^{i-1} = \frac{c\big(1-(1-c)^n\big)}{1-(1-c)} = 1-(1-c)^n.$$

Then with $b_{n+1} = 1 - c(1 - e_{n+1})$,

$$\begin{aligned} a_{n+1} - b_{n+1}a_n &= c(1-c)^n(e_{n+1}-1) + c(1-e_{n+1})a_n \\ &= c(1-e_{n+1})\Big(1-(1-c)^n + c\sum_{i=1}^{n}(1-c)^{i-1}(e_i-1)\Big) \\ &= c(1-e_{n+1})c\sum_{i=1}^{n}(1-c)^{i-1}(1+e_i-1) \\ &= c^2\sum_{i=1}^{n}(1-c)^{i-1}(e_i - e_{n+1}e_i) =: -d_{n+1}. \end{aligned}$$

Therefore

$$\big\|a_{n+1} - b_{n+1}a_n\big\| = \|d_{n+1}\| \le c^2\sum_{i=1}^{n}(1-c)^{i-1}\varepsilon_{n+1} < c\varepsilon_{n+1} < \frac{1}{2\|a_n^{-1}\|}.$$

Since Lemma 8.5.3 shows that $\|a_n^{-1}b_{n+1}^{-1}\| < 2\|a_n^{-1}\|$, we see that a_{n+1} is invertible. Moreover

$$\begin{aligned} a_{n+1}^{-1} &= (b_{n+1}a_n - d_{n+1})^{-1} = a_n^{-1}b_{n+1}^{-1}\big(1 - d_{n+1}a_n^{-1}b_{n+1}^{-1}\big)^{-1} \\ &= a_n^{-1}b_{n+1}^{-1}\big(1 + d_{n+1}a_n^{-1}b_{n+1}^{-1}\big(1 - d_{n+1}a_n^{-1}b_{n+1}^{-1}\big)^{-1}\big) \end{aligned}$$

Thus

$$\|a_{n+1}^{-1} - a_n^{-1}b_{n+1}^{-1}\| \le \frac{\|d_{n+1}\|\|a_n^{-1}b_{n+1}^{-1}\|^2}{1 - \|d_{n+1}a_n^{-1}b_{n+1}^{-1}\|} < \frac{c\varepsilon_{n+1}4\|a_n^{-1}\|^2}{1 - 2c\varepsilon_{n+1}\|a_n^{-1}\|} < \frac{\varepsilon}{2^{n+2}K}.$$

Finally, by Lemma 8.5.3, $\|b_{n+1}^{-1}x - x\| \le \|e_{n+1}x - x\| < \varepsilon_{n+1}$. Hence

$$\begin{aligned} \|a_{n+1}^{-1}x - a_n^{-1}x\| &\le \|a_n^{-1}b_{n+1}^{-1}x - a_n^{-1}x\| + \|(a_{n+1}^{-1} - a_n^{-1}b_{n+1}^{-1})x\| \\ &\le \|a_n^{-1}\|\|b_{n+1}^{-1}x - x\| + K\frac{\varepsilon}{2^{n+2}K} \\ &< \|a_n^{-1}\|\varepsilon_{n+1} + \frac{\varepsilon}{2^{n+2}} < \frac{\varepsilon}{2^{n+1}}. \end{aligned}$$

Therefore (‡) is satisfied. ■

If we let $X = \overline{\mathfrak{A}a}$ and apply Cohen's factorization theorem, we obtain the following.

8.5.5. COROLLARY. *If $\mathfrak{A}$ is a Banach algebra with bounded left approximate identity, then $\mathfrak{A}^2 = \mathfrak{A}$. Moreover, if $a \in \mathfrak{A}$ and $\varepsilon > 0$, there are $b, c \in \mathfrak{A}$ with $c \in \overline{\mathfrak{A}a}$, $\|c - a\| < \varepsilon$, $\|b\| \le K + 2$ and $a = bc$, where the b.l.a.i is bounded by K.*

Exercises for Chapter 8

8.1 If $\mathfrak{A}$ is a Banach algebra, define $\mathfrak{A}^{op}$ to be the same Banach space with a product defined by $a \cdot b := ba$ for $a, b \in \mathfrak{A}$. Show that $\mathfrak{A}^{op}$ is a Banach algebra.

8.2 Let X be a Banach space, and let $\mathcal{K}(X)$ be the space of compact operators.
(a) Find all minimal closed left ideals of $\mathcal{K}(X)$.
(b) Find all maximal modular left ideals of $\mathcal{K}(X)$.

8.3 Let $\mathfrak{A}$ be a Banach algebra with left $\mathfrak{A}$-module X. For $x \in X \setminus \{0\}$, let $I_x = \{a \in \mathfrak{A} : ax = 0\}$ and $id(x) = \{e \in \mathfrak{A} : ex = x\}$.
(a) Show that every $e \in id(x)$ is a right modular unit for I_x.
(b) If $\mathfrak{A}x = X$, show that $id(x)$ is non-empty and hence I_x is a modular left ideal.

8.4 Let $\mathfrak{A}, \mathfrak{B}$ be (non-commutative) Banach algebras, and let $\theta : \mathfrak{A} \to \mathfrak{B}$ be a surjective homomorphism.
(a) Show that $\theta(\mathrm{rad}(\mathfrak{A})) \subset \mathrm{rad}(\mathfrak{B})$.
(b) If $\mathfrak{B}$ is semisimple, prove that $\ker\theta$ is closed.

8.5 Let $\mathfrak{A}$ be a (non-commutative) Banach algebra. Let $\mathfrak{J}$ be a closed ideal of $\mathfrak{A}$. Show that $\text{rad}(\mathfrak{J}) = \mathfrak{J} \cap \text{rad}(\mathfrak{A})$.

8.6 (a) Let X be an infinite dimensional Banach space. Show that $\mathcal{B}(X)$ is semisimple and $\{0\}$ is a primitive ideal.
(b) Show every minimal left ideal of $\mathcal{B}(X)$ has the form $X\varphi = \{x\varphi : x \in X\}$ for some $\varphi \in X^*$, where $(x\varphi)(y) = \varphi(y)x$.
(c) Show every non-zero 2-sided ideal of $\mathcal{B}(X)$ contains all finite rank operators.
(d) Show that $\mathcal{B}(X)$ contains a non-zero nilpotent element A. Find an element B so that AB is not quasinilpotent.

8.7 Show that if P is a maximal modular 2-sided ideal in a Banach algebra $\mathfrak{A}$, then it is primitive.

8.8 Show that if P is a primitive ideal, and L_1, L_2 are left ideals such that $L_1 L_2 \subset P$, then either $L_1 \subset P$ or $L_2 \subset P$.
HINT: $P = I : \mathfrak{A}$ for a maximal modular left ideal I. If $L_2 \not\subset P$, show that $L_2\mathfrak{A} + I = \mathfrak{A}$.

8.9 Show that if $J \lhd \mathfrak{A}$ is a primitive ideal such that $\mathfrak{A}/J$ is finite dimensional, then J is a maximal (2-sided) ideal.

8.10 Show that if J is an ideal of a Banach algebra $\mathfrak{A}$, and $a \in \mathfrak{A}$, then

$$\sigma_{\mathfrak{A}/J}(a + J) \subset \sigma_{\mathfrak{A}}(a).$$

8.11 (*Wedderburn theorem*) Let $\mathfrak{A}$ be a finite dimensional semisimple Banach algebra.
(a) Use Exercise **8.9** to show that the primitive ideals are precisely the maximal ideals.
(b) Show that there is a minimal finite set of primitive ideals $P_1, \dots, P_m$ with $\bigcap_{i=1}^m P_i = \{0\}$.
(c) Define $J_i = \bigcap_{j \neq i} P_j$. Show that $\mathfrak{A} \simeq P_i \oplus J_i$, and deduce that there is an integer n_i so that $J_i \simeq M_{n_i}$.
(d) Prove that $\mathfrak{A} \simeq J_1 \oplus J_2 \oplus \cdots \oplus J_m$.
(e) Prove that $P_1, \dots, P_m$ is the complete list of all primitive ideals.

8.12 Let $\mathfrak{A}$ be a Banach algebra and let $\Pi_{\mathfrak{A}}$ denote the set of primitive ideals. For $J \subset \mathfrak{A}$, define $\text{hull}\, J = \{P \in \Pi_{\mathfrak{A}} : P \supset J\}$; and for and $E \subset \Pi_{\mathfrak{A}}$, set

$$\ker E = \bigcap \{P : P \in E\} \quad \text{and} \quad \overline{E} = \text{hull} \ker E.$$

(a) If $E = \text{hull}(J)$, show that $\overline{E} = E$.
(b) Show that if $E_1, E_2 \subset \Pi_{\mathfrak{A}}$, then $\overline{E_1 \cup E_2} = \overline{E_1} \cup \overline{E_2}$.
(c) Verify the Kuratowski closure axioms (Exercise **1.7**) to put the *hull-kernel topology* on $\Pi_{\mathfrak{A}}$.
(d) Show that if J is a modular left ideal, then $\text{hull}(J)$ is compact.

8.13 Let $\mathfrak{A}$, $\mathfrak{B}$ be Banach algebras with $\mathfrak{B}$ semisimple. If $\pi : \mathfrak{A} \to \mathfrak{B}$ is a surjective homomorphism, show that $\operatorname{rad}(\mathfrak{A}) \subset \ker \pi$.

8.14 Let $\mathfrak{A}$ be the Banach algebra of infinite sequences $a = (a_k)$ with $a_k \in M_2$ such that $\lim_{k\to\infty} a_k = d$ exists and $d \in \mathfrak{D}_2$, the algebra of diagonal matrices with respect to a fixed orthonormal basis e_1, e_2. The norm is $\|a\| = \sup_{k\geq 1} \|a_k\|$. Define $\varphi_i(a) = \lim_{k\to\infty} \langle a_k e_i, e_i \rangle = d_{ii}$ for $i = 1, 2$.

(a) Show that the primitive ideals are $P_k = \{a : a_k = 0\}$ for $k \geq 1$ and $P_{\infty,i} = \ker \varphi_i$ for $i = 1, 2$.
(b) Compute the hull-kernel topology on $\Pi_{\mathfrak{A}}$.
(c) Show that this topology is not Hausdorff.

8.15 (*Eidelheit*) Let X and Y be Banach spaces, and let $\theta : \mathcal{B}(X) \to \mathcal{B}(Y)$ be an isomorphism. Prove that there is an isomorphism $S : X \to Y$ such that $\theta(A) = SAS^{-1}$.
HINT: Show that θ is continuous. Fix $x_0 \in X$, $\varphi_0 \in X^*$ with $\varphi_0(x_0) = 1$. Show that $\theta(x_0\varphi_0) = y_0\psi_0$ and $\psi_0(y_0) = 1$. For $x \in X$, define $S : X \to Y$ by

$$\theta(x\varphi_0) = \theta((x\varphi_0)(x_0\varphi_0)) = (Sx)\psi_0.$$

Then S is invertible. For $A \in \mathcal{B}(X)$ and $x \in X$,

$$\theta(A)Sx = \theta(A(x_0\varphi_0))y_0 = SAx.$$

8.16 Let $\mathfrak{A} = l_2 \oplus \mathbb{C}r$ with $\|(a_n) + \alpha r\| = \|(a_n)\|_2 + |\alpha|$. Define

$$\big((a_n) + \alpha r\big)\big((b_n) + \beta r\big) = (a_n b_n).$$

(a) Find $\operatorname{rad}\mathfrak{A}$.
(b) Show that $\mathfrak{A}^2 = l_1$.
(c) Pick any non-zero linear functional ψ on l_2 which is zero on l_1. Define a new norm $|||(a_n)+\alpha r||| = \|(a_n)\|_2 + |\psi((a_n)) - \alpha|$. Show that $(\mathfrak{A}, |||\cdot|||)$ is complete, making $|||\cdot|||$ a Banach algebra norm on $\mathfrak{A}$ which is not equivalent to the original.

8.17 Let $\mathfrak{A}$ be a Banach algebra with bounded left approximate identity. Suppose that $a_n \in \mathfrak{A}$ and $\lim\limits_{n\to\infty} a_n = 0$. Show that there are $b, c_n \in \mathfrak{A}$ with $\lim\limits_{n\to\infty} c_n = 0$ and $a_n = bc_n$.
HINT: let $X = c_0(\mathfrak{A}) = \{(a_n)_{n\geq 1} : \lim\limits_{n\to\infty} a_n = 0\}$.

8.18 Let $\mathfrak{A}$ be a Banach algebra.

(a) If $a_0 \in \operatorname{rad}\mathfrak{A}$, show that $x^2 - x + a_0 = 0$ has a solution $r \in \operatorname{rad}\mathfrak{A} \cap \{a_0\}''$; i.e. r commutes with the commutant of a_0.
HINT: $x = -\frac{1}{2}\sum_{k\geq 1} \binom{1/2}{k}(-4a_0)^k$.
(b) Show that if $a \in \mathfrak{A}$ and $a^2 - a = q \in \operatorname{rad}\mathfrak{A}$, then there is an $r \in \operatorname{rad}\mathfrak{A}$ such that $p = a + r = p^2$.
HINT: look for $r \in \operatorname{rad}\mathfrak{A} \cap \{a\}'$. Reduce this to solving $x^2 + (2a-1)x + q = 0$. Substitute $y = (2a-1)x$. Show that $(2a-1)^2$ is invertible.

8.19 Suppose that $\mathfrak{A}$ is a Banach algebra satisfying the second sentence of Corollary 8.5.5. Prove that $\mathfrak{A}$ has a bounded left approximate identity.
HINT: Given $a \in \mathfrak{A}$, write $a = b_n c_n$ with $\|b_n\| \leq K + 2$ and $\|a - c_n\| < 2^{-n}$. Estimate $\|a - b_n a\|$. Apply Exercise **6.27**.

8.20 Let X be a Banach left $\mathfrak{A}$-module. Define a right module action of $\mathfrak{A}$ on X^* by $fa(x) = f(ax)$ for $f \in X^*$, $x \in X$ and $a \in \mathfrak{A}$.
(a) Show that X^* is a right Banach $\mathfrak{A}$-module.
(b) Suppose that Y is a closed submodule of X. Show that X/Y is also a Banach left $\mathfrak{A}$-module. Then show that as Banach right $\mathfrak{A}$-modules, Y^* and $X^*/Y^\perp$ are isometrically isomorphic. Similarly show that $(X/Y)^*$ and $Y^\perp$ are isometrically isomorphic as Banach right $\mathfrak{A}$-modules.

8.21 (*Arens product*) Let $\mathfrak{A}$ be a Banach algebra. Define a left action of $\mathfrak{A}^{**}$ on $\mathfrak{A}^*$ by $(Af)(a) = A(fa)$ for $A \in \mathfrak{A}^{**}$, $f \in \mathfrak{A}^*$, $a \in \mathfrak{A}$ and fa comes from the right module action of $\mathfrak{A}$ on $\mathfrak{A}^*$ as in Exercise **8.20**. Define the *Arens product* of $A, B \in \mathfrak{A}^{**}$ by $(A\Box B)(f) = A(Bf)$ for $f \in \mathfrak{A}^*$.
(a) Prove that this makes $\mathfrak{A}^{**}$ into a Banach algebra and that the injection of $\mathfrak{A}$ into $\mathfrak{A}^{**}$ is a homomorphism.
(b) Show that if $(A_\lambda)_\Lambda$ is a net in $\mathfrak{A}^{**}$ converging weak-$*$ to A, then $(A_\lambda B)_\Lambda$ converges weak-$*$ to AB for $B \in \mathfrak{A}^{**}$.
(c) If $(A_\lambda)_\Lambda$ converges weak-$*$ to A and $(B_\mu)_{\Lambda'}$ converges weak-$*$ to B, show that $A\Box B = \lim_\Lambda \lim_{\Lambda'} A_\lambda \Box B_\mu$. In particular, we can take A_λ and B_μ in $\mathfrak{A}$.
(d) The same construction for $\mathfrak{A}^{op}$ (see Exercise **8.1**) yields another product $\Diamond$ on $\mathfrak{A}^{**}$, namely $(\mathfrak{A}^{op\,**})^{op}$. Say $\mathfrak{A}$ is *regular* if the two products agree; i.e. $\mathfrak{A}^{op\,**} = \mathfrak{A}^{**\,op}$. Show that if $(A_\lambda)_\Lambda$ converges weak-$*$ to A and $(B_\mu)_{\Lambda'}$ converges weak-$*$ to B, then $A\Diamond B = \lim_{\Lambda'} \lim_\Lambda A_\lambda \Diamond B_\mu$.

8.22 Show that c_0 is regular, and that the Arens product on $l_\infty = c_0^{**}$ is the usual product.

8.23 Let $\mathfrak{A}$ be a Banach algebra and let $\mathfrak{A}^{**}$ have the Arens product.
(a) Suppose that $\{e_\lambda\}_\Lambda$ is a bounded right approximate identity in $\mathfrak{A}$. Let E be a weak-$*$ cluster point of the net $\{\iota(e_\lambda)\}_\Lambda$ in $\mathfrak{A}^{**}$. Prove that E is a right identity.
HINT: $\iota(e_\lambda)(fa) = (fa)(e_\lambda) = f(ae_\lambda)$.
(b) Suppose that $\mathfrak{A}^{**}$ has a right identity E. Prove that $\mathfrak{A}$ has a bounded right approximate identity.
HINT: by Goldstine's theorem, there is a net $\{e_\lambda\}_\Lambda$ in $\mathfrak{A}$ such that $\{\iota(e_\lambda)\}_\Lambda$ converges weak-$*$ to E.
(c) If $\mathfrak{A}$ is regular, show that $\mathfrak{A}^{**}$ has an identity element if and only if $\mathfrak{A}$ has a bounded (2-sided) approximate identity.
HINT: Exercise **6.25**

8.24 Let (X, π) be a Banach left $\mathfrak{A}$-module which is topologically transitive; i.e. the only closed submodules are $\{0\}$ and X.

(a) Show that if $0 \neq x \in X$, then $\overline{\mathfrak{A}x} = X$.
(b) Show that $\ker \pi$ is a prime ideal; i.e. if J_1, J_2 are ideals and $J_1 J_2 \subset \ker \pi$, then $J_1 \subset \ker \pi$ or $J_2 \subset \ker \pi$.

8.25 (*Nagata–Higman*). A $\mathbb{C}$-algebra $\mathfrak{A}$ is *nil* if every element is nilpotent, and *nilpotent* if there is an $n \in \mathbb{N}$ such that the product of any n elements of $\mathfrak{A}$ is 0. Show that if every element of $\mathfrak{A}$ is nilpotent of order n, then $\mathfrak{A}$ is nilpotent of order $2^n - 1$ as follows.
(a) Show $f_n(a,b) = \sum_{i=0}^{n-1} a^i b a^{n-1-i} = 0$ for $a, b \in \mathfrak{A}$; and $f_n(a,1) = na^{n-1}$.
HINT: The coefficient of λ in $0 = (a + \lambda b)^n$ is $f_n(a,b)$.
(b) Show that $a^{n-1} c b^{n-1} = 0$ for all $a, b, c \in \mathfrak{A}$ by evaluating in two ways

$$\sum_{i,j=0}^{n-1} a^i c b^j a^{n-1-i} b^{n-1-j} = \sum_{j=0}^{n-1} f_n(a, cb^j) b^{n-1-j} = \sum_{i=0}^{n-1} a^i c f_n(b, a^{n-1-i}).$$

(c) Let J be the ideal generated by $\{a^{n-1} : a \in \mathfrak{A}\}$. Then in $\mathfrak{A}/J$, every element is nilpotent of order $n-1$. If the result is known for $n-1$, show that for $M = 2^{n-1} - 1$, $\mathfrak{A}^M \subset J$ and hence $\mathfrak{A}^{2M+1} \subset J\mathfrak{A}J = \{0\}$.
(d) Complete the proof by induction.

8.26 (*Grabiner*). Show that every nil Banach algebra is nilpotent.
HINT: let $A_n = \{a \in \mathfrak{A} : a^n = 0\}$, and note that $\mathfrak{A} = \bigcup_{n \geq 1} A_n$. Show that $\mathfrak{A} = A_{n_0}$ for some n_0. Then apply Exercise **8.25**.

Notes on Chapter 8

The notion of a modular ideal is due to I. Segal [282], but the terminology is due to Jacobson [159], who characterized the radical [158] in 1945. Jacobson [157] proved the density theorem, also in 1945, to extend classical work of Wedderburn on finite dimensional algebras. Barry Johnson [162] established the uniqueness of norm for a Banach algebra. The Cohen factorization theorem is from [59].

See Bonsall and Duncan [37] for this material and more.

CHAPTER 9

C*-algebras

The study of self-adjoint subalgebras of $\mathcal{B}(H)$ is an essential tool in the study of linear operators on Hilbert space. However the subject of C*-algebras has developed into a separate discipline of its own, with a more algebraic character. In this chapter, we develop their basic properties culminating in the GNS construction, showing that every abstract C*-algebra can be faithfully represented on a Hilbert space. Then we establish the spectral theorem for normal operators. After that, we take a look at a variety of examples. Special attention is paid to C*-algebras of groups and the property of amenability.

Recall from Section 7.6 on commutative C*-algebras that a C*-algebra $\mathfrak{A}$ is a Banach $*$-algebra satisfying the C*-norm condition

$$\|a^*a\| = \|a\|^2 \quad \text{for all} \quad a \in \mathfrak{A}.$$

The prototype example is a closed self-adjoint subalgebra of $\mathcal{B}(H)$ for a Hilbert space H. In this case, the C*-norm condition is straightforward:

$$\begin{aligned} \|T\|^2 &= \sup_{\|x\|\le 1} \|Tx\|^2 = \sup_{\|x\|\le 1} \langle Tx, Tx\rangle \\ &= \sup_{\|x\|\le 1} \langle T^*Tx, x\rangle \le \|T^*T\| \le \|T^*\|\,\|T\| = \|T\|^2. \end{aligned}$$

A C*-subalgebra of $\mathcal{B}(H)$ is called a *concrete C*-algebra*. One of our goals is to show that every C*-algebra can be represented as a concrete C*-algebra.

9.1. Positive elements

In this section, we develop basic properties of positive elements in C*-algebras. In particular, we show that non-unital C*-algebras always have a positive contractive approximate identity.

Recall that an element a of a C*-algebra is positive if $a = a^*$ and $\sigma(a) \subset [0, \infty)$. We need a couple of easy consequences of the continuous functional calculus.

9.1.1. LEMMA. *If $a = a^* \in \mathfrak{A}$, a C*-algebra, then the following are equivalent:*

(1) $a \ge 0$.

(2) $a = b^2$ *for some* $b = b^* \in \mathfrak{A}$.

K. R. Davidson, *Functional Analysis and Operator Algebras*, CMS/CAIMS Books in Mathematics 13,
https://doi.org/10.1007/978-3-031-63665-3_9

(3) $\|t - a\| \le t$ *for all* $t \ge \|a\|$ *in* $\mathbb{R}$.

(4) *there exists* $t \ge \|a\|$ *in* $\mathbb{R}$ *so that* $\|t - a\| \le t$.

PROOF. If $a \ge 0$, then $\sigma(a) \subset [0, \|a\|]$. The function $f(x) = \sqrt{x}$ lies in $C_0(\sigma(a) \setminus \{0\})$. By the Continuous functional calculus 7.6.5, $b = f(a) \in \mathfrak{A}^+$. Thus, $b = b^*$ since f is real valued; and $b^2 = f^2(a) = x(a) = a$.

If (2) holds, let $g(x) = x^2$ on $\sigma(b)$. Now $\sigma(b) \subset \mathbb{R} \cap \overline{b_{\|b\|}(0)} = [-\|b\|, \|b\|]$. For $t \ge \|b\|^2 = \|a\|$, the function $t - g(x)$ has range in $[t - \|b\|^2, t]$. Therefore $\|t - a\| = \|(t - g)(b)\| = \|t - g\|_\infty \le t$.

Clearly (3) implies (4). So suppose that $\|t - a\| \le t$ for some $t \ge \|a\|$. Then $\sigma(t-a) \subset \mathbb{R} \cap \overline{b_t(0)} = [-t, t]$. Let $h(x) = t - x$. Then $a = h(t-a)$ is self-adjoint with $\sigma(a) \subset \operatorname{Ran}(h) = [0, 2t]$. Therefore $a \ge 0$. ∎

9.1.2. COROLLARY. *The set* $\mathfrak{A}^+$ *of positive elements in* $\mathfrak{A}$ *forms a closed cone; i.e. if* $a, b \in \mathfrak{A}^+$, *then* $a + b \in \mathfrak{A}^+$ *and* $ta \in \mathfrak{A}^+$ *for* $t \ge 0$.

PROOF. Take $r > \|a\|$ and $s > \|b\|$. Then

$$\|(r + s) - (a + b)\| \le \|r - a\| + \|s - b\| \le r + s.$$

Therefore $a + b \ge 0$ by Lemma 9.1.1. Clearly $ta \ge 0$ if $t \ge 0$. Finally, if $a = \lim_{n\to\infty} a_n$ is a limit of positive elements, let $t > \sup_{n\ge1} \|a_n\|$. By Lemma 9.1.1,

$$\|t - a\| = \lim_{n\to\infty} \|t - a_n\| \le \lim_{n\to\infty} t = t.$$

So by Lemma 9.1.1 again, a is positive. ∎

9.1.3. LEMMA. *If* $a \in \mathfrak{A}_{sa} = \{x \in \mathfrak{A} : x = x^*\}$, *then there are* $a_+, a_- \in \mathfrak{A}^+$ *such that*

$$a = a_+ - a_- \quad \textit{and} \quad a_+ a_- = 0.$$

PROOF. Let $f(x) = \max\{x, 0\}$ and $g(x) = \max\{-x, 0\}$. Set $a_+ = f(a)$ and $a_- = g(a)$ using the continuous functional calculus on $C^*(a)$. Because both f, g are positive, $a_\pm$ are both positive. Since $f - g = x$ and $fg = 0$, the result follows from the continuous functional calculus. ∎

If $T \in \mathcal{B}(H)$, then $\langle T^*Tx, x\rangle = \langle Tx, Tx\rangle = \|Tx\|^2 \ge 0$. By Remark 5.4.2(2), $\sigma(T^*T) \subset [0, \infty)$. So T^*T is a positive operator.

This is not obvious in an abstract C*-algebra. Indeed, Gelfand and Naimark originally had this as an additional axiom. This condition was removed by Kaplansky a few years later. They also had split the axiom $\|a^*a\| = \|a\|^2$ into two parts: $\|a^*a\| = \|a^*\|\,\|a\|$ and $\|a^*\| = \|a\|$. We showed in Section 7.6 that the latter is an easy consequence of the C*-norm condition. It is actually sufficient to just assume that $\|a^*a\| = \|a^*\|\,\|a\|$, but this was not shown until 1960 [126].

9.1.4. THEOREM. *Let a belong to a C*-algebra $\mathfrak{A}$. Then a^*a is positive.*

PROOF. Let $b = a^*a$. Then $b = b^*$, so we can write $b = b_+ - b_-$ for $b_\pm \in \mathfrak{A}^+$ and $b_+b_- = 0$. Let $f(x) = \sqrt{x}$ on $[0, \|b_-\|]$. Set $c = f(b_-)$, so that $c \geq 0$ and $c^2 = b_-$. There are polynomials $p_n(x)$ with $p_n(0) = 0$ converging uniformly to $f(x)$ on $[0, \|b_-\|]$. Hence $cb_+ = \lim_{n\to\infty} p_n(b_-)b_+ = 0$. Let $d = ac$. Then

$$d^*d = ca^*ac = c(b_+ - b_-)c = -cb_-c = -b_-^2 \leq 0.$$

Write $d = x + iy$ where $x = \frac{1}{2}(d + d^*)$ and $y = \frac{1}{2i}(d - d^*)$. Then

$$d^*d + dd^* = (x^2 + ixy - iyx + y^2) + (x^2 - ixy + iyx + y^2) = 2x^2 + 2y^2 \geq 0.$$

Hence by Corollary 9.1.2,

$$dd^* = (d^*d + dd^*) - d^*d = x^2 + y^2 + b_-^2 \geq 0.$$

By Lemma 8.2.1,

$$\sigma(dd^*) \cup \{0\} = \sigma(d^*d) \cup \{0\} \subset (-\infty, 0] \cap [0, \infty) = \{0\}.$$

Since d^*d is self-adjoint, $\|d^*d\| = \operatorname{spr}(d^*d) = 0$. Therefore $0 = d^*d = -b_-^2$; and hence $b_- = 0$. That is, $a^*a = b_+ \geq 0$. ■

9.1.5. COROLLARY. *C*-algebras are semisimple.*

PROOF. Suppose that $a \in \operatorname{rad}\mathfrak{A}$ for a C*-algebra $\mathfrak{A}$. Then a^*a is positive and $\|a^*a\| = \|a\|^2 = \operatorname{spr}(a^*a) = 0$. Hence $a = 0$. ■

9.1.6. DEFINITION. If $a, b \in \mathfrak{A}_{sa}$, we say $a \leq b$ if $b - a \geq 0$.

9.1.7. PROPOSITION. *Let $a, b \in \mathfrak{A}_{sa}$.*

(1) *If $a \leq b$ and $x \in \mathfrak{A}$, then $x^*ax \leq x^*bx$.*

(2) *If $0 \leq a \leq b$, then $\|a\| \leq \|b\|$.*

(3) *If $a \geq 0$, then a is invertible if and only if $a \geq \varepsilon 1$ for some $\varepsilon > 0$.*

(4) *If $0 \leq a \leq b$ and a is invertible, then $b^{-1} \leq a^{-1}$.*

PROOF. (1). Factor $b - a = c^*c$. Then

$$x^*bx - x^*ax = x^*(c^*c)x = (cx)^*(cx) \geq 0.$$

Thus $x^*ax \leq x^*bx$.

(2). Working in $\widetilde{\mathfrak{A}}$ if necessary, we have $0 \leq a \leq b \leq \|b\|1$. Now a and 1 belong to $C^*(a)$ and applying the Gelfand map yields $0 \leq \hat{a} \leq \|b\|$ in $C(\sigma(a))$. Thus $\|a\| = \|\hat{a}\|_\infty \leq \|b\|$.

(3). Now a is invertible if and only if $0 \notin \sigma(a)$. Since $a \geq 0$, $\sigma(a) \subset [0, \infty)$. Hence a is invertible if and only if there is an $\varepsilon > 0$ so that $\sigma(a) \subset [\varepsilon, \infty)$, which occurs if and only if there is an $\varepsilon > 0$ so that $\sigma(a - \varepsilon 1) \subset [0, \infty)$. This happens if and only if $a - \varepsilon 1 \geq 0$, or equivalently, $a \geq \varepsilon 1$.

(4). If $0 \leq a \leq b$ and a is invertible, then $b \geq a \geq \varepsilon 1$ for some $\varepsilon > 0$; and thus b is also invertible by (3). Clearly, a^{-1} and b^{-1} are self-adjoint. Moreover,

$$\sigma(a^{-1}) = \{\lambda^{-1} : \lambda \in \sigma(a)\} \subset (0, \infty).$$

So a^{-1} is positive, and has a positive square root $a^{-1/2}$. Similarly b^{-1} has a positive square root $b^{-1/2}$. Therefore by (1), $0 \leq b^{-1/2}ab^{-1/2} \leq b^{-1/2}bb^{-1/2} = 1$. Hence

$$\|a^{1/2}b^{-1/2}\|^2 = \|b^{-1/2}ab^{-1/2}\| \leq 1$$

by (2). The adjoint map is isometric, so

$$\|a^{1/2}b^{-1}a^{1/2}\| = \|b^{-1/2}a^{1/2}\|^2 \leq 1.$$

Hence $a^{1/2}b^{-1}a^{1/2} \leq 1$. Using (1) again, we get $b^{-1} \leq a^{-1/2}1a^{-1/2} = a^{-1}$. ■

9.1.8. DEFINITION. An *approximate identity* or *approximate unit* for a C*-algebra is a contractive approximate identity consisting of positive elements.

When a C*-algebra is non-unital, there is always an approximate identity.

9.1.9. THEOREM. *Every C*-algebra has an approximate identity.*

PROOF. Let $\Lambda = \{a \in \mathfrak{A}^+ : \|a\| < 1\}$ ordered by the order $\leq$ on $\mathfrak{A}_{sa}$. We first show that Λ is directed. Let $a, b \in \Lambda$. Define

$$f(t) = \frac{t}{1-t} \text{ on } [0,1) \quad \text{and} \quad g(t) = \frac{t}{1+t} = 1 - \frac{1}{1+t} \text{ on } [0, \infty).$$

Then $g(f(t)) = t$. Let $x = f(a) + f(b)$ and $c = g(x)$. Note that $f(0) = 0 = g(0)$, so $f(a)$, $f(b)$ and $g(x)$ all belong to $\mathfrak{A}$. Now $c \geq 0$. Since g maps $\sigma(x)$ onto a compact subset of $[0, 1)$, we have $\|c\| < 1$. Thus $c \in \Lambda$. Now $y = f(a) \leq x$ implies $1 + y \leq 1 + x$, which are both invertible. Hence $(1 + x)^{-1} \leq (1 + y)^{-1}$. Therefore

$$a = g(y) = 1 - (1 + y)^{-1} \leq 1 - (1 + x)^{-1} = g(x) = c.$$

Similarly $b \leq c$. Therefore Λ is directed.

Suppose that $a \leq b$ in Λ and $x \in \mathfrak{A}$. Note that $(1 - b)^2 \leq 1 - b$. Therefore

$$\begin{aligned}\|x - bx\|^2 &= \|x^*(1-b)^2x\| \leq \|x^*(1-b)x\| \\ &\leq \|x^*(1-a)x\| \leq \|x\|\,\|x - ax\|.\end{aligned}$$

Now if $x \in \mathfrak{A}^+$, we can choose $a_n = g(nx) = \dfrac{nx}{1+nx}$. Then

$$\|x - a_n x\| = \left\|\frac{x}{1+nx}\right\| \leq \left\|\frac{t}{1+nt}\right\|_\infty < \frac{1}{n}.$$

Hence for $b \geq a_n$ in Λ, we have

$$\|x - bx\|^2 \leq \|x\| \, \|x - ax\| < \frac{\|x\|}{n}.$$

Therefore this converges to 0. For general x, we have

$$\|x - bx\|^2 = \|(1-b)xx^*(1-b)\| \leq \|xx^* - bxx^*\|.$$

Since $xx^* \in \mathfrak{A}^+$, the RHS converges to 0, and therefore $\|x - bx\| \to 0$ for all $x \in \mathfrak{A}$. On the right, $\|x - xb\| = \|bx^* - x^*\| \to 0$. Hence Λ is a contracive approximate identity of positive elements. ∎

9.1.10. REMARKS.
(1) If $\mathfrak{A}$ is separable, there is an approximate identity which is an increasing sequence. See Exercise **9.1**.

(2) In the proof, we bounded $(1-b)^2 \leq 1 - b \leq 1 - a$ when $0 \leq a \leq b$ in Λ. This was done because $0 \leq a \leq b$ does not imply $a^2 \leq b^2$. A simple example occurs in M_2. Let

$$A = \begin{bmatrix} 1 & 1 \\ 1 & 1 \end{bmatrix} \quad \text{and} \quad B = \begin{bmatrix} 1 & 1 \\ 1 & 2 \end{bmatrix}.$$

Then $0 \leq A \leq B$, but $A^2 \not\leq B^2$ because

$$B^2 - A^2 = \begin{bmatrix} 2 & 3 \\ 3 & 5 \end{bmatrix} - \begin{bmatrix} 2 & 2 \\ 2 & 2 \end{bmatrix} = \begin{bmatrix} 0 & 1 \\ 1 & 3 \end{bmatrix}$$

which has a negative determinant.

(3) If $0 \leq a \leq b$ and $ab = ba$, then $a^2 \leq b^2$. To see this, we can work in the abelian algebra $C^*(a, b, 1)$ which is $*$-isomorphic to $C(X)$ for some compact space X. Then with $f = \hat{a}$ and $g = \hat{b}$, we have $0 \leq f \leq g$ and hence $f^2 \leq g^2$; so that $a^2 \leq b^2$.

9.1.11. DEFINITION. A function f on $[a, b]$ is *operator monotone* if whenever $A \leq B \in \mathcal{B}(H)$, both with spectrum in $[a, b]$, then $f(A) \leq f(B)$.

9.1.12. PROPOSITION. *For* $0 < p \leq 1$, $f(x) = x^p$ *is operator monotone on* $[0, \infty)$.

PROOF. If $0 \leq A \leq B$, then $I + sA \leq I + sB$ for $s > 0$; whence

$$(I + sB)^{-1} \leq (I + sA)^{-1}; \quad \text{and so} \quad I - (I + sA)^{-1} \leq I - (I + sB)^{-1}.$$

That is, $1-(1+sx)^{-1} = \frac{sx}{1+sx}$ is operator monotone. Let $f_s(x) = \frac{x}{1+sx}$. Compute, substituting $u = sx$

$$\int_0^\infty f_s(x)s^{-p}\,ds = \int_0^\infty \frac{xs^{-p}}{1+sx}\,ds = \int_0^\infty \frac{x^{1+p}u^{-p}}{1+u}\frac{du}{x}$$
$$= x^p \int_0^\infty \frac{u^{-p}}{1+u}\,du = \gamma_p x^p.$$

For $0 \le x \le M$, this formula is valid uniformly. Thus given $\varepsilon > 0$, there is a Riemann sum which approximates x^p within ε on $[0, M]$:

$$\Big| x^p - \frac{1}{\gamma_p}\frac{M}{n}\sum_{i=1}^{n} f_{Mi/n}(x)\big(\tfrac{Mi}{n}\big)^{-p}\Big| < \varepsilon.$$

Since the Riemann sum is operator monotone, we get $B^p - A^p > -2\varepsilon I$. But ε is arbitrary, so $A^p \le B^p$. ∎

9.2. Ideals and Quotients

In this section, we show that the quotient of a C*-algebra by a closed ideal is a C*-algebra. Then we deduce some fundamental properties of homomorphisms of C*-algebras.

9.2.1. Definition. An *ideal* of a C*-algebra is a closed 2-sided ideal. Left and right ideals are assumed to be closed unless otherwise stated. A subalgebra $\mathfrak{B}$ of a C*-algebra $\mathfrak{A}$ is *hereditary* if $b \in \mathfrak{B}^+$ and $0 \le a \le b$ for $a \in \mathfrak{A}$ implies $a \in \mathfrak{B}$.

9.2.2. Theorem. *Let J be an ideal of a C*-algebra $\mathfrak{A}$. Then J is self-adjoint. Let $\{e_\lambda\}_\Lambda$ be an approximate unit for J. Then for $a \in \mathfrak{A}$,*

$$\|a + J\| = \inf_J \|a - j\| = \lim_\Lambda \|a - ae_\lambda\|.$$

The quotient $\mathfrak{A}/J$ is a C-algebra.*

Proof. Let $\mathfrak{B} = J \cap J^* \supset JJ^*$. Then $\mathfrak{B}$ is a C*-algebra. Let $\{e_\lambda\}$ be an approximate identity for $\mathfrak{B}$. Fix $a \in J$. Then

$$\begin{aligned}\lim_\Lambda \|a^* - a^*e_\lambda\|^2 &= \lim_\Lambda \|(a - e_\lambda a)(a^* - a^*e_\lambda)\| \\ &= \lim_\Lambda \|aa^* - e_\lambda aa^* - aa^*e_\lambda + e_\lambda aa^* e_\lambda\| \\ &\le \lim_\Lambda \|aa^* - e_\lambda aa^*\| + \|aa^* - e_\lambda aa^*\|\|e_\lambda\| = 0\end{aligned}$$

because $aa^* \in \mathfrak{B}$. Now $e_\lambda \in J$ and so $a^*e_\lambda \in J$; whence $a^* \in J$. Thus $J = J^*$.

Since $J = J^* = \mathfrak{B}$, it is a C*-algebra, and thus $\{e_\lambda\}_\Lambda$ is an approximate unit. Let $a \in \mathfrak{A}$, and let $\dot{a} = a + J$. For any $\lambda \in \Lambda$, $\|a - ae_\lambda\| \geq \|\dot{a}\|$. Given $\varepsilon > 0$, pick $j \in J$ so that $\|a - j\| < \|\dot{a}\| + \varepsilon$. Then

$$\begin{aligned}\|\dot{a}\| &\leq \liminf_{\Lambda} \|a - ae_\lambda\| \leq \limsup_{\Lambda} \|a - ae_\lambda\| \\ &\leq \limsup_{\Lambda} \|(a-j)(1-e_\lambda)\| + \|j - je_\lambda\| \leq \|a-j\| < \|\dot{a}\| + \varepsilon.\end{aligned}$$

Since $\varepsilon > 0$ was arbitrary, we have $\|\dot{a}\| = \lim_\Lambda \|a - ae_\lambda\|$.

Because $J = J^*$ is an ideal, $\mathfrak{A}/J$ is a Banach $*$-algebra and also $\|\dot{a}^*\| = \|\dot{a}\|$. To verify the C*-norm condition,

$$\begin{aligned}\|\dot{a}^*\dot{a}\| &= \lim_{\Lambda} \|a^*a(1 - e_\lambda)\| \geq \lim_{\Lambda} \|(1 - e_\lambda)a^*a(1 - e_\lambda)\| \\ &= \lim_{\Lambda} \|a(1 - e_\lambda)\|^2 = \|\dot{a}\|^2 = \|\dot{a}^*\|\,\|\dot{a}\| \geq \|\dot{a}^*\dot{a}\|.\end{aligned}$$

Hence $\|\dot{a}^*\dot{a}\| = \|\dot{a}\|^2$. Therefore $\mathfrak{A}/J$ is a C*-algebra. ■

9.2.3. COROLLARY. *Every ideal of a C*-algebra is hereditary.*

PROOF. Suppose that $0 \leq a \leq b$ and $b \in J \triangleleft \mathfrak{A}$. Choose c, d in $\mathfrak{A}^+$ so that $c^2 = a$ and $d^2 = b - a$. Then in $\mathfrak{A}/J$, we have

$$\dot{a} = \dot{c}^2 \geq \dot{0} \quad \text{and} \quad -\dot{a} = \dot{b} - \dot{a} = \dot{d}^2 \geq \dot{0}.$$

Thus $\dot{a} = \dot{0}$; whence $a \in J$. ■

Now we come to a crucial fact about C*-algebras which distinguishes them from other Banach algebras. It is analogous to the first isomorphism theorem for rings but with a stronger conclusion.

9.2.4. THEOREM. *Let $\mathfrak{A}$ and $\mathfrak{B}$ be C*-algebras, and let $\pi : \mathfrak{A} \to \mathfrak{B}$ be a $*$-homomorphism. Let $J = \ker \pi$. Then $\|\pi\| = 1$ unless $\pi = 0$, J is closed, $\pi(\mathfrak{A})$ is a C*-algebra, and there is a factorization through the quotient $\mathfrak{A}/J$:*

$$\mathfrak{A} \xrightarrow{\ q\ } \mathfrak{A}/J \xrightarrow{\ \dot{\pi}\ } \pi(\mathfrak{A}) \subset \mathfrak{B}.$$

The map $\dot{\pi}$ is an isometric $*$-isomorphism.

PROOF. If $a = a^* \in \mathfrak{A}$, then since $\sigma(\pi(a)) \subset \sigma(a)$,

$$\|\pi(a)\| = \operatorname{spr}(\pi(a)) \leq \operatorname{spr}(a) = \|a\|.$$

In general,

$$\|\pi(a)\|^2 = \|\pi(a^*a)\| \leq \|a^*a\| = \|a\|^2.$$

Therefore $\|\pi\| \leq 1$. So π is continuous, and hence $J = \ker \pi$ is a closed ideal. It follows that π factors as $\pi = \dot{\pi}q$ where $q : \mathfrak{A} \to \mathfrak{A}/J$ is the quotient map. Hence $\dot{\pi}$ is an injective $*$-homomorphism and $\|\dot{\pi}\| \leq 1$.

To see that $\dot{\pi}$ is isometric, suppose there is an $a \in \mathfrak{A}/J$ so that

$$r = \|\dot{\pi}(a)\| < \|a\| = s.$$

Observe that $s^2 = \|a^*a\| = \mathrm{spr}(a^*a)$, and thus $s^2 \in \sigma(a^*a) \subset [0, s^2]$. Let

$$f(t) = \begin{cases} 0 & \text{if } 0 \le t \le r^2 \\ \frac{x-r^2}{s^2-r^2} & \text{if } r^2 \le x \le s^2 \end{cases} \quad \text{and} \quad b = f(a^*a).$$

Note that $f \in C_0\big((0, \|a^*a\|]\big)$, and thus $b \in \mathfrak{A}$. Moreover $0 \le b \le 1$ and

$$\|b\| = \sup\{|f(t)| : t \in \sigma(a^*a)\} = f(s^2) = 1.$$

Because $\sigma(\dot{\pi}(a^*a)) \subset [0, r^2]$, by Corollary 7.6.6 we have

$$\dot{\pi}(b) = \dot{\pi}(f(a^*a)) = f(\dot{\pi}(a^*a)) = 0.$$

Hence $\dot{\pi}$ is not injective, a contradiction. Therefore $\dot{\pi}$ is isometric.

It follows that $\pi(\mathfrak{A}) = \dot{\pi}(\mathfrak{A}/J)$ is closed and $*$-isomorphic to $\mathfrak{A}/J$. In particular, $\pi(\mathfrak{A})$ is a C*-algebra. ■

This corollary is analogous to the second isomorphism theorem for rings, again with a stronger conclusion.

9.2.5. COROLLARY. *Let $\mathfrak{A}$ be a C*-algebra. If J is an ideal of $\mathfrak{A}$ and $\mathfrak{B}$ is a C*-subalgebra of $\mathfrak{A}$, then $\mathfrak{B} + J$ is a C*-algebra, and there is a canonical isometric isomorphism*

$$(\mathfrak{B} + J)/J \simeq \mathfrak{B}/(\mathfrak{B} \cap J).$$

PROOF. Let $q : \mathfrak{A} \to \mathfrak{A}/J$ be the quotient map. Then $q(\mathfrak{B}) = (\mathfrak{B} + J)/J$ is a closed C*-subalgebra of $\mathfrak{A}/J$ by Theorem 9.2.4. Hence $q^{-1}(q(\mathfrak{B})) = \mathfrak{B} + J$ is a closed C*-subalgebra of $\mathfrak{A}$. The kernel of $q|_{\mathfrak{B}}$ is $\mathfrak{B} \cap J$, so again Theorem 9.2.4 shows that $q(\mathfrak{B}) \simeq \mathfrak{B}/(\mathfrak{B} \cap J)$. ■

9.2.6. COROLLARY. *Let $\mathfrak{A}$ be a C*-algebra, and let $\mathfrak{B}$ be a C*-subalgebra. For $b \in \mathfrak{B}$, $\sigma_{\mathfrak{B}}(b) = \sigma_{\mathfrak{A}}(b)$; i.e. C*-algebras are* inverse closed.

PROOF. We need to show that if $b \in \mathfrak{B}$ has an inverse $b^{-1} \in \mathfrak{A}$, then $b^{-1} \in \mathfrak{B}$. First suppose that $b = b^*$. Then $\mathfrak{C} = C^*(b, b^{-1})$ is a commutative subalgebra of $\mathfrak{A}$. Since b is invertible in $\mathfrak{C}$, $0 \notin \sigma_{\mathfrak{C}}(b)$. Thus there are polynomials $p_n \in \mathbb{C}[x]$ so that $p_n(0) = 0$ and $p_n(x)$ converge uniformly to x^{-1} on $\sigma_{\mathfrak{C}}(b)$. By the functional calculus, $p_n(b)$ belong to $\mathfrak{B}$ and they converge to b^{-1} in $\mathfrak{C}$. Hence $b^{-1} \in \mathfrak{B}$.

For general $b \in \mathfrak{B}$ with inverse in $\mathfrak{A}$, note that $(b^*b)^{-1} = b^{-1}b^{-1*}$ is invertible in $\mathfrak{A}$. By the first paragraph, $(b^*b)^{-1}$ lies in $\mathfrak{B}$. Therefore $b^{-1} = (b^*b)^{-1}b^*$ belongs to $\mathfrak{B}$. ■

9.2.7. DEFINITION. A representation π of a C*-algebra is *faithful* if it is injective.

Theorem 9.2.4 shows that faithful representations are isometric.

9.3. Von Neumann Algebras

Chapter 10 is entirely devoted to von Neumann algebras. Here we develop some basic ideas needed to do some operator theory. This will be needed, for example, when we discuss the spectral theorem.

9.3.1. DEFINITION. A *von Neumann algebra* is an unital C*-subalgebra of $\mathcal{B}(H)$ which is closed in the weak operator topology. A von Neumann algebra is *maximal abelian* if it is abelian and is not contained in any larger abelian algebra.

9.3.2. DEFINITION. If $\mathcal{S} \subset \mathcal{B}(H)$, its *commutant* is

$$\mathcal{S}' = \{T \in \mathcal{B}(H) : TS = ST \text{ for all } S \in \mathcal{S}\}.$$

By $\mathcal{S}''$, we mean $(\mathcal{S}')'$. This is the *double commutant* of $\mathcal{S}$.

The commutant of $\mathcal{S}$ is always a unital algebra which is closed in the weak operator topology. Indeed, if $A, B \in \mathcal{S}'$, then $A+\lambda B$ and $AB \in \mathcal{S}'$; and if $T_\lambda \in \mathcal{S}'$ and $T_\lambda \xrightarrow{\text{WOT}} T$, then

$$TS = \text{WOT-}\lim_{\Lambda} T_\lambda S = \text{WOT-}\lim_{\Lambda} ST_\lambda = ST.$$

Also if $\mathcal{S} = \mathcal{S}^*$, then $\mathcal{S}'$ is self-adjoint; and therefore is a von Neumann algebra.

9.3.3. EXAMPLE. If $\mathfrak{A}$ is a C*-subalgebra of $\mathcal{B}(H)$, then $\mathfrak{A}'$ and $\mathfrak{A}''$ are von Neumann algebras. If $\mathfrak{A}$ is unital, then $\overline{\mathfrak{A}}^{\text{WOT}}$ is a von Neumann algebra. We shall see that as long as $\mathfrak{A}$ is non-degenerate, meaning that $\overline{\mathfrak{A}H} = H$, then $\overline{\mathfrak{A}}^{\text{WOT}}$ is a von Neumann algebra.

If $T \in \mathcal{B}(H)$, let $W^*(T) = \overline{C^*(T)}^{\text{WOT}}$. This is the von Neumann algebra generated by T.

9.3.4. PROPOSITION. *Let (X, μ) be a σ-finite measure. Let $H = L^2(\mu)$. Consider the representation $\pi : L^\infty(\mu) \to \mathcal{B}(H)$ by $\pi(f) = M_f$. Then $\pi(L^\infty(\mu))$ is a maximal abelian subalgebra of $\mathcal{B}(L^2(\mu))$. The weak operator topology on $\pi(L^\infty(\mu))$ coincides with the weak-$*$ topology on $L^\infty(\mu)$ induced by the dual pairing with $L^1(\mu)$. Thus π is continuous from the weak-$*$ topology on $L^\infty(\mu)$ onto $\pi(L^\infty(\mu))$ with the* WOT *topology.*

PROOF. First, observe that if $h, k \in L^2(\mu)$, then

$$\langle M_f h, k\rangle = \int f(h\bar{k})\,d\mu.$$

The function $h\bar{k}$ belongs to $L^1(\mu)$, and conversely, every function in $L^1(\mu)$ factors as the product of two functions in $L^2(\mu)$. It follows that the weak operator topology on $\pi(L^\infty(\mu))$ coincides with the weak-$*$ topology on $L^\infty(\mu)$ arising from the duality $L^1(\mu)^* = L^\infty(\mu)$. Indeed, if $\{f_\lambda\}_\Lambda \subset L^\infty(\mu)$ converge weak-$*$ to f, then

$$\langle M_f h, k\rangle = \int f(h\bar{k})\,d\mu = \lim_\Lambda \int f_\lambda(h\bar{k})\,d\mu = \lim_\Lambda \langle M_{f_\lambda} h, k\rangle$$

for $h, k \in L^2(\mu)$. Hence π is continuous from the weak-$*$ topology on $L^\infty(\mu)$ onto $\pi(L^\infty(\mu))$ with the WOT topology.

Clearly $\pi(L^\infty(\mu))$ is an abelian subalgebra of $\mathcal{B}(L^2(\mu))$. Suppose first that μ is finite. Let $A \in \pi(L^\infty(\mu))'$. Then $A1 = h \in L^2(\mu)$. Then for any $f \in L^\infty(\mu)$,

$$Af = AM_f 1 = M_f A1 = M_f h = hf.$$

For $\varepsilon > 0$, if $Y = \{x \in X : |h(x)| > \|A\| + \varepsilon\}$ has positive measure, we let $f = \operatorname{sign}(\bar{h})\mu(Y)^{-1/2}\chi_Y$, which is a unit vector. Thus

$$\|A\| \ge \|Af\| = \|hf\|_2 = \mu(Y)^{-1/2}\|\,|h|\chi_Y\|_2 > \|A\| + \varepsilon.$$

This contradiction shows that $\|h\|_\infty \le \|A\|$. Therefore A and M_h are both continuous, and agree on the dense subset $L^\infty(\mu)$. Thus $A = M_h$. We deduce that $\pi(L^\infty(\mu))' = \pi(L^\infty(\mu))$. Hence $\pi(L^\infty(\mu))$ is WOT-closed, and so is a von Neumann algebra. It does not commute with any other operators, and therefore it is maximal abelian.

Now suppose that μ is σ-finite, and let $A \in \pi(L^\infty(\mu))'$. We can write X as a countable disjoint union $X = \dot{\bigcup}_{n\ge1} X_n$ where $\mu(X_n) < \infty$. Now $M_{\chi_{X_n}}$ lies in $\pi(L^\infty(\mu))$, and commutes with A. Thus $A|_{L^2(X_n)}$ commutes with $\pi(L^\infty(\mu))|_{L^2(X_n)}$. By the previous paragraph, there is a function $h_n \in L^\infty(X_n)$ with $\|h_n\|_\infty \le \|A\|$ so that $A|_{L^2(X_n)} = M_{h_n}$. Therefore, letting $h \in L^\infty(\mu)$ be given by $h|_{X_n} = h_n$ for $n \ge 1$, we see that $A = M_h$. Again, it now follows that $\pi(L^\infty(\mu))$ is a maximal abelian von Neumann algebra. ■

In von Neumann algebras, one can often detect WOT-continuity by considering increasing nets; i.e. a net in $\mathcal{N}_{sa}$ such that $a_\lambda \le a_\mu$ when $\lambda \le \mu$. The interplay between the order and the topology plays an important role in von Neumann algebras. See Definition 9.3.8 and later in Section 10.3. We need a lemma.

9.3.5. LEMMA. *Suppose that $\{A_\lambda : \lambda \in \Lambda\}$ is an increasing net of self-adjoint operators in $\mathcal{B}(H)$ which is bounded above by M; i.e. $A_\lambda \le A_\mu \le M$ if $\lambda \le \mu$. Then* SOT–$\lim_\Lambda A_\lambda = A =: \sup A_\lambda$ *exists, and is the least upper bound for $\{A_\lambda\}$. In particular, if each A_λ is a projection, then A is the projection onto $\overline{\bigcup_\Lambda A_\lambda H}$.*

PROOF. For $\xi \in H$, the net $\langle A_\lambda \xi, \xi \rangle$ is an increasing net of real numbers bounded above by $M\|\xi\|^2$. Hence $\Gamma(\xi) = \lim_\Lambda \langle A_\lambda \xi, \xi \rangle$ exists. Use the polarization identity to define a sesquilinear form

$$\begin{aligned}
[\xi, \eta] &= \frac{1}{4}\big(\Gamma(\xi+\eta) - \Gamma(\xi-\eta) + i\Gamma(\xi+i\eta) - i\Gamma(\xi - i\eta)\big) \\
&= \lim_\Lambda \frac{1}{4}\big(\langle A_\lambda \xi + \eta, \xi + \eta\rangle - \langle A_\lambda \xi - \eta, \xi - \eta\rangle \\
&\qquad + i\langle A_\lambda \xi + i\eta, \xi + i\eta\rangle - i\langle A_\lambda \xi - i\eta, \xi - i\eta\rangle\big) \\
&= \lim_\Lambda \langle A_\lambda \xi, \eta\rangle.
\end{aligned}$$

By Proposition 4.2.2, this defines a bounded linear operator A so that

$$\langle A\xi, \eta\rangle = [\xi, \eta] = \lim_\Lambda \langle A_\lambda \xi, \eta\rangle.$$

This shows that $A = \text{WOT–}\lim_\Lambda A_\lambda$. In particular, $A = A^*$ and $\langle A\xi, \xi\rangle \geq \langle A_\lambda \xi, \xi\rangle$ for all $\xi \in H$ and $\lambda \in \Lambda$. Hence $A \geq A_\lambda$; whence A is an upper bound for the net. On the other hand, if $B \geq A_\lambda$ for all $\lambda \in \Lambda$, then

$$\langle (B - A)\xi, \xi\rangle = \lim_\Lambda \langle (B - A_\lambda)\xi, \xi\rangle \geq 0.$$

Thus $B \geq A$. Therefore A is the least upper bound for this net.

If $D \geq 0$, the sesquilinear form $[\xi, \eta] = \langle D\xi, \eta\rangle$ is positive semidefinite. Thus by the Cauchy–Schwarz inequality, we have

$$\begin{aligned}
\|D\xi\|^2 = [\xi, D\xi] &\leq [\xi, \xi]^{1/2}[D\xi, D\xi]^{1/2} \\
&= \langle D\xi, \xi\rangle^{1/2}\langle D^3\xi, \xi\rangle^{1/2} \\
&\leq \langle D\xi, \xi\rangle^{1/2}\|D\|^{3/2}\|\xi\|.
\end{aligned}$$

Apply this to $A - A_\lambda$ to obtain

$$\|(A - A_\lambda)\xi\|^2 \leq \langle (A - A_\lambda)\xi, \xi\rangle^{1/2}\|(A - A_\lambda)\|^{3/2}\|\xi\|.$$

For $\lambda \geq \lambda_0$, the RHS is bounded by

$$\langle (A - A_\lambda)\xi, \xi\rangle^{1/2}\|(A - A_{\lambda_0})\|^{3/2}\|\xi\|,$$

which converges to 0. That is, $A_\lambda \xi \to A\xi$; i.e. $A = \text{SOT–}\lim_\Lambda A_\lambda$.

Finally, if A_λ is the projection onto a subspace M_λ, then $M_\lambda \subset M_\mu$ if $\lambda \leq \mu$. Let $M = \overline{\bigcup_\Lambda M_\lambda}$. If $\xi \perp M$, then $A\xi = \lim_\Lambda A_\lambda \xi = 0$. If $x \in \bigcup_\Lambda M_\lambda$, say $x \in M_{\lambda_0}$, then for $\lambda \geq \lambda_0$, $A_\lambda \xi = \xi$; and thus $A\xi = \xi$. Since A is continuous, $A = P_M$. ■

The next result shows that a von Neumann algebra $\mathcal{N}$ always has many projections. The projections are the extreme points of $[0, I] = \{A \in \mathcal{N} : 0 \leq A \leq I\}$. See Exercise **9.5**.

9.3.6. THEOREM. *Let $0 \leq A \leq I$ belong to a von Neumann algebra $\mathcal{N} \subset \mathcal{B}(H)$. Then A belongs to the norm-closed convex hull of the projections in $\mathcal{N}$.*

PROOF. We may as well let $\mathcal{N} = W^*(A)$ be the (commutative) von Neumann algebra generated by A. Let $U \subset \sigma(A)$ be open, and consider

$$\mathcal{F}_U = \{f \in C(\sigma(A)) : 0 \le f \le \chi_U\}.$$

We consider this as a partially ordered set with the usual order on $C(\sigma(A))$. This is directed because $f, g \le f \vee g = \max\{f, g\}$. Define an increasing net $f(A)$ using the continuous functional calculus. Let $P_U = \text{SOT--}\lim_{f \in \mathcal{F}_U} f(A)$, which exists in $\mathcal{N}$ by Lemma 9.3.5. We claim that this is a projection. We have $0 \le f \le 1$ for $f \in \mathcal{F}_U$, and hence $0 \le P_U \le I$. For any $f \in \mathcal{F}_U$, $f^{1/2} \in \mathcal{F}_U$; so that $P_U \ge f(A)^{1/2}$. Since P_U commutes with $f(A)$, Remark 9.1.10 (3) shows that $P_U^2 \ge f(A)$. This holds for all $f \in \mathcal{F}_U$, and therefore $P_U \le P_U^2 \le I$. However, since $P_U \le I$, we have $P_U^2 \le P_U$ (again by the functional calculus). Thus $P_U = P_U^2$ is a projection.

Now $\sigma(A) \subset [0, 1]$. For $n \ge 1$, let $U_{j,n} = (\frac{j}{2^n}, 1]$ for $1 \le j \le 2^n$; and let $P_{j,n} = P_{U_{j,n}}$. Define

$$A_n = \frac{1}{2^n} \sum_{j=1}^{2^n} P_{j,n} = \sup\{f(A) : f \in C[0, 1],\ f(t) \le 2^{-n} \lfloor 2^n t \rfloor\}.$$

Observe that each $P_{j,n}$ is the sup of $f_j(A)$ where $f_j(t) \le \chi_{U_{j,n}}$, and hence A_n is the sup of $f(A)$ where $f \in C[0, 1]$ satisfy

$$f(t) \le g_n(t) = 2^{-n} \sum_{j=1}^{2^n} \chi_{U_{j,n}} = 2^{-n} \lfloor 2^n t \rfloor.$$

Also observe that $t - 2^{-n} \le g_n(t) \le t$ on $[0, 1]$. It follows that

$$A - 2^{-n} I \le A_n \le A.$$

Thus $A = \lim_{n \to \infty} A_n$ in norm. Evidently each A_n is a convex combination of projections in $\mathcal{N}$. ■

Recall that a *symmetry* is a self-adjoint unitary S. Thus $\sigma(S) = \{1, -1\}$ except for $S = \pm I$. Since S is normal, $C^*(S) \simeq C(\sigma(S)) = C(\{-1, 1\})$. By the continuous functional calculus, $P = \chi_{\{1\}}(S)$ is a projection and $\chi_{\{-1\}}(S) = P^\perp$. So $S = P - P^\perp = 2P - I$. This formula is also valid for $S = \pm I$.

9.3.7. COROLLARY. *If $\mathcal{N}$ is a von Neumann algebra, then $\mathcal{N}$ is the norm closed span of its projections. The convex hull of the symmetries in $\mathcal{N}$ is the set $\{A \in \mathcal{N} : -I \le A \le I\}$.*

PROOF. Every self-adjoint operator $A = A^*$ in $\mathcal{N}$ can be scaled and translated so that $0 \le rA + sI \le I$. By Theorem 9.3.6, this operator is approximated in norm by a convex combination of projections. Therefore A can be approximated in norm by a linear combination of projections.

Given $-I \leq A \leq I$, form $B = \frac{1}{2}(A + I)$. This is in the closed convex hull of projections. Thus $A = 2B - I$ is in the closed convex hull of the symmetries. ∎

9.3.8. DEFINITION. If $\mathfrak{A}$ and $\mathfrak{B}$ are C*-algebras, then $\rho : \mathfrak{A} \to \mathfrak{B}$ is a *positive linear map* if $\varphi(\mathfrak{A}^+) \subset \mathfrak{B}^+$. A positive linear map ρ from a von Neumann algebra $\mathcal{M}$ into a von Neumann algebra $\mathcal{N}$ is *normal* if whenever $\{a_\lambda\}_\Lambda$ is a bounded increasing net in $\mathcal{M}_{sa}$, then $\rho(\sup_\Lambda a_\lambda) = \sup_\Lambda \rho(a_\lambda)$.

9.3.9. PROPOSITION. *A $*$-isomorphism π from a von Neumann algebra $\mathcal{M}$ onto a von Neumann algebra $\mathcal{N}$ is always normal.*

PROOF. A homomorphism between C*-algebras is positive because if $a \geq 0$, then $\pi(a) = \pi(a^{1/2})^*\pi(a^{1/2}) \geq 0$. If $\{a_\lambda\}_\Lambda$ is an increasing net in $\mathcal{M}$, let $a = \sup_\Lambda a_\lambda$. Then $\pi(a) \geq \pi(a_\lambda)$ for all λ, and thus $\pi(a) \geq b := \sup_\Lambda \pi(a_\lambda)$. Now π^{-1} is an isomorphism of $\mathcal{N}$ onto $\mathcal{M}$, so that similarly, $a \leq \pi^{-1}(b)$. Therefore $\pi(a) \geq b \geq \pi(a)$, whence $\pi(a) = b$; i.e. π is normal. ∎

9.4. Density Theorems

In this section, we prove two important results about a C*-subalgebra of $\mathcal{B}(H)$ and the von Neumann algebra that it generates. The first, due to von Neumann, shows that the SOT-closure of a non-degenerate C*-subalgebra equals the double commutant. The second, due to Kaplansky, shows that the unit ball of $\mathfrak{A}$ is SOT-dense in the unit ball of $\mathfrak{A}''$.

Recall that a C*-subalgebra of $\mathcal{B}(H)$ is *non-degenerate* if $\overline{\mathfrak{A}H} = H$.

9.4.1. VON NEUMANN DOUBLE COMMUTANT THEOREM.
For every non-degenerate C-subalgebra $\mathfrak{A}$ of $\mathcal{B}(H)$, $\mathfrak{A}'' = \overline{\mathfrak{A}}^{\text{WOT}} = \overline{\mathfrak{A}}^{\text{SOT}}$.*

PROOF. Clearly $\overline{\mathfrak{A}}^{\text{SOT}} \subset \overline{\mathfrak{A}}^{\text{WOT}} \subset \mathfrak{A}''$. The latter holds because $\mathfrak{A}''$ is WOT-closed and contains $\mathfrak{A}$. Fix $T \in \mathfrak{A}''$. A basic SOT-open neighbourhood of T is given by vectors $\xi_j \in H$ of the form

$$U(T, \xi_1, \ldots, \xi_n) = \{A \in \mathcal{B}(H) : \sum_{j=1}^{n} \|(A - T)\xi_j\|^2 < 1\}.$$

To prove that $\mathfrak{A}'' \subset \overline{\mathfrak{A}}^{\text{SOT}}$, it suffices to show that each of these neighbourhoods intersects $\mathfrak{A}$.

Case $n = 1$. Notice that $M = \overline{\mathfrak{A}\xi_1}$ is invariant for $\mathfrak{A}$. Thus $AP_M = P_M A P_M$ for $A \in \mathfrak{A}$. Since $\mathfrak{A}$ is self-adjoint,

$$P_M A = (A^* P_M)^* = (P_M A^* P_M)^* = P_M A P_M = AP_M;$$

i.e. $P_M \in \mathfrak{A}'$. Thus $TP_M = P_M T$. Hence $T\xi_1 \in M$. So we may pick $A \in \mathfrak{A}$ so that $\|A\xi_1 - T\xi_1\| < 1$.

Case $n \geq 2$. Let $H^{(n)} = H_1 \oplus \cdots \oplus H_n$ be the Hilbert space direct sum of n Hilbert spaces $H_j \simeq H$. Let $\mathfrak{A}^{(n)} = \{A^{(n)} : A \in \mathfrak{A}\}$ where $A^{(n)}(\xi_1, \dots, \xi_n) = (A\xi_1, \dots, A\xi_n)$. Now $\mathcal{B}(H^{(n)})$ consists of all $n \times n$ matrices B with coefficients $B_{ij} \in \mathcal{B}(H_j, H_i)$. One readily checks that

$$[B_{ij}]A^{(n)} - A^{(n)}[B_{ij}] = [B_{ij}A - AB_{ij}].$$

Thus $B = [B_{ij}] \in \mathfrak{A}^{(n)\prime}$ if and only if each $B_{ij} \in \mathfrak{A}'$; that is, $\mathfrak{A}^{(n)\prime} = M_n(\mathfrak{A}')$. Similarly, $\mathfrak{A}^{(n)\prime\prime} = \{T^{(n)} : T \in \mathfrak{A}''\}$.

Now let $T \in \mathfrak{A}''$ and $\xi = (\xi_1, \dots, \xi_n)$. By Case $n = 1$, there is $A \in \mathfrak{A}$ so that

$$1 > \|(A^{(n)} - T^{(n)})\xi\|^2 = \sum_{j=1}^{n} \|(A - T)\xi_j\|^2.$$

Hence $A \in U(T, \xi_1, \dots, \xi_n)$. Therefore $\mathfrak{A}'' = \overline{\mathfrak{A}}^{\text{WOT}} = \overline{\mathfrak{A}}^{\text{SOT}}$. ∎

9.4.2. COROLLARY. *Let $\mathfrak{A}$ be a non-degenerate C*-subalgebra of $\mathcal{B}(H)$. Then $\mathfrak{A}''_{sa} = \overline{\mathfrak{A}_{sa}}^{\text{WOT}} = \overline{\mathfrak{A}_{sa}}^{\text{SOT}}$.*

PROOF. If $T = T^* \in \mathfrak{A}''_{sa}$, choose a net $A_\lambda \in \mathfrak{A}$ so that $A_\lambda \xrightarrow{\text{WOT}} T$. Since the adjoint is WOT continuous, $A_\lambda^* \xrightarrow{\text{WOT}} T$ as well. Hence $\operatorname{Re} A_\lambda = \frac{1}{2}(A_\lambda + A_\lambda^*) \xrightarrow{\text{WOT}} T$. The weak operator topology and strong operator topology have the same continuous linear functionals and thus have the same closed convex sets. See Exercise **3.4**. Therefore there is a net in $\mathfrak{A}_{sa}$ converging SOT to T as well. ∎

It is frequently necessary to control the norm of the net converging SOT to an element T. This does not follow immediately from the double commutant theorem. A trick is required.

9.4.3. LEMMA. *Let $f(t) = \dfrac{2t}{1+t^2} \in C_0(\mathbb{R})$. If $A_\lambda = A_\lambda^*$ is a net with SOT–$\lim_\Lambda A_\lambda = S$, then SOT–$\lim_\Lambda f(A_\lambda) = f(S)$.*

PROOF. Note that $\|f\|_\infty = 1$, so $\|f(A_\lambda)\| \leq 1$ and $\|f(S)\| \leq 1$. Compute

$$\begin{aligned}
f(A_\lambda) - f(S) &= 2A_\lambda(I + A_\lambda^2)^{-1} - 2S(I + S^2)^{-1} \\
&= 2(I + A_\lambda^2)^{-1}\big(A_\lambda(I + S^2) - (I + A_\lambda^2)S\big)(I + S^2)^{-1} \\
&= 2(I + A_\lambda^2)^{-1}\big((A_\lambda - S) + A_\lambda(S - A_\lambda)S\big)(I + S^2)^{-1} \\
&= 2(I + A_\lambda^2)^{-1}(A_\lambda - S)(I + S^2)^{-1} - \tfrac{1}{2}f(A_\lambda)(A_\lambda - S)f(S).
\end{aligned}$$

Given $\xi \in H$, let $\eta = (I+S^2)^{-1}\xi$ and $\zeta = f(S)\xi$. Then

$$\begin{aligned}\lim_\Lambda \|(f(A_\lambda)-f(S))\xi\| &= \lim_\Lambda \|2(I+A_\lambda^2)^{-1}(A_\lambda-S)\eta - \tfrac{1}{2}f(A_\lambda)(A_\lambda-S)\zeta\| \\ &\le \lim_\Lambda 2\|(I+A_\lambda^2)^{-1}\|\|(A_\lambda-S)\eta\| + \tfrac{1}{2}\|f(A_\lambda)\|\|(A_\lambda-S)\zeta\| \\ &\le 2\lim_\Lambda \|(A_\lambda-S)\eta\| + \tfrac{1}{2}\lim_\Lambda \|(A_\lambda-S)\zeta\| = 0.\end{aligned}$$

Therefore $f(A_\lambda) \xrightarrow{\text{SOT}} f(S)$. ■

9.4.4. Kaplansky Density Theorem. *The unit ball of $\mathfrak{A}_{sa}$ is* SOT *dense in $\mathfrak{A}''_{sa}$, and the unit ball of $\mathfrak{A}$ is* SOT *dense in the unit ball of $\mathfrak{A}''$.*

Proof. Let $f(t) = \dfrac{2t}{1+t^2} \in C_0(\mathbb{R})$. Then $f'(t) = \dfrac{2(1-t^2)}{(1+t^2)^2} \ge 0$; so f is strictly increasing on $[-1,1]$ and $f(\pm 1) = \pm 1$. Hence f is a homeomorphism of $[-1,1]$ onto itself. Let $g : [-1,1] \to [-1,1]$ be the inverse function.

Fix $T = T^* \in \overline{b_1(\mathfrak{A}''_{sa})}$. Define $S = g(T)$, so that $T = f(S)$. By Corollary 9.4.2, there is a net A_λ in $\mathfrak{A}_{sa}$ converging SOT to S. By Lemma 9.4.3, $f(A_\lambda) \xrightarrow{\text{SOT}} f(S) = T$. Since $\|f\|_\infty = 1$, we have $\|f(A_\lambda)\| \le 1$.

For an arbitrary $T \in \overline{b_1(\mathfrak{A}'')}$, consider $\widetilde{T} = \left[\begin{smallmatrix} 0 & T \\ T^* & 0\end{smallmatrix}\right] \in \overline{b_1(M_2(\mathfrak{A}''))_{sa}}$. Now $M_2(\mathfrak{A}'') = M_2(\mathfrak{A})''$. So by the first part of this proof, there is a net in the unit ball of $M_2(\mathfrak{A})_{sa}$ converging SOT to $\widetilde{T}$. The $1,2$ entry of this net converges SOT to T and has a norm at most 1. ■

In section 5.4, we developed a polar decomposition for compact operators. Now we can extend that to any von Neumann algebra, but specifically to $\mathcal{B}(H)$. Recall that a partial isometry is an operator $U = UU^*U$. Such an operator is isometric from $U^*H = U^*UH$ onto $UH = UU^*H$ and is zero on $(U^*H)^\perp$; and U^*U and UU^* are projections onto $\operatorname{Ran} U$ and $(\ker U)^\perp$, respectively.

9.4.5. Polar Decomposition. *Let $T \in \mathcal{B}(H)$. Then $|T| = (T^*T)^{1/2}$ lies in $C^*(T)$. There is a unique partial isometry $U \in W^*(T)$ so that $T = U|T|$ and* $\ker U = \ker T$. *Also UU^* projects onto* $\overline{\operatorname{Ran} T}$ *and U^*U projects onto* $\overline{\operatorname{Ran} T^*}$.

Proof. First $f(x) = \sqrt{x} \in C_0(0, \|T^*T\|]$, so that $|T|$ lies in the non-unital C*-algebra generated by T^*T; and in particular lies in $C^*(T)$. Calculate

$$\big\||T|\xi\big\|^2 = \langle |T|\xi, |T|\xi\rangle = \langle |T|^2\xi, \xi\rangle = \langle T^*T\xi, \xi\rangle = \langle T\xi, T\xi\rangle = \|T\xi\|^2.$$

In particular, $\ker |T| = \ker T = (\operatorname{Ran} T^*)^\perp$. Since $|T|$ is self-adjoint,

$$\overline{\operatorname{Ran} |T|} = (\ker |T|)^\perp = \overline{\operatorname{Ran} T^*}.$$

There is a unique operator U defined on $\operatorname{Ran} |T|$ given by $U(|T|\xi) = T\xi$. The calculation above shows that U is isometric on $\operatorname{Ran} |T|$ with range $\operatorname{Ran} T$, and thus norm 1. Therefore it extends to an isometry from $\overline{\operatorname{Ran}|T|}$ onto $\overline{\operatorname{Ran} T}$. Define $U = 0$ on $\ker T = (\operatorname{Ran}|T|)^\perp$, and extend by linearity. This is a partial isometry with $\ker U = \ker T$. By construction, it satisfies $U|T| = T$.

Finally we show that $U \in C^*(T)'' = W^*(T)$. Let $X \in C^*(T)'$. If $\xi \in \ker T$, then $0 = XT\xi = TX\xi$, so that $X \ker T \subset \ker T$. Therefore $UX\xi = 0 = XU\xi$. Next consider $\xi \in \operatorname{Ran}|T|$, say $\xi = |T|\eta$. Then

$$UX\xi = UX|T|\eta = U|T|X\eta = TX\eta = XT\eta = XU|T|\eta = XU\xi.$$

By continuity, UX and XU agree on $\overline{\operatorname{Ran}|T|}$. By linearity, $UX = XU$. Thus $U \in C^*(T)'' = W^*(T)$ by the Double Commutant theorem. ■

9.4.6. COROLLARY. *If $f \in C_0(\sigma(|T|) \setminus \{0\})$, then $Uf(|T|)$ belongs to $C^*(T)$. If T is invertible, then U is a unitary in $C^*(T)$,*

PROOF. Any $f \in C_0(0, \|T\|]$ can be approximated by a sequence of polynomials with $p_n(0) = 0$. Say $p_n(t) = tq_n(t)$. Then

$$Uf(|T|) = \lim_{n\to\infty} Up_n(|T|) = \lim_{n\to\infty} U|T|q_n(|T|) = \lim_{n\to\infty} Tq_n(|T|).$$

Thus $Uf(|T|)$ belongs to $C^*(T)$.

If T is invertible, so is $|T|$. Thus $U = T|T|^{-1}$ lies in $C^*(T)$ and is invertible. Now $U^*U = P_{\overline{\operatorname{Ran} T^*}} = I$ and similarly $UU^* = P_{\overline{\operatorname{Ran} T}} = I$. So U is unitary. ■

As an application, we show that in every unital C*-algebra, the closed convex hull of the unitary elements is the whole unit ball. This may be surprising since there are often many other extreme points, and the unitaries may not even be dense in the set of extreme points. (See Exercise **9.5**.) This is in contrast to Milman's Theorem 3.5.8.

9.4.7. RUSSO–DYE THEOREM. *Let $\mathfrak{A}$ be a unital C*-algebra. The convex hull of the set $\mathcal{U}(\mathfrak{A})$ of unitaries in $\mathfrak{A}$ contains the open unit ball of $\mathfrak{A}$. Thus $\overline{\operatorname{conv}(\mathcal{U})} = \overline{b_1(\mathfrak{A})}$.*

PROOF. Suppose that $a \in \mathfrak{A}$ with $\|a\| < 1$ is invertible. Then the polar decomposition of a is $a = up$ where $u \in \mathcal{U}$ and $p = |a|$. Then $v_\pm := p \pm i(1-p^2)^{1/2}$ are unitary and $a = \frac{1}{2}(uv_+ + uv_-)$. Thus $a \in \operatorname{conv}(\mathcal{U})$.

If $\|a\| < 1$ and $u \in \mathcal{U}$, then $\frac{1}{2}(a+u) = \frac{u}{2}(1 + u^*a)$ is invertible. Hence $\frac{1}{2}(a+u) \in \operatorname{conv}(\mathcal{U})$. This holds for all $u \in \mathcal{U}$ and thus taking convex combinations shows that $\frac{1}{2}(a+x) \in \operatorname{conv}(\mathcal{U})$ whenever $x \in \operatorname{conv}(\mathcal{U})$.

Given $a \in b_1(\mathfrak{A})$, define $a_1 = \frac{1}{2}(a+1)$ and $b_1 = \frac{1}{2}(a-1)$; and for $n \geq 2$, set

$$a_n = \frac{1}{2}(a + a_{n-1}) = (1-2^{-n})a + 2^{-n}1$$

and

$$b_n = \frac{1}{2}(a + b_{n-1}) = (1 - 2^{-n})a - 2^{-n}1.$$

By induction, a_n and b_n lie in $\operatorname{conv}(\mathcal{U})$ for $n \geq 1$. Hence so does $(1 - 2^{-n})a$. If $c \in \mathfrak{A}$ satisfies $\|c\| < 1 - 2^{-n}$, apply this to $a = (1 - 2^{-n})^{-1}c$ to get $c \in \operatorname{conv}(\mathcal{U})$. Hence $b_1(\mathfrak{A}) \subset \operatorname{conv}(\mathcal{U})$ and $\overline{\operatorname{conv}(\mathcal{U})} = \overline{b_1(\mathfrak{A})}$. ■

9.5. Transitivity

When we talk about a representation of a C*-algebra, we always mean a $*$-representation into $\mathcal{B}(H)$ for some Hilbert space H. Since $\mathcal{B}(H)$ is itself a C*-algebra, representations are always contractive by Theorem 9.2.4. We saw in Examples 8.2.5 and 8.2.6 that for Banach algebras, there is an important distinction between irreducible representations and topologically irreducible representations. For C*-algebras, the two notions coincide.

9.5.1. Lemma. *Let $\mathfrak{A}$ be a C*-algebra. A representation $\pi : \mathfrak{A} \to \mathcal{B}(H)$ is topologically irreducible if and only if $\pi(\mathfrak{A})' = \mathbb{C}I$.*

Proof. A representation is not topologically irreducible if and only if H has a proper closed submodule; i.e. a closed invariant subspace M for $\pi(\mathfrak{A})$. Since $\pi(\mathfrak{A}) = \pi(\mathfrak{A})^*$, this subspace is reducing, or equivalently, $P_M \in \pi(\mathfrak{A})'$ (see the proof of the double commutant theorem). Therefore $\pi(\mathfrak{A})' \neq \mathbb{C}I$.

Conversely, if $\pi(\mathfrak{A})' \neq \mathbb{C}I$, then this von Neumann algebra contains a non-trivial projection P by Theorem 9.3.6. So $M = PH$ is a proper closed submodule for $\pi(\mathfrak{A})$; whence $\pi(\mathfrak{A})$ is not topologically irreducible. ■

9.5.2. Lemma. *Let $\mathfrak{A}$ be a C*-algebra, and let $\pi : \mathfrak{A} \to \mathcal{B}(H)$ be a topologically irreducible representation. If $T \in \mathcal{B}(H)$, $\varepsilon > 0$ and M is a finite dimensional subspace of H, then there is $a \in \mathfrak{A}$ so that $\|(\pi(a) - T)|_M\| < \varepsilon$ and $\|a\| \leq \|T|_M\|$.*

Proof. Let $S = TP_M$. By scaling, we may suppose that $\|S\| = 1$. By Lemma 9.5.1, $\pi(\mathfrak{A})'' = (\mathbb{C}I)' = \mathcal{B}(H)$. By Kaplansky's density theorem, there is $b \in \mathfrak{A}$ so that $\|\pi(b)\| \leq 1$ and $\|(\pi(b) - S)|_M\| < \varepsilon/2$. Now

$$\|\pi(b)\| = \|b + \ker \pi\| = \inf\{\|b - j\| : j \in \ker \pi\}.$$

Pick j so that $\|b - j\| < 1 + \varepsilon/2$. Set $a = (1 + \varepsilon/2)^{-1}(b - j)$. Then $\|a\| < 1$ and

$$\|(\pi(a) - S)|_M\| \leq \|\pi(a - b)\| + \|(\pi(b) - S)|_M\| \leq \frac{\varepsilon/2}{1 + \varepsilon/2} + \frac{\varepsilon}{2} < \varepsilon. \quad ■$$

9.5.3. Kadison Transitivity Theorem. *Let $\mathfrak{A}$ be a C*-algebra, and let $\pi : \mathfrak{A} \to \mathcal{B}(H)$ be a topologically irreducible representation. Then π is irreducible. For any $T \in \mathcal{B}(H)$, $\varepsilon > 0$ and a finite dimensional subspace $M \subset H$, there is $a \in \mathfrak{A}$ with $\|a\| < \|T\| + \varepsilon$ so that $\pi(a)|_M = T|_M$.*

Proof. We use Lemma 9.5.2 repeatedly. First choose a_0 with so that

$$\|a_0\| \le \|T\| \quad \text{and} \quad \|(\pi(a_0) - T)|_M\| < \frac{\varepsilon}{2}.$$

Recursively select $a_n \in \mathfrak{A}$ so that

$$\|a_n\| < 2^{-n}\varepsilon \quad \text{and} \quad \Big\|\Big(\sum_{k=0}^{n} \pi(a_k) - T\Big)\Big|_M\Big\| < 2^{-n-1}\varepsilon.$$

Let $a = \sum_{k\ge 0} a_k$. Then $\|a\| < \|T\| + \sum_{k\ge 1} 2^{-n}\varepsilon = \|T\| + \varepsilon$; and

$$\Big\|\big(\pi(a) - T\big)\big|_M\Big| = \lim_{n\to\infty} \Big\|\Big(\sum_{k=0}^{n} \pi(a_k) - T\Big)\Big|_M\Big\| = 0.$$

In particular, for any non-zero vector ξ and vector η, there is an $a \in \mathfrak{A}$ so that $\pi(a)\xi = \eta$. Hence π is irreducible. ∎

9.5.4. Remark. Once one knows that π is irreducible, Jacobson's density theorem applies to show that $\pi(\mathfrak{A})$ is strictly dense. Kadison's theorem adds optimal norm control.

9.6. States

In this section, we study positive linear functionals on C*-algebras. They will be used to build $*$-representations in the following section.

9.6.1. Definition. A linear functional φ on a C*-algebra is a *positive linear functional* if $a \ge 0$ implies that $\varphi(a) \ge 0$. A *state* is a positive linear functional of norm 1.

9.6.2. Example. If $\pi : \mathfrak{A} \to \mathcal{B}(H)$ is a non-degenerate representation, then for any $\xi \in H$,

$$\varphi(a) = \langle \pi(a)\xi, \xi \rangle$$

is positive. If $\|\xi\| = 1$, this is a state. It is clear that $\|\varphi\| \le \|\xi\|^2$. Suppose that $\{e_\lambda\}_\Lambda$ is an approximate identity. Since π is non-degenerate,

$$\lim_{\Lambda} \pi(e_\lambda)\pi(a)\eta = \pi(a)\eta \quad \text{for all} \quad a \in \mathfrak{A} \text{ and } \eta \in H.$$

It follows that $\pi(e_\lambda) \xrightarrow{\text{SOT}} I$. Thus $\|\varphi\| = \|\xi\|^2$.

9.6.3. LEMMA. *A positive linear functional φ on a C^*-algebra $\mathfrak{A}$ satisfies the Cauchy–Schwarz inequality*

$$|\varphi(b^*a)|^2 \le \varphi(a^*a)\varphi(b^*b) \quad \textit{for} \quad a.b \in \mathfrak{A}.$$

PROOF. Define $[a,b] = \varphi(b^*a)$. Check that this is a positive semidefinite sesquilinear form. Thus it satisfies the Cauchy–Schwarz inequality. ∎

9.6.4. LEMMA. *A positive linear functional φ on a C^*-algebra $\mathfrak{A}$ is continuous. If $\mathfrak{A}$ is unital, then $\|\varphi\| = \varphi(1)$. If $\mathfrak{A}$ is not unital, and $\{e_\lambda\}_\Lambda$ is an approximate identity, then $\|\varphi\| = \lim_\Lambda \varphi(e_\lambda)$.*

PROOF. Suppose that $\mathfrak{A}$ is unital. If $0 \le a \le 1$, then $0 \le \varphi(a) \le \varphi(1)$. For general a with $\|a\| \le 1$,

$$|\varphi(a)|^2 = |\varphi(1^*a)|^2 \le \varphi(a^*a)\varphi(1) \le \varphi(1)^2.$$

Hence $\|\varphi\| = \varphi(1)$.

If $\mathfrak{A}$ is not unital, we first verify continuity on $\mathfrak{A}^+$. If it isn't continuous, then there are $a_n \ge 0$ with $\|a_n\| \le 2^{-n}$ and $\varphi(a_n) > 1$. Let $a = \sum_{n\ge1} a_n$, which is positive and $\|a\| \le 1$. However

$$\varphi(a) = \sum_{n=1}^{N} \varphi(a_n) + \varphi\Big(\sum_{n>N} a_n\Big) > N$$

for all $N \ge 1$. This is absurd. Hence there is a constant C so that $0 \le \varphi(a) \le C\|a\|$ for $a \ge 0$. A general element of a can be written as

$$a = \operatorname{Re} a + i \operatorname{Im} a = a_1 - a_2 + ia_3 - ia_4 \quad \text{for} \quad 0 \le a_i \le \|a\|,\ 1 \le i \le 4.$$

Now $0 \le \varphi(a_i) \le C\|a\|$. Therefore

$$|\varphi(a)|^2 = \big(\varphi(a_1) - \varphi(a_2)\big)^2 + \big(\varphi(a_3) - \varphi(a_4)\big)^2 \le 2C^2\|a\|^2.$$

Hence φ is continuous with $\|\varphi\| \le \sqrt{2}C$.

Now let $\{e_\lambda\}_\Lambda$ be an approximate identity for $\mathfrak{A}$. Observe that

$$\varphi(e_\mu) = \lim_{\lambda\in\Lambda} \varphi(e_\lambda e_\mu e_\lambda) \le \liminf_{\Lambda} \varphi(e_\lambda^2) \le \liminf_{\Lambda} \varphi(e_\lambda).$$

Therefore

$$\limsup_{\mu\in\Lambda} \varphi(e_\mu) \le \sup_{\mu\in\Lambda} \varphi(e_\mu) \le \liminf_{\Lambda} \varphi(e_\lambda).$$

Hence $M := \lim_\Lambda \varphi(e_\lambda) \le \|\varphi\|$ exists. For $a \in \mathfrak{A}$ with $\|a\| \le 1$, using Cauchy–Schwarz,

$$|\varphi(a)|^2 = \lim_{\Lambda} |\varphi(e_\lambda a)|^2 \le \limsup_{\Lambda} |\varphi(a^*a)|\varphi(e_\lambda^2) \le \|\varphi\| M.$$

Now take the supremum over the unit ball to get

$$\|\varphi\|^2 \le \|\varphi\| M \quad \text{or} \quad \|\varphi\| \le M.$$

Thus $\|\varphi\| = M$. ∎

9.6.5. LEMMA. *If $\mathfrak{A}$ is unital and φ is a functional such that $\|\varphi\| = \varphi(1) = 1$, then φ is a state. If $\mathfrak{A}$ is not unital, and $\{e_\lambda\}_\Lambda$ is an approximate unit such that $\|\varphi\| = \lim_\Lambda \varphi(e_\lambda) = 1$, then φ is a state.*

PROOF. First suppose that $\mathfrak{A}$ is unital. If $a = a^*$, write $\varphi(a) = x + iy \in \mathbb{C}$. Then for $t \in \mathbb{R}$,

$$|\varphi(a + it)|^2 = |x + i(y + t)|^2 = x^2 + y^2 + 2yt + t^2$$

is bounded by

$$\|a + it\|^2 = \|(a - it)(a + it)\| = \|a^2 + t^2\| \le \|a\|^2 + t^2.$$

Therefore $2yt \le \|a\|^2 - x^2 - y^2$ for all real t, which forces $y = 0$. Hence $\varphi(a)$ is real. Now if $0 \le a \le 1$, then $-1 \le 2a - 1 \le 1$. So $\|2a - 1\| \le 1$, whence $-1 \le 2\varphi(a) - 1 \le 1$. This forces $0 \le \varphi(a) \le 1$. Thus φ is positive.

In the non-unital case, let $\widetilde{\varphi}$ be any Hahn–Banach extension of φ to $\widetilde{\mathfrak{A}}$. Let $\zeta = \widetilde{\varphi}(1)$. Then $|\zeta| \le \|1\| = 1$. Also $\|1 - 2e_\lambda\| \le 1$, so

$$1 \ge \lim_\Lambda |\varphi(1 - 2e_\lambda)| = \lim_\Lambda |\zeta - 2\varphi(e_\lambda)| = |\zeta - 2|.$$

Thus $\zeta \in \overline{\mathbb{D}} \cap \overline{\mathbb{D} + 2} = \{1\}$. Therefore $\|\widetilde{\varphi}\| = \widetilde{\varphi}(1) = 1$. Hence $\widetilde{\varphi}$ is a state on $\widetilde{\mathfrak{A}}$; so φ is positive, and thus is a state on $\mathfrak{A}$. ∎

9.6.6. DEFINITION. If $\mathfrak{A}$ is a C*-algebra, let $\mathcal{S}(\mathfrak{A})$ denote the set of all states on $\mathfrak{A}$; and let $\mathcal{Q}(\mathfrak{A}) = \{\varphi : \mathfrak{A} \to \mathbb{C} : \varphi \ge 0,\ \|\varphi\| \le 1\}$ be the set of *quasistates*.

9.6.7. REMARK. Let's write $\mathfrak{A}^\dagger$ to denote the dual space of $\mathfrak{A}$, to avoid confusion with the adjoint. If $\mathfrak{A}$ is unital, then by Lemma 9.6.5,

$$\mathcal{S}(\mathfrak{A}) = \{\varphi \in \mathfrak{A}^\dagger : \|\varphi\| \le 1 = \varphi(1)\} = \overline{b_1(\mathfrak{A}^\dagger)} \cap \{\varphi : \varphi(1) = 1\}.$$

By the Banach–Alaoglu theorem, this is the intersection of a weak-$*$ compact convex set with a weak-$*$ closed convex set. Hence $\mathcal{S}(\mathfrak{A})$ is compact and convex.

If $\mathfrak{A}$ is not unital, the state space need not be compact. However

$$\mathcal{Q}(\mathfrak{A}) = \overline{b_1(\mathfrak{A}^\dagger)} \cap \bigcap_{a \ge 0} \{\varphi : \varphi(a) \ge 0\}.$$

This is a weak-$*$ compact, convex set because the set $\{\varphi : \varphi(a) \ge 0\}$ is weak-$*$ closed and convex.

By the Krein–Milman theorem, the weak-$*$ closed convex hull of the extreme points is the whole weak-$*$ compact, convex set. So extreme points will be important.

9.6.8. DEFINITION. If $\mathfrak{A}$ is a C*-algebra, a *pure state* on $\mathfrak{A}$ is a state φ such that $0 \le \psi \le \varphi$ for a (positive) linear functional ψ implies that $\psi \in \mathbb{R}^+\varphi$.

9.6.9. PROPOSITION. *A state φ on a C*-algebra $\mathfrak{A}$ is pure if and only if it is an extreme point of $\mathcal{S}(\mathfrak{A})$.*

PROOF. If φ is not extreme in $\mathcal{S}(\mathfrak{A})$, then $\varphi = \frac{1}{2}(\psi_1 + \psi_2)$ where $\psi_1 \ne \psi_2$ are distinct states. Then $0 \le \frac{1}{2}\psi_1 \le \varphi$ and $\frac{1}{2}\psi_1$ is not a scalar multiple of φ.

Conversely, if φ is not pure, suppose that $0 \le \psi \le \varphi$ for a positive linear functional ψ which is not a multiple of φ. Let $t = \psi(1)$. Then $\psi_1 = t^{-1}\psi$ is a state, as is $\psi_2 = (1-t)^{-1}(\varphi - \psi)$. Neither equals φ, but $\varphi = t\psi_1 + (1-t)\psi_2$. So φ is not extreme. ■

9.6.10. PROPOSITION. *Let $\mathfrak{A}$ be a C*-algebra. Then*

$$\operatorname{ext}(\mathcal{Q}(\mathfrak{A})) = \{0\} \cup \operatorname{ext}(\mathcal{S}(\mathfrak{A})).$$

Moreover $\overline{\operatorname{conv}\operatorname{ext}(\mathcal{S}(\mathfrak{A}))}^{\text{weak-}*}$ *contains $\mathcal{S}(\mathfrak{A})$, with equality in the unital case.*

PROOF. If $\varphi \in \mathcal{Q}(\mathfrak{A})$ has $0 < \|\varphi\| < 1$, then it evidently is not extreme. So $\operatorname{ext}(\mathcal{Q}(\mathfrak{A}) \subset \{0\} \cup \operatorname{ext}(\mathcal{S}(\mathfrak{A}))$. Also a convex combination of elements in $\mathcal{Q}(\mathfrak{A})$ can have norm 1 or 0 only if these elements are norm 1 or 0, respectively. Thus 0 and $\operatorname{ext}(\mathcal{S}(\mathfrak{A}))$ are extreme in $\mathcal{Q}(\mathfrak{A})$.

Since $\mathcal{Q}(\mathfrak{A})$ is weak-$*$ compact and convex, the Krein–Milman theorem shows that $\mathcal{Q}(\mathfrak{A}) = \overline{\operatorname{conv}\operatorname{ext}(\mathcal{Q}(\mathfrak{A}))}^{\text{weak-}*}$. But 0 is of no help to get a state in the closed convex hull, so $\overline{\operatorname{conv}\operatorname{ext}(\mathcal{S}(\mathfrak{A}))}^{\text{weak-}*} \supset \mathcal{S}(\mathfrak{A})$. In the unital case, $\mathcal{S}(\mathfrak{A})$ is compact and convex, so we get equality by the Krein–Milman theorem. ■

9.6.11. EXAMPLE. Let $\mathfrak{A} = C(X)$ for a compact Hausdorff space X. The Reisz–Markov theorem shows that the positive linear functionals on $C(X)$ have the form $\varphi(f) = \int f\,d\mu$ where μ is a finite positive regular Borel measure on X. The states are given by the set of probability measures. A measure is extreme if and only if it is a point mass δ_x for $x \in X$. See Example 3.5.6 for the case of $X = [0, 1]$.

Now suppose that X is a non-compact, locally compact Hausdorff space and $\mathfrak{A} = C_0(X)$. The Reisz–Markov theorem shows that the positive linear functionals on $C_0(X)$ have the form $\varphi(f) = \int f\,d\mu$ where μ is a finite positive Radon measure on X, i.e. a finite regular Borel measure. In this case however, the weak-$*$ closed convex hull of the extreme points δ_x of $\mathcal{S}(C_0(X))$ is all of $\mathcal{Q}(C_0(X))$ because 0 is a weak-$*$ limit of δ_x as x goes to "infinity".

The next goal is to show that there are enough pure states to determine the norm on $\mathfrak{A}_{sa}$.

9.6.12. LEMMA. *If $a = a^*$ in $\mathfrak{A}$, then there is a pure state φ on $\mathfrak{A}$ so that $|\varphi(a)| = \|a\|$.*

PROOF. We may assume that $\mathfrak{A}$ is unital. In $C^*(a) \simeq C(\sigma(a))$, we have $\|a\| = \text{spr}(a) = \|\hat{a}\|_\infty$. Choose a multiplicative linear functional $\varphi \in \mathcal{M}_{C^*(a)}$ such that $|\varphi(a)| = \|a\|$. Extend this by Hahn–Banach to a norm 1 functional on $\mathfrak{A}$. Then $1 = \|\varphi\| = \varphi(1)$. Thus φ is a state by Lemma 9.6.5.

Let $F = \{\psi \in \mathcal{S}(\mathfrak{A}) : \psi(a) = \varphi(a)\}$. This is a closed convex subset of $\mathcal{S}(\mathfrak{A})$. Moreover F is a face because if $\frac{1}{2}(\psi_1 + \psi_2) \in F$ for two states ψ_i, then $\varphi(a) = \frac{1}{2}\big(\psi_1(a) + \psi_2(a)\big)$. Because $\varphi(a) = \pm\|a\|$ and $|\psi_i(a)| \le \|a\|$, we must have $\varphi(a) = \psi_1(a) = \psi_2(a)$. Since F is a compact convex set, it has an extreme point ψ. Since F is a face, ψ is extreme in $\mathcal{S}(\mathfrak{A})$. By Lemma 9.6.9, ψ is pure. ∎

9.6.13. DEFINITION. A linear functional φ on a C*-algebra $\mathfrak{A}$ is a *self-adjoint functional* if $a = a^*$ implies that $\varphi(a) \in \mathbb{R}$. Let $\mathfrak{A}_{sa}^\dagger$ denote the set of all self-adjoint linear functionals on $\mathfrak{A}_{sa}$, which is the real dual of this real Banach space.

In measure theory, the Jordan decomposition splits a real-valued measure into a difference of two positive measures which are mutually singular. We can think of this as a result about self-adjoint functionals on an abelian C*-algebra. The analogue of this for arbitrary C*-algebras remains true.

9.6.14. THEOREM. *The unit ball $\overline{b_1(\mathfrak{A}_{sa}^\dagger)} = \text{conv}(\mathcal{S}(\mathfrak{A}) \cup -\mathcal{S}(\mathfrak{A}))$. Let φ be a self-adjoint linear functional on a C*-algebra $\mathfrak{A}$. Then there are positive linear functionals ψ_1 and ψ_2 so that $\varphi = \psi_1 - \psi_2$ and $\|\varphi\| = \|\psi_1\| + \|\psi_2\|$.*

PROOF. Since 0 belongs to $\text{conv}(\mathcal{S}(\mathfrak{A}) \cup -\mathcal{S}(\mathfrak{A}))$, we have

$$\text{conv}(\mathcal{S}(\mathfrak{A}) \cup -\mathcal{S}(\mathfrak{A})) = \text{conv}(\mathcal{Q}(\mathfrak{A}) \cup -\mathcal{Q}(\mathfrak{A})).$$

But $\mathcal{Q}(\mathfrak{A})$ is weak-$*$ compact, and thus so is this convex hull. Clearly this is contained in $\overline{b_1(\mathfrak{A}_{sa}^\dagger)}$, which is also weak-$*$ compact by the Banach–Alaoglu theorem. If these two sets are not equal, choose $\varphi \in b_1(\mathfrak{A}_{sa}^\dagger) \setminus \overline{\text{conv}(\mathcal{S}(\mathfrak{A}) \cup -\mathcal{S}(\mathfrak{A}))}$. By the Hahn–Banach theorem, there is an $a = a^* \in \mathfrak{A}_{sa}$ so that

$$\varphi(a) > \sup\{\psi(a) : \psi \in \mathcal{S}(\mathfrak{A}) \cup -\mathcal{S}(\mathfrak{A})\}.$$

However $\varphi(a) \le \|a\|$, and Lemma 9.6.12 shows that there is a state ψ so that $\pm\psi(a) = \|a\|$. Hence this separation is not possible. Therefore we have equality.

Let $\varphi \in \mathfrak{A}^\dagger_{sa}$ with $\|\varphi\| = 1$. Write $\varphi = t\psi_1 - (1-t)\psi_2$ for $\psi_i \in \mathcal{S}(\mathfrak{A})$. Then

$$\|\varphi\| = 1 = t + (1-t) = \|t\psi_1\| + \|(1-t)\psi_2\|.$$

For general φ, just scale. ■

In fact, this decomposition is unique. See Theorem 10.2.10 and Corollary 10.5.6.

9.6.15. COROLLARY. *Every functional on $\mathfrak{A}$ is a linear combination of four states.*

PROOF. Given $\varphi \in \mathfrak{A}^\dagger$, define $\varphi^*(a) := \overline{\varphi(a^*)}$. It is easy to check that this is also linear. Moreover

$$\operatorname{Re}\varphi := \tfrac{1}{2}(\varphi + \varphi^*) \quad \text{and} \quad \operatorname{Im}\varphi := \tfrac{1}{2i}(\varphi - \varphi^*)$$

are self-adjoint functionals such that $\varphi = \operatorname{Re}\varphi + i\operatorname{Im}\varphi$. By the theorem above, these can each be written as the difference of two positive functionals. ■

9.7. Constructing Representations

We now come to the fundamental construction of representations of an abstract C*-algebra which will lead to a spatial representation.

9.7.1. GELFAND–NAIMARK–SEGAL CONSTRUCTION. *Let φ be a state of a C*-algebra $\mathfrak{A}$. Then there is a representation π_φ on a Hilbert space H_φ and a unit vector ξ_φ which is a cyclic vector for π_φ such that*

$$\varphi(a) = \langle \pi_\varphi(a)\xi_\varphi, \xi_\varphi\rangle \quad \textit{for} \quad a \in \mathfrak{A}.$$

PROOF. Define a sesquilinear form $[a, b] = \varphi(b^*a)$ as before. Let

$$L = \{a \in \mathfrak{A} : [a, a] = 0\}.$$

We claim that L is a closed left ideal. It is closed since φ is continuous. If $a \in L$ and $b \in \mathfrak{A}$, then $|\varphi(b^*a)|^2 \le \varphi(a^*a)\varphi(b^*b) = 0$. Thus

$$L = \{a \in \mathfrak{A} : \varphi(b^*a) = 0 \text{ for all } b \in \mathfrak{A}\}.$$

In particular, L is a subspace invariant under left multiplication by $\mathfrak{A}$, and hence is a closed left ideal.

Therefore $\mathfrak{A}/L$ has a positive definite inner product $\langle \xi_x, \xi_y\rangle = [x, y]$ for x, y in $\mathfrak{A}$ and $\xi_x = x + L$. It is well defined because if $m, n \in L$, then

$$[x + m, y + n] = [x, b] + [x, n] + [m, y + n] = [x, y] + \overline{[n, x]} + 0 = [x, y].$$

The completion in the corresponding norm is a Hilbert space H_φ.

Define $\pi_0(a)\xi_x = \xi_{ax}$ on $\mathfrak{A}/L$. This is well defined because L is a left ideal.

$$\begin{aligned}\langle \pi_0(a^*)\xi_x, \xi_y\rangle &= \varphi(y^*a^*x) = \varphi((ay)^*x) \\ &= \langle \xi_x, \xi_{ay}\rangle = \langle \xi_x, \pi_0(a)\xi_y\rangle = \langle \pi_0(a)^*\xi_x, \xi_y\rangle.\end{aligned}$$

Hence $\pi_0(a^*) = \pi_0(a)^*$. Also

$$\begin{aligned}\|\pi(a)\|^2 &= \sup_{\|\xi_x\|\le 1} \|\pi(a)\xi_x\|^2 = \sup_{\|\xi_x\|\le 1} \varphi(x^*a^*ax) \\ &\le \sup_{\|\xi_x\|\le 1} \|a^*a\|\varphi(x^*x) = \|a\|^2.\end{aligned}$$

Thus $\|\pi_0\| \le 1$. It follows that there is a continuous extension π_φ of π_0 to a $*$-representation on H_φ.

If $\mathfrak{A}$ is unital, we set $\xi_\varphi = \xi_1$; and $\|\xi_\varphi\|^2 = \varphi(1) = 1$. If it isn't, let $\{e_\lambda\}_\Lambda$ be an approximate identity; and let $\xi_\lambda = \xi_{e_\lambda}$. We will show that $\{\xi_\lambda\}_\Lambda$ is Cauchy in H_φ. By Lemma 9.6.4, $1 = \|\varphi\| = \lim_\Lambda \varphi(e_\lambda)$. Given $\varepsilon > 0$, choose λ_0 so that $\varphi(e_{\lambda_0}) > 1 - \varepsilon$. Then for $\mu \ge \lambda_0$, we can select $\lambda \ge \mu$ so that $\|e_\lambda e_\mu - e_\mu\| < \varepsilon$. Then

$$\operatorname{Re}\varphi(e_\lambda e_\mu) = \varphi(e_\mu) + \operatorname{Re}\varphi(e_\mu - e_\lambda e_\mu) > 1 - \varepsilon - \varepsilon = 1 - 2\varepsilon.$$

Therefore

$$\|\xi_\lambda - \xi_\mu\|^2 = \varphi(e_\lambda^2) - 2\operatorname{Re}\varphi(e_\lambda e_\mu) + \varphi(e_\mu^2) \le 1 - 2(1-2\varepsilon) + 1 = \left(2\sqrt{\varepsilon}\right)^2.$$

Now if $\mu, \nu \ge \lambda_0$, we can choose λ as before so that $\|e_\lambda e_\nu - e_\nu\| < \varepsilon$ as well. Then

$$\|\xi_\mu - \xi_\nu\| \le \|\xi_\mu - \xi_\lambda\| + \|\xi_\lambda - \xi_\nu\| \le 4\sqrt{\varepsilon}.$$

It follows that $\{\xi_\lambda\}_\Lambda$ is Cauchy. Let $\xi_\varphi = \lim_\Lambda \xi_\lambda$. Then $\|\xi_\varphi\|^2 = \lim_\Lambda \varphi(e_\lambda^2) \le 1$. But if $\varphi(e_\mu) > 1 - \varepsilon$, then

$$\|\xi_\varphi\|^2 = \lim_\Lambda \varphi(e_\lambda^2) \ge \lim_\Lambda \varphi(e_\lambda e_\mu e_\lambda) = \varphi(e_\mu) > 1 - \varepsilon.$$

So ξ_φ is a unit vector. In both the unital and non-unital case, we obtain

$$\langle \pi(a)\xi_\varphi, \xi_\varphi\rangle = \lim_\Lambda \langle \pi(a)\dot{e}_\lambda, \dot{e}_\lambda\rangle = \lim_\Lambda \varphi(e_\lambda a e_\lambda) = \varphi(a).$$

Lastly, we show that ξ_φ is cyclic. For $x \in \mathfrak{A}$,

$$\pi(a)\xi_\varphi = \lim_\Lambda \pi(a)\xi_\lambda = \lim_\Lambda \xi_{ae_\lambda} = \xi_a.$$

Thus $\pi(\mathfrak{A})\xi_\varphi$ is dense in H. So ξ_φ is cyclic. ■

9.7.2. COROLLARY. *If $\mathfrak{A}$ is not unital, then every non-degenerate representation $\pi : \mathfrak{A} \to \mathcal{B}(H)$ has a unique extension to a representation $\widetilde{\pi} : \widetilde{\mathfrak{A}} \to \mathcal{B}(H)$; and every state φ on $\mathfrak{A}$ has a unique extension to a state on $\widetilde{\mathfrak{A}}$.*

PROOF. Since $\pi(\mathfrak{A})H$ is dense, this forces $\widetilde{\pi}(1) = I$. Conversely, it is easy to check that this defines a $*$-representation.

Uniqueness of an extension of φ is clear because we require $\widetilde{\varphi}(1) = 1$. For existence, extend the representation π_φ to $\widetilde{\pi}_\varphi$ on $\widetilde{\mathfrak{A}}$. Then we can define

$$\widetilde{\varphi}(a+\lambda) = \langle \widetilde{\pi}(a+\lambda)\xi_\varphi, \xi_\varphi\rangle = \langle (\pi_\varphi(a) + \lambda I)\xi_\varphi, \xi_\varphi\rangle = \varphi(a) + \lambda.$$

This is a state on $\widetilde{\mathfrak{A}}$ extending φ. ■

9.7.3. DEFINITION. Two representations $\pi_i : \mathfrak{A} \to \mathcal{B}(H_i)$ are *unitarily equivalent* ($\pi_1 \simeq \pi_2$) if there is a unitary $U \in \mathcal{B}(H_1, H_2)$ so that $\pi_2(a) = U\pi_1(a)U^*$ for all $a \in \mathfrak{A}$.

9.7.4. PROPOSITION. *Suppose that $\pi : \mathfrak{A} \to \mathcal{B}(H)$ is a representation with a unit cyclic vector ξ. Set $\varphi(a) = \langle \pi(a)\xi, \xi\rangle$. Then π is unitarily equivalent to π_φ via a unitary $U \in \mathcal{B}(H_\varphi, H)$ such that $U\xi_\varphi = \xi$.*

PROOF. We may assume that $\mathfrak{A}$ is unital by setting $\pi(1) = I$ if necessary. Define $U\pi_\varphi(a)\xi_\varphi = \pi(a)\xi$ for $a \in \mathfrak{A}$. Then for $a, b \in \mathfrak{A}$,

$$\begin{aligned}\langle U\pi_\varphi(a)\xi_\varphi, U\pi_\varphi(b)\xi_\varphi\rangle &= \langle \pi(a)\xi, \pi(b)\xi\rangle = \langle \pi(b^*a)\xi, \xi\rangle = \varphi(b^*a)\\ &= \langle \pi_\varphi(b^*a)\xi_\varphi, \xi_\varphi\rangle = \langle \pi_\varphi(a)\xi_\varphi, \pi_\varphi(b)\xi_\varphi\rangle.\end{aligned}$$

This shows that U preserves inner products, and hence is isometric. Thus it extends by continuity to an isometry from $\overline{\mathfrak{A}\xi_\varphi} = H_\varphi$ onto $\overline{\mathfrak{A}\xi} = H$. That is, U is unitary.

Now

$$U\pi_\varphi(a)\,\pi_\varphi(b)\xi_\varphi = U\pi_\varphi(ab)\xi_\varphi = \pi(ab)\xi = \pi(a)U\pi_\varphi(b)\xi_\varphi.$$

Thus $U\pi_\varphi(a)$ and $\pi(a)U$ agree on a dense set, and hence they are equal. Therefore $\pi(a) = U\pi_\varphi(a)U^*$; and $U\xi_\varphi = U\pi_\varphi(1)\xi_\varphi = \pi(1)\xi = \xi$. ■

Next we wish to show that pure states correspond to irreducible representations. This starts with a lemma.

9.7.5. LEMMA. *Let $\pi : \mathfrak{A} \to \mathcal{B}(H)$ be a representation of $\mathfrak{A}$ with a unit cyclic vector ξ; and let $\varphi(a) = \langle \pi(a)\xi, \xi\rangle$ be the corresponding state. For any $\psi \in \mathfrak{A}^\dagger$, $0 \le \psi \le \varphi$ if and only if there is $0 \le T \le I$ in $\pi(\mathfrak{A})'$ so that $\psi(a) = \langle \pi(a)\xi, T\xi\rangle$.*

PROOF. Given $T \in \pi(\mathfrak{A})'$ with $0 \le T \le I$, the functional

$$\psi(a) := \langle \pi(a)\xi, T\xi\rangle = \langle T^{1/2}\pi(a)\xi, T^{1/2}\xi\rangle = \langle \pi(a)T^{1/2}\xi, T^{1/2}\xi\rangle$$

is positive. Also $\varphi - \psi$ has the same form using $I - T$; so it is also positive. Thus $0 \le \psi \le \varphi$.

Conversely, suppose that $0 \le \psi \le \varphi$. Let $L = \{a \in \mathfrak{A} : \varphi(a^*a) = 0\}$. Define a positive semidefinite sesquilinear form on $\mathfrak{A}/L$ by $[\xi_a, \xi_b] = \psi(b^*a)$. Check that this is well defined because $L \subset L_1 = \{a \in \mathfrak{A} : \psi(a^*a) = 0\}$. This form is bounded because $[\xi_a, \xi_a] = \psi(a^*a) \le \varphi(a^*a) = \|\xi_a\|^2$. Hence by Cauchy–Schwarz,

$$|[\xi_a, \xi_b]| \le [\xi_a, \xi_a]^{1/2}[\xi_b, \xi_b]^{1/2} \le \|\xi_a\| \, \|\xi_b\|.$$

Thus it extends to a positive semidefinite sesquilinear form on H by continuity. By Proposition 4.2.2, there is a bounded operator T with $\|T\| \le 1$ so that $[\eta, \zeta] = \langle \eta, T\zeta \rangle$. Since this is positive semidefinite, $\langle \eta, T\eta \rangle = [\eta, \eta] \ge 0$. Hence $T \ge 0$; so $0 \le T \le I$. For $c \in \mathfrak{A}$,

$$\begin{aligned}\langle (\pi(c)T - T\pi(c))\xi_a, \xi_b \rangle &= \langle \xi_a, T\pi(c^*)\xi_b \rangle - \langle \pi(c)\xi_a, T\xi_b \rangle \\ &= [\xi_a, \xi_{c^*b}] - [\xi_{ca}, \xi_b] = \psi(b^*ca) - \psi(b^*ca) = 0.\end{aligned}$$

Since the vectors ξ_a for $a \in \mathfrak{A}$ are dense in H, this shows that $\pi(c)T = T\pi(c)$. Therefore $T \in \pi(\mathfrak{A})'$. ∎

9.7.6. THEOREM. *Let $\pi : \mathfrak{A} \to \mathcal{B}(H)$ be a representation with cyclic vector ξ, and set $\varphi(a) = \langle \pi(a)\xi, \xi \rangle$. Then π is irreducible if and only if φ is pure.*

PROOF. If φ is not pure, there is a positive functional ψ with $0 \le \psi \le \varphi$ which is not a scalar multiple of φ. By Lemma 9.7.5, there is a positive operator $T \in \pi(\mathfrak{A})'$ and $0 \le T \le I$ so that $\psi(a) = \langle \pi(a)\xi, T\xi \rangle$. Now T is not a scalar multiple of I, and thus $\pi(\mathfrak{A})' \ne \mathbb{C}I$. By Lemma 9.5.1, π is not irreducible.

Conversely if π is not irreducible, then $\pi(\mathfrak{A})' \ne \mathbb{C}I$. Therefore there is a non-scalar $T \in \pi(\mathfrak{A})'$ with $0 \le T \le I$. Set $\psi = \langle \pi(a)\xi, T\xi \rangle$. Then $0 \le \psi \le \varphi$, and ψ is not a multiple of φ. Hence φ is not pure. ∎

We have reached the main goal of this section.

9.7.7. GELFAND–NAIMARK THEOREM. *Let $\mathfrak{A}$ be a C*-algebra. Then $\mathfrak{A}$ is isometrically $*$-isomorphic to a concrete C*-algebra. Moreover this representation may be taken to be a direct sum of irreducible representations. If $\mathfrak{A}$ is separable, then this representation can be taken on a separable Hilbert space.*

PROOF. Let $\pi = \bigoplus_{\varphi \in \mathrm{ext}(\mathcal{S}(\mathfrak{A}))} \pi_\varphi$ be a representation on $H = \bigoplus_{\varphi \in \mathrm{ext}(\mathcal{S}(\mathfrak{A}))} H_\varphi$. This is a direct sum of irreducible representations by Lemma 9.6.9 and Theorem 9.7.6. By Lemma 9.6.12, $\|\pi(a)\| = \|a\|$ for all $a = a^* \in \mathfrak{A}$. Thus

$$\|\pi(b)\|^2 = \|\pi(b^*b)\| = \|b^*b\| = \|b\|^2 \quad \text{for all} \quad b \in \mathfrak{A}.$$

Hence π is isometric.

Suppose that $\mathfrak{A}$ is separable, and $\{a_i : i \ge 1\}$ is a dense subset of $b_1(\mathfrak{A}_{sa})$. For each a_i, choose an irreducible representation π_i so that $\|\pi_i(a_i)\| = \|a_i\|$. Each H_i is separable as it is a completion of $\mathfrak{A}/L_\varphi$, which is separable; and thus $H = \bigoplus_{i \ge 1} H_i$ is separable. ∎

9.8. Representations and Ideals

In this section, we analyze the relationship between a representation on $\mathfrak{A}$ and an ideal.

9.8.1. THEOREM. *Let $J \lhd \mathfrak{A}$ be an ideal, and let $\pi : J \to \mathcal{B}(H)$ be a non-degenerate representation. There is a unique representation $\widetilde{\pi} = \operatorname{Ind}\pi : \mathfrak{A} \to \mathcal{B}(H)$ such that $\widetilde{\pi}|_J = \pi$. Moreover, $\overline{\pi(J)}^{\text{SOT}} = \overline{\widetilde{\pi}(\mathfrak{A})}^{\text{SOT}}$ and $\pi(J)' = \widetilde{\pi}(\mathfrak{A})'$. Also $\widetilde{\pi}$ is irreducible if and only if π is irreducible.*

PROOF. Let $\{e_\lambda\}_\Lambda$ be an approximate identity for J. Then

$$\lim_\Lambda \pi(e_\lambda)\pi(j)\xi = \lim_\Lambda \pi(e_\lambda j)\xi = \pi(j)\xi$$

Since $\{\pi(j)\xi : j \in J, \xi \in H\}$ is dense in H and $\|\pi(e_\lambda)\| \le 1$, this shows that SOT–$\lim_\Lambda \pi(e_\lambda) = I$.

Define $\widetilde{\pi}(a)\pi(j)\xi = \pi(aj)\xi$ for $\pi(j)\xi \in H$. It is clear that this is necessary if $\widetilde{\pi}$ exists, and since $\overline{\pi(J)H} = H$, this establishes uniqueness. We will show that this is well defined. If $\pi(j_1)\xi_1 = \pi(j_2)\xi_2$, then

$$\begin{aligned}\widetilde{\pi}(a)\pi(j_1)\xi_1 = \pi(aj_1)\xi_1 &= \lim_\Lambda \pi(ae_\lambda j_1)\xi_1 = \lim_\Lambda \pi(ae_\lambda)\pi(j_1)\xi_1\\ &= \lim_\Lambda \pi(ae_\lambda)\pi(j_2)\xi_2 = \lim_\Lambda \pi(ae_\lambda j_2)\xi_2 = \pi(aj_2)\xi_2 = \widetilde{\pi}(a)\pi(j_2)\xi_2.\end{aligned}$$

This also shows that

$$\|\widetilde{\pi}(a)\pi(j)\xi\| = \lim_\Lambda \|\pi(ae_\lambda)\pi(j)\xi\| \le \|a\|\,\|\pi(j)\xi\|.$$

Therefore $\widetilde{\pi}$ is continuous, and so extends to be defined on all of H by continuity. Thus $\widetilde{\pi}(a) =$ SOT–$\lim_\Lambda \pi(ae_\lambda)$; whence $\overline{\pi(J)}^{\text{SOT}} = \overline{\widetilde{\pi}(\mathfrak{A})}^{\text{SOT}}$. Hence $\pi(J)' = \widetilde{\pi}(\mathfrak{A})'$.

Next we show that $\widetilde{\pi}$ is a $*$-representation.

$$\begin{aligned}\langle \widetilde{\pi}(a^*)\pi(j)\xi_1, \pi(j_2)\xi_2\rangle &= \langle \pi(a^*j_1)\xi_1, \pi(j_2)\xi_2\rangle = \langle \pi(j_2^*)\pi(a^*j_1)\xi_1, \xi_2\rangle\\ &= \langle \pi(j_2^*a^*j_1)\xi_1, \xi_2\rangle = \langle \pi(j_2^*a^*)\pi(j_1)\xi_1, \xi_2\rangle\\ &= \langle \pi(j_1)\xi_1, \pi(aj_2)\xi_2\rangle = \langle \pi(j_1)\xi_1, \widetilde{\pi}(a)\pi(j_2)\xi_2\rangle\\ &= \langle \widetilde{\pi}(a)^*\pi(j)\xi_1, \pi(j_2)\xi_2\rangle.\end{aligned}$$

Therefore $\widetilde{\pi}(a^*) = \widetilde{\pi}(a)^*$. And

$$\widetilde{\pi}(ab)\pi(j)\xi = \pi(abj)\xi = \widetilde{\pi}(a)\pi(bj)\xi = \widetilde{\pi}(a)\widetilde{\pi}(b)\pi(j)\xi.$$

Hence $\widetilde{\pi}$ is multiplicative, and linearity is obvious.

Since $\widetilde{\pi}(\mathfrak{A})' = \pi(J)'$, it is irreducible exactly when this commutant is $\mathbb{C}I$ by Lemma 9.5.1 and Theorem 9.5.3. So $\widetilde{\pi}$ is irreducible if and only if π is. ∎

9.8.2. THEOREM. *Let $J \lhd \mathfrak{A}$ be ideal, and let $\pi : \mathfrak{A} \to \mathcal{B}(H)$ be a non-degenerate representation. The projection P onto $\overline{\pi(J)H}$ belongs to $\pi(\mathfrak{A})' \cap \pi(\mathfrak{A})''$.*

Let $\sigma(a) = \pi(a)|_{PH}$ *and* $\tau(a) = \pi(a)|_{P^\perp H}$. *Then* $\sigma = \operatorname{Ind}(\sigma|_J)$ *and* $\tau|_J = 0$. *If* π *is irreducible, then either* $\pi|_J = 0$ *or* $\pi|_J$ *is irreducible and* $\pi = \operatorname{Ind}\pi|_J$.

PROOF. Note that $\pi(\mathfrak{A})\overline{\pi(J)H} = \overline{\pi(J)H}$; so PH is invariant for $\pi(\mathfrak{A})$. Thus $P \in \pi(\mathfrak{A})'$. If $\{e_\lambda\}_\Lambda$ is an approximate identity for J, then SOT–$\lim_\Lambda \pi(e_\lambda) = P$. Thus $P \in \overline{\pi(J)}^{\text{SOT}} \subset \mathfrak{A}''$. Therefore P belongs to $\pi(\mathfrak{A})' \cap \pi(\mathfrak{A})''$.

The restriction of π to PH is a representation σ; and $\sigma|_J$ is non-degenerate. Thus by Theorem 9.8.1, $\operatorname{Ind}(\sigma|_J) = \sigma$. By construction, $\tau|_J = 0$. If π is irreducible, then either σ is vacuous and $\pi = \tau$ annihilates J or $\pi = \sigma = \operatorname{Ind}(\sigma|_J)$. In the latter case, $\sigma|_J$ is irreducible by Theorem 9.8.1. ■

9.9. The Spectral Theorem

The spatial structure of normal operators is an important strengthening of the more straightforward continuous functional calculus that we established in Theorem 7.6.5. The main property that we use is that $C^*(n)$ is abelian, so we will deal with the more general situation of representations of abelian C*-algebras.

9.9.1. THEOREM. *Let* X *be an LCH space, and let* $\pi : C_0(X) \to \mathcal{B}(H)$ *be a representation with cyclic vector* ξ. *Then there is a regular Borel probability measure* μ *on* X *so that* $\pi \simeq \sigma_\mu$, *where* $\sigma_\mu : C_0(X) \to \mathcal{B}(L^2(\mu))$ *by* $\sigma_\mu(f) = M_f$.

PROOF. We may suppose that $\|\xi\| = 1$. Define a state by $\varphi(f) = \langle \pi(f)\xi, \xi\rangle$. By Proposition 9.7.4, $\pi \simeq \pi_\varphi$. By the Riesz–Markov theorem, there is a finite regular Borel measure μ on X so that $\varphi(f) = \displaystyle\int f\,d\mu$ for $f \in C_0(X)$. Also $1 = \|\varphi\| = \mu(X)$; so μ is a probability measure.

Define $U : C_0(X) \to H$ by $Uf = \pi(f)\xi$. Then

$$\|Uf\|_2^2 = \langle \pi(f)\xi, \pi(f)\xi\rangle = \langle \pi(|f|^2)\xi, \xi\rangle = \int |f|^2\,d\mu = \|f\|^2_{L^2(\mu)}.$$

Thus U is an isometry from a dense subset of $L^2(\mu)$ onto $\pi(C_0(X))\xi \subset H$. Extend it by continuity to an isometry U of $L^2(\mu)$ onto $\overline{\pi(C_0(X))\xi} = H$. Hence U is unitary. Moreover, for $f, g \in C_0(X)$,

$$\pi(f)Ug = \pi(f)\pi(g)\xi = \pi(fg)\xi = Ufg = UM_f g.$$

Hence $U^*\pi(f)U$ and M_f agree on a dense set and thus are equal by continuity. That is, $\pi(f) = UM_fU^*$. ■

We can also identify the double commutant.

9.9.2. THEOREM. *Let X be a LCH space, and let $\pi : C_0(X) \to \mathcal{B}(H)$ be a representation with cyclic vector ξ, and let μ be a regular Borel probability measure μ on X so that $\pi \simeq \sigma_\mu$. Then $\pi(C_0(X))''$ is a maximal abelian algebra unitarily equivalent to $L^\infty(\mu)$ acting on $L^2(\mu)$. There is a unique normal extension $\widetilde{\pi}$ of $L^\infty(\mu)$ onto $\pi(C_0(X))''$ extending π.*

PROOF. In Proposition 9.3.4, the representation $\widetilde{\sigma}_\mu : L^\infty(\mu) \to \mathcal{B}(L^2(\mu))$ given by $\widetilde{\sigma}_\mu(f) = M_f$ was shown to be an isomorphism onto a maximal abelian subalgebra of $\mathcal{B}(L^2(\mu))$, and this map is also weak-$*$ to WOT continuous. This algebra clearly contains $\sigma_\mu(C_0(X))$. Since it is a von Neumann algebra, it is spanned by its projections, which are of the form $\widetilde{\sigma}_\mu(\chi_E)$ for some Borel set E.

Given E, the regularity of μ provides a compact set K_n and open set U_n such that $K_n \subset E \subset U_n$ and $\mu(U \setminus K) < \frac{1}{n}$. By Urysohn's lemma for the LCH case, there is a function $f_n \in C_0(X)$ so that $\chi_{K_n} \le f_n \le \chi_{U_n}$. Then $f_n \to \chi_E$ a.e.(μ). Thus for any $h \in L^1(\mu)$, $|f_n h| \le |h|$, so by the Lebesgue Dominated Convergence theorem,

$$\lim_{n\to\infty} \int h f_n \, d\mu = \int h \chi_E \, d\mu.$$

Hence χ_E lies in $\overline{C_0(X)}^{\text{weak-}*}$. This shows that $C_0(X)$ is weak-$*$ dense in $L^\infty(\mu)$. Since $\widetilde{\sigma}_\mu$ is weak-$*$ to WOT-continuous, the extension of σ_μ from $C_0(X)$ to $L^\infty(\mu)$ is unique. It follows that $\sigma_\mu(C_0(X))$ is WOT-dense in $\widetilde{\sigma}(L^\infty(\mu))$. By the Double Commutant theorem, we have $\sigma_\mu(C_0(X))'' = \widetilde{\sigma}(L^\infty(\mu))$. It is normal because if f_n are increasing to $f \in L^\infty(\mu)$, then $f_n \xrightarrow{w^*} f$ and thus $\widetilde{\sigma}(f_n) \xrightarrow{\text{WOT}} \widetilde{\sigma}(f)$; and thus $\widetilde{\sigma}(f) = \sup \widetilde{\sigma}(f_n)$.

Let $U : L^2(\mu) \to H$ be the unitary provided by Theorem 9.9.1 such that $\pi = U\sigma_\mu U^*$. Then

$$\pi(C_0(X))'' = U\sigma_\mu(C_0(X))''U^* = U\widetilde{\sigma}(L^\infty(\mu))U^*.$$

It follows that $\widetilde{\pi}(f) = U\widetilde{\sigma}_\mu(f)U^*$ is a normal $*$-isomorphism of $L^\infty(\mu)$ onto $\pi(C_0(X))''$. Since $\widetilde{\sigma}(L^\infty(\mu))$ is maximal abelian by Proposition 9.3.4, so is $\pi(C_0(X))''$. ■

The following easy result is valid for all C*-algebras.

9.9.3. LEMMA. *Let $\mathfrak{A}$ be a C*-algebra and let $\pi : \mathfrak{A} \to \mathcal{B}(H)$ be a non-degenerate representation. Then π is a direct sum of cyclic representations.*

PROOF. If $\xi \ne 0$, $H_\xi = \overline{\pi(\mathfrak{A})\xi}$ is invariant for $\pi(\mathfrak{A})$, and hence it reduces $\pi(\mathfrak{A})$. If $\eta \perp H_\xi$, then

$$\langle \pi(a)\eta, \pi(b)\xi \rangle = \langle \eta, \pi(a^*b)\xi \rangle = 0.$$

Thus $H_\eta \perp H_\xi$. By Zorn's lemma, there is a maximal family of vectors $\{\xi_\lambda\}_\Lambda$ such that $H_{\xi_\lambda} \perp H_{\xi_\mu}$ for $\lambda \ne \mu \in \Lambda$. Let $M = \sum_\Lambda^\oplus H_{\xi_\lambda}$. If $M \ne H$, select

$0 \neq \eta \in M^\perp$. Then $H_\eta \perp H_{\xi_\lambda}$ for all $\lambda \in \Lambda$, contradicting the maximality of the set. Hence $H = \sum_\Lambda^\oplus H_{\xi_\lambda}$. Let π_λ be the restriction of π to the reducing subspace H_{ξ_λ}. Then ξ_λ is a cyclic vector for π_λ by construction, and $\pi = \sum_\Lambda^\oplus \pi_\lambda$. ■

9.9.4. COROLLARY. *Let X be a LCH space, and let $\pi : C_0(X) \to \mathcal{B}(H)$ be a representation. Then there is a family of regular Borel probability measures $\{\mu_\lambda\}_\Lambda$ on X so that $\pi \simeq \sum_\Lambda^\oplus \sigma_{\mu_\lambda}$ on $\sum_\Lambda^\oplus L^2(\mu_\lambda)$.*

9.9.5. COROLLARY. *Let $N \in \mathcal{B}(H)$ be a normal operator. Then N is unitarily equivalent to a multiplication operator.*

PROOF. Since $C^*(N) \simeq C(\sigma(N))$, we can apply the previous result to the representation arising from the continuous functional calculus, $\pi(f) = f(N)$ with $\pi(z) = N$. The unitary equivalence between π and $\sum_\Lambda^\oplus L^2(\mu_\lambda)$ identifies N with a direct sum of multiplication operators M_z on $L^2(\mu_\lambda)$, where each μ_λ is a regular Borel probability measure on $\sigma(N)$. ■

When H is separable, we can obtain some additional information.

9.9.6. THEOREM. *Let X be an LCH space, and let $\pi : C_0(X) \to \mathcal{B}(H)$ be a representation of a separable Hilbert space H. There is a unit vector ξ such that the restriction to $M = \overline{\pi(C_0(X))\xi}$ is faithful on $\pi(C_0(X))''$. Thus there is a regular Borel probability measure μ on X and a normal isomorphism $\tilde{\pi}$ of $L^\infty(\mu)$ onto $\pi(C_0(X))''$ which extends π.*

PROOF. By Lemma 9.9.3, π is a direct sum of cyclic representations with cyclic vector ξ_i, say $M_i = \overline{\pi(C_0(X))\xi_i}$ and $H = \sum^\oplus M_i$. Moreover, there are regular Borel probability measures μ_i so that these subrepresentations $\pi_i(f) = \pi(f)|_{M_i}$, are unitarily equivalent to σ_{μ_i}, say via $\operatorname{Ad} U_i$. Since H is separable, there are at most countably many such measures, which we enumerate as $\{\mu_i : 1 \le i \le n \text{ or } i < \infty\}$ for some ordinal $n \in \mathbb{N} \cup \{\omega\}$. We will deal with the countably infinite case, and the reader can easily modify the argument if n is finite. So $\pi \simeq \sigma = \sum_{i\ge1}^\oplus \sigma_{\mu_i}$.

Define $\mu = \sum\limits_{i\ge1} 2^{-i}\mu_i$. This is a regular Borel probability measure such that $\mu_i \ll \mu$ and $\mu(A) > 0$ implies that $\mu_i(A) > 0$ for some i. For each i, there is a Borel set E_i so that μ_i is equivalent to $\chi_{E_i}\mu$. Define a unit vector $\xi = \sum_{i\ge1} 2^{-i/2}\xi_i$. Since the cyclic subspaces for each ξ_i are pairwise orthogonal, for any $f \in C_0(X)$,

$$\langle \pi(f)\xi, \xi\rangle = \sum_{i\ge1} 2^{-i}\langle \pi(f)\xi_i, \xi_i\rangle = \sum_{i\ge1} 2^{-i} \int f\, d\mu_i = \int f\, d\mu.$$

Let $M = \overline{\pi(C_0(X))\xi}$, and let $\tau(A) = A|_M$ be defined on $\pi(C_0(X))''$. Since M is reducing for $\pi(C_0(X))$, it is reducing for $\pi(C_0(X))''$, and thus this is a normal

$*$-homomorphism. By Theorem 9.9.2, $\tau\pi(C_0(X))''$ is unitarily equivalent to $L^\infty(\mu)$ acting on $L^2(\mu)$.

Consider the map $q_i : L^\infty(\mu) \to L^\infty(\mu_i)$ by $q_i(f) = f$, which makes sense since two functions which agree a.e.(μ) must also agree a.e.(μ_i). The kernel of this map is

$$J_i = \ker q_i = \{f \in L^\infty(\mu) : f = 0\,\text{a.e.}(\mu_i)\} = \chi_{E_i^c} L^\infty(\mu).$$

This ideal is weak-$*$ closed, and q_i is weak-$*$ continuous.

Let $\rho_i = \widetilde{\sigma}_{\mu_i} q_i$. This takes $L^\infty(\mu)$ onto $\widetilde{\sigma}_{\mu_i}(L^\infty(\mu_i))$, and it is normal and weak-$*$ to WOT-continuous. The map $\rho = \sum_{i\geq 1}^{\oplus} \rho_i$ maps $L^\infty(\mu)$ into $\sum_{i\geq 1}^{\oplus} \widetilde{\sigma}_{\mu_i}$ $(L^\infty(\mu_i))$ by $\rho(f) = \sum_{i\geq 1}^{\oplus} M_{f,\mu_i}$ where M_{f,μ_i} is multiplication by f on $L^2(\mu_i)$. The kernel of this map is

$$\bigcap_{i\geq 1} J_i = \chi_{(\bigcup_{i\geq 1} E_i)^c} L^\infty(\mu) = \{0\}.$$

Thus ρ is an isomorphism. It is clear that the range of ρ is a von Neumann algebra that contains $\sigma(C_0(X))$. Moreover, if $f \in L^\infty(\mu)$, let $f_n \in C_0(X)$ such that $f_n \to f$ a.e.(μ) and $\|f_n\|_\infty \leq \|f\|_\infty$. Then

$$\rho(f) = \text{WOT–lim}\,\rho(f_n) = \text{WOT–lim}\,\sigma(f_n).$$

That means that

$$\sigma(C_0(X))'' = \overline{\sigma(C_0(X))}^{\text{WOT}} = \rho(L^\infty(\mu)).$$

Finally there is a unitary U such that $\pi = U\sigma U^*$. Therefore

$$\pi(C_0(X))'' = U\sigma(C_0(X))''U^* = U\rho(L^\infty(\mu))U^*;$$

and $\widetilde{\pi}(f) = U\rho(f)U^*$ is the desired extension. ∎

We apply this to a normal operator to get a much larger, more powerful functional calculus. Note that the measure used depends on the operator, not just the spectrum.

9.9.7. L^∞ Functional Calculus. *Let $N \in \mathcal{B}(H)$ be a normal operator on a separable Hilbert space. There is a regular Borel probability measure μ on $\sigma(N)$ and a weak-$*$–WOT-continuous isomorphism of $L^\infty(\mu)$ onto $W^*(N)$ that extends the continuous functional calculus.*

We briefly explain how we can develop an operator valued integration theory for a normal operator. The idea is to use the L^∞ functional calculus to define a projection valued measure.

9.9.8. Definition. A *spectral measure* on a measure space $(X, \mathcal{B})$ (where $\mathcal{B}$ is a σ-algebra of subsets of X) is a map $E : \mathcal{B} \to \Pr(H)$, the set of orthogonal projections in $\mathcal{B}(H)$, such that

(1) $E(\varnothing) = 0$ and $E(X) = I$, and

(2) countable additivity: if A_i are disjoint elements of $\mathcal{B}$ for $i \geq 1$, then

$$E\Big(\bigcup_{i\geq 1} A_i\Big) = \text{SOT-}\sum_{i\geq 1} E(A_i).$$

If μ is a scalar measure on $(X, \mathcal{B})$, we say that $\mu \ll E$ if $E(A) = 0$ implies $\mu(A) = 0$; and $E \ll \mu$ if $\mu(A) = 0$ implies that $E(A) = 0$.

We define the integral as it is done in measure theory. First if $f = \sum_{i=1}^n \lambda_i \chi_{A_i}$ is a simple function for disjoint sets $A_i \in \mathcal{B}$, we set

$$\int f\, dE = \sum_{i=1}^n \lambda_i E(A_i).$$

Observe that

$$\Big\| \int f\, dE \Big\| \leq \max\{|\lambda_i| : 1 \leq i \leq n\} = \|f\|_\infty.$$

Every bounded measurable function is a uniform limit of simple functions of no larger norm. Thus one readily checks that we can extend the integral by continuity to all bounded measurable functions on X.

9.9.9. Example. Let N be a normal operator on a separable Hilbert space. Using the $L^\infty(\mu)$ functional calculus obtained in Theorem 9.9.7, we define a spectral measure

$$E_N(A) = \chi_A(N) \quad \text{for all} \quad A \in \text{Bor}(\sigma(N)).$$

If A_i are disjoint Borel sets, and $A = \dot{\bigcup}_{i\geq 1} A_i$, then the weak-$*$ to WOT continuity shows that

$$w^*\text{-}\lim_{n\to\infty} \sum_{i=1}^n \chi_{A_i} = \chi_A \quad \text{and} \quad \text{SOT-}\sum_{i\geq 1} E_n(A_i) = E_N(A).$$

Thus E_N is countably additive. Therefore we can write

$$N = z(N) = \int z\, dE_N.$$

More generally, suppose that X is a LCH space, and $\pi : C_0(X) \to \mathcal{B}(H)$ is a representation on a separable Hilbert space H. By Theorem 9.9.6, there is a finite regular Borel measure on X and a WOT-continuous $*$-isomorphism $\tilde{\pi}$ of $L^\infty(\mu)$ onto $\pi(C_0(X))''$. Define a spectral measure $E_\pi(A) = \tilde{\pi}(\chi_A)$ for Borel sets $A \subset X$. Then as in the singly generated case, we get

$$\pi(f) = \int f\, dE_\pi \quad \text{for} \quad f \in L^\infty(\mu).$$

9.9.10. PROPOSITION. *Let X be an LCH space, and let $\pi : C_0(X) \to \mathcal{B}(H)$ be a representation on a separable Hilbert space H. If ν is a regular Borel probability measure on X such that $\nu \ll E_\pi$, then there is a unit vector $\eta \in H$ so that*

$$\int f\,d\nu = \langle \pi(f)\eta, \eta\rangle \quad \textit{for} \quad f \in C_0(X).$$

PROOF. Let μ be the probability measure obtained from Theorem 9.9.6; and let ξ be a unit vector so that $\int f\,d\mu = \langle \pi(f)\xi, \xi\rangle$. As $\nu \ll E_\pi \simeq \mu$, the Radon–Nikodym theorem provides a positive function $h \in L^1(\mu)$ of norm 1 such that $\nu = h\mu$. Then

$$\int f\,d\nu = \int fh\,d\mu = \langle M_f h^{1/2}, h^{1/2}\rangle_{L^2(\mu)} \quad \text{for} \quad f \in C_0(X).$$

The restriction of π to the cyclic subspace $M = \overline{\pi(C_0(X))\xi}$ is unitarily equivalent to σ_μ, say via U. Thus $\eta = Uh^{1/2}$ is the desired vector. ∎

9.10. Direct Limits

In this section, we explain another construction of C*-algebras. In particular, we define UHF algebras and classify them up to $*$-isomorphism.

9.10.1. DEFINITION. Suppose that $\{(\mathfrak{A}_n, \pi_{n,n+1}) : n \geq 1\}$ is a sequence of C*-algebras with $*$-homomorphisms $\pi_{n,n+1} : \mathfrak{A}_n \to \mathfrak{A}_{n+1}$. The *direct limit* (also called *inductive limit*) is a C*-algebra $\mathfrak{A}$ together with homomorphisms $\pi_n : \mathfrak{A}_n \to \mathfrak{A}$ such that $\pi_{n+1}\pi_{n,n+1} = \pi_n$ for $n \geq 1$ with the universal property: if $\mathfrak{B}$ is a C*-algebra and $\rho_n : \mathfrak{A}_n \to \mathfrak{B}$ satisfy $\rho_{n+1}\pi_{n,n+1} = \rho_n$ for $n \geq 1$, then there is a unique $*$-homomorphism $\rho : \mathfrak{A} \to \mathfrak{B}$ such that $\rho\pi_n = \rho_n$ for $n \geq 1$. We have a commuting diagram:

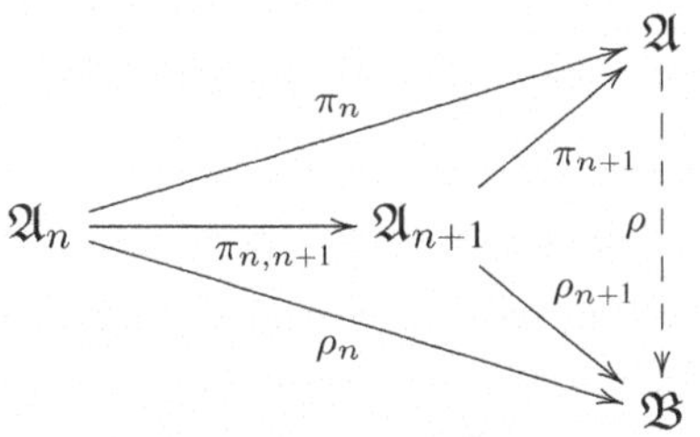

We write $\mathfrak{A} = \varinjlim(\mathfrak{A}_n, \pi_{n,n+1})$ or just $\mathfrak{A} = \varinjlim \mathfrak{A}_n$ if the maps are understood.

The situation is easier when each $\pi_{n,n+1}$ is injective. In this case, we can consider $\mathfrak{A}_n$ as a C*-subalgebra of $\mathfrak{A}_{n+1}$. The maps π_n also turn out to be injective, and $\mathfrak{A} = \overline{\bigcup_{n\geq 1} \pi_n(\mathfrak{A}_n)}$ is the closure of a nested sequence of subalgebras.

9.10.2. THEOREM. *Suppose that $\{(\mathfrak{A}_n, \pi_{n,n+1}) : n \geq 1\}$ is a sequence of C*-algebras with $*$-homomorphisms $\pi_{n,n+1} : \mathfrak{A}_n \to \mathfrak{A}_{n+1}$. Then there is a unique direct limit $\mathfrak{A} = \varinjlim(\mathfrak{A}_n, \pi_{n,n+1})$. If each $\pi_{n,n+1}$ is injective, then each π_n is injective. Also if each $\mathfrak{A}_n$ is unital and the $\pi_{n,n+1}$ are unital, then $\mathfrak{A}$ is unital and the π_n are unital.*

PROOF. Uniqueness is straightforward. If $\mathfrak{A}$ and $\mathfrak{B}$ are both direct limits, then in the diagram above, there is also a unique $*$-homomorphism $\pi : \mathfrak{B} \to \mathfrak{A}$ so that $\pi\rho_n = \pi_n$. Then since the diagram commutes, $\pi\rho\pi_n$ could replace π_n. By uniqueness, $\pi\rho = id_{\mathfrak{A}}$. Similarly $\rho\pi = id_{\mathfrak{B}}$, so that $\mathfrak{A}$ and $\mathfrak{B}$ are isomorphic via maps compatible with the diagram. Thus the direct limit is unique.

For existence, let

$$\mathfrak{D} = l_\infty(\{\mathfrak{A}_n\}) = \{(a_1, a_2, a_3, \dots) : a_n \in \mathfrak{A}_n,\ \sup \|a_n\| < \infty\} \quad \text{and}$$
$$\mathfrak{E}_0 = \{(a_1, a_2, a_3, \dots) \in \mathfrak{D} : \exists_N \pi_{n,n+1}(a_n) = a_{n+1} \text{ for } n \geq N\}.$$

Let $\mathfrak{E} = \overline{\mathfrak{E}_0}$ and let $I = c_0(\{\mathfrak{A}_n\}) = \{(a_1, a_2, a_3, \dots) \in \mathfrak{D} : \lim a_n = 0\}$. Then I is an ideal in $\mathfrak{E}$ since the sequences which are eventually 0 belong to $\mathfrak{E}_0$. Set $\mathfrak{A} = \mathfrak{E}/I$. Let $\pi_{n,m} = \pi_{m-1,m} \cdots \pi_{n,n+1}$ for $n < m$. Define

$$\pi_n(a_n) = (0, \dots, 0, a_n, \pi_{n,n+1}(a_n), \dots, \pi_{n,m}(a_n), \dots) + I$$

where the a_n is the nth coordinate. It is routine to verify that $\pi_{n+1}\pi_{n,n+1} = \pi_n$. Observe that $\pi_n(\mathfrak{A}_n) \subset \pi_{n+1}(\mathfrak{A}_{n+1})$ and $\bigcup_{n\geq 1} \pi_n(\mathfrak{A}_n)$ is a dense subalgebra of $\mathfrak{A}$.

Suppose that $\mathfrak{B}$ is another C*-algebra with maps $\rho_n : \mathfrak{A}_n \to \mathfrak{B}$ such that $\rho_{n+1}\pi_{n,n+1} = \rho_n$ for $n \geq 1$. Define ρ on $\pi_n(\mathfrak{A}_n)$ by $\rho(\pi_n(a_n)) = \rho_n(a_n)$. We need to check that this is well defined. First, if $a_n, a_n' \in \mathfrak{A}_n$ such that $\pi_n(a_n) = \pi_n(a_n')$, then $\lim_{m\to\infty} \pi_{n,m}(a_n - a_n') = 0$. Thus

$$\rho_n(a_n - a_n') = \lim_{m\to\infty} \rho_m(\pi_{n,m}(a_n - a_n')) = 0.$$

Hence $\rho(\pi_n(a_n')) = \rho_n(a_n') = \rho_n(a_n) = \rho(\pi_n(a_n))$. We also need to check that if $\pi_n(a_n) = \pi_m(a_m')$ for $n < m$, then the definition of ρ is the same. Let $a_m = \pi_{n,m}(a_n)$. Then $\pi_m(a_m) = \pi_m(a_m')$, and thus $\rho(\pi_m(a_m')) = \rho(\pi_m(a_m))$. Therefore,

$$\rho(\pi_n(a_n)) = \rho_n(a_n) = \rho_m(\pi_{n,m}(a_n)) = \rho(\pi_m(a_m')).$$

It is easy to check that ρ is contractive, self-adjoint, linear, and multiplicative on a dense subalgebra of $\mathfrak{A}$. Hence it extends uniquely to a $*$-homomorphism $\rho : \mathfrak{A} \to \mathfrak{B}$ such that $\rho\pi_n = \rho_n$ for $n \geq 1$. Therefore $\mathfrak{A} = \varinjlim(\mathfrak{A}_n, \pi_{n,n+1})$.

If each $\pi_{n,n+1}$ is injective, then they are all isometric. Hence $\|\pi_n(a)\| = \limsup_{m\to\infty} \|\pi_{m,n}(a)\| = \|a\|$ for all $a \in \mathfrak{A}_n$; whence π_n is injective.

If each $\pi_{n,n+1}$ is unital, then $e := (1, 1, 1, \dots) + I$ belongs to $\mathfrak{A}$ and $\pi_n(1) = e$. Now $e\pi_m(a) = \pi_m(a) = \pi_m(a)e$ for all $a \in \mathfrak{A}_m$ for $m \geq 1$. Since the union of the $\pi_m(\mathfrak{A}_m)$ is dense in $\mathfrak{A}$, e is the unit for $\mathfrak{A}$, so that it is unital, as are the π_n. ∎

9.10.3. EXAMPLE. Let $\pi_{n,n+1} : M_n \to M_{n+1}$ by

$$\pi_{n,n+1}(a) = a \oplus 0 = \left[\begin{array}{ccc|c} & & & 0 \\ & a & & 0 \\ & & & 0 \\ \hline 0 & 0 & 0 & 0 \end{array}\right]$$

This maps M_n onto the $n \times n$ corner of M_{n+1}. Then $\varinjlim(M_n, \pi_{n,n+1}) = \mathfrak{K}$, the compact operators via the maps π_n sending $a \in M_n$ to $a \oplus 0$ in $\mathfrak{K}$. These maps are all injective, and so we can think of the M_n as a nested sequence of subalgebras of $\mathfrak{K}$. Any sequence of maps $\rho_n : M_n \to \mathfrak{B}$ uniquely defines $\rho(a)$ if a lies in $\bigcup_{n\geq 1} \pi_{n,n+1}(M_n)$. In particular, the matrix units $E_{ij} = e_i e_j^*$ for the standard basis are sent to $f_{ij} = \rho(E_{ij})$ satisfying $f_{ij} f_{kl} = \delta_{jk} f_{il}$ and $f_{ji} = f_{ij}^*$ and $\|f_{ij}\| \leq 1$. If some $\rho_n \neq 0$, then each ρ_n is injective because M_n is simple. So $\{f_{ij} : i, j \geq 1\}$ is a set of matrix units in $\mathfrak{B}$ which spans a subalgebra isomorphic to $\mathfrak{K}$, and is the range of the completion ρ of $\bigcup_{n\geq 1} \rho_n$ (defined on the dense union).

9.10.4. EXAMPLE. If $n, k \in \mathbb{N}$, we can define a *unital* embedding $\pi : M_n \to M_{kn}$ by $\pi(a) = a \otimes I_k = a^{(k)}$. Conversely, suppose that $\pi : M_n \to M_m$ is a *unital* homomorphism. Note that $\pi(E_{11})$ is a projection, and let $k = \operatorname{rank} \pi(E_{11})$. Since $E_{ii} = E_{i1} E_{11} E_{1i}$ and $E_{11} = E_{1i} E_{ii} E_{i1}$, it follows that $\operatorname{rank} \pi(E_{ii}) = k$ as well. Thus

$$m = \operatorname{rank}(I_m) = \operatorname{rank} \pi(I_n) = \operatorname{rank} \sum_{i=1}^{n} \pi(E_{ii}) = kn$$

because $\pi(E_{ii})$ are pairwise orthogonal projections. Hence n divides m.

Let $\rho : M_n \to M_{kn}$ be another unital homomorphism. Let $\xi_1^1, \ldots, \xi_k^1$ be an orthonormal basis for $\operatorname{Ran} \pi(E_{11})$ and let $\eta_1^1, \ldots, \eta_k^1$ be an orthonormal basis for $\operatorname{Ran} \rho(E_{11})$. Define $\xi_s^i = \pi(E_{i1})\xi_s^1$ and $\eta_s^i = \rho(E_{i1})\eta_s^1$ for $1 \leq i \leq n$ and $1 \leq s \leq k$. Then these form an orthonormal basis for $\operatorname{Ran} \pi(E_{ii})$ and $\operatorname{Ran} \rho(E_{ii})$, respectively. Define a unitary $U \in M_{kn}$ by setting $U\xi_s^i = \eta_s^i$. Then it is easy to check that $U\pi(E_{ij})U^* = \rho(E_{ij})$ for $1 \leq i, j \leq n$. It follows that $\rho \simeq \pi$. This says that there is a unique unital homomorphism of M_n into M_{kn} up to unitary equivalence.

9.10.5. EXAMPLE. Let $k_i \geq 2$ be integers and set $m_n = \prod_{i=1}^{n} k_i$. Define $\pi_{n,n+1} : M_{m_n} \to M_{m_{n+1}}$ be the unital imbeddings $\pi_{n,n+1}(a) = a \otimes I_{k_{n+1}}$. By Example 9.10.4, we see that these imbeddings are unique up to unitary equivalence. Thus the direct limit does not depend on any choices. We let $\mathfrak{A} = \varinjlim M_{m_n}$, where the imbeddings are understood to be unital. Then since the maps are all injective, we may identify M_{m_n} with $\pi_n(M_{m_n})$ as a nested sequence of unital subalgebras of $\mathfrak{A}$ with dense union. The limit is called a *uniformly hyperfinite algebra* or *UHF algebra*.

This next lemma shows that ideals in a direct limit are *inductive*, meaning that they satisfy the conclusion of the lemma.

9.10.6. LEMMA. *Suppose that* $\mathfrak{A} = \overline{\bigcup_{n\geq 1} \mathfrak{A}_n}$ *is a C*-algebra which is the closure of the union of a nested sequence of C*-subalgebras* $\mathfrak{A}_n$. *Let* I *be an ideal of* $\mathfrak{A}$. *Then* $I = \overline{\bigcup_{n\geq 1} I \cap \mathfrak{A}_n}$.

PROOF. Let $I_n = I \cap \mathfrak{A}_n$. This is a closed ideal in $\mathfrak{A}_n$. By Corollary 9.2.5, $\mathfrak{A}_n/I_n \simeq (\mathfrak{A}_n + I)/I$. Let $a \in I$ and choose $a_n \in \mathfrak{A}_n$ with $a = \lim_n a_n$. Then

$$0 = \|a + I\| = \lim_n \|a_n + I\| = \lim_n \|a_n + I_n\|.$$

Select $b_n \in I_n$ so that $\|a_n - b_n\| \leq 2\|a_n + I_n\|$. Then $a = \lim_n b_n$. Hence $I = \overline{\bigcup_{n\geq 1} I_n}$. ∎

9.10.7. DEFINITION. Two projections p, q in a C*-algebra $\mathfrak{A}$ are *Murray-von Neumann equivalent* or just *equivalent projections* if there is a partial isometry $u \in \mathfrak{A}$ so that $u^*u = p$ and $uu^* = q$. We write $p \sim q$.

9.10.8. LEMMA. *Suppose that* $\mathfrak{A} = \overline{\bigcup_{n\geq 1} \mathfrak{A}_n}$ *is a C*-algebra which is the closure of the union of a nested sequence of C*-subalgebras* $\mathfrak{A}_n$. *Let* p *be a projection in* $\mathfrak{A}$. *Then* p *is equivalent in* $\mathfrak{A}$ *to a projection in* $\bigcup_{n\geq 1} \mathfrak{A}_n$.

PROOF. Let $0 < \varepsilon \leq \frac{1}{6}$. Select an $a = a^* \in \mathfrak{A}_n$ for n large so that $\|p - a\| < \varepsilon$. If $\lambda \in \mathbb{R}$ satisfies $|\lambda| \geq \varepsilon$ and $|1 - \lambda| \geq \varepsilon$, then $\|(p - \lambda)^{-1}\| \leq \varepsilon^{-1}$; so that $a - \lambda$ is also invertible. Therefore $\sigma(a) \subset [-\varepsilon, \varepsilon] \cup [1 - \varepsilon, 1 + \varepsilon]$. Let $f = \chi_{[1-\varepsilon,1+\varepsilon]}$ in $C(\sigma(a))$, and set $q = f(a) \in \mathfrak{A}_n$. Then $q = q^* = q^2$ is a projection. Moreover

$$\|p - q\| \leq \|p - a\| + \|a - q\| < \varepsilon + \|x - f\|_{\sigma(a)} < 2\varepsilon.$$

Note that $\sigma(pqp) \subset \{0\} \cup [1 - 2\varepsilon, 1]$ because

$$p \geq pqp = ppp - p(p - q)p \geq (1 - 2\varepsilon)p.$$

Define a function on $\sigma(pqp)$ by

$$g(x) = \begin{cases} 0 & \text{if } x = 0 \\ x^{-1/2} & \text{if } 1 - 2\varepsilon \leq x \leq 1 \end{cases}$$

Then $b = g(pqp)$ satisfies $(pqp)^{1/2}b = b(pqp)^{1/2} = p$.

Write $qp = u|qp|$ be the polar decomposition, where $|qp| = (pqp)^{1/2}$ and $u = qpb$. Then,

$$u^*u = b(pqp)b = p \quad \text{and} \quad uu^*u = up = u$$

shows that u is a partial isometry in $\mathfrak{A}$. Now $uu^* = qpb^2pq \le q$ and

$$\|q - uu^*\| = \|q(q - p + p - b^2)q\| \le 2\varepsilon + \left|1 - \frac{1}{1-2\varepsilon}\right| < 5\varepsilon < 1.$$

Since $uu^* \le q$ is a projection, we obtain $uu^* = q$. Hence p and q are equivalent. ■

9.10.9. DEFINITION. A *trace* or *tracial state* on a C*-algebra is a state τ such that $\tau(ab) = \tau(ba)$ for all $a, b \in \mathfrak{A}$.

9.10.10. THEOREM. *A UHF algebra $\mathfrak{A}$ is simple and has a unique trace. The trace on the projections in $\mathfrak{A}$ takes all values $\{\frac{k}{m_n} : 0 \le k \le m_n,\ n \ge 1\}$.*

PROOF. Write $\mathfrak{A} = \overline{\bigcup_{n\ge1} \mathfrak{A}_n}$ where $\mathfrak{A}_n = M_{m_n}$. Let I be a proper ideal of $\mathfrak{A}$. Then $I_n = I \cap \mathfrak{A}_n$ does not contain the identity, and so $I_n = \{0\}$ because M_{m_n} is simple. By Lemma 9.10.6, $I = \{0\}$; whence $\mathfrak{A}$ is simple.

Note that $\tau_n(a) = \frac{1}{m_n}\operatorname{Tr}_{m_n}(a)$ is a tracial state on $\mathfrak{A}_n = M_{m_n}$. This trace is unique because if τ is a tracial state on M_{m_n}, then

$$\tau(E_{ii}) = \tau(E_{i1}E_{1i}) = \tau(E_{1i}E_{i1}) = \tau(E_{11})$$

and

$$1 = \tau(I) = \sum_{i=1}^{m_n} \tau(E_{ii}) = m_n\tau(E_{11}).$$

Thus $\tau(E_{ii}) = \frac{1}{m_n}$. Likewise any rank one projection $\xi\xi^*$ is equivalent to E_{11} via the partial isometry ξe_1^*, so $\tau(\xi\xi^*) = 1$. The projections in M_{m_n} span M_{m_n}, so τ is uniquely determined.

We can define $\tau(a) = \tau_n(a)$ for $a \in \mathfrak{A}_n$ and $n \ge 1$. This is well defined since for $a \in \mathfrak{A}_n$,

$$\tau_{n+1}(\pi_{n,n+1}(a)) = \frac{1}{m_{n+1}}\operatorname{Tr}_{m_{n+1}}(a \otimes I_{k_{n+1}}) = \frac{k_{n+1}}{m_{n+1}}\operatorname{Tr}_{m_n}(a) = \tau_n(a).$$

This defines a positive norm 1 linear functional on $\bigcup_{n\ge1}\mathfrak{A}_n$, and thus it extends to $\mathfrak{A}$ by continuity. The tracial property also extends by continuity. So τ is a trace. Conversely any trace must restrict to τ_n on $\mathfrak{A}_n$, so τ is unique.

If $p \in \mathfrak{A}_n$ is a projection of rank k in M_{m_n}, then $\tau(p) = \tau_n(p) = \frac{k}{m_n}$. Each value, for $0 \le k \le m_n$, is attained by some projection. On the other hand, if p is a projection in $\mathfrak{A}$, Lemma 9.10.8 shows that there is a projection $q \in \mathfrak{A}_n$ for large n which is equivalent to p, say $p = u^*u$ and $q = uu^*$. Then

$$\tau(p) = \tau(u^*u) = \tau(uu^*) = \tau(q) \in \left\{\frac{k}{m_n} : 0 \le k \le m_n,\ n \ge 1\right\}.$$ ■

9.10.11. DEFINITION. Given $k_i \geq 2$ and $m_n = \prod_{i=1}^n k_i$, we associate a *supernatural number* which is a formal product $\prod_{p \in P} p^{\alpha_p}$ where P is the set of prime numbers and

$$\alpha_p = \sup\{k : p^k | m_n \text{ for some } n \geq 1\}.$$

It is perhaps easier to consider $\alpha : P \to \mathbb{N}_0 \cup \{\infty\}$ as a function on P.

9.10.12. GLIMM THEOREM. *Two UHF algebras are isomorphic if and only if they have the same supernatural number.*

PROOF. Theorem 9.10.10 shows that the set of values taken by the trace on the set of projections is an isomorphism invariant. These tracial values determine the supernatural number associated to $\{m_n\}$ because α_p is the sup of all k such that p^{-k} is the trace of some projection in $\mathfrak{A}$, and conversely the supernatural number determines this set of traces. So isomorphism implies equal supernatural numbers.

Conversely, suppose that $\mathfrak{A} = \overline{\bigcup_{n \geq 1} M_{a_n}}$ and $\mathfrak{B} = \overline{\bigcup_{n \geq 1} M_{b_n}}$ are two UHF algebras with the same supernatural number. Let $\pi_{m,n} : \mathfrak{A}_n \to \mathfrak{A}_m$ and $\widetilde{\pi}_{m,n} : \mathfrak{B}_n \to \mathfrak{B}_m$ be the unital imbeddings. If n is given, every prime power dividing a_n will divide b_m for m large enough, so there is some m so that $a_n | b_m$. Likewise given b_m, there is some large n so that $b_m | a_n$. Hence we can find two subsequences $n_1 < n_2 < n_3 < \dots$ and $m_1 < m_2 < m_3 < \dots$ so that $a_{n_i} | b_{m_i}$ and $b_{m_i} | a_{n_{i+1}}$.

Let $\rho_i' : \mathfrak{A}_{n_i} \to \mathfrak{B}_{m_i}$ and $\sigma_i' : \mathfrak{B}_{m_i} \to \mathfrak{A}_{n_{i+1}}$ be unital imbeddings. Then $\sigma_i' \rho_i'$ is an unital imbedding of $\mathfrak{A}_{n_i}$ into $\mathfrak{A}_{n_{i+1}}$. But $\pi_i := \pi_{n_i, n_{i+1}}$ is another unital imbedding. By Example 9.10.4, there is a unitary $u_2 \in \mathfrak{A}_{n_2}$ so that

$$\sigma_1 := \mathrm{Ad}(u_2)\sigma_1'(b) := u_2 \sigma_i'(b) u_2^*$$

is an imbedding of $\mathfrak{B}_{m_1}$ into $\mathfrak{A}_{n_2}$ such that $\pi_1 = \sigma_1 \rho_1$. Similarly there is a unitary $v_2 \in \mathfrak{B}_{m_2}$ so that $\rho_2 = \mathrm{Ad}(v_2)\rho_2'$ satisfies $\rho_2 \sigma_1 = \widetilde{\pi}_1 := \widetilde{\pi}_{m_1, m_2}$. There is no need to adjust the first imbedding $\rho_1 = \rho_1'$. Recursively, we obtain unitaries $u_i \in \mathfrak{A}_{n_i}$ and $v_i \in \mathfrak{B}_{m_i}$ satisfying $\sigma_i = \mathrm{Ad}(u_{i+1})\sigma_i'$ and $\rho_i = \mathrm{Ad}(v_{i+1})\rho_i'$ so that $\pi_i = \sigma_i \rho_i$ and $\widetilde{\pi}_i = \rho_{i+1}\sigma_i$ for $i \geq 1$. We obtain the following commutative diagram:

$$\begin{array}{ccccccccc}
\mathfrak{A}_{n_1} & \xrightarrow{\pi_1} & \mathfrak{A}_{n_2} & \xrightarrow{\pi_2} & \mathfrak{A}_{n_3} & \xrightarrow{\pi_3} & \mathfrak{A}_{n_4} & \xrightarrow{\pi_4} & \cdots \\
\rho_1 \downarrow & \overset{\sigma_1}{\nearrow} & \rho_2 \downarrow & \overset{\sigma_2}{\nearrow} & \rho_3 \downarrow & \overset{\sigma_3}{\nearrow} & \rho_4 \downarrow & \overset{\sigma_4}{\nearrow} & \\
\mathfrak{B}_{m_1} & \xrightarrow[\widetilde{\pi}_1]{} & \mathfrak{B}_{m_2} & \xrightarrow[\widetilde{\pi}_2]{} & \mathfrak{B}_{m_3} & \xrightarrow[\widetilde{\pi}_3]{} & \mathfrak{B}_{m_4} & \xrightarrow[\widetilde{\pi}_4]{} & \cdots
\end{array}$$

We define ρ to be the continuous extension of $\bigcup_{i \geq 1} \rho_i$, which is an injective homomorphism of $\mathfrak{A}$ into $\mathfrak{B}$ and σ to be the continuous extension of $\bigcup_{i \geq 1} \sigma_i$, which is an injective homomorphism of $\mathfrak{B}$ into $\mathfrak{A}$. By commutativity, $\sigma\rho = id_{\mathfrak{A}}$ and $\rho\sigma = id_{\mathfrak{B}}$. Hence $\mathfrak{A}$ and $\mathfrak{B}$ are isomorphic. ∎

9.11. Cuntz Algebras

In this section, we give an interesting and important family of C*-algebras defined by a set of relations. We will not examine general conditions under which a universal C*-algebra exists. Rather we shall illustrate the idea with the Cuntz algebras.

9.11.1. DEFINITION. The *universal C*-algebra* generated by a set of relations is a C*-algebra with generators satisfying these relations which has every other example of such a C*-algebra as a canonical quotient, meaning that generators are sent to the corresponding generators.

The universal C*-algebra generated by $d \geq 2$ isometries $\mathfrak{s}_1, \ldots, \mathfrak{s}_d$ such that $\sum_{i=1}^{d} \mathfrak{s}_i \mathfrak{s}_i^* = I$ is the *Cuntz algebra* $\mathcal{O}_d$.

9.11.2. PROPOSITION. *The Cuntz algebras exist.*

PROOF. First note that examples of C*-algebras satisfying the Cuntz relations exist. Let H be a separable Hilbert space with orthonormal basis $\{\xi_n : n \geq 0\}$ Define $S_i \in \mathcal{B}(H)$ by

$$S_i \xi_n = \xi_{dn+i-1} \quad \text{for} \quad 1 \leq i \leq d \text{ and } n \geq 0.$$

It is easy to check that each S_i is an isometry and that their ranges are orthogonal and span the whole space.

We can make the condition on the isometries into algebraic relations. The property that S_i is an isometry is $S_i^* S_i = I$. The condition $\sum_{i=1}^{d} S_i S_i^* = I$ forces the ranges of the S_i's to be pairwise orrthogonal. Hence $S_j^* S_i = 0$ if $i \neq j$. So we obtain the relations

$$\mathfrak{s}_j^* \mathfrak{s}_i = \delta_{ij} I \text{ for } 1 \leq i, j \leq d \quad \text{and} \quad \sum_{i=1}^{d} \mathfrak{s}_i \mathfrak{s}_i^* = I.$$

Fix a separable Hilbert space H, and let

$$\mathcal{C} = \Big\{ S = (S_1, \ldots, S_d) \in \mathcal{B}(H)^d : S_i^* S_j = \delta_{ij} I,\ 1 \leq i, j \leq d, \text{ and } \sum_{i=1}^{d} S_i S_i^* = I \Big\}.$$

Then define $\mathfrak{s}_i = \bigoplus_{S \in \mathcal{C}} S_i$ for $1 \leq i \leq d$, and let $\mathcal{O}_d = C^*(\mathfrak{s}_1, \ldots, \mathfrak{s}_d)$. This d-tuple satisfies the Cuntz relations.

Observe that if $\mathfrak{A} = C^*(T_1, \ldots, T_d)$ is a C*-algebra satisfying the Cuntz relations, then any non-zero quotient of $\mathfrak{A}$ also satisfies these relations. Now $\mathfrak{A}$ can be expressed as a direct sum of cyclic subrepresentations, and since $\mathfrak{A}$ is finitely generated, these subrepresentations act on a separable Hilbert space. This space cannot be finite dimensional since the isometries are proper. So each one is unitarily equivalent to some $S \in \mathcal{C}$. Hence there are cardinals α_S for $S \in \mathcal{C}$ so that

$T_i \simeq \sum_{S\in C}^{\oplus} S_i^{(\alpha_S)}$. From this, it follows that there is a $*$-homomorphism of $\mathcal{O}_d$ onto $\mathfrak{A}$ sending $\mathfrak{s}_i$ to T_i for $1 \le i \le d$. Therefore $\mathcal{O}_d$ is universal for these relations. ∎

We will next derive some basic properties of $\mathcal{O}_d$. Let $\mathbb{F}_d^+$ denote the *free semigroup on d letters*, meaning the set of all words $w = i_1 i_2 \dots i_k$ with $i_j \in \{1, \dots, d\}$. This is a monoid with multiplication given by concatenation, so if $v = j_1 \dots j_l$, then $wv = i_1 i_2 \dots i_k j_1 \dots j_l$, and the empty word $\varnothing$ is the identity. Let $|w| = k$ denote the length of the word w. We will write $\mathfrak{s}_w = \mathfrak{s}_{i_1} \dots \mathfrak{s}_{i_k}$. Observe that if $|w| = |v| = k$, then

$$\mathfrak{s}_v^* \mathfrak{s}_w = \mathfrak{s}_{j_k}^* \dots \mathfrak{s}_{j_2}^* (\mathfrak{s}_{j_1}^* \mathfrak{s}_{i_1}) \mathfrak{s}_{i_2} \dots \mathfrak{s}_{i_k} = 0 \quad \text{if} \quad i_1 \ne j_1.$$

To get a non-zero product, we must have repeated cancellation of all k terms, and then the product is 1. Thus $\mathfrak{s}_v^* \mathfrak{s}_w = \delta_{v,w} 1$ when $|w| = |v|$. Hence any product of elements from $\{\mathfrak{s}_1, \dots, \mathfrak{s}_d, \mathfrak{s}_1^*, \dots, \mathfrak{s}_d^*\}$ will reduce to a shorter word unless all of the terms $\mathfrak{s}_i$ are to the left of all the adjoint terms $\mathfrak{s}_j^*$'s. Thus $\mathcal{O}_d$ is the norm-closed span of $\{\mathfrak{s}_v \mathfrak{s}_w^* : v, w \in \mathbb{F}_d^+\}$.

9.11.3. LEMMA. *For $k \ge 1$, $\mathcal{F}_d(k) = \operatorname{span}\{\mathfrak{s}_v \mathfrak{s}_w^* : |v| = |w| = k\}$ is isomorphic to M_{d^k} and $\mathcal{F}_d = \overline{\bigcup_{k\ge 1} \mathcal{F}_d(k)}$ is a UHF algebra of type d^∞.*

PROOF. First observe that the terms $\{\mathfrak{s}_v \mathfrak{s}_w^* : |v| = |w| = k\}$ form a set of matrix units because if $|v_i| = |w_i| = k$, then

$$\mathfrak{s}_{v_1} \mathfrak{s}_{w_1}^* \mathfrak{s}_{v_2} \mathfrak{s}_{w_2}^* = \delta_{w_1, v_2} \mathfrak{s}_{v_1} \mathfrak{s}_{w_2}^*.$$

In particular $\{\mathfrak{s}_w \mathfrak{s}_w^* : |w| = k\}$ is a set of d^k pairwise orthogonal projections summing to 1. A term $\mathfrak{s}_v \mathfrak{s}_w^*$ is a partial isometry with initial space $\mathfrak{s}_w \mathfrak{s}_w^*$ and range space $\mathfrak{s}_v \mathfrak{s}_v^*$. Hence $\mathcal{F}_d(k)$ is isomorphic to M_{d^k}. Now

$$\mathfrak{s}_v \mathfrak{s}_w^* = \mathfrak{s}_v \Big(\sum_{i=1}^{d} \mathfrak{s}_i^* \mathfrak{s}_i \Big) \mathfrak{s}_w^* = \sum_{i=1}^{d} \mathfrak{s}_{vi} \mathfrak{s}_{wi}^*.$$

Thus $\mathcal{F}_d(k) \subset \mathcal{F}_d(k+1)$ is a unital embedding. Therefore $\mathcal{F}_d = \overline{\bigcup_{k\ge 1} \mathcal{F}_d(k)}$ is a UHF algebra of type d^∞. ∎

An *expectation* of a C*-algebra onto a subalgebra is a norm 1 projection. See section 10.6 for properties of these maps.

9.11.4. THEOREM. *There is a faithful expectation Φ of $\mathcal{O}_d$ onto $\mathcal{F}_d$.*

PROOF. For $\lambda \in \mathbb{T}$, the elements $\lambda \mathfrak{s}_1, \dots, \lambda \mathfrak{s}_d$ satisfy the same relations. By the universal property, there is a $*$-homomorphism ρ_λ of $\mathcal{O}_d$ into itself such that $\rho_\lambda(\mathfrak{s}_i) = \lambda \mathfrak{s}_i$ for $1 \le i \le d$. Clearly $\rho_{\overline{\lambda}} = \rho^{-1}$, so ρ_λ is a $*$-automorphism. Observe that

$$\rho_\lambda(s_v s_w^*) = \lambda^{|v|} \mathfrak{s}_v \overline{\lambda}^{|w|} \mathfrak{s}_w^* = \lambda^{|v|-|w|} \mathfrak{s}_v \mathfrak{s}_w^* \quad \text{for} \quad v, w \in \mathbb{F}_d^+.$$

The function $t \to \rho_{e^{2\pi it}}(X)$ for $X \in \mathcal{O}_d$ is norm continuous because this is true for the algebraic span of words $s_v s_w^*$, which are dense in $\mathcal{O}_d$, and the fact that $*$-homomorphisms are contractive. Define

$$\Phi(X) = \int_0^1 \rho_{e^{2\pi it}}(X)\, dt \quad \text{for} \quad X \in \mathcal{O}_d.$$

This makes sense as a Riemann integral. Note that

$$\|\Phi(X)\| \leq \int_0^1 \|\rho_{e^{2\pi it}}(X)\|\, dt = \int_0^1 \|X\|\, dt = \|X\|.$$

So Φ is contractive. Moreover, we have

$$\Phi(s_v s_w^*) = \int_0^1 e^{2\pi i(|v|-|w|)t} \mathfrak{s}_v \mathfrak{s}_w^*\, dt = \begin{cases} \mathfrak{s}_v \mathfrak{s}_w^* & \text{if } |v| = |w| \\ 0 & \text{if } |v| \neq |w|. \end{cases}$$

So Φ maps the dense algebraic span of words $s_v s_w^*$ into $\mathcal{F}_d$, and is the identity on $\bigcup_{k\geq 1} \mathcal{F}_d(k)$. By continuity, Φ is an expectation onto $\mathcal{F}$.

Suppose that $X \gneq 0$. Then $\rho_\lambda(X) \gneq 0$ for all $\lambda \in \mathbb{T}$; and thus $\Phi(X) \geq 0$. Choose some vector ξ so that $\langle X\xi, \xi\rangle > 0$. By continuity, $\langle \rho_{e^{2\pi it}}(X)\xi, \xi\rangle > 0$ on some neighbourhood of $t = 0$, and is non-negative for all t. Hence $\langle \Phi(X)\xi, \xi\rangle > 0$. Therefore Φ is faithful. ■

9.11.5. LEMMA. *For $k \geq 1$, there is an isometry $W_k \in \mathcal{O}_d$ such that for $X \in$ span$\{\mathfrak{s}_v\mathfrak{s}_w^* : |v| \leq k,\ |w| \leq k\}$, we have $\Phi(X) = W_k^* X W_k$.*

PROOF. Let $W_k = \sum_{|x|=k} \mathfrak{s}_x \mathfrak{s}_1^k \mathfrak{s}_2 \mathfrak{s}_x^*$. Compute for $|v| \leq k$ and $|w| \leq k$:

$$W_k^* \mathfrak{s}_v \mathfrak{s}_w^* W_k = \sum_{|x|=|y|=k} \mathfrak{s}_x \mathfrak{s}_2^* \mathfrak{s}_1^{*k} \mathfrak{s}_x^* (\mathfrak{s}_v \mathfrak{s}_w^*) \mathfrak{s}_y \mathfrak{s}_1^k \mathfrak{s}_2 \mathfrak{s}_y^*$$

For a term in this sum to be non-zero, we need $y = wy'$ for $|y'| = k - |w|$ and $x = vx'$ for $|x'| = k - |v|$, in which case we obtain a term in the middle of the form $\mathfrak{s}_{x'1^k2}^* \mathfrak{s}_{y'1^k2}$. Again for this to be non-zero, we must get perfect cancellation. If $|v| \neq |w|$, we eventually get down to a factor of the form $\mathfrak{s}_2^*\mathfrak{s}_1 = 0$ or $\mathfrak{s}_1^*\mathfrak{s}_2 = 0$. Otherwise, when $|v| = |w|$, we require $x' = y'$ to get a non-zero product of 1. The sum is now reduced to

$$W_k^* \mathfrak{s}_v \mathfrak{s}_w^* W_k = \sum_{|x'|=k-|v|} \mathfrak{s}_{vx'} \mathfrak{s}_{wx'}^* = \mathfrak{s}_v \Big(\sum_{|x'|=k-|v|} \mathfrak{s}_{x'} \mathfrak{s}_{x'}^* \Big) \mathfrak{s}_w^* = \mathfrak{s}_v \mathfrak{s}_w^*.$$

Hence $W_k^* \mathfrak{s}_v \mathfrak{s}_w^* W_k = \Phi(\mathfrak{s}_v \mathfrak{s}_w^*)$ in all cases. In particular, taking $v = w = \varnothing$, we get $W_k^* W_k = 1$, so that W_k is an isometry. By linearity, we get $\Phi(X) = W_k^* X W_k$ for $X \in$ span$\{\mathfrak{s}_v \mathfrak{s}_w^* : |v| \leq k,\ |w| \leq k\}$. ■

9.11.6. THEOREM. *If $0 \neq X \in \mathcal{O}_d$, there are $A, B \in \mathcal{O}_d$ such that $AXB = 1$.*

PROOF. Now $X^*X \gneq 0$, so that $\Phi(X^*X) \neq 0$. Normalize so $\|\Phi(X^*X)\| = 1$. Pick an element $Y = Y^* \in \operatorname{span}\{\mathfrak{s}_v\mathfrak{s}_w^* : |v| \leq k,\ |w| \leq k\}$ for some large k such that $\|Y - X^*X\| \leq \frac{1}{4}$. Then $\|\Phi(Y)\| > \frac{3}{4}$ and its spectrum lies in $[-\frac{1}{4}, \frac{5}{4}]$. Let W_k be the isometry from Lemma 9.11.5 such that $\Phi(Y) = W_k^*YW_k \in \mathcal{F}_d(k)$. Since $\mathcal{F}_d(k) \simeq M_{d^k}$, $\Phi(Y)$ can be diagonalized, and it has a positive eigenvalue $t > \frac{3}{4}$. That is, there is a minimal projection $E \in \mathcal{F}_d(k)$ such that $E\Phi(Y) = \Phi(Y)E = tE$. This projection E is equivalent in $\mathcal{F}_d(k)$ to $E_1 = \mathfrak{s}_1^k\mathfrak{s}_1^{*k}$; whence there is a partial isometry $U \in \mathcal{F}_d(k)$ so that $U^*EU = E_1$. Therefore

$$(t^{-1}\mathfrak{s}_1^{*k}U^*EW_k^*)Y(W_kEU\mathfrak{s}_1^k) = \mathfrak{s}_1^{*k}U^*EU\mathfrak{s}_1^k = 1.$$

Let $A = t^{-1/2}\mathfrak{s}_1^{*k}U^*EW_k^* = B^*$. Then $AYB = 1$, so

$$\begin{aligned}\|AX^*XB - I\| &= \|AX^*XB - AYB\| = \|A(X^*X - Y)B\| \\ &\leq \|A\|\,\|X^*X - Y\|\,\|B\| < \frac{1}{4}\|A\|^2 = \frac{1}{4t} < \frac{1}{3}.\end{aligned}$$

Hence AX^*XB is invertible, so that $(AX^*)X(B(AX^*XB)^{-1}) = 1$. ■

9.11.7. COROLLARY. *For $d \geq 2$, $\mathcal{O}_d$ is simple. Thus if $S = (S_1, \ldots, S_d)$ satisfies the Cuntz relations, then $C^*(S_1, \ldots, S_d) \simeq \mathcal{O}_d$.*

PROOF. Let $\mathcal{J}$ be a non-zero ideal in $\mathcal{O}_d$, and let $0 \neq X \in \mathcal{J}$. Then there are $A, B \in \mathcal{O}_d$ so that $AXB = 1$, and thus $\mathcal{J} = \mathcal{O}_d$. ■

9.11.8. EXAMPLE. Let $\mathbb{F}_d^+$ be the free semigroup on d letters and consider isometries on $F_d^2 = \ell_2(\mathbb{F}_d^+)$, with orthonormal basis $\{\xi_w : w \in \mathbb{F}_d^+\}$, defined by

$$L_i\xi_w = \xi_{iw} \quad \text{for} \quad 1 \leq i \leq d \text{ and } w \in \mathbb{F}_d^+.$$

Notice that $L_1, \ldots, L_d$ are isometries with ranges $\operatorname{Ran} L_i = \operatorname{span}\{\xi_{iw} : w \in \mathbb{F}_d^+\}$, which are pairwise orthogonal. However $\xi_\varnothing$ is orthogonal to all of them. So

$$I - \sum_{i=1}^{d} L_iL_i^* = \xi_\varnothing\xi_\varnothing^*.$$

This is the rank 1 projection onto $\mathbb{C}\xi_\varnothing$. Let $\mathcal{E}_d = C^*(L_1, \ldots, L_d)$.

We first show that $\mathcal{K}(F_d^2) \subset \mathcal{E}_d$. Indeed, $L_v\xi_\varnothing\xi_\varnothing^*L_w^* = \xi_v\xi_w^*$ for $v, x \in \mathbb{F}_d^+$ form a set of matrix units for the compact operators. Let $q : \mathcal{E}_d \to \mathcal{E}_d/\mathcal{K}(F_d^2)$ be the quotient map. Then

$$\sum_{i=1}^{d} q(L_i)q(L_i)^* = q(I - \xi_\varnothing\xi_\varnothing^*) = 1 \quad \text{and} \quad q(L_i^*)q(L_j) = \delta_{ij}q(I) = \delta_{ij}.$$

Thus $\mathcal{E}_d/\mathcal{K}(F_d^2)$ is isomorphic to $\mathcal{O}_d$. Therefore

$$0 \longrightarrow \mathcal{K}(F_d^2) \longrightarrow \mathcal{E}_d \longrightarrow \mathcal{O}_d \longrightarrow 0.$$

The algebra $\mathcal{E}_d$ is called the *Cuntz–Toeplitz algebra*.

Let π be a representation of $\mathcal{E}_d$. By Theorem 9.8.2, there is a decomposition $\pi = \pi_a \oplus \pi_c$, where π_a is induced from a representation of $\mathcal{K}(F_d^2)$ and π_c factors through $\mathcal{O}_d$. But every representation of $\mathcal{K}(F_d^2)$ is a multiple of the identity representation. Therefore $\pi \simeq id^{(\alpha)} \oplus \pi_c$.

9.11.9. EXAMPLE. Let $u = i_k i_{k-1} \dots i_1 \in \mathbb{F}_d^+$ with $k \geq 1$ and let $\lambda \in \mathbb{T}$. Let $K = \mathbb{C}^k \oplus H^{(k(d-1))}$ where $H = F_d^2$. Fix a basis $\eta_1, \dots, \eta_k$ for $\mathbb{C}^k$. For each $1 \leq j \leq k$, index $d-1$ copies of F_d^2 as $H_{j,s}$ for $1 \leq s \leq d$, $s \neq i_j$. Let $H_{j,s}$ have orthonormal basis $\xi_{j,s,w}$ for $w \in \mathbb{F}_d^+$. Define isometries $S_1, \dots, S_d$ by

$$\begin{aligned} S_i \eta_j &= \begin{cases} \eta_{j+1} & \text{if } 1 \leq j < k \text{ and } i = i_j \\ \lambda \eta_1 & \text{if } j = k \text{ and } i = i_k \\ \xi_{j,i,\varnothing} & \text{if } i \neq i_j \end{cases} \\ S_i \xi_{j,s,w} &= \ \xi_{j,s,iw} \qquad \text{for } 1 \leq j \leq k,\ 1 \leq s \leq d \text{ and } w \in \mathbb{F}_d^+. \end{aligned}$$

A little thought shows that each S_i is an isometry and their ranges are spanned by disjoint subsets of the basis and every basis vector is in the range of some S_i. This shows that $S_1, \dots, S_d$ generate a copy of $\mathcal{O}_d$. For future reference, we call this representation $\sigma_{u,\lambda}$.

9.11.10. EXAMPLE. Let $H = L^2(0,1)$ and fix $\lambda \in L^\infty(0,1)$ with $|\lambda(t)| = 1$ a.e. Define isometries $S_1, \dots, S_d$ in $\mathcal{B}(H)$ by

$$S_i f(t) = \begin{cases} \lambda(t)\sqrt{d}\, f(dt + 1 - i) & \text{if } \frac{i-1}{d} \leq t \leq \frac{i}{d} \\ 0 & \text{otherwise.} \end{cases}$$

It is easy to check that these are isometries with $\operatorname{Ran} S_i = L^2(\frac{i-1}{d}, \frac{i}{d})$. This shows that $S_1, \dots, S_d$ generate a copy of $\mathcal{O}_d$.

9.12. Group C*-algebras

In this section, let G be a discrete group. We will see how we can use C*-algebras to study their unitary representations. There is a version of this theory for locally compact groups, but we will not study it here.

9.12.1. DEFINITION. If G is a discrete group, a *unitary representation* is a homomorphism $\sigma : G \to \mathcal{U}(H)$, where $\mathcal{U}(H)$ is the group of unitary operators acting on a Hilbert space H. The *left regular representation* is the map

$$\lambda : G \to \mathcal{U}(l_2(G)) \quad \text{given by} \quad \lambda(g)\delta_h = \delta_{gh} \quad \text{for} \quad g, h \in G$$

where $\{\delta_h : h \in G\}$ is the standard orthonormal basis for $l_2(G)$. The *right regular representation* is the map $\rho : G \to \mathcal{U}(l_2(G))$ given by $\rho(g)\delta_h = \delta_{hg^{-1}}$.

We have already seen in the Banach algebra chapter that one can form a natural $*$-algebra from G, at least for LCA groups.

9.12.2. DEFINITION. If G is a discrete group, the *group algebra* $\mathbb{C}G$ is the set of all *finite* sums $\sum_G a_g g$, for $a_g \in \mathbb{C}$, with G forming a basis, and multiplication extending the multiplication on G:

$$\Big(\sum_{g\in G} a_g g\Big)\Big(\sum_{h\in G} b_h h\Big) = \sum_{g\in G}\sum_{h\in G} a_g b_h gh.$$

Make it into a $*$-algebra by setting $\big(\sum_{g\in G} a_g g\big)^* = \sum_{g\in G} \overline{a_g} g^{-1}$. Similarly we define $l_1(G)$ to be the Banach $*$-algebra obtained as the completion of $\mathbb{C}G$ in the norm

$$\Big\|\sum_{g\in G} a_g g\Big\|_1 = \sum_{g\in G} |a_g|.$$

9.12.3. PROPOSITION. *Let H be a Hilbert space. There is a natural bijection (by restriction) between*

(1) *unitary representations of G into $\mathcal{U}(H)$,*

(2) *$*$-representations of $\mathbb{C}G$ into $\mathcal{B}(H)$, and*

(3) *$*$-representations of $l_1(G)$ into $\mathcal{B}(H)$.*

Moreover every $*$-representation of $l_1(G)$ has norm 1.

PROOF. Clearly a $*$-representation of $l_1(G)$ restricts to a $*$-representation of $\mathbb{C}G$. If σ is a $*$-representation of $\mathbb{C}G$, then $\sigma(g)^{-1} = \sigma(g^{-1}) = \sigma(g)^*$, which shows that $\sigma(g)$ is unitary. Thus it restricts to a unitary representation of G. Finally, if σ is a unitary representation of G, then

$$\sigma\Big(\sum_{g\in G} a_g g\Big) = \sum_{g\in G} a_g \sigma(g)$$

is readily seen to be a $*$-representation of $l_1(G)$. Moreover $\|\sigma(g)\| = 1 = \|g\|$ and

$$\Big\|\sigma\Big(\sum_{g\in G} a_g g\Big)\Big\| \le \sum_{g\in G} |a_g|\|\sigma(g)\| = \sum_{g\in G} |a_g| = \Big\|\sum_{g\in G} a_g g\Big\|_1.$$

Thus σ has norm 1. ∎

9.12.4. DEFINITION. If G is a discrete group, the *universal C*-algebra of G*, $C^*(G)$, is the completion of $\mathbb{C}G$ in the norm

$$\Big\|\sum_{g\in G} a_g g\Big\|_u = \sup_\sigma \Big\|\sigma\Big(\sum_{g\in G} a_g g\Big)\Big\|,$$

where the supremum is taken over all $*$-representations of $\mathbb{C}G$ into $\mathcal{B}(H)$. The preceding proposition shows that this supremum is at most $\big\|\sum_{g\in G} a_g g\big\|_1$.

The *reduced C*-algebra* of G, $C_\lambda(G)$, is the completion of $\mathbb{C}G$ in the norm

$$\Big\|\sum_{g\in G} a_g g\Big\|_\lambda = \Big\|\sum_{g\in G} a_g \lambda(g)\Big\|.$$

Similarly, we define $C^*_\rho(G)$. More generally we can define $C^*_\sigma(G)$ for any unitary representation σ of G.

The *von Neumann algebra of* G is $\mathcal{L}(G) = C^*_\lambda(G)''$ and similarly $\mathcal{R}(G) = C^*_\rho(G)''$.

9.12.5. Remark. Any $*$-representation of $\mathbb{C}G$ splits as a direct sum of cyclic representations, and these will act on a Hilbert space of dimension at most $|G|$. Thus in order to ensure that we are taking the supremum over a set, we consider representations into a fixed collection of Hilbert spaces H_n for $n \le |G|$ with $\dim H_n = n$. If we then take the direct sum σ_u of all of the representations into these spaces, we obtain a universal representation of $\mathbb{C}G$ such that $\big\|\sum_{g\in G} a_g g\big\|_u = \big\|\sigma_u\big(\sum_{g\in G} a_g g\big)\big\|$. Thus $C^*(G) \simeq C^*(\sigma_u(G))$.

9.12.6. Proposition. *Let G be a discrete group. Then for any unitary representation σ of G, there is a unique map of $C^*(G)$ onto $C^*_\sigma(G)$ taking the generators $\sigma_u(g)$ onto $\sigma(g)$. In particular, $C^*_\lambda(G)$ is a quotient of $C^*(G)$.*

The norm on $C^*(G)$ is obtained by taking the supremum over all irreducible unitary representations.

Proof. By definition, the norm $\|\cdot\|_u$ dominates $\|\cdot\|_\sigma$. Thus the identity map on $\mathbb{C}G$ considered as a subspace of $C^*(G)$ to $C^*_\sigma(G)$ is contractive. Hence it extends by continuity to a unique $*$-homomorphism of $C^*(G)$ onto the C*-algebra generated by $\sigma(G)$, namely $C^*_\sigma(G)$.

A representation of $C^*(G)$ is irreducible if and only if the restriction to the generators, G, is irreducible. By the Gelfand–Naimark theorem 9.7.7, it suffices to take the supremum over the irreducible representations to recover the norm. ∎

9.12.7. Remark. $\mathcal{R}(G) \subset \mathcal{L}(G)'$ and $\mathcal{L}(G) \subset \mathcal{R}(G)'$. Indeed,

$$\lambda(g)\rho(h)\delta_k = \delta_{gkh} = \rho(h)\lambda(g)\delta_k \quad \text{for} \quad g,h,k \in G.$$

By linearity and continuity, $C^*_\sigma(G)$ and $C^*_\rho(G)$ commute. Then extending this to the WOT closures completes the claim. We will show in Theorem 10.10.12 that actually $\mathcal{R}(G) = \mathcal{L}(G)'$ and $\mathcal{L}(G) = \mathcal{R}(G)'$.

The reduced C*-algebra $C^*_\lambda(G)$ and the group von Neumann algebra $\mathcal{L}(G)$ always have a trace, known as the *standard trace*, defined in the following proposition.

9.12.8. PROPOSITION. *Let G be a discrete group. Then $C^*_\lambda(G)$ has a faithful trace τ defined on $\mathbb{C}G$ by*

$$\tau\big(\lambda(\sum a_g g)\big) = \big\langle \sum a_g \lambda(g)\delta_e, \delta_e \big\rangle = a_e.$$

Moreover, τ extends to a WOT*-continuous faithful trace on $\mathcal{L}(G)$.*

PROOF. Now $\tau(T) = \langle T\delta_e, \delta_e\rangle$ is a WOT-continuous state on $\mathcal{B}(l_2(G))$. In particular, it is a state on $C^*_\lambda(G)$ and a WOT-continuous state on $\mathcal{L}(G)$.

For $a = \sum a_g g$ in $\mathbb{C}G$, it is easy to verify that $\langle \sum a_g\lambda(g)\delta_e, \delta_e\rangle = a_e$. Let $b = \sum b_g g$ be another element of $\mathbb{C}G$. We have

$$\tau(ab) = \tau\big(\sum_{g,h\in G} a_g b_h gh\big) = \sum_{g\in G} a_g b_{g^{-1}}$$

and

$$\tau(ba) = \tau\big(\sum_{g,h\in G} b_h a_g hg\big) = \sum_{g\in G} b_{g^{-1}} a_g = \tau(ab).$$

Thus τ is a trace on $\mathbb{C}G$. The trace property extends by continuity to $C^*_\lambda(G)$.

If $b \in \mathcal{L}(G)$, Kaplansky's density theorem shows that there is a bounded net $(b_\lambda)_\Lambda$ in $C^*_\lambda(G)$ converging WOT to b. If $a \in C^*_\lambda(G)$, then

$$\tau(ab) = \lim_\Lambda \tau(ab_\lambda) = \lim_\Lambda \tau(b_\lambda a) = \tau(ba).$$

Now we can repeat the argument, taking a WOT limit of $(a_\mu)_{\Lambda'}$ in $C^*_\lambda(G)$ converging WOT to $a \in \mathcal{L}(G)$. It follows that $\tau(ab) = \tau(ba)$ for $a, b \in \mathcal{L}(G)$.

It is enough to check that τ is faithful on $\mathcal{L}(G)$. Let $b \gneq 0$ in $\mathcal{L}(G)$ and factor $b = a^*a$. If $a\delta_e = 0$, then

$$a\delta_g = a\rho(g)\delta_e = \rho(g)a\delta_e = 0 \quad \text{for all} \quad g \in G;$$

and thus $a = 0$. Since $b \neq 0$, we must have $a\delta_e \neq 0$. Therefore

$$\tau(b) = \langle a^*a\delta_e, \delta_e\rangle = \|a\delta_e\|^2 > 0.$$

That is, τ is faithful. ∎

For the case of abelian G, we can build on what we learned about $L^1(G)$ in Chapter 7.

9.12.9. THEOREM. *Let G be a discrete abelian group. Then $C^*(G) = C^*_\lambda(G) = C(\widehat{G})$, where $\widehat{G}$ is the (compact) dual group. Moreover $\mathcal{L}(G) = \mathcal{R}(G) = L^\infty(m)$ acting on $L^2(m)$, where m is the Haar measure on $\widehat{G}$.*

PROOF. Since G is abelian, $C^*(G)$ is a commutative C*-algebra. Now $C^*(G)$ is determined by its irreducible representations, which are the one-dimensional representations associated to the characters of G, namely $\widehat{G}$. By Theorem 7.5.1, $\widehat{G}$ is a compact group. Theorem 7.4.5 shows that the Gelfand transform of $l_1(G)$ is

$A(\widehat{G}) \subset C(\widehat{G})$. The algebra $A(\widehat{G})$ is self-adjoint and separates points, and thus the completion in this norm is $C(\widehat{G})$ by Stone–Weierstrass. Therefore $C^*(G) \simeq C(\widehat{G})$.

Let m be the Haar measure on $\widehat{G}$, normalized so that $m(\widehat{G}) = 1$. Define a map $U : \mathbb{C}G \to L^2(m)$ by $Uf = \hat{f}$, the Fourier transform of f. We consider $\mathbb{C}G$ as a (dense) subspace of $l_2(G)$. Then for any $g, h \in G$ and $\gamma_1 \in \widehat{G}$, using translation invariance of m,

$$\begin{aligned}\langle U\delta_g, U\delta_h\rangle &= \int_{\widehat{G}} \hat{g}(\gamma)\overline{\hat{h}}(\gamma)\,dm(\gamma) = \int_{\widehat{G}} \widehat{gh^{-1}}(\gamma)\,dm(\gamma)\\ &= \int_{\widehat{G}} \widehat{gh^{-1}}(\gamma\gamma_1)\,dm(\gamma\gamma_1) = \int_{\widehat{G}} \widehat{gh^{-1}}(\gamma\gamma_1)\,dm(\gamma)\\ &= \widehat{gh^{-1}}(\gamma_1)\int_{\widehat{G}} \widehat{gh^{-1}}(\gamma)\,dm(\gamma) = \widehat{gh^{-1}}(\gamma_1)\langle U\delta_g, U\delta_h\rangle.\end{aligned}$$

If $g = h$, then we obtain $\|U\delta_g\|^2 = 1$. If $g \neq h$, there is some character $\gamma_1 \in \widehat{G}$ such that $\widehat{gh^{-1}}(\gamma_1) \neq 1$; so it follows that $\langle U\delta_g, U\delta_h\rangle = 0$. Thus U carries an orthonormal basis to an orthonormal set in $L^2(m)$. Hence U extends by continuity to an isometry onto the closure of $A(\widehat{G})$ in $L^2(m)$. Now $A(\widehat{G})$ is uniformly dense in $C(\widehat{G})$, and $C(\widehat{G})$ is dense in $L^2(m)$. Hence U is a unitary operator taking $l_2(G)$ onto $L^2(m)$. This is a special case of the Plancherel theorem for LCA groups.

Now calculate for $g, h \in G$,

$$U\lambda(g)U^*\hat{h} = U\lambda(g)\delta_h = U\delta_{gh} = \widehat{gh} = M_{\hat{g}}\hat{h}.$$

Therefore $U\lambda(g)U^*$ agrees with the multiplication operator $M_{\hat{g}}$ on an orthonormal basis. Since both are continuous, they must be equal. Hence by linearity, we get $U\lambda(f)U^* = M_{\hat{f}}$ for all $f \in \mathbb{C}G$. Since $A(\widehat{G})$ is dense in $C(\widehat{G})$, conjugation by U carries $C^*_\lambda(G)$ onto $\{M_f : f \in C(\widehat{G})\}$. Passing to the WOT closure, we see that $U\mathcal{L}(G)U^* = L^\infty(m)$ acting on $L^2(m)$. The support of m is $\widehat{G}$ because every open set has a positive measure. Hence $UC^*_\lambda(G)U^* = C(\widehat{G})$ acting by multiplication. Therefore the quotient map from $C^*(G)$ onto $C^*_\lambda(G)$ is an isomorphism. ∎

9.12.10. EXAMPLE. Let S_∞ be the group of finite permutations of $\mathbb{N}$. We can write $S_\infty = \bigcup_{n\geq 1} S_n$ where S_n is the permutation group on the first n elements of $\mathbb{N}$, considered as fixing the remaining elements. Thus $\mathbb{C}S_\infty = \bigcup_{n\geq 1} \mathbb{C}S_n$ is an increasing union of finite dimensional subalgebras.

Since $\mathbb{C}S_n$ is a finite dimensional C*-algebra, it has a unique C*-algebra norm. In particular, $C^*(S_n) = C^*_\lambda(S_n)$ because the representation λ is faithful on the group ring. Therefore

$$C^*(S_\infty) = \varinjlim_{n\to\infty} C^*(S_n) = \varinjlim_{n\to\infty} C^*_\lambda(S_n) = C^*_\lambda(S_\infty).$$

Moreover, this C*-algebra is evidently AF (the closure of an increasing union of finite dimensional subalgebras).

9.12.11. DEFINITION. The *trivial representation* τ_0 of $C^*(G)$ is the 1-dimensional representation determined by the character $\tau_0(g) = 1$ for all $g \in G$.

9.12.12. EXAMPLE. Now we look at a noncommutative example in detail. Let $\mathbb{F}_S$ be the free group on a set of generators S of cardinality $\kappa \geq 2$, possibly infinite. This group consists of all words in $S \cup S^{-1}$ subject only to the relations $ss^{-1} = e = s^{-1}s$. A word is called *reduced* if there are no adjacent pairs ss^{-1} or $s^{-1}s$ for $s \in S$. Every word is equivalent to a unique reduced word. The *length* $|g|$ of a group element counts the number of characters in the reduced word. The empty word $\varnothing$ is the identity element e.

The free groups have many unitary representations. To see this, note first that for any unitaries $U_s \in \mathcal{B}(H)$ for $s \in S$ and H of any dimension, there is a unitary representation σ with $\sigma(s) = U_s$. This is because there are no relations to satisfy in the group. In particular, there are irreducible representations on any space of dimension at most countable. For example, in dimension $n < \infty$, let

$$U_1 e_j = e^{2j\pi i/n} e_j \quad \text{and} \quad U_2 e_j = e_{j+1 (\text{mod } n)} \quad \text{for} \quad 1 \leq j \leq n.$$

The eigenvalues of U_1 are the basis vectors e_j, and so its reducing subspaces are spanned by a subset of this basis. However, powers of U_2 move each basis vector to all of the others. So $\{U_1, U_2\}$ is irreducible. We can define the image of the remaining generators arbitrarily. On $l_2(\mathbb{Z})$, we can take $U_1 e_j = e^{ij} e_j$ and $U_2 e_j = e_{j+1}$ for $j \in \mathbb{Z}$. The argument for irreducibility is similar.

The finite dimensional unitary representations suffice to norm $C^*(\mathbb{F}_S)$. To establish this, let σ be a unitary representation of $\mathbb{F}_S$ on $\mathcal{B}(H)$, for an infinite dimensional H. Let $a \in \mathbb{C}\mathbb{F}_S$ and $\xi \in H$ with $\|\xi\| = 1$. It suffices to construct a finite dimensional representation π on a finite dimensional subspace $M \subset H$ containing ξ so that $\|\pi(a)\xi\| = \|\sigma(a)\xi\|$. Write $a = \sum a_g g$ where $a_g \neq 0$ for only finitely many words in $S \cup S^{-1}$. Then there is a finite subset $F \subset S$ so that these are words in $F \cup F^{-1}$ of length at most n. Let $N = \operatorname{span}\{\sigma(g)\xi : g \in \mathbb{F}_F,\ |g| \leq n\}$. Let N' be a subspace orthogonal to N of the same dimension; and let V be a partial isometry such that $V^*V = P_N$ and $VV^* = P_{N'}$. For each $s \in S$, let $A_s = P_N \sigma(s)|_N$. Define operators on $M = N \oplus N'$ by

$$U_s = \begin{bmatrix} A_s & (I_N - A_s A_s^*)^{1/2} V^* \\ V(I_N - A_s^* A_s)^{1/2} & -V A_s V^* \end{bmatrix}$$

where this 2×2 matrix acts on $N \oplus N'$. A simple calculation shows that these are unitary matrices. Hence $\pi(s) = U_s$ extends to a unitary representation of $\mathbb{F}_S$ on M. By construction, for every word $g = s_k \dots s_1$ in $F \cup F^{-1}$ of length at most n, $\sigma(g)\xi \in N$. Now ξ and $\sigma(s_1)\xi \in N$, and $\|\sigma(s_1)\xi\| = \|\xi\| = 1$. Thus $\pi(s_1)\xi = \sigma(s_1)\xi$ since $\pi(s_1)$ is isometric and restricts to σ on N. Then $\sigma(s_2)\sigma(s_1)\xi \in N$ and $\pi(s_2)\pi(s_1)\xi = \sigma(s_2)\sigma(s_1)\xi$. By induction, we see that

$\pi(g)\xi = \sigma(g)\xi$. Therefore $\pi(a)\xi = \sigma(a)\xi$. Thus

$$\|\sigma(a)\| = \sup_{\|\xi\|=1} \|\sigma(a)\xi\| \le \sup_{\pi \text{ f.d.}} \|\pi(a)\|.$$

In particular, if $s_1, \dots, s_k \in S \cup S^{-1}$, then $\|s_1 + \cdots + s_k\| = k$ because this norm is obtained from the trivial representation τ_0, and is clearly an upper bound.

The character of $C^*_\lambda(\mathbb{F}_S)$ is quite different. The following is known as *Powers' averaging technique.*

9.12.13. Lemma. *Let $u \in S$. Then for $g \in \mathbb{F}_S$,*

$$\lim_{n\to\infty} \frac{1}{2n+1} \sum_{i=-n}^{n} \lambda(u^i)\lambda(g)\lambda(u^{-i}) = \begin{cases} \lambda(g) & \text{if } g = u^k \text{ for } k \in \mathbb{Z} \\ 0 & \text{otherwise.} \end{cases}$$

Proof. Let $M = \operatorname{span}\{\delta_g : g = e \text{ or } g = u^{\pm 1}h \text{ in reduced form}\}$. For any $k \ne 0$, $\lambda(u^k)M^\perp \subset M$; while if $v, w \in S \cup S^{-1} \setminus \{u, u^{-1}\}$ and $g = vhw$ or $g = v$ in reduced form, for any $h \in \mathbb{F}_S$, then $\lambda(g)M \subset M^\perp$. In matrix form with respect to $H = M \oplus M^\perp$, we have

$$\lambda(u^k) = \begin{bmatrix} * & * \\ * & 0 \end{bmatrix} \quad \text{and} \quad \lambda(vhw) = \begin{bmatrix} 0 & * \\ * & * \end{bmatrix}.$$

If $A = \begin{bmatrix} 0 & 0 \\ * & * \end{bmatrix}$ and $C = \begin{bmatrix} * & * \\ 0 & 0 \end{bmatrix}$, then $A^*C = C^*A = 0$, so that

$$\|A + C\|^2 = \|A^*A + C^*C\| \le \|A\|^2 + \|C\|^2.$$

Now

$$\lambda(u^k)A\lambda(u^{-k}) = \begin{bmatrix} * & * \\ * & 0 \end{bmatrix} \begin{bmatrix} 0 & 0 \\ * & * \end{bmatrix} \begin{bmatrix} * & * \\ * & 0 \end{bmatrix} = \begin{bmatrix} * & * \\ 0 & 0 \end{bmatrix}.$$

Thus for $m < n$,

$$\Big\| \sum_{i=m}^{n} \lambda(u^i)A\lambda(u^{-i}) \Big\|^2 = \Big\| \sum_{i=0}^{n-m} \lambda(u^i)A\lambda(u^{-i}) \Big\|^2 = \Big\| A + \sum_{i=1}^{n-m} \lambda(u^i)A\lambda(u^{-i}) \Big\|^2$$
$$\le \|A\|^2 + \Big\| \sum_{i=1}^{m-n} \lambda(u^i)A\lambda(u^{-i}) \Big\|^2$$

because the sum has the form of C. By induction, we deduce that

$$\Big\| \sum_{i=m}^{n} \lambda(u^i)A\lambda(u^{-i}) \Big\|^2 \le (n - m + 1)\|A\|^2.$$

By taking adjoints, the same holds for B of the form $B = A^* = \begin{bmatrix} 0 & * \\ 0 & * \end{bmatrix}$. Now $\lambda(vhw)$ can be written as the sum of two contractions, one of the form of A and the other of the form of B. We deduce that

$$\Big\|\frac{1}{2n+1}\sum_{i=-n}^{n}\lambda(u^i)\lambda(vhw)\lambda(u^{-i})\Big\| \le \frac{2\sqrt{2n+1}}{2n+1} = \frac{2}{\sqrt{2n+1}}.$$

Finally, note that for $\lambda(u^k vhwu^l)$, factoring out the $\lambda(u^k)$ and $\lambda(u^l)$ yields

$$\Big\|\frac{1}{2n+1}\sum_{i=-n}^{n}\lambda(u^k)\lambda(u^i)\lambda(vhw)\lambda(u^{-i})\lambda(u^l)\Big\| \le \frac{2}{\sqrt{2n+1}}.$$

Therefore

$$\lim_{n\to\infty}\frac{1}{2n+1}\sum_{i=-n}^{n}\lambda(u^i)\lambda(g)\lambda(u^{-i}) = 0 \quad \text{for} \quad g \notin \{u^k : k \in \mathbb{Z}\}.$$

Clearly $\frac{1}{2n+1}\sum_{i=-n}^{n}\lambda(u^i)\lambda(u^k)\lambda(u^{-i}) = \lambda(u^k)$ for $k \in \mathbb{Z}$. So the result follows. ∎

9.12.14. REMARK. In $C^*(\mathbb{F}_S)$ for $|S| \ge 2$, $\|\sum_{i=-2}^{2} u^i v u^{-i}\| = 5$ follows from the trivial representation τ_0. This lemma shows that

$$\Big\|\sum_{i=-2}^{2}\lambda(u^i)\lambda(v)\lambda(u^{-i})\Big\| \le 2\sqrt{5} < 5.$$

Hence $C^*_\lambda(\mathbb{F}_S)$ is a proper quotient of $C^*(\mathbb{F}_S)$.

In the next section, we study when $C^*(G) = C^*_\lambda(G)$.

We finish this section with another result that clearly distinguishes $C^*_\lambda(\mathbb{F}_S)$ from $C^*(\mathbb{F}_S)$. Recall that $C^*(\mathbb{F}_S)$ is far from simple, since it has irreducible representations of all finite and countable dimensions.

9.12.15. THEOREM. *For $|S| \ge 2$, $C^*_\lambda(\mathbb{F}_S)$ is simple and has a unique trace.*

PROOF. Suppose that $a = \sum a_g g \in \mathbb{C}\mathbb{F}_S$. Fix two generators $u, v \in S$. Then by the lemma,

$$\lim_{n\to\infty}\frac{1}{2n+1}\sum_{i=-n}^{n}\lambda(u^i)\lambda(a)\lambda(u^{-i}) = \sum_k a_{u^k}\lambda(u^k).$$

Therefore

$$\lim_{m\to\infty}\lim_{n\to\infty}\frac{1}{(2m+1)(2n+1)}\sum_{j=-m}^{m}\sum_{i=-n}^{n}\lambda(v^j)\lambda(u^i)\lambda(a)\lambda(u^{-i})\lambda(v^{-j}) = a_e I.$$

These averages are contractive operations, and thus this formula extends by continuity to $C^*_\lambda(\mathbb{F}_S)$ to provide a contractive linear map $\tau : C^*_\lambda(\mathbb{F}_S) \to \mathbb{C}$. This is readily seen to coincide with the standard trace $\tau(\lambda(a)) = \langle \lambda(a)\delta_e, \delta_e\rangle$.

If τ' is a trace, then for $a \in \mathbb{CF}_S$,

$$\tau'(\lambda(a)) = \frac{1}{(2m+1)(2n+1)} \sum_{j=-m}^{m} \sum_{i=-n}^{n} \tau'\big(\lambda(v^j)\lambda(u^i)\lambda(a)\lambda(u^{-i})\lambda(v^{-j})\big)$$

converges to $\tau'(a_e I) = a_e$. Hence $\tau' = \tau$, and the trace is unique.

Suppose that J is a non-zero ideal. Let $0 \neq a \in J$. Then $0 \lneq a^*a$. Since τ is faithful by Proposition 9.12.8, $\tau(a^*a) = \alpha > 0$. By the averaging technique above, we obtain elements

$$\frac{1}{(2m+1)(2n+1)} \sum_{j=-m}^{m} \sum_{i=-n}^{n} \lambda(v^j)\lambda(u^i)\lambda(a^*a)\lambda(u^{-i})\lambda(v^{-j})$$

in the ideal J converging in norm to αI. Thus $J = C^*_\lambda(\mathbb{F}_S)$. Therefore $C^*_\lambda(\mathbb{F}_S)$ is simple. ■

9.13. Amenability

There is a property of certain groups which seems to epitomize good behaviour. In this section and later in Theorem 10.10.20, we show that it characterizes certain nice properties of the group C*-algebra and von Neumann algebra.

9.13.1. Definition. A discrete group G is *amenable* if there is a left translation invariant state on $l^\infty(G)$; that is, a state m such that $m({}_g f) = m(f)$ for all f in $l^\infty(G)$ and $g \in G$, where ${}_g f(h) = f(g^{-1}h)$. This is a *left invariant mean* for G. Similarly, m is a *right invariant mean* if $m(f_g) = m(f)$ for $f \in l^\infty(G)$ and $g \in G$, where $f_g(h) = f(hg^{-1})$.

Define $\check{f}(g) = f(g^{-1})$ for $f \in l^\infty(G)$. If m is a left invariant mean, then $\tilde{m}(f) = m(\check{f})$ is a right invariant mean. We can do better!

9.13.2. Proposition. *If G is amenable, there is a state on $l^\infty(G)$ which is both left and right invariant.*

Proof. Let m be a left invariant mean on G. Define $T \in \mathcal{B}(l^\infty(G))$ by $(Tf)(h) = m(f_h)$. Then for $g \in G$,

$$T({}_g f)(h) = m({}_g f_h) = m(f_h) = Tf(h);$$

and

$$T(f_g)(h) = m(f_{gh}) = Tf(gh) = {}_{g^{-1}}Tf(h).$$

Define a state on $l^\infty(G)$ by $\mu(f) = m(Tf)$. By these formulae,

$$\mu(f) = \mu({}_g f) = \mu(f_g).$$

Therefore μ is both left and right invariant. ■

You can build new amenable groups from old ones using various standard constructions.

9.13.3. PROPOSITION. *Subgroups, quotients, extensions, and direct limits of amenable groups are amenable.*

PROOF. Suppose that G is amenable and H is a subgroup. Let m be an invariant mean on G. By the Axiom of Choice, select $A \subset G$ so that A intersects each *right* coset Hg in a single point. Define a $*$-homomorphism of $l^\infty(H)$ into $l^\infty(G)$ by $\pi(f)(ha) = f(h)$ for all $h \in H$ and $a \in A$. Compute for $h, h' \in H$ and $a \in A$,

$$\pi({}_hf)(h'a) = {}_hf(h') = f(h^{-1}h') = \pi(f)(h^{-1}h'a) = {}_h\pi(f)(h'a).$$

Then $\widetilde{m} = m\pi$ is a state on $l^\infty(H)$, and

$$\widetilde{m}({}_hf) = m(\pi({}_hf)) = m({}_h\pi(f)) = m(\pi(f)) = \widetilde{m}(f).$$

Hence $\widetilde{m}$ is an invariant mean for H.

If N is a normal subgroup, define $\pi : l^\infty(G/N) \to l^\infty(G)$ by $\pi(f)(g) = f(gN)$. An easy calculation shows that $\widetilde{m} = m\pi$ is an invariant mean for G/N.

Now suppose that N and G/N are amenable. Let m_N be a left invariant mean on N, and let $\widetilde{m}$ be a left invariant mean on G/N. Define a unital positive linear map $P : l^\infty(G) \to l^\infty(G/N)$ by $(Pf)(gN) = m_N({}_{g^{-1}}f|_N)$. This is well defined because if $gN = hN$, then $h = gn$ for $n \in N$ and thus ${}_{h^{-1}}f|_N = {}_{n^{-1}}({}_{g^{-1}}f|_N)$. Hence $m_N({}_{h^{-1}}f|_N) = m_N({}_{g^{-1}}f|_N)$, and therefore P is well defined. Observe that for $g \in G$,

$$P({}_gf)(hN) = m_N({}_{h^{-1}g}f|_N) = Pf(g^{-1}hN) = {}_{gN}(Pf)(hN).$$

Now define a state m on $l^\infty(G)$ by $m(f) = \widetilde{m}(Pf)$. Then m is left invariant.

Suppose that G is a direct limit of amenable groups. Since quotients are also amenable, we may assume that the connecting maps are injective. Thus we can consider $G = \bigcup_\Lambda G_\lambda$ where $G_\lambda \subset G_\mu$ if $\lambda \le \mu$ in the net Λ. Let m_λ be a left invariant mean for G_λ, $\lambda \in \Lambda$. Let $\rho_\lambda : l^\infty(G) \to l^\infty(G_\lambda)$ be the restriction map $\rho_\lambda(f) = f|_{G_\lambda}$, and set $\widetilde{m}_\lambda(f) = m_\lambda(\rho_\lambda f)$. Since the state space of $l^\infty(G)$ is weak-$*$ compact, we can obtain a weak-$*$ convergent cofinal subnet with limit m. Then if $g \in G_\mu$, then $m_\lambda({}_gf) = m_\lambda(f)$ for all $\lambda \ge \mu$; whence $m({}_gf) = m(f)$. Therefore m is a left invariant mean. ■

9.13.4. EXAMPLES.

(1) Every finite group G is amenable. The state $m(f) = \frac{1}{|G|}\sum_G f(g)$ is evidently left and right translation invariant.

(2) S_∞ is amenable because it is an inductive limit of finite groups.

(3) $\mathbb{F}_S$ is not amenable if $|S| \geq 2$. Let $u, v \in S$ be distinct. Define

$$A_0 = \{u^{2n}g : n \in \mathbb{Z},\ g = e \text{ or } g = s^{\pm 1}h \text{ in reduced form for } s \in S \setminus \{u\}\}.$$

Then $A_1 = uA_0$ is disjoint from A_0 and $A_0 \cup A_1 = \mathbb{F}_S$. Hence

$$\chi_{A_1} = {}_u\chi_{A_0} \quad \text{and} \quad \chi_{A_0} + \chi_{A_1} = 1.$$

Similarly, let

$$B_0 = \{v^{3n}g : n \in \mathbb{Z},\ g = e \text{ or } g = s^{\pm 1}h \text{ in reduced form for } s \in S \setminus \{v\}\}.$$

Then B_0, $B_1 = vB_0$ and $B_2 = v^2 B_0$ are disjoint and $B_0 \cup B_1 \cup B_2 = \mathbb{F}_S$. Thus

$$\chi_{B_1} = {}_{v^{-1}}\chi_{B_0}, \quad \chi_{B_2} = {}_{v^{-2}}\chi_{B_0} \quad \text{and} \quad \chi_{B_0} + \chi_{B_1} + \chi_{B_2} = 1.$$

If there were a left invariant mean m on $\mathbb{F}_S$, then $m(\chi_{A_1}) = \frac{1}{2}$ and $m(\chi_{B_1}) = \frac{1}{3}$. But $vA_1 \subset B_1$, so that ${}_{v^{-1}}\chi_{A_1} \leq \chi_{B_1}$. Therefore $\frac{1}{2} \leq \frac{1}{3}$, a contradiction. Hence $\mathbb{F}_S$ is not amenable.

(4) For $\mathbb{Z}$, one can take any Banach limit on $l^\infty(\mathbb{Z})$ (Example 2.6.22).

In fact, every abelian group G is amenable.

9.13.5. Proposition. *If $C^*_\lambda(G)$ has a character, then G is amenable.*

Proof. Suppose that τ is a $*$-homomorphism of $C^*_\lambda(G)$ into $\mathbb{C}$. Extend it to a state on $\mathcal{B}(l_2(G))$ by Hahn–Banach. If $g \in G$,

$$|\tau(\lambda(g))|^2 = \tau(\lambda(g^{-1}))\tau(\lambda(g)) = \tau\big(\lambda(g)\lambda(g^{-1})\big) = \tau(1).$$

By Exercise **9.20**, $\tau(A\lambda(g)) = \tau(A)\tau(\lambda(g)) = \tau(\lambda(g)A)$ for all A in $\mathcal{B}(l_2(G))$. Thus $\tau(\lambda(g)A\lambda(g^{-1})) = \tau(A)$. Imbed $l^\infty(G)$ into $\mathcal{B}(l_2(G))$ as diagonal operators $M_f\delta_g = f(g)\delta_g$. Define m as the restriction of τ to $l^\infty(G)$. This is a state such that

$$m({}_g f) = \tau(\lambda(g)M_f\lambda(g)^*) = m(f).$$

Therefore m is an invariant mean; and so G is amenable. ∎

Recall that a group G is solvable if there is a subnormal series $\{e\} = G_0 \lhd G_1 \lhd \cdots \lhd G_n = G$ such that G_{i-1} is a normal subgroup of G_i and G_i/G_{i-1} is abelian for $1 \leq i \leq n$.

9.13.6. Corollary. *Solvable groups are amenable.*

Proof. If G is abelian, then $C^*(G)$ is commutative and thus has a character. So abelian groups are amenable. By Proposition 9.13.3, an extension of an amenable group by an amenable group is amenable. By induction, all solvable groups are amenable. ∎

9.13.7. Definition. A bounded representation of a group G is a homomorphism π of G into $\mathcal{B}(H)^{-1}$ such that $\|\pi\| = \sup_{g\in G} \|\pi(g)\| < \infty$.

One early application of amenability is the following:

9.13.8. Dixmier Theorem. *Let G be a discrete amenable group. Suppose that $\pi : G \to \mathcal{B}(H)^{-1}$ is a bounded representation. Then there is an invertible operator $S \in \mathcal{B}(H)$ with $\|S\|\,\|S^{-1}\| \le \|\pi\|^2$ such that $\sigma(g) = S\pi(g)S^{-1}$ is a unitary representation.*

Proof. Let m be a left invariant mean for G. For $\xi, \eta \in H$, define

$$F_{\xi,\eta}(g) = \langle \pi(g^{-1})\xi, \pi(g^{-1})\eta\rangle \quad \text{and} \quad [\xi,\eta] = m(F_{\xi,\eta}).$$

This is a sesquilinear form on H which is bounded since

$$|[\xi,\eta]| \le \sup_{g\in G} \|\pi(g^{-1})\|^2 \|\xi\|\,\|\eta\| = \|\pi\|^2 \|\xi\|\,\|\eta\|;$$

and is positive definite since

$$[\xi,\xi] = m(\|\pi(g^{-1})\xi\|^2) \ge \inf_{g\in G} \|\pi(g)\|^{-2}\|\xi\|^2 = \|\pi\|^{-2}\|\xi\|^2.$$

By Proposition 4.2.2, there is a unique operator $T \in \mathcal{B}(H)$ so that $[\xi,\eta] = \langle \xi, T\eta\rangle$. Moreover

$$\|\pi\|^{-2}\|\xi\|^2 \le [\xi,\xi] = \langle \xi, T\xi\rangle \le \|\pi\|^2\|\xi\|^2$$

shows that $T \ge 0$ and $\|\pi\|^{-2}I \le T \le \|\pi\|^2 I$. Define $S = T^{1/2}$. Then

$$[\xi,\eta] = \langle \xi, T\eta\rangle = \langle S\xi, S\eta\rangle \quad \text{and} \quad \|S\|\,\|S^{-1}\| \le \|\pi\|^2.$$

For $h \in G$ and $\xi, \eta \in H$,

$${}_hF_{\xi,\eta}(g) = F_{\xi,\eta}(h^{-1}g) = \langle \pi(g^{-1}h)\xi, \pi(g^{-1}h)\eta\rangle = F_{\pi(h)\xi,\pi(h)\eta}(g).$$

Thus

$$\begin{aligned}\langle S\pi(h)S^{-1}\xi, S\pi(h)S^{-1}\eta\rangle &= [\pi(h)S^{-1}\xi, \pi(h)S^{-1}\eta] = m(F_{\pi(h)S^{-1}\xi,\pi(h)S^{-1}\eta}) \\ &= m({}_hF_{S^{-1}\xi,S^{-1}\eta}) = [S^{-1}\xi, S^{-1}\eta] = \langle \xi,\eta\rangle.\end{aligned}$$

Hence $\sigma(h) := S\pi(h)S^{-1}$ preserves the inner product and is invertible. Therefore $\sigma(h)$ is unitary for every $h \in G$. ∎

We now define some other averaging notions which will be shown to be equivalent to amenability.

9.13.9. Definition. Let $\mathrm{FPr}(G)$ be the set of *finitely supported probability measures* on G, which we consider as the norm 1 elements of $l_1(G)^+ \cap \mathbb{C}G$. These measures act on $l^\infty(G)$ as linear functionals. We define $(a \cdot \mu)(f) = \mu({}_af)$ for $a \in G$; so $(a \cdot \mu)(g) = \mu(ag)$. Say G has an *approximate invariant mean* if for $A \subset G$ finite and $\varepsilon > 0$, there is $\mu \in \mathrm{FPr}(G)$ with $\|\mu - a \cdot \mu\|_1 < \varepsilon$ for $a \in A$.

Say G satisfies the *Følner condition* if for $A \subset G$ finite and $\varepsilon > 0$, there is a finite $F \subset G$ so that $|F \triangle aF| < \varepsilon |F|$ for $a \in A$.

Say that τ_0 is *weakly contained* in λ if $l^2(G)$ has *almost invariant vectors*: there are unit vectors $\{\xi_\alpha\}_\Lambda \subset l^2(G)$ so that $\lim_\Lambda \langle \lambda(g)\xi_\alpha, \xi_\alpha\rangle = \tau_0(g) = 1$ for $g \in G$.

9.13.10. REMARKS. Elements of $\mathrm{FPr}(G)$ are weak-$*$ continuous on $l^\infty(G)$.

If G is countable, we can get a *Følner sequence* consisting of finite subsets $\{F_n : n \geq 1\}$ of G such that $\lim\limits_{n\to\infty} \frac{|F_n \triangle gF_n|}{|F_n|} = 0$ for each $g \in G$.

If ξ_n are unit vectors such that $\langle \lambda(g)\xi_n, \xi_n\rangle \to 1$, then

$$\|\lambda(g)\xi_n - \xi_n\|_2^2 = 2 - 2\,\mathrm{Re}\langle \lambda(g)\xi_n, \xi_n\rangle \to 0.$$

We are ready to establish several useful equivalent conditions to amenability.

9.13.11. THEOREM. *Let G be a discrete group. The following are equivalent:*

(1) *G is amenable.*

(2) *G has an approximate invariant mean.*

(3) *G satisfies the Følner condition.*

(4) *τ_0 is weakly contained in λ.*

(5) *$\|\sum_{g\in F} \lambda(g)\| = |F|$ for every finite $F \subset G_0$ where $e \in G_0$ generates G.*

(6) *$C^*_\lambda(G)$ has a character.*

PROOF. Assume (1): G is amenable with left-invariant mean m. Let $A = \{a_1, \dots, a_n\} \subset G$. Consider

$$X = \{(\mu - a_1\cdot\mu, \dots, \mu - a_n\cdot\mu) : \mu \in \mathrm{FPr}(G)\}.$$

This is a bounded convex subset of $l_1(G)^n$ since $\mathrm{FPr}(G)$ is convex. We claim that $0 \in \overline{X}$. If this were not the case, then by the Hahn–Banach theorem, there is a linear functional $f = (f_1, \dots, f_n) \in l^\infty_{\mathbb{R}}(G)^n$ and $\varepsilon > 0$ such that

$$\varepsilon \leq \sum_{i=1}^n (\mu - a_i\cdot\mu)(f_i) = \mu\Big(\sum_{i=1}^n f_i - {}_{a_i}f_i\Big) \quad \text{for all} \quad \mu \in \mathrm{FPr}(G).$$

Take $\mu = \delta_g$ for $g \in G$ to get $\sum_{i=1}^n f_i(g) - {}_{a_i}f_i(g) \geq \varepsilon$. Thus $\sum_{i=1}^n f_i - {}_{a_i}f_i \geq \varepsilon 1$. Applying m yields $0 \geq \varepsilon$, a contradiction. Hence $0 \in \overline{X}$. So given $\varepsilon > 0$, we can find $\mu \in \mathrm{FPr}(G)$ so that $\|(\mu - a_1\cdot\mu, \dots, \mu - a_n\cdot\mu)\| = \sum_{i=1}^n \|\mu - a_i\cdot\mu\|_1 < \varepsilon$. Therefore (2) holds.

Assume (2). Given A and ε, let $\mu \in \mathrm{FPr}(G)$ with $\sum_{i=1}^n \|\mu - a_i\cdot\mu\|_1 < \varepsilon$. Let $\mathrm{Ran}(\mu) = \{0 = t_0 < t_1 < \dots < t_k\}$. Set $F_j = \{g : \mu(g) \geq t_j\}$ and set $c_j = t_j - t_{j-1}$ for $1 \leq j \leq k$. Then $\mu = \sum_{j=1}^k c_j \chi_{F_j}$ and $F_j \supset F_{j+1}$. In particular, $1 = \|\mu\| = \sum_{j=1}^k c_j|F_j|$. Also $a_i\cdot\mu = \sum_{j=1}^k c_j \chi_{a_i^{-1}F_j}$. If $g \in G$ and $\mu(g) = t_l$

and $(a_i \cdot \mu)(g) = \mu(_{a_i}g) = t_m$, then $g \in F_j \triangle a_i^{-1} F_j$ for $l < j \leq m$ or $m < j \leq l$. Therefore

$$|\mu(g) - a_i \cdot \mu(g)| = |t_l - t_m| = \sum \{c_j : g \in F_j \triangle a_i^{-1} F_j\}.$$

Hence

$$\|\mu - a_i \cdot \mu\|_1 = \sum_{j=1}^{k} c_j |F_j \triangle a_i^{-1} F_j|.$$

Summing over i yields

$$\sum_{j=1}^{k} c_j \sum_{i=1}^{n} |F_j \triangle a_i^{-1} F_j| = \sum_{i=1}^{n} \|\mu - a_i \cdot \mu\|_1 < \varepsilon = \varepsilon \sum_{j=1}^{k} c_j |F_j|.$$

Therefore there is some j_0 so that $\sum_{i=1}^{n} |F_{j_0} \triangle a_i^{-1} F_{j_0}| \leq \varepsilon |F_{j_0}|$. This is (3).

Assume (3). Given $A \subset G$ finite, take $\varepsilon = \frac{1}{|A|}$, and let F be a corresponding Følner set. Define a unit vector $\xi_A = \frac{1}{|F|^{1/2}} \chi_F \in l^2(G)$. Then

$$\langle \lambda(g)\xi_A, \xi_A \rangle = \frac{|F \cap g^{-1}F|}{|F|} = 1 - \frac{|F \setminus g^{-1}F|}{|F|} > 1 - \frac{1}{|A|} \quad \text{for} \quad g \in A.$$

Build a net based on the finite subsets of G ordered by inclusion. By the remarks preceding the theorem, this forces $\|\lambda(g)\xi_A - \xi_A\| \to 0$ for all $g \in G$. So (4) holds.

Assume (4) and let $\{\xi_\alpha\}_\Lambda$ be a net of almost invariant unit vectors. Then $\|\sum_{g \in F} \lambda(g)\xi_\alpha\| \to |F|$ for any finite $F \subset G$; whence $\|\sum_{g \in F} \lambda(g)\| = |F|$. Conversely, if $e \in F \subset G_0$ and $\|\sum_{g \in F} \lambda(g)\| = |F|$, then there are unit vectors ξ_k such that $\|\sum_{g \in F} \lambda(g)\xi_k\| \to |F|$. It follows that $\|\lambda(g)\xi_k - \xi_k\| \to 0$ for $g \in F$ because if the sum of n unit vectors has norm close to n, then the n vectors are close to each other; and this sum includes $\lambda(e)\xi_k = \xi_k$. This also holds for $\lambda(g^{-1})\xi_k$ since $\langle \lambda(g^{-1})\xi_k, \xi_k \rangle = \langle \xi_k, \lambda(g)\xi_k \rangle \to 1$. Clearly this extends to finite products of elements of $G_0 \cup G_0^{-1}$. Hence if G_0 generates G, then τ_0 is weakly contained in λ. Therefore (4) and (5) are equivalent.

Assume (4) again. The net of states $\varphi_\alpha(a) = \langle a\xi_\alpha, \xi_\alpha \rangle$ on $C^*_\lambda(G)$ converges on $\lambda(\mathbb{C}G)$ to (the linear extension of) τ_0. Since this net has norm 1, it converges on $C^*_\lambda(G)$ as well. This shows that τ_0 is the weak-$*$ limit of $\{\varphi_\alpha\}_\Lambda$, and hence is a state on $C^*_\lambda(G)$. It is multiplicative on G, and hence on $\mathbb{C}G$. So it is multiplicative on $C^*_\lambda(G)$ by continuity. Therefore $C^*_\lambda(G)$ has a character.

If (6) holds, then G is amenable by Proposition 9.13.5. ■

9.13.12. DEFINITION. Suppose that G is a countable group with a finite set $S = S^{-1}$ of generators. Each element $g \in G$ is a finite product of elements of S, and the number required for a minimal representation is called $|g|$. Let $B_S(n) = \{g : |g| \leq n\}$. A countable group G has *subexponential growth* if $\lim_{n \to \infty} |B_S(n)|^{1/n} = 1$.

Note that $|B_S(n+m)| \le |B_S(n)|\,|B_S(m)|$ is a submultiplicative sequence, and hence $\lim_{n\to\infty} |B_S(n)|^{1/n}$ exists. (See Exercise **4.24**.) It is an easy exercise (Exercise **9.35**) to show that this limit is independent of the choice of symmetric generating set S. The example of Exercise **9.40** has polynomial growth.

9.13.13. PROPOSITION. *If G is a countable group with subexponential growth, then G is amenable.*

PROOF. We will construct a Følner sequence for G. As $\lim_{m\to\infty} |B_S(m)|^{1/m} = 1$, we can choose m_n so that

$$1 + \tfrac{1}{n} > |B_S(m_n)|^{1/m_n} \ge |B_S(m)|^{1/m} \quad \text{for} \quad m > m_n.$$

Let $F_n = B_S(m_n)$. Hence $|B_S(m_n+k)| \le |B_S(m_n)|^{1+\frac{k}{m_n}}$. If $g \in G$ and $|g| = k$, then $gF_n \subset B_S(m_n+k)$. Thus $gF_n \setminus F_n \subset B_S(m_n+k) \setminus B_S(m_n)$. Therefore

$$\begin{aligned} \frac{|F_n \triangle gF_n|}{|F_n|} &\le \frac{2|B_S(m_n+k) \setminus B_S(m_n)|}{|B_S(m_n)|} \\ &\le 2\frac{|B_S(m_n)|^{1+k/m_n} - |B_S(m_n)|}{|B_S(m_n)|} < 2\big((1+\tfrac{1}{n})^k - 1\big) \to 0 \end{aligned}$$

as $n \to \infty$. Therefore G is amenable by Theorem 9.13.11. ■

An important tool for determining the norm on a C*-algebra is the set of states. For $C^*(G)$ and $C^*_\lambda(G)$, this can be transferred to certain functions on the group.

9.13.14. DEFINITION. A function $\varphi : G \to \mathbb{C}$ is *positive definite* if for every $F \subset G$ finite, the matrix $\big[\varphi(g^{-1}h)\big]_{g,h\in F} \ge 0$. Let $\mathcal{P}(G)$ denote the set of all positive definite functions on G such that $\varphi(e) = 1$, endowed with the topology of pointwise convergence.

9.13.15. THEOREM. *If $\widetilde{\varphi}$ is a state on $C^*(G)$, then $\varphi := \widetilde{\varphi}|_G$ is positive definite. Conversely, every $\varphi \in \mathcal{P}(G)$ arises in this manner. Moreover this is a homeomorphism from $(\mathcal{S}(C^*(G)), w^*)$ onto $\mathcal{P}(G)$.*

If φ has finite support, then φ extends to a state on $C^*_\lambda(G)$.

PROOF. If $\widetilde{\varphi}$ is a state on $C^*(G)$, let π be the GNS representation with unit vector ξ so that $\widetilde{\varphi}(a) = \langle \pi(a)\xi, \xi\rangle$. Let $P = \xi\xi^*$ be the projection onto $\mathbb{C}\xi$, and note that $\varphi(g) = \langle \sigma(g)\xi, \xi\rangle = P\sigma(g)P|_{\mathbb{C}\xi}$. Then for $F = \{g_1, \dots, g_n\} \subset G$,

$$0 \le \begin{bmatrix} P\sigma(g_1)^* \\ \vdots \\ P\sigma(g_n)^* \end{bmatrix} \begin{bmatrix} \sigma(g_1)P & \dots & \sigma(g_n)P \end{bmatrix} = \begin{bmatrix} \varphi(g_1^{-1}g_1) & \dots & \varphi(g_1^{-1}g_n) \\ \vdots & \ddots & \vdots \\ \varphi(g_n^{-1}g_1) & \dots & \varphi(g_n^{-1}g_n) \end{bmatrix}.$$

So φ is positive definite, and $\varphi(e) = \widetilde{\varphi}(I) = 1$.

Conversely, let $\varphi \in \mathcal{P}(G)$. Put an inner product on $\mathbb{C}G$ by

$$\langle \sum_{g\in G} a_g g, \sum_{h\in G} b_h h\rangle = \sum_{g\in G}\sum_{h\in G} a_g \overline{b_h}\varphi(h^{-1}g).$$

This is easily seen to be sesquilinear, and if $F = \{g : a_g \neq 0\} = \{g_1, \dots, g_n\}$, set $\zeta = [a_1 \ \dots \ a_n]^t$. Then

$$\langle \sum_{g\in G} a_g g, \sum_{g\in G} a_g g\rangle = \langle [\,\varphi(h^{-1}g)\,]\zeta, \zeta\rangle \geq 0.$$

Thus this form is positive semidefinite. Let H be the Hilbert space completion. Let ξ be the image of e; so $\|\xi\|^2 = \varphi(e) = 1$. Define a unitary representation on H by $\sigma(g)\sum_h b_h h = \sum_h b_h gh$. This map is an isometry because in the inner product, we sum terms $a_{h_i}\overline{a_{h_j}}\varphi(h_j^{-1}g^{-1}gh_i) = a_{h_i}\overline{a_{h_j}}\varphi(h_j^{-1}h_i)$. Since H is spanned by $\mathbb{C}G$, it is easy to see that $\sigma(g)$ has a dense range, and thus is unitary. The definition makes it clear that this is a homomorphism of G. Therefore by the universal property of $C^*(G)$, σ extends to a representation σ of $C^*(G)$. The state $\widetilde{\varphi}(a) = \langle\sigma(a)\xi, \xi\rangle$ extends φ.

If a net $\{\widetilde{\varphi}_\lambda\}_\Lambda$ converges weak-$*$ to $\widetilde{\varphi}$, then restricting to G yields pointwise convergence. Conversely, if $\{\varphi_\lambda\}_\Lambda$ in $\mathcal{P}(G)$ converges pointwise to φ, then the linear extension converges pointwise on $\mathbb{C}G$. Thus this net is bounded by 1 and converges weak-$*$ on a dense subalgebra of $C^*(G)$. Therefore it converges weak-$*$.

Suppose $\varphi \in \mathcal{P}(G)$ has finite support, i.e. $F = \{h \in G : \varphi(h) \neq 0\}$ is finite. Define $T \in \mathcal{B}(l_2(G))$ by

$$T = \sum_{h\in F} \varphi(h)\rho(h) \quad \text{and} \quad \zeta = T\delta_e = \sum_{h\in F}\varphi(h)\delta_{h^{-1}} = \sum_{h\in G}\overline{\varphi(h)}\delta_h.$$

Then $\|T\| \leq \sum_{h\in F}|\varphi(h)| < \infty$ and $T \in \mathcal{L}(G)'$. Also $\langle\delta_g, \zeta\rangle = \varphi(g)$ for $g \in G$. For a finitely supported vector $\xi = \sum_G a_g\delta_g \in l_2(G) \cap \mathbb{C}G$,

$$\begin{aligned}\langle\xi, T\xi\rangle &= \Big\langle \sum_{g\in G} a_g\delta_g, \sum_{h\in G} a_h\lambda(h)T\delta_e\Big\rangle = \sum_{g\in G}\sum_{h\in G}\overline{a_h}a_g\langle\delta_g, \lambda(h)\zeta\rangle \\ &= \sum_{g\in G}\sum_{h\in G}\overline{a_h}a_g\langle\delta_{h^{-1}g}, \zeta\rangle = \sum_{g\in G}\sum_{h\in G}\overline{a_h}a_g\varphi(h^{-1}g) \geq 0.\end{aligned}$$

Hence T is positive. Let $\eta = T^{1/2}\delta_e$. Then

$$\langle\lambda(g)\eta, \eta\rangle = \langle\lambda(g)T^{1/2}\delta_e, T^{1/2}\delta_e\rangle = \langle\lambda(g)\delta_e, T\delta_e\rangle = \langle\delta_g, \zeta\rangle = \varphi(g).$$

Therefore $\widetilde{\varphi}(\lambda(a)) = \langle\lambda(a)\eta, \eta\rangle$ is a state on $C^*_\lambda(G)$ extending φ. ■

9.13.16. LEMMA. *The pointwise product of two positive definite functions on G is positive definite.*

PROOF. This is immediate from the fact that the Schur product (entrywise product of matrix entries) of two positive definite matrices is positive definite. To see this, note that the spectral theorem for matrices shows that every positive matrix is the sum of rank 1 positive matrices of the form $\big[a_i\overline{a_j}\big]$. The Schur product of $\big[a_i\overline{a_j}\big]$ and $\big[b_i\overline{b_j}\big]$ is $\big[a_ib_i\overline{a_jb_j}\big]$, which is positive. The result follows. ∎

We are now ready to characterize when $C^*_\lambda(G) = C^*(G)$. The proof relies on positive definite functions.

9.13.17. THEOREM. *Let G be a discrete group. The following are equivalent:*

(1) *G is amenable.*

(2) *There is a net $\{\varphi_\alpha\}_\Lambda$ of finitely supported positive definite functions on G converging pointwise to* 1.

(3) $C^*_\lambda(G) = C^*(G)$.

PROOF. Suppose that G is amenable with left-invariant mean m. By Theorem 9.13.11, τ_0 is weakly contained in λ. Let $\{\xi_\alpha\}_\Lambda \subset l^2(G)$ be a net of unit vectors so that $\lim_\Lambda \langle \lambda(g)\xi_\alpha, \xi_\alpha\rangle = 1$ for $g \in G$. With an asymptotically small change, we may suppose that each ξ_α is finitely supported. Define states $\varphi_\alpha(a) = \langle \lambda(a)\xi_\alpha, \xi_\alpha\rangle$. Then the restriction of φ_α to G is positive definite, and converges to $\tau_0(g) = 1$ for all $g \in G$. Thus (2) holds.

Assume (2). Let $\widetilde{\varphi}$ be a state on $C^*(G)$; and let $\varphi = \widetilde{\varphi}|_G$. By Lemma 9.13.16, $\psi_\alpha := \varphi\varphi_\alpha$ are positive definite with finite support. By Theorem 9.13.15, there are states $\widetilde{\psi}_\alpha$ on $C^*_\lambda(G)$ extending each ψ_α. The functions ψ_α converge pointwise to φ. Thus by Theorem 9.13.15 again, $\widetilde{\psi}_\alpha$ converge weak-$*$ to $\widetilde{\varphi}$. If $a \in \mathbb{C}G$, then

$$\|a\|_u^2 = \|a^*a\|_u = \sup_{\mathcal{S}(C^*(G))} \widetilde{\varphi}(a^*a) = \sup_{\mathcal{S}(C^*(G))} \lim_\Lambda \widetilde{\psi}_\alpha(a^*a) \le \|\lambda(a^*a)\| = \|a\|_\lambda^2.$$

Therefore the universal norm equals the reduced norm, and thus $C^*(G) = C^*_\lambda(G)$.

Suppose that (3) holds. The trivial representation τ_0 is a character on $C^*(G)$, and therefore on $C^*_\lambda(G)$. So G is amenable by Proposition 9.13.5. ∎

9.14. Crossed Products

A group can act on a C*-algebra by $*$-automorphisms. The encoding of such actions leads to an interesting construction of C*-algebras. In this section, we again restrict our attention to discrete groups.

9.14.1. DEFINITION. A *C*-algebra dynamical system* $(G, \mathfrak{A}, \alpha)$ of a (discrete) group G on a C*-algebra $\mathfrak{A}$ is given by a homomorphism $\alpha : G \to \mathrm{Aut}(\mathfrak{A})$. A *covariant representation* (u, π) of $(G, \mathfrak{A}, \alpha)$ is a unitary representation $u : G \to \mathcal{B}(H)$ and a $*$-representation $\pi : \mathfrak{A} \to \mathcal{B}(H)$ such that

$$u_g\pi(a)u_g^* = \pi\alpha_g(a) \quad \text{for} \quad g \in G \text{ and } a \in \mathfrak{A}.$$

9.14.2. EXAMPLE. Covariant representations always exist. Here we define the family of *regular representations*. Let $\pi : \mathfrak{A} \to \mathcal{B}(H)$ be a non-degenerate representation. Form a Hilbert space $K = H \otimes l_2(G) = \sum_G^{\oplus} H_g$ consisting of vectors $\xi = \sum_G \xi_g \otimes \delta_g$, for $\xi_g \in H$, such that

$$\|\xi\|_2^2 = \sum_{g \in G} \|\xi_g\|^2 < \infty.$$

Define $\pi_g = \pi\alpha_g^{-1}$ and $\widetilde{\pi} = \sum_G^{\oplus} \pi_g$, so that $\widetilde{\pi}(a)(\xi \otimes \delta_g) = (\pi_g(a)\xi) \otimes \delta_g$. Also define a unitary representation $(1 \otimes \lambda)(g)\, \xi \otimes \delta_h := (1 \otimes \lambda_g)\, \xi \otimes \delta_h = \xi \otimes \delta_{gh}$. Then

$$\begin{aligned} \big((1 \otimes \lambda_g)\widetilde{\pi}(a)(1 \otimes \lambda_g^*)\big)(\xi \otimes \delta_h) &= (1 \otimes \lambda_g)\widetilde{\pi}(a)(\xi \otimes \delta_{g^{-1}h}) \\ &= (1 \otimes \lambda_g)(\pi_{g^{-1}h}(a)\xi) \otimes \delta_{g^{-1}h} \\ &= (\pi\alpha_{h^{-1}g}(a)\xi) \otimes \delta_h = (\pi\alpha_{h^{-1}}\alpha_g(a)\xi) \otimes \delta_h \\ &= (\pi_h\alpha_g(a)\xi) \otimes \delta_h = \widetilde{\pi}\alpha_g(a)(\xi \otimes \delta_h). \end{aligned}$$

Therefore $(1 \otimes \lambda_g)\widetilde{\pi}(1 \otimes \lambda_g^*) = \widetilde{\pi}\alpha_g$. We will denote this covariant pair by $(1 \otimes \lambda, \widetilde{\pi})$ associated to π.

Given a C*-dynamical system $(G, \mathfrak{A}, \alpha)$, form the $*$-algebra $\mathfrak{A}G$ of finitely supported elements $x = \sum_G x_g e_g$ with $x_g \in \mathfrak{A}$ and convolution product

$$(xe_g)(ye_h) = x(e_g y e_g^{-1})e_{gh} = x\alpha_g(y)e_{gh}$$

and adjoint

$$(x_g e_g)^* = e_{g^{-1}} x_g^* = (e_{g^{-1}} x_g^* e_g)e_{g^{-1}} = \alpha_g^{-1}(x_g^*)e_{g^{-1}}.$$

Observe that if (u, π) is a covariant representation, then we can define a representation $\pi \rtimes u : \mathfrak{A}G \to \mathcal{B}(H)$ by

$$\pi \rtimes u\Big(\sum_G x_g e_g\Big) = \sum_G \pi(x_g)u_g.$$

Note that

$$\begin{aligned} \big(\pi \rtimes u(xe_g)\big)\big(\pi \rtimes u(ye_h)\big) &= \pi(x)u_g\pi(y)u_h = \pi(x)(u_g\pi(y)u_g^*)u_{gh} \\ &= \pi(x)\pi\alpha_g(y)u_{gh} = \pi(x\alpha_g(y))u_{gh} = \pi \rtimes u\big((xe_g)(ye_h)\big) \end{aligned}$$

and

$$\begin{aligned} \pi \rtimes u\big((x_g e_g)^*\big) &= \pi \rtimes u\big(\alpha_g^{-1}(x_g^*)e_{g^{-1}}\big) = \pi\alpha_g^{-1}(x_g^*)u_{g^{-1}} = \big(u_g\pi\alpha_g^{-1}(x_g)\big)^* \\ &= \big((u_g\pi\alpha_g^{-1}(x_g)u_g^*)u_g\big)^* = \big(\pi\alpha_g\alpha_g^{-1}(x_g)u_g\big)^* \\ &= \big(\pi(x_g)u_g\big)^* = \big(\pi \rtimes u(x_g e_g)\big)^*. \end{aligned}$$

Therefore $\pi \rtimes u$ is a $*$-representation of $\mathfrak{A}G$.

9.14.3. DEFINITION. The *C*-algebra crossed product* $\mathfrak{A} \rtimes_\alpha G$ is the completion of $\mathfrak{A}G$ in the norm

$$\|x\| = \sup_{(u,\pi)\text{ covariant}} \|\pi \rtimes u(x)\| = \sup_{(u,\pi)\text{ covariant}} \Big\| \sum_G \pi(x_g) u_g \Big\|.$$

The *reduced C*-algebra crossed product* $\mathfrak{A} \rtimes_{\alpha r} G$ is the completion with respect to the norm $\|x\|_r$ obtained as the supremum over all regular representations.

9.14.4. REMARK. If $\mathfrak{A}$ has a unit, then $\mathfrak{A} \rtimes_\alpha G$ contains a copy of $\mathfrak{A}$ and a copy of a unitary representation of G. This will be a universal covariant representation. When $\mathfrak{A}$ is not unital, the unitaries do not lie in the crossed product, but au_g and $u_g a$ do exist there for any $a \in \mathfrak{A}$.

9.14.5. EXAMPLE. The simplest example is to take $\mathfrak{A} = \mathbb{C}$ and $\alpha_g = id$. Then $\mathbb{C} \rtimes_\alpha G = C^*(G)$ and $\mathbb{C} \rtimes_{\alpha r} G = C^*_\lambda(G)$.

9.14.6. REMARK. Note that $\|x\|_r \le \|x\| \le \|x\|_1 = \sum_G \|x_g\| < \infty$ for $x \in \mathbb{C}G$. For a regular representation $(1 \otimes \lambda, \widetilde{\pi})$ associated to a faithful representation $\pi : \mathfrak{A} \to \mathcal{B}(H)$,

$$\begin{aligned}
\|x\|_r &\ge \sup_{\|\xi\|=1} \Big\| \sum_G \pi\alpha_g^{-1}(x_g) 1 \otimes \lambda_g(\xi \otimes \delta_e) \Big\| \\
&= \sup_{\|\xi\|=1} \Big\| \sum_G (\pi\alpha_g^{-1}(x_g)\xi) \otimes \delta_g \Big\| \\
&\ge \sup_{g\in G} \|\pi\alpha_g^{-1}(x_g)\| = \sup_{g\in G} \|x_g\| = \|x\|_\infty.
\end{aligned}$$

Thus the map from $\mathfrak{A}G$ into $\mathfrak{A} \rtimes_{\alpha r} G$ and $\mathfrak{A} \rtimes_\alpha G$ is injective. In particular, $\mathfrak{A} \to \mathfrak{A}e$ is a $*$-homomorphism, and thus $\mathfrak{A}$ is contained in $\mathfrak{A} \rtimes_{\alpha r} G$ and $\mathfrak{A} \rtimes_\alpha G$.

The identity map on $\mathfrak{A}G$ extends to a contractive map of $\mathfrak{A} \rtimes_\alpha G$ onto $\mathfrak{A} \rtimes_{\alpha r} G$, and thus $\mathfrak{A} \rtimes_{\alpha r} G$ is a quotient of $\mathfrak{A} \rtimes_\alpha G$. Likewise for every covariant pair (u, π), the map $\pi \rtimes u$ extends by continuity to a $*$-representation of $\mathfrak{A} \rtimes_\alpha G$, which we also denote by $\pi \rtimes u$.

Next we show that the left regular representations for faithful representations of $\mathfrak{A}$ all determine the same norm.

9.14.7. PROPOSITION. *Let $(G, \mathfrak{A}, \alpha)$ be a C*-dynamical system. Then the norm on $\mathfrak{A}G$ arising from a regular representation $(1 \otimes \lambda, \widetilde{\pi})$ associated to a* faithful *representation π is independent of the choice of π. Hence $\|x\|_r = \|\widetilde{\pi} \rtimes (1 \otimes \lambda)(x)\|$ for $x \in \mathfrak{A}G$.*

PROOF. Let $P_g = P_{H_g}$ and $P_F = \sum_{g \in F} P_g$ for $F \Subset P$ finite. Consider an element of $\mathfrak{A}G$, say $x = \sum_G x_g e_g$. Then if $\gamma = \widetilde{\pi} \rtimes (1 \otimes \lambda)$,

$$\|\gamma(x)\| = \sup_{F \Subset P} \|P_F \gamma(x) P_F\|.$$

We consider $\gamma(x)$ as an infinite matrix with matrix entries $X_{g,h} \in \mathcal{B}(H_h, H_g)$, where

$$\begin{aligned}(X_{g,h}\xi) \otimes \delta_g &= P_g \gamma(x) P_h(\xi \otimes \delta_h) \\ &= P_g \widetilde{\pi}(x_{gh^{-1}})(1 \otimes \lambda)(gh^{-1})(\xi \otimes \delta_h) \\ &= (\pi \alpha_g^{-1}(x_{gh^{-1}})\xi) \otimes \delta_g.\end{aligned}$$

Thus $X_{g,h} = \pi \alpha_g^{-1}(x_{gh^{-1}})$. Therefore to get $P_F \gamma(x) P_F$, where $|F| = n$, we apply $\pi^{(n)}$ to $Y_F \in M_n(\mathfrak{A})$ where the g, h entry is $Y_{g,h} = \alpha_g^{-1}(x_{gh^{-1}})$ for $g, h \in F$. Since π is faithful, so is $\pi^{(n)}$, and hence it is isometric. That is, $\|P_F X P_F\| = \|Y_F\|$ is independent of π. Taking the supremum over all finite F shows that $\|\gamma(x)\|$ is independent of the choice of π, provided that it is faithful. ■

The following computational lemma shows that we can "absorb" any unitary representation of G into the left regular representation by tensoring.

9.14.8. LEMMA (FELL ABSORPTION PRINCIPLE). *Let (u, π) be any covariant representation of a C*-dynamical system $(G, \mathfrak{A}, \alpha)$. Then $(u \otimes \lambda, \pi \otimes 1)$ is unitarily equivalent to the regular covariant representation $(1 \otimes \lambda, \widetilde{\pi})$.*

PROOF. Define a unitary U on $H \otimes l_2(G)$ by $U(\xi \otimes \delta_h) = u_h \xi \otimes \delta_h$ for $h \in G$. Then

$$\begin{aligned}U(1 \otimes \lambda_g)U^*(\xi \otimes \delta_h) &= U(1 \otimes \lambda_g)(u_h^* \xi \otimes \delta_h) = U(u_h^* \xi \otimes \delta_{gh}) \\ &= u_{gh} u_{h^{-1}} \xi \otimes \delta_{gh} = (u_g \otimes \lambda_g)(\xi \otimes \delta_h)\end{aligned}$$

and since $u_h \pi(a) u_h^* = \pi \alpha_h(a)$,

$$\begin{aligned}U\widetilde{\pi}(a)U^*(\xi \otimes \delta_h) &= U\widetilde{\pi}(a)(u_h^* \xi \otimes \delta_h) = U(\pi \alpha_h^{-1}(a) u_h^* \xi) \otimes \delta_h \\ &= (u_h \pi \alpha_h^{-1}(a) u_h^* \xi) \otimes \delta_h = \pi(a)\xi \otimes \delta_h \\ &= (\pi \otimes 1)(a)(\xi \otimes \delta_h).\end{aligned}$$

Thus the regular representation $(1 \otimes \lambda, \widetilde{\pi})$ is unitarily equivalent to the covariant representation $(u \otimes \lambda, \pi \otimes 1)$. ■

Now using our knowledge of amenable groups, we can show that for amenable groups, the reduced crossed product always coincides with the full crossed product.

9.14.9. THEOREM. *Let $(G, \mathfrak{A}, \alpha)$ be a C*-dynamical system with G amenable. Then the canonical map of $\mathfrak{A} \rtimes_\alpha G$ onto $\mathfrak{A} \rtimes_{\alpha r} G$ is an isomorphism.*

PROOF. Let (u,π) be a covariant representation of $(G,\mathfrak{A},\alpha)$. We can define $\gamma' = (\pi\otimes 1)\rtimes(u\otimes\lambda)$. Then Lemma 9.14.8 shows that $\|\gamma'(x)\| \le \|x\|_r$ for $x \in \mathfrak{A}G$. Let $x = \sum_{g\in F} x_g e_g \in \mathfrak{A}G$ for $F \Subset G$ finite. It will suffice to show that $\|\pi\rtimes u(x)\| \le \|\gamma'(x)\|$.

Since G is amenable, Theorem 9.13.11(4) provides unit vectors $\eta_n \in l_2(G)$ such that $\|\lambda_g\eta_n - \eta_n\| < \frac{1}{n}$ for all $g \in F$. Therefore for any unit vector $\xi \in H$,

$$\begin{aligned}\|\gamma'(x)(\xi\otimes\eta_n)\| &= \Big\|\sum_{g\in F}\pi(x_g)u_g\xi\otimes\lambda_g\eta_n)\Big\| \\ &\ge \Big\|\Big(\sum_{g\in F}\pi(x_g)u_g\xi\Big)\otimes\eta_n\Big\| - \sum_F \|x_g\|\|\lambda_g\eta_n-\eta_n\| \\ &> \Big\|\sum_{g\in F}\pi(x_g)u_g\xi\Big\| - \frac{1}{n}\sum_F\|x_g\| = \big\|(\pi\rtimes u)(x)\xi\big\| - \frac{\|x\|_1}{n}.\end{aligned}$$

Letting $n\to\infty$ and taking the supremum over ξ yields $\|\pi\rtimes u(x)\| \le \|\gamma'(x)\|$. Therefore $\|x\| = \|x\|_r$. ■

This is another situation in which there is a canonical expectation of a subalgebra.

9.14.10. PROPOSITION. *There is a faithful expectation of $\mathfrak{A}\rtimes_{\alpha r} G$ onto $\mathfrak{A}$ extending the map $\mathcal{E}(x) = x_e$ for $x \in \mathfrak{A}G$.*

PROOF. Let (u,π) be a covariant representation with faithful π. Represent $\mathfrak{A}\rtimes_{\alpha r} G$ on $K = H\otimes l_2(G)$ by $\gamma = (\pi\otimes 1)\rtimes(u\otimes\lambda)$. By Lemma 9.14.8 and Proposition 9.14.7, γ is faithful. For $x = \sum x_g e_g \in \mathfrak{A}G$, $\gamma(x) = \sum \pi(x_g)u_g\otimes\lambda_g$ lies in the C*-algebra $\mathfrak{T} = (1\otimes\rho)(G)'$, where ρ is the right regular representation of G. The operators $T\in\mathfrak{T}$ have the property that

$$TP_g = T(1\otimes\rho_g)P_e(1\otimes\rho_g^*) = (1\otimes\rho_g)TP_e(1\otimes\rho_g^*).$$

In particular, if $TP_e = 0$, then $T = 0$.

Let $x = \sum x_g e_g \in \mathfrak{A}G$. Then

$$P_e\gamma(x)P_e = P_e\sum\pi(x_g)u_g\otimes\lambda_g P_e = (\pi(x_e)\otimes I)P_e.$$

So $\|(\pi\otimes 1)(x_e)\| = \|(\pi(x_e)\otimes I)P_e\| \le \|\gamma(x)\|$. Thus $\mathcal{E}(\gamma(x)) = \pi\otimes 1(x_e)$ is a contractive projection of $\gamma(\mathfrak{A}G)$ onto $\pi\otimes 1(\mathfrak{A})$. Hence it extends to an expectation of $\gamma(\mathfrak{A}\rtimes_{\alpha r} G)$ onto $\pi\otimes 1(\mathfrak{A})$. To see that it is faithful, suppose that $X \gneqq 0$ in $\gamma(\mathfrak{A}\rtimes_{\alpha r} G)$. Factor it as $X = T^*T$. As observed above, $TP_e \ne 0$. Therefore

$$P_eXP_e = P_eT^*TP_e = (TP_e)^*(TP_e) \ne 0;$$

and thus $\mathcal{E}(X) \ne 0$. ■

9.14.11. REMARK. Let G be a discrete abelian group, and let $(G, \mathfrak{A}, \alpha)$ be a C*-dynamical system. The dual group $\widehat{G}$ of characters of G is a compact group by Theorem 7.5.1 with Haar measure m. Let (u, π) be a faithful covariant representation of $\mathfrak{A} \rtimes_\alpha G$. Observe that if $\gamma \in \widehat{G}$, then $\gamma u(g) := \gamma(g) u_g$ defines a unitary representation and $(\gamma u, \pi)$ is also covariant. The universal property of the crossed product shows that the map σ_γ from $\pi \times u(x)$ to $\pi \times \gamma u(x)$ is a $*$-homomorphism. Since $\sigma_{\gamma^{-1}} = \sigma_\gamma^{-1}$, these maps are $*$-automorphisms of $\mathfrak{A} \rtimes_\alpha G$. It is not hard to see that

$$\mathcal{E}(x) = \int_{\widetilde{G}} \sigma_\gamma(x)\, dm(\gamma).$$

Indeed one checks on terms of the form au_g

$$\int_{\widetilde{G}} \sigma_\gamma(au_g)\, dm(\gamma) = au_g \int_{\widetilde{G}} \gamma(g)\, dm(\gamma) = \begin{cases} a & \text{if } g = 0 \\ 0 & \text{if } g \neq 0 \end{cases} = \mathcal{E}(au_g).$$

By continuity and linearity, this extends to $\mathfrak{A} \rtimes_\alpha G = \mathfrak{A} \rtimes_{\alpha r} G$.

9.14.12. EXAMPLE. Let G be any discrete group, and let G act on itself by $g \cdot x = gx$ for $x \in G$. This is a homeomorphism of G (as a set with the discrete topology), and thus it induces an automorphism of $C_0(G)$ by $\alpha_g(f)(x) = f(g^{-1}x)$. The functions $\{\delta_h : h \in G\}$ span $C_0(G)$ and $\alpha_g(\delta_h) = \delta_{gh}$.

We may suppose that $C_0(G) \rtimes_\alpha G$ is faithfully represented on $\mathcal{B}(H)$ by a covariant representation (u, π). Then $\{\pi(\delta_h) : h \in G\}$ is a set of pairwise orthogonal projections; and $u_g \pi(\delta_h) u_g^* = \pi(\delta_{gh})$. Thus $u_g \pi(\delta_h) = \pi(\delta_{gh}) u_g$. Define

$$E_{g,h} = u_{gh^{-1}} \pi(\delta_h) = \pi(\delta_g) u_{gh^{-1}} = \pi(\delta_g) u_{gh^{-1}} \pi(\delta_h).$$

Then

$$E_{g_1,h_1} E_{g_2,h_2} = \pi(\delta_{g_1}) u_{g_1 h_1^{-1}} \pi(\delta_{h_1}) \pi(\delta_{g_2}) u_{g_2 h_2^{-1}} \pi(\delta_{h_2}) = \begin{cases} E_{g_1,h_2} & \text{if } h_1 = g_2 \\ 0 & \text{otherwise} \end{cases}.$$

The collection $\{E_{g,h} : g, h \in G\}$ forms a set of matrix units, and thus they generate a C*-algebra isomorphic to the compact operators $\mathcal{K} = \mathcal{K}(l_2(G))$. This C*-algebra contains $\pi(C_0(G)) = \overline{\operatorname{span}}\{\pi(\delta_h) : h \in G\}$ and it is generated by $\pi(C_0(G))$ and multiplications by u_g since $u_g \pi(\delta_h) = E_{gh,h}$ and $\pi(\delta_h) u_g = E_{h,hg^{-1}}$. Therefore $C_0(G) \rtimes_\alpha G \simeq \mathcal{K}(l_2(G))$.

The significance of this example is that all information about G is lost except for its cardinality. So in this situation, the crossed product provides limited new information.

9.14.13. EXAMPLE (*Irrational rotation algebras*). Let θ be an irrational real number. Define a homeomorphism of the unit circle $\mathbb{T}$ by $h(z) = e^{2\pi\theta i} z$. This induces an automorphism of $C(\mathbb{T})$ by $\alpha_1(f) = f \circ h^{-1}$. Thus we obtain a dynamical system $(\mathbb{Z}, C(\mathbb{T}), \alpha)$ where $\alpha_n = \alpha_1^n$. Since $C(\mathbb{T})$ is unital, $C(\mathbb{T}) \rtimes_\alpha \mathbb{Z}$ is generated by two unitaries, u which induces α_1 and v corresponding to the function z. Thus

$$uvu^* = (z \circ h^{-1})(v) = e^{-2\pi\theta i}v \quad \text{or} \quad vu = e^{2\pi\theta i}uv.$$

Conversely, if (U, V) are unitaries satisfying $VU = e^{2\pi\theta i}UV$, then we obtain a representation $\pi : C(\mathbb{T}) \to \mathcal{B}(H)$ by $\pi(f) = f(V)$ using the continuous functional calculus. Then

$$Uf(V)U^* = f(UVU^*) = f(e^{-2\pi\theta i}V) = f \circ h^{-1}(V).$$

Thus (U, π) is a covariant pair. We will say that (U, V) is a covariant pair of unitaries. The algebra $\mathcal{A}_\theta = C(\mathbb{T}) \rtimes_\alpha \mathbb{Z}$ is called an *irrational rotation algebra.*

It is not difficult to exhibit explicit covariant representations. Let $H = L^2(\mathbb{T}, m)$ where m is normalized Lebesgue measure on $\mathbb{T}$. Define $V = M_z$ and $Uf = f \circ h^{-1}$. Note that U is unitary because h is measure preserving. Then for $w \in \mathbb{T}$,

$$UVf(w) = U(zf)(w) = (z \circ h^{-1})(w)(f \circ h^{-1})(w) = \big(e^{-2\pi\theta i}z(f \circ h^{-1})\big)(w))$$

and

$$VUf(w) = Vf \circ h^{-1}(w) = \big(z(f \circ h^{-1})\big)(w) = e^{2\pi\theta i}UVf(w).$$

Another example is given by the bilateral shift u and $v = \operatorname{diag}(e^{2\pi n\theta i}, n \in \mathbb{Z})$ on $l_2(\mathbb{Z})$.

We can obtain a universal covariant pair $(\mathfrak{u}, \mathfrak{v})$ by taking a direct sum over the set of all covariant pairs of unitaries in a fixed separable Hilbert space. Since polynomials in z and z^{-1} are dense in $C(\mathbb{T})$ by Weierstrass's theorem, we see that polynomials in $\mathfrak{u}$, $\mathfrak{u}^*$, $\mathfrak{v}$ and $\mathfrak{v}^*$ are dense in $\mathcal{A}_\theta$; so that $\mathcal{A}_\theta = C^*(\mathfrak{u}, \mathfrak{v})$. Moreover, because of the commutation relation between $\mathfrak{u}$ and $\mathfrak{v}$, $\operatorname{span}\{\mathfrak{u}^m\mathfrak{v}^n : m, n \in \mathbb{Z}\}$ is dense in $\mathcal{A}_\theta$.

Observe that if (U, V) is a covariant pair, then so is $(\lambda U, \mu V)$ for any $\lambda, \mu \in \mathbb{T}$. By the universal property of the crossed products, there is a homomorphism

$$\beta_{\lambda,\mu} : \mathcal{A}_\theta \to \mathcal{A}_\theta \quad \text{such that} \quad \beta_{\lambda,\mu}(\mathfrak{u}) = \lambda\mathfrak{u} \quad \text{and} \quad \beta_{\lambda,\mu}(\mathfrak{v}) = \mu\mathfrak{v}.$$

Then $\beta_{\overline{\lambda},\overline{\mu}}\beta_{\lambda,\mu} = id$, so $\beta_{\lambda,\mu}$ is an automorphism. The map β on $\mathbb{T}^2$ sending λ, μ to $\beta_{\lambda,\mu}$ is point-norm continuous, meaning that for each $A \in \mathcal{A}_\theta$,

$$(\lambda, \mu) \to \beta_{\lambda,\mu}(A)$$

is continuous. We can see this first for a polynomial $A = \sum a_{m,n}\mathfrak{u}^m\mathfrak{v}^n$ (finite sum) that

$$\|\beta_{\lambda,\mu}(A) - \beta_{\lambda',\mu'}(A)\| \le \sum |a_{m,n}||\lambda^m\mu^n - \lambda'^m\mu'^n|$$

converges to 0 as $(\lambda', \mu') \to (\lambda, \mu)$. Since automorphisms are isometric, this extends to the closure.

Now we can explicitly exhibit the expectation of $\mathcal{A}_\theta$ onto $C(\mathbb{T})$. In fact, we will define expectations onto $C^*(\mathfrak{v})$ and $C^*(\mathfrak{u})$,

$$\mathcal{E}_1(A) = \int_0^1 \beta_{e^{2\pi it},1}(A)\,dt \quad \text{and} \quad \mathcal{E}_2(A) = \int_0^1 \beta_{1,e^{2\pi it}}(A)\,dt.$$

By the previous paragraph, these are Riemann integrals of continuous functions and thus are well defined. Moreover, since automorphisms are positive and contractive, $\mathcal{E}_1$ and $\mathcal{E}_2$ are positive and contractive. Compute

$$\mathcal{E}_1(\mathfrak{u}^m\mathfrak{v}^n) = \int_0^1 e^{2\pi imt}\mathfrak{u}^m\mathfrak{v}^n\,dt = \begin{cases} \mathfrak{v}^n & \text{if } m=0 \\ 0 & \text{if } m\neq 0. \end{cases}$$

By linearity, we see that $\mathcal{E}_1\big(\sum_{m,n} a_{m,n}\mathfrak{u}^m\mathfrak{v}^n\big) = \sum_n a_{0,n}\mathfrak{v}^n$. By continuity, we see that $\mathcal{E}_1(f(\mathfrak{v})) = f(\mathfrak{v})$ for $f \in C(\mathbb{T})$ and $\operatorname{Ran}\mathcal{E}_1 = C^*(\mathfrak{v})$. Therefore this is an expectation. It is faithful because if $A \gneq 0$, then there is a unit vector ξ so that $\langle A\xi,\xi\rangle > 0$. It follows that $\langle \beta_{e^{2\pi it},1}(A)\xi,\xi\rangle > 0$ on an interval around $t=0$ and non-negative everywhere. Thus $\langle \mathcal{E}_1(A)\xi,\xi\rangle > 0$. Similarly $\mathcal{E}_2$ is a faithful expectation of $\mathcal{A}_\theta$ onto $C^*(\mathfrak{u})$.

The composition $\tau = \mathcal{E}_1\mathcal{E}_2 = \mathcal{E}_2\mathcal{E}_1$ is a faithful positive map of $\mathcal{A}_\theta$ onto $\mathbb{C}1$. To see that $\mathcal{E}_1$ and $\mathcal{E}_2$ commute, observe that

$$\mathcal{E}_1\mathcal{E}_2(\mathfrak{u}^m\mathfrak{v}^n) = \mathcal{E}_2\mathcal{E}_1(\mathfrak{u}^m\mathfrak{v}^n) = \begin{cases} 1 & \text{if } m=n=0 \\ 0 & \text{otherwise.} \end{cases}$$

We claim that τ is a trace. It is enough to check the trace property on monomials

$$\tau\big((\mathfrak{u}^k\mathfrak{v}^l)(\mathfrak{u}^m\mathfrak{v}^n)\big) = \tau\big(e^{2\pi\theta lmi}\mathfrak{u}^{k+m}\mathfrak{v}^{l+n}\big) = \begin{cases} e^{2\pi\theta lmi} & \text{if } k+m=l+n=0 \\ 0 & \text{otherwise} \end{cases}.$$

When $k+m=l+n=0$, we have $lm=kn$ and thus

$$\tau\big((\mathfrak{u}^k\mathfrak{v}^l)(\mathfrak{u}^m\mathfrak{v}^n)\big) = \tau\big((\mathfrak{u}^m\mathfrak{v}^n)(\mathfrak{u}^k\mathfrak{v}^l)\big).$$

By linearity and continuity, $\tau(AB)=\tau(BA)$ for all $A, B \in \mathcal{A}_\theta$.

We claim that the trace on $\mathcal{A}_\theta$ is unique. Let τ' be a trace on $\mathcal{A}_\theta$. Since τ' is a state, we have $\tau'(1)=1$. If $m\neq 0$,

$$\tau'(\mathfrak{u}^m\mathfrak{v}^n) = \tau'(\mathfrak{v}\mathfrak{u}^m\mathfrak{v}^n\mathfrak{v}^*) = e^{2\pi\theta mi}\tau'(\mathfrak{u}^m\mathfrak{v}^n).$$

Thus $\tau'(\mathfrak{u}^m\mathfrak{v}^n) = 0$. Similarly $\tau'(\mathfrak{u}^m\mathfrak{v}^n) = 0$ if $n\neq 0$. So τ' agrees with τ on a spanning set, whence they are equal.

In fact, there is a useful formula for the trace. Observe that

$$\begin{aligned} \frac{1}{2k+1}\sum_{j=-k}^{k} \mathfrak{v}^j(\mathfrak{u}^m\mathfrak{v}^n)\mathfrak{v}^{-j} &= \frac{1}{2k+1}\sum_{j=-k}^{k} e^{2\pi\theta jmi}(\mathfrak{u}^m\mathfrak{v}^n) \\ &= \begin{cases} \mathfrak{v}^n & \text{if } m=0 \\ \frac{1}{2k+1}\Big(\frac{\sin(2k+1)\pi m\theta}{\sin \pi m\theta}\Big)\mathfrak{u}^m\mathfrak{v}^n & \text{if } m\neq 0 \end{cases}. \end{aligned}$$

The Dirichlet kernel turns up in this calculation (see section 2.5). This converges to $\delta_{m0}\mathfrak{u}^m\mathfrak{v}^n = \mathcal{E}_1(\mathfrak{u}^m\mathfrak{v}^n)$ for all $m, n \in \mathbb{Z}^2$. Since these averages are contractive, we see that

$$\mathcal{E}_1(A) = \lim_{k\to\infty} \frac{1}{2k+1} \sum_{j=-k}^{k} \mathfrak{v}^j A \mathfrak{v}^{-j} \quad \text{for} \quad A \in \mathcal{A}_\theta.$$

Similarly,

$$\mathcal{E}_2(A) = \lim_{k\to\infty} \frac{1}{2k+1} \sum_{j=-k}^{k} \mathfrak{u}^j A \mathfrak{u}^{-j} \quad \text{for} \quad A \in \mathcal{A}_\theta.$$

It follows that

$$\tau(A) = \lim_{k\to\infty} \frac{1}{(2k+1)^2} \sum_{j=-k}^{k} \sum_{l=-k}^{k} \mathfrak{u}^l \mathfrak{v}^j A \mathfrak{v}^{-j} \mathfrak{u}^{-l} \quad \text{for} \quad A \in \mathcal{A}_\theta.$$

Now we can show that $\mathcal{A}_\theta$ is simple. For if J is a non-zero ideal in $\mathcal{A}_\theta$, let $A \gneq 0$ belong to J. Since τ is faithful, $\tau(A) > 0$. But

$$\tau(A) = \lim_{k\to\infty} \frac{1}{(2k+1)^2} \sum_{j=-k}^{k} \sum_{l=-k}^{k} \mathfrak{u}^l \mathfrak{v}^j A \mathfrak{v}^{-j} \mathfrak{u}^{-l} \in J.$$

So J contains a non-zero scalar, and thus $J = \mathcal{A}_\theta$. This implies that any covariant pair (U, V) generates an isomorphic copy of $\mathcal{A}_\theta$.

9.14.14. EXAMPLE. Let $\theta = \frac{m}{n}$ with $\gcd(m, n) = 1$. Define a homeomorphism of $\mathbb{T}$ by $h(z) = e^{2\pi\theta i}z$. Let $\mathbb{Z}/n\mathbb{Z}$ act on $C(\mathbb{T})$ by $\alpha_s(f) = f \circ h^{-s}$. Arguing as in the previous example, $C(\mathbb{T}) \rtimes_\alpha \mathbb{Z}/n\mathbb{Z}$ is the universal C*-algebra generated by two unitaries u, v such that

$$vu = e^{2\pi i\theta} uv \quad \text{and} \quad u^n = 1.$$

We will show that $C(\mathbb{T}) \rtimes_\alpha \mathbb{Z}/n\mathbb{Z} \simeq C(\mathbb{T}, M_n) \simeq M_n(C(\mathbb{T}))$.

Let u be the backward shift in M_n given by $ue_j = e_{j-1 \bmod n}$ for a basis $e_1, \dots, e_n$. This unitary has a skew-adjoint logarithm ai by the functional calculus, so $u = e^{ai}$ with $a = a^*$. Since $\gcd(m, n) = 1$, there is an integer $p \in [1, n-1]$ such that $pm \equiv 1 \pmod{n}$. Thus $h^p(z) = e^{2\pi\theta pi}z = e^{2\pi i/n}z$. Observe that

$$u \operatorname{diag}(d_0, d_1, \dots, d_{n-1})u^* = \operatorname{diag}(d_1, d_2, \dots, d_{n-1}, d_0).$$

Therefore with subscripts computed modulo n,

$$u^p \operatorname{diag}(d_0, d_1, \dots, d_{n-1})u^{*p} = \operatorname{diag}(d_p, d_{1+p}, \dots, d_{n-1+p}).$$

Imbed $C(\mathbb{T})$ into $C(\mathbb{T}, M_n)$ by

$$\begin{aligned}\pi(f)(e^{2\pi i t}) &= e^{ptai} \operatorname{diag}\big(\alpha_0(f)(e^{\frac{2\pi i t}{n}}), \alpha_1(f)(e^{\frac{2\pi i t}{n}}), \dots, \alpha_{n-1}(f)(e^{\frac{2\pi i t}{n}})\big)e^{-ptai} \\ &= e^{ptai} \operatorname{diag}\big(f(e^{\frac{2\pi i t}{n}}), f(e^{\frac{2\pi i t}{n}+\theta}), \dots, f(e^{\frac{2\pi i t}{n}+(n-1)\theta})\big)e^{-ptai}\end{aligned}$$

for $0 \le t \le 1$. It is clear that this function is continuous for $0 < t < 1$. Using $e^{pai} = u^p$, we check that

$$\begin{aligned}\pi(f)(e^{2\pi i 1}) &= u^p \operatorname{diag}\big(\alpha_0(f)(e^{\frac{2\pi i}{n}}), \alpha_1(f)(e^{\frac{2\pi i}{n}}), \dots, \alpha_{n-1}(f)(e^{\frac{2\pi i}{n}})\big)u^{*p} \\ &= \operatorname{diag}\big(\alpha_p(f)(e^{\frac{2\pi i}{n}}), \alpha_{1+p}(f)(e^{\frac{2\pi i}{n}}), \dots, \alpha_{n-1+p}(f)(e^{\frac{2\pi i}{n}})\big) \\ &= \operatorname{diag}\big(\alpha_0(f)h^{-p}(e^{\frac{2\pi i}{n}}), \alpha_1(f)h^{-p}(e^{\frac{2\pi i}{n}}), \dots, \alpha_{n-1}(f)h^{-p}(e^{\frac{2\pi i}{n}})\big) \\ &= \operatorname{diag}\big(\alpha_0(f)(1), \alpha_1(f)(1), \dots, \alpha_{n-1}(f)(1)\big) = \pi(f)(1).\end{aligned}$$

We have succeeded in wrapping the function $f(z)$ continuously around a circle in M_n.

Let $U_1(z) = u$ be a constant function. Note that $u = e^{ai}$ commutes with e^{ptai}. Compute

$$\begin{aligned}&U_1\pi(f)U_1^*(e^{2\pi it}) \\ &\quad = ue^{ptai} \operatorname{diag}\big(\alpha_0(f)(e^{\frac{2\pi it}{n}}), \alpha_1(f)(e^{\frac{2\pi it}{n}}), \dots, \alpha_{n-1}(f)(e^{\frac{2\pi it}{n}})\big)e^{-ptai}u^* \\ &\quad = e^{ptai} \operatorname{diag}\big(\alpha_1(f)(e^{\frac{2\pi it}{n}}), \alpha_2(f)(e^{\frac{2\pi it}{n}}), \dots, \alpha_0(f)(e^{\frac{2\pi it}{n}})\big)e^{-ptai} \\ &\quad = \pi\alpha_1(f)(e^{2\pi it}).\end{aligned}$$

Thus $s \to U_s = U_1^s$ is a unitary representation of $\mathbb{Z}/n\mathbb{Z}$ (since $u^n = 1$) such that $U_s\pi(f)U_s^* = \pi\alpha_s(f)$.

Consider $\mathfrak{B} = C^*(\pi(C(\mathbb{T})), \{U_s\}) \subset C(\mathbb{T}, M_n)$. At any point $z = e^{2\pi it}$, notice that $\pi(f)(z) = e^{ptai}De^{-ptai}$ is unitarily equivalent to the diagonal operator D; and $U_s(z) = u^s = e^{ptai}u^se^{-ptai}$ is unchanged by this unitary equivalence. To recognize this, let $\xi_j(z) = e^{ptai}e_j$ for $0 \le j < n$ be an orthonormal basis for M_n. Let $X = \sum_{l=0}^{n-1} \pi(f_l)U^l \in \mathfrak{B}$. Compute the matrix entries of $X(z)$ with respect to the basis $\{\xi_j(z)\}$, and observe that only $l = j-k$ yields a non-zero term in

$$\begin{aligned}X(z)_{j,k} &= \Big\langle \sum_{l=0}^{n-1} \pi(f_l)(z)u^l\, \xi_j(z), \xi_k(z)\Big\rangle \\ &= \big\langle \pi(f_{j-k})\xi_k(z), \xi_k(z)\big\rangle = f_{j-k}\big(e^{\frac{2\pi it}{n}+k\theta}\big).\end{aligned}$$

Now suppose that $G = \big[g_{j,k}\big] \in C(\mathbb{T}, M_n)$. To see that $G \in \mathfrak{B}$, we have to select the element X so that

$$f_l\big(e^{\frac{2\pi it}{n}+k\theta}\big) = g_{l+k,k}(e^{2\pi it}) \quad \text{for} \quad 0 \le l, k < n.$$

As above, it is clear that these functions are continuous at $e^{2\pi it}$ for $0 < t < 1$. The same calculation now shows that one obtains the equal values at $t = 0$ and $t = 1$; whence f_l are continuous on $\mathbb{T}$. Hence $\mathfrak{B}$ is all of $C(\mathbb{T}, M_n)$.

Now let

$$V(e^{2\pi it}) = e^{ptai} \operatorname{diag}(1, e^{2\pi i\theta}, e^{2\pi i2\theta}, \dots, e^{2\pi i(n-1)\theta})e^{-ptai}e^{-\frac{2\pi it}{n}}.$$

Evidently this is continuous for $t \in (0,1)$. Compute

$$\begin{aligned}V(1) &= e^{pai}\,\mathrm{diag}(1, e^{2\pi i\theta}, e^{2\pi i 2\theta}, \dots, e^{2\pi i(n-1)\theta})e^{-pai}e^{-\frac{2\pi i}{n}} \\ &= u^p\,\mathrm{diag}(1, e^{2\pi i\theta}, e^{2\pi i 2\theta}, \dots, e^{2\pi i(n-1)\theta})u^{-p}e^{-\frac{2\pi i}{n}} \\ &= e^{2\pi i p\theta}e^{-\frac{2\pi i}{n}}\,\mathrm{diag}(1, e^{2\pi i\theta}, e^{2\pi i 2\theta}, \dots, e^{2\pi i(n-1)\theta}) \\ &= \mathrm{diag}(1, e^{2\pi i\theta}, e^{2\pi i 2\theta}, \dots, e^{2\pi i(n-1)\theta}) = V(0).\end{aligned}$$

Notice that $VUV^*(e^{2\pi it}) = e^{-2\pi i\theta}U$. Next we calculate for $F = \sum_{l=0}^{n-1}\pi(f_l)U^l$,

$$\begin{aligned}\frac{1}{n}\sum_{k=0}^{n-1}V^kFV^{*k} - \frac{1}{n}\sum_{k=0}^{n-1}\sum_{l=0}^{n-1}V^k\pi(f_l)U^lV^{*k} &= \frac{1}{n}\sum_{k=0}^{n-1}\sum_{l=0}^{n-1}\pi(f_l)V^kU^lV^{*k} \\ &= \frac{1}{n}\sum_{k=0}^{n-1}\sum_{l=0}^{n-1}\pi(f_l)U^le^{-2\pi ikl\theta} = \sum_{l=0}^{n-1}\pi(f_l)U^l\frac{1}{n}\sum_{k=0}^{n-1}e^{-2\pi ikl\theta} \\ &= \pi(f_0) = \mathcal{E}_1(F).\end{aligned}$$

This is the expectation of $\pi \rtimes U(C(\mathbb{T}) \rtimes_\alpha \mathbb{Z}/n\mathbb{Z})$ onto $\pi(C(\mathbb{T}))$ that picks out the coordinate f_0. Note that in this case, $C(\mathbb{T})\mathbb{Z}/n\mathbb{Z} = C(\mathbb{T}) \rtimes_\alpha \mathbb{Z}/n\mathbb{Z}$ as a set, so no completion is necessary. In Proposition 9.14.10, we showed that the expectation $\mathcal{E}$ of $C(\mathbb{T}) \rtimes_\alpha \mathbb{Z}/n\mathbb{Z}$ onto $C(\mathbb{T})$ obtained by selecting the coordinate f_0 is faithful. It is clear that $\mathcal{E} = \mathcal{E}_1(\pi \rtimes U)$, and thus $\pi \rtimes U$ is faithful. Therefore

$$C(\mathbb{T}) \rtimes_\alpha \mathbb{Z}/n\mathbb{Z} \simeq C(\mathbb{T}, M_n) \simeq M_n(C(\mathbb{T})).$$

This C*-algebra has many irreducible representations. They correspond to evaluation at points of $\mathbb{T}$. Evaluation at $z = e^{2\pi it}$ corresponds to the generators $U(z) = u$, the backward shift on $\mathbb{C}^n$, and

$$v = \pi(z)(e^{2\pi it}) = e^{ptai}\,\mathrm{diag}\left(e^{\frac{2\pi it}{n}}, e^{\frac{2\pi it}{n}+\theta}, \dots, e^{\frac{2\pi it}{n}+(n-1)\theta}\right)e^{-ptai}.$$

But we can conjugate by e^{-ptai}, which commutes with u, and get

$$ue_j = e_{j-1 \bmod n} \quad \text{and} \quad v = \mathrm{diag}\left(e^{\frac{2\pi it}{n}}, e^{\frac{2\pi it}{n}+\theta}, \dots, e^{\frac{2\pi it}{n}+(n-1)\theta}\right).$$

Note that the spectrum of u is the set of nth roots of 1, while the spectrum of v is the set of nth roots of $e^{2\pi it}$.

Exercises for Chapter 9

9.1 Show that if $\mathfrak{A}$ is a separable C*-algebra, then it has an approximate unit consisting of an increasing sequence.

9.2 Show that if $0 \le a$, $0 \le b$ and $\|a\| < 1$ and $\|b\| < 1$, then

$$c = b + (1-b)^{1/2}[(1-b)^{-1/2}(a-b)(1-b)^{-1/2}]_+(1-b)^{1/2}$$

satisfies $b \le c$, $a \le c$ and $\|c\| < 1$.

9.3 Show that if J is a closed ideal of a C*-algebra $\mathfrak{A}$ and $a \in \mathfrak{A}$, then there is an element $j \in J$ such that $\|a - j\| = \text{dist}(a, J)$.
HINT: write $|a| - \|a + J\|1$ as $b_+ - b_-$.

9.4 Show that if $0 \le a \le b$ in a C*-algebra $\mathfrak{A}$, then $0 \le a^{1/2} \le b^{1/2}$.
HINT: Use the fact that inverses reverse order to show that $\|a^{1/2}(b+\varepsilon I)^{-1/2}\| \le 1$. Then show that $\text{spr}((b+\varepsilon I)^{-1/4} a^{1/2} (b+\varepsilon I)^{-1/4}) \le 1$.

9.5 (a) For a von Neumann algebra $\mathcal{N}$, show that the extreme points of $[0, I] = \{A \in \mathcal{N} : 0 \le A \le I\}$ are the projections.
(b) For a von Neumann algebra $\mathcal{N}$, show that the extreme points of $\mathcal{N}_{sa1}$ are precisely the symmetries, i.e. elements $s = s^*$ such that $s^2 = 1$.
(c) Show that for $\mathcal{N} = \mathcal{B}(H)$, the set of extreme points of the unit ball is not contained in the closure of the set of unitaries.

9.6 Let $\mathcal{K}(H)$ denote the C*-algebra of compact operators on a separable Hilbert space H. Fix an orthonormal basis $\{e_i : i \ge 1\}$ and let $E_{ij} = e_i e_j^*$ for $1 \le i, j < \infty$. Let $\pi : \mathcal{K}(H) \to \mathcal{B}(K)$ be a $*$-representation.
(a) Let $K_1 = \text{Ran}(\pi(E_{11}))$ and let $\{f_{1,\lambda} : \lambda \in \Lambda\}$ be an orthonormal basis for K_1. Define $f_{i,\lambda} = \pi(E_{i1}) f_{1,\lambda}$. Show that $\{f_{i,\lambda} : \alpha \in \Lambda\}$ is an orthonormal basis for $K_i := \text{Ran}(\pi(E_{ii}))$; and that K is the Hilbert space direct sum of the K_i's.
(b) Let $H_\lambda = \overline{\text{span}\{f_{i,\lambda} : i \ge 1\}}$. Show that H_λ is a reducing subspace for $\pi(\mathcal{K}(H))$, and that $\pi_\lambda(K) = \pi(K)|_{H_\lambda}$ is unitarily equivalent to the identity representation.
(c) Deduce that every representation of $\mathcal{K}(H)$ is a direct sum of representations unitarily equivalent to the identity representation. In particular, $\mathcal{K}(H)$ has only one irreducible representation up to unitary equivalence.

9.7 *The projections in a von Neumann algebra form a complete lattice.*
(a) Let $\text{Pr}(\mathcal{M})$ denote the projections of a von Neumann algebra $\mathcal{M}$. Show that the ranges of these projections are precisely the invariant subspaces of $\mathcal{M}'$.
(b) Hence show that if $P, Q \in \text{Pr}(\mathcal{M})$, then the projections $P \vee Q$ onto $\overline{PH + QH}$ and $P \wedge Q$ onto $PH \cap QH$ belong to $\mathcal{M}$.
(c) Show that $(P \vee Q)^\perp = P^\perp \wedge Q^\perp$ and $(P \wedge Q)^\perp = P^\perp \vee Q^\perp$.
(d) Show that if $\{P_\lambda\}_\Lambda \subset \text{Pr}(\mathcal{M})$, then $\bigvee_\Lambda P_\lambda$, the projection onto $\overline{\bigcup_\Lambda P_\lambda H}$, belongs to $\mathcal{M}$, as does $\bigwedge_\Lambda P_\lambda$, the projection onto $\overline{\bigcap_\Lambda P_\lambda H}$.
(e) State and prove the analogue of (c) for infinite families.

9.8 The strong-$*$ topology on $\mathcal{B}(\mathcal{H})$ is the weakest topology such that $T \to Tx$ for $x \in \mathcal{H}$ and $T \to T^*$ are continuous maps. Show that the linear functionals on $\mathcal{B}(\mathcal{H})$ which are strong-$*$ continuous coincide with the WOT-continuous functionals.
HINT: compare with Exercise **3.4**.

9.9 (a) The σ-weak (or ultraweak) topology is the weakest topology on $\mathcal{B}(H)$ such that the functionals $\omega(T) = \sum_{n\geq 1}\langle Tx_n, y_n\rangle$ are continuous for all families $\{x_n, y_n : n \geq 1\}$ of vectors such that $\sum_{n\geq 1} \|x_n\|\,\|y_n\| < \infty$. Likewise the σ-strong topology is the topology determined by seminorms of the form $\rho(T) = \big(\sum_{n\geq 1} \|Tx_n\|^2\big)^{1/2}$ where $\sum_{n\geq 1} \|x_n\|^2 < \infty$. Prove that these two topologies have the same continuous linear functionals.

(b) Find a linear functional on $\mathcal{B}(H)$ which is σ-weak continuous but not WOT continuous. Hence find a subspace of $\mathcal{B}(H)$ such that its σ-weak closure is properly contained in its WOT-dense closure.

(c) Rework the proof of von Neumann's Double Commutant theorem to show that if $\mathfrak{A}$ is a non-degenerate C*-subalgebra of $\mathcal{B}(H)$, then $\mathfrak{A}''$ equals the σ-strong closure of $\mathfrak{A}$.

9.10 Let $\mathfrak{A} \subset \mathcal{B}(H)$ be a concrete C*-algebra. Suppose that $A \in \mathfrak{A}$ has a polar decomposition $A = U|A|$. Show that $Uf(|A|)$ belongs to $\mathfrak{A}$ provided that $f \in \mathrm{C}(\sigma(|A|))$ and $f(0) = 0$.

9.11 Let $A \in \mathcal{B}(H)$, and suppose that A is bounded below. Show that in the polar decomposition $A = U|A|$, U is an isometry in $C^*(A)$.

9.12 Show that if $T \in \mathcal{B}(H)$ is not compact, then there is an infinite dimensional subspace on which T is bounded below.
HINT: consider spectral projections of $|T|$ for $[\varepsilon, \|T\|]$ for $\varepsilon > 0$. If all are finite rank, show $|T|$ is compact.

9.13 (*Kadison–Pedersen*). If $\mathfrak{A}$ is a unital C*-algebra and $a \in \mathfrak{A}$ satisfies $\|a\| < 1 - 2^{-n}$, how many unitaries are needed in our proof of the Russo–Dye theorem to express a as a convex combination?

9.14 Compute the extreme points of the state space of $\mathcal{K}(H)$ for a separable Hilbert space H, and find the norm and weak-$*$ closure of the convex hull of these extreme points.

9.15 Two positive functionals ψ_1, ψ_2 on a unital C*-algebra $\mathfrak{A}$ are *orthogonal* if $\|\psi_1 - \psi_2\| = \|\psi_1\| + \|\psi_2\|$. Show that this property holds if and only if for every $\varepsilon > 0$, there is some $a \in \mathfrak{A}$, $0 \leq a \leq 1$, such that $\psi_1(a) < \varepsilon$ and $\psi_2(1 - a) < \varepsilon$.
HINT: ($\Rightarrow$) choose $x = x^*$ with $\|x\| = 1$ and $\psi_1 - \psi_2(x) > \|\psi_1\| + \|\psi_2\| - \varepsilon$. Use $a = \frac{1}{2}(1 + x)$. ($\Leftarrow$) evaluate at $2a - 1$.

9.16 (a) (*Fuglede theorem*) If X commutes with a normal operator N, show that it commutes with N^*.
HINT: $f(z) = e^{zN^*}Xe^{-zN^*} = e^{zN^*}e^{-\bar{z}N}Xe^{\bar{z}N}e^{-zN^*}$ is bounded and entire.

(b) (*Fuglede–Putnam theorem*) Show that if M and N are normal operators and $MX = XN$, then $M^*X = XN^*$.
HINT: $\left[\begin{smallmatrix} M & 0 \\ 0 & N \end{smallmatrix}\right]$ commutes with $\left[\begin{smallmatrix} 0 & X \\ 0 & 0 \end{smallmatrix}\right]$.

(c) Show that if M and N are normal and $MX = XN$ and X is injective with dense range, then M and N are unitarily equivalent.
HINT: use the polar decomposition of X.

9.17 Say that σ is a *subrepresentation* of $\pi : \mathfrak{A} \to \mathcal{B}(H)$ if there is a projection $P \in \pi(\mathfrak{A})'$ so that σ is unitarily equivalent to the restriction of $\pi(\mathfrak{A})$ to PH. Suppose that σ_i are *irreducible* subrepresentations of π corresponding to projections $P_i \in \pi(\mathfrak{A})'$. Show that if $P_1P_2 \neq 0$, then σ_1 and σ_2 are unitarily equivalent.
HINT: use the polar decomposition of P_1P_2.

9.18 Let π be a non-degenerate representation of $C_0(X)$ on a separable Hilbert space H, where X is LCH. Write $\pi \simeq \sum_i^\oplus \pi_i$ where $\pi_i \simeq \sigma_{\mu_i}$ are cyclic representations and μ_i are regular Borel probability measures of X.
(a) Let $\mu = \sum_i 2^{-i}\mu_i$. Show that there are Borel sets $E_i \subset X$ so that μ_i is mutually absolutely to $\chi_{E_i}\mu$.
(b) Show that there are Borel sets $F_j \subset X$ with $F_{j+1} \subset F_j$ and measures $\nu_j = \frac{1}{\mu(F_j)}\chi_{F_j}\mu$ so that $\pi \simeq \sum_j^\oplus \sigma_j$ where $\sigma_j \simeq \sigma_{\nu_j}$. Show that σ_{j+1} is equivalent to a subrepresentation of σ_j.
(c) Define $G_j = F_j \setminus F_{j+1}$ and $G_\infty = \bigcap_j F_j$. Let $\lambda_j = \frac{1}{\mu(G_j)}\chi_{G_j}\mu$ including $j = \infty$. Let $\tau_j = \sigma_{\lambda_j}$. Show that $\pi \simeq \sum_{j\geq 1}^\oplus \tau_j^{(j)} \oplus \tau_\infty^{(\infty)}$ and τ_j has no non-zero subrepresentation equivalent to any subrepresentation of τ_k for $j \neq k$.
(d) Describe the commutant $\pi(C_0(X))'$ explicitly.

9.19 Let $\mathfrak{A}$ be a finite dimensional C*-algebra.
(a) Let π be an irreducible representation of $\mathfrak{A}$ on $\mathcal{B}(H)$. Prove that $\dim H < \infty$, that $\pi(\mathfrak{A}) = \mathcal{B}(H)$ and that $\ker \pi$ is a maximal ideal.
(b) Show that $\mathfrak{A}$ has a faithful finite dimensional representation.
(c) Prove that $\mathfrak{A}$ is $*$-isomorphic to the direct sum of a finite number of full matrix algebras.

9.20 Let φ be a state on a C*-algebra $\mathfrak{A}$. Suppose that $a \in \mathfrak{A}$ satisfies $|\varphi(a)|^2 = \varphi(a^*a)$. Show that $\varphi(ba) = \varphi(b)\varphi(a)$ for all $b \in \mathfrak{A}$.
HINT: use Lemma 9.6.3 to show that $|\varphi(za+b)|^2 \leq \varphi((za+b)^*(za+b))$ for $z \in \mathbb{C}$. Expand.

9.21 (a) Let $\mathfrak{A}$ be a unital C*-algebra and let π be a representation of $M_n(\mathfrak{A})$ on K. Show that there is a representation σ of $\mathfrak{A}$ such that $\pi\big[a_{ij}\big] \simeq \big[\sigma(a_{ij})\big]$.
HINT: let E_{ij} be a set of matrix units for M_n. Show that $F_{ij} = \pi(E_{ij})$ is a set of matrix units in $\pi(M_n(\mathfrak{A}))$. Define $\sigma(a) = F_{11}\pi(a \oplus 0_{n-1})|_{F_{11}K}$.
(b) Show that σ is irreducible if and only if π is.
(c) Extend this result to non-unital C*-algebras.

9.22 Let S be the unilateral shift on l_2 and let $\mathfrak{T} = C^*(S)$.
(a) Show that $\mathfrak{T}$ contains all compact operators.
(b) Identify $\mathfrak{T}/\mathcal{K}(l_2)$.

(c) Find all irreducible representations of $\mathfrak{T}$.

9.23 (a) Let φ be a state on $C(X)$. Prove that $|\varphi(f)| \le \varphi(|f|)$.
HINT: : Reisz Representation theorem.
(b) Show that this fails for non-abelian C*-algebras.
HINT: try M_2.

9.24 Let $\mathfrak{A}$ be a C*-algebra with ideal J, and let φ be a state on $\mathfrak{A}$. Show that there is a unique decomposition $\varphi = \varphi_1 + \varphi_2$ as a sum of positive linear functionals such that the GNS representation π_1 for $\varphi_1/\|\varphi_1\|$ is induced from its restriction to J and the GNS representation π_2 for $\varphi_2/\|\varphi_2\|$ annihilates J. Moreover $\pi_\varphi \simeq \pi_1 \oplus \pi_2$.
HINT: Apply Theorem 9.8.2 to the GNS representation π_φ.

9.25 Let $\mathfrak{A} \subset \mathfrak{B}$ be an inclusion of C*-algebras, and let φ be a state on $\mathfrak{A}$.
(a) First assume that $\mathfrak{A}$ contains an approximate identity for $\mathfrak{B}$. Show that there is a state ψ on $\mathfrak{B}$ extending φ.
(b) If $\mathfrak{B} \subset \mathcal{B}(H)$ and $(e_\lambda)_\Lambda$ is an approximate identity for $\mathfrak{A}$, show that e_λ converge SOT to the projection P onto $\overline{\mathfrak{A}H}$. Use (a) to extend φ to a state on $\mathcal{B}(PH)$, and then extend it to $\mathcal{B}(H)$.
(c) Show that there is a state ψ on $\mathfrak{B}$ extending φ.

9.26 Let $\mathfrak{A} \subset \mathfrak{B}$ be an inclusion of C*-algebras
(a) Show that if π is a cyclic representation of $\mathfrak{A}$ on H, there is a Hilbert space $K \supset H$ and a representation σ of $\mathfrak{B}$ on K so that H reduces $\sigma(\mathfrak{A})$ and $\sigma(a)|_H = \pi(a)$ for $a \in \mathfrak{A}$.
HINT: Exercise **9.25**.
(b) Extend (a) to arbitrary representations.

9.27 Let $\mathfrak{A}$ be a C*-algebra, and let f be a continuous linear functional on $\mathfrak{A}$. Show that there is a representation π of $\mathfrak{A}$ on a Hilbert space H and vectors $\xi, \eta \in H$ so that $f(a) = \langle \pi(a_\xi, \eta\rangle$ for $a \in \mathfrak{A}$.

9.28 (*Akemann–Pedersen, Arveson*). A *quasicentral approximate unit* for $J \lhd \mathfrak{A}$ is an approximate identity $\{f_\lambda\}_\Lambda$ for J such that $\lim_\Lambda \|f_\lambda a - a f_\lambda\| = 0$ for all $a \in \mathfrak{A}$.
(a) Begin with a convex approximate unit $\{e_\mu\}_{\Lambda'}$ for J. If $\mu_0 \in \Lambda'$ and $a_1, \dots, a_n \in \mathfrak{A}$, let $a = a_1 \oplus \cdots \oplus a_n$. Show that $0^{(n)} \in \overline{\mathrm{conv}}\{a e_\mu^{(n)} - e_\mu^{(n)a} : \mu \ge \mu_0 \in \Lambda'\}$.
HINT: use Hahn–Banach and Exercises **9.27** and **9.25**(b).
(b) Build a quasicentral approximate unit for J.

9.29 (a) Let $\mathfrak{s}_1, \mathfrak{s}_2$ be the generators of $\mathcal{O}_2$. Show that $C^*(\mathfrak{s}_1, \mathfrak{s}_2\mathfrak{s}_1, \mathfrak{s}_2^2) \simeq \mathcal{O}_3$.
(b) Show that for $n, k \ge 2$, there is a subalgebra of $\mathcal{O}_n$ isomorphic to $\mathcal{O}_{k(n-1)+1}$.

9.30 (a) Show that the representation $\sigma_{u,\lambda}$ of Example 9.11.9 is irreducible if and only if $u \ne v^p$ for some $p \ge 2$ (i.e. u is primitive).
(b) Show that if $\lambda_1 \ne \lambda_2$ or u_2 is not a cyclic permutation of u_1, then the two representations σ_{u_1,λ_1} and σ_{u_2,λ_2} are not unitarily equivalent.

9.31 Let $\mathcal{O}_\infty$ denote the universal C*-algebra generated by a countable family $\{\mathfrak{s}_i : i \geq 1\}$ satisfying the relations $\mathfrak{s}_i^* \mathfrak{s}_j = \delta_{ij}$ for all $i, j \geq 1$.
(a) Show that $\mathcal{O}_\infty$ exists.
(b) Show that $\mathcal{O}_\infty$ is simple.

9.32 (a) Show that $M_3(\mathcal{O}_2) \simeq \mathcal{O}_2$.
HINT: consider $T_1 = \begin{bmatrix} 0 & 0 & 0 \\ I & 0 & 0 \\ 0 & \mathfrak{s}_1 & \mathfrak{s}_2 \end{bmatrix}$ and $T_2 = \begin{bmatrix} \mathfrak{s}_1 & \mathfrak{s}_2\mathfrak{s}_1 & \mathfrak{s}_2^2 \\ I & 0 & 0 \\ 0 & 0 & 0 \end{bmatrix}$.
(b) Show that $M_n(\mathcal{O}_2) \simeq \mathcal{O}_2$ for $n \geq 2$.

9.33 Show that if $T \in \mathcal{B}(H)$ is invertible and $\sup\{\|T^n\| : n \in \mathbb{Z}\} < \infty$, prove that T is similar to a unitary operator.
HINT: Dixmier's theorem.

9.34 Let $\mathfrak{A} = \varinjlim(\mathfrak{A}_n, \pi_{n,n+1})$ and $\mathfrak{B} = \varinjlim(\mathfrak{B}_n, \rho_{n,n+1})$ be direct limits. Suppose that there are $*$-isomorphisms $\sigma_n : \mathfrak{A}_n \to \mathfrak{B}_n$ such that $\rho_{n,n+1}\sigma_n = \sigma_{n+1}\rho_{n,n+1}$ for $n \geq 1$. Prove that $\mathfrak{A}$ and $\mathfrak{B}$ are isomorphic.

9.35 Let G be a countable discrete group with two finite generating sets S and T. Show that G has subexponential growth with respect to S if and only if it has subexponential growth with respect to T.

9.36 Let $G = \mathbb{Z}_2 * \mathbb{Z}_3$. This is the group generated by $\{u, v\}$ such that $u^2 = 1 = v^3$.
(a) Prove that G is not amenable.
(b) Prove that $C^*_\lambda(G)$ is simple.

9.37 Let $u_{00} = \begin{bmatrix} 1 & 0 \\ 0 & 1 \end{bmatrix}$, $u_{10} = \begin{bmatrix} 1 & 0 \\ 0 & -1 \end{bmatrix}$, $u_{01} = \begin{bmatrix} 0 & 1 \\ 1 & 0 \end{bmatrix}$ and $u_{11} = \begin{bmatrix} 0 & 1 \\ -1 & 0 \end{bmatrix}$; and define $\alpha_{ij} \in \mathrm{Aut}(M_2)$ by $\alpha_{ij} = \mathrm{ad}\, u_{ij}$ for $i, j \in \{0, 1\}$. Prove $M_2 \rtimes_\alpha (\mathbb{Z}/2\mathbb{Z})^2 \simeq M_4$.
HINT: u is not a group representation of $(\mathbb{Z}/2\mathbb{Z})^2$. Let $\pi(A) = A \oplus A$ represent M_2 on $\mathbb{C}^4$. Find a unitary representation of $(\mathbb{Z}/2\mathbb{Z})^2$ on M_4 which implements α.

9.38 Let $\mathfrak{A}$ be the universal C*-algebra generated by two projections.
(a) Show that every irreducible representation of $\mathfrak{A}$ is 1 or 2 dimensional.
(b) Prove that $\mathfrak{A}$ is isomorphic to the subalgebra of $C([0, 1], M_2)$ of functions F such that $F(0)$ and $F(1)$ are diagonal with respect to a fixed basis for $\mathbb{C}^2$.

9.39 Let X be a finite set with n elements, and let σ be a cyclic permutation of elements of X. Prove that $C(X) \rtimes_\sigma \mathbb{Z}/n\mathbb{Z} \simeq M_n(C(\mathbb{T}))$.

9.40 The *discrete Heisenberg group* H_3 is the multiplicative subset of 3×3 matrices of the form $\begin{bmatrix} 1 & a & c \\ 0 & 1 & b \\ 0 & 0 & 1 \end{bmatrix}$ for $a, b, c \in \mathbb{Z}$. Let $u = \begin{bmatrix} 1 & 1 & 0 \\ 0 & 1 & 0 \\ 0 & 0 & 1 \end{bmatrix}$ and $v = \begin{bmatrix} 1 & 0 & 0 \\ 0 & 1 & 1 \\ 0 & 0 & 1 \end{bmatrix}$.
(a) Show that $w = uvu^{-1}v^{-1}$ generates the centre and H_3 is solvable.
(b) Let π be an irreducible representation of $C^*(H_3)$. Show that $\pi(w) = e^{2\pi\theta i} I$.
(c) If θ is irrational, show that $\pi(C^*(H_3)) \simeq \mathcal{A}_\theta$.
(d) If $\theta = \frac{m}{n}$ with $\gcd(m, n) = 1$, show that $\pi(C^*(H_3)) \simeq M_n$.

9.41 (a) Show that every separable C*-algebra is isomorphic to a C*-subalgebra of the *Calkin algebra* $\mathcal{B}(H)/\mathcal{K}(H)$.

(b) Show that the Calkin algebra has no representation on a separable Hilbert space. HINT: show that it contains a copy of l_∞/c_0.

9.42 Let S be the unilateral shift and let $\mathfrak{s} = q(S)$, where $q : \mathcal{B}(H) \to \mathcal{B}(H)/\mathcal{K}(H)$ is the quotient map. Show that $C^*(\mathfrak{s})$ is abelian, but there is no abelian C*-subalgebra $\mathfrak{A} \subset \mathcal{B}(H)$ such that $q(\mathfrak{A}) = C^*(\mathfrak{s})$.
HINT: Fredholm index.

Notes on Chapter 9

The first C*-algebras studied were the von Neumann algebras, beginning with the work of von Neumann [316] in 1936, where he proved the double commutant theorem. See the notes in the next chapter for more detail. The original theory of C*-algebras was developed by Gelfand and Naimark [120] in 1943. They had an additional axiom that $1 + a^*a$ is invertible for $a \in \mathfrak{A}$, but conjectured that this was unnecessary. This condition was removed by Kaplansky (unpublished) and Fukamiya [117] in 1952. Kaplansky [173] proved his density theorem in 1951. Segal [283] expanded and clarified the GNS construction in 1947, and established the existence of approximate identities. Gelfand and Naimark [121] overlaps with Segal. The spectral theorem for normal operators is due to Stone [290] and von Neumann [314]. Hilbert had established a weaker version for self-adjoint operators.

Glimm [125] defined and classified the UHF algebras. Cuntz [65] defined the algebras $\mathcal{O}_n$ and developed their basic properties. These have both proven to be extremely important examples. Powers [248] showed that $C^*_\lambda(\mathbb{F}_2)$ is simple with unique trace. Invariant means of discrete groups were introduced by von Neumann [312] in 1929. Their study has a long history. The characterization of amenability in terms of the left regular representation is due to Hulanicki [155]. A version of the crossed product was introduced in 1958 by Turumaru [306]. The modern definition was studied extensively a decade later by Zeller-Meier [334]. The roots of the idea go back to the Murray and von Neumann group measure construction [211].

There are a number of books on C*-algebras. The author is keen on his own book [70]. Pedersen's book [234] is excellent, and has a lot of material on C*-algebras of groups and group actions on operator algebras. There is Dixmier's classic book [95], which is also very interested in the connection to group representations. Murphy's book [210] is another reasonable source. Crossed products are covered in Williams [326] and also [124]. Amenability in various contexts is studied in Runde [270]. For some more advanced topics, also with a strong view of groups, is the wonderful book by Brown and Ozawa [41]. Those interested in the modern theory of C*-algebras will want to learn about K-theory, and the book by Rørdam, Larsen and Lausten [263] is a very readable source. Lin [191] provides an introduction to the classification program.

CHAPTER 10

Von Neumann Algebras

In this chapter, we study von Neumann algebras in detail. We begin by showing that $\mathcal{B}(H)$, and hence every von Neumann algebra, is a dual space. Then we study the tight relationship between normal and weak-$*$ continuous homomorphisms. We show that every C*-algebra sits inside a universal von Neumann algebra. Then we characterize abstract von Neumann algebras as C*-algebras which are dual spaces. The last half of the chapter is devoted to the classification of von Neumann algebras into types, and exhibiting examples.

10.1. Trace Class Operators

In this section, we will show that the dual of $\mathcal{K}(H)$ is $\mathfrak{S}_1$ and the dual of $\mathfrak{S}_1$ is $\mathcal{B}(H)$. Hence $\mathcal{B}(H)$ is a dual space. The space of compact operators is not a dual space (Exercise **5.33**).

Recall that the singular decomposition of a compact operator K has the form $K = \sum_{n\ge 1} s_n\, \eta_n \xi_n^*$ for two orthonormal families $\{\xi_n\}$ and $\{\eta_n\}$ and a decreasing sequence (s_n) in c_0^+. In Definition 5.4.7, we defined the *trace class operators*

$$\mathfrak{S}_1 = \Big\{K \in \mathcal{K}(H) : \|K\|_1 = \sum_{n\ge 1} s_n < \infty\Big\}.$$

10.1.1. LEMMA. *Let $T \in \mathcal{K}(H)$ and let $\{e_\lambda\}_\Lambda$ be any orthonormal basis for H. Then*

$$\|T\|_1 = \sum_{\Lambda} \langle |T| e_\lambda, e_\lambda\rangle \in [0, +\infty].$$

Let $\mathcal{E} = \{e_\lambda\}_\Lambda$ and $\mathcal{F} = \{f_\lambda\}_\Lambda$ denote arbitrary orthonormal families in H of the same cardinality. Then

$$\|T\|_1 = \sup_{|\mathcal{E}|=|\mathcal{F}|} \sum_{\Lambda} |\langle T e_\lambda, f_\lambda\rangle| = \sup_{|\mathcal{E}|=|\mathcal{F}|<\infty} \sum_{\Lambda} |\langle T e_\lambda, f_\lambda\rangle|.$$

PROOF. Let $\{e_\lambda\}_\Lambda$ and $\{f_\mu\}_{\Lambda'}$ be orthonormal bases. By the Parseval identity

K. R. Davidson, *Functional Analysis and Operator Algebras*, CMS/CAIMS Books in Mathematics 13,
https://doi.org/10.1007/978-3-031-63665-3_10

$$\sum_{\Lambda} \|Te_\lambda\|^2 = \sum_{\Lambda}\sum_{\Lambda'} |\langle Te_\lambda, f_\mu\rangle|^2 = \sum_{\Lambda'}\sum_{\Lambda} |\langle Te_\lambda, f_\mu\rangle|^2$$
$$= \sum_{\Lambda'}\sum_{\Lambda} |\langle T^* f_\mu, e_\lambda\rangle|^2 = \sum_{\Lambda'} \|T^* f_\mu\|^2.$$

In particular, the first sum must be independent of the choice of basis.

Write $T = \sum_{n\ge1} s_n \eta_n \xi_n^*$ for orthonormal sets $\{\xi_n\}_{n\ge1}$ and $\{\eta_n\}_{n\ge1}$. Then $|T| = \sum_{n\ge1} s_n \xi_n \xi_n^*$ and $|T|^{1/2} = \sum_{n\ge1} s_n^{1/2} \xi_n \xi_n^*$. Let $\mathcal{E} = \{e_\lambda\}_\Lambda$ be an orthonormal basis including $\{\xi_n\}_{n\ge1}$. Thus $\||T|^{1/2}\xi_n\|^2 = s_n$ and $|T|^{1/2}e_\lambda = 0$ if $e_\lambda \in \mathcal{E} \setminus \{\xi_n\}_{n\ge1}$. Therefore

$$\|T\|_1 = \sum_{n\ge1} s_n = \sum_{\Lambda} \||T|^{1/2} e_\lambda\|^2 = \sum_{\Lambda'} \||T|^{1/2} f_\mu\|^2$$
$$= \sum_{\Lambda'} \langle |T|^{1/2} f_\mu, |T|^{1/2} f_\mu\rangle = \sum_{\Lambda'} \langle |T| f_\mu, f_\mu\rangle.$$

Again this sum is independent of the choice of basis.

Next we calculate

$$\begin{aligned}\sum_{\Lambda} |\langle Te_\lambda, f_\lambda\rangle| &= \sum_{\Lambda} \Big| \sum_{n\ge1} \langle s_n \langle e_\lambda, \xi_n\rangle \eta_n, f_\lambda\rangle \Big| \\ &\le \sum_{\Lambda}\sum_{n\ge1} s_n |\langle e_\lambda, \xi_n\rangle|\, |\langle \eta_n, f_\lambda\rangle| \\ &= \sum_{n\ge1} s_n \sum_{\Lambda} |\langle \xi_n, e_\lambda\rangle|\, |\langle \eta_n, f_\lambda\rangle| \\ &\le \sum_{n\ge1} s_n \Big(\sum_{\Lambda} |\langle \xi_n, e_\lambda\rangle|^2 \sum_{\Lambda} |\langle \eta_n, f_\lambda\rangle|^2 \Big)^{1/2} \\ &\le \sum_{n\ge1} s_n \|\xi_n\|\, \|\eta_n\| = \sum_{n\ge1} s_n = \|T\|_1.\end{aligned}$$

Thus the supremum is at most $\|T\|_1$. On the other hand, if we use $\mathcal{E}_n = \{\xi_1, \dots, \xi_n\}$ and $\mathcal{F}_n = \{\eta_1, \dots, \eta_n\}$, then

$$\sum_{i=1}^{n} |\langle T\xi_i, \eta_i\rangle| = \sum_{i=1}^{n} s_i.$$

Taking the sup over $n \ge 1$ yields $\|T\|_1$. ∎

We collect various elementary properties of $\mathfrak{S}_1$.

10.1.2. THEOREM.

(1) *If* $T \in \mathfrak{S}_1$, *then* $T^* \in \mathfrak{S}_1$ *and* $\|T^*\|_1 = \|T\|_1$. *Also* $\|T\| \le \|T\|_1$.

(2) *If* $S, T \in \mathfrak{S}_1$, *then* $S + T \in \mathfrak{S}_1$ *and* $\|S + T\|_1 \le \|S\|_1 + \|T\|_1$.

(3) $\mathfrak{S}_1$ *is a Banach space.*

(4) *Let* $T \in \mathfrak{S}_1$, *and let* $\{e_\lambda\}_\Lambda$ *and* $\{f_\mu\}_{\Lambda'}$ *be orthonormal bases for* H. *Then* $\sum\limits_\Lambda \langle Te_\lambda, e_\lambda\rangle = \sum\limits_{\Lambda'} \langle Tf_\mu, f_\mu\rangle$. *This defines a norm 1 linear functional on* $\mathfrak{S}_1$ *denoted* $\operatorname{Tr}(T)$; *and* $\|T\|_1 = \operatorname{Tr}(|T|)$.

(5) *If* $T \in \mathfrak{S}_1$ *and* $A, B \in \mathcal{B}(H)$, *then* $ATB \in \mathfrak{S}_1$ *and*

$$\|ATB\|_1 \leq \|A\|\,\|T\|_1\|B\|.$$

Thus $\mathfrak{S}_1$ *is a (non-closed) 2-sided ideal of* $\mathcal{B}(H)$.

(6) *If* $T \in \mathfrak{S}_1$ *and* $A \in \mathcal{B}(H)$, *then* $\operatorname{Tr}(AT) = \operatorname{Tr}(TA)$.

PROOF. If $T = \sum_{n\geq 1} s_n \eta_n \xi_n^*$, then $T^* = \sum_{n\geq 1} s_n \xi_n \eta_n^*$. So $\|T^*\|_1 = \|T\|_1$ by Lemma 10.1.1. Moreover, $\|T\| = s_1 \leq \|T\|_1$. Hence (1) holds.

For (2), if $\{e_\lambda\}_\Lambda$ and $\{f_\lambda\}_\Lambda$ are orthonormal families in H, Lemma 10.1.1 yields

$$\sum_\Lambda |\langle (S+T)e_\lambda, f_\lambda\rangle| \leq \sum_\Lambda |\langle Se_\lambda, f_\lambda\rangle| + |\langle Te_\lambda, f_\lambda\rangle| \leq \|S\|_1 + \|T\|_1.$$

Taking the supremum yields $\|S+T\|_1 \leq \|S\|_1 + \|T\|_1$. Hence $\mathfrak{S}_1$ is a normed vector space.

For completeness, consider a Cauchy sequence $\{T_n\}$ in $\mathfrak{S}_1$. By (1), this is Cauchy in $\mathcal{K}(H)$. Let T be the limit in $\mathcal{K}(H)$. Note that $\langle Te, f\rangle = \lim\limits_{n\to\infty} \langle T_n e, f\rangle$ for $e, f \in H$. Also $\|T_n\|_1$ is Cauchy, so let $L = \lim_{n\to\infty} \|T_n\|_1$. If $\{e_1, \dots, e_k\}$ and $\{f_1, \dots, f_k\}$ are finite orthonormal sets of the same size,

$$\sum_{i=1}^k |\langle Te_i, f_i\rangle| = \lim_{n\to\infty} \sum_{i=1}^k |\langle T_n e_i, f_i\rangle| \leq \lim_{n\to\infty} \|T_n\|_1 = L.$$

By Lemma 10.1.1, $\|T\|_1 \leq L$. So $T \in \mathfrak{S}_1$. Given $\varepsilon > 0$, choose N so large that $\|T_n - T_m\|_1 < \varepsilon$ for $N \leq m \leq n$. Then for $N \leq m$, applying the same argument to $\{T_n - T_m\}_{n\geq m}$ yields $\|T - T_m\|_1 \leq \varepsilon$. Therefore $\lim\limits_{n\to\infty} T_n = T$ in $\mathfrak{S}_1$.

For (4), calculate

$$\sum_\Lambda \langle Te_\lambda, e_\lambda\rangle = \sum_\Lambda \sum_{n\geq 1} s_n \langle e_\lambda, \xi_n\rangle\langle \eta_n, e_\lambda\rangle.$$

This sum, with absolute values, turns up in the calculation of Lemma 10.1.1 with $f_\lambda = e_\lambda$. This sum is bounded by $\|T\|_1$, so this sum is absolutely convergent. Therefore

$$\sum_\Lambda \langle Te_\lambda, e_\lambda\rangle = \sum_{n\geq 1} s_n \sum_\Lambda \langle \eta_n, e_\lambda\rangle\overline{\langle \xi_n, e_\lambda\rangle} = \sum_{n\geq 1} s_n \langle \eta_n, \xi_n\rangle.$$

The last expression is independent of the basis $\{e_\lambda\}_\Lambda$. Therefore there is a well-defined linear functional $\operatorname{Tr}(T) = \sum_\Lambda \langle Te_\lambda, e_\lambda\rangle$ on $\mathfrak{S}_1$ which is independent of the basis. By Lemma 10.1.1, $|\operatorname{Tr}(T)| \leq \operatorname{Tr}(|T|) = \|T\|_1$. So $\|\operatorname{Tr}\| = 1$.

Let $A \in \mathcal{B}(H)$. Note that

$$|AT| = (T^* A^* AT)^{1/2} \le (T^* \|A\|^2 T)^{1/2} = \|A\|\,|T|.$$

Then by Lemma 10.1.1,

$$\|AT\|_1 = \sum_\Lambda \langle |AT| e_\lambda, e_\lambda\rangle \le \sum_\Lambda \|A\| \langle |T| e_\lambda, e_\lambda\rangle = \|A\|\,\|T\|_1.$$

By taking adjoints and using (1), we get $\|TB\| \le \|T\|_1 \|B\|$. Hence $\mathfrak{S}_1$ is a 2-sided ideal of $\mathcal{B}(H)$. It isn't closed in $\mathcal{B}(H)$ because $\mathfrak{S}_1$ is dense in $\mathcal{K}(H)$.

For the trace property (6), we extend $\{\xi_n\}$ and $\{\eta_n\}$ to orthonormal bases,

$$\begin{aligned}\mathrm{Tr}(AT) &= \sum_{n\ge1} \langle AT\xi_n, \xi_n\rangle = \sum_{n\ge1} s_n \langle A\eta_n, \xi_n\rangle \\ &= \sum_{n\ge1} \langle A\eta_n, T^*\eta_n\rangle = \sum_{n\ge1} \langle TA\eta_n, \eta_n\rangle = \mathrm{Tr}(TA).\end{aligned}$$ ∎

Now we can obtain our duality result. Throughout this chapter, we will write $\mathfrak{A}^\dagger$ for the dual space of a C*-algebra $\mathfrak{A}$ to avoid confusion with the adjoint.

10.1.3. THEOREM. *The dual of $\mathcal{K}(H)$ is $\mathfrak{S}_1$ and the dual of $\mathfrak{S}_1$ is $\mathcal{B}(H)$ with the pairing $(T, A) = \mathrm{Tr}(TA)$ for $T \in \mathfrak{S}_1$ and $A \in \mathcal{B}(H)$.*

PROOF. For $T \in \mathfrak{S}_1$, define $\tau_T(A) = \mathrm{Tr}(TA)$. This is linear on $\mathcal{B}(H)$, and $|\mathrm{Tr}(TA)| \le \|T\|_1\|A\|$, so that $\|\tau_T\| \le \|T\|_1$. On the other hand, if $T = \sum_{n\ge1} s_n\eta_n\xi_n^*$, let $K_p = \sum_{n=1}^p \xi_n\eta_n^*$. Then $\|K_p\| = 1$, K_p is finite rank, and

$$\tau_T(K_p) = \mathrm{Tr}(TK_p) = \mathrm{Tr}\Big(\sum_{n=1}^p s_n\eta_n\eta_n^*\Big) = \sum_{n=1}^p s_n.$$

Hence $\|\tau_T|_{\mathcal{K}(H)}\| = \|T\|_1$. So $\mathfrak{S}_1$ is isometric to a subspace of $\mathcal{K}(H)^\dagger$.

Now let $\varphi \in \mathcal{K}(H)^\dagger$. Define a sesquilinear form on H by $[\eta, \xi] = \varphi(\eta\xi^*)$. Then

$$\big|[\eta, \xi]\big| \le \|\varphi\|\,\|\eta\xi^*\| = \|\varphi\|\,\|\eta\|\,\|\xi^*\|.$$

By Proposition 4.2.2, there is a bounded operator $T \in \mathcal{B}(H)$ so that

$$\varphi(\eta\xi^*) = [\eta, \xi] = \langle T\eta, \xi\rangle = \mathrm{Tr}(T\eta\xi^*).$$

By linearity, $\varphi(F) = \mathrm{Tr}(TF)$ for all finite rank operators F. If P is a finite rank projection, then $TP \in \mathfrak{S}_1$ and

$$\|TP\|_1 = \|\tau_{TP}\| = \|\varphi|_{P\mathcal{K}(H)}\| \le \|\varphi\|.$$

If T is not compact, then T is bounded below on an infinite dimensional subspace by Exercise **9.12**; so this inequality will be violated for a large finite dimensional subspace. If T is compact, say $T = \sum_{n\ge1} s_n\eta_n\xi_n^*$, then using the same operators K_p as before, we obtain $\sum_{n=1}^p s_n \le \|\varphi\|$ for all $p \ge 1$. Therefore $T \in \mathfrak{S}_1$ and $\varphi = \tau_T$; whence $\mathcal{K}(H)^\dagger = \mathfrak{S}_1$.

Similarly, if $A \in \mathcal{B}(H)$, then $\psi_A(T) = \operatorname{Tr}(AT)$ is bounded on $\mathfrak{S}_1$ with $|\psi_A(T)| \le \|A\|\,\|T\|_1$. If ξ is a unit vector and $A\xi = \lambda\eta$ for a unit vector η, let $T = \xi\eta^*$. Then $\|T\|_1 = 1$ and $|\psi_A(T)| = |\operatorname{Tr}(\lambda\eta\eta^*)| = |\lambda|$. Taking the supremum over the unit ball shows that $\|\psi_A\| = \|A\|$.

Now if $\psi \in \mathfrak{S}_1^\dagger$, we again define $[\eta, \xi] = \psi(\eta\xi^*)$ and use Proposition 4.2.2 to obtain a bounded operator $A \in \mathcal{B}(H)$ so that

$$\psi(\eta\xi^*) = [\eta, \xi] = \langle A\eta, \xi\rangle = \operatorname{Tr}(A\eta\xi^*).$$

By linearity, $\psi(F) = \operatorname{Tr}(AF) = \psi_A(F)$ for all finite rank operators. But ψ and ψ_A are continuous and agree on a dense subspace. Thus $\psi = \psi_A$. Therefore $\mathcal{B}(H) = \mathfrak{S}^\dagger$ isometrically. ∎

10.1.4. Remark. The functional τ_T on $\mathcal{B}(H)$ is positive if and only if $T \ge 0$. Indeed, if $\tau_T \ge 0$, then $0 \le \tau_T(\xi\xi^*) = \langle T\xi, \xi\rangle$ for all $\xi \in H$, whence $T \ge 0$. Conversely, if $T \ge 0$, say $T = \sum_{n\ge1} s_n\xi_n\xi_n^*$, then $\tau_T(A) = \sum_{n\ge1} s_n\langle A\xi_n, \xi_n\rangle$ is a positive linear combination of vector states, and hence is positive.

When $T \ge 0$ and $\|T\|_1 = 1$, τ_T is a weak-$*$ continuous state. Every weak-$*$ continuous linear functional τ_T is a linear combination of weak-$*$ continuous states obtained by writing $T = T_1 - T_2 + iT_3 - iT_4$ where $T_i \ge 0$, and scaling the T_i's.

10.2. The Weak* Topology and Normality

Theorem 10.1.3 provides a weak-$*$ topology on $\mathcal{B}(H)$. As a consequence, we will show that all von Neumann algebras are dual spaces. We relate the notions of normal maps, defined in terms of the order structure, and weak-$*$ continuous maps.

In the literature, the weak-$*$ topology is also called the *σ-weak topology* or ultraweak topology. The basic open neighbourhoods of 0 are given by a finite set of trace class operators $T_1, \dots, T_n$:

$$\{A \in \mathcal{B}(H) : |\operatorname{Tr}(AT_i))| < 1,\ 1 \le i \le n\}.$$

The finite rank operators yield functionals which are WOT-continuous, but there are more functionals in $\mathfrak{S}_1$. Thus there are more open sets. This topology is stronger than the weak operator topology.

Another way to understand the weak-$*$ topology is to consider the infinite ampliation of operators in $\mathcal{B}(H)$, by analogy to the technique used in the proof of the Double Commutant theorem. Let $H^{(\infty)} = \sum_{n\ge1}^{\oplus} H_n$ where each $H_n = H$. This is the Hilbert space direct sum and consists of vectors $\xi = (\xi_1, \xi_2, \dots)$ such that

$$\|\xi\| = \Big(\sum_{n\ge1} \|\xi_n\|^2\Big)^{1/2} < \infty.$$

For $A \in \mathcal{B}(H)$, define $A^{(\infty)} \in \mathcal{B}(H^{(\infty)})$ by $A^{(\infty)}(\xi_1, \xi_2, \dots) = (A\xi_1, A\xi_2, \dots)$. The weak operator topology on $\mathcal{B}(H^{(\infty)})$ determines the σ-weak topology on $\mathcal{B}(H)$, which is the weakest topology such that the functionals

$$\varphi(A) = \langle A^{(\infty)}\xi, \eta\rangle = \sum_{n\geq 1} \langle A\xi_n, \eta_n\rangle$$

are continuous for all ξ and η in $H^{(\infty)}$. Define $T = \sum_{n\geq 1} \xi_n \eta_n^*$. Then

$$\|T\|_1 \leq \sum_{n\geq 1} \|\xi_n\| \, \|\eta_n\| \leq \Big(\sum_{n\geq 1} \|\xi_n\|^2\Big)^{1/2} \Big(\sum_{n\geq 1} \|\eta_n\|^2\Big)^{1/2} = \|\xi\| \, \|\eta\|.$$

Thus $T \in \mathfrak{S}_1$ and $\varphi = \tau_T$. Hence this is a weak-$*$ continuous functional.

Conversely, if $T = \sum_{n\geq 1} s_n \xi_n \eta_n^*$ is trace class with orthonormal sets $\{\xi_n\}$ and $\{\eta_n\}$, define

$$\xi = (\sqrt{s_1}\xi_1, \sqrt{s_2}\xi_2, \dots) \quad \text{and} \quad \eta = (\sqrt{s_1}\eta_1, \sqrt{s_2}\eta_2, \dots).$$

Then $\|\xi\| = \|\eta\| = \big(\sum_{n\geq 1} s_n\big)^{1/2} = \|T\|_1^{1/2}$ and

$$\langle A^{(\infty)}\xi, \eta\rangle = \sum_{n\geq 1} s_n \langle A\xi_n, \eta_n\rangle = \operatorname{Tr}(AT).$$

Thus this functional is σ-weak continuous. Therefore these two topologies coincide.

This suggests another topology, the *σ-strong topology*, which is the weakest topology such that the seminorms

$$A \to \|A^{(\infty)}\xi\| = \Big(\sum_{n\geq 1} \|A\xi_n\|^2\Big)^{1/2}$$

are continuous for all $\xi \in H^{(\infty)}$, or equivalently $\sum_{n\geq 1} \|\xi_n\|^2 < \infty$. This is the SOT-topology on $\mathcal{B}(H^{(\infty)})$ restricted to $\mathcal{B}(H)^{(\infty)}$. Since the σ-weak topology on $\mathcal{B}(H)$ corresponds to the WOT topology on $\mathcal{B}(H)^{(\infty)}$, Exercise **3.4** shows that the σ-strong topology has the same continuous linear functionals as the σ-weak topology, namely the weak-$*$ continuous functionals.

The WOT topology is weaker than the SOT topology and the weak-$*$ topology, and both of these are weaker than the σ-SOT topology, which is weaker than the norm topology. If we return to the proof of the Double Commutant theorem, it is easy to see that the proof also works for the case $n = \infty$, showing that if $\mathfrak{A} \subset \mathcal{B}(H)$ is a non-degenerate C*-algebra, then $\mathfrak{A}'' \subset \overline{\mathfrak{A}}^{\sigma\text{-SOT}}$. We obtain a strengthening of the Double Commutant theorem.

10.2.1. THEOREM. *Let $\mathfrak{A} \subset \mathcal{B}(H)$ be a non-degenerate C*-algebra. Then*

$$\mathfrak{A}'' = \overline{\mathfrak{A}}^{\text{WOT}} = \overline{\mathfrak{A}}^{\text{SOT}} = \overline{\mathfrak{A}}^{\textit{weak-}*} = \overline{\mathfrak{A}}^{\sigma\text{-SOT}}.$$

Thus a unital weak-$*$ closed C-subalgebra of $\mathcal{B}(H)$ is a von Neumann algebra.*

The weak-$*$ closed subspaces of a dual space are dual spaces by Corollary 2.6.17. Since $\mathcal{B}(H)$ is a dual space, every von Neumann algebra has a predual and a corresponding weak-$*$ topology.

10.2.2. COROLLARY. *Let $\mathcal{M} \subset \mathcal{B}(H)$ be a von Neumann algebra. Then $\mathcal{M}$ has a predual $\mathcal{M}_*$.*

We will frequently use the following easy result.

10.2.3. PROPOSITION. *The* WOT *and weak-$*$ topologies coincide on $\overline{b_1(\mathcal{B}(H))}$.*

PROOF. Since the WOT topology is weaker than the weak-$*$ topology, it suffices to show that if $\{A_\lambda\}_\Lambda$ is a bounded net in $\mathcal{B}(H)$ (say by M) such that WOT–$\lim_\Lambda A_\lambda = A$, them w^*–$\lim_\Lambda A_\lambda = A$. Let $T \in \mathfrak{S}_1$, $\varepsilon > 0$ and pick a finite rank F so that $\|T - F\|_1 < \varepsilon$. Since $\tau_F(A) = \operatorname{Tr}(AF)$ is WOT-continuous, there is a λ_0 so that $|\operatorname{Tr}((A - A_\lambda)F)| < \varepsilon$ for all $\lambda \geq \lambda_0$. Hence

$$|\operatorname{Tr}((A - A_\lambda)T)| \leq |\operatorname{Tr}((A - A_\lambda)F)| + \|A - A_\lambda\| \|T - F\|_1 < (2M + 1)\varepsilon.$$

Thus $\operatorname{Tr}((A - A_\lambda)T) \to 0$. So w^*–$\lim_\Lambda A_\lambda = A$. ∎

10.2.4. DEFINITION. A bounded linear functional φ on a von Neumann algebra $\mathcal{M}$ is *normal* if whenever $(A_\lambda)_\Lambda$ is a monotone increasing bounded net of self-adjoint operators with $A = \sup_\Lambda A_\lambda = $ SOT–$\lim_\Lambda A_\lambda$, we have $\varphi(A) = \lim_\Lambda \varphi(A_\lambda)$.

A bounded linear functional φ on $\mathcal{M}$ is *completely additive* if whenever $\{E_\lambda\}_\Lambda$ are pairwise orthogonal projections with $E = $ SOT–$\sum_\Lambda E_\lambda = \sup_\Lambda\{E_\lambda\}$, we have $\varphi(E) = \sum_\Lambda \varphi(E_\lambda)$.

When φ is positive, this definition agrees with Definition 9.3.8. In this case, $\varphi(A_\lambda)$ is monotone increasing and less than $\varphi(A)$. So normality means that φ respects suprema. This is a property of the von Neumann algebra and is independent of how it is represented on some Hilbert space. Note that normality is in general easier to check than weak-$*$ continuity.

Countable additivity is even easier to verify. One defines the sum to be the limit over the net of finite subsets of Λ ordered by inclusion. If this sum converges, then only countably many of the terms $\varphi(E_\lambda)$ are non-zero. If φ is completely additive, the sum $\sum_\Lambda \varphi(E_\lambda)$ must be unconditional, and hence absolutely convergent. It is clear that normal maps are completely additive. Observe that $\varphi^*(A) = \overline{\varphi(A^*)}$ is also completely additive. So $\operatorname{Re}\varphi = \frac{1}{2}(\varphi + \varphi^*)$ and $\operatorname{Im}\varphi = \frac{1}{2i}(\varphi - \varphi^*)$ are completely additive, and self-adjoint; i.e. they take real values on $\mathcal{M}_{sa}$.

10.2.5. LEMMA. *Let $\mathcal{M}$ be a von Neumann algebra, and let φ be a completely additive self-adjoint linear functional on $\mathcal{M}$. Suppose that E is a projection in $\mathcal{M}$. Then there is a projection $P \le E$ such that $\varphi|_{P\mathcal{M}P}$ is positive and $\varphi(P) \ge \varphi(E)$.*

PROOF. Consider families $\{E_\lambda\}$ of pairwise orthogonal projections such that $E_\lambda \le E$ and $\varphi(E_\lambda) < 0$. The union of a nested collection of such families is a larger such family. Hence by Zorn's lemma, there is a maximal family of such projections $\{E_\lambda\}_\Lambda$. Let

$$E_0 = \text{SOT-}\sum_\Lambda E_\lambda = \sup_\Lambda\{E_\lambda\}.$$

By complete additivity,

$$\varphi(E_0) = \sum_\Lambda \varphi(E_\lambda) < 0$$

unless $E_0 = 0$. Let $P = E - E_0$. Then $\varphi(P) = \varphi(E) - \varphi(E_0) \ge \varphi(E)$, with equality only when $P = E$. If $0 \le Q \le P$ is a projection such that $\varphi(Q) < 0$, this would violate the maximality of the family $\{E_\lambda\}$. Hence $\varphi(Q) \ge 0$. If $0 \le A \le P$, by Theorem 9.3.6 A is the norm limit of convex combinations of projections less than P. Therefore $\varphi(A) \ge 0$; whence $\varphi|_{P\mathcal{M}P} \ge 0$. ∎

Next we identify the normal states. Recall that positive linear functionals on C*-algebras are automatically norm continuous, with $\|\varphi\| = \varphi(1)$, by Lemma 9.6.4.

10.2.6. THEOREM. *Let φ be a positive linear functional on a von Neumann algebra $\mathcal{M}$. The following are equivalent:*

(1) *φ is weak-$*$ continuous.*

(2) *φ is normal.*

(3) *φ is completely additive.*

(4) *φ is SOT-continuous on the unit ball.*

PROOF. If φ is non-zero, we can normalize so that $\varphi(I) = 1$, making φ a state.

(1) $\Rightarrow$ (2). Suppose that φ is weak-$*$ continuous, and let $\{A_\lambda\}_\Lambda$ be an increasing net in $\mathcal{M}_{sa}$ with $\sup_\Lambda A_\lambda = A$. By restricting to $\{\lambda \ge \lambda_0\}$, we get $A_{\lambda_0} \le A_\lambda \le A$, making this is a bounded net. By Lemma 9.3.5, $A = \text{SOT}\text{–}\lim_\Lambda A_\lambda$, and hence $A = \text{WOT}\text{–}\lim_\Lambda A_\lambda$. The WOT and weak-$*$ topologies agree on closed balls. Hence $A = w^*\text{–}\lim_\Lambda A_\lambda$. Therefore $\varphi(A) = \lim_\Lambda \varphi(A_\lambda)$. Thus φ is normal.

(2) $\Rightarrow$ (3) is clear.

(3) $\Rightarrow$ (4). Let φ be a completely additive state. If $E \ne 0$ is a projection in $\mathcal{M}$, choose a vector $\xi \in H$ so that $\varphi(E) < \langle E\xi, \xi\rangle =: \psi(E)$. Clearly ψ is weak-$*$ continuous, and thus countably additive. Apply Lemma 10.2.5 to $\psi - \varphi$ to obtain a projection $P \le E$ so that

$$(\psi - \varphi)|_{P\mathcal{M}P} \ge 0 \quad \text{and} \quad (\psi - \varphi)(P) \ge (\psi - \varphi)(E) > 0.$$

Thus $P \neq 0$. By the Cauchy–Schwarz inequality, for any $B \in \mathcal{M}$,

$$|\varphi(BP)|^2 \leq \varphi(PB^*BP)\varphi(I) \leq \psi(PB^*BP) = \langle PB^*BP\xi, \xi\rangle = \|BP\xi\|^2.$$

Therefore $\varphi_P(B) := \varphi(BP)$ is SOT continuous.

Use Zorn's lemma to obtain a maximal family $\{P_\lambda\}_\Lambda$ of pairwise orthogonal projections such that φ_{P_λ} is SOT continuous. Let $P_0 = \sup\{P_\lambda\}_\Lambda$. If $P_0 < I$, the previous paragraph yields a projection $0 \neq P \leq P_0^\perp$ such that φ_P is SOT continuous, contradicting the maximality of $\{P_\lambda\}_\Lambda$. Thus $P_0 = I$. For $F \Subset \Lambda$ finite, let $P_F = \sum_{\lambda\in F} P_\lambda$. By complete additivity,

$$1 = \varphi(I) = \sup_{F \Subset \Lambda} \varphi(P_F) = \lim_{F \Subset \Lambda} \varphi(P_F).$$

Let $B \in \mathcal{M}$ with $\|B\| \leq 1$. Then

$$|\varphi(BP_F^\perp)|^2 \leq \varphi(B^*B)\varphi(P_F^\perp) \leq 1 - \varphi(P_F).$$

Thus $\lim_{F\Subset\Lambda} \varphi(BP_F^\perp) = 0$ *uniformly* over the unit ball. Therefore

$$\varphi(B) = \lim_{F\Subset\Lambda} \sum_{\lambda\in F} \varphi(BP_\lambda) = \lim_{F\Subset\Lambda} \varphi(BP_F).$$

is a uniform limit of SOT continuous functionals, and hence is SOT continuous on the unit ball.

(4) $\Rightarrow$ (1). Since φ is SOT continuous on the unit ball, if K is a closed convex subset of $\mathbb{C}$, then $\varphi^{-1}(K) \cap \overline{b_1(\mathcal{M})}$ is an SOT-closed convex set. Hence it is also WOT closed, because these two topologies have the same continuous linear functionals (Exercise **3.4**), whence the same closed convex sets. Therefore φ is WOT continuous on the ball. The WOT and weak-$*$ topologies agree on closed balls by Proposition 10.2.3. Thus it is weak-$*$ continuous on the closed ball. By the Krein–Smulian theorem 3.6.2, φ is weak-$*$ continuous. ∎

10.2.7. EXAMPLE. Let $T = \sum_{n\geq 1} 2^{-n}\xi_n\xi_n^*$ for an orthonormal set $\{\xi_n : n \geq 1\}$. Then $\tau_T(A) = \mathrm{Tr}(AT) = \mathrm{Tr}(T^{1/2}AT^{1/2})$ is a positive, weak-$*$ continuous functional on $\mathcal{B}(H)$, and $\|T\|_1 = 1$. Thus τ_T is a normal state. However since T is not finite rank, it is not WOT continuous, and thus is not SOT continuous.

We can see this explicitly. Assume that $\{\xi_n : n \geq 1\}$ is a basis, so that τ_T is faithful. Let Λ be the net of finite dimensional subspaces F of H, and let $T_F = \tau_T(P_F^\perp)^{-1}P_F^\perp$. This is an unbounded net. However for $\xi \in H$, $T_F\xi = 0$ for $F \geq \operatorname{span}\xi$. Thus T_F converges SOT to 0, and hence also WOT. However $\tau_T(T_F) = 1$ for all F, so it does not converge to $\tau_T(0) = 0$.

Now we extend this theorem to apply to arbitrary functionals on $\mathcal{M}$. Statement (4) regarding SOT-continuity on the unit ball has to wait until Corollary 10.2.11.

10.2.8. THEOREM. *Let φ be a bounded linear functional on a von Neumann algebra $\mathcal{M}$. The following are equivalent:*

(1) *φ is weak-$*$ continuous.*

(2) *φ is normal.*

(3) *φ is completely additive.*

PROOF. Clearly weak-$*$ continuous implies normal implies completely additive; so (1) $\Rightarrow$ (2) $\Rightarrow$ (3). So assume that φ is completely additive. Then $\operatorname{Re}\varphi$ and $\operatorname{Im}\varphi$ are completely additive. So we may suppose that φ is self-adjoint and can scale so that $\|\varphi\| \leq 1$.

Define $\alpha = \sup\{\varphi(A) : 0 \leq A \leq I\} \geq 0$. Then $\alpha \leq 1$. Fix $0 < \varepsilon \leq \frac{1}{2}$. Choose $A \in \mathcal{M}$ with $0 \leq A \leq 1$ so that $\varphi(A) > \alpha - \varepsilon$. By Theorem 9.3.6, we can replace A by a convex combination of projections, say $A = \sum_{i=1}^n t_i E_i$, where $t_i > 0$ and $\sum_i t_i = 1$. Then for some i, $\varphi(E_i) > \alpha - \varepsilon$; so we have a projection E with $\varphi(E) > \alpha - \varepsilon$. By Lemma 10.2.5, there is a projection $P_1 \leq E$ such that $\varphi|_{P_1\mathcal{M}P_1} \geq 0$ and $\varphi(P_1) \geq \varphi(E) > \alpha - \varepsilon$. By Theorem 10.2.6, $\varphi|_{P_1\mathcal{M}P_1}$ is weak-$*$ continuous. Let $P_2 = P_1^\perp$ and define $\varphi_{ij}(A) = \varphi(P_iAP_j)$ for $1 \leq i, j \leq 2$.

Observe that if $X \in \mathcal{M}$ with $\|X\| \leq 1$, then $P_1 + P_1XP_2 + P_2X^*P_1 + P_2 \geq 0$. Indeed if $\xi = P_1\xi$ and $\eta = P_2\eta$, then

$$\begin{aligned}\langle (P_1 + P_1XP_2 + P_2X^*P_1 + P_2)\xi + \eta, \xi + \eta\rangle &= \|\xi\|^2 + 2\operatorname{Re}\langle X\eta, \xi\rangle + \|\eta\|^2 \\ &\geq \|\xi\|^2 - 2\operatorname{Re}\|X\|\,\|\eta\|\,\|\xi\| + \|\eta\|^2 \geq 0.\end{aligned}$$

Multiplying on both sides by $(1-\varepsilon)^{1/2}P_1 + \varepsilon^{1/2}P_2$ or by $\varepsilon^{1/2}P_1 - (1-\varepsilon)^{1/2}P_2$ yields

$$0 \leq (1-\varepsilon)P_1 + (\varepsilon - \varepsilon^2)^{1/2}(P_1XP_2 + P_2X^*P_1) + \varepsilon P_2 =: B$$

and

$$0 \leq \varepsilon P_1 - (\varepsilon - \varepsilon^2)^{1/2}(P_1XP_2 + P_2X^*P_1) + (1-\varepsilon)P_2 = I - B.$$

Thus $0 \leq B \leq I$, so that

$$\begin{aligned}\alpha \geq \varphi(B) &= (1-\varepsilon)\varphi(P_1) + (\varepsilon - \varepsilon^2)^{1/2}\varphi(P_1XP_2 + P_2X^*P_1) + \varepsilon\varphi(P_2) \\ &> (1-\varepsilon)(\alpha - \varepsilon) + 2(\varepsilon - \varepsilon^2)^{1/2}\operatorname{Re}\varphi_{12}(X) - \varepsilon.\end{aligned}$$

Therefore

$$\operatorname{Re}\varphi_{12}(X) < \frac{\varepsilon(\alpha + 2 - \varepsilon)}{2\sqrt{\varepsilon(1-\varepsilon)}} \leq \sqrt{\varepsilon}\frac{3}{2\sqrt{1/2}} < 3\sqrt{\varepsilon}.$$

Since we can multiply X by any $e^{i\theta}$, we deduce that $\|\varphi_{12}\| < 3\sqrt{\varepsilon}$; and $\varphi_{21} = \varphi_{12}^*$ has the same estimate. If $F \leq P_2$ is a projection, then

$$\varphi_{22}(F) = \varphi(F) = \varphi(F + P_1) - \varphi(P_1) \leq \alpha - (\alpha - \varepsilon) = \varepsilon.$$

Now let $\psi = -\varphi_{22}$ on $P_2\mathcal{M}P_2$. If $F \leq P_2$ is a projection, then $\psi(F) \geq -\varepsilon$. Repeat the argument of the previous two paragraphs using ψ. We obtain a projection $Q_1 \leq P_2$ so that $\psi|_{Q_1\mathcal{M}Q_1} \geq 0$, Setting $Q_2 = P_2 - Q_1$ and $\psi_{ij}(X) = \psi(Q_iXQ_j)$

for $1 \le i, j \le 2$, we obtain $\|\psi_{12}\| = \|\psi_{21}\| < 3\sqrt{\varepsilon}$. We also get that $\psi_{22}(F) \le \varepsilon$ if $F \le Q_2$ is a projection. Therefore $|\psi_{22}(F)| < \varepsilon$. In the algebra $Q_2\mathcal{M}Q_2$, the differences of two complementary projections of the form $F - (Q_2 - F)$ are the extreme points of $\overline{b_1(Q_2\mathcal{M}Q_2)_{sa}}$. Thus $\|\psi_{22}\| \le 2\|\psi_{22}|_{(Q_2\mathcal{M}Q_2)_{sa}}\| \le 4\varepsilon$.

The completely additive positive functionals φ_{11} and ψ_{11} are weak-$*$ continuous by Theorem 10.2.6. Therefore

$$\begin{aligned}\operatorname{dist}(\varphi, \mathcal{M}_*) &\le \|\varphi - \varphi_{11} + \psi_{11}\| \\ &\le \|\varphi_{12}\| + \|\varphi_{21}\| + \|\psi_{12}\| + \|\psi_{21}\| + \|\psi_{22}\| < 12\sqrt{\varepsilon} + 4\varepsilon.\end{aligned}$$

The predual $\mathcal{M}_*$ is a Banach space and the canonical injection into $\mathcal{M}^\dagger$ is isometric, so $\mathcal{M}_*$ is closed in $\mathcal{M}^\dagger$. As $\varepsilon > 0$ was arbitrary, $\varphi \in \mathcal{M}_*$ is weak-$*$ continuous. ∎

There is a version of polar decomposition for weak-$*$ continuous linear functionals. We first need the notion of support for a normal positive linear functional.

10.2.9. PROPOSITION. *Let φ be a normal positive linear functional on $\mathcal{M}$. Then there is the largest projection F in* $\ker\varphi$. *Define the* support *of φ to be $s(\varphi) = F^\perp$. Then $\varphi(A) = \varphi(F^\perp A F^\perp)$ and φ is faithful on $F^\perp \mathcal{M} F^\perp$.*

PROOF. Suppose that P and Q are projections such that $\varphi(P) = 0 = \varphi(Q)$. The kernel of $P + Q$ is $(P^\perp \wedge Q^\perp)H = (P \vee Q)^\perp H$. As $0 \le P + Q \le 2I$, the spectral projection

$$E_{(0,2]}(P+Q) = \big(E_{\{0\}}(P+Q)\big)^\perp = P \vee Q,$$

which has range $\overline{PH + QH}$. (See Exercise **9.7**.) Let $E_n := E_{[1/n,2]}(P + Q)$ for $n \ge 1$. Since $(P + Q)E_n \ge \frac{1}{n} E_n$, we have $E_n \le n(P + Q)$ and so

$$0 \le \varphi(E_n) \le n\varphi(P + Q) = 0.$$

Now $P \vee Q = \bigvee_{n \ge 1} E_n = \text{SOT–}\lim_{n\to\infty} E_n$. By normality,

$$\varphi(P \vee Q) = \sup_n \varphi(E_n) = 0.$$

Thus the set of projections in $\ker\varphi$ is upward directed, and so has a supremum F. Again $\varphi(F) = 0$ by normality. Let $s(\varphi) = F^\perp$.

If $X = F^\perp X F$ and $\|X\| \le 1$, then for any $\varepsilon > 0$ and $k \in \{0, 1, 2, 3\}$,

$$A_k = \varepsilon F^\perp + i^k X + (i^k X)^* + \varepsilon^{-1} F = \begin{bmatrix} \varepsilon^{1/2} \\ i^k \varepsilon^{-1/2} \end{bmatrix}^* \begin{bmatrix} F & X \\ X^* & F^\perp \end{bmatrix} \begin{bmatrix} \varepsilon^{1/2} \\ i^k \varepsilon^{-1/2} \end{bmatrix} \ge 0.$$

Thus $0 \le \varphi(A_k) = \varepsilon\varphi(F^\perp) + 2\operatorname{Re}(i^k\varphi(X))$. As ε is arbitrary, we deduce that $\varphi(X) = 0$. Hence φ vanishes on $F^\perp \mathcal{M} F$; and similarly on $F\mathcal{M}F^\perp$. If $0 \le A \le F$, then $\varphi(A) = 0$, so φ vanishes on $F\mathcal{M}F$. Therefore $\varphi(A) = \varphi(F^\perp A F^\perp)$.

Suppose that $P \le F^\perp$ is a non-zero projection. Then P is not in $\ker\varphi$, so $\varphi(P) > 0$. Now if $A = F^\perp A F^\perp > 0$, then there is some non-zero spectral projection $P \leq F^\perp$ of A so that $A \geq \varepsilon P$ for $\varepsilon > 0$. Thus $\varphi(A) \geq \varepsilon\varphi(P) > 0$. That is, φ is faithful on $F^\perp \mathcal{M} F^\perp$. ∎

Note that $\mathcal{M}_*$ becomes a bimodule over $\mathcal{M}$ if we define

$$(A\varphi)(B) = \varphi(BA) \quad \text{and} \quad (\varphi A)(B) = \varphi(AB) \quad \text{for } \varphi \in \mathcal{M}_* \text{ and } A, B \in \mathcal{M}.$$

It is routine to check that $\|A\varphi B\| \leq \|A\|\,\|\varphi\|\,\|B\|$.

10.2.10. Polar Decomposition for Normal Functionals. *Let $\varphi \in \mathcal{M}_*$ be a normal functional on a von Neumann algebra $\mathcal{M}$. Then there are a unique positive normal functional ψ and a partial isometry V with $\varphi = V\psi$ and $V^*V = s(\psi)$. Moreover $\psi = V^*\varphi$ and $\|\psi\| = \|\varphi\| = \varphi(V^*)$.*

If $\varphi = \varphi^*$, then $\varphi = \varphi_+ - \varphi_-$, where $\varphi_\pm$ are positive normal functionals such that $\|\varphi\| = \|\varphi_+\| + \|\varphi_-\| = \|\varphi_+ + \varphi_-\|$, in a unique way.

Proof. Scale so that $\|\varphi\| = 1$. As $\mathcal{M}$ is the dual of $\mathcal{M}_*$, the Hahn–Banach theorem provides some $A \in \mathcal{M}$ such that $\|A\| = 1$ and $\varphi(A) = 1$. The non-empty set

$$F = \{A \in \mathcal{M} : \|A\| \leq 1 \text{ and } \varphi(A) = 1\}$$

is convex, bounded and weak-$*$ closed; whence it is weak-$*$ compact by the Banach–Alaoglu theorem. By the Krein–Milman theorem, F has an extreme point U. It is easy to check that F is a face of the unit ball, and thus U is also an extreme point of the unit ball of $\mathcal{M}$. By Exercise **10.7**, U is a partial isometry.

Define $\psi = U\varphi$. Then

$$1 = \|\varphi\| \geq \|\psi\| \geq \psi(I) = \varphi(U) = 1.$$

By Lemma 9.6.5, ψ is positive. As $\psi \in \mathcal{M}_*$, it is normal. Hence $P := s(\psi)$ is defined, and we set $V = U^*P$. Note that since U is a partial isometry,

$$\psi(I - UU^*) = \varphi(U - UU^*U) = \varphi(0) = 0.$$

Hence $I - UU^*$ is orthogonal to P, so that $P \leq UU^*$. Therefore $V^*V = PUU^*P = P$.

We claim that $\varphi = V\psi = (U^*PU)\varphi$. If $(I - U^*U)\varphi \neq 0$, fix $A \in \mathcal{M}$ with $\|A\| \leq 1$ such that $\varphi\big(A(I - U^*U)\big) = \alpha > 0$. Now for $t > 0$,

$$\|A(I - U^*U) + tU\|^2 = \|A(I - U^*U)A^* + t^2UU^*\| \leq 1 + t^2.$$

However $\varphi(A(I - U^*U) + tU) = \alpha + t > \sqrt{1 + t^2}$ for large t, contradicting $\|\varphi\| = 1$. So $(I - U^*U)\varphi = 0$ and thus $\varphi = U^*U\varphi = U^*\psi = U^*P\psi = V\psi$. Now $\psi = P\psi = (PU)\varphi = V^*\varphi$. Therefore $1 = \|\varphi\| = \|\psi\| = \psi(I) = \varphi(V^*)$.

Now consider uniqueness. Suppose that $\varphi = W\omega$ for a positive normal functional ω and partial isometry W such that $W^*W = s(\omega)$. Observe that

$$\psi(A) = \psi(AV^*V) = \varphi(AV^*) = \omega(AV^*W) = \omega(s(\omega)AV^*W).$$

Setting $A = s(\omega)^\perp$ yields that $\psi(s(\omega)^\perp) = 0$, so that $s(\psi) \leq s(\omega)$. Reversing the roles shows that $s(\omega) = s(\psi) = P$.

It follows that $W^*V \in P\mathcal{M}P$. Write $W^*V = A + iB$ for $A, B \in (P\mathcal{M}P)_{sa}$. Then

$$\psi(A) + i\psi(B) = \psi(W^*V) = \varphi(W^*) = \omega(W^*W) = \omega(P) = 1.$$

Hence $\psi(A) = 1$ and $\psi(B) = 0$. Since $\|A\| \leq \|W^*V\| \leq 1$, $A \leq P$ and $\psi(P - A) = 0$. By Proposition 10.2.9, ψ is faithful on $P\mathcal{M}P$; whence $A = P$. Since $1 \geq \|A+iB\|^2 = \|P+B^2\| = 1+\|B\|^2$, we get $B = 0$. That is, $W^*V = P$ and $W = WP = (WW^*)V =: EV$ where $E = WW^*$ is a projection. Similarly $V^*W = P$, so $V = (VV^*)W = FW$ where $F = VV^*$. Hence

$$E = WW^* = (EV)(V^*E) = EFE \quad \text{and} \quad F = VV^* = (FW)(W^*F) = FEF.$$

So $EF^{\perp}E = 0$ implies that $E \leq F$ and similarly $F \leq E$; whence $E = F$. Therefore $W = EV = VV^*V = V$; and hence $\omega = \psi$.

Now suppose that $\varphi = \varphi^*$. We have $\psi \in \mathcal{M}_*^+$ and an isometry V with $\varphi = V\psi$ and $V^*V = s(\psi)$. Let $\widetilde{\psi} = V\psi V^*$, which is also positive. Observe that for $A \in \mathcal{M}$,

$$\begin{aligned}\varphi(A) &= \overline{\varphi(A^*)} = \overline{\psi(A^*V)} = \psi(V^*A) \\ &= \psi(V^*AV^*V) = \widetilde{\psi}(AV^*) = (V^*\widetilde{\psi})(A).\end{aligned}$$

By uniqueness of the polar decomposition, $V^* = V$ and $\widetilde{\psi} = \psi$. Therefore $\psi = V^*\varphi = V\varphi = (V\varphi)^* = \varphi V^* = \varphi V$. A self-adjoint partial isometry has spectrum contained in $\{\pm 1, 0\}$; whence $V = E - F$ for projections E, F with $EF = 0$ and $P = VV^* = E + F$. Now for $A \in \mathcal{M}$,

$$\begin{aligned}\varphi(A) &= ((E-F)\psi)\big((E+F)A(E+F)\big) = \psi(EAE-EAF+FAE-FAF) \\ &= (\psi(E-F))\big((E+F)A(E+F)\big) = \psi(EAE+EAF-FAE-FAF).\end{aligned}$$

We deduce that $\psi(EAF) = 0 = \psi(FAE)$; whence $\varphi(A) = \psi(EAE - FAF)$. Set $\varphi_+ = E\psi E$ and $\varphi_- = F\psi F$. Clearly $\varphi_\pm \in \mathcal{M}_*^+$ and $\varphi = \varphi_+ - \varphi_-$. Moreover $s(\varphi_+) = E$ and $s(\varphi_-) = F$. Thus

$$\begin{aligned}\|\varphi_+ + \varphi_-\| &\leq \|\varphi_+\| + \|\varphi_-\| = \varphi_+(E) + \varphi_-(F) = \varphi(V) \\ &= \|\varphi\| = (\varphi_+ + \varphi_-)(E + F) \leq \|\varphi_+ + \varphi_-\|.\end{aligned}$$

Suppose that $\varphi = \varphi_1 - \varphi_2$ for $\varphi_i \in \mathcal{M}_*^+$ with $\|\varphi\| = \|\varphi_1\| + \|\varphi_2\|$. Then

$$\|\varphi_+\| = \varphi(E) = (\varphi_1 - \varphi_2)(E) \leq \varphi_1(E) \leq \|\varphi_1\|.$$

Similarly, $\|\varphi_-\| \leq \varphi_2(F) \leq \|\varphi_2\|$. But $\|\varphi_+\| + \|\varphi_-\| = \|\varphi_1\| + \|\varphi_2\|$, forcing $\|\varphi_+\| = \varphi_1(E) = \|\varphi_1\|$ and $\|\varphi_1\| = \varphi_2(F) = \|\varphi_2\|$. Therefore $s(\varphi_1) \leq E$ and $s(\varphi_2) \leq F$, so they have orthogonal supports. Therefore

$$\varphi = (s(\varphi_1) - s(\varphi_2))(\varphi_1 + \varphi_2) = (E - F)(\varphi_+ + \varphi_-).$$

By the uniqueness of the polar decomposition, we have $s(\varphi_1) = E$, $s(\varphi_2) = F$ and $\varphi_1 + \varphi_2 = \varphi_+ + \varphi_-$. Thus $\varphi_1 = E(\varphi_+ + \varphi_-) = \varphi_+$ and similarly $\varphi_2 = \varphi_-$. ∎

The positive functional ψ in the polar decomposition is called $|\varphi|$. The first part of the following is immediate. The rest is in Exercise **10.8**.

10.2.11. COROLLARY. *Every $\varphi \in \mathcal{M}_*$ is a linear combination of four normal states. Thus a functional on $\mathcal{M}$ is weak-$*$ continuous if and only if it is* SOT *continuous on the unit ball.*

A positive normal linear functional on $\mathcal{B}(H)$ has the form

$$\tau_T(A) = \operatorname{Tr}(AT) = \sum_{n\geq 1} s_n \langle A\xi_n, \xi_n\rangle,$$

where $T \geq 0$ lies in $\mathfrak{S}_1$ and has the form $T = \sum_{n\geq 1} s_n \xi_n \xi_n^*$ for some orthonormal set $\{\xi_n : n \geq 1\}$, where $\sum_{n\geq 1} s_n = \|T\|_1$. However it is not immediately clear that a positive normal functional φ on a von Neumann algebra $\mathcal{M}$ has this form. Since $\mathcal{M}_* = \mathfrak{S}_1/\mathcal{M}_\perp$ is a quotient of $\mathfrak{S}_1$, we can always find some $T \in \mathfrak{S}_1$ so that $\varphi = \tau_T|_{\mathcal{M}}$. However it isn't clear that $T \geq 0$, and indeed there are many choices for this extension. Nor is it clear that this extension can be made in a norm-preserving manner.

10.2.12. PROPOSITION. *Let $\mathcal{M} \subset \mathcal{B}(H)$ be a von Neumann algebra and let φ be a normal state in $\mathcal{M}_*^+$. Then there is a positive $T \in \mathfrak{S}_1^+$ with $\|T\|_1 = 1$ so that $\varphi = \tau_T|_{\mathcal{M}}$. If $\varphi \in \mathcal{M}_*$, there is a $T \in \mathfrak{S}_1$ with $\|T\|_1 = \|\varphi\|$ so that $\varphi = \tau_T|_{\mathcal{M}}$.*

PROOF. As noted above, there is some $S \in \mathfrak{S}_1$ so that $\varphi = \tau_S|_{\mathcal{M}}$. The discussion at the beginning of this section shows that there are vectors $\xi, \eta \in H^{(\infty)}$ such that $\varphi(A) = \langle A^{(\infty)}\xi, \eta\rangle$. If $A \geq 0$, then $\varphi(A) \geq 0$ and so

$$\begin{aligned} 4\varphi(A) &= 2\langle A^{(\infty)}\xi, \eta\rangle + 2\langle \xi, A^{(\infty)}\eta\rangle = 2\langle A^{(\infty)}\xi, \eta\rangle + 2\langle A^{(\infty)}\eta, \xi\rangle \\ &= \langle A^{(\infty)}(\xi+\eta), \xi+\eta\rangle - \langle A^{(\infty)}(\xi-\eta), \xi-\eta\rangle \leq \langle A^{(\infty)}(\xi+\eta), \xi+\eta\rangle. \end{aligned}$$

Let $\zeta = \frac{1}{2}(\xi+\eta)$. Then for $A \in \mathcal{M}$, $\varphi(A^*A) \leq \langle (A^*A)^{(\infty)}\zeta, \zeta\rangle = \|A^{(\infty)}\zeta\|^2$.

Let $K_0 = \mathcal{M}^{(\infty)}\zeta$ and $K = \overline{K_0}$. Map $\mathcal{M}$ onto the dense subspace K_0 of K by $\lambda(A) = A^{(\infty)}\zeta$. As in the GNS construction, $[A, B] = \varphi(B^*A)$ defines a sesquilinear form on $\mathcal{M}$. By the Cauchy–Schwarz inequality,

$$|[A, B]| \leq \varphi(A^*A)^{1/2}\varphi(B^*B)^{1/2} \leq \|A^{(\infty)}\zeta\|_2 \|B^{(\infty)}\zeta\|_2.$$

Thus we obtain a sesquilinear form on K_0 by $[\![A^{(\infty)}\zeta, B^{(\infty)}\zeta]\!] = \varphi(B^*A)$. This is well defined and contractive, and thus it extends to a bounded sesquilinear form on K. By Proposition 4.2.2, there is an operator $X \in \mathcal{B}(K)$ with $\|X\| \leq 1$ so that

$$\varphi(B^*A) = [A^{(\infty)}\zeta, B^{(\infty)}\zeta] = \langle A^{(\infty)}\zeta, XB^{(\infty)}\zeta\rangle.$$

Since φ is positive, $\langle \kappa, X\kappa\rangle \geq 0$ for $\kappa \in K_0$; and hence $X \geq 0$. For $A, B, C \in \mathcal{M}$,

$$\begin{aligned} \langle A^{(\infty)}\zeta, X(BC)^{(\infty)}\zeta\rangle &= \varphi(C^*B^*A) = \langle (B^*A)^{(\infty)}\zeta, XC^{(\infty)}\zeta\rangle \\ &= \langle A^{(\infty)}\zeta, B^{(\infty)}XC^{(\infty)}\zeta\rangle. \end{aligned}$$

Therefore X commutes with $B^{(\infty)}|_K$. Let $Y = X^{1/2}$. Then

$$\varphi(B^*A) = \langle A^{(\infty)}\zeta, XB^{(\infty)}\zeta\rangle = \langle A^{(\infty)}Y\zeta, B^{(\infty)}Y\zeta\rangle.$$

In particular, $\varphi(A) = \langle A^{(\infty)}Y\zeta, Y\zeta\rangle$. If $Y\zeta = \nu = (\nu_1, \nu_2, \dots)$, then

$$\varphi(A) = \langle A^{(\infty)}\nu, \nu\rangle = \sum_{n\geq 1}\langle A\nu_n, \nu_n\rangle.$$

Now φ is a state, so that

$$1 = \varphi(I) = \|\nu\|^2 = \sum_{n\geq 1}\|\nu_n\|^2.$$

The trace class operator $T = \sum_{n\geq 1}\nu_n\nu_n^*$ is positive and $\|T\|_1 = 1$. Therefore $\varphi(A) = \mathrm{Tr}(AT)$. So τ_T is a norm-preserving positive normal extension of φ to $\mathcal{B}(H)$.

Now suppose that $\varphi \in \mathcal{M}_*$. By the polar decomposition, $\varphi = V\psi$ for a positive normal functional ψ and a partial isometry V with $V^*V = s(\psi)$. Let T be the positive trace class operator such that $\psi(A) = \mathrm{Tr}(AT)$. Then $\varphi(A) = \psi(AV) = \mathrm{Tr}(AVT)$. Moreover $\|VT\|_1 \leq \|V\|\,\|T\|_1 = \|\psi\| = \|\varphi\| \leq \|VT\|_1$. Therefore τ_{VT} is a norm-preserving weak-$*$ continuous extension of φ to $\mathcal{B}(H)$. ■

Another interesting consequence of Theorem 10.2.6 is that the weak-$*$ topology on a von Neumann algebra is determined by the order structure. Therefore, if a von Neumann algebra is represented as a von Neumann algebra on another Hilbert space, the weak-$*$ topology there must be the same. This is not the case for the weak operator topology. For example, for the infinite ampliation $\mathcal{B}(H)^{(\infty)}$, the WOT topology coincides with the weak-$*$ topology; while the weak operator topology on $\mathcal{B}(H)$ is strictly weaker than the weak-$*$ topology.

In fact, a stronger statement is true. A von Neumann algebra has a unique predual. This means that we can talk about *the* predual of a von Neumann algebra.

10.2.13. THEOREM. *Every von Neumann algebra $\mathcal{M}$ has a unique predual.*

PROOF. Let a closed subspace $Y \subset \mathcal{M}^\dagger$ be a predual of $\mathcal{M}$. By Proposition 3.6.10, the main properties of Y are that it norms $\mathcal{M}$ and the unit ball of $\mathcal{M}$ is τ_Y compact.

We first show that $\mathcal{M}_{sa}$ is τ_Y closed. Suppose that $\{A_\lambda\}_\Lambda$ is a net in $\overline{b_1(\mathcal{M}_{sa})}$ which converges τ_Y to $A = C+iD$, where $C, D \in \mathcal{M}_{sa}$. Let $d \in \sigma(D)\cap\{\pm\|D\|\}$. For any $t > 0$, the net $A_\lambda + itdI$ converges τ_Y to $A + itdI = C + i(D + tdI)$. Now

$$\|A_\lambda + itdI\|^2 = \|(A_\lambda + itdI)^*(A_\lambda + itdI)\| = \|A_\lambda\|^2 + t^2d^2 \leq 1 + t^2d^2.$$

Therefore

$$(t+1)^2d^2 = \|D + tdI\|^2 \leq \|A + itdI\|^2 \leq 1 + t^2d^2 \quad \text{for all} \quad t > 0.$$

Hence $d = 0$. Thus $\overline{b_1(\mathcal{M}_{sa})}$ is τ_Y closed. By scaling, $\overline{b_r(\mathcal{M}_{sa})}$ is τ_Y-compact for all $r > 0$. By the Krein–Smulian theorem 3.6.2, $\mathcal{M}_{sa}$ is τ_Y closed. Next consider $\mathcal{M}^+$, which is a convex cone. Note that

$$\mathcal{M}^+ \cap \overline{b_1(\mathcal{M})} = \left(\overline{b_1(\mathcal{M}_{sa})} + I\right) \cap \overline{b_1(\mathcal{M})}$$

is τ_Y compact. Again the Krein–Smulian theorem shows that $\mathcal{M}^+$ is τ_Y closed.

Let $\{A_\lambda\}_\Lambda$ be an increasing net with $A = \sup A_\lambda$. This is a bounded net, so by τ_Y compactness, there is a cofinal subnet Λ' converging τ_Y to $B = B^*$. Since $\mathcal{M}^+$ is τ_Y closed, for any $\mu \in \Lambda'$,

$$B - A_\mu = \tau_Y - \lim_{\lambda \geq \mu,\, \lambda \in \Lambda'} A_\lambda - A_\mu \geq 0$$

and hence $B \geq A$. However,

$$B - A = \tau_Y - \lim_{\Lambda'} A_\lambda - A \leq 0.$$

Therefore $B = A$. This holds for any cofinal convergent subnet Λ', and thus there is a unique cluster point; whence $\tau_Y - \lim_\Lambda A_\lambda = A$.

Now suppose that $\varphi \in Y$. It follows that $\varphi(A) = \lim_\Lambda \varphi(A_\lambda)$. That is, φ is normal. By Theorem 10.2.8, φ belongs to $\mathcal{M}_*$. Therefore $Y \subseteq \mathcal{M}_*$. By the Hahn–Banach theorem, since Y norms $\mathcal{M}$, we have $Y = \mathcal{M}_*$. ■

In section 10.6, we will show that this leads to an abstract characterization of von Neumann algebras: every C*-algebra which is a dual space has a faithful weak-$*$ continuous representation as a von Neumann algebra.

10.3. Normal Homomorphisms

We begin this section by showing that normal homomorphisms of von Neumann algebras are weak-$*$ continuous. This leads us to a characterization of normal homomorphisms as the composition of a few very special maps.

10.3.1. THEOREM. *Let $\pi : \mathcal{M} \to \mathcal{N}$ be an unital $*$-homomorphism of one von Neumann algebra into another. The following are equivalent:*

(1) *π is normal.*

(2) *π is completely additive.*

(3) *π is weak-$*$ to weak-$*$ continuous.*

(4) *π is* WOT *to* WOT *continuous on the unit ball.*

(5) *π is* SOT *to* SOT *continuous on the unit ball.*

(6) *$\pi(\mathcal{M})$ is a von Neumann algebra and* $\ker \pi = P\mathcal{M}$ *for a projection $P \in \mathcal{Z}(\mathcal{M}) := \mathcal{M} \cap \mathcal{M}'$.*

PROOF. Clearly (1) implies (2). Suppose that π is completely additive. Let φ be a weak-$*$ continuous state on $\mathcal{N}$. Then φ is normal. Hence $\varphi\pi$ is a completely additive state on $\mathcal{M}$. By Theorem 10.2.6, $\varphi\pi$ is weak-$*$ continuous. Every weak-$*$ continuous functional on $\mathcal{N}$ is a linear combination of weak-$*$ continuous states by Remark 10.1.4. Thus $\varphi\pi$ is weak-$*$ continuous for all weak-$*$ continuous φ. Therefore π is weak-$*$ to weak-$*$ continuous. So (2) implies (3).

If π is weak-$*$ to weak-$*$ continuous, then it is weak-$*$ to weak-$*$ continuous on the unit ball. However the weak-$*$ topology and the WOT topologies agree on the ball, so π is WOT to WOT continuous on the unit ball. Hence (3) implies (4).

Assume (4). It is enough to show SOT-SOT continuity at 0 for bounded nets in $\mathcal{M}$. Suppose that $\sup_\Lambda \|A_\lambda\| \le 1$ and $\text{SOT–}\lim_\Lambda A_\lambda = 0$. Then $\||A_\lambda|\xi\| = \|A_\lambda\xi\| \to 0$ for $\xi \in H_\mathcal{M}$; so $\text{SOT–}\lim_\Lambda |A_\lambda| = 0$. Now

$$\|A_\lambda^* A_\lambda \xi\| = \||A_\lambda|^2\xi\| \le \||A_\lambda|\xi\| \to 0;$$

so $\text{SOT–}\lim_\Lambda A_\lambda^* A_\lambda = 0$, and thus $\text{WOT–}\lim_\Lambda A_\lambda^* A_\lambda = 0$. Therefore

$$\text{WOT–}\lim_\Lambda \pi(A_\lambda)^*\pi(A_\lambda) = 0.$$

If $\eta \in H_\mathcal{N}$,

$$0 = \lim_\Lambda \langle \pi(A_\lambda)^*\pi(A_\lambda)\eta, \eta\rangle = \lim_\Lambda \|\pi(A_\lambda)\eta\|^2.$$

Hence $\text{SOT–}\lim_\Lambda \pi(A_\lambda) = 0$. So (5) holds.

If (5) holds, let $\{A_\lambda\}_\Lambda$ be an increasing net in $\mathcal{M}_{sa}$ with $\sup_\Lambda A_\lambda = A$. Then by Lemma 9.3.5, $A = \text{SOT–}\lim_\Lambda A_\lambda$. We can assume that this net is bounded by restricting to $\lambda \ge \lambda_0$ so that $A_{\lambda_0} \le A_\lambda \le A$. Thus $\pi(A) = \text{SOT–}\lim_\Lambda \pi(A_\lambda)$. Hence π is normal. Therefore (1) holds.

Suppose that (3) holds. The closed unit ball of $\mathcal{M}$ is weak-$*$ compact by the Banach–Alaoglu theorem. Since π is weak-$*$ to weak-$*$ continuous, the image is weak-$*$ compact. A $*$-homomorphism factors as a quotient map and an isometric isomorphism. Thus the open unit ball is mapped onto the open unit ball of the image. Let T belong to the open unit ball of $\pi(\mathcal{M})''$. By Kaplansky's theorem, there is a net $\{\pi(A_\lambda)\}_\Lambda$ in the open unit ball of $\pi(\mathcal{M})$ converging WOT to T; and thus it converges weak-$*$. By the previous observation, the preimage A_λ may be selected in the open unit ball of $\mathcal{M}$. By compactness, there is a cofinal subnet converging weak-$*$ to some A in the closed unit ball of $\mathcal{M}$. By weak-$*$ continuity, $\pi(A) = T$. Therefore $\pi(\mathcal{M}) = \pi(\mathcal{M})''$ is a von Neumann algebra.

Let $\mathcal{J} = \ker\pi$, and let $\{E_\lambda\}_\Lambda$ be an approximate identity for $\mathcal{J}$. The bounded net E_λ converges SOT to the projection P onto $\overline{\mathcal{J}H}$ because $E_\lambda H \subset \overline{\mathcal{J}H}$ and $\lim_\Lambda E_\lambda(J\xi) = J\xi$ for all $J \in \mathcal{J}$ and $\xi \in H$. Hence it converges weak-$*$. Since π is weak-$*$ continuous, $\pi(P) = 0$; so $P \in J$. Thus P is a unit for $\mathcal{J}$; and hence $P\mathcal{J}P = \mathcal{J}$. Since $\mathcal{J}$ is an ideal, for any $A \in \mathcal{M}$, AP and PA belong to $\mathcal{J}$, and thus equal PAP. Therefore $AP = PA$, so that $P \in \mathcal{M}\cap\mathcal{M}'$. Clearly $\mathcal{J} = P\mathcal{M} = \mathcal{M}P$ is weak-$*$ closed. Thus (6) holds.

Now if (6) holds, the map $q : \mathcal{M} \to \mathcal{M}/\ker\pi = P^\perp\mathcal{M}|_{P^\perp H} =: \widetilde{M}$ is the quotient map onto the von Neumann algebra $\widetilde{M}$. Thus we can factor $\pi = \tilde{\pi}q$

where $\widetilde{\pi} : \widetilde{M} \to \pi(\mathcal{M})$ is an isomorphism of von Neumann algebras. By Proposition 9.3.9, $\widetilde{\pi}$ is normal. Now q is also normal, since if $\{A_\lambda\}_\Lambda$ is an increasing net with $\sup A_\lambda = A$, then $\text{SOT–}\lim_\Lambda P^\perp A_\lambda = P^\perp A$. Hence $q(A) = P^\perp A = \sup q(A_\lambda)$. Therefore π is normal. So (1) holds. ∎

10.3.2. REMARK. Condition (6) cannot be weakened to just say that $\pi(\mathcal{M})$ is a von Neumann algebra. Consider a multiplicative linear functional on l_∞. This is by definition a $*$-homomorphism onto the von Neumann algebra $\mathbb{C}$. The point evaluations δ_n are normal, but all the other multiplicative linear functionals are not.

The von Neumann algebra $\mathcal{Z}(\mathcal{M}) = \mathcal{M} \cap \mathcal{M}'$ played a role in our proof. This algebra arises in various contexts throughout this chapter.

10.3.3. DEFINITION. The *centre* of a von Neumann algebra $\mathcal{M}$ is the abelian algebra $\mathcal{Z}(\mathcal{M}) = \mathcal{M} \cap \mathcal{M}'$. A von Neumann algebra $\mathcal{M}$ is a *factor* if $\mathcal{Z}(\mathcal{M}) = \mathbb{C}I$.
If $A \in \mathcal{M}_+$, the *central cover* of A is $c(A) = \inf\{Z \in \mathcal{Z}(\mathcal{M}) : A \le Z\}$.

Firstly, this infimum makes sense since $\mathcal{Z}(\mathcal{M})$ is abelian. We always have $A \le \|A\| I$, so $\|c(A)\| = \|A\|$. If $P \in \mathcal{M}$ is a projection, then $P = P^2 \le c(P)^2$ because P and $c(P)$ commute. Thus $c(P) \le c(P)^2 \le I$, so that $c(P)$ is a projection, namely the smallest projection in $\mathcal{Z}(\mathcal{M})$ greater than P.

10.3.4. LEMMA. *If $P \in \mathcal{M}$ is a projection, then $c(P)$ is the projection onto $\overline{\mathcal{M}PH}$. Also if $\mathcal{G}$ is any group of unitaries which generates $\mathcal{M}$ as a von Neumann algebra, then $c(P) = \bigvee_{U \in \mathcal{G}} U^*PU$.*

PROOF. Clearly $\overline{\mathcal{M}PH}$ is invariant for $\mathcal{M}$ and $\mathcal{M}'$, and thus the projection onto it lies in $\mathcal{Z}(\mathcal{M})$, and it dominates P. Conversely the range of $c(P)$ contains PH and is $\mathcal{M}$ invariant, so contains $\overline{\mathcal{M}PH}$. Thus these projections are equal.
Note that $U^*PU \le U^*c(P)U^* = c(P)$; and thus

$$c(P) \ge \bigvee_{U \in \mathcal{G}} U^*PU =: Q.$$

Now $U^*QU = Q$ for $U \in \mathcal{G}$, so Q commutes with $\mathcal{G}'' = \mathcal{M}$, and hence $Q \in \mathcal{Z}(\mathcal{M})$. Since $P \le Q \le c(P)$, we have $Q = c(P)$. ∎

We will now provide a complete description of normal homomorphisms between von Neumann algebras.

10.3.5. DEFINITION. If $\mathcal{M}$ is a von Neumann algebra on H and α is a cardinal number, then the map $\varphi_\alpha : \mathcal{M} \to \mathcal{B}(H^{(\alpha)})$ by $\pi(A) = A^{(\alpha)}$ is an *ampliation*.

If E is a projection in $\mathcal{M}'$, the homomorphism $\pi : \mathcal{M} \to \mathcal{B}(EH)$ given by $\pi_E(A) = EA|_{EH}$ is a *compression*. The image is a von Neumann algebra $\mathcal{M}_E = E\mathcal{M}E|_{EH}$. We also write $\mathcal{M}'_E$ for $E\mathcal{M}'E|_{EH}$.

It is clear that an ampliation is a normal isomorphism. However in general, an infinite ampliation is not WOT–WOT continuous as noted in the discussion prior to Theorem 10.2.13. Also if $U \in \mathcal{B}(H, K)$ is unitary, then the unitary conjugation $\operatorname{Ad} U(A) = UAU^*$ is a normal isomorphism (in fact a unitary equivalence) of $\mathcal{M}$ onto $U\mathcal{M}U^*$. Note that $\mathcal{M}'_E$ is a von Neumann algebra because $E\mathcal{M}'E$ is a WOT-closed subalgebra of $\mathcal{M}'$ which leaves EH invariant. However the compression of $\mathcal{M}'$ to EH is not multiplicative.

10.3.6. LEMMA. *Let $\mathcal{M}$ be a von Neumann algebra, and let E be a projection in $\mathcal{M}'$. The compression map π_E is a normal homomorphism. It is an isomorphism if and only if $c(E) = I$. Moreover, $(\mathcal{M}_E)' = \mathcal{M}'_E$.*

PROOF. Since $E \in \mathcal{M}'$, EH is a reducing subspace for $\mathcal{M}$, so the compression map is a normal homomorphism. Note that $\ker \pi_E = \mathcal{M}P$ for the largest projection $P \in \mathcal{Z}(\mathcal{M}) = \mathcal{Z}(\mathcal{M}')$ orthogonal to E, namely $P = c(E)^\perp$. (See Exercise **10.3**.) Hence π_E is an isomorphism if and only if $c(E) = I$. Clearly $\mathcal{M}_E$ commutes with $E\mathcal{M}'E|_{EH}$. Conversely if $A \in (\mathcal{M}_E)'$, then $EAE \in \mathcal{M}'$. So $(\mathcal{M}_E)' = \mathcal{M}'_E$. ∎

With these three basic normal isomorphisms, we can obtain all of them. The following result provides three somewhat different views of their structure.

10.3.7. THEOREM. *Let $\pi : \mathcal{M}_1 \to \mathcal{M}_2$ be a normal isomorphism. Then*

(1) *$\pi = \operatorname{Ad} U \circ \pi_E \circ \varphi_\alpha$ where φ_α is an ampliation of cardinality α, π_E is a compression for a projection $E \in \varphi_\alpha(\mathcal{M})'$ with $c(E) = I$, and $\operatorname{Ad} U$ is a unitary conjugation.*

(2) *There is a von Neumann algebra $\mathcal{N}$ and projections $P_i \in \mathcal{N}'$ with $c(P_1) = c(P_2) = I$ such that $\mathcal{M}_i$ is unitarily equivalent to $P_i\mathcal{N}$ and $\pi = \pi_{P_2}\pi_{P_1}^{-1}$.*

(3) *There are projections $Q_i \in \mathcal{M}'_i$ with $c(Q_i) = I$ so that $Q_i\mathcal{M}_i$ are unitarily equivalent via U so that $\pi = \pi_{Q_2}^{-1} \circ \operatorname{Ad} U \circ \pi_{Q_1}$.*

PROOF. (1) Decompose H_2 into a direct sum of cyclic subspaces for $\mathcal{M}_2$. This is accomplished by using Zorn's lemma to find a maximal family of pairwise orthogonal reducing subspaces $H_{\xi_\lambda} = \overline{\mathcal{M}_2\xi_\lambda}$ for $\lambda \in \Lambda$ with unit cyclic vector ξ_λ. Define the normal state $\varphi_\lambda(A) = \langle A\xi_\lambda, \xi_\lambda\rangle$. Then $\psi_\lambda = \varphi_\lambda\pi$ is a normal state on $\mathcal{M}_1$. Thus it is weak-$*$ continuous, so there are vectors $\eta_{\lambda,n} \in H_1$ so that

$$\psi_\lambda(A) = \sum_{n\geq 1}\langle A\eta_{\lambda,n}, \eta_{\lambda,n}\rangle = \langle A^{(\infty)}\eta_\lambda, \eta_\lambda\rangle$$

where $\eta_\lambda = (\eta_{\lambda,1}, \eta_{\lambda,2}, \eta_{\lambda,3}, \dots) \in \underline{H_1^{(\infty)}}$ is a unit vector. By Proposition 9.7.4, the restriction of $\mathcal{M}_1^{(\infty)}$ to $H_{\eta_\lambda} = \overline{\mathcal{M}_1^{(\infty)}\eta_\lambda}$ is unitarily equivalent to $\mathcal{M}_2|_{H_{\xi_\lambda}}$, say via $\operatorname{Ad} U_\lambda$. If $E_\lambda = P_{H_{\eta_\lambda}}$, then this restriction map is just π_{E_λ}.

Now take a direct sum $\mathcal{N}$ of $\alpha = |\Lambda|$ copies of $\mathcal{M}_1^{(\infty)}$ on $H = (H_1^{(\infty)})^{(\alpha)}$. There is an ampliation map φ_α of $\mathcal{M}_1$ onto $\mathcal{N}$. Let $E = \sum_\Lambda^\oplus E_\lambda$. Then by the previous paragraph, $\pi = \operatorname{Ad} U \circ \pi_E \circ \varphi_\alpha$, where $U = \sum_\Lambda^\oplus U_\lambda$. Since this map is an isomorphism, $c(E) = I$.

For (2), take $\mathcal{N}$ as above. Let $P_2 = E$ and let P_1 be the projection onto the first copy of H_1 in the first copy of $H_1^{(\infty)}$ in H. By the previous paragraph, $\mathcal{N}P_2$ is unitarily equivalent to $\mathcal{M}_2$ via π_{P_2} and clearly $\mathcal{N}P_1$ is unitarily equivalent to $\mathcal{M}_1$ via π_{P_1}. So $\pi = \pi_{P_2}\pi_{P_1}^{-1}$. We have noted already that $c(P_2) = I$, and clearly $c(P_1) = I$ as well.

For (3), using the setup from (2), the unit ball of $P_1\mathcal{N}'P_2$ is WOT-compact. By the Krein–Milman theorem, it has an extreme point U. Then U must be a partial isometry. Indeed, if it has polar decomposition VA for a partial isometry $V \in P_1\mathcal{N}'P_2$ and a positive operator $0 \le A \le V^*V$ in $P_2\mathcal{N}'P_2$. But then

$$U = \frac{1}{2}(V + V(2A - V^*V))$$

is the average of two elements of $P_1\mathcal{N}'P_2$. Since U is extreme, $U = V$.

Next we let $Q_1 = UU^*$ and $Q_2 = U^*U$ be the range and initial projections of U in $\mathcal{N}'$. We claim that $(P_1 - Q_1)\mathcal{N}'(P_2 - Q_2) = \{0\}$. For otherwise, there is an element $B = (P_1 - Q_1)B(P_2 - Q_2) \in P_1\mathcal{N}'P_2$ with $\|B\| = 1$. However then $U \pm B \in P_1\mathcal{N}'P_2$ and

$$\|U \pm B\| = \|Q_1UQ_2 + Q_1^\perp BQ_2^\perp\| = \max\{\|U\|, \|B\|\} = 1.$$

Hence U is not extreme, a contradiction. Thus $(P_1 - Q_1)\mathcal{N}'(P_2 - Q_2) = \{0\}$.

By Lemma 10.3.4, this means that $c(P_1 - Q_1)$ is orthogonal to $c(P_2 - Q_2)$. Now

$$I = c(P_1) = c(Q_1) \vee c(P_1 - Q_1) \le c(Q_1) + c(P_1 - Q_1)$$

Similarly, $I \le c(Q_2) + c(P_2 - Q_2)$. Therefore

$$I \le c(Q_2) + c(P_2 - Q_2) \le c(Q_2) + c(P_1 - Q_1)^\perp \le c(Q_2) + c(Q_1) = 2c(Q_1).$$

The last step follows because Q_1 and Q_2 are Murray-von Neumann equivalent, and thus $c(Q_1) = c(Q_2)$ by Lemma 10.3.4 again. This means that $c(Q_1) = c(Q_2) = I$. The partial isometry $U \in \mathcal{N}'$ now satisfies for $B \in \mathcal{N}$:

$$U^*(Q_1B)U = U^*(Q_1BQ_1)U = Q_2(U^*BU)Q_2 = Q_2B.$$

This is a unitary equivalence between $Q_1\mathcal{N} = Q_1\mathcal{M}_1$ and $Q_2\mathcal{N} = Q_2\mathcal{M}_2$. Since $\mathcal{N}_{P_i} \simeq \mathcal{M}_i$ and $c(Q_i) = I$, we have $\mathcal{N}_{Q_i} \simeq \mathcal{M}_i$. Thus $\pi = \pi_{Q_2}^{-1} \circ \operatorname{Ad} U \circ \pi_{Q_1}$. ∎

10.3.8. COROLLARY. *Let $\pi : \mathcal{M}_1 \to \mathcal{M}_2$ be a normal homomorphism. Then $\pi = \operatorname{Ad} U \circ \pi_E \circ \varphi$ where φ is an ampliation, π_E is a compression and $\operatorname{Ad} U$ is a unitary equivalence.*

PROOF. Since π is normal, $\ker \pi$ is weak-$*$ closed, and thus has the form $\mathcal{M}_1 F^\perp$ for some $F \in \mathcal{Z}(\mathcal{M}_1)$ by Theorem 10.3.1(5). Thus $\pi = \widetilde{\pi}\pi_F$ and $\widetilde{\pi}$ is a normal isomorphism of $\mathcal{M}_1 F$ onto $\mathcal{M}_2$. Therefore $\widetilde{\pi} = \operatorname{Ad} U \circ \pi_E \circ \widetilde{\varphi}$ where U is unitary, $E \in (\mathcal{M}_1 F)'$ and $\widetilde{\varphi}$ is an ampliation of $\mathcal{M}_1 F$. Let φ be the corresponding ampliation of $\mathcal{M}_1$. Then $\widetilde{\varphi}\pi_F = \pi_{\varphi(F)}\varphi$. Consider E as an element of $\varphi(\mathcal{M}_1)'$. Since $E \leq \varphi(F)$, we have

$$\pi_E \widetilde{\varphi} \pi_F = \pi_E \pi_{\varphi(F)} \varphi = \pi_E \varphi.$$

Hence $\pi = \operatorname{Ad} U \circ \pi_E \circ \varphi$. ∎

10.4. Abelian von Neumann Algebras

In this section, we determine the structure of abelian von Neumann algebras on a separable Hilbert space. Then we characterize when such an algebra is maximal abelian. We begin by showing that abelian von Neumann algebras on a separable Hilbert space are singly generated.

10.4.1. THEOREM. *Let $\mathcal{M} \subset \mathcal{B}(H)$ be an abelian von Neumann algebra on separable Hilbert space. There is an element $A = A^* \in \mathcal{M}$ so that $\mathcal{M} = C^*(A)''$.*

PROOF. The closed unit ball B of $\mathcal{M}$ is WOT-closed, and thus it is compact in the WOT topology. Since H is separable, the WOT topology on B is metrizable by Lemma 3.6.6. Thus there is a countable set of projections $\{P_i : i \geq 1\}$ in $\mathcal{M}$ which is WOT-dense in the set of all projections in $\mathcal{M}$. Since the projections in $\mathcal{M}$ span the algebra by Corollary 9.3.7, we have

$$\mathcal{M} = \overline{\operatorname{span}\{P_i : i \geq 1\}}^{\text{WOT}} = \overline{C^*(\{P_i : i \geq 1\})}^{\text{WOT}} = C^*(\{P_i : i \geq 1\})''.$$

Define $A = \sum_{i \geq 1} 3^{-i} P_i$. Clearly $A = A^* \in \mathcal{M}$ and $0 \leq A \leq \frac{1}{2} I$. We claim that $C^*(A) = C^*(\{P_i : i \geq 1\})$. Observe that

$$0 \leq P_1^\perp A = \sum_{i \geq 2} 3^{-i} P_1^\perp P_i \leq \sum_{i \geq 2} 3^{-i} P_1^\perp = \frac{1}{6} P_1^\perp$$

and

$$P_1 A = \frac{1}{3} P_1 + \sum_{i \geq 2} 3^{-i} P_1 P_i \geq \frac{1}{3} P_1.$$

Thus with $H = P_1H \oplus P_1^\perp H$, which splits H into reducing subspaces for A, $A = A_1 \oplus A_2$ where $\frac{1}{3} \le A_1 \le \frac{1}{2}$ and $0 \le A_2 \le \frac{1}{6}$. Thus $\sigma(A_1) \subset [\frac{1}{3}, \frac{1}{2}]$ and $\sigma(A_2) \subset [0, \frac{1}{6}]$; and $\sigma(A) = \sigma(A_1) \cup \sigma(A_2)$. The function $\chi_{[\frac{1}{3},\frac{1}{2}]}$ is continuous on $\sigma(A)$ and $\chi_{[\frac{1}{3},\frac{1}{2}]}(A) = E_A([\frac{1}{3}, \frac{1}{2}]) = P_1$. Hence $P_1 \in C^*(A)$.

Now $3A - P_1 = \sum_{i\ge1} 3^{-i}P_{i+1}$ belongs to $C^*(A)$. The same analysis shows that $P_2 \in C^*(A)$. Recursively we show that $P_i \in C^*(A)$ for all $i \ge 1$. Therefore

$$C^*(A)'' = C^*(\{P_i : i \ge 1\})'' = \mathcal{M}. \qquad \blacksquare$$

By applying the spectral theorem to the positive generator A, we obtain our first characterization.

10.4.2. COROLLARY. *Let $\mathcal{M} \subset \mathcal{B}(H)$ be an abelian von Neumann algebra on separable Hilbert space. Then there is a regular Borel probability measure μ on $[0, 1]$ and a normal isomorphism of $L^\infty(\mu)$ onto $\mathcal{M}$.*

PROOF. By Theorem 10.4.1, we may assume that $\mathcal{M} = C^*(A)''$ for some $0 \le A = A^* \le I$. Apply Theorem 9.9.6 to the map $\pi : C(\sigma(A)) \to \mathcal{B}(H)$ obtained from the continuous functional calculus of A. $\blacksquare$

10.4.3. DEFINITION. If $\mathfrak{A}$ is a subalgebra of $\mathcal{B}(H)$, a vector $\xi \in H$ is a *separating vector* for $\mathfrak{A}$ if $A\xi = 0$ for $A \in \mathfrak{A}$ implies that $A = 0$.

10.4.4. EXAMPLES.

(1) Consider $L^\infty(\mu)$ acting on $L^2(\mu)$. If μ is a finite measure, the function 1 is both cyclic and separating. When μ is σ-finite, any function $h \in L^2(\mu)$ such that $h \ne 0$ a.e.(μ) is both cyclic and separating.

(2) Let $\mathcal{K}(H)$ denote the compact operators on a Hilbert space H. Then any non-zero vector is cyclic for $\mathcal{K}(H)$. However there is no separating vector.

Next we obtain two useful lemmas about cyclic and separating vectors.

10.4.5. LEMMA. *Let $\mathfrak{A} \subset \mathcal{B}(H)$ be a C^*-algebra. Then $\xi \in H$ is a cyclic vector for $\mathfrak{A}$ if and only if it is separating for $\mathfrak{A}'$.*

PROOF. Suppose that ξ is a cyclic vector for $\mathfrak{A}$. If $T \in \mathfrak{A}'$ satisfies $T\xi = 0$, then $TA\xi = AT\xi = 0$. Thus T vanishes on $\overline{\mathfrak{A}\xi} = H$; whence $T = 0$. That is, ξ is a separating vector for $\mathfrak{A}'$.

Conversely, if ξ is a separating vector for $\mathfrak{A}'$, let $M = \overline{\mathfrak{A}\xi}$. This is invariant for $\mathfrak{A}$, and hence is reducing. Therefore the orthogonal projection $P = P_M$ belongs to

$\mathfrak{A}'$. Now $P^\perp \xi = 0$. Since ξ is separating for $\mathfrak{A}'$, this means that $P^\perp = 0$. Therefore $M = H$ and ξ is cyclic for $\mathfrak{A}$. ■

10.4.6. LEMMA. *Let $\mathcal{M} \subset \mathcal{B}(H)$ be an abelian von Neumann algebra on a separable Hilbert space H. Then $\mathcal{M}$ has a separating vector.*

PROOF. Let $\mathcal{M} = C^*(A)''$ for some $A = A^*$. By Corollary 9.9.4, the representation $\pi : C(\sigma(A)) \to \mathcal{B}(H)$ via the functional calculus for A decomposes as a finite or countable direct sum of cyclic representations, $\pi \simeq \sum_{i\geq 1}^{\oplus} \pi_i$ on $H = \sum_{i\geq 1}^{\oplus} H_i$, where ξ_i is a unit cyclic vector for π_i. Then $\overline{\mathcal{M}\xi_i} = \overline{\pi(C(\sigma(A)))\xi_i} = H_i$. Define $\xi = \sum_{i\geq 1} 2^{-i}\xi_i$.

Note that since $\mathcal{M}$ is abelian, $\mathcal{M}' \supset \mathcal{M}$. Also since each H_i reduces $\mathcal{M}$, the projections P_i onto H_i lie in $\mathcal{M}'$. Therefore $\overline{\mathcal{M}'\xi}$ contains $2^i P_i \xi = \xi_i$, and hence contains $\overline{\mathcal{M}\xi_i} = H_i$. Therefore $\overline{\mathcal{M}'\xi} = H$. That is, ξ is cyclic for $\mathcal{M}'$. By the previous lemma, ξ is separating for $\mathcal{M}$. ■

10.4.7. DEFINITION. A C*-subalgebra $\mathfrak{D}$ of a C*-algebra $\mathfrak{A}$ is a *maximal abelian self-adjoint algebra* (masa) in $\mathfrak{A}$ if it is abelian and not contained in any larger abelian algebra. If $\mathfrak{A} = \mathcal{B}(H)$, we say $\mathfrak{D}$ is a masa.

We have seen that $L^\infty(\mu)$ acting on $L^2(\mu)$ is maximal abelian in Proposition 9.3.4. Now we wish to characterize all masas in $\mathcal{B}(H)$. Later in Corollary 10.9.9, we will obtain the fine structure of an arbitrary abelian von Neumann algebra on a separable space.

10.4.8. THEOREM. *Let $\mathcal{M}$ be an abelian von Neumann algebra in $\mathcal{B}(H)$ for separable H. The following are equivalent:*

(1) *$\mathcal{M}$ has a cyclic vector.*

(2) *$\mathcal{M}'$ is abelian.*

(3) *$\mathcal{M}$ is maximal abelian.*

(4) *$\mathcal{M}$ is unitarily equivalent to $L^\infty(\mu)$ acting on $L^2(\mu)$ for a regular Borel probability measure on $[0, 1]$.*

PROOF. Theorem 9.9.2 shows that (1) implies (4). Clearly (4) implies (3), which implies (2). Finally if (2) holds, $\mathcal{M}'$ has a separating vector by Lemma 10.4.6. Thus $\mathcal{M}$ has a cyclic vector by Lemma 10.4.5. So (1) holds. ■

10.4.9. COROLLARY. *Let $\mathcal{M}$ be an abelian von Neumann algebra in $\mathcal{B}(H)$ for separable H. There is a subspace N so that the restriction map $\rho(A) = A|_N$ is a faithful normal isomorphism such that $\rho(\mathcal{M})$ is maximal abelian in $\mathcal{B}(N)$.*

PROOF. By Lemma 10.4.6, $\mathcal{M}$ has a separating vector ξ. Let $N = \overline{\mathcal{M}\xi}$, which is invariant for $\mathcal{M}$; whence $P_N \in \mathcal{M}'$. Therefore the restriction to N is a faithful normal representation of $\mathcal{M}$. But $\mathcal{M}|_N$ has a cyclic vector ξ, so it is maximal abelian by Theorem 10.4.8. ∎

10.5. The Universal Representation

We already made use of the direct sum of all irreducible representations of a C*-algebra to draw some important conclusions. The direct sum over all cyclic representations has very nice properties as well. This universal representation π_u has the important property that the von Neumann algebra $\pi_u(\mathfrak{A})''$ is isomorphic and weak-* homeomorphic to the double dual $\mathfrak{A}^{\dagger\dagger}$ of $\mathfrak{A}$. We continue to use † for dual spaces to avoid confusion with the adjoint.

10.5.1. DEFINITION. Let $\mathfrak{A}$ be a C*-algebra. The *universal representation* of $\mathfrak{A}$ is $\pi_u = \bigoplus_{\varphi \in \mathcal{S}(\mathfrak{A})} \pi_\varphi$. The *universal enveloping von Neumann algebra* of $\mathfrak{A}$ is $\mathcal{M}_u = \overline{\pi_u(\mathfrak{A})}^{\text{WOT}} = \pi_u(\mathfrak{A})''$.

10.5.2. THEOREM. *If $\mathfrak{A}$ is a C*-algebra, let $\mathcal{M}_u$ be its universal enveloping von Neumann algebra. Then the predual $\mathcal{M}_*$ of $\mathcal{M}_u$ is isometrically isomorphic to $\mathfrak{A}^\dagger$. Thus $\mathcal{M}_u$ is isomorphic as a Banach space to $\mathfrak{A}^{\dagger\dagger}$ via an isometric, weak-$*$–weak-$*$ continuous isomorphism that fixes $\mathfrak{A}$.*

PROOF. Let $\mathcal{M}_*$ denote the predual of $\mathcal{M}_u$. Each state $\varphi \in \mathcal{S}(\mathfrak{A})$ is given by $\varphi(a) = \langle \pi_\varphi(a)\xi_\varphi, \xi_\varphi \rangle$. Since H_φ is a summand of H_u, this extends to a weak-$*$ continuous functional on $\mathcal{M}_u$. Every linear functional in $\mathfrak{A}^\dagger$ is a linear combination of (four) states, and thus also extends to a weak-$*$ continuous functional on $\mathcal{M}_u$. This yields a canonical injection of $\mathfrak{A}^\dagger$ into $\mathcal{M}_*$.

By Kaplansky's density theorem, $b_1(\mathfrak{A})$ is weak-$*$ dense in $b_1(\mathcal{M}_u)$. Thus this injection of $\mathfrak{A}^\dagger$ into $\mathcal{M}_*$ is isometric, and hence closed. If $\mathfrak{A}^\dagger \neq \mathcal{M}_*$, then any $\psi \in \mathcal{M}_* \setminus \mathfrak{A}^\dagger$ would agree on $\mathfrak{A}$ with $\varphi = \psi|_{\mathfrak{A}}$. Thus $0 \neq \psi - \varphi \in \mathfrak{A}^\perp \subset \mathcal{M}_*$. By the Hahn–Banach theorem, there is some $A \in \mathcal{M}_u$ so that $(\psi - \varphi)(A) \neq 0$. This contradicts the weak-$*$ density of $\mathfrak{A}$ in $\mathcal{M}_u$. Hence $\mathfrak{A}^\dagger = \mathcal{M}_*$. Therefore $\mathcal{M}_u = \mathfrak{A}^{\dagger\dagger}$ as a Banach space via a map which fixes the canonical copies of $\mathfrak{A}$ in both of them.

Since we have an isometric identification between $\mathcal{M}_*$ and $\mathfrak{A}^\dagger$, the dual maps are isometric and are continuous if we endow both $\mathcal{M}_u$ and $\mathfrak{A}^{\dagger\dagger}$ with their weak-$*$ topologies. Therefore it is a weak-$*$–weak-$*$ homeomorphism. ∎

10.5.3. COROLLARY. *If $\mathfrak{A}$ is a C*-algebra, a multiplication can be defined on $\mathfrak{A}^{\dagger\dagger}$ for $A, B \in \mathfrak{A}^{\dagger\dagger}$ by*

$$AB = \underset{\Lambda}{w^*\text{–lim}}\ \underset{\Gamma}{w^*\text{–lim}}\ a_\lambda b_\gamma = \underset{\Gamma}{w^*\text{–lim}}\ \underset{\Lambda}{w^*\text{–lim}}\ a_\lambda b_\gamma$$

for any nets $\{a_\lambda\}_\Lambda$ and $\{b_\gamma\}_\Gamma$ in $\mathfrak{A}$ such that $w^*\text{–lim}_\Lambda a_\lambda = A$ and $w^*\text{–lim}_\Gamma b_\gamma = B$. The adjoint is defined by $A^* = w^*\text{–lim}_\Lambda a_\lambda^*$, making $\mathfrak{A}^{\dagger\dagger}$ into a von Neumann algebra.

PROOF. To define a multiplication on $\mathfrak{A}^{\dagger\dagger}$ directly, we use the identification with the von Neumann algebra $\mathcal{M}_u$. Thus if $A, B \in \mathfrak{A}^{\dagger\dagger}$ and $w^*\text{–lim}_\Lambda a_\lambda = A$ and $w^*\text{–lim}_\Gamma b_\gamma = B$, we know from the product on $\mathcal{M}_u$ that

$$\underset{\Lambda}{w^*\text{–lim}}\ \underset{\Gamma}{w^*\text{–lim}}\ a_\lambda b_\gamma = \underset{\Lambda}{w^*\text{–lim}}\ a_\lambda B = AB$$

and

$$\underset{\Gamma}{w^*\text{–lim}}\ \underset{\Lambda}{w^*\text{–lim}}\ a_\lambda b_\gamma = \underset{\Gamma}{w^*\text{–lim}}\ A b_\gamma = AB.$$

So these limits exist and are equal, independent of the choice of nets. Similarly, A^* is defined as $A^* = w^*\text{–lim}_\Lambda a_\lambda^*$ independent of the net. Finally, since $\mathcal{M}_u$ is a C*- algebra, $\mathfrak{A}^{\dagger\dagger}$ is a C*-algebra. Moreover, that $\mathfrak{A}^{\dagger\dagger}$ is a von Neumann algebra is exhibited by the faithful $*$-representation onto $\mathcal{M}_u$. ■

10.5.4. REMARK. If $\mathfrak{B}$ is any Banach algebra, one can define a product on $\mathfrak{B}^{**}$ by either formula: given nets $w^*\text{–lim}_\Lambda a_\lambda = A$ and $w^*\text{–lim}_\Gamma b_\gamma = B$ in $\mathfrak{B}$, let

$$A \Box B = \underset{\Lambda}{w^*\text{–lim}}\ \underset{\Gamma}{w^*\text{–lim}}\ a_\lambda b_\gamma \quad \text{or} \quad A \Diamond B = \underset{\Gamma}{w^*\text{–lim}}\ \underset{\Lambda}{w^*\text{–lim}}\ a_\lambda b_\gamma.$$

These are called *Arens products*. See Exercise **8.21**. In general, they are different. A Banach algebra is *Arens regular* if these two products on the double dual agree. So C*-algebras are Arens regular.

The importance of the universal representation lies in the fact that representations of $\mathfrak{A}$ extend to $\mathcal{M}_u$.

10.5.5. THEOREM. *Let $\mathfrak{A}$ be a C*-algebra. Then each representation π of $\mathfrak{A}$ extends to a unique normal homomorphism $\widetilde{\pi}$ of $\mathcal{M}_u$ onto $\pi(\mathfrak{A})''$. Conversely, if σ is a faithful representation of $\mathfrak{A}$ such that for each representation π of $\mathfrak{A}$, $\pi\sigma^{-1}$ extends to a unique normal homomorphism of $\sigma(\mathfrak{A})''$ onto $\pi(\mathfrak{A})''$, then $\widetilde{\sigma}$ is a $*$-isomorphism of $\mathcal{M}_u$ onto $\sigma(\mathfrak{A})''$.*

PROOF. Every representation π is a direct sum of cyclic representations $(\pi_\lambda, \xi_\lambda)$. We may reduce to the case where the cyclic representations arise from distinct states $\varphi_\lambda(a) = \langle \pi_\lambda(a)\xi_\lambda, \xi_\lambda \rangle$, since this only affects multiplicity. Thus there is a direct summand ρ_λ of π_u unitarily equivalent to π_λ. Let $N = \sum_\Lambda^\oplus H_{\rho_\lambda}$. Then the restriction of π_u to N is unitarily equivalent to π, say via $\operatorname{Ad} u$ for some unitary u.

The compression ρ_N of $\mathcal{M}_u$ to N is clearly a normal homomorphism into $\mathcal{B}(N)$. Thus $\widetilde{\pi} = \operatorname{Ad} u\, \rho_N$ is normal. By Theorem 10.3.1, $\widetilde{\pi}(\mathcal{M}_u)$ is a von Neumann algebra and $\widetilde{\pi}$ is weak-$*$–weak-$*$ continuous. Since $\pi_u(\mathfrak{A})$ is weak-$*$ dense in $\mathcal{M}_u$, it follows that the range of $\widetilde{\pi}$ equals $\overline{\pi(\mathfrak{A})}^{\text{weak-}*} = \pi(\mathfrak{A})''$. Uniqueness follows from the weak-$*$ density of $\pi_u(\mathfrak{A})$ in $\mathcal{M}_u$.

If $\sigma(\mathfrak{A})''$ has this same universal property, then applying the property to *id* shows that $\widetilde{\sigma}$ must be faithful. ■

10.5.6. COROLLARY. *If $\mathfrak{A}$ is a C*-algebra and $\varphi \in \mathfrak{A}^\dagger_{sa}$ is a self-adjoint linear functional, then the Jordan decomposition $\varphi = \varphi_+ - \varphi_-$ with $\|\varphi\| = \|\varphi_+\| + \|\varphi_-\|$ is unique.*

PROOF. The existence of this decomposition is in Theorem 9.6.14. We consider $\mathfrak{A}^\dagger$ as the predual of $\mathfrak{A}^{\dagger\dagger}$. By the uniqueness of the polar decomposition for normal functionals, Theorem 10.2.10, this decomposition is unique. ■

10.5.7. DEFINITION. Two representations π and σ of a C*-algebra $\mathfrak{A}$ are *quasi-equivalent* if there is a normal isomorphism $\rho : \pi(\mathfrak{A})''$ onto $\sigma(\mathfrak{A})''$ such that $\sigma = \rho\pi$. We write $\pi \sim \sigma$. Two representations π and σ are *disjoint* if no subrepresentation of σ is unitarily equivalent to a subrepresentation of π.

The *central cover* of a representation π of $\mathfrak{A}$ is the projection $c(\pi)$ in the centre $\mathcal{Z}(\mathcal{M}_u) = \mathcal{M}_u \cap \mathcal{M}_u'$ such that $\ker \widetilde{\pi} = c(\pi)^\perp \mathcal{M}_u$. (See Exercise **10.3**.)

Quasi-equivalence is much stronger than just $\ker \pi = \ker \sigma$ but is a bit weaker than unitary equivalence. The reason is that it ignores multiplicity; for example,

$$(\pi \oplus \pi)(\mathfrak{A})'' = \pi(\mathfrak{A})''^{(2)} \simeq \pi(\mathfrak{A})''.$$

This is the only difference, as we shall see. It is easy to verify that quasi-equivalence is an equivalence relation.

10.5.8. THEOREM. *Let π and σ be representations of a C*-algebra $\mathfrak{A}$. The following are equivalent:*

(1) *π and σ are quasi-equivalent.*

(2) $c(\pi) = c(\sigma)$.

(3) *There are subrepresentations π_1 of π and σ_1 of σ which are unitarily equivalent and $c(\pi_1) = c(\pi) = c(\sigma) = c(\sigma_1)$.*

(4) *There is a cardinal α so that $\pi^{(\alpha)}$ and $\sigma^{(\alpha)}$ are unitarily equivalent.*

PROOF. Given a projection $P \in \mathcal{Z}(\mathcal{M}_u)$, let $\pi_P(A) = PA|_{PH_u}$ for $A \in \mathcal{M}_u$. Notice that $\ker \pi_P = P^\perp \mathcal{M}_u$. If $c(\pi) = P$, then $\ker \widetilde{\pi} = P^\perp \mathcal{M}_u$. Therefore $\widetilde{\pi}$ factors through π_P, say $\widetilde{\pi} = \rho \pi_P$, where ρ is an isomorphism of $P\mathcal{M}_u$ onto $\pi(\mathfrak{A})''$ such that $\rho(\pi_P \pi_u(a)) = \pi(a)$ for $a \in \mathfrak{A}$. It follows that $\pi \sim \pi_P \pi_u$.

If $\pi \sim \sigma$, $c(\pi) = P$ and $c(\sigma) = Q$, then $\pi_P \pi_u \sim \pi_Q \pi_u$. Let $\rho : P\mathcal{M}_u \to Q\mathcal{M}_u$ be a normal isomorphism such that $\pi_Q \pi_u = \rho \pi_P \pi_u$. By the normality of the maps π_P, π_Q and ρ and the weak-$*$ density of $\pi_u(\mathfrak{A})$ in $\mathcal{M}_u$, we deduce that $\pi_P = \rho \pi_Q$. Therefore

$$P^\perp \mathcal{M}_u = \ker \pi_P = \ker \rho \pi_Q = Q^\perp \mathcal{M}_u.$$

Hence $P = Q$ and $c(\pi) = c(\sigma)$. So (1) implies (2).

Conversely, if $c(\pi) = c(\sigma) = P$, then $\pi \sim \pi_P \pi_u \sim \sigma$. So (2) implies (1).

Suppose that $\pi \sim \sigma$, and let $P = c(\pi) = c(\sigma)$. Let $\rho : \pi(\mathfrak{A})'' \to \sigma(\mathfrak{A})''$ be a normal isomorphism such that $\widetilde{\sigma} = \rho \widetilde{\pi}$. By Theorem 10.3.7(3), there are projections $E \in \pi(\mathfrak{A})'$ and $F \in \sigma(\mathfrak{A})'$ with $c(E) = I$ and $c(F) = I$ so that there is a unitary equivalence implementing a map of $E\pi(\mathfrak{A})''$ onto $F\sigma(\mathfrak{A})''$:

$$\pi_1(a) := \pi_E \pi(a) \to \sigma_1(a) := \pi_F \sigma(a) \quad \text{for} \quad a \in \mathfrak{A}.$$

Hence we have $c(\pi_1) = c(\pi) = P = c(\sigma) = c(\sigma_1)$. Thus (1) implies (3).

Conversely, (3) immediately implies (2).

Next, (4) implies that there is a spatial isomorphism between $\sigma^{(\alpha)}(\mathfrak{A})''$ and $\pi^{(\alpha)}(\mathfrak{A})''$ intertwining the two representations. Hence there is a normal isomorphism between $\sigma(\mathfrak{A})''$ and $\pi(\mathfrak{A})''$ intertwining the two representations. So (1) holds.

Conversely, by Theorem 10.3.7, (1) implies that both σ and π are unitarily equivalent to a subrepresentation of $\varphi(\sigma) \simeq \sigma^{(\alpha)}$, and likewise, to subrepresentations of $\pi^{(\beta)}$. We may assume that $\alpha = \beta$ is infinite by increasing them if necessary. Thus $\sigma^{(\alpha)}$ is a subrepresentation of $\pi^{(\alpha\alpha)} = \pi^{(\alpha)}$ and $\pi^{(\alpha)}$ is a subrepresentation of $\sigma^{(\alpha\alpha)} = \sigma^{(\alpha)}$. By Exercise **10.15**, $\sigma^{(\alpha)}$ and $\pi^{(\alpha)}$ are unitarily equivalent. So (1) holds. ■

This shows that the lattice of projections in $\mathcal{Z}(\mathcal{M}_u)$ contains a great deal of information about the representations of $\mathfrak{A}$. One easy observation is that if $c(\pi)$ is a minimal projection, then π is a *factor representation*, meaning that $\pi(\mathfrak{A})''$ is a factor. This is because

$$\mathcal{P}(\mathcal{Z}(\pi(\mathfrak{A})'')) = \{p = p^* = p^2 \in \mathcal{Z}(\mathcal{M}_u) : 0 \le p \le c(\pi)\} = \{0, c(\pi)\}.$$

In particular, this includes irreducible representations, since then $\pi(\mathfrak{A})'' = \mathcal{B}(H_\pi)$ is a factor. Since an irreducible representation has no proper subrepresentations, we get an immediate corollary. It also follows from Exercise **10.10**.

10.5.9. COROLLARY. *Let π and σ be irreducible representations of a C*-algebra $\mathfrak{A}$. Then $\pi \sim \sigma$ if and only if $\pi \simeq \sigma$.*

10.5.10. COROLLARY. *Let π and σ be representations of a C*-algebra $\mathfrak{A}$. If $c(\sigma) \le c(\pi)$, then σ is quasi-equivalent to a subrepresentation of π.*

PROOF. Let $\rho = \pi_{c(\sigma)}\pi$. Then $c(\rho) = c(\sigma)$; whence $\rho \sim \sigma$. ■

We finish this section with a characterization of disjoint representations.

10.5.11. PROPOSITION. *Let π and σ be representations of a C*-algebra $\mathfrak{A}$. The following are equivalent:*

(1) *π and σ are disjoint.*

(2) $c(\pi) \perp c(\sigma)$.

(3) $(\pi \oplus \sigma)(\mathfrak{A})'' = \pi(\mathfrak{A})'' \oplus \sigma(\mathfrak{A})''$.

(4) $(\pi \oplus \sigma)(\mathfrak{A})' = \pi(\mathfrak{A})' \oplus \sigma(\mathfrak{A})'$.

PROOF. If (2) fails, then $c(\pi)c(\sigma) = E \neq 0$. Let $\rho = \pi_E\pi$; so that $c(\rho) = E$. Then by Corollary 10.5.10, ρ is quasi-equivalent to a subrepresentation of σ. By Theorem 10.5.8, there is a subrepresentation of ρ unitarily equivalent to a subrepresentation of σ. And thus π and σ are not disjoint. So (1) fails.

Conversely, if (1) fails, there is a representation ρ which is unitarily equivalent to a subrepresentation of both σ and π. Then $c(\sigma)c(\pi) \ge c(\rho) \neq 0$. So (2) fails.

Observe that

$$\ker \widetilde{\pi} \oplus \widetilde{\sigma} = \ker \widetilde{\pi} \cap \ker \widetilde{\sigma} = c(\pi)^\perp \mathcal{M}_u \cap c(\sigma)^\perp \mathcal{M}_u = (c(\pi) \vee c(\sigma))^\perp \mathcal{M}_u.$$

Thus $c(\pi \oplus \sigma) = c(\pi) \vee c(\sigma)$. If (2) holds, then $c(\pi \oplus \sigma) = c(\pi) + c(\sigma)$. Therefore

$$(\pi \oplus \sigma)(\mathfrak{A})'' \simeq (c(\pi) + c(\sigma))\mathcal{M}_u = c(\pi)\mathcal{M}_u \oplus c(\sigma)\mathcal{M}_u \simeq \pi(\mathfrak{A})'' \oplus \sigma(\mathfrak{A})''.$$

So (3) holds. While if (2) fails, say $c(\pi)c(\sigma) = E \neq 0$, then we can write $c(\pi) = P + E$ and $c(\sigma) = Q + E$ where E, P, Q are pairwise orthogonal. Applying the argument above to $\pi = \pi_P\pi \oplus \pi_E\pi$ and $\sigma = \pi_E\sigma \oplus \pi_Q\sigma$, we have

$$\pi(\mathfrak{A})'' \oplus \sigma(\mathfrak{A})'' \simeq P\mathcal{M}_u \oplus E\mathcal{M}_u \oplus E\mathcal{M}_u \oplus Q\mathcal{M}_u,$$

while $c(\pi \oplus \sigma) = c(\pi) \vee c(\sigma) = P + E + Q$, so that

$$(\pi \oplus \sigma)(\mathfrak{A})'' \simeq P\mathcal{M}_u \oplus E\mathcal{M}_u \oplus Q\mathcal{M}_u.$$

Thus (3) fails.

If $\mathcal{M}$ and $\mathcal{N}$ are von Neumann algebras, then $(\mathcal{M} \oplus \mathcal{N})' = \mathcal{M}' \oplus \mathcal{N}'$ because operators commuting with $I \oplus 0$ have the form $A \oplus B$. Since $\mathfrak{A}''' = \mathfrak{A}'$ for any C*-algebra, (3) and (4) are equivalent. ■

10.6. Abstract von Neumann Algebras

In this section, we establish Sakai's abstract characterization of von Neumann algebras as dual C*-algebras. Our proof relies on the structure of expectations. Thus we begin with information about them.

10.6.1. DEFINITION. If $\mathfrak{A}$ is a C*-algebra and $\mathfrak{B}$ is a C*-subalgebra of $\mathfrak{A}$, an *expectation* of $\mathfrak{A}$ onto $\mathfrak{B}$ is a norm 1 idempotent $\mathcal{E} \in \mathcal{B}(\mathfrak{A})$ with range $\mathfrak{B}$.

10.6.2. TOMIYAMA THEOREM. *Let $\mathfrak{A}$ be a C*-algebra with C*-subalgebra $\mathfrak{B}$. Suppose that $\mathcal{E}$ is an expectation of $\mathfrak{A}$ onto $\mathfrak{B}$. Then*

(1) *$\mathcal{E}$ is positive.*

(2) *$\mathcal{E}(b_1ab_2) = b_1\mathcal{E}(a)b_2$ for $a \in \mathfrak{A}$ and $b_1, b_2 \in \mathfrak{B}$. (bimodule property)*

(3) *$\mathcal{E}(a)^*\mathcal{E}(a) \le \mathcal{E}(a^*a)$ for $a \in \mathfrak{A}$. (Schwarz inequality)*

PROOF. We can reduce to the von Neumann algebra setting by considering $\mathcal{E}^{**} : \mathfrak{A}^{\dagger\dagger} \to \mathfrak{B}^{\dagger\dagger}$. By Theorem 10.5.2, $\mathfrak{B}^{\dagger\dagger}$ is a weak-$*$ closed subalgebra of the von Neumann algebra $\mathfrak{A}^{\dagger\dagger}$. Once the result is established for $\mathcal{E}^{**}$, it follows for $\mathcal{E}$ by restriction since $\mathcal{E}^{**}|_{\mathfrak{A}} = \mathcal{E}$. So for the remainder of the proof, suppose that $\mathcal{E} : \mathcal{M} \to \mathcal{N}$ is a norm 1 projection of a von Neumann algebra $\mathcal{M}$ onto a weak-$*$ closed subalgebra $\mathcal{N}$. Since von Neumann algebras are unital, we have identities $1 = 1_{\mathcal{M}}$ and $1_{\mathcal{N}}$. One possible issue is that $1_{\mathcal{N}}$ may be a proper projection in $\mathcal{M}$.

Let e be a projection in $\mathcal{M}$ such that either e or $e^\perp$ is in $\mathcal{N}$. Then $e\mathcal{N} \subset \mathcal{N}$. For $a, b \in \mathcal{M}$,

$$\begin{aligned}\|ea + e^\perp b\|^2 &= \|(ea + e^\perp b)^*(ea + e^\perp b)\| = \|a^*ea + b^*e^\perp b\| \\ &\le \|a^*ea\| + \|b^*e^\perp b\| = \|ea\|^2 + \|e^\perp b\|^2.\end{aligned}$$

Thus if $t > 0$,

$$\|e^\perp\mathcal{E}(ea + e^\perp t\mathcal{E}(ea))\|^2 \le \|ea + e^\perp t\mathcal{E}(ea)\|^2 \le \|ea\|^2 + t^2\|e^\perp\mathcal{E}(ea)\|^2.$$

However $\mathcal{E}(ea) \in \mathcal{N}$, so that $e^\perp t\mathcal{E}(ea) \in \mathcal{N}$ and hence

$$e^\perp\mathcal{E}(ea + e^\perp t\mathcal{E}(ea)) = e^\perp\mathcal{E}(ea) + e^\perp t\mathcal{E}(ea) = (1+t)e^\perp\mathcal{E}(ea).$$

Therefore

$$(1+t)^2\|e^\perp\mathcal{E}(ea)\|^2 \le \|ea\|^2 + t^2\|e^\perp\mathcal{E}(ea)\|^2.$$

As $t \to \infty$, we obtain $e^\perp\mathcal{E}(ea) = 0$. Reversing roles yields $e\mathcal{E}(e^\perp a) = 0$. Thus

$$\mathcal{E}(ea) = e\mathcal{E}(ea) + e^\perp\mathcal{E}(ea) = e\big(\mathcal{E}(a) - \mathcal{E}(e^\perp a)\big) = e\mathcal{E}(a).$$

The projections in $\mathcal{N}$ span $\mathcal{N}$, and thus by linearity and continuity, we obtain $\mathcal{E}(ca) = c\mathcal{E}(a)$ for $c \in \mathcal{N}$ and $a \in \mathcal{M}$. In particular,

$$\mathcal{E}(1) = 1_{\mathcal{N}}\mathcal{E}(1) = \mathcal{E}(1_{\mathcal{N}}) = 1_{\mathcal{N}}.$$

Let $a \in \mathcal{M}_+$ and let ψ be a state on $\mathcal{N}$. Define $\varphi = \psi\mathcal{E}$ in $\mathcal{M}^\dagger$. Then $\varphi(1) = \psi(1_\mathcal{N}) = 1$ and $\|\varphi\| \leq \|\psi\| \, \|\mathcal{E}\| \leq 1$. By Lemma 9.6.5, φ is a state on $\mathcal{M}$; whence $0 \leq \varphi(a) = \psi(\mathcal{E}(a))$. This holds for all states on $\mathcal{N}$, and hence $\mathcal{E}(a) \geq 0$; i.e. $\mathcal{E}$ is positive. Every self-adjoint element $b = b^*$ is the difference of two positive elements, and hence $\mathcal{E}(b) = \mathcal{E}(b)^*$. Therefore $\mathcal{E}$ is self-adjoint, meaning $\mathcal{E}(x^*) = \mathcal{E}(x)^*$ for $x \in \mathcal{M}$.

Now we obtain that $\mathcal{E}(ae) = \mathcal{E}(ea^*)^* = (e\mathcal{E}(a^*))^* = \mathcal{E}(a)e$. As above, we obtain $\mathcal{E}(ac) = \mathcal{E}(a)c$ for all $a \in \mathcal{M}$ and $c \in \mathcal{N}$. Therefore $\mathcal{E}$ is an $\mathcal{N}$-bimodule map.

Finally, if $\mathcal{E}(a) = b$ for $a \in \mathcal{M}$,

$$0 \leq \mathcal{E}((a-b)^*(a-b)) = \mathcal{E}(a^*a) - b^*\mathcal{E}(a) - \mathcal{E}(a)^*b + b^*b = \mathcal{E}(a^*a) - b^*b.$$

Hence we have the inequality $\mathcal{E}(a)^*\mathcal{E}(a) \leq \mathcal{E}(a^*a)$. ∎

10.6.3. Remark. It follows from the theorem above that when $1_\mathcal{N} < 1_\mathcal{M}$,

$$\mathcal{E}(a) = 1_\mathcal{N}\mathcal{E}(a)1_\mathcal{N} = \mathcal{E}(1_\mathcal{N} a 1_\mathcal{N}).$$

Thus $\mathcal{E}$ annihilates $1_\mathcal{N}^\perp \mathcal{M} + \mathcal{M} 1_\mathcal{N}^\perp$.

The following is an important method for obtaining expectations.

10.6.4. Proposition. *If $\mathcal{M}$ is a von Neumann algebra for which there is an amenable group G of unitary operators such that $G' = \mathcal{M}$, then there is an expectation of $\mathcal{B}(H)$ onto $\mathcal{M}$.*

Proof. Let m be an invariant mean on $l^\infty(G)$. For each $T \in \mathcal{B}(H)$, define $f_T(u) = uTu^*$ for $u \in G$. Construct a form on $H \times H$ by

$$[\xi, \eta] = m(\langle f_T(u)\xi, \eta\rangle) \quad \text{for} \quad \xi, \eta \in H.$$

Then since $T \to f_T$ is a linear isometric map, it is easy to see that $[\xi, \eta]$ is a sesquilinear form on H which is bounded by $\|T\|$. By Proposition 4.2.2, there is a unique operator $\mathcal{E}(T)$ such that

$$\langle \mathcal{E}(T)\xi, \eta\rangle = [\xi, \eta],$$

and $\|\mathcal{E}(T)\| \leq \|T\|$. It is clear that $\mathcal{E}$ is linear, so $\mathcal{E}$ is a contractive linear map.

If $A \in \mathcal{M}$, then $f_A(u) = u^*Au = A$ is constant and thus $\mathcal{E}(A) = A$. Conversely, if $v \in G$, then

$$(v^* f_T v)(u) = v^*uTu^*v = f_T(v^*u) = {}_v f_T(u)$$

is a left translate of f_T. Therefore

$$\begin{aligned}\langle (v^*\mathcal{E}(T)v)\xi,\eta\rangle &= \langle \mathcal{E}(T)v\xi, v\eta\rangle = [v\xi, v\eta]\\ &= m(\langle (v^* f_T(u)v)\xi,\eta\rangle) = m(\langle {}_v f_T(u)\xi,\eta\rangle)\\ &= m(\langle f_T(u)\xi,\eta\rangle) = \langle \mathcal{E}(T)\xi,\eta\rangle.\end{aligned}$$

Hence $v^*\mathcal{E}(T)v = \mathcal{E}(T)$, and so $\mathcal{E}(T) \in G' = \mathcal{M}$. We deduce that $\mathcal{E}$ is an expectation of $\mathcal{B}(H)$ onto $\mathcal{M}$. ■

10.6.5. Example. If $\mathcal{M}$ is a von Neumann algebra with abelian commutant, there is an expectation onto $\mathcal{M}$. In this case, we can take G to be the unitary group of $\mathcal{M}'$.

Consider $\mathcal{M} = L^\infty(0,1)$ acting on $L^2(0,1)$. Since $\mathcal{M}$ is maximal abelian, there is an expectation onto $\mathcal{M}$. It turns out that there are many such expectations. (See Exercise **10.16**.)

However there is no normal expectation. To see this, suppose that K is a compact operator on $L^2(0,1)$; and let $P_{[a,b]} = M_{\chi_{[a,b]}}$. For any $t \in [0,1]$, the sequence $P_{[t-\delta,t+\delta]}$ converges SOT to 0. By Exercise **5.8**, $\lim_{\delta\to 0} \|P_{[t-\delta,t+\delta]}K\| = 0$. Hence by compactness of $[0,1]$, for any $\varepsilon > 0$, we can find a finite partition $0 = t_0 < t_1 < \cdots < t_n = 1$ so that

$$\|P_{[t_{i-1},t_i]}K\| < \varepsilon \quad \text{for} \quad 1 \le i \le n.$$

By the bimodule property of expectations, we have

$$\|P_{[t_{i-1},t_i]}\mathcal{E}(K)\| = \|\mathcal{E}(P_{[t_{i-1},t_i]}K)\| \le \|P_{[t_{i-1},t_i]}K\| < \varepsilon.$$

Now $P_{[t_{i-1},t_i]}\mathcal{E}(K) = P_{[t_{i-1},t_i]}\mathcal{E}(K)P_{[t_{i-1},t_i]}$ because $\mathcal{M}$ is abelian. Since $\mathcal{E}(K)$ is an orthogonal direct sum of these terms for $1 \le i \le n$, we deduce that $\|\mathcal{E}(K)\| < \varepsilon$. As $\varepsilon > 0$ is arbitrary, we see that $\mathcal{E}(K) = 0$. But $\mathcal{K}(H)$ is SOT-dense in $\mathcal{B}(L^2(0,1))$, so this map is not normal.

The same argument works for any non-atomic σ-finite measure μ and abelian subalgebra $L^\infty(\mu) \subset \mathcal{B}(H)$.

On the other hand, if $\mathcal{M} = l^\infty$ acting on l^2, then the expectation is normal and unique. Let P_i be the projection onto $\mathbb{C}e_i$ where $\{e_i : i \ge 1\}$ is the standard basis. Then for $T \in \mathcal{B}(l^2)$ and any expectation $\mathcal{E}$ onto l^∞,

$$P_i\mathcal{E}(T) = P_i\mathcal{E}(T)P_i = \mathcal{E}(P_iTP_i) = t_{ii}E_{ii}.$$

Therefore $\mathcal{E}(T) = \sum_{i\ge 1} t_{ii}E_{ii}$ is the diagonal of T. This is easily seen to be normal.

We use Tomiyama's theorem to provide a nice proof of Sakai's abstract characterization of von Neumann algebras.

10.6.6. Sakai Theorem. *A C*-algebra $\mathfrak{A}$ has a faithful representation π as a von Neumann algebra if and only if $\mathfrak{A}$ is the dual of a Banach space X. In this* *case, the representation is a homeomorphism between* $\mathfrak{A}$ *with the weak-$*$ topology* τ_X *and* $\pi(\mathfrak{A})$ *with its weak-$*$ topology.*

PROOF. Recall from Corollary 10.2.2 that every von Neumann algebra is a dual space. Indeed the predual is unique by Theorem 10.2.13. Consider the converse.

Let X be a predual of $\mathfrak{A}$, and let j be the canonical embedding of X into $X^{**} = \mathfrak{A}^\dagger$. Then $\mathcal{E} = j^* : \mathfrak{A}^{\dagger\dagger} \to \mathfrak{A} = X^*$ is a contraction onto $\mathfrak{A}$ such that $\mathcal{E}(a) = a$ for $a \in \mathfrak{A}$. By Theorem 10.5.2, $\mathfrak{A}^{\dagger\dagger}$ is isomorphic and weak-$*$ homeomorphic to the universal von Neumann algebra of $\mathfrak{A}$. Thus $\mathcal{E}$ is an expectation of $\mathfrak{A}^{\dagger\dagger}$ onto $\mathfrak{A}$. By Tomiyama's theorem, $\mathcal{E}$ is a positive $\mathfrak{A}$-bimodule map. Since $\mathcal{E} = j^*$, $\mathcal{E}$ is $\tau_{\mathfrak{A}^\dagger}$–$\tau_X$ continuous. In particular, $\mathcal{J} = \ker \mathcal{E}$ is $\tau_{\mathfrak{A}^\dagger}$ closed.

The bimodule property yields $\mathfrak{A}\mathcal{J}\mathfrak{A} \subset \mathcal{J}$. Since $\mathfrak{A}$ is $\tau_{\mathfrak{A}^\dagger}$-dense in $\mathfrak{A}^{\dagger\dagger}$, it follows that $\mathcal{J}$ is a weak-$*$ closed ideal of $\mathfrak{A}^{\dagger\dagger}$. Hence by Exercise **10.3**, there is a central projection $P \in \mathcal{Z}(\mathfrak{A}^{\dagger\dagger})$ so that $\mathcal{J} = P^\perp \mathfrak{A}^{\dagger\dagger}$. Then

$$\mathfrak{A} \simeq \mathfrak{A}^{\dagger\dagger}/\mathcal{J} = P\mathfrak{A}^{\dagger\dagger} = \pi_P(\mathfrak{A}^{\dagger\dagger}) =: \mathcal{M}.$$

The linear isomorphism ε of $\mathfrak{A}^{\dagger\dagger}/\mathcal{J}$ onto $\mathfrak{A}$ arises from the factorization $\mathcal{E} = \varepsilon q$, where $q : \mathfrak{A}^{\dagger\dagger} \to \mathfrak{A}^{\dagger\dagger}/\mathcal{J}$ is the quotient map, and ε is the induced isomorphism. However if $a, b \in \mathfrak{A}^{\dagger\dagger}$, then $a - \mathcal{E}(a)$ and $b - \mathcal{E}(b)$ belong to the ideal $\mathcal{J}$. Thus using the bimodule property,

$$\begin{aligned} 0 &= \mathcal{E}\big((a - \mathcal{E}(a))(b - \mathcal{E}(b))\big) \\ &= \mathcal{E}(ab) - \mathcal{E}\big(a\mathcal{E}(b)\big) - \mathcal{E}\big(\mathcal{E}(a)b\big) + \mathcal{E}\big(\mathcal{E}(a)\mathcal{E}(b)\big) \\ &= \mathcal{E}(ab) - \mathcal{E}(a)\mathcal{E}(b). \end{aligned}$$

That is, $\mathcal{E}$ is multiplicative, and hence is a $*$-homomorphism. It follows that $\pi := \pi_P \varepsilon^{-1}$ is a $*$-isomorphism of $\mathfrak{A}$ onto $\mathcal{M}$. Therefore $\mathfrak{A}$ is realized faithfully as a von Neumann algebra.

Moreover the map $\pi^{-1} = \mathcal{E}|_{\mathcal{M}}$ of $\mathcal{M}$ onto $\mathfrak{A}$ is a $\tau_{\mathcal{M}_*}$–τ_X continuous bijection. It is isometric, and so carries the closed unit ball onto the closed unit ball. These balls are compact in the $\tau_{\mathcal{M}_*}$ and τ_X topologies, respectively. Thus this restriction is a $\tau_{\mathcal{M}_*}$–τ_X homeomorphism. By the Krein–Smulian theorem, $\mathcal{E}$ is a $\tau_{\mathcal{M}_*}$–τ_X homeomorphism. Therefore π is a τ_X–$\tau_{\mathcal{M}_*}$ homeomorphism. ■

10.7. Types for von Neumann Algebras

The classification of von Neumann algebras begins with an analysis of the lattice of projections $\mathcal{P}(\mathcal{M})$ in $\mathcal{M}$. A projection $P \in \mathcal{M}$ corresponds to the invariant subspace PH for $\mathcal{M}'$. Thus $\mathcal{P}(\mathcal{M}) = \operatorname{Lat} \mathcal{M}'$ is the lattice of invariant subspaces of $\mathcal{M}'$. This shows that $\mathcal{P}(\mathcal{M})$ is a complete lattice, meaning that is closed under arbitrary meets, $\bigwedge_\Lambda P_\lambda$ is the projection onto $\bigcap_\Lambda P_\lambda H$, and joins, $\bigvee_\Lambda P_\lambda$ is the projection onto $\overline{\operatorname{span}}\big(\bigcup_\Lambda P_\lambda H\big)$. It is also closed under orthogonal complements.

See the discussion following Definition 5.5.1. We defined (Murray-von Neumann) equivalence of projections in Definition 9.10.7, but we repeat it here for emphasis.

10.7.1. DEFINITION. Two projections P, Q in a von Neumann algebra $\mathcal{M}$ are *(Murray-von Neumann) equivalent* if there is a partial isometry $U \in \mathcal{M}$ such that $P = U^*U$ and $Q = UU^*$. We write $P \sim Q$. If there is a projection $Q_1 \leq Q$ such that $P \sim Q_1$, we write $P \preceq Q$. The notation $P \prec F$ means that $P \preceq F$ but $P \not\sim Q$. If N is a (not necessarily closed) subspace of H, let $[N]$ denote the orthogonal projection onto $\overline{N}$.

Since we frequently wish to sum infinite families of orthogonal projections or related partial isometries, we will suppress the fact that the sum converges SOT in our notation. Recall Definition 10.3.3 of the central cover $c(P)$.

10.7.2. LEMMA. *Suppose that P, Q are projections in $\mathcal{M}$.*

(1) *If Z is a projection in $\mathcal{Z}(\mathcal{M})$ and $P \sim Q$, then $ZP \sim ZQ$.*

(2) *If $A \in \mathcal{M}$, the partial isometry U in the polar decomposition yields an equivalence between $[AH]$ and $[A^*H]$.*

(3) *If $c(P)c(Q) \neq 0$, then there are non-zero projections $P_1 \leq P$ and $Q_1 \leq Q$ such that $P_1 \sim Q_1$.*

(4) *Suppose that $\{P_\lambda\}_\Lambda$ and $\{Q_\lambda\}_\Lambda$ are two families of pairwise orthogonal projections in $\mathcal{M}$. If $P_\lambda \sim Q_\lambda$ for all $\lambda \in \Lambda$, then $\sum_\Lambda P_\lambda \sim \sum_\Lambda Q_\lambda$.*

PROOF. If $U^*U = P$ and $UU^* = Q$, then ZU implements the equivalence between ZP and ZQ. (2) is immediate from Theorem 9.4.5. By Lemma 10.3.4, $c(Q) = \bigvee W^*QW$ as W runs over the unitary group of $\mathcal{M}$. Since $c(P)c(Q) \neq 0$, there is a unitary W so that $QWP \neq 0$. Let U be the partial isometry in the polar decomposition of QWP. Then $Q_1 = UU^* \leq Q$ and $U^*U = P_1 \leq P$ and $P_1 \sim Q_1$. If partial isometries U_λ implement the equivalence $P_\lambda \sim Q_\lambda$, then $U = \sum_\Lambda U_\lambda$ implements an equivalence of their sums. ∎

The following result is not difficult, but it is so useful, we call it a theorem.

10.7.3. COMPARISON THEOREM. *Suppose that P and Q are projections in $\mathcal{M}$. Then there are projections $P_0 \leq P$ and $Q_0 \leq Q$ such that $P_0 \sim Q_0$ and $c(P - P_0) \perp c(Q - Q_0)$. Hence there is a projection $Z \in \mathcal{Z}(\mathcal{M})$ such that*

$$ZP \sim ZQ_0 \leq ZQ \quad \text{and} \quad Z^\perp Q \sim Z^\perp P_0 \leq Z^\perp P.$$

PROOF. Consider collections $\{P_\lambda\}_\Lambda$ and $\{Q_\lambda\}_\Lambda$ of pairwise orthogonal projections less than P and Q respectively such that $P_\lambda \sim Q_\lambda$ for each $\lambda \in \Lambda$. The union of a nested family of such sets is another such pair. Therefore by Zorn's lemma, there is a maximal family with this property. By Lemma 10.7.2(4), $P_0 := \sum_\Lambda P_\lambda \sim \sum_\Lambda Q_\lambda =: Q_0$. By maximality, there is no subprojection of $P - P_0$ which is equivalent to a subprojection of $Q - Q_0$. By Lemma 10.7.2(3), $c(P - P_0)c(Q - Q_0) = 0$. Let $Z = c(Q - Q_0)$. Then $ZP = ZP_0 \sim ZQ_0 \leq ZQ$ and $Z^\perp Q = Z^\perp Q_0 \sim Z^\perp P_0 \leq Z^\perp P$. ∎

10.7.4. COROLLARY. *Let $\mathcal{M}$ be a factor, and let P, Q be projections in $\mathcal{M}$. Then either $P \prec Q$, $P \sim Q$ or $Q \prec P$.*

PROOF. The projection Z provided by Theorem 10.7.3 is either 0 or I. So either $Q \preceq P$ or $P \preceq Q$. ∎

We now identify some useful properties of projections in von Neumann algebras that measure their "size" in various ways.

10.7.5. DEFINITION. A projection P in $\mathcal{M}$ is *finite* if $P \sim Q \leq P$ implies that $P = Q$. Otherwise it is an *infinite projection*. Say that P is *purely infinite* if $0 \neq Q \leq P$ is always infinite. A projection P is *semifinite* if for every central projection Z such that $ZP \neq 0$, there is a finite projection $0 \neq Q \leq ZP$. And P is *properly infinite* if for every central projection Z such that $ZP \neq 0$, ZP is infinite. A projection P is *abelian* if $\mathcal{M}_P = P\mathcal{M}P|_{PH}$ is abelian.

A von Neumann algebra is finite, infinite, purely infinite, semifinite or properly infinite if I has this property.

A projection P is *σ-finite* if every collection of pairwise orthogonal non-zero projections less than P is at most countable. A von Neumann algebra is *σ-finite* (also called *countably decomposable*) if I is σ-finite.

10.7.6. REMARKS.
(1) All of these properties are preserved by equivalence of projections.

(2) A projection P is *minimal* if there are no smaller non-zero projections, and then $\mathcal{M}_P = \mathbb{C}P$; so P is abelian. Abelian implies finite, which implies semifinite. These are all measures of smallness, and it follows from the first remark and part (1) of the next proposition that if $Q \preceq P$ and P is abelian, finite or semifinite, then so is Q.

(3) The notion of σ-finiteness is a measure of smallness. No infinite von Neumann algebra is separable, but one can ask whether the predual is separable. A von Neumann algebra acting on a separable Hilbert space is σ-finite and has a separable predual, since $\mathcal{M}_*$ is a quotient of $\mathcal{B}(H)_*$. Conversely, the property that a von Neumann algebra has a faithful representation on a separable Hilbert space is equivalent to separable predual and also to σ-finiteness plus a countable SOT-dense subset. We will not prove these results here.

The next proposition collects various easy consequences of these definitions. More useful properties will be established later in Section 10.11.

10.7.7. PROPOSITION. *Let $\mathcal{M}$ be a von Neumann algebra.*

(1) *Suppose that $0 \neq P_1 \leq P$ and P is abelian, finite, semifinite or purely infinite. Then so is P_1.*

(2) *If $\{P_\lambda\}_\Lambda$ are all purely infinite (semifinite) projections, then $P = \bigvee_\Lambda P_\lambda$ is purely infinite (semifinite). Thus there is a largest purely infinite (semifinite) projection, and it lies in $\mathcal{Z}(\mathcal{M})$.*

(3) *If P is a properly infinite projection, then it decomposes uniquely as a sum $P = P_s + P_p$ of a semifinite projection P_s and a purely infinite projection P_p; and $c(P_s) \perp c(P_p)$.*

(4) *If $\{P_\lambda\}_\Lambda$ are abelian (finite) projections such that $c(P_\lambda) \perp c(P_{\lambda'})$ for $\lambda \neq \lambda' \in \Lambda$, then $P = \sum_\Lambda P_\lambda$ is an abelian (finite) projection. There is a smallest central projection $Z \in \mathcal{Z}(\mathcal{M})$ that dominates all abelian (finite) projections, and there is an abelian (finite) projection P with $c(P) = Z$. Also there is a largest finite central projection.*

(5) *If E, F are abelian projections such that $c(E) = c(F)$, then $E \sim F$.*

PROOF. Statement (1) is clear if P is abelian or purely infinite. Suppose that P is finite, $P_2 \leq P_1 \leq P$ and $P_2 \sim P_1$. Then $P_2 + (P - P_1) \leq P$ and

$$P_2 + (P - P_1) \sim P_1 + (P - P_1) = P.$$

Since P is finite, $P_2 + (P - P_1) = P$. Hence $P_2 = P_1$; so that P_1 is finite.

Now suppose that P is semifinite, and let $0 < P_1 \leq P$. Set $Z = c(P_1)$. There is a finite projection $0 \neq Q \leq ZP$. Recall that Z is the join of all projections equivalent to P_1 by Lemma 10.3.4. Hence there is a projection $Q_1 \sim Q$ such that $P_1 Q_1 \neq 0$. By Lemma 10.7.2(2), there is a projection $0 \neq Q_2 \leq Q_1$ with $Q_2 \sim P_2 \leq P_1$. By the finite case, Q_2 and thus P_2 is finite as desired.

(2) Suppose that each P_λ is purely infinite and Q is a finite projection with $Q \leq P$. Then for some λ, $QP_\lambda \neq 0$. By Lemma 10.7.2, there is a projection $0 \neq Q_1 \leq Q$ with $Q_1 \sim P_1 \leq P_\lambda$. By (1), Q_1 is finite, and hence P_1 is finite. This contradicts the fact that P_λ is purely infinite. So P is purely infinite. The set of all purely infinite projections is invariant under unitary conjugation. Thus by Lemma 10.3.4, the join Q of all purely infinite projections satisfies $Q = UQU^*$ for all $U \in \mathcal{M}$ unitary. Thus $Q \in \mathcal{Z}(\mathcal{M})$, it is purely infinite, and by definition, it dominates all purely infinite projections.

Now suppose that each P_λ is semifinite. Let $Q \leq P$. Again for some λ, $QP_\lambda \neq 0$; and so there is a projection $0 \neq Q_1 \leq Q$ with $Q_1 \sim P_1 \leq P_\lambda$. Since P_λ is semifinite, so is P_1. Thus there is a finite projection $0 \neq P_2 \leq P_1$. The equivalence $Q_1 = UU^*$ and $P_1 = U^*U$ shows that $Q_2 := UP_2U^* \leq Q_1 \leq Q$ is finite.

Thus P is semifinite. Again the set of all semifinite projections is invariant under unitary conjugation; and thus the span of this collection is central and dominates all semifinite projections.

(3) Suppose that P is properly infinite. Let P_s be the span of all finite projections less than P. Then P_s is semifinite. Moreover there are no finite projections less than $P_p = P - P_s$; so it is purely infinite. Finally by Lemma 10.3.4, $c(P_s)$ is spanned by projections $Q_1 \sim P_1 \leq P$ which are finite. If $c(P_s)c(P_p) \neq 0$, then there is a finite projection Q_1 with $Q_1 P_p \neq 0$. By Lemma 10.7.2, there is a projection $0 \neq Q_2 \leq Q_1$ with $Q_2 \sim P_2 \leq P_\lambda$. This contradicts the fact that P_p is purely infinite. Therefore $c(P_s)c(P_p) = 0$.

(4) If each P_λ is abelian with pairwise orthogonal central support projections, then $P\mathcal{M}P = \sum_\Lambda c(P_\lambda)P_\lambda\mathcal{M}P_\lambda$ is a direct sum of abelian algebras, and hence is abelian. By Zorn's lemma, we can find a maximal family of such projections. If there were an abelian projection $Q \leq c(P)^\perp$, then the collection would not be maximal. Hence $Z = c(P)$ dominates every abelian projection and is the smallest central projection with this property.

Now suppose that each P_λ is finite, and let $Q \leq P$ such that $Q \sim P$. Then $Q_\lambda = Qc(P_\lambda) \leq P_\lambda$ and $Q_\lambda \sim P_\lambda$ by Lemma 10.7.2. Since they are finite, $Q_\lambda = P_\lambda$ for all $\lambda \in \Lambda$. Therefore $Q = P$ and P is finite. Again we could take a maximal family of finite projections P_λ with pairwise orthogonal central supports. In this case, $Z = c(P)$ must dominate every finite projection Q, as otherwise $QZ^\perp$ will be finite, contradicting the maximality of the family. Since P is finite, it is clear that Z is the smallest central projection with this property.

Let $\{Z_\lambda\}_\Lambda$ be a maximal family of pairwise orthogonal finite central projections. Then $Z = \sum_\Lambda Z_\lambda$ is finite and central. If there is a finite central projection P such that $P \not\leq Z$, then $Z^\perp P$ is a non-zero finite central projection orthogonal to Z. This would contradict the maximality of our family. Thus Z is the largest finite central projection.

(5) We may suppose that E and F are abelian projections with $c(E) = c(F) = I$. By the Comparison theorem 10.7.3, there is a central projection Z such that $ZE \sim ZF_0 \leq ZF$ and $Z^\perp F \sim Z^\perp E_0 \leq Z^\perp E$. Note that $c(ZE) = c(ZF_0) = c(ZF) = Z$. If $ZF_0 < ZF$, by Lemma 10.3.4 there is an $A \in \mathcal{M}$ so that

$$\begin{aligned} 0 \neq Z(F - F_0)AZF_0 &= Z(F - F_0)(ZFAZF)ZF_0 \\ &= Z(F - F_0)ZF_0(ZFAZF) = 0, \end{aligned}$$

where we are using the fact that $ZF\mathcal{M}ZF$ is abelian. Therefore $ZF_0 = ZF$. Similarly $Z^\perp E_0 = Z^\perp E$. Therefore $E = ZE + Z^\perp E \sim ZF + Z^\perp F = F$. ■

10.7.8. Corollary. *Let P be an abelian projection in a von Neumann algebra $\mathcal{M}$. Then $c(P)\mathcal{Z}(\mathcal{M}) \simeq \mathcal{M}_P$ via the map $c(P)Z \to PZ$.*

PROOF. By Lemma 10.3.6, $(\mathcal{M}'_P)' = \mathcal{M}_P$. Hence by the double commutant theorem, $(\mathcal{M}_P)' = \mathcal{M}'_P$. Since P is an abelian projection,

$$\mathcal{M}_P = \mathcal{Z}(\mathcal{M}_P) = \mathcal{M}_P \cap \mathcal{M}'_P = P(\mathcal{M} \cap \mathcal{M}')P = P\mathcal{Z}(\mathcal{M}).$$

Now $Z \to PZ$ is a $*$-homomorphism with kernel $c(P)^\perp \mathcal{Z}(\mathcal{M})$. So this yields an isomorphism with $c(P)\mathcal{Z}(\mathcal{M})$. ∎

We are now ready to define the *type* of a von Neumann algebra.

10.7.9. DEFINITION. A von Neumann algebra is *type* I if every non-zero projection dominates a non-zero abelian projection. A von Neumann algebra is *type* II_1 if it is finite and there are no non-zero abelian projections. A von Neumann algebra is *type* II_∞ if it is semifinite, properly infinite and there are no non-zero abelian projections. A von Neumann algebra is *type* III if it is purely infinite.

A von Neumann algebra is *atomic* if every non-zero projection dominates a minimal projection. A von Neumann algebra with no minimal projections is *diffuse*.

10.7.10. REMARK. Note that the type of a von Neumann algebra depends only on the internal structure, not on its spatial representation. Thus the type is an isomorphism invariant.

10.7.11. THEOREM. *Every von Neumann algebra has unique central projections Z_{I}, Z_{II_1}, Z_{II_∞} and Z_{III} summing to I such that $\mathcal{M}Z_{\mathrm{I}}$, $\mathcal{M}Z_{\mathrm{II}_1}$, $\mathcal{M}Z_{\mathrm{II}_\infty}$ and $\mathcal{M}Z_{\mathrm{III}}$ are type* I, II_1, II_∞ *and* III *respectively.*

PROOF. By Proposition 10.7.7(4), there is an abelian projection P so that the central projection $Z_{\mathrm{I}} := c(P)$ dominates all abelian projections. To see that $\mathcal{M}Z_{\mathrm{I}}$ is type I, let $0 \neq Q \leq Z_{\mathrm{I}}$. Then $c(Q)c(P) = c(Q) \neq 0$. By Lemma 10.7.2, there are projections $0 \neq Q_1 \leq Q$ and $P_1 \leq P$ with $Q_1 \sim P_1$. Then P_1 is an abelian projection, and thus so is Q_1. So $\mathcal{M}Z_{\mathrm{I}}$ is type I. Moreover there are no abelian projections in $\mathcal{M}Z_{\mathrm{I}}^\perp$.

Let Z_f be the largest finite central projection, and define $Z_{\mathrm{II}_1} = Z_{\mathrm{I}}^\perp Z_f$. Then Z_{II_1} is the largest finite central projection which does not dominate an abelian projection. Thus it is type II_1. Moreover, since $(Z_{\mathrm{I}} + Z_{\mathrm{II}_1})^\perp$ is properly infinite, there is no finite central projection in $\mathcal{M}(Z_{\mathrm{I}} + Z_{\mathrm{II}_1})^\perp$.

Decompose $(Z_{\mathrm{I}} + Z_{\mathrm{II}_1})^\perp$ into its semifinite part Z_{II_∞} and its purely infinite part Z_{III}. Then $\mathcal{M}Z_{\mathrm{II}_\infty}$ is semifinite, has no abelian projections and no finite central projections; so it is type II_∞. Finally $\mathcal{M}Z_{\mathrm{III}}$ is purely infinite, so it is type III. The uniqueness of these projections yields uniqueness of the decomposition. ∎

10.7.12. COROLLARY. *A von Neumann algebra is type* I *if and only if there is an abelian projection P with $c(P) = I$.*

10.7.13. COROLLARY. *A von Neumann algebra factor has type* I, II_1, II_∞ *or* III.

10.7.14. EXAMPLE. It will take us some work to identify examples of type II and III. Observe that every abelian von Neumann algebra is type I, since the identity is an abelian projection. Also $\mathcal{B}(H)$ is type I for any Hilbert space H. This is because every non-zero projection $P \in \mathcal{B}(H)$ dominates a rank one projection. This is a minimal projection, and thus is abelian.

10.8. Tensor products

We now give a very brief introduction to tensor products. Tensor products of C*-algebras are a big subject because there are often different possible C*-algebra norms that can be put on the algebraic tensor product. For von Neumann algebras, only one tensor product is used. In this book, we will only make use of the von Neumann algebras $\mathcal{M} \mathbin{\overline{\otimes}} \mathbb{C}I_K$ and $\mathcal{M} \mathbin{\overline{\otimes}} \mathcal{B}(K)$ which are needed to discuss multiplicity in type I von Neumann algebras.

Recall that for any two linear spaces X and Y, the *algebraic tensor product*, $X \odot Y$, is the algebraic span of *elementary tensors* $x \otimes y$ for $x \in X$ and $y \in Y$ subject to the relations: for $\lambda_i \in \mathbb{C}$, $x, x_i \in X$ and $y, y_i \in Y$,

- $\lambda_1(x_1 \otimes y) + \lambda_2(x_2 \otimes y) = (\lambda_1 x_1 + \lambda_2 x_2) \otimes y$.
- $\lambda_1(x \otimes y_1) + \lambda_2(x \otimes y_2) = x \otimes (\lambda_1 y_1 + \lambda_2 y_2)$.

We reserve the notation $X \otimes Y$ for appropriate completions of $X \odot Y$, depending on the structure on X and Y. If A and B are algebras, then $A \odot B$ becomes an algebra with multiplication defined by $(a_1 \otimes b_1)(a_2 \otimes b_2) = a_1 a_2 \otimes b_1 b_2$ on elementary tensors, and extending to the linear span using the distributive law.

We will need the tensor product of Hilbert spaces, which is straightforward. Proposition 10.8.2 provides several different ways of representing it.

10.8.1. DEFINITION. If H and K are Hilbert spaces, put an inner product on $H \odot K$ by

$$\langle \xi_1 \otimes \eta_1, \xi_2 \otimes \eta_2 \rangle = \langle \xi_1, \xi_2 \rangle_H \langle \eta_1, \eta_2 \rangle_K \quad \text{for} \quad \xi_i \in H,\ \eta_i \in K$$

and extending by linearity in the first variable and conjugate linearity in the second. The *Hilbert space tensor product* $H \otimes K$ is the completion of $H \odot K$ with respect to the norm induced by this inner product.

10.8.2. PROPOSITION. *Let H and K be Hilbert spaces with orthonormal bases $\{\xi_i : i \in I\}$ and $\{\eta_j : j \in J\}$ respectively.*

(1) *$\{\xi_i \otimes \eta_j : i \in I,\ j \in J\}$ is an orthonormal basis for $H \otimes K$.*

(2) *If* $|J| = \alpha$, *then* $H \otimes K \simeq H^{(\alpha)} = \sum_J^{\oplus} H_j$ *where* $H_j \simeq H$ *for* $j \in J$.

(3) *If* $H = L^2(\mu)$ *for a* σ*-finite regular Borel measure* (X, μ) *and* K *is separable, let* $L^2(X, \mu, K)$ *be the space of Borel functions* $f : X \to K$ *such that*

$$\|f\|_2^2 = \int \|f(x)\|^2 \, d\mu(x) < \infty$$

modulo the subspace of null vectors. Then $H \otimes K \simeq L^2(X, \mu, K)$.

PROOF. It is readily checked that $\{\xi_i \otimes \eta_j : i \in I, \ j \in J\}$ is an orthonormal set. The span of these elementary tensors contains all vectors of the form $\xi \otimes \eta$ provided ξ is a finite linear combination of the basis $\{\xi_i : i \in I\}$ and likewise η is a finite linear combination of the basis $\{\eta_j : j \in J\}$. Thus the completion equals the completion of $H \odot K$. Therefore $\{\xi_i \otimes \eta_j : i \in I, \ j \in J\}$ is an orthonormal basis for $H \otimes K$.

We can think of $H \otimes K$ as $\sum_J^{\oplus} H \otimes \mathbb{C}\eta_j = \sum_J^{\oplus} H_j$ where $H_j = H \otimes \mathbb{C}\eta_j \simeq H$ via the unitary taking $\xi \otimes \eta_j$ to ξ. This is an l_2-direct sum of $|J| = \alpha$ copies of H, which is denoted by $H^{(\alpha)}$.

Define maps $Y_j : L^2(X, \mu, K) \to L^2(\mu)$ by $Y_j f(x) = \langle f(x), \eta_j \rangle$ a.e.(μ). Then define $Y : L^2(X, \mu, K) \to \sum_J^{\oplus} H_j$ by $Yf = \sum_J^{\oplus} Y_j f$. Compute using Tonelli's theorem

$$\begin{aligned} \|Yf\|_2^2 &= \sum_{j \in J} \|Y_j f\|_2^2 = \sum_{j \in J} \int |\langle f(x), \xi_j \rangle|^2 \, d\mu \\ &= \int \sum_{j \in J} |\langle f(x), \xi_j \rangle|^2 \, d\mu = \int \|f(x)\|^2 \, d\mu = \|f\|_2^2. \end{aligned}$$

Thus Y is an isometry. The range is easily seen to be dense, since if $h \in L^2(\mu)$, then $f = h\eta_j$ is mapped to the vector with h in the jth coordinate and zeros elsewhere. Therefore Y is a unitary map of $L^2(X, \mu, K)$ onto $H \otimes K$. ∎

10.8.3. PROPOSITION. *If* $A \in \mathcal{B}(H)$ *and* $B \in B(K)$, *then there is a unique operator* $A \otimes B \in \mathcal{B}(H \otimes K)$ *such that* $(A \otimes B)(\xi \otimes \eta) = A\xi \otimes B\eta$ *for* $\xi \in H$ *and* $\eta \in K$. *Moreover,* $\|A \otimes B\| = \|A\| \, \|B\|$.

PROOF. Consider the ampliation $A \otimes I$. Fix an orthonormal basis $\{\eta_j : j \in J\}$ for K. Then a vector ω in $H \otimes K$ can be written as $\omega = \sum_J^{\oplus} \zeta_j \otimes \eta_j$ for vectors $\zeta_j \in H$ where $\|\omega\|^2 = \sum_J \|\zeta_j\|^2 < \infty$. Then $(A \otimes I)\omega = \sum_J^{\oplus} A\zeta_j \otimes \eta_j$. Thus

$$\|(A \otimes I)\omega\|^2 = \Big\| \sum_{j \in J}^{\oplus} A\zeta_j \otimes \eta_j \Big\|^2 = \sum_{j \in J} \|A\zeta_j\|^2 \le \|A\|^2 \sum_{j \in J} \|\zeta_j\|^2 = (\|A\| \|\omega\|)^2.$$

It follows that $\|A \otimes I\| = \|A\|$. Now if $\xi \in H$ and $\eta \in K$, write $\eta = \sum_j b_j \eta_j$ with $\sum_j |b_j|^2 = \|\eta\|^2$. Then $\xi \otimes \eta = \sum_j b_j \xi \otimes \eta_j$; whence

$$(A \otimes I)\xi \otimes \eta = \sum_j b_j A\xi \otimes \eta_j = A\xi \otimes \sum_j b_j \eta_j = A\xi \otimes \eta.$$

Thus $A \otimes I$ is bounded and uniquely defined on the linear span of vectors $\xi \otimes \eta$, which is dense in $H \otimes K$. In particular, this operator is independent of the choice of basis for K.

Similarly we define $I \otimes B$. Then set $A \otimes B = (A \otimes I)(I \otimes B)$. Therefore

$$(A \otimes B)(\xi \otimes \eta) = (A \otimes I)(I \otimes B)(\xi \otimes \eta) = (A \otimes I)(\xi \otimes B\eta) = A\xi \otimes B\eta.$$

Clearly $\|A \otimes B\| \leq \|A \otimes I\| \, \|I \otimes B\| = \|A\| \, \|B\|$. This is an equality since $\|(A \otimes B)(\xi \otimes \eta)\| = \|A\xi\| \, \|B\eta\|$; and choosing unit vectors ξ and η appropriately approximates $\|A\| \, \|B\|$. Finally we have a bounded operator which is uniquely defined on a dense set, and so $A \otimes B$ is well defined. ∎

10.8.4. DEFINITION. If $\mathfrak{A}$ and $\mathfrak{B}$ are concrete C*-algebras on Hilbert spaces H and K, respectively, the *spatial tensor product* $\mathfrak{A} \otimes \mathfrak{B}$ is the norm closure of $\mathfrak{A} \odot \mathfrak{B}$ in $\mathcal{B}(H \otimes K)$. When $\mathcal{M}$ and $\mathcal{N}$ are von Neumann algebras, the *von Neumann algebra tensor product* $\mathcal{M} \overline{\otimes} \mathcal{N}$ is the WOT-closure of $\mathcal{M} \odot \mathcal{N}$ in $\mathcal{B}(H \otimes K)$. When $\mathcal{N} = \mathbb{C}I$, the algebra $\mathcal{M} \overline{\otimes} \mathbb{C}I = \mathcal{M}^{(\alpha)}$ for $\alpha = |\dim K|$ is an *ampliation* of $\mathcal{M}$.

Note that the map $\pi : \mathcal{M} \to \mathcal{M} \overline{\otimes} \mathbb{C}I$ by $\pi(A) = A \otimes I$ is a normal isomorphism.

10.8.5. PROPOSITION. *Let $\mathcal{M}$ be a von Neumann algebra on H, and let K be a Hilbert space with orthonormal basis $\{\eta_j : j \in J\}$. Then $\mathcal{M} \overline{\otimes} \mathcal{B}(K)$ can be identified with all infinite matrices $\big[M_{ij}\big]_{J \times J}$ with $M_{ij} \in \mathcal{M}$ that represent bounded operators on $H^{(\alpha)}$, where $\alpha = |J|$, with norm*

$$\big\| [M_{ij}] \big\| = \sup_{F \Subset J} \big\| [M_{ij}]_{F \times F} \big\|$$

where the supremum is taken over all finite subsets F of J.

Suppose that $H = L^2(\mu)$ for a σ-finite regular Borel measure (X, μ), $\mathcal{M}$ is $L^\infty(\mu)$ acting on $L^2(\mu)$, and K is separable. Then $\mathcal{M} \overline{\otimes} \mathcal{B}(K)$ can be identified with $L^\infty(X, \mu, \mathcal{B}(K))$, the space of bounded Borel functions into $\mathcal{B}(K)$ modulo the set of null functions, i.e. $\{f : f = 0 \text{ a.e.}(\mu)\}$.

PROOF. Let $Q_j = \eta_j \eta_j^*$ be the rank 1 projection of K onto $\mathbb{C}\eta_j$. An operator A on $H \otimes K \simeq \sum_J^{\oplus} H_j$ is determined by the infinite matrix $\big[A_{ij}\big]_{J \times J}$ where $A_{ij} = (I \otimes Q_i)A(I \otimes Q_j)$ considered as an element of $\mathcal{B}(H_j, H_i)$ for $i, j \in J$. If

$A \in \mathcal{M} \odot \mathcal{B}(K)$, then there is a finite sum $A = \sum M_k \otimes B_k$. Thus

$$A_{ij} = (I \otimes Q_i) \sum M_k \otimes B_k (I \otimes Q_j) = \sum M_k \otimes Q_i B_k Q_j.$$

Since $Q_i B_k Q_j$ is a multiple of $\eta_i \eta_j^*$, this sum has the form $M_{ij} \otimes (\eta_i \eta_j^*)$ for $M_{ij} \in \mathcal{M}$. So A is represented by the matrix $\big[M_{ij}\big]_{J \times J}$. If $(A_\lambda)_\Lambda$ is a WOT-convergent net in $\mathcal{M} \odot \mathcal{B}(K)$ with limit A, then

$$A_{ij} = (I \otimes Q_i) A (I \otimes Q_j) = \text{WOT-}\lim_{\Lambda} (I \otimes Q_i) A_\lambda (I \otimes Q_j) \in \mathcal{M} \otimes \mathbb{C}(\eta_i \eta_j^*).$$

Therefore every element of $\mathcal{M} \overline{\otimes} \mathcal{B}(K)$ has such a matrix form.

Conversely, suppose that $\big[M_{ij}\big]$ is an infinite matrix with coefficients in $\mathcal{M}$ such that the operators $A_F = \sum_{i,j \in F} A_{ij} \otimes (\eta_i \eta_j^*)$ for finite subsets $F \Subset J$ are uniformly bounded, i.e. $\sup_{F \Subset J} \|A_F\| < \infty$. Evidently $A_F \in \mathcal{M} \odot \mathcal{B}(K)$. By compactness, there is a cofinal subnet (of the net of finite subsets of J) which converges weak-$*$ to an operator $A \in \mathcal{M} \overline{\otimes} \mathcal{B}(K)$. It follows that $Q_i A Q_j = M_{ij}$. Set $Q_F = \sum_{j \in F} Q_j$. Then we have $Q_F A Q_F = A_F$. Thus, in fact, the net A_F converges weak-$*$ to A. Therefore every infinite matrix over $\mathcal{M}$ represents an element of $\mathcal{M} \overline{\otimes} \mathcal{B}(K)$.

When $H = L^2(\mu)$, we identify $H \otimes K$ with $L^2(X, \mu, K)$. Using the notation from the previous paragraph, each A_F is an $|F| \times |F|$ matrix with coefficients from $\mathcal{M} = L^\infty(\mu)$. We can think of this as a Borel function from X into $M_{|F|}$, which we then identify with $Q_F \mathcal{B}(K) Q_F$. So each A_F is identified with an element of $L^\infty(X, \mu, Q_F \mathcal{B}(K) Q_F) \subset L^\infty(X, \mu, \mathcal{B}(K))$. The right-hand side is WOT-closed, so taking limits identifies A with an element of $L^\infty(X, \mu, \mathcal{B}(K))$. Conversely, if $f \in L^\infty(X, \mu, \mathcal{B}(K))$, then $f_F(x) = Q_F f(x) Q_F$ belongs to $L^\infty(X, \mu, Q_F \mathcal{B}(K) Q_F)$, which we identify with an operator A_F by reversing the correspondence above. Taking WOT limits identifies f with an operator $A \in \mathcal{M} \overline{\otimes} \mathcal{B}(K)$. ■

There is a general theorem due to Tomita identifying $(\mathcal{M} \otimes \mathcal{N})'$ with $\mathcal{M}' \otimes \mathcal{N}'$. See [300, Theorem IV.5.9]. We will only establish a very special case.

10.8.6. PROPOSITION. *Let $\mathcal{M}$ be a von Neumann algebra on H, and let K be a Hilbert space. Then*

$$(\mathcal{M} \overline{\otimes} \mathbb{C} I_K)' = \mathcal{M}' \overline{\otimes} \mathcal{B}(K) \quad \textit{and} \quad (\mathcal{M} \overline{\otimes} \mathcal{B}(K))' = \mathcal{M}' \overline{\otimes} \mathbb{C} I_K.$$

PROOF. It is clear that if $M \in \mathcal{M}$, $A \in \mathcal{M}'$ and $B \in \mathcal{B}(K)$, then $M \otimes I$ commutes with $A \otimes B$. Hence every element of $(\mathcal{M} \overline{\otimes} \mathbb{C} I_K)$ commutes with $A \otimes B$. Thus $(\mathcal{M} \overline{\otimes} \mathbb{C} I_K)'$ contains $\mathcal{M}' \overline{\otimes} \mathcal{B}(K)$. On the other hand, we identify an element A of $\mathcal{B}(H \otimes K)$ with an infinite matrix $\big[A_{ij}\big]_{J \times J}$. The operator $M \otimes I$ corresponds to the matrix with $M_{jj} = M$ and $M_{ij} = 0$ if $i \neq j$. Thus $[M \otimes I, A]$ has the matrix $\big[M A_{ij} - A_{ij} M\big]_{J \times J}$. This is 0 for all $M \in \mathcal{M}$ if and only if each $A_{ij} \in \mathcal{M}'$. Therefore $(\mathcal{M} \otimes \mathbb{C} I_K)' \simeq \mathcal{M}' \otimes \mathcal{B}(K)$. By the Double Commutant theorem, $(\mathcal{M} \overline{\otimes} \mathcal{B}(K))' \simeq \mathcal{M}' \overline{\otimes} \mathbb{C} I_K$. ■

10.8.7. DEFINITION. A set of *matrix units* is a system of partial isometries $\{E_{ij} : i, j \in \Lambda\}$ such that for $i, j, k, l \in \Lambda$,

- $E_{ji} = E_{ij}^*$,
- $E_{ij}E_{kl} = \delta_{jk}E_{il}$, and
- $\sum E_{ii} = I$

Note that E_{ii} form a set of pairwise orthogonal projections summing SOT to I. Let K be a Hilbert space with $\dim K = |\Lambda|$. If $\{\eta_j : j \in \Lambda\}$ is an orthonormal basis for K, the rank 1 operators $\{\eta_i\eta_j^* : i, j \in \Lambda\}$ form a set of matrix units with these same relations. If $|\Lambda| = n < \infty$, these matrix units span $\mathcal{B}(K) \simeq M_n$; so $\operatorname{span}\{E_{ij}\} \simeq M_n$. When Λ is infinite, the norm-closed span $\overline{\operatorname{span}}\{\eta_i\eta_j^* : i, j \in \Lambda\}$ is the space of compact operators on K; and the WOT closure is $\mathcal{B}(K)$. Hence the WOT closure of $\operatorname{span}\{E_{ij} : i, j \in \Lambda\}$ is isomorphic to $\mathcal{B}(K)$.

The existence of a set of matrix units in a von Neumann algebra leads to a decomposition as a tensor product.

10.8.8. PROPOSITION. *Suppose that $\mathcal{M}$ is a von Neumann algebra on H containing a set of matrix units $\{E_{ij} : i, j \in \Lambda\}$. Let K be a Hilbert space with $\dim K = |\Lambda|$; and let $P = E_{j_0j_0}$ for some $j_0 \in \Lambda$. Then $\mathcal{M}$ is unitarily equivalent to $\mathcal{M}_P \overline{\otimes} \mathcal{B}(K)$.*

PROOF. Let $\{\eta_j : j \in \Lambda\}$ be an orthonormal basis for K. Define a linear map $U : PH \otimes K \to H$ by

$$U \sum_{k\in\Lambda} \xi_k \otimes \eta_k = \sum_{k\in\Lambda} E_{kj_0}\xi_k.$$

This is easily seen to be an isometry onto $\overline{\sum_{k\in\Lambda} E_{kk}H} = H$, so U is unitary. An element of $\mathcal{M}_P \overline{\otimes} \mathcal{B}(K)$ is given by a matrix $A = \left[M_{ij}\right]_{\Lambda\times\Lambda}$ for M_{ij} in $\mathcal{M}_P$ which determines a bounded operator in the sense that the net of operators

$$A_F = \sum_{i,j\in F} M_{ij} \otimes (\eta_i\eta_j^*) \quad \text{for} \quad F \Subset \Lambda$$

is bounded and converges SOT in $\mathcal{B}(PH \otimes K)$. Note that

$$\begin{aligned} UA_F \sum_{k\in\Lambda} \xi_k \otimes \eta_k &= \sum_{i,j\in F} UM_{ij} \otimes (\eta_i\eta_j^*) \sum_{k\in\Lambda} \xi_k \otimes \eta_k \\ &= \sum_{i,j\in F} UM_{ij}\xi_j \otimes \eta_i = \sum_{i,j\in F} E_{ij_0} M_{ij}\xi_j \\ &= \sum_{i,j\in F} (E_{ij_0} M_{ij} E_{j_0 j}) E_{jj_0}\xi_j = \sum_{i,j\in F} (E_{ij_0} M_{ij} E_{j_0 j}) \sum_{k\in\Lambda} E_{kj_0}\xi_k \\ &= \sum_{i,j\in F} (E_{ij_0} M_{ij} E_{j_0 j}) U \sum_{k\in\Lambda} \xi_k \otimes \eta_k . \end{aligned}$$

Therefore we obtain a net

$$UA_F U^* = \sum_{i,j\in F} E_{ij_0} M_{ij} E_{j_0 j} \in \mathcal{M},$$

bounded by $\|A\|$, and it converges SOT to $\sum_{i,j\in\Lambda} E_{ij_0} M_{ij} E_{j_0 j}$ in $\mathcal{M}$.

On the other hand, if $B \in \mathcal{M}$, we let $M_{ij} = E_{j_0 i} B E_{jj_0} \in \mathcal{M}_P$. Then setting $Q_F = \sum_{i\in F} E_{ii}$, we see that

$$Q_F B Q_F = \sum_{i,j\in F} E_{ij_0} M_{ij} E_{j_0 j};$$

and this converges SOT to B for the net of finite subsets $F \Subset \Lambda$. It follows that $A_F = U^*(Q_F B Q_F)U$ in $\mathcal{M}_P \overline{\otimes} \mathcal{B}(K)$ converge SOT to $A = U^*BU$. Hence $UAU^* = B$. Therefore this determines a unitary equivalence of von Neumann algebras. ∎

10.8.9. COROLLARY. *If $\mathcal{M}$ is a von Neumann algebra and $\{E_j\}_\Lambda$ is a family of equivalent projections such that $\sum_\Lambda E_j = I$, then $\mathcal{M}$ is unitarily equivalent to $\mathcal{M}_{E_{j_0}} \overline{\otimes} \mathcal{B}(K)$ where* $\dim K = |\Lambda|$.

PROOF. Since $\sum_\Lambda E_j = I$, it follows that $\{E_j\}_\Lambda$ is a family of pairwise orthogonal projections. Fix some $j_0 \in \Lambda$ and select partial isometries E_{ij_0} for $i \in \Lambda$ such that $E_{ij_0} E_{ij_0}^* = E_i$ and $E_{ij_0}^* E_{ij_0} = E_{j_0}$. Then $E_{ij} := E_{ij_0} E_{jj_0}^*$ for $i, j \in \Lambda$ form a set of matrix units in $\mathcal{M}$. Hence $\mathcal{M}$ is unitarily equivalent to $\mathcal{M}_{E_{j_0}} \overline{\otimes} \mathcal{B}(K)$ where $\dim K = |\Lambda|$. ∎

10.8.10. COROLLARY. *If $\pi : \mathcal{M} \overline{\otimes} \mathcal{B}(K) \to \mathcal{B}(H)$ is a normal representation, then there is a normal representation $\rho : \mathcal{M} \to \mathcal{B}(H_0)$ so that π is unitarily equivalent to $\rho \otimes$ id.*

PROOF. $\pi(\mathbb{C}I \overline{\otimes} \mathcal{B}(K))$ is an isomorphic copy of $\mathcal{B}(K)$ because $\mathcal{B}(K)$ has no non-trivial WOT-closed ideals. In particular, it contains a set of matrix units

$\{E_{ij} : i, j \in \Lambda\}$ spanning the image of $\mathcal{B}(K)$. Let $H_0 = E_{j_0 j_0} H$ and let $\rho = \pi_{E_{j_0 j_0}}$. The identifications in Proposition 10.8.8 yield the unitary equivalence. ∎

10.9. Type I von Neumann Algebras

In this section, we completely classify the type I von Neumann algebras. In Example 10.7.14, we showed that every abelian von Neumann algebra is type I.

10.9.1. Lemma. *A von Neumann algebra with abelian commutant is type* I.

Proof. Suppose that $\mathcal{M}'$ is abelian, and let P be a projection in $\mathcal{M}$. Pick a unit vector $\xi = P\xi$, and define $M = \overline{\mathcal{M}'\xi}$. The projection $E = P_M \in \mathcal{M}$ and $EH = \overline{\mathcal{M}'\xi} \subset \overline{\mathcal{M}'PH} = PH$; so $E \leq P$. By Lemma 10.3.6, π_E is a normal homomorphism of $\mathcal{M}'$ onto $\mathcal{M}'_E$. Thus $\mathcal{M}'_E$ is abelian, and by construction, ξ is a cyclic vector. By Theorem 10.4.8, $\mathcal{M}'_E$ is a masa. By Lemma 10.3.6 again, $\mathcal{M}_E = (\mathcal{M}'_E)' = \mathcal{M}'_E$ is abelian; i.e. $E \leq P$ is an abelian projection for $\mathcal{M}$. Therefore $\mathcal{M}$ is type I. ∎

10.9.2. Theorem. *Let $\mathcal{M}$ be a von Neumann algebra. The following are equivalent:*

(1) *$\mathcal{M}$ is type* I.

(2) *$\mathcal{M}'$ is type* I.

(3) *$\mathcal{M}$ is isomorphic to a von Neumann algebra with abelian commutant.*

(4) *$\mathcal{M}'$ is isomorphic to a von Neumann algebra with abelian commutant.*

Proof. Assume that $\mathcal{M}$ is type I. Let E be an abelian projection in $\mathcal{M}$ with $c(E) = I$. Since $E \in \mathcal{M}''$ and $c(E) = I$, the compression map $\pi_E : \mathcal{M}' \to \mathcal{M}'_E$ by $\pi(A) = EAE|_{EH}$ is a normal isomorphism, and $(\mathcal{M}'_E)' = E\mathcal{M}E|_{EH}$ by Lemma 10.3.6. Hence $\mathcal{M}'_E$ has abelian commutant. So (1) implies (4). Thus (2) implies (3) as well. If (4) holds, Lemma 10.9.1 shows $\mathcal{M}'$ is type I. So (4) implies (2). Similarly (3) implies (1). So (1) $\Rightarrow$ (4) $\Rightarrow$ (2) $\Rightarrow$ (3) $\Rightarrow$ (1). ∎

10.9.3. Definition. In a type I von Neumann algebra $\mathcal{M}$, a central projection Z is *γ-homogeneous* for a cardinal number γ if there is a family $\{E_i : i \in \gamma\}$ of equivalent abelian projections such that $\sum_{i\in\gamma} E_i = Z$.

10.9.4. Remark. Suppose that $\mathcal{M}$ is γ-homogeneous. When projections sum to a projection, they must be pairwise orthogonal. Since these are equivalent projections, Corollary 10.8.9 shows that $\mathcal{M}_Z \simeq \mathcal{M}_{E_{i_0}} \overline{\otimes} \mathcal{B}(K)$ where $\dim K = |\gamma|$. As E_{i_0} is an abelian projection, $\mathcal{M}_{E_{i_0}}$ is an abelian von Neumann algebra.

Conversely, if $\mathcal{M} \simeq \mathcal{N} \overline{\otimes} \mathcal{B}(K)$ for an abelian von Neumann algebra $\mathcal{N}$, then $\mathcal{M}' \simeq \mathcal{N}' \overline{\otimes} \mathbb{C}I_K$. Therefore $\mathcal{Z}(M) \simeq \mathcal{N} \overline{\otimes} \mathbb{C}I_K$. Let P_0 be a rank 1 projection in $\mathcal{B}(K)$. Then $P = 1_{\mathcal{N}} \otimes P_0$ is an abelian projection in $\mathcal{M}$ with $c(P) = I$. By Corollary 10.7.12, $\mathcal{M}$ is type I.

10.9.5. THEOREM. *Let $\mathcal{M}$ be a type* I *von Neumann algebra. Then there is a unique family of central projections $\{Z_\gamma\}_\Gamma$ in $\mathcal{M}$ indexed by distinct cardinal numbers such that $\sum_\Gamma Z_\gamma = I$ and there are families $\{E_{\gamma,i} : i \in \gamma\}$ of equivalent abelian projections such that $\sum_{i\in\gamma} E_{\gamma,i} = Z_\gamma$.*

PROOF. We claim that there is a largest γ-homogeneous central projection Z_γ. Use Zorn's lemma to find a maximal family $\{P_\lambda\}_\Lambda$ of pairwise orthogonal γ-homogeneous central projections. By Exercise **10.23**, $Z_\gamma = \sum_\Lambda P_\lambda$ is γ-homogeneous. If there were a central γ-homogeneous projection $P \not\leq Z_\gamma$, then $PZ_\gamma^\perp$ would be γ-homogeneous by Exercise **10.23** again, contradicting the maximality of our maximal family. So Z_γ is the desired projection.

Next we show that $Z_\gamma Z_\delta = 0$ if $\gamma \neq \delta$. If not, we restrict to $P = Z_\gamma Z_\delta$; so we may suppose that $P = I$. Thus there are two families $\{E_i : i \in \gamma\}$ and $\{F_j : j \in \delta\}$ of pairwise orthogonal abelian projections with $c(E_i) = c(F_j) = I$ and $\sum_\gamma E_i = \sum_\delta F_j = I$. By Proposition 10.7.7 (5), $E_i \sim E_{i'} \sim F_j \sim F_{j'}$ for $i, i' \in \gamma$ and $j, j' \in \delta$. Thus $\mathcal{M}_{E_1} \simeq \mathcal{M}_{F_1}$ is isomorphic to $\mathcal{Z}$, where $\mathcal{Z}$ is the centre of $\mathcal{M}$, by Corollary 10.7.8. By Corollary 10.8.9, there are unitary equivalences

$$\mathcal{M} \simeq \mathcal{M}_{E_1} \overline{\otimes} \mathcal{B}(K_1) \simeq \mathcal{M}_{F_1} \overline{\otimes} \mathcal{B}(K_2)$$

where $\dim K_1 = \gamma$ and $\dim K_2 = \delta$.

First suppose that δ is finite. If φ is any multiplicative linear functional on $\mathcal{M}_{F_1}$, there is a $*$-homomorphism of $\mathcal{M}$ onto $\mathcal{B}(K_2) \simeq M_\delta$ obtained by identifying $\mathcal{M}$ with $\mathcal{M}_{F_1} \overline{\otimes} \mathcal{B}(K_2) \simeq M_\delta(\mathcal{M}_{F_1})$ and applying φ to each matrix entry. The restriction to $\mathcal{B}(K_1) \simeq \mathbb{C}I \overline{\otimes} \mathcal{B}(K_1) \subset \mathcal{M}$ is a unital $*$-homomorphism of $\mathcal{B}(K_1)$ into M_δ. This shows that $\gamma = \dim K_1$ is also finite, since $\mathcal{B}(H)$ has no finite dimensional quotients if H is infinite dimensional. Moreover, since M_γ is simple, it forces $\gamma \leq \delta$. Reversing roles shows that $\gamma = \delta$.

Now we can suppose that both γ and δ are infinite. Let φ be a normal state on $\mathcal{Z}$ and let $Z_0 = s(\varphi)$. By Proposition 10.2.9, φ is faithful on $Z_0 \mathcal{Z} Z_0$.

For $i \in \gamma$, by Corollary 10.7.8, there is an isomorphism $\pi_i : \mathcal{Z} \simeq \mathcal{M}_{E_i}$ via $Z \to ZE_i$. So we may define normal states on $\mathcal{M}$ by $\varphi_i(X) = \varphi\pi_i^{-1}(E_i X E_i)$. Then $\varphi_i(Z_0 E_i) = 1 > \varphi_i(Z' E_i)$ for any central projection $Z' < Z_0$ in $\mathcal{Z}$. Let

$$A_i = \{j \in \delta : \varphi_i(F_j) \neq 0\} \quad \text{for} \quad i \in \gamma.$$

Since $\{F_j\}$ are pairwise orthogonal and φ_i is completely additive, each A_i must be a countable set. On the other hand, if $\varphi_i(F_j) = 0$, then $E_i F_j E_i = 0$, as otherwise it will dominate εP for a projection $P = ZE_i \le E_i$ for some central projection Z. But then $\varphi_i(F_j) \ge \varepsilon\varphi(Z) > 0$. If $E_iF_jE_i = 0$ for all $i \in \gamma$, then F_j is a projection in $\mathcal{M} \simeq \mathcal{M}_{E_1} \overline{\otimes} \mathcal{B}(K_1)$ with zero diagonal, and thus must be 0. This isn't the case, and hence there is some i_0 so that $j \in A_{i_0}$. Therefore

$$\bigcup_{i\in\gamma} A_i = \delta \quad \text{and hence} \quad \delta \le \aleph_0\gamma = \gamma.$$

Reversing roles shows that $\gamma = \delta$.

We claim that $Z_\Gamma := \sum_\Gamma Z_\gamma = I$. If not, then we restrict to $Z_\Gamma^\perp H$ and reach a situation in which there are no homogeneous central projections for any cardinal. Since $\mathcal{M}$ is type I, there is an abelian projection E_1 with $c(E_1) = I$. Let $\{E_j\}_J$ be a maximal family of pairwise orthogonal abelian projections with $c(E_j) = I$. Let $Q = \sum_J E_j$ and let $Z = c(Q^\perp)$. If $Z = I$, then since $Q^\perp \mathcal{M} Q^\perp = (\mathcal{M}'_{Q^\perp})'$ is type I, it has an abelian projection E with $c(E) = I$. But then E is an abelian projection for $\mathcal{M}$ orthogonal to Q, and hence $\{E, E_j\}_J$ is a larger family, contradicting maximality. Therefore $Z \ne I$. It follows that $\{E_jZ^\perp\}_J$ a maximal family of pairwise orthogonal abelian projections with $c(E_jZ^\perp) = Z^\perp$ such that $\sum_J E_jZ^\perp = Z^\perp$. By Proposition 10.7.7 (5), $E_iZ^\perp \sim E_jZ^\perp$ for all $i, j \in J$. Therefore $Z^\perp$ is $|J|$-homogeneous. This contradiction shows that $Z_\Gamma = I$.

Uniqueness follows since the maximal homogeneous central projections are pairwise orthogonal and sum to the identity. ∎

10.9.6. DEFINITION. If $\mathcal{M}_\lambda$ are von Neumann algebras on H_λ for $\lambda \in \Lambda$, the *infinite product* $\mathcal{M} = \prod_\Lambda \mathcal{M}_\lambda$ denotes the von Neumann algebra on $H = \sum_\Lambda^\oplus H_\lambda$ consisting of operators $A = (A_\lambda)_\Lambda$ with $A_\lambda \in \mathcal{M}_\lambda$ and $\|A\| = \sup_\Lambda \|A_\lambda\| < \infty$.

The issue here is that an infinite direct sum is by definition the norm closure of the finitely non-zero sequences, while the infinite product uses all bounded sequences. For a finite set, direct sum and product are equivalent.

10.9.7. DEFINITION. A type I von Neumann algebra $\mathcal{M}$ which is unitarily equivalent to $\mathcal{N} \overline{\otimes} \mathcal{B}(H) \overline{\otimes} \mathbb{C}I_K$, where $\mathcal{N}$ is abelian, $\dim H = \alpha$ and $\dim K = \beta$, is *homogeneous of degree* α with *multiplicity* β. Such an algebra is called *type* I_α.

Thus a factor of type I_n is isomorphic to M_n if $n < \infty$ and to $\mathcal{B}(H_\alpha)$ with $\dim H_\alpha = \alpha$ if α is infinite. Multiplicity is used for abelian von Neumann algebras in the context of the spectral theorem. See Exercise **10.30**.

Finally we can characterize the spatial structure of type I von Neumann algebras.

10.9.8. THEOREM. *Let $\mathcal{M}$ be a type* I *von Neumann algebra. Then there are families of pairwise orthogonal central projections $\{Z_\gamma\}_\Gamma$ and $\{Z'_\delta\}_\Delta$ in $\mathcal{Z}(\mathcal{M})$*

and maximal abelian von Neumann algebras $\mathcal{A}_{\gamma\delta}$ *for* $\gamma \in \Gamma$ *and* $\delta \in \Delta$ *so that*

$$Z_\gamma Z'_\delta \mathcal{M} \simeq \mathcal{A}_{\gamma\delta} \overline{\otimes} \mathcal{B}(H_\gamma) \overline{\otimes} \mathbb{C} I_{K_\delta},$$

where $\dim H_\gamma = \gamma$ *and* $\dim K_\delta = \delta$*. Thus we obtain, up to unitary equivalence, that*

$$\mathcal{M} \simeq \prod_{\Gamma \times \Delta} \mathcal{A}_{\gamma\delta} \overline{\otimes} \mathcal{B}(H_\gamma) \overline{\otimes} \mathbb{C} I_{K_\delta}$$

and

$$\mathcal{M}' \simeq \prod_{\Gamma \times \Delta} \mathcal{A}_{\gamma\delta} \overline{\otimes} \mathbb{C} I_{H_\gamma} \overline{\otimes} \mathcal{B}(K_\delta).$$

PROOF. By Theorem 10.9.5, there are central projections $\{Z_\gamma\}_\Gamma$ in $\mathcal{Z}$ indexed by distinct cardinal numbers such that $\sum_\Gamma Z_\gamma = I$ and families $\{E_{\gamma,i} : i \in \gamma\}$ of equivalent abelian projections such that $\sum_{i\in\gamma} E_{\gamma,i} = Z_\gamma$. Similarly, applying the theorem to $\mathcal{M}'$, there are central projections $\{Z'_\delta\}_\Delta$ in $\mathcal{Z}$ indexed by distinct cardinal numbers such that $\sum_\Delta Z'_\delta = I$ and $\mathcal{M}' Z'_\delta$ is homogeneous of degree δ.

We let $\mathcal{M}_{\gamma,\delta}$ be the restriction of $\mathcal{M}$ to the Hilbert space $H_{\gamma,\delta} = Z_\gamma Z'_\delta H$. Let $E_i = E_{\gamma,i} Z'_\delta$, and fix some $i_0 \in \gamma$. By Corollary 10.8.9, $\mathcal{M}_{\gamma,\delta} \simeq \mathcal{M}_{E_{i_0}} \overline{\otimes} \mathcal{B}(H_\gamma)$ and $\mathcal{M}'_{\gamma,\delta} \simeq \mathcal{M}'_{E_{i_0}} \overline{\otimes} \mathbb{C} I_{H_\gamma}$. Now there are pairwise orthogonal abelian projections $\{F_j : j \in \delta\}$ in $\mathcal{M}'$ with $\sum_{j\in\delta} F_j = Z_\gamma Z'_\delta$. Fix some $j_0 \in \delta$. Another application of Proposition 10.8.8 and Corollary 10.8.9 yields the more precise structure $\mathcal{M}'_{\gamma,\delta} \simeq \mathcal{M}'_{F_{j_0} E_{i_0}} \overline{\otimes} \mathcal{B}(H_\delta) \overline{\otimes} \mathbb{C} I_{H_\gamma}$. This finer decomposition shows that $\mathcal{M}_{\gamma,\delta} \simeq \mathcal{M}_{F_{j_0} E_{i_0}} \overline{\otimes} \mathbb{C} I_{H_\delta} \overline{\otimes} \mathcal{B}(H_\gamma)$. Now $\mathcal{M}'_{F_{j_0} E_{i_0}}$ and $\mathcal{M}_{F_{j_0} E_{i_0}}$ are both abelian and are mutual commutants. Therefore they are equal and are maximal abelian, so let $\mathcal{A}_{\gamma,\delta} = \mathcal{M}_{F_{j_0} E_{i_0}}$. Then

$$\mathcal{M}_{\gamma,\delta} \simeq \mathcal{A}_{\gamma,\delta} \overline{\otimes} \mathbb{C} I_{H_\delta} \overline{\otimes} \mathcal{B}(H_\gamma) \quad \text{and} \quad \mathcal{M}'_{\gamma,\delta} \simeq \mathcal{A}_{\gamma,\delta} \overline{\otimes} \mathcal{B}(H_\delta) \overline{\otimes} \mathbb{C} I_{H_\gamma}.$$

Finally putting everything together by taking a product over $\Gamma \times \Delta$ yields the final form. ■

The structure of maximal abelian von Neumann algebras on separable spaces, Theorem 10.4.8, yields the following special case.

10.9.9. COROLLARY. *Let* $\mathcal{M}$ *be an abelian von Neumann algebra on a separable Hilbert space. Then there is a finite regular Borel measure* (X, μ) *on a metric space* X *and a partition* $X = \bigsqcup_{n=1}^{\infty} X_n \sqcup X_\infty$ *into disjoint Borel sets and measures* $\mu_n = \mu|_{X_n}$ *such that* $\mathcal{M}$ *is unitarily equivalent to*

$$\prod_{n=1}^{\infty} L^\infty(\mu_n) \overline{\otimes} \mathbb{C} I_n \,\oplus\, L^\infty(\mu_\infty) \overline{\otimes} \mathbb{C} I_\infty$$

where I_α *is the identity of a Hilbert space* H_α *with* $\dim H_\alpha = \alpha$ *for* $1 \le \alpha \le \infty$.

The following is an immediate consequence of Theorem 10.9.8.

10.9.10. COROLLARY. *Every type* I *von Neumann algebra factor* $\mathcal{M}$ *is isomorphic* $\mathcal{B}(H)$ *for some Hilbert space* H. *More precisely, there are Hilbert spaces* H, K *so that* $\mathcal{M}$ *is unitarily equivalent to* $\mathcal{B}(H)\overline{\otimes}\mathbb{C}I_K$.

10.10. II_1 von Neumann Algebras

Some books like [94] and [234] take the definition of a II_1 von Neumann algebra as being a von Neumann algebra with a faithful family of traces and no type I summand. Other books define it as we have done and prove that there are traces. We will put off the proof until Section 10.12, but the examples we construct will have a trace. We explore the consequences of having a normal trace on a von Neumann algebra. Several examples are studied, including $\mathcal{L}(G)$ for a discrete group G.

10.10.1. REMARK. A von Neumann algebra is finite if and only if it does not contain a proper isometry. Indeed, if U is a proper isometry, then

$$I > UU^* \sim U^*U = I$$

shows that $\mathcal{M}$ is infinite. On the other hand, if $Q < P \sim Q$ and $UU^* = Q$ and $U^*U = P$, then $V = U + P^\perp$ is a proper isometry.

A von Neumann algebra with a faithful trace is finite because if $U^*U = I$, then $UU^* \le I$ and

$$\tau(I - UU^*) = \tau(I - U^*U) = 0.$$

Thus $I - UU^* = 0$ and U is unitary.

10.10.2. DEFINITION. A von Neumann algebra $\mathcal{M}$ is *hyperfinite* if there is a nested sequence $\mathfrak{A}_n$ of finite dimensional C*-subalgebras so that $\mathcal{M} = \overline{\bigcup_{n\ge1}\mathfrak{A}_n}^{\text{WOT}}$.

Our first example comes from UHF algebras, which we showed have a unique trace in Theorem 9.10.10.

10.10.3. THEOREM. *Let* $\mathfrak{A}$ *be a UHF algebra with trace* τ. *Let* $\mathcal{R} = \pi_\tau(\mathfrak{A})''$. *Then* $\mathcal{R}$ *is a hyperfinite* II_1 *factor and has a unique faithful, normal trace* τ. *Two projections in* $\mathcal{R}$ *are equivalent if and only if they have the same trace.*

PROOF. By the Double Commutant theorem, $\mathcal{R} = \overline{\pi_\tau(\mathfrak{A})}^{\text{WOT}}$, so it is hyperfinite. The trace on $\mathfrak{A}$ is given by $\tau(a) = \langle \pi_\tau(a)\xi_\tau, \xi_\tau\rangle$. Extend the definition of τ to a WOT-continuous state on $\mathcal{B}(H_\tau)$ by $\tau(A) = \langle A\xi_\tau, \xi_\tau\rangle$. This is tracial on $\pi_\tau(\mathfrak{A})$. By WOT-continuity, the tracial property extends to $\mathcal{R}$. Hence this is a normal trace on $\mathcal{R}$. In fact, it is the unique normal trace because the restriction to $\pi_\tau(\mathfrak{A})$ is unique by Theorem 9.10.10, and $\pi_\tau(\mathfrak{A})$ is WOT dense in $\mathcal{R}$. Since τ maps the projections in $\mathfrak{A}$ onto a dense subset of $[0,1]$ and τ is normal, τ maps the projections in $\mathcal{R}$ onto $[0,1]$.

Next we show that ξ_τ is a separating vector for $\mathcal{R}$. Indeed, suppose that $N \in \mathcal{R}$ such that $N\xi_\tau = 0$. Then for all $A \in \pi_\tau(\mathfrak{A})$, we have

$$\begin{aligned}\|NA\xi_\tau\|^2 &= \langle A^*N^*NA\xi_\tau, \xi_\tau\rangle = \tau(A^*N^*NA) = \tau(NAA^*N^*)\\ &\le \|A\|^2\tau(NN^*) = \|A\|^2\tau(N^*N) = \|A\|^2\langle N\xi_\tau, N\xi_\tau\rangle = 0.\end{aligned}$$

Therefore N vanishes on a dense subspace, and thus $N = 0$. In particular, if $N > 0$, we have $\tau(N) = \|N^{1/2}\xi_\tau\|^2 > 0$. So τ is faithful. Since ξ_τ is cyclic and separating for $\mathcal{R}$, it is cyclic and separating for $\mathcal{R}'$ by Lemma 10.4.5.

Suppose that $0 < P < I$ is a projection in $\mathcal{Z}(\mathcal{R})$. Then $0 < \tau(P) < 1$ Observe that $\tau_P(A) = \tau(P)^{-1}\tau(AP)$ is a normal tracial state, and $\tau_P \ne \tau$ because $\tau_P(P^\perp) = 0 \ne \tau(P^\perp) = 1 - \tau(P)$. The uniqueness of the normal tracial state shows that $\mathcal{R}$ is a factor.

Corollary 10.7.4 shows that if P, Q are projections in $\mathcal{R}$, then $P \prec Q$, $P \sim Q$ or $P \succ Q$. If $P \prec Q$, then $P \sim Q_1 < Q$ with $UU^* = P$ and $U^*U = Q_1$ for some partial isometry U. Then

$$\tau(P) = \tau(UU^*) = \tau(U^*U) = \tau(Q_1) < \tau(Q).$$

The inequality follows since τ is faithful. Similarly, $P \succ Q$ implies $\tau(P) > \tau(Q)$. Hence $P \sim Q$ if and only if $\tau(P) = \tau(Q)$.

If $P \ne 0$, let Q be a projection with $0 < \tau(Q) \le \frac{1}{2}\tau(P)$. Then $Q \sim Q_1 \le P$ and $Q \sim Q_2 \le P - Q_1$. So $Q_1 \sim Q_2$, and thus there is a partial isometry U with $U^*U = Q_1$ and $UU^* = Q_2$. Hence $P\mathcal{R}P$ is not abelian. So $\mathcal{R}$ is a II_1 factor. ∎

10.10.4. Remark. We could have concluded the proof in a different way. The only finite factors are of type I_n for $n < \infty$ and II_1. But the type I_n factor is M_n, which is finite dimensional. So any infinite dimensional finite factor is II_1. M_n also has minimal (hence abelian) projections, and II_1 factors do not.

The type I_n factors have the usual normalized trace τ_n on M_n. It is faithful, and is normal because M_n is finite dimensional. If $\mathcal{M} \simeq M_n(L^\infty(\mu)) \simeq L^\infty(X, \mu, M_n)$ for a probability measure μ, then

$$\tau(A) = \int \tau_n(A(x))h(x)\,d\mu(x)$$

is a faithful normal trace on $\mathcal{M}$ for any $h \in L^1(\mu)$ such that $h > 0$ a.e.(μ) and $\|h\|_1 = 1$. Combining this idea for each finite n shows that every finite type I von Neumann algebra on a separable Hilbert space has a faithful normal trace.

In Theorem 10.12.10, we will show that every type II_1 von Neumann algebra acting on a separable Hilbert space has a faithful normal trace. More generally, every type II_1 von Neumann algebra has a faithful family of normal traces.

10.10.5. DEFINITION. Let $\mathcal{M}$ be a von Neumann algebra with a faithful normal trace τ. Put a 2-norm on $\mathcal{M}$ by $\|A\|_2 = \tau(A^*A)^{1/2}$. Denote the completion as $L^2(\mathcal{M},\tau)$. The *standard form* of $\mathcal{M}$ is $\pi_\tau(\mathcal{M})$, which acts on $L^2(\mathcal{M},\tau)$.

We first show that π_τ is a faithful normal representation.

10.10.6. PROPOSITION. *Let $\mathcal{M}$ be a von Neumann algebra with a faithful normal trace τ. Then π_τ is a normal $*$-isomorphism of $\mathcal{M}$ onto its image.*

If $\mathfrak{A} \subset \mathcal{M}$ is a C-subalgebra which is* WOT *dense in $\mathcal{M}$ with trace $\tau' = \tau|_{\mathfrak{A}}$, then $\pi_{\tau'} = \pi_\tau|_{\mathfrak{A}}$ and $\pi_\tau(\mathcal{M}) = \pi_{\tau'}(\mathfrak{A})''$.*

PROOF. First observe that $H_\tau = L^2(\mathcal{M},\tau)$ because the set of vectors of 2-norm 0 is just $\{0\}$. Therefore H_τ is the completion of $\mathcal{M}$ in $\|\cdot\|_2$. For $A \in \mathcal{M}$, we let $\xi_A = A$ as an element of $L^2(\mathcal{M},\tau)$. In particular, $\xi_1 = \xi_\tau$.

Suppose that $\{A_\lambda\}_\Lambda$ is an increasing net in $\mathcal{M}_{sa}$ with $A = \sup_\Lambda A_\lambda$. Then for any $B \in \mathcal{M}$, $B^*AB = \text{SOT--}\lim_\Lambda B^*A_\lambda B$; whence $\tau(B^*AB) = \lim_\Lambda \tau(B^*A_\lambda B)$. This shows that

$$\lim_\Lambda \langle \pi_\tau(A_\lambda)\xi_B, \xi_B\rangle = \lim_\Lambda \tau(B^*A_\lambda B) = \tau(B^*AB) = \langle \pi_\tau(A)\xi_B, \xi_B\rangle.$$

Thus $\pi_\tau(A) = \sup_\Lambda \pi_\tau(A_\lambda)$. Therefore π_τ is normal. By Theorem 10.3.1, $\pi_\tau(\mathcal{M})$ is a von Neumann algebra. This map is faithful because if $A \neq 0$,

$$\|\pi_\tau(A)\| \geq \|\pi_\tau(A)\xi_1\|_2 = \|A\|_2 > 0.$$

If $\mathfrak{A}$ is a WOT dense subalgebra, then $M = \{\xi_A : A \in \mathfrak{A}\}$ is dense in $L^2(\mathcal{M},\tau)$. Indeed, if $A \in \mathcal{M}$, Kaplansky's density theorem provides a bounded net $A_\lambda \in \mathfrak{A}$ converging SOT to A. Then

$$\|\xi_A - \xi_{A_\lambda}\|_2^2 = \tau\big((A-A_\lambda)^*(A-A_\lambda)\big) = \|\pi_\tau(A-A_\lambda)\xi_1\|^2 \to 0.$$

It follows that $H_{\tau'} = L^2(\mathcal{M},\tau)$ and $\pi_{\tau'} = \pi_\tau|_{\mathfrak{A}}$. Thus $\pi_\tau(\mathcal{M}) = \pi_{\tau'}(\mathfrak{A})''$. ■

10.10.7. DEFINITION. Let $\mathcal{M}$ be a von Neumann algebra with a faithful normal trace τ. Define a conjugate linear map J on $L^2(\mathcal{M},\tau)$ by $J\xi_A = \xi_{A^*}$ and extending by continuity to the whole space.

Observe that

$$\|J\xi_A\|^2 = \|\xi_{A^*}\|^2 = \tau(AA^*) = \tau(A^*A) = \|\xi_A\|^2.$$

Thus J is an isometry and carries the dense subspace $\{\xi_A : A \in \mathcal{M}\}$ onto itself. Hence J extends to a *conjugate linear* isometry of $L^2(\mathcal{M},\tau)$ onto itself. Also $J^2 = I$ since this is clearly true on this dense subspace.

10.10.8. DEFINITION. Let $\mathfrak{A}$ be a C*-algebra. Define a C*-algebra $\mathfrak{A}^{\text{op}}$ to be $\mathfrak{A}$ as a Banach algebra with involution *, but with the product $a \circ b := ba$. We say that $\mathfrak{A}$ is *anti-isomorphic* to $\mathfrak{B}$ if $\mathfrak{A}$ is $*$-isomorphic to $\mathfrak{B}^{\text{op}}$.

Note that $\mathfrak{A}^{\text{op}}$ satisfies the C*-norm condition, and hence is a C*-algebra.

The operator J is used to identify the commutant $\mathcal{M}'$ in the standard representation. A more general version of this, valid for arbitrary von Neumann algebras, is called Tomita–Takesaki theory.

10.10.9. THEOREM. *Let $\mathcal{M}$ be a von Neumann algebra with a faithful normal trace τ in standard form in $\mathcal{B}(L^2(\mathcal{M},\tau))$. Then $\mathcal{M}$ is anti-isomorphic to $\mathcal{M}' = J\mathcal{M}J$. Moreover $\tau(T) = \langle T\xi_1, \xi_1\rangle$ is also a faithful normal trace on $\mathcal{M}'$.*

PROOF. For $A, B, C \in \mathcal{M}$,

$$(JAJ)B\xi_C = JA\xi_{C^*B^*} = \xi_{BCA^*} = BJ\xi_{AC^*} = B(JAJ)\xi_C.$$

Since $\{\xi_C : C \in \mathcal{M}\}$ is dense in $L^2(\mathcal{M},\tau)$, $JAJ \in \mathcal{M}'$; whence $J\mathcal{M}J \subset \mathcal{M}'$.

Next we show that if $T \in \mathcal{M}'$, then $JT\xi_1 = T^*\xi_1$. Since ξ_1 is cyclic for $\mathcal{M}$, choose $A_n \in \mathcal{M}$ so that $T\xi_1 = \lim_{n\to\infty} \xi_{A_n}$. Then for $B \in \mathcal{M}$,

$$\begin{aligned}\langle JT\xi_1, \xi_B\rangle &= \lim_{n\to\infty} \langle J\xi_{A_n}, \xi_B\rangle = \lim_{n\to\infty} \langle \xi_{A_n^*}, \xi_B\rangle = \lim_{n\to\infty} \tau(B^*A_n^*) \\ &= \lim_{n\to\infty} \tau(A_n^*B^*) = \lim_{n\to\infty} \langle \xi_{B^*}, \xi_{A_n}\rangle = \langle \xi_{B^*}, T\xi_1\rangle \\ &= \langle T^*B^*\xi_1, \xi_1\rangle = \langle B^*T^*\xi_1, \xi_1\rangle = \langle T^*\xi_1, B\xi_1\rangle = \langle T^*\xi_1, \xi_B\rangle.\end{aligned}$$

Since $\{\xi_B : B \in \mathcal{M}\}$ is dense in $L^2(\mathcal{M},\tau)$, we have $JT\xi_1 = T^*\xi_1$.

Now we can show that $J\mathcal{M}'J \subset \mathcal{M}$. Let $R, S, T \in \mathcal{M}'$. Then

$$\begin{aligned}(JTJ)S(R\xi_1) &= JT(R^*S^*\xi_1) = J(TR^*S^*)\xi_1 \\ &= SRT^*\xi_1 = SRJT\xi_1 = SR(JTJ)\xi_1\end{aligned}$$

Taking $S = I$, we have $(JTJ)R\xi_1 = R(JTJ)\xi_1$. Plugging this in, we obtain

$$(JTJ)S(R\xi_1) = SR(JTJ)\xi_1 = S(JTJ)R\xi_1.$$

Since ξ_1 is a separating (and cyclic) vector for $\mathcal{M}$, it is cyclic (and separating) for $\mathcal{M}'$. This shows that $JTJ \in \mathcal{M}'' = \mathcal{M}$. Hence $\mathcal{M}' \subset J\mathcal{M}J$. Therefore $\mathcal{M}' = J\mathcal{M}J$.

Define $\rho(A) = JA^*J$ for $A \in \mathcal{M}$. This maps $\mathcal{M}$ onto $\mathcal{M}'$. Observe that

$$\rho(A)\xi_B = JA^*J\xi_B = J\xi_{A^*B^*} = \xi_{BA}.$$

It follows that $\rho(A)$ is a linear map that multiplies on the right, as opposed to A which multiplies on the left. This makes it clear why they commute. Also

$$\rho(A_1A_2) = J(A_1A_2)^*J = JA_2^*JJA_1^*J = \rho(A_2)\rho(A_1);$$

i.e. ρ reverses multiplication. Clearly ρ is normal. Thus ρ is a normal anti-isomorphism.

Next we observe that $\tau(T) = \langle T\xi_1, \xi_1\rangle$ is a normal trace on $\mathcal{M}'$ as well. Indeed, it is clearly a WOT continuous state and

$$\tau(\rho(AB)) = \langle \rho(AB)\xi_1, \xi_1\rangle = \langle \xi_{AB}, \xi_1\rangle = \tau(AB).$$

Similarly, $\tau(\rho(BA)) = \tau(BA) = \tau(AB) = \tau(\rho(AB))$. Hence τ is also a trace on $\mathcal{M}'$. Finally, if $T > 0$ in $\mathcal{M}'$, then $T = \rho(A)$ for some $A > 0$ in $\mathcal{M}$. Thus

$$\tau(T) = \tau(\rho(A)) = \tau(A) > 0.$$

Therefore τ is faithful on $\mathcal{M}'$. ■

10.10.10. COROLLARY. *Let $\mathcal{M}$ be a type II_1 von Neumann algebra with a faithful normal trace τ in standard form in $\mathcal{B}(L^2(\mathcal{M}, \tau))$. Then $\mathcal{M}'$ is type II_1.*

PROOF. Since J is an anti-isomorphism, $JPJ \sim JQJ$ if and only if $P \sim Q$. It follows that $\operatorname{Ad} J$ takes finite projections to finite projections and abelian projections to abelian projections. Since $\mathcal{M}$ is II_1, it has no abelian projections and I is finite. Thus $\mathcal{M}'$ has no abelian projections and I is finite. That is, $\mathcal{M}$ is type II_1. ■

10.10.11. COROLLARY. *Let $\mathcal{M}$ be a hyperfinite von Neumann algebra with a faithful normal trace τ in standard form in $\mathcal{B}(L^2(\mathcal{M}, \tau))$. Then $\mathcal{M}'$ is hyperfinite.*

PROOF. Let $\mathfrak{A} = \bigcup_{n\geq 1} \mathfrak{A}_n$ be a nested sequence of finite dimensional C*-algebras $\mathfrak{A}_n$ which is WOT dense in $\mathcal{M}$. Let ρ be the isomorphism of $\mathcal{M}$ onto $\mathcal{M}'^{\mathrm{op}}$ just constructed. Then $\rho(\mathfrak{A}) = \bigcup_{n\geq 1} \rho(\mathfrak{A}_n)$ is a nested sequence of finite dimensional C*-algebras $\rho(\mathfrak{A}_n)$ which is WOT dense in $\mathcal{M}'^{\mathrm{op}}$. Therefore $\rho(\mathfrak{A})^{\mathrm{op}} = \bigcup_{n\geq 1} \rho(\mathfrak{A}_n)^{\mathrm{op}}$ is a nested sequence of finite dimensional C*-algebras which is WOT dense in $\mathcal{M}'$. Thus $\mathcal{M}'$ is hyperfinite. ■

Next we look at the group von Neumann algebras. Let G be a discrete group. Consider the group von Neumann algebras $\mathcal{L}(G) = C^*_\lambda(G)''$ and $\mathcal{R}(G) = C^*_\rho(G)''$, both acting on $l_2(G)$. This will provide us with other examples of II_1 factors.

We showed in Proposition 9.12.8 that $\tau(A) = \langle A\xi_e, \xi_e\rangle$ is a faithful normal trace on $\mathcal{L}(G)$. It is likewise a faithful normal trace on $\mathcal{R}(G)$. We also observed in Remark 9.12.7 that $\mathcal{R}(G) \subset \mathcal{L}(G)'$. We make this more precise.

10.10.12. THEOREM. *Let G be a discrete group. Then $\mathcal{R}(G) = \mathcal{L}(G)'$ is isomorphic and anti-isomorphic to $\mathcal{L}(G)$.*

PROOF. Observe that $\lambda \simeq \pi_\tau$ follows from Proposition 9.7.4. It follows that $\mathcal{L}(G)$ is a von Neumann algebra with a faithful normal trace in standard form.

By Theorem 10.10.9, $\mathcal{L}(G)' = J\mathcal{L}(G)J$ where $J(A\xi_e) = A^*\xi_e$ for $A \in \mathcal{L}(G)$. In particular, $J\xi_g = J\lambda(g)\xi_e = \xi_{g^{-1}}$. A computation shows that

$$J\lambda(g)J\xi_h = J\lambda(g)\xi_{h^{-1}} = J\xi_{gh^{-1}} = \xi_{hg^{-1}} = \rho(g)\xi_h.$$

Therefore $JC^*_\lambda(G)J = C^*_\rho(G)$, and thus $J\mathcal{L}(G)J = \mathcal{R}(G)$. Therefore by Theorem 10.10.9, $\mathcal{R}(G) = \mathcal{L}(G)'$ is anti-isomorphic to $\mathcal{L}(G)$. However ρ is unitarily equivalent to λ via $U\xi_g = \xi_{g^{-1}}$, which follows as in the computation above replacing the conjugate unitary map J with the unitary map U. Hence $\mathcal{R}(G)$ and $\mathcal{L}(G)$ are unitarily equivalent and thus isomorphic. ■

10.10.13. DEFINITION. We say that G is an *ICC group* or an *infinite conjugacy class group* if for every $e \neq g \in G$, the conjugacy class $\mathrm{Cl}(g) = \{hgh^{-1} : h \in G\}$ is infinite.

10.10.14. PROPOSITION. *Let G be a discrete group. Then $\mathcal{L}(G)$ is a factor if and only if G is ICC.*

PROOF. Suppose that there is some $g \neq e$ in G so that $\mathrm{Cl}(g) = \{g_1, \dots, g_n\}$ is finite. Let $A = \sum_{i=1}^n \lambda(g_i)$. This lies in $\mathcal{L}(G)$, even $C^*_\lambda(G)$, and is not scalar since $A\xi_e = \sum_{i=1}^n \xi_{g_i}$ is not a multiple of ξ_e. If $h \in G$,

$$\lambda(h)A\lambda(h)^* = \sum_{i=1}^n \lambda(hg_ih^{-1}) = A$$

because $\{hg_ih^{-1} : g_i \in \mathrm{Cl}(g)\}$ is just a permutation of $\mathrm{Cl}(g)$. Hence $\lambda(h)A = A\lambda(h)$ for every $h \in G$. It follows that A commutes with $\mathcal{L}(G)$, and so lies in the centre $\mathcal{Z}(G)$. Therefore $\mathcal{L}(G)$ is not a factor.

Now suppose that G is ICC, and let $A \in \mathcal{Z}(G) = \mathcal{L}(G) \cap \mathcal{R}(G)$. Let $A\xi_e = \sum_G a_g\xi_g$. Then for $h \in G$,

$$A\xi_e = A\lambda(h)\rho(h)\xi_e = \lambda(h)\rho(h)A\xi_e = \sum_G a_g\xi_{hgh^{-1}}.$$

It follows that $a_{hgh^{-1}} = a_g$ for all h; so a_g is constant on $\mathrm{Cl}(g)$, which is infinite. The vector $A\xi_e$ lies in $l_2(G)$, and hence $\sum_G |a_g|^2 < \infty$. So for each conjugacy class other than $\{e\}$, we must have $a_g = 0$. Hence $A\xi_e = a_e\xi_e$. Since ξ_e is a separating vector for $\mathcal{L}(G)$, $A = a_eI$. Thus $\mathcal{Z}(G) = \mathbb{C}I$, and $\mathcal{L}(G)$ is a factor. ■

10.10.15. EXAMPLE. Let S_∞ be the group of finite permutations of $\mathbb{N}$. See Example 9.12.10. We can write $S_\infty = \bigcup_{n\geq 1} S_n$ where S_n is the permutation group on the first n elements of $\mathbb{N}$, considered as fixing the remaining elements. Thus $\mathbb{C}S_\infty = \bigcup_{n\geq 1} \mathbb{C}S_n$ is an increasing union of finite dimensional subalgebras. We showed that

$$C^*(S_\infty) = \varinjlim_{n\to\infty} C^*(S_n) = \varinjlim_{n\to\infty} C^*_\lambda(S_n) = C^*_\lambda(S_\infty).$$

This C*-algebra is evidently AF (the closure of an increasing union of finite dimensional subalgebras). Hence $\mathcal{L}(S_\infty)$ is hyperfinite.

We claim that S_∞ is an ICC group. Let $e \neq g \in S_\infty$. Then $g \in S_n$ for some large n, and in its reduced form, contains a cycle $(i_1 \dots i_k)$ for $k \geq 2$. For each $h = (i_1 m)$ for $m > n$, hgh^{-1} contains the cycle $(mi_2 \dots i_k)$. These are distinct elements. Hence $\mathcal{L}(S_\infty)$ is a factor. Therefore this is a hyperfinite II_1 factor.

10.10.16. Example. See Example 9.12.12. Observe that the free group $\mathbb{F}_S$ for $|S| \geq 2$ is an ICC group. Indeed if $e \neq g = a_1 \dots a_k$ is an element of the group in reduced form with $a_i \in S \cup S^{-1}$, choose $b \in S \cup S^{-1} \setminus \{a_1^{-1}, a_k\}$ and choose j so that $s_j \notin \{b, b^{-1}\}$. Set $t_n = s_j^n b$ and note that $t_n g t_n^{-1} = s_j^n b g b^{-1} s_j^{-n}$ is in reduced form. These words are distinct, so $\mathbb{F}_S$ is ICC. Therefore $\mathcal{L}(\mathbb{F}_S)$ is a factor.

By Theorem 9.12.15, there is a unique trace on $C^*_\lambda(\mathbb{F}_S)$. It follows that $\mathcal{L}(\mathbb{F}_S)$ has a unique normal trace. This must be the trace $\tau(A) = \langle A\xi_e, \xi_e\rangle$, and this trace is faithful. Moreover, we have $\lambda = \pi_\tau$ is the standard representation and $\mathcal{L}(\mathbb{F}_S)' = \mathcal{R}(\mathbb{F}_S)$. Therefore $\mathcal{L}(\mathbb{F}_S)$ is a factor with a unique faithful normal trace. Because $\mathcal{L}(\mathbb{F}_S)$ is infinite dimensional, it is a II_1 factor. We will see shortly in Corollary 10.10.21 that it is not hyperfinite.

A curious pathology occurs when S is uncountable. In this case, the algebra acts on a non-separable Hilbert space; and its predual is not separable. Even so, it has a unique faithful trace. Thus a collection of pairwise orthogonal non-zero projections in $\mathcal{L}(\mathbb{F}_S)$ must be countable; (i.e. it is *countably decomposable*). Compare this with Exercise **10.36**.

Now let's consider the relationship between amenability of G and the group von Neumann algebra.

10.10.17. Definition. A von Neumann algebra $\mathcal{M} \subset \mathcal{B}(H)$ is called *injective* if there is an expectation of $\mathcal{B}(H)$ onto $\mathcal{M}$.

10.10.18. Remark. The word injective suggests a categorical meaning. The category is the collection of operator systems with unital completely positive maps as morphisms. In particular, being injective is a property of $\mathcal{M}$ and not a property of the imbedding into $\mathcal{B}(H)$. See Definition 14.7.12.

Here is an easy way to establish injectivity of a von Neumann algebra.

10.10.19. Proposition. *If $\mathcal{M} \subset \mathcal{B}(H)$ and $\mathcal{M}'$ is hyperfinite, then $\mathcal{M}$ is injective.*

Proof. Write $\mathcal{M}'$ as the WOT-closure of an increasing net of finite dimensional C*-subalgebras $\mathfrak{A}_\lambda$. Then $\mathcal{M} = \bigcap_\Lambda \mathfrak{A}'_\lambda$.

We first construct an expectation $\mathcal{E}_\lambda$ onto $\mathfrak{A}'_\lambda$. By the Wedderburn theorem for semisimple finite dimensional algebras (see Exercise **9.19**) each $\mathfrak{A}_\lambda$ is a direct sum of full matrix algebras $\mathfrak{A}_\lambda \simeq \bigoplus_{i=1}^k \mathcal{M}_{n_i}$. Choose a finite subgroup G_i of the unitary group $\mathcal{U}_{n_i}$ which spans $\mathcal{M}_{n_i}$, such as the group generated by the diagonal matrices with entries ± 1 and the permutation matrices with respect to a fixed orthonormal basis. Then $G = G_1 \times \cdots \times G_k$ is a finite subgroup of the unitary group of $\mathfrak{A}_\lambda$ which spans $\mathfrak{A}_\lambda$. Define

$$\mathcal{E}_\lambda(a) = \frac{1}{|G|} \sum_{g \in G} gag^*.$$

It is clear that $\|\mathcal{E}_\lambda\| \le 1$. For $h \in G$,

$$h\mathcal{E}_\lambda(a) = \frac{1}{|G|} \sum_{g \in G} hgag^* = \frac{1}{|G|} \sum_{g \in G} (hg)a(hg)^*h = \mathcal{E}_\lambda(a)h.$$

This shows that the range of $\mathcal{E}_\lambda$ is contained in $\mathfrak{A}'_\lambda$. On the other hand, if $a \in \mathfrak{A}'_\lambda$, then it is clear that $\mathcal{E}_\lambda(a) = a$. Hence $\mathcal{E}_\lambda$ is an expectation onto $\mathfrak{A}'_\lambda$.

Consider the maps $\mathcal{E}_\lambda$ as a net in $\mathcal{B}(\mathcal{B}(H))$ endowed with the topology of pointwise WOT convergence. Since they are all norm 1, we can consider them as elements in the product space

$$\prod_{A \in \mathcal{B}(H)} \left(\overline{b_{\|A\|}(\mathcal{B}(H))}, \text{WOT}\right) \quad \text{by } \mathcal{E} \to (\mathcal{E}(A))_{A \in \mathcal{B}(H)}.$$

This is a product of compact Hausdorff spaces, and thus it is compact by Tychonoff's theorem 1.8.7. Therefore there is a cofinal subnet which converges WOT at each point $A \in \mathcal{B}(H)$. Let $\mathcal{E}$ denote the limit of this subnet. It is clear that $\mathcal{E}$ is a linear map of norm at most 1 which takes positives to positives. The range of $\mathcal{E}$ belongs to $\bigcap_\Lambda \mathfrak{A}'_\lambda = \mathcal{M}$. Moreover, since $\mathcal{E}_\mu(b) = b$ for $\mu \ge \Lambda$ and $b \in \mathfrak{A}'_\lambda \supset \mathcal{M}$, $\mathcal{E}(b) = b$ for $b \in \mathcal{M}$. Therefore $\mathcal{E}$ is an expectation onto $\mathcal{M}$. ∎

We now provide another characterization of the amenability of G, this time in terms of $\mathcal{L}(G)$.

10.10.20. Theorem. *Let G be a discrete group. Then G is amenable if and only if $\mathcal{L}(G)$ is injective.*

Proof. Suppose that G is amenable and m is a left-invariant mean on $l^\infty(G)$. For $A \in \mathcal{B}(l_2(G))$, let $\widehat{A}(g) = \rho(g)A\rho(g)^*$. Define a sesquilinear form on $l_2(G)$ by

$$[\xi, \eta]_A = m(\langle \widehat{A}(g)\xi, \eta \rangle) \quad \text{for} \quad \xi, \eta \in l_2(G).$$

This is clearly bounded by $\|A\|$, linear in the first variable, and conjugate linear in the second. By Proposition 4.2.2, there is a unique operator $\mathcal{E}(A)$ such that

$$\langle \mathcal{E}(A)\xi, \eta \rangle = [\xi, \eta]_A.$$

Since $\|\mathcal{E}(A)\| \leq \|A\|$, this is a contractive linear map on $\mathcal{B}(H)$. If $A \in \mathcal{L}(G)$, then $\rho(g)A\rho(g)^* = A$ is constant, and thus $\mathcal{E}(A) = A$.

Let $h \in G$. By left invariance,

$$\begin{aligned}\langle \rho(h)\mathcal{E}(A)\rho(h)^*\xi, \eta \rangle &= [\rho(h)^*\xi, \rho(h)^*\eta]_A = m(\langle \rho(g)A\rho(g)^*\rho(h)^*\xi, \rho(h)^*\eta \rangle) \\ &= m(\langle \rho(hg)A\rho(hg)^*\xi, \eta \rangle) = m(\langle \widehat{A}(hg)\xi, \eta \rangle) \\ &= m({}_{h^{-1}}\langle \widehat{A}(g)\xi, \eta \rangle) = m(\langle \widehat{A}(g)\xi, \eta \rangle) = \langle \mathcal{E}(A)\xi, \eta \rangle.\end{aligned}$$

Thus $\rho(h)\mathcal{E}(A)\rho(h)^* = \mathcal{E}(A)$. Therefore $\mathcal{E}(a) \in \mathcal{R}(G)' = \mathcal{L}(G)$. Hence the range of $\mathcal{E}$ is $\mathcal{L}(G)$ and $\mathcal{E}$ is a norm 1 projection. So $\mathcal{L}(G)$ is injective.

Conversely, suppose that $\mathcal{L}(G)$ is injective, and let $\mathcal{E}$ be an expectation of $\mathcal{B}(l_2(G))$ onto $\mathcal{L}(G)$. Imbed $l^\infty(G)$ into $\mathcal{B}(l_2(G))$ as the diagonal operators $M_f\xi_g = f(g)\xi_g$. Let τ be the canonical trace on $\mathcal{L}(G)$ given by $\tau(A) = \langle A\xi_e, \xi_e \rangle$. Define $m(f) = \tau(\mathcal{E}(M_f))$ for $f \in l^\infty(G)$. This is a positive linear functional such that $m(I) = 1$, so it is a state. If $f \in l^\infty(G)$, then $\lambda(g)M_f\lambda(g^{-1}) = M_{{}_gf}$. Therefore by the bimodule property of $\mathcal{E}$ and the trace property of τ,

$$m({}_gf) = \tau(\mathcal{E}(\lambda(g)M_f\lambda(g^{-1}))) = \tau(\lambda(g)\mathcal{E}(M_f)\lambda(g^{-1})) = \tau(\mathcal{E}(M_g)) = m(f).$$

Thus m is a left-invariant mean. ■

10.10.21. COROLLARY. *If $|S| \geq 2$, $\mathcal{L}(\mathbb{F}_S)$ is a non-hyperfinite* II_1 *factor.*

PROOF. In Example 10.10.16, it is shown that $\mathcal{L}(\mathbb{F}_S)$ is a II_1 factor. In Example 9.13.4(3), we saw that $\mathbb{F}_S$ is not amenable. Hence by Theorem 10.10.20, $\mathcal{L}(\mathbb{F}_S)$ is not injective. Finally, by Proposition 10.10.19, $\mathcal{L}(\mathbb{F}_S)$ is not hyperfinite. ■

It is a long-standing problem whether $\mathcal{L}(\mathbb{F}_n)$ are isomorphic to each other for $2 \leq n < \infty$. It is known that either they are all different or all isomorphic.

10.11. More about Equivalence of Projections

We develop a few more results about the equivalence of projections. They will be used in our analysis of type II_∞ and III von Neumann algebras, as well as in the construction of a trace for II_1 algebras.

The following is a variation on the proof of the Schroeder–Bernstein theorem.

10.11.1. PROPOSITION. *Let P, Q be projections in $\mathcal{M}$. If $P \preceq Q \preceq P$, then $P \sim Q$.*

PROOF. We may suppose that $VV^* \leq P = U^*U$ and $UU^* \leq Q = V^*V$ for partial isometries U and V in $\mathcal{M}$. Let $E_0 = P - VV^*$ and $F_0 = Q - UU^*$. If either $E_0 = 0$ or $F_0 = 0$, then V or U provides the equivalence. Otherwise, we recursively define

$$E_n = VF_{n-1}V^* \quad \text{and} \quad F_n = UE_{n-1}U^* \quad \text{for} \quad n \geq 1.$$

Note that $E_0 \perp VV^*$ is orthogonal to $E_n \leq VV^*$ and similarly $F_0 \perp F_n$ for $n \geq 1$. Now U maps orthogonal spaces to orthogonal spaces, and hence $F_1 = UE_0U^*$ is orthogonal to $F_n = UE_nU^*$ for $n \geq 2$. It now follows that $E_1 = VF_0V^*$ is orthogonal to $E_n = VF_{n-1}V^*$, $n \geq 2$. Recursively we see that $\{E_n : n \geq 0\}$ and $\{F_n : n \geq 0\}$ are sets of pairwise orthogonal projections. Moreover, by construction, $E_n \sim F_{n\pm 1}$.

Let $E_\infty = P - \sum_{n\geq 0} E_n$ and $F_\infty = Q - \sum_{n\geq 0} F_n$. Then

$$\sum_{n\geq 0} UE_nU^* = \sum_{n\geq 1} F_n \quad \text{and} \quad \sum_{n\geq 0} VF_nV^* = \sum_{n\geq 1} E_n.$$

Hence

$$UE_\infty U^* = UU^* - \sum_{n\geq 0} UE_nU^* = (Q - F_0) - \sum_{n\geq 1} F_n = F_\infty.$$

Therefore $E_\infty \sim F_\infty$; and so by Proposition 10.7.2(4),

$$P = E_\infty + \sum_{n\geq 0} E_{2n} + \sum_{n\geq 0} E_{2n+1} \sim F_\infty + \sum_{n\geq 0} F_{2n+1} + \sum_{n\geq 0} F_{2n} = Q. \qquad \blacksquare$$

The next lemma shows that properly infinite projections live up to their name in a strong sense.

10.11.2. LEMMA. *Let P be a properly infinite projection in $\mathcal{M}$. There is a family $\{E_i : i \geq 1\}$ of pairwise orthogonal projections in $\mathcal{M}$ such that $E_i \sim E_j \sim P$ for $i, j \geq 1$ and $\sum_{i\geq 1} E_i = P$.*

PROOF. We can work in $P\mathcal{M}P$ and assume that $P = I$. Since $\mathcal{M}$ is infinite, it contains a proper isometry U. Let

$$P_n = U^{n-1}U^{*n-1} - U^nU^{*n} \quad \text{for} \quad n \geq 1.$$

Note that $(UP_n)(UP_n)^* = P_{n+1}$ and $(UP_n)^*(UP_n) = P_n$, so $P_n \sim P_{n+1}$ for $n \geq 1$. By Zorn's lemma, we can extend $\{P_n\}_{n\geq 1}$ to a maximal family $\{P_\lambda\}_\Lambda$ of projections with $P_\lambda \sim P_1$. Let $Q = \left(\sum_\Lambda P_\lambda\right)^\perp$. By the comparison Theorem 10.7.3 applied to P_1 and Q, there is a central projection Z so that $QZ \precsim P_1Z$ and $P_1Z^\perp \precsim QZ^\perp$. If $Z = 0$, then $P_1 \sim Q_1 \leq Q$. But then $\{P_\lambda\} \cup \{Q_1\}$ would be a larger family of equivalent orthogonal projections; and thus $Z \neq 0$. Let $Q_0 = QZ \sim Q_1 \leq P_1Z$; and observe that $Q_0 \sim Q_n = U^{n-1}Q_1U^{*n-1} \leq P_nZ$. Define $F_n = P_nZ - Q_n + Q_{n-1}$ for $n \geq 1$ and $F_\lambda = P_\lambda Z$ for $\lambda \in \Lambda \setminus \mathbb{N}$. Then $\{F_\lambda\}_\Lambda$ are pairwise orthogonal equivalent projections with $\sum_\Lambda F_\lambda = Z$. Since Λ

is infinite, we can partition it into a countably infinite collection of disjoint subsets Λ_i with $|\Lambda_i| = |\Lambda|$ for $i \geq 1$. Define $E_i = \sum_{\Lambda_i} F_\lambda$. Then by Lemma 10.7.2(4), $E_i \sim E_j$ for $i, j \in \mathbb{N}$; and by pairing Λ_i with Λ, we have $E_i \sim \sum E_i = Z$.

Now let $\{Z_\lambda\}_\Lambda$ be a maximal collection of pairwise orthogonal central projections such that there are pairwise orthogonal equivalent projections $\{E_i^\lambda\}_{i\geq 1}$ such that $E_i^\lambda \sim \sum_{\mathbb{N}} E_i^\lambda = Z_\lambda$. It must be the case that $Z = \sum_\Lambda Z_\lambda = I$, for otherwise we apply the previous paragraph to $Z^\perp \mathcal{M}$. Finally we define $E_i = \sum_\Lambda E_i^\lambda$, and observe that $E_i \sim E_j \sim I$ and $\sum_{\mathbb{N}} E_i = I$. ■

10.11.3. REMARK. In the proof, we use an infinite projection to produce a countable family of pairwise orthogonal equivalent projections P_n. Conversely, if there are such projections in $\mathcal{M}$, then $P = \sum_{n\geq 1} P_n$ is infinite because it is equivalent to $P - P_1$. Thus this cannot occur in a finite von Neumann algebra.

The following consequence is straightforward—just partition $\mathbb{N}$ into n infinite subsets.

10.11.4. COROLLARY. *Let P be a properly infinite projection in $\mathcal{M}$, and let $n \geq 2$. Then there is a family $\{E_i : 1 \leq i \leq n\}$ of pairwise orthogonal projections in $\mathcal{M}$ such that $E_i \sim E_j \sim P$ and $\sum_{i=1}^n E_i = P$.*

10.11.5. COROLLARY. *Let $\mathcal{M}$ be a properly infinite von Neumann algebra, and let P be a properly infinite projection. If Q is a σ-finite projection with $c(Q) \leq c(P)$, then $Q \preceq P$. If $\mathcal{M}$ is σ-finite, Q is properly infinite and $c(Q) = c(P)$, then $Q \sim P$. In particular, in a σ-finite type* III *factor, any two non-zero projections are equivalent.*

PROOF. Use comparison for P and Q. There is a central projection Z so that $ZP \preceq ZQ$ and $Z^\perp Q \preceq Z^\perp P$. In $Z^\perp \mathcal{M}$ there is nothing to prove. So we can assume that $Z = I$, $P \leq Q$ and $c(P) = I$; and restrict to $\mathcal{M}_Q$, which is σ-finite. Write $P = \sum_{i\geq 1} E_i$ with $E_i \sim E_j \sim P$. Let $\{F_j : j \in J\}$ be a maximal family of pairwise orthogonal projections including $\{E_i : i \geq 1\}$ such that $0 \neq F_j \preceq P$ for each $j \in J$. Since $\mathcal{M}_Q$ is σ-finite, J is countable and infinite; so we may suppose that $J = \mathbb{N}$. Let $R = \sum F_j$. If $R^\perp \neq 0$, then $c(R^\perp)c(P) = c(R^\perp) \neq 0$. But then by Lemma 10.7.2(3), there is a projection $0 \neq F \leq R^\perp$ with $F \preceq P$, contradicting maximality of $\{F_j\}$. Hence $R = I$. Therefore

$$I = \sum_{j\in\mathbb{N}} F_j \preceq \sum_{j\in\mathbb{N}} E_j = P \leq I.$$

By Proposition 10.11.1, $P \sim I$. Hence in our original situation, $ZQ \sim ZP$.

If $\mathcal{M}$ or P is σ-finite, Q is properly infinite and $c(P) = c(Q)$, then the first part yields $Q \preceq P \preceq Q$. Hence $Q \sim P$ by Proposition 10.11.1. If $\mathcal{M}$ is a σ-finite type

III factor, every non-zero projection is properly infinite, σ-finite and has central support I. So they are equivalent. ■

In many cases, a von Neumann algebra is not simple as a C*-algebra. See Exercise **10.32**. We shall see there are two cases where a factor is simple.

10.11.6. COROLLARY. *A σ-finite type* III *factor $\mathcal{M}$ is a simple C*-algebra.*

PROOF. Let $\mathcal{J}$ be a non-zero norm-closed ideal in $\mathcal{M}$, and let $0 \neq A \in \mathcal{J}_+$. All of the spectral projections of A belong to $\{A\}'' \subset \mathcal{M}$. There is some $t > 0$ so that the spectral projection $P = E_A[t, \|A\|] \neq 0$. Then $tP \leq A$. Since ideals are hereditary, $P \in \mathcal{J}$. By Corollary 10.11.5, $P \sim I$. So there is an isometry $U \in \mathcal{M}$ such that $UU^* = P$ and $U^*U = I$. Thus $I = U^*PU \in \mathcal{J}$, and hence $\mathcal{J} = \mathcal{M}$. ■

This next result is known as the "*halving lemma*". In view of Corollary 10.11.4, this is of especial interest in the finite case.

10.11.7. LEMMA. *Let $\mathcal{M}$ be a von Neumann algebra with no type* I *summand. Let P be a projection in $\mathcal{M}$. Then there is a projection $Q \leq P$ so that $Q \sim P - Q$. This projection is unique up to equivalence, i.e. if $P > R \sim P - R$, then $R \sim Q$.*

PROOF. We again assume that $P = I$ by restricting to $P\mathcal{M}P$. Consider collections $\{(Q_\lambda, R_\lambda)\}_\Lambda$ of projections so that $Q_\lambda \sim R_\lambda$ and $\{Q_\lambda, R_\lambda : \lambda \in \Lambda\}$ consists of pairwise orthogonal projections. It is an easy application of Zorn's lemma to obtain a maximal collection. Now $Q_0 = \sum_\Lambda Q_\lambda$ and $R_0 = \sum_\Lambda R_\lambda$ are equivalent and orthogonal.

We will show that $Q_0 + R_0 = I$. If not, let $P_0 = (Q_0 + R_0)^\perp \neq 0$. Then $P_0\mathcal{M}P_0$ is not abelian since there is no type I summand. So there is a projection $P_1 \leq P_0$ which is not in the centre of $P_0\mathcal{M}P_0$. Let $Q_1 = P_0 - P_1$, and note that $c(P_1)c(Q_1) \neq 0$. By Lemma 10.7.2, there are equivalent projections $0 < P_2 \leq P_1$ and $0 < Q_2 \leq Q_1$. This contradicts the maximality of the original family. Thus we must have $Q_0 + R_0 = I$.

Suppose that R is another projection with $R \sim R^\perp$. By the Comparison theorem 10.7.3, there is a central projection Z so that $ZQ \preceq ZR$ and $Z^\perp Q \succeq Z^\perp R$. Hence

$$ZQ \preceq ZR \sim ZR^\perp \preceq ZQ^\perp \sim ZQ.$$

By Proposition 10.11.1, $ZQ \sim ZR$. Similarly we get that $Z^\perp Q \sim Z^\perp R$, and therefore $Q \sim R$. ■

10.11.8. COROLLARY. *Let $\mathcal{M}$ be a type* II_1 *von Neumann algebra. For each $n \geq 1$, there is a projection $P_{2^{-n}}$ and equivalent pairwise orthogonal projections*

$P_i^n \sim P_{2^{-n}}$ for $1 \le i \le 2^n$ such that $\sum_{i=1}^{2^n} P_i^n = I$. Moreover, they can be chosen so that $\sum_{n\ge1} P_{2^{-n}} = I$.

PROOF. The existence of $P_{1/2}$ is provided by the halving lemma. Recursively construct $P_{2^{-n}} \le P_{2^{1-n}}$ such that $P_{2^{-n}} \sim P_{2^{1-n}} - P_{2^{-n}}$. Define $P_1^0 = I$, $P_1^1 = P_{1/2}$ and $P_2^1 = P_{1/2}^\perp$. Assume that we have defined $P_i^n \sim P_{2^{-n}}$ for $1 \le i \le 2^n$ so that $P_{2i-1}^n + P_{2i}^n = P_i^{n-1}$ for $1 \le i \le 2^{n-1}$. Now

$$P_{2^{-n}} = P_{2^{-n-1}} + (P_{2^{-n}} - P_{2^{-n-1}}) \sim P_i^n;$$

and this identification splits $P_i^n = P_{2i-1}^{n+1} + P_{2i}^{n+1}$ where $P_{2i-1}^{n+1} \sim P_{2i}^{n+1} \sim P_{2^{-n-1}}$. Thus the projections P_i^n are defined recursively.

If we identify $P_{2^{-n}}$ with $P_{2^n-1}^n$, we have pairwise orthogonal projections with $\sum_{n=1}^N P_{2^{-n}} = I - P_{2^N}^N$. Clearly $P_{2^n}^n$ forms a decreasing sequence of projections. Let Q be the limit. Observe that $Q \le P_{2^n}^n \sim P_{2^{-n}}$ for $n \ge 1$. Hence for each $n \ge 1$, we can find equivalent projections $Q \sim Q_n \le P_{2^{-n}}$ The projections Q_n are pairwise orthogonal. By Remark 10.11.3, $Q = 0$. Therefore $\sum_{n\ge1} P_{2^{-n}} = I$. ■

10.11.9. DEFINITION. A projection P in a finite von Neumann algebra $\mathcal{M}$ is called a *monic projection of order* n if there are n pairwise orthogonal projections $P = P_1 \sim \cdots \sim P_n$ such that $\sum_{i=1}^n P_i = c(P)$.

10.11.10. LEMMA. *Every projection P in a finite von Neumann algebra $\mathcal{M}$ is the sum of monic projections.*

PROOF. Let $0 \ne Z \le c(P)$ be a central projection. We will show that there is a monic projection $Q \le ZP$. By Corollary 10.11.8, there are monic projections $P_{2^{-n}}$, each equivalent to half of the previous one, which sum to Z. By comparison applied to ZP and $ZP_{2^{-n}}$, there is a central projection $Z_n \le Z$ so that $Z_nP \preceq Z_nP_{2^{-n}}$ and $(Z - Z_n)P_{2^{-n}} \preceq (Z - Z_n)P$. If $Z_n \ne Z$, then $(Z - Z_n)P_{2^{-n}}$ is a monic projection equivalent to a projection $Q \le (Z - Z_n)P$; whence Q is monic. Suppose that $Z_n = Z$ for all $n \ge 1$. Then we obtain $ZP \sim Q_n \le ZP_{2^{-n}}$. Arguing as in Corollary 10.11.8, we deduce that $ZP = 0$, contrary to the fact that $c(ZP) = Z$. Hence, for some n, there is a monic projection Q dominated by $(Z - Z_n)P$.

Let $\{P_j\}$ be a maximal family of pairwise orthogonal monic projections dominated by P, and let $Q = \sum P_j$. If $Q \ne P$, apply the previous paragraph to $P - Q$ to contradict maximality. ■

10.11.11. COROLLARY. *A type II_1 factor $\mathcal{M}$ is a simple C*-algebra.*

PROOF. Let $\mathcal{J}$ be a non-zero norm-closed ideal in $\mathcal{M}$, and let $0 \ne A \in \mathcal{J}_+$. All of the spectral projections of A belong to $\{A\}'' \subset \mathcal{M}$. There is some $t > 0$ so that the spectral projection $P = E_A[t, \|A\|] \ne 0$. Then $tP \le A$. Since ideals are hereditary, $P \in \mathcal{J}$. By Lemma 10.11.10, there is a monic projection $Q \le P$. Hence there are pairwise orthogonal projections $Q = Q_1 \sim \cdots \sim Q_n$ so that $\sum_{i=1}^n Q_i = c(Q)$. As $\mathcal{M}$ is a factor, $c(Q) = I$. Thus there are partial isometries $U_i \in \mathcal{M}$ such that $U_i U_i^* = Q_i$ and $U_i^* U_i = Q$. Therefore $Q_i = U_i^* Q U_i \in \mathcal{J}$, and hence $I \in \mathcal{J}$; whence $\mathcal{J} = \mathcal{M}$. ∎

The following is mostly about finite projections.

10.11.12. THEOREM. *Let P, Q be projections in $\mathcal{M}$.*

(1) $(P \vee Q) - P \sim Q - (P \wedge Q)$. (Kaplansky identity)

(2) *If P and Q are finite, then $P \vee Q$ is finite.*

(3) *If P and Q are finite and $P \sim Q$, then $P^\perp \sim Q^\perp$. Hence P and Q are unitarily equivalent.*

(4) *If P and Q are finite and $P \preceq Q$, then $Q^\perp \preceq P^\perp$.*

(5) *If P is finite, and that (Q_n) is an increasing sequence of projections with $Q = \sup_n Q_n$ such that $Q_n \preceq P$ for $n \ge 1$, then $Q \preceq P$.*

PROOF. (1) Let $A = P^\perp Q$. Then $[\ker A] = Q^\perp + (P \wedge Q)$ and similarly

$$[\ker A^*] = P + (P^\perp \wedge Q^\perp) = P + (P \vee Q)^\perp.$$

Hence $[A^* H] = [\ker A]^\perp = Q - (P \wedge Q)$ and

$$[AH] = [\ker A^*]^\perp = P^\perp (P \vee Q) = (P \vee Q) - P.$$

The equivalence now follows from Proposition 10.7.2(2).

(2) Suppose that P and Q are finite. Now $P \vee Q = P + ((P \vee Q) - P)$. By (1), we have $(P \vee Q) - P \sim Q - (P \wedge Q) \preceq Q$. By Proposition 10.7.7(1), $(P \vee Q) - P$ is finite, $((P \vee Q) - P) \perp P$ and $P + ((P \vee Q) - P) = P \vee Q$. Hence we may assume that $P \perp Q$. By restriction, we may assume that $P + Q = I$.

If $P + Q = I$ is infinite, there is a non-zero properly infinite central projection Z. We may restrict to $\mathcal{N} = Z\mathcal{M}$, which is properly infinite, but has finite projections $P_1 = ZP$ and $Q_1 = ZQ = P_1^\perp$. By Corollary 10.11.4, there is a projection $E \in \mathcal{N}$ so that $E \sim E^\perp \sim I$. Apply comparison to $E \wedge P_1$ and $E^\perp \wedge P_1^\perp$. There is a central projection Z_1 so that

$$Z_1 E \wedge Z_1 P_1 = Z_1(E \wedge P_1) \preceq Z_1(E^\perp \wedge P_1^\perp) = Z_1 E^\perp \wedge Z_1 P_1^\perp$$

and

$$Z_1^\perp E \wedge Z_1^\perp P_1 = Z_1^\perp (E \wedge P_1) \succeq Z_1^\perp (E^\perp \wedge P_1^\perp) = Z_1^\perp E^\perp \wedge Z_1^\perp P_1^\perp.$$

We compute first using (1) and the fact $(F \wedge G)^\perp = F^\perp \vee G^\perp$,

$$\begin{aligned} Z_1 E &= (Z_1 E \wedge Z_1 P_1) + (Z_1 E - (Z_1 E \wedge Z_1 P_1)) \\ &\sim (Z_1 E \wedge Z_1 P_1) + ((Z_1 E \vee Z_1 P_1) - Z_1 P_1) \end{aligned}$$

$$\begin{aligned} &\precsim (Z_1E^\perp \wedge Z_1P_1^\perp) + (Z_1E \vee Z_1P_1)Z_1P_1^\perp \\ &= (Z_1E^\perp \wedge Z_1P_1^\perp)Z_1P_1^\perp + (Z_1E^\perp \wedge Z_1P_1^\perp)^\perp Z_1P_1^\perp = Z_1P_1^\perp. \end{aligned}$$

But Z_1E is properly infinite and $Z_1P_1^\perp$ is finite, which is a contradiction unless $Z_1 = 0$. Similarly, $Z_1^\perp E^\perp \precsim Z_1^\perp P_1$; which leads to $Z_1^\perp = 0$. Hence $Z = 0$ and $P + Q$ is finite.

(3) It suffices to establish this in $\mathcal{M}_{P\vee Q}$, since an equivalence of $(P \vee Q) - P$ and $(P \vee Q) - Q$ yields the equivalence of $P^\perp$ and $Q^\perp$. By (2), $P \vee Q$ is finite; whence $\mathcal{M}_{P\vee Q}$ is finite. By the Comparison theorem applied to $P^\perp$ and $Q^\perp$, there is a central projection $Z \in \mathcal{Z}(\mathcal{M}_{P\vee Q})$ so that

$$ZP^\perp \sim E \le ZQ^\perp \quad \text{and} \quad Z^\perp Q^\perp \sim F \le Z^\perp P^\perp.$$

Thus

$$Z = ZP + ZP^\perp \sim ZQ + E \le ZQ + ZQ^\perp = Z.$$

Since Z is finite, $E = ZQ^\perp$. Similarly $F = Z^\perp P^\perp$. Hence

$$P^\perp = ZP^\perp + Z^\perp P^\perp \sim ZQ^\perp + Z^\perp Q^\perp = Q^\perp.$$

If $U^*U = P$, $UU^* = Q$, $V^*V = P^\perp$ and $VV^* = Q^\perp$, then $W = U + V$ is unitary and $WPW^* = UPU^* = Q$.

(4) As $P \precsim Q$, there is a projection $Q_1 \le Q$ with $P \sim Q_1$. Therefore by (3), $Q^\perp \le Q_1^\perp \sim P^\perp$.

(5) Let $E_1 = Q_1$ and $E_n = Q_n - Q_{n-1}$ for $n \ge 2$. Construct pairwise orthogonal projections $F_n \sim E_n$ with $F_n \le P$ as follows. Since $E_1 \prec P$, there is some $F_1 \sim E_1$ with $F_1 \le P$. Suppose that $F_1, \dots, F_n$ have been constructed, and let $P_n = \sum_{i=1}^n F_i$. Then $Q_n = \sum_{i=1}^n E_i \sim P_n$ and $Q_{n+1} = Q_n + E_{n+1} \precsim P$. Thus $Q_{n+1} \sim R \le P$ and this equivalence identifies $Q_n \sim R_n < R$ and $E_{n+1} \sim R - R_n \le P - R_n$ By (3), working in $P\mathcal{M}P$ and using $P_n \sim R_n$, we obtain $E_{n+1} \precsim P - P_n$. Thus there is a projection $F_{n+1} \sim E_{n+1}$ with $F_{n+1} \le P - P_n$. Recursively, we construct the projections (F_n). By Lemma 10.7.2(4), $Q \sim \sum_n F_n \le P$; whence $Q \precsim P$. ∎

10.12. The Centre-Valued Trace

In this section, we will establish the existence of traces on II_1 von Neumann algebras. We show that every finite von Neumann algebra $\mathcal{M}$ has a unique tracial expectation onto $\mathcal{Z}(\mathcal{M})$, and it is normal and faithful. We make use of the Eberlein–Smulian theorem 3.7.3 and the Ryll-Nardzewski fixed point theorem 3.8.13. This leads to a representation of type II_1 von Neumann algebras in which the commutant is also type II_1.

10.12.1. Definition. A *centre-valued state* on a finite von Neumann algebra $\mathcal{M}$ is an expectation $\Phi : \mathcal{M} \to \mathcal{Z}(\mathcal{M})$. It is *faithful* if $0 \lneq A$ implies $\Phi(A) \ne 0$. A *centre-valued trace* on a finite von Neumann algebra $\mathcal{M}$ is a centre-valued state Φ such that $\Phi(AB) = \Phi(BA)$ for $A, B \in \mathcal{M}$.

If $\mathcal{M}$ has a faithful centre-valued trace, then $\mathcal{M}$ must be finite. Indeed, otherwise $\mathcal{M}$ contains a proper isometry S. The trace property shows that

$$0 = \Phi(S^*S - SS^*) = \Phi(I - SS^*)$$

and since $I - S^*S \gneq 0$, Φ is not faithful.

We require a characterization of weakly compact subsets of $\mathcal{M}_*$.

10.12.2. LEMMA. *Let $\mathcal{M}$ be a von Neumann algebra with predual $\mathcal{M}_*$. A bounded subset $K \subset \mathcal{M}_*$ has weakly compact weak closure $\overline{K}^w$ if and only if*

($\blacklozenge$) *if (E_n) are pairwise orthogonal projections in $\mathcal{M}$, then $\omega(E_n) \to 0$ uniformly for $\omega \in K$.*

PROOF. Suppose that ($\blacklozenge$) holds. Let $K_1 = \overline{\iota(K)}^{w^*}$ be the weak-$*$ closure of K in $\mathcal{M}^\dagger$. This is $\tau_{\mathcal{M}}$-compact by the Banach–Alaoglu theorem. If we show that $K_1 \subset \iota(\mathcal{M}_*)$, then $\overline{K}^w = \iota^{-1}(K_1)$ is $\tau_{\mathcal{M}}$-compact in $\mathcal{M}_*$; i.e. weakly compact.

Let $(E_\lambda)_\Lambda$ be a set of pairwise orthogonal projections in $\mathcal{M}$, and let $E = \sum_\Lambda E_\lambda$. By normality of $\omega \in \mathcal{M}_*$, we have $\omega(E) = \lim_{J \Subset \Lambda} \sum_{\lambda \in J} \omega(E_\lambda)$ over the net of finite subsets of Λ ordered by inclusion. We will show that this convergence is uniform on K. If it is not uniform, then there are a $\delta > 0$, functionals $\omega_n \in K$ and pairwise disjoint finite sets $J_n \Subset \Lambda$ such that $\bigl|\sum_{\lambda \in J_n} \omega_n(E_\lambda)\bigr| \geq \delta$ for $n \geq 1$. Define $F_n = \sum_{\lambda \in J_n} E_\lambda$. This is a pairwise orthogonal sequence of projections in $\mathcal{M}$ such that $|\omega_n(F_n)| \geq \delta$. As this contradicts ($\blacklozenge$), we obtain uniform convergence of these sums over K. That is, for any $\delta > 0$, there is a finite subset $J_0 \Subset \Lambda$ so that if $J' \cap J_0 = \varnothing$, then $\bigl|\sum_{\lambda \in J'} \omega(E_\lambda)\bigr| < \delta$ for every $\omega \in K$. By normality, for any finite $J \supset J_0$, $\bigl|\omega\bigl(\sum_{\lambda \in \Lambda \setminus J} E_\lambda\bigr)\bigr| \leq \delta$. Therefore if $\varphi \in K_1$, we get $\bigl|\varphi\bigl(\sum_{\lambda \in \Lambda \setminus J} E_\lambda\bigr)\bigr| \leq \delta$. Hence

$$\Bigl|\varphi(E) - \sum_{\lambda \in J} \varphi(E_\lambda)\Bigr| \leq \delta \quad \text{for} \quad J \supset J_0,\ J \Subset \Lambda.$$

This shows that $\lim_{J \Subset \Lambda} \sum_{\lambda \in J} \varphi(E_\lambda) = \varphi(E)$. That is, φ is completely additive, and hence weak-$*$ continuous by Theorem 10.2.6. Therefore $K_1 \subset \iota(\mathcal{M}_*)$, and so $\overline{K}^w$ is weakly compact.

Conversely, suppose that $\overline{K}^w$ is weakly compact, but ($\blacklozenge$) fails. This means that there are $\delta > 0$, pairwise orthogonal projections (E_n) in $\mathcal{M}$ and $\omega_n \in K$ such that $|\omega_n(E_n)| \geq \delta$. By the Eberlein–Smulian theorem, we may drop to a weakly convergent subsequence, still denoted (ω_n), with weak limit ω. Let $\mathcal{N}$ be the abelian von Neumann algebra generated by $\{E_n : n \geq 1\}$, which is isomorphic to l_∞. Let $\varphi_n = \omega_n|_{\mathcal{N}}$ and $\varphi = \omega|_{\mathcal{N}}$. The restriction map from $\mathcal{M}_*$ to $\mathcal{N}_*$ is weakly continuous, so that $\varphi, \varphi_n \in \mathcal{N}_* \simeq l_1$, and the sequence (φ_n) converges weakly in

l_1 to φ. By the Schur property (Exercise **3.34**(b)), φ_n converges to φ in norm. By normality, $\varphi(E_n) \to 0$, so there is some N so that for $n \geq N$, $|\varphi(E_n)| < \delta/2$ and $\|\varphi - \varphi_n\| < \delta/2$. Thus $|\varphi_n(E_n)| < \delta$, a contradiction. Therefore (♦) holds. ■

For each unitary $U \in \mathcal{U}(\mathcal{M})$, define a linear isometry on $\mathcal{M}_*$ by $\mathrm{Ad}_U\,\omega(A) = (U\omega U^*)(A) = \omega(U^*AU)$. Observe that

$$\mathrm{Ad}_U\,\mathrm{Ad}_V\,\omega(A) = (\mathrm{Ad}_V\,\omega)(U^*AU) = \omega(V^*U^*AUV) = \mathrm{Ad}_{UV}\,\omega(A).$$

So $U \to \mathrm{Ad}\,U$ is a group homomorphism of $\mathcal{U}(\mathcal{M})$ into the group of surjective isometries of $\mathcal{M}_*$. For $\omega \in \mathcal{M}_*$, let $K_\omega = \{\mathrm{Ad}_U\,\omega : U \in \mathcal{U}(\mathcal{M})\}$ and let $L_\omega = \overline{\mathrm{conv}}(K_\omega)$.

10.12.3. Lemma. *Let $\mathcal{M}$ be a finite von Neumann algebra. For any $\omega \in \mathcal{M}_*$, the convex set L_ω is weakly compact.*

Proof. We first show that $\overline{K_\omega}^w$ is weakly compact. We apply Lemma 10.12.2. Suppose that (♦) is not satisfied. Then there are $\delta > 0$, a sequence of pairwise orthogonal projections (E_n) in $\mathcal{M}$ and elements $\omega_n = \mathrm{Ad}_{U_n}\,\omega \in K$ such that $|\omega_n(E_n)| \geq \delta$. Let $F_n = U_n^* E_n U_n$; so $F_n \sim E_n$ and $|\omega(F_n)| \geq \delta$. Define

$$P_{n,m} = \bigvee_{i=n}^{m} F_i, \quad P_n = \bigvee_{i\geq n} F_i = \bigvee_{m\geq n} P_{n,m} \quad \text{and} \quad P = \bigwedge_{n\geq 1} P_n.$$

Note that $P_{n,m} \leq P_{n,m+1}$ for $m \geq n$. Define projections $Q_{n,n} := P_{n,n}$ and $Q_{n,m} := P_{n,m} - P_{n,m-1}$ for $m > n$. Then $Q_{n,n} = F_n \sim E_n$, and by Kaplansky's identity, Proposition 10.11.12(1), for $m > n$,

$$Q_{n,m} = (P_{n,m-1} \vee F_m) - P_{n,m-1} \sim F_m - (P_{n,m-1} \wedge F_m) \preceq F_m \sim E_m.$$

Since $\{Q_{n,m} : m \geq n\}$ and $\{E_m : m \geq n\}$ are pairwise orthogonal families of projections, $P_n = \sum_{m\geq n} Q_{n,m} \preceq \sum_{m\geq n} E_m$. By Proposition 10.11.12(4), $I - \sum_{m\geq n} E_m \preceq P_n^\perp \leq P^\perp$. The left-hand side is an increasing sequence converging strongly to I, so by Proposition 10.11.12(5), $I \preceq I - P$. Since $\mathcal{M}$ is finite, this forces $P = 0$. Therefore P_n decreases strongly to 0. Since $F_n \leq P_n$, (F_n) converges strongly to 0. As ω is normal, we have $\omega(F_n) \to 0$, contrary to our earlier assumption that (♦) failed. Thus by Lemma 10.12.2, $\overline{K_\omega}^w$ is weakly compact. Hence L_ω is weakly compact by Theorem 3.7.5 of Krein and Smulian. ■

10.12.4. Theorem. *Let $\mathcal{M}$ be a finite von Neumann algebra with centre $\mathcal{Z}$. For each $\varphi \in \mathcal{Z}_*$, there is a unique $\widetilde{\varphi} \in \mathcal{M}_*$ such that $\widetilde{\varphi}|_{\mathcal{Z}} = \varphi$ and $\widetilde{\varphi}(U^*AU) = \widetilde{\varphi}(A)$ for all $A \in \mathcal{M}$ and $U \in \mathcal{U}(\mathcal{M})$. Moreover the map $S\varphi = \widetilde{\varphi}$ is a positive linear isometry; i.e. $\|\widetilde{\varphi}\| = \|\varphi\|$ and if $\varphi \geq 0$, then $\widetilde{\varphi} \geq 0$.*

PROOF. Let $\varphi \in \mathcal{Z}_*$. Since $\mathcal{Z} \subset \mathcal{M} \subset \mathcal{B}(H)$ with the induced weak-$*$ topology, we know that $\mathcal{Z}_* \simeq \mathfrak{S}_1(H)/\mathcal{Z}_\perp$. In particular, φ extends to a weak-$*$ continuous functional $\omega \in \mathcal{M}_*$. By Lemma 10.12.3, the set L_ω is weakly compact. The group $\{\mathrm{Ad}_U : U \in \mathcal{U}(\mathcal{M})\}$ consists of linear isometries which map K_ω into itself, and thus by weak and norm continuity, they map L_ω into itself. By Corollary 3.8.14 of the Ryll-Nardzewski fixed point theorem, there is an element $\widetilde{\varphi} \in L_\omega$ such that $\mathrm{Ad}_U\, \widetilde{\varphi} = \widetilde{\varphi}$ for all $U \in \mathcal{U}(\mathcal{M})$. Hence $\widetilde{\varphi} \in \mathcal{M}_*$ and $\widetilde{\varphi}(U^*AU) = \widetilde{\varphi}(A)$ for all $A \in \mathcal{M}$ and $U \in \mathcal{U}(\mathcal{M})$. If $Z \in \mathcal{Z}$ and $U \in \mathcal{U}(\mathcal{M})$,

$$\mathrm{Ad}_U\, \omega(Z) = \omega(U^*ZU) = \omega(Z) = \varphi(Z).$$

Hence $\psi(Z) = \varphi(Z)$ for all $\psi \in L_\omega$; whence $\widetilde{\varphi}|_{\mathcal{Z}} = \varphi$.

Now let $\psi \in \mathcal{M}_*$ such that $\psi|_{\mathcal{Z}} = \varphi$ and $\mathrm{Ad}_U\, \psi = \psi$ for all $U \in \mathcal{U}(\mathcal{M})$. By the polar decomposition for normal functionals, Theorem 10.2.10, we write $\psi = V|\psi|$. Then $\psi = \mathrm{Ad}_U(V|\psi|) = (UVU^*)(\mathrm{Ad}_U\, |\psi|)$. By the uniqueness of the polar decomposition, $UVU^* = V$ for all $U \in \mathcal{U}(\mathcal{M})$; whence $V \in \mathcal{M}' \cap \mathcal{M} = \mathcal{Z}$. Then $\|\varphi\| \le \|\psi\| = \psi(V^*) = \varphi(V^*) \le \|\varphi\|$; so we have equality. If $\varphi \ge 0$, then $\psi(I) = \varphi(I) = \|\varphi\| = \|\psi\|$, so ψ is positive by Lemma 9.6.5.

Now if $\varphi = 0$, we obtain $\psi = 0$. Thus in general, if $\widetilde{\varphi}, \psi$ are two invariant extensions of φ, then $\widetilde{\varphi} - \psi$ is an invariant extension of 0, and thus $\psi = \widetilde{\varphi}$; i.e. this extension is unique. Therefore the map $S\varphi = \widetilde{\varphi}$ is well defined, positive and isometric. Note that if $\varphi_i \in \mathcal{Z}_*$, then $\alpha_1\widetilde{\varphi}_1 + \alpha_2\widetilde{\varphi}_2$ is an invariant extension of $\alpha_1\varphi_1 + \alpha_2\varphi_2$. By uniqueness, S is linear. ■

We now have the necessary ingredients to establish the existence and uniqueness of a centre-valued trace.

10.12.5. THEOREM. *Let $\mathcal{M}$ be a finite von Neumann algebra. Then $\mathcal{M}$ has a faithful normal centre-valued trace. This is the unique centre-valued trace.*

PROOF. Let $S : \mathcal{Z}_* \to \mathcal{M}_*$ be the positive linear isometry provided by Theorem 10.12.4. Define $\Phi = S^* : \mathcal{M} \to \mathcal{Z}$. Then Φ is weak-$*$ to weak-$*$ continuous, and thus is normal; and $\|\Phi\| = 1$. If $Z \in \mathcal{Z}$ and $\varphi \in \mathcal{Z}_*$, then $\Phi(Z)(\varphi) = Z(S\varphi) = Z(\varphi)$; whence $\Phi(Z) = Z$ and so $\Phi^2 = \Phi$. By Tomiyama's theorem, the expectation Φ is a positive $\mathcal{Z}$-bimodule map.

If $A \in \mathcal{M}, U \in \mathcal{U}(\mathcal{M})$ and $\varphi \in \mathcal{Z}_*$,

$$\Phi(U^*AU)(\varphi) = (S\varphi)(U^*AU) = (S\varphi)(A) = \Phi(A)(\varphi).$$

Therefore $\Phi(U^*AU) = \Phi(A)$. Setting $B = U^*A$, we have $\Phi(BU) = \Phi(UB)$ for any $B \in \mathcal{M}$ and $U \in \mathcal{U}(\mathcal{M})$. Therefore $\Phi(BC) = \Phi(CB)$ for all C in the weak-$*$ closed linear span of $\mathcal{U}(\mathcal{M})$. However the span of $\mathcal{U}(\mathcal{M})$ is all of $\mathcal{M}$ (see the Russo–Dye theorem 9.4.7). Therefore Φ is a centre-valued trace.

Consider the set $\mathcal{J} = \{A \in \mathcal{M} : \Phi(A^*A) = 0\}$. By weak-$*$ continuity, $\mathcal{J}$ is weak-$*$ closed. If $A, B \in \mathcal{J}$, then $0 \le (A+B)^*(A+B) \le 2(A^*A + B^*B)$. So

$\mathcal{J}$ is a subspace. For $B \in \mathcal{M}$ and $A \in \mathcal{J}$,

$$0 \le \Phi(A^*B^*BA) \le \|B\|^2\Phi(A^*A) = 0.$$

Thus $\mathcal{M}\mathcal{J} \subset \mathcal{J}$. The trace property shows that $\mathcal{J} = \{A \in \mathcal{M} : \Phi(AA^*) = 0\}$. Hence $\mathcal{J}\mathcal{M} \subset \mathcal{J}$. Therefore $\mathcal{J}$ is a weak-$*$ closed 2-sided ideal. So we obtain a central projection $P \in \mathcal{Z}$ so that $\mathcal{J} = P\mathcal{M}$. However $0 = \Phi(P) = P$; whence $P = 0$. Therefore Φ is faithful.

Now let Ψ be a centre-valued trace which is not assumed to be normal or faithful. Suppose that P is a monic projection, say $P = P_1 \sim \cdots \sim P_n$ and $\sum_{i=1}^n P_i = c(P)$. If $P \sim Q$, there is a partial isometry U so that $Q = UPU^*$. Then $\Psi(P) = \Psi(U^*UP) = \Psi(UPU^*) = \Psi(Q)$. Thus

$$c(P) = \Psi(c(P)) = \sum_{i=1}^{n} \Psi(P_i) = n\Psi(P).$$

The same argument holds for Φ, and thus $\Psi(P) = \Phi(P)$ for every monic projection.

By Lemma 10.11.10, every projection in $\mathcal{M}$ is the sum of a sequence monic projections; say $P = \sum_{n\ge1} P_n$. By Tomiyama's theorem, Ψ is a positive map. Therefore

$$\Psi(P) \ge \lim_{n\to\infty} \sum_{i=1}^{n} \Psi(P_i) = \lim_{n\to\infty} \sum_{i=1}^{n} \Phi(P_i) = \Phi(P).$$

Similarly, $\Psi(P^\perp) \ge \Phi(P^\perp)$. But $\Psi(P) + \Psi(P^\perp) = \Psi(I) = I = \Phi(P) + \Phi(P^\perp)$. Therefore $\Psi(P) = \Phi(P)$ for all projections, and hence they agree on the norm-closed linear span of the projections. By Corollary 9.3.7, this is all of $\mathcal{M}$. Therefore Φ is the unique centre-valued trace. ∎

10.12.6. COROLLARY. *Let $\mathcal{M}$ be a finite von Neumann algebra with centre-valued trace Φ. If P and $Q \in \mathcal{M}$ are projections, then $P \preceq Q$ if and only if $\Phi(P) \le \Phi(Q)$. So $P \sim Q$ if and only if $\Phi(P) = \Phi(Q)$.*

PROOF. Let P, Q be projections in $\mathcal{M}$ such that $P \preceq Q$. Then $P \sim Q_1 \le Q$. Thus $\Phi(P) = \Phi(Q_1) \le \Phi(Q)$. Since Φ is faithful, this is an equality only if $Q_1 = Q$, so that $P \sim Q$.

Conversely, suppose that $\Phi(P) \le \Phi(Q)$. By the Comparison theorem 10.7.3, there is a central projection Z so that $ZP \sim ZQ_0 \le ZQ$ and $Z^\perp Q \sim Z^\perp P_0 \le Z^\perp P$. Then

$$Z^\perp\Phi(Q) = Z^\perp\Phi(P_0) \le Z^\perp\Phi(P) \le Z^\perp\Phi(Q).$$

Since this is an equality and Φ is faithful, $Z^\perp Q \sim Z^\perp P_0 = Z^\perp P$. Therefore $P = ZP + Z^\perp P \sim ZQ_0 + Z^\perp Q \le Q$, so $P \preceq Q$. ∎

10.12.7. COROLLARY. *If $\mathcal{M}$ is a II_1 factor, then it has a unique trace τ and it is normal. The range of τ on the projections is $[0, 1]$.*

PROOF. In the case of a factor, $\mathcal{Z}(\mathcal{M}) = \mathbb{C}I$. So the unique centre-valued trace Φ is normal, faithful and takes values in $\mathbb{C}I$. Let τ be the corresponding scalar-valued trace. Conversely, any scalar-valued trace induces a centre-valued trace and hence equals τ.

By Corollary 10.11.8, there is a monic projection P of order 2^n, say with pairwise orthogonal equivalent projections $P \sim P_i$ for $1 \le i \le 2^n$. Then $\tau\big(\sum_{i=1}^{k} P_i\big) = \frac{k}{2^n}$. So the range of τ on $\mathcal{P}(\mathcal{M})$ is dense in $[0,1]$. We can construct pairwise orthogonal projections P_i such that P_i is monic of order 2^i. Given a real number $t = \sum_{i\ge1} \varepsilon_i 2^{-i}$ where $\varepsilon_i \in \{0,1\}$, form $P = \sum_{i\ge1} \varepsilon_i P_i$. Then

$$\tau(P) = \sum_{i\ge1} \varepsilon_i \tau(P_i) = \sum_{i\ge1} \varepsilon_i 2^{-i} = t.$$

Hence the range of τ on the set of projections is $[0,1]$. ■

10.12.8. REMARK. If $\mathcal{M}$ is a finite von Neumann algebra with non-trivial centre, then there is more than one trace on $\mathcal{M}$. Indeed, if ψ is any normal state on $\mathcal{Z}(\mathcal{M})$, then $\psi\Phi$ is a normal trace on $\mathcal{M}$. If τ is any normal trace on a finite von Neumann algebra $\mathcal{M}$, one can show as in the proof of Theorem 10.12.5 that $\mathcal{J} = \{A \in \mathcal{M} : \tau(A^*A) = 0\}$ is a weak-$*$ closed 2-sided ideal, and thus has the form $P\mathcal{M}$ for a projection $P \in \mathcal{Z}$. Therefore $s(\tau) = P^\perp$ is a central projection.

10.12.9. DEFINITION. A collection $\{\tau_\lambda\}_\Lambda$ of normal traces on $\mathcal{M}$ is a *faithful family of traces* if for every $0 \lneq A \in \mathcal{M}$, there is some λ so that $\tau_\lambda(A) > 0$.

10.12.10. THEOREM. *Every finite von Neumann algebra $\mathcal{M}$ has a faithful family of normal traces. Moreover, there is a family $\{\tau_\lambda\}_\Lambda$ such that the support projections Z_{τ_λ} are pairwise orthogonal and $\sum_\Lambda Z_{\tau_\lambda} = I$.*

If $\mathcal{M}$ acts on a separable Hilbert space, then it has a faithful normal trace.

PROOF. By Remark 10.12.8, every finite von Neumann algebra has normal traces. Let $\{\tau_\lambda\}_\Lambda$ be a maximal family of normal traces such that the support projections $Z_{\tau_\lambda} = s(\tau_\lambda)$ are pairwise orthogonal. Let $Z = \big(\sum_\Lambda Z_{\tau_\lambda}\big)^\perp$. If $Z \ne 0$, then for any normal state ψ on $Z\mathcal{Z}$, $\tau = \psi\Phi$ is another normal trace with support $Z_\tau \le Z$. This contradicts the maximality of the original family, and thus $\sum_\Lambda Z_{\tau_\lambda} = I$. Each τ_λ is faithful on $Z_\lambda\mathcal{M}$ by Proposition 10.2.9. Thus $\{\tau_\lambda\}_\Lambda$ is a faithful family of normal traces.

If $\mathcal{M}$ acts on a separable Hilbert space, then $\mathcal{Z}(\mathcal{M})$ is countably decomposable, and hence the set Λ is countable. A convex combination τ of $\{\tau_\lambda\}_\Lambda$ with all non-zero coefficients is a faithful normal trace. ■

10.12.11. THEOREM. *Let $\mathcal{M}$ be a type* II_1 *von Neumann algebra. Then $\mathcal{M}$ has a faithful normal representation π such that $\pi(\mathcal{M})'$ is anti-isomorphic to $\mathcal{M}$ and has type* II_1.

PROOF. Let $\{\tau_\lambda\}_\Lambda$ be a faithful family of traces with pairwise orthogonal supports provided by Theorem 10.12.10. Let π_{τ_λ} be the normal representation of $\mathcal{M}$ on $L^2(\mathcal{M}, \tau_\lambda)$. Clearly π_{τ_λ} factors through $Z_{\tau_\lambda}\mathcal{M}$, and it is faithful on $Z_{\tau_\lambda}\mathcal{M}$ by Proposition 10.10.6. Hence $\pi = \sum_\Lambda^\oplus \pi_{\tau_\lambda}$ is a faithful normal representation of $\mathcal{M}$.

Note that $\pi(\mathcal{M}) = \prod_\Lambda \pi_{\tau_\lambda}(Z_{\tau_\lambda}\mathcal{M})$. Since the $\{Z_{\tau_\lambda}\}$ are pairwise orthogonal, the representations $\{\pi_{\tau_\lambda}\}_\Lambda$ are disjoint. Therefore $\pi(\mathcal{M})' = \prod_\Lambda \pi_{\tau_\lambda}(Z_{\tau_\lambda}\mathcal{M})'$. By Corollary 10.10.10, each $\pi_{\tau_\lambda}(Z_{\tau_\lambda}\mathcal{M})'$ is anti-isomorphic to $\pi_{\tau_\lambda}(Z_{\tau_\lambda}\mathcal{M})$, and thus is type II_1. Therefore $\pi(\mathcal{M})'$ is anti-isomorphic to $\mathcal{M}$, and is type II_1. ∎

10.13. Infinite von Neumann Algebras

In this section, we obtain structural results about type II_∞ algebras. We then show that the type I, II or III is preserved by commutants. Also we show that every faithful normal representation of a type III algebra is a unitary equivalence.

We first make use of the projection lemmas to get some easy consequences.

10.13.1. PROPOSITION. *Let $\mathcal{M}$ be a properly infinite von Neumann algebra. Then $\mathcal{M}$ is unitarily equivalent to $\mathcal{M} \overline{\otimes} \mathcal{B}(H)$ for H a separable infinite dimensional Hilbert space and to $\mathcal{M} \overline{\otimes} M_n$ for $n \geq 2$.*

PROOF. By Lemma 10.11.2. there is a family $\{E_i : i \geq 1\}$ of equivalent pairwise orthogonal projections such that $\sum_{i\geq 1} E_i = I$ and $E_i \sim I$. Hence there are partial isometries $E_{i1} = U_i E_1$ so that $E_{i1}E_{i1}^* = E_i$ and $E_{i1}^* E_{i1} = E_1$. Then it is easy to check that $E_{ij} = E_{i1}E_{j1}^*$ is a set of matrix units for a separable Hilbert space H. By Proposition 10.8.8, $\mathcal{M}$ is unitarily equivalent to $\mathcal{M}_{E_1} \overline{\otimes} \mathcal{B}(H)$. However, since $E_1 \sim I$, there is an isometry U so that $UU^* = E_1$. Thus U implements a unitary equivalence of $\mathcal{M}$ onto $\mathcal{M}_{E_1}$. Therefore $\mathcal{M}$ is unitarily equivalent to $\mathcal{M} \overline{\otimes} \mathcal{B}(H)$.

A similar argument works for $\mathcal{M} \overline{\otimes} M_n$ using Corollary 10.11.4. ∎

10.13.2. THEOREM. *Let $\mathcal{M}$ be a von Neumann algebra, and let K be any Hilbert space. Then $\mathcal{M}$ has type* I, II *or* III *if and only if $\mathcal{M} \overline{\otimes} \mathcal{B}(K)$ has the same type.*

PROOF. If $\mathcal{M}$ is type I, then it has an abelian projection E with $c(E) = I$. Then $E \otimes E_{11}$ is an abelian projection in $\mathcal{M} \overline{\otimes} \mathcal{B}(K)$ and $c(E \otimes E_{11}) = I$; so $\mathcal{M} \overline{\otimes} \mathcal{B}(K)$ is type I. Conversely, if $\mathcal{M} \overline{\otimes} \mathcal{B}(K)$ is type I and P is a projection in $\mathcal{M}$, then $P \otimes E_{11}$ is a projection in $\mathcal{M} \overline{\otimes} \mathcal{B}(K)$. Hence it dominates an abelian projection, which must have the form $E \otimes E_{11}$. Therefore E is an abelian projection in $\mathcal{M}$ with $E \leq P$; so that $\mathcal{M}$ is type I.

Suppose that $\mathcal{M} \otimes \mathcal{B}(K)$ is type III, and let $P \neq 0$ be a projection in $\mathcal{M}$. Then $P \otimes E_{11}$ is a projection in $\mathcal{M} \otimes \mathcal{B}(K)$. Thus there is a projection $Q \lneq P$ such that $Q \sim P \otimes E_{11}$, say via a partial isometry U. Then Q, U^*U and UU^* are less than $I \otimes E_{11}$, and thus there is a projection $E \in \mathcal{M}$ and a partial isometry $V \in \mathcal{M}$ so that $Q = E \otimes E_{11}$ and $U = V \otimes E_{11}$. Thus $E \lneq P$ and $E \sim P$, so that P is infinite. Hence $\mathcal{M}$ is type III.

Conversely, suppose that $\mathcal{M}$ is type III. Then by Proposition 10.13.1, $\mathcal{M} \simeq \mathcal{M} \overline{\otimes} \mathcal{B}(H)$ for a separable infinite dimensional Hilbert space H. If $\dim K \leq \aleph_0$, then $\mathcal{M} \overline{\otimes} \mathcal{B}(K) \simeq \mathcal{M}$ is type III. Suppose that P is a projection in $\mathcal{M} \overline{\otimes} \mathcal{B}(K)$, where $\dim K$ is some uncountable cardinal. Let $\xi = P\xi$ be a unit vector. Choosing an orthonormal basis $\{\eta_i\}$ for K, we can write $\xi = \sum \xi_i \otimes \eta_i$. Only countably many of these terms are non-zero, and thus there is a separable subspace $K_0 \subset K$ so that $\xi \in H \otimes K_0$. Let R_0 be the projection of K onto K_0. The projection $Q = P \wedge (I \otimes R_0)$ is a non-zero projection in $\mathcal{M} \overline{\otimes} \mathcal{B}(K_0)$, which is a corner of $\mathcal{M} \otimes \mathcal{B}(K)$. This corner is isomorphic to $\mathcal{M}$, and thus there is a projection $Q_1 \lneq Q$ such that $Q_1 \sim Q$. Hence Q and thus P is infinite. Therefore $\mathcal{M} \otimes \mathcal{B}(K)$ is type III.

Finally suppose that $\mathcal{M}$ is type II. Then $\mathcal{M} \otimes \mathcal{B}(K)$ cannot have a central summand of type I or III, so it must be type II. The converse is the same. ■

The following provides an easy way to obtain II_∞ von Neumann algebras from II_1 algebras.

10.13.3. COROLLARY. *Let $\mathcal{M}$ be a II_1 von Neumann algebra, and let H be a separable infinite dimensional Hilbert space. Then $\mathcal{M} \overline{\otimes} \mathcal{B}(H)$ is type II_∞.*

PROOF. Let $\mathcal{N} = \mathcal{M} \overline{\otimes} \mathcal{B}(H)$. This is type II. Therefore $\mathcal{N}$ is semifinite but properly infinite. Thus it is type II_∞. ■

10.13.4. EXAMPLE. Let $\mathcal{R}$ be the hyperfinite II_1 factor of Theorem 10.10.3. Then $\mathcal{R}_\infty := \mathcal{R} \overline{\otimes} \mathcal{B}(H)$ is a II_∞ factor since

$$\mathcal{Z}(\mathcal{R}_\infty) = \mathcal{R}_\infty \cap \mathcal{R}'_\infty = (\mathcal{R} \overline{\otimes} \mathcal{B}(H)) \cap (\mathcal{R}' \overline{\otimes} \mathbb{C} I_H) = \mathcal{Z}(\mathcal{R}) \overline{\otimes} \mathbb{C} I_H = \mathbb{C} I_{H \otimes H}.$$

Also it is hyperfinite. Indeed suppose that $\mathfrak{A}_n$ is a sequence of finite dimensional C*-algebras with union weak-$*$ dense in $\mathcal{R}$. Write $H = \overline{\bigcup_{n \geq 1} H_n}$ where $H_n \subset H_{n+1}$ is n-dimensional. Then $\mathfrak{A}_n \otimes \mathcal{B}(H_n)$ is finite dimensional and $\overline{\bigcup_{n \geq 1} \mathfrak{A}_n \otimes \mathcal{B}(H_n)}$ is weak-$*$ dense in $\mathcal{R}_\infty$.

10.13.5. EXAMPLE. Let $n \geq 2$. Then $\mathcal{L}(\mathbb{F}_n)$ is a II_1 factor which is not injective by Corollary 10.10.21. Thus $\mathcal{M} = \mathcal{L}(\mathbb{F}_n) \overline{\otimes} \mathcal{B}(H)$ is a II_∞ factor. Note that $\mathcal{M}$ is not injective, for if $\mathcal{E}$ were a norm 1 projection of $\mathcal{B}(l_2(\mathbb{F}_n) \otimes H)$ onto $\mathcal{M}$, then define $\mathcal{E}_1 : \mathcal{B}(l_2(\mathbb{F}_n)) \to \mathcal{L}(\mathbb{F}_n)$ by $\mathcal{E}_1 = P\mathcal{E}J$ where

$$J(A) = A \otimes E_{11} \quad \text{and} \quad P(X) = (I \otimes E_{11})X(I \otimes E_{11})|_{l_2(\mathbb{F}_n)\otimes\mathbb{C}\xi_1}.$$

Then $\mathcal{E}_1$ would show that $\mathcal{L}(\mathbb{F}_n)$ was injective, contrary to fact. It follows that $\mathcal{M}$ is not isomorphic to $\mathcal{R}_\infty$ of the previous example.

The next result shows that this method actually produces all type II_∞ algebras.

10.13.6. Theorem. *Let $\mathcal{M}$ be a II_∞ von Neumann algebra. Then there is a II_1 von Neumann algebra $\mathcal{N}$ and central projections $\{Z_\alpha\}$ indexed by infinite cardinals with $\sum Z_\alpha = I$ such that $\mathcal{M} \simeq \prod_\alpha Z_\alpha \mathcal{N} \overline{\otimes} \mathcal{B}(H_\alpha)$, where $\dim H_\alpha = \alpha$. In particular, if $\mathcal{M}$ acts on a separable Hilbert space, then $\mathcal{M} \simeq \mathcal{N} \overline{\otimes} \mathcal{B}(H)$ where H is separable.*

Proof. There is a finite projection $0 \neq F \leq Z$ for any central projection Z of $\mathcal{M}$. By Zorn's lemma, there is a maximal family $\{F_i\}_\mathcal{I}$ of finite projections such that $\{c(F_i)\}_\mathcal{I}$ are pairwise orthogonal. It must be the case that $F = \sum_\mathcal{I} F_i$ has $c(F) = I$, for otherwise we find another finite projection dominated by $c(F)^\perp$. By Proposition 10.7.7(4), F is finite. Therefore $\mathcal{N} = F\mathcal{M}F$ is finite; so it is type II_1.

Let $\{F_\lambda\}_\Lambda$ be a maximal family of pairwise orthogonal projections equivalent to F. This collection is infinite because the sum of finitely many finite projections is finite by Proposition 10.11.12(2). Let $P = \sum_\Lambda F_\lambda$. By the Comparison theorem applied to $P^\perp$ and F, there is central projection Z such that $ZP^\perp \sim F_0 \leq ZF$ and $Z^\perp F \prec Z^\perp P^\perp$. The maximality of this family shows that $Z \neq 0$. Now let $\{F_i : i \geq 1\}$ be a countable subset of $\{F_\lambda\}_\Lambda$ Then each ZF_i can be decomposed as $ZF_i = F_{i,0} + F_{i,1}$ where $F_{i0} \sim F_0$. Let $G_1 = ZP^\perp + F_{1,1}$, $G_i = F_{i-1,0} + F_{i,1}$ for $i \geq 2$, and $G_\lambda = ZF_\lambda$ otherwise. Then $G_\lambda \sim ZF$ and $\sum_\Lambda G_\lambda = Z$.

Let $\{Z_\mu\}_{\Lambda'}$ be a maximal family of pairwise orthogonal central projections such that there are infinite families $\{G_{\mu,\lambda} : \lambda \in \Lambda_\mu\}$ with $G_{\mu,\lambda} \sim Z_\mu F$ and $\sum_{\Lambda_\mu} G_{\mu,\lambda} = Z_\mu$. The argument above applied to $\big(\sum_{\Lambda'} Z_\mu\big)^\perp \mathcal{M}$ would provide another central projection with the same properties unless $\sum_{\Lambda'} Z_\mu = I$. Hence this is true. By Corollary 10.8.9, $Z_\mu\mathcal{M} \simeq Z_\mu\mathcal{N} \overline{\otimes} \mathcal{B}(H_\mu)$ where $\dim H_\mu = |\Lambda_\mu|$. It is a simple matter to collect all sets of the same cardinality into one term. In particular, if $\mathcal{M}$ acts on a separable Hilbert space, then each set is countably infinite, and $\mathcal{M} \simeq \mathcal{N} \overline{\otimes} \mathcal{B}(H)$ where H is separable. ∎

10.13.7. Lemma. *Let $\mathcal{M}$ be a von Neumann algebra such that $\mathcal{M}'$ is type* I, II *or* III. *Let π be a faithful normal representation of $\mathcal{M}$ on a Hilbert space H. Then $\pi(\mathcal{M})'$ is the same type as $\mathcal{M}'$.*

PROOF. This is an easy application of Theorem 10.3.7 on the structure of isomorphisms. Factor $\pi = \operatorname{Ad} U \circ \pi_E \circ \varphi_\alpha$ where φ_α is an ampliation, π_E is a compression by a projection in $\varphi_\alpha(\mathcal{M})'$ with $c(E) = I$ and $\operatorname{Ad} U$ is a unitary equivalence. Now $\varphi_\alpha(\mathcal{M})' = \pi(\mathcal{M})' \overline{\otimes} \mathcal{B}(H_\alpha)$ by Proposition 10.8.6. This does not change the type by Theorem 10.13.2.

By Lemma 10.3.6, $\pi_E \circ \varphi_\alpha(\mathcal{M})' = E\pi(\mathcal{M})' \overline{\otimes} \mathcal{B}(H_\alpha)E$. If $\pi(\mathcal{M})'$ is type I, then every projection $P \in E\pi(\mathcal{M})' \overline{\otimes} \mathcal{B}(H_\alpha)E$ is a projection in $\pi(\mathcal{M})' \overline{\otimes} \mathcal{B}(H_\alpha)$ with $P \le E$. Hence it dominates a non-zero abelian projection. Thus it is type I. Similarly if $\pi(\mathcal{M})'$ is type II, every such P dominates a finite projection and thus is semifinite. It contains no abelian projections, and hence has type II. Lastly if $\pi(\mathcal{M})'$ is type III, then $\pi(\mathcal{M})' \overline{\otimes} \mathcal{B}(H_\alpha)$ has no non-zero finite projections, and thus neither does $\pi_E \circ \varphi_\alpha(\mathcal{M})'$. So it is type III.

Finally the type is clearly preserved by unitary equivalence. ∎

We can now show that type is preserved by the commutant.

10.13.8. THEOREM. *A von Neumann algebra $\mathcal{M}$ is type* I, II *or* III *if and only if $\mathcal{M}'$ is the same type.*

PROOF. By Theorem 10.9.2, $\mathcal{M}$ is type I if and only if $\mathcal{M}'$ is type I.

Suppose that $\mathcal{M}$ is type II_1. By Theorem 10.12.11, $\mathcal{M}$ has a faithful representation π such that $\pi(\mathcal{M})'$ is type II_1. By Lemma 10.13.7, $\mathcal{M}'$ is type II. If $\mathcal{M}$ is type II_∞, then by Theorem 10.13.6, there is a II_1 algebra $\mathcal{N}$ so that $\mathcal{M} \simeq \prod_\alpha Z_\alpha \mathcal{N} \overline{\otimes} \mathcal{B}(H_\alpha)$. Then

$$\mathcal{M}' \simeq \prod_\alpha Z_\alpha \mathcal{N}' \overline{\otimes} \mathbb{C} I_{H_\alpha}.$$

Since $\mathcal{N}'$ is type II, so is $\mathcal{M}'$.

Finally if $\mathcal{M}'$ is type III, then $\mathcal{M}'$ must be type III because any central summand of other type would yield a central summand of $\mathcal{M}$ of the same type. ∎

10.13.9. THEOREM. *Let $\mathcal{M}$ be a type* III *von Neumann algebra on a separable Hilbert space. Suppose that π is a normal representation on a separable Hilbert space. Then $\pi(\mathcal{M})$ is unitarily equivalent to $\mathcal{M}$.*

PROOF. By Corollary 10.11.6, $\mathcal{M}$ is simple and hence π is faithful. By Theorem 10.3.7 again, factor $\pi = \operatorname{Ad} U \circ \pi_E \circ \varphi_\alpha$. Since $\mathcal{M}$ and $\pi(\mathcal{M})$ are separably acting, we can take $\alpha \le \aleph_0$. By Proposition 10.13.1, $\mathcal{M}$ is unitarily equivalent to $\varphi_\alpha(\mathcal{M})$. Since $c(E) = I$, Corollary 10.11.4 shows that $E \sim I$, say U is an isometry with $UU^* = E$. Then $\pi_E^{-1}(A) = U^*AU$ is a unitary equivalence. Thus $\pi(\mathcal{M})$ is unitarily equivalent to $\mathcal{M}$. ∎

There is a useful characterization of semifinite von Neumann algebras in terms of traces on the positive part. This provides a method of recognizing type III algebras.

10.13.10. Definition. A *tracial weight* on a von Neumann algebra $\mathcal{M}$ is a map on the positive cone into the extended positive reals, $\tau : \mathcal{M}_+ \to [0, \infty]$, such that

(1) $\tau(A + B) = \tau(A) + \tau(B)$ for all $A, B \in \mathcal{M}_+$.

(2) $\tau(tA) = t\tau(A)$ for $t \geq 0$ and $A \in \mathcal{M}_+$. (convention: $0 \cdot \infty = 0$.)

(3) $\tau(A^*A) = \tau(AA^*)$ for all $A \in \mathcal{M}$.

Say that τ is a *faithful* if $A \gneq 0$ implies that $\tau(A) > 0$; and τ is a *finite* if $\tau(I) < \infty$; and *semifinite* if $A \gneq 0$ implies that there is $0 \leq B \leq A$ such that $0 < \tau(B) < \infty$; and *normal* if for every increasing net $\{A_\lambda\}_\Lambda$ in $\mathcal{M}_+$, $\tau(\sup_\Lambda A_\lambda) = \sup_\Lambda \tau(A_\lambda)$.

In particular, if $P \sim Q$ are equivalent projections, then $\tau(P) = \tau(Q)$. A map with properties (1) and (2) is called a weight, and these exist on all von Neumann algebras including type III. They play an important role in more advanced topics.

10.13.11. Example. $\mathcal{B}(H)$ has a tracial weight that extends the trace Tr on $\mathfrak{S}_{1+}$ by setting $\operatorname{Tr}(A) = \infty$ if $A \geq 0$ is not trace class. It is straightforward to check that this tracial weight is semifinite, normal, and faithful.

10.13.12. Proposition. *A von Neumann algebra $\mathcal{M}$ is semifinite if and only if it has a faithful, normal, semifinite tracial weight.*

Proof. If $\mathcal{M}$ has a central summand $Z\mathcal{M}$ of type III, then for every non-zero projection $P \in Z\mathcal{M}$, there is a projection $Q < P$ such that $P \sim Q \sim P - Q$ by Corollary 10.11.4. Then $\tau(P) = \tau(Q) + \tau(P - Q) = 2\tau(P)$. Thus $\tau(P) \in \{0, \infty\}$. Since τ is faithful, $\tau(P) = \infty$. If $A \gneq 0$, there is a non-zero projection P and $t > 0$ so that $A \geq tP$, and hence $\tau(A) = \infty$. Therefore τ is not semifinite on $Z\mathcal{M}$. It follows that $\mathcal{M}$ must be a semifinite von Neumann algebra.

Conversely, suppose that $\mathcal{M}$ is semifinite. By the structure of type I and II algebras in Theorem 10.9.8 and Theorem 10.13.6, any semifinite von Neumann algebra can be written as a product $\mathcal{M} = \prod_\Lambda Z_\lambda \mathcal{M}$ for a partition $\{Z_\lambda\}_\Lambda$ of central projections with $\sum_\Lambda Z_\lambda = I$ so that each summand $Z_\lambda \mathcal{M} \simeq \mathcal{N}_\lambda \overline{\otimes} \mathcal{B}(H_\lambda)$ where each N_λ is either abelian or type II_1 and H_λ is a Hilbert space. Theorem 10.12.10 shows that we can further decompose each $\mathcal{N}_\lambda$ into central summands, each with a faithful normal trace. We may assume that this has already been done. Call it τ_λ.

Let $\{E_{i,j} : i, j \in \mathcal{I}_\lambda\}$ be a set of matrix units for $\mathcal{B}(H_\lambda)$. Define a tracial weight on $Z_\lambda \mathcal{M}$ by identifying $\mathcal{N}_\lambda$ with $\mathcal{N}_\lambda \otimes \mathbb{C}E_{11}$ and setting

$$\widetilde{\tau}_\lambda(A) = \sum_{i \in \mathcal{I}_\lambda} \tau_\lambda(E_{1i} A E_{i1}) \quad \text{for} \quad A \in Z_\lambda \mathcal{M}_+.$$

If the series diverges, the value ∞ is assigned. The additivity and positive homogeneity of $\widetilde{\tau}_\lambda$ follows from the linearity of τ_λ. The trace property follows from normality and the fact that a sum of positive terms is unconditional: if $A \in Z_\lambda\mathcal{M}$,

$$\begin{aligned}\widetilde{\tau}_\lambda(A^*A) &= \widetilde{\tau}_\lambda\Big(\sum_{j\in\mathcal{I}} A^*E_{j1}E_{1j}A\Big) = \sum_{i,j\in\mathcal{I}} \tau_\lambda\big((E_{1i}A^*E_{j1})(E_{1j}AE_{i1})\big)\\ &= \sum_{i,j\in\mathcal{I}} \tau_\lambda\big((E_{1j}AE_{i1})(E_{1i}A^*E_{j1})\big) = \sum_{j\in\mathcal{I}} \tau_\lambda(E_{1j}AA^*E_{j1})\\ &= \widetilde{\tau}_\lambda(AA^*).\end{aligned}$$

If $\{A_\mu\}_{\Lambda'}$ is an increasing net with $\sup A_\mu = A \in Z_\lambda\mathcal{M}$, then normality of τ_λ shows that

$$\tau(E_{1i}AE_{i1}) = \sup_{\mu\in\Lambda'} \tau(E_{1i}A_\mu E_{i1}).$$

From this, it follows that $\sup_{\mu\in\Lambda'} \widetilde{\tau}_\lambda(A_\mu) = \widetilde{\tau}_\lambda(A)$. Hence $\widetilde{\tau}_\lambda$ is normal. It is semifinite because if $A \gneq 0$, then there is some i so that $E_{1i}AE_{i1} \neq 0$. Since τ_λ is faithful, $\tau_\lambda(E_{1i}AE_{i1}) > 0$ and is finite. Now $A^{1/2}E_{ii}A^{1/2} \leq A$ and

$$\widetilde{\tau}_\lambda(A^{1/2}E_{ii}A^{1/2}) = \widetilde{\tau}_\lambda(E_{1i}A^{1/2}A^{1/2}E_{i1}) \in (0,\infty).$$

Therefore $\widetilde{\tau}_\lambda$ is semifinite and faithful.

Finally we define $\tau(A) = \sum_\Lambda \widetilde{\tau}_\lambda(Z_\lambda A)$ for $A \in \mathcal{M}_+$. It is easy to check as in the previous paragraph that τ is a faithful, normal, semifinite tracial weight. ■

10.13.13. REMARK. If $\mathcal{N}$ is a finite von Neumann algebra which is not countably decomposable, then Theorem 10.12.10 will provide an uncountable faithful family $\{\tau_\lambda\}_\Lambda$ of normal traces with pairwise orthogonal supports Z_λ. As in the proof above, we can define $\tau(A) = \sum_\Lambda \tau_\lambda(Z_\lambda A)$. This is a faithful, normal, semifinite tracial weight on $\mathcal{N}$.

10.13.14. COROLLARY. *If $\mathcal{M}$ is a semifinite factor, the semifinite tracial weight τ on $\mathcal{M}$ is unique up to a scalar multiple.*

PROOF. If $\mathcal{M}$ is II_∞, then $\mathcal{M} \simeq \mathcal{N}\overline{\otimes}\mathcal{B}(H)$ where $\mathcal{N}$ is a II_1 factor. If τ is a semifinite tracial weight on $\mathcal{M}$, then by restriction to $\mathcal{N} \otimes \mathbb{C}E_{11}$, we obtain a trace τ_1 on $\mathcal{N}$. If we multiply it by $\tau(I \otimes E_{11})^{-1}$, we can suppose that it is unital. The uniqueness of the trace of $\mathcal{N}$ shows that τ is unique up to a scalar as $\tau(A) = \sum \tau_1(E_{1i}AE_{i1})$.

A similar argument works for $\mathcal{M} = \mathcal{B}(H)$. ■

10.14. The Group Measure Space Construction

The purpose of this section is to provide another general construction of von Neumann algebras, some of which will be type III. Here we discuss an analogue of the reduced crossed product for von Neumann algebras.

Let X be a separable metric space and let μ be a σ-finite Borel measure on X. Then $\mathcal{A} = L^\infty(\mu)$ acting on $L^2(\mu)$ is a separably acting abelian von Neumann algebra. Let G be a countable discrete group, and let $(G, \mathcal{A}, \alpha)$ be a C*-dynamical system. By Theorem 10.3.1, every automorphism of a von Neumann algebra is normal.

An automorphism α of $L^\infty(\mu)$ induces an automorphism of the space of measurable subsets of X modulo null sets by its action on projections. In fact, there is a Borel measurable map $T : X \to X$ such that $\alpha(\chi_E) = \chi_{T^{-1}(E)}$. We will not need this fact, so we will not prove it.

Recall from section 9.14 how the reduced crossed product was defined. Let $\pi_g = \alpha_g^{-1}$ and define a representation of $\mathcal{A}$ on $L^2(\mu) \otimes l_2(G) = \sum_G^{\oplus} L^2(\mu) \otimes \mathbb{C}\delta_g$ by $\widetilde{\pi} = \sum_G^{\oplus} \pi_g$. Next define a unitary representation of G by $u_g = 1 \otimes \lambda_g$, where $(1 \otimes \lambda_g)\, \xi \otimes \delta_h = \xi \otimes \delta_{gh}$. As in Example 9.14.2, $(1 \otimes \lambda, \widetilde{\pi})$ is a covariant representation of $(G, \mathcal{A}, \alpha)$. The *von Neumann algebra crossed product* $\mathcal{A} \rtimes_\alpha G$ is the WOT-closed algebra $\overline{\mathcal{A} \rtimes_{\alpha r} G}^{\text{WOT}}$. Since $\mathcal{A}$ is unital, this is the WOT-closed algebra generated by $\widetilde{\pi}(\mathcal{A})$ and $u(G)$. To simplify notation, we will identify $\widetilde{\pi}(\mathcal{A})$ with $\mathcal{A}$.

The reader will notice that we are using the same notation for a von Neumann crossed product as for the full crossed product of a C*-algebra. There should be no confusion because this is the crossed product used in the von Neumann algebra context.

Proposition 9.14.10 showed that there is a faithful expectation of $\mathfrak{A} \rtimes_{\alpha r} G$ onto $\mathfrak{A}$ for any C*-algebra dynamical system $(G, \mathfrak{A}, \alpha)$. The same proof extends to yield a faithful expectation $\mathcal{E}$ of $\mathcal{A} \rtimes_\alpha G$ onto $\mathcal{A}$. Indeed, $\mathcal{E}$ is obtained by a compression to $L^2(\mu) \otimes \mathbb{C}\delta_e$ followed by $\widetilde{\pi}$, and thus $\mathcal{E}$ is normal. If $X \in \mathcal{A}G$, so that $X = \sum x_g u_g$ is a finite sum with $x_g \in \mathcal{A}$, then $\mathcal{E}(X) = x_e$. If $Y = \sum y_h u_h \in \mathcal{A}G$ also, then

$$XY = \sum_{g,h \in G} x_g u_g y_h u_h = \sum_{g,h \in G} x_g (u_g y_h u_g^*) u_g u_h = \sum_{g,h \in G} x_g \alpha_g(y_h) u_{gh}.$$

Hence $\mathcal{E}(XY) = \sum_{g \in G} x_g \alpha_g(y_{g^{-1}})$.

Remark 9.14.11 now also applies to yield a unique Fourier expansion for any element $X \in \mathfrak{A} \rtimes_{\alpha r} G$ given by $x_g = \mathcal{E}(X u_g^*)$. The map taking X to the sequence $(x_g)_G$ is injective. (As usual, the Fourier series $\sum x_g u_g$ generally does not converge.) However, if $X, Y \in \mathcal{A} \rtimes_\alpha G$ have Fourier coefficients x_g and y_g, respectively, then the formula in the last paragraph extends

$$(XY)_g = \mathcal{E}(XY u_g^*) = \sum_{h \in G} \mathcal{E}\big(x_h u_h y_{h^{-1}g} u_{h^{-1}}\big) = \sum_{h \in G} x_h \alpha_h(y_{h^{-1}g}).$$

10.14.1. DEFINITION. An automorphism α on a von Neumann algebra $\mathcal{M}$ acts *freely* if there is a set $\{P_i\}_{\mathcal{I}}$ of projections in $\mathcal{M}$ such that $\sum_{\mathcal{I}} P_i = I$ and $P_i \alpha(P_i) = 0$ for $i \in \mathcal{I}$. Say that an action of a discrete group G on $\mathcal{M}$ is free if α_g acts freely for all $g \neq e$.

10.14.2. REMARK. When $\mathcal{A}$ is abelian, α acts freely if and only if whenever P is a non-zero projection in $\mathcal{A}$, then $\alpha|_{P\mathcal{A}} \neq$ *id*. Indeed, if α acts freely, take projections $\{P_i\}_{\mathcal{I}}$ in $\mathcal{A}$ such that $\sum_{\mathcal{I}} P_i = I$ and $P_i\alpha(P_i) = 0$ for $i \in \mathcal{I}$. If $P \neq 0$ is a projection, there is some j so that $Q = PP_j \neq 0$. It follows that $Q\alpha(Q) = PP_j\alpha(P_j)\alpha(P) = 0$. Hence $\alpha|_{P\mathcal{A}} \neq$ *id*.

Conversely, if α does not act freely, let $\{P_i\}_{\mathcal{I}}$ be a maximal family of pairwise orthogonal projections in $\mathcal{A}$ such that $P_i\alpha(P_i) = 0$ for $i \in \mathcal{I}$. Then define $P = \big(\sum_{\mathcal{I}} P_i\big)^{\perp} \neq 0$. We claim that $\alpha|_{P\mathcal{A}} =$ *id*. Indeed, if this is not the case, then there is a projection $Q \leq P$ such that $Q \neq \alpha(Q)$. If $Q \not\leq \alpha(Q)$, let $P_0 = Q\alpha(Q)^{\perp}$ and note that $P_0\alpha(P_0) = 0$, contrary to the maximality of the family $\{P_i\}_{\mathcal{I}}$. Similarly if $Q \lneq \alpha(Q)$, let $P_0 = \alpha^{-1}(\alpha(Q)Q^{\perp}) = Q\alpha^{-1}(Q)^{\perp}$. Then again $P_0\alpha(P_0) = \alpha^{-1}(Q)^{\perp}QQ^{\perp}\alpha(Q) = 0$.

10.14.3. PROPOSITION. *The algebra $\widetilde{\pi}(\mathcal{A})$ is maximal abelian in $\mathcal{A} \rtimes_\alpha G$ if and only if G acts freely on $\mathcal{A}$.*

PROOF. Suppose that $X \in \mathcal{A} \rtimes_\alpha G$ commutes with $\widetilde{\pi}(\mathcal{A})$ and has Fourier series $X \sim \sum x_g u_g$, where $x_g = \mathcal{E}(Xu_g^*)$. Then for all $a \in \mathcal{A}$,

$$0 = \mathcal{E}\big((aX - Xa)u_g^*\big) = a\mathcal{E}(Xu_g^*) - \mathcal{E}(Xu_g^*)(u_g a u_g^*) = x_g(a - \alpha_g(a)).$$

If $g \neq e$ and α_g acts freely on $\mathcal{A}$, let $\{P_i\}$ be projections in $\mathcal{A}$ with $\sum P_i = I$ and $P_i\alpha(P_i) = 0$. Taking $a = P_i$ yields $x_g P_i = x_g(P_i - \alpha_g(P_i))P_i = 0$ and thus $x_g = \sum x_g P_i = 0$. It follows that $X = x_e \in \mathcal{A}$.

Conversely, if G does not act freely, then there is some $g \in G \setminus \{e\}$ such that α_g does not act freely. By Remark 10.14.2, there is a non-zero projection $P \in \mathcal{A}$ so that $\alpha_g|_{P\mathcal{A}} =$ *id*. Hence $Pu_g = \alpha_g(P)u_g = u_g P u_g^* u_g = u_g P$. Then for $a \in \mathcal{A}$,

$$\begin{aligned} a(Pu_g) - (Pu_g)a &= (aP - u_g P a u_g^*)u_g \\ &= (aP - \alpha_g(Pa))u_g = (aP - aP)u_g = 0. \end{aligned}$$

Thus $\widetilde{\pi}(\mathcal{A})$ is not maximal abelian. ∎

10.14.4. DEFINITION. A group action $\alpha : G \to \mathrm{Aut}(\mathcal{A})$ on an abelian von Neumann algebra $\mathcal{A}$ is *ergodic* if whenever P is a projections such that $\alpha_g(P) = P$ for all $g \in G$, then $P \in \{0, I\}$.

10.14.5. COROLLARY. *If $(G, \mathcal{A}, \alpha)$ is a free action, then $\mathcal{A} \rtimes_\alpha G$ is a factor if and only if G acts ergodically on $\mathcal{A}$.*

PROOF. Since G acts freely, $\mathcal{A}$ is maximal abelian in $\mathcal{A} \rtimes_\alpha G$. Therefore, the centre of $\mathcal{A} \rtimes_\alpha G$ is contained in $\mathcal{A}$ and is spanned by those projections $P \in \mathcal{A}$ such that $P = u_g P u_g^* = \alpha_g(P)$ for all $g \in G$. This is just $\{0, I\}$ if and only if G acts ergodically. ∎

10.14.6. DEFINITION. Let $\alpha : G \to \mathrm{Aut}(\mathcal{A})$ be a group action on an abelian von Neumann algebra $\mathcal{A}$. A semifinite normal tracial weight τ on $\mathcal{A}$ is *G-invariant* if $\tau \circ \alpha_g^{-1} = \tau$ for every $g \in G$.

We can now explain how to determine the type of the crossed product in certain cases.

10.14.7. THEOREM. *Let $\mathcal{A} = L^\infty(\mu)$ where μ is a non-atomic σ-finite Borel measure on a separable metric space X. Let G be a countably infinite discrete group such that $(G, \mathcal{A}, \alpha)$ is a free, ergodic action. Then*

(1) *$\mathcal{A} \rtimes_\alpha G$ is a type* II_1 *factor if and only if there is a faithful finite G-invariant normal trace on $\mathcal{A}$.*

(2) *$\mathcal{A} \rtimes_\alpha G$ is a type* II_∞ *factor if and only if there is a faithful semifinite and infinite G-invariant normal tracial weight on $\mathcal{A}$.*

(3) *$\mathcal{A} \rtimes_\alpha G$ is a type* III *factor if and only if there is no semifinite G-invariant normal tracial weight on $\mathcal{A}$.*

PROOF. By Proposition 10.14.3 and Corollary 10.14.5, $\mathcal{A} \rtimes_\alpha G$ is a factor, $\mathcal{A}$ is a masa and $\mathcal{E}$ is a normal expectation onto $\mathcal{A}$.

If $\mathcal{A} \rtimes_\alpha G$ is a type I factor, then by Corollary 10.9.10, it is isomorphic to $\mathcal{B}(H)$, where H is an infinite dimensional separable Hilbert space because $\mathcal{A}$ is infinite dimensional. Moreover there is a normal expectation $\mathcal{E}$ of $\mathcal{A} \rtimes_\alpha G$ onto $\mathcal{A}$. By Example 10.6.5, this is not possible. Therefore $\mathcal{A} \rtimes_\alpha G$ is either type II or III.

Suppose that $\mathcal{A}$ has a faithful normal G-invariant semifinite tracial weight φ. Define a tracial weight on $\mathcal{A} \rtimes_\alpha G$ by $\tau(X) = \varphi(\mathcal{E}(X))$. This is normal and faithful since φ and $\mathcal{E}$ are normal and faithful, and it is positive, additive, and positive homogeneous since both have these properties. Let $X \in \mathcal{A} \rtimes_\alpha G$ have Fourier series $\sum_{g\in G} x_g u_g$. Then X^* has Fourier series

$$\Big(\sum_{g\in G} x_g u_g\Big)^* = \sum_{g\in G} u_g^* x_g^* u_g u_g^* = \sum_{g\in G} \alpha_g^{-1}(x_g)^* u_{g^{-1}}.$$

Thus $(X^*)_{g^{-1}} = \alpha_g^{-1}(x_g)^*$ and so $(X^*)_g = \alpha_g(x_{g^{-1}})^*$. Therefore

$$\mathcal{E}(X^*X) = \sum_{g\in G} \alpha_g(x_{g^{-1}})^* \alpha_g(x_{g^{-1}}) = \sum_{g\in G} \alpha_g\big((x_{g^{-1}})^* x_{g^{-1}}\big),$$

and similarly,

$$\mathcal{E}(XX^*) = \sum_{g\in G} x_g \alpha_g\big(\alpha_g^{-1}(x_g)^*\big) = \sum_{g\in G} x_g x_g^*.$$

We compute

$$\tau(X^*X) = \sum_{g\in G} \varphi \circ \alpha_g\big((x_{g^{-1}})^* x_{g^{-1}}\big) = \sum_{g\in G} \varphi\big((x_{g^{-1}})^* x_{g^{-1}}\big)$$
$$= \sum_{g\in G} \varphi\big(x_{g^{-1}}(x_{g^{-1}})^*\big) = \varphi(\mathcal{E}(XX^*)) = \tau(XX^*).$$

Hence τ is a tracial weight.

Since $\mathcal{A}$ is separably acting and φ is semifinite, there is a partition $\{P_i : i \geq 1\}$ of projections in $\mathcal{A}$ with $\varphi(P_i) < \infty$ and $\sum_{i\geq 1} P_i = I$. If $Y = X^*X \gneq 0$ in $\mathcal{A} \rtimes_\alpha G$, then

$$\tau(Y) = \sum_{i\geq 1} \tau\big(X^*(P_i \otimes I)X\big)$$

and

$$\tau\big(X^*(P_i \otimes I)X\big) = \tau\big((P_i \otimes I)XX^*(P_i \otimes I)\big)$$
$$\leq \|X\|^2 \tau(P_i \otimes I) = \|Y\|\varphi(P_i) < \infty.$$

Since τ is faithful, at least one of these terms is non-zero and finite. Observe that $X^*(P_i \otimes I)X \leq Y$, and hence τ is semifinite. By Proposition 10.13.12, $\mathcal{A} \rtimes_\alpha G$ is semifinite.

Conversely, suppose that $\mathcal{A} \rtimes_\alpha G$ is semifinite. Let τ be a faithful semifinite normal tracial weight on $\mathcal{A} \rtimes_\alpha G$. Then the restriction $\varphi = \tau|_{\mathcal{A}}$ is also a faithful semifinite normal tracial weight. Moreover, for $a \in \mathcal{A}$ and $g \in G$,

$$\varphi \circ \alpha_g(a) = \tau(u_g a u_g^*) = \tau(u_g^* u_g a) = \varphi(a).$$

Therefore φ is G-invariant. This establishes (3) by elimination, and it remains to distinguish the II_1 case from II_∞. By Corollary 10.13.14, the semifinite tracial weight τ is unique up to a scalar factor. In the II_1 case, τ and hence φ is finite and in the II_∞ case, it is infinite. ■

10.14.8. EXAMPLE. Let $\mathcal{A} = L^\infty(\mathbb{T}, m)$ acting on $L^2(\mathbb{T}, m)$, and let θ be an irrational real number. Let $\mathbb{Z}$ act on $L^2(\mathbb{T}, m)$ by rotation through the angle $2\pi\theta$ as in Example 9.14.13. The action is free and ergodic, and m is invariant. The von Neumann algebra $L^\infty(\mathbb{T}, m) \rtimes_\alpha \mathbb{Z})$ is the WOT-closure of the same representation applied to the irrational rotation algebra $C(\mathbb{T}) \rtimes_\alpha \mathbb{Z}$. Integration against the Lebesgue measure is a finite faithful normal $\mathbb{Z}$-invariant trace on $\mathcal{A}$. Hence this is a II_1 factor.

10.14.9. EXAMPLE. Let $\mathcal{A} = L^\infty(\mathbb{R}, m)$. Let $\mathbb{Q}$ act on $L^\infty(\mathbb{R}, m)$ by translation. The action is free and ergodic, and m is invariant. This measure is infinite, so integration against m is semifinite but infinite, faithful, normal, and $\mathbb{Q}$-invariant. Hence $L^\infty(\mathbb{R}, m) \rtimes_\alpha \mathbb{Q})$ is type II_∞.

10.14.10. Example. Let $\mathcal{A} = L^\infty(\mathbb{R}, m)$. Let $\mathbb{Q}$ act on $L^\infty(\mathbb{R}, m)$ by translation and let $\mathbb{Q}^*$ act by multiplication on $\mathbb{R}$. The group $\mathbb{Q} \rtimes \mathbb{Q}^*$ is represented as the automorphisms $\alpha_{r,s}(f)(x) = f(sx + r)$. The group operation is $(r_1, s_1)(r_2, s_2) = (r_1 s_2 + r_2, s_1 s_2)$ and the group is a semidirect product where the subgroup $N = \{(r, 1) : r \in \mathbb{Q}\}$ is normal. The group $\mathbb{Q}$ acts by m-preserving homeomorphisms, and m is the unique $\mathbb{Q}$-translation invariant measure (up to a scalar multiple) equivalent to m. However, the action of $\mathbb{Q}^*$ does not preserve the measure. Hence there is no $\mathbb{Q} \rtimes \mathbb{Q}^*$-invariant semifinite tracial weight. Therefore $\mathcal{R}(L^\infty(\mathbb{R}, m) \rtimes_\alpha \mathbb{Q} \rtimes \mathbb{Q}^*)$ is type III.

A related example is given by the subgroup $\mathbb{Q} \rtimes \mathbb{Z}$ where $\mathbb{Z}$ is identified with the subgroup $\{a^n : n \in \mathbb{Z}\}$ for any $a \in \mathbb{Q} \setminus \{0, \pm 1\}$. This also yields a type III factor for any a.

Exercises for Chapter 10

10.1 Let $\mathfrak{S}_2 = \{K \in \mathcal{K}(H) : \|K\|_2 = \big(\sum_{n \geq 1} s_n^2\big)^{1/2} < \infty\}$.
(a) Show that $\|T\|_2 = \big(\sum_\Lambda \|Te_\lambda\|^2\big)^{1/2}$ for any orthonormal basis $\{e_\lambda\}_\Lambda$.
(b) If $T \in \mathfrak{S}_2$, then $T^* \in \mathfrak{S}_2$ and $\|T^*\|_2 = \|T\|_2$.
(c) $\mathfrak{S}_2$ is a Banach space.
(d) If $S, T \in \mathfrak{S}_2$, then $S^*T \in \mathfrak{S}_1$ and $\|ST\|_1 \leq \|S\|_2 \|T\|_2$.
Hint: Use Lemma 10.1.1 and Cauchy–Schwarz.
(e) $\langle S, T \rangle = \mathrm{Tr}(T^*S)$ is an inner product on $\mathfrak{S}_2$, making it a Hilbert space.
(f) If $S, T \in \mathfrak{S}_2$, then $\mathrm{Tr}(ST) = \mathrm{Tr}(TS)$.

10.2 Provide another proof that $\sum_\Lambda |\langle Te_\lambda, f_\lambda \rangle| \leq \|T\|_1$.
Hint: Note that $\sum_\Lambda |\langle Te_\lambda, f_\lambda \rangle| = \sum_\Lambda |\langle |T|^{1/2} e_\lambda, |T|^{1/2} U^* f_\lambda \rangle|$, where $T = U|T|$ in polar form. Use Cauchy–Schwarz and $\|T^*\|_1 = \|T\|_1$.

10.3 Let L be a weak-$*$ closed left ideal of a von Neumann algebra $\mathcal{M}$.
(a) Show that there is a projection $P \in \mathcal{M}$ such that $L = \mathcal{M}P$.
(b) Show that L is a 2-sided ideal if and only if $P \in \mathcal{Z}(\mathcal{M})$.

10.4 Use the fact that $(\overline{b_1(\mathcal{B}(H))}, w^*)$ is compact and $(\overline{b_1(\mathcal{B}(H))}, \textsc{wot})$ is Hausdorff to provide another proof that the identity map from one to the other is a homeomorphism.

10.5 Let H be a Hilbert space with $\dim H = \alpha > \aleph_0$. Prove that the closed ideals of $\mathcal{B}(H)$ have the form $J_1 = \mathcal{K}(H)$ and $\mathcal{J}_\beta = \{T : \dim \overline{TH} \leq \beta\}$ for cardinals $\aleph_0 \leq \beta \leq \alpha$.

10.6 Show that the only normal representations of $\mathcal{B}(H)$ are ampliations.

10.7 (*Kadison*). Let $\mathcal{M}$ be a von Neumann algebra.

(a) Show that the extreme points of the unit ball of $\mathcal{M}$ are the partial isometries $U \in \mathcal{M}$ such that $(I - UU^*)\mathcal{M}(I - U^*U) = \{0\}$.
HINT: look at the proof of Theorem 10.3.7(3).
(b) Show that there is a central projection $Z \in \mathcal{Z}(\mathcal{M})$ such that $(I - UU^*) \leq Z$ and $(I - U^*U) \leq Z^\perp$. Hence UZ is an isometry and $UZ^\perp$ is a co-isometry.
(c) Show that the extreme points of $b_1(\mathcal{M}_{sa})$ are the symmetries $U = P - P^\perp$ for projections $P \in \mathcal{M}$.

10.8 Show that a linear functional on a von Neumann algebra $\mathcal{M}$ is weak-$*$ continuous if and only if it is SOT-continuous on the unit ball of $\mathcal{M}$.
HINT: use Corollary 10.2.11 and the proof of Theorem 10.2.6.

10.9 Let $\pi : \mathcal{M} \to \mathcal{N}$ be a $*$-homomorphism between two von Neumann algebras. Prove that if π is completely additive, then it is normal.

10.10 Show that a $*$-isomorphism between $\mathcal{B}(H)$ and $\mathcal{B}(K)$ is given by a unitary conjugation.

10.11 Let $\mathcal{M}$ be an abelian von Neumann algebra with a cyclic vector ξ.
(a) Let $T \in \mathcal{M}'$ and choose $A_n \in \mathcal{M}$ so that $T\xi = \lim A_n\xi$. Prove that the sequence $A_n^*\xi$ is Cauchy. HINT: elements of $\mathcal{M}$ are normal.
(b) Prove that $T^*\xi = \lim A_n^*\xi =: \eta$. HINT: compute $\langle \eta, B\xi \rangle$ for $B \in \mathcal{M}$.
(c) Prove that $\|T^*B\xi\| = \|TB\xi\|$ for $B \in \mathcal{M}$. Deduce that T is normal.
(d) Prove that $\mathcal{M}$ is maximal abelian.

10.12 Let φ, ψ be pure states on a C*-algebra $\mathfrak{A}$, and let $\mathcal{M}_u = \pi_u(\mathfrak{A})''$.
(a) Let $\ker \widetilde{\pi_\varphi} = \mathcal{M}_u Z^\perp$ where Z is a central projection. (Exercise **10.3**.) Show that $c(\pi_\varphi) = Z$; and that $Z^\perp$ is minimal in $\mathcal{Z}(\mathcal{M}_u)$.
(b) Show that if $\|\varphi - \psi\| < 2$, then $\pi_\varphi \simeq \pi_\psi$. HINT: show $c(\pi_\varphi) = c(\pi_\psi)$.

10.13 Let $\mathfrak{B}$ be a C*-subalgebra of a C*-algebra $\mathfrak{A}$. Show that the weak-$*$ closure of $\pi_u(\mathfrak{B})$ in $\mathcal{M}_u = \mathfrak{A}^{\dagger\dagger}$ is isomorphic to $\mathfrak{B}^{\dagger\dagger}$.

10.14 Show that every ideal of a C*-algebra is an M-ideal as follows:
(a) Say $J \lhd \mathfrak{A}$. Show that there is a central projection Z in $\mathfrak{A}^{\dagger\dagger}$ so that $J^{\perp\perp} = Z\mathfrak{A}^{\dagger\dagger}$.
(b) Show that $Z\mathfrak{A}^{\dagger\dagger}$ and $Z^\perp\mathfrak{A}^{\dagger\dagger}$ are complementary M-summands of $\mathfrak{A}^{\dagger\dagger}$.
(c) Show that $\mathfrak{A}^\dagger = Z\mathfrak{A}^\dagger \oplus_1 Z^\perp\mathfrak{A}^\dagger$, $Z\mathfrak{A}^\dagger \simeq J^\dagger$ and $Z^\perp\mathfrak{A}^\dagger = J^\perp$.

10.15 Show that if σ and π are representations of a C*-algebra $\mathfrak{A}$, each unitarily equivalent to a subrepresentation of the other, then they are unitarily equivalent.
HINT: suppose u and v are isometries so that $\sigma(a) = u^*\pi(a)u$ and $\pi(a) = v^*\sigma(a)v$ for $a \in \mathfrak{A}$. Let $q = \text{SOT-}\sum_{n \geq 0}(uv)^n(1 - uu^*)(uv)^{*n}$ and let $w = qv^* + q^\perp u$.

10.16 (*Arveson*). Let $\mathcal{M} = L^\infty(\mathbb{T})$ acting on $H = L^2(\mathbb{T})$. Let $e_n = e^{in\theta}$ for $n \in \mathbb{Z}$ be the standard orthonormal basis for H, and let $U = M_z$ be the bilateral shift.

(a) Let L be a Banach limit on l_∞ (Example 2.6.22). For $A \in \mathcal{B}(H)$, let $f_A(\xi, \eta) = \big(\langle U^{*n} A U^n \xi, \eta\rangle\big)_{n \ge 1} \in l_\infty$. Define a sesquilinear form $[\xi, \eta] = L(f_A(\xi, \eta))$. Use Proposition 4.2.2 to obtain an operator $\mathcal{E}(A)$ such that $[\xi, \eta] = \langle \mathcal{E}(A)\xi, \eta\rangle$. Show that $\mathcal{E}$ is an expectation of $\mathcal{B}(H)$ onto $\mathcal{M}$. Let P_n be the projection onto $\overline{\operatorname{span}}\{e_k : k \ge n\}$. Show that $\mathcal{E}(P_n) = I$ for all $n \in \mathbb{Z}$.
HINT: show that $\mathcal{E}(A) \in \{U\}' = \mathcal{M}$.

(b) Construct an expectation $\mathcal{F}$ onto $\mathcal{M}$ such that $\mathcal{F}(P_n) = 0$ for all $n \in \mathbb{Z}$.

10.17 Use Sakai's theorem to give another proof of the uniqueness of the predual of a von Neumann algebra.
HINT: Let X_1 and X_2 be preduals of $\mathcal{M}$, and let π_1, π_2 be faithful representations as in Sakai's theorem. Apply Proposition 9.3.9 to $\pi_2\pi_1^{-1}$.

10.18 Let $\mathcal{M}$ be a finite von Neumann algebra with a faithful normal trace τ, and let $\mathcal{N} \subset \mathcal{M}$ be a von Neumann subalgebra.

(a) Define $T : \mathcal{N} \to \mathcal{N}_*$ by $(Ta)(b) = \tau(ab)$ for $a, b \in \mathcal{N}$. Show that $\|Ta\| = \tau(|a|)$ and that $\operatorname{Ran} T$ is dense in $\mathcal{N}_*$.
HINT: polar decomposition and Hahn–Banach.

(b) Define $\mathcal{E} : \mathcal{M} \to \mathcal{N}$ by $\mathcal{E}(x)(Ta) = \tau(xa)$ for $x \in \mathcal{M}$ and $a \in \mathcal{N}$. Prove that $\mathcal{E}$ is a normal expectation of $\mathcal{M}$ onto $\mathcal{N}$ which preserves the trace: $\tau \circ \mathcal{E} = \tau$.

(c) Prove that $\mathcal{E}$ is the unique trace preserving expectation of $\mathcal{M}$ onto $\mathcal{N}$.

10.19 (a) Let $\mathcal{Z}$ be an *abelian* von Neumann algebra. Suppose that $\{P_\lambda\}_\Lambda$ are projections with $P = \bigvee P_\lambda$. Find projections $P'_\lambda \le P_\lambda$ so that $P = \sum_\Lambda P'_\lambda$.
HINT: well order Λ and set $P'_\lambda = P_\lambda\big(\bigvee_{\mu<\lambda} P_\mu\big)^\perp$.

(b) Show that this result fails in the 2×2 matrices.

10.20 Show that if P, Q and R are finite projections in $\mathcal{M}$ with $P \le R$, then the modular law holds: $P \vee (Q \wedge R) = (P \vee Q) \wedge R$.

10.21 Suppose that P_1, P_2, Q_1, Q_2 are finite projections such that $P_1P_2 = 0 = Q_1Q_2$. Suppose that $P_1 \sim Q_1$ and $P_1 + P_2 \sim Q_1 + Q_2$. Show that $P_2 \sim Q_2$.

10.22 (a) Show that an abelian projection in a factor is minimal.

(b) Show that an abelian projection cannot be halved as in Lemma 10.11.7.

10.23 Let $\mathcal{M}$ be a type I von Neumann algebra.

(a) Show that if $P, Q \in \mathcal{Z}(\mathcal{M})$ with $PQ \ne 0$, and P is γ-homogeneous, then PQ is γ-homogeneous.

(b) Show that if $\{Z_\lambda\}_\Lambda$ is a set of pairwise orthogonal γ-homogeneous projections, then $Z = \sum_\Lambda Z_\lambda$ is γ-homogeneous.

10.24 Let $\mathcal{M}$ be a von Neumann algebra.

(a) Show that if $P, Q \in \mathcal{P}(\mathcal{M})$ are unitarily equivalent, then $P \sim Q$.

(b) Show that if $P \sim Q$, then there are projections P_λ, Q_λ for $\lambda \in \Lambda$ such that $P = \sum_\Lambda P_\lambda$, $Q = \sum_\Lambda Q_\lambda$ and $P_\lambda \simeq Q_\lambda$ for $\lambda \in \Lambda$.

10.25 Let $\{E_n : n \geq 1\}$ and $\{F_n : n \geq 1\}$ be increasing sequences of projections with $\sup E_n = I = \sup F_n$.

(a) If $E_n F_n = F_n E_n$ for $n \geq 1$, show that $\sup E_n \wedge F_n = I$.

(b) Construct an example such that $E_n \wedge F_n = 0$ for all $n \geq 1$.
HINT: $H = L^2(0,1)$ and E_n is the space of polynomials of degree at most m.

10.26 Show that if $\pi : \mathcal{M} \to \mathcal{N}$ is a (normal) isomorphism, then $\mathcal{M}$ and $\mathcal{N}$ have the same type.

10.27 Let $\mathcal{M}$ be a von Neumann algebra.

(a) Suppose that $S \in \mathcal{M}$ is a proper isometry. Let $P_n = S^n S^{*n} - S^{n+1} S^{*n+1}$ for $n \geq 0$ and $P_\infty = \left(\sum_{n \geq 0} P_n\right)^\perp$. Show that $P_i \sim P_j$ for $0 \leq i, j < \infty$ and that $P_\infty S = S P_\infty$ is unitary on $P_\infty H$.

(b) Prove that there is a sequence of unitaries U_n converging SOT to S.

(c) Show that the unitary group of $\mathcal{M}$ is SOT-closed if and only if $\mathcal{M}$ is finite.

10.28 Show that every (separably acting) type I von Neumann algebra is injective.

10.29 Suppose that $A \in \mathcal{B}(H)$ and $B \in \mathcal{B}(K)$. If $A \oplus A$ is unitarily equivalent to $B \oplus B$, show that A is unitarily equivalent to B.
HINT: suppose that $U(A \oplus A)U^* = B \oplus B$ for U unitary. Then $\varphi = \operatorname{Ad} U$ is a unitary equivalence between $\mathcal{M} = C^*(A \oplus A)''$ and $\mathcal{N} = C^*(B \oplus B)''$ and also between $\mathcal{M}'$ and $\mathcal{N}'$. Let $E = I_H \oplus 0_H \in \mathcal{M}'$ and $F = I_K \oplus 0_K \in \mathcal{N}'$. Show that $\varphi(E) \sim \varphi(E)^\perp$ in $\mathcal{N}'$ and hence $\varphi(E) \sim F$.

10.30 Let N be a normal operator on a separable Hilbert space H with scalar spectral measure μ on $\sigma(N)$. Prove that there are pairwise disjoint Borel subsets $X_\infty, X_1, X_2, \ldots$ of $\sigma(N)$ so that $N \simeq \bigoplus_{n \geq 1} (M_z \otimes I_n) \oplus (M_z \otimes I_\infty)$ on the Hilbert space $H = \sum_{n \geq 1}^{\oplus} L^2(X_n, \mu) \otimes H_n \oplus L^2(X_\infty, \mu) \otimes H_\infty$, where $\dim H_n = n$. The function $f : \sigma(N) \to \mathbb{N}_0 \cup \{\infty\}$ with $f^{-1}(n) = X_n$ and $f^{-1}(0) = \sigma(N) \setminus \left(\bigcup_{n \geq 1} X_n \cup X_\infty\right)$ is called the *multiplicity function* of N.

10.31 Let $\mathcal{M}$ be a type II_1 von Neumann algebra with centre-valued trace Φ.

(a) Let $0 \neq P \in \mathcal{M}$ be a projection and let $\varepsilon > 0$. Show that there is a projection $0 \neq Q \leq P$ such that $\Phi(Q) \leq \varepsilon \Phi(P)$.
HINT: halving lemma.

(b) If $Z \gneq 0$ in $\mathcal{Z}(\mathcal{M})$, show that there is a central projection E and $\varepsilon > 0$ such that $ZE \geq \varepsilon E$.

(c) Let $Z \in \mathcal{Z}(\mathcal{M})$ with $0 \leq Z \leq I$. Let $\mathcal{C}$ be a maximal chain of projections (in the usual order) such that $\Phi(P) \leq Z$. Let $P_1 = \sup \mathcal{C}$. Prove that $\Phi(P_1) = Z$. Hence Φ maps $\mathcal{P}(\mathcal{M})$ onto $[0, I]$ in $\mathcal{Z}(\mathcal{M})$.

10.32 (a) Show that if $\mathcal{M}$ is not a factor, then $\mathcal{M}$ is not a simple C*-algebra.

(b) Show that if $\mathcal{M}$ is not σ-finite, then the norm-closed ideal generated by σ-finite projections is proper.

(c) Show that if $\mathcal{M}$ is a II_∞ factor, then the norm-closed ideal generated by finite projections is proper.
HINT: use Theorem 10.13.6 and show that the ideal is isomorphic to $\mathcal{N} \otimes \mathcal{K}(H)$.

10.33 (a) Let $\mathcal{M}$ be a factor and let $\mathcal{J}$ be a norm-closed ideal. Show that $\mathcal{J}$ is determined by its projections, and that if $P \in \mathcal{J}$ and $Q \preceq P$, then $Q \in \mathcal{J}$.
(b) Show that if $\mathcal{I}$ is another norm-closed ideal, then either $\mathcal{I} \subset \mathcal{J}$ or $\mathcal{J} \subset \mathcal{I}$.
(c) Prove that there is a largest proper norm-closed ideal, possibly $\{0\}$.
(d) If $\mathcal{M}$ is a separably acting II_∞ factor, then there is a unique non-trivial norm-closed ideal, and it is generated by the finite projections.

10.34 Let $\mathcal{M} \subset \mathcal{B}(H)$ be a finite von Neumann algebra with faithful trace τ acting on separable H. Show that the metric $d_2(A, B) = \tau((A - B)^*(A - B))^{1/2}$ yields the same topology on $\overline{b_1(\mathcal{M})}$ as the SOT topology.

10.35 Show that if $\mathcal{M}$ has no type I summand and $n \geq 2$, then there is a von Neumann algebra $\mathcal{N}$ so that $\mathcal{M} \simeq \mathcal{N} \overline{\otimes} M_n$.

10.36 Let $\mathcal{M}$ be a II_1 factor with faithful normal trace τ. Prove that the following are equivalent.
(1) $L^2(\mathcal{M}, \tau)$ is separable.
(2) $\mathcal{M}_*$ is separable.
(3) $\mathcal{M}$ contains a separable C*-subalgebra $\mathfrak{A}$ which is WOT dense.

10.37 Let $\mathcal{N} = l_\infty([0, 1])$ be the von Neumann algebra of all bounded functions on $[0, 1]$ acting on $l_2([0, 1])$. Show that $\mathcal{N}$ does not have a separable predual, but it does have a separable, weak-$*$ dense subalgebra, $C[0, 1]$.

10.38 Let $\mathcal{M}$ be a II_1 von Neumann algebra with centre-valued trace Φ. If P, Q are projections in $\mathcal{M}$, show that $P \preceq Q$ if and only if $\Phi(P) \leq \Phi(Q)$.
HINT: use the Comparison theorem.

10.39 Let $\mathcal{M}$ be a type II_∞ von Neumann algebra with faithful normal semifinite tracial weight τ. Let $\mathfrak{p}_\tau = \{A \in \mathcal{M}_+ : \tau(A) < \infty\}$, $\mathfrak{n}_\tau = \{X \in \mathcal{M} : \tau(X^*X) < \infty\}$ and $\mathfrak{m}_\tau = \{\sum_{i=1}^n X_i Y_i : X_i, Y_i \in \mathfrak{n}_\tau,\ n \geq 1\}$. Prove that there is a linear functional $\tilde{\tau}$ on $\mathfrak{m}_\tau$ such that
(a) $\mathfrak{n}_\tau$ and $\mathfrak{m}_\tau$ are (non-closed) ideals of $\mathcal{M}$.
(b) $\mathfrak{m}_\tau \cap \mathcal{M}_+ = \mathfrak{p}_\tau$ and $\tilde{\tau}|_{\mathfrak{p}_\tau} = \tau|_{\mathfrak{p}_\tau}$.
(c) $\tilde{\tau}(AX) = \tilde{\tau}(XA)$ for all $A \in \mathcal{M}$ and $X \in \mathfrak{m}_\tau$.
(d) $\tilde{\tau}(XY) = \tilde{\tau}(YX)$ for all $X, Y \in \mathfrak{n}_\tau$.

10.40 Take the setup of the previous exercise.
(a) Show that $\|X\|_1 := \tau(|X|)$ is a norm on $\mathfrak{m}_\tau$.
(b) If $X \in \mathfrak{m}_\tau$, then $\tau_X(A) = \tilde{\tau}(AX)$ is an element of $\mathcal{M}_*$ with $\|\tau_X\| = \|X\|_1$.
(c) This map extends to an isometric map of $L^1(\mathcal{M}, \tau)$, the completion of $\mathfrak{m}_\tau$ in $\|\cdot\|_1$, onto $\mathcal{M}_*$.

10.41 Let $\mathcal{A}$ be an abelian von Neumann algebra, and let α be an automorphism of $\mathcal{A}$. Prove that α is free if and only if $XA = A\alpha(X)$ for all $X \in \mathcal{A}$ implies that $A = 0$.

Notes on Chapter 10

Von Neumann introduced *rings of operators*, now known as von Neumann algebras, in [316], and proved the double commutant theorem. In a series of deep papers [211–213, 317] from the mid-1930s to the mid-1940s, von Neumann, often in collaboration with Murray, developed the theory of von Neumann algebras. Mainly they dealt with factors on separable Hilbert space. They considered not just equivalence, but relative size, of projections. Type I factors have a discrete set of sizes, either $n < \infty$ for factors isomorphic to M_n or ∞ for factors isomorphic to $\mathcal{B}(H)$. Type II_1 factors have sizes order isomorphic to $[0, 1]$ and II_∞ factors to $[0, \infty]$. They use this to construct a trace on a II_1 factor. Lastly in type III factors, all non-zero projections are equivalent, yielding the set $\{0, \infty\}$. They constructed examples of all types, using the group measure construction to construct a type III algebra. They showed that there is a unique hyperfinite II_1 factor, and found the examples $\mathcal{L}(G)$ including non-hyperfinite examples. So a huge portion of the material in this chapter is due to Murray and von Neumann.

The dual and double dual of the compacts were analyzed by Dixmier [92]. He introduced the weak-$*$ topology and observed that von Neumann algebras are dual spaces. In 1949, Dixmier [91] proved the existence of the centre-valued trace, extending the Murray–von Neumann results on traces for finite factors. He also introduced normal states and homomorphisms in [93]. The proof of the existence of the centre-valued trace given here is due to Yeadon [333].

Sherman [285] announced the result that the second dual of a C*-algebra was isomorphic to the universal von Neumann algebra in 1950, and Takeda [299] published a proof in 1954. The uniqueness of the predual of a von Neumann algebra was established by Dixmier [93]. Sakai [273] shows that a C*-algebra which is a dual space can be faithfully represented as a von Neumann algebra. Tomiyama [305] gave a new proof using expectations, which he introduced in 1957.

Takesaki's three-volume treatise on von Neumann algebras is a thorough treatment of the subject. The material covered here is contained in the first volume [300]. Dixmier [94], Sakai [273], Kadison and Ringrose [167] and Stratila and Zsido [291] are other comprehensive books covering this material and more. More recent theory is covered in Sunder [292]. See Kadison's article [166] for an overview of the early history of operator algebras. Some modern directions in the field include free probability (see Nica and Speicher [218]) and the theory of subfactors (see Jones and Sunder [163]).

Part III

Operator Theory

Chapter 11

Hardy Spaces

The Hardy spaces are spaces of analytic functions on the unit disc $\mathbb{D}$. The most important are H^1, H^2, H^∞ and the disc algebra $A(\mathbb{D})$, which also fits into this scheme. We develop a general factorization theorem which helps identify important structures in these spaces. At the end, we apply this information to obtain interesting algebraic information about the Banach algebras $A(\mathbb{D})$ and H^∞.

11.1. Harmonic Functions

In this section, we quickly develop properties of harmonic functions on $\mathbb{D}$. We relate these functions to their boundary values when they belong to various spaces. There is a clear connection with Fourier analysis. In particular, we establish a theorem of Fatou about nontangential limits at the boundary.

11.1.1. Definition. A real- or complex-valued C^2 function u on an open subset of $\mathbb{R}^2$ is *harmonic* if $\Delta u := u_{xx} + u_{yy} = 0$ everywhere.

We will identify $\mathbb{R}^2$ with $\mathbb{C}$ and complex numbers $z = x + iy$. In this chapter, we only deal with harmonic functions on $\mathbb{D} = \{z : |z| < 1\}$. It is clear that if u is harmonic, then $\operatorname{Re} u$ and $\operatorname{Im} u$ are harmonic.

11.1.2. Example. If f is analytic on an open set, then $f = u + iv$ is harmonic, as are u and v. This follows from the Cauchy–Riemann equations: $v_x = -u_y$ and $v_y = u_x$. Since u and v are C^∞,

$$\Delta u = u_{xx} + u_{yy} = v_{yx} - v_{xy} = 0.$$

So u and similarly v are harmonic; thus f is harmonic.

This example has a well-known converse which is often taught in a first course in complex analysis. The fact that $\mathbb{D}$ is simply connected is important, but hidden by the fact that we use the convexity of the disc instead.

K. R. Davidson, *Functional Analysis and Operator Algebras*, CMS/CAIMS Books in Mathematics 13,
https://doi.org/10.1007/978-3-031-63665-3_11

11.1.3. PROPOSITION. *Let u be a real harmonic function on $\mathbb{D}$. Then there is a unique real harmonic function v on $\mathbb{D}$ with $v(0) = 0$ such that $f = u + iv$ is analytic.*

PROOF. We will use Green's theorem: *if Ω is open in $\mathbb{C}$ with C^1 boundary $\partial\Omega$, and $f, g \in C^1(\overline{\Omega})$, then* $\int_{\partial\Omega} f\,dx + g\,dy = \iint_\Omega g_x - f_y\,dxdy$. Let Γ be the piecewise linear path from $(0,0)$ to $(0,x)$ to (x,y). We use the Cauchy–Riemann equations to define:

$$v(x,y) = \int_\Gamma v_x\,dx + v_y\,dy = \int_\Gamma -u_y\,dx + u_x\,dy.$$

If Γ' is another piecewise C^1 curve in $\mathbb{D}$ from $(0,0)$ to (x,y), then $\Gamma - \Gamma'$ is a closed curve bounding a region Ω with $\overline{\Omega} \subset \mathbb{D}$. Thus by Green's theorem,

$$\int_{\Gamma-\Gamma'} -u_y\,dx + u_x\,dy = \iint_\Omega u_{xx} + u_{yy}\,dxdy = 0.$$

Hence v is defined independent of the path used.

To compute $v_x(x_0, y_0)$, evaluate

$$\begin{aligned} v_x(x_0,y_0) &= \lim_{h\to 0} \frac{v(x_0+h, y_0) - v(x_0,y_0)}{h} \\ &= \lim_{h\to 0} \frac{1}{h} \int_{x_0}^{x_0+h} -u_y(t,y_0)\,dt = -u_y(x_0,y_0). \end{aligned}$$

Similarly, $v_y(x_0,y_0) = u_x(x_0,y_0)$. Therefore $f = u+iv$ is analytic by the Cauchy–Riemann equations. If $g = u+iv_1$ is also analytic, then $f-g = i(v-v_1)$ is real and analytic, hence constant. Since $v(0) = 0$ is specified, the function is unique. ■

11.1.4. COROLLARY. *Harmonic functions are C^∞.*

If u is a real harmonic function on $\mathbb{D}$, write $u = \operatorname{Re} f(z)$ where f is analytic on $\mathbb{D}$. Then there is a power series $f(z) = \sum_{n\geq 0} a_n z^n$ which converges on $\mathbb{D}$, and uniformly on $\overline{\mathbb{D}_r}$ for $r < 1$. Hence if we switch to polar coordinates,

$$u = \operatorname{Re} f = \operatorname{Re} a_0 + \tfrac{1}{2}\sum_{n\geq 1} a_n z^n + \overline{a}_n \overline{z}^n = \sum_{n=-\infty}^{\infty} b_n r^{|n|} e^{in\theta}$$

where $b_0 = \operatorname{Re} a_0$, $b_n = \frac{a_n}{2}$ and $b_{-n} = \frac{\overline{a}_n}{2}$ for $n \geq 1$. So this series converges absolutely, and uniformly on each $\overline{\mathbb{D}_r}$ for $r < 1$. This formula extends to complex-valued functions. Conversely every convergent series of this type is harmonic.

The harmonic function v constructed in Proposition 11.1.3 is called the *harmonic conjugate* of u. It has the form

$$v = \operatorname{Im}(f - f(0)) = \frac{1}{2i} \sum_{n \geq 1} a_n z^n - \overline{a}_n \overline{z}^n = \sum_{n=-\infty}^{\infty} -i \operatorname{sign}(n) b_n r^{|n|} e^{in\theta}.$$

11.1.5. The Poisson formula. We develop an integral formula for the harmonic extension of a continuous function on $\mathbb{T}$.

Assume that u is harmonic on $\mathbb{D}_R$ for $R > 1$. Then $u = \sum_{n=-\infty}^{\infty} b_n r^{|n|} e^{in\theta}$ and this series converges uniformly on $\overline{\mathbb{D}}$. In particular, $u|_{\mathbb{T}} = \sum_{n=-\infty}^{\infty} b_n e^{in\theta}$ is an absolutely convergent Fourier series, and

$$b_n = \frac{1}{2\pi} \int_{-\pi}^{\pi} u(e^{it}) e^{-int} \, dt \quad \text{for} \quad n \in \mathbb{Z}.$$

Therefore, we can derive the *Poisson formula*

$$\begin{aligned} u(re^{i\theta}) &= \sum_{n=-\infty}^{\infty} b_n r^{|n|} e^{in\theta} = \sum_{n=-\infty}^{\infty} \frac{1}{2\pi} \int_{-\pi}^{\pi} u(e^{it}) e^{-int} \, dt \, r^{|n|} e^{in\theta} \\ &= \frac{1}{2\pi} \int_{-\pi}^{\pi} u(e^{it}) \sum_{n=-\infty}^{\infty} r^{|n|} e^{in(\theta - t)} \, dt =: \frac{1}{2\pi} \int_{-\pi}^{\pi} u(e^{it}) P_r(\theta - t) \, dt \end{aligned}$$

where, by summing two geometric series, we have

$$\begin{aligned} P_r(t) &= \sum_{n=-\infty}^{\infty} r^{|n|} e^{int} = \frac{1}{1 - re^{it}} + \frac{re^{-it}}{1 - re^{-it}} \\ &= \frac{1 - re^{-it} + re^{-it} - r^2}{(1 - re^{it})(1 - re^{-it})} = \frac{1 - r^2}{1 - 2r \cos t + r^2}. \end{aligned}$$

The function $P_r(t)$ is known as the *Poisson kernel*.

A similar calculation shows that the harmonic conjugate is given by

$$v(re^{i\theta}) = \frac{1}{2\pi} \int_{-\pi}^{\pi} u(e^{it}) Q_r(\theta - t) \, dt$$

where the *conjugate kernel* is

$$Q_r(\theta) = \frac{1}{i} \sum_{n \neq 0} \operatorname{sign}(n) r^{|n|} e^{in\theta} = \frac{2r \sin \theta}{1 - 2r \cos t + r^2}.$$

(See Exercise **11.4**.) Combining them, we get

$$f(z) - i \operatorname{Im} f(0) = \frac{1}{2\pi} \int_{-\pi}^{\pi} u(e^{it}) H_r(\theta - t) \, dt$$

where the *Herglotz kernel* is

$$H_r(e^{i\theta}) = P_r(\theta) + iQ_r(\theta) = \frac{1 - r^2 + 2ir \sin \theta}{1 - 2r \cos t + r^2} = \frac{1 + re^{i\theta}}{1 - re^{i\theta}}.$$

Here are some basic properties of the Poisson kernel.

11.1.6. PROPOSITION. *For* $0 \le r < 1$ *and* $t \in \mathbb{R}$,

(1) P_r *is* 2π*-periodic.*

(2) $P_r(t) > 0$.

(3) $P_r(-t) = P_r(t)$.

(4) $P_r(t)$ *is monotone decreasing on* $[0, \pi]$.

(5) $\|P_r\|_1 = \frac{1}{2\pi}\displaystyle\int_{-\pi}^{\pi} P_r(t)\,dt = 1$.

(6) *For* $\delta > 0$, $\lim\limits_{r\to 1^-} \max_{\delta\le|t|\le\pi} P_r(t) = 0$.

PROOF. The first four properties are easy. For (5), take $u = 1$. By the Poisson formula,

$$1 = u(r) = \frac{1}{2\pi}\int_{-\pi}^{\pi} P_r(-t)\,dt = \frac{1}{2\pi}\int_{-\pi}^{\pi} P_r(t)\,dt.$$

For (6), $\max_{\delta\le|t|\le\pi} P_r(t) = P_r(\delta) = \frac{1-r^2}{1-2r\cos\delta+r^2} \to 0$ as $r \to 1^-$. ■

Two important properties of harmonic functions now follow easily.

11.1.7. MEAN VALUE PROPERTY. *If* u *is harmonic on a ball* $b_r(z_0)$ *and continuous on* $\overline{b_r(z_0)}$, *then*

$$u(z_0) = \frac{1}{2\pi}\int_{-\pi}^{\pi} u(z_0 + re^{it})\,dt.$$

PROOF. A linear change of variables makes $z_0 = 0$ and $r = 1$, which does not affect harmonicity. On any disc of radius $s < 1$, the Poisson formula shows that

$$u(z_0) = \frac{1}{2\pi}\int_{-\pi}^{\pi} u(z_0 + se^{it})\,dt.$$

Let $s \to 1^-$ and take the limit. ■

11.1.8. MAXIMUM PRINCIPLE. *If* u *is harmonic on a bounded domain* Ω *and continuous on the closure, then* u *attains its maximum modulus on the boundary. If it attains its maximum modulus at an interior point, then* u *is constant.*

PROOF. If $z_0 \in \Omega$, then z_0 is the centre of a disc $\overline{b_r(z_0)} \subset \Omega$. Then

$$|u(z_0)| = \left|\frac{1}{2\pi}\int_{-\pi}^{\pi} u(z_0 + re^{it})\,dt\right| \le \frac{1}{2\pi}\int_{-\pi}^{\pi} |u(z_0 + re^{it})|\,dt.$$

If $|u(z_0)| = \|u\|_\infty$, then u is constant on the circle, and hence on the whole disc by Poisson's formula. Then since Re u and Im u are the real parts of analytic functions, they will be constant on the whole connected component, namely, Ω. Otherwise u cannot attain its maximum modulus at an interior point. Since u is continuous on $\overline{\Omega}$, which is compact, u attains it maximum modulus on the boundary. ■

11.1.9. DEFINITION. Let μ be a regular complex Borel measure on $\mathbb{T}$, i.e. $\mu \in M(\mathbb{T})$. (For convenience, we can consider μ as a measure on $(-\pi, \pi]$.) The *Fourier coefficients* of μ are

$$\hat{\mu}(n) = \int_{-\pi}^{\pi} e^{-int}\, d\mu(t) \quad \text{for} \quad n \in \mathbb{Z}.$$

If $f \in L^1(\mathbb{T})$, we associate it to the measure fm, where $m = \frac{1}{2\pi}\, dt$ is normalized Haar measure on $\mathbb{T}$. So we get the familiar definition of the Fourier series of an $L^1(\mathbb{T})$-function:

$$\hat{f}(n) = \frac{1}{2\pi} \int_{-\pi}^{\pi} f(e^{it}) e^{-int}\, dt.$$

The *Fourier series* of μ is the *formal* series $\mu \sim \sum_{n=-\infty}^{\infty} \hat{\mu}(n) e^{in\theta}$.

It is easy to see that $|\hat{\mu}(n)| \leq \|\mu\|$, whence $\big(\hat{\mu}(n)\big)_{n\in\mathbb{Z}}$ is a bounded sequence. Thus

$$\widetilde{\mu}(re^{i\theta}) = \sum_{n=-\infty}^{\infty} \hat{\mu}(n) r^{|n|} e^{in\theta} = \int P_r(\theta - t)\, d\mu(t)$$

converges absolutely, and uniformly on $\overline{\mathbb{D}_r}$ if $r < 1$. This is called the *harmonic extension* of μ to $\mathbb{D}$. In particular, we write $\widetilde{f}$ for the harmonic extension of a function $f \in L^1(\mathbb{T})$. The next result explains the relationship between μ and $\widetilde{\mu}$ in various cases of interest.

11.1.10. THEOREM. *Let $f \in L^1(\mathbb{T})$ or $\mu \in M(\mathbb{T})$. Let $f_r(e^{i\theta}) = \widetilde{f}(re^{i\theta})$.*

(1) **(Poisson theorem)** *If $f \in C(\mathbb{T})$, then f_r converges uniformly to f. Thus $\widetilde{f}$ extends to a continuous function on $\overline{\mathbb{D}}$ with boundary value f.*

(2) *If $f \in L^p(\mathbb{T})$ for $1 \leq p < \infty$, then f_r converges to f in $L^p(\mathbb{T})$.*

(3) *If $f \in L^\infty(\mathbb{T})$, then f_r converges weak-$*$ to f in $L^\infty(\mathbb{T})$.*

(4) *If $\mu \in M(\mathbb{T})$, then $f_r\, dm = \widetilde{\mu}(re^{i\theta}) \frac{dt}{2\pi}$ converges weak-$*$ to μ in $M(\mathbb{T})$. Moreover, $\mu \geq 0$ if and only if $\widetilde{\mu} \geq 0$.*

In all cases, the norm of f_r is increasing to the norm of the limit function/measure.

PROOF. (1) First assume that f is continuous. Then

$$\begin{aligned}
\big|f(e^{i\theta})-f_r(e^{i\theta})\big| &= \Big|f(\theta)\frac{1}{2\pi}\int_{-\pi}^{\pi} P_r(\theta-t)\,dt - \frac{1}{2\pi}\int_{-\pi}^{\pi} f(t)P_r(\theta-t)\,dt\Big| \\
&\le \frac{1}{2\pi}\int_{-\pi}^{\pi} |f(\theta)-f(t)|P_r(\theta-t)\,dt \\
&= \tfrac{1}{2\pi}\int_{-\delta}^{\delta}|f(\theta)-f(t)|P_r(\theta-t)\,dt + \tfrac{1}{2\pi}\int_{\delta\le|t|\le\pi}|f(\theta)-f(t)|P_r(\theta-t)\,dt.
\end{aligned}$$

Let $\varepsilon > 0$. By uniform continuity, we can choose $\delta > 0$ so that $|f(\theta)-f(t)| < \varepsilon$ if $|\theta - t| \le \delta$. By Proposition 11.1.6(6), choose $R < 1$ so that $P_R(\delta) < \frac{\varepsilon}{2\|f\|_\infty}$. Then for $r \ge R$,

$$\begin{aligned}
\big|f(e^{i\theta})-f_r(e^{i\theta})\big| &\le \frac{1}{2\pi}\int_{-\delta}^{\delta} \varepsilon P_r(\theta-t)\,dt + \frac{1}{2\pi}\int_{\delta\le|t|\le\pi} 2\|f\|_\infty \frac{\varepsilon}{2\|f\|_\infty}\,dt \\
&< \varepsilon + \varepsilon = 2\varepsilon.
\end{aligned}$$

Therefore f_r converges uniformly to f. Hence

$$u(re^{i\theta}) = \begin{cases} \widetilde{f}(re^{i\theta}) & \text{if } 0 \le r < 1 \\ f(e^{i\theta}) & \text{if } r = 1 \end{cases}$$

is continuous on the closed disc. The supremum over $\overline{\mathbb{D}_r}$ always occurs on the boundary by the maximum principle, and thus $\|f_r\|_\infty$ increases to $\|f\|_\infty$.

(2) Now take $f \in L^p(\mathbb{T})$ for $1 \le p < \infty$. By Hölder's inequality for the measure $P_r(\theta - t)\,dt$ and Tonelli's theorem,

$$\begin{aligned}
\|f_r\|_p^p &= \frac{1}{2\pi}\int_{-\pi}^{\pi}\Big|\frac{1}{2\pi}\int_{-\pi}^{\pi} f(t)P_r(\theta-t)\,dt\Big|^p\,d\theta \\
&\le \frac{1}{2\pi}\int_{-\pi}^{\pi}\Big(\frac{1}{2\pi}\int_{-\pi}^{\pi}|f(t)|^pP_r(\theta-t)\,dt\Big)\Big(\frac{1}{2\pi}\int_{-\pi}^{\pi}1^qP_r(\theta-t)\,dt\Big)^{\frac{p}{q}}\,d\theta \\
&= \frac{1}{2\pi}\int_{-\pi}^{\pi}|f(t)|^p\,dt = \|f\|_p^p.
\end{aligned}$$

By Poisson's theorem, if $g \in C(\mathbb{T})$, then g_r converges uniformly to g, and hence converges in L^p. As $C(\mathbb{T})$ is dense in $L^p(\mathbb{T})$, for $f \in L^p(\mathbb{T})$ and $\varepsilon > 0$, there is $g \in C(\mathbb{T})$ with $\|f-g\|_p < \varepsilon$. We obtain the estimate

$$\|f-f_r\|_p \le \|f-g\|_p + \|g-g_r\|_p + \|(g-f)_r\|_p < 2\varepsilon + \|g-g_r\|_p.$$

Thus f_r converges to f in $L^p(\mathbb{T})$. If $0 < s < 1$, then $\|f_{rs}\|_p = \|(f_r)_s\|_p \le \|f_r\|_p$. Therefore $\|f_r\|_p$ is increasing to $\|f\|_p$.

(3) Next take $f \in L^\infty(\mathbb{T})$. Then $\widetilde{f}$ is bounded on $\mathbb{D}$. If $g \in L^1(\mathbb{T})$, then by Fubini's theorem,

$$\begin{aligned}\langle f_r, g\rangle &= \frac{1}{2\pi}\int_{-\pi}^{\pi}\frac{1}{2\pi}\int_{-\pi}^{\pi} f(t)P_r(\theta - t)\,dt\,g(\theta)\,d\theta \\ &= \frac{1}{2\pi}\int_{-\pi}^{\pi} f(t)\frac{1}{2\pi}\int_{-\pi}^{\pi} P_r(\theta - t)g(\theta)\,d\theta\,dt \\ &= \frac{1}{2\pi}\int_{-\pi}^{\pi} f(t)g_r(t)\,dt = \langle f, g_r\rangle.\end{aligned}$$

As $r \to 1^-$, g_r converges to g in L^1, and so $\lim\limits_{r\to 1^-}\langle f_r, g\rangle = \lim\limits_{r\to 1^-}\langle f, g_r\rangle = \langle f, g\rangle$. The maximum principle shows that $\|f_r\|_\infty$ increases to $\|f\|_\infty$.

(4) Finally let $\mu \in M(\mathbb{T})$. For $g \in C(\mathbb{T})$, again by Fubini's theorem,

$$\begin{aligned}\langle f_r, g\rangle &= \frac{1}{2\pi}\int_{-\pi}^{\pi}\int_{-\pi}^{\pi} P_r(\theta - t)\,d\mu(t)\,g(\theta)\,d\theta \\ &= \int_{-\pi}^{\pi}\frac{1}{2\pi}\int_{-\pi}^{\pi} P_r(\theta - t)g(\theta)\,d\theta\,d\mu(t) = \int_{-\pi}^{\pi} g_r(t)\,d\mu(t).\end{aligned}$$

By Poisson's theorem, g_r converges uniformly to g, so

$$\lim_{r\to 1^-}\langle f_r, g\rangle = \lim_{r\to 1^-}\int_{-\pi}^{\pi} g_r(t)\,d\mu(t) = \int_{-\pi}^{\pi} g\,d\mu.$$

That is, $f_r\frac{dt}{2\pi}$ converges weak-$*$ to μ. If $\mu \geq 0$, the Poisson formula shows that $\widetilde{\mu} \geq 0$. Conversely if $\widetilde{\mu} \geq 0$, then $f_r\frac{d\theta}{2\pi} \geq 0$. They converge weak-$*$ to μ, whence $\mu \geq 0$. As for $L^1(\mathbb{T})$, $\|f_r\|_1$ is increasing to $\|\mu\|$. ∎

11.1.11. DEFINITION. For $1 \leq p \leq \infty$, let

$$h_p = \{u \text{ harmonic on } \mathbb{D} : \|u\|_p := \sup_{0\leq r<1}\|u_r\|_p < \infty\}$$

where $u_r(e^{i\theta}) = u(re^{i\theta})$.

This next result complements the previous one, and ties down the relationship between the L^p spaces and harmonic functions.

11.1.12. THEOREM.

(1) *If* $1 < p \leq \infty$ *and* $u \in h_p$, *there is a unique* $f \in L^p(\mathbb{T})$ *such that* $u = \widetilde{f}$.

(2) *If* $u \in h_1$, *there is a unique measure* $\mu \in M(\mathbb{T})$ *such that* $u = \widetilde{\mu}$.

PROOF. If $1 < p \leq \infty$, then $L^p(\mathbb{T}) = L^q(\mathbb{T})^*$ where $\frac{1}{p} + \frac{1}{q} = 1$. By assumption, $\sup_{0\leq r<1}\|u_r\|_p = R < \infty$. By the Banach–Alaoglu theorem, $\overline{b_R(L^p(\mathbb{T}))}$ is weak-$*$ compact. Thus there is a subsequence $r_k \to 1$ such that $\underset{k\to\infty}{w^*\text{-lim}}\, u_{r_k} = f$ exists in $L^p(\mathbb{T})$. Then

$$\widetilde{f}(re^{i\theta}) = \lim_{k\to\infty} \frac{1}{2\pi} \int_{-\pi}^{\pi} u_{r_k}(e^{it}) P_r(\theta - t)\, dt = \lim_{k\to\infty} u(r_k re^{i\theta}) = u(re^{i\theta}).$$

For $p = 1$, the argument does not work because $L^1(\mathbb{T})$ is not a dual space. But $M(\mathbb{T}) = C(\mathbb{T})^*$ is a dual space. The proof runs the same way using measures $u_r dm = \frac{1}{2\pi} u_r(e^{it})\, dt$. Again a cofinal subsequence converges weak-$*$ to a measure μ, and the same argument shows that $u = \widetilde{\mu}$. ■

One can also ask about pointwise a.e. convergence, and that works as well. Indeed, since any sequence f_{r_k} converges to f in L^p for any $f \in L^p$, it follows by standard measure theory that some subsequence converges to f a.e.(m). For many purposes, that is sufficient.

11.1.13. DEFINITION. A function u on $\mathbb{D}$ *converges nontangentially* to L at $e^{i\theta}$ if whenever $z_n = r_n e^{it_n} \in \mathbb{D}$ converge to $e^{i\theta}$ and $|t_n - \theta| \le C(1 - r_n)$ for $n \ge 1$ and some $C < \infty$, then $\lim_{n\to\infty} u(z_n) = L$. If this happens a.e.$(m)$ on $\mathbb{T}$ with limit $u(e^{i\theta})$, we write n. t. $\lim_{z\to e^{i\theta}} u(z) = u(e^{i\theta})$ a.e.(m).

Geometrically, this says that the limit exists a.e.(m) provided that one approaches $e^{i\theta}$ from within a wedge that bounds the region away from the tangent line at $e^{i\theta}$. One could also consider the *radial limits* $\lim_{r\to 1^-} u(re^{i\theta})$. One reason it is worth some extra effort to establish nontangential convergence is that this property is conformally invariant. So if τ is a conformal map of $\mathbb{D}$ onto itself and $u(z)$ converges nontangentially to $u(e^{i\theta})$ a.e.(m), then $u(\tau(z))$ converges nontangentially to $u(\tau(e^{i\theta}))$ a.e.(m). See Exercise **11.5**.

Perhaps it isn't so obvious what to expect for nontangential limits of the harmonic extension of a measure. If $\mu \in M(\mathbb{T})$, we can as usual consider this as a measure on $(-\pi, \pi]$. It is determined by the function $F(t) = \mu(-\pi, t]$ of bounded variation. Indeed, $\mu = dF$ is the Lebesgue–Stieltjes measure of F. Then $\dfrac{d\mu}{dm} = 2\pi F'(t)$, where $F'(t)$ exists a.e.(m) by Lebesgue's differentiation theorem. Thus the Radon–Nikodym theorem says that $\mu = 2\pi F'(t)\, dm + \nu$, where $\nu \perp m$ is singular.

11.1.14. FATOU THEOREM. *Let $\mu \in M(\mathbb{T})$. Write $\mu = dF = 2\pi F'\, dm + \nu$, where $\nu \perp m$. Then $\widetilde{\mu}(z) \to \dfrac{d\mu}{dm}(e^{i\theta}) = 2\pi F'(e^{i\theta})$ a.e.(m) nontangentially.*

PROOF. Without loss of generality, we can consider $\theta = 0$ and a real measure μ. Suppose that $F'(0)$ exists. Replace μ by $\mu - 2\pi F'(0)m$. This changes $\widetilde{\mu}$ by the constant function $2\pi F'(0)$ and changes the derivative to $F' - F'(0)$. Also replacing

F by $F - F(0)$ has no effect at all on the measure dF. So we have normalized things to that $F'(0) = F(0) = 0$. Now $F'(0) = 0$ implies that for any $\varepsilon > 0$, there is a $\delta > 0$ so that $|t| < \delta$ implies that $|F(t)| < \varepsilon|t|$. We consider $z = re^{i\theta}$ where $|\theta| \leq C(1-r)$ and r is close to 1, so that $|\theta| \leq \delta/2$. If we integrate by parts,

$$\begin{aligned}\widetilde{\mu}(re^{i\theta}) &= \int_{-\pi}^{\pi} P_r(\theta - t)\,dF(t) = P_r(\theta - t)F(t)\Big|_{-\pi}^{\pi} + \int_{-\pi}^{\pi} F(t)P_r'(\theta - t)\,dt \\ &= P_r(\theta - \pi)\big(F(\pi) - F(-\pi)\big) + \int_{-\pi}^{\pi} F(t)P_r'(\theta - t)\,dt.\end{aligned}$$

For the first term, since $|\theta| \leq \frac{\pi}{2}$, it is bounded by

$$P_r(\theta - \pi)|\mu(\mathbb{T})| \leq P_r(\tfrac{\pi}{2})\|\mu\| = \frac{1-r^2}{1+r^2}\|\mu\|.$$

We split the integral at $\pm\delta$:

$$\begin{aligned}\Big|\int_{-\pi}^{-\delta} F(t)P_r'(\theta - t)\,dt\Big| &\leq \|F\|_\infty \int_{-\pi}^{|\theta|-\delta} P_r'(t)\,dt \\ &\leq \|\mu\|\big(P_r(|\theta| - \delta) - P_r(-\pi)\big) \leq \|\mu\|P_r(\delta/2).\end{aligned}$$

Similarly, $\big|\int_\delta^\pi F(t)P_r'(\theta - t)\,dt\big| \leq \|\mu\|P_r(\delta/2)$. These three terms tend to 0 as $r \to 1^-$, so will each be less than ε if $R \leq r < 1$. Since $P_r(t)$ is increasing on $[-\pi, 0]$ and then decreasing, we have $tP_r'(t) < 0$. Using $|F(t)| \leq \varepsilon|t|$ in $[-\delta, \delta]$, we estimate the remaining term

$$\begin{aligned}\Big|\int_{-\delta}^{\delta} F(\theta - t)P_r'(t)\,dt\Big| &\leq \int_{-\delta}^{\delta} \varepsilon|tP_r'(\theta - t)|\,dt \leq \int_{-\pi}^{\pi} \varepsilon(|\theta| + |s|)|P_r'(s)|\,ds \\ &= 2\varepsilon|\theta| \int_{-\pi}^{0} P_r'(s)\,ds + \varepsilon \int_{-\pi}^{\pi} -sP_r'(s)\,ds \\ &\leq 2\varepsilon|\theta|P_r(0) + \varepsilon(-s)P_r(s)\Big|_{-\pi}^{\pi} + \varepsilon\int_{-\pi}^{\pi} P_r(s)\,ds \\ &\leq 2\varepsilon C(1-r)\frac{1+r}{1-r} + 0 + \varepsilon \leq (4C+1)\varepsilon.\end{aligned}$$

Hence for r sufficiently close to 1 and $|\theta| \leq C(1-r)$, accounting for the other three terms, we get $|\widetilde{\mu}(re^{i\theta})| < (4C+4)\varepsilon$. But $\varepsilon > 0$ was arbitrary, and thus $\text{n.t.}\lim_{z\to 1}\widetilde{\mu}(z) = 0 = F'(0)$.

After translating, we obtain $\text{n.t.}\lim_{z\to e^{i\theta}}\widetilde{\mu}(z) = \dfrac{d\mu}{dm}(e^{i\theta})$ whenever the derivative exists, which is a.e.(m). ■

11.1.15. Remark. If $\mu \geq 0$ is a non-zero positive measure singular to Lebesgue measure, then $\widetilde{\mu} \to 0$ a.e.(m). However, it converges $\widetilde{\mu} \to +\infty$ a.e.(μ). To see this, note that $\dfrac{dm}{d\mu} = 0$ a.e.(μ) and thus $\dfrac{d\mu}{dm} = +\infty$ a.e.(μ). That is, for a.a. (μ) $e^{i\theta} \in \mathbb{T}$,

$\lim\limits_{h\to 0} \dfrac{\mu(\{e^{it} : |t-\theta| < h\})}{2h} = +\infty$. Taking $r = 1 - h$, we have

$$\min_{|t|\le h} P_r(t) = P_r(h) = \frac{1-r^2}{(1-r)^2 + 4r\sin^2\frac{h}{2}} \ge \frac{2h-h^2}{h^2(2-h)} = \frac{1}{h},$$

since $\sin^2\frac{h}{2} \le \frac{h^2}{4}$. Therefore as $r \to 1^-$,

$$\widetilde{\mu}(re^{i\theta}) \ge \int_{\theta-h}^{\theta+h} P_r(\theta - t)\, d\mu(t) \ge \frac{\mu(\{e^{it} : |t-\theta| < h\})}{h} \to +\infty \text{ a.e.}(\mu).$$

This is also valid nontangentially.

11.2. Hardy spaces H^p

The Hardy spaces H^p are the analytic counterpart of the spaces h^p of harmonic functions. Transferring the results of the previous section to H^p for $1 < p < \infty$ is straightforward. However, the theory for H^1 requires considerably more work. The key is the F.&M. Riesz theorem.

11.2.1. Definition. The *Hardy spaces* for $1 \le p \le \infty$ are

$$H^p = \{f \in L^p(\mathbb{T}) : \hat{f}(n) = 0 \text{ for } n < 0\}$$

and

$$H^p(\mathbb{D}) = \{f(z) \in \mathcal{O}(\mathbb{D}) : \sup_{0<r<1} \|f_r\|_p < \infty\},$$

where $\mathcal{O}(\mathbb{D})$ is the space of analytic functions on $\mathbb{D}$, and $\|f\|_p = \sup_{0<r<1} \|f_r\|_p$. Let H^p_0 and $H^p_0(\mathbb{D})$ denote the subspace of functions with $\hat{f}(0) = 0$ and $f(0) = 0$, respectively, and let $A_0 = \{f \in A(\mathbb{D}) : f(0) = 0\}$.

Some of the results of the previous section can be restated as follows:

11.2.2. Corollary. *If $1 < p \le \infty$, then $H^p(\mathbb{D}) = h_p \cap \mathcal{O}(\mathbb{D})$. If $f \in H^p$, then $\widetilde{f} \in H^p(\mathbb{D})$ and f_r converges to f in L^p if $1 < p < \infty$, and weak-$*$ when $p = \infty$. Conversely, if $h \in H^p(\mathbb{D})$, then there is an $f \in H^p$ such that $h = \widetilde{f}$.*

Proof. Each $h \in h_p$ has a power series $h = \sum_{n\in\mathbb{Z}} \hat{h}(n) r^{|n|} e^{in\theta}$ converging uniformly on $\overline{\mathbb{D}_r}$ for $r < 1$. Thus it follows that

$$h_p \cap \mathcal{O}(\mathbb{D}) = \{h \in h_p : \hat{h}(n) = 0 \text{ for } n < 0\} = H^p(\mathbb{D}).$$

The rest follows immediately from Theorems 11.1.10 and 11.1.12. ∎

11.2.3. REMARKS.

(1) When $p = 1$, every $f \in H^1$ has harmonic extension in $H^1(\mathbb{D})$. Some hard work is still required to get the other direction. Theorem 11.1.12 tells us that if $f \in H^1(\mathbb{D})$, there is a measure $\mu \in M(\mathbb{T})$ such that $f = \widetilde{\mu}$; so $\hat{\mu}(n) = 0$ for $n < 0$. We have the set inclusions

$$A(\mathbb{D}) \subset H^\infty(\mathbb{D}) \subset H^p(\mathbb{D}) \subset H^r(\mathbb{D}) \subset H^1(\mathbb{D}) \quad \text{for} \quad \infty > p > r > 1.$$

It is evident that $H^\infty(\mathbb{D})$ is an algebra. Since $(fg)_r = f_r g_r$, we see that the weak-$*$ limit of the product is the product of the weak-$*$ limits. So the multiplication in H^∞ is the restriction of the multiplication in $L^\infty(\mathbb{T})$.

(2) All of the spaces H^p for $1 \le p < \infty$ are separable because $L^p(\mathbb{T})$ is separable. Note that H^∞ is a weak-$*$ closed subalgebra of L^∞ because the Fourier coefficients $\hat{f}(n) = \frac{1}{2\pi}\int_{-\pi}^{\pi} f(e^{it})e^{-int}\,dt$ for $n < 0$ are weak-$*$ continuous functionals on L^∞.

(3) The only real functions in H^p are the constants. Indeed, if $h \in H^1$ and $\bar{h} = h$, then $\hat{h}(n) = \overline{\hat{h}(-n)} = 0$ for $n \ge 1$. Thus $\hat{h} = a\delta_0$, so that $h = a$ is constant.

(4) The space H^2 is a Hilbert space with inner product coming from $L^2(\mathbb{T})$. We let $\zeta_n(e^{i\theta}) = e^{in\theta}$ for $n \ge 0$ be the standard orthonormal basis.

We now address the issue of $H^1(\mathbb{D})$ versus H^1. Theorem 11.1.12 tells us that a function $f \in H^1(\mathbb{D})$ is the harmonic extension of a measure $\mu \in M(\mathbb{T})$ such that $\hat{\mu}(n) = 0$ for $n < 0$. Our goal is to show that $\mu \ll m$, and hence $\mu = f\,dm$ for some $f \in H^1$. This is a result of F.&M. Riesz. We use a Hilbert space approach due to Helson and Lowdenslager.

11.2.4. LEMMA. *Let μ be a finite positive Borel measure on $\mathbb{T}$ with Lebesgue decomposition $\mu = w\,dm + \nu$ where $\nu \perp m$. Let $H^2(\mu)$ denote the closure of $A(\mathbb{D})$ in $L^2(\mu)$, and let $H^2_0(\mu)$ denote the closure of A_0. Set $h = P_{H^2_0(\mu)}1$. If $1 \notin H^2_0(\mu)$, then*

(1) *$|1-h|^2\,d\mu = c\,dm$ for $c = \mathsf{d}(\mu) := \|P^\perp_{H^2_0(\mu)}1\|^2_{L^2(\mu)} > 0$. So $m \ll \mu$.*

(2) *$(1-h)^{-1} \in H^2$ and $(1-h)w \in L^2$.*

(3) *$\mathsf{d}(w\,dm) = \mathsf{d}(\mu)$ and $\mathsf{d}(\nu) = 0$.*

PROOF. Note that $H^2_0(\mu) = M_z H^2(\mu)$ where $M_z \in \mathcal{B}(L^2(\mu))$ is multiplication by $z(e^{i\theta}) = e^{i\theta}$. Now $1 - h$ is in $H^2(\mu)$ orthogonal to $H^2_0(\mu)$. However for $n > 0$, $(1-h)z^n \in H^2_0(\mu)$. Thus setting $\sigma = |1-h|^2\,d\mu$, we have

$$0 = \langle 1-h, (1-h)z^n\rangle = \int |1-h|^2 e^{-in\theta}\,d\mu = \widehat{\sigma}(n).$$

Since σ is a real measure, we have $\widehat{\sigma}(n) = 0$ for $n < 0$ as well. A measure can be recovered from its Fourier series by Theorem 11.1.10(4), and hence this map is injective. Therefore, there is a constant $c \neq 0$ so that $|1-h|^2\,d\mu = c\,dm$. Moreover

$$c = \int c\,dm = \int |1-h|^2\,d\mu = \|1-h\|^2_{L^2(\mu)} = \|P^{\perp}_{H^2_0(\mu)} 1\|^2_{L^2(\mu)} = \mathsf{d}(\mu) > 0.$$

Now

$$c\,dm = |1-h|^2\,d\mu = |1-h|^2 w\,dm + |1-h|^2\,d\nu;$$

so we deduce that $|1-h|^2 w = c$ a.e.(m) and $h = 1$ a.e.(ν). Therefore $m \ll \mu$.

We obtain that $c^{-1} w\,dm = |1-h|^{-2}\,dm$ is a finite measure, and thus $(1-h)^{-1}$ belongs to $L^2(m)$. Moreover, for $n \geq 1$,

$$\begin{aligned}\int (1-h)^{-1} e^{in\theta}\,dm &= c^{-1} \int (1-h)^{-1} e^{in\theta} |1-h|^2 w\,dm \\ &= c^{-1} \int e^{in\theta} (1-\bar{h}) w\,dm = c^{-1} \int e^{in\theta}(1-\bar{h})\,d\mu \\ &= c^{-1} \langle z^n, 1-h\rangle_{L^2(\mu)} = 0,\end{aligned}$$

because $1-h = 0$ a.e.(ν) and $1-h$ is orthogonal to $H^2_0(\mu)$ in $L^2(\mu)$. Hence, the negative Fourier coefficients of $(1-h)^{-1}$ all vanish, whence $(1-h)^{-1} \in H^2$. Now $|1-h|^2 w = c$ a.e.(m) implies that $|1-h| w = c|1-h|^{-1} \in L^2$.

We can decompose $L^2(\mu) = L^2(w\,dm) \oplus L^2(\nu)$. This decomposes each $f \in L^2(\mu)$ as $f = f_a \oplus f_s$. Since $h_s = 1_s$, we have $\mathsf{d}(\nu) = 0$. In particular, $1 - h = 1_a - h_a \oplus 0$ is orthogonal to every $p = p_a \oplus p_s \in A_0$. It follows that $1_a - h_a$ is orthogonal to A_0 in $L^2(w\,dm)$. Moreover h is a limit in $L^2(\mu)$ of sequence $p_k \in A_0$, from which it follows that h_a lies in $H^2_0(w\,dm)$. Since $1_a = (1_a - h_a) + h_a$ is an orthogonal decomposition, we deduce that

$$\mathsf{d}(w\,dm) = \|1_a - h_a\|^2 = \|1-h\|^2 = c = \mathsf{d}(\mu). \qquad \blacksquare$$

Now we can show that measures of analytic type are absolutely continuous.

11.2.5. F. & M. Riesz Theorem. *If $\mu \in M(\mathbb{T})$ is analytic, i.e. $\hat{\mu}(n) = 0$ for $n < 0$, then μ is absolutely continuous to Lebesgue measure, and $d\mu = f\,dm$ for some $f \in H^1$.*

Proof. By the Lebesgue decomposition, we write $\mu = w\,dm + \nu$ where $\nu \perp m$. Define $\sigma = (1+|w|)\,dm + |\nu|$. Thus $\mu \ll \sigma$. Since $\widehat{1-p}(0) = 1$ for $p \in A_0$,

$$\mathsf{d}(\sigma) = \inf_{p \in A_0} \|1-p\|^2_{L^2(\sigma)} \geq \inf_{p \in A_0} \|1-p\|^2_{L^2(m)} = 1.$$

Let $h = P_{H^2_0(\sigma)} 1$. There is a sequence $p_k \in A_0$ such that $h = \lim p_k$ in $L^2(\sigma)$. Since μ is analytic, $\int z^n\,d\mu = 0$ for $n > 0$. Therefore

$$\int (1-h) z^n\,d\mu = \lim_{k\to\infty} \int (1-p_k) z^n\,d\mu = 0.$$

By Lemma 11.2.4, $1 - h = 0$ a.e.$(|\nu|)$, and hence

$$\int (1-h) z^n w \, dm = 0 \quad \text{for all} \quad n > 0.$$

The same lemma shows that $(1-h)w \in L^2$, and thus it lies in H^2. Likewise $(1-h)^{-1} \in H^2$, so there is a sequence $q_k \in A(\mathbb{D})$ converging to $(1-h)^{-1} \in H^2$. Hence for all $n > 0$,

$$\begin{aligned} \hat{w}(-n) &= \int z^n w \, dm = \int (1-h) w \, z^n (1-h)^{-1} \, dm \\ &= \lim_{k \to \infty} \int (1-h) z^n q_k w \, dm = 0. \end{aligned}$$

Therefore $w \in H^1$.

By subtraction, we also obtain that $\hat{\nu}(n) = 0$ for all $n < 0$. However, Lemma 11.2.4 shows that $1 \in H_0^2(|\nu|)$. Thus there is a sequence $q_k \in A_0$ converging to 1 in $L^2(|\nu|)$. Hence

$$\hat{\nu}(0) = \int 1 \, d\nu = \lim_{k \to \infty} \int q_k \, d\nu = 0.$$

It follows that $\bar{z}\nu$ is a singular analytic measure. Thus $\hat{\nu}(1) = \widehat{\bar{z}\nu}(0) = 0$. We can now show recursively that $\hat{\nu}(n) = 0$ for all $n \in \mathbb{Z}$. Therefore $\nu = 0$ and $\mu \ll m$.

By the Radon–Nikodym theorem, $d\mu = f \, dm$ for some $f \in L^1$. Moreover $\hat{f}(n) = \hat{\mu}(n) = 0$ for $n < 0$. Therefore $f \in H^1$. ■

Our main application of the F.&M. Riesz theorem is the identification between $H^1(\mathbb{D})$ and H^1.

11.2.6. THEOREM. *Let $f \in H^1(\mathbb{D})$. Then $f^*(e^{i\theta}) :=$ n. t. $\lim_{r \to 1^-} f(re^{i\theta})$ exists* a.e.(m), *$f^* \in H^1$, f is the Poisson integral of f^* and $\|f^*\|_1 = \|f\|_1$.*

PROOF. Since $H^1(\mathbb{D}) \subset h_1$, Theorems 11.1.10 and 11.1.12 show that f_r converge weak-$*$ to a measure μ and f is the Poisson integral of μ. Since f is analytic, we have

$$\hat{\mu}(n) = \lim_{r \to 1^-} \hat{f}_r(n) = 0 \quad \text{for} \quad n < 0.$$

By the F.&M. Riesz theorem, $\mu = w \, dm$ for some $w \in L^1$ with $\hat{w}(n) = 0$ for $n < 0$. That is, $w \in H^1$. Fatou's theorem shows that there is a nontangential limit f^* a.e.(m), and that $f^* = w$. Therefore, f is the Poisson integral of f^*, and f_r converges to f^* in L^1. Finally, $\|f_r\|_1$ is monotone increasing, and so

$$\|f^*\|_1 = \lim_{r \to 1^-} \|f_r\|_1 = \sup_{0<r<1} \|f_r\|_1 = \|f\|_1.$$

Thus this pairing is isometric. ■

Now we collect a few other results about the Hardy spaces.

11.2.7. PROPOSITION. *If $1 \le p < \infty$, then $H^{p*} \simeq L^q(\mathbb{T})/H^q_0$ where $\frac{1}{p}+\frac{1}{q}=1$. If $1 < p < \infty$, then H^p is reflexive. Also $H^\infty = \big(L^1(\mathbb{T})/H^1_0\big)^*$.*

PROOF. Since $H^p \subset L^p(\mathbb{T})$, $H^{p*} \simeq L^p(\mathbb{T})^*/H^{p\perp}$. Now $L^p(\mathbb{T})^* = L^q(\mathbb{T})$ if $p < \infty$. The pairing is $\langle f, g\rangle = \frac{1}{2\pi}\int f(e^{it})g(e^{it})\,dt$. Since H^p is spanned by $\zeta_n = e^{in\theta}$ for $n \ge 0$,

$$H^{p\perp} = \{g \in H^q(\mathbb{T}) : \langle g, \zeta_n\rangle = \hat{g}(-n) = 0 \text{ for } n \ge 0\} = H^q_0.$$

Therefore $H^{p*} \simeq L^q(\mathbb{T})/H^q_0$.

If $1 < p < \infty$, H^p is a subspace of a reflexive space, and thus is reflexive.

Since H^∞ is weak-$*$ closed, it is a dual space $H^\infty \simeq \big(L^1(\mathbb{T})/H^\infty_\perp\big)^*$. Since z^n for $n \ge 0$ span $A(\mathbb{D})$, which is weak-$*$ dense in H^∞, the argument of the previous paragraph shows that $H^\infty_\perp = H^1_0$. Therefore $H^\infty = \big(L^1(\mathbb{T})/H^1_0\big)^*$. ■

It turns out that H^1 is also a dual space.

11.2.8. THEOREM. *$\big(C(\mathbb{T})/A_0\big)^* \simeq H^1$ and $H^{1*} \simeq L^\infty/H^\infty_0$ are dual pairings, and also $\big(C(\mathbb{T})/A(\mathbb{D})\big)^* \simeq H^1_0$ and $(H^1_0)^* \simeq L^\infty/H^\infty$.*

PROOF. The last step of the following computation follows from the F.&M. Riesz theorem.

$$\begin{aligned}\big(C(\mathbb{T})/A_0\big)^* &\simeq \{\varphi \in C(\mathbb{T})^* : \varphi \perp A_0\}\\ &= \{\mu \in M(\mathbb{T}) : \mu(z^n) = 0 \text{ for } n > 0\}\\ &= \{\mu \in M(\mathbb{T}) : \hat{\mu}(n) = 0 \text{ for } n < 0\} = H^1.\end{aligned}$$

Since $H^1 \subset L^1$, we have $H^{1*} \simeq L^\infty/(H^1)^\perp$. And

$$\begin{aligned}(H^1)^\perp &= \{f \in L^\infty : 0 = \frac{1}{2\pi}\int f(e^{it})e^{int}\,dt \text{ for } n \ge 0\}\\ &= \{f \in L^\infty : 0 = \hat{f}(-n) \text{ for } n \ge 0\} = H^\infty_0.\end{aligned}$$

The second pair of statements are proven in the same manner. ■

One advantage of $H^p(\mathbb{D})$ is that analytic functions can be evaluated at points in $\mathbb{D}$. These are continuous functionals. We can transfer this to H^p.

11.2.9. PROPOSITION. *For $1 \le p \le \infty$ and $w = re^{i\theta} \in \mathbb{D}$, the functional $\varepsilon_w(f) = \widetilde{f}(w)$ is continuous on H^p, and is given by integration against $P_r(\theta - t)$ or alternatively $g_w(e^{it}) = \dfrac{1}{1 - we^{-it}}$. Moreover,*

$$\|\varepsilon_w\|_{H^{p*}} \le \frac{1}{1-|w|}, \quad \|\varepsilon_w\|_{H^{2*}} = \frac{1}{\sqrt{1-|w|^2}} \quad \text{and} \quad \|\varepsilon_w\|_{H^\infty_*} = 1.$$

For $p = 2$, the function $k_w(z) = \dfrac{1}{1 - z\overline{w}} = \sum_{n\ge 0} \bar{w}^n \zeta_n \in H^2$ yields point evaluation by the inner product $\varepsilon_w(f) = \langle f, k_w\rangle$.

The span of $\{\varepsilon_w : w \in \mathbb{D}\}$ is dense in H^{p*} if $1 < p < \infty$, is weak-$*$ dense in H^{1*}, and is norm dense in H^∞_*.

PROOF. We have already seen that evaluation at w is achieved by integration against $P_r(\theta - t)$, which is continuous on $\mathbb{D}$ and so belongs to L^p for all p. A calculation shows that

$$\varepsilon_w(f) = \sum_{n\ge 0} \hat{f}(n) w^n = \frac{1}{2\pi}\int_{-\pi}^{\pi} f(e^{it}) \sum_{n\ge 0} w^n e^{-int}\, dt = \frac{1}{2\pi}\int_{-\pi}^{\pi} f(e^{it}) \frac{1}{1-we^{-it}}\, dt.$$

Since $\left|\frac{1}{1-we^{-it}}\right| \le \frac{1}{1-|w|}$, it follows that

$$\|\varepsilon_w\|_{H^{p*}} \le \left\|\frac{1}{1 - we^{-it}}\right\|_q \le \frac{1}{1 - |w|}.$$

For $p = 2$, we use the conjugate linear pairing between L^2/H^2_0 and H^2 to get

$$k_w(z) = \sum_{n\ge 0} \overline{w}^n z^n = \frac{1}{1 - z\overline{w}}.$$

Then $\varepsilon_w(h) = \langle h, k_w\rangle$ for $h \in H^2$ and $\|k_w\|_2^2 = \sum_{n\ge 0} |w^2|^n = \frac{1}{1-|w|^2}$. For $p = \infty$, $\|\varepsilon_w\| \le \|P_r(\theta - t)\|_1 = 1$ and $\varepsilon_w(1) = 1$. So $\|\varepsilon_w\|_{H^\infty_*} = 1$.

If $f \in H^p$ annihilates $\{\varepsilon_w : w \in \mathbb{D}\}$, then $f = 0$. By the Hahn–Banach theorem, span$\{\varepsilon_w : w \in \mathbb{D}\}$ is weak-$*$ dense in H^{p*}. Hence for $1 < p < \infty$, which are reflexive spaces, the span is norm dense in H^{p*}. Likewise, the span is norm dense in H^∞_*. ■

The set of functions $\{k_w : w \in \mathbb{D}\}$ in H^2 that perform point evaluations is called the *Szegö kernel*, which is the *reproducing kernel* for H^2. For the general theory, see Section 15.9.

This proposition has consequences for the maximal ideal space $\mathcal{M}_{H^\infty}$.

11.2.10. THEOREM. *The embedding $w \mapsto \varepsilon_w$ is a homeomorphism of $\mathbb{D}$ onto the open subset $\{\varphi \in \mathcal{M}_{H^\infty} : |\varphi(z)| < 1\}$.*

An $h \in H^\infty$ is invertible if and only if $\inf_{z\in\mathbb{D}} |h(z)| > 0$. So $\sigma_{H^\infty}(h) = \overline{h(\mathbb{D})}$.

PROOF. Since H^∞ functions are continuous on $\mathbb{D}$, the map $w \mapsto \varepsilon_w$ is continuous from $\mathbb{D}$ into $\mathcal{M}_{H^\infty}$. Clearly the range lies in $\{\varphi \in \mathcal{M}_{H^\infty} : |\varphi(z)| < 1\}$.

Suppose that φ is a multiplicative functional on H^∞ such that $\varphi(z) = w \in \mathbb{D}$. If $h \in H^\infty$ and $h(w) = a$, then

$$h(z) = h(a) + \frac{h(z) - a}{z - w}(z - w) \quad \text{and} \quad g(z) = \frac{h(z) - a}{z - w}.$$

Thus g is a bounded analytic function (after fixing the removable singularity at w). Hence

$$\varphi(h) = \varphi(a) + \varphi(g)\varphi(z-w) = a = \varepsilon_w(h).$$

Therefore, there is a unique character φ with $\varphi(z) = w$. Thus, the disc is mapped onto $\mathcal{D} = \{\varphi \in \mathcal{M}_{H^\infty} : |\varphi(z)| < 1\}$, which is evidently open in the weak-$*$ topology on $\mathcal{M}_{H^\infty}$. The inverse map $\varepsilon_w \to w$ from $\mathcal{D}$ to $\mathbb{D}$ is given by $\varepsilon_w \to \varepsilon_w(z)$, which is continuous. Thus, the two sets are homeomorphic.

If h is not bounded away from 0 on $\mathbb{D}$, then the Gelfand transform vanishes on $\overline{\mathcal{D}}$ and h is not invertible. Conversely, if h is bounded away from 0, then $g(z) = 1/h(z)$ is a bounded analytic functions and $gh = 1$. Hence $\sigma_{H^\infty}(h) = \overline{h(\mathbb{D})}$. ∎

The last part of the proof would be evident if $\mathcal{D}$ were dense in $\mathcal{M}_{H^\infty}$. It is a deep result of Carleson known as the *corona theorem* that this is the case. From now on, we identify $\mathbb{D}$ with $\mathcal{D}$ as a subset of $\mathcal{M}_{H^\infty}$.

We record one more technical lemma for future use.

11.2.11. LEMMA. *$H^2 \cdot H^2 \subset H^1$ and $\varepsilon_w(fg) = \varepsilon_w(f)\varepsilon_w(g)$ for $w \in \mathbb{D}$ and $f, g \in H^2$.*

PROOF. This is valid for $f, g \in A(\mathbb{D})$. Let $f, g \in H^2$ and let $\widetilde{f}, \widetilde{g}$ be their analytic extensions to $\mathbb{D}$. Then for $r < 1$, $f_r, g_r \in A(\mathbb{D})$ and converge to f and g, respectively, in L^2. Clearly $fg \in L^1(\mathbb{T})$. Therefore

$$\begin{aligned}\|fg - f_r g_r\|_1 &\le \|(f - f_r)g\|_1 + \|f_r(g - g_r)\|_1 \\ &\le \|f - f_r\|_2\|g\|_2 + \|f_r\|_2\|g - g_r\|_2 \to 0.\end{aligned}$$

Thus $f_r g_r = (fg)_r$ converges in L_1 to fg. Hence $\widetilde{fg} = \widetilde{f}\,\widetilde{g}$. So $fg \in H^1$ and

$$\varepsilon_w(fg) = \lim_{r\to 1^-} \varepsilon_w(f_r g_r) = \lim_{r\to 1^-} \varepsilon_w(f_r)\varepsilon_w(g_r) = f(w)g(w).$$ ∎

11.3. Consequences of the Jensen Formula

In this section, we establish Jensen's formula, which provides information about the zeros of analytic functions. This is then extended to H^1.

11.3.1. DEFINITION. For $0 \ne a \in \mathbb{D}$, the *Blaschke factor* $b_a(z) = \dfrac{\bar{a}}{|a|}\dfrac{a - z}{1 - \bar{a}z}$ is a Möbius map that takes $\mathbb{D}$ conformally onto itself, $b_a(\mathbb{T}) = \mathbb{T}$, $b_a(a) = 0$ and $b_a(0) = |a| > 0$. Let $b_0(z) = z$. A *finite Blaschke product* is a product of a finite number of Blaschke factors, possibly with repeated zeros, and a constant $\lambda \in \mathbb{T}$.

We have $|b_a(e^{i\theta})| = 1$, and thus $|B(e^{i\theta})| = 1$ for every finite Blaschke product B. Observe that $e^{i\theta} \mapsto b_a(e^{i\theta})$ carries the unit circle $\mathbb{T}$ homeomorphically onto itself with the same orientation. Hence, it has winding number 1 around every point in the unit disc. It follows that a product B of n Blaschke factors maps $\mathbb{T}$ onto itself with winding number n around every point in the unit disc. By the argument principle, B takes each value in $\mathbb{D}$ exactly n times (counting multiplicity).

We begin this section with a useful formula involving the zeros of an analytic function.

11.3.2. JENSEN FORMULA. *Suppose that f is analytic on a neighbourhood of $\overline{\mathbb{D}}$, $f(0) \neq 0$ and f does not vanish on $\mathbb{T}$. Let $\{a_1, \dots, a_k\}$ be the zeros of f in $\mathbb{D}$, repeated according to multiplicity. Then*

$$\log|f(0)| = \frac{1}{2\pi}\int_{-\pi}^{\pi} \log|f(e^{it})|\,dt + \sum_{j=1}^{k} \log|a_j|.$$

PROOF. Let B be the finite Blaschke product with zeros $\{a_1, \dots, a_k\}$. Then $g(z) = f(z)/B(z)$ has removable singularities at each a_j, and thus is analytic and non-zero on a neighbourhood of $\overline{\mathbb{D}}$. Hence, there is an analytic branch of $\log g$ on a neighbourhood of $\overline{\mathbb{D}}$. Note that $|g(e^{i\theta})| = |f(e^{i\theta})|$. By Cauchy's theorem,

$$\begin{aligned}\log|g(0)| &= \operatorname{Re}\log g(0) = \operatorname{Re}\frac{1}{2\pi}\int_{-\pi}^{\pi} \log g(e^{it})\,dt \\ &= \frac{1}{2\pi}\int_{-\pi}^{\pi} \log|g(e^{it})|\,dt = \frac{1}{2\pi}\int_{-\pi}^{\pi} \log|f(e^{it})|\,dt.\end{aligned}$$

Therefore we obtain

$$\log|f(0)| = \log|g(0)| + \log|B(0)| = \frac{1}{2\pi}\int_{-\pi}^{\pi} \log|f(e^{it})|\,dt + \sum_{j=1}^{k} \log|a_j|. \blacksquare$$

11.3.3. COROLLARY. *Suppose that f is analytic on a neighbourhood of $\overline{\mathbb{D}}$ and f does not vanish on $\mathbb{T}$. Then*

$$\log|f(z)| \le \frac{1}{2\pi}\int_{-\pi}^{\pi} \log|f(e^{it})| P_r(\theta - t)\,dt \quad \textit{for} \quad z = re^{i\theta} \in \mathbb{D}.$$

PROOF. Let B be the finite Blaschke product with zeros corresponding to the zeros of f including multiplicity. Then $f(z) = B(z)g(z)$ for $g(z)$ analytic and non-zero in a neighbourhood of $\overline{\mathbb{D}}$. Hence $\log|g(z)| = \operatorname{Re}\log g(z)$ is harmonic.

Since $|f(e^{it})| = |g(e^{it})|$,

$$\begin{aligned}\log|f(z)| &= \log|g(z)| + \log|B(z)| \le \log|g(z)| \\ &= \frac{1}{2\pi}\int_{-\pi}^{\pi} \log|g(e^{it})|P_r(\theta - t)\,dt \\ &= \frac{1}{2\pi}\int_{-\pi}^{\pi} \log|f(e^{it})|P_r(\theta - t)\,dt.\end{aligned}$$ ■

Jensen's formula has some important consequences for the H^p spaces.

11.3.4. THEOREM. *Let $0 \neq f \in H^1$ with zeros $\{a_i\}$, repeated according to multiplicity. Then*

(1) $\log|f(e^{i\theta})| \in L^1$.

(2) $\log|f(z)| \le \dfrac{1}{2\pi}\displaystyle\int_{-\pi}^{\pi} \log|f(e^{it})|P_r(\theta - t)\,dt$ *for* $z = re^{i\theta} \in \mathbb{D}$.

(3) $\sum_{i\ge 1} 1 - |a_i| < \infty$.

PROOF. We consider f both as an element of H^1 and as an analytic function in $H^1(\mathbb{D})$. If f has a zero at 0 of order n, factor out the z^n. This has no effect on the first and third statements, and increases the LHS in the second statement. So we may assume that $f(0) \neq 0$. Choose $r_k \to 1^-$ such that f does not vanish on $r_k\mathbb{T}$. By Theorem 11.1.10, $f_{r_k}(e^{i\theta})$ converges to $f(e^{i\theta})$ in the L^1 norm. By Fatou's theorem, $f_{r_k}(e^{i\theta}) \to f(e^{i\theta})$ a.e.(m) as well.

Let $\{a_1, \dots, a_{n_k}\}$ be the zeros of f in $r_k\mathbb{D}$. Then f_{r_k} has zeros $\{\frac{a_1}{r_k}, \dots, \frac{a_{n_k}}{r_k}\}$. By Jensen's formula,

$$\begin{aligned}\log|f(0)| = \log|f_{r_k}(0)| &= \tfrac{1}{2\pi}\int_{-\pi}^{\pi} \log|f(r_k e^{it})|\,dt + \sum_{i=1}^{n_k} \log\frac{|a_i|}{r_k} \\ &\le \tfrac{1}{2\pi}\int_{-\pi}^{\pi} \log|f(r_k e^{it})|\,dt.\end{aligned}$$

Set $\log^+ x = \max\{0, \log x\}$ and $\log^- x = \max\{0, -\log x\}$ for $x > 0$. Then

$$\log^+|f(r_k e^{it})| \le |f(r_k e^{it})| \in L^1.$$

Thus $\big\|\log^+|f_{r_k}|\big\|_1 \le \|f_{r_k}\|_1 \le \|f\|_1$. And by Jensen's formula,

$$\big\|\log^-|f(r_k e^{it})|\big\|_1 \le -\log|f(0)| + \big\|\log^+|f(r_k e^{it})|\big\|_1 \le -\log|f(0)| + \|f\|_1.$$

Therefore

$$\big\|\log|f_{r_k}|\big\|_1 \le -\log|f(0)| + 2\|f\|_1 < \infty.$$

Since $\big|\log|f_{r_k}|\big| \to \big|\log|f|\big|$ a.e.(m), Fatou's lemma shows that $\log|f| \in L^1$.

To get the second statement requires more delicacy. Let $\varepsilon > 0$. By Corollary 11.3.3 for $z = re^{i\theta}$ and $r_k > r$,

$$\log|f(z)| = \log|f_{r_k}(z/r_k)| \le \tfrac{1}{2\pi}\int_{-\pi}^{\pi} \log|f(r_k e^{it})| P_{r/r_k}(\theta - t)\,dt$$
$$\le \tfrac{1}{2\pi}\int_{-\pi}^{\pi} \log(|f(r_k e^{it})| + \varepsilon) P_{r/r_k}(\theta - t)\,dt.$$

If $a, b > 0$, $|\log(a+\varepsilon) - \log(b+\varepsilon)| \le \log\big(1 + \frac{|b-a|}{\varepsilon}\big)$ by Exercise **11.7**. Thus

$$\big|\log(|f(r_k e^{it})| + \varepsilon) - \log(|f(e^{it})| + \varepsilon)\big| \le \log\left(1 + \frac{\big||f(r_k e^{it)})| - |f(e^{it})|\big|}{\varepsilon}\right)$$
$$\le \frac{|f(r_k e^{it)}) - f(e^{it})|}{\varepsilon}.$$

Therefore

$$\big\|\log(|f_{r_k}| + \varepsilon) - \log(|f| + \varepsilon)\big\|_1 \le \frac{\|f_{r_k} - f\|_1}{\varepsilon}.$$

Hence

$$\big\|\log(|f(r_k e^{it})|+\varepsilon) P_{r/r_k}(\theta-t) - \log(|f(e^{it})|+\varepsilon) P_r(\theta-t)\big\|_1$$
$$\le \big\|\log(|f_{r_k}|+\varepsilon) - \log(|f|+\varepsilon)\big\|_1 \|P_{r/r_k}\|_\infty + \big\|\log(|f|+\varepsilon)\big\|_1 \|P_r - P_{r/r_k}\|_\infty.$$

Letting $k \to \infty$ yields

$$\tfrac{1}{2\pi}\int_{-\pi}^{\pi} \log(|f(e^{it})|+\varepsilon) P_r(\theta-t)\,dt = \lim_{k\to\infty} \tfrac{1}{2\pi}\int_{-\pi}^{\pi} \log(|f(r_k e^{it})|+\varepsilon) P_{\frac{r}{r_k}}(\theta-t)\,dt$$
$$\ge \log|f(z)|.$$

Finally, since

$$\log|f| \le \log(|f| + \varepsilon) \le \log(|f| + 1) \le |f|$$

and the extremes lie in L^1, the Lebesgue dominated convergence theorem shows

$$\log|f(z)| \le \lim_{\varepsilon\to 0^+} \tfrac{1}{2\pi}\int_{-\pi}^{\pi} \log(|f(e^{it})| + \varepsilon) P_r(\theta - t)\,dt$$
$$= \tfrac{1}{2\pi}\int_{-\pi}^{\pi} \log|f(e^{it})| P_r(\theta - t)\,dt.$$

Now we consider the zeros of f, say $\{a_i : i \ge 1\}$ ordered so that $|a_i| \le |a_{i+1}|$. From Jensen's formula, we have for $k \ge K$,

$$\sum_{i=1}^{n_K} \log\frac{r_k}{|a_i|} \le \sum_{i=1}^{n_k} \log\frac{r_k}{|a_i|} = \tfrac{1}{2\pi}\int_{-\pi}^{\pi} \log|f(r_k e^{it})|\,dt - \log|f(0)|$$
$$\le \big\|\log^+|f_{r_k}|\big\|_1 - \log|f(0)| \le \|f\|_1 - \log|f(0)|.$$

Fixing K and letting $k \to \infty$ yields

$$\sum_{i=1}^{n_K} \log\frac{1}{|a_i|} \le \|f\|_1 - \log|f(0)|.$$

Now let $K \to \infty$ to get $\sum_{i=1}^{\infty} \log \frac{1}{|a_i|} < \infty$. For $\frac{1}{2} \le x < 1$, we have

$$2(x-1) \le \log x \le x - 1 \quad \text{so} \quad 1 - x \le \log \frac{1}{x} \le 2(1-x).$$

As there are only finitely many zeros with $|a_i| < \frac{1}{2}$, we see that

$$\sum_{i=1}^{\infty} 1 - |a_i| \le \sum_{|a_i|<1/2} 1 - |a_i| + \sum_{|a_i|\ge 1/2} \log \frac{1}{|a_i|} < \infty. \qquad \blacksquare$$

11.3.5. REMARK. The inequalities at the end of the proof also show that

$$\sum_{|a_i|\ge 1/2} \log \frac{1}{|a_i|} \le 2 \sum_{|a_i|<1/2} 1 - |a_i|.$$

Thus $\sum_{i\ge 1} 1 - |a_i|$ converges if and only if $\sum_{i\ge 1} \log \frac{1}{|a_i|}$ does. Hence for large N,

$$\prod_{i>N} \frac{1}{|a_i|} \le e^{2\sum_{i>N} 1-|a_i|} \quad \text{and} \quad \prod_{i>N} |a_i| \ge e^{-2\sum_{i>N} 1-|a_i|}.$$

11.3.6. COROLLARY. *If $0 \ne f \in H^1$, then $f(e^{i\theta}) \ne 0$ a.e.(m). If μ is an analytic measure, then μ is mutually absolutely continuous with Lebesgue measure.*

PROOF. Since $\log |f|$ is integrable, $f \ne 0$ a.e.(m). If μ is an analytic measure, then the F.&M. Riesz theorem shows that $d\mu = f\,dm$ for some $f \in H^1$. Therefore $m \ll \mu$ as well. $\blacksquare$

11.4. Inner and Outer Functions

This section examines the structure of inner and outer functions, which are defined here. We start by building an infinite Blaschke product from the zeros of f. The outer functions are those functions which are extremal for Jensen's inequality. When we factor out an appropriate Blaschke product and outer function from f, what remains is a singular inner function.

First we wish to show that the condition $\sum_{i\ge 1} 1 - |a_i| < \infty$ is sufficient to be the zero set of an H^∞ function.

11.4.1. LEMMA. *If u_n are continuous functions on a domain Ω and $\sum_{i\ge 1} |u_i|$ converges uniformly on compact subsets, then $\prod_{i\ge 1} 1 + u_i$ converges uniformly on compact subsets. Moreover, the product is non-zero at z unless $1 + u_i(z) = 0$ for some i.*

PROOF. Let $f_N(z) = \prod_{i=1}^{N} 1 + u_i(z)$ for $N \geq 1$. Let $K \subset \Omega$ be compact, and let $\varepsilon > 0$. Choose N so that $\sum_{n>N} |u_n(z)| < \varepsilon$ for all $z \in K$, and set $M(K) = \sup_{z\in K} \sum_{i\geq 1} |u_i(z)|$. Then for $N \leq m < n$,

$$\begin{aligned}|f_n(z) - f_m(z)| &\leq |f_m(z)|\Big(\prod_{i=m+1}^{n} (1 + |u_i(z)|) \;-1\Big)\\ &\leq \prod_{i=1}^{m}(1 + |u_i(z)|)\Big(\prod_{i=m+1}^{n} e^{|u_i|} \;-1\Big)\\ &\leq e^{\sum_{i\geq 1}|u_i|}\Big(e^{\sum_{i>m}|u_i|} \;-1\Big)\\ &\leq e^{M(K)}\big(e^{\varepsilon} - 1\big) \leq e^{M(K)+1}\varepsilon.\end{aligned}$$

Thus, $\{f_n|_K\}$ is Cauchy for each K, and hence the product converges u.c.c. on Ω. The tail of the product is close to 1, so the infinite product vanishes at z only if some $1 + u_i(z) = 0$. ∎

11.4.2. DEFINITION. A function $\omega \in H^\infty$ is *inner* if $|\omega(e^{i\theta})| = 1$ a.e.(m). The inner function constructed in the next theorem is called an (*infinite*) *Blaschke product*. An inner function with no zeros in $\mathbb{D}$ is a *singular inner function.*

11.4.3. THEOREM. *Suppose that $a_i \in \mathbb{D}$ for $i \geq 1$ and $\sum_{i\geq 1} 1 - |a_i| < \infty$. Then $B(z) = \prod_{i\geq 1} b_{a_i}(z)$ converges absolutely in $\mathbb{D}$, and uniformly on $\overline{r\mathbb{D}}$ for $r < 1$. Moreover $B \in H^\infty$, $|B(e^{i\theta})| = 1$ a.e.(m), i.e. B is inner, and the zero set of B is $\{a_i : i \geq 1\}$ including multiplicity.*

PROOF. Let's write $b_{a_i}(z) = 1 - u_i(z)$. If $a_i = 0$, then $|u_i(z)| = |z - 1| \leq 2$. For $a_i \neq 0$,

$$u_i(z) = 1 - \frac{\bar{a}_i(a_i - z)}{|a_i|(1 - \bar{a}_i z)} = \frac{|a_i| - |a_i|\bar{a}_i z - |a_i|^2 + \bar{a}_i z}{|a_i|(1 - \bar{a}_i z)} = \Big(\frac{|a_i| + \bar{a}_i z}{|a_i|}\Big)\frac{1 - |a_i|}{1 - \bar{a}_i z}.$$

Therefore

$$|u_i(z)| \leq 2\frac{1 - |a_i|}{1 - r} \quad \text{for} \quad |z| \leq r < 1.$$

Hence if $|z| \leq r < 1$, then

$$\sum_{i\geq 1} |u_i(z)| \leq \frac{2}{1 - r}\sum_{i\geq 1} 1 - |a_i| < \infty.$$

Thus $\sum_{i\geq 1} |u_i|$ converges uniformly on compact subsets of $\mathbb{D}$. By Lemma 11.4.1, the infinite product converges u.c.c. to a function $B(z)$ which is analytic on $\mathbb{D}$. Clearly $\|B\|_\infty \leq 1$ since each $\|b_{a_i}\|_\infty = 1$. The zero set of B is just the union of zeros of each b_{a_i}, namely, $\{a_i : i \geq 1\}$ including multiplicity.

Let $B_n(z) = \prod_{i=1}^n b_{a_i}(z)$. We will show that this sequence of finite Blaschke products converges to $B(z)$ in $H^2(\mathbb{D})$. For $m < n$, we first compute in H^2:

$$\begin{aligned}\|B_n - B_m\|_2^2 &= \frac{1}{2\pi}\int |B_n|^2 - 2\,\mathrm{Re}\, B_n\overline{B_m} + |B_m|^2\,dt \\ &= 2 - 2\,\mathrm{Re}\,\frac{1}{2\pi}\int \frac{B_n}{B_m}\,dt \\ &= 2\Big(1 - \mathrm{Re}\,\frac{B_n(0)}{B_m(0)}\Big) = 2\big(1 - \prod_{i=m+1}^{n} |a_i|\big)\end{aligned}$$

By Remark 11.3.5, the RHS converges to 0 as $m \to \infty$. Therefore $\{B_n\}$ is a Cauchy sequence in H^2. The pairing between H^2 and $H^2(\mathbb{D})$ is isometric so this sequence converges in $\mathbb{D}$ in the norm $\sup_{0<r<1} \|f_r\|_2$. However, the functions $B_n(z)$ converge u.c.c. to $B(z)$ in $\mathbb{D}$. In particular, $(B_n)_r$ converges uniformly to B_r, which implies convergence in the 2-norm. Hence, $B_n(z)$ converges to $B(z)$ in $H^2(\mathbb{D})$, and so also in H^2. Now $|B_n(e^{i\theta})| = 1$ everywhere. There is a subsequence of $B_n(e^{i\theta})$ which converges to $B(e^{i\theta})$ a.e.(m). Therefore $|B(e^{i\theta})| = 1$ a.e.(m). So B is inner. ∎

11.4.4. REMARK. Suppose that J is an arc of $\mathbb{T}$ which is bounded away from $\overline{\{a_i : i \geq 1\}}$, say by $\delta > 0$. Then we also obtain $|u_i(z)| \leq 2\frac{1-|a_i|}{1-\delta}$ on a neighbourhood of J. It follows that the product defining $B(z)$ converges on J and extends to be analytic across J.

On the other hand, if $e^{i\theta} \in \overline{\{a_i : i \geq 1\}} \cap \mathbb{T}$, then 0 is a cluster point of $B(z)$ at $e^{i\theta}$. But $B(re^{it})$ has radial limits in $\mathbb{T}$ a.e., and so on a dense subset of $\mathbb{T}$. Thus 1 is also a cluster point of $|B(z)|$ as $z \to e^{i\theta}$ in $\mathbb{D}$. Therefore, B cannot extend to be even continuous at any point in $\overline{\{a_i : i \geq 1\}} \cap \mathbb{T}$.

11.4.5. DEFINITION. A function $F \in H^1(\mathbb{D})$ is *outer* if there is $0 \leq k \in L^1$ with $\log k \in L^1$ and $\lambda \in \mathbb{T}$ so that

$$F(z) = \lambda \exp\Big(\frac{1}{2\pi}\int_{-\pi}^{\pi} \frac{e^{it}+z}{e^{it}-z}\log k(e^{it})\,dt\Big).$$

We first recall a useful inequality about convex functions. A function φ on $[a, b]$ is convex if $\{(x, y) : y \geq \varphi(x), x \in [a, b]\}$ is a convex set. By the Hahn–Banach theorem for $\mathbb{R}^2$, this says that

$$\varphi(x_0) = \sup_{\alpha x+\beta \leq \varphi} \alpha x_0 + \beta.$$

11.4.6. Jensen Inequality. *Let φ be a convex function on $\mathbb{R}$, and let μ be a probability measure. For $h \in L^1_{\mathbb{R}}(\mu)$,*

$$\varphi\Big(\int h\,d\mu\Big) \le \int \varphi(h)\,d\mu.$$

Proof. Let $x_0 = \displaystyle\int h\,d\mu$. Then

$$\varphi(x_0) = \sup_{\alpha x+\beta\le\varphi} \alpha x_0 + \beta = \sup_{\alpha x+\beta\le\varphi} \int \alpha h + \beta\,d\mu \le \int \varphi(h)\,d\mu. \qquad \blacksquare$$

11.4.7. Proposition. *If $0 \le k \in L^1$ such that $\log k \in L^1$, then*

$$F(z) = \lambda \exp\Big(\frac{1}{2\pi}\int_{-\pi}^{\pi} \frac{e^{it}+z}{e^{it}-z}\log k(e^{it})\,dt\Big)$$

belongs to H^1 and $|F(e^{i\theta})| = k(e^{i\theta})$ a.e.(m).

Proof. Recall that $H_r(t) = \frac{e^{it}+z}{e^{it}-z} = P_r(\theta-t) + iQ_r(\theta-t)$ for $z = re^{i\theta}$ is the Herglotz kernel. The real harmonic function

$$u(z) = u(re^{i\theta}) = \frac{1}{2\pi}\int_{-\pi}^{\pi} P_r(\theta-t)\log k(e^{it})\,dt$$

belongs to h_1 and u_r converges to $\log k(e^{it})$ in L^1. The function $v(z)$ is the harmonic conjugate of u, and

$$g(z) = u(z) + iv(z) = \frac{1}{2\pi}\int_{-\pi}^{\pi} \frac{e^{it}+z}{e^{it}-z}\log k(e^{it})\,dt$$

is analytic in $\mathbb{D}$. Thus $F(z) = \lambda e^{g(z)}$ is also analytic in $\mathbb{D}$. Moreover,

$$|F(z)| = e^{\operatorname{Re} g(z)} = e^{u(z)}.$$

Therefore by Fatou's theorem, $|F(re^{i\theta})|$ converges nontangentially to $e^{\log k} = k(e^{i\theta})$ a.e.(m).

We let $\varphi(x) = e^x$, which is convex, and $\mu = P_r(\theta-t)\,\frac{dt}{2\pi}$, which is a probability measure. Then by Jensen's inequality

$$\begin{aligned}|F(z)| = e^{u(z)} &= \exp\Big(\int \log k(e^{it})P_r(\theta-t)\,\frac{dt}{2\pi}\Big)\\ &\le \int k(e^{it})P_r(\theta-t)\,\frac{dt}{2\pi} = k_r(e^{i\theta}).\end{aligned}$$

Therefore $\|F_r\|_1 \le \|k_r\|_1 \le \|k\|_1$. So $F(z) \in H^1(\mathbb{D})$, and there is an L^1 limit $\lim_{r\to 1^-} F_r = F$. Thus by Fatou's theorem,

$$|F(e^{i\theta})| = \text{n.t.}\lim_{r\to1^-}|F_r(e^{i\theta})| = k(e^{i\theta}) \text{ a.e.}(m). \qquad \blacksquare$$

We can now establish some basic properties of outer functions.

11.4.8. THEOREM. *For $F \in H^1$, the following are equivalent:*

(1) *F is outer.*

(2) $F(z) = \lambda \exp\Big(\frac{1}{2\pi}\displaystyle\int_{-\pi}^{\pi} \frac{e^{it}+z}{e^{it}-z} \log|F(e^{it})|\,dt\Big)$ *for some $\lambda \in \mathbb{T}$.*

(3) $\log|F(z_0)| = \frac{1}{2\pi}\displaystyle\int_{-\pi}^{\pi} \log|F(e^{it})|P_r(\theta - t)\,dt$ *for some $z_0 = re^{i\theta} \in \mathbb{D}$.*

(4) *For all $f \in H^1$ such that $|f(e^{i\theta})| = |F(e^{i\theta})|$ a.e.(m), $|f(z)| \le |F(z)|$ for all $z \in \mathbb{D}$.*

PROOF. The equivalence of (1) and (2) is contained in the previous theorem. Thus $|F(z)| = e^{u(z)}$ where u is the Poisson integral of $\log|F|$. Hence for all $z = re^{i\theta} \in \mathbb{D}$,

$$\log|F(z)| = u(z) = \frac{1}{2\pi}\int_{-\pi}^{\pi} \log|F(e^{it})|P_r(\theta - t)\,dt.$$

If $f \in H^1$ and $|f(e^{i\theta})| = |F(e^{i\theta})|$ a.e.(m), then by Theorem 11.3.4,

$$\log|f(z)| \le \frac{1}{2\pi}\int_{-\pi}^{\pi} \log|f(e^{it})|P_r(\theta - t)\,dt = \log|F(z)|.$$

Hence (2) implies (3) for all $z \in \mathbb{D}$ and (4).

If (3) or (4) holds, we set $G(z) = \exp\Big(\frac{1}{2\pi}\displaystyle\int_{-\pi}^{\pi} \frac{e^{it}+z}{e^{it}-z} \log|F(e^{it})|\,dt\Big)$. Applying the previous paragraph, we obtain $|F(z)| \le |G(z)|$ for all $z \in \mathbb{D}$. Hence $g(z) = \dfrac{F(z)}{G(z)}$ is analytic on $\mathbb{D}$ and $\|g\|_\infty \le 1$. Now (3) implies that $|g(z_0)| = 1$ while (4) implies that $|g(z)| = 1$ everywhere. By the maximum principle, $g = \lambda \in \mathbb{T}$ is constant. So $F(z) = \lambda G(z)$, which proves (2). ∎

The following useful corollary is immediate. This holds, in particular, if $f(e^{i\theta}) \ge \varepsilon > 0$ a.e.(m).

11.4.9. COROLLARY. *If $0 \le f \in L^p$ for $1 \le p \le \infty$, then there is an outer function $h \in H^p$ such that $|h| = f$ a.e.(m) if and only if $\displaystyle\int_{-\pi}^{\pi} \log f(e^{it})\,dt > -\infty$.*

Now we do a similar analysis to describe singular inner functions.

11.4.10. PROPOSITION. *Let $S(z)$ be a singular inner function. Then there is a positive measure $\mu \in M_+(\mathbb{T})$ which is singular to m such that*

$$S(z) = \lambda \exp\Big(-\int_{-\pi}^{\pi} \frac{e^{it}+z}{e^{it}-z}\,d\mu\Big)$$

for some $\lambda \in \mathbb{T}$. Every function of this form is inner.

PROOF. Since S does not vanish on $\mathbb{D}$, there is an analytic branch of $\log S(z)$. Thus $u(z) = \log|S(z)| = \operatorname{Re}\log S(z)$ is harmonic. Now $\|S\|_\infty = 1$, so that $u \le 0$. Moreover for $0 < r < 1$,

$$\frac{1}{2\pi}\int_{-\pi}^{\pi} |u(re^{i\theta})|\, d\theta = \frac{-1}{2\pi}\int_{-\pi}^{\pi} u(re^{i\theta})\, d\theta = -u(0) = -\log|S(0)| < \infty.$$

Therefore $u \in h_1$.

By Theorem 11.1.12, there is a measure $\mu = -w^*\text{–}\lim_{r\to 1^-} u_r dm$ which is positive since $u_r \le 0$. However $|S_r|$ converges to 1 a.e.(m), so u_r converges to 0 a.e.(m). By Fatou's theorem, this shows that $\mu \perp m$. Finally, the analytic function with u as real part is obtained by integration against the Herglotz kernel. Thus

$$F(z) = \exp\Big(-\int_{-\pi}^{\pi} \frac{e^{it}+z}{e^{it}-z}\, d\mu\Big)$$

has $|F(z)| = e^u = |S(z)|$. Therefore $S = \lambda F$ for some $\lambda \in \mathbb{T}$.

Conversely, suppose that $\mu \in M_+(\mathbb{T})$ and $\mu \perp m$. Then $|S(z)| = e^{-\widetilde{\mu}(z)}$. By Fatou's theorem, $|S_r| \to e^0 = 1$ a.e.(m) as $r \to 1^-$. Hence S is inner. ■

11.4.11. EXAMPLE. $S(z) = e^{-\frac{1+z}{1-z}}$ is the singular inner function corresponding to the point mass δ_1.

11.4.12. REMARK. In a certain sense, a singular inner function has zeros on the boundary. We observed in Remark 11.1.15 that if μ is a positive singular measure, then $\widetilde{\mu}(re^{i\theta}) \to \infty$ a.e.(μ). This translates into saying that

$$\lim_{r\to 1^-} |S(re^{i\theta})| = \lim_{r\to 1^-} e^{-\widetilde{\mu}(re^{i\theta})} = 0 \text{ a.e.}(\mu).$$

Since $|S(e^{i\theta})| = 1$ a.e.(m), which holds on a dense set, S cannot extend to be continuous at any point of $\overline{\operatorname{supp}\mu}$.

Also note that the integral formula for $S(z)$ is valid for any $z \in \mathbb{C} \setminus \overline{\operatorname{supp}\mu}$, and so $S(z)$ is analytic there. In particular, if J is an open arc of $\mathbb{T}$ disjoint from the support of μ, then $S(z)$ continues analytically across J. Thus the restriction to this arc is C^∞ and modulus 1 everywhere.

This leads to an important factorization theorem for H^p functions. The result is the *inner-outer factorization*.

11.4.13. THEOREM. *Let $f \in H^p$ for $1 \le p \le \infty$. Then there is a factorization $f = BSF$ where*

(1) *B is the Blaschke product for the zeros of f including multiplicity.*

(2) *S is a singular inner function.*

(3) *F is an outer function with $|F(e^{it})| = |f(e^{it})|$ a.e.(m).*

These terms are unique up to a scalar in $\mathbb{T}$.

PROOF. Let $F(z) = \exp\Big(\frac{1}{2\pi}\int_{-\pi}^{\pi} \frac{e^{it}+z}{e^{it}-z} \log|F(e^{it})|\,dt\Big)$. Then F is outer and $|f(e^{i\theta})| = |F(e^{i\theta})|$ a.e.(m). Therefore $F \in H^1(\mathbb{D})$ and $F(e^{i\theta}) \in L^p$, so $F \in H^1 \cap L^p = H^p$. By Theorem 11.4.8, $|f(z)| \le |F(z)|$ for $z \in \mathbb{D}$. Hence $g(z) = f(z)/F(z)$ is analytic in $\mathbb{D}$ with $\|g\|_\infty = 1$. Now f_r and F_r converge to f and F in L^p and g_r converges weak-$*$ to g in L^∞. By Fatou's theorem, $f_r \to f$ a.e.(m) and $F_r \to F$ a.e.(m). Thus $g(e^{i\theta}) = f(e^{i\theta})/F(e^{i\theta}) \in \mathbb{T}$ a.e.(m). Hence g is inner. Now $S(z) = g(z)/B(z)$ is analytic in $\mathbb{D}$ because it has removable singularities at the zeros of f. Let B_n be the finite Blaschke product from the first n zeros of f. Then B_n converges u.c.c. to $B(z)$, so $S(z)$ is the u.c.c. limit of $s_n(z) := g(z)/B_n(z)$. Now $\|s_n\|_\infty = 1$, and therefore $\|S\|_\infty \le 1$. Arguing as above, we see that $|S(e^{i\theta})| = 1$ a.e.(m). Since S has no zeros, it is a singular inner function. Since F and B are unique up to a scalar in $\mathbb{T}$, so is S. ∎

11.4.14. COROLLARY. *If $f \in H^1$, then there are functions $f_i \in H^2$ so that $f = f_1 f_2$ and $\|f_1\|_2^2 = \|f_2\|_2^2 = \|f\|_1$.*

PROOF. Let

$$G(z) = \exp\Big(\frac{1}{2\pi}\int_{-\pi}^{\pi} \frac{e^{it}+z}{e^{it}-z}\,\frac{\log|F(e^{it})|}{2}\,dt\Big).$$

Then $G(z)^2 = F(z)$ is the outer factor of f, so that $f = \omega F$ where ω is inner. Let $f_1 = \omega G$ and $f_2 = G$. ∎

The set of inner functions forms a semigroup. Say that ω_1 divides ω_2 if $\omega_2 = \omega_1 h$ for some $h \in H^\infty$. But clearly this forces $|h| = 1$ a.e.(m), so that h is inner. The following proposition is left as an exercise.

11.4.15. PROPOSITION. *For two inner functions $\omega_i = B_i S_i$ where B_i are Blaschke products and S_i are singular with measures μ_i, the following are equivalent:*

(1) *ω_1 divides ω_2.*

(2) *The zeros of B_1 counting multiplicity is a subset of the zero set of B_2 and $\mu_1 \le \mu_2$.*

(3) *$\omega_1 H^2 \supset \omega_2 H^2$.*

The inner function defined by the next proposition is called the *greatest common divisor* (gcd) of a family of inner functions.

11.4.16. PROPOSITION. *If $\mathcal{S} \subset H^\infty$, there is a unique inner function (up to a scalar) which is the largest inner function that divides every $h \in \mathcal{S}$.*

PROOF. For each $h \in \mathcal{S}$, factor $h = BSF$. The greatest common divisor of the inner factors is what we want. Take the intersection of the zero sets (including multiplicity) of all the Blaschke products. For the singular parts, take the largest measure smaller than all of the measures for each function. More precisely, given two positive measures μ and ν, write $\nu = f\mu + \nu_s$ where $f \in L^\infty(\mu)$ and $\nu_s \perp \mu$. Then $\min\{1, f\}\mu$ is the minimum $\mu \wedge \nu$ of μ and ν. Then given a set of measures $\{\mu_\alpha\}$, let $\mathcal{M} = \{\mu_{\alpha_1} \wedge \cdots \wedge \mu_{\alpha_n} : n < \infty\}$. Define $\mu_0(A) = \inf_{\mu \in \mathcal{M}} \mu(A)$ for Borel sets A. Check that this is a measure. ∎

11.5. The Disc Algebra $A(\mathbb{D})$

We have already seen the disc algebra as a standard example several times. In particular, in Examples 2.2.7, 6.2.7, 7.1.14 and 7.8.4, we have considered $A(\mathbb{D})$ as a Banach algebra, examined the spectrum of individual elements, shown that it is semisimple and found the Shilov boundary. In this section, we will primarily study its ideal structure and its relation to interpolation.

Recall that the maximal ideal space of $A(\mathbb{D})$ is $\overline{\mathbb{D}}$, so that we can consider $A(\mathbb{D})$ as a subalgebra of $C(\overline{\mathbb{D}})$. We showed that the Shilov boundary of $A(\mathbb{D})$ is $\mathbb{T}$. One direction follows from the maximum modulus principle. The other direction was established using the functions $z + e^{i\theta}$ which peak at a single point. So the restriction map of $A(\mathbb{D})$ into $C(\mathbb{T})$ is isometric. Hence, we can also consider $A(\mathbb{D})$ as a subalgebra of $C(\mathbb{T})$.

11.5.1. DEFINITION. A uniform algebra A is a *Dirichlet algebra* if $\operatorname{Re} A$ is dense in $C_{\mathbb{R}}(\partial_A)$.

The prototype for Dirichlet algebras is $A(\mathbb{D})$.

11.5.2. PROPOSITION. *$A(\mathbb{D})$ is a Dirichlet algebra.*

PROOF. Let $u \in C_{\mathbb{R}}(\mathbb{T})$. Let $u(z)$ be the harmonic extension of u, and let $v(z)$ be its harmonic conjugate. Then $h(z) = u(z) + iv(z)$ is an analytic function on $\mathbb{D}$. Thus $h_r = u_r + iv_r$ belong to $A(\mathbb{D})$ for $0 < r < 1$, and so $u_r \in \operatorname{Re} A(\mathbb{D})$. By Poisson's theorem, u_r converges to u uniformly as $r \to 1^-$. Therefore $A(\mathbb{D})$ is a Dirichlet algebra. ∎

Here is a result of an unusual type.

11.5.3. Wermer Maximality Theorem. *Let B be a closed subalgebra of $C(\mathbb{T})$ such that $A(\mathbb{D}) \subset B$. Then either $B = A(\mathbb{D})$ or $B = C(\mathbb{T})$.*

PROOF. Consider the ideal $J = zB$. This is closed because multiplication by z is an isometry on $C(\mathbb{T})$.

Case 1: $J = B$. In this case, there is some $f \in B$ so that $zh = 1$, whence $\bar{z} = \frac{1}{z} \in B$. But then $B \supset \overline{\mathrm{Alg}\{z, \bar{z}\}} = C(\mathbb{T})$ by the Weierstrass theorem.

Case 2: J is a proper ideal. By Zorn's lemma, J is contained in a maximal ideal M of B. Let $\varphi : B \to B/M \simeq \mathbb{C}$ be the multiplicative linear functional on B associated to M. Use the Hahn–Banach theorem to extend φ to a functional of norm 1 on $C(\mathbb{T})$. By the Riesz representation theorem, there is a measure $\mu \in M(\mathbb{T})$ such that

$$\varphi(f) = \int_{\mathbb{T}} f \, d\mu \quad \text{for all} \quad f \in B.$$

Now

$$1 = \varphi(1) = \int_{\mathbb{T}} 1 \, d\mu \leq \int_{\mathbb{T}} d|\mu| = \|\mu\| = \|\varphi\| = 1.$$

Thus μ is a probability measure on $\mathbb{T}$. Since $1 \in B$, we have $z^n \in J \subset M$ for $n \geq 1$. Hence

$$0 = \varphi(z^n) = \int_{\mathbb{T}} z^n \, d\mu = \hat{\mu}(-n) \quad \text{for} \quad n \geq 1.$$

By positivity, this means that $\hat{\mu}(n) = 0$ for $n \geq 1$ as well. Thus $\mu \in \mathbb{C}m$ and is a probability measure, so $\mu = m$. Now if $f \in B$, then $z^n f \in J \subset M$ for $n \geq 1$. Thus

$$0 = \varphi(z^n f) = \int_{\mathbb{T}} f z^n \, d\mu = \hat{f}(-n) \quad \text{for} \quad n \geq 1.$$

Therefore $f \in A(\mathbb{D})$; so $B = A(\mathbb{D})$. ∎

11.5.4. Definition. Let $\mathcal{A}$ be a uniform algebra on X. A closed subset $E \subset X$ is a *peak set* if there is an $f \in \mathcal{A}$ such that $f|_E = 1$ and $|f(x)| < 1$ for all $x \in X \backslash E$. A *peak point* is a peak set consisting of a single point.

Say that E is a *peak interpolation set* if for every $0 \neq g \in C(E)$, there is an $h \in \mathcal{A}$ such that $h|_E = g$ and $|h(x)| < \|g\|_\infty$ for every $x \in X \setminus E$.

It is clear that every peak point must belong to the Shilov boundary. More generally, the Shilov boundary must intersect every peak set. For a proper closed subset of $\overline{\mathbb{D}}$ to be a peak set for $A(\mathbb{D})$, it must be contained in the unit circle. This is because if f attains its maximum modulus at an interior point, then it is constant. Also if $f|_{\mathbb{T}} = 1$, then $f = 1$, so $\mathbb{T}$ is not a peak set. Moreover, if $E \subset \mathbb{T}$ is a peak set for f, then $1 - f$ vanishes on E and nowhere else. Thus Corollary 11.3.6 shows that $m(E) = 0$.

Peak interpolation sets are peak sets, but $\overline{\mathbb{D}}$ is a peak set because $1 \in A(\mathbb{D})$, but it is not an interpolation set. Typically this is a stronger condition than peak set, but for $A(\mathbb{D})$, they coincide if we ignore the trivial case of $\overline{\mathbb{D}}$.

11.5.5. LEMMA. *$A(\mathbb{D})^* = L^1/H_0^1 \oplus_1 M_s(\mathbb{T})$ where $M_s(\mathbb{T})$ is the space of measures which are singular to Lebesgue measure.*

PROOF. Since $A(\mathbb{D}) \subset C(\mathbb{T})$, we have $A(\mathbb{D})^* \simeq C(\mathbb{T})^*/A(\mathbb{D})^\perp$. The Riesz representation theorem states that $C(\mathbb{T})^* = M(\mathbb{T})$. If $\mu \in A(\mathbb{D})^\perp$, then

$$\hat{\mu}(-n) = \int_{\mathbb{T}} z^n \, d\mu = 0 \quad \text{for} \quad n \geq 0.$$

By the F.&M. Riesz Theorem 11.2.5, $\mu = hm$ for some $h \in H^1$, and $\hat{\mu}(0) = 0$ means that $h \in H_0^1$.

The space of measures decomposes as $M(\mathbb{T}) \simeq L^1 \oplus_1 M_s(\mathbb{T})$ where $f \in L^1$ corresponds to the measure $fm \ll m$ and $M_s(\mathbb{T})$ is the space of measures singular to m. This decomposition is a consequence of the Radon–Nikodym theorem. Since $A(\mathbb{D})^\perp = H_0^1 \subset L^1$, we obtain $A(\mathbb{D})^* = L^1/H_0^1 \oplus_1 M_s(\mathbb{T})$. ■

Since measures play a major role in the next proof, we will write $\mu(f)$ instead of the more cumbersome $\int_{\mathbb{T}} f \, d\mu$. M-ideals (Definition 2.7.7) play a key role in our proof.

11.5.6. RUDIN–CARLESON THEOREM. *If $E \subset \mathbb{T}$ is a closed set with $m(E) = 0$, then E is a peak interpolation set for $A(\mathbb{D})$.*

PROOF. Consider the restriction map $\rho : A(\mathbb{D}) \to C(E)$. It is clear that $\ker \rho = I(E) = \{f \in A(\mathbb{D}) : f|_E = 0\}$. Our first step is to show that $I(E)$ is an M-ideal. By Lemma 11.5.5,

$$A(\mathbb{D})^* = L^1/H_0^1 \oplus_1 M_s(\mathbb{T}) = L^1/H_0^1 \oplus_1 M_s(\mathbb{T} \setminus E) \oplus_1 M(E).$$

At the second step, we decompose $M_s(\mathbb{T}) = M_s(\mathbb{T} \setminus E) \oplus_1 M_s(E)$. But since $m(E) = 0$, E does not support any non-zero measures absolutely continuous to m, so $M_s(E) = M(E) = C(E)^*$. By Lemma 4.1.7, $I(E)^\perp = (\ker \rho)^\perp = \overline{\operatorname{Ran} \rho^*}$. But for $\mu \in M(E)$, $\rho^*(\mu)(f) = \mu(f|_E) = \mu(f)$. That is, $\rho^*(\mu) = \mu$ where we consider a measure on E as a measure on $\mathbb{T}$ which has measure 0 on $\mathbb{T} \setminus E$. Thus we have exhibited the L-summand $M(E)$ in $A(\mathbb{D})^*$. Hence $I(E)$ is an M-ideal. This argument contains additional information. The map ρ^* is an isometry, and hence ρ is surjective and a co-isometry, i.e. ρ is a quotient map. This means that $A(\mathbb{D})/I(E) \simeq C(E)$.

Now we can apply Theorem 2.7.11. This shows that ρ maps the closed unit ball of $A(\mathbb{D})$ onto the closed unit ball of $C(E)$. Thus every $g \in C(E)$ can be interpolated by some function $h \in A(\mathbb{D})$ with $\|h\|_\infty = \|g\|_\infty$. To complete the proof, it suffices to show that E is a peak set, for if f peaks on E, then fh interpolates g and if $z \in \overline{\mathbb{D}} \setminus E$, then $|f(z)h(z)| < |h(z)| \le \|g\|_\infty$.

First consider $F_1 = E \cup \{\lambda\}$ for some $\lambda \in \mathbb{T} \setminus E$. Since F_1 is closed and $m(F_1) = 0$, there is some $h \in A(\mathbb{D})$ with $\|h\| = 1 = h|_E$ and $h(\lambda) = 0$. Let $f_1 = \frac{1}{2}(1+h)$. Then $\operatorname{Ran} f_1 \cap \mathbb{T} = \{1\}$, so it attains its maximum modulus only on $f_1^{-1}(1) =: F_2$. Now f_1 is not constant since $f_1(\lambda) = \frac{1}{2}$. Since $f_1 - 1$ vanishes on F_2, Corollary 11.3.6 shows that $m(F_2) = 0$. Let g_2 be a continuous function on F_2 so that $0 \le g_2 \le 1$ and $g_2^{-1}(1) = E$. By the previous paragraph, there is $f_2 \in A(\mathbb{D})$ with $\|f_2\| = 1$ so that $f_2|_{F_2} = g_2$. Then $f = f_1 f_2$ peaks precisely at E. ■

11.5.7. Remark. Note that if f peaks on E, then $|f(z)| < 1$ for all $z \in \mathbb{D}$ as well. This is because if $|f(z)| = 1$, then f would be constant by the maximum modulus principle.

Next we will describe all of the closed ideals in $A(\mathbb{D})$. This needs some preparation.

11.5.8. Definition. If $\omega = BS$ is inner, the *support* of ω is

$$\operatorname{supp}(\omega) := \overline{\operatorname{supp}(\mu)} \cup \overline{Z(B)} \cap \mathbb{T},$$

where $Z(B)$ is the set of zeros of B in $\mathbb{D}$, μ is the singular measure for S and $\overline{\operatorname{supp}(\mu)}$ is the complement of all open sets which are μ-null.

11.5.9. Theorem. *Let $0 \ne f \in A(\mathbb{D})$ have inner-outer factorization $f = \omega F$. Then $K = \operatorname{supp}(\omega)$ has $m(K) = 0$, $F \in A(\mathbb{D})$ and $F|_K = 0$.*

Proof. Let $\omega = BS$. If $e^{i\theta} \in \overline{Z(B)} \cap \mathbb{T}$, there are zeros z_i of B such that $z_i \to e^{i\theta}$. Thus $f(e^{i\theta}) = \lim_{i\to\infty} f(z_i) = 0$. Let μ be the singular measure for S. For μ-almost all $e^{i\theta} \in \operatorname{supp}(\mu)$, $\lim_{r\to 1^-} \widetilde{S}(re^{i\theta}) = 0$ by Remark 11.4.12. Any isolated point of $\operatorname{supp}\mu$ must correspond to a point mass in μ. Therefore by continuity, $f(e^{i\theta}) = 0$ on $\overline{\operatorname{supp}(\mu)}$. Hence f vanishes on K. By Corollary 11.3.6, $m(K) = 0$.

Both B and S are continuous on $\mathbb{T} \setminus K$, and hence F is also continuous there. For $e^{i\theta} \in K$, we have $f(e^{i\theta}) = 0$. Given $\varepsilon > 0$, there is $\delta > 0$ so that $|f(e^{it})| < \varepsilon$ for $t \in (\theta - \delta, \theta + \delta)$. Let $|\theta' - \theta| < \frac{\delta}{2}$. Since F is outer, by Theorem 11.4.8

$$\begin{aligned}\log|F(re^{i\theta'})| &= \frac{1}{2\pi}\int_{-\pi}^{\pi} \log|f(e^{it})| P_r(\theta' - t)\,dt \\ &\le \frac{1}{2\pi}\int_{-\delta/2}^{\delta/2} \log\varepsilon\; P_r(t)\,dt + \frac{1}{2\pi}\int_{\frac{\delta}{2}\le|t|\le\pi} \log^+|f(e^{it})| P_r(\tfrac{\delta}{2})\,dt\end{aligned}$$

$$\leq \log \varepsilon \, \tfrac{1}{2\pi} \int_{-\delta/2}^{\delta/2} P_r(t)\, dt + \| \log^+ |f| \|_1 P_r(\tfrac{\delta}{2}) \to \log \varepsilon \quad \text{as } r \to 1^-.$$

Hence $|F(e^{i\theta'})| \leq \varepsilon$ on $(\theta - \frac{\delta}{2}, \theta + \frac{\delta}{2})$. Therefore $F(e^{i\theta}) = 0$ and F is continuous at $e^{i\theta}$. So $F \in A(\mathbb{D})$ and $F|_K = 0$. ∎

If $\{0\} \neq J \subset A(\mathbb{D})$ is an ideal, let

$$Z(J) = \{e^{i\theta} : f(e^{i\theta}) = 0 \text{ for all } f \in J\}.$$

This is a closed set of measure 0. Likewise if $K \subset \mathbb{T}$, we let

$$I(K) = \{f \in A(\mathbb{D}) : f|_K = 0\}.$$

For organizational reasons, we discuss the invariant subspaces of the shift operator in the next chapter. It could be done in this chapter, and books on H^p spaces do it before this section. We need Corollary 12.2.15 which shows that the weak-$*$ closed ideals of H^∞ have the form ωH^∞ for ω inner.

11.5.10. THEOREM. *Let $K \subset \mathbb{T}$ be a closed set with Lebesgue measure 0. Let ω be an inner function such that* $\operatorname{supp}(\omega) \subset K$. *Then $I(K, \omega) = \omega I(K)$ is a closed ideal of $A(\mathbb{D})$. Every non-zero closed ideal of $A(\mathbb{D})$ has this form.*

PROOF. Clearly $I(K)$ is a closed ideal in $A(\mathbb{D})$. Since multiplication by ω is an isometry, $\omega I(K)$ is closed in H^∞. If $F \in I(K)$, then ωF is continuous on K because ω is bounded, and it is continuous on $\mathbb{T} \setminus K$ because both F and ω are continuous there. So $I(K, \omega) \subset A(\mathbb{D})$, and it is clearly an ideal.

Conversely, let J be a closed ideal of $A(\mathbb{D})$. Let $K = Z(J)$. By Proposition 11.4.16, the gcd of the inner factors of elements of J is defined, say ω. Then by Theorem 11.5.9, $\operatorname{supp}(\omega) \subset K$. Hence $J \subset I(K, \omega)$. Let

$$J' = \{F \in A(\mathbb{D}) : \omega F \in J\}.$$

This is a closed ideal because multiplication by ω is isometric. By Theorem 11.5.9, $J' \subset I(K)$ and $J = \omega J'$. Therefore it suffices to show that $J' = I(K)$.

Let φ be a linear functional on $A(\mathbb{D})$ which annihilates J'. By the Hahn–Banach and Riesz representation theorems, there is a measure μ on $\mathbb{T}$ so that $\varphi(f) = \int f \, d\mu$. By the Radon–Nikodym theorem, $\mu = w \, dm + \nu$ where $\nu \perp m$ is singular. For any $f \in J'$,

$$0 = \varphi(f z^n) = \int z^n f \, d\mu = \widehat{f\mu}(-n) \quad \text{for} \quad n \geq 0.$$

By the F.&M. Riesz theorem, $f\mu = fw \, dm + f\nu \ll m$ and hence $f\nu = 0$. Thus $\operatorname{supp}(\nu) \subset Z(f)$ for all $f \in J'$. But $Z(J') = Z(J) = K$, whence $\operatorname{supp}(\nu) \subset K$. Therefore ν annihilates $I(K)$ and, in particular, J'. Hence $w \, dm$ annihilates J', i.e. $\int fw \, dm = 0$ for all $f \in J'$. As $w \in L^1$, it follows that $w \, dm$ annihilates $M = \overline{J'}^{\text{weak-}*}$ in H^∞. This is a weak-$*$ closed subspace of H^∞ which is invariant under multiplication by $\overline{A(\mathbb{D})}^{\text{weak-}*} = H^\infty$, i.e. M is a weak-$*$ closed ideal of H^∞.

Therefore, by Corollary 12.2.15, there is an inner function u so that $M = uH^\infty$. This u must divide every element of J', but the elements of J' have no common inner factor. Hence $u = 1$ and $M = H^\infty$. Since $w\,dm \perp H^\infty$, in particular, $w\,dm \perp I(K)$. Consequently φ annihilates $I(K)$. By the Hahn–Banach theorem, $J' = I(K)$, and so $J = I(K, \omega)$. ■

11.5.11. COROLLARY. *Every closed ideal of $A(\mathbb{D})$ is singly generated as a closed ideal.*

PROOF. Let $J = I(K, \omega)$. By the Rudin–Carleson theorem, there is a function $f \in A(\mathbb{D})$ which peaks precisely on K. So $f|_K = 1$ and $|f(e^{i\theta})| < 1$ for $e^{i\theta} \in \mathbb{T} \setminus K$. Factor $1 - f = uF$ where F is outer and u is inner. By Theorem 11.5.9, $F \in A(\mathbb{D})$; and $Z(F) = Z(1 - f) = K$ by construction. Consider the closed ideal J' generated by ωF. Then the gcd of the inner factors of the generating set is ω and $Z(f) = K$, so $J' = I(K, \omega)$. ■

11.5.12. COROLLARY. *Suppose that $f_1, \dots, f_n \in A(\mathbb{D})$ have no common zeros in $\overline{\mathbb{D}}$. Then there are $g_1, \dots, g_n \in A(\mathbb{D})$ so that $\sum_{i=1}^n f_i g_i = 1$.*

PROOF. The conclusion is equivalent to saying that the ideal J generated by $f_1, \dots, f_n$ is all of $A(\mathbb{D})$. The zero set of this ideal is $\mathbb{T} \cap \bigcap_{i=1}^n Z(f_i) = \varnothing$. By Theorem 11.5.10, $J = I(\varnothing, B) = BA(\mathbb{D})$ where B is an inner function with empty support, namely, a finite Blaschke product. However, if the f_i had a common Blaschke factor, then they would have a common zero. So $B = 1$ and $J = A(\mathbb{D})$. ■

11.5.13. REMARK. This is an unnecessarily difficult way to prove this corollary. The maximal ideal space of $A(\mathbb{D})$ is $\overline{\mathbb{D}}$. If $f_1, \dots, f_n$ have no common zeros in $\overline{\mathbb{D}}$, then they do not lie inside any maximal ideal. Hence, they must generate $A(\mathbb{D})$ as a closed ideal. But any element $\sum_{i=1}^n f_i g_i$ of the ideal $\langle f_1, \dots, f_n \rangle$ in the ball $b_1(1)$ is invertible, and thus closure is not required.

11.6. The Maximal Ideal Space of H^∞

The maximal ideal space of H^∞ is a very large, complicated object. In this section, we obtain a few details about it. In particular, we identify $\mathcal{M}_{L^\infty}$ as its Shilov boundary.

Recall Theorem 11.2.10 which showed that the unit disc $\mathbb{D}$ sits in $\mathcal{M}_{H^\infty}$ homeomorphically. We showed that $h \in H^\infty$ is invertible if and only if $|h(z)|$ is bounded below on $\mathbb{D}$. Here is another useful characterization.

11.6.1. PROPOSITION. *A function $h \in H^\infty$ is invertible if and only if it is outer and is invertible in L^∞, i.e. $|h(e^{i\theta})| \geq \varepsilon > 0$ a.e.(m).*

PROOF. If h is invertible in H^∞, then the inner factor of h is constant because inner functions are not bounded below on $\mathbb{D}$. Clearly the inverse is also an inverse in L^∞. So $|h|$ is bounded below a.e. Conversely, if h is outer and bounded below a.e.(m) on $\mathbb{T}$, then $\log|h(e^{i\theta})|$ is bounded. Thus

$$\frac{1}{h(z)} = \exp\Big(\frac{-1}{2\pi}\int_{-\pi}^{\pi} \frac{e^{it}+z}{e^{it}-z} \log|h(e^{it})|dt\Big)$$

belongs to H^∞. So h is invertible. ∎

Now $\mathbb{D}$ is identified with the open set $\{\varphi \in \mathcal{M}_{H^\infty} : |\varphi(z)| < 1\}$ as in Theorem 11.2.10. These characters are also special in that they are weak-$*$ continuous. One direction is easy, since $\varepsilon_{re^{i\theta}}$ is obtained by integration against the L^1 function $P_r(\theta - t)\,dm(t)$.

11.6.2. LEMMA. *A bounded sequence $h_n \in H^\infty$ converges weak-$*$ to h if and only if $h_n(z)$ converges u.c.c. to $h(z)$ on $\mathbb{D}$.*

PROOF. If h_n converges weak-$*$ to h, then for $w = re^{i\theta} \in \mathbb{D}$, then

$$\lim_{n\to\infty} h_n(w) = \lim_{n\to\infty} \frac{1}{2\pi}\int_{-\pi}^{\pi} h_n(e^{it})P_r(\theta-t)\,dt = \frac{1}{2\pi}\int_{-\pi}^{\pi} h(e^{it})P_r(\theta-t)\,dt = h(w).$$

Let $M = \sup_{n\geq 1}\|h_n\|$. For $R < 1$, the set $\mathcal{D}_R = \{\varepsilon_w : w \in \overline{\mathbb{D}_R}\}$ is compact in L^1. Given $\delta > 0$, select an δ-net $\{\varepsilon_{w_i} : 1 \leq i \leq k\}$ for $\mathcal{D}_R$. Then choose N so that $|h_n(w_i) - h(w_i)| < \delta$ for $1 \leq i \leq k$ and all $n \geq N$. For $|w| \leq R$, pick i so that $\|\varepsilon_w - \varepsilon_{w_i}\|_1 < \delta$. Then

$$\begin{aligned}|h_n(w) - h(w)| &\leq |h_n(w) - h_n(w_i)| + |h_n(w_i) - h(w_i)| + |h(w_i) - h(w)| \\ &\leq 2M\|\varepsilon_w - \varepsilon_{w_i}\|_1 + \delta < (2M+1)\delta.\end{aligned}$$

Hence, the convergence is uniform on $\overline{\mathbb{D}_R}$.

Conversely, if a bounded sequence h_n converges u.c.c. to h, let $f \in L^1$. It suffices to prove convergence on a dense subset of L^1, so consider f_r for $r < 1$. Then

$$\lim_{n\to\infty}\langle h_n, f_r\rangle = \lim_{n\to\infty}\langle (h_n)_r, f\rangle = \langle h_r, f\rangle = \langle h, f_r\rangle.$$

Therefore h_n converges weak-$*$ to h. ∎

11.6.3. PROPOSITION. *The weak-$*$ continuous characters of H^∞ are just the evaluation functionals $\{\varepsilon_w : w \in \mathbb{D}\}$.*

PROOF. A linear functional is weak-$*$ continuous precisely when its kernel is weak-$*$ closed. In the case of ε_w, then kernel is $b_w H^\infty$ which is weak-$*$ closed. If

φ is another character, then $\varphi(z) = \lambda \in \mathbb{T}$. Hence $\ker\varphi$ contains $(\lambda - z)H^\infty$. Let $k_n(z) = \dfrac{\lambda - z}{(1+\frac{1}{n})\lambda - z}$. Then

$$|k_n(z)| = \left|1 - \frac{1}{n}\,\frac{1}{(1+\frac{1}{n})\lambda - z}\right| \le 1 + \frac{1}{n\frac{1}{n}} = 2 \quad \text{and} \quad |k_n(z) - 1| \le \frac{1}{n(1-|z|)}.$$

Hence $\|k_n\|_\infty \le 2$ and k_n converges u.c.c. to 1. Therefore, if $h \in H^\infty$, then $hk_n \in (\lambda - z)H^\infty$ and hk_n converges u.c.c. to h. By the lemma above, $(\lambda - z)H^\infty$ is weak-$*$ dense in H^∞. So φ is not weak-$*$ continuous. ■

11.6.4. DEFINITION. For $\lambda \in \mathbb{T}$, the *fibre over* λ is

$$\mathcal{M}_\lambda = \{\varphi \in \mathcal{M}_{H^\infty} : \varphi(z) = \lambda\}.$$

If $h \in H^\infty$, let the *cluster set* of h at $\lambda \in \mathbb{T}$ be

$$\mathrm{Cl}(h, \lambda) = \{w \in \mathbb{C} : \exists z_n \in \mathbb{D},\ z_n \to \lambda,\ h(z_n) \to w\}.$$

11.6.5. LEMMA. *If $h \in H^\infty$ and $\lambda \in \mathbb{T}$, then* $\mathrm{Cl}(h, \lambda) \subset \hat{h}(\mathcal{M}_\lambda)$.

PROOF. Suppose that $z_n \in \mathbb{D}$ converge to λ and $h(z_n) \to w$. Let

$$J = \{g \in H^\infty : \lim_{n\to\infty} g(z_n) = 0\}.$$

Clearly J is an ideal of H^∞, and it is proper since $1 \notin J$. By Zorn's lemma, there is a maximal ideal $M \subset J$. Let φ be the corresponding character. Since $z - \lambda \in J$, $0 = \varphi(z) - \lambda$, whence $\varphi \in \mathcal{M}_\lambda$. Since $h - w \in J$, we have $\hat{h}(\varphi) = \varphi(h) = w$. So $\mathrm{Cl}(H, \lambda) \subset \hat{h}(\mathcal{M}_\lambda)$. ■

11.6.6. COROLLARY. *Let $h \in H^\infty$ and $\lambda \in \mathbb{T}$. Then $\hat{h}$ is constant on $\mathcal{M}_\lambda$ if and only if h is continuous at λ.*

PROOF. The lemma above shows that if $\hat{h}$ is constant on $\mathcal{M}_\lambda$, then $\mathrm{Cl}(h, \lambda)$ is a singleton, and thus h is continuous at λ.

Conversely, suppose that h is continuous at λ. After scaling and translation by a constant, we may suppose that $h(\lambda) = 0$ and $\|h\| = 1$. Let $k(z) = (1 + \bar{\lambda}z)/2$ in $A(\mathbb{D})$. Note that $|k(z)| < 1 = k(\lambda)$ for $z \in \overline{\mathbb{D}} \setminus \{\lambda\}$. Hence k^n converges u.c.c. to 0 on $\overline{\mathbb{D}} \setminus \{\lambda\}$. We claim that $h(1 - k^n)$ converges to h in norm. Let $\varepsilon > 0$. By continuity at λ, there is a ball $b_\delta(\lambda) \cap \mathbb{D}$ on which $|h(z)| < \varepsilon$. There is an N so that $|k^n(z)| < \varepsilon$ on $\overline{\mathbb{D}} \setminus b_\delta(\lambda)$ for $n \ge N$. Thus for $n \ge N$,

$$\|h - h(1 - k^n)\| = \|hk^n\| \le \max\{\|h\|\|k^n\|_{\overline{\mathbb{D}}\setminus b_\delta(\lambda)}, \varepsilon\|k^n\|_{b_\delta(\lambda)\cap\overline{\mathbb{D}}}\} \le \varepsilon.$$

Hence if $\varphi \in \mathcal{M}_\lambda$,

$$\hat{h}(\varphi) = \lim_{n\to\infty} \varphi(h(1 - k^n)) = \lim_{n\to\infty} \varphi(h)\varphi(1 - k^n) = 0.$$

Therefore $\hat{h}$ is constant on $\mathcal{M}_\lambda$. ∎

Using the finer structure of inner and outer functions, we can identify $\hat{h}(\mathcal{M}_\lambda)$. This is another result which would be straightforward if we knew the corona theorem.

11.6.7. THEOREM. *For $h \in H^\infty$ and $\lambda \in \mathbb{T}$, $\hat{h}(\mathcal{M}_\lambda) = \mathrm{Cl}(h, \lambda)$.*

PROOF. It is enough to show that if $0 \notin \mathrm{Cl}(h, \lambda)$, then $\varphi(h) \neq 0$ for all $\varphi \in \mathcal{M}_\lambda$. In this case, there is an open set $N = \mathbb{D} \cap b_r(\lambda)$ on which $|h(z)| \geq \delta > 0$. Factor $h = \omega F$ where $\omega = BS$ is inner and F is outer. Then $|\omega(z)| \geq \delta/\|h\|$ on N. We observed in the proof of Theorem 11.5.9 that $0 \subset \mathrm{Cl}(\omega, \lambda)$ if λ belongs to $\mathrm{supp}(\omega)$. Hence the arc $J = b_r(\lambda) \cap \mathbb{T}$ is disjoint from $\mathrm{supp}(\omega)$. By Remarks 11.4.4 and 11.4.12, ω extends to be analytic across J, and, in particular, it is continuous there. By Corollary 11.6.6, $\hat{\omega}$ is constant on $\mathcal{M}_\lambda$ and has modulus 1. As $|\omega(z)| < 1$, we have $|F(z)| \geq \delta$ on N. Since F_r converges weak-$*$ to F, we have $|F| \geq \delta$ a.e.(m) on J. We have

$$F(z) = \exp\Big(\frac{1}{2\pi}\int_{\mathbb{T}\setminus J} \frac{e^{it}+z}{e^{it}-z} \log|F(e^{it})|dt\Big) \exp\Big(\frac{1}{2\pi}\int_{J} \frac{e^{it}+z}{e^{it}-z} \log|F(e^{it})|dt\Big).$$

Call this factorization $F =: F_1F_2$. Observe that $|F_2| \geq e^{\log\delta} = \delta$ a.e.(m) on $\mathbb{T}$ and F_2 is outer, so it is invertible with $\|F_2^{-1}\| \leq \delta^{-1}$ by Proposition 11.6.1. Since F_1 is obtained from an integral over $\mathbb{T} \setminus J$, it continues analytically across J as well, and thus is constant of modulus 1 on $\mathcal{M}_\lambda$. Hence for $\varphi \in \mathcal{M}_\lambda$,

$$|\hat{f}(\varphi)| = |\varphi(\omega)|\,|\varphi(F_1)|\,|\varphi(F_2)| = |\varphi(F_2^{-1})^{-1}| \geq \frac{1}{\|F_2^{-1}\|} \geq \delta.$$

Therefore, $\hat{f}$ is bounded away from 0 on $\mathcal{M}_\lambda$. ∎

Next we investigate the relationship between $\mathcal{M}_{H^\infty}$ and $\mathcal{M}_{L^\infty}$. Since L^∞ is a commutative C*-algebra, we have $L^\infty \simeq C(\mathcal{M}_{L^\infty})$ by Theorem 7.6.3.

11.6.8. PROPOSITION. *The topology on $\mathcal{M}_{L^\infty}$ is generated by a collection of clopen sets.*

PROOF. The topology on $\mathcal{M}_{L^\infty}$ is the weakest topology that makes each $\hat{f}$ continuous. Since the simple functions are norm dense in L^∞, this is the weakest topology which makes each $\widehat{\chi_A}$ continuous for all Borel sets $A \subset \mathbb{T}$. Now $\chi_A^2 = \chi_A$, and hence $\widehat{\chi_A}^2 = \widehat{\chi_A}$. Thus $\widehat{\chi_A}$ is the characteristic function of

$$\{\varphi \in \mathcal{M}_{L^\infty} : \varphi(\chi_A) = 1\} = \{\varphi \in \mathcal{M}_{L^\infty} : \varphi(\chi_A) \neq 0\}.$$

Hence, this set is both closed and open in $\mathcal{M}_{L^\infty}$. If all of these sets are clopen, then every simple function is continuous. Hence, these sets determine the topology. ∎

We need a weakening of the notion of a Dirichlet algebra.

11.6.9. DEFINITION. A uniform algebra $A \subset C(X)$ is called *logmodular* if $\{\log|f| : f \in A^{-1}\}$ is norm dense in $C_{\mathbb{R}}(X)$. Say that A is *strongly logmodular* if this is an equality.

11.6.10. PROPOSITION. *H^∞ is strongly logmodular in L^∞.*

PROOF. Let $u \in L^\infty_{\mathbb{R}}$. By Proposition 11.4.7 applied to e^u, the function $\widetilde{u}$ has a harmonic conjugate $\widetilde{v}$ so that

$$F_u(z) = e^{\widetilde{u}+i\widetilde{v}} = \exp\Big(\frac{1}{2\pi}\int_{-\pi}^{\pi} \frac{e^{it}+z}{e^{it}-z}\, u(e^{it})\,dt\Big)$$

is an H^∞ function with $\log|F_u| = u$ a.e.(m), i.e. $\log|F_u| = u$ in L^∞. ■

11.6.11. THEOREM. *The restriction map $\rho : \mathcal{M}_{L^\infty} \to \mathcal{M}_{H^\infty}$ is a homeomorphism onto the Shilov boundary of H^∞.*

PROOF. Clearly ρ is continuous and since $\mathcal{M}_{L^\infty}$ is compact, $X = \rho(\mathcal{M}_{L^\infty})$ is a compact subset of $\mathcal{M}_{H^\infty}$. To see that ρ is injective, suppose that $\varphi \neq \psi \in \mathcal{M}_{L^\infty}$. Then there is an $u \in L^\infty_{\mathbb{R}}$ such that $\varphi(u) \neq \psi(u)$. Now $\log|F_u| = u$ and so

$$|F_u| = e^u = \sum_{n\geq 0} \frac{u^n}{n!} \quad\text{and thus}\quad \varphi(|F_u|) = \sum_{n\geq 0} \frac{\varphi(u^n)}{n!} = e^{\varphi(u)}.$$

Write $F_u = a + ib$ where $a, b \in L^\infty_{\mathbb{R}}$. Then $\varphi(F_u) = \varphi(a) + i\varphi(b)$, and $\varphi(a)$ and $\varphi(b)$ are real because φ is a state of L^∞. Hence

$$\varphi(|F_u|)^2 = \varphi(|F_u|^2) = \varphi(a^2+b^2) = |\varphi(a)+i\varphi(b)|^2 = |\varphi(F_u)|^2.$$

Similarly,

$$|\psi(F_u)| = \psi(|F_u|) = e^{\psi(u)} \neq e^{\varphi(u)} = \varphi(|F_u|) = |\varphi(F_u)|.$$

Therefore ρ is one to one, and thus is a homeomorphism onto X.

Clearly X is a boundary because $\|\hat{h}\|_X = \|h\|_{L^\infty} = \|h\|$. There is no smaller boundary because if $Y \subset X$ is a proper closed subset, there is an $u \in L^\infty_{\mathbb{R}} = C_{\mathbb{R}}(X)$ so that $\|u\|_Y < \|u\|_\infty$. Then F_u does not attain its norm $e^{\|u\|}$ on Y. So X is the Shilov boundary. ■

11.6.12. COROLLARY. *The C*-algebra generated by H^∞ inside L^∞ is L^∞.*

PROOF. By the previous proof, H^∞ separates points of $X = \mathcal{M}_{L^\infty}$. As H^∞ is a subalgebra of L^∞ containing the constants, the result follows from the Stone–Weierstrass theorem. ■

11.6.13. PROPOSITION. *H^∞ is not a Dirichlet algebra.*

PROOF. Suppose that H^∞ were Dirichlet in L^∞. Let $X = \mathcal{M}_{L^\infty}$. If $A \subset \mathbb{T}$ is Borel with $0 < m(A) < 1$, choose $h \in H^\infty$ with $\|\chi_A - \operatorname{Re} h\| < \frac{1}{3}$. Then

$$\hat{h}(X) \subset \{z : \operatorname{Re} z \le \tfrac{1}{3},\ |z| \le \|h\|\} \cup \{z : \operatorname{Re} z \ge \tfrac{2}{3},\ |z| \le \|h\|\} =: K_1 \dot{\cup} K_2.$$

By Runge's theorem, χ_{K_2} is a norm limit of a sequence of polynomials p_n on $K_1 \cup K_2$. Therefore, $p_n(\hat{h})$ converges uniformly to χ_A in L^∞. This shows that $\chi_A \in H^\infty$. But non-zero functions in H^∞ cannot vanish on a set of positive measure by Corollary 11.3.6. Also, since the characteristic functions χ_A span L^∞, this would show that $H^\infty = L^\infty$, which is absurd. Thus H^∞ is not Dirichlet. ∎

This next result uses the logmodularity property to show that characters on H^∞ have unique Hahn–Banach extensions to L^∞. This is another tight connection between H^∞ and L^∞.

11.6.14. GLEASON–WHITNEY THEOREM. *Let $\varphi \in \mathcal{M}_{H^\infty}$. Then φ has a unique Hahn–Banach extension to L^∞ given by $L(u) = \log|\varphi(F_u)|$ for $u \in L^\infty_{\mathbb{R}}$.*

PROOF. For $u, v \in L^\infty_{\mathbb{R}}$, observe that $F_{u+v} = F_u F_v$ and $F_{tu} = F_u^t$ for $t \in \mathbb{R}$. Also $F_1 = e$ is constant. Define $L(u) = \log|\varphi(F_u)|$ for $u \in L^\infty_{\mathbb{R}}$. Then

$$L(u+v) = \log|\varphi(F_{u+v})| = \log|\varphi(F_u)\varphi(F_v)| = L(u) + L(v)$$

and

$$L(tu) = \log|\varphi(F_{tu})| = \log|\varphi(F_u^t)| = tL(u).$$

Hence L is real linear on $L^\infty_{\mathbb{R}}$. Also if $u \ge 0$, then $|F_{-u}| = e^{-u} \le 1$, whence $|\varphi(F_{-u})| \le \|F_{-u}\|_\infty \le 1$ and so $L(-u) \le 0$. Therefore L is positive. Since $L(1) = \log|\varphi(F_1)| = \log e = 1$, L is a state, and thus $\|L\| = 1$. Now complexify L to a functional on L^∞ by $L(u + iv) = L(u) + iL(v)$.

If $h \in H^\infty$, write $h = u + iv$. The harmonic conjugate of u is $v - v(0)$. Hence

$$\begin{aligned} L(u) &= \log|\varphi(F_u)| = \log|\varphi(e^{u+iv-iv(0)})| \\ &= \log|e^{-iv(0)}\varphi(e^h)| = \log|e^{\varphi(h)}| = \operatorname{Re}\varphi(h). \end{aligned}$$

Similarly $L(v) = \operatorname{Im}\varphi(h)$, and thus $L(h) = \varphi(h)$. Therefore L is a Hahn–Banach extension of φ.

Suppose that L_2 is another Hahn–Banach extension of φ. By Exercise **9.23**, $|L_2(f)| \le L_2(|f|)$ for $f \in L^\infty$. Let $u \in L^\infty_{\mathbb{R}}$. Then for $t \in \mathbb{R}$,

$$|\varphi(F_{-tu})| = |L_2(F_{-tu})| \le L_2(|F_{-tu}|) = L_2(e^{-tu}).$$

Hence

$$\begin{aligned} 1 &= |\varphi(F_{tu})\varphi(F_{-tu})| \le L(|F_{tu}|)L_2(|F_{-tu}|) = L(e^{tu})L_2(e^{-tu}) \\ &= \sum_{m\ge 0} \frac{L(u^m)t^m}{m!} \sum_{n \ge 0} \frac{L_2(u^n)(-t)^n}{n!} =: f(t). \end{aligned}$$

The function $f(t)$ is real analytic, and attains its minimum at $f(0) = 1$. Thus

$$0 = f'(0) = L(u)L_2(1) - L(1)L_2(u) = L(u) - L_2(u).$$

Therefore $L_2 = L$. ∎

Here is another characterization of the points in $\mathcal{M}_{L^\infty}$.

11.6.15. THEOREM. *For $\varphi \in \mathcal{M}_{H^\infty}$, the following are equivalent:*

(1) $\varphi \in X = \rho(\mathcal{M}_{L^\infty})$.

(2) $\varphi(h) \in \sigma_{L^\infty}(h)$ *for all* $h \in H^\infty$.

(3) $|\varphi(\omega)| = 1$ *for all inner functions* ω.

(4) $\varphi(\omega) \neq 0$ *for all inner functions* ω.

PROOF. If $\varphi \in X$, then $\varphi(h) \in \sigma_{L^\infty}(h)$ by Gelfand's theorem. If ω is inner, then $\bar{\omega} \in L^\infty$ and $\omega\bar{\omega} = 1$. So ω is a unitary in L^∞, whence $\sigma_{L^\infty}(\omega) \subset \mathbb{T}$. Since φ is a state on L^∞, it is self-adjoint. Thus $(1) \Rightarrow (2) \Rightarrow (3) \Rightarrow (4)$.

If $h \in H^\infty$ is invertible in L^∞, then $|h|$ is bounded below; so $u = \log|h| \in L^\infty$. Therefore, $h = \omega F_u$ where ω is inner. Now F_u is invertible so that $\varphi(F_u) \neq 0$. If (4) holds, then $\varphi(\omega) \neq 0$, and hence $\varphi(h) \neq 0$. Now if $\varphi(h) = \lambda$, then $\varphi(h - \lambda) = 0$ and so $h - \lambda$ is not invertible in L^∞. Thus $0 \in \sigma_{L^\infty}(h - \lambda) = \sigma_{L^\infty}(h) - \lambda$. Therefore (4) implies (2).

Suppose that (2) holds and let L be the Hahn–Banach extension of φ to L^∞ provided by the Gleason–Whitney theorem. Let $u \in L^\infty_{\mathbb{R}}$, and let $\widehat{F_u}$ denote the Gelfand transform of F_u on $\mathcal{M}_{L^\infty}$. Then $u = \log|F_u|$ and $\hat{u} = \log|\widehat{F_u}|$. By assumption, $\varphi(F_u) \in \sigma_{L^\infty}(F_u) = \operatorname{Ran}\widehat{F_u}$. Thus $L(u) = \log|\varphi(F_u)| \in \operatorname{Ran}\hat{u}$.

If L is not a character of L^∞, then for each $x \in X$, there is some $f \in L^\infty$ with $L(f) \neq f(x)$. The simple functions are dense in L^∞, and so we may suppose that $f = \chi_{A_x}$ for some Borel set $A_x \subset \mathbb{T}$. Replace A_x by $\mathbb{T} \setminus A_x$ if necessary so that $\widehat{\chi_{A_x}}(x) = 1$. Then we must have $L(\chi_{A_x}) = 0$ since $\operatorname{Ran}\widehat{\chi_{A_x}} \in \{0, 1\}$. Now $E_x = \{y \in X : \widehat{\chi_{A_x}}(y) = 1\}$ is a clopen neighbourhood of x in X. The collection of sets $\{E_x : x \in X\}$ is an open cover; so there is a finite subcover, say $E_{x_1}, \dots, E_{x_n}$. Thus $f = \sum_{i=1}^n \chi_{A_{x_i}}$ has $\hat{f} \geq 1$. But $L(f) = 0$, contradicting the previous paragraph. Therefore $L \in \mathcal{M}_{L^\infty}$. ∎

11.7. More about H^∞

In this section, we explore some properties of H^∞ as an algebra. First we provide the analogue of the Wermer maximality theorem. It is not as dramatic, as there are many algebras between H^∞ and L^∞. They are known as *Douglas algebras*, and have a very interesting structure [118]. The following theorem shows that the smallest Douglas algebra is $H^\infty + C$. We first show that $H^\infty + C$ is indeed a norm-closed algebra.

11.7.1. LEMMA. *$H^\infty + C$ is a closed subalgebra of L^∞.*

PROOF. By Theorem 11.2.8, there is a natural isometric imbedding

$$C(\mathbb{T})/A(\mathbb{D}) \xrightarrow{\ i\ } \big(C(\mathbb{T})/A(\mathbb{D})\big)^{**} = (H^1_0)^* = L^\infty/H^\infty$$

where $i(f + A(\mathbb{D})) = f + H^\infty$; so $i(C/A(\mathbb{D})) = (H^\infty + C)/H^\infty$. The image is closed because i is isometric, and hence its preimage in L^∞, namely, $H^\infty + C$, is closed.

Note that $H^\infty + C = \overline{\bigcup_{n\geq 1} \bar{z}^n H^\infty}$. Since $\bar{z}^n H^\infty \cdot \bar{z}^m H^\infty \subset \bar{z}^{n+m} H^\infty$, the union is an algebra, whence $H^\infty + C$ is an algebra. ∎

11.7.2. HOFFMAN–SINGER THEOREM. *Suppose that A is a closed subalgebra of L^∞ containing H^∞. Then $A = H^\infty$ or $A \supset H^\infty + C$, where $C = C(\mathbb{T})$.*

PROOF. By Theorem 11.6.14, we can consider $\mathcal{M}_{L^\infty} \subset \mathcal{M}_A \subset \mathcal{M}_{H^\infty}$ because each character on H^∞ extends uniquely to a norm 1 functional on A, and the subset which are multiplicative on A forms its maximal ideal space. Since H^∞ separates points of $\mathcal{M}_{L^\infty}$, so does A, and $\mathcal{M}_{L^\infty}$ is its Shilov boundary.

If $\varepsilon_0 \notin \mathcal{M}_A$, then $\sigma_A(z) = \hat{z}(\mathcal{M}_A) \not\ni 0$. Thus z is invertible in A, and hence $A \supset \mathrm{Alg}\{z, \bar{z}\} = C$. Therefore $A \supset H^\infty + C$.

In the remaining case, $\varepsilon_0 \in \mathcal{M}_A$. Now ε_0 has a unique extension to L^∞ given by $L(f) = \frac{1}{2\pi}\int_{-\pi}^{\pi} f(e^{it})\,dt$. Let $f \in A$. For $n > 0$,

$$\hat{f}(-n) = \frac{1}{2\pi}\int_{-\pi}^{\pi} f(e^{it})e^{int}\,dt = \varepsilon_0(fz^n) = \varepsilon_0(f)\varepsilon_0(z)^n = 0.$$

Therefore $f \in H^\infty$. So $A = H^\infty$. ∎

Since C is weak-$*$ dense in L^∞, the following consequence is immediate.

11.7.3. COROLLARY. *There is no weak-$*$ closed subalgebra $A \subset L^\infty$ sitting properly between H^∞ and L^∞.*

Let Ω denote the set of all inner functions. Some theory of conformal maps is required for the next proof.

11.7.4. DOUGLAS–RUDIN THEOREM. *The set $U = \{u\bar{v} : u, v \in \Omega\}$ of quotients of inner functions is dense in the set of all unimodular functions in L^∞. Thus $C^*(\Omega) = L^\infty$.*

PROOF. Since Ω is a semigroup, the self-adjoint set U is also closed under multiplication. The simple unimodular functions are dense in the set of unimodular functions in L^∞. Thus, it suffices to approximate each unimodular function of the form: $e^{i\alpha}\chi_E + e^{i\beta}\chi_{E^c}$ for $E \subset \mathbb{T}$ Borel and $0 < m(E) < 1$, since finite products of these functions yield all unimodular simple functions.

Let $\varepsilon > 0$ and let I and J be disjoint closed arcs of $\mathbb{T}$ of length at most ε centred at $e^{i\alpha}$ and $e^{i\beta}$. Let $S^2 = \mathbb{C} \cup \{\infty\}$ be the Riemann sphere. The region $\mathbb{S} = S^2 \setminus (I \cup J)$ is a doubly connected region. By the theory of conformal maps, there are an $r \in (0,1)$, an annulus $\mathbb{A} = \{z \in \mathbb{C} : r < |z| < \frac{1}{r}\}$ and a conformal map $F : \mathbb{A} \to \mathbb{S}$. It extends to be continuous from $\overline{\mathbb{A}}$ onto $\overline{\mathbb{S}}$, taking the two boundary curves $r\mathbb{T}$ and $\frac{1}{r}\mathbb{T}$ onto I and J, respectively. (Replace $F(w)$ by $F(\frac{1}{w})$ if necessary to flip the two boundary circles.) The map F is analytic except for a simple pole at $p = F^{-1}(\infty)$.

The map $G(w) = w + \frac{1}{w}$ is analytic on a neighbourhood of $\overline{\mathbb{A}}$. Note that $G(e^{i\theta}) = e^{i\theta} + e^{-i\theta} = 2\cos\theta$; hence, it maps the unit circle onto $[-2,2]$. Also

$$G(te^{i\theta}) = G(t^{-1}e^{-i\theta}) = te^{i\theta} + t^{-1}e^{-i\theta} \quad \text{for} \quad r < t < 1.$$

The reader can check that this is a continuous path of ellipses, and G maps $\mathbb{A}$ onto a solid ellipse $\mathbb{E}$. G is continuous on $\overline{\mathbb{A}}$ and maps $\partial\mathbb{A}$ onto $\partial\mathbb{E}$.

Let H be a conformal map of $\mathbb{E}$ onto $\mathbb{D}$ taking $G(p)$ to 0. Then $\varphi = H \circ G$ maps $\mathbb{A}$ onto $\mathbb{D}$, it is continuous on $\overline{\mathbb{A}}$, and maps $\partial\mathbb{A}$ onto $\mathbb{T}$. Also $\varphi(p) = 0$. Now $\psi(w) = F(w)\varphi(w)$ is analytic on $\mathbb{A}$ and continuous on $\overline{\mathbb{A}}$ since p is now a removable singularity. Both F and φ map $\partial\mathbb{A}$ into $\mathbb{T}$, and thus so does ψ. Since ψ is continuous on $\overline{\mathbb{A}}$, the maximum modulus principle shows that ψ maps $\mathbb{A}$ into $\mathbb{D}$.

Let h be an outer function with $|h| = r\chi_E + \frac{1}{r}\chi_{E^c}$ a.e.(m); so h is invertible in H^∞. Since $\|h\| = \|h^{-1}\| = \frac{1}{r}$, we have $r < |h(z)| < \frac{1}{r}$ for $z \in \mathbb{D}$. However on the circle, $|h(e^{it})| \in \{r, \frac{1}{r}\}$ a.e.(m). This shows that h maps $\mathbb{D}$ into the annulus $\mathbb{A}$ and it takes $\mathbb{T}$ into $\partial\mathbb{A}$ a.e.(m). Therefore $u = \psi \circ h$ is analytic from $\mathbb{D}$ into $\mathbb{D}$ and $|u(e^{it})| = 1$ a.e.(m). So u is inner. Similarly, $v = \varphi \circ h$ is analytic from $\mathbb{D}$ into $\mathbb{D}$ and $|v(e^{it})| = 1$ a.e.(m). So v is inner. Finally, $\frac{u}{v} = F \circ h$ and $F \circ h(e^{it}) \in I$ when $|h(e^{it})| = r$ and $F \circ h(e^{it}) \in J$ when $|h(e^{it})| = \frac{1}{r}$. Thus

$$\|e^{i\alpha}\chi_E + e^{i\beta}\chi_{E^c} - \tfrac{u}{v}\|_\infty \le \max\{\operatorname{diam}(I)/2, \operatorname{diam}(J)/2\} < \varepsilon.$$

Since $\varepsilon > 0$ is arbitrary, $e^{i\alpha}\chi_E + e^{i\beta}\chi_{E^c} \in \overline{U}$. So U is dense in the unimodular functions.

Since $\chi_E = \frac{1}{2}(\chi_E - \chi_{E^c}) + \frac{1}{2}$, it follows that $C^*(\Omega) = L^\infty$. ∎

11.7.5. COROLLARY. *The algebra $\bigcup_{\omega\in\Omega} \overline{\omega} H^\infty$ is norm dense in L^∞, where Ω is the semigroup of inner functions.*

The following nice observation is due to Bernard.

11.7.6. PROPOSITION. *Let A be a uniform algebra on X which is generated by its unimodular functions U. Then $\overline{\operatorname{conv}}(U) = \overline{b_1(A)}$.*

PROOF. Since U is a semigroup, $\overline{\operatorname{span}}\, U$ is a closed algebra, and thus equals A by hypothesis. Let $f \in \operatorname{span} U$ with $\|f\| < 1$, say $f = \sum_{i=1}^n \lambda_i u_i$ for $u_i \in U$. Set $u = \prod_{j=1}^n u_j$. Then $\bar{f}u = \sum_{i=1}^n \overline{\lambda_i} v_i$ belongs to A, where $v_i = \prod_{j \neq i} u_j \in U$. Now

$$|f - e^{it}u| = |fe^{-it}\bar{u} - 1| = |1 - e^{it}\bar{f}u|.$$

Since $\|e^{it}\bar{f}u\| = \|f\|_\infty < 1$, $1 - e^{it}\bar{f}u$ is invertible in A. Therefore for any $t \in \mathbb{R}$,

$$\frac{f - e^{it}u}{1 - e^{it}\bar{f}u}$$

is unimodular in A, so lies in U. Since $g(z) = \frac{f - zu}{1 - z\bar{f}u}$ is analytic on $\mathbb{D}$, the mean value property yields

$$\frac{1}{2\pi}\int_{-\pi}^{\pi} \frac{f - e^{it}u}{1 - e^{it}\bar{f}u}\,dt = \frac{f - 0}{1 - 0} = f.$$

This sum makes sense as a Riemann sum, and thus f is approximated by

$$\frac{1}{n}\sum_{j=1}^{n} \frac{f - e^{2\pi ij/n}u}{1 - e^{2\pi ij/n}\bar{f}u}.$$

These Riemann sums belong to $\operatorname{conv} U$ as required. ■

For an immediate application to $A(\mathbb{D})$, see Exercise **11.19**.

For the next result, we require a result from the next chapter, Corollary 12.4.14, which is established using operator theory methods.

11.7.7. MARSHALL THEOREM. *H^∞ is the closed span of the inner functions.*

PROOF. Let $J = \overline{\operatorname{span}\,\Omega}$ and let $N = \{f \in H^\infty : \exists u \in \Omega,\ \bar{f}u \in H^\infty\}$. We first show that $\overline{N} = J$. As we have seen, if $f = \sum_{i=1}^n \lambda u_i \in J$, then $\bar{f}u \in H^\infty$ for $u = \prod_{j=1}^n u_j$. Thus $f \in N$. Conversely, if $\bar{f}u \in N$ and $\|f\| < 1$, the proof of Bernard's proposition 11.7.6 shows that $f \in \overline{\operatorname{conv}\,\Omega}$. Thus $J = \overline{N}$.

Now let $h \in H^\infty$ and let $\varepsilon > 0$. By the Douglas–Rudin Theorem, there are inner functions $u_1, \dots, u_n, v$ and scalars λ_i so that

$$\Big\|h - \sum_{i=1}^{n} \lambda_i u_i \bar{v}\Big\| < \varepsilon.$$

Let $f = \sum_{i=1}^n \lambda_i u_i \in J$. Then

$$\operatorname{dist}(\varepsilon^{-1} f\bar{v}, H^\infty) \le \varepsilon^{-1}\|f\bar{v} - h\| < 1.$$

By Corollary 12.4.14, there is a unimodular function $\mu \in L^\infty$ in $\varepsilon^{-1} f\bar{v} + H^\infty$, say $\mu = \varepsilon^{-1} f\bar{v} + g$ for $g \in H^\infty$. Then, $\mu v = \varepsilon^{-1} f + gv \in H^\infty$, so μv is inner. Hence

$gv = \mu v - \varepsilon^{-1} f \in \operatorname{span} \Omega \subset N$. Thus, there is another inner function w so that $\overline{gvw} \in H^\infty$. Therefore $\bar{g}w \in H^\infty$, and so $g \in N$. Finally,

$$\begin{aligned} \|h + \varepsilon g\| &= \|hv + \varepsilon gv\| \le \|hv - f\| + \|f + \varepsilon gv\| \\ &= \|h - f\bar{v}\| + \varepsilon \|\mu v\| < 2\varepsilon. \end{aligned}$$

As $\varepsilon > 0$ is arbitrary, $h \in N$. ■

As an immediate consequence of Marshall's theorem and the previous proposition, we obtain

11.7.8. COROLLARY. *The closed convex hull of the set of inner functions is* $\overline{\operatorname{conv} \Omega} = \overline{b_1(H^\infty)}$.

Exercises for Chapter 11

11.1 Suppose that $u(z) \ge 0$ is harmonic on $\mathbb{D}$. Prove that for $0 \le r < 1$ and $\theta \in \mathbb{R}$,

$$\frac{1-r}{1+r} u(0) \le u(re^{i\theta}) \le \frac{1+r}{1-r} u(0).$$

11.2 (a) Show that if $u \in h_p$ for $p \ge 1$, then $|u(re^{i\theta})| \le \left(\dfrac{1+r}{1-r}\right)^{1/p} \|u\|_p$.

(b) Show that there is a universal constant C_p so that if $f \in H^p$ for $p \ge 1$, then

$$|f'(re^{i\theta})| \le C_p (1-r)^{-1-1/p} \|f\|_p.$$

HINT: Consider $g(z) = \frac{f(z) - f(z_0)}{z - z_0}$ for $z_0 = re^{i\theta}$.

11.3 (a) Show that the space h_2 of harmonic functions on $\mathbb{D}$ is a Hilbert space with inner product $\langle u, v \rangle = \lim_{r \to 1^-} \frac{1}{2\pi} \int_{-\pi}^{\pi} u(re^{it}) \overline{v(re^{it})}\, dt$.

(b) Show that for $w \in \mathbb{D}$, the linear functional $\varepsilon_w(u) = u(w)$ is continuous. Find the function $k_w \in h_2$ so that $\varepsilon_w(u) = \langle u, k_w \rangle$.

11.4 Sum the series $Q_r(\theta) = \frac{1}{i} \sum_{n \ne 0} \operatorname{sign}(n) r^{|n|} e^{in\theta}$ to get a formula.

11.5 (a) Show that if τ is a conformal map of $\mathbb{D}$ onto itself and $u(z)$ converges nontangentially to $u(e^{i\theta})$ a.e.(m), then

$$\text{n. t.} \lim_{z \to e^{i\theta}} u(\tau(z)) = u(\tau(e^{i\theta})) \text{ a.e.}(m).$$

HINT: The conformal maps of $\mathbb{D}$ are $\tau(z) = \lambda \frac{z-a}{1-\bar{a}z}$ for $a \in \mathbb{D}$ and $\lambda \in \mathbb{T}$.

(b) If u is inner, show that $u \circ \tau$ is inner.

11.6 (a) Prove Proposition 11.4.15.

(b) Give necessary and sufficient conditions for a set S of inner functions to have a least common multiple.

11.7 If $a, b > 0$ and $\varepsilon > 0$, $|\log(a+\varepsilon) - \log(b+\varepsilon)| \le \log\left(1 + \frac{|b-a|}{\varepsilon}\right)$.

11.8 Let $f(z) \in H^1(\mathbb{D})$.
(a) Use Corollary 11.4.14 to factor $f = f_1 f_2$ for $f_i \in H^2(\mathbb{D})$. Use the identification between $H^2(\mathbb{D})$ and H^2 to show that f_r converges to f^* in L^1.
(b) Give another proof of the F.&M. Riesz theorem.
HINT: Show that f_r converges weak-$*$ to $f^*\,dm$.

11.9 (a) Construct a Blaschke product such that $\overline{Z(B)}$ contains $\mathbb{T}$.
(b) Show that $\lim_{z \to e^{i\theta},\ z \in \mathbb{D}} B(z)$ does not exist a.e. on $\mathbb{T}$. This is in contrast to nontangential limits.

11.10 If $g, h \in H^p$, h is outer and $|g| \le |h|$ a.e.(m), show that $g = fh$ for some $f \in H^\infty$.

11.11 Show that $(1-z)^\alpha$ belongs to H^p if and only if $p\alpha > -1$, and it is outer.

11.12 If $f(z)$ and $1/f(z)$ are both in $H^1(\mathbb{D})$, prove that f is outer.

11.13 If $f \in H^1(\mathbb{D})$ and $\operatorname{Re} f(z) > 0$ on $\mathbb{D}$, prove that f is outer.
HINT: $f+\varepsilon$ is outer. Split $\int \log|f+\varepsilon|\,dt$ over $E = \{t : |f(e^{it})| \ge 1/2\}$ and $\mathbb{T} \setminus E$.

11.14 Let $\omega(z) = e^{\frac{z+1}{z-1}}$ be the singular inner function for the unit point mass at 1.
(a) Describe the level lines of $|\omega|$, and hence describe the possible limits of $\lim \omega(z_k)$ for a sequence z_k converging in $\mathbb{D}$ to 1. What is the nontangential limit?
(b) $f(z) = 1/\omega(z)$ is analytic on $\mathbb{D}$, continuous on $\overline{\mathbb{D}} \setminus \{1\}$, and has $|f(e^{i\theta})| = 1$ on $\mathbb{T} \setminus \{1\}$, yet $|f(z)| > 1$ everywhere on $\mathbb{D}$. What does this mean for the maximum modulus theorem? If an analytic function on $\mathbb{D}$ has radial limits a.e.(m) in L^∞, is it necessarily in H^∞?

11.15 If $f, g \in H^1$, f is outer and $h = g/f \in L^1$, show that $h \in H^1$.
HINT: Reduce to the case g outer and use the integral formula.

11.16 (*Smirnov*). Prove that if $f \in H^p$, $p < r$ and $f \in L^r(\mathbb{T})$, then $f \in H^r$.
HINT: Show that $f \in h_r \cap \mathcal{O}(\mathbb{D})$.

11.17 Show that the unit disc $\mathbb{D}$ is dense in $\mathcal{M}_{H^\infty}$ if and only if
(†) if $f_1, \ldots, f_n \in H^\infty$ and $\inf_{z \in \mathbb{D}} \sum_{i=1}^n |f_i(z)| = \delta > 0$, then there are $g_1, \ldots, g_n$ in H^∞ such that $\sum_{i=1}^n f_i g_i = 1$.

11.18 (*Hoffman*). Suppose that $f \in H^\infty$ and $\inf_{z \in \mathbb{D}} |1-z| + |f(z)| = \delta > 0$. Show there are $g, h \in H^\infty$ such that $(1-z)g(z) + f(z)h(z) = 1$.
HINT: Factor f as in the proof of Theorem 11.6.7. Find $h \in H^\infty$ such that fh extends to be analytic on a neighbourhood of 1 and $(fh)(1) = 1$.

11.19 What is the closed convex hull of the finite Blaschke products?

11.20 (*Kakutani*). Let α be an automorphism of H^∞.

(a) For $f \in H^\infty$, show that $\overline{(\alpha f)(\mathbb{D})} = \overline{f(\mathbb{D})}$.
(b) Show that $\tau = \alpha z$ maps $\mathbb{D}$ into itself.
(c) For $f \in H^\infty$ and $\lambda \in \mathbb{D}$, factor $f - f(\tau(\lambda)) = (z - \tau(\lambda))g$. Use this to show that $\alpha f(\lambda) = f(\tau(\lambda))$.
(d) Show that τ is a conformal map of $\mathbb{D}$ onto itself, and $\alpha f = f \circ \tau$ for $f \in H^\infty$.

11.21 Show that the Gleason–Whitney theorem identifies $\mathcal{M}_{H^\infty}$ with a weak-$*$ closed compact subset of $L^{\infty *}$. Hence, the harmonic extension $\widetilde{f}(z)$ of any L^∞ function extends to a continuous function on $\mathcal{M}_{H^\infty}$.

11.22 If $f \in H^\infty$ and $\|f\|_\infty \le 1$, show that f is a Blaschke product if and only if

$$\lim_{r\to 1} \frac{1}{2\pi} \int_{-\pi}^{\pi} \log|f(re^{it})|\, dt = 0.$$

11.23 (*Frostman Theorem*). Show that if $\omega(z)$ is inner, then $b_a(\omega(z))$ is a Blaschke product for almost all $a \in \mathbb{D}$. Deduce that the Blaschke products are norm dense in the set of inner functions. Follow this outline:
(a) Show that $G(z) = \displaystyle\int_{\mathbb{D}} \log|b_a(z)|dA(a)$ is continuous on $\overline{\mathbb{D}}$ and 0 on $\mathbb{T}$, where $dA(a)$ is normalized area measure on $\mathbb{D}$.
(b) Show (carefully verify limits, etc.) that

$$0 = \lim_{r\to 1} \frac{1}{2\pi} \int_{-\pi}^{\pi} G(\omega(re^{it}))\, dt = \int_{\mathbb{D}} \lim_{r\to 1} \frac{1}{2\pi} \int_{-\pi}^{\pi} \log|b_a(\omega(re^{it}))|\, dt.$$

(c) Apply Exercise **11.22**.

11.24 (*Sarason*). Show that $\mathcal{M}_{H^\infty + C} = \mathcal{M}_{H^\infty} \setminus \mathbb{D}$.

11.25 Let A be a closed subalgebra of L^∞ containing H^∞.
(a) Show that $\mathcal{M}_A$ can be identified with a subset of $\mathcal{M}_{H^\infty}$.
HINT: Gleason–Whitney theorem.
(b) Show that $\mathcal{M}_{L^\infty}$ is the Shilov boundary.

11.26 Let A be a closed subalgebra of L^∞ containing H^∞. Let

$$\mathcal{U}_A = \{u \in A^{-1} : |u| = 1 \text{ a.e.}(m)\}.$$

(a) Show that A is generated by H^∞ and $\mathcal{U}_A$.
HINT: If $f \in A^{-1}$, find an invertible $g \in H^\infty$ such that $|f| = |g|$ a.e.(m).
(b) Show that $\mathcal{M}_A = \{\varphi \in \mathcal{M}_{H^\infty} : |\varphi(u)| = 1 \text{ for all } u \in \mathcal{U}_A\}$.
HINT: Represent φ as an integral against a probability measure μ on $\mathcal{M}_{L^\infty}$. Show that each $u \in \mathcal{U}_A$ is constant on $Y = \operatorname{supp}(\mu)$. Hence $A|_Y = H^\infty|_Y$. Deduce that φ is multiplicative on A.

11.27 (*Axler*). Exact factorization in L^∞ using $H^\infty + C$.
(a) If $f \in L^\infty$, find Blaschke products B_n so that $\lim_{n\to\infty} \operatorname{dist}(B_n f, H^\infty) = 0$.
HINT: Use the Douglas–Rudin and Frostman theorems.

(b) Find finite Blaschke products F_n so that $B_n = F_n B'_n$ and $B = \prod_n B'_n \neq 0$.
HINT: Make the sum of $1 - |a_i|$ for the zeros of B'_n less than 2^{-n}.
(c) Show that $Bf \in H^\infty + C$. Hence $f = \overline{B}h$ for some $h \in H^\infty + C$.
(d) A measurable set $E \subset \mathbb{T}$ is a *set of uniqueness* for an algebra $A \subset L^\infty$ if $f \in A$ and $f|_E = 0$ implies that $f = 0$. Show that E is a set of uniqueness for $H^\infty + C$ if and only if $m(\mathbb{T} \setminus E) = 0$.

11.28 (a) If $f \in H^1$ and $f \sim \sum_{n \geq 0} a_n e^{in\theta}$, show that there is a function $F \in H^1$ such that $F \sim \sum_{n \geq 0} A_n e^{in\theta}$, $\|F\|_1 \leq \|f\|_1$ and $|a_n| \leq A_n$ for $n \geq 0$.
HINT: Factor f in $H^2 \cdot H^2$, and replace the Fourier coefficients with absolute values for thc H^2 functions.

(b) Compute $\displaystyle\lim_{r \to 1} \int_0^{2\pi} (\pi - \theta) \operatorname{Im} F(re^{i\theta})\, d\theta$.

(c) Hence establish *Hardy's inequality*: $\displaystyle\sum_{n \geq 1} \frac{1}{n} |a_n| \leq \pi \|f\|_1$.

Notes on Chapter 11

The study of harmonic and analytic functions goes way back. The Dirichlet problem for a bounded domain Ω in $\mathbb{C}$ asks for a harmonic function on Ω which extends to a continuous function on $\overline{\Omega}$ with a prescribed continuous boundary function. This problem goes back to Green in 1828, and the result for the disc is due to Dirichlet in 1850. The relation between various classes of functions, their Fourier series and their boundary values goes back to work of Herglotz for positive measures in 1911, Young for L^p for $1 < p < \infty$ in 1913 and Gross for L^1 in 1915. Fatou established his boundary limit theorem in 1906, not long after Lebesgue introduced his new measure theory. Hardy's analysis in 1915 of the behaviour of the p-norms of f_r for analytic functions on the disc is seen as the beginning of Hardy spaces. The Riesz brothers wrote only one joint paper [259], and it has been extremely influential. We use the approach of Helson and Lowdenslager [145].

Jensen's formula goes back to 1899. Blaschke introduced his products [31] in 1915. F. Riesz factored H^p functions as $f = Bg$ for a Blaschke product and a function with no zeros in [258]. The refinement into the inner-outer factorization is due to Smirnov [287].

Wermer established his maximality theorem in [322]. Rudin [266] and Carleson [45] independently proved Theorem 11.5.6 in 1957. The description of the ideals of $A(\mathbb{D})$ was first done by Beurling (unpublished) and by Rudin in [267].

Hoffman published a major treatise on the structure of $\mathcal{M}_{H^\infty}$ in [151]. He identified many analytic discs in $\mathcal{M}_{H^\infty} \setminus \mathbb{D}$. The Shilov boundary was found in 1960 by I.J. Schark [278], a pseudonym for I. Kaplansky, J. Wermer, S. Kakutani, R.

C. Buck, H. Royden, A. Gleason, R. Arens and K. Hoffman. The Gleason–Whitney theorem appears in [123].

Hoffman and Singer [152] give a new proof of Wermer's theorem which yields what we call the Hoffman–Singer theorem. They actually obtained Corollary 11.7.3. Douglas and Rudin [100] establish Theorem 11.7.4. Douglas [98] asked whether every closed subalgebra of L^∞ containing H^∞ is generated by H^∞ and the inverses of the inner functions invertible in it. This was eventually verified by results of Chang [47] and Marshall [199]. See [118] for a treatment. Marshall [200] shows that the inner functions generate H^∞ and the corollary about their convex hull.

A very nice treatment of the ideas in this chapter is contained in Hoffman's book [150]. Duren [106] also treats H^p spaces very thoroughly from a more function-theoretic perspective. Another nice treatment is the book by Koosis [182]. Garnett's wonderful book [118] assumes more prerequisites, and covers a lot of material about H^∞ including Carleson's results on interpolating sequences, Carleson's corona theorem, Douglas algebras and much more.

CHAPTER 12

Isometries and Toeplitz Operators

Isometries are the simplest non-normal, non-compact operators. They have a lot of interesting features, and we investigate their structure. Toeplitz operators are formed by the compression of multiplication operators on L^2 to H^2. Their analysis requires a nice blending of ideas from operator theory and function theory.

12.1. Isometries

There are two basic examples of isometries on Hilbert space, the unitary operators and the unilateral shift S. These are the building blocks for an arbitrary isometry.

12.1.1. DEFINITION. If V is an isometry, say that a subspace $W \subset H$ is a *wandering space* if $\{V^n W : n \geq 0\}$ are pairwise orthogonal.

Observe that if W is wandering, $M = \sum_{n\geq 0}^{\oplus} V^n W$ is invariant for V. Let $\alpha = \dim W$, and let $\{\xi_j : j \in \alpha\}$ be an orthonormal basis for W. Then the subspaces $M_j = \operatorname{span}\{V^n \xi_j : n \geq 0\}$ are pairwise orthogonal invariant subspaces which span M. The restrictions $V|_{M_j} \simeq S$. Thus $V|_M \simeq S^{(\alpha)}$.

12.1.2. WOLD DECOMPOSITION. *Let $V \in \mathcal{B}(H)$ be an isometry on a Hilbert space H. Let $H_0 = \bigcap_{n\geq 1} V^n H$ and $W = \ker V^*$. Then H_0 reduces V, $U := V|_{H_0}$ is unitary, W is wandering for V, $H_0^\perp = H_1 := \sum_{n\geq 0}^{\oplus} V^n W$ and $V \simeq U \oplus S^{(\alpha)}$ where $\alpha = \dim W$.*

PROOF. First $W = \ker V^*$ is a wandering subspace. Indeed, W is orthogonal to $\operatorname{Ran} V \supset V^n W$ for $n \geq 1$ since if $\zeta \in W$,

$$\langle \zeta, V\eta \rangle = \langle V^* \zeta, \eta \rangle = 0.$$

Since V is an isometry, $V^m W \perp V^{m+n} W$ for $n \geq 1$. Thus by the remarks preceding the theorem, $H_1 := \sum_{n\geq 0}^{\oplus} V^n W$ is invariant and $V|_{H_1} \simeq S^{(\alpha)}$ where

K. R. Davidson, *Functional Analysis and Operator Algebras*, CMS/CAIMS Books in Mathematics 13,
https://doi.org/10.1007/978-3-031-63665-3_12

$\alpha = \dim W$. Observe that

$$\operatorname{Ran} V^k = (\ker V^{*k})^\perp = (W \oplus \cdots \oplus V^{k-1}W)^\perp = H_1^\perp \oplus \sum_{n\geq k}^{\oplus} V^n W.$$

Therefore

$$H_0 = \bigcap_{n\geq 1} V^n H = H_1^\perp.$$

Since $VH = W^\perp = H_0 \oplus VH_1$, we have $VH_0 = H_0$. Hence $V|_{H_0}$ is unitary. ∎

12.1.3. DEFINITION. An isometry is *pure* if it is has no unitary summand.

The following easy proposition is left as an exercise.

12.1.4. PROPOSITION. *An isometry V is pure if and only if* $\operatorname{SOT-}\lim_{n\to\infty} V^{*n} = 0$.

12.1.5. DEFINITION. Let $\pi : \mathcal{B}(H) \to \mathcal{B}(H)/\mathcal{K}(H)$ be the quotient by the compact operators onto the Calkin algebra. If $T \in \mathcal{B}(H)$, its *essential spectrum* is

$$\sigma_e(T) = \sigma_{\mathcal{B}(H)/\mathcal{K}(H)}(\pi(T)) = \{\lambda \in \mathbb{C} : \lambda I - T \text{ is not Fredholm}\}.$$

Also the *essential norm* is $\|T\|_e = \|\pi(T)\|$.

The equivalence of these two definitions of essential spectrum is the content of Atkinson's theorem 5.3.8.

12.1.6. EXAMPLE. Let S be the unilateral shift with respect to an orthonormal basis $\{\xi_n : n \geq 0\}$. We have seen that $\sigma(S) = \overline{\mathbb{D}}$. Also S is Fredholm with $\operatorname{ind} S = -1$, and $S^*S = I$ and $SS^* = I - \xi_0\xi_0^*$, so that $\pi(S)^{-1} = \pi(S^*) = \pi(S)^*$. Thus $\pi(S)$ is unitary, so $\sigma_e(S) \subset \mathbb{T}$. One way to see that $\sigma_e(S) = \mathbb{T}$ is that $\mathbb{D}$ is in the resolvent of $\pi(S)$, and the Fredholm index is a locally constant function, so $\operatorname{ind} \lambda I - S = -1$ on $\mathbb{D}$. But $\lambda I - S$ is invertible, and thus $\operatorname{ind} \lambda I - S = 0$ for $|\lambda| > 1$. So these must be separate components of $\mathbb{C} \setminus \sigma_e(S)$, and hence $\mathbb{T} \subset \sigma_e(S)$. Alternatively, we can observe that $S \simeq \lambda S$ for $|\lambda| = 1$, so that $\sigma_e(S)$ must have circular symmetry.

We record a couple of elementary facts about the essential norm for future use.

12.1.7. PROPOSITION. *Let $T \in \mathcal{B}(H)$. Suppose that $P_n \leq P_{n+1}$ for $n \geq 1$ is an increasing sequence of finite rank projections converging strongly to I. Then $\|T\|_e = \lim_{n\to\infty} \|TP_n^\perp\|$. If ξ_n are unit vectors in H such that $\xi_n \xrightarrow{w} 0$ weakly. then $\|T\|_e \geq \limsup \|T\xi_n\|$.*

PROOF. Since $TP_n^\perp$ is a compact perturbation of T, $\|TP_n^\perp\| \geq \|T\|_e$ for all $n \geq 1$, and this sequence is decreasing. If K is compact, then

$$\lim_{n\to\infty} \|TP_n^\perp\| \leq \lim_{n\to\infty} \|(T+K)P_n^\perp\| + \|KP_n^\perp\| \leq \|T+K\|.$$

Taking the infimum over $K \in \mathcal{K}(H)$ yields $\lim_{n\to\infty} \|TP_n^\perp\| = \|T\|_e$.

If $\xi_n \xrightarrow{w} 0$, then $\|K\xi_n\| \to 0$. This is clear if K is finite rank, and the general case follows by approximation. Therefore

$$\|T+K\| \geq \limsup \|(T+K)\xi_n\| = \limsup \|T\xi_n\|.$$

Again taking the infimum over $K \in \mathcal{K}(H)$ yields $\|T\|_e \geq \limsup \|T\xi_n\|$. ∎

Now we can compute the C*-algebra of an isometry. If U is unitary, we have already seen that $C^*(U) \simeq C(\sigma(U))$ in Corollary 7.6.4. An ideal of a C*-algebra is *essential* if it has non-zero intersection with every other non-zero ideal.

12.1.8. COBURN THEOREM. *Let S be the unilateral shift. Then $C^*(S)$ contains the compact operators $\mathfrak{K}$ as an essential ideal, and there is a short exact sequence*

$$0 \longrightarrow \mathfrak{K} \longrightarrow C^*(S) \xrightarrow{\pi} C(\mathbb{T}) \longrightarrow 0.$$

If V is any isometry, there is a $*$-homomorphism of $C^*(S)$ onto $C^*(V)$ taking S to V. If V is a proper isometry, this is a $*$-isomorphism.

PROOF. Let S be the unilateral shift with respect to an orthonormal basis $\{\xi_n : n \geq 0\}$. Then $C^*(S)$ contains $S^*S - SS^* = \xi_0\xi_0^*$. Hence, it contains $S^i\xi_0\xi_0^*S^{*j} = \xi_i\xi_j^*$ for $i, j \geq 0$. These matrix units span $\mathcal{K}(H)$. Thus $\mathcal{K}(H)$ is an ideal in $C^*(S)$. It is an essential ideal because if $0 \neq J \lhd C^*(S)$, pick $0 \neq A \in J$. Then there is some basis vector ξ_i so that $A\xi_i \neq 0$. Hence $0 \neq A(\xi_i\xi_i^*) \in J \cap \mathcal{K}(H)$.

The quotient $C^*(S)/\mathcal{K}(H)$ is generated by $\pi(S)$, which is unitary by Example 12.1.6. Moreover $\sigma(\pi(S)) = \sigma_e(S) = \mathbb{T}$. Therefore, $C^*(\pi(S)) \simeq C(\mathbb{T})$ via a map such that $\pi(S) = z$. Thus there is a short exact sequence

$$0 \longrightarrow \mathfrak{K} \longrightarrow C^*(S) \xrightarrow{\pi} C(\mathbb{T}) \longrightarrow 0.$$

If V is a proper isometry, then by the Wold decomposition, $V \simeq S^{(\alpha)} \oplus U$ where $\alpha \geq 1$ and U is unitary. Define $\rho : C^*(S) \to C^*(V)$ by $\rho(a) = a^{(\alpha)} \oplus \pi(a)(U)$, where $\pi(a) \in C(\mathbb{T})$ is evaluated at U by the continuous functional calculus. Finally

$$\rho(S) = S^{(\alpha)} \oplus z(U) = S^{(\alpha)} \oplus U = V.$$

This is a $*$-isomorphism. If V is unitary, compose the quotient map π with the map of $C(\mathbb{T})$ onto $C^*(V)$ via the functional calculus. This sends S to z to V. ∎

Recall that if $f : \mathbb{T} \to \mathbb{C}$ is a continuous curve and $z_0 \notin \operatorname{Ran} f$, then the *winding number* of the oriented curve $f(\mathbb{T})$ around z_0 is an integer. Think of $\mathbb{T}$ as $\{e^{i\theta} : 0 \leq \theta \leq 2\pi\}$. Choose a logarithm $w(0)$ for $f(1) - z_0$, and then there is a unique continuous choice of a logarithm $w(\theta)$ for $f(e^{i\theta})$ for $0 \le \theta \le 2\pi$ beginning with $w(0)$. Then

$$\operatorname{ind}_{z_0}(f) = \frac{w(2\pi) - w(0)}{2\pi i}.$$

Two logarithms of $f(1) - z_0$ have the same real part, $\log |f(1) - z_0|$, and the imaginary part is the argument, which is defined only up to a multiple of 2π. So winding number measures the number of times the curve circles z_0 in a counterclockwise direction.

12.1.9. THEOREM. *Let S be the unilateral shift, and let $\pi : C^*(S) \to C(\mathbb{T})$ be the quotient map. If $A \in C^*(S)$, then $\sigma_e(A) = \operatorname{Ran} \pi(A)$. If $\lambda \notin \sigma_e(A)$, then*

$$\operatorname{ind} \lambda I - A = -\operatorname{ind}_\lambda(\pi(A)).$$

PROOF. The spectrum of $\pi(A)$ in $C^*(S)/\mathcal{K}(H)$ and in $\mathcal{B}(H)/\mathcal{K}(H)$ is the same since C*-algebras are inverse closed by Corollary 9.2.6. Therefore $\sigma_e(A) = \sigma(\pi(A))$. Let $f = \pi(A)$ and let $\lambda \notin \sigma_e(A)$. The Fredholm index is invariant under compact perturbations by Corollary 5.3.9, and thus it is a function of the image $f(\pi(S))$. The index is a continuous locally constant function on $\mathcal{F}(H)$. Therefore, it is a homotopy invariant, meaning that $\operatorname{ind}_\lambda f_t(\pi(S))$ does not change along a continuous path $t \to f_t$ in $C(\mathbb{T})$ for $0 \le t \le 1$ providing that $f_t(z) \ne \lambda$ for any $t \in [0, 1]$ and $z \in \mathbb{T}$. It is a standard fact that if $\operatorname{ind}_\lambda(f) = n$, then f is homotopic in $\mathbb{C} \setminus \{\lambda\}$ to $f_1(z) = \lambda + z^n$. Therefore

$$\operatorname{ind} \lambda I - A = \operatorname{ind} \lambda I - (\lambda I + S^n) = \operatorname{ind} S^n = -n = -\operatorname{ind}_\lambda \pi(A). \qquad \blacksquare$$

For deeper facts about shifts, we need a function-theoretic perspective.

12.2. Invariant Subspaces for Shifts

A few invariant subspaces for the shift are apparent from our usual view as the unilateral shift. A deeper answer requires the viewpoint of analytic function theory.

We start by representing the bilateral shift on $L^2(\mathbb{T})$ as M_z. Recall that $L^2(\mathbb{T})$ has an orthonormal basis $\zeta_n(e^{i\theta}) = e^{in\theta}$ for $n \in \mathbb{Z}$. Clearly, $M_z\zeta_n = \zeta_{n+1}$ for $n \in \mathbb{Z}$; so M_z is unitarily equivalent to the bilateral shift. The subspace H^2 is invariant for M_z, so we can define $T_z = M_z|_{H^2}$. Evidently, T_z is unitarily equivalent to the unilateral shift S. Let m be normalized Lebesgue measure on $\mathbb{T}$.

12.2.1. DEFINITION. An invariant subspace for an operator T is called *doubly invariant* if it is also invariant for T^*, and hence is reducing. If it is invariant for T but contains no non-zero reducing subspace, then it is called *simply invariant*.

12.2.2. THEOREM. *Let $\mu \in M^+(\mathbb{T})$ be a finite regular Borel measure on $\mathbb{T}$. Let $N \subset L^2(\mu)$ be invariant for M_μ, multiplication by z on $L^2(\mu)$. Then*

(1) *If N is reducing, there is a measurable set $E \subset \mathbb{T}$ so that $N = L^2(E, \mu)$.*

(2) *If N is simply invariant, then there is a function $\varphi \in L^2(\mu)$ so that $|\varphi|^2\,d\mu = m$, and there is a unitary map $U : H^2 \to N$ given by $Uf = \varphi f$ so that $N = \varphi H^2$ and $M_\mu|_N \simeq S$. The function φ is unique up to multiplication by a scalar in $\mathbb{T}$.*

(3) *If $M_\mu|_N$ has both a unitary and shift part, then $N = L^2(E,\mu) \oplus \varphi H^2$ where $m(E) = 0$ and φ is as in (2).*

PROOF. Clearly $L^2(E,\mu)$ is a reducing subspace for any measurable $E \subset \mathbb{T}$. Conversely, we know that $\{M_\mu\}'' = L^\infty(\mu)$ by the L^∞ functional calculus, and this is maximal abelian by Theorem 9.9.2. If N is a reducing subspace, then $P_N \in \{M_\mu\}' = L^\infty(\mu)$. Hence there is a measurable $E \subset \mathbb{T}$ so that $P_N = M_{\chi_E}$.

If $M_\mu N \subsetneq N$, let $\varphi \in N \ominus M_\mu N$ be a unit vector. Then

$$0 = \langle \varphi, z^n \varphi\rangle = \int e^{-int} |\varphi|^2\,d\mu = \widehat{|\varphi|^2\mu}(n) \quad \text{for} \quad n \geq 1.$$

Hence, this also holds for $n \leq -1$, as $|\varphi|^2\,d\mu$ is a positive measure. Since $\|\varphi\|_2 = 1$, we have

$$\widehat{|\varphi|^2\mu}(n) = \widehat{m}(n) \quad \text{for all} \quad n \in \mathbb{Z}.$$

The map that takes a measure to its Fourier series is injective because this determines its harmonic extension, which determines the measure. Therefore $|\varphi|^2\mu = m$. Define a map $U : H^2 \to L^2(\mu)$ by $Uh = \varphi h$. Then

$$\langle Uh, Uk\rangle_{L^2(\mu)} = \langle \varphi h, \varphi k\rangle_{L^2(\mu)} = \int h\bar{k}|\varphi|^2\,d\mu = \int h\bar{k}\,dm = \langle h, k\rangle.$$

Thus U is an isometry. Moreover $UT_z h = \varphi z h = M_\mu U h$. Since $U1 = \varphi \in N$ and N is invariant, $UH_2 \subset N$.

If $\psi \in N \ominus M_\mu N$ is another unit vector orthogonal to φ, then $|\psi|^2\mu = m$. In particular, $|\psi| = |\varphi|$ a.e.(μ). However,

$$0 = \langle \varphi, z^n \psi\rangle = \langle \psi, z^n\varphi\rangle \quad \text{for} \quad n \geq 0.$$

Arguing as above, we see that $\widehat{\varphi\overline{\psi}\mu} = 0$. Hence $\varphi\overline{\psi} = 0$ a.e.(μ). This is contradictory, and therefore $\dim N \ominus M_\mu N = 1$.

Hence if $N \neq UH^2$, then $N \ominus UH^2$ must be reducing. By the first part, it has the form $L^2(E,\mu)$. In addition, $L^2(E,\mu)$ is invariant under M_μ^*. So $L^2(E,\mu)$ is orthogonal to $\overline{\bigcup_{n\geq 1} M_\mu^{-n} UH^2} = L^2(|\varphi|^2\mu)$. Therefore $\chi_E\mu$ and $|\varphi|^2\mu = m$ are mutually singular; so that $m(E) = 0$. ∎

12.2.3. COROLLARY. *The invariant subspaces of M_z on $L^2(\mathbb{T})$ have the form $L^2(E)$, which are doubly invariant, or uH^2 where $u \in L^\infty$ is* unimodular, *i.e.* $|u| = 1$ a.e.(m), *which are simply invariant.*

PROOF. In the simply invariant case, $N = uH^2$ and $|u|^2 m = m$, or $|u| = 1$ a.e.(m). The third case cannot occur since $m(E) = 0$ makes $L^2(E, m) = \{0\}$. ∎

Note that multiplication by any inner function ω is an isometry on H^p for all p. In particular, ωH^2 is a closed subspace of H^2 which is invariant for T_z. We can now give a complete description of the invariant subspaces of the unilateral shift.

12.2.4. BEURLING THEOREM. *Let N be a non-zero invariant subspace for $T_z \in \mathcal{B}(H^2)$. Then there is an inner function $\omega \in H^\infty$ such that $N = \omega H^2$. The inner function is unique up to a scalar multiple.*

PROOF. Invariant subspaces for T_z in H^2 are invariant for M_z as subspaces of L^2. No non-zero subspace of the form $L^2(E)$ is contained in H^2 because they contain non-constant real functions. So $N = \omega H^2$ for some unimodular function $\omega \in H^2 \cap L^\infty = H^\infty$. Therefore ω is inner. If $\omega H^2 = \chi H^2$ for another inner function χ, then both $\overline{\omega}\chi$ and $\omega\overline{\chi}$ are in H^2. As these functions are conjugates, they must be constants. ∎

12.2.5. EXAMPLE. Let b_a be a Blaschke factor for $a \in \mathbb{D}$. Since $|a| < 1$, $I - \bar{a}T_z$ is invertible. Thus

$$b_a H^2 = (aI - T_z)(I - \bar{a}T_z)^{-1} H^2 = (a - z)H^2 = \{h \in H^2 : h(a) = 0\}.$$

Indeed, $(b_a h)(a) = 0$. Conversely, if $h(a) = 0$, then $g(z) = \dfrac{h(z)}{b_a(z)}$ has a removable singularity at a. Since $|b_a(e^{i\theta})| = 1$, we have $g \in H^2$ and $h = b_a g$. Note that $b_a H^2 = \{k_a\}^\perp$ because $\langle h, k_a \rangle = h(a)$. This subspace has codimension 1. In fact, since $|b_a(e^{i\theta})| = 1$, T_{b_a} is an isometry. Thus $\operatorname{Ran}(T_{b_a})^\perp = \mathbb{C}k_a$, and so $\operatorname{ind} T_{b_a} = -1$.

Next observe that $T_{b_a}^n$ has range $(a - z)^n H^2$. This is the space of functions h in H^2 such that

$$0 = h(a) = h'(a) = \cdots = h^{(n-1)}(a).$$

Therefore $\bigcap_{n \geq 1} \operatorname{Ran} T_{b_a}^n = \{0\}$ because any analytic function on $\mathbb{D}$ with a zero of infinite order at a must be 0. Hence by the Wold decomposition, we see that T_{b_a} is unitarily equivalent to the unilateral shift.

If B is a finite Blaschke product of n Blaschke factors (including z), then BH^2 has codimension n. So $\operatorname{ind} T_B = -\deg(B) = -n$. In particular, if the zeros $a_1, \dots, a_n$ are distinct, then $BH^2 = \{k_{a_1}, \dots, k_{a_n}\}^\perp$.

12.2.6. Remark. On the other hand, if $\omega = BS$ is inner but not a finite Blaschke product, then either B is an infinite Blaschke product or the singular part $S = S(\mu)$ corresponds to a non-zero singular measure μ. In the first case, the range is orthogonal to $\operatorname{span}\{k_{a_n} : B(a_n) = 0\}$ and so has infinite codimension. And if there is a non-trivial singular part, then $S(2^{-n}\mu)$ is a proper isometry for all $n \geq 1$, and hence again $S(\mu) = \prod_{n\geq 1} S(2^{-n}\mu)$ has range with infinite codimension.

12.2.7. Proposition. *A vector $f \in H^2$ is cyclic for T_z, i.e.*

$$\overline{\operatorname{span}}\{fz^n : n \geq 0\} = H^2,$$

if and only if it is outer.

Proof. Let $N = \overline{\operatorname{span}}\{fz^n : n \geq 0\}$. Clearly this is invariant for T_z. By Beurling's theorem, there is an inner function ω, unique up to a scalar multiple in $\mathbb{T}$, such that $N = \omega H^2$. If f is not cyclic, i.e. $N \neq H^2$, then ω is not constant and $f = \omega h$ for some $h \in H^2$. As f has a proper inner factor, it is not outer.

Conversely, if f is not outer, then $f = \omega h$ for some proper inner factor ω. Then $N \subset \omega H^2$ is a proper invariant subspace. So f is not cyclic. ∎

12.2.8. Corollary. *A vector $f \in L^2$ generates a simply invariant subspace for M_z if and only if $\log|f|$ is integrable.*

Proof. If the subspace $N = \operatorname{span}\{fz^n : n \geq 0\}$ is simply invariant, then $N = uH^2$ for a unimodular function u. Thus $f = uh$ for $h \in H^2$. Therefore $|f| = |h|$, so that $\log|f| = \log|h| \in L^1$ by Theorem 11.3.4.

Conversely, if $\log|f| \in L_1$, by Proposition 11.4.7 and Theorem 11.4.8, there is an outer function $h \in H^2$ with $|h| = |f|$. Hence there is a unimodular function u so that $f = uh$. Therefore $N = u\operatorname{span}\{hz^n : n \geq 0\} = uH^2$ is simply invariant. ∎

12.2.9. Remark. This corollary requires the full treatment that we give of inner and outer functions. In some cases, one can find a simpler argument. For example, suppose that $|f| \geq \varepsilon > 0$. Then

$$\operatorname{dist}(f, zN) = \inf_{p\in\mathbb{C}[z]} \|f - fzp\|_2 \geq \varepsilon \inf_{p\in\mathbb{C}[z]} \|1 - zp\|_2 = \varepsilon.$$

Hence $zN \neq N$, so N is simply invariant.

The following corollary explaining the remaining cases is left as Exercise **12.7**.

12.2.10. Corollary. *If $f \in L^2$ and $f^{-1}(0) = E$ has $m(E) > 0$, then $\overline{\operatorname{span}}\{fz^n : n \geq 0\} = L^2(\mathbb{T} \setminus E)$.*
A vector $f \in L^2$ is cyclic for M_z if and only if $\log|f|$ is not integrable and $m(f^{-1}(0)) = 0$.

Next we extend the invariant subspace result to L^p. We will write M_z for the multiplication operator by $z(e^{i\theta}) = e^{i\theta}$ on L^p for any $p \in [1, \infty]$.

12.2.11. LEMMA. *If $1 \le p < \infty$ and $N \subset L^p$ is a closed invariant subspace for M_z, then $N = H^\infty N$. If $N \subset L^\infty$ is a weak-$*$ closed invariant subspace for M_z, then $N = H^\infty N$. Moreover, if $N \subset H^\infty$, then it is an ideal.*

PROOF. If $f \in N$, then $pf \in N$ for any polynomial in z. If $h \in A(\mathbb{D})$, there are polynomials p_n converging to h uniformly. Hence

$$\|hf - p_n f\|_p \le \|h - p_n\|_\infty \|f\|_p \to 0;$$

whence $hf \in N$. Now if $h \in H^\infty$, then $h_r \in A(\mathbb{D})$, $\|h_r\|_\infty \le \|h\|_\infty$ and h_r converges weak-$*$ to h as $r \to 1^-$. Either by dropping to a subsequence or applying Fatou's theorem, we may find $r_k \to 1$ so that $h_{r_k} \to h$ a.e.(m). For $p < \infty$, the Lebesgue dominated convergence theorem applies to

$$\lim_{k\to\infty} \|hf - h_{r_k} f\|_p^p = \lim_{k\to\infty} \int_{\mathbb{T}} |h - h_{r_k}|^p |f|^p \, dm = 0.$$

Thus $hf \in N$. Finally, if $p = \infty$ and $g \in L^1$, the LDCT again shows

$$\lim_{k\to\infty} \langle hf - h_{r_k} f, g\rangle = \lim_{k\to\infty} \int_{\mathbb{T}} (h - h_{r_k}) fg \, dm = 0.$$

Thus $hf = w^*\text{–}\lim h_{r_k} f \in N$. If $N \subset H^\infty$, then N is an ideal of H^∞. ∎

The heavy lifting is done in the next lemma.

12.2.12. LEMMA. *If $1 \le p < r \le \infty$ and $N \subset L^p$ is a closed invariant subspace for M_z, then $M = N \cap L^r$ is a closed invariant subspace of L^r and N is the closure of M in L^p. If $r = \infty$, then M is weak-$*$ closed.*

PROOF. Clearly M is invariant for M_z. Suppose that a sequence $f_n \in M$ converges to $f \in L^r$. Then since $\|f - f_n\|_p \le \|f - f_n\|_r$, it also converges in L^p. So $f \in N \cap L^r = M$. Thus M is closed. Now if $r = \infty$, suppose that $\{f_\lambda\}_\Lambda \subset M$ is a net converging weak-$*$ to $f \in L^\infty$, i.e. $\lim_\Lambda \langle f_\lambda, g\rangle = \langle f, g\rangle$ for all $g \in L^1$. In particular, this holds for $g \in L^q$ where $\frac{1}{p} + \frac{1}{q} = 1$. So $\{f_\lambda\}_\Lambda$ converges weakly in L^p. But the Hahn–Banach theorem shows that norm-closed subspaces are weakly closed, and hence $f \in N \cap L^\infty = M$. Therefore M is weak-$*$ closed in L^∞.

The second statement is trickier. Suppose that $f \in N$. Let

$$k_n(e^{i\theta}) = \begin{cases} 0 & \text{if } |f(e^{i\theta})| \le n \\ \log n - \log|f(e^{i\theta})| & \text{if } |f(e^{i\theta})| \ge n. \end{cases}$$

Observe that if $0 < x$, then $0 \le \log^+ x \le \frac{1}{p} x^p$. Thus

$$0 \ge k_{n+1} \ge k_n \ge -\log^+|f| \ge -\tfrac{1}{p}|f|^p \in L^1 \quad \text{and} \quad \lim_{n\to\infty} k_n = 0 \text{ a.e.}(m).$$

Let $u_n(z) = \widetilde{k_n}(z)$ and let $v_n(z)$ be its harmonic conjugate. Define

$$h_n(z) = e^{u_n(z)+iv_n(z)} = \exp\Big(\frac{1}{2\pi}\int_{-\pi}^{\pi} \frac{e^{it}+z}{e^{it}-z} k_n(e^{it})\, dt\Big).$$

Then $|h_n| = e^{k_n} \leq 1$ a.e.(m), so that $h_n \in H^\infty$ and $\|h_n\|_\infty \leq 1$. Notice that $\|h_n f\|_\infty \leq n$ and so $h_n f \in L^r$. Also $h_n f \in H^\infty N = N$ by Lemma 12.2.11. Thus $h_n f \in M$. Observe that

$$h_n(0) = e^{u_n(0)} = \exp\Big(\frac{1}{2\pi}\int_{-\pi}^{\pi} k_n(e^{it})\, dt\Big).$$

This converges to $e^0 = 1$ by the LDCT. Now

$$\|1 - h_n\|_2^2 = \|1\|_2^2 + \|h_n\|_2^2 - 2\,\mathrm{Re}\langle h_n, 1\rangle \leq 2(1 - h_n(0)) \to 0.$$

Thus h_n converges to 1 in L^2; so some subsequence converges to 1 a.e.(m). Hence $h_{n_i} f$ converges to f in L^p by LDCT. Therefore N is the closure of M in L^p. ∎

12.2.13. REMARK. There is an interesting and useful fact established in this proof. If $f \in L^p$, there is a sequence $h_n \in H^\infty$ such that $h_n f \in L^\infty$ and converge to f in L^p.

12.2.14. THEOREM. *If $1 \leq p < \infty$ and $N \subset L^p$ is a closed invariant subspace for M_z or if $N \subset L^\infty$ is a weak-$*$ closed invariant subspace for M_z, then either N is also invariant for $M_{\bar z}$ and $N = L^p(E)$ for some measurable $E \subset \mathbb{T}$ or N is simply invariant and there is a unimodular function $u \in L^\infty$ so that $N = uH^p$.*

PROOF. First suppose that $1 \leq p < 2$. Then $M = N \cap L^2$ is norm closed and M_z invariant. If M is doubly invariant, then $M = L^2(E)$ for some Borel $E \subset \mathbb{T}$. Hence N is the closure of $L^2(E)$ in L^p, namely, $L^p(E)$. If M is simply invariant, then there is a unimodular function $u \in L^\infty$ so that $M = uH^2$. Again we conclude that N is the closure of M in L^p, namely, uH^p.

Now suppose that $2 < p$. Let q be the dual exponent, $\frac{1}{p} + \frac{1}{q} = 1$. Set $M = N_\perp = \{g \in L^q : \langle f, g\rangle = 0 \text{ for all } f \in N\}$. Then

$$\langle f, gz\rangle = \langle fz, g\rangle = 0 \quad \text{for all} \quad f \in N \text{ and } g \in M.$$

Hence M is M_z invariant. Now $N = (N_\perp)^\perp = M^\perp$ because N is weak-$*$ closed (which is the same as norm closed if $p < \infty$). Thus if $M = L^q(E)$ is doubly invariant, then $N = L^q(E)^\perp = L^p(\mathbb{T} \setminus E)$ is also doubly invariant. And if $M = uH^q$, then

$$N = (uH^q)^\perp = \bar{u}H_0^p = (\bar{u}z)H^p$$

because $H^{q\perp} = H_0^p$ and $|\bar{u}z| = 1$ a.e.(m) is unimodular. ∎

We obtain an immediate consequence, again because non-zero H^p functions are non-zero a.e.(m) on $\mathbb{T}$. So $H^p \cap L^p(E) = \{0\}$ for any proper $E \subset \mathbb{T}$.

12.2.15. COROLLARY. *The closed invariant subspaces for M_z on H^p for $1 \le p < \infty$ are the spaces of the form ωH^p for ω inner. The weak-$*$ closed ideals in H^∞ are ωH^∞ for ω inner.*

We now derive another proof of the F. & M. Riesz theorem which does not rely on the deeper material about inner and outer functions of the last chapter. Recall that the theorem says: *If $\nu \in M(\mathbb{T})$ is analytic, i.e.* $\hat{\nu}(n) = 0$ for $n < 0$, *then $|\nu|$ is mutually absolutely continuous to Lebesgue measure and $\nu = hm$ for some $h \in H^1(\mathbb{T})$.*

PROOF OF F. & M. RIESZ THEOREM. Let $\mu = |\nu|$. By the Radon–Nikodym theorem, $\nu = k\mu$ for some k with $|k| = 1$ a.e.(μ). Let $N = \overline{\operatorname{span}}\{kz^n : n \ge 0\}$ in $L^2(\mu)$ be the closed invariant subspace for M_μ generated by k. Then in $L^2(\mu)$,

$$\langle kz^n, \overline{z}\rangle = \int z^{n+1} k \, d\mu = \hat{\nu}(-n-1) = 0 \quad \text{for} \quad n \ge 0.$$

Therefore $N \perp \overline{z}$, so that N is a proper subspace. Now N cannot be reducing, for then it would have the form $L^2(E, \mu)$, and since $|k| = 1$ a.e.(μ) and $k \in N$, this would mean that $N = L^2(\mu)$, a contradiction. Also if N contains a reducing subspace $M \ni g$, then $z^n g \in M \subset N$ for all $n \in \mathbb{Z}$. Hence

$$0 = \langle gz^n, \overline{z}\rangle = \int gz^{n+1} \, d\mu = \widehat{gz\mu}(n) \quad \text{for} \quad n \in \mathbb{Z}.$$

This means that $gz\mu = 0$, and hence $g = 0$. Therefore $M_\mu|_N$ is a pure isometry. By Theorem 12.2.2, there is a $\varphi \in L^2(\mu)$ so that $|\varphi|^2\mu = m$. Since $k \in N$ and $k \neq 0$ a.e.(μ), we have $\varphi \neq 0$ a.e.(μ). Hence μ and m are mutually absolutely continuous. Finally $\nu = k\mu = hm$ where $h = k|\varphi|^{-2}$. Since this is a finite measure, $h \in L^1(\mathbb{T})$, and $\hat{h}(n) = \hat{\nu}(n) = 0$ for $n < 0$. Thus $h \in H^1$. ∎

12.3. Toeplitz Operators

In this section, we define the Toeplitz operators and find out a few things about their spectra.

12.3.1. DEFINITION. If $f \in L^\infty(\mathbb{T})$, we define a *Toeplitz operator* $T_f \in \mathcal{B}(H^2)$ by $T_f = P_{H^2} M_f|_{H^2}$. If $A \subset L^\infty(\mathbb{T})$ is a closed subalgebra, let $\mathcal{T}(A)$ denote the norm-closed algebra generated by $\{T_f : f \in A\}$.

We begin by deriving a number of elementary facts about Toeplitz operators. For the remainder, we will write L^∞ for $L^\infty(\mathbb{T})$ and L^2 for $L^2(\mathbb{T})$, and we write P for P_{H^2}.

12.3.2. Proposition. *For $f \in L^\infty$, $T_f^* = T_{\bar{f}}$ and $\|T_f\| = \|T_f\|_e = \|f\|_\infty$.*

Proof. For $h, k \in H^2$,

$$\begin{aligned} \langle T_f^* h, k\rangle &= \langle h, T_f k\rangle = \langle h, Pfk\rangle = \langle h, fk\rangle \\ &= \langle \bar{f}h, k\rangle = \langle P\bar{f}h, k\rangle = \langle T_{\bar{f}}h, k\rangle. \end{aligned}$$

Thus $T_f^* = T_{\bar{f}}$.

Now $\|T_f\|_e \le \|T_f\| \le \|M_f\| = \|f\|_\infty$. For $\varepsilon > 0$, choose a trigonometric polynomial $p = \sum_{jk=-N}^{N} a_j e^{ij\theta} \in L^2$ with $\|p\|_2 = 1$ so that $\|fp\|_2 \ge \|f\|_\infty - \varepsilon$. Write $fp = \sum_{j=-\infty}^{\infty} b_j e^{ij\theta}$. Now choose M so large that

$$\Big\| \sum_{j=-M}^{\infty} b_j e^{ij\theta} \Big\|_2 > \|f\|_\infty - \varepsilon.$$

Then $z^{N+M+k}p \in H^2$ and

$$\|T_f z^{N+M+k}p\|_2 \ge \Big\| \sum_{j=N+k}^{\infty} b_{j-N-M-k} e^{ij\theta} \Big\|_2 > \|f\|_\infty - \varepsilon.$$

Since $z^{N+M+k}p$ converges weakly to 0 in H^2, Proposition 12.1.7 shows that

$$\|T_f\|_e \ge \limsup \|T_f z^{N+M+k}p\| > \|f\|_\infty - \varepsilon.$$

Now let ε go to 0. ∎

12.3.3. Definition. If $f \in L^\infty$, the *Hankel operator* $H_f \in \mathcal{B}(H^2, H^{2\perp})$ is $H_f = P^\perp M_f P|_{H^2}$.

12.3.4. Proposition. *If $f \in L^\infty$, then $H_f = 0$ if and only if $f \in H^\infty$. For $f, g \in L^\infty$, $T_{fg} - T_f T_g = H_{\bar{f}}^* H_g$.*

Proof. Note that $H_f = 0$ exactly when H^2 is invariant for M_f. If H^2 is invariant, then $f = M_f 1 \in H^2 \cap L^\infty = H^\infty$. Conversely, if $h \in H^\infty$, H^2 is invariant. If $f, g \in L^\infty$,

$$\begin{aligned} T_{fg} - T_f T_g &= PM_f M_g P - PM_f PM_g P|_{H^2} = PM_f P^\perp M_g P|_{H^2} \\ &= (P^\perp M_{\bar{f}} P)^*(P^\perp M_g P)|_{H^2} = H_{\bar{f}}^* H_g. \end{aligned}$$

∎

12.3.5. Corollary. *For $f \in L^\infty$ and $g, h \in H^\infty$, $T_{\bar{g}} T_f T_h = T_{\bar{g}fh}$.*

Proof. $T_{fh} - T_f T_h = H_{\bar{f}}^* H_h = 0$. Taking adjoints, we get $T_{\bar{g}f} = T_{\bar{g}} T_f$. ∎

12.3.6. Proposition. *For $f \in L^\infty$,* $\operatorname{rank}(T_f T_z - T_z T_f) \le 1$.

PROOF. By the previous proposition, $T_f T_z - T_z T_f = T_{zf} - T_z T_f = H_{\bar{z}}^* H_f$. However $H_{\bar{z}} = \zeta_{-1}\zeta_0^*$ is rank 1. The result follows. ∎

12.3.7. COROLLARY. *For $f \in L^\infty$ and $g \in C(\mathbb{T})$, $T_f T_g - T_{fg}$ and $T_{fg} - T_g T_f$ are compact.*

PROOF. By the previous proposition,

$$T_{z^n} T_f - T_{fz^n} = T_z^n T_f - T_f T_z^n = \sum_{k=0}^{n-1} T_z^{n-1-k}(T_z T_f - T_f T_z) T_z^k$$

is finite rank for $n \geq 1$. Taking adjoints, $T_{\bar{z}^n} T_f - T_f T_{\bar{z}^n} = T_{\bar{z}^n f} - T_f T_{\bar{z}^n}$ are also finite rank. Now taking linear combinations, we have $T_f T_p - T_{fp}$ and $T_{fp} - T_p T_f$ are finite rank if p is a polynomial in z and $\bar{z}$. By the Weierstrass theorem, these trigonometric polynomials are dense in $C(\mathbb{T})$. Hence $T_f T_g - T_{fg}$ and $T_{fg} - T_g T_f$ are compact for $g \in C(\mathbb{T})$. ∎

12.3.8. REMARK. This provides another view of the C*-algebra of the unilateral shift. Recall Coburns Theorem 12.1.8. Let's write C for $C(\mathbb{T})$. We have

$$C^*(T_z) = \mathcal{T}(C) = \{T_f + K : f \in C,\ K \in \mathcal{K}(H^2)\}.$$

Indeed, taking $f, g \in C(\mathbb{T})$, we have $T_f T_g - T_{fg} \in \mathcal{K}(H^2)$. We know that $C^*(T_z)$ contains all compact operators, and in the Calkin algebra, $\pi(T_f)\pi(T_g) = \pi(T_{fg})$. In particular, $\pi(T_z)$ is unitary and if p is a trig polynomial, $p(\pi(T_z)) = \pi(T_p)$. Hence $f(\pi(T_z)) = \pi(T_f)$ for all $f \in C(\mathbb{T})$. We obtain a positive isometric linear section from $C(\mathbb{T})$ to $C^*(T_z)$ by sending f to T_f.

12.3.9. THEOREM. *The following are equivalent for $A \in \mathcal{B}(H^2)$:*

(1) *A is a Toeplitz operator.*

(2) *$A = T_z^* A T_z$.*

(3) *$\langle A\zeta_{n+1}, \zeta_{m+1}\rangle = \langle A\zeta_n, \zeta_m\rangle$ for $n, m \geq 0$.*

PROOF. If $A = T_f$, then $T_z^* T_f T_z = T_{\bar{z}fz} = T_f$ by Corollary 12.3.5. If $A = T_z^* A T_z$, then

$$\langle A\zeta_n, \zeta_m\rangle = \langle T_z^* A T_z \zeta_n, \zeta_m\rangle = \langle A\zeta_{n+1}, \zeta_{m+1}\rangle.$$

This says that the matrix entries are constant on diagonals.

Finally suppose that this latter condition holds. Consider the bounded sequence $A_n = M_{z^n}^* PAPM_{z^n}$ in $\mathcal{B}(L^2)$. Observe that if $n < m$, then $A_n = P_{-n} A_m P_{-n}$, where P_{-n} is the projection onto $z^{-n}H^2 = \operatorname{span}\{\zeta_k : k \geq -n\}$. Since $\bigcup_{n\geq 1} z^{-n}H^2$ is dense in L^2, it follows that A_n is converging in the WOT topology to the operator B such that

$$\langle B\zeta_{n+1}, \zeta_{m+1}\rangle = \langle B\zeta_n, \zeta_m\rangle = \langle A\zeta_{n+|n|+|m|}, \zeta_{m+|n|+|m|}\rangle \quad \text{for} \quad n, m \in \mathbb{Z}.$$

In particular, $B = M_z^* B M_z$. Since M_z is unitary, this implies that $M_z B = B M_z$, i.e. $B \in \{M_z\}' = \{M_f : f \in L^\infty\}$. Thus there is some $f \in L^\infty$ so that $B = M_f$. By construction, $A = PBP|_{H_2} = T_f$ is Toeplitz. ∎

Now we will obtain some information about the spectrum of Toeplitz operators.

12.3.10. LEMMA. *If* $0 \neq f \in L^\infty$, *then either* $\ker T_f = \{0\}$ *or* $\ker T_f^* = \{0\}$. *Hence if* T_f *is Fredholm and* $\operatorname{ind} T_f = 0$, *then* T_f *is invertible.*

PROOF. Suppose to the contrary that $h, k \in H^2$ are non-zero functions such that $T_f h = 0 = T_{\bar{f}} k$. Then fh and $\bar{f}k$ lie in $H^{2\perp} = \overline{H_0^2}$. Hence $g = \overline{fhk} \in H_0^1$ and $\bar{g} = \overline{\bar{f}kh} \in H_0^1$ as well by Lemma 11.2.11. Hence $\hat{g}(n) = 0$ for all $n \in \mathbb{Z}$, whence $g = 0$. Since h and k are non-zero a.e.(m) by Corollary 11.3.6, we have $f = 0$.

If T_f is Fredholm and index 0, we have $\dim\ker T_f = \dim\ker T_f^* = 0$. Thus T_f is injective and has closed range $(\ker T) f^*)^\perp = H_2$. Therefore, it is invertible by the Banach isomorphism theorem. ∎

12.3.11. COROLLARY. *If* $f \in C(\mathbb{T})$, *then* $\sigma_e(T_f) = \operatorname{Ran} f$ *and*

$$\sigma(T_f) = \operatorname{Ran} f \cup \{\lambda : \operatorname{ind} f - \lambda \neq 0\}.$$

PROOF. We identify T_z with the unilateral shift so that we can apply Theorem 12.1.9. By Remark 12.3.8, the quotient $\pi : C^*(T_z) \to C(\mathbb{T})$ satisfies $\pi(T_f) = f$. Therefore $\sigma_e(T_f) = \sigma_{C(\mathbb{T})}(f) = \operatorname{Ran} f$. Moreover, if $\lambda \in \mathbb{C} \setminus \operatorname{Ran} f$, then T_f is Fredholm and

$$\operatorname{ind}(T_f - \lambda I) = -\operatorname{ind}(f - \lambda).$$

Lemma 12.3.10 shows that $T_f - \lambda I$ is invertible precisely when this index is 0. This occurs on the unbounded component of $\mathbb{C} \setminus \operatorname{Ran} f$ and possibly on some of the bounded components. The components on which $f - \lambda$ has non-zero winding number are in the spectrum. ∎

12.3.12. LEMMA. *If* $f \in L^\infty$ *and* T_f *is left or right Fredholm, then* f *is invertible in* L^∞.

PROOF. First suppose that T_f is left invertible, which means that it is bounded below, say by $\varepsilon > 0$. Let $\xi = \sum_{k=-N}^{N} a_k \zeta_k$ be a finitely supported vector in L^2. Then

$$\|M_f \xi\| = \|M_f z^N \xi\| \geq \|T_f z^N \xi\| \geq \varepsilon \|z^N \xi\|_2 = \varepsilon \|\xi\|_2.$$

These vectors are dense in L^2, and thus M_f is bounded below. Hence, f is essentially bounded below, i.e. $|f| \geq \varepsilon$ a.e.(m), so that f is invertible in L^∞.

If T_f is right invertible, then $T_f^* = T_{\bar{f}}$ is left invertible. Thus $\bar{f}$ is invertible in L^∞; so f is also invertible.

Next suppose that T_f is left Fredholm, meaning that there is an operator A so that $AT_f = I + K$ for some $K \in \mathcal{K}(H^2)$, or equivalently that $\ker T_f$ is finite dimensional and T_f has closed range. By Lemma 12.3.10, either $\ker T_f = \{0\}$, which implies that T_f is left invertible or $\ker T_f^* = \{0\}$, so that T_f is surjective and thus right invertible. Either way, f is invertible in L^∞. The case of right Fredholm operators follows by taking adjoints. ∎

In the following result, the first inclusion is due to Hartman and Wintner, and the second inclusion is due to Brown and Halmos.

12.3.13. THEOREM. *If $f \in L^\infty$, then*

$$\operatorname{ess\,ran}(f) \subset \sigma_e(T_f) \quad \textit{and} \quad \sigma(T_f) \subset \overline{\operatorname{conv}(\operatorname{ess\,ran} f)}.$$

PROOF. If $\lambda \in \operatorname{ess\,ran} f$, then $f - \lambda$ is not invertible in L^∞. Hence by Lemma 12.3.12, $T_f - \lambda I$ is not Fredholm, so $\lambda \in \sigma_e(T_f)$.

Suppose that $f \in L^\infty$ and $\operatorname{Re} f \geq \varepsilon > 0$. Then

$$\operatorname{ess\,ran} f \subset C = \overline{b_{\|f\|_\infty}(0)} \cap \{z : \operatorname{Re} z \geq \varepsilon\}.$$

If r is sufficiently large (e.g. $r > \frac{1}{\varepsilon}\|f\|_\infty$), then $C \subset b_r(r)$ and thus $\|r - f\|_\infty < r$. Hence $\|T_{f/r} - I\| < 1$, so that $\frac{1}{r}T_f$ is invertible. The geometric interpretation of this is that if H is a closed half plane containing $\operatorname{ess\,ran} f$, then $\sigma(T_f) \subset H$. Therefore, $\sigma(T_f)$ is contained in the intersection of all closed half planes containing $\operatorname{ess\,ran} f$, which is $\overline{\operatorname{conv}(\operatorname{ess\,ran} f)}$. ∎

12.3.14. THEOREM. *If $f = \bar{f} \in L^\infty$ is real valued, then*

$$\sigma(T_f) = \sigma_e(T_f) = \big[\operatorname{ess\,inf} f, \operatorname{ess\,sup} f\big].$$

PROOF. Since f is real valued, $T_f = T_f^*$ and $\sigma(T_f) \subset \mathbb{R}$. If $T_f - rI$ is Fredholm for $r \in \mathbb{R}$, then it has index 0 as r lies in the unbounded component of $\rho_e(T_f)$. Hence by Lemma 12.3.10, T_f is invertible. Thus $\sigma(T_f) = \sigma_e(T_f)$.

Suppose that $r \in \mathbb{R}$ and $T_f - rI$ is invertible. There is some $h \in H^2$ so that $T_{f-r}h = 1$. So $(f - r)h = 1 + \bar{g}$ for some $g \in H_0^2$. Therefore

$$(f - r)|h|^2 = \overline{(f - r)h}\, h = (1 + g)h \in H^2 \cdot H^2 \subset H^1.$$

But $(f - r)|h|^2$ is real valued, and hence it is constant by Remark 11.2.3(3); say $(f - r)|h|^2 = s \in \mathbb{R}$. Since $|h|^2 > 0$ a.e.(m), $\operatorname{sign}(f - r)$ is constant. Therefore, either $r > \operatorname{ess\,sup} f$ or $r < \operatorname{ess\,inf} f$. ∎

It is a non-trivial result of Widom that the essential spectrum $\sigma_e(T_f)$ is always connected. See [99, Theorem 7.45].

12.4. Analytic Toeplitz Operators

In this section, we study the operator algebra $\mathcal{T}(H^\infty)$. This is a very tractable situation. We also collect some results which can be applied more generally.

12.4.1. Proposition. $\{T_z\}' = \mathcal{T}(H^\infty) = \{T_h : h \in H^\infty\}$.

Proof. We have seen that if $h \in H^\infty$, then $T_h T_z = T_{hz} = T_z T_h$. Since all analytic Toeplitz operators commute, and the map from f to T_f is isometric, $\mathcal{T}(H^\infty) = \{T_h : h \in H^\infty\}$ is a norm-closed algebra.

Suppose that $A \in \mathcal{B}(H^2)$ and $T_z A = A T_z$. Then $A = T_z^* T_z A = T_z^* A T_z$, so that $A = T_f$ for $f \in L^\infty$ by Theorem 12.3.9. If $f \notin H^\infty$, then $\hat{f}(-n) \neq 0$ for some $n \geq 1$. Thus

$$\langle T_f T_z^n \zeta_0, \zeta_0 \rangle = \langle T_f \zeta_n, \zeta_0 \rangle = \hat{f}(-n) \neq 0,$$

while

$$\langle T_z^n T_f \zeta_0, \zeta_0 \rangle = \langle T_z^{n-1} T_f \zeta_0, T_z^* \zeta_0 \rangle = 0.$$

So T_f does not commute with T_z. ■

12.4.2. Corollary. *The* WOT *closed algebra $W(T_z)$ generated by T_z in $\mathcal{B}(H^2)$ is $\mathcal{T}(H^\infty)$.*

Proof. By the proposition, $\mathcal{T}(H^\infty)$ is a commutant, and thus is a WOT closed subalgebra of $\mathcal{B}(H^2)$ containing T_z. On the other hand, if $h \in H^\infty$, then $h_r \in A(\mathbb{D})$ for $r < 1$ and $h_r \xrightarrow{w^*} h$. Now h_r has an absolutely convergent Fourier series, so T_{h_r} belong to the norm-closed algebra generated by T_z. If $f, g \in H^2$, then $f\bar{g} \in L^1$. Thus

$$\begin{aligned} \lim_{r \to 1^-} \langle T_{h_r} f, g \rangle &= \lim_{r \to 1^-} \frac{1}{2\pi} \int_{-\pi}^{\pi} h_r(e^{it}) f(e^{it}) \overline{g(e^{it})}\, dt \\ &= \frac{1}{2\pi} \int_{-\pi}^{\pi} h(e^{it}) f(e^{it}) \overline{g(e^{it})}\, dt = \langle T_h f, g \rangle. \end{aligned}$$

Therefore, T_{h_r} converge to T_h in the WOT topology. Hence $W(T_z) = \mathcal{T}(H^\infty)$. ■

12.4.3. Theorem. *If $f \in H^\infty$, then $\sigma(T_f) = \sigma_{H^\infty}(f) = \overline{f(\mathbb{D})}$.*

Proof. First, $\sigma_{H^\infty}(f) = \overline{f(\mathbb{D})}$ by Theorem 11.2.10. If f is invertible in H^∞, then $g = 1/f \in H^\infty$. So $T_f T_g = T_1 = T_g T_f$ and T_f is invertible.

Suppose that T_f is invertible with inverse A. Then

$$T_f(T_z A - AT_z) = T_z(T_f A) - (T_f A)T_z = 0.$$

But T_f is invertible, so $A \in \{T_z\}' = \mathcal{T}(H^\infty)$. Therefore, f is invertible in H^∞. ∎

12.4.4. COROLLARY. *If $f \in H^\infty$, then the following are equivalent:*

(1) *T_f is Fredholm.*

(2) *$f = BF$ where F is invertible in H^∞ and B is a finite Blaschke product.*

(3) *$\hat{f}$ does not vanish on $\mathcal{M}_{H^\infty} \setminus \mathbb{D}$.*

(4) *there is an $r < 1$ and $\varepsilon > 0$ so that $|f(z)| \geq \varepsilon$ for $r \leq |z| < 1$.*

Moreover $\operatorname{ind}(T_f) = -\lim_{r\to 1^-} \operatorname{ind} f_r = -\deg B$.

PROOF. Assume that T_f is Fredholm. Since $f \in H^\infty$, $\ker T_f = \{0\}$. And $\overline{\operatorname{Ran} T_f}$ is invariant for T_z, and so has the form ωH^2 for some inner function ω. Hence

$$-\operatorname{ind} T_f = \dim(H^2 \ominus \omega H^2) < \infty.$$

This implies that $\omega = B$ is a finite Blaschke product of degree $-\operatorname{ind} T_f$ by Example 12.2.5 and Remark 12.2.6. We can factor $f = BF$ and note that $\operatorname{ind} T_F = \operatorname{ind} T_f - \operatorname{ind} T_B = 0$. Hence, T_F is invertible by Lemma 12.3.10. Thus by Theorem 12.4.3, F is invertible in H^∞. So (2) holds.

Now (2) shows that $\hat{f} = \hat{B}\hat{F}$. Since $|\hat{F}|$ is bounded below on $\mathcal{M}_{H^\infty}$ and $|\hat{B}| = 1$ on $\mathcal{M}_{H^\infty} \setminus \mathbb{D}$, it follows that $\hat{f}$ is bounded below on $\mathcal{M}_{H^\infty} \setminus \mathbb{D}$. Since $\mathcal{M}_{H^\infty} \setminus \mathbb{D}$ is compact, this is equivalent to $\hat{f}$ being non-zero there, i.e. (3) holds.

By continuity, this implies that $|f(z)|$ is bounded away from 0 on an annulus $\{z : r \leq |z| < 1\}$, so (3) implies (4). Condition (4) shows that f has only finitely many zeros. Let B be the corresponding finite Blaschke product. Then $F = f/B$ is never 0 and is bounded away from 0 on $\mathbb{D}$ and hence is invertible. Thus T_F is invertible. Therefore $T_f = T_B T_F$ is Fredholm, which is (1).

As F is invertible, it has a logarithm and so $\operatorname{ind} F_r = 0$ for all $r < 1$. By the continuous case,

$$\operatorname{ind} T_B = -\operatorname{ind} B = -\lim_{r\to 1^-} \operatorname{ind} B_r.$$

Since $f_r = B_r F_r$, $\operatorname{ind} f_r = \operatorname{ind} B_r$. Hence

$$\operatorname{ind}(T_f) = \operatorname{ind} T_B = -\lim_{r\to 1^-} \operatorname{ind} f_r = -\deg B.$$ ∎

12.4.5. REMARK. Beurling's theorem shows that the invariant subspace of T_z, or equivalently of $\mathcal{T}(H^\infty)$, has the form ωH^2 for ω inner. Since $|\omega| = 1$, T_ω is an isometry. Conversely, if $T_f \in \mathcal{T}(H^\infty)$ is an isometry, then $I = T_f^* T_f = T_{\overline{f}f}$ implies that $\overline{f}f = 1$. Therefore f is inner.

By Proposition 12.2.7, a function $f \in H^\infty$ is outer if and only if f is cyclic for T_z. This is equivalent to saying that $\operatorname{Ran} T_f$ is dense in H^2.

12.4.6. DEFINITION. If $\mathcal{A} \subset \mathcal{B}(H)$, the *lattice of invariant subspaces*

$$\operatorname{Lat} \mathcal{A} = \{P \in \operatorname{Pr}(H) : P^\perp A P = 0 \text{ for all } A \in \mathcal{A}\}.$$

If $\mathcal{L}$ is a set of projections on H, the *invariant algebra*

$$\operatorname{Alg}(\mathcal{L}) = \{A \in \mathcal{B}(H) : P^\perp A P = 0 \text{ for all } P \in \mathcal{L}\}.$$

An algebra $\mathcal{A}$ of operators is *reflexive* if $\mathcal{A} = \operatorname{Alg} \operatorname{Lat} \mathcal{A}$. An operator T is called *reflexive* if $W(T)$, the unital WOT-closed algebra generated by T, is reflexive.

12.4.7. REMARK. The collection of invariant subspaces for a set $\mathcal{A}$ is a *complete lattice*. Indeed, if N_α are invariant for $\mathcal{A}$, it is easy to check that $\bigwedge N_\alpha = \bigcap N_\alpha$ is invariant, and $\bigvee N_\alpha = \overline{\operatorname{span} \bigcup N_\alpha}$ is invariant, and that $\{0\}$ and H are in $\operatorname{Lat} \mathcal{A}$. We identify an invariant subspace N with the projection P_N. Then N is invariant if and only if $AP_N = P_N A P_N$, or equivalently $P_N^\perp A P_N = 0$.

If $\mathcal{L}$ is a set of projections, then $\operatorname{Alg}(\mathcal{L})$ is a WOT closed unital operator algebra. Indeed, it is easy to see that it is a subspace. If $S, T \in \operatorname{Alg}(\mathcal{L})$ and $P \in \mathcal{L}$, then

$$P^\perp S T P = (P^\perp S P) T P + P^\perp S (P^\perp T P) = 0.$$

So $\operatorname{Alg} \mathcal{L}$ is an algebra containing I. If $T_\alpha \in \operatorname{Alg} \mathcal{L}$ converge WOT to T, take $\xi \in PH$ and $\eta \in P^\perp H$. Then

$$\langle T\xi, \eta\rangle = \lim \langle T_\alpha \xi, \eta\rangle = 0.$$

Thus $P^\perp T P = 0$, whence $\operatorname{Alg} \mathcal{L}$ is WOT closed.

This use of reflexive is not related to the notion of a reflexive Banach space.

12.4.8. LEMMA. *Let $k_w = \sum_{n \geq 0} \overline{w}^n \zeta_n$ be the kernel function for evaluation at $w \in \mathbb{D}$ in H^2. Then for $f \in H^\infty$, $T_f^* k_w = \overline{f(w)} k_w$ and $\big\langle T_f \frac{k_w}{\|k_w\|}, \frac{k_w}{\|k_w\|} \big\rangle = f(w)$.*

PROOF. This follows from the kernel property: for $h \in H^2$,

$$\begin{aligned} \langle h, T_f^* k_w\rangle &= \langle T_f h, k_w\rangle = \langle fh, k_w\rangle \\ &= f(w) h(w) = f(w) \langle h, k_w\rangle = \langle h, \overline{f(w)} k_w\rangle. \end{aligned}$$

Therefore $T_f^* k_w = \overline{f(w)} k_w$. Hence

$$\Big\langle T_f \frac{k_w}{\|k_w\|}, \frac{k_w}{\|k_w\|} \Big\rangle = \Big\langle \frac{k_w}{\|k_w\|}, T_f^* \frac{k_w}{\|k_w\|} \Big\rangle = \Big\langle \frac{k_w}{\|k_w\|}, \overline{f(w)} \frac{k_w}{\|k_w\|} \Big\rangle = f(w). \qquad \blacksquare$$

12.4.9. REMARK. Since k_w is an eigenvector for $\mathcal{T}(H^\infty)^*$, the one-dimensional subspace $\mathbb{C}k_w$ is invariant for $\mathcal{T}(H^\infty)^*$. Therefore

$$\{k_w\}^\perp = M_w = \{h \in H^2 : h(w) = 0\}$$

is invariant for $\mathcal{T}(H^\infty)$. By Beurling's theorem, there is an inner function with this range. The function is b_w. See Example 12.2.5. This has a simple zero at w and is a conformal map of $\mathbb{D}$ onto itself, and maps the circle $\mathbb{T}$ bijectively onto itself.

12.4.10. THEOREM. *$\mathcal{T}(H^\infty)$ is reflexive; so the unilateral shift is a reflexive operator.*

PROOF. Suppose that $A \in \operatorname{Alg\,Lat} T_z$. Then $AM_w \subset M_w$ for $w \in \mathbb{D}$. Therefore $A^*\mathbb{C}k_w \subset \mathbb{C}k_w$, say $A^*k_w = a_w k_w$. Then

$$A^*T_z^*k_w = A^*\bar{w}k_w = a_w\bar{w}k_w = T_z^*a_wk_w = T_z^*A^*k_w.$$

Thus $A^*T_z^* - T_z^*A^*$ vanishes on $\overline{\operatorname{span}}\{k_w : w \in \mathbb{D}\} = H^2$ by Proposition 11.2.9. Hence A^* commutes with T_z^*, so A commutes with T_z. Therefore $A \in \mathcal{T}(H^\infty)$. ■

Here is a useful formula for the distance to H^∞ in L^∞.

12.4.11. NEHARI THEOREM. *Let $f \in L^\infty$. Then* $\operatorname{dist}(f, H^\infty) = \|H_f\|$.

PROOF. By Proposition 11.2.7, the preannihilator of H^∞ in L^∞ is H_0^1. By Corollary 11.4.14, each $k \in H^1$ can be factored as $k = g_1g_2$ with $g_1, g_2 \in H^2$ and $\|g_1\|_2^2 = \|g_2\|_2^2 = \|k\|_1$. So if $h = zk \in H_0^1$, we get $h = g_1(zg_2)$ where $g_1 \in H^2$ and $zg_2 \in H_0^2$. Compute

$$\begin{aligned}
\operatorname{dist}(f, H^\infty) &= \sup_{\substack{h\in H_\perp^\infty \\ \|h\|_1=1}} \langle f, h\rangle = \sup_{\substack{h\in H_0^1 \\ \|h\|_1=1}} \frac{1}{2\pi}\int f(e^{it})h(e^{it})\,dt \\
&= \sup_{\substack{g_1\in H^1,\, g_2\in H_0^2 \\ \|g_1\|_2=\|g_2\|_2=1}} \frac{1}{2\pi}\int f(e^{it})g_1(e^{it})g_2(e^{it})\,dt \\
&= \sup_{\substack{g_1\in H^1,\, \overline{g_2}\in H^{2\perp} \\ \|g_1\|_2=\|\overline{g_2}\|_2=1}} \langle M_f g_1, \overline{g_2}\rangle = \|H_f\|.
\end{aligned}$$

■

We will make use of Nehari's theorem to prove a result of Adamjan, Arov and Krein that is useful in the proof of Marshall's theorem in the previous chapter.

12.4.12. LEMMA. *Suppose that $T \in \mathcal{B}(H)$ and Q is a rank 1 projection such that $1 = \|T\| > \|TQ^\perp\|$. Then there is a unique one-dimensional subspace $\mathbb{C}\xi$ such that $1 = \|\xi\| = \|T\xi\|$.*

PROOF. Let $\|TQ^\perp\| = r < 1$. Let $|T| = (T^*T)^{1/2}$ be the absolute value, and recall that $\||T|\xi\| = \|T\xi\|$ for $\xi \in H$. The positive operator $|T|$ attains its norm precisely on the spectral subspace $M = E_{|T|}(\{1\}) = \ker I - |T|$. Let $N = E_{|T|}(r, 1]$. If $\dim N \geq 2$, then N contains a unit vector $\eta = Q^\perp \eta$. Thus $\|TQ^\perp\| \geq \|T\eta\| > r$, a contradiction. Since $\||T|\| = 1$, $E_{|T|}[1-\varepsilon, 1] \neq 0$ for all $\varepsilon > 0$. These spaces must all be one dimensional for $0 < \varepsilon < 1 - r$, say $\mathbb{C}\xi$ for $\|\xi\| = 1$. Therefore $\|T\xi\| = \||T|\xi\| \geq 1 - \varepsilon$ for all $\varepsilon > 0$; so $\|T\xi\| = 1$. Thus $\dim M = 1$, and M is the only subspace on which T attains its norm. ■

12.4.13. ADAMJAN–AROV–KREIN THEOREM. *Suppose that $f \in L^\infty$ and $1 = \operatorname{dist}(f, H^\infty) > \operatorname{dist}(zf, H^\infty)$. Then there is a unique unimodular function $u \in f + H^\infty$.*

PROOF. By Nehari's theorem, $1 = \|H_f\| > \|H_{zf}\|$. Since H^∞ is weak-$*$ closed, there is some $h \in H^\infty$ so that $\|f - h\|_\infty = 1$. However,

$$\|H_{zf}\| = \|P^\perp M_f M_z P\| = \|P^\perp M_f P_{zH^2} M_z\| = \|H_f P_0^\perp\|$$

where $P_0 = \zeta_0 \zeta_0^*$ is the projection onto $\mathbb{C}\zeta_0$. By Lemma 12.4.12, there is a unique unit vector $k \in H^2$ such that

$$\begin{aligned} 1 = \|H_f k\|_2 = \|H_{f-h} k\|_2 = \|P^\perp (f-h)k\|_2 \\ \leq \|(f-h)k\|_2 \leq \|f-h\|_\infty \|k\|_2 = 1. \end{aligned}$$

Thus both inequalities must be equalities. This means that $|f - h| = 1$ a.e.(m) on $\{e^{i\theta} : k(e^{i\theta}) \neq 0\}$, which is a set of full measure by Corollary 11.3.6. That is, $u = f - h$ is a unimodular function in $f + H^\infty$.

Conversely, if $v \in f + H^\infty$ is unimodular, then $h_1 = f - v$ is another closest point, $\|f - h_1\|_\infty = 1$. But then $\|f - \frac{1}{2}(h + h_1)\|_\infty = 1$. By the previous paragraph, this difference is unimodular:

$$1 = \left|\frac{1}{2}\big((f-h) + (f-h_1)\big)\right| \text{ a.e.}(m) \leq \frac{1}{2}\big(|f-h| + |f-h_1|\big) \leq 1.$$

Thus $f - h_1 = f - h$ a.e.(m), so that $h_1 = h$ is the unique closest point. Therefore u is also unique. ■

12.4.14. COROLLARY. *Suppose that $f \in L^\infty$ and $\|f\| < 1$. Then there is a unimodular function $u \in f + H^\infty$.*

PROOF. We can choose $a \in \mathbb{C}$ so that $\operatorname{dist}\big((f + a)\bar{z}, H^\infty\big) = 1$. Then $g = (f + a)\bar{z}$ satisfies

$$1 = \operatorname{dist}(g, H^\infty) > \|f\| \geq \operatorname{dist}(f, H^\infty) = \operatorname{dist}(zg, H^\infty).$$

By Theorem 12.4.13, there is a unimodular function $u \in g + H^\infty$. Therefore $zu \in zg + H^\infty = f + H^\infty$. ■

12.5. Isometries are Reflexive

It is of interest to identify classes of operators which are reflexive. In this section, we show that normal operators and isometries are reflexive.

12.5.1. DEFINITION. A weak-$*$ closed operator algebra $\mathcal{A} \subset \mathcal{B}(H)$ has property $\mathbb{A}_1$ (or $\mathbb{A}_1(r)$) if for every weak-$*$ continuous functional $\varphi \in \mathcal{A}_*$, there are vectors ξ, η such that $\varphi(A) = \langle A\xi, \eta \rangle$ for $A \in \mathcal{A}$ (and if $\varepsilon > 0$, one can choose them so that $\|\xi\| \, \|\eta\| < (r + \varepsilon)\|\varphi\|$).

12.5.2. THEOREM. *Let $\mathcal{M}$ be an abelian von Neumann algebra which is countably generated. Then $\mathcal{M}$ has $\mathbb{A}_1(1)$.*

PROOF. If $\mathcal{M}$ is generated as a von Neumann algebra by $T_1, T_2, \dots$, let $\mathfrak{A} = C^*(T_1, T_2, \dots)$ which is separable and abelian. The identity representation decomposes as a direct sum of cyclic representations of $\mathfrak{A}$ on pairwise orthogonal separable subspaces $\{(H_\lambda, \xi_\lambda)\}_\Lambda$. Suppose that $\varphi \in \mathcal{M}_*$ with $\|\varphi\| = 1$. Since $\mathcal{M}_*$ is a quotient of $\mathcal{B}(H)_*$, there is a trace class operator K such that $\varphi(A) = \operatorname{Tr}(AK)$ for $A \in \mathcal{M}$. There are at most countably many λ such that $P_{H_\lambda} K \neq 0$ or $KP_{H_\lambda} \neq 0$, say $\lambda \in F$. Let $N = \bigvee_{\lambda \in F} H_\lambda$, and replace K by $P_N K P_N$. Now $\mathcal{M}|_N$ is an abelian von Neumann algebra on a separable Hilbert space. Thus, if we can solve the problem in the case of separable Hilbert space, we are done.

So suppose that H is separable. By Corollary 10.4.9, there is an invariant subspace N_0 so that the restriction to N_0 is faithful, and $\mathcal{M}|_{N_0}$ is a masa. By Theorem 10.4.8, $\mathcal{M}|_{N_0}$ is unitarily equivalent to $L^\infty(\mu)$ acting on $L^2(\mu)$. Now $\varphi \in \mathcal{M}_* \simeq L^1(\mu)$, so there is a function $k \in L^1(\mu)$ with $\|k\|_1 = 1$ so that $\varphi(M_h) = \int hk \, d\mu$. Factor $k = f\bar{g}$ where $|f| = |g| = |k|^{1/2}$, so that $f, g \in L^2(\mu)$ and $\|f\|_2 = \|g\|_2 = 1$ and

$$\varphi(M_h) = \int hk \, d\mu = \int hf\bar{g} \, d\mu = \langle M_h f, g \rangle.$$

Using the unitary equivalence, we obtain unit vectors $\eta, \zeta \in N$ so that $\varphi(A) = \langle A\eta, \zeta \rangle$ and $\|\eta\| \, \|\zeta\| = \|\varphi\| = 1$. ∎

The following result is due to Sarason.

12.5.3. THEOREM. *Let N be a normal operator on H. Then N is reflexive, i.e.* $\operatorname{Alg}\operatorname{Lat} N = W(N)$.

PROOF. Let $\mathcal{M} = W^*(N)$. This is an abelian von Neumann algebra. By Exercise **12.5**, $\mathcal{M}$ is a reflexive operator algebra. By Theorem 12.5.2, $\mathcal{M}$ has $\mathbb{A}_1(1)$. Now $\operatorname{Alg}\operatorname{Lat} N \subset \operatorname{Alg}\operatorname{Lat} \mathcal{M} = \mathcal{M}$. Suppose that $T \in W^*(N) \setminus W(N)$. Since $W(N)$ is weakly closed, there is a WOT-continuous linear functional φ so that

$\varphi|_{W(N)} = 0$ and $\varphi(T) \neq 0$. By property $\mathbb{A}_1$, there are vectors $\xi, \eta \in H$ so that $\varphi(A) = \langle A\xi, \eta\rangle$ for all $A \in \mathcal{M}$.

Let $M = \overline{W(N)\xi}$. Clearly $M \in \operatorname{Lat} N$. Since φ annihilates $W(N)$, we have $0 = \varphi(A) = \langle A\xi, \eta\rangle$ for all $A \in W(N)$. This shows that $\eta \perp M$. However $0 \neq \varphi(T) = \langle T\xi, \eta\rangle$, and therefore $T\xi \notin M$. So $T \notin \operatorname{Alg Lat} N$. Therefore $\operatorname{Alg Lat} N = W(N)$. ■

12.5.4. DEFINITION. An operator T is *reductive* if every invariant subspace is reducing. Let $W(T)$ be the WOT-closed unital algebra generated by T.

12.5.5. REMARK. If there is a net of polynomials $\{p_\lambda\}_\Lambda$ such that $p_\lambda(T)$ converges WOT to T^*, then $T^* \in W(T) \subset \operatorname{Alg Lat} T$. And if $T^* \in \operatorname{Alg Lat} T$, then T is reductive.

The problem of whether every reductive operator is normal is equivalent to the invariant subspace problem. Of course, not all normal operators are reductive. For example, the bilateral shift has invariant subspaces which are not reducing.

12.5.6. COROLLARY. *Let N be a normal operator on H. If N is reductive, then $W(N) = W^*(N)$ is a von Neumann algebra.*

PROOF. If N is reductive, $\operatorname{Lat} N$ is complemented, and thus $N^* \in \operatorname{Alg Lat} N$. By Theorem 12.5.3, $W(N) = \overline{C^*(N)}^{\text{WOT}} = W^*(N)$ is a von Neumann algebra. ■

Every unitary is the direct sum of unitaries U_λ with a cyclic vector. By the spectral theorem, these operators are unitarily equivalent to multiplication operators M_λ by the coordinate function z on $L^2(\mu_\lambda)$ for Borel probability measures on $\mathbb{T}$. Decompose the measure $\mu_\lambda = \mu_{\lambda,a} + \mu_{\lambda,s}$ where $\mu_{\lambda,a} \ll m$ and $\mu_{\lambda,s} \perp m$ are the absolutely continuous and singular parts. So each $U_\lambda \simeq U_{\lambda,a} \oplus U_{\lambda,s}$. We then obtain the decomposition $U \simeq U_a \oplus U_s$ where $U_a = \sum_\Lambda^\oplus U_{\lambda,a}$ and $U_s = \sum_\Lambda^\oplus U_{\lambda,s}$ are the absolutely continuous and singular parts of U. Each $\mu_{\lambda,a} \approx \chi_{A_\lambda} m$ for some Borel set A_λ. A standard measure-theoretic argument shows that

$$r = \sup\Big\{m\Big(\bigcup_{\lambda\in F} A_\lambda\Big) : F \subset \Lambda \text{ countable}\Big\}$$

is attained by some countable set F_0. Setting $A = \bigcup_{\lambda\in F_0} A_\lambda$ yields a Borel set such that $m(A_\lambda \setminus A) = 0$ for all $\lambda \in \Lambda$, and the spectral measure E_a is equivalent to $\mu_a = \chi_A m$. If H is separable, then there is a singular measure $\mu_s \approx E_s$. But if H is not separable, there is in general no such measure. For example, consider U on $l^2(\mathbb{T})$ with basis $\{\delta_z : z \in \mathbb{T}\}$ and $U\delta_z = z\delta_z$.

12.5.7. LEMMA. *Let $U \simeq U_a \oplus U_s$ be a unitary operator split into its absolutely continuous and singular parts on $H = H_a \oplus H_s$. If $U_a X = X U_s$, then $X = 0$. In particular, $P_{H_s} \in W^*(U)$ and $W^*(U) = W^*(U_a) \oplus W^*(U_s)$.*

PROOF. Suppose that $U_a X = XU_s$. By the Fuglede–Putnam theorem (Exercise **9.16**), $X^* U_a = U_s X^*$. Hence $U_s X^* X = X^* X U_s$. This implies that $\ker X^* X = \ker X$ reduces U_s. Similarly $U_a X X^* = X X^* U_a$, so that $\ker XX^* = (\operatorname{Ran} X)^\perp$ reduces U_a. Hence, X is an injective operator from $(\ker X)^\perp$ onto $\operatorname{Ran} X$, so that

$$U_a|_{\overline{\operatorname{Ran} X}} X = XU_s|_{(\ker X)^\perp}.$$

By Exercise **9.16** $U_a|_{\overline{\operatorname{Ran} X}}$ and $U_s|_{(\ker X)^\perp}$ are unitarily equivalent. This is impossible, since unitary equivalence preserves the measure class of the spectral measure. Thus $X = 0$. Similarly, there is no operator Y such that $YU_a = U_s Y$. Therefore P_{H_s} commutes with $W^*(U)'$, and so it lies in $W^*(U)'' = W^*(U)$ by the double commutant theorem. It clearly also belongs to $W^*(U)'$, and hence P_{H_s} lies in the centre of $W^*(U)$. Thus $W^*(U)$ decomposes as $W^*(U_a) \oplus W^*(U_s)$. ■

Much of the following theorem is due to Wermer in 1952. The rest is due to Deddens.

12.5.8. THEOREM. *Let $U \simeq U_a \oplus U_s$ be a unitary operator split into its absolutely continuous and singular parts on $H = H_a \oplus H_s$. Then*

(1) $\operatorname{Lat} U = \operatorname{Lat} U_a \oplus \operatorname{Lat} U_s$, *and every $M \in \operatorname{Lat} U_s$ is reducing.*

(2) *U is reductive if and only if $m \not\ll \mu_a$.*

(3)
$$W(U) \simeq \begin{cases} L^\infty(\mu_a) \oplus W^*(U_s) & \text{if } m \not\ll \mu_a \\ H^\infty \oplus W^*(U_s) & \text{if } m \ll \mu_a. \end{cases}$$

If H is separable, then $E_{U_s} \approx \mu_s$ for a σ-finite singular regular Borel measure on $\mathbb{T}$, and $W(U_s) \simeq L^\infty(\mu_s)$.

PROOF. Let M be a reducing subspace of U. Then P_M commutes with U and U^*. By Lemma 11.5.5, it commutes with P_{H_s}. Hence, $M = M_a \oplus M_s$ where $M_a = M \cap H_a$ and $M_s = M \cap H_s$. Clearly $M_a \in \operatorname{Lat} U_a$ and $M_s \in \operatorname{Lat} U_s$.

Now let M be an arbitrary invariant subspace for U. Then $U|_M$ is an isometry, so we can apply the Wold decomposition. The subspace $M_0 = M \ominus UM$ is a wandering space, meaning that the subspaces $M_n = U^n M_0$ are pairwise orthogonal for $n \geq 0$. Moreover, $M = \sum_{n\geq 0}^{\oplus} M_n \oplus L$ where L reduces U. It is easy to see that $M_n = U^n M_0$ for $n \leq -1$ are also pairwise orthogonal. The space $N = \sum_{n\in\mathbb{Z}}^{\oplus} M_n$ reduces U and $U|_N \simeq W^{(\alpha)}$ where W is the bilateral shift and $\alpha = \dim M_0$. The bilateral shift is absolutely continuous with spectral measure equivalent to m. Thus $N \subset H_a$. The reducing subspace L decomposes as $L = L_a \oplus L_s$ where $L_a = L \cap H_a$ and $L_s = L \cap H_s$. Thus if $M_a = \sum_{n\geq 0}^{\oplus} M_n \oplus L_a$, we have $M = M_a \oplus L_s$, where $M_a \in \operatorname{Lat} U_a$ and $L_s \in \operatorname{Lat} U_s$. In particular, every invariant subspace for U_s is reducing.

This also shows that if U is not reductive, then there is a reducing subspace N such that $U|_N \simeq W$ is a bilateral shift. Thus $m \ll E_U$. Hence if $m \not\ll \mu_a$, then

$m \not\ll E_U$ and U is reductive. Conversely, if $m \ll \mu_a$, by Proposition 9.9.10, there is a vector $\xi \in H$ so that

$$\langle f(U)\xi, \xi\rangle = \int f\, dm \quad \text{for} \quad f \in C(\mathbb{T}).$$

The cyclic reducing subspace $N = \overline{C^*(U)\xi}$ has $U|_N \simeq M_z$ on $L^2(m)$, which is unitarily equivalent to the bilateral shift. Therefore, there is an orthonormal basis $\{\zeta_n : n \in \mathbb{Z}\}$ for N so that $U\zeta_n = \zeta_{n+1}$ for $n \in \mathbb{Z}$. Thus, $M = \operatorname{span}\{\zeta_n : n \geq 0\}$ is an invariant subspace which doesn't reduce U. Hence U is not reductive.

By Lemma 12.5.7, $W^*(U) = W^*(U_a) \oplus W^*(U_s)$. When $m \not\ll \mu_a$, U is reductive. Hence by Corollary 12.5.6,

$$W(U) = W^*(U) = W^*(U_a) \oplus W^*(U_s) \simeq L^\infty(\mu_a) \oplus W^*(U_s).$$

In general, U_s is reductive so that $W(U_s) = W^*(U_s)$.

Now suppose that $m \ll \mu_a \approx E_{U_a}$. Then $W^*(U_a) \simeq L^\infty(m)$ and there is a subspace N so that $W = U_a|_N$ is a bilateral shift, say with respect to a basis $\{\zeta_n : n \in \mathbb{Z}\}$. Then $M = \operatorname{span}\{\zeta_n : n \geq 0\}$ is invariant. If $f \in L^\infty$ and $f(U_a) \in \operatorname{Alg}\operatorname{Lat} U_a$, then

$$\hat{f}(n) = \langle f(U_a)\zeta_0, \zeta_n\rangle = 0 \quad \text{for} \quad n \leq -1.$$

Therefore $f \in H^\infty$. Conversely, if $f \in H^\infty$, then $f_r(x) = f(rz)$ belongs to $A(\mathbb{D})$ and so $f_r(U_a) \in W(U_a)$. Moreover, $f_r(U_a) \xrightarrow{\text{WOT}} f(U_a)$ by the L^∞ functional calculus. We deduce that $W(U_a) = H^\infty(U_a) \simeq H^\infty$. Finally, by Lemma 12.5.7 and Theorem 12.5.3,

$$\begin{aligned} W(U_a \oplus U_s) &= \operatorname{Alg}\operatorname{Lat}(U_a \oplus U_s) = \operatorname{Alg}(\operatorname{Lat} U_a \oplus \operatorname{Lat}(U_s)) \\ &= \operatorname{Alg}(\operatorname{Lat} U_a) \oplus \operatorname{Alg}(\operatorname{Lat}(U_s)) = H^\infty(U_a) \oplus W^*(U_s). \end{aligned}$$

If H is separable, then $E_s \approx \mu_s$ for a σ-finite regular Borel measure on $\mathbb{T}$, and $W(U_s) \simeq L^\infty(\mu_s)$. ■

Next we want to extend this result to all isometries.

12.5.9. Deddens Theorem. *Let S be a proper isometry. Then S is reflexive. Write $S \simeq T_z^{(\alpha)} \oplus U_a \oplus U_s$ for $\alpha \geq 1$. Then*

$$W(S) = W(T_z^{(\alpha)} \oplus U_a) \oplus W^*(U_s) \simeq H^\infty \oplus W^*(U_s),$$

where $h \in H^\infty$ is mapped to $T_h^{(\alpha)} \oplus h(U_a)$ and $h(U_a)$ is evaluated using the L^∞ functional calculus.

Proof. Let $S_a = T_z^{(\alpha)} \oplus U_a$. There is a bilateral shift W (of multiplicity α) and an invariant subspace L so that $W|_L \simeq T_z^{(\alpha)}$. Let $U = W \oplus U_a \oplus U_s$. Any $M \in \operatorname{Lat} S$ can be considered as an element of $\operatorname{Lat} U$. By the Wermer–Deddens theorem, $M = M_a \oplus M_s$ for $M_a \subset L \oplus H_a$ and $M_s \subset H_s$, and M_s reduces U_s. Therefore $\operatorname{Lat}(S) = \operatorname{Lat}(S_a) \oplus \operatorname{Lat} U_s$ and

$$\operatorname{Alg}\operatorname{Lat} S = \operatorname{Alg}\operatorname{Lat} S_a \oplus \operatorname{Alg}\operatorname{Lat} U_s = \operatorname{Alg}\operatorname{Lat} S_a \oplus W^*(U_s).$$

This reduces the analysis to a study of S_a.

By Corollary 12.4.2 and Theorem 12.4.10,

$$W(T_z) = \operatorname{Alg}\operatorname{Lat} T_z = \mathcal{T}(H^\infty) = \{T_h : h \in H^\infty\}.$$

For any $h \in H^\infty$, $h_r \in A(\mathbb{D})$ and $h_r(S_a) = T_{h_r}^{(\alpha)} \oplus h_r(U_a)$. This converges WOT to $T_h^{(\alpha)} \oplus h(U_a)$, so $H^\infty(S_a) \subset W(S_a)$. On the other hand,

$$\operatorname{Alg}\operatorname{Lat} S_a \subset \operatorname{Alg}\operatorname{Lat} T_z^{(\alpha)} \oplus \operatorname{Alg}\operatorname{Lat} U_a \subset \mathcal{T}(H^\infty)^{(\alpha)} \oplus L^\infty(\mu_a).$$

There is a Borel set $A \subset \mathbb{T}$ so that $\mu_a \approx \chi_A m$. By Proposition 9.9.10, there is a cyclic subspace $L_a \subset H_a$ so that $U_a|_{L_a} \simeq M_z$ on $L^2(A, m)$. We identify L_a with $L^2(A, m)$. We also pick a copy L_+ of H^2 in $H^{2(\alpha)}$. Define

$$M_0 = \{(h, \chi_A h) \in H^2 \oplus L^2(A, m) : h \in H^2\}.$$

This is readily seen to be closed since

$$\|h\|_2^2 \le \|h\|_2^2 + \|\chi_A h\|_2^2 \le 2\|h\|_2^2.$$

So convergence in the first coordinate is equivalent to convergence in M_0. It is easy to see that M_0 is invariant for S_a. Hence, if $T = T_f^{(\alpha)} \oplus g(U_a) \in \operatorname{Alg}\operatorname{Lat} S_a$ for $f \in H^\infty$ and $g \in L^\infty(A, m)$, then $T(1, \chi_A) = (f, g) \in M_0$, and thus $(f, g) = (f, \chi_A f)$. Therefore $g = f\chi_A$. That is, $T = f(S_a)$. Hence

$$\operatorname{Alg}\operatorname{Lat} S_a \subset H^\infty(S_a) = W(S_a) \subset \operatorname{Alg}\operatorname{Lat} S_a.$$

Thus $\operatorname{Alg}\operatorname{Lat} S_a = W(S_a) \simeq H^\infty$. ■

12.6. Invertibility of Toeplitz Operators

Now that we understand the analytic Toeplitz operators, we return to the general problem. Our goal is the Widom–Devinatz theorem characterizing invertibility of Toeplitz operators.

12.6.1. Lemma. *Suppose that $f \in (L^\infty)^{-1}$. Then T_f is invertible if and only if T_u is invertible, where $u = \dfrac{f}{|f|}$. Also T_f is Fredholm if and only if T_u is Fredholm, and they have the same index.*

PROOF. Since $|f|$ has a bounded logarithm, there is an outer function $g \in H^\infty$ such that $|g| = |f|^{-1/2}$. Thus $g^{-1} \in H^\infty$ and so T_g is invertible. But $\bar{g}gf = u$, hence

$$T_u = T_{\bar{g}fg} = T_g^* T_f T_g.$$

Therefore T_f is invertible precisely when T_u is. Likewise T_f is Fredholm if and only if T_u is Fredholm, and

$$\operatorname{ind}(T_u) = \operatorname{ind}(T_g^*) + \operatorname{ind}(T_f) + \operatorname{ind}(T_g) = \operatorname{ind}(T_f)$$

because $\operatorname{ind}(T_g) = \operatorname{ind}(T_g^*) = 0$. ■

Next we consider the invertibility of T_u when u is unimodular.

12.6.2. PROPOSITION. *If $u \in L^\infty$ is unimodular, then T_u is left invertible if and only if* $\operatorname{dist}(u, H^\infty) < 1$. *Moreover, T_u is invertible if and only if there is an invertible function $h \in H^\infty$ with $\|u - h\|_\infty < 1$.*

PROOF. Suppose that $h \in H^\infty$ with $\|u - h\|_\infty < 1$. Then $\|1 - \bar{u}h\|_\infty < 1$, so that $1 > \|I - T_{\bar{u}h}\| = \|I - T_u^* T_h\|$. Therefore $T_u^* T_h$ is invertible, and so T_u^* has a right inverse, whence T_u is left invertible. Also T_h is left invertible, so it is injective and has closed range. If h is invertible in H^∞, then T_h is invertible. Therefore T_u is also invertible.

Conversely, suppose that T_u is left invertible. Then T_u is bounded below, say by $\varepsilon > 0$. Since u is unimodular, M_u is unitary, and thus

$$M_u|_{H^2} = \begin{bmatrix} T_u \\ H_u \end{bmatrix}$$

is an isometry. Since T_u is bounded below by ε, $\|H_u\| \le \sqrt{1 - \varepsilon^2} < 1$. By Nehari's theorem, $\operatorname{dist}(u, H^\infty) < 1$. Let $\|u - h\|_\infty < 1$ for some $h \in H^\infty$. Then as in the first paragraph, $T_u^* T_h$ is invertible. If T_u is invertible, then T_h is invertible too. By Theorem 12.4.3, h is invertible in H^∞. ■

We now have the tools that we need to characterize invertibility.

12.6.3. WIDOM–DEVINATZ THEOREM. *Let $f \in L^\infty$ be invertible, and let $u = \frac{f}{|f|}$. Then the following are equivalent:*

(1) *T_f is invertible.*

(2) $\operatorname{dist}(u, H^\infty) < 1$ *and* $\operatorname{dist}(\bar{u}, H^\infty) < 1$.

(3) *There is an invertible function $h \in H^\infty$ with $\|u - h\|_\infty < 1$.*

(4) *$f = he^{a+ib}$ where $h \in (H^\infty)^{-1}$, $a, b \in L^\infty_{\mathbb{R}}$ and $\|b\|_\infty < \frac{\pi}{2}$.*

PROOF. Lemma 12.6.1 shows that T_f is invertible if and only if T_u is invertible. This is the same as being both left and right invertible, or, equivalently, both T_u

and $T_u^* = T_{\bar u}$ being left invertible. By Proposition 12.6.2, this is equivalent to $\operatorname{dist}(u, H^\infty) < 1$ and $\operatorname{dist}(\bar u, H^\infty) < 1$ and also equivalent to the existence of an invertible function $h \in H^\infty$ with $\|u - h\|_\infty < 1$.

Now suppose that $h \in (H^\infty)^{-1}$ and $\|u - h\| = 1 - \delta < 1$. Then

$$1 - \delta = \|u - h\| = \|1 - \bar u h\| = \|1 - u\bar h\|.$$

This implies that there is a $\theta < \frac{\pi}{2}$ so that

$$\operatorname{Ran}(u\bar h) \subset b_{1-\delta}(1) \subset \{z : |\arg z| < \theta\}.$$

Hence $u\bar h = e^{a+ib}$ for $a, b \in L^\infty_{\mathbb{R}}$ and $\|b\| \le \theta < \frac{\pi}{2}$. For $z \in \mathbb{D}$,

$$|h(z)| = |h^{-1}(z)|^{-1} \ge \|h^{-1}\|^{-1} \ge \delta.$$

Thus there is an analytic branch $k(z) = \log h(z)$ on $\mathbb{D}$ with $k \in H^\infty$. Therefore

$$u = (u\bar h)h \frac{1}{h\bar h} = e^{a+ib} h e^{-k-\bar k} = h e^{c+ib}$$

where $c = a - 2\operatorname{Re} k$. So (4) holds.

Finally, if (4) holds, then $T_f = T_g T_h$ where $g = e^{a+ib}$ and $\|b\| < \frac{\pi}{2}$. Now $\operatorname{Re} g \ge e^{-\|a\|} \cos \|b\| > 0$. By Theorem 12.3.13, T_g is invertible. Since T_h is also invertible, T_f is invertible. ■

The main problem with this theorem is that these conditions are generally too difficult to check.

12.7. The algebra $H^\infty + C$

Recall from Lemma 11.7.1 that $H^\infty + C$ is a closed subalgebra of L^∞. This will be useful for obtaining a criterion for T_f to be Fredholm for $f \in H^\infty + C$, and computing the index.

12.7.1. Hartman Theorem. *If $f \in L^\infty$, $\operatorname{dist}(f, H^\infty + C) = \|H_f\|_e$.*

Proof. We apply Proposition 12.1.7 using the projections $P_n^\perp = M_{z^n} P M_{z^n}^*$ onto $z^n H^2$.

$$\begin{aligned}\|H_f\|_e &= \lim_{n\to\infty} \|H_f P_n^\perp\| = \lim_{n\to\infty} \|P^\perp M_f M_{z^n} P M_{z^n}^*\| \\ &= \lim_{n\to\infty} \|P^\perp M_{fz^n} P\| = \lim_{n\to\infty} \|H_{fz^n}\| = \lim_{n\to\infty} \operatorname{dist}(fz^n, H^\infty) \\ &= \lim_{n\to\infty} \operatorname{dist}(f, \bar z^n H^\infty) = \operatorname{dist}(f, H^\infty + C).\end{aligned}$$

In the middle, we applied Nehari's theorem. ■

The first corollary follows from the fact that $H^\infty + C$ is closed.

12.7.2. COROLLARY. *For $f \in L^\infty$, H_f is compact if and only if $f \in H^\infty + C$.*

12.7.3. COROLLARY. *If $f \in L^\infty$, then $T_g T_f - T_{gf}$ is compact for all $g \in L^\infty$ if and only if $f \in H^\infty + C$.*

PROOF. Since $T_g T_f - T_{gf} = H_{\bar g}^* H_f$, this is clearly compact if $f \in H^\infty + C$ independent of g. Conversely, taking $g = \bar f$, we see that if $T_{\bar f} T_f - T_{\bar f f} = H_{\bar f}^* H_f$ is compact, then so is H_f, and hence $f \in H^\infty + C$. ∎

Now $\mathcal{T}(H^\infty + C) \supset \mathcal{T}(C) \supset \mathcal{K}(H^2)$. By the preceeding result, we see that T_f and T_g commute modulo compacts if $f, g \in H^\infty + C$. We obtain

12.7.4. COROLLARY. *$\mathcal{T}(H^\infty + C) = \{T_f + K : f \in H^\infty + C,\ K \in \mathcal{K}(H^2)\}$. Thus $\mathcal{T}(H^\infty + C)/\mathcal{K}(H^2) \simeq H^\infty + C$.*

When u is unimodular, we can apply Proposition 12.6.2 to obtain a condition for being left Fredholm.

12.7.5. PROPOSITION. *If $u \in L^\infty$ is unimodular, then T_u is left Fredholm if and only if* $\operatorname{dist}(u, H^\infty + C) < 1$.

PROOF. Since $H^\infty + C = \overline{\bigcup_{n \geq 0} z^{-n} H^\infty}$, we have

$$\begin{aligned} \operatorname{dist}(u, H^\infty + C) < 1 &\iff \exists n \geq 0 \ s.t.\ \operatorname{dist}(uz^n, H^\infty) = \operatorname{dist}(u, z^{-n} H^\infty) < 1 \\ &\iff \exists n \geq 0 \ s.t.\ T_{uz^n} = T_u T_z^n \text{ is left invertible.} \end{aligned}$$

Let A be a left inverse for $T_u T_z^n$. Then in the Calkin algebra, by Proposition 12.3.6,

$$\pi(I) = \pi(A)\pi(T_u)\pi(T_z^n) = \pi(A)\pi(T_z^n)\pi(T_u).$$

So $\pi(T_u)$ is left invertible, i.e. T_u is left Fredholm.

Conversely, if T_u is left Fredholm, then T_u has closed range and finite dimensional kernel. Thus, there is a large integer n so that $\ker T_u \cap z^n H^2 = \{0\}$. Therefore, T_{uz^n} is injective, and has closed range, and so is left invertible. Hence $\operatorname{dist}(uz^n, H^\infty) < 1$. This completes the equivalence. ∎

We can apply this to obtain information about the spectrum of Toeplitz operators with symbol in subalgebras of L^∞. Our particular interest is in $H^\infty + C$.

12.7.6. THEOREM. *If $H^\infty \subset A \subset L^\infty$ is a closed algebra, then for $f \in A$,*

$$\sigma(T_f) = \sigma_{\mathcal{T}(A)}(T_f).$$

PROOF. Clearly $\sigma(T_f) \subset \sigma_{\mathcal{T}(A)}(T_f)$. So suppose that T_f is invertible in $\mathcal{B}(H^2)$. By Lemma 12.3.12, $1/f \in L^\infty$, and, in particular, f is bounded below. By

Remark 12.2.9, there is an outer function $h \in H^2$ with $|f| = |h|$ a.e.(m). Thus, $h \in H^\infty$ is invertible in L^∞ and is outer, so is invertible in H^∞ by Proposition 11.6.1. Write $f = uh$ for a unimodular function u. Since $T_f = T_u T_h$ and both T_f and T_h are invertible, we see that T_u is invertible. Taking adjoints, $T_{\bar{u}}$ is also invertible. By Proposition 12.6.2, there is an invertible function $g \in H^\infty$ so that

$$1 > \|\bar{u} - g\|_\infty = \|1 - ug\|_\infty.$$

Hence $\|I - T_{ug}\| < 1$ and thus $T_{ug} = T_u T_g$ is invertible in $\mathcal{T}(A)$. Now T_g is invertible in $\mathcal{T}(H^\infty)$ and thus in $\mathcal{T}(A)$. Therefore, T_u is invertible in $\mathcal{T}(A)$. Finally, $T_f = T_u T_h$ is invertible in $\mathcal{T}(A)$. ■

12.7.7. COROLLARY. *If $f \in H^\infty + C$, then $\sigma_e(T_f) = \sigma_{H^\infty + C}(f)$, i.e. T_f is Fredholm if and only if f is invertible in $H^\infty + C$.*

PROOF. If f is invertible in $H^\infty + C$, then by Corollary 12.7.4,

$$\pi(T_f)\pi(T_{1/f}) = \pi(I) = \pi(T_{1/f})\pi(T_f).$$

So T_f is Fredholm.

Conversely, if T_f is Fredholm and $\operatorname{ind} T_f = n$, then

$$\operatorname{ind} T_{fz^n} = \operatorname{ind} T_f T_{z^n} = \operatorname{ind} T_f + n \operatorname{ind} T_z = 0.$$

By Lemma 12.3.10, T_{fz^n} is invertible. By Theorem, 12.7.6, T_{fz^n} is invertible in $\mathcal{T}(H^\infty + C)$. By Corollary 12.7.4, $H^\infty + C$ is a quotient of $\mathcal{T}(H^\infty + C)$. Hence fz^n is invertible in $H^\infty + C$. Because z^n is invertible in $H^\infty + C$, so is f. ■

Since $H^\infty + C$ is a commutative Banach algebra, we can test invertibility on its maximal ideal space.

12.7.8. THEOREM. $\mathcal{M}_{H^\infty + C} = \mathcal{M}_{H^\infty} \setminus \mathbb{D}$.

PROOF. There is a natural map ρ from $\mathcal{M}_{H^\infty + C}$ to $\mathcal{M}_{H^\infty}$ given by restriction. Since z is invertible in $H^\infty + C$ and $\|z\| = \|z^{-1}\| = 1$, $\psi(z) \in \mathbb{T}$ for any $\psi \in \mathcal{M}_{H^\infty + C}$. So ρ maps into $\mathcal{M}_{H^\infty} \setminus \mathbb{D}$.

To see that this map is a bijection, we show that each $\varphi \in \mathcal{M}_{H^\infty} \setminus \mathbb{D}$ has a unique Hahn–Banach extension to $H^\infty + C$, and it is multiplicative. Take any such extension ψ. Then $\psi|_C$ is norm 1 and $\psi(1) = 1$, so it is a state on C by Lemma 9.6.5. Let $\psi(z) = \lambda \in \mathbb{T}$. Then $\psi(z^n) = \lambda^n$ for $n \geq 0$. By positivity, $\psi(z^n) = \lambda^n$ for $n < 0$. Thus $\psi(f) = f(\lambda)$ for all $f \in C$. This shows that the extension is unique.

Let $h \in H^\infty$, and write $h = p + z^n k$ for p a polynomial of degree $n - 1$ and $k \in H^\infty$. Then

$$\psi(z^{-n}h) = \psi(z^{-n}p) + \psi(k) = \lambda^{-n}\big(\psi(p) + \psi(z^n k)\big) = \psi(z^{-n})\psi(h).$$

By linearity, $\psi(fh) = \psi(f)\psi(h)$ for $f \in C$ and $h \in H^\infty$. Since ψ is already multiplicative on both C and H^∞, another application of linearity shows that ψ is multiplicative on $H^\infty + C$. Thus ρ is a continuous bijection onto $\mathcal{M}_{H^\infty} \setminus \mathbb{D}$. Both spaces are compact and Hausdorff, so ρ is a homeomorphism. ∎

The following lemma is easy, and is only long because of notation.

12.7.9. LEMMA. *The harmonic extension of functions in $H^\infty + C$ is approximately multiplicative in the sense: if $f, g \in H^\infty + C$ and $\varepsilon > 0$, there is an $R < 1$ so that*

$$\big|\widetilde{fg}(re^{i\theta}) - \tilde{f}(re^{i\theta})\tilde{g}(re^{i\theta})\big| < \varepsilon \quad \textit{for} \quad R \le r < 1 \textit{ and } \theta \in \mathbb{R}.$$

PROOF. It suffices to establish this on a dense subset of $H^\infty + C$, so take

$$f(z) = \sum_{j=0}^{m} a_j \bar{z}^j + h(z) \quad \text{and} \quad g(z) = \sum_{j=0}^{n} b_j \bar{z}^j + k(z)$$

for $h, k \in H^\infty$. Then

$$\tilde{f}(re^{i\theta}) = \sum_{j=0}^{m} a_j r^j e^{-ij\theta} + h(re^{i\theta}) \quad \text{and} \quad \tilde{g}(re^{i\theta}) = \sum_{j=0}^{n} b_j r^j e^{-ij\theta} + k(re^{i\theta}).$$

Now $z^m f = F$ and $z^n g = G$ belong to H^∞ and

$$\big|\tilde{f}(re^{i\theta}) - z^{-m}F(re^{i\theta})\big| = \Big|\sum_{j=0}^{m} a_j(r^j - r^{-j})e^{-ij\theta}\Big| \le \sum_{j=0}^{m} |a_j|(r^{-j} - r^j).$$

Similarly we have

$$\big|\tilde{g}(re^{i\theta}) - z^{-m}G(re^{i\theta})\big| \le \sum_{j=0}^{n} |b_j|(r^{-j} - r^j).$$

Write $fg = \sum_{j=0}^{m+n} c_j r^j e^{-ij\theta} + l(re^{i\theta})$ for $l \in H^\infty$ and $z^{m+n} fg = FG$. Then

$$\big|\widetilde{fg}(re^{i\theta}) - z^{-m-n}FG(re^{i\theta})\big| \le \sum_{j=0}^{m+n} |c_j|(r^{-j} - r^j)$$

All of these quantities can be made arbitrarily small by choosing r sufficiently close to 1. Therefore $\big|\tilde{f}(re^{i\theta})\tilde{g}(re^{i\theta}) - \widetilde{fg}(re^{i\theta})\big|$ is bounded by

$$\big|\tilde{f}(re^{i\theta}) - z^{-m}F(re^{i\theta})\big|\big|\tilde{g}(re^{i\theta})\big| + |z^{-m}F(re^{i\theta})|\big|\tilde{g}(re^{i\theta}) - z^{-m}G(re^{i\theta})\big|,$$

which tends to 0 as $r \to 1$ independent of θ. ∎

Finally, we can characterize invertibility of $f \in H^\infty + C$, and compute the Fredholm index of T_f.

12.7.10. DOUGLAS THEOREM. *For $f \in H^\infty + C$, the following are equivalent:*

(1) *f is invertible in $H^\infty + C$.*

(2) *There are* $R < 1$ *and* $\varepsilon > 0$ *so* $|\widetilde{f}(re^{i\theta})| \geq \varepsilon$ *for* $R \leq r < 1$ *and* $\theta \in \mathbb{R}$.

(3) T_f *is Fredholm.*

Moreover $\operatorname{ind} T_f = -\lim_{r\to 1^-} \operatorname{ind} \widetilde{f}(r\mathbb{T})$.

PROOF. If $g \in H^\infty + C$ such that $fg = 1$, then Lemma 12.7.9 shows that

$$|\widetilde{f}(re^{i\theta})\widetilde{g}(re^{i\theta}) - 1| < \frac{1}{2} \quad \text{for} \quad R \leq r < 1.$$

Since $|\widetilde{g}(re^{i\theta})| \leq \|g\|_\infty$, we get $|\widetilde{f}(re^{i\theta})| \geq \frac{1}{2\|g\|_\infty}$ for $R \leq r < 1$.

Conversely, suppose first that $f \in H^\infty$ and $|f(re^{i\theta})| \geq \varepsilon$ for $R \leq r < 1$. Then the only zeros of f lie in $R\overline{\mathbb{D}}$, so there are only finitely many counting multiplicity, say $\alpha_1, \dots, \alpha_n$. Let $b = \prod_{j=1}^n b_{\alpha_j}$ be the finite Blaschke product with the same zeros as f including multiplicity. Then $h = f/b$ is bounded and analytic on $\mathbb{D}$ and has no zeros. Since b is continuous on $\overline{\mathbb{D}}$ and $|b(e^{i\theta})| = 1$, we see that $|h(re^{i\theta})| \geq \varepsilon$ for $R' \leq r < 1$. Therefore $1/h$ is a bounded analytic function, and h is invertible in H^∞. Now $T_f = T_b T_h$ is Fredholm because T_h is invertible, and T_b is Fredholm by Theorem 12.1.9. Thus

$$\operatorname{ind} T_f = \operatorname{ind} T_b + \operatorname{ind} T_h = -\operatorname{ind} b|_{\mathbb{T}}.$$

For $r \geq R$, the argument principle shows that $\operatorname{ind} f_r|_{\mathbb{T}}$ and $\operatorname{ind} b_r|_{\mathbb{T}}$ count the number of zeros inside $r\mathbb{T}$, namely, n. So

$$\operatorname{ind} T_f = -\operatorname{ind} b|_{\mathbb{T}} = -\operatorname{ind} b_r|_{\mathbb{T}} = -\operatorname{ind} f_r|_{\mathbb{T}} \quad \text{for} \quad R \leq r < 1.$$

Next suppose that $f = \bar{z}^n h$ for $h \in H^\infty$ and $|f(re^{i\theta})| \geq \varepsilon$ for $R \leq r < 1$. Then by Lemma 12.7.9, $|h(re^{i\theta})| \geq \varepsilon/2$ for $R' \leq r < 1$. By the previous paragraph, T_h is Fredholm, so $T_f = T_{\bar{z}^n} T_h$ is Fredholm and

$$\begin{aligned} \operatorname{ind} T_f &= \operatorname{ind} T_{\bar{z}^n} + \operatorname{ind} T_h = -\operatorname{ind} \bar{z}^n|_{\mathbb{T}} - \operatorname{ind} h_r|_{\mathbb{T}} \\ &= -\operatorname{ind} r^n \bar{z}^n h_r \quad \text{for} \quad R' \leq r < 1. \end{aligned}$$

By Lemma 12.7.9, for $R'' \leq r < 1$, this index coincides with $-\operatorname{ind} \widetilde{f}|_{r\mathbb{T}}$.

For general $f \in H^\infty + C$, we approximate f by $\bar{z}^n h$ with $\|f - \bar{z}^n h\|_\infty < \frac{\varepsilon}{3}$. Then one sees that $\widetilde{\bar{z}^n h}$ is bounded below by $\frac{2\varepsilon}{3}$ for $R \leq r < 1$, and hence it is Fredholm. Thus $|h|$ is bounded below by $\frac{\varepsilon}{2}$ on some smaller annulus, so that T_h^{-1} is bounded by $\frac{2}{\varepsilon}$. Therefore

$$\|\pi(T_{\bar{z}^n h})^{-1}\| = \|\pi(T_z^n)\pi(T_{1/h})\| = \|1/h\|_\infty < \frac{2}{\varepsilon}.$$

Since $\|T_f - T_{\bar{z}^n h}\| < \frac{\varepsilon}{3}$, it follows that T_f is Fredholm with the same index. Since the index of a curve is a homotopy invariant, and we can see that f and $\bar{z}^n h$ can be connected by a straight line maintaining the condition on the harmonic extension on an annulus, it also follows that f and $\bar{z}^n h$ have the same winding number on $r\mathbb{T}$ for r close to 1.

Finally, (3) implies (1) by Corollary 12.7.7. ■

12.7.11. REMARK. Unfortunately this result does not extend to L^∞. There are Toeplitz operators which are Fredholm, even invertible, but the harmonic extension is not bounded away from 0 on an annulus. See Exercise **12.17**. Also Wolff constructed a function whose harmonic extension is bounded away from 0 on an annulus, but the Toeplitz operator is not Fredholm. Apparently this was done in 1983, but was published posthumously in [332].

Exercises for Chapter 12

12.1 Show that an isometry V is pure if and only if SOT–$\lim_{n\to\infty} V^{*n} = 0$.

12.2 Show that if $f \in L^\infty$, then $\operatorname{spr}(T_f) = \|f\|_\infty$.

12.3 (a) Show that an operator $A \in \mathcal{B}(H^2, H^{2\perp})$ is a Hankel operator if and only if $AT_z = P^\perp M_z A$.
(b) Show that the matrix for a Hankel operator with respect to the standard basis $\{\zeta_n : n \in \mathbb{Z}\}$ for $L^2(\mathbb{T})$ satisfies $a_{i,-j} = a_{1+i,1-j}$ for $i \geq 0$ and $j \geq 1$.
(c) Show that $\ker H_h$ is invariant for T_z.

12.4 (a) Let $f \in L^\infty$. Show that $P^\perp f$ is a rational function if and only if there are $h \in H^2$ and a finite Blaschke product B such that $f = \overline{B}h$.
(b) (*Kronecker*). Show that a Hankel operator H_f is finite rank if and only if $P^\perp f$ is a rational function.
HINT: Exercise **12.3**(c).

12.5 Show that every von Neumann algebra is reflexive.
HINT: Invariant subspaces correspond to projections in the commutant.

12.6 Let $\omega(z)$ be an inner function and let $M = H^2 \ominus \omega H^2$. Let $A_\omega = P_M T_z|_M$ be the compression of the shift to M.
(a) Find all of the invariant subspaces of A.
(b) Show that the map taking T_h to $h(A) := P_M T_h|_M$ for $h \in A(\mathbb{D})$ is an algebra homomorphism from $\mathcal{T}(A(\mathbb{D}))$ into $\mathrm{Alg}(A)$.
(c) Show that this map extends to a WOT-continuous homomorphism from $\mathcal{T}(H^\infty)$ into $W(A)$.
(d) Show that the kernel is $T_\omega \mathcal{T}(H^\infty) = \{T_{\omega h} : h \in H^\infty\}$.

12.7 Let M_z be the bilateral shift on L^2.
(a) If $f \in L^2$ and $f^{-1}(0) = E$ has $m(E) > 0$, show $\overline{\operatorname{span}}\{f z^n : n \geq 0\} = L^2(E)$.

(b) Show that a vector $f \in L^2$ is cyclic for M_z if and only if $\log|f|$ is not integrable and $m(f^{-1}(0)) = 0$.

12.8 Take the fact that the outer functions are the cyclic vectors in H^p, i.e. $H^p = \overline{\text{span}\{fz^n : n \geq 0\}}$, as the definition of outer function. Use operator theory to show:
(a) If $f \in L^2$ and $|f| \geq \varepsilon > 0$ a.e.(m), show that there is an outer function $h \in H^2$ such that $|f| = |h|$ a.e.(m).
HINT: Let $M = \overline{\text{span}}\{fz^n : n \geq 0\}$. Show that $\text{dist}(f, zM) > 0$.
(b) Let $h \in H^\infty$. Show that h is invertible in H^∞ if and only if h is outer and invertible in L^∞.
HINT: Consider $\text{Ran}\, T_h$.
(c) If $f \in H^1$, find $h_1, h_2 \in H^2$ so that $f = h_1 h_2$ and $|h_1|^2 = |h_2|^2 = |f|$ a.e.(m).
HINT: Let $g = |f|^{1/2} \in L^2$, and let $M = \overline{\text{span}}\{gz^n : n \geq 0\}$. Show that M is simply invariant. Find outer $h \in H^2$ so that $|g| = |h|$. Select polynomials p_n in z so that $hp_n \to 1$ in H^2. Show that fp_n is Cauchy in L^1.

12.9 For $f \in L^2$, define $T_f h = P(fh)$ for $h \in H^\infty$, so this is defined on a dense subspace of H^2. Prove that the following are equivalent:
(1) $f \in L^\infty$.
(2) T_f extends to a bounded operator on H^2.
(3) $\|T_f k_z\| \leq C\|k_z\|$ for all $z \in \mathbb{D}$.
(4) $\sup_{z\in\mathbb{D}} \left|\left\langle T_f \frac{k_z}{\|k_z\|_2}, \frac{k_z}{\|k_z\|_2}\right\rangle\right| < \infty$.

12.10 Use Theorem 12.2.2 to give another proof of the F.&M. Riesz theorem. HINT: If μ is an analytic measure, consider $M = \overline{\text{span}}\{z^n : n > 0\}$ in $L^2(|\mu|)$. Write $d\mu = h\, d|\mu|$. Show that $\bar{h} \perp M$ and M has no reducing part.

12.11 (*Gohberg–Krupnik theorem*). Let PC denote the algebra generated by the piecewise continuous functions on $\mathbb{T}$ (i.e. a finite number of discontinuities, and left and right limits exist at each point). Show that $\mathcal{T}(PC)/\mathcal{K}(H^2)$ is abelian, and compute its maximal ideal space.
HINT: Use a partition of unity argument to show that if $f, g \in PC$ do not have common discontinuities, then $T_f T_g - T_{fg}$ is compact. Then identify the maximal ideal space with $\mathbb{T} \times [0, 1]$ with a strange topology.

12.12 Let $f \in H^\infty + C$. Suppose that $\lambda \in \sigma(T_f) \setminus \sigma_{H^\infty+C}(f)$. Prove that either λ is an eigenvalue for T_f or $\overline{\lambda}$ is an eigenvalue for $T_{\bar{f}}$.

12.13 Let $\text{Aut}(\mathcal{T}(C))$ denote the group of automorphisms of $\mathcal{T}(C)$.
(a) Let $\alpha \in \text{Aut}(\mathcal{T}(C))$. Show that $\alpha(\mathcal{K}) = \mathcal{K}$.
(b) Hence show that α induces an automorphism $\widetilde{\alpha}$ of C. By considering the action on maximal ideals, show that there is a homeomorphism τ of $\mathbb{T}$ such that $\widetilde{\alpha}(f) = f(\tau)$.
(c) Show that τ is order preserving. Let $\text{Homeo}_+(\mathbb{T})$ denote the group of all these order preserving homeomorphisms.

(d) Show that the map from $\mathrm{Aut}(\mathcal{T}(C))$ into $\mathrm{Homeo}_+(\mathbb{T})$ is surjective.
HINT: You may use the fact that if S is a proper isometry, then there is a compact operator K such that $S+K$ is a shift (i.e. no unitary part in the Wold decomposition).

12.14 Let $QC = (H^\infty + C) \cap \overline{(H^\infty + C)}$. Let $\pi : \mathcal{B}(H) \to \mathcal{B}(H)/\mathcal{K}(H)$ be the quotient map.
(a) Show that if $f \in L^\infty$, then $M_f P - PM_f$ is compact on $L^2(\mathbb{T})$ if and only if $f \in QC$.
(b) Show that $A \in \mathcal{T}(L^\infty)$ commutes modulo the compact operators with every Toeplitz operator if and only if $A = T_f + K$ for some $f \in QC$ and $K \in \mathcal{K}$.
HINT: Write $A = T_f + K$ where K belongs to the commutator ideal, and consider $T_u^*(T_u A - AT_u)$ for inner functions u.

12.15 More about QC.
(a) Show that $\pi(T_f)$ is a unitary element of the Calkin algebra if and only if f is a unimodular function in QC.
(b) For $f \in QC$, show that T_f is Fredholm if and only if $|f| \geq \varepsilon$ a.e.(m) for some $\varepsilon > 0$.

12.16 A function in QC. Let $\Omega = \{z = x+iy : 0 < x < 1 \text{ and } -2 < y < \sin\frac{1}{x}\}$. Let f be a conformal map of $\mathbb{D}$ onto Ω. Show that $f \in H^\infty$ and $\mathrm{Re}\, f \in C$, but $\mathrm{Im}\, f$ is not in C. Thus $\mathrm{Im}\, f \in QC \setminus (H^\infty \cup C)$.

12.17 (*Douglas*) Invertibility and the harmonic extension.
(a) Find a trigonometric polynomial p such that T_p is invertible but the Poisson extension satisfies $\widetilde{p}(0) = 0$.
HINT: Approximate a continuous function with these properties.
(b) Let B be an infinite Blaschke product. Show that $T_{p(B)}$ is unitarily equivalent to a direct sum of infinitely many copies of T_p, and thus it is invertible.
(c) Show that $\widetilde{p(B)}$ vanishes on the zeros of B, a sequence of points approaching the boundary.

12.18 (*Sarason*) For $f, g \in L^\infty$ and $\lambda \in \mathbb{T}$, define

$$d_\lambda(f,g) = \operatorname{ess.}\limsup_{z\to\lambda} |f(z) - g(z)| \quad \text{and} \quad d_\lambda(f, H^\infty) = \inf_{g\in H^\infty} d_\lambda(f,g).$$

(a) Show that $d_\lambda(f, H^\infty)$ is an upper semicontinuous function of λ.
(b) Show that $\mathrm{dist}(f, H^\infty + C) = \max_{\lambda\in\mathbb{T}} d_\lambda(f, H^\infty)$.
HINT: Let M be the RHS and $\varepsilon > 0$. Find open arcs $J_1, \dots, J_n$ covering $\mathbb{T}$ and $h_1, \dots, h_n \in H^\infty$ so that $d_\lambda(f, h_i) < M + \varepsilon$ if $\lambda \in J_i$. Let g_i be a partition of unity for this cover and set $h = \sum_{i=1}^n g_i h_i$. Show that $h \in H^\infty + C$ and $\|f - h\| < M + \varepsilon$.
(c) Prove that if $f, g \in L^\infty$ and $\min\{d_\lambda(\overline{f}, H^\infty), d_\lambda(g, H^\infty)\} = 0$ for every $\lambda \in \mathbb{T}$, then $T_f T_g - T_{fg}$ is compact.
HINT: $\varepsilon > 0$, $A = \{\lambda : d_\lambda(\overline{f}, H^\infty) \geq \varepsilon\}$ and $B = \{\lambda : d_\lambda(g, H^\infty) \geq \varepsilon\}$ are

disjoint closed sets. Let u be continuous with $u|_A = 0$, $u|_B = 1$ and $0 \le u \le 1$. Then $T_f T_g = T_f T_u T_g + T_f T_{1-u} T_g = T_{fu} T_g + T_f T_{(1-u)g} + K_1$, with K_1 is compact. Compute $\operatorname{dist}(\overline{f}u, H^\infty + C)$ and $\operatorname{dist}((1-u)g, H^\infty + C)$.

12.19 (*Douglas–Sarason*) Suppose that $f \in L^\infty$ and for each $\lambda \in \mathbb{T}$, there is an arc $J \ni \lambda$ and $g \in L^\infty$ such that $f - g|_J = 0$ a.e.(m) and T_g is left Fredholm. Prove that T_f is left Fredholm.
HINT: Use part (b) of the previous exercise.

12.20 (*Arveson*) A norm 1 projection of $\mathcal{B}(H^2)$ onto the Toeplitz operators.

(a) Let Λ be a Banach limit on l_∞ (Example 2.6.22). For $A \in \mathcal{B}(L^2)$, define a sesquilinear form on L^2 by $[\xi, \eta] = \Lambda(\langle M_z^n \xi, A M_z^n \eta \rangle)$. This defines an operator $\mathcal{E}(A)$ by $[\xi, \eta] = \langle \xi, \mathcal{E}(A)\eta \rangle$. Prove that $\mathcal{E}$ is an expectation of $\mathcal{B}(L^2)$ onto $\{M_f : f \in L^\infty\}$.
(b) Show that if P_n is the projection onto $\overline{\operatorname{span}}\{\zeta_k : k \ge n\}$, then $\mathcal{E}(P_n) = I$.
(c) Show that if A is lower triangular with respect to the basis $\{\zeta_k : k \in \mathbb{Z}\}$, then $\mathcal{E}(A)$ is lower triangular, so lies in $\{M_f : f \in H^\infty\}$.
(d) Define $\mathcal{F} : \mathcal{B}(H^2) \to \mathcal{B}(H^2)$ by $\mathcal{F}(A) = P\mathcal{E}(PAP)P|_{H^2}$. Prove that $\mathcal{F}$ is a norm 1 projection onto the space of Toeplitz operators.
(e) Show that $\mathcal{F}(T_f A) = \mathcal{F}(A T_f) = \mathcal{F}(A) T_f$ for $A \in \mathcal{B}(H^2)$ and $f \in H^\infty$.
(f) Prove that if A is lower triangular, then $\mathcal{F}(A)$ lies in $\mathcal{T}(H^\infty)$.

Notes on Chapter 12

An early version of the Wold decomposition was due to von Neumann [313] in 1929. The Wold decomposition is from Wold's 1938 thesis, which was published as a book, reprinted in [331]. Coburn analysed the C*-algebra of an isometry in [58]. Beurling's theorem [28] about invariant subspaces for the bilateral and unilateral shift dates to 1949. Helson and Lowdenslager analysed shifts on $L^2(\mu)$ [146]; see also [144]. Theorem 12.5.8 is mostly due to Wermer [321] in 1952. Deddens proved that isometries are reflexive in [87].

Some of the basic results on Toeplitz operators are from Brown–Halmos [39]. Lemma 12.3.10 is due to Coburn [57]. The spectrum of analytic Toeplitz operators is due to Wintner [328] in 1929. The Hartman–Wintner paper is [142]. Nehari's theorem [216] dates from 1957. Hartman's theorem [141] is from 1958, and determines which Hankel operators are compact. Adamjan–Arov–Krein [1] establish Theorem 12.4.13 in [1]. Widom [324] proved a version of Theorem 12.6.3, strengthened by Devinatz [88]. Douglas's theorem for Toeplitz operators with symbol in $H^\infty + C$ is [97].

This chapter was strongly influenced by Douglas's book [99]. See also Nikolski [220]. More encyclopedic treatments of Toeplitz operators can be found in Böttcher–Silbermann [38] and Nikolski [219].

CHAPTER 13

Nest Algebras

The study of nest algebras is the study of triangular forms for operators and operator algebras. In this chapter, we first study some basic properties of nest algebras. The bulk of the chapter is devoted to a proof of the Similarity theorem, which describes precisely when two nest algebras acting on a separable Hilbert space are similar. This is followed by a number of applications.

13.1. Nests: definitions and examples

We recall from Definition 12.4.6 and Remark 12.4.7 that if $\mathcal{A}$ is an algebra of operators, then $\operatorname{Lat}\mathcal{A}$ is a complete lattice of its invariant subspaces. Conversely, if $\mathcal{L}$ is a collection of subspaces, then $\operatorname{Alg}\mathcal{L}$ is the unital WOT-closed operator algebra consisting of all operators leaving each element of $\mathcal{L}$ invariant.

13.1.1. DEFINITION. A *nest* is a complete lattice $\mathcal{N}$ which is totally ordered by containment. A *nest algebra* is the algebra $\mathcal{T}(\mathcal{N}) = \operatorname{Alg}\mathcal{N}$ for a nest $\mathcal{N}$.

13.1.2. EXAMPLES.

(1) Let $\mathcal{P} = \{0 = P_0 \subset P_1 \subset P_k \subset P_{k+1} \subset \cdots \subset H\}$ where $\bigvee_{k\geq 1} P_k = H$. This nest is order isomorphic to $\omega = \mathbb{N}_0 \cup \{\infty\}$. Then $\mathcal{T}(\mathcal{P})$ consists of all operators which are *block upper triangular* with respect to the decomposition $H = \sum_{k\geq 1}^{\oplus} M_k$ where $M_k = P_k \ominus P_{k-1}$. Another such nest $\mathcal{P}'$ is unitarily equivalent to $\mathcal{P}$ if and only if $\dim M_k' = \dim M_k$ for $k \geq 1$.

(2) $\mathcal{P}^\perp = \{\{0\}, H, P_k^\perp : k \geq 1\}$. This nest is order isomorphic to $\{-\infty\} \cup -\mathbb{N}_0$. Then $\mathcal{T}(\mathcal{P}^\perp) = \mathcal{T}(\mathcal{P})^*$ is the algebra of all block lower triangular operators with respect to the decomposition $H = \sum_{k\geq 1}^{\oplus} M_k$.

(3) Let $H = L^2(0,1)$ and for $0 \leq t \leq 1$, let

$$N_t = L^2(0,t) = \{f \in L^2(0,1) : f(x) = 0 \text{ a.e. for } t < x < 1\}.$$

This is the *Volterra nest*, and is order isomorphic to $[0,1]$. The nest algebra $\mathcal{T}(\mathcal{N})$ has a continuous upper triangular form.

K. R. Davidson, *Functional Analysis and Operator Algebras*, CMS/CAIMS Books in Mathematics 13,
https://doi.org/10.1007/978-3-031-63665-3_13

(4) Let $H = L^2((0,1)^2)$ and let $\mathcal{M} = \{M_t : 0 \le t \le 1\}$ where for $0 \le t \le 1$,

$$M_t = N_t \otimes L^2(0,1) = \{f(x,y) : f(x,y) = 0 \text{ a.e.}(m^2) \text{ for } t < x < 1\}.$$

This nest is also order isomorphic to $[0,1]$. However, it is not unitarily equivalent to $\mathcal{N}$, as will be seen later.

(5) Let $H = l^2(\mathbb{Q}) = \operatorname{span}\{\delta_r : r \in \mathbb{Q}\}$ and set $Q_t = \operatorname{span}\{e_r : r < t\}$ for $t \in \mathbb{R}$ and $Q_t^+ = \operatorname{span}\{e_r : r \le t\}$. The *Cantor nest* is $\{\{0\}, Q_t, Q_t^+, H : t \in \mathbb{R}\}$. Note that if $r \in \mathbb{Q}$, then $Q_r^+ = Q_r \oplus \mathbb{C}\delta_r$. But if $t \in \mathbb{R} \setminus \mathbb{Q}$, then $Q_t^+ = Q_t$. Then $Q_s \le Q_s^+ < Q_t \le Q_t^+$ for $s < t \in \mathbb{R}$. The order type of this nest can be identified with $[-\infty, \infty] \cup \{r^+ : r \in \mathbb{Q}\}$ with total order $s < r < r^+ < t$ if $s < r < t$ and $r \in \mathbb{Q}$. Topologically, we can identify this with the Cantor set. This will be made more precise later.

(6) Let $H = l^2(\mathbb{Q}) \oplus L^2(\mathbb{R})$ and let

$$\mathcal{R} = \{R_t = Q_t \oplus N_t, R_r^+ = Q_r^+ \oplus N_r : t \in \mathbb{R},\ r \in \mathbb{Q}\}$$

where $N_t = \{f \in L^2(\mathbb{R}) : f(x) = 0 \text{ a.e. for } x > t\}$. This has the same order type as $\mathcal{Q}$.

13.1.3. Definition. Let $\mathcal{N}$ be a nest. For $N \in \mathcal{N}$, define

$$N_- = \bigvee\{M \in \mathcal{N} : M < N\} \quad \text{and} \quad N_+ = \bigwedge\{M \in \mathcal{N} : M > N\}.$$

By convention, $0_- = 0$ and $H_+ = H$. When $N \neq N_+$, the subspace $A = N_+ \ominus N$ is an *atom* of $\mathcal{N}$. If there are no atoms, then $\mathcal{N}$ is a *non-atomic nest* or a *continuous nest*. If $\bigvee\{A : \text{atoms of } \mathcal{N}\} = H$, we say that $\mathcal{N}$ is an *atomic nest*.

In the order on $\mathcal{N}$, N_+ is the immediate successor of N if it exists, and otherwise $N_+ = N$. Similarly, N_- is the immediate predecessor of N if it exists, and otherwise $N_- = N$. When $N \neq N_+$, then $(N_+)_- = N$. Thus the atoms are also $\{N \ominus N_- : N \in \mathcal{N}\}$.

13.1.4. Examples. We consider the examples of Examples 13.1.2 again.

(1) In the nest $\mathcal{P}$, we have $P_{k+} = P_{k+1}$ and $P_{k-} = P_{k-1}$. Also $H_- = H$. The atoms of $\mathcal{P}$ are the subspaces $M_k = P_k \ominus P_{k-1}$. Since $\bigvee_{k \ge 1} M_k = H$, this is an atomic nest.

(2) In $\mathcal{P}^\perp$, $(P_k^\perp)_- = P_{k+1}^\perp$ and $(P_k^\perp)_+ = P_{k-1}^\perp$. Here $0_+ = 0$. This nest has the same atoms as $\mathcal{P}$, and so it is atomic.

(3) The Volterra nest has no atoms, so it is a continuous nest.

(4) $\mathcal{M}$ is also a continuous nest.

(5) In the Cantor nest, $Q_{t-} = Q_t$ for all $t \in \mathbb{R}$, while $Q_{t+} = Q_t$ when $t \notin \mathbb{Q}$ and $Q_{r+} = Q_r^+$ when $r \in \mathbb{Q}$. Also $(Q_r^+)_- = Q_r$. The atoms are $A_r = Q_r^+ \ominus Q_r = \mathbb{C}\delta_r$ for $r \in \mathbb{Q}$. These span $l^2(\mathbb{Q})$, so this is an atomic nest. There are only countably many atoms even though $\mathcal{Q}$ is an uncountable nest.

(6) The nest R has the same atoms as $\mathcal{Q}$, but now they do not span the whole space. This nest is neither atomic nor continuous.

The next lemma is the simple observation that large "corners" of $\mathcal{B}(H)$ belong to a nest algebra.

13.1.5. LEMMA. *Let $N \in \mathcal{N}$. Then $\mathcal{T}(\mathcal{N})$ contains $P_{N_+}\mathcal{B}(H)P_N^\perp$.*

PROOF. Let $M \in \mathcal{N}$. If $M \leq N$, then $P_{N_+}\mathcal{B}(H)P_N^\perp M = 0 \subseteq M$. So M is invariant. Otherwise $M \geq N_+$, so that $P_{N_+}\mathcal{B}(H)P_N^\perp M \subseteq N_+ \subseteq M$. ■

This equivalently shows that $\mathcal{T}(\mathcal{N})$ contains $P_N\mathcal{B}(H)P_{N_-}^\perp$.
The nest can be recovered from the nest algebra.

13.1.6. THEOREM. *If $\mathcal{N}$ is a nest, then* $\operatorname{Lat}\mathcal{T}(\mathcal{N}) = \mathcal{N}$, *i.e. nests are reflexive.*

PROOF. Clearly $\mathcal{N} \subset \operatorname{Lat}\mathcal{T}(\mathcal{N})$. Let $M \in \operatorname{Lat}\mathcal{T}(\mathcal{N})$. Set

$$N_0 = \bigwedge\{N \in \mathcal{N} : M \subseteq N\}.$$

So $M \subseteq N_0$, but if $N < N_0$ in $\mathcal{N}$, then there is a vector $\xi \in M \setminus N$. Hence $P_N^\perp\xi \neq 0$. Therefore by the lemma,

$$M \supset P_{N_+}\mathcal{B}(H)P_N^\perp\xi = N_+.$$

Therefore

$$M \supseteq \bigvee\{N_+ : N < N_0\} = \begin{cases} N_0 & \text{if } N_0 = N_{0-} \\ (N_{0-})_+ = N_0 & \text{if } N_{0-} < N_0. \end{cases}$$

So $M = N_0 \in \mathcal{N}$. Thus $\operatorname{Lat}\mathcal{T}(\mathcal{N}) = \mathcal{N}$. ■

13.1.7. PROPOSITION. *If $\mathcal{N}$ is a nest on H, then $\mathcal{T}(\mathcal{N})' = \mathbb{C}I$.*

PROOF. Let A commute with $\mathcal{T}(\mathcal{N})$. If $A \neq 0$, then it doesn't vanish on the dense subspace $\bigcup_{N<H} N_+$. Pick $N_0 < H \in \mathcal{N}$ and $\xi_0 \in N_{0+}$ with $A\xi_0 \neq 0$. For any $\eta \in N_0^\perp$, we have $\xi_0\eta^* \in \mathcal{T}(\mathcal{N})$. Thus

$$0 \neq (A\xi_0)\eta^* = A(\xi_0\eta^*) = (\xi_0\eta^*)A = \xi_0(A^*\eta)^*.$$

Therefore $A\xi_0 \in \mathbb{C}\xi_0$; say $A\xi_0 = \lambda\xi_0$. It follows that $A^*\eta = \bar{\lambda}\eta$ for all $\eta \in N_0^\perp$, i.e. $A^*|N_0^\perp = \bar{\lambda}I|N_0^\perp$. Repeating the calculation for an arbitrary $\xi \in N_{0+}$ yields

$A\xi = \lambda\xi$, i.e. $A|_{N_{0+}} = \lambda I|_{N_{0+}}$. Now we can repeat this for any $N < H$ with $N_+ > 0$ and get the same scalar λ because $N_+ \cap N_{0+} \neq \{0\}$ and $N^\perp \cap N_0^\perp \neq \{0\}$. Hence $A = \lambda I$ on $\bigcup_{N<H} N_+$, which is dense in H. By continuity, $A = \lambda I$. ■

We will frequently identify $\mathcal{N}$ with the set $\{P_N : N \in \mathcal{N}\}$ of orthogonal projections onto elements of $\mathcal{N}$. So $\mathcal{N}'$ and $\mathcal{N}''$ are von Neumann algebras. By the Double Commutant theorem, $\mathcal{N}'' = \overline{C^*(\mathcal{N})}^{\text{WOT}}$ is abelian.

13.1.8. PROPOSITION. *Let $\mathcal{N}$ be a nest. Then $\mathcal{T}(\mathcal{N}^\perp) = \mathcal{T}(\mathcal{N})^*$ and*

$$\mathcal{T}(\mathcal{N}) \cap \mathcal{T}(\mathcal{N})^* = \mathcal{N}' := \{T \in \mathcal{B}(H) : TP_N = P_N T \text{ for all } N \in \mathcal{N}\}.$$

PROOF. Note that $T \in \mathcal{T}(\mathcal{N}^\perp)$ if and only if $P_N T P_N^\perp = 0$ for all $N \in \mathcal{N}$, which is equivalent to $P_N^\perp T^* P_N = 0$ for all $N \in \mathcal{N}$, i.e. $T^* \in \mathcal{T}(\mathcal{N})$.

Now if $T \in \mathcal{T}(\mathcal{N}) \cap \mathcal{T}(\mathcal{N})^*$, then

$$TP_N = P_N T P_N = (P_N T^* P_N)^* = (T^* P_N)^* = P_N T.$$

Conversely, if $T \in \mathcal{N}'$, then

$$P_N^\perp T P_N = P_N^\perp P_N T = 0 = P_N T P_N^\perp.$$

Thus $T \in \mathcal{T}(\mathcal{N}) \cap \mathcal{T}(\mathcal{N})^*$. ■

Recall that the *order topology* on a totally ordered set Ω has a subbase of open intervals $(\alpha, \infty) = \{\gamma \in \Omega : \gamma > \alpha\}$ and $(-\infty, \beta) = \{\gamma \in \Omega : \gamma < \beta\}$ for $\alpha, \beta \in \Omega$. Including the collection $(\alpha, \beta) = (\alpha, \infty) \cap (-\infty, \beta)$ for $\alpha < \beta$ yields a base of open sets. Let $(\mathcal{N}, <)$ denote the nest $\mathcal{N}$ with the order topology. When we identify $\mathcal{N}$ with the corresponding set of projections, we can consider the WOT and SOT topologies on $\mathcal{N}$.

13.1.9. THEOREM. *If $\mathcal{N}$ is a nest, then the identity map is a homeomorphism between $(\mathcal{N}, <)$, $(\mathcal{N}, \text{SOT})$ and $(\mathcal{N}, \text{WOT})$. They are compact and Hausdorff.*

If H is separable, then there is a unit vector $\xi \in H$ so that $N \to \|P_N\xi\|^2$ is an order homeomorphism of $(\mathcal{N}, <)$ onto a compact subset of $[0, 1]$.

PROOF. Recall from Lemma 9.3.5 that if N_λ is an increasing net, then

$$\text{SOT-}\lim_{\Lambda} P_{N_\lambda} = P_N \quad \text{where } N = \bigvee N_\lambda.$$

A similar statement holds for decreasing nets.

Suppose that $N_\lambda \in \mathcal{N}$ and $\text{WOT-}\lim_\Lambda P_{N_\lambda} = E$. Then $0 \le E \le I$ and $E \in \mathcal{N}''$. Define $N_0 = \bigwedge\{N \in \mathcal{N} : E \le P_N\}$. Suppose that $N_1 < N_0 < N_2$ in $\mathcal{N}$. Then $P_{N_1} \not\ge E$, so there is a vector $\xi \in N_1^\perp$ such that $E\xi \neq 0$. Now

$$\langle P_{N_\lambda}\xi, \xi\rangle \to \langle E\xi, \xi\rangle = \|E^{1/2}\xi\|^2 > 0$$

because $E^{1/2}\xi = 0$ implies that $E\xi = 0$. Hence, there is some $\lambda_1 \in \Lambda$ so that $P_{N_\lambda}\xi \neq 0$ for $\lambda \geq \lambda_1$. Thus $N_\lambda > N_1$, so $N_\lambda \geq N_{1+}$ for $\lambda \geq \lambda_1$. Similarly WOT–$\lim_\Lambda P_{N_\lambda}^\perp = I - E$, and there is a $\lambda_2 \in \Lambda$ so that $N_\lambda \leq N_{2-}$ for $\lambda \geq \lambda_2$. Combining these two facts, we see that $P_{N_{1+}} \leq E \leq P_{N_{2-}}$ for $\lambda \geq \lambda_0 \geq \{\lambda_1, \lambda_2\}$. By Lemma 9.3.5, the increasing net $\{P_{N_+} : N < N_0\}$ converges SOT to P_M where $M = \bigvee_{N<N_0} N_+ = N_0$ as shown in the proof of Theorem 13.1.6. Similarly the decreasing net $\{P_{N_-} : N > N_0\}$ converges SOT to P_{N_0}. Since the net P_{N_λ} is sandwiched in between these two nets, we deduce that $E = P_{N_0}$. Thus $(\mathcal{N}, \text{WOT})$ is closed. Moreover, this argument shows that P_{N_λ} converge SOT to P_{N_0}. Since the order interval (N_1, N_2) about N_0 was arbitrary, this also shows that N_λ converge to N_0 in the order topology. That is, the identity map is continuous from $(\mathcal{N}, \text{WOT})$ into $(\mathcal{N}, \text{SOT})$ and $(\mathcal{N}, <)$.

The set $\{P_N : N \in \mathcal{N}\}$ is contained in the WOT-compact set $\overline{b_1(\mathcal{B}(H)})$. Therefore, it is compact in the WOT-topology. The other two sets are clearly Hausdorff. Thus, the identity maps from $(\mathcal{N}, \text{WOT})$ into $(\mathcal{N}, \text{SOT})$ and $(\mathcal{N}, <)$ are homeomorphisms.

If H is separable, Lemma 10.4.6 shows that the abelian von Neumann algebra $\mathcal{N}''$ has a separating vector, say ξ. We may suppose that $\|\xi\| = 1$, so that $\varphi(A) = \langle A\xi, \xi\rangle$ is a faithful normal state on $\mathcal{N}''$. Then $\Phi(N) = \varphi(P_N) = \|P_N\xi\|^2$ is continuous from $(\mathcal{N}, \text{SOT})$ into $[0, 1]$. So the range is compact, $\Phi(0) = 0$ and $\Phi(H) = 1$. Since φ is faithful, this map is injective and preserves the order. Therefore, Φ is an order homeomorphism onto its range. ∎

13.1.10. COROLLARY. *If H is separable, there is a Borel probability measure on $\omega = \Phi(\mathcal{N})$ so that $\mathcal{N}'' \simeq L^\infty(\mu)$. Thus, there is a (countably additive) spectral measure $\mathcal{E} : \mathrm{Bor}(\omega) \to \mathrm{Pr}(\mathcal{N}'')$ such that $\mathcal{E}((a, b]) = \Phi^{-1}(b) - \Phi^{-1}(a)$.*

PROOF. The state φ is defined on $\mathcal{N}'' \supset C^*(\mathcal{N}) \simeq C(\Phi(\mathcal{N}))$. By the Riesz Representation theorem, there is a Borel probability measure on $\Phi(\mathcal{N})$ so that $\varphi(A) = \displaystyle\int f\,d\mu$ where $f \in C(\Phi(\mathcal{N}))$ corresponds to $A \in C^*(\mathcal{N})$. The state comes from a faithful normal state on $\mathcal{N}''$. By Corollary 10.4.2, $\mathcal{N}'' \simeq L^\infty(\mu)$.

The inverse of this map is a $*$-isomorphism of $L^\infty(\mu)$ onto $\mathcal{N}''$. When restricted to Borel characteristic functions, it yields a (countably additive) spectral measure defined on the set of Borel subsets $\mathrm{Bor}(\omega)$ onto the set of projections in $\mathcal{N}''$. If $M < N$, $\Phi(M) = a$ and $\Phi(N) = b$, then clearly $\mathcal{E}((a, b]) = P_N - P_M$. ∎

13.1.11. EXAMPLE. Let $\mathcal{Q}$ be the Cantor nest on $l^2(\mathbb{Q})$. Let $\xi = \sum_{r\in\mathbb{Q}} a_r\delta_r$ be a unit vector such that $a_r \neq 0$ for all $r \in \mathbb{Q}$. Then ξ is a separating vector for $\mathcal{N}$. Let $\Phi(N) = \langle P_N\xi, \xi\rangle$. The range is a compact subset of $[0, 1]$. For any two elements $s < t$ in $\mathbb{R}$, there is $r \in \mathbb{Q}$ so that $Q_s < Q_r < Q_r^+ < Q_t$. Thus $\Phi(\mathcal{Q}) \cap [\Phi(Q_s), \Phi(Q_t)]$ omits the interval $(\Phi(Q_r), \Phi(Q_r^+))$. It follows that $\Phi(\mathcal{Q})$

is a compact nowhere dense subset of $[0, 1]$. It is also perfect, meaning that there are no isolated points. It is a result from real analysis that a compact nowhere dense perfect subset of a metric space is homeomorphic to the Cantor set. Since it is a subset of $[0, 1]$, it is possible to arrange this in an order preserving manner. Indeed, the points $\Phi(Q_r)$ and $\Phi(Q_r^+)$ are sent to the endpoints of the intervals which are deleted in the construction of the Cantor set. The order structure on those intervals is countable and dense, in that there are intervals in between any two others. So one can match the intervals $(\Phi(Q_r), \Phi(Q_r^+))$ to the intervals in the complement of the Cantor set in an order preserving way recursively.

13.1.12. DEFINITION. An *interval of a nest* $\mathcal{N}$ is a projection $E = P_{N_2} - P_{N_1}$ for $N_1 < N_2$ in $\mathcal{N}$. Define $\Phi_E(T) = ETE|_{EH}$ for $T \in \mathcal{T}(\mathcal{N})$.

For the next result, we require an easy result, Proposition 14.1.2, from the next chapter.

13.1.13. PROPOSITION. *Let E be an interval of a nest $\mathcal{N}$. Then Φ_E is a surjective,* WOT*-continuous, contractive homomorphism of $\mathcal{T}(\mathcal{N})$ onto $\mathcal{T}(E\mathcal{N})$, where $E\mathcal{N}$ is considered as a nest on EH. Conversely, if E is a projection such that Φ_E is a homomorphism, then E is an interval of $\mathcal{N}$.*

PROOF. By Proposition 14.1.2, the compression to EH is a homomorphism if and only if EH is the difference of two invariant subspaces of $\mathcal{T}(\mathcal{N})$. So by Theorem 13.1.6, this occurs precisely when $EH = N_2 \ominus N_1$ is an interval of $\mathcal{N}$. Moreover Φ_E is WOT-continuous and contractive.

To see that it is surjective, let $A \in \mathcal{T}(E\mathcal{N})$. Define $T = EAE \in \mathcal{B}(H)$. If $N \le N_1$, then $TN = \{0\}$. If $N \ge N_2$, then $TN \subset EH \subset N$. If $N_1 < N < N_2$, then $N = N_1 + EN$ and $TN = AEN \subset EN \subset N$. Thus $T \in \mathcal{T}(\mathcal{N})$ and $\Phi_E(T) = A$. ■

If A is an atom of $\mathcal{N}$, then $A\mathcal{N} = \{0, AH\}$. Hence we obtain:

13.1.14. COROLLARY. *If A is an atom of the nest $\mathcal{N}$, then Φ_A is a homomorphism of $\mathcal{T}(\mathcal{N})$ onto $\mathcal{B}(AH)$.*

13.1.15. LEMMA. *Let $0 = N_0 \subset N_1 \subset \cdots \subset N_k = H$ be a finite subnest of $\mathcal{N}$. Let $E_i = P_{N_i} - P_{N_{i-1}}$ for $1 \le i \le k$. Let $T \in \mathcal{T}(\mathcal{N})$. Then T is invertible in $\mathcal{T}(\mathcal{N})$ if and only if $\Phi_{E_i}(T)$ is invertible in $\mathcal{T}(E_i\mathcal{N})$ for $1 \le i \le k$, and moreover $\Phi_{E_i}(T^{-1}) = \Phi_{E_i}(T)^{-1}$.*

PROOF. If T is invertible in $\mathcal{T}(\mathcal{N})$, then $\Phi_{E_i}(T^{-1}) = \Phi_{E_i}(T)^{-1}$ are invertible for $1 \le i \le k$ because Φ_{E_i} are homomorphisms.

Conversely, suppose that $\Phi_{E_i}(T)$ is invertible in $\mathcal{T}(E_i\mathcal{N})$ for $1 \le i \le k$. Let $T_{ij} = E_iTE_j$ be considered as an element of $\mathcal{B}(E_jH, E_iH)$. Then $T_{ij} = 0$ if $i > j$ and $T_{ii} = \Phi_{E_i}(T)$. We can consider T as a $k \times k$ upper triangular matrix. Let $D = \sum_{i=1}^k E_iTE_i$ be the $k \times k$ diagonal matrix with entries T_{ii} on the diagonal. Observe that D is invertible in $\mathcal{T}(\mathcal{N})$ with diagonal inverse with i,i entry $\Phi_{E_i}(T)^{-1}$. Therefore, $D^{-1}T$ is upper triangular with diagonal entries E_i. So $D^{-1}T - I = A$ is strictly upper triangular with respect to this finite block decomposition. Now $A^k = 0$, and thus $(I+A)^{-1} = \sum_{i=0}^{k-1}(-1)^iA^i$ in $\mathcal{T}(\mathcal{N})$. This is upper triangular with diagonal I. Therefore $T^{-1} = (I+A)^{-1}D^{-1}$ belongs to $\mathcal{T}(\mathcal{N})$. ∎

13.1.16. Remark. It can happen that $T \in \mathcal{T}(\mathcal{N})$ is invertible in $\mathcal{B}(H)$ but not in $\mathcal{T}(\mathcal{N})$. A simple example is $\mathcal{N} = \{\{0\}, l^2(-\mathbb{N}), l^2(\mathbb{Z})\}$. Let U be the bilateral shift. Then $U^* \in \mathcal{T}(\mathcal{N})$ but U is not. Indeed, U is lower triangular.

13.2. Compact Operators in Nest Algebras

One very nice result about compact operators in nest algebras is Ringrose's Theorem 5.5.8. In this section, we will show that the compact operators in $\mathcal{T}(\mathcal{N})$ contain a contractive approximate identity. Recall that if $\zeta, \eta \in H$, then $\zeta\eta^*$ is the rank 1 operator $\zeta\eta^*(\xi) = \langle \xi, \eta\rangle\zeta$.

13.2.1. Lemma. *A rank 1 operator $\zeta\eta^*$ belongs to $\mathcal{T}(\mathcal{N})$ if and only if there is an $N \in \mathcal{N}$ so that $\zeta \in N$ and $\eta \in N_-^\perp$.*

Proof. The if direction is immediate from Lemma 13.1.5.

Conversely, suppose that $\zeta\eta^* \in \mathcal{T}(\mathcal{N})$. Let $N = \overline{\mathcal{T}(\mathcal{N})\zeta}$. Then N is the smallest element of $\operatorname{Lat}\mathcal{T}(\mathcal{N}) = \mathcal{N}$ containing ζ. If $M \in \mathcal{N}$, $M < N$ and $\xi \in M$, then $\zeta\eta^*\xi = \langle\xi,\eta\rangle\zeta \in M$. Since $\zeta \notin M$, $\xi \perp \eta$. So $\eta \in M^\perp$. Thus

$$\eta \in \bigwedge_{M<N} M^\perp = \Big(\bigvee_{M<N} M\Big)^\perp = N_-^\perp. \qquad \blacksquare$$

13.2.2. Remark. Alternatively, we could say that $\zeta\eta^*$ belongs to $\mathcal{T}(\mathcal{N})$ if and only if there is an $N \in \mathcal{N}$ so that $\zeta \in N_+$ and $\eta \in N^\perp$.

An operator algebra can contain finite rank elements which are not the sum of rank one elements in the algebra. Also a finite rank element F could be a sum of rank ones, but require more than rank F terms. (See Exercises **13.19** and **13.18**.) This is not the case in a nest algebra.

13.2.3. THEOREM. *Let $\mathcal{N}$ be a nest, and let $F \in \mathcal{T}(\mathcal{N})$ have finite rank n. Then F is the sum of n rank 1 elements of $\mathcal{T}(\mathcal{N})$.*

PROOF. We will establish this by induction on n. If $n = 1$ or 0, there is nothing to prove. Suppose the result is established for rank less than n, and suppose that $F \in \mathcal{T}(\mathcal{N})$ has rank n. The function $d(N) = \dim(N \cap \operatorname{Ran} F)$ is increasing, and $d(H) = n$. Let n_0 be the smallest non-zero value taken by d. The set of subspaces $\mathcal{S} = \{N \in \mathcal{N} : d(N) = n_0\}$ is non-empty, and $N \cap \operatorname{Ran} F = Z$ is constant on $\mathcal{S}$. Thus $N_0 = \bigwedge\{N \in \mathcal{S}\}$ belongs to $\mathcal{S}$. Let $\zeta \in Z$ be a unit vector and let $\eta = F^*\zeta$. We will show that $\zeta\eta^* \in \mathcal{T}(\mathcal{N})$ and $\operatorname{rank}(F - \zeta\eta^*) = n - 1$. Then the result will follow from the induction hypothesis.

If $N < N_0$, then $FP_N = P_N F P_N$ has range in $\operatorname{Ran} F \cap N = \{0\}$. Thus $N \subset \ker F$, whence $N_{0-} \subset \ker F$. Now $\eta \in \operatorname{Ran} F^* = (\ker F)^\perp$, so $\eta \in N_{0-}^\perp$. Therefore $\zeta\eta^* \in P_{N_0}\mathcal{B}(H)P_{N_{0-}}^\perp \subset \mathcal{T}(\mathcal{N})$. Finally,

$$\operatorname{rank}(F - \zeta\eta^*) = \operatorname{rank}(F - \zeta\eta^*)^* = \operatorname{rank}(F^* - \eta\zeta^*).$$

Since $\ker F^* = (\operatorname{Ran} F)^\perp$ and $\zeta \in \operatorname{Ran} F \cap \ker(F^* - \eta\zeta^*)$, it follows that

$$\ker F^* + \mathbb{C}\zeta \subset \ker(F^* - \eta\zeta^*).$$

So $\operatorname{rank}(F^* - \eta\zeta^*) \le \operatorname{rank} F^* - 1 = n - 1$. The result follows. ∎

Our next goal is to build a contractive approximate identity of finite rank operators in $\mathcal{T}(\mathcal{N})$. It is easy to deal with atoms of the nest. So we first explain what happens with a continuous nest.

13.2.4. LEMMA. *Let $\mathcal{N}$ be a continuous nest, let ξ be a unit vector, and let $H_\xi = \overline{\operatorname{span}\{P_N\xi : N \in \mathcal{N}\}}$. Then $P = P_{H_\xi} \in \mathcal{N}'$ and there are finite rank operators $R_n \in P\mathcal{T}(\mathcal{N})P$ with $\|R_n\| \le 1$ such that* SOT*–$\lim_{n\to\infty} R_n = P$.

PROOF. Note that $H_\xi = \overline{C^*(\mathcal{N})\xi} = \overline{\mathcal{N}''\xi}$ is invariant for $\mathcal{N}''$. Hence $P \in \mathcal{N}'$. Since $\mathcal{N}' \subset \mathcal{T}(\mathcal{N})$, $P\mathcal{T}(\mathcal{N})P \subset \mathcal{T}(\mathcal{N})$. Moreover $P\mathcal{T}(\mathcal{N})P|_{H_\xi} = \mathcal{T}(P\mathcal{N}) \subset \mathcal{B}(H_\xi)$. Clearly $P\mathcal{N}$ is also a continuous nest. So we may suppose that H is separable and ξ is a cyclic vector for $\mathcal{N}''$.

The state $\varphi(A) = \langle A\xi, \xi\rangle$ maps $\mathcal{N}$ onto $[0, 1]$. Since ξ is a cyclic vector for $\mathcal{N}''$, it is a separating vector for $\mathcal{N}'$ by Lemma 10.4.5. Thus φ is a faithful normal state on $\mathcal{N}'$ and so φ is injective on $\mathcal{N}$. For $n \ge 1$ and $0 \le k \le 2^n$, let $N_{k,n} = \varphi|_{\mathcal{N}}^{-1}\left(\frac{k}{2^n}\right)$. Define unit vectors $\xi_{k,n} = 2^{n/2}(P_{N_{k,n}} - P_{N_{k-1,n}})\xi$ for $1 \le k \le 2^n$, and set

$$R_n = \sum_{k=2}^{2^n} \xi_{k-1,n}\xi_{k,n}^*.$$

Observe that if $n < m$, then

$$\xi_{k,n} = 2^{(n-m)/2} \sum_{j=1}^{2^{m-n}} \xi_{2^{m-n}(k-1)+j,m}.$$

Therefore

$$\begin{aligned}\xi_{k,n} - R_m\xi_{k,n} &= 2^{(n-m)/2}\Big(\sum_{j=1}^{2^{m-n}} \xi_{2^{m-n}(k-1)+j,m} - \sum_{j=0}^{2^{m-n}-1} \xi_{2^{m-n}(k-1)+j,m}\Big) \\ &= 2^{(n-m)/2}(\xi_{2^{m-n}k,m} - \xi_{2^{m-n}(k-1),m}).\end{aligned}$$

Hence $\lim_{m\to\infty} R_m\xi_{k,n} = \xi_{k,n}$ and similarly $\lim_{m\to\infty} R_m^*\xi_{k,n} = \xi_{k,n}$. This extends to any vector in

$$\overline{\operatorname{span}\{\xi_{k,n} : 1 \le k \le 2^n,\ n \ge 1\}} = H_\xi.$$

Since $R_m = PR_mP$, we obtain SOT*–$\lim_{n\to\infty} R_n = P$. ■

We can now complete the argument for arbitrary nests.

13.2.5. ERDOS THEOREM. *Let $\mathcal{N}$ be a nest. Then there is a net $\{F_\lambda\}_\Lambda$ of finite rank contractions in $\mathcal{T}(\mathcal{N})$ which converge* SOT* *to I. If H is separable, there is a sequence.*

PROOF. Let $\{A_i : i \in \mathcal{I}\}$ be the atoms of $\mathcal{N}$. Set $Q =$ SOT–$\sum_{i\in\mathcal{I}} A_i$. If $\dim A_i < \infty$, then P_{A_i} is a finite rank identity for $\mathcal{B}(A_i)$. But if $\dim A_i$ is infinite, there is a net of finite rank projections in $\mathcal{B}(A_i)$ converging SOT* to A_i. If A_i is separable, this can be taken to be a sequence. If H is separable, then $\mathcal{I}$ is finite or countable.

Then $Q \in \mathcal{N}''$ and $Q^\perp\mathcal{N}$ is a continuous nest on $H_c = Q^\perp H$. By Zorn's lemma, there is a maximal family of pairwise orthogonal subspaces $\{H_{\xi_j} : j \in \mathcal{J}\}$ in H_c so that $H_{\xi_j} = \overline{\operatorname{span}\{P_N\xi_j : N \in \mathcal{N}\}}$. Let P_j be the projection onto H_{ξ_j}. Note that SOT–$\sum_{\mathcal{J}} P_j = Q^\perp$. By the lemma, on each H_{ξ_j}, there is a sequence of finite rank contractions $R_{j,m}$ in $\mathcal{T}(\mathcal{N})$ converging strong-$*$ to P_j. If H is separable, then $\mathcal{J}$ is finite or countable.

Now it is routine to build a net that combines all of these nets into one with the desired property. In the separable case, one can build a sequence. ■

13.2.6. COROLLARY. *The net $\{F_\lambda\}_\Lambda$ is a contractive approximate identity for $\mathcal{T}(\mathcal{N}) \cap \mathcal{K}(H)$. The unit ball of $\mathcal{T}(\mathcal{N}) \cap \operatorname{Fin}(H)$ is* SOT* *dense in the unit ball of $\mathcal{T}(\mathcal{N})$.*

PROOF. If $T \in \mathcal{T}(\mathcal{N})$ and $\|T\| \le 1$, then TF_λ is finite rank, has norm at most 1, and converges SOT* to T. If $K \in \mathcal{T}(\mathcal{N}) \cap \mathcal{K}(H)$, then $F_\lambda K$ converges to K in norm. See Exercise **5.8**. Similarly KF_λ converges to K in norm because $F_\lambda^* K^*$

converges to K^* in norm. Therefore $\{F_\lambda\}_\Lambda$ is a contractive approximate identity for $\mathcal{T}(\mathcal{N}) \cap \mathcal{K}(H)$. ■

13.3. The Distance Formula

In this short section, we establish a formula for the distance to a nest algebra. A precise distance formula like this is rare. However, there is a much larger class of operator algebras with such a formula which is valid up to a constant.

13.3.1. PARROTT LEMMA. *Let $H = H_1 \oplus H_2$ and $K = K_1 \oplus K_2$ be Hilbert spaces. Suppose that $A \in \mathcal{B}(H_1, K_1)$, $B \in \mathcal{B}(H_2, K_1)$, $C \in \mathcal{B}(H_1, K_2)$. Then*

$$\inf_{X \in \mathcal{B}(H_2, K_2)} \left\| \begin{bmatrix} A & B \\ C & X \end{bmatrix} \right\| = \max \left\{ \| [A \ \ B] \|, \left\| \begin{bmatrix} A \\ C \end{bmatrix} \right\| \right\}$$

and the infimum is attained.

PROOF. Scale so that the RHS is 1. Then

$$I \geq [A \ \ B] \, [A \ \ B]^* = AA^* + BB^* \quad \text{and} \quad I \geq \begin{bmatrix} A \\ C \end{bmatrix}^* \begin{bmatrix} A \\ C \end{bmatrix} = A^*A + C^*C.$$

Thus $C^*C \leq I - A^*A = D_A^2$ where $D_A = (I - A^*A)^{1/2}$. Therefore

$$\|C\xi\|^2 = \langle C^*C\xi, \xi\rangle \leq \langle (I - A^*A)\xi, \xi\rangle = \|D_A\xi\|^2.$$

We define $L \in \mathcal{B}(H_1, K_2)$ by $LD_A\xi = C\xi$ on $\operatorname{Ran} D_A$. Since $\|L\| \leq 1$, this extends to $\overline{\operatorname{Ran} D_A}$ by continuity. Define L to be 0 on $(\operatorname{Ran} D_A)^\perp = \ker(I - A^*A)$ and extend it by linearity to an element of $\mathcal{B}(H_1, K_2)$. Then $L \in \mathcal{B}(H_1, K_2)$ is a contraction such that $C = LD_A$. This factorization trick is known as the *Douglas lemma*. Similarly, $BB^* \leq I - AA^*$, so there is a contraction $M \in \mathcal{B}(K_1, K_2)$ such that $B = D_{A^*}M$.

Now let $X_0 = -LA^*M$. Then

$$\begin{bmatrix} A & B \\ C & X_0 \end{bmatrix} = \begin{bmatrix} A & D_{A^*}M \\ LD_A & -LA^*M \end{bmatrix} = \begin{bmatrix} I & 0 \\ 0 & L \end{bmatrix} \begin{bmatrix} A & D_{A^*} \\ D_A & -A^* \end{bmatrix} \begin{bmatrix} I & 0 \\ 0 & M \end{bmatrix}.$$

The first and third terms are contractions, and the middle term U is unitary because

$$UU^* = \begin{bmatrix} A & D_{A^*} \\ D_A & -A^* \end{bmatrix} \begin{bmatrix} A & D_{A^*} \\ D_A & -A^* \end{bmatrix}^* = \begin{bmatrix} I & AD_A - D_{A^*}A \\ D_AA^* - A^*D_{A^*} & I \end{bmatrix}.$$

Now there is a sequence of polynomials $p_n(x)$ converging to $\sqrt{x}$ uniformly on $[0, 1]$. Thus $D_A = \lim_{n\to\infty} p_n(I - A^*A)$ and $D_{A^*} = \lim_{n\to\infty} p_n(I - AA^*)$. Since $A(I - A^*A)^k = (I - AA^*)^k A$ for $k \geq 0$, we deduce that $AD_A = D_{A^*}A$. Thus $UU^* = I$, and similarly $U^*U = I$. Therefore, the product is a contraction, so the infimum is at most 1. However, the RHS is clearly a lower bound, so the minimum is 1 and is attained with X_0. ■

13.3.2. COROLLARY. *Let $\mathcal{N}$ be a finite nest on H. If $A \in \mathcal{B}(H)$, then*

$$\operatorname{dist}(A, \mathcal{T}(\mathcal{N})) = \max_{N \in \mathcal{N}} \|P_N^\perp A P_N\|.$$

PROOF. Let $\mathcal{N}$ consist of $0 = N_0 \subset N_1 \subset \cdots \subset N_n = H$. Write $H = \sum_{1 \le i \le n}^{\oplus} M_i$ where $M_i = N_i \ominus N_{i-1}$; and let $P_j = P_{M_j}$. If $A \in \mathcal{B}(H)$, we let $A_{ij} = P_i A P_j$ considered as an element of $\mathcal{B}(H_j, H_i)$. Then A is represented by the $n \times n$ matrix $\left[A_{ij}\right]$. In computing $\operatorname{dist}(A, \mathcal{T}(\mathcal{N})) = \inf_{T \in \mathcal{T}(\mathcal{N})} \|A - T\|$, the lower triangular entries A_{ij} for $i > j$ cannot be changed. However any choice of $T_{ij} \in \mathcal{B}(H_j, H_i)$ for $i \le j$ and $T_{ij} = 0$ for $i > j$ yields an operator $T = \left[T_{ij}\right]$ in $\mathcal{T}(\mathcal{N})$. Thus, the distance is given by the infimum of the norm of an $n \times n$ matrix with entries A_{ij} for $i > j$ and arbitrary entries X_{ij} for $i < j$.

The quantities $\|P_{N_i}^\perp A P_{N_i}\|$ are the rectangular corners in the lower left of the matrix with entries consisting of A_{ij} that cannot be modified. So the maximum of these norms is a lower bound for the distance. There is no harm in setting $X_{1j} = 0 = X_{in}$. The minimum norm is achieved by a repeated application of Parrott's lemma. First look at the first two columns. All of the entries are specified except for X_{22}. It can be completed using Parrot's lemma with norm no greater

$$\left[\begin{array}{c|c} 0 & 0 \\ A_{21} & X_{22} \\ \hline A_{31} & A_{32} \\ \vdots & \vdots \\ A_{n1} & A_{n2} \end{array}\right], \quad \left[\begin{array}{cc|c} 0 & 0 & 0 \\ A_{21} & X_{22} & X_{23} \\ A_{31} & A_{32} & X_{33} \\ \hline A_{41} & A_{42} & A_{43} \\ \vdots & \vdots & \vdots \\ A_{n1} & A_{n2} & A_{n3} \end{array}\right] \quad \cdots \quad \left[\begin{array}{ccc|c} 0 & \cdots & 0 & 0 \\ A_{21} & \cdots & X_{2,n-2} & X_{2,n-1} \\ \vdots & \cdots & \vdots & \vdots \\ A_{n-1,1} & \cdots & A_{n-1,n-2} & X_{n-1,n-1} \\ \hline A_{n1} & \cdots & A_{n,n-2} & A_{n,n-1} \end{array}\right]$$

than $\max\{\|P_{N_1}^\perp A P_{N_1}\|, \|P_{N_2}^\perp A P_{N_2}\|\}$. Then in the next matrix, since X_{22} is now defined, we can use Parrott's lemma again to define X_{23} and X_{33} to complete the matrix with norm no greater than $\max\{\|P_{N_1}^\perp A P_{N_1}\|, \|P_{N_2}^\perp A P_{N_2}\|, \|P_{N_3}^\perp A P_{N_3}\|\}$. Eventually one reaches the last matrix where the entries $X_{2,n-1}, \dots, X_{n-1,n-1}$ are filled in to attain at matrix of norm $\max_{N \in \mathcal{N}} \|P_N^\perp A P_N\|$. ∎

This distance formula extends to all nest algebras by a routine limit procedure and WOT-compactness.

13.3.3. ARVESON DISTANCE FORMULA. *Let $\mathcal{N}$ be a nest on a Hilbert space H. For $A \in \mathcal{B}(H)$,*

$$\operatorname{dist}(A, \mathcal{T}(\mathcal{N})) = \sup_{N \in \mathcal{N}} \|P_N^\perp A P_N\|.$$

PROOF. The supremum is a lower bound for the distance since for $T \in \mathcal{T}(\mathcal{N})$,

$$\|A - T\| \ge \|P_N^\perp (A - T) P_N\| = \|P_N^\perp A P_N\|.$$

Let $\{\mathcal{N}_\lambda\}_\Lambda$ be the collection of finite subnests of $\mathcal{N}$ ordered by inclusion. By the Corollary 13.3.2, for each λ, there is $T_\lambda \in \mathcal{T}(\mathcal{N}_\lambda)$ such that

$$\|A - T_\lambda\| = \operatorname{dist}(A, \mathcal{T}(\mathcal{N}_\lambda)) = \max_{N \in \mathcal{N}_\lambda} \|P_N^\perp A P_N\|.$$

Now $\|T_\lambda\| \le 2\|A\|$, so that $\{T_\lambda\}_\Lambda$ is a bounded net in $\mathcal{B}(H)$. Since the closed ball in $\mathcal{B}(H)$ is compact in the WOT topology, there is a cofinal subnet $\{T_\gamma\}_\Gamma$ which converges WOT to an operator T. Then

$$\|A - T\| \le \limsup \|A - T_\gamma\| \le \sup_{N \in \mathcal{N}} \|P_N^\perp A P_N\|.$$

For each $N \in \mathcal{N}$, once $\mathcal{N}_\gamma$ contains N, we have $P_N^\perp T_\gamma P_N = 0$. Thus $P_N^\perp T P_N = 0$. Therefore $T \in \mathcal{T}(\mathcal{N})$ and achieves the desired distance. ■

13.3.4. COROLLARY. *Let $K \in \mathcal{K}(H)$. Then*

$$\operatorname{dist}(K, \mathcal{T}(\mathcal{N}) \cap \mathcal{K}(H)) = \operatorname{dist}(K, \mathcal{T}(\mathcal{N})).$$

PROOF. Let $T \in \mathcal{T}(\mathcal{N})$ with $\|K - T\| = \operatorname{dist}(K, \mathcal{T}(\mathcal{N}))$. Let $\{F_\lambda\}_\Lambda$ be a contractive approximate identity for $\mathcal{T}(\mathcal{N}) \cap \mathcal{K}(H)$. Then

$$\|K - F_\lambda T\| \le \|K - F_\lambda K\| + \|F_\lambda(K - T)\| \le \|K - F_\lambda K\| + \operatorname{dist}(K, \mathcal{T}(\mathcal{N})).$$

Since $F_\lambda T \in \mathcal{T}(\mathcal{N}) \cap \mathcal{K}(H)$ and $\|K - F_\lambda K\| \to 0$, the result follows. ■

13.3.5. EXAMPLE. In general, the distance to $\mathcal{T}(\mathcal{N}) \cap \mathcal{K}(H)$ is not attained. Let $\mathcal{N}$ be the Volterra nest in $\mathcal{B}(L^2(0,1)$, and let $P_t = P_{L^2(0,t)}$. Let $K = 11^*$ be the rank 1 projection onto $\mathbb{C}1$. It is easy to compute

$$\|P_t^\perp 11^* P_t\| = \|\chi_{(t,1)}\chi_{(0,t)}^*\| = \sqrt{(1-t)t}.$$

The supremum is attained at $t = \frac{1}{2}$, namely, $\frac{1}{2}$. Let $T \in \mathcal{T}(\mathcal{N})$ with $\|K - T\| = \frac{1}{2}$. Let $\xi = \sqrt{2}\chi_{(0,1/2)}$, so $\|\xi\| = 1$. Then $K\xi = \frac{1}{\sqrt{2}}$ and

$$P_{1/2}^\perp K P_{1/2}\xi = \frac{1}{\sqrt{2}}\chi_{(1/2,1)} =: \eta,$$

where $\|\eta\| = \frac{1}{2}$. Thus $\|(K - T)\xi\|^2 = \|P_{1/2}(K - T)\xi\|^2 + \|\eta\|^2 \ge \frac{1}{4}$. Therefore

$$T\xi = P_{1/2}K\xi = \frac{1}{\sqrt{2}}\chi_{(0,1/2)} = \frac{1}{2}\xi.$$

Then for $t < \frac{1}{2}$,

$$\frac{1}{2}P_t^\perp\xi = P_t^\perp T\xi = P_t^\perp T P_t^\perp \xi.$$

Let $\zeta_t = (\frac{1}{2} - t)^{-1/2}P_t^\perp\xi = (\frac{1}{2} - t)^{-1/2}\chi_{(t,1/2)}$. Then $\|\zeta_t\| = 1$ and $P_t^\perp T\zeta_t = \frac{1}{2}\zeta_t$. Therefore $\|T\zeta_t\| \ge \frac{1}{2}$. However, ζ_t converges weakly to 0 as $t \to \frac{1}{2}$, so that $C\zeta_t \to 0$ in norm for any compact operator C. Thus the closest operators cannot be compact.

13.4. Isomorphisms of Nest Algebras

The goal of this section is a result of Ringrose that isomorphisms between nest algebras are spatial. Every nest algebra contains a masa containing $\mathcal{N}$. This is one major tool. The other is an algebraic characterization of rank 1 elements.

First, we note that a similarity carries one nest algebra onto another.

13.4.1. Lemma. *Let $\mathcal{N}$ be a nest in H, and let $A \in \mathcal{B}(H)$ be invertible. Then* $\operatorname{Ad} A(\mathcal{T}(\mathcal{N})) = A\mathcal{T}(\mathcal{N})A^{-1} = \mathcal{T}(A\mathcal{N})$.

Proof. If $T \in \mathcal{T}(\mathcal{N})$ and $N \in \mathcal{N}$, then

$$(ATA^{-1})AN = ATN \subset AN.$$

Thus $ATA^{-1} \in \mathcal{T}(A\mathcal{N})$. Conversely, if $S \in \mathcal{T}(A\mathcal{N})$, then $A^{-1}SA \in \mathcal{T}(\mathcal{N})$ and $\operatorname{Ad} A(A^{-1}SA) = S$. Thus $\operatorname{Ad} A(\mathcal{T}(\mathcal{N})) = \mathcal{T}(A\mathcal{N})$. ∎

Now we focus on maps between two masas.

13.4.2. Theorem. *Let $\mathcal{A}$ and $\mathcal{B}$ be two masas in $\mathcal{B}(H)$ acting on separable Hilbert spaces. Let $\alpha : \mathcal{A} \to \mathcal{B}$ be an algebra isomorphism. Then there is a unitary operator U such that $\alpha = \operatorname{Ad} U$.*

Proof. Since elements of $\mathcal{A}$ and $\mathcal{B}$ are normal and α preserves spectrum,

$$\|a\| = \operatorname{spr}(a) = \operatorname{spr}(\alpha(a)) = \|\alpha(a)\| \quad \text{for} \quad a \in \mathcal{A}.$$

So α is isometric. If $p = p^2$ is a projection in $\mathcal{A}$, then $\alpha(p)^2 = \alpha(p)$, whence $\alpha(p)$ is also a projection. By Theorem 9.3.6, the span of the projections in $\mathcal{A}$ is norm dense in $\mathcal{A}$. Observe that

$$\alpha\Big(\sum_{i=1}^{n} \overline{a_i} p_i\Big) = \sum_{i=1}^{n} \overline{a_i}\alpha(p_i) = \Big(\sum_{i=1}^{n} a_i \alpha(p_i)\Big)^* = \alpha\Big(\sum_{i=1}^{n} a_i p_i\Big)^*.$$

By the density of these terms, we see that α is a $*$-isomorphism. By Proposition 9.3.9, α is a normal isomorphism.

By Corollary 10.4.2, there are Borel probability measures (X, μ) and (Y, ν) on separable metric spaces X and Y so that $\mathcal{A} \simeq L^\infty(\mu)$ acting on $L^2(\mu)$ and $\mathcal{B} \simeq L^\infty(\nu)$ acting on $L^2(\nu)$. We make these identifications and consider $\alpha : L^\infty(\mu) \to L^\infty(\nu)$. Let $\tau(f) = \displaystyle\int f\,d\mu$ be the normal state on $L^\infty(\mu)$ extending the map $\chi_A \to \mu(A)$. Define a normal state on $L^\infty(\nu)$ by $\varphi(g) = \tau\alpha^{-1}(g)$. This determines a probability measure $\widetilde{\mu}$ on Y by

$$\widetilde{\mu}(B) = \varphi(\chi_B) = \tau\alpha^{-1}(\chi_B) = \mu(A) \quad \text{where} \quad \alpha(\chi_A) = \chi_B.$$

Indeed, normality implies that $\widetilde{\mu}$ is countably additive. Then $\varphi(g) = \displaystyle\int g\,d\widetilde{\mu}$.

Now $\nu(B) = 0$ implies that $\alpha^{-1}(\chi_B) = 0$, so that $\widetilde{\mu}(B) = 0$. Hence $\widetilde{\mu} \ll \nu$. Conversely, $\nu(B) \neq 0$ implies that $\alpha^{-1}(\chi_B) \neq 0$, so that $\widetilde{\mu}(B) > 0$. Thus $\nu \ll \widetilde{\mu}$. By the Radon–Nikodym theorem, there is a strictly positive function $h \in L^1(\nu)$ so that $\widetilde{\mu} = h\nu$. Define $U \in \mathcal{B}(L^2(\mu), L^2(\nu))$ by first setting

$$Uf = h^{1/2}\alpha(f) \quad \text{for} \quad f \in L^\infty(\mu).$$

Then

$$\begin{aligned}\|Uf\|^2_{L^2(\nu)} &= \|h^{1/2}\alpha(f)\|_2^2 = \int |\alpha(f)|^2 h \, d\nu \\ &= \int \alpha(|f|^2)\, d\widetilde{\mu} = \varphi(\alpha(|f|^2)) = \tau(|f|^2) = \|f\|^2_{L^2(\mu)}.\end{aligned}$$

Hence U is isometric and extends to all of $L^2(\mu)$ by continuity. However $\operatorname{Ran} U$ contains all functions in $L^2(\nu)$ of the form $h^{1/2}g$ where $g \in L^\infty(\nu)$. Since $h > 0$, this set is dense in $L^2(\nu)$. Therefore U is unitary. Finally, if $f, g \in L^\infty(\mu)$, then

$$(UM_gU^*)Uf = UM_gf = U(gf) = h^{1/2}\alpha(g)\alpha(f) = M_{\alpha(g)}Uf.$$

Since UM_gU^* and $M_{\alpha(g)}$ are bounded operators that agree on a dense set, we have $\alpha = \operatorname{Ad} U$. ■

We need an easy technical lemma.

13.4.3. LEMMA. *Let $\mathcal{D}$ be a masa in $\mathcal{B}(H)$. Suppose that $\xi_n \in H$ for $n \geq 1$ and $\sum_{n\geq 1} \|\xi_n\| < \infty$. Then there is a vector $\eta \in H$ and $D_n \in \mathcal{D}$ with $\|D_n\| \leq 1$ so that $D_n\eta = \xi_n$ for $n \geq 1$.*

PROOF. Let H_0 be the smallest $\mathcal{D}$-invariant subspace containing the vectors $\{\xi_n : n \geq 1\}$. This space is separable and is the range of a projection $P \in \mathcal{D}$. By Theorem 10.4.8, $\mathcal{D}P|_{H_0}$ is unitarily equivalent to $L^\infty(\mu)$ acting on $L^2(\mu)$ for some probability measure μ. Then ξ_n correspond to functions $f_n \in L^2(\mu)$. Let $g = \sum_{n\geq 1} |f_n|$. Then $\|g\|_2 \leq \sum_{n\geq 1} \|\xi_n\| < \infty$. Define $h_n = f_n/g$. Then $\|h_n\|_\infty \leq 1$ and $M_{h_n}g = f_n$. By the unitary equivalence, there are elements $D_n = D_nP \in \mathcal{D}$ with $\|D_n\| \leq 1$ and a vector η such that $D_n\eta = \xi_n$. ■

This allows us to show that all linear maps which commute with a masa are bounded.

13.4.4. LEMMA. *Let $\mathcal{D}$ be a masa in $\mathcal{B}(H)$. Suppose that T is a linear map defined on all of H such that $TD = DT$ for all $D \in \mathcal{D}$. Then $T \in \mathcal{D}$.*

PROOF. Since $\mathcal{D}$ is maximal abelian in $\mathcal{B}(H)$, it suffices to show that T is continuous. If T is not continuous, there are vectors $\xi_n \in H$ with $\|\xi_n\| \leq 2^{-n}$ and $\|T\xi_n\| > n$. By Lemma 13.4.3, there is a vector $\eta \in H$ and $D_n \in \mathcal{D}$ with

$\|D_n\| \le 1$ so that $D_n\eta = \xi_n$ for $n \ge 1$. Compute:

$$n < \|T\xi_n\| = \|TD_n\eta\| = \|D_nT\eta\| \le \|T\eta\|.$$

This is a contradiction for large n. Hence T is continuous, and so belongs to $\mathcal{D}$. ∎

The next step is to characterize the rank 1 elements of a nest algebra. The following lemma is very easy for $\mathcal{B}(H)$.

13.4.5. Lemma. *Let $\mathcal{N}$ be a nest, and let $T \in \mathcal{T}(\mathcal{N})$. Then T has rank 1 if and only if $ATB = 0$ for $A, B \in \mathcal{T}(\mathcal{N})$ implies that either $AT = 0$ or $TB = 0$.*

Proof. If $T = \xi\eta^*$, then $ATB = (A\xi)(B^*\eta)^*$. This is 0 only if $A\xi = 0$ or $B^*\eta = 0$, in which case either $AT = 0$ or $TB = 0$, respectively.

Suppose that T has rank at least 2. Let $N_0 = \bigvee\{N \in \mathcal{N} : \operatorname{rank} TP_N \le 1\}$. Then there are $N_k \in \mathcal{N}$ such that $\operatorname{rank} TP_{N_k} \le 1$ and $P_{N_0} = \text{SOT–}\lim P_{N_k}$. Thus $TP_{N_0} = \text{SOT–}\lim TP_{N_k}$. Since the rank 1 operators are SOT-closed, TP_{N_0} has rank at most 1. Pick $N_1 > N_0$ with $N_1 < H$ unless $H = N_{0+}$, in which case $N_1 = H$ and $H_- = N_0$. Then TP_{N_1} has rank at least 2, and $N_{1-} \ne H$ in all cases. Similarly, there is an $N_2 \in \mathcal{N}$ such that $\operatorname{rank}(P_{N_2}^\perp TP_{N_1}) \ge 2$ and $N_{2+} \ne 0$.

Pick $\xi \in N_1$ with $P_{N_2}^\perp TP_{N_1}\xi = \eta \ne 0$. Pick a unit vector ζ in $\operatorname{Ran} P_{N_2}^\perp TP_{N_1}$ orthogonal to η. Let $v_1 \in N_{1-}^\perp$ and $v_2 \in N_{2+}$ be unit vectors. Then by Lemma 13.2.1, $A = v_2\zeta^*$ and $B = \xi v_1^*$ belong to $\mathcal{T}(\mathcal{N})$. Then

$$TB = (T\xi)v_1^* = \eta v_1^* \ne 0 \quad \text{and} \quad AT = v_2\zeta^*T = v_2(T^*P_{N_2}^\perp\zeta)^* \ne 0.$$

The latter holds because $\ker(T^*P_{N_2}^\perp) = (\operatorname{Ran} P_{N_2}^\perp T)^\perp$ and $\zeta \in \operatorname{Ran} P_{N_2}^\perp T$. However $ATB = v_2\zeta^*\eta v_1^* = 0$ because $\langle \eta, \zeta\rangle = 0$. ∎

The next lemma reduces the situation to one which fixes a masa. Note that if $\mathcal{N}$ is a nest, then $\mathcal{N}''$ is an abelian von Neumann algebra. Zorn's lemma shows that there is a masa $\mathcal{D} \supset \mathcal{N}''$. Then $\mathcal{D} = \mathcal{D}' \subset \mathcal{N}' = \mathcal{T}(\mathcal{N}) \cap \mathcal{T}(\mathcal{N})^*$.

13.4.6. Lemma. *Let $\mathcal{N}, \mathcal{M}$ be nests in H_1 and H_2, respectively, and let $\mathcal{D}$ be a masa in $\mathcal{T}(\mathcal{N})$. Suppose that $\alpha : \mathcal{T}(\mathcal{N}) \to \mathcal{T}(\mathcal{M})$ is an algebra isomorphism, not assumed to be continuous. Then there is an invertible operator $S \in \mathcal{B}(H_1, H_2)$ and $\beta \in \operatorname{Aut}(\mathcal{T}(\mathcal{N}))$ such that $\beta|_{\mathcal{D}} = \mathrm{id}_{\mathcal{D}}$ and $\alpha = \operatorname{Ad} S \circ \beta$, i.e. $\alpha(T) = S\beta(T)S^{-1}$.*

Proof. Observe that $\alpha(\mathcal{D})$ is a maximal abelian subalgebra of $\mathcal{T}(\mathcal{M})$ (not necessarily self-adjoint). Maximality implies that it is WOT-closed, and, in particular, it is norm closed. If $D \in \mathcal{D}$, then

$$\|D\| = \operatorname{spr}(D) = \operatorname{spr}(\alpha(D)) \le \|\alpha(D)\|.$$

Therefore $\alpha^{-1}|_{\alpha(\mathcal{D})}$ has norm at most 1. By Banach's Isomorphism theorem, $\alpha|_{\mathcal{D}}$ is norm continuous.

Let $\mathcal{U}$ be the unitary group of $\mathcal{D}$. Then $\alpha(\mathcal{U})$ is a bounded abelian group in $\mathcal{B}(H_2)$. Abelian groups are amenable by Corollary 9.13.6. Therefore by Dixmier's Theorem 9.13.8, there is an invertible operator $B \in \mathcal{B}(H_2)$ so that $\mathcal{V} = \operatorname{Ad} B(\alpha(\mathcal{U}))$ is a group of unitary operators. It follows that $\mathcal{B} = \mathcal{V}''$ is maximal abelian and self-adjoint in $\operatorname{Ad} B(\mathcal{T}(\mathcal{M})) = \mathcal{T}(B\mathcal{M})$ by Lemma 13.4.1. Therefore

$$(B\mathcal{M})'' = \{P_{BM} : M \in \mathcal{M}\}'' \subset \mathcal{B} \subset (B\mathcal{M})' = \mathcal{T}(B\mathcal{M}) \cap \mathcal{T}(B\mathcal{M})^*.$$

But any masa in $\mathcal{B}(H_2)$ containing $\mathcal{B}$ will be contained in $(B\mathcal{M})'$, and so $\mathcal{B}$ is maximal as a subalgebra of $\mathcal{B}(H_2)$, i.e. $\mathcal{B}$ is a masa.

The map $\gamma = \operatorname{Ad} B \circ \alpha$ is an algebra isomorphism of $\mathcal{T}(\mathcal{N})$ onto $\mathcal{T}(B\mathcal{M})$ which takes $\mathcal{D}$ onto $\mathcal{B}$. Applying Theorem 13.4.2 to $\gamma^{-1}|_{\mathcal{B}}$ yields a unitary operator $U \in \mathcal{B}(H_2, H_1)$ so that $\beta = \operatorname{Ad}(UB)\alpha$ satisfies $\beta|_{\mathcal{D}} = id_{\mathcal{D}}$. Now

$$\beta(\mathcal{T}(\mathcal{N})) = \operatorname{Ad}(UB)\mathcal{T}(\mathcal{M}) = \mathcal{T}(UB\mathcal{M})$$

is a nest algebra containing $\mathcal{D}$. So if $L \in UB\mathcal{M}$, then $P_L \in (UB\mathcal{M})'' \subset \mathcal{D}$. It follows that $\beta(P_N) = P_N$ for $N \in \mathcal{N}$ and $\beta^{-1}(P_L) = P_L$ for $L \in UB\mathcal{M}$. If $T \in \mathcal{T}(\mathcal{N})$, then

$$0 = \beta(P_N^\perp T P_N) = P_N^\perp \beta(T) P_N.$$

Hence $\mathcal{T}(UB\mathcal{M}) = \beta(\mathcal{T}(\mathcal{N})) \subset \mathcal{T}(\mathcal{N})$ and so $\mathcal{N} \subset \operatorname{Lat} \mathcal{T}(UB\mathcal{M}) = UB\mathcal{M}$ by Theorem 13.1.6. Conversely, if $L \in UB\mathcal{M}$ and $T \in \mathcal{T}(\mathcal{N})$, then

$$P_L^\perp T P_L = \beta^{-1}(P_L^\perp \beta(T) P_L) = 0.$$

Hence $L \in \mathcal{M}$ by Theorem 13.1.6. Thus $UB\mathcal{M} = \mathcal{N}$, so that $\beta(\mathcal{T}(\mathcal{N})) = \mathcal{T}(\mathcal{N})$.

Finally, setting $S = B^{-1}U^*$, we have $S\mathcal{N} = SUB\mathcal{M} = \mathcal{M}$. Thus $\operatorname{Ad} S$ is an isomorphism of $\mathcal{T}(\mathcal{N})$ onto $\mathcal{T}(\mathcal{M})$. Clearly $\operatorname{Ad} S\, \beta = \operatorname{Ad}(SUB)\alpha = \alpha$. ■

We are now ready to prove the isomorphism theorem.

13.4.7. Ringrose Isomorphism Theorem. *Let $\mathcal{N}, \mathcal{M}$ be nests in H_1 and H_2, respectively, and let $\alpha : \mathcal{T}(\mathcal{N}) \to \mathcal{T}(\mathcal{M})$ be an algebra isomorphism. Then there is an invertible operator $S \in \mathcal{B}(H_1, H_2)$ such that $\alpha = \operatorname{Ad} S$.*

Proof. Let $\mathcal{D}$ be a masa in $\mathcal{T}(\mathcal{N})$. The previous lemma reduces the problem to establishing the result for $\beta \in \operatorname{Aut}(\mathcal{T}(\mathcal{N}))$ such that $\beta|_{\mathcal{D}} = id_{\mathcal{D}}$. Lemma 13.4.5 shows that β carries the set of rank 1 elements of $\mathcal{T}(\mathcal{N})$ onto itself. The proof amounts to exploiting these two facts.

Let $0 \neq N_0 \in \mathcal{N}$ such that $N_{0-} \neq H_1$. If $\xi_0 \in N_0$ and $\eta_0 \in N_{0-}^\perp$ are non-zero, then $\xi_0\eta_0^* \in \mathcal{T}(\mathcal{N})$ and has rank 1. Thus

$$\beta(\xi_0\eta_0^*) = \beta(P_{N_0}\xi_0\eta_0^* P_{N_{0-}}^\perp) = P_{N_0}\, \beta(\xi_0\eta_0^*)\, P_{N_{0-}}^\perp$$

has rank 1. Therefore $\beta(\xi_0\eta_0^*) = u_0v_0^*$ for $u_0 \in N_0$ and $v_0 \in N_{0-}^\perp$. The vectors u_0 and v_0 are unique up to a scalar.

Now suppose that $\xi_1, \xi_2 \in N_0$. By Lemma 13.4.3, there are $D_0, D_1, D_2 \in \mathcal{D}$ and a vector $\xi \in N_0$ so that $D_i\xi = \xi_i$ for $i = 0, 1, 2$. Suppose that $\beta(\xi\eta_0^*) = uv^*$. Then

$$\beta\Big(\sum_{i=0}^{2} \lambda_i \xi_i \eta_0^*\Big) = \beta\Big(\sum_{i=0}^{2} \lambda_i D_i \xi \eta_0^*\Big) = \sum_{i=0}^{2} \lambda_i D_i u v^*.$$

For $\lambda_0 = 1$ and $\lambda_1 = \lambda_2 = 0$, we have $u_0 v_0^* = (D_0 u)v^*$. So v and v_0 are collinear. After rescaling u and v, we may suppose that $v = v_0$. Then $D_0 u = u_0$ and $\beta(\xi_i \eta_0) = D_i u v_0^*$ for $i = 1, 2$. If ξ' is a different vector in N_0 and $D_i' \in \mathcal{D}$ satisfied $D_i'\xi' = \xi_i$, then $\beta(\xi'\eta_0) = u'v_0^*$ and $u_i = D_i'u'$. That is, we obtain the same vector u_i independent of the choice of ξ; even $\xi = \xi_i$ will work.

Define a map $S : N_0 \to N_0$ by $\beta(\xi\eta_0^*) = (S\xi)v_0^*$. This is linear because by the previous paragraph,

$$S(\lambda_1\xi_1 + \lambda_2\xi_2) = \lambda_1 u_1 + \lambda_2 u_2 = \lambda_1 S\xi_1 + \lambda_2 S\xi_2.$$

If $D \in \mathcal{D}P_{N_0}$, then

$$(SD\xi)v_0^* = \beta(D\xi\eta^*) = D\beta(\xi\eta^*) = (DS\xi)v_0^*.$$

Thus $SD = DS$. Hence S commutes with $\mathcal{D}|_{N_0}$, which is a masa. By Lemma 13.4.4, S is continuous and belongs to $\mathcal{D}|_{N_0}$.

Similarly, there is an operator $T \in \mathcal{D}|_{N_{0-}^\perp}$ so that $\beta(\xi_0\eta^*) = u_0(T\eta)^*$.

Now if $\eta_1 \in N_{0-}^\perp$, then there is a vector $\eta \in N_{0-}^\perp$ and $E_0, E_1 \in \mathcal{D}P_{N_{0-}}^\perp$ such that $E_i\eta = \eta_i$ for $i = 0, 1$. Let $\beta(\xi\eta^*) = uv^*$. Then setting $v_1 = T\xi_1$,

$$\begin{aligned}
\beta\big((\lambda_0\xi_0 + \lambda_1\xi_1)(\mu_0\eta_0 + \mu_1\eta_1)^*\big) &= \beta\big((\lambda_0 D_0 + \lambda_1 D_1)\xi\eta^*(\mu_0 E_0 + \mu_1 E_1)^*\big) \\
&= (\lambda_0 D_0 + \lambda_1 D_1)\,\beta(\xi\eta^*)\,(\mu_0 E_0 + \mu_1 E_1)^* \\
&= (\lambda_0 D_0 u + \lambda_1 D_1 u)(\mu_0 E_0 v + \mu_1 E_1 v)^* \\
&= \lambda_0\overline{\mu_0}\,u_0 v_0^* + \lambda_1\overline{\mu_0}\,(D_1 u)v_0^* + \lambda_0\overline{\mu_1}\,u_0(E_1 v)^* + \lambda_1\overline{\mu_1}\,(D_1 u)(E_1 v)^*.
\end{aligned}$$

In particular, looking at $\beta(\xi_1\eta_0^*)$ and $\beta(\xi_0\eta_1^*)$, we have $D_1 u = S\xi_1$ and $E_1 v = T\eta_1$. Thus for $\xi_1\eta_1^* \in P_{N_0}\mathcal{B}(H_1)P_{N_{0-}}^\perp$,

$$\beta(\xi_1\eta_1^*) = (S\xi_1)(T\eta_1)^* = S(\xi_1\eta_1^*)T^*.$$

Once we fixed the choice of u_0 and v_0, the definitions of S and T were determined. The only freedom was to replace u_0 and v_0 by γu_0 and $\gamma^{-1}v_0$, which has the effect of changing S, T to γS and $\overline{\gamma}^{-1}T$, i.e. they are unique up to a scalar. Because β is one to one, S and T are injective, and because β must map the rank 1 elements of $P_{N_0}\mathcal{B}(H_1)P_{N_{0-}}^\perp$ onto itself, S and T must also be surjective. As they are continuous, they also have continuous inverses.

If $0 \neq N \in \mathcal{N}$ and $N_- \neq H_1$, we can similarly obtain operators $S_N \in \mathcal{D}|_N$ and $T_N \in \mathcal{D}|_{N_-^\perp}$ so that $\beta(\xi_1\eta_1^*) = S_N(\xi_1\eta_1^*)T_N^*$ for rank 1 elements of $P_N\mathcal{B}(H_1)P_{N_-}^\perp$. By considering $\xi \in N_0 \wedge N$ and $\eta \in (N_{0-} \vee N_-)^\perp$, we see that $S\xi\eta T^* = \beta(\xi\eta^*) = S_N\xi\eta^* T_N^*$. The only flexibility is to multiply S_N and T_N by scalars. Thus choosing this scaling appropriately, we get $S_N P_{N_0\wedge N} = SP_{N_0\wedge N}$

and $T_N P^{\perp}_{N_{0-} \vee N_-} = T P^{\perp}_{N_{0-} \vee N_-}$. Similarly S_{N_1} and S_{N_2} will now agree on $N_1 \wedge N_2$ and T_{N_1} and T_{N_2} will agree on $(N_{1-} \vee N_{2-})^{\perp}$. Therefore, there is a common operator S defined on $\bigcup_{N_- \neq H_1} N$ and operator T defined on $\bigcup_{N \neq 0} N_-^{\perp}$ so that $\beta(\xi\eta^*) = S(\xi\eta^*)T^*$ for all rank 1 operators in $\mathcal{T}(\mathcal{N})$. Moreover, each S_N and T_N is invertible.

Next we use multiplicativity: $(\xi_1\eta_1^*)(\xi_2\eta_2^*) = \langle \xi_2, \eta_1 \rangle \xi_1\eta_2^*$. Hence applying β,

$$\langle \xi_2, \eta_1 \rangle S\xi_1\eta_2^*T^* = (S\xi_1\eta_1^*T^*)(S\xi_2\eta_2^*T^*) = \langle T^*S\xi_2, \eta_1 \rangle S\xi_1\eta_2^*T^*.$$

Therefore, if we take arbitrary $\eta_1, \xi_2 \in N_{1-}^{\perp} \cap N_2$ for small $N_1 \neq 0$ and large N_2 with $N_{2-} \neq H_1$, we find that $T^*S = I$ on $N_{1-}^{\perp} \cap N_2$. Hold N_2 fixed. Then $T^*S_{N_2} = I$ on $\bigcup_{0 \neq N_1,\, N_{1-} \neq H_1} N_{1-}^{\perp} \cap N_2$. This is dense in N_2 if $0 = 0_+$ and if $0 = (0_+)_-$. It follows that $T^*|_{N_{1-}^{\perp} \cap N_2}$ is bounded below by $\|S_{N_2}\|^{-1}$; and since S_{N_2} is invertible, $\|T^*|_{N_{1-}^{\perp} \cap N_2}\| \leq \|S_{N_2}^{-1}\|^{-1}$. These estimates hold uniformly on the union, and hence $S_{N_2}^{*-1}$ is the continuous extension of T to N_2. Since T is continuous and invertible on $N_{2-}^{\perp}$, we have extended the definition of T to an operator on H_1, and it belongs to $\mathcal{D}^{-1}$. Similarly, S extends by continuity to an invertible element of $\mathcal{D}$ and $S^{-1} = T^*$.

We now can conclude that $\beta(\xi\eta^*) = S\,\xi\eta^*\,S^{-1}$ for all rank one elements in $\mathcal{T}(\mathcal{N})$. Let $A \in \mathcal{T}(\mathcal{N})$. If $\xi\eta^* \in \mathcal{T}(\mathcal{N})$, then

$$\begin{aligned} \beta(A)S(\xi\eta^*)S^{-1} &= \beta(A)\beta(\xi\eta^*) = \beta(A(\xi\eta^*)) = \beta((A\xi)\eta^*) \\ &= S(A\xi)\eta^*S^{-1} = SAS^{-1}S(\xi\eta^*)S^{-1}. \end{aligned}$$

Therefore $\beta(A)S\xi = (SAS^{-1})S\xi$. This holds for any $\xi \in N$ provided $N_- \neq H_1$, which is a dense set of vectors. Since $SN_2 = N_2$, the set of vectors $S\xi$ is also dense in H_1. Now both $\beta(A)$ and SAS^{-1} are continuous, and thus they are equal. That is, $\beta = \operatorname{Ad} S$. ■

13.5. The Similarity Theorem

Ringrose raised the natural question of when two nests are similar. In particular, he wondered whether similar nests are unitarily equivalent. The first question has a very nice answer when H is separable, and it shows that the second question has a negative answer. In the remainder of this chapter, H will always be a separable Hilbert space.

We need to establish a number of definitions in order to state the Similarity theorem.

13.5.1. Definition. An order preserving bijection of a nest $\mathcal{N}$ onto a nest $\mathcal{M}$ is an *order isomorphism*. An order isomorphism $\theta : \mathcal{N} \to \mathcal{M}$ *preserves dimension* if $\dim\big(\theta(N_2) \ominus \theta(N_1)\big) = \dim(N_2 \ominus N_1)$ for all $N_1 < N_2 \in \mathcal{N}$.

13.5.2. REMARK. In the separable case, which is our concern in this section, intervals are either finite dimensional or separably infinite dimensional. Any interval E of $\mathcal{N}$ such that $E\mathcal{N}$ is infinite has infinite dimension. So the finite case only arises when $E\mathcal{N}$ is finite. In this case, E is a finite sum of atoms. Thus, dimension preserving reduces to the simpler statement that θ preserves the dimension of atoms. That is, if $N^+ > N$, then $\dim\big(\theta(N_+) \ominus \theta(N)\big) = \dim(N_+ \ominus N)$.

13.5.3. DEFINITION. If S is an invertible operator such that $S\mathcal{N} = \mathcal{M}$, we write θ_S for the order isomorphism $\theta_S(N) = SN$. An order isomorphism $\theta : \mathcal{N} \to \mathcal{M}$ is *implemented by a similarity* if there is an invertible operator S such that $\theta = \theta_S$. We write $\theta \sim id_{\mathcal{N}}$. In this case, $\operatorname{Ad} S$ is an isomorphism of $\mathcal{T}(\mathcal{N})$ onto $\mathcal{T}(\mathcal{M})$.

13.5.4. DEFINITION. An order isomorphism $\theta : \mathcal{N} \to \mathcal{M}$ is *implemented by an approximate unitary equivalence* (write $\theta \sim_a id_{\mathcal{N}}$) if there is a sequence of unitaries U_n for $n \geq 1$ so that

$$\lim_{n\to\infty} \sup_{N\in\mathcal{N}} \|P_{\theta(N)} - U_n P_N U_n^*\| = 0.$$

This is an *approximate unitary equivalence of nests mod* $\mathcal{K}$ if, in addition,

$$f(N) = P_{\theta(N)} - U_n P_N U_n^*$$

is a norm-continuous function of $\mathcal{N}$ into $\mathcal{K}(H)$. We write $\theta \sim_{\mathcal{K}} id_{\mathcal{N}}$.

Note that $f(N) = P_{\theta(N)} - U_n P_N U_n^*$ is always SOT* continuous, since both $N \to P_N$ and $N \to P_{\theta(N)}$ are SOT*-continuous by Theorem 13.1.9.

13.5.5. DEFINITION. If X is a compact metric space, let $C_{s*}(X, \mathcal{B}(H))$ denote the C*-algebra of all SOT* continuous functions from X into $\mathcal{B}(H)$. Also let $C_n(X, \mathcal{K})$ denote the C*-subalgebra of norm-continuous functions into the compact operators.

13.5.6. PROPOSITION. *Let X be a compact metric space. Then $C_n(X, \mathcal{K})$ is an ideal in $C_{s*}(X, \mathcal{B}(H))$.*

PROOF. Since X is compact and $f \in C_{s*}(X, \mathcal{B}(H))$, then $\operatorname{Ran} f$ is SOT*-compact. In particular, it is bounded, say by M. Let $g \in C_n(X, \mathcal{K})$ and fix $x_0 \in X$ and $\varepsilon > 0$. Clearly $f(x)g(x)$ takes values in $\mathcal{K}(H)$. By the norm continuity, there is a $\delta_1 > 0$ so that $d(x, x_0) < \delta_1$ implies $\|g(x) - g(x_0)\| < \varepsilon$. Choose a finite rank operator F with $\|g(x_0) - F\| < \varepsilon$, and hence $\|g(x) - F\| < 2\varepsilon$ if $d(x, x_0) < \delta_1$. Clearly $f(x)F$ converges in norm to $f(x_0)F$ as $x \to x_0$, so that

$\|(f(x) - f(x_0))F\| < \varepsilon$ for $d(x, x_0) < \delta_2$. Thus

$$\begin{aligned}\|f(x)g(x)-f(x_0)g(x_0)\| &\le \|f(x)\|\|g(x)-F\|+\|(f(x)-f(x_0))F\| \\ &\quad + \|f(x_0)\|\|F-g(x_0)\| \\ &< M(2\varepsilon) + \varepsilon + M\varepsilon = (3M+1)\varepsilon\end{aligned}$$

for $d(x, x_0) < \min\{\delta_1, \delta_2\}$. Hence fg is norm continuous. Taking adjoints, we obtain that gf is also norm continuous. Hence $C_n(X, \mathcal{K})$ is an ideal. ∎

The constant function $f(x) = K \in \mathcal{K}(H)$ is norm continuous, so we obtain a useful consequence.

13.5.7. Corollary. *If $f \in C_{S*}(X, \mathcal{B}(H))$ and K is compact, then fK and Kf are in $C_n(X, \mathcal{K})$.*

If $\theta : \mathcal{N} \to \mathcal{M}$ is an order isomorphism, we will also consider θ as the function $\theta(N) = P_{\theta(N)}$. Clearly this belongs to $C_{s*}(\mathcal{N}, \mathcal{B}(H))$. If θ_1 and θ_2 are order isomorphisms onto (generally different) nests on the same Hilbert space H, we define

$$\|\theta_1 - \theta_2\| = \sup_{N\in\mathcal{N}} \|P_{\theta_1(N)} - P_{\theta_2(N)}\|.$$

With this convention, we see that $\theta \sim_a id_{\mathcal{N}}$ means that for all $\varepsilon > 0$, there is a unitary U so that $\|\theta - \theta_U\| < \varepsilon$. Likewise $\theta \sim_{\mathcal{K}} id_{\mathcal{N}}$ means that for all $\varepsilon > 0$, there is a unitary U so that $\theta - \theta_U \in C_n(\mathcal{N}, \mathcal{K})$ and $\|\theta - \theta_U\| < \varepsilon$.

13.5.8. Davidson Similarity Theorem. *Let $\theta : \mathcal{N} \to \mathcal{M}$ be an order isomorphism of nests in a separable Hilbert space. The following are equivalent:*

(1) *θ preserves dimension.*

(2) *$\theta = \theta_{S_n}$ for a sequence of invertible operators such that $S_n = U_n + K_n$ where U_n are unitary and $\|K_n\| \to 0$.*

(2′) *$\theta = \theta_{S_n}$ for a sequence of invertible operators such that $S_n = U_n + K_n$ where U_n are unitary, K_n are compact and $\|K_n\| \to 0$.*

(3) *θ is implemented by an approximate unitary equivalence.*

(3′) *θ is implemented by an approximate unitary equivalence mod $\mathcal{K}$.*

(4) *$\theta = \theta_S$ for an invertible operator S.*

In this case, $\mathcal{T}(\mathcal{M})$ is similar to $\mathcal{T}(\mathcal{N})$.

13.5.9. Overview. Clearly $(2') \Rightarrow (2) \Rightarrow (4) \Rightarrow (1)$ and $(3') \Rightarrow (3)$. Showing $(2') \Rightarrow (3')$ and $(2) \Rightarrow (3)$ are straightforward. Also $(3') \Rightarrow (2')$ and $(3) \Rightarrow (2)$ aren't too bad. They will be established in this section. Then we will deal with $(1) \Rightarrow (2')$.

13.5.10. LEMMA (2) $\Rightarrow$ (3). *If $S = U + K$ for U unitary and $\|K\| < \varepsilon < 1$, then $\|\theta_S - \theta_U\| < \frac{\varepsilon}{1-\varepsilon}$.*

PROOF. Nothing is changed if we consider U^*S and $U^*U = I$ instead. So we assume that $U = I$. Then $S = I + K$ and $S^{-1} = I + L$ for

$$L = (I + K)^{-1} - I = \sum_{n\geq 1}(-1)^n K^n.$$

So $\|L\| < \sum_{n\geq 1} \varepsilon^n = \frac{\varepsilon}{1-\varepsilon}$.

If $P = P_N$ and $Q = P_{SN}$, then

$$\|Q-P\| = \|QP^\perp - Q^\perp P\| = \max\{\|QP^\perp\|, \|Q^\perp P\|\} = \max\{\|P^\perp Q\|, \|Q^\perp P\|\}$$

because $QP^\perp$ and $Q^\perp P$ have orthogonal ranges and orthogonal domains. If $\xi = Q\xi$ is a unit vector in SN, then $\eta = (I + L)\xi \in S^{-1}SN = N$. Thus

$$\|P^\perp Q\xi\| = \|P^\perp((I + L)\xi - L\xi)\| = \|P^\perp L\xi\| \leq \|L\| < \frac{\varepsilon}{1-\varepsilon}.$$

Hence $\|QP^\perp\| = \|P^\perp Q\| \leq \|L\|$. Similarly if $\xi = P\xi$, then $\eta = (I+K)\xi$ belongs to SN. Thus

$$\|Q^\perp P\xi\| = \|Q^\perp(I + K)\xi - Q^\perp K\xi\| \leq \|K\| < \varepsilon.$$

Hence $\|Q^\perp P\| < \varepsilon$. Therefore $\|\theta_S - \theta_U\| \leq \max\{\varepsilon, \|L\|\} < \frac{\varepsilon}{1-\varepsilon}$. ■

Only a bit more work is needed if K is compact to get the difference in $C_n(\mathcal{N}, \mathcal{K})$.

13.5.11. LEMMA (2′) $\Rightarrow$ (3′). *If $S = U + K$ for U unitary, K compact and $\|K\| < \varepsilon < 1$ and $S\mathcal{N} = \mathcal{M}$, then $f(N) = P_{\theta_S} - UP_NU^* \in C_n(\mathcal{N}, \mathcal{K})$ and $\|f\| < \frac{\varepsilon}{1-\varepsilon}$.*

PROOF. Again we may suppose that $U = I$. Write $S^{-1} = I + L$ as before. By the previous lemma, we have $\|f\| < \frac{\varepsilon}{1-\varepsilon}$. For $N \in \mathcal{N}$, let $Q_N = SP_NS^{-1}$. This is idempotent with range SN. Hence with respect to $H = SN \oplus (SN)^\perp$, we have

$$Q_N = \begin{bmatrix} I & B_N \\ 0 & 0 \end{bmatrix} \quad \text{and} \quad Q_N - Q_N^* = \begin{bmatrix} 0 & B_N \\ -B_N^* & 0 \end{bmatrix}$$

where $B_N = QP_{SN}^\perp$. Thus $B_N = P_{SN}(Q_N - Q_N^*)$. Let $g(N) = Q_N - Q_N^*$ and $h(N) = P_N$. Compute

$$\begin{aligned} g(N) = Q_N - Q_N^* &= (I + K)P_N(I + L) - (I + L^*)P_N(I + K^*) \\ &= KP_N + P_NL + KP_NL - L^*P_N - P_NK^* - L^*P_NK^* \\ &= (K - L^*)h(N) + h(N)(L - K^*) + Kh(N)L - L^*h(N)K^*. \end{aligned}$$

By Corollary 13.5.7, $g \in C_n(\mathcal{N}, \mathcal{K})$. Hence $f(N) = P_{SN}g(N) \in C_n(\mathcal{N}, \mathcal{K})$ by Proposition 13.5.6. ■

13.5.12. Lemma. *Suppose $\theta : \mathcal{N} \to \mathcal{M}$ is an order isomorphism such that $\theta - id_{\mathcal{N}} \in C_n(\mathcal{N}, \mathcal{K})$ and $\|\theta - id_{\mathcal{N}}\| = \gamma < \frac{1}{2}$. Then given $0 < \varepsilon < \gamma$, there is a unitary $U \in I + \mathcal{K}(H)$ with $\|\theta - \theta_U\| < \varepsilon$ and $\theta - \theta_U \in C_n(\mathcal{N}, \mathcal{K})$.*

Proof. By Theorem 13.1.9, we can identify $\mathcal{N}$ with a compact subset of $[0, 1]$ via an order preserving homeomorphism φ. Extend θ to a map on the intervals of $\mathcal{N}$ by setting $\Theta(N_2 \ominus N_1) = P_{\theta(N_2)} - P_{\theta(N_1)}$ for $N_1 < N_2$ in $\mathcal{N}$. Then it is easy to see that $\|\Theta - id\| \le 2\|\theta - id_{\mathcal{N}}\| = 2\gamma$ and $\Theta(E) - E$ is compact.

Let $\varepsilon > 0$. By the uniform continuity of $f(N) = P_{\theta(N)} - P_N$, there is a $\delta > 0$ so that $|\varphi(N) - \varphi(M)| < \delta$ for $N, M \in \mathcal{N}$ implies that $\|f(N) - f(M)\| < \frac{\varepsilon}{5}$. Form a partition

$$0 = N_0 < N_1 < \cdots < N_k = H$$

of $\mathcal{N}$ so that either $\varphi(N_i) - \varphi(N_{i-1}) < \delta$ or $E_i = P_{N_i} - P_{N_{i-1}}$ is an atom. Note that if $N_{i-1} < N < N_i$, then E_i is not an atom. Thus with $E = P_N - P_{N_{i-1}}$,

$$\|\Theta(E) - E\| = \|f(N) - f(N_{i-1})\| < \frac{\varepsilon}{5}.$$

When E_i is an atom, we have $\|\Theta(E_i) - E_i\| \le 2\gamma < 1$.

Let $S = \sum_{i=1}^{k} \Theta(E_i)E_i$. This belongs to $I + \mathcal{K}(H)$. Compute

$$I \ge S^*S = \sum_{i=1}^{k} E_i\Theta(E_i)E_i = I + \sum_{i=1}^{k} E_i(\Theta(E_i) - E_i)E_i \ge (1 - 2\gamma)I.$$

Hence S^*S is invertible and $S^*S - I \in \mathcal{K}(H)$. Similarly SS^* is invertible, so S is invertible. The polar decomposition of S is $S = U|S|$ where S, $|S|$ and U lie in $I + \mathcal{K}(H)$. Since $|S|E_iH = E_iH$ and $SE_iH = \Theta(E_i)H$, we have $UE_iH = \Theta(E_i)H$. In particular, $UN_i = \theta(N_i)$ for $0 \le i \le k$.

Suppose $N_{i-1} < N < N_i$ in $\mathcal{N}$, and $E = P_N - P_{N_{i-1}}$. Then

$$\|P_{\theta(N)} - P_{UN}\| = \|P_{\theta(N_{i-1})} + \Theta(E) - P_{\theta(N_{i-1})} - UEU^*\| = \|\Theta(E) - UEU^*\|.$$

As E_i is not an atom, $\|\Theta(E_i) - E_i\| < \frac{\varepsilon}{5}$. So $E_i \ge S^*SE_i \ge (1 - \frac{\varepsilon}{5})E_i$, whence $E_i \le |S^*S|^{-1/2}E_i < \left(1 + \frac{\varepsilon}{5}\right)E_i$. Thus $\left\||S|^{-1}E_i - E_i\right\| < \frac{\varepsilon}{5}$. Also $\|SE_i - E_i\| = \|(\Theta(E_i) - E_i)E_i\| < \frac{\varepsilon}{5}$. Therefore

$$UE_i - E_i = S|S|^{-1}E_i - SE_i + SE_i - E_i = S(|S|^{-1}E_i - E_i) + (SE_i - E_i)$$

is compact and has norm at most $\frac{2\varepsilon}{5}$. It follows that

$$\begin{aligned}\|\Theta(E) - UEU^*\| &\le \|\Theta(E) - E\| + \|E - UE\| + \|U(E - EU^*)\| \\ &< \frac{\varepsilon}{5} + \frac{2\varepsilon}{5} + \frac{2\varepsilon}{5} = \varepsilon.\end{aligned}$$

Therefore $\|\theta - \theta_U\| < \varepsilon$. Since $U \in I + \mathcal{K}$, $\theta_U - id \in C_n(\mathcal{N}, \mathcal{K})$, and therefore $\theta - \theta_U \in C_n(\mathcal{N}, \mathcal{K})$. ∎

Next we use the distance formula to show that close, approximately unitarily equivalent nests are similar via an invertible close to I.

13.5.13. LEMMA. *Let $\theta : \mathcal{N} \to \mathcal{M}$ be an order isomorphism so that $\theta \sim_a \mathrm{id}_{\mathcal{N}}$ and $\|\theta - \mathrm{id}_{\mathcal{N}}\| = \gamma$. Given $\varepsilon > 0$, there is an invertible T with $\|T - I\| \leq 12\gamma$ such that $\|\theta - \theta_T\| < \varepsilon$. Moreover if $\theta \sim_{\mathcal{K}} \mathrm{id}_{\mathcal{N}}$ and $\gamma < \frac{1}{2}$, we may choose T so that $T - I \in \mathcal{K}(H)$.*

PROOF. We may suppose that $\varepsilon < \gamma$. Choose a U unitary so that $\|\theta - \theta_U\| < \varepsilon$. In the second case, we apply Lemma 13.5.12 to obtain $U \in I + \mathcal{K}(H)$ as well. If $\gamma \geq \frac{1}{6}$, then $\|U - I\| \leq 2 \leq 12\gamma$; so we may take $T = U$.

Otherwise, $\gamma < \frac{1}{6}$. Compute using the Arveson distance formula

$$\begin{aligned}\mathrm{dist}(U, \mathcal{T}(\mathcal{N})) &= \sup_{\mathcal{N}} \|P_N^\perp U P_N\| = \sup_{\mathcal{N}} \|P_N^\perp P_{UN}\| \leq \sup_{\mathcal{N}} \|P_{UN} - P_N\| \\ &= \|\theta_U - \mathrm{id}_{\mathcal{N}}\| \leq \|\theta_U - \theta\| + \|\theta - \mathrm{id}_{\mathcal{N}}\| < \varepsilon + \gamma < 2\gamma.\end{aligned}$$

Take an $A \in \mathcal{T}(\mathcal{N})$ with $\|U - A\| < 2\gamma$. In the case in which $U = I + K$ for $K \in \mathcal{K}$, by Corollary 13.3.4 we have

$$\mathrm{dist}(U, \mathcal{T}(\mathcal{N})) = \mathrm{dist}(K, \mathcal{T}(\mathcal{N})) = \mathrm{dist}(K, \mathcal{T}(\mathcal{N}) \cap \mathcal{K}(H)).$$

In this case, we can take $A = I + L$ for $L \in \mathcal{T}(\mathcal{N}) \cap \mathcal{K}(H)$.

Now A is invertible in $\mathcal{B}(H)$. We will show that $A^{-1} \in \mathcal{T}(\mathcal{N})$. It is an easy exercise to show that this holds if and only if $AN = N$ for all $N \in \mathcal{N}$. If this fails for some N, then AN is a proper closed subspace of N. Therefore $\|P_{AN} - P_N\| = 1$. However by Lemma 13.5.10, $\|\theta_A - \theta_U\| < \frac{2\gamma}{1-2\gamma} < 3\gamma$. So

$$\|P_{AN} - P_N\| \leq \|\theta_A - \mathrm{id}_{\mathcal{N}}\| \leq \|\theta_A - \theta_U\| + \|\theta_U - \mathrm{id}_{\mathcal{N}}\| < 3\gamma + 2\gamma < 1.$$

Hence $A^{-1} \in \mathcal{T}(\mathcal{N})$. Define $T = UA^{-1}$. Then

$$\|T - I\| = \|(U - A)A^{-1}\| \leq \|U - A\|\|A^{-1}\| < 2\gamma \frac{1}{1 - 2\gamma} < 3\gamma.$$

Now $\theta_T(N) = UA^{-1}N = UN = \theta_U(N)$. Thus $\|\theta_T - \mathrm{id}_{\mathcal{N}}\| < \varepsilon$. Moreover, if $U = I + K$ and $A = I + L$ with K, L compact, then $T - I$ is also compact. ■

We can now show that (3) and (3′) imply (2) and (2′), respectively. The two proofs are essentially the same, so we just keep track of the additional features in the second case.

13.5.14. THEOREM. (3) $\Rightarrow$ (2) *and* (3′) $\Rightarrow$ (2′).

PROOF. Assume (3) (respectively (3′)). Let $\theta : \mathcal{N} \to \mathcal{M}$ be an order isomorphism such that $\theta \sim_a id_{\mathcal{N}}$ (respectively $\theta \sim_{\mathcal{K}} id_{\mathcal{N}}$). Fix $\frac{1}{2} \geq \varepsilon > 0$ and choose $\varepsilon_n > 0$ so that $\prod_{n \geq 1}(1 + 12\varepsilon_n) < 1 + \varepsilon$. Select a unitary $U = U_{n_0}$ so that $\|\theta - \theta_U\| < \varepsilon_1$ (and belongs to $C_n(\mathcal{N}, \mathcal{K})$). Use Lemma 13.5.13 to recursively select T_n invertible operators with $\|T_n - I\| < 12\varepsilon_n$ (and also $T_n - I \in \mathcal{K}$) so that $\|\theta_{T_nT_{n-1}\dots T_1U} - \theta\| < \varepsilon_{n+1}$. Then $T = \lim_{n\to\infty} T_nT_{n-1}\dots T_1U$ converges in norm (and $T - U \in \mathcal{K}$), and

$$\|T - U\| = \Big\| \prod_{n\geq 1} T_n \; - I \Big\| \leq \prod (1 + 12\varepsilon_n) \; - 1 < \varepsilon.$$

Finally, by Lemma 13.5.10 applied to $\theta_{T_nT_{n-1}\dots T_1U}$, we get

$$\|\theta_T - \theta_{T_nT_{n-1}\dots T_1U}\| \leq 2\|T - T_nT_{n-1}\dots T_1U\|.$$

Therefore

$$\begin{aligned}\|\theta_T - \theta\| &\leq \|\theta_T - \theta_{T_nT_{n-1}\dots T_1U}\| + \|\theta_{T_nT_{n-1}\dots T_1U} - \theta\| \\ &\leq 2\|T - T_nT_{n-1}\dots T_1U\| + \varepsilon_{n+1} \to 0.\end{aligned}$$

Hence $\theta_T = \theta$. ∎

It remains to establish $(1) \Rightarrow (2')$.

13.6. Continuous Nests

The goal of this section is to establish a special case of $(1) \Rightarrow (2')$. By Theorem 13.5.14, it suffices to show $(1) \Rightarrow (3')$.

13.6.1. ANDERSEN THEOREM. *Let $\mathcal{N}$ and $\mathcal{M}$ be continuous nests on a separable Hilbert space. Let $\theta : \mathcal{N} \to \mathcal{M}$ be an order isomorphism, and let $\varepsilon > 0$. Then there is a unitary W so that $\theta - \theta_W \in C_n(\mathcal{N}, \mathcal{K})$ and $\|\theta - \theta_W\| < \varepsilon$. Hence $\theta \sim_{\mathcal{K}} \mathrm{id}_{\mathcal{N}}$.*

Let $\mathcal{N} = \{N_t : 0 \leq t \leq 1\}$ and similarly parameterize $\mathcal{M} = \{M_t{:}0 \leq t \leq 1\}$ so that $\theta(N_t) = M_t$. Let ξ_1 be a unit separating vector for $\mathcal{N}''$, so that $\varphi_1(t) = \|P_{N_t}\xi_1\|^2$ is a strictly increasing function of $\mathcal{N}$ onto $[0, 1]$ as in Theorem 13.1.9. Then we can reparameterize $\mathcal{N}$ and $\mathcal{M}$ so that $\varphi_1(t) = t$. Then the measure in Corollary 13.1.10 is just Lebesgue measure and ξ_1 is the constant function 1. So $\mathcal{N}|_{H_{\xi_1}}$ is unitarily equivalent to the Volterra nest.

By Zorn's lemma, there is a finite or countable family of vectors $\{\xi_j : j \in J\}$ of norm 1 such that $H_{\xi_j} = \overline{\mathcal{N}''\xi_j}$ are pairwise orthogonal and sum to $H = \sum_{j\in J}^{\oplus} H_{\xi_j}$. Define $\varphi_j(t) = \|P_{N_t}\xi_j\|^2$.

13.6.2. LEMMA. *There is a countable family $\{\eta_n : n \geq 1\}$ of unit vectors in H so that $H = \sum_{n\geq 1}^{\oplus} H_{\eta_n}$ and $\psi_n(t) = \|P_{N_t}\eta_n\|^2$ are all strictly increasing.*

PROOF. Let $X_j = \mathrm{supp}(\varphi_j) = [0, 1] \setminus \bigcup\{(a, b) \subset [0, 1] : \varphi_j(a) = \varphi_j(b)\}$. If $J \subset \mathbb{N}$ is finite, we let $X_n = \varnothing$ for $n \in \mathbb{N} \setminus J$. Let $\{A_n : n \geq 1\}$ be a partition of $[0, 1]$ into disjoint Borel sets with the property that $m(A_n \cap (a, b)) > 0$ for all

$n \geq 1$ and all $0 \leq a < b \leq 1$. The existence of such sets is a standard construction in measure theory. Let $B_n = A_n \setminus X_n$ for $n \geq 2$ and $B_1 = [0,1] \setminus \bigcup_{n\geq 2} B_n \supset A_1$. Define $\eta_n = \xi_n \oplus M\chi_{B_n}\xi_1$ for $n \geq 2$ and $\eta_1 = M\chi_{B_1}\xi_1$, where we have identified $\mathcal{N}''|_{H_{\xi_1}}$ with $L^\infty(0,1)$ acting as multiplication operators on $L^2(0,1)$. Define $\psi_n(t) = \|P_{N_t}\eta_n\|^2$.

Note that for $0 \leq a < b \leq 1$,

$$\psi_1(b) - \psi_1(a) = \|M\chi_{(a,b)} M_{B_1} 1\|^2 = m\big((a,b) \cap B_1\big) > 0.$$

For $n \geq 2$,

$$\begin{aligned}\psi_n(b) - \psi_n(a) &= \varphi_n(b) - \varphi_n(a) + \|M\chi_{(a,b)} M_{B_n} 1\|^2 \\ &= \varphi_n(b) - \varphi_n(a) + m\big((a,b) \cap B_n\big) > 0.\end{aligned}$$

This is positive because if $\varphi_n(b) = \varphi_n(a)$, then $(a,b) \cap X_n = \varnothing$ and hence $m((a,b) \cap B_n) > 0$. Therefore all ψ_n are strictly increasing.

Next, note that

$$H_{\eta_n} = \overline{\mathcal{N}''\eta_n} = \overline{\mathcal{N}''\xi_n} \oplus \overline{\mathcal{N}''M\chi_{B_n}\xi_1} = H_{\xi_n} \oplus L^2(B_n).$$

These sets are pairwise orthogonal and

$$\sum_{n\geq 1} \oplus H_{\eta_n} = L^2(B_1) \oplus \sum_{n\geq 2} \oplus H_{\xi_n} \oplus L^2(B_n) = L^2(0,1) \oplus \sum_{n\geq 2} \oplus H_{\xi_n} = H.$$

We can now normalize the $\{\eta_n\}$ without affecting this. ∎

We need an elementary estimate.

13.6.3. Lemma. *Suppose that η_i, ζ_i are vectors in H with $\|\eta_i\| \leq 1$, $\|\zeta_i\| \leq 1$ for $1 \leq i \leq L$ and*

$$\max\{|\langle \eta_i, \eta_j\rangle|,\ |\langle \zeta_i, \zeta_j\rangle| : 1 \leq i < j \leq L\} < \frac{1}{L^2}.$$

Then $\big\| \sum_{i=1}^L \eta_i \zeta_j^* \big\| < 2$.

Proof. Let $\xi_1, \dots, \xi_L$ be an orthonormal set in $\mathbb{C}^L$. Define $A = \sum_{i=1}^L \eta_i \xi_i^*$ and $B = \sum_{i=1}^L \xi_i \zeta_i^*$. So $AB = \sum_{i=1}^L \eta_i \zeta_j^*$. Compute

$$\begin{aligned}\|A\|^2 = \|A^*A\| &= \Big\| \sum_{i,j=1}^L \langle \eta_j, \eta_i\rangle \xi_i \xi_j^* \Big\| \\ &\leq \Big\| \sum_{i=1}^L \langle \eta_i, \eta_i\rangle \xi_i \xi_i^* \Big\| + \Big\| \sum_{1\leq i\neq j\leq L} \langle \eta_j, \eta_i\rangle \xi_i \xi_j^* \Big\| \\ &\leq \max\{\|\eta_i\|^2\} + L(L-1)\max\{|\langle \eta_i, \eta_j\rangle| : i \neq j\} < 1 + 1 = 2.\end{aligned}$$

Similarly $\|B\|^2 < 2$. Hence $\|AB\| < 2$. ∎

In Exercise **9.28**, we defined the notion of a quasicentral approximate unit for an ideal of a C*-algebra. We require a similar notion here.

13.6.4. DEFINITION. A *quasicentral approximate unit for $\mathcal{K}$ relative to $\mathcal{N}$* is a sequence F_n of positive finite rank operators with $\|F_n\| \le 1$ and $\text{SOT--}\lim\limits_{n\to\infty} F_n = I$ such that

$$\lim_{n\to\infty} \sup_{0\le t\le 1} \|P_{N_t} F_n - F_n P_{N_t}\| = 0.$$

We will use the commutator notation: $[A, B] = AB - BA$. Note that if P is a projection, then $[P, A] = PA - AP = PAP^\perp - P^\perp AP$ and so (Exercise **4.22**)

$$\|[P, A]\| = \max\{\|PAP^\perp\|, \|P^\perp AP\|\}.$$

13.6.5. LEMMA. *Let $\mathcal{N} = \{N_t : 0 \le t \le 1\}$ be a continuous nest with a cyclic unit vector ξ, i.e. $\overline{\mathcal{N}''\xi} = H$. For $n \ge 1$ and $1 \le k \le 2^n$, let*

$$E_{k,n} = P_{N_{2^{-n}k}} - P_{N_{2^{-n}(k-1)}} \quad \textit{and} \quad \xi_{k,n} = E_{k,n}\xi / \|E_{k,n}\xi\|.$$

Let $P_n = \sum_{k=1}^{2^n} \xi_{k,n}\xi_{k,n}^*$ be the projection onto $\operatorname{span}\{\xi_{k,n} : 1 \le k \le 2^n\}$*. Given $\varepsilon > 0$ and n, there is a positive finite rank operator F with $P_n \le F \le P_m$, for some $m > n$, so that* $\sup_{0\le t\le 1} \|[P_{N_t}, F]\| < \varepsilon$.

PROOF. Let $\varphi(t) = \|P_{N_t}\xi\|^2$. Then φ is strictly increasing because ξ is cyclic, and is continuous since it is SOT-continuous on $\mathcal{N}$. Hence $\xi_{k,n}$ is defined for all k and n. By uniform continuity,

$$\lim_{n\to\infty} \max_{1\le k\le 2^n} \|E_{k,n}\xi\|^2 \le \lim_{n\to\infty} \sup_{|s-t|\le 2^{-n}} |\varphi(s) - \varphi(t)| = 0.$$

If $m = n + p$, we can write $\xi_{k,n} = \sum_{2^p(k-1)<j\le 2^p k} a_j \xi_{j,m}$ where

$$a_j = \langle \xi_{k,n}, \xi_{j,m}\rangle = \frac{\langle E_{k,n}\xi, E_{j,m}\xi\rangle}{\|E_{k,n}\xi\|\,\|E_{j,m}\xi\|} = \frac{\|E_{j,m}\xi\|}{\|E_{k,n}\xi\|}.$$

Therefore with n fixed,

$$\delta(n, m) := \max_{1\le j\le 2^m} |a_j| \to 0 \quad \text{as } m \to \infty.$$

Let $L = 2\lceil 1 + \frac{1}{\varepsilon}\rceil$. Select $n = m_1 < \cdots < m_L$ so that $\delta(m_i, m_{i+1}) < \frac{1}{L^2}$ for $1 \le i < L$. Define $F = \frac{1}{L}\sum_{i=1}^{L} P_{m_i}$. Then F is positive, finite rank and $P_n \le F \le P_{m_L}$. Fix $0 < t < 1$. For each $1 \le i \le L$, there is a unique integer k_i so that $k_i - 1 < 2^{m_i}t \le k_i$. Then

$$[P_{N_t}, \xi_{k,m_i}\xi_{k,m_i}^*] = (P_{N_t}\xi_{k,m_i})(P_{N_t}^\perp \xi_{k,m_i})^* - (P_{N_t}^\perp \xi_{k,m_i})(P_{N_t}\xi_{k,m_i})^*$$

However if $k < k_i$, then $P_{N_t}^\perp \xi_{k,m_i} = 0$ and if $k > k_i$, then $P_{N_t}\xi_{k,m_i} = 0$. Thus

$$[P_{N_t}, P_{m_i}] = [P_{N_t}, \xi_{k_i,m_i}\xi_{k_i,m_i}^*] = \eta_i\zeta_i^* - \zeta_i\eta_i^*$$

where $\eta_i = P_{N_t}\xi_{k_i,m_i}$ and $\zeta_i = P_{N_t}^{\perp}\xi_{k_i,m_i}$. Now

$$|\langle \eta_i, \eta_j\rangle| + |\langle \zeta_i, \zeta_j\rangle| = \langle \xi_{k_i,m_i}, \xi_{k_j,m_j}\rangle \le \delta(m_i, m_j) < \frac{1}{L^2}.$$

Therefore by Lemma 13.6.3,

$$\begin{aligned}\|P_{N_t}F - FP_{N_t}\| &= \max\{\|P_{N_t}FP_{N_t}^{\perp}\|, \|P_{N_t}^{\perp}FP_{N_t}\|\}\\ &= \max\Big\{\frac{1}{L}\Big\|\sum_{i=1}^{L}\eta_i\zeta_i^*\Big\|, \frac{1}{L}\Big\|\sum_{i=1}^{L}\zeta_i\eta_i^*\Big\|\Big\} < \frac{2}{L} < \varepsilon.\end{aligned}$$

This estimate is independent of t, so the result follows. ■

From this result and Lemma 13.6.2, the following corollary is straightforward. We will not need this explicitly, so we leave the details as Exercise **13.2**. Instead, we use the result in the cyclic case to build unitaries intertwining two continuous nests which are close on finite dimensional subspaces, and then we use the quasicentral approximate unit to splice these maps together into a single map with small intertwining error. The argument is modelled on the proof of Voiculescu's Weyl–von Neumann theorem [311] (see also [70], §II.5).

13.6.6. COROLLARY. *If $\mathcal{N}$ is a continuous nest on a separable Hilbert space, then there is a quasicentral approximate unit for $\mathcal{K}$ relative to $\mathcal{N}$.*

13.6.7. LEMMA. *Let $\mathcal{N} = \{N_t : 0 \le t \le 1\}$ be a continuous nest with cyclic unit vector ξ. Let $\mathcal{M} = \{M_t : 0 \le t \le 1\}$ be another continuous nest with cyclic unit vector η. Given n_0, let P_{n_0} be defined as in Lemma 13.6.5. Given $\varepsilon > 0$, there is a unitary $U \in \mathcal{B}(H_{\mathcal{N}}, H_{\mathcal{M}})$ so that*

$$\sup_{0\le t\le 1} \|(P_{M_t}U - UP_{N_t})P_{n_0}\| < \varepsilon.$$

PROOF. We adopt the notation of Lemma 13.6.5. Similarly for $n \ge 1$ and $1 \le k \le 2^n$, we define

$$F_{k,n} = P_{M_{2^{-n}k}} - P_{M_{2^{-n}(k-1)}} \quad \text{and} \quad \eta_{k,n} = F_{k,n}\eta/\|F_{k,n}\eta\|.$$

Choose $n > n_0$ so that $\delta(n_0, n) < \varepsilon/2$. Define $U\xi_{k,n} = \eta_{k,n}$ for $1 \le k \le 2^n$, and extend the definition of U on each $E_{k,n}H_{\mathcal{N}}$ so that $UE_{k,n}H_{\mathcal{N}} = F_{k,n}H_{\mathcal{M}}$.

We verify that this works. Fix $0 < t < 1$. There are unique integers k_0 and k_1 so that $k_0 - 1 < 2^{n_0}t \le k_0$ and $k_1 - 1 < 2^n t \le k_1$. If $k < k_1$, then

$$(P_{M_t}U - UP_{N_t})\xi_{k,n} = P_{M_t}\eta_{k,n} - U\xi_{k,n} = 0$$

and if $k > k_1$, then

$$(P_{M_t}U - UP_{N_t})\xi_{k,n} = P_{M_t}\eta_{k,n} - 0 = 0.$$

While

$$\|(P_{M_t}U - UP_{N_t})\xi_{k_1,n}\| = \|P_{M_t}\eta_{k_1,n} - UP_{N_t}\xi_{k,n}\| \le 2.$$

Hence $(P_{M_t}U - UP_{N_t})\xi_{k,n_0} = 0$ if $k \neq k_0$, and

$$\|(P_{M_t}U - UP_{N_t})\xi_{k_0,n_0}\| = |a_{k_1}|\|P_{M_t}\eta_{k_1,n} - UP_{N_t}\xi_{k_1,n}\| < \varepsilon.$$

Therefore $(P_{M_t}U - UP_{N_t})P_{n_0}$ has rank 1, and norm at most ε. ∎

We require a commutator inequality, although one can get away with somewhat less. See Exercise **13.15**. It is suspected that the correct constant is 1, but the best known constant is $\frac{5}{4}$.

13.6.8. Pedersen Inequality. *Let $E \geq 0$, $T \in \mathcal{B}(H)$ and $0 < p < 1$. Then*

$$\|TE^p - E^pT\| \leq \tfrac{5}{4}\|TE - ET\|^p\|T\|^{1-p}.$$

Proof. Both sides are homogeneous of degree 1, so we may normalize so that $\|T\| = 1$, and then ignore the factor $\|T\|^{1-p}$. First suppose that U is unitary. Then since x^p is operator monotone by Proposition 9.1.12 and $(1+x)^p \leq 1 + x^p$,

$$\begin{aligned} UE^pU^* &= (UEU^*)^p = (UEU^* - E + E)^p \\ &\leq \big(\|UEU^* - E\| + E\big)^p \leq \|UEU^* - E\|^p + E^p. \end{aligned}$$

Therefore

$$\|UE^p - E^pU\| = \|UE^pU^* - E^p\| \leq \|UEU^* - E\|^p = \|UE - EU\|^p.$$

Next assume that $T = T^*$. Define a unitary $U = (I + \frac{i}{2}T)(I - \frac{i}{2}T)^{-1}$. Compute

$$\begin{aligned} UE - EU &= (I - \tfrac{i}{2}T)^{-1}\big((I + \tfrac{i}{2}T)E(I - \tfrac{i}{2}T) - (I - \tfrac{i}{2}T)E(I + \tfrac{i}{2}T)\big)(I - \tfrac{i}{2}T)^{-1} \\ &= (I - \tfrac{i}{2}T)^{-1}i(TE - ET)(I - \tfrac{i}{2}T)^{-1}. \end{aligned}$$

Hence

$$TE - ET = -i(I - \tfrac{i}{2}T)(UE - EU)(I - \tfrac{i}{2}T)$$

and similarly

$$TE^p - E^pT = -i(I - \tfrac{i}{2}T)(UE^p - E^pU)(I - \tfrac{i}{2}T).$$

Therefore since $(I - \frac{i}{2}T)^{-1}$ is a contraction,

$$\begin{aligned} \|TE^p - E^pT\| &\leq \|I - \tfrac{i}{2}T\|^2\|UE^p - E^pU\| \leq (1 + \tfrac{1}{4}\|T\|^2)\|UE - EU\|^p \\ &= \tfrac{5}{4}\|(I - \tfrac{i}{2}T)^{-1}i(TE - ET)(I - \tfrac{i}{2}T)^{-1}\|^p \leq \tfrac{5}{4}\|TE - ET\|^p. \end{aligned}$$

Finally if T is not self-adjoint, consider

$$\widetilde{T} = \begin{bmatrix} 0 & T \\ T^* & 0 \end{bmatrix} \quad \text{and} \quad \widetilde{E} = \begin{bmatrix} E & 0 \\ 0 & E \end{bmatrix}.$$

Then $\widetilde{T} = \widetilde{T}^*$, $\|\widetilde{T}\| = \|T\| = 1$, and

$$\|\widetilde{T}\widetilde{E}^p - \widetilde{E}^p\widetilde{T}\| = \left\| \begin{bmatrix} 0 & TE^p - E^pT \\ -(TE^p - E^pT)^* & 0 \end{bmatrix} \right\| = \|TE^p - E^pT\|.$$

The same holds for E, and thus $\|TE^p - E^pT\| \leq \frac{5}{4}\|TE - ET\|^p$. ∎

Now we splice these unitary intertwiners together using a quasicentral approximate unit.

13.6.9. LEMMA. *Let $\mathcal{N} = \{N_t : 0 \le t \le 1\}$ be a continuous nest with cyclic unit vector ξ. Let $\mathcal{M} = \{M_t : 0 \le t \le 1\}$ be another continuous nest. Given $\varepsilon > 0$, there is an isometry $U \in \mathcal{B}(H_\mathcal{N}, H_\mathcal{M})$ so that*

$$f(t) = P_{M_t}U - UP_{N_t} \in C_n(\mathcal{N}, \mathcal{K}) \quad \textit{and} \quad \|f\| \le \varepsilon.$$

PROOF. By Lemma 13.6.2, we may suppose that $H_\mathcal{M} \simeq \sum_{j\ge1}^{\oplus} H_j$ where each H_j has a cyclic unit vector η_j for $\mathcal{M}''|_{H_j}$ and $\operatorname{supp}(\eta_j) = [0,1]$ (i.e. $\|P_{M_t}\eta_j\|^2$ is strictly increasing).

We adopt the notation of Lemma 13.6.5. Using Lemmas 13.6.5 and 13.6.7, recursively select an increasing sequence n_j, positive finite rank contractions F_j, unitary operators $U_j \in \mathcal{U}(H_\mathcal{N}, H_j)$ such that

(i) $P_{n_j} \le F_j \le P_{n_{j+1}}$.

(ii) $\sup_{0\le t\le1} \left\|P_{N_t}F_j - F_jP_{N_t}\right\| < \dfrac{\varepsilon^2}{32\cdot 4^j}$.

(iii) $\sup_{0\le t\le1} \left\|(P_{M_t}U_j - U_jP_{N_t})P_{n_{j+1}}\right\| < \dfrac{\varepsilon}{2^{j+1}}$.

We initialize the construction with $n_0 = 0$, $n_1 = 1$ and $F_0 = 0$. Suppose that we have F_{j-1} and n_j. By Lemma 13.6.5, we can find n_{j+1} and F_j satisfying (i) and (ii). Then we use Lemma 13.6.7 to construct the unitary U_j mapping $H_\mathcal{N}$ onto H_j satisfying (iii).

Define operators $E_j = (F_j - F_{j-1})^{1/2}$ for $j \ge 1$ and $U = \text{SOT-}\sum_{j\ge1} U_jE_j$. Observe that

$$U^*U = \text{SOT-}\sum_{j\ge1} E_j^2 = \text{SOT-}\sum_{j\ge1} F_j - F_{j-1} = I.$$

Thus U is an isometry of $H_\mathcal{N}$ into $H_\mathcal{M}$. By Pedersen's inequality 13.6.8,

$$\begin{aligned}
\|[P_{N_t}, E_j]\| &\le \frac{5}{4}\|[P_{N_t}, F_j - F_{j-1}]\|^{1/2} \\
&\le \frac{5}{4}\sqrt{\frac{\varepsilon^2}{32\cdot 4^j} + \frac{\varepsilon^2}{32\cdot 4^{j-1}}} = \frac{\varepsilon}{4\cdot 2^{j-1}}\sqrt{\frac{125}{128}} < \frac{\varepsilon}{2^{j+1}}.
\end{aligned}$$

Therefore

$$\begin{aligned}
\|P_{M_t}U - UP_{N_t}\| &= \Big\|\sum_{j\ge1} P_{M_t}U_jE_j - U_jE_jP_{N_t}\Big\| \\
&\le \Big\|\sum_{j\ge1}(P_{M_t}U_j - U_jP_{N_t})E_j\Big\| + \Big\|\sum_{j\ge1} U_jP_{N_t}E_j - U_jE_jP_{N_t}\Big\| \\
&\le \sum_{j\ge1} \left\|(P_{M_t}U_j - U_jP_{N_t})P_{n_{j+1}}\right\| + \|[P_{N_t}, E_j]\|
\end{aligned}$$

$$< \sum_{j\geq 1} \frac{\varepsilon}{2^{j+1}} + \frac{\varepsilon}{2^{j+1}} = \varepsilon.$$

This is a uniformly convergent sequence for $t \in [0, 1]$ of finite rank, norm-continuous functions. Hence, the limit lies in $C_n(\mathcal{N}, \mathcal{K})$ and has norm at most ε. ■

We are ready to complete the proof of Theorem 13.6.1.

PROOF OF ANDERSEN'S THEOREM. We may suppose that $\theta(N_t) = M_t$ for $0 \leq t \leq 1$. Let $\mathcal{N}^{(\infty)}$ denote the nest $\{N_t^{(\infty)} : 0 \leq t \leq 1\}$ on $H = H_{\mathcal{N}}^{(\infty)}$. By applying Lemma 13.6.2, we can write $H = \sum_{i\geq 1}^{\oplus} H_i$ so that there are unit vectors ξ_i such that $\overline{\mathcal{N}^{(\infty)\prime\prime}\xi_i} = H_i$ and $\mathrm{supp}(\xi_i) = [0, 1]$ for $i \geq 1$. Likewise, we write $H_{\mathcal{M}} = \sum_{i\geq 1}^{\oplus} K_i$ where there are unit vectors η_i such that $\overline{\mathcal{M}'\eta_i} = K_i$ and $\mathrm{supp}(\eta_i) = [0, 1]$ for $i \geq 1$.

For each $i \geq 1$, apply Lemma 13.6.9 to obtain isometries $U_i \in \mathcal{B}(H_i, K_i)$ so that $f_i(t) = P_{M_t}U_i - U_iP_{N_t^{(\infty)}} \in C_n(\mathcal{N}, \mathcal{K})$ and $\|f_i\| < 2^{-i}\varepsilon$. Then $U = \sum_{i\geq 1}^{\oplus} U_i$ is an isometry of H into $H_{\mathcal{M}}$ such that

$$f(t) = P_{M_t}U - UP_{N_t^{(\infty)}} = \sum_{i\geq 1}^{\oplus} f_i(t)$$

is a norm convergent sum, and thus $f \in C_n(\mathcal{N}, \mathcal{K})$ and $\|f\| < \sum_{i\geq 1} 2^{-i}\varepsilon = \varepsilon$. Let $P = UU^*$ be the range projection of U. Then

$$g(t) = [P_{M_t}, P] = (P_{M_t}U - UP_{N_t^{(\infty)}})U^* - U(P_{M_t}U - UP_{N_t^{(\infty)}})^*.$$

This belongs to $C_n(\mathcal{N}, \mathcal{K})$ and $\|g\| \leq 2\|f\| < 2\varepsilon$.

This shows that with a small $C_n(\mathcal{N}, \mathcal{K})$ perturbation, $\mathcal{M}$ has a direct summand unitarily equivalent to $\mathcal{N}^{\infty}$. The idea is to split off one copy of $\mathcal{N}$ leaving something very close to $\mathcal{M}$ behind. To make this precise, let $S \in \mathcal{B}(H_{\mathcal{N}}^{(\infty)})$ be the shift $S(\zeta_1, \zeta_2, \zeta_3, \ldots) = (0, \zeta_1, \zeta_2, \ldots)$. Also define $J \in \mathcal{B}(H_{\mathcal{N}}, H_{\mathcal{N}}^{(\infty)})$ by $J\zeta = (\zeta, 0, 0, \ldots)$. Note that

$$SP_{N_t^{(\infty)}} = P_{N_t^{(\infty)}}S \quad \text{and} \quad JP_{N_t} = P_{N_t^{(\infty)}}J \quad \text{and} \quad SS^* + JJ^* = I_{H_{\mathcal{N}}^{(\infty)}}.$$

Define $W \in \mathcal{B}(H_{\mathcal{M}} \oplus H_{\mathcal{N}}, H_{\mathcal{M}})$ by $W = \begin{bmatrix} P^{\perp} + USU^* & UJ \end{bmatrix}$. Then W is unitary and

$$h(t) := P_{M_t}W - W(P_{M_t} \oplus P_{N_t}) = \begin{bmatrix} X(t) & Y(t) \end{bmatrix}$$

where

$$\begin{aligned} X(t) &= P_{M_t}(P^{\perp} + USU^*) - (P^{\perp} + USU^*)P_{M-t} \\ &= [P_{M_t}, P^{\perp}] + (P_{M_t}U - UP_{N_t^{(\infty)}})SU^* - US(P_{M_t}U - UP_{N_t^{(\infty)}})^* \end{aligned}$$

and

$$Y(t) = P_{M_t}UJ - UJP_{N_t} = (P_{M_t}U - UP_{N_t^{(\infty)}})J.$$

Both $X(t), Y(t) \in C_n(\mathcal{N},\mathcal{K})$ and $\|X(t)\| < 4\varepsilon$ and $\|Y(t)\| < \varepsilon$. Hence $\|h\| < 5\varepsilon$.

Reversing the roles of $\mathcal{M}$ and $\mathcal{N}$, we obtain a unitary $V \in \mathcal{B}(H_{\mathcal{M}} \oplus H_{\mathcal{N}}, H_{\mathcal{N}})$ such that $P_{N_t}V - V(P_{M_t} \oplus P_{N_t})$ belongs to $C_n(\mathcal{N},\mathcal{K})$ with norm less than 5ε. It follows that $WV^* \in \mathcal{B}(H_{\mathcal{N}}, H_{\mathcal{M}})$ is unitary such that $P_{M_t}WV^* - WV^*P_{N_t}$ is in $C_n(\mathcal{N},\mathcal{K})$ with norm less than 10ε. Since $\varepsilon > 0$ is arbitrary, $\theta \sim_{\mathcal{K}} id_{\mathcal{N}}$. ∎

By Theorem 13.5.14, it follows that the similarity theorem is valid for continuous nests.

13.6.10. Larson Theorem. *Let $\mathcal{N}$ and $\mathcal{M}$ be continuous nests on a separable Hilbert space, let $\theta : \mathcal{N} \to \mathcal{M}$ be an order isomorphism, and let $\varepsilon > 0$. Then there is an invertible operator $T = U + K$ with U unitary, K compact and $\|K\| < \varepsilon$ such that $T\mathcal{N} = \mathcal{M}$ and $\theta_T = \theta$.*

13.7. The general case

The remaining problem can be pictured by considering Examples 13.1.2(5,6), the Cantor nest $\mathcal{Q} = \{\{0\}, Q_t, Q_r^+, H : t \in \mathbb{R},\ r \in \mathbb{Q}\}$ on $l^2(\mathbb{Q})$ and the nest $\mathcal{R}$ on $l^2(\mathbb{Q}) \oplus L^2(\mathbb{R})$ given by $R_t = Q_t \oplus L^2(-\infty, t)$ and $R_r^+ = Q_r^+ \oplus L^2(-\infty, r)$. The map $\theta(Q_t) = R_t$ is an order isomorphism. They have the same one-dimensional atoms $Q_r^+ \ominus Q_r = \mathbb{C}\delta_r = R_r^+ \ominus R_r$ for $r \in \mathbb{Q}$. However, $\mathcal{R}$ has a large continuous part while $\mathcal{Q}$ is atomic. The idea of the remainder of the proof is to apply a similarity to $\mathcal{Q}$ to introduce a large continuous part so that we can then apply the results of the previous section.

13.7.1. Definition. Let $\mathcal{N}$ be a nest which is order isomorphic via θ to a compact set $\omega \subset [0,1]$. Define $N_t = \theta^{-1}(t)$ for $t \in \omega$ and intervals $E_{\mathcal{N}}(a,b] = P_{N_b} - P_{N_a}$. Say that an invertible operator T *acts discontinuously* on $\mathcal{N}$ if there are disjoint intervals $(a_n, b_n]$ for $n \geq 1$ such that SOT-$\sum_{n\geq 1} E_{\mathcal{N}}(a_n, b_n] = I$ but SOT-$\sum_{n\geq 1} E_{T\mathcal{N}}(a_n, b_n] \neq I$.

Say that an interval E of $\mathcal{N}$ has *non-atomic part* if E is not the sum of the atoms that it dominates.

We show that an uncountable nest can be split into countably many uncountable intervals, leaving an uncountable piece that doesn't support any non-zero operator in $\mathcal{T}(\mathcal{N})$.

13.7.2. Lemma. *Let $\mathcal{N}$ be an uncountable nest order isomorphic to a compact set $\omega \subset [0,1]$. Then there are disjoint intervals $(a_n, b_n]$ for $n \geq 1$ such that* SOT-$\sum_{n\geq 1} E_{\mathcal{N}}(a_n, b_n] = I$ *and $\omega \setminus \bigcup_{n\geq 1}(a_n, b_n)$ and $\omega \cap (a_n, b_n)$ are all uncountable.*

PROOF. Let $\xi_1 \in H_\mathcal{N}$ be a unit separating vector for $\mathcal{N}''$. Expand this to a finite or countable set of unit vectors $\{\xi_i\}$ so that $H_{\xi_i} = \overline{\mathcal{N}''\xi_i}$ are pairwise orthogonal and $H_\mathcal{N} = \sum_i^\oplus H_{\xi_i}$. (For example, see the discussion preceding Lemma 13.6.2.) Choose $c_i > 0$ so that $\sum_i c_i = 1$. Define $\varphi(t) = \sum_i c_i \|P_{N_t}\xi_i\|^2$ for $t \in \omega$. This is a strictly increasing, continuous function on ω which can be extended uniquely to a monotone increasing, right continuous function on $[0, 1]$ which is locally constant on $[0, 1]\setminus\omega$. Let $\mu = d\varphi$ be the corresponding Lebesgue–Stieltjes measure on $[0, 1]$, and observe that $\operatorname{supp}(\mu) = \omega$.

Since ω is an uncountable compact metric space, one can construct a Cantor set $C \subset \omega$ so that $\mu(C) = 0$. Then construct a Cantor subset $D \subset C$ so that if the components of $[0, 1] \setminus D$ are (a_n, b_n) for $n \geq 1$, then $(a_n, b_n) \cap C$ is uncountable. The idea is to take a monotone increasing continuous function f of C onto $[0, 1]$ such as the Cantor function, and then define D to be the inverse image of the usual middle thirds Cantor set in $[0, 1]$. Thus $\mu(D) = 0$ and D is uncountable, and $(a_n, b_n) \cap \omega$ is uncountable for each component of $[0, 1] \setminus D$.

To complete the proof, we consider $P = \text{SOT-}\sum_{n\geq 1} E_\mathcal{N}(a_n, b_n]$. Then

$$\begin{aligned} 1 \geq \sum_i c_i \|P\xi_i\|^2 &= \sum_i \sum_{n\geq 1} c_i \|E_\mathcal{N}(a_n, b_n]\xi_i\|^2 \\ &= \sum_{n\geq 1} \varphi(b_n) - \varphi(a_n) = \mu(\omega \setminus D) = 1. \end{aligned}$$

Therefore $\xi_i \in PH_\mathcal{N}$ for all i, so that

$$PH_\mathcal{N} = \mathcal{N}''PH_\mathcal{N} \supset \text{span}\,\Big\{\bigcup_i \mathcal{N}''\xi_i\Big\} = H_\mathcal{N}.$$

Therefore $P = I$ as claimed. ∎

An uncountable nest could be atomic, such as the Cantor nest, Example 13.1.2(5). The next lemma shows that after a similarity close to I, the new nest can be forced to have a non-atomic part.

13.7.3. LEMMA. *Let $\mathcal{N}$ be an uncountable nest and let $\varepsilon > 0$. Then there is an invertible operator $T = I + K$ where K is compact and $\|K\| < \varepsilon$ which acts discontinuously on $\mathcal{N}$. In particular, if $\mathcal{N}$ is atomic, then $T\mathcal{N}$ is not.*

PROOF. Use the previous lemma to partition $\mathcal{N}$ into pairwise orthogonal intervals $E_n = E_\mathcal{N}(a_n, b_n]$ for $n \geq 1$ so that $\omega \cap (a_n, b_n)$ is uncountable for $n \geq 1$,

$$D = \omega \setminus \bigcup_{n\geq 1} (a_n, b_n) \text{ is uncountable} \quad \text{and} \quad \text{SOT-}\sum_n E_n = I.$$

Since $(a_n, b_n) \cap \omega$ is uncountable, the nest $E_n\mathcal{N}|E_nH_\mathcal{N}$ is uncountable, whence E_n is infinite dimensional. We identify $E_nH_\mathcal{N}$ with $L^2(a_n, b_n)$, so that

$$H_\mathcal{N} = \sum_{n\geq 1}^{\oplus} E_nH_\mathcal{N} \simeq \sum_{n\geq 1}^{\oplus} L^2(a_n, b_n) = L^2(\mathcal{O}) =: H_1 \quad \text{where} \quad \mathcal{O} = \bigcup_{n\geq 1}(a_n, b_n).$$

Since $[0,1] \setminus \mathcal{O} = D$ is nowhere dense, $\mathcal{O}$ is dense in $[0,1]$. Let W be a unitary implementing the equivalence between $H_\mathcal{N}$ and H_1.

Let ν be a non-atomic probability measure on D. Let $H_2 = H_1 \oplus L^2(\nu)$. Form two continuous nests $\mathcal{L}$ and $\mathcal{M}$ on H_1 and H_2, respectively: for $0 \leq t \leq 1$, set

$$L_t = L^2(\mathcal{O} \cap (0,t)) \quad \text{and} \quad M_t = L_t \oplus L^2(D \cap (0,t), \nu).$$

Then these are both continuous nests, and the unitary equivalence between $H_\mathcal{N}$ and H_1 shows that for $d \in D$, we have $N_d \simeq L_d$. By Larson's theorem, there is an operator $S = U + K$ where U is unitary, K is compact and $\|K\| < \varepsilon$ such that $SL_t = M_t$ for $0 \leq t \leq 1$. Note that S acts discontinuously on $\mathcal{L}$ with respect to the intervals $F_n = L^2(a_n, b_n)$ since $\text{SOT-}\sum_{n\geq 1} F_n = I$ but

$$\begin{aligned} \text{SOT-}\sum_{n\geq 1} P_{M_{b_n}} - P_{M_{a_n}} &= \text{SOT-}\sum_{n\geq 1} L^2(a_n, b_n) \oplus L^2((a_n, b_n), \nu) \\ &= L^2(\mathcal{O}) \oplus \{0\} \neq H_2. \end{aligned}$$

Let $S = V|S|$ be the polar decomposition, and let $T = W^*|S|W$. Then $T - I$ is compact and norm less than ε. Since V and W are unitary, they preserve countable partitions of the identity. Thus, T must similarly act discontinuously on $\mathcal{N}$ with respect to the partition $\{E_n\}$. ∎

In the next lemma, we iterate this construction. We are careful not to destroy any continuous part which is introduced.

13.7.4. Lemma. *Let $\mathcal{N}$ be an uncountable nest and let $\varepsilon > 0$. Then there is an invertible operator $T = I + K$ where K is compact and $\|K\| < \varepsilon$ such that $T\mathcal{N}$ has non-atomic part on every uncountable interval.*

Proof. Choose $\varepsilon_n > 0$ such that $\prod_{n\geq 1} 1 + \varepsilon_n < 1 + \varepsilon$. Identify $\mathcal{N}$ with a compact set $\omega \subset [0,1]$. In the following construction, we will find invertible operators $S_n = I + K_n$ where K_n are compact and $\|K_n\| < \varepsilon_n$, invertible operators $T_n = S_nT_{n-1}$ and nests $\mathcal{N}_n = T_n\mathcal{N}$, where $T_0 = I$.

Assume that T_{n-1} and $\mathcal{N}_{n-1}$ are defined. Let $E_{k,n} = E_{\mathcal{N}_{n-1}}\left(\frac{k-1}{2^n}, \frac{k}{2^n}\right]$ for $1 \leq k \leq 2^n$. If $E_{k,n}\mathcal{N}_{n-1}$ is either countable or has a non-atomic part, then let $S_{k,n} = E_{k,n}$. Otherwise, $E_{k,n}\mathcal{N}_{n-1}$ is uncountable and atomic. Use the previous lemma to define $S_{k,n} = E_{k,n}S_{k,n}E_{k,n} = E_{k,n} + K_{k,n}$, where $K_{k,n}$ is compact, $\|K_{k,n}\| < \varepsilon_n$ and $S_{k,n}$ acts discontinuously on $E_{k,n}\mathcal{N}_{n-1}$. Then $S_{k,n}E_{k,n}\mathcal{N}_{n-1}$ has a non-trivial non-atomic part. Define $S_n = \sum_{k=1}^{2^n} S_{k,n}$. This is invertible, and

$S_n - I = K_n := \sum_{k=1}^{2^n} K_{k,n}$ is compact with norm $\|K_n\| = \max_k\{\|K_{k,n}\|\} < \varepsilon_n$. Then define $T_n = S_n T_{n-1}$ and $\mathcal{N}_n = S_n(T_{n-1}\mathcal{N}) = T_n\mathcal{N}$. By construction, every diadic interval, $E_{\mathcal{N}_n}(\frac{k-1}{2^n}, \frac{k}{2^n}]\mathcal{N}_n$ for $1 \leq k \leq 2^n$ is either a countable nest or has a non-atomic part.

The operator $T = \lim_{n\to\infty} T_n$ exists and $T - I$ is compact of norm less than ε. By construction, the non-atomic part of $\mathcal{N}_n$ is never changed in later iterations. Thus every diadic interval $E_{T\mathcal{N}}(\frac{k-1}{2^n}, \frac{k}{2^n}]T\mathcal{N}$ of $T\mathcal{N}$ for $1 \leq k \leq 2^n$ and $n \geq 1$ is either a countable nest or it has a non-atomic part. Since every interval of $\mathcal{N}$ which is uncountable must contain a diadic interval which is uncountable, we see that it has a non-atomic part. ■

We can now complete the proof of $(1) \Rightarrow (2')$ of the Similarity theorem.

COMPLETION OF THE PROOF OF THE SIMILARITY THEOREM 13.5.8. Let $\theta : \mathcal{N} \to \mathcal{M}$ be a dimension preserving order isomorphism, and let $\varepsilon < \frac{1}{2}$. Apply Lemma 13.7.4 to both $\mathcal{N}$ and $\mathcal{M}$. There are invertible operators T_1 and T_2 on $H_{\mathcal{N}}$ and $H_{\mathcal{M}}$ with $K_i = T_i - I$ compact and $\|K_i\| < \varepsilon/4$ so that $\mathcal{N}_1 = T_1\mathcal{N}$ and $\mathcal{M}_1 = T_2\mathcal{M}$ have non-atomic part on every uncountable interval. Then $\theta_1 = \theta_{T_2}\theta\theta_{T_1}^{-1}$ is the corresponding order isomorphism between $\mathcal{N}_1$ and $\mathcal{M}_1$.

Let P and Q be the sums of all atoms of $T_1\mathcal{N}$ and $T_2\mathcal{M}$, respectively. Since θ_1 preserves dimension, there is a unitary operator $U_a : PH_{\mathcal{N}_1} \to QH_{\mathcal{M}_1}$ such that $\theta_{U_a} = \theta_1|_{P\mathcal{N}_1}$, i.e. it matches up the atoms via θ_1. Now $P^\perp\mathcal{N}_1$ and $Q^\perp\mathcal{M}_2$ are continuous nests with the same support ω_0. If $\operatorname{supp}(\mathcal{N}) = \omega$, then

$$\omega_0 = \{t \in \omega : (t-\delta, t+\delta) \cap \omega \text{ is uncountable for all } \delta > 0\}.$$

By Larson's theorem, there is an invertible operator $S_c = U_c + K$ with U_c unitary, K compact and $\|K\| < \varepsilon/4$ so that $\theta_{S_c} = \theta_1|_{P^\perp\mathcal{N}_1}$. Then $S = T_2^{-1}(U_a + S_c)T_1$ implements $\theta_S = \theta$, and $U = (U_a + U_c)$ is unitary. Note that $T_2^{-1} = (I + K_2)^{-1} = I - L_2$ where L_2 is compact and

$$\|L_2\| = \|K_2 \sum_{k\geq 0}(-K_2)^k\| \leq \|K_2\|(1 - \|K_2\|)^{-1} < \frac{\varepsilon/4}{1-\varepsilon/4} < \frac{\varepsilon}{3}.$$

Thus

$$\begin{aligned} S - U &= (I - L_2)(U + K)(I + K_1) - U \\ &= UK_1 + K + KK_1 + L_2U + L_2UK_1 + L_2K + L_2KK_1. \end{aligned}$$

This is compact and since $\varepsilon < \frac{1}{2}$,

$$\|S - U\| < \frac{\varepsilon}{4} + \frac{\varepsilon}{4} + \frac{\varepsilon}{4}\frac{\varepsilon}{4} + \frac{\varepsilon}{3} + \frac{\varepsilon}{3}\frac{\varepsilon}{4} + \frac{\varepsilon}{3}\frac{\varepsilon}{4} + \frac{\varepsilon}{3}\frac{\varepsilon}{4}\frac{\varepsilon}{4} < \varepsilon.$$

This establishes $(1) \Rightarrow (2')$ and completes the proof. ■

13.8. Consequences of the Similarity Theorem

In this section, we collect a number of direct consequences of the Similarity theorem. We first point out that similar nests need not be unitarily equivalent. We discuss a property called factorization, and show that only countable nests have this property. We show that the group of outer automorphisms of $\mathcal{T}(\mathcal{N})$ is isomorphic to the group of dimension preserving order automorphisms of $\mathcal{N}$. Lastly, we prove a theorem of Larson and Pitts that characterizes equivalent idempotents in $\mathcal{T}(\mathcal{N})$.

One immediate observation is that there are similar nests which are not unitarily equivalent. If $U\mathcal{N} = \mathcal{M}$ with U unitary, then $\mathcal{T}(\mathcal{M}) = U\mathcal{T}(\mathcal{N})U^*$. Therefore $\mathcal{M}'' = U\mathcal{N}''U^*$. In particular, if $\mathcal{N}$ is atomic, then so is $\mathcal{M}$. However, we showed that the Cantor nest $\mathcal{Q}$ and the order isomorphic nest $\mathcal{R}$ of Examples 13.1.2 (5) and (6) are similar. Also the Volterra nest $\mathcal{N}$ and the continuous nest $\mathcal{M}$ on $L^2((0,1)^2)$ of Examples 13.1.2 (3) and (4) are similar. To see that they are not unitarily equivalent, note that $\mathcal{N}'' = L^\infty(0,1)$ is maximal abelian. However, $\mathcal{M}''$ is not because it commutes with the unitary operators $U_s f(x,y) = f(x, y+s)$ (where addition is modulo 1) which shift the y variable. Since $\mathcal{M}''$ is spanned by $P_{M_t} = M_{\chi_{(0,t)\times(0,1)}}$ which only involve the x variable, these unitaries lie in $\mathcal{M}'$ but not $\mathcal{M}''$.

13.8.1. Definition. A nest $\mathcal{N}$ is *multiplicity free* if $\mathcal{N}''$ is maximal abelian. An operator $T \in \mathcal{B}(H)$ is *hyperintransitive* if $\operatorname{Lat} T$ contains a multiplicity free nest.

13.8.2. Example. The Volterra nest $\mathcal{N}$ on $L^2(0,1)$ is multiplicity free. The Volterra operator on $L^2(0,1)$ given by $Vf(x) = \displaystyle\int_x^1 f(t)\,dt$ leaves $N_t \in \mathcal{N}$ invariant. In fact, $\operatorname{Lat} V = \mathcal{N}$ [69, chapter 5]. Let S be an invertible operator such that $S\mathcal{N} = \mathcal{M}$ mentioned above. Then $\operatorname{Lat} SVS^{-1} = S \operatorname{Lat} V = \mathcal{M}$. So V is hyperintransitive but SVS^{-1} is not.

13.8.3. Proposition. *Every compact operator in $\mathcal{B}(H)$ is similar to a hyperintransitive operator.*

Proof. Let K be a compact operator on H. By Corollary 5.5.6, $\operatorname{Lat} K$ contains a maximal subnest $\mathcal{N}$ of $\mathcal{B}(H)$. This just means that every atom is one dimensional. The continuous part may have multiplicity, which would mean that K is not hyperintransitive. Let $\mathcal{M}$ be another maximal nest order isomorphic to $\mathcal{N}$ which is multiplicity free. Let S be invertible such that $S\mathcal{N} = \mathcal{M}$. Then SKS^{-1} is hyperintransitive. ∎

13.8.4. Definition. Say that an algebra $\mathcal{A} \subset \mathcal{B}(H)$ has *factorization* if every invertible positive operator $Q \in \mathcal{B}(H)$ factors as $Q = A^*A$ for some A which is invertible in $\mathcal{A}$.

13.8.5. Lemma. *Let $T \in \mathcal{B}(H)$ be invertible, and let $\mathcal{N}$ be a nest on H. The following are equivalent:*

(1) *There is an $A \in \mathcal{T}(\mathcal{N})^{-1}$ such that $T^*T = A^*A$.*

(2) *There is an $A \in \mathcal{T}(\mathcal{N})^{-1}$ and a unitary $U \in \mathcal{B}(H)$ such that $T = UA$.*

(3) *There is a unitary $U \in \mathcal{B}(H)$ such that $TN = UN$ for all $N \in \mathcal{N}$.*

Proof. If (1) holds, use polar decomposition to write $T = U|T| = U|A|$ and $A = V|A|$. Since T and A are invertible, U and V are unitary. Hence $T = (UV^*)A$.

If (2) holds, $TN = UAN = UN$ for all $N \in \mathcal{N}$. So (3) holds.

If (3) holds, then $A = U^*T$ satisfies $AN = N$ for $N \in \mathcal{N}$. Since A is invertible in $\mathcal{B}(H)$, this shows that $A^{-1}N = N$ as well. Thus $A, A^{-1} \in \mathcal{T}(\mathcal{N})$. Finally, $A^*A = T^*UU^*T = T^*T$. So (1) holds. ∎

13.8.6. Proposition. *A nest $\mathcal{N}$ on a separable Hilbert space is unitarily equivalent to $T\mathcal{N}$ for every invertible operator T if and only if $\mathcal{N}$ is countable.*

Proof. Suppose that $\mathcal{N}$ is countable, and let P be the sum of all of the atoms $\{A \in \mathbb{A}\}$ of $\mathcal{N}$. Then if $P \neq I$, $P^\perp\mathcal{N}$ would be a non-atomic nest, and hence uncountable. Therefore $P = I$. If T is invertible, then $T\mathcal{N}$ is also a countable nest, so its atoms $\{B \in \mathbb{B}\}$ sum to $Q = I$ as well. For each $A \in \mathbb{A}$, there is $N \in \mathcal{N}$ so that $A = P_{N_+} - P_N$. Then $B = P_{TN_+} - P_{TN}$ has the same dimension. The map θ_T induces a bijection $\Theta_T : \mathbb{A} \to \mathbb{B}$ taking A to B. Thus, there is a partial isometry $U_A \in \mathcal{B}(H)$ such that $U_A A U_A^* = B = \Theta_T(A)$ for each $A \in \mathbb{A}$. Hence $U = \sum_{\mathbb{A}} U_A$ maps $PH = H$ isometrically onto $QH = H$, so it is unitary. If $N \in \mathcal{N}$, we can write $P_N = \sum_{A \leq P_N} A$ as a sum of atoms. Then

$$UP_NU^* = \sum_{A \leq P_N} U_A A U_A^* = \sum_{A \leq P_N} \Theta_T(A) = P_{\theta(N)}.$$

Hence $\theta_U = \theta_T$.

On the other hand, if $\mathcal{N}$ is uncountable, then it is similar to a nest with non-atomic part. That part can be multiplicity free and it can have higher multiplicity. As we have seen, these similar nests are not unitarily equivalent. ∎

We now obtain a characterization of which nest algebras have the factorization property.

13.8.7. Corollary. *Let $\mathcal{N}$ be a nest on a separable Hilbert space. Then $\mathcal{T}(\mathcal{N})$ has factorization if and only if $\mathcal{N}$ is countable. If $\mathcal{N}$ is uncountable and $\varepsilon > 0$, there is a compact operator $K = K^*$ with $\|K\| < \varepsilon$ such that $Q = I + K$ does not factor as A^*A for $A \in \mathcal{T}(\mathcal{N})^{-1}$.*

PROOF. The first part is immediate from Proposition 13.8.6 and the equivalence of (1) and (3) in Lemma 13.8.5.

If $\mathcal{N}$ is uncountable, then there is a similar nest $\mathcal{M}$ which is not unitarily equivalent. By the Similarity theorem, there is an invertible operator $T = U + L$ which implements the similarity, where U is unitary, L is compact and $\|L\| < \varepsilon/3$. By Lemma 13.8.5, $T^*T = I + K$ does not factor as A^*A for $A \in \mathcal{T}(\mathcal{N})^{-1}$. Here $K = U^*L + L^*U + L^*L$ is compact and $\|K\| < \varepsilon$. ■

13.8.8. DEFINITION. Let $\operatorname{Aut}\mathfrak{A}$ denote the group of algebraic automorphisms of $\mathfrak{A}$. Also $\operatorname{Inn}\mathfrak{A}$ is the group of *inner automorphisms* $\operatorname{Ad}A$ for $A \in \mathfrak{A}^{-1}$, and $\operatorname{Out}\mathfrak{A}$ is the group $\operatorname{Aut}\mathfrak{A}/\operatorname{Inn}\mathfrak{A}$ of *outer automorphisms.*

If $\mathcal{N}$ is a nest, let $\operatorname{Aut}\mathcal{N}$ denote the group of dimension preserving order automorphisms of $\mathcal{N}$.

13.8.9. THEOREM. *Let $\mathcal{N}$ be a nest on a separable Hilbert space. Every automorphism of $\mathcal{T}(\mathcal{N})$ is spatial, $\alpha = \operatorname{Ad}S$ for $S \in \mathcal{B}(H)^{-1}$, and hence is norm and* WOT*-continuous. If $\|\alpha - id\| < 1$, then α is inner, so* $\operatorname{Inn}\mathcal{T}(\mathcal{N})$ *is an open and closed normal subgroup. The quotient* $\operatorname{Out}\mathcal{T}(\mathcal{N})$ *is naturally isomorphic to* $\operatorname{Aut}\mathcal{N}$.

PROOF. By Ringrose's isomorphism theorem 13.4.7, every automorphism is spatial. Norm and WOT-continuity are automatic. If $\alpha \in \operatorname{Aut}\mathcal{T}(\mathcal{N})$, then $\alpha = \operatorname{Ad}S$ for an invertible operator S. Since $\mathcal{T}(\mathcal{N})' = \mathbb{C}I$ by Proposition 13.1.7, the operator S is unique up to a scalar multiple. Now α induces an order isomorphism of $\mathcal{N}$ by $\Theta(\alpha) = \theta_S$. If $\alpha \in \ker\Theta$, then $SN = N$ for all $N \in \mathcal{N}$, so that S and S^{-1} are in $\mathcal{T}(\mathcal{N})$ and hence α is inner. So far, we have not needed separability.

Conversely, if $\theta \in \operatorname{Aut}(\mathcal{N})$, the Similarity theorem shows that $\theta = \operatorname{Ad}S$ for some invertible S. Hence Θ maps $\operatorname{Aut}\mathcal{T}(\mathcal{N})$ onto $\operatorname{Aut}\mathcal{N}$. Therefore Θ induces an isomorphism of $\operatorname{Out}\mathcal{T}(\mathcal{N})$ onto $\operatorname{Aut}\mathcal{N}$.

Suppose that $\|\alpha - id\| < 1$, and let $\alpha = \operatorname{Ad}S$. If $\theta(S) \neq id_{\mathcal{N}}$, there is some $N \in \mathcal{N}$ such that $SN \neq N$. Then

$$\|\alpha(P_N) - P_N\| = \|SP_NS^{-1} - P_N\| \geq 1.$$

Indeed, if $SN \subsetneq N$, take a unit vector $x \in N \ominus SN$ and note that

$$\|(SP_NS^{-1} - P_N)x\| = \|P_{SN}y - x\| \geq \|x\| = 1,$$

and if $N \subsetneq SN$, a similar argument applies. Therefore $\theta_S = id_{\mathcal{N}}$ and hence α is inner. So *id* is in the interior of $\operatorname{Inn}\mathcal{T}(\mathcal{N})$, and hence $\operatorname{Inn}\mathcal{T}(\mathcal{N})$ is open. The union of all other cosets of $\operatorname{Inn}\mathcal{T}(\mathcal{N})$ is also open, so $\operatorname{Inn}\mathcal{T}(\mathcal{N})$ is closed. It is the kernel of a homomorphism, and thus is normal. ■

The following definition of equivalent idempotents is the ring-theoretic analogue of the Murray–von Neumann equivalence of projections in a von Neumann algebra.

13.8.10. DEFINITION. In a unital Banach algebra $\mathfrak{A}$, two idempotents e and f (i.e. $e^2 = e$) are *algebraically equivalent* if there are $u, v \in \mathfrak{A}$ such that $e = uv$ and $f = vu$. Write $e \sim f$. We also write $e \approx f$ if e is *similar* to f, i.e. there is an $s \in \mathfrak{A}^{-1}$ such that $f = s^{-1}es$.

13.8.11. PROPOSITION. *Algebraic equivalence of idempotents is an equivalence relation.*

Two idempotents $e, f \in \mathfrak{A}$ are similar if and only if $e \sim f$ and $1 - e \sim 1 - f$.

PROOF. Suppose that $e \sim f$, say $e = uv$ and $f = vu$. Let $x = euf$ and $y = fve$. Then $xy = eufve = euvuve = e^4 = e$ and similarly $yx = f$. So we may suppose that $u = euf$ and $v = fve$.

Suppose that $e \sim f$ and $f \sim g$, say $e = uv$, $f = vu = xy$ and $g = yx$ where $u = euf$, $v = fve$, $x = fxg$ and $y = gyx$. Then

$$(ux)(yv) = ufv = e \quad \text{and} \quad (yv)(ux) = yfx = g.$$

So $e \sim g$. The other properties of an equivalence relation are clear.

If $f = s^{-1}es$, then $(1 - f) = s^{-1}(1 - e)s$ and

$$(s^{-1}e)(es) = f \quad \text{and} \quad (es)(s^{-1}e)(es) = e.$$

Thus $f \sim e$ and likewise $1 - f \sim 1 - e$. Conversely, suppose that $e = uv$, $f = vu$, $1 - e = xy$ and $1 - f = yx$ where $u = euf$, $v = fve$, $x = (1 - e)x(1 - f)$ and $y = (1 - f)y(1 - e)$. Let $s = u + x$ and $t = v + y$. Then

$$st = uv + uf(1 - f)y + x(1 - f)fv + xy = e + (1 - e) = 1$$

and

$$ts = vu + ve(1 - e)x + y(1 - e)eu + yx = f + (1 - f) = 1.$$

Hence s is invertible in $\mathfrak{A}$. Finally,

$$s^{-1}es = (v + y)e(u + x) = vex = f.$$

Therefore $f \approx e$. ∎

Now we can characterize equivalence of idempotents in a nest algebra.

13.8.12. LARSON–PITTS THEOREM. *Let $\mathcal{N}$ be a nest on a separable Hilbert space, and let $E, F \in \mathcal{T}(\mathcal{N})$ be idempotents.*

(1) *$E \sim F$ if and only of* $\operatorname{rank}(QEQ) = \operatorname{rank}(QFQ)$ *for all intervals Q of $\mathcal{T}(\mathcal{N})$.*

(2) *$E \approx F$ if and only if for all intervals Q of $\mathcal{T}(\mathcal{N})$,*

$$\operatorname{rank}(QEQ) = \operatorname{rank}(QFQ) \textit{ and } \operatorname{rank}(Q(I - E)Q) = \operatorname{rank}(Q(I - F)Q).$$

PROOF. Suppose that $E \sim F$, say $E = UV$ and $F = VU$ for $U, V \in \mathcal{T}(\mathcal{N})$. By Proposition 13.1.13, the compression map $\Phi_Q(T) = QTQ$ is a homomorphism. Hence

$$\Phi_Q(E) = \Phi_Q(U)\Phi_Q(F)\Phi_Q(V) \quad \text{and} \quad \Phi_Q(F) = \Phi_Q(V)\Phi_Q(E)\Phi_Q(U).$$

These identities show that $\operatorname{rank}(\Phi_Q(E)) \le \operatorname{rank}(\Phi_Q(F)) \le \operatorname{rank}(\Phi_Q(E))$. Hence they are equal.

For the converse, observe that $E\mathcal{N}$ is a nest on EH. Note $E|_N = \Phi_N(E)$ is a bounded idempotent, and thus has closed range. Indeed, $EN = N \cap EH$ because $EN \subset N \cap EH$ and if $x = Ey \in N$, then $x = Ex \in EN$. Since multiplication by E is SOT-continuous, $E\mathcal{N}$ is a complete nest. The same goes for $F\mathcal{N}$.

Define $\theta : E\mathcal{N} \to F\mathcal{N}$ by $\theta(EN) = FN$. If $N_1 < N_1$ and $Q = P_{N_2} - P_{N_1}$, we have by hypothesis that $\operatorname{rank}(QEQ) = \operatorname{rank}(QFQ)$, whence

$$\dim(EN_2 \ominus EN_1) = \dim(FN_2 \ominus FN_1).$$

This shows that θ preserves dimension, and also shows that it is well defined. By the Similarity theorem, there is an invertible operator $A \in \mathcal{B}(EH, FH)$ such that $AEN = FN$ for $N \in \mathcal{N}$. Setting $B = A^{-1}$, we have $BFN = EN$ as well. We may consider $U = FAE$ and $V = EBF$ as elements of $\mathcal{B}(H)$. Then $UN = FN \subset N$ and $VN = EN \subset N$ for $N \in \mathcal{N}$, so $U, V \in \mathcal{T}(\mathcal{N})$. Clearly $UV = F$ and $VU = E$, so $E \sim F$.

The second statement follows from the first part combined with the second part of Proposition 13.8.11. ∎

13.9. The Distance to a MASA

We establish a distance estimate from any operator in $\mathcal{B}(H)$ to a masa. This will be utilized in the next section on perturbations.

13.9.1. DEFINITION. If $\mathcal{A}$ is a reflexive subalgebra of $\mathcal{B}(H)$, define a seminorm on $\mathcal{B}(H)$ by

$$\beta(T) = \sup_{L \in \operatorname{Lat} \mathcal{A}} \|P_L^\perp T P_L\| \quad \text{for} \quad T \in \mathcal{B}(H).$$

Say that $\mathcal{A}$ is *hyperreflexive* if there is a constant C so that $\operatorname{dist}(T, \mathcal{A}) \le C\beta(T)$ for all $T \in \mathcal{B}(H)$. The optimal constant C is the *distance constant.*

For any $A \in \mathcal{A}$ and $L \in \operatorname{Lat} \mathcal{A}$, $P_L^\perp A P_L = 0$, whence

$$\beta(T) = \sup_{L \in \operatorname{Lat} \mathcal{A}} \|P_L^\perp (T - A) P_L\| \le \inf_{A \in \mathcal{A}} \|T - A\| = \operatorname{dist}(T, \mathcal{A}).$$

On the other hand, $\beta(T) = 0$ means that $T \in \operatorname{Alg} \operatorname{Lat} \mathcal{A} = \mathcal{A}$. We have seen that nest algebras are hyperreflexive with an exact formula, i.e. $C = 1$.

Let $\mathcal{D}$ be a maximal abelian subalgebra of $\mathcal{B}(H)$. It has several properties which can be used to estimate the distance of an operator T to $\mathcal{D}$. Since $\mathcal{D} = \mathcal{D}'$, the invariant subspaces are the ranges of projections in $\mathcal{D}$, i.e. $\operatorname{Lat}\mathcal{D} = \operatorname{Pr}(\mathcal{D})$, the set of projections in $\mathcal{D}$.

Define a derivation $\delta_T(A) = AT - TA$ from $\mathcal{D}$ into $\mathcal{B}(H)$. Then $\delta_T = 0$ precisely when $T \in \mathcal{D}' = \mathcal{D}$. If $D \in \mathcal{D}$ with $\|T - D\| = \operatorname{dist}(T, \mathcal{D})$, then for $A \in \mathcal{D}$,

$$\|\delta_T(A)\| = \|\delta_{T-D}(A)\| \le \|A\|\,\|T-D\| + \|T-D\|\,\|A\| \le 2\operatorname{dist}(T,\mathcal{D})\|A\|.$$

Now the self-adjoint part of the unit ball of $\mathcal{D}$ is the closed convex hull of the symmetries $2P - I = P - P^\perp$ by Corollary 9.3.7. Therefore

$$\begin{aligned}\|\delta_T|_{\mathcal{D}_{sa}}\| &= \sup_{P\in\operatorname{Pr}(\mathcal{D})} \|\delta_T(2P-1)\| = \sup_{P\in\operatorname{Pr}(\mathcal{D})} 2\|TP - PT\| \\ &= 2 \sup_{P\in\operatorname{Pr}(\mathcal{D})} \max\{\|P^\perp TP\|, \|PTP^\perp\| = 2\beta(T).\end{aligned}$$

But $\|\delta_T|_{\mathcal{D}_{sa}}\| \le \|\delta_T\| \le 2\|\delta_T|_{\mathcal{D}_{sa}}\|$. Hence

$$2\beta(T) \le \|\delta_T\| \le 4\beta(T).$$

The purpose of this section is to establish the following more precise estimate.

13.9.2. THEOREM. *Let $\mathcal{D}$ be a maximal abelian subalgebra of $\mathcal{B}(H)$. Then for $T \in \mathcal{B}(H)$,*

$$\beta(T) = \sup_{P\in\operatorname{Pr}(\mathcal{D})} \|P^\perp TP\| \le \operatorname{dist}(T,\mathcal{D}) \le \|\delta_T|_{\mathcal{D}_{sa}}\| = 2\beta(T).$$

Thus a masa is hyperreflexive with distance constant at most 2.

PROOF. Note that $\operatorname{Pr}(\mathcal{D}) = \operatorname{Lat}\mathcal{D}$ is a commutative lattice with $P \wedge Q = PQ$ and $P \vee Q = P + Q - PQ$. Also

$$(2P - I)(2Q - I) = 2(2PQ - P - Q) + I = 2(PQ + P^\perp Q^\perp) - I.$$

This shows that $G = \{2P - I : P \in \operatorname{Pr}(\mathcal{D})\}$ is an abelian group. Since $\operatorname{Pr}(\mathcal{D})$ spans $\mathcal{D}$, we have $G' = \mathcal{D}$. By Proposition 10.6.4, there is an expectation $\mathcal{E}$ of $\mathcal{B}(H)$ onto $\mathcal{D}$, using a function $f_T(u) = uTu^*$ for $u \in G$ and $T \in \mathcal{B}(H)$, given by

$$\langle \mathcal{E}(T)\xi, \eta\rangle = m(\langle f_T(u)\xi, \eta\rangle) \quad \text{for} \quad \xi, \eta \in H,$$

where m is an invariant mean on $l^\infty(G)$. We estimate

$$\begin{aligned}\|T - \mathcal{E}(T)\| &= \sup_{\|\xi\|=\|\eta\|=1} |\langle T - \mathcal{E}(T)\xi, \eta\rangle| = \sup_{\|\xi\|=\|\eta\|=1} m(\langle (T - uTu^*)\xi, \eta\rangle) \\ &\le \sup_{u\in G} \|Tu - uT\| = \|\delta_T|_{\mathcal{D}_{sa}}\| = 2\beta(T).\end{aligned}$$

■

13.10. Perturbations

In this section, we consider nests which are close in a certain sense, and show that this implies that the nest algebras are similar and close. Conversely, if two nest algebras are close, then so are the nests, and the algebras are similar.

13.10.1. DEFINITION. Two subspaces M and N are in *generic position* if

$$M \cap N = \{0\}, \quad M \cap N^{\perp} = \{0\}, \quad M^{\perp} \cap N = \{0\}, \quad \text{and} \quad M^{\perp} \cap N^{\perp} = \{0\}.$$

Note that $\begin{bmatrix} \cos^2\theta & \cos\theta\sin\theta \\ \cos\theta\sin\theta & \sin^2\theta \end{bmatrix}$ is a projection. In the following, the letter C suggests cosine and S suggests sine.

13.10.2. PROPOSITION. *If subspaces M and N are in generic position, then there are positive contractions C and $S = (I - C^2)^{1/2}$ with* $\ker C = \ker S = \{0\}$ *and a simultaneous unitary equivalence*

$$P_M \simeq \begin{bmatrix} I & 0 \\ 0 & 0 \end{bmatrix} \quad \textit{and} \quad P_N \simeq \begin{bmatrix} C^2 & CS \\ CS & S^2 \end{bmatrix}.$$

PROOF. Decompose $H = M \oplus M^{\perp}$. Then we get

$$P_M \simeq \begin{bmatrix} I & 0 \\ 0 & 0 \end{bmatrix} \quad \text{and} \quad P_N \simeq \begin{bmatrix} X & Y \\ Y^* & Z \end{bmatrix}.$$

Then X and Z are positive contractions. Since $M \cap N = \{0\}$, 1 is not an eigenvalue of X. Similarly, 0 is not an eigenvalue of X, and 0, 1 are not eigenvalues of Z. Since $P_N^2 = P_N$,

$$X^2 + YY^* = X, \quad Z^2 + Y^*Y = Z \quad \text{and} \quad XY + YZ = Y.$$

Since 1 is not an eigenvalue of X, Y is injective. Similarly Y^* is injective, so that Y has dense range in M. By the polar decomposition, we see that $Y = U|Y|$ where U is a partial isometry that must map $M^{\perp}$ isometrically onto M. If we use U as the identification between $M^{\perp}$ and M, we can assume that Y is positive. Let $C = X^{1/2}$ and $S = Z^{1/2}$. Our identities become

$$Y^2 = C^2 - C^4 = S^2 - S^4 \quad \text{and} \quad C^2Y + YS^2 = Y.$$

Thus Y commutes with C and S and $Y(C^2 + S^2) = Y$. Since Y is injective, $C^2 + S^2 = I$. Therefore $S = (I - C^2)^{1/2}$. Finally $Y^2 = C^2(I - C^2) = (CS)^2$, whence $Y = CS$. ■

13.10.3. LEMMA. *Suppose that P and Q are orthogonal projections such that* $\|P - Q\| = \sin\theta < 1$ *for* $0 \le \theta < \frac{\pi}{2}$. *Let U be the partial isometry in the polar decomposition of QP. Then* $\|U - P\| = 2\sin\frac{\theta}{2} < \sqrt{2}$.

PROOF. If $P \wedge Q = R \neq 0$, we can replace P and Q by $P - R$ and $Q - R$. This does not affect the problem unless $P = Q$, which is a trivial case. Likewise, we may suppose that $P \vee Q = I$. Then PH and QH are in generic position.

Let $C = (PQP)^{1/2}$ and $S = (1 - C^2)^{1/2}$. By Proposition 13.10.2, we may suppose that P, Q and U are unitarily equivalent to

$$P = \begin{bmatrix} I & 0 \\ 0 & 0 \end{bmatrix}, \quad Q = \begin{bmatrix} C^2 & CS \\ CS & S^2 \end{bmatrix} \quad \text{and} \quad U = \begin{bmatrix} C & 0 \\ S & 0 \end{bmatrix}.$$

Then since the two columns of $P - Q$ have orthogonal ranges,

$$\|P - Q\| = \left\| \begin{bmatrix} S^2 & -CS \\ -CS & -S^2 \end{bmatrix} \right\| = \|(S^4 + C^2S^2)\|^{1/2} = \|S\|.$$

So $\|S\| = \sin\theta$, and thus $C = (1 - S^2)^{1/2} \geq \cos\theta\, I$, and $\cos\theta$ is optimal. Therefore

$$\|U - P\| = \left\| \begin{bmatrix} C - I & 0 \\ S & 0 \end{bmatrix} \right\| = \|(C - I)^2 + S^2\|^{1/2} = \sqrt{\|2(I - C)\|}.$$

Hence $\|U - P\| = \sqrt{2(1 - \cos\theta)} = 2\sin\frac{\theta}{2}$. ■

This lemma allows us to establish the following result only assuming that $\|P_i - Q_i\| < 1$, instead of requiring a smaller bound.

13.10.4. THEOREM. *Let P_1 and P_2 (respectively Q_1 and Q_2) be pairwise orthogonal projections on a separable Hilbert space. Suppose that $\|P_1 - Q_1\| < 1$ and $\|P_2 - Q_2\| < 1$. Then*

$$\operatorname{rank}(P_1 + P_2)^\perp = \operatorname{rank}(Q_1 + Q_2)^\perp.$$

PROOF. We may suppose that $\operatorname{rank}(P_1 + P_2)^\perp \leq \operatorname{rank}(Q_1 + Q_2)^\perp$. If both are infinite, there is nothing to prove, so we assume that $\operatorname{rank}(P_1 + P_2)^\perp < \infty$. Form the polar decomposition $Q_iP_i = U_i(P_iQ_iP_i)^{1/2}$ for $i = 1, 2$. Then $U = U_1 + U_2$ is a partial isometry of $(P_1 + P_2)H$ onto $(Q_1 + Q_2)H$. Thus, U is left semi-Fredholm since it has finite nullity and closed range. However by the previous lemma,

$$\begin{aligned} \|U - (P_1 + P_2)\|^2 &= \big\|(U_1 - P_1)P_1 + (U_2 - P_2)P_2\big\|^2 \\ &= \big\|\big((U_1 - P_1)P_1 + (U_2 - P_2)P_2\big)\big(P_1(U_1^* - P_1) + P_2(U_2^* - P_2)\big)\big\| \\ &\leq \|(U_1 - P_1)P_1(U_1^* - P_1)\| + \|(U_2 - P_2)P_2(U_2^* - P_2)\| \\ &= \|U_1 - P_1\|^2 + \|U_2 - P_2\|^2 < 4. \end{aligned}$$

Thus $\|U - (P_1 + P_2)\| < 2$. If $\operatorname{ind} U \neq 0$, then in the Calkin algebra, $\pi(U)$ would be the quotient of a proper isometry. Thus if $\operatorname{rank}(Q_1 + Q_2)^\perp < \infty$, then $\sigma_e(U) = \mathbb{T}$ or if $\operatorname{rank}(Q_1 + Q_2)^\perp = \infty$, then $\sigma_e(U) = \overline{\mathbb{D}}$. Either would imply that

$$\|U - (P_1 + P_2)\|_e = \|\pi(U - I)\| = 2.$$

Hence $\operatorname{ind} U = 0$, which means that $\operatorname{rank}(P_1 + P_2)^\perp = \operatorname{rank}(Q_1 + Q_2)^\perp$. ■

Our first perturbation result shows that close nests are similar.

13.10.5. THEOREM. *Let $\theta : \mathcal{N} \to \mathcal{M}$ be an order isomorphism such that $\|\theta - id_{\mathcal{N}}\| = \gamma < 1$. Then there is an invertible operator S such that $\theta_S = \theta$. If $\gamma < \frac{1}{2}$, we can arrange that $\|S - I\| < 2\gamma$.*

PROOF. If $N_1 < N_2 \in \mathcal{N}$, let $M_i = \theta(N_i)$. Then $P_1 = P_{N_1}$ and $P_2 = P_{N_2}^{\perp}$ are orthogonal projections, as are $Q_1 = P_{M_1}$ and $Q_2 = P_{M_2}^{\perp}$. By hypothesis, we have $\|P_1 - Q_1\| < 1$ and $\|P_2 - Q_2\| = \|P_{M_2} - P_{N_2}\| < 1$. By Theorem 13.10.4,

$$\dim(N_2 \ominus N_1) = \operatorname{rank}(P_1 + P_2)^{\perp} = \operatorname{rank}(Q_1 + Q_2)^{\perp} = \dim(M_2 \ominus M_1).$$

Therefore θ preserves dimension.

By the Similarity theorem, there is an invertible operator $T = U + K$ where U is unitary, K is compact, and $\|K\| < \delta$ so that $\theta_T = \theta$. Take δ so small that

$$\gamma + \gamma\delta + \delta < 1 \quad \text{and} \quad \gamma < \frac{(1-\delta)^2}{1+\delta} \quad \text{such as} \quad \delta \leq \frac{1-\gamma}{3}.$$

When $\gamma < \frac{1}{2}$, we also want $\dfrac{1+\delta}{1-\gamma-\gamma\delta-\delta} < 2$ such as $\delta \leq \frac{1-2\gamma}{4}$. Use Arveson's distance formula,

$$\begin{aligned} \operatorname{dist}(T, \mathcal{T}(\mathcal{N})) &= \sup_{N \in \mathcal{N}} \|P_N^{\perp} T P_N\| = \sup_{N \in \mathcal{N}} \|P_N^{\perp} P_{\theta(M)} T P_N\| \\ &\leq \sup_{N \in \mathcal{N}} \|P_N - P_{\theta(N)}\| \|T\| < \gamma(1+\delta). \end{aligned}$$

Choose $V \in \mathcal{T}(\mathcal{N})$ with $\|T - V\| < \gamma(1+\delta)$. Then

$$\|U - V\| \leq \|U - T\| + \|T - V\| < \delta + \gamma + \gamma\delta < 1;$$

so V is invertible.

We claim that $V^{-1} \in \mathcal{T}(\mathcal{N})$. Since $VN \subseteq N$, this is equivalent to showing that $VN = N$, i.e. $\theta_V = id_{\mathcal{N}}$. Suppose contrarily that for some N, $VN \subsetneq N$. Let $M = \theta(N)$. Let $P_1 = P_{VN}$, $P_2 = P_N^{\perp}$, $Q_1 = P_M$ and $Q_2 = P_M^{\perp}$. Since

$$\operatorname{rank}(P_1 + P_2)^{\perp} \geq 1 \neq 0 = \operatorname{rank}(Q_1 + Q_2)^{\perp} \quad \text{and} \quad \|P_2 - Q_2\| \leq \gamma < 1,$$

Theorem 13.10.4 shows that

$$1 = \|P_1 - Q_1\| = \max\left\{\|P_1 Q_1^{\perp}\|, \|P_1^{\perp} Q_1\|\right\} \leq \max\left\{\|P_N Q_1^{\perp}\|, \|P_1^{\perp} Q_1\|\right\}.$$

Since $\|P_N Q_1^{\perp}\| \leq \|P_N - P_M\| \leq \gamma$, we have $\|P_1^{\perp} Q_1\| = 1$. Thus, there is a unit vector $\eta = P_M \eta$ so that $\|P_1^{\perp}\eta\| > 1 - \delta$. Let $\xi = \|T^{-1}\eta\|^{-1} T^{-1}\eta$. This is a unit vector in N. Since $\|T^{-1}\| \leq (1-\delta)^{-1}$, $T\xi = t\eta$ where $t \geq 1 - \delta$. Then

$$\begin{aligned} \gamma(1+\delta) &> \|(T - V)\xi\| = \|t\eta - V\xi\| \\ &\geq \|P_1^{\perp}(t\eta - V\xi)\| \geq (1-\delta)\|P_1^{\perp}\eta\| > (1-\delta)^2. \end{aligned}$$

This contradicts the initial choice of δ, so $VN = N$ and $V \in \mathcal{T}(\mathcal{N})^{-1}$.

Thus $S = TV^{-1}$ implements θ. Now $\|U - V\| < \gamma + \gamma\delta + \delta$ implies that $\|V^{-1}\| \le (1 - \gamma - \gamma\delta - \delta)^{-1}$. If $\gamma < \frac{1}{2}$, then our choice of δ ensures that

$$\|S - I\| = \|(T - V)V^{-1}\| \le \frac{\gamma(1+\delta)}{1 - \gamma - \gamma\delta - \delta} < 2\gamma.$$

If $\gamma \ge \frac{1}{2}$, we need a bigger constant such as $\|T - I\| \le 2 + \delta < 5\gamma$. ■

The following corollary is immediate from Theorem 13.10.5 and the Similarity theorem.

13.10.6. Corollary. *Let $\theta : \mathcal{N} \to \mathcal{M}$ be an order isomorphism between nests on separable Hilbert space. Then $\mathcal{N}$ and $\mathcal{M}$ are similar if and only if there is a unitary operator U such that $\|\theta - \theta_U\| < 1$.*

13.10.7. Definition. If L and M are subspaces of a Banach space X, we define a distance between them:

$$d(L, M) = \max\{\sup_{\xi \in b_1(L)} \|\xi + M\|, \sup_{\eta \in b_1(M)} \|\eta + L\|\}.$$

This distance is comparable to the Hausdorff distance between the unit balls. If L and M are C*-algebras, these two quantities are equal. But for nonself-adjoint operator algebras, it will affect constants.

We are interested in showing that two nest algebras which are close to one another are similar. We begin with the converse. Close nests yield close nest algebras because of the distance formula. This result is only of interest when $\gamma < \frac{1}{2}$ since the distance between two subspaces is always at most 1.

13.10.8. Proposition. *Let $\theta : \mathcal{N} \to \mathcal{M}$ be an order isomorphism such that $\|\theta - id_{\mathcal{N}}\| = \gamma < \frac{1}{2}$. Then $d(\mathcal{T}(\mathcal{N}), \mathcal{T}(\mathcal{M})) \le 2\gamma$.*

Proof. Let $A \in \mathcal{T}(\mathcal{M})$ with $\|A\| \le 1$. Then with $M = \theta(N)$,

$$\begin{aligned} \operatorname{dist}(A, \mathcal{T}(\mathcal{N})) &= \sup_{N \in \mathcal{N}} \|P_N^\perp A P_N\| = \sup_{N \in \mathcal{N}} \|P_N^\perp A P_N - P_M^\perp A P_M\| \\ &= \sup_{N \in \mathcal{N}} \|P_N^\perp A (P_N - P_M) + (P_N^\perp - P_M^\perp) A P_M\| \\ &\le \sup_{N \in \mathcal{N}} 2\|P_N - P_M\| = 2\gamma. \end{aligned}$$

Reversing the roles of $\mathcal{M}$ and $\mathcal{N}$ yields $d(\mathcal{T}(\mathcal{N}), \mathcal{T}(\mathcal{M})) \le 2\gamma$. ■

13.10.9. Lemma. *Let $\mathcal{B}$ be a unital Banach algebra. Suppose that $B \in \mathcal{B}$ with $\|B - B^2\| = \varepsilon < \frac{1}{7}$ and $\|B\| \le \beta$. Then there is an idempotent $F = F^2 \in \mathcal{B}$ such that $\|B - F\| \le (4\beta + 2)\varepsilon$.*

PROOF. Use the power series expansion $(1-x)^{-1/2} = \sum_{n\geq 0} \binom{2n}{n}\left(\frac{x}{4}\right)^n$, which has radius of convergence 1, to define

$$A = \left(I - 4(B - B^2)\right)^{-1/2} = \sum_{n\geq 0} \binom{2n}{n} (B - B^2)^n.$$

This converges absolutely to an element of $\mathcal{B}$. Moreover

$$\begin{aligned} \|A - I\| &\leq \sum_{n\geq 1} \binom{2n}{n} \varepsilon^n = (1 - 4\varepsilon)^{-1/2} - 1 \\ &= \frac{4\varepsilon}{1 - 4\varepsilon + \sqrt{1 - 4\varepsilon}} < \frac{4\varepsilon}{\frac{3}{7} + \sqrt{\frac{3}{7}}} < 4\varepsilon. \end{aligned}$$

Since A comes from the functional calculus of $B - B^2$, it commutes with B and

$$A^2(I - 2B)^2 = (I - 4(B - B^2))^{-1}(I - 4B + 4B^2) = I.$$

Set $I - 2F = A(I - 2B)$; so that $I - 4(F - F^2) = I$ and thus $F = F^2$. Finally

$$\begin{aligned} \|F - B\| &= \frac{1}{2}\|(I - 2B) - (I - 2F)\| = \frac{1}{2}\|(I - A)(I - 2B)\| \\ &< \frac{1}{2} 4\varepsilon(2\beta + 1) = (4\beta + 2)\varepsilon. \end{aligned}$$

■

We can now show that any operator algebra which is close to a nest algebra is itself a nest algebra similar to the first one.

13.10.10. THEOREM. *Let $\mathcal{N}$ be a nest on a separable Hilbert space H, and let $\mathcal{B}$ be a norm-closed subalgebra of $\mathcal{B}(H)$ such that $d(\mathcal{T}(\mathcal{N}), \mathcal{B}) = \gamma < \frac{1}{20}$. Then $\mathcal{B}$ is a nest algebra similar to $\mathcal{T}(\mathcal{N})$ via an invertible operator S with $\|S - I\| < 4\gamma$.*

PROOF. There is a masa $\mathcal{D}$ with $\mathcal{N}'' \subset \mathcal{D} \subset \mathcal{N}'$. Let $L \in \mathcal{L} = \operatorname{Lat} \mathcal{B}$ and let $P = P_L$. For $D \in b_1(\mathcal{D})$, select $B, C \in \mathcal{B}$ so that $\|D - B\| \leq \gamma$ and $\|D^* - C\| \leq \gamma$. Then

$$\begin{aligned} \|PD - DP\| &= \max\{\|P^\perp DP\|, \|PDP^\perp\|\} \\ &= \max\{\|P^\perp(D - B)P\|, \|P(D - C^*)P^\perp\|\} \leq \gamma. \end{aligned}$$

Hence $\|\delta_P|_{\mathcal{D}}\| \leq \gamma$. By Theorem 13.9.2, $\operatorname{dist}(P, \mathcal{D}) \leq \gamma$.

Let $D \in \mathcal{D}$ with $\|P - D\| \leq \gamma$. Since $P = P^*$, this remains valid if we replace D with $\frac{1}{2}(D + D^*)$, so we can assume that $D = D^*$. Now if $\operatorname{dist}(\lambda, \{0, 1\}) > \gamma$, we have

$$\|(\lambda I - P)^{-1}\| = \frac{1}{\operatorname{dist}(\lambda, \sigma(P))} < \frac{1}{\gamma}.$$

Since $\|(\lambda I - P) - (\lambda I - D)\| \leq \gamma$, $\lambda I - D$ is invertible. Therefore

$$\sigma(D) \subset [-\gamma, \gamma] \cup [1 - \gamma, 1 + \gamma].$$

Using the continuous functional calculus, $Q := \chi_{[1-\gamma,1+\gamma]}(D)$ is a projection in $\mathcal{D}$ with

$$\|P - Q\| \leq \|P - D\| + \|D - Q\| \leq 2\gamma < \frac{1}{10} < \frac{1}{2}.$$

Next we show that $Q = P_N$ for some $N \in \mathcal{N}$. For if N isn't in $\mathcal{N}$, there is an element $T \in \mathcal{T}(\mathcal{N})$ with $A = Q^\perp T Q \neq 0$. We may normalize so that $\|A\| = 1$. Choose $B \in \mathcal{B}$ with $\|A - B\| \leq \gamma$. Then

$$\begin{aligned} 1 = \|A\| &= \|Q^\perp A Q - P^\perp B P\| \\ &\leq \|(Q^\perp - P^\perp) A Q\| + \|P^\perp (A - B) Q\| + \|P^\perp B (Q - P)\| \\ &\leq 2\gamma + \gamma + 2\gamma = 5\gamma < 1. \end{aligned}$$

This contradiction shows that Q belongs to $\mathcal{N}$. The projection Q is unique, for if Q' is any other projection in $\mathcal{D}$, then

$$\|P - Q'\| \geq \|Q - Q'\| - \|P - Q\| \geq 1 - 2\gamma > \frac{9}{10} > \frac{1}{2}.$$

Thus we can define a map $\theta : \mathcal{L} \to \mathcal{N}$ by $\theta(L) = N$ where $Q = P_N$ for the unique $N \in \mathcal{N}$ with $\|P - Q\| < \frac{1}{2}$.

Conversely, let $Q = P_N$ for some $N \in \mathcal{N}$. Since $\|2Q - I\| = 1$, we can choose $B \in \mathcal{B}$ with $\|(2Q - I) - (2B - I)\| \leq \gamma$, whence $\|Q - B\| \leq \frac{\gamma}{2}$. So $\|B\| \leq 1 + \frac{\gamma}{2}$. Compute

$$\begin{aligned} \|B - B^2\| &= \|(B - B^2) - (Q - Q^2)\| = \|B - Q - B(B - Q) - (B - Q)Q\| \\ &\leq \|(B - Q)Q^\perp\| + \|B(B - Q)\| \leq (1 + \|B\|)\frac{\gamma}{2} \leq \gamma + \tfrac{1}{4}\gamma^2. \end{aligned}$$

Apply Lemma 13.10.9 to obtain an idempotent $F \in \mathcal{B}$ with

$$\|F - B\| \leq (4\|B\| + 2)(\gamma + \tfrac{1}{4}\gamma^2) < 6.2\gamma.$$

Hence $\|F - Q\| < 7\gamma$.

Let $M = FH$. If M is not invariant for $\mathcal{B}$, there is an element $C \in \mathcal{B}$ such that $D = (I - F)CF \neq 0$. Normalize so that $\|D\| = 1$. Select $T \in \mathcal{T}(\mathcal{N})$ with $\|T - D\| \leq \gamma$. Then $\|F\| < 1 + 7\gamma$ and $\|T\| \leq 1 + \gamma$, whence

$$\begin{aligned} 1 = \|D\| &= \|(I - F)DF - Q^\perp T Q\| \\ &\leq \|(I - F - Q^\perp)DF + Q^\perp(D - T)F + Q^\perp T(F - Q)\| \\ &< 7\gamma\|F\| + \gamma\|F\| + \|T\|7\gamma < (15 + 63\gamma)\gamma < .91 < 1. \end{aligned}$$

This contradiction shows that M is invariant.

Let $P = P_M$. Note that $PF = F$ and $FP = P$. Thus

$$\begin{aligned} \|Q - P\| &= \|P^\perp(Q - P) + P(Q - P)\| = \|P^\perp Q - PQ^\perp\| \\ &= \max\left\{\|P^\perp Q\|, \|PQ^\perp\|\right\} = \max\left\{\|P^\perp(Q - F)\|, \|P(Q^\perp - (I - F))\|\right\} \\ &= \max\left\{\|P^\perp(Q - F)\|, \|P(F - Q))\|\right\} \leq \|Q - F\| < 7\gamma < \frac{1}{2}. \end{aligned}$$

Therefore $\theta(M) = N$. Hence θ is surjective.

Now we show that θ is a bijection. Let $N \in \mathcal{N}$ and let $M \in \mathcal{L}$ be the invariant subspace for $\mathcal{B}$ just found which is the range of an idempotent $F \in \mathcal{B}$ with $\|F - P_N\| < 7\gamma$. Suppose that $L \in \mathcal{L}$ also has $\theta(L) = N$. From the early part of the proof, we have $\|P_N - P_L\| \le 2\gamma$. Since $F \in \mathcal{B}$, it leaves L invariant. Write $H = L \oplus L^\perp$. With respect to this decomposition, F has a matrix $F = \begin{bmatrix} F_{11} & F_{12} \\ 0 & F_{22} \end{bmatrix}$. So F_{11} and F_{22} are idempotents. Since $P_L = \begin{bmatrix} I & 0 \\ 0 & 0 \end{bmatrix}$ and $\|P_L - F\| < 9\gamma < 1$, we have $\|F_{11} - I\| < 1$ and $\|F_{22}\| < 1$. This forces $F_{11} = I$ and $F_{22} = 0$. Thus $L = \operatorname{Ran} F = M$.

Next we show that $\mathcal{L}$ is a nest and θ is an order isomorphism. Suppose that $N_1 < N_2$ in $\mathcal{N}$, and let $L_1, L_2 \in \mathcal{L}$ with $\theta(L_i) = N_i$. We have $\|P_{L_i} - P_{N_i}\| \le 2\gamma$. There are idempotents $F_1, F_2 \in \mathcal{B}$ with $\operatorname{Ran} F_i = L_i$ such that $\|P_{N_i} - F_i\| \le 7\gamma$. Observe that since L_2 is invariant for F_1,

$$(F_1F_2)^2 = F_1(F_2F_1F_2) = F_1(F_1F_2) = F_1F_2$$

is an idempotent whose range equals $L_1 \cap L_2$. Compute

$$\begin{aligned}\|P_{L_1} - F_1F_2\| &\le \|P_{L_1} - P_{N_1}\| + \|P_{N_1}(P_{N_2} - F_2)\| + \|(P_{N_1} - F_1)F_2\| \\ &\le 2\gamma + 7\gamma + 7\gamma(1 + 7\gamma) < 19\gamma < 1.\end{aligned}$$

As the range of F_1F_2 is contained in L_1, this forces $\operatorname{Ran} F_1F_2 = L_1$. Therefore $L_1 \cap L_2 = L_1$, i.e. $L_1 < L_2$.

We have an order isomorphism of nests $\theta : \mathcal{L} \to \mathcal{N}$ with $\|\theta - id\| \le 2\gamma < \frac{1}{10}$. By Theorem 13.10.5, there is an invertible S with $\|S - I\| < 4\gamma$ so that $S\mathcal{L} = \mathcal{N}$. Moreover, by Proposition 13.10.8, $d(\mathcal{T}(\mathcal{N}), \mathcal{T}(\mathcal{L})) \le 4\gamma$. However $\mathcal{B} \subseteq \mathcal{T}(\mathcal{L})$ and

$$d(\mathcal{B}, \mathcal{T}(\mathcal{L})) \le d(\mathcal{B}, \mathcal{T}(\mathcal{N})) + d(\mathcal{T}(\mathcal{N}), \mathcal{T}(\mathcal{L})) \le 5\gamma < 1.$$

A proper subspace is distance 1, so $\mathcal{B} = \mathcal{T}(\mathcal{L})$. ■

When $\mathcal{B}$ is known to be a nest algebra, we can get improved constants by using just part of the proof in both directions.

13.10.11. LANCE THEOREM. *Let $\mathcal{N}$ and $\mathcal{M}$ be nests on a separable Hilbert space H. Suppose that $d(\mathcal{T}(\mathcal{N}), \mathcal{T}(\mathcal{M})) = \gamma < \frac{1}{5}$. Then $\mathcal{T}(\mathcal{M})$ is a nest algebra similar to $\mathcal{T}(\mathcal{N})$ via an invertible operator S with $\|S - I\| < 4\gamma$.*

PROOF. The first three paragraphs of the proof of Theorem 13.10.10 show that there is a map $\theta : \mathcal{M} \to \mathcal{N}$ such that $\|\theta - id_{\mathcal{M}}\| \le 2\gamma$. (Note that we needed $5\gamma < 1$ in the proof.) Reversing the role of $\mathcal{M}$ and $\mathcal{N}$ yields a map $\eta : \mathcal{N} \to \mathcal{M}$ such that $\|\eta - id_{\mathcal{N}}\| \le 2\gamma$. Therefore $\|\eta\theta - id_{\mathcal{M}}\| \le 4\gamma$ and $\|\theta\eta - id_{\mathcal{N}}\| \le 4\gamma$. Since $\|P_{M_1} - P_{M_2}\| = 1$ if $M_1 \ne M_2$, this forces $\eta\theta = id_{\mathcal{M}}$. Similarly $\theta\eta = id_{\mathcal{N}}$. Therefore θ is an order isomorphism. By Theorem 13.10.5, $\mathcal{N}$ and $\mathcal{M}$ are similar via an invertible operator S with $\|S - I\| \le 4\gamma$. ■

13.11. Interpolation Theorem

In this section, we will establish a result of Orr on factoring certain elements in continuous nest algebras. It begins with an analysis of Larson's ideal $\mathcal{R}^\infty(\mathcal{N})$.

13.11.1. Definition. Let $\mathcal{N}$ be a nest on a separable Hilbert space H. A *partition* of $\mathcal{N}$ is a countable family $\{E_n : n \geq 1\}$ of intervals of $\mathcal{N}$ such that SOT-$\sum_n E_n = I$. Let $\mathcal{R}^\infty(\mathcal{N})$ denote the set of elements $T \in \mathcal{T}(\mathcal{N})$ such that for each $\varepsilon > 0$, there is a partition $\{E_n : n \geq 1\}$ so that $\|E_n T E_n\| < \varepsilon$ for $n \geq 1$.

For $N \in \mathcal{N} \setminus \{0\}$, define $i_N^-(T) = \inf_{M<N} \|(P_N - P_M)T(P_N - P_M)\|$. For $N \in \mathcal{N} \setminus H$, define $i_N^+(T) = \inf_{N<M} \|(P_M - P_N)T(P_M - P_N)\|$.

13.11.2. Proposition. *Let $\mathcal{N}$ be a nest on a separable Hilbert space H. Then $\mathcal{R}^\infty(\mathcal{N})$ is a norm-closed two-sided ideal of $\mathcal{T}(\mathcal{N})$.*

Proof. Suppose that $\mathcal{P} = \{E_n : n \geq 1\}$ is a partition of $\mathcal{N}$. Define $\Delta_\mathcal{P} : \mathcal{T}(\mathcal{N}) \to \mathcal{T}(\mathcal{N})$ by $\Delta_\mathcal{P}(T) = \sum_n E_n T E_n$. It is readily checked that this is a unital contractive homomorphism. Hence $\|\Delta_\mathcal{P}(T)\|$ is a continuous seminorm on $\mathcal{T}(\mathcal{N})$. Now $\mathcal{R}^\infty(\mathcal{N}) = \{T \in \mathcal{T}(\mathcal{N}) : \inf_\mathcal{P} \|\Delta_\mathcal{P}(T)\| = 0\}$. Therefore $\mathcal{R}^\infty(\mathcal{N})$ is a norm-closed subspace. If $A, B \in \mathcal{T}(\mathcal{N})$, $T \in \mathcal{R}^\infty(\mathcal{N})$ and $\|\Delta_\mathcal{P}(T)\| < \varepsilon$, then

$$\|\Delta_\mathcal{P}(ATB)\| = \|\Delta_\mathcal{P}(A)\Delta_\mathcal{P}(T)\Delta_\mathcal{P}(B)\| \leq \|A\|\|\Delta_\mathcal{P}(T)\|\|B\| < \|A\|\|B\|\varepsilon.$$

It follows that $\mathcal{R}^\infty(\mathcal{N})$ is a two-sided ideal. ■

13.11.3. Lemma. *Let $\mathcal{N}$ be a nest on a separable Hilbert space H order isomorphic to a closed set $\omega \subset [0,1]$. If $T \in \mathcal{T}(\mathcal{N})$ and $0 \neq N \in \mathcal{N}$, then*

$$\limsup_{M \uparrow N} i_M^-(T) \leq i_N^-(T),$$

i.e. $f_T(N) := i_N^-(T)$ is left upper semicontinuous. The set $\mathcal{S} = \{N : i_N^-(T) < a\}$ is open in the topology on ω generated by the half-open intervals $(x, y]$. Thus, $\mathcal{S}$ corresponds to a subset $S \subset \omega$ which is the disjoint union of an open set U (in the metric topology) and a countable set C so that $U \cup C = S \subset \overline{U}$.

Proof. If $i_N^-(T) = a$ and $\varepsilon > 0$, there is an $L < N$ such that

$$a \leq \|(P_N - P_L)T(P_N - P_L)\| < a + \varepsilon.$$

If $L < M \leq N$, then $i_M^-(T) \leq \|(P_M - P_L)T(P_M - P_L)\| < a + \varepsilon$. Therefore $\limsup_{M\uparrow N} i_M^-(T) \leq i_N^-(T)$. It follows that the set $\{N : i_N^-(T) \geq a\}$ is closed under increasing limits.

If $i_N^-(T) < a$ and $\varepsilon < a - i_N^-(T)$, then again there is $L < N$ so that if $L < M \leq N$, then

$$\|(P_M - P_L)T(P_M - P_L)\| < \|(P_N - P_L)T(P_N - P_L)\| < i_N^-(T) + \varepsilon < a.$$

Thus $(L, N] \subset \mathcal{S}$. Hence, the corresponding subset $S \subset \omega$ is a union of half-open intervals $(x, y]$. Such a set is an arbitrary open set in the topology on ω generated by such intervals. It is easy to see that the components of S must be either half-open intervals or open intervals. As they are disjoint and have interior, there are at most countably many. So the set C, consisting of those right endpoints which are included for some components, is at most countable. Hence $S = U \cup C$ where $U = \operatorname{int} S$ and $C = S \setminus U$. Clearly $S \subset \overline{U}$. ∎

The previous proposition shows that we can define the support of $T \in \mathcal{T}(\mathcal{N})$, which is always Borel.

13.11.4. DEFINITION. For $T \in \mathcal{T}(\mathcal{N})$, the *support* of T is

$$\operatorname{supp}(T) = \{N \in \mathcal{N} \setminus \{0\} : i_N^-(T) > 0\}.$$

13.11.5. LEMMA. *Let $\mathcal{N}$ be a nest on a separable Hilbert space H, and let $\mathcal{M} = S\mathcal{N}$ for an invertible operator S. Set $C = \|S\|\,\|S^{-1}\|$. If $T \in \mathcal{T}(\mathcal{N})$ and $N \in \mathcal{N}$, then*

$$C^{-1} i_{SN}^-(STS^{-1}) \le i_N^-(T) \le C i_{SN}^-(STS^{-1}).$$

Hence $\operatorname{supp}(STS^{-1}) = S \operatorname{supp}(T)$.

PROOF. Let $i_N^-(T) = a$ and $\varepsilon > 0$, and find $L < N$ such that

$$\|(P_N - P_L)T(P_N - P_L)\| < a + \varepsilon.$$

Let $M = SN$ and $K = SL$. Then $P_K^\perp S = P_K^\perp S P_L^\perp$ and $S^{-1}P_M = P_N S^{-1} P_M$. Therefore

$$\begin{aligned}\|(P_M - P_K)STS^{-1}(P_M - P_K)\| &= \|P_M P_K^\perp S P_L^\perp T P_N S^{-1} P_M P_K^\perp\| \\ &\le \|S\|\,\|P_L^\perp T P_N\|\,\|S^{-1}\| = \|S\|\,\|P_N P_L^\perp T P_N P_L^\perp\|\,\|S^{-1}\| < C(a+\varepsilon).\end{aligned}$$

Hence $i_{SN}^-(STS^{-1}) \le C i_N^-(T)$. Reversing roles of $\mathcal{M}$ and $\mathcal{N}$ yield the other inequality. The conclusion for the supports is immediate. ∎

We will use the similarity theorem to show that there are many idempotents in $\mathcal{R}^\infty(\mathcal{N})$ for a continuous nest $\mathcal{N}$. Here are the main things that we will require.

13.11.6. PROPOSITION. *Let $\mathcal{N}$ be a continuous nest on separable H.*

(1) *If Q is an idempotent in $\mathcal{T}(\mathcal{N})$, $N \in \mathcal{N}$, then $i_N^-(Q) = 0$ or $i_N^-(Q) \ge 1$.*

(2) *If Q is an idempotent in $\mathcal{R}^\infty(\mathcal{N})$, there is a partition $\mathcal{P} = \{E_n : n \ge 1\}$ such that $\Delta_{\mathcal{P}}(Q) = 0$.*

(3) *If* $\operatorname{supp}(Q)$ *is empty, then $Q = 0$.*

(4) *If $0 \neq N \in \mathcal{N}$, there is an idempotent $Q \in \mathcal{R}^\infty(\mathcal{N})$ with $\|Q\| \le 2$ and* $N \in \operatorname{supp}(Q)$.

PROOF. (1) If E is an interval of $\mathcal{N}$, then since the map $T \to ETE$ is a homomorphism, EQE is idempotent, and hence is either 0 or has norm at least 1. This passes to $i_N^-(Q)$. (2) Thus if $Q \in \mathcal{R}^\infty(\mathcal{N})$, there is a partition $\mathcal{P}$ with $\|\Delta_{\mathcal{P}}(Q)\| < 1$, whence $\Delta_{\mathcal{P}}(Q) = 0$.

(3) If supp(Q) is empty, then $i_I^-(Q) = 0$ and thus there is $N < I$ with $P_N^\perp Q = P_N^\perp Q P_N^\perp = 0$. Consider $\mathcal{S} = \{N \in \mathcal{N} : P_N^\perp Q = 0\}$. This set is upward directed. Let $N_0 = \bigwedge \mathcal{S}$. Then there are $N_i \in \mathcal{S}$ with $N_i \to N_0$. Hence $P_{N_0}^\perp Q = \text{SOT–}\lim_i P_{N_i}^\perp Q = 0$. So N_0 belongs to $\mathcal{S}$. If $N_0 > 0$, then since $i_{N_0}^-(Q) = 0$, there is an $M < N_0$ with $(P_{N_0} - P_M)Q(P_{N_0} - P_M) = 0$. But then $P_M^\perp Q = (P_{N_0} - P_M)QP_{N_0}^\perp$ is nilpotent. But is also idempotent, and hence $P_M^\perp Q = 0$. This shows that $M \in \mathcal{S}$, contradicting the definition of N_0. Hence $0 \in \mathcal{S}$ and $Q = 0$.

(4) As in the discussion preceding Lemma 13.6.2, we may parameterize $\mathcal{N}$ as $\{N_t : 0 \le t \le 1\}$ in such a way that the spectral measure for $\mathcal{N}$ is equivalent to Lebesgue measure on $[0, 1]$. Let $N = N_{t_0} > 0$ be given. Choose a non-atomic probability measure ν on $[0, 1]$ supported on a compact set X of Lebesgue measure 0 such that $\nu((s, t_0]) > 0$ for all $s < t_0$. Consider the Hilbert space $H' = H \oplus L^2(\nu)$ and the continuous nest $M_t = N_t \oplus L^2((0, t), \nu)$ for $0 \le t \le 1$. By the similarity theorem, there is an invertible $S \in \mathcal{B}(H, H')$ with $\|S\|\,\|S^{-1}\| < 2$ such that $SN_t = M_t$ for $0 \le t \le 1$.

Let P be the orthogonal projection onto $L^2(\nu)$. Now supp(P) contains M_{t_0} as for any $s < t_0$, $P_{M_s}^\perp P P_{M_{t_0}}$ is the projection onto $L^2((s, t_0], \nu)$, which is non-zero. Thus $Q = S^{-1}PS$ is an idempotent in $\mathcal{T}(\mathcal{N})$ with $\|Q\| \le \|S\|\,\|S^{-1}\| < 2$. By Lemma 13.11.5, $t_0 \in$ supp(Q). We will show that $Q \in \mathcal{R}^\infty(\mathcal{N})$. Since X is compact, $[0, 1]\backslash X$ can be written as a disjoint union of open intervals (x_n, y_n). Since $m(X) = 0$, we have $\sum_{n\ge1} y_n - x_n = 1$. Thus the intervals $E_n = P_{N_{y_n}} - P_{N_{x_n}}$ form a partition of $\mathcal{N}$. Let $F_n = P_{M_{y_n}} - P_{M_{x_n}}$. Then by construction $F_n P F_n = 0$. Hence as in Lemma 13.11.5, we calculate

$$E_n Q E_n = P_{N_{x_n}}^\perp (S^{-1}PS) P_{N_{y_n}} = P_{N_{x_n}}^\perp S^{-1}(P_{M_{x_n}}^\perp P P_{M_{y_n}}) S P_{N_{y_n}} = 0.$$

Therefore $Q \in \mathcal{R}^\infty(\mathcal{N})$. ∎

13.11.7. COROLLARY. *Let $\mathcal{N}$ be a continuous nest on separable H. There are pairwise orthogonal projections $F_i \in \mathcal{N}''$ and idempotents $Q_i = F_i Q_i F_i$ in $\mathcal{R}^\infty(\mathcal{N})$ with $\|Q_i\| \le 2$ such that $Q = \text{SOT-}\sum_i Q_i$ is an idempotent in $\mathcal{T}(\mathcal{N})$ algebraically equivalent to I. So there are $U_0, V_0 \in \mathcal{T}(\mathcal{N})$ such that $U_0 Q V_0 = I$.*

PROOF. As before, we can identity $\mathcal{N}$ with $[0, 1]$ in such a way that the spectral measure is Lebesgue measure. It is a standard construction in measure theory that there exist pairwise disjoint measurable subsets X_i such that $m(X_i \cap (a, b)) > 0$ for any $0 \le a < b \le 1$. Let F_i be the projection in $\mathcal{N}''$ corresponding to χ_{X_i} via the spectral measure of Corollary 13.1.10. Note that $F_i\mathcal{T}(\mathcal{N})F_i = \mathcal{T}(F_i\mathcal{N})$ when

$F_i\mathcal{N}$ is considered as a nest on F_iH. Then $F_i\mathcal{N}$ is a continuous nest with the same support because the support of X_i is the whole interval.

Let $\{r_i : i \geq 1\}$ be a dense subset of $[0, 1]$. For each $i \geq 1$, use Proposition 13.11.6(4) to obtain $Q_i \in \mathcal{R}^\infty(F_i\mathcal{N})$ with $r_i \in \operatorname{supp}(Q_i)$ and $\|Q_i\| \leq 2$. We identify Q_i with $F_iQ_iF_i \in \mathcal{T}(\mathcal{N})$, and observe that $Q_i \in R^\infty(\mathcal{N})$; indeed, the same partitions work. Let $Q = \text{SOT-}\sum_i Q_i$. This is clearly an idempotent in $\mathcal{T}(\mathcal{N})$ with $\|Q\| \leq 2$. Note that $\operatorname{supp}(Q) = (0, 1]$ because for any $t > 0$ and $0 \leq s < t$, there is some $r_i \in (s, t)$ and hence

$$\|(P_{N_t} - P_{N_s})Q(P_{N_t} - P_{N_s})\| \geq |(P_{N_t} - P_{N_s})Q_i(P_{N_t} - P_{N_s})\| \geq 1.$$

It follows that $(P_{N_t} - P_{N_s})Q(P_{N_t} - P_{N_s})$ is a non-zero idempotent, necessarily of infinite rank, for any $s < t$. By the Larson–Pitts Theorem 13.8.12, Q is algebraically equivalent to I, i.e. there are $U_0, V_0 \in \mathcal{T}(\mathcal{N})$ such that $V_0U_0 = Q$ and $U_0V_0 = I$. Therefore $U_0QV_0 = I$. ∎

It is now possible to state the interpolation theorem. Recall that every nest has a spectral measure $\mathcal{E}$ by Corollary 13.1.10. The function $i^-_N(X)$ makes sense if $X \in \mathcal{B}(H)$, although we are generally interested in $T \in \mathcal{T}(\mathcal{N})$. So we state it for that case.

A technical lemma is needed.

13.11.8. Lemma. *Let $\mathcal{N}$ be a continuous nest on a separable H, and let $T \in \mathcal{T}(\mathcal{N})$. Let G_k be intervals of $\mathcal{N}$ such that for each $k \geq 1$, there is $N_k \in \mathcal{N}$ with $i^-_{N_k}(G_kTG_k) > 1$. Let $\{\xi_k : k \geq 1\}$ and $\{\eta_k : k \geq 1\}$ be two orthonormal families in H. Then there are orthogonal families $\{\alpha_k : k \geq 1\}$ and $\{\beta_k : k \geq 1\}$ with $\alpha_k, \beta_k \in \overline{b_1(G_kH)}$ such that $A = \sum_{k\geq1} \xi_k\alpha_k^*$ and $B = \sum_{k\geq1} \beta_k\eta_k^*$ satisfy $ATB = \sum_{k\geq1} \xi_k\eta_k^*$.*

Proof. The idea is to construct α_k and β_k so that $\langle T\beta_k, \alpha_l\rangle = \delta_{k,l}$. Once this is accomplished,

$$ATB = \sum_{k\geq1}\sum_{l\geq1} \xi_k\langle T\beta_l, \alpha_k\rangle\eta_l^* = \sum_{k\geq1} \xi_k\eta_k^*.$$

These vectors are defined recursively. Suppose that α_i and β_i are defined for $1 \leq i < k$. Define a finite dimensional subspace

$$L_k = \operatorname{span}\{\beta_i,\ T^*\alpha_i, T^*G_k\alpha_i,\ T^*G_kT\beta_i : 1 \leq i < k\}.$$

The existence of N_k means that for any $M < N_k$,

$$\|(P_{N_k} - P_M)G_kTG_k(P_{N_k} - P_M)\| > 1.$$

As M increases to N_k, $P_{N_k} - P_M$ converges to 0 SOT. It follows that the essential norm $\|G_kTG_k\|_e > 1$. Thus there is a unit vector $\beta_k \in G_kH$ orthogonal to L_k with $\|G_kT\beta_k\| > 1$. Define $\alpha_k = \|G_kT\beta_k\|^{-2}G_kT\beta_k$. So $\|\alpha_k\| = \|G_kT\beta_k\|^{-1} < 1$.

Note that β_k is orthogonal to β_i for $i < k$, whence $\{\beta_i : i \geq 1\}$ is an orthonormal family. Compute

$$\langle \alpha_k, \alpha_i \rangle = \|G_k T \beta_k\|^{-2} \langle \beta_k, T^* G_k \alpha_i \rangle = 0 \quad \text{for} \quad i < k.$$

So $\{\alpha_i : i \geq 1\}$ is an orthogonal family. Also $\langle T\beta_k, \alpha_i \rangle = \langle \beta_k, T^* \alpha_i \rangle = 0$ for $i < k$, and

$$\langle T\beta_i, \alpha_k \rangle = \|G_k T \beta_k\|^{-2} \langle T^* G_k T \beta_i, \beta_k \rangle = 0.$$

Finally, $\langle T\beta_k, \alpha_k \rangle = \|G_k T\beta_k\|^{-2} \langle T\beta_k, G_k T\beta_k \rangle = 1$. ∎

13.11.9. Orr Interpolation Theorem. *Let $\mathcal{N}$ be a continuous nest on separable H. Suppose that $T \in \mathcal{T}(\mathcal{N})$, $a > 0$ and $\mathcal{S} = \{N \in \mathcal{N} : i^-_N(T) \geq a\}$. Then there are $A, B \in \mathcal{T}(\mathcal{N})$ such that $ATB = \mathcal{E}(\mathcal{S})$.*

Proof. First assume that $\mathcal{E}(\mathcal{S}) = I$. There is no loss if we scale T so that $a > 1$. Let $Q_i = F_i Q_i F_i$ be the idempotents in $\mathcal{R}^\infty(\mathcal{N})$ provided by Corollary 13.11.7; and $Q = \text{SOT-}\sum_i Q_i$. For each $i \geq 1$, let $\mathcal{P}_i = \{E_n^{(i)} : n \geq 1\}$ be a partition of $\mathcal{N}$ such that $E_n^{(i)} Q_i E_n^{(i)} = 0$ for $n \geq 1$, as provided by Proposition 13.11.6(2). For each n, split the interval $E_n^{(i)}$ into countably many consecutive intervals $E_{n,j}^{(i)}$ ordered by $j \in \mathbb{Z}$. The projections $E_{n,j}^{(i)} F_i$ are infinite dimensional, and thus there are partial isometries $U_{n,j}^{(i)}$ mapping $E_{n,j}^{(i)} F_i H$ onto $E_{n,j-1}^{(i)} F_i H$. Let $U_i = \sum_{n \geq 1} \sum_{j \in \mathbb{Z}} U_{n,j}^{(i)}$ and $U = \sum_{i \geq 1} U_i$. Observe that each U_i is a partial isometry that maps $F_i H$ onto itself. Moreover, U_i belongs to $\mathcal{T}(\mathcal{N})$ since each $U_{n,j}^{(i)}$ is strictly upper triangular. Therefore U is unitary and lies in $\mathcal{T}(\mathcal{N})$. Moreover, U_i is block diagonal with respect to the partition $\mathcal{P}_i$.

Observe that since Q_i is strictly upper triangular with respect to the partition $\mathcal{P}_i$ that $U_i^{*2} Q_i$ is also strictly upper triangular with respect to $\mathcal{P}_i$, and thus belongs to $\mathcal{T}(\mathcal{N})$. Since $U_i = F_i U_i F_i$, we deduce that $U^{*2} Q$ belongs to $\mathcal{T}(\mathcal{N})$ as well.

Choose an orthonormal basis for each $E_{n,j}^{(i)} F_i H$, for $i, n \geq 1$ and $j \in \mathbb{Z}$, and list them all as $\{\eta_k : k \geq 1\}$, which is an orthonormal basis for H. Define $\xi_k = U^2 \eta_k$. This is another orthonormal basis for H because U^2 is unitary. For each k, there is some i, n and j so that $\eta_k \in E_{n,j}^{(i)} F_i H$ and so $\xi_k \in E_{n,j-2}^{(i)} F_i H$. We apply Lemma 13.11.8 using $G_k = E_{n,j-1}^{(i)}$. Since $\mathcal{E}(\mathcal{S}) = I$, there are elements N_k of the nest such that $i^-_{N_k}(G_k T G_k) > 1$. Hence, there are orthogonal families $\{\alpha_k : k \geq 1\}$ and $\{\beta_k : k \geq 1\}$ with $\alpha_k, \beta_k \in \overline{b_1(G_k H)}$ such that $A = \sum_{k \geq 1} \xi_k \alpha_k^*$ and $B = \sum_{k \geq 1} \beta_k \eta_k^*$ satisfy $ATB = \sum_{k \geq 1} \xi_k \eta_k^* = U^2$. Note that since $\alpha_k \in E_{n,j-1}^{(i)}$ and $\xi_k \in E_{n,j-2}^{(i)}$ that $\xi_k \alpha_k^* \in \mathcal{T}(\mathcal{N})$, whence $A \in \mathcal{T}(\mathcal{N})$. Similarly $B \in \mathcal{T}(\mathcal{N})$. Let $U_0, V_0 \in \mathcal{T}(\mathcal{N})$ as in Corollary 13.11.7. Finally, we have that

$$(U_0 A) T \big(B(U^{*2} Q) V_0\big) = U_0 (ATB) U^{*2} Q V_0 = U_0 U^2 U^{*2} Q V_0 = I.$$

Now we deal with the general case. Let $P = E(\mathcal{S})$. Since $P \in \mathcal{N}''$, define $\mathcal{N}_1 = P\mathcal{N}|_{PH}$. This is a continuous nest. As before, we obtain an idempotent Q and unitary U in $\mathcal{T}(\mathcal{N}_1)$ such that $U^{*2}Q \in \mathcal{T}(\mathcal{N}_1)$. As in the previous paragraph, Q is algebraically equivalent to P, the identity of $\mathcal{T}(\mathcal{N}_1)$. So there are $U_0, V_0 \in \mathcal{T}(\mathcal{N}_1)$ such that $U_0QV_0 = P$. Moreover, there are intervals $E_{n,j}^{(i)}$ of $\mathcal{N}$ such that $E_{n,j}^{(i)}F_iP$ are pairwise orthogonal and sum to P. Then PQP, PUP and $PU^{*2}QP$ belong to $\mathcal{T}(\mathcal{N})$. Arguing as above, we see that there are elements $A, B \in \mathcal{T}(\mathcal{N})$ such that $ATB = PU^2P$ and so

$$\begin{aligned}(PU_0P)ATB(PU^{*2}QP)(PV_0P) &= PU_0PU^2PU^{*2}QPV_0P \\ &= PU_0QV_0P = P = \mathcal{E}(\mathcal{S}).\end{aligned}$$ ∎

13.12. Connectedness of the Invertibles

It is a long-standing problem whether the set of invertible operators which are upper triangular with respect to a fixed orthonormal basis $\{\xi_n : n \geq 1\}$ are connected. However, it is known that they are connected in many nest algebras, and the general problem reduces to this special case. We will show here that the invertibles in a continuous nest algebra are connected. For information about further developments, see the notes at the end of the chapter.

Say that a WOT-closed unital operator algebra $\mathcal{A}$ has *infinite multiplicity* if it is continuously isomorphic to $\mathcal{A}\overline{\otimes}\mathcal{B}(l_2)$, the WOT-closed algebra of bounded operators on $H \otimes l_2$ which have an infinite matrix with entries in $\mathcal{A}$. See Proposition 10.8.5 for the von Neumann algebra version. It is a consequence of the similarity theorem that if a nest (on separable Hilbert space) has no finite rank atoms, then it has infinite multiplicity. In particular, this is the case for a continuous nest.

If $\mathcal{A}$ has infinite multiplicity, then it is also isomorphic to $M_n(\mathcal{A})$ for any $n \geq 2$. Some matrix tricks will simplify the problem for us.

13.12.1. PROPOSITION. *Let $\mathcal{A}$ be a* WOT*-closed unital operator algebra with infinite multiplicity.*

(1) *If $A \in \mathcal{A}^{-1}$, then $A \oplus A^{-1}$ is connected to I_2 in $M_2(\mathcal{A})$.*

(2) *If $A \in \mathcal{A}^{-1}$, then $A \oplus I$ is connected to I_2 in $M_2(\mathcal{A})$.*

(3) *If $\left[\begin{smallmatrix} I & B \\ C & D \end{smallmatrix}\right]$ is invertible in $M_2(\mathcal{A})$, then it is connected to I_2 in $M_2(\mathcal{A})$.*

(4) *If $T = \left[\begin{smallmatrix} A & B \\ C & D \end{smallmatrix}\right]$ is invertible in $M_2(\mathcal{A})$ and one of the matrix coefficients is 0, then T is connected to I_2 in $M_2(\mathcal{A})$.*

(5) *If there are $S, T \in \mathcal{A}$ such that $SAT = I$ and $\left[\begin{smallmatrix} A & B \\ C & D \end{smallmatrix}\right]$ is invertible in $M_2(\mathcal{A})$, then it is connected to I_2 in $M_2(\mathcal{A})$.*

PROOF. (1) This works in any Banach algebra. The path

$$\begin{bmatrix} \cos\theta & \sin\theta \\ -\sin\theta & \cos\theta \end{bmatrix} \begin{bmatrix} A & 0 \\ 0 & I \end{bmatrix} \begin{bmatrix} \cos\theta & -\sin\theta \\ \sin\theta & \cos\theta \end{bmatrix} \begin{bmatrix} I & 0 \\ 0 & A^{-1} \end{bmatrix}$$

for $0 \le \theta \le \frac{\pi}{2}$ does the job.

(2) Since $\mathcal{A}$ has infinite multiplicity, let $\Phi : \mathcal{A} \to \mathcal{A}\overline{\otimes}\mathcal{B}(l_2)$ be a continuous isomorphism, and let $\widetilde{A} = \Phi(A)$. Then we can similarly identify $A \oplus I$ with $\widetilde{A} \oplus \widetilde{I}$, and $\widetilde{I} = I \otimes I$. So there is a unitary equivalence between $\widetilde{A} \oplus \widetilde{I}$ and $\widetilde{A} \oplus (\widetilde{I} \oplus \widetilde{I}) \oplus (\widetilde{I} \oplus \widetilde{I}) \oplus \dots$. By (1), there is a continuous path from this to

$$\widetilde{A} \oplus (\widetilde{A}^{-1} \oplus \widetilde{A}) \oplus (\widetilde{A}^{-1} \oplus \widetilde{A}) \oplus \cdots = (\widetilde{A} \oplus \widetilde{A}^{-1}) \oplus (\widetilde{A} \oplus \widetilde{A}^{-1}) \oplus (\widetilde{A} \oplus \widetilde{A}^{-1}) \oplus \dots$$

The second arrangement connects back to $\widetilde{I} \oplus \widetilde{I} \oplus \widetilde{I} \oplus \dots$, which is equivalent to $\widetilde{I} \oplus \widetilde{I}$. Pulling back via Φ^{-1} yields a continuous path from $A \oplus I$ to I_2.

(3) One can factor

$$\begin{bmatrix} I & B \\ C & D \end{bmatrix} = \begin{bmatrix} I & 0 \\ C & I \end{bmatrix} \begin{bmatrix} I & 0 \\ 0 & D - CB \end{bmatrix} \begin{bmatrix} I & B \\ 0 & I \end{bmatrix}.$$

The two outer factors connect to I_2 within $M_2(\mathcal{A})^{-1}$ via a straight line segment. Then the middle term connects to I_2 by (2).

(4) First suppose that $A = 0$. The set of invertible elements is open, so $T_t = \left[\begin{smallmatrix} tI & B \\ C & D \end{smallmatrix}\right]$ is a straight line segment of invertible elements for $0 \le t \le \varepsilon$ for some small $\varepsilon > 0$. Apply (3). Clearly the same works for $D = 0$. If $B = 0$,

$$\begin{bmatrix} A & 0 \\ C & D \end{bmatrix} \begin{bmatrix} 0 & I \\ -I & 0 \end{bmatrix} = \begin{bmatrix} 0 & A \\ -D & C \end{bmatrix}$$

This connects to I_2, and multiplying this path by $R = \left[\begin{smallmatrix} 0 & -I \\ I & 0 \end{smallmatrix}\right]$ yields a path from T to R. The path $R_\theta = \left[\begin{smallmatrix} \cos\theta & -\sin\theta \\ \sin\theta & \cos\theta \end{smallmatrix}\right]$ connects R to I_2. Again $C = 0$ is the same.

(5) If $UV = I$, then there are invertible matrices

$$\begin{bmatrix} U & 0 \\ I - VU & V \end{bmatrix} \begin{bmatrix} V & I - VU \\ 0 & U \end{bmatrix} = I_2 = \begin{bmatrix} V & I - VU \\ 0 & U \end{bmatrix} \begin{bmatrix} U & 0 \\ I - VU & V \end{bmatrix}.$$

Thus $\left[\begin{smallmatrix} S & 0 \\ I-ATS & AT \end{smallmatrix}\right]$ and $\left[\begin{smallmatrix} T & I-TSA \\ 0 & SA \end{smallmatrix}\right]$ are invertible in $M_2(\mathcal{A})$. They are connected to I_2 by (4). So we can connect $\left[\begin{smallmatrix} A & B \\ C & D \end{smallmatrix}\right]$ to

$$\begin{bmatrix} S & 0 \\ I - ATS & AT \end{bmatrix} \begin{bmatrix} A & B \\ C & D \end{bmatrix} \begin{bmatrix} T & I - TSA \\ 0 & SA \end{bmatrix} = \begin{bmatrix} I & X \\ Y & Z \end{bmatrix}.$$

This connects to I_2 by (3). ∎

13.12.2. DEFINITION. Say that an element A of a unital algebra $\mathcal{A}$ is *interpolating* is there are $S, T \in \mathcal{A}$ such that $SAT = I$.

13.12.3. THEOREM. *Let $\mathcal{N}$ be a continuous nest on separable H. The interpolating elements are dense in $\mathcal{T}(\mathcal{N})$. Hence, the set of invertible elements $\mathcal{T}(\mathcal{N})^{-1}$ is path connected.*

PROOF. Let $A \in \mathcal{T}(\mathcal{N})$ and let $\varepsilon > 0$. Set $\mathcal{S} = \{N \in \mathcal{N} : i_N^-(A) < \varepsilon/2\}$. Define $T = A + \varepsilon\mathcal{E}(\mathcal{S})$. By Lemma 13.11.3, $\mathcal{S}$ corresponds to a set

$$S = U \cup C \subset \overline{U} \subset [0,1],$$

where U is open and C is countable. In particular, $i_N^-(\mathcal{E}(\mathcal{S})) = 1$ for $N \in U$ and $i_N^-(\mathcal{E}(\mathcal{S})) = 0$ for $N \in [0,1] \setminus \overline{U}$. Since i_N^- is a seminorm, for $N \in U$, $i_N^-(T) \geq \varepsilon - i_N^-(A) > \varepsilon/2$, and for $N \in [0,1] \setminus \overline{U}$, $i_N^-(T) \geq i_N^-(A) \geq \varepsilon/2$. Since $T \mapsto i_N^-(T)$ is left upper semicontinuous, $i_N^-(T) \geq \varepsilon/2$ for all $N > 0$. By the interpolation theorem, T is an interpolating element. Therefore, interpolating elements are dense in $\mathcal{T}(\mathcal{N})$.

Now if $X = \left[\begin{smallmatrix} A & B \\ C & D \end{smallmatrix}\right]$ is invertible in $M_2(\mathcal{T}(\mathcal{N}))$, this shows that a perturbation of the 1,1 entry by less than $\|X^{-1}\|^{-1}$ yields an element $Y \in M_2(\mathcal{T}(\mathcal{N}))$ with an interpolating coefficient. These two invertible elements are connected in $M_2(\mathcal{T}(\mathcal{N}))^{-1}$ by a line segment. By the similarity theorem, continuous nest algebras have infinite multiplicity. By Proposition 13.12.1(5), Y is connected to the identity. Hence, the invertibles of $M_2(\mathcal{T}(\mathcal{N}))^{-1}$ are connected. But $\mathcal{T}(\mathcal{N})$ is isomorphic to $M_2(\mathcal{T}(\mathcal{N}))$, so the invertibles are connected. ■

Now we can extend this to any nest algebra of infinite multiplicity, namely, a nest in which every atom is infinite dimensional.

13.12.4. COROLLARY. *Let $\mathcal{N}$ be a nest on separable H with only infinite dimensional atoms. Then the invertibles in $\mathcal{T}(\mathcal{N})$ are connected.*

PROOF. Let $\{A_j : j \in J\}$ be the atoms of $\mathcal{N}$, and let $\Delta(T) = \sum_{j\in J} A_j T A_j$ be the diagonal of $T \in \mathcal{T}(\mathcal{N})$. Then the range of Δ is the von Neumann algebra $\prod_{j\in J} \mathcal{B}(A_j H)$. The invertibles in a von Neumann algebra are connected (Exercise **13.24**). So there is a continuous path in $\mathcal{T}(\mathcal{M})^{-1}$ from T to an invertible element S such that $\Delta(S) = I$.

Build a continuous nest on each atom A_j, and in this way form a continuous nest $\mathcal{M}$ on H containing $\mathcal{N}$. Then it is easy to see that S is invertible in $\mathcal{T}(\mathcal{M})$. By Theorem 13.12.3, there is a path in $\mathcal{T}(\mathcal{M})^{-1}$ connecting S to I. Since $\mathcal{T}(\mathcal{M})$ is contained in $\mathcal{T}(\mathcal{N})$, this is also a path in $\mathcal{T}(\mathcal{N})^{-1}$. So the set of invertibles is connected. ■

Exercises for Chapter 13

13.1 Show that a nest $\mathcal{N}$ is maximal if and only if every atom is one dimensional.

13.2 (a) Provide the details for Corollary 13.6.6 building a quasicentral approximate unit for $\mathcal{K}$ relative to a continuous nest.

(b) Prove that every nest $\mathcal{N}$ on a separable Hilbert space has a contractive quasicentral approximate unit for the compact operators, i.e. there are compact contractions $E_n \in \mathcal{T}(\mathcal{N})$ such that $\lim_{n\to\infty} \sup_{N\in\mathcal{N}} \|E_n P_N - P_N E_n\| = 0$ and SOT–$\lim_n E_n = I$.

HINT: Deal with the atoms first and reduce to the continuous nest case.

13.3 Show that every compact quasinilpotent operator Q is a norm limit of compact nilpotent operators.
HINT: Let $\mathcal{N}$ be a maximal nest with $Q \in \mathcal{T}(\mathcal{N})$. Use Ringrose's theorem and the finite rank approximate identity.

13.4 (*Deddens*). Let $A \in \mathcal{B}(H)$ be positive and invertible. Let $\mathcal{N}$ be the nest generated by the spectral subspaces $E_A[0,t)$ for $t \ge 0$. Show that $\mathcal{T}(\mathcal{N})$ consists of all operators $T \in \mathcal{B}(H)$ such that $\sup_{n\ge 0} \|A^n T A^{-n}\| < \infty$.

13.5 (*Lidskii*). Let K be a trace class operator, and let $\{\lambda_i, i \in I\}$ be a list of the eigenvalues of K including multiplicity. Prove that $\operatorname{Tr} K = \sum_{i\in I} \lambda_i$.
HINT: Let $\mathcal{N}$ be a maximal nest with $K \in \mathcal{T}(\mathcal{N})$. Let $\{A_j : j \in J\}$ be the (rank 1) atoms of $\mathcal{N}$. Show that $\varphi(A) = \sum_J \Phi_{A_j}(A)$ is continuous on $\mathfrak{S}_1 \cap \mathcal{T}(\mathcal{N})$. Show that $\varphi(F) = \operatorname{Tr}(F)$ when F is finite rank.

13.6 (*Erdos–Power*). Let $\mathcal{J}$ be a WOT-closed ideal of $\mathcal{T}(\mathcal{N})$. Define a map $\varphi : \mathcal{N} \to \mathcal{N}$ by $\varphi(N) = \overline{\mathcal{J}N}$. Prove that $\mathcal{J} = \{T \in \mathcal{T}(\mathcal{N}) : AN \subset \varphi(N)\}$.
HINT: Show that $\mathcal{J}$ contains every rank 1 operator in this set.

13.7 (*Longstaff*). Let $\mathcal{L}$ be a subspace lattice. For $L \in \mathcal{L}$, define

$$L_- = \bigvee\{M \in \mathcal{L} : M \not\geq L\}.$$

Show that the rank 1 operator $\xi\eta^*$ is in $\operatorname{Alg}\mathcal{L}$ if and only if there is an $L \in \mathcal{L}$ such that $\xi \in L$ and $\eta \in L_-^\perp$.

13.8 (*Lance*). Let $\mathcal{N}$ be a nest on H.

(a) If $T \in \mathcal{T}(\mathcal{N})$ and $T\xi = \eta$, show that $\sup_{N\in\mathcal{N}} \frac{\|P_N^\perp \eta\|}{\|P_N^\perp \xi\|} \le \|T\|$.

(b) If $\mathcal{F}$ is a finite nest, let $A_1, \dots, A_n$ be the atoms in increasing order. If $\xi, \eta \in H$, let $\xi_j = A_j\xi$ and $\eta_j = A_j\eta$. If $\sum_{j=k}^n \|\eta_j\|^2 \le \sum_{j=k}^n \|\xi_j\|^2$ for $1 \le k \le n$, show that there is a finite rank $T \in \mathcal{T}(\mathcal{F})$ with $T\xi = \eta$ and $\|T\| \le 1$.
HINT: Use induction. If $T = A_1^\perp T$ has $TA_1^\perp\xi = A_1^\perp\eta$, show that $\|\eta_1\| \le \|(I - T^*T)^{1/2}\eta\|$.

(c) Show that if $\sup_{N\in\mathcal{N}} \frac{\|P_N^\perp \eta\|}{\|P_N^\perp \xi\|} = M < \infty$, there is a $T \in \mathcal{T}(\mathcal{N})$ with $T\xi = \eta$.

13.9 (*Davidson–Power*). Let $\mathcal{N} = \{0\} \cup \{N_k : k \in \mathbb{Z}\} \cup H$ be a nest on H. If K is compact, show that there is a $T \in \mathcal{T}(\mathcal{N}) \cap \mathcal{K}(H)$ with $\|K - T\| = \operatorname{dist}(K, \mathcal{T}(\mathcal{N}))$.
HINT: Reduce the problem to a finite nest by finding integers k_1, k_2 so that $\|KP_{N_{k_1}}\| \leq \frac{1}{2}\operatorname{dist}(K, \mathcal{T}(\mathcal{N}))$ and $\|P_{N_{k_2}}^{\perp} K\| \leq \frac{1}{2}\operatorname{dist}(K, \mathcal{T}(\mathcal{N}))$.

13.10 Let $\mathcal{A}$ be a hyperreflexive algebra with property $\mathbb{A}_1(r)$. Prove that every weak-$*$ closed unital subalgebra $\mathcal{B}$ of $\mathcal{A}$ is hyperreflexive.
HINT: First find a closest element to T in $\mathcal{A}$. Then $\operatorname{dist}(A, \mathcal{B}) = \sup_{\varphi \in b_1(\mathcal{B}_\perp)} |\varphi(A)|$.

13.11 (a) Let $\mathfrak{A}$ be a finite dimensional unital C*-algebra in $\mathcal{B}(H)$. Show that $\mathfrak{A}'$ is hyperreflexive.
HINT: Average uTu^* over the unitary group of $\mathfrak{A}$.
(b) Let $\mathcal{M}$ be a von Neumann algebra such that $\mathcal{M} = \mathfrak{A}'$ where $\mathfrak{A}$ is an AF C*-algebra (the closure of an increasing union of finite dimensional C*-algebra). Show that $\mathcal{M}$ is hyperreflexive.
(c) Show that $\mathcal{B}(H)^{(\infty)} = \mathcal{B}(H) \overline{\otimes} \mathbb{C} I_H$ is hyperreflexive in $\mathcal{B}(H \otimes H)$.

13.12 (*Arveson*). Let $\mathcal{A}$ be a weak-$*$-closed operator algebra on H, and let $\mathcal{A}^{(\infty)} = \{A^{(\infty)} \in \mathcal{B}(H^{(\infty)}) : A \in \mathcal{A}\}$ be its infinite ampliation.
(a) Prove that $\mathcal{A}^{(\infty)}$ has $\mathbb{A}_1(1)$.
(b) If $T \in \mathcal{B}(H)$, then $\operatorname{dist}(T, \mathcal{A}) = \sup\{\|P^\perp T^{(\infty)} P\| : P \in \operatorname{Lat} \mathcal{A}^{(\infty)}\}$.
(c) Show that $\mathcal{A}^{(\infty)}$ is hyperreflexive.
HINT: Use Exercises **13.11** and **13.10**.

13.13 Let $\mathcal{T}_\infty \subset \mathcal{B}(H)$ be the algebra of upper triangular operators with respect to a fixed orthonormal basis $\xi_1, \xi_2, \ldots$.
(a) Show that the map taking an operator $A \in \mathcal{B}(H)$ to its upper triangular part is unbounded even on the finite rank operators.
HINT: Find the Fourier series of the function $f(\theta) = \theta - \pi$ for $0 \leq \theta \leq 2\pi$. Use the Toeplitz operator T_f, compressed to the initial $n \times n$ corner.
(b) Hence show that $\overline{\mathcal{T}_\infty + \mathcal{T}_\infty^*} \neq \mathcal{B}(H)$.
(c) Show that $C^*(\mathcal{T}_\infty) = \mathcal{B}(H)$.
HINT: If A is positive and invertible, write $A^{1/2} = [\eta_1\ \eta_2\ \ldots]$ where $\eta_i = A^{1/2}\xi_i$. Apply the Gram–Schmidt process to $\eta_1, \eta_2, \ldots$, and hence construct a unitary U so that $UA^{1/2} \in \mathcal{T}_\infty$.

13.14 (*Ringrose*). *The radical of a nest algebra.* Let $\mathcal{N}$ be a nest.
(a) If $\mathcal{F} = \{0\} = N_0 < N_1 < \cdots < N_n = H$ is a finite subnest of $\mathcal{N}$, let $\mathcal{T}_0(\mathcal{F})$ be the subalgebra of $\mathcal{T}(\mathcal{F})$ with zero diagonal, i.e. $E_i T E_i = 0$ for $E_i = P_{N_i} - P_{N_{i-1}}$ for $1 \leq i \leq n$. Show that $\mathcal{T}_0(\mathcal{F})$ is a nil ideal, i.e. $T_1 T_2 \cdots T_n = 0$ for any $T_i \in \mathcal{T}_0(\mathcal{F})$.
(b) Show that $\mathcal{J} = \overline{\bigcup_{\mathcal{F} \subset \mathcal{N}} \mathcal{T}_0(\mathcal{F})}$ is a closed ideal contained in $\operatorname{rad} \mathcal{T}(\mathcal{N})$.
(c) Show that $T \in \mathcal{J}$ if and only if for any $\varepsilon > 0$, there is a finite nest $\mathcal{F} \subset \mathcal{N}$ such that $\max_i \|E_i T E_i\| < \varepsilon$.

(d) Show that $T \in \mathcal{T}(\mathcal{N})$ is not in $\mathcal{J}$ if and only if there is some $N_0 \in \mathcal{N}$ so that either (i) $N_0 < H$ and $\inf_{N>N_0} \|(P_N - P_{N_0})T(P_N - P_{N_0})\| > 0$ or (ii) $N_0 > 0$ and $\inf_{N>N_0} \|(P_{N_0} - P_N)T(P_{N_0} - P_N)\| > 0$.
(e) Use (d)(i) to construct a sequence $N_1 > N_2 > \dots$ with $N_k \to N_0$ and unit vectors $\xi_i = E_i\xi$, where $E_i = P_{N_i} - P_{N_{i+1}}$, such that $E_iT\xi_i = \eta_i$ and $\inf \|\eta_i\| = \delta > 0$. Let $S = \sum \xi_{i+1}\eta_i^*/\|\eta_i\|^2$. Show that $S \in \mathcal{T}(\mathcal{N})$ and ST is not quasinilpotent.
(f) Show that $\mathcal{J} = \operatorname{rad} \mathcal{T}(\mathcal{N})$.
(g) Show that $\mathcal{J} = \{T \in \mathcal{T}(\mathcal{N}) : i_N^-(T) = i_N^+(T) = 0 \text{ for all } N \in \mathcal{N}\}$.

13.15 This is a "soft" substitute for Pedersen's inequality 13.6.8. Let $E \geq 0$.
(a) Show that for $T \in \mathcal{B}(H)$, $\|E^n, T]\| \leq n\|E\|^{n-1}\|[E, T]\|$.
(b) Show that if p is a polynomial, there is a constant C so that for $T \in \mathcal{B}(H)$, $\|p(E), T]\| \leq C\|[E, T]\|$.
(c) Show that if $f \in C(\sigma(E))$, there is a constant C so that for $T \in \mathcal{B}(H)$, $\|f(E), T]\| \leq C\|[E, T]\|$.

13.16 The *commutator ideal* of a Banach algebra $\mathfrak{A}$ is the closed ideal generated by the elements $\{ab - ba : a, b \in \mathfrak{A}\}$. Prove that the commutator ideal of a continuous nest algebra is the whole algebra.
HINT: Do it for the algebra of 2×2 matrices over the Volterra nest algebra.

13.17 (a) Let $\mathcal{N}$ be a continuous nest on H, and let $A \in \mathcal{B}(H)$ be invertible. Show that there is $T \in \mathcal{T}(\mathcal{N})$ and U unitary such that $A = UT$.
HINT: Split $\mathcal{N}$ into consecutive intervals ordered like $\mathbb{Z}$, $\{E_k : k \in \mathbb{Z}\}$. Construct a unitary U^* so that $U^*AE_k = E_{k-1}U^*A$.
(b) Generalize this to arbitrary nests.

13.18 Let $\xi_1, \xi_2, \eta_1, \eta_2, \eta_3$ be an orthonormal basis for $\mathbb{C}_5$. Let $M_1 = \operatorname{span}\{\xi_1, \eta_1\}$, $M_2 = \operatorname{span}\{\xi_2, \eta_2\}$ and $M_3 = \operatorname{span}\{\xi_1 + \xi_2, \eta_3\}$. Set

$$\mathcal{A} = \operatorname{Alg}\{\mathbb{C}\xi_1, \mathbb{C}\xi_2, \mathbb{C}(\xi_1 + \xi_2), M_1, M_2, M_3\}.$$

Show that $F = \xi_1\eta_1^* + \xi_2\eta_2^* + (\xi_1 + \xi_2)\eta_3^*$ belongs to $\mathcal{A}$ but is not the sum of two rank one elements of $\mathcal{A}$.

13.19 (*Hopenwasser–Moore*). Let H have an orthonormal basis $\{\xi_i, \eta_i : i \geq 1\}$. Let $\mathcal{D}$ be the algebra of all diagonal operators for this basis. Define $\mathcal{A} = \mathcal{D} + \mathcal{J}$ where $\mathcal{J}$ consists of all operators T such that $T\xi_i = 0$ and $T\eta_i \in \operatorname{span}\{\xi_j : j \neq i\}$.
(a) Find $\operatorname{Lat} \mathcal{A}$ and prove that $\mathcal{A}$ is reflexive.
(b) Show that the span of the rank one elements of $\mathcal{A}$ is SOT-dense in $\mathcal{A}$.
(c) Let $\theta_1 = \sum_{i\geq 1} \frac{1}{i}\xi_i$, $\theta_2 = \sum_{i\geq 1} \frac{1}{i^2}\xi_i$, $\zeta_1 = \sum_{i\geq 1} \frac{1}{i}\eta_i$ and $\zeta_2 = \sum_{i\geq 1} \frac{1}{i^2}\eta_i$. Let $F = \zeta_1\theta_2^* - \zeta_2\theta_1^*$. Show that $F \in \mathcal{A}$ has rank 2 but it is not the sum of finitely many rank one elements of $\mathcal{A}$.
HINT: Show that the range of a rank one element of $\mathcal{J}$ has non-zero entries on a full rectangle $X \times Y \subset \mathbb{N}^2$ which is disjoint from the diagonal D. (Here X

is identified with a subset of $\{\xi_i : i \geq 1\}$ and Y with a subset of $\{\eta_i : i \geq 1\}$.)
Show that a finite union of such sets cannot equal $\mathbb{N}^2 \setminus D$.

13.20 Let $\mathcal{D}_3$ be the 3×3 diagonal algebra in M_3. Let $T = \begin{bmatrix} 1&1&1\\1&1&1\\1&1&1 \end{bmatrix}$. Show that $\beta_{\mathcal{D}_3}(T) < \operatorname{dist}(T, \mathcal{D}_3)$.
HINT: If $D \in \mathcal{D}_3$ is closest to T, any permutation of D is equally close. Average.

13.21 Let $\mathcal{N}$ be a continuous nest. Construct an idempotent $Q \in \mathcal{T}(\mathcal{N})$ such that $Q, I - Q$ and I are algebraically equivalent.

13.22 (a) Show that $\mathcal{R}^\infty(\mathcal{N}) \cap \mathcal{N}' = \{0\}$.
(b) Show that $\mathcal{R}^\infty(\mathcal{N})$ is not invariant under similarity.

13.23 (*Orr*). Let $\mathcal{N}$ be a nest on a separable Hilbert space H with scalar spectral measure μ. Prove that $\mathcal{R}^\infty(\mathcal{N})$ is equal to the set of all $T \in \mathcal{T}(\mathcal{N})$ such that $\mathcal{E}(\{N : i_N^-(T) > 0\}) = 0$, i.e. $i_N^-(T) = 0$ a.e.(μ).
HINT: Lemma 13.11.3.

13.24 Show that the invertibles in any von Neumann algebra $\mathcal{M}$ are connected.
HINT: Use polar decomposition, reduce to the unitary case, and use the spectral theorem to find a logarithm bounded by π.

13.25 (*Davidson–Orr–Pitts*). Let $\mathcal{A} \subset \mathcal{B}(H)$ be a unital WOT-closed algebra.
(a) An operator S is a *symmetry* if $S^2 = I$. Show that every symmetry in any Banach algebra is connected to I within the invertibles. Hence, a product of symmetries in connected to I within the invertibles.
(b) Show that $\begin{bmatrix} I & B \\ 0 & I \end{bmatrix}$ is a product of two symmetries in $M_2(\mathcal{A})$.
HINT: $\begin{bmatrix} I & B \\ 0 & -I \end{bmatrix}$ is a symmetry.
(c) If $U, V \in \mathcal{A}$ satisfy $UV = I \neq VU$, then $\begin{bmatrix} V & I-VU \\ 0 & U \end{bmatrix}$ is a product of six symmetries in $M_2(\mathcal{A})$.
HINT: $\begin{bmatrix} I & 0 \\ U & I \end{bmatrix} \begin{bmatrix} 0 & I \\ I & 0 \end{bmatrix} \begin{bmatrix} I & 0 \\ V & I \end{bmatrix} \begin{bmatrix} I & U \\ 0 & I-2VU \end{bmatrix}$.
(d) If $\mathcal{A}$ has infinite multiplicity, show that $T = A \oplus I$ is a product of 4 symmetries.
HINT: Consider $T \simeq A \oplus I \oplus I \oplus \ldots$, and set

$$S_1 = \begin{bmatrix} 0 & A \\ A^{-1} & 0 \end{bmatrix} \oplus \begin{bmatrix} 0 & A \\ A^{-1} & 0 \end{bmatrix} \oplus \ldots, \qquad S_2 = \begin{bmatrix} 0 & I \\ I & 0 \end{bmatrix} \oplus \begin{bmatrix} 0 & I \\ I & 0 \end{bmatrix} \oplus \ldots,$$
$$S_3 = I \oplus \begin{bmatrix} 0 & A^{-1} \\ A & 0 \end{bmatrix} \oplus \begin{bmatrix} 0 & A^{-1} \\ A & 0 \end{bmatrix} \oplus \ldots, \qquad S_4 = I \oplus S_2.$$

(e) Show that if $\mathcal{N}$ is a continuous nest, then every $T \in \mathcal{T}(\mathcal{N})^{-1}$ is a product of 28 symmetries.
HINT: Go through the proof of connectedness and keep track.

Notes on Chapter 13

Nest algebras were defined by Ringrose [260] in which he established Theorem 5.5.8. He establishes the reflexivity of nests in [261] in which he characterizes the radical, Exercise **13.14**. The results of section 13.2 are taken from Erdos [112]. Arveson's distance formula was taken from [16]; the proof follows Power [246]. Parrott's lemma is taken from [223]. The Ringrose isomorphism theorem is in [262].

Andersen's theorem was established in [8]. The proof in this book comes from [67]. Larson [187] then established Theorem 13.6.10. The full similarity theorem is due to Davidson [66]. That proof utilized the compact perturbations of nest algebras, which the proof here avoids. Pedersen's inequality is taken from [233].

Proposition 13.8.3 and Corollary 13.8.7 are also taken from Larson [187]. The structure of automorphisms, Theorem 13.8.9, is taken from Davidson–Wagner [85]. The Larson–Pitts paper on idempotents in nest algebras is [189]. Perturbations of nests and nest algebras were studied by Lance [188] using cohomological methods without the Similarity theorem. Theorem 13.10.10 is taken from Choi–Davidson [51] and our treatment of perturbations is from [68] and [**?**].

The seminorms i_N^- were introduced by Ringrose [261] mentioned above. Larson [187] defined the ideal $\mathcal{R}^\infty(\mathcal{N})$ and established its basic properties. Orr [221] proves the interpolation theorem. One nice consequence is that $R^\infty(\mathcal{N})$ is the largest ideal which has non-trivial intersection with the diagonal $\mathcal{N}'$. Davidson and Orr [79] prove that nests of infinite multiplicity have a connected set of invertibles. Then Davidson, Orr and Pitts [80] extend this to any nest for which there is a finite bound on the number of atoms in a finite rank interval of the nest. This reduced the general problem to deciding whether the invertibles are connected in the upper triangular operators with respect to an orthonormal basis $\{\xi_n : n \geq 1\}$. Ji and Zhang [161] show that if $\mathcal{N}$ is order isomorphic to $\mathbb{N} \cup \{\infty\}$ and the rank of the atoms grows sufficiently quickly, then the invertibles are connected.

The main text source for this material is the author's book [69]. See also [17].

CHAPTER 14

Dilation Theory

The theory of dilations, both for single operators and operator algebras, is a powerful tool for analyzing operators and representations. It is a method of extending an operator (or a representation of an operator algebra) to another acting on a larger space which has better properties. This can be exploited to provide structural results, such as an extended functional calculus.

We first study the Sz.-Nagy dilation of a contraction to an isometry and then to a unitary. This yields an H^∞ functional calculus for all contractions with no unitary summand. We then dilate commuting pairs of contractions, and provide several examples showing that problems can arise for three or more commuting contractions.

Then we study completely positive maps and a general view of dilation theory due to Arveson. This leads us to define the minimal C*-algebra or C*-envelope of an operator algebra or operator system. We take a brief look at some similarity problems. We finish the chapter with an abstract characterization of operator systems and operator spaces.

14.1. Dilation of a Contraction

14.1.1. DEFINITION. A *(power) dilation* of an operator $T \in \mathcal{B}(H)$ is an operator $S \in \mathcal{B}(K)$ on a Hilbert space $K \supset H$ such that $P_H S^n|_H = T^n$ for $n \geq 0$. If H is invariant for S, then the dilation is called an *extension*. If $H^\perp$ is invariant for S, then the dilation is called a *co-extension*. They have the matrix forms:

$$\begin{bmatrix} * & 0 \\ * & T \end{bmatrix} \qquad \begin{bmatrix} T & 0 \\ * & * \end{bmatrix} \qquad \begin{bmatrix} * & 0 & 0 \\ * & T & 0 \\ * & * & * \end{bmatrix}$$

extension   coextension   dilation

The third matrix form will be justified shortly.

It is a *unitary dilation* if S is unitary, and a *normal dilation* if S is normal. It is an *isometric dilation* if S is an isometry and H is coinvariant for S.

An isometric dilation is *minimal* if $K = \overline{\mathcal{A}(S)H}$, where $\mathcal{A}(S)$ is the closed, unital algebra generated by S. A unitary dilation U is minimal if $K = \overline{C^*(U)H}$.

K. R. Davidson, *Functional Analysis and Operator Algebras*, CMS/CAIMS Books in Mathematics 13,
https://doi.org/10.1007/978-3-031-63665-3_14

If $M \subset N$ are invariant subspaces for an operator A or an algebra of operators $\mathcal{A}$, then $N \ominus M$ is called a *semi-invariant subspace.*

If $H = N \ominus M$ is semi-invariant, the compression of S to $T = P_H S|_H$ preserves powers. Hence it preserves polynomials. Since it is contractive, it extends to a contractive homomorphism from $\mathcal{A}(S)$ into the closed algebra $\mathcal{A}(T)$. The converse is also true.

14.1.2. PROPOSITION. *Let $\mathcal{A}$ be an algebra of operators on a Hilbert space K, and let H be a closed subspace. Then the compression map $\Psi_H(A) = P_H A|_H$ is a homomorphism if and only if H is semi-invariant. Moreover, Ψ_H is contractive and* WOT*-continuous.*

PROOF. Suppose $K_1 \subset K_2$ are invariant subspaces for $\mathcal{A}$ and $H = K_2 \ominus K_1$. Decompose $K = K_2^\perp \oplus H \oplus K_1$. Then every $A \in \mathcal{A}$ has the form:

$$A = \begin{bmatrix} * & 0 & 0 \\ * & \Psi_H(A) & 0 \\ * & * & * \end{bmatrix}.$$

From this it is clear that Ψ_H is multiplicative; so it is a homomorphism. This map is evidently contractive and WOT-continuous.

Conversely, suppose that Ψ_H is multiplicative. Define $K_2 = \overline{\mathcal{A}H}$. This is invariant for $\mathcal{A}$ and contains H. Define $K_1 = K_2 \cap H^\perp$. Then $K_2 \ominus K_1 = H$. Let $P = P_H$. If $A, B \in \mathcal{A}$, then

$$0 = \Psi_H(AB) - \Psi_H(A)\Psi_H(B) = PABP - PAPBP|_H = PAP^\perp BP|_H.$$

Suppose that $\xi \in K_1$ but $A \in \mathcal{A}$ and $A\xi \notin K_1$. Since $\xi \in K_2$, $A\xi \in K_2 \setminus K_1$. Therefore $\xi = P^\perp \xi$ and $PAP^\perp \xi \neq 0$. On the other hand, since $K_2 = \overline{\mathcal{A}H}$, there is some $\eta \in H$ and $B \in \mathcal{A}$ so that $B\eta \approx \xi$. Thus

$$PAP^\perp BP\eta \approx PAP^\perp \xi \neq 0,$$

which is a contradiction. Thus $AK_1 \subset K_1$. Therefore K_1 is invariant for $\mathcal{A}$, and $H = K_2 \ominus K_1$ is semi-invariant. ∎

14.1.3. COROLLARY. *A dilation S of an operator T has the form:*

$$S = \begin{bmatrix} * & 0 & 0 \\ * & T & 0 \\ * & * & * \end{bmatrix}.$$

Every contraction has a minimal isometric dilation. It is unique up to the equivalence described in the following theorem.

14.1.4. Sz.-Nagy Dilation Theorem. *Let $T \in \mathcal{B}(H)$ with $\|T\| \le 1$. Then there is a minimal isometric dilation S on a Hilbert space $K \supseteq H$. If S_1 is another minimal isometric dilation on $K_1 \supseteq H$, there is a unitary operator $W \in \mathcal{B}(K, K_1)$ such that $W|_H = I_H$ and $WSW^* = S_1$.*

Proof. Let $D_T = (I - T^*T)^{1/2}$ be the *defect operator* of T. Let $M = \overline{\operatorname{Ran} D_T}$, and let $P = P_M$. Let $K = H \oplus M^{(\infty)} = H \oplus \sum_{n\ge1}^{\oplus} M_n$ where $M_n \simeq M$ for $n \ge 1$. Define $S \in \mathcal{B}(K)$ by

$$S(\xi, \eta_1, \eta_2, \eta_3, \dots) = (T\xi, PD_T\xi, \eta_1, \eta_2, \dots),$$

where $x \in H$ and $y_n \in M_n$. As a matrix,

$$S = \begin{bmatrix} T & 0 & 0 & \dots \\ PD_T & 0 & 0 & \ddots \\ 0 & I & 0 & \ddots \\ 0 & 0 & I & \ddots \\ \vdots & \ddots & \ddots & \ddots \end{bmatrix}$$

The first column is isometric with range in $H \oplus M_1$ because

$$\begin{aligned} \|T\xi\|^2 + \|PD_T\xi\|^2 &= \langle T\xi, T\xi\rangle + \langle (I - T^*T)^{1/2}\xi, (I - T^*T)^{1/2}\xi\rangle \\ &= \langle (T^*T + I - T^*T)\xi, \xi\rangle = \|\xi\|^2. \end{aligned}$$

Every other column is a shift from M_n onto M_{n+1}. Thus S is an isometry that is a co-extension of T.

Observe that $H \vee SH = H \oplus \overline{PD_T H} = H \oplus M_1$. Since $S^n M_1 = M_{n+1}$ for $n \ge 1$, we have $\overline{\mathcal{A}(S)H} = K$. Therefore S is a minimal isometric dilation of T.

Suppose that S_1 is another minimal isometric dilation on $K_1 \supseteq H$. Define a map $W : K \to K_1$ by

$$W \sum_{i=0}^{n} S^i \xi_i = \sum_{i=0}^{n} S_1^i \xi_i \quad \text{for} \quad n \ge 0 \text{ and } \xi_i \in H.$$

We will check that this is a well-defined unitary operator. Compute

$$\begin{aligned} \Big\| \sum_{i=0}^{n} S^i \xi_i \Big\|^2 &= \sum_{i,j=0}^{n} \langle S^i \xi_i, S^j \xi_j\rangle \\ &= \sum_{0\le j\le i\le n} \langle S^{i-j}\xi_i, \xi_j\rangle + \sum_{0\le i<j\le n} \langle \xi_i, S^{j-i}\xi_j\rangle \\ &= \sum_{0\le j\le i\le n} \langle T^{i-j}\xi_i, \xi_j\rangle + \sum_{0\le i<j\le n} \langle \xi_i, T^{j-i}\xi_j\rangle. \end{aligned}$$

The computation of $\big\| \sum_{i=0}^{n} S_1^i \xi_i \big\|^2$ yields the identical result. In particular, whenever $\sum_{i=0}^{n} S^i \xi_i = 0$, it follows that $\sum_{i=0}^{n} S_1^i \xi_i = 0$. Hence, the map W is well defined and isometric. It carries a dense subspace of K onto a dense subspace of K_1 by minimality. Thus, W extends uniquely to a unitary operator from K onto K_1. Finally, by construction, $S_1 W = WS$ on this dense subspace, and hence on K by continuity. Thus $S_1 = WSW^*$. Note that $W\xi = \xi$ for $\xi \in H$. ∎

There are several nice ways to extend this to a unitary dilation.

14.1.5. Corollary. *Let $T \in \mathcal{B}(H)$ with $\|T\| \le 1$. Then there is a minimal unitary dilation U on a Hilbert space $L \supseteq H$. If U_1 is another minimal unitary dilation on $L_1 \supseteq H$, there is a unitary operator $W \in \mathcal{B}(L, L_1)$ such that $W|_H = I_H$ and $WUW^* = U_1$.*

Proof. *First proof.* Let S be a minimal isometric dilation of T on K. By the Wold decomposition 12.1.2, $S \simeq T_z^{(\alpha)} \oplus V$ on $K \simeq H^{2(\alpha)} \oplus K_0$ where T_z is the unilateral shift and V is unitary. Identify K with the obvious subspace of $L_0 := L^{2(\alpha)} \oplus K_0$ and define $U_0 = M_z^{(\alpha)} \oplus V$. Then it is evident that U_0 is a unitary dilation of S, and hence of T. It may not be minimal even if S is, so we define $L = \overline{C^*(U_0)H}$ and set $U = U_0|_L$. This is now a minimal unitary dilation of T.

Second proof. Let S be a minimal isometric dilation of T on K. Then S^* is a contraction, so it has a minimal isometric dilation U_0^*. A bit of thought about the structure of S shows that this is the same U_0 as in the previous paragraph.

Uniqueness is proven in the same manner as the uniqueness of the minimal isometric dilation. ∎

For a third proof, we require a lemma.

14.1.6. Lemma. *Suppose that $T \in \mathcal{B}(H)$ is a contraction. Let P be the projection onto $M = \overline{\operatorname{Ran} D_T}$ and let Q be the projection onto $N = \overline{\operatorname{Ran} D_{T^*}}$. The operator $V = \begin{bmatrix} D_{T^*} & T \\ -T^* & D_T \end{bmatrix}$ is unitary on $H \oplus H$. The operator T maps $M^\perp$ isometrically onto $N^\perp$, and $T^*|_{N^\perp} = (T|_{M^\perp})^*$ maps $N^\perp$ isometrically back to $M^\perp$. So*

$$V_0 = (I \oplus P)V(Q \oplus I)|_{N \oplus H} = \begin{bmatrix} D_{T^*}Q & T \\ -PT^*Q & PD_T \end{bmatrix}$$

is a unitary operator mapping $N \oplus H$ onto $H \oplus M$.

Proof. Check that $T(I - T^*T)^k = (I - TT^*)^k T$ for $k \ge 1$. If $p_n(x)$ are polynomials which converge to $\sqrt{x}$ uniformly on [0, 1], then

$$TD_T = \lim_{n\to\infty} Tp_n(I - T^*T) = \lim_{n\to\infty} p_n(I - TT^*)T = D_{T^*}T.$$

Now a straightforward calculation shows that V is unitary. Observe that

$$N^\perp = (\operatorname{Ran} D_{T^*})^\perp = \ker(I - TT^*) = \{\xi \in H : \|T^*\xi\| = \|\xi\|\}.$$

On this space, T^* is isometric and it maps $N^\perp$ onto

$$M^\perp = (\operatorname{Ran} D_T)^\perp = \ker(I - T^*T) = \{\xi \in H : \|T\xi\| = \|\xi\|\}.$$

Similarly, T maps $M^\perp$ isometrically onto $N^\perp$, and $T^*|_{N^\perp} = (T|_{M^\perp})^*$. Therefore, the unitary V maps $N^\perp \oplus \{0\}$ onto $\{0\} \oplus M^\perp$.

We deduce that T^* maps N into M, since to be norm 1 requires that TN be orthogonal to $TN^\perp = M^\perp$. Therefore $T^*Q = PT^* = PT^*Q$. Similarly, TM is contained in N. Hence, V maps $N \oplus H$ into $H \oplus M$. By the previous paragraph and the fact that V is unitary, it follows that V is an isometry of $N \oplus H$ onto $H \oplus M$. ■

Third proof. Let $N_n \simeq N$ for $n \leq -1$ and $M_n \simeq M$ for $n \geq 1$, and form

$$K = H \oplus M^{(\infty)} \oplus N^{(\infty)} = \sum_{n=-1}^{-\infty} N_n \oplus H \oplus \sum_{n \geq 1}^{\oplus} M_n.$$

Define

$$U = \left[\begin{array}{cccc|c|ccc}
\ddots & \ddots & \ddots & \ddots & \ddots & \ddots & \ddots & \ddots \\
\ddots & I & 0 & 0 & \ddots & \ddots & \ddots & \ddots \\
\ddots & 0 & I & 0 & 0 & \ddots & \ddots & \ddots \\
\hline
\ddots & 0 & 0 & D_{T^*}Q & T & 0 & \ddots & \ddots \\
\hline
\ddots & \ddots & 0 & -PT^*Q & PD_T & 0 & 0 & \ddots \\
\ddots & \ddots & \ddots & 0 & 0 & I & 0 & \ddots \\
\ddots & \ddots & \ddots & \ddots & 0 & 0 & I & \ddots \\
\ddots & \ddots & \ddots & \ddots & \ddots & \ddots & \ddots & \ddots
\end{array}\right].$$

By the Lemma, we see that the 2×2 matrix in the middle is V_0, which is a unitary mapping $N_{-1} \oplus H$ onto $H \oplus M_1$. Since the restriction to each N_k for $k \leq -2$ is a shift onto N_{k+1} and the restriction to each M_n for $n \geq 1$ is a shift onto M_{n+1}, one sees that U is unitary.

It is evident that U dilates T since H is semi-invariant and $P_H U|_H = T$. It is easy to see that $U|_{K^+}$ is a minimal isometric dilation of T on $K_+ = H \oplus M^{(\infty)}$ and $U^*|_{K_-}$ is a minimal isometric dilation of T^* on $K_- = N^{(\infty)} \oplus H$. Thus, we see that $\bigvee_{n \geq 0} U^n H = K_+$ and $\bigvee_{n \leq 0} U^n H = K_-$. Therefore $\bigvee_{n \in \mathbb{Z}} U^n H = K$, whence U is a minimal dilation. ■

We collect some more information about the minimal unitary dilation. Most of this is evident from the matrix form of U. Our proof will not assume the matrix form.

14.1.7. Corollary. *Let $T \in \mathcal{B}(H)$ be a contraction on H, and let $U \in \mathcal{B}(K)$ be its minimal unitary dilation. Then*

$$M_1 = (UH \vee H) \cap H^\perp \quad \text{and} \quad N_{-1} = (U^*H \vee H) \cap H^\perp$$

are wandering subspaces, meaning $U^n M_1$ are pairwise orthogonal for $n \in \mathbb{Z}$ and similarly $U^n N_{-1}$ are pairwise orthogonal for $n \in \mathbb{Z}$. Moreover,

$$K_+ = H \oplus \sum_{n\geq 0}^{\oplus} U^n M_1 \quad \text{and} \quad K_- = H \oplus \sum_{n\leq 0}^{\oplus} U^n N_1$$

are the smallest invariant subspace for U containing H and the smallest invariant subspace for U^* containing H, respectively, and

$$K = (K_- \ominus H) \oplus H \oplus (K_+ \ominus H) = \sum_{n\leq 0}^{\oplus} U^n N_{-1} \oplus H \oplus \sum_{n\geq 0}^{\oplus} U^n M_1.$$

Proof. Clearly K_+ is the smallest invariant subspace containing H. By Proposition 14.1.2, H is a semi-invariant subspace for U. From its proof, $K_+^0 := K_+ \ominus H$ is also invariant. Note that $M_1 = UH \cap K_+^0$ and $M_0 = U^* M_1 \subset H$. Therefore for $n \geq 1$,

$$M_n = U^n M_0 \subset U^{n-1} K_+^0 \subset K_+^0 \subset H^\perp.$$

Hence $M_0 \perp M_n$ for $n \geq 1$. Since U is unitary,

$$M_k = U^k M_0 \perp U^k M_n = M_{k+n} \quad \text{for} \quad k \in \mathbb{Z} \text{ and } n \geq 1.$$

This shows that M_0 (or M_1) is wandering. Thus $M = \sum_{n\in\mathbb{Z}}^{\oplus} M_n$ reduces U and $U|_M$ is a bilateral shift of multiplicity $\dim M_0$.

Observe that U^* is the minimal unitary dilation of T^*. Repeating the argument of the first paragraph for T^* shows that K_- is the smallest invariant subspace for U^* containing H, N_{-1} is wandering and $N_0 := UN_{-1} \subset H$. Define $N = \sum_{n\in\mathbb{Z}}^{\oplus} N_n$. Again N reduces U and $U|_N$ is a bilateral shift of multiplicity $\dim N_0$.

Let $K_-^0 = K_- \cap H^\perp$. Thus $L := K_-^0 \oplus H \oplus K_+^0$ is an orthogonal decomposition. Since $UK_-^0 \subset K_- \subset L$ and K_+ is invariant for U, $L \in \operatorname{Lat} U$. Similarly $L \in \operatorname{Lat} U^*$, so that L is a reducing subspace containing H. By minimality, $L = H$. ∎

The following result due to von Neumann, which he proved using facts about Blaschke products, has an easy proof using dilations.

14.1.8. von Neumann Inequality. *Let $T \in \mathcal{B}(H)$ with $\|T\| \leq 1$. If $p \in \mathbb{C}[z]$ or $p \in A(\mathbb{D})$, then*

$$\|p(T)\| \leq \sup_{z\in\overline{\mathbb{D}}} |p(z)| =: \|p\|_\infty.$$

PROOF. Let U be a unitary (power) dilation of T on K. Then since H is semi-invariant, compression to H is a homomorphism. Thus by the continuous functional calculus,

$$\|p(T)\| = \|P_H p(U) P_H\| \le \|p(U)\| = \sup_{z\in\sigma(U)} |p(z)| \le \sup_{z\in\mathbb{T}} |p(z)| = \|p\|_\infty. \quad \blacksquare$$

14.1.9. COROLLARY. *The contractive representations of $A(\mathbb{D})$ are in bijective correspondence with the set of contractions via $\rho \to \rho(z)$.*

PROOF. Clearly a contractive representation of $A(\mathbb{D})$ is determined by $T = \rho(z)$, and $\|T\| \le \|z\| = 1$. Conversely, von Neumann's inequality shows that there is a continuous extension of the polynomial calculus to $A(\mathbb{D})$, and yields a contractive representation. ■

In fact, there is a stronger inequality which is proved in much the same manner.

14.1.10. THEOREM. *Let $T \in \mathcal{B}(H)$ with $\|T\| \le 1$. If $p = [p_{ij}] \in M_n(\mathbb{C}[z])$ or $p \in M_n(A(\mathbb{D}))$, then*

$$\|p(T)\| = \big\|\big[p_{ij}(T)\big]\big\| \le \sup_{z\in\overline{\mathbb{D}}} \big\|\big[p_{ij}(z)\big]\big\| =: \|p\|_\infty.$$

PROOF. Arguing as before, let U be a unitary dilation of T. Since T is a direct sum of contractions on separable spaces, it suffices to assume that U acts on a separable space. Using the Spectral theorem, we represent U as a multiplication operator M_u on $L^2(\mu)$ with $|u| = 1$a.e.(μ). Then

$$\|p(T)\| \le \|p(U)\| = \big\|\big[M_{p_{ij}(u)}\big]\big\|_\infty \le \sup_{z\in\mathbb{T}} \big\|\big[p_{ij}(z)\big]\big\| = \|p\|_\infty. \quad \blacksquare$$

14.2. The H^∞ Functional Calculus

We have seen in several places that a more powerful functional calculus is a very useful tool. For normal operators, we have an L^∞ functional calculus. But for arbitrary operators, we only have the Riesz–Dunford functional calculus. In this section, we use dilation theory to significantly improve on this.

14.2.1. DEFINITION. A contraction T on H is *completely non-unitary* (c.n.u.) if there is no reducing subspace M for T such that $T|_M$ is unitary.

A unitary operator U is *absolutely continuous* if the spectral measure $E_U \ll m$.

14.2.2. PROPOSITION. *Let $T \in \mathcal{B}(H)$ with $\|T\| \le 1$. Then there is a unique reducing subspace N such that $T|_N$ is unitary and $T|_{N^\perp}$ is completely non-unitary.*

PROOF. First note that if $U = P_N T|_N$ is unitary, then N reduces T because with respect to $N \oplus N^\perp$,

$$T = \begin{bmatrix} U & X \\ Y & Z \end{bmatrix}, \quad T^*T = \begin{bmatrix} I + Y^*Y & * \\ * & * \end{bmatrix} \quad \text{and} \quad TT^* = \begin{bmatrix} I + XX^* & * \\ * & * \end{bmatrix}.$$

Since $T^*T \le I$ and $TT^* \le I$, we require $X = Y = 0$, so that $T \simeq U \oplus Z$.

Next suppose that M and N are reducing subspaces with $T|_M$ and $T|_N$ both unitary. Let $L = (M \vee N) \cap M^\perp$. Since M and N are reducing, so are $M \vee N$ and L. If $\xi = \eta + \zeta \in N$ with $\eta \in M$ and $\zeta \in L$, we have $T\eta \in M, T\zeta \in L$ and

$$\|\eta\|^2 + \|\zeta\|^2 = \|\xi\|^2 = \|T\xi\|^2 = \|T\eta\|^2 + \|T\zeta\|^2 = \|\eta\|^2 + \|T\zeta\|^2.$$

The set of vectors ζ that arise in this way, $P_L N$, is dense in L. Thus T is isometric on L. Applying the same argument to T^* shows that $T^*|_L$ is isometric also. Hence $T|_L$ is unitary. Therefore T is unitary on $M \oplus L = M \vee N$.

Let $\{M_\lambda\}_\Lambda$ be a maximal collection of pairwise orthogonal subspaces of H which reduce T and $T|_{M_\lambda}$ are unitary. Let $M = \sum_\Lambda^\oplus M_\lambda$. Clearly M is reducing and $T|_M$ is unitary. Suppose that there was another reducing subspace $N \not\subset M$ such that $T|_N$ is unitary. Then $L = (M \vee N) \cap M^\perp$ is reducing, $T|_L$ is unitary, and $L \perp M_\lambda$ for all λ. This contradicts the maximality of the collection. Thus M contains every reducing subspace on which T is unitary. It follows that $T|_{M^\perp}$ is completely non-unitary and that M is uniquely defined. ∎

See the discussion in Section 12.5 about how to decompose a unitary $U \simeq U_a \oplus U_s$ into its absolutely continuous and singular parts.

14.2.3. COROLLARY. *Let $T \in \mathcal{B}(H)$ with $\|T\| \le 1$, and suppose that T is c.n.u. Then the minimal unitary dilation U of T is absolutely continuous.*

PROOF. By Corollary 14.1.7, $K_+^0 = \sum_{k\ge1}^\oplus M_k \subset \sum_{k\in\mathbb{Z}}^\oplus M_k = M$. Since M_0 is wandering, $U|_M$ is a bilateral shift, and hence absolutely continuous. Similarly, setting $N = \sum_{k\in\mathbb{Z}}^\oplus N_k$, we have that $U|_N$ is absolutely continuous. But $M \vee N \supset K_-^0 \oplus K_+^0 = H^\perp$. Therefore the singular part of U is contained in H. If it is non-zero, then T would not be c.n.u. Hence U is absolutely continuous. ∎

The L^∞ functional calculus for unitary operators is powerful, and well understood. So we can safely restrict our attention to c.n.u. contractions in trying to expand the Riesz–Dunford functional calculus for contractions.

14.2.4. SZ.-NAGY–FOIAŞ FUNCTIONAL CALCULUS. *Let $T \in \mathcal{B}(H)$ be a c.n.u. contraction. There is a contractive, weak-$*$-WOT-continuous homomorphism $\Phi_T : H^\infty \to \mathcal{W}(T)$ such that $h(T) = \Phi_T(h)$ coincides with the Riesz–Dunford functional calculus when h is analytic in a neighbourhood of $\overline{\mathbb{D}}$.*

If $h \in H^\infty$, let $\widetilde{h}(z) = \overline{h(\bar{z})}$. Then $h(T^*) = \widetilde{h}(T)^*$.

PROOF. By Corollary 14.2.3, the minimal unitary dilation U of T is absolutely continuous. Therefore, it has an $L^\infty(m)$ functional calculus which is contractive and weak-$*$–WOT-continuous. If $h \in H^\infty$, we set $h_r(z) = h(rz)$ for $0 < r < 1$. These functions are analytic in a neighbourhood of $\overline{\mathbb{D}}$, and thus belong to $A(\mathbb{D})$. Moreover h_r converges weak-$*$ to h in L^∞.

By Corollary 14.1.9, we can evaluate $h_r(T)$. Now $p(T) = P_H p(U)|_H$ for any polynomial, and thus for any element of $A(\mathbb{D})$ by norm continuity. Since the L^∞ functional calculus Φ_U for U is weak-$*$–WOT-continuous, we see $h_r(U)$ converges in the WOT topology to $h(U)$. In particular, K_+ and K_+^0 remain invariant for $h(U)$; so H is semi-invariant. We define $\Phi_T(h) = P_H h(U)|_H$ for $h \in H^\infty$. The compression map $\Psi_H : \mathcal{B}(K) \to \mathcal{B}(H)$ given by $\Psi_H(A) = P_H A|_H$ is WOT-continuous, and restricted to $\Phi_U(H^\infty)$, it is a contractive homomorphism. Thus $\Phi_T = \Psi_H \Phi_U|_{H^\infty}$ is a composition of two contractive homomorphisms, and it is weak-$*$–WOT-continuous. This agrees with the Riesz–Dunford functional calculus when both are defined.

Since T^* is also a c.n.u. contraction, it also has an H^∞-functional calculus. For $h \in H^\infty$ with Taylor series $h(z) = \sum_{n\ge 0} a_n z^n$, the series $h_r(z) = \sum_{n\ge0} a_n r^n z^n$ converges absolutely on $\overline{\mathbb{D}}$ for $0 < r < 1$. Thus $\widetilde{h_r}(z) = \sum_{n\ge0} \overline{a_n} r^n z^n$ and

$$h_r(T^*) = \sum_{n\ge0} a_n r^n T^{*n} = \Big(\sum_{n\ge0} \overline{a_n} r^n T^n\Big)^* = \widetilde{h_r}(T)^*.$$

Letting $r \to 1$ and using the WOT-continuity, we obtain $h(T^*) = \widetilde{h}(T)^*$. ■

Recall that $h \in H^2$ is outer if and only if h is a cyclic vector for the shift T_z.

14.2.5. PROPOSITION. *If $T \in \mathcal{B}(H)$ is a c.n.u. contraction and $h \in H^\infty$ is outer, then $h(T)$ is injective with dense range.*

PROOF. Suppose that $\xi \in \ker h(T)$ with $\|\xi\| = 1$. Let U be the minimal unitary dilation of T on K. Then $\varphi(f) = \langle f(U)\xi, \xi\rangle$ is a state on $C(\mathbb{T})$. By the Riesz–Markov representation theorem, there is a Borel probability measure μ so that $\varphi(f) = \int f\,d\mu$. By Corollary 14.2.3, U is absolutely continuous, and thus $\mu \ll m$; so that $d\mu = k\,dm$ for $k \in L^1_+(m)$. Compute for $n \ge 0$,

$$0 = \langle T^n h(T)\xi, \xi\rangle = \langle (z^n h)(U)\xi, \xi\rangle = \frac{1}{2\pi}\int e^{int} h(e^{it}) k(e^{it})\,dt.$$

Since $\overline{\operatorname{span}}\{z^n h : n \ge 0\} = H^2$, we deduce that $\frac{1}{2\pi}\int k(e^{it})\,dt = 0$. But $k \ge 0$, so $k = 0$, a contradiction. Hence $\ker h(T) = \{0\}$ and $h(T)$ is injective.

Similarly $\widetilde{h}(T^*) = h(T)^*$ is injective, and therefore $h(T)$ has dense range. ■

We will apply this theory to study a class of operators called C_0 operators.

14.2.6. EXAMPLE. If T is a c.n.u contraction such that Φ_T has kernel, then this kernel is a non-trivial weak-$*$ closed ideal of H^∞. By Corollary 12.2.15, there is an inner function ω such that $\ker \Phi_T = \omega H^\infty$. This is possible for any inner function.

Since ωH^2 is invariant for T_z, then the subspace $K_\omega = H^2 \ominus \omega H^2$ is coinvariant for T_z. Let $S_\omega = P_{K_\omega} T_z|_{K_\omega}$. Observe that the minimal unitary dilation of S_ω is M_z acting on L^2. Therefore $h(S_\omega) = P_{K_\omega} M_h|_{K_\omega}$. Hence $h \in \ker \Phi_{S_\omega}$ if and only if $hK_\omega \subset \omega H^2$. Since $h(\omega H^2) \subset \omega H^2$, this is the same as $hH^2 \subset \omega H^2$, which occurs exactly when ω divides h, namely, $h \in \omega H^\infty$.

14.2.7. DEFINITION. A contraction T such that $\ker \Phi_T = \omega H^\infty$ is called a C_0 *operator*, and $m_T = \omega$ is the *minimal function* of T.

14.2.8. LEMMA. *Suppose that T is a C_0 contraction with $m_T = \omega$. If $\omega = \omega_1\omega_2$ is a proper factorization, set $M = \ker \omega_1(T)$. With respect to $H = M \oplus M^\perp$, we have*

$$T \simeq \begin{bmatrix} T_1 & X \\ 0 & T_2 \end{bmatrix}.$$

Then T_i are C_0 and $m_{T_i} = \omega_i$ for $i = 1, 2$. In particular, $\{0\} \neq M \neq H$.

PROOF. Let $h \in H^\infty$. Observe that $h(T)$ has the form $\left[\begin{smallmatrix} h(T_1) & Y \\ 0 & h(T_2) \end{smallmatrix}\right]$. This follows since it is true for polynomials, and this form is preserved by WOT limits. But the $(1,1)$ entry of $\omega_1(T)$ is 0, and thus $\omega_1(T_1) = 0$. Also $0 = \omega(T) = \omega_1(T)\omega_2(T)$ shows that $\operatorname{Ran}(\omega_2(T)) \subset M$. Thus $\omega_2(T_2) = 0$. Hence $\theta_i = m_{T_i}$ divides ω_i. However, we have

$$\theta_1\theta_2(T) = \theta_1(T)\theta_2(T) = \left[\begin{smallmatrix} 0 & Y \\ 0 & \theta_1(T_2) \end{smallmatrix}\right] \left[\begin{smallmatrix} \theta_2(T_1) & Z \\ 0 & 0 \end{smallmatrix}\right] = \left[\begin{smallmatrix} 0 & 0 \\ 0 & 0 \end{smallmatrix}\right].$$

Hence ω divides $\theta_1\theta_2$; so that $\theta_i = \omega_i$ for $i = 1, 2$. This shows that $M \neq \{0\}$, for then $\omega = m_T = \omega_2$, and similarly $M \neq H$. ∎

14.2.9. COROLLARY. *Suppose that T is a C_0 contraction which is not scalar. Then T has proper hyperinvariant subspaces.*

PROOF. Since T is not scalar, m_T is not a Blaschke factor b_λ. Thus it has a proper factorization $m_T = \omega_1\omega_2$. By the lemma, $M = \ker \omega_1(T)$ is a proper subspace. If $AT = TA$, then $A\omega_1(T) = \omega_1(T)A$. Thus if $\xi \in \ker \omega_1(T)$,

$$\omega_1(T)A\xi = A\omega_1(T)\xi = 0.$$

So $AM \subset M$, and M is hyperinvariant. ∎

Recall that if ω is inner, then $\omega = BS$ for a Blaschke product B and singular inner function S with singular measure μ. Then $Z(B)$ is the zero set of B in $\mathbb{D}$, and $\operatorname{supp}(\omega) = \operatorname{supp}(\mu) \cup (\overline{Z(B)} \cap \mathbb{T})$. Moreover, $\operatorname{supp}(\omega)$ is the smallest closed subset

C of $\mathbb{T}$ such that ω can be continued analytically across any arc in $\mathbb{T} \setminus C$, and ω is non-zero there. Let $s_\omega = Z(B) \cup \operatorname{supp}(\omega)$.

14.2.10. THEOREM. *Suppose that T is a C_0 contraction with $m_T = \omega$. Then $\sigma(T) = s_\omega$.*

PROOF. Suppose that $\lambda \in \overline{\mathbb{D}} \setminus s_\omega$. Then ω extends to be analytic in a neighbourhood of λ. Let $h(z) = \dfrac{\omega(\lambda) - \omega(z)}{\lambda - z}$. Then h is bounded and analytic on $\mathbb{D}$, and so lies in H^∞. Thus

$$h(T)(\lambda I - T) = \omega(\lambda)I - \omega(T) = \omega(\lambda)I \neq 0.$$

Hence $(\lambda I - T)^{-1} = \omega(\lambda)^{-1}h(T)$. Therefore $\sigma(T) \subset s_\omega$.

Suppose that $\lambda \in Z(B)$. Then the Blaschke factor $b_\lambda(z) = \dfrac{\lambda - z}{1 - \bar{\lambda} z}$ is a non-scalar factor of ω. By Lemma 14.2.8, $\ker(\lambda I - T) = \ker b_\lambda(T) \neq \{0\}$, whence $\lambda \in \sigma(T)$. So $\sigma(T) \supset \overline{Z(B)}$.

Suppose that $\operatorname{supp}(\mu) \not\subset \sigma(T)$. Then there is a closed ball C with $\mu(C) > 0$ disjoint from $\sigma(T)$. By the previous paragraph, $\overline{Z(B)} \cap C = \varnothing$. Split $\mu = \mu_1 + \mu_2$ where μ_1 has support in $C \cap \mathbb{T}$ and μ_2 has support in $\mathbb{T} \setminus C$. Let S_i be the singular inner function for the measure μ_i. Set $\omega_1 = S_1$ and $\omega_2 = BS_2$, so $\omega = \omega_1\omega_2$. Apply Lemma 14.2.8: $M = \ker \omega_1(T)$, $H = M \oplus M^\perp$ and

$$T \simeq \begin{bmatrix} T_1 & X \\ 0 & T_2 \end{bmatrix}.$$

If $\lambda \in C$, then $(\lambda I - T)^{-1}$ commutes with T, and hence leaves M invariant. Since $m_{T_2} = \omega_2$, $\sigma(T_2) \subset s_{\omega_2}$ is disjoint from C. Hence, we have

$$\begin{aligned} I = (\lambda I - T)^{-1}(\lambda I - T) &= \begin{bmatrix} A_1 & Y \\ 0 & A_2 \end{bmatrix} \begin{bmatrix} \lambda I - T_1 & -X \\ 0 & \lambda I - T_2 \end{bmatrix} \\ &= \begin{bmatrix} A_1(\lambda I - T_1) & Z \\ 0 & A_2(\lambda I - T_2) \end{bmatrix}. \end{aligned}$$

Hence A_1 is a left inverse for $\lambda I - T_1$. Similarly A_1 is a right inverse for $\lambda I - T_1$. We conclude that $\sigma(T_1)$ is disjoint from C. But $\sigma(T_1) \subset s_{T_1} = \operatorname{supp}(\mu_1) \subset C$. This shows that $\sigma(T_1)$ is empty, a contradiction. Therefore $\sigma(T) = s_\omega$. ∎

14.3. Commuting Pairs of Operators

It is a natural question whether the results for one contraction can be extended, in a variety of possible ways, to two or more variables. In this section, we will deal with two commuting contractions.

14.3.1. ANDO THEOREM. *Let $T_1, T_2 \in \mathcal{B}(H)$ be contractions such that $T_1T_2 = T_2T_1$. Then there is a Hilbert space $K \supset H$ and isometries V_i dilating T_i such that $V_1V_2 = V_2V_1$ and $P_H V_1^m V_2^n|_H = T_1^m T_2^n$ for $m, n \geq 0$.*

PROOF. Let S_1 and S_2 be isometric dilations of T_1 and T_2 on the same Hilbert space K. If the minimal dilation of one has a smaller dimension for $K \ominus H$, then it will be necessary to add another isometric summand such as the identity operator on the orthogonal complement. Then

$$S_i = \begin{bmatrix} T_i & 0 \\ * & * \end{bmatrix}, \quad S_1S_2 = \begin{bmatrix} T_1T_2 & 0 \\ * & * \end{bmatrix} \quad \text{and} \quad S_2S_1 = \begin{bmatrix} T_2T_1 & 0 \\ * & * \end{bmatrix}.$$

Thus both S_1S_2 and S_2S_1 are isometric dilations of T_1T_2. Let S_0 be the minimal isometric dilation of T_1T_2 on $K_0 \supset H$. We can identify K_0 as a subspace of K so that $K = K_0 \oplus K_1$ and $S_1S_2 \simeq S_0 \oplus R_1$ for an isometry $R_1 \in \mathcal{B}(K_1)$. Similarly there is a unitary map of K_0 onto a subspace K_0' of K, which is the identity on H, so that $K = K_0' \oplus K_2$ and $S_2S_1 \simeq S_0' \oplus R_2$ for $S_0' \simeq S_0$ via a unitary that fixes H, and an isometry $R_2 \in \mathcal{B}(K_2)$. Define isometries on $K \oplus (K_1 \oplus K_2)^{(\infty)}$ by

$$W_1 = S_1 \oplus (I_{K_1} \oplus I_{K_2})^{(\infty)} \quad \text{and} \quad W_2 = S_2 \oplus (R_1 \oplus R_2)^{(\infty)}.$$

These are also isometric dilations of T_1 and T_2, but now

$$W_1W_2 \simeq S_0 \oplus R_1 \oplus (R_1 \oplus R_2)^{(\infty)} \simeq S_0' \oplus R_2 \oplus (R_1 \oplus R_2)^{(\infty)} \simeq W_2W_1.$$

Moreover, the unitary U implementing this equivalence intertwines S_0 and S_0', acts as the identity on H, and matches up copies of R_i. We have $W_2W_1 = UW_1W_2U^*$. Define $V_1 = UW_1$ and $V_2 = W_2U^*$. Then V_i dilates T_i for $i = 1, 2$, and

$$V_1V_2 = UW_1W_2U^* = W_2W_1 = (W_2U^*)(UW_1) = V_2V_1.$$

Finally since V_i have the same matrix form as the matrices for S_i above, it is evident that $P_H V_1^m V_2^n|_H = T_1^m T_2^n$ for $m, n \geq 0$. ∎

We can extend these commuting isometries to commuting unitaries.

14.3.2. THEOREM. *Let $S_1, \dots, S_n$ be commuting isometries in $\mathcal{B}(H)$. Then there is a Hilbert space $K = K_- \oplus H$ and commuting unitaries $U_1, \dots, U_n$ which have the form:*

$$U_i = \begin{bmatrix} * & 0 \\ * & S_i \end{bmatrix} \quad \textit{for} \quad 1 \leq i \leq n,$$

and so simultaneously dilate the S_i.

PROOF. Take a minimal unitary dilation $U_1 = \begin{bmatrix} * & 0 \\ * & S_1 \end{bmatrix}$ on $K_1 = \bigvee_{k\geq 0} U_1^{*k} H$. Define operators

$$V_i \sum_{k=0}^{m} U_1^{*k} \xi_k = \sum_{k=0}^{m} U_1^{*k} S_i \xi_k \quad \text{for} \quad \xi_k \in H,\ m \in \mathbb{N}, \text{ and } 2 \leq i \leq n.$$

This defines linear maps on the algebraic span of $\bigcup_{k\geq 0} U_1^{*k} H$ such that

$$\begin{aligned} \Big\| V_i \sum_{k=0}^{m} U_1^{*k} \xi_k \Big\| &= \Big\| \sum_{k=0}^{m} U_1^{*k} S_i \xi_k \Big\| = \Big\| U_1^{m} \sum_{k=0}^{m} U_1^{*k} S_i \xi_k \Big\| \\ &= \Big\| \sum_{k=0}^{m} U_1^{m-k} S_i \xi_k \Big\| = \Big\| \sum_{k=0}^{m} S_1^{m-k} S_i \xi_k \Big\| \\ &= \Big\| S_i \sum_{k=0}^{m} S_1^{m-k} \xi_k \Big\| = \Big\| U_1^{m} \sum_{k=0}^{m} U_1^{*k} \xi_k \Big\| = \Big\| \sum_{k=0}^{m} U_1^{*k} \xi_k \Big\|. \end{aligned}$$

This shows that each V_i is isometric. In particular, if $\sum_{k=0}^{m} U_1^{*k} \xi_k = 0$, then V_i sends it to 0, and thus V_i is well defined. Hence, each V_i extends to be an isometry on $\mathcal{K}_1$ by continuity. You can easily check that this formula is also correct for $i = 1$.

Moreover, note that if S_i is already unitary, then the vectors $\eta_k = S_i \xi_k$ are arbitrary vectors in H, and thus the range of V_i contains all vectors of the form $\sum_{k=0}^{m} U_1^{*k} \eta_k$, which are dense in $\mathcal{K}_1$. Hence, in this case, V_i is surjective, and thus is also unitary.

Suppose that $i \neq j$. Compute

$$\begin{aligned} V_i V_j \sum_{k=0}^{m} U_1^{*k} \xi_k &= V_i \sum_{k=0}^{m} U_1^{*k} S_j \xi_k = \sum_{k=0}^{m} U_1^{*k} S_i S_j \xi_k \\ &= \sum_{k=0}^{m} U_1^{*k} S_j S_i \xi_k = V_j \sum_{k=0}^{m} U_1^{*k} S_i \xi_k = V_j V_i \sum_{k=0}^{m} U_1^{*k} \xi_k. \end{aligned}$$

Therefore $V_i V_j = V_j V_i$. Note that each V_i agrees with S_i on each $\xi_0 \in H$, so they have the matrix form $V_i = \left[\begin{smallmatrix} * & 0 \\ * & S_i \end{smallmatrix}\right]$.

Now we repeat this construction for each $1 \leq i \leq n$. Eventually each S_i is extended to a unitary, and from that point on, it remains unitary. At each stage, the isometries commute, so we end up with commuting unitaries $U_1, \dots, U_n$. The matrix form of U_i remains the same after each iteration. Finally, this is a joint unitary dilation because H is an invariant subspace for all of them. ∎

Applying this to the isometric dilation of Ando's theorem, we obtain a result which is also often called Ando's theorem:

14.3.3. COROLLARY. *Let $T_1, T_2 \in \mathcal{B}(H)$ be two commuting contractions. Then there is a Hilbert space $K \supset H$ and unitaries $U_i \in \mathcal{B}(K)$ dilating T_i such that $U_1 U_2 = U_2 U_1$ and*

$$U_i = \begin{bmatrix} * & 0 & 0 \\ * & T_i & 0 \\ * & * & * \end{bmatrix} \quad \textit{for} \quad i = 1, 2.$$

Hence $P_H U_1^m U_2^n|_H = T_1^m T_2^n$ *for* $m, n \geq 0$.

This immediately yields a two-variable von Neumann inequality.

14.3.4. COROLLARY. *Let* $T_1, T_2 \in \mathcal{B}(H)$ *be two commuting contractions. For a matrix polynomial* $[p_{ij}] \in M_n(\mathbb{C}[z_1, z_2])$,

$$\| [p_{ij}(T_1, T_2)] \| \leq \sup_{(z_1,z_2)\in\overline{\mathbb{T}}^2} \| [p_{ij}(z_1, z_2)] \|.$$

PROOF. Dilate T_1, T_2 to commuting unitaries U_1, U_2. Note that U_1 also commutes with U_2^*, so $C^*(U_1, U_2)$ is commutative, and is isomorphic to $C(X)$ where X is the joint spectrum of U_1, U_2. Since $\sigma(U_i) \subset \mathbb{T}$, we have $X \subset \mathbb{T}^2$. Define a $*$-representation ρ of $C(\mathbb{T}^2)$ by the continuous functional calculus $\rho(f) = f(U_1, U_2)$. Then $\rho^{(n)} : M_n(C(\mathbb{T}^2)) \to M_n(\mathcal{B}(K)) \simeq \mathcal{B}(K^{(n)})$ by

$$\rho^{(n)} [f_{ij}] = [f_{ij}(U_1, U_2)]$$

is also a $*$-representation. Now $*$-representations of C*-algebras are contractive. When we restrict the matrix entries to be polynomials, the 3x3 matrix form of U_1 and U_2 shows that the $2, 2$ entry of $[p_{ij}(U_1, U_2)]$ is $[p_{ij}(T_1, T_2)]$. Therefore

$$\| [p_{ij}(T_1, T_2)] \| \leq \| [p_{ij}(U_1, U_2)] \| \leq \sup_{(z_1,z_2)\in\overline{\mathbb{T}}^2} \| [p_{ij}(z_1, z_2)] \|. \qquad \blacksquare$$

Ando's theorem has another very useful formulation due to Sz.-Nagy and Foiaş.

14.3.5. COMMUTANT LIFTING THEOREM. *Let* T *be a contraction on* H *and let* $X \in \mathcal{B}(H)$ *commute with* T. *Also let* $U \in \mathcal{B}(K)$, *for* $K \supset H$, *be a unitary dilation of* T. *Then there is a simultaneous dilation of* X *to an operator* $Y \in \mathcal{B}(K)$ *such that* $\|Y\| = \|X\|$, $UY = YU$ *and* $P_H U^m Y^n|_H = T^m X^n$ *for* $m, n \geq 0$.

PROOF. First we normalize so that $\|X\| = 1$. By Ando's theorem, there is a pair of commuting unitaries U_1, U_2 on a Hilbert space K' that simultaneously dilate T and X.

$$U_1 = \begin{bmatrix} * & 0 & 0 \\ * & T & 0 \\ * & * & * \end{bmatrix} \quad \text{and} \quad U_2 = \begin{bmatrix} * & 0 & 0 \\ * & X & 0 \\ * & * & * \end{bmatrix}.$$

There is a subspace $H \subset K_0' \subset K'$ which reduces U_1, and $U_1|_{K_0'}$ is a minimal unitary dilation of T. Recall that

$$K_0' = \bigvee_{n\in\mathbb{Z}} U^n H = \Big(\big(\bigvee_{n\le 0} U^n H\big) \ominus H\Big) \oplus H \oplus \Big(\big(\bigvee_{n\ge 0} U^n H\big) \ominus H\Big).$$

Thus we obtain

$$U_{10} := U_1|_{K_0'} = \begin{bmatrix} * & 0 & 0 \\ * & T & 0 \\ * & * & * \end{bmatrix} \quad \text{and} \quad Y_0 := P_{K_0'} U_2|_{K_0'} = \begin{bmatrix} * & 0 & 0 \\ * & X & 0 \\ * & * & * \end{bmatrix}.$$

Then U_{10} is unitary, but Y_0 will generally not be unitary. If we write $K' = K_0' \oplus K_1'$, then

$$U_1 = \begin{bmatrix} U_{10} & 0 \\ 0 & U_{11} \end{bmatrix} \quad \text{and} \quad U_2 = \begin{bmatrix} Y_0 & * \\ * & * \end{bmatrix}.$$

Since U_1 and U_2 commute, we see that $U_{10}Y_0 = Y_0 U_{10}$.

Now consider the given unitary dilation U of T on $\mathcal{B}(K)$. There is, similarly, a subspace $H \subset K_0 \subset K$ which reduces U, and $U|_{K_0}$ is a minimal unitary dilation of T. By uniqueness of the minimal unitary dilation, there is a unitary $W : K_0' \to K_0$ with $W|_H = I_H$ such that $U|_{K_0} = W U_{10} W^*$. Write $U = U|_{K_0} \oplus U|_{K_0^\perp}$. Define

$$Y = W Y_0 W^* \oplus 0.$$

Then Y is a contraction which commutes with U and maintains the triangular form

$$U = \begin{bmatrix} * & 0 & 0 \\ * & T & 0 \\ * & * & * \end{bmatrix} \quad \text{and} \quad Y = \begin{bmatrix} * & 0 & 0 \\ * & X & 0 \\ * & * & * \end{bmatrix}.$$

From this, we see that U and Y simultaneously dilate T and X, $UY = YU$ and U is the original unitary. ∎

We can do the same for an isometric dilation simply by dilating the isometry to a unitary, applying the commutant lifting theorem, and restricting back down.

14.3.6. Corollary. *Let T be a contraction on H and let $X \in \mathcal{B}(H)$ commute with T. Also let $S \in \mathcal{B}(K)$, for $K \supset H$, be an isometric co-extension of T. Then there is a simultaneous dilation of X to an operator Y such that $\|Y\| = \|X\|$, $UY = YU$ and $P_H S^m Y^n|_H = T^m X^n$ for $m, n \geq 0$.*

14.4. Nevanlinna–Pick Interpolation

In this section, we provide an application of commutant lifting with an operator-theoretic proof of an old result in function theory due to Pick and independently to Nevanlinna. This theorem has been turned into a property that certain Hilbert spaces of functions can have. These ideas have had a significant influence on the development of function-theoretic operator theory. See Section 15.12.

The problem is simple to state: let $z_1, \dots, z_n$ be distinct points in $\mathbb{D}$ and let $w_1, \dots, w_n \in \mathbb{C}$. Is there a function $f \in H^\infty$ such that

$$f(z_i) = w_i \quad \text{for } 1 \le i \le n \quad \text{and} \quad \|f\|_\infty \le 1?$$

Pick's solution was a necessary and sufficient condition that a certain matrix be positive semidefinite. Nevanlinna actually parameterized the set of all solutions by different methods.

This problem was solved using operator theory by Sarason by establishing a special case of the commutant lifting theorem. Recall that for $f \in H^\infty$, the Toeplitz operator T_f on H^2 is just multiplication by f. It has norm $\|T_f\| = \|f\|_\infty$ by Proposition 12.3.2. Also the kernel functions $k_w(z) = \frac{1}{1-z\overline{w}}$ for $w \in \mathbb{D}$ satisfy

$$\langle T_f h, k_w \rangle = f(w)h(w) \quad \text{for } w \in \mathbb{D} \quad \text{and} \quad T_f^* k_w = \overline{f(w)} k_w.$$

See Proposition 11.2.9 and Lemma 12.4.8.

14.4.1. THEOREM. *Given distinct points $z_1, \dots, z_n$ in $\mathbb{D}$ and $w_1, \dots, w_n \in \mathbb{C}$, there is a function $f \in H^\infty$ so that $f(z_i) = w_i$ for $1 \le i \le n$ and $\|f\|_\infty \le 1$ if and only if*

$$\left[\frac{1 - w_i \overline{w_j}}{1 - z_i \overline{z_j}} \right]_{k \times k} \ge 0.$$

PROOF. First we show that this condition is necessary. Suppose that f is a solution with $\|f\|_\infty \le 1$. Let $M = \operatorname{span}\{k_{z_1}, \dots, k_{z_n}\} = H^2 \ominus BH^2$ where $B = \prod_{i=1}^n b_{z_i}$ is the finite Blaschke product with simple zeros at $z_1, \dots, z_n$. See Example 12.2.5. Then $A = P_M T_f|_M$ has $\|A\| \le \|f\|_\infty \le 1$, or equivalently $I - AA^* \ge 0$. We can test this against the linear map $K : \mathbb{C}^n \to M$ by $Ke_i = k_{z_i}$, which has range M. Thus $I - AA^* \ge 0$ if and only if $K^*(I - AA^*)K \ge 0$. Compute

$$\begin{aligned}
\langle K^*(I - AA^*)Ke_j, e_i \rangle &= \langle (I - AA^*)k_{z_j}, k_{z_i} \rangle \\
&= \langle k_{z_j}, k_{z_i} \rangle - \langle T_f^* k_{z_j}, T_f^* k_{z_i} \rangle \\
&= \langle k_{z_j}, k_{z_i} \rangle - \langle \overline{f(z_j)} k_{z_j}, \overline{f(z_i)} k_{z_i} \rangle \\
&= \frac{1}{1 - z_i \overline{z_j}} - \langle \overline{w_j} k_{z_j}, \overline{w_i} k_{z_i} \rangle = \frac{1 - w_i \overline{w_j}}{1 - z_i \overline{z_j}}.
\end{aligned}$$

Therefore $\|A\| \le 1$ if and only if

$$\left[\frac{1 - w_i \overline{w_j}}{1 - z_i \overline{z_j}} \right]_{n \times n} \ge 0.$$

For the converse, note that BH^2 is invariant for T_z, and hence the subspace M is coinvariant. Let $S = P_M T_z|_M$, and observe that S dilates to the isometry T_z. Define $A \in \mathcal{B}(M)$ by setting $A^* k_{z_i} = \overline{w_i} k_{z_i}$ for $1 \le i \le n$, which specifies A^* on a basis. The computation above, suitably rearranged, shows that $I - AA^* \ge 0$ and thus $\|A\| \le 1$. We claim that $SA = AS$. We have $S^* k_{z_i} = T_z^* k_{z_i} = \overline{z_i} k_{z_i}$. Hence

$$(S^*A^* - A^*S^*)k_{z_i} = \overline{z_i}\overline{w_i} k_{z_i} - \overline{w_i}\overline{z_i} k_{z_i} = 0.$$

Therefore $(AS - SA)^* = 0$.

By the commutant lifting theorem, there is an operator $T \in \mathcal{B}(H^2)$ such that T dilates A, so that $P_M T|_M = A$, $\|T\| = \|A\| \le 1$ and T commutes with T_z. The latter condition means that $T = T_f$ for some $f \in H^\infty$ by Proposition 12.4.1. Since $\|T_f\| \le 1$, we have $\|f\|_\infty \le 1$. Finally,

$$\overline{f(z_i)} k_{z_i} = T_f^* k_{z_i} = A^* k_{z_i} = \overline{w_i} k_{z_i} \quad \text{for} \quad 1 \le i \le n.$$

Thus $f(z_i) = w_i$ interpolates the values w_i as desired. ∎

This proof is also valid for matrix-valued functions. That is, we can think of $M_k(H^\infty)$, the $k \times k$ matrices with coefficients in H^∞, as bounded matrix-valued analytic functions by identifying $F = [f_{ij}]$ with

$$F(z) = [f_{ij}(z)] \text{ for } z \in \mathbb{D} \quad \text{and} \quad \|F\|_\infty = \sup_{z \in \mathbb{D}} \|F(z)\|.$$

We make use of the fact that $M_k(H^\infty)$ can be represented as matrices with Toeplitz operator entries acting on $H^{2(k)} = H^2 \otimes \mathbb{C}^k$. A vector in $H^{2(k)}$ can be given as a k-tuple $h(z) = (h_1(z), \ldots, h_k(z))$ or we can think of h as an analytic function with values in $\mathbb{C}^k$; so that $h(z) = \sum_{i=1}^k h_i(z) e_i$ where $e_1, \ldots, e_k$ is the standard orthonormal basis for $\mathbb{C}^k$. Then

$$T_F \begin{bmatrix} h_1 \\ \vdots \\ h_k \end{bmatrix} = \Big[\sum_{j=1}^k f_{ij} h_j \Big]_{i=1}^k \quad \text{for} \quad h_1, \ldots, h_k \in H^2.$$

We can also define multiplication operators M_F on $L^{2(k)}$ by a similar formula. We leave it to the reader to check that $\|M_F\| = \|T_F\| = \|F\|_\infty$ by essentially the same argument as in the scalar case (Exercise **14.15**).

The proof of the matrix result is virtually identical except for the technicalities.

14.4.2. THEOREM. *Given points $z_1, \ldots, z_n$ in $\mathbb{D}$ and matrices $W_1, \ldots, W_n$ in M_k, there is a function $F \in M_k(H^\infty)$ so that*

$$F(z_s) = [f_{ij}(z_s)] = W_s \ \textit{for } 1 \le s \le n \quad \textit{and} \quad \|F\|_\infty \le 1$$

if and only if

$$\left[\frac{I_k - W_s W_t^*}{1 - z_s \overline{z_t}} \right]_{n \times n} \ge 0.$$

PROOF. First we show that this condition is necessary. Suppose that F is a solution with $\|F\|_\infty \le 1$. Let $\widetilde{M} = M \otimes \mathbb{C}^k$ where $M = \operatorname{span}\{k_{z_s} : 1 \le s \le n\} = H^2 \ominus BH^2$ as above. Note that $BH^2 \otimes \mathbb{C}^k = \{h = (Bh_1, \dots, Bh_k) : h_i \in H^2\}$ is invariant for T_F for any $F \in M_k(H^\infty)$ because the coefficients remain divisible by B. Thus $\widetilde{M}$ is coinvariant. Define $A = P_{\widetilde{M}} T_F|_{\widetilde{M}}$. Then $\|A\| \le \|F\|_\infty \le 1$. Since $A^* = T_F^*|_{\widetilde{M}}$, we have

$$A^* k_{z_s} \otimes e_j = k_{z_s} \otimes F(z_s)^* e_j \quad \text{for} \quad 1 \le s \le n,\ 1 \le j \le k.$$

Set $\widetilde{K} : \mathbb{C}^n \otimes \mathbb{C}^k \to \widetilde{M}$ by

$$\widetilde{K} e_s \otimes e_j = k_{z_s} \otimes e_j \quad \text{for} \quad 1 \le s \le n,\ 1 \le j \le k.$$

This maps onto $\widetilde{M}$, and hence $I - AA^* \ge 0$ if and only if $K^*(I - AA^*)K \ge 0$. We compute

$$\begin{aligned}
\langle K^*(I - AA^*)Ke_t \otimes e_j, e_s \otimes e_i\rangle &= \langle (I - AA^*)k_{z_t} \otimes e_j, k_{z_s} \otimes e_i\rangle \\
&= \langle k_{z_t}, k_{z_s}\rangle \langle e_j, e_i\rangle - \langle k_{z_t}, k_{z_s}\rangle \langle F(z_t)^* e_j, F(z_s)^* e_i\rangle \\
&= \langle k_{z_t}, k_{z_s}\rangle \langle (I_k - F(z_s)F(z_t)^*) e_j, e_i\rangle \\
&= \frac{\langle (I_k - W_s W_t^*) e_j, e_i\rangle}{1 - z_s \overline{z_t}}.
\end{aligned}$$

Therefore, we have $\|A\| \le 1$ if and only if

$$0 \le K^*(I - AA^*)K = \left[\frac{I_k - W_s W_t^*}{1 - z_s \overline{z_t}}\right]_{n \times n}.$$

For the converse, we define A on $\widetilde{M}$ by

$$A^* k_{z_s} \otimes e_j = k_{z_s} \otimes W_s^* e_j \quad \text{for} \quad 1 \le s \le n,\ 1 \le j \le k.$$

Running the computation above in the other direction, we deduce that $I - AA^* \ge 0$ and thus $\|A\| \le 1$. Let $S = P_{\widetilde{M}} T_z^{(n)}|_{\widetilde{M}} = \big(T_z^{*\,(n)}|_{\widetilde{M}}\big)^*$ and observe that since $\widetilde{M}$ is coinvariant for $T_z^{(n)}$, we get that S is dilated by $T_z^{(n)}$. Next we check that $AS = SA$ as before because

$$S^* A^* (k_{z_s} \otimes e_j) = S^* (k_{z_s} \otimes W_s^* e_j) = \overline{z_s} k_{z_s} \otimes W_s^* e_j = A^* S^* (k_{z_s} \otimes e_j).$$

By the commutant lifting theorem, we can find $T \in \mathcal{B}(H^{2(k)})$ which dilates A, so that $T^*|_{\widetilde{M}} = A^*$, $\|T\| \le 1$ and T commutes with $T_z^{(n)}$. To commute with $T_z^{(n)}$, which is a $k \times k$ diagonal matrix with T_z in the diagonal entries, the operator T must be a matrix with coefficients in $\{T_z\}' = \mathcal{T}(H^\infty)$. That is, $T = T_F$ for some $F \in M_k(H^\infty)$ with $\|F\|_\infty = \|T\| \le 1$. Then $A^* = T_F^*|_{\widetilde{M}}$, and hence as in the first part, we obtain

$$k_{z_s} \otimes F(z_s)^* e_j = A^* k_{z_s} \otimes e_j = k_{z_s} \otimes W_s^* e_j \quad \text{for} \quad 1 \le s \le n,\ 1 \le j \le k.$$

That is, $F(z_s) = W_s$ for $1 \le s \le n$ as desired. ∎

14.5. Counterexamples for more variables

In this section, we will look at some counterexamples that show the limitations of dilating three or more contractions. The existence of such counterexamples was first established by Varopoulos. We present some minimal examples with different features.

14.5.1. EXAMPLE (Holbrook). Consider matrices

$$T_1 = \begin{bmatrix} 0 & 1 & 0 & 0 \\ 0 & 0 & 0 & 1 \\ 0 & 0 & 0 & 0 \\ 0 & 0 & 0 & 0 \end{bmatrix}, \ T_2 = \begin{bmatrix} 0 & -\frac{1}{2} & \frac{\sqrt{3}}{2} & 0 \\ 0 & 0 & 0 & -\frac{1}{2} \\ 0 & 0 & 0 & \frac{\sqrt{3}}{2} \\ 0 & 0 & 0 & 0 \end{bmatrix}, \ T_3 = \begin{bmatrix} 0 & -\frac{1}{2} & -\frac{\sqrt{3}}{2} & 0 \\ 0 & 0 & 0 & \frac{1}{2} \\ 0 & 0 & 0 & -\frac{\sqrt{3}}{2} \\ 0 & 0 & 0 & 0 \end{bmatrix}.$$

Then an easy computation shows that $\|T_i\| = 1$ and

$$T_iT_j = \begin{bmatrix} 0 & 0 & 0 & x_{ij} \\ 0 & 0 & 0 & 0 \\ 0 & 0 & 0 & 0 \\ 0 & 0 & 0 & 0 \end{bmatrix} \quad \text{where} \quad x_{ij} = \begin{cases} 1 & \text{if } i = j \\ -\frac{1}{2} & \text{if } i \neq j \end{cases}.$$

In particular, these three contractions commute. Let

$$p(z_1, z_2, z_3) = z_1^2 + z_2^2 + z_3^2 - 2(z_1z_2 + z_2z_3 + z_3z_1).$$

An easy computation shows that

$$p(T_1, T_2, T_3) = \begin{bmatrix} 0 & 0 & 0 & 6 \\ 0 & 0 & 0 & 0 \\ 0 & 0 & 0 & 0 \\ 0 & 0 & 0 & 0 \end{bmatrix}$$

and hence $\|p(T_1, T_2, T_3)\|$=6. However, we claim $\|p\|_\infty = \sup\limits_{z_i \in \mathbb{T}} |p(z_1, z_2, z_3)| = 5$.

To see this, we rewrite

$$p(z_1, z_2, z_3) = (z_1 + z_2 - z_3)^2 - 4z_1z_2 = z_1z_2\big((s + \bar{s} - w)^2 - 4\big)$$

where $s^2 = z_1/z_2$ and $w = z_3\overline{z_1}s$. Note that $s, w \in \mathbb{T}$, and so $t = s + \bar{s} \in [-2, 2]$. There are two square roots of z_1/z_2, but changing the sign changes the sign of w, and so the RHS is unchanged. Thus we need to calculate

$$\|p\|_\infty = \sup_{\substack{t \in [-2,2] \\ |w|=1}} |(t - w)^2 - 4| = \sup_{z \in \Omega} |z^2 - 4|$$

where, setting $z = x + iy$, we have

$$\begin{aligned} \Omega &= \{z = t - w : t \in [-2, 2] \text{ and } |w| = 1\} \\ &= \{z : |x| \le 2,\ |y| \le 1\} \cup \{z : |x - 2|^2 + y^2 \le 1\} \cup \{z : |x + 2|^2 + y^2 \le 1\}. \end{aligned}$$

By the maximum modulus principle, it suffices to take the supremum over

$$\partial\Omega \subset \{t \pm i : t \in [-2,2]\} \cup \{\pm 2 + w : |w| = 1\}.$$

Thus a routine calculation using calculus (on $x = t^2$) shows that

$$\begin{aligned}\|p\|_\infty &= \max\Big\{ \sup_{-2\le t\le 2} |(t+i)^2 - 4|,\ \sup_{|w|=1} |(2-w)^2 - 4| \Big\} \\ &= \max\Big\{ \sup_{0\le x\le 4} \sqrt{(x-5)^2 + 4x},\ \sup_{|w|=1} |w^2 - 4w| \Big\} = 5.\end{aligned}$$

The reader should verify this. This shows that T_1, T_2, T_3 do not satisfy von Neumann's inequality. If they had a common commuting unitary dilation, then as in the two-variable case, von Neumann's inequality would hold.

14.5.2. Example (Parrott). Let

$$U = \begin{bmatrix} 1 & 0 \\ 0 & -1 \end{bmatrix} \quad\text{and}\quad V = \begin{bmatrix} 0 & 1 \\ 1 & 0 \end{bmatrix}.$$

Note that $VU = -UV$ and $U = U^*$ and $V = V^*$. Define 4×4 matrices

$$T_1 = \begin{bmatrix} 0_2 & I_2 \\ 0_2 & 0_2 \end{bmatrix}, \quad T_2 = \begin{bmatrix} 0_2 & U \\ 0_2 & 0_2 \end{bmatrix}, \quad\text{and}\quad T_3 = \begin{bmatrix} 0_2 & V \\ 0_2 & 0_2 \end{bmatrix}.$$

Then $T_iT_j = 0$ for all $1 \le i, j \le 3$. In particular, they commute.

We first show that this triple satisfies the scalar von Neumann inequality. Let

$$p(z_1, z_2, z_3) = a_0 + a_1 z_1 + a_2 z_2 + a_3 z_3 + \text{h.o.t.}$$

(h.o.t. means higher order terms). Choose $\omega_i \in \mathbb{T}$ so that $a_0 = \omega_0 |a_0|$ and $a_i\omega_i = \omega_0 |a_i|$ for $1 \le i \le 3$. Then

$$h(z) := \overline{\omega_0} p(\omega_1 z, \omega_2 z, \omega_3 z) = |a_0| + (|a_1| + |a_2| + |a_3|)z + \text{h.o.t.}$$

Therefore

$$\begin{aligned}\|p(T_1,T_2,T_3)\| &= \left\| \begin{bmatrix} a_0 I_2 & a_1 I_2 + a_2 U + a_3 V \\ 0_2 & a_0 I_2 \end{bmatrix} \right\| \\ &\le \left\| \begin{bmatrix} |a_0| & |a_1| + |a_2| + |a_3| \\ 0_2 & |a_0| \end{bmatrix} \right\| \le \|T_h\| = \|h\|_\infty = \|p\|_\infty.\end{aligned}$$

However, the matrix-valued von Neumann inequality fails. Indeed we have

$$\begin{aligned}\left\| \begin{bmatrix} z_1 & z_2 & 0 \\ z_3 & 0 & z_2 \\ 0 & z_3 & -z_1 \end{bmatrix} \right\| &= \left\| \begin{bmatrix} z_1 & 0 & 0 \\ 0 & z_3 & 0 \\ 0 & 0 & z_1\overline{z_2}z_3 \end{bmatrix} \begin{bmatrix} 1 & 1 & 0 \\ 1 & 0 & 1 \\ 0 & 1 & -1 \end{bmatrix} \begin{bmatrix} 1 & 0 & 0 \\ 0 & \overline{z_1}z_2 & 0 \\ 0 & 0 & z_2\overline{z_3} \end{bmatrix} \right\| \\ &= \left\| \begin{bmatrix} 1 & 1 & 0 \\ 1 & 0 & 1 \\ 0 & 1 & -1 \end{bmatrix} \right\| = \sqrt{3}.\end{aligned}$$

However

$$\left\| \begin{bmatrix} T_1 & T_2 & 0 \\ T_3 & 0 & T_2 \\ 0 & T_3 & -T_1 \end{bmatrix} \right\| = \left\| \begin{bmatrix} 0 & X \\ 0 & 0 \end{bmatrix} \right\| = \|X\|$$

where

$$X = \begin{bmatrix} I & U & 0 \\ V & 0 & U \\ 0 & V & -I \end{bmatrix} = \begin{bmatrix} I & 0 & 0 \\ 0 & V & 0 \\ 0 & 0 & VU^* \end{bmatrix} \begin{bmatrix} I & I & 0 \\ I & 0 & I \\ 0 & I & I \end{bmatrix} \begin{bmatrix} I & 0 & 0 \\ 0 & U & 0 \\ 0 & 0 & V^*U \end{bmatrix}.$$

Hence

$$\|X\| = \left\| \begin{bmatrix} 1 & 1 & 0 \\ 1 & 0 & 1 \\ 0 & 1 & 1 \end{bmatrix} \right\| = 2.$$

Again the argument for the matrix von Neumann inequality in three variables would follow from the existence of a simultaneous unitary dilation. So no such dilation exists.

14.5.3. EXAMPLE (Choi–Davidson). Let $H = \mathbb{C}^3$ have an orthonormal basis f, e_1, e_2. For $i = 1, 2$, pick $\theta_i \in (0, \pi/2)$ and set $c_i = \cos\theta_i$ and $s_i = \sin\theta_i$. Define

$$A_1 = e_1 f^* = \begin{bmatrix} 0 & 0 & 0 \\ 1 & 0 & 0 \\ 0 & 0 & 0 \end{bmatrix} \quad A_2 = e_2 f^* = \begin{bmatrix} 0 & 0 & 0 \\ 0 & 0 & 0 \\ 1 & 0 & 0 \end{bmatrix} \quad A_3 = (c_1e_1 + s_1e_2) f^* = \begin{bmatrix} 0 & 0 & 0 \\ c_1 & 0 & 0 \\ s_1 & 0 & 0 \end{bmatrix} \quad A_4 = (c_2e_1 + is_2e_2) f^* = \begin{bmatrix} 0 & 0 & 0 \\ c_2 & 0 & 0 \\ is_2 & 0 & 0 \end{bmatrix}.$$

Observe that $A_iA_j = 0$ for all $1 \le i, j \le 4$; so they commute. We will show that these matrices do not have simultaneous coextensions to commuting isometries.

Suppose that the A_i dilate to commuting isometries of the form

$$S_i = \begin{bmatrix} A_i & 0 \\ D_i & V_i \end{bmatrix} \quad \text{for} \quad i = 1, 2, 3, 4$$

on a Hilbert space K. We may assume that K is the smallest invariant subspace for the S_i containing H. If we write $S^k = S_1^{k_1} S_2^{k_2} S_3^{k_3} S_4^{k_4}$ for $k = (k_1, k_2, k_3, k_4) \in \mathbb{N}_0^4$, then $K = \bigvee_{k \in \mathbb{N}_0^4} S^k H$.

Observe that since $\|A_i f\| = 1$, we have $S_i f = A_i f$ for $i = 1, 2, 3, 4$. Hence

$$(c_1S_1 + s_1S_2 - S_3) f = 0.$$

Therefore

$$0 = S^k (c_1S_1 + s_1S_2 - S_3) f = (c_1S_1 + s_1S_2 - S_3) S^k f.$$

Since $S_i f = e_i$ for $i = 1, 2$,

$$\ker(c_1S_1 + s_1S_2 - S_3) \supset \bigvee_{k \in \mathbb{N}^4} S^k \operatorname{span}\{f, S_1 f, S_2 f\} = K.$$

So $S_3 = c_1S_1 + s_1S_2$. Similarly $S_4 = c_2S_1 + is_2S_2$.

Since S_3 is an isometry,

$$I = S_3^*S_3 = c_1^2 S_1^*S_1 + s_1^2 S_2^*S_2 + c_1 s_1 (S_2^*S_1 + S_1^*S_2) \\ = I + c_1 s_1 (S_2^*S_1 + S_1^*S_2).$$

It follows that $S_2^*S_1 + S_1^*S_2 = 0$. Likewise, using the fact that S_4 is an isometry,

$$0 = (iS_2)^*S_1 + S_1^*(iS_2) = -i(S_2^*S_1 - S_1^*S_2).$$

Hence $S_2^*S_1 - S_1^*S_2 = 0$. Therefore $S_2^*S_1 = 0$. This implies that S_1 and S_2 have pairwise orthogonal ranges, and therefore do not commute. This contradiction establishes the result.

We will not prove it, but these matrices do satisfy the scalar von Neumann inequality, but fail the inequality for 1×2 matrix polynomials.

14.5.4. REMARK. It is known that any family of commuting 2×2 contractions dilate to commuting isometries. It remains an open question about three 3×3 commuting contractions. However, it has been shown by Knese [180] that the scalar von Neumann inequality holds for any commuting family of 3×3 contractions.

14.5.5. REMARK. We show in the next chapter that if $T_1, \dots, T_d$ are contractions such that $\left\| \begin{bmatrix} T_1 & \dots & T_d \end{bmatrix} \right\| \le 1$, then the T_i's simultaneously dilate to isometries S_i such that $\left\| \begin{bmatrix} S_1 & \dots & S_d \end{bmatrix} \right\| = 1$. If the T_i's commute, we obtain a dilation to a certain commuting family with special properties.

14.6. Completely Positive Maps

A critical tool to extend dilation theory to a wider context is the notion of completely positive and completely bounded maps. These ideas are also crucial in the general study of operator algebras.

14.6.1. DEFINITION. A (concrete) *operator space* is a closed subspace $\mathcal{A}$ of a C*-algebra $\mathfrak{A}$ together with a family of norms on $M_n(\mathcal{A})$ induced from the unique C*-algebra norm on $M_n(\mathfrak{A})$. A (concrete) *operator system* is a closed self-adjoint subspace $\mathcal{S}$ of a unital C*-algebra $\mathfrak{A}$ containing the unit 1.

If $\varphi : \mathcal{A} \to \mathfrak{B}$ is a continuous linear map into a C*-algebra $\mathfrak{B}$, we define

$$\varphi_n : M_n(\mathcal{A}) \to M_n(\mathfrak{B}) \text{ by } \varphi_n\big[a_{ij}\big] = \big[\varphi(a_{ij})\big].$$

Say that φ is *completely bounded* (CB) if $\|\varphi\|_{cb} := \sup_{n\ge1} \|\varphi_n\| < \infty$. It is a *complete contraction* if $\|\varphi\|_{cb} \le 1$.

If $\mathcal{S}$ is an operator system, we say that φ is n-positive if φ_n is a positive map. If φ_n is positive for all $n \ge 1$, we say that φ is *completely positive* (CP). If φ is also unital, we call it UCP.

14.6.2. REMARKS. An operator system is spanned by its positive elements. Indeed, if $S \in \mathcal{S}$, we can write $S = X + iY$ where X, Y are self-adjoint elements of $\mathcal{S}$. But $X = \frac{1}{2}(\|X\|I + X) - \frac{1}{2}(\|X\|I - X)$ is in the span of positive elements of $\mathcal{S}$. Also $\mathcal{S}_+ = \{S \in \mathcal{S} : S \geq 0\}$ is a closed cone such that $\mathcal{S}_+ \cap (-\mathcal{S}_+) = \{0\}$ and $\mathcal{S}_+ - \mathcal{S}_+ = \mathcal{S}_{sa}$.

There is an interesting abstract characterization of operator systems and operator spaces. This will be treated in Section 14.14.

14.6.3. EXAMPLES.

(1) A $*$-homomorphism $\pi : \mathfrak{A} \to \mathfrak{B}$ between C*-algebras is always completely positive. This is because $\pi_n : M_n(\mathfrak{A}) \to M_n(\mathfrak{B})$ is also a $*$-homomorphism, and thus π_n is positive and $\|\pi_n\| = 1$. Therefore $\|\pi\|_{cb} = 1$, so that π is also completely contractive.

(2) If $V : H \to K$ is a linear map and $\pi : \mathfrak{A} \to \mathcal{B}(H)$ is a representation, define $\varphi(a) = V^*\pi(a)V \in \mathcal{B}(H)$. Clearly this is positive. Also $\|\varphi(a)\| \leq \|V\|^2\|a\|$, and hence $\|\varphi\| \leq \|V\|^2$. If $\mathfrak{A}$ is unital, then $\varphi(1) = V^*V$ so $\|\varphi\| = \|V\|^2$. If $\mathfrak{A}$ is not unital, a similar argument works using an approximate unit. The key point is that

$$\varphi_n\big[a_{ij}\big] = \big[\varphi(a_{ij})\big] = \big[V^*\pi(a)V\big] = V^{*\,(n)}\pi_n(a)V^{(n)}.$$

By the previous argument, we see that φ_n is positive and $\|\varphi_n\| = \|V\|^2$. Thus φ is completely positive and completely bounded. If φ is unital, then $V^*V = I$, i.e. V is an isometry. Then φ is completely contractive.

(3) The prototypical positive unital map that is not completely positive and not completely contractive is the transpose map on M_n for $n \geq 2$:

$$\tau\big[a_{ij}\big] = \big[a_{ji}\big].$$

This map is self-adjoint and $\tau(AB) = \tau(B)\tau(A)$. Thus if $B = A^*A \geq 0$, then

$$\tau(B) = \tau(A)\tau(A)^* \geq 0.$$

Hence the transpose is a positive map.

Consider the standard matrix units E_{ij} in M_n with respect to the basis $\xi_1, \dots, \xi_n$. Then $A = \begin{bmatrix} E_{11} & E_{12} \\ E_{21} & E_{22} \end{bmatrix} \geq 0$ in $M_2(M_n)$. Set $\eta = \begin{bmatrix} \xi_2 \\ -\xi_1 \end{bmatrix}$. Then

$$\langle \tau_2(A)\eta, \eta\rangle = \begin{bmatrix} \xi_2^* & -\xi_1^* \end{bmatrix} \begin{bmatrix} E_{11} & E_{21} \\ E_{12} & E_{22} \end{bmatrix} \begin{bmatrix} \xi_2 \\ -\xi_1 \end{bmatrix} = -2.$$

Thus τ is not 2-positive. Moreover, $\tau_n\big[E_{ji}\big] = \big[E_{ij}\big]$. The first matrix has a single 1 in each row and column, and thus has norm 1. But $\big\|\big[E_{ij}\big]\big\| = n$. Hence $\|\tau_n\| = n$ as the upper bound is elementary.

14.6.4. PROPOSITION. *If $\varphi : \mathcal{S} \to \mathfrak{B}$ is a completely positive map on an operator system, then $\|\varphi\|_{cb} = \|\varphi\| = \|\varphi(1)\|$.*

PROOF. Clearly, $\|\varphi(I)\| \le \|\varphi\| \le \|\varphi\|_{cb}$. Let $P = \varphi(1)$. We make use of the easy fact that

$$\|a\| \le 1 \quad \text{if and only if} \quad \begin{bmatrix} 1 & a \\ a^* & 1 \end{bmatrix} \ge 0.$$

Let $S = \begin{bmatrix} s_{ij} \end{bmatrix} \in M_n(\mathcal{S})$ with $\|S\| \le 1$. Then $\begin{bmatrix} I_n & S \\ S^* & I_n \end{bmatrix} \ge 0_{2n}$ in $M_{2n}(\mathcal{S})$. Thus

$$0_{2n} \le \varphi_{2n}\Big(\begin{bmatrix} I_n & S \\ S^* & I_n \end{bmatrix}\Big) \le \begin{bmatrix} P^{(n)} & \varphi_n(S) \\ \varphi_n(S^*) & P^{(n)} \end{bmatrix} \le \begin{bmatrix} \|P\|I_n & \varphi_n(S) \\ \varphi_n(S)^* & \|P\|I_n \end{bmatrix}.$$

This implies that $\|\varphi_n(S)\| \le \|P\|$. Thus $\|\varphi_n\| = \|P\| = \|\varphi(1)\|$ for all $n \ge 1$. ∎

14.6.5. REMARK. This proof shows that if φ is 2-positive, then $\|\varphi\| = \|\varphi(1)\|$. If φ is only known to be positive, the best estimate is $\|\varphi\| \le 2\|\varphi(1)\|$. If $a = a^*$ and $\|a\| \le 1$, then $-1 \le a \le 1$ and thus $-\varphi(1) \le \varphi(a) \le \varphi(1)$. So $\|\varphi(a)\| \le \|\varphi(1)\|$. In general, $x = a + ib$ for $a = a^*$ and $b = b^*$, so that

$$\|\varphi(x)\| \le \|\varphi(a)\| + \|\varphi(b)\| \le 2\|\varphi(1)\| \, \|x\|.$$

Consider the operator system $\mathcal{S} = \operatorname{span}\{1, z, \overline{z}\} \subset C(\mathbb{T})$. Define

$$\varphi(a + bz + c\overline{z}) = \begin{bmatrix} a & 2b \\ 2c & a \end{bmatrix}.$$

To check that this is positive, note that $a + bz + \overline{bz} = a + 2\operatorname{Re}(bz) \ge 0$ on $\mathbb{T}$ only when $a \ge 2|b|$. But then

$$\varphi(a + bz + \overline{bz}) = \begin{bmatrix} a & 2b \\ 2\overline{b} & a \end{bmatrix} \ge \begin{bmatrix} 2|b| & 2b \\ 2\overline{b} & 2|b| \end{bmatrix} \ge 0.$$

However $\|\varphi(z)\| = 2$, and hence $\|\varphi\| = 2 = 2\|\varphi(1)\|$. By Proposition 14.6.4, φ is not completely positive.

14.6.6. PROPOSITION. *If $\mathcal{A}$ is an operator space and $\varphi : \mathcal{A} \to \mathbb{C}$ is a continuous linear functional, then $\|\varphi\|_{cb} = \|\varphi\|$. If $\mathcal{S}$ is an operator system and $\varphi : \mathcal{S} \to \mathbb{C}$ is positive, then φ is completely positive.*

PROOF. Let $A = \begin{bmatrix} a_{ij} \end{bmatrix} \in M_n(\mathcal{A})$ and let $\xi = (\xi_1, \dots, \xi_n)$ and $\eta = (\eta_1, \dots, \eta_n)$ be unit vectors in $\mathbb{C}^n$. Then

$$|\langle \varphi_n(A)\xi, \eta\rangle| = |\langle [\varphi(a_{ij})]\xi, \eta\rangle| = \Big|\sum_{i,j=1}^{n} \varphi(a_{ij})\xi_j\overline{\eta_i}\Big|$$
$$= \big|\varphi(\eta^*\begin{bmatrix} a_{ij} \end{bmatrix}\xi)\big| \le \|\varphi\| \, \|\eta\| \, \big\|\begin{bmatrix} a_{ij} \end{bmatrix}\big\| \, \|\xi\| = \|\varphi\| \, \|A\|.$$

Thus $\|\varphi_n\| = \|\varphi\|$ for all $n \ge 1$.

If φ is positive on an operator system, a similar computation for $A \geq 0$ and $\eta = \xi$ yields

$$\langle \varphi_n(A)\xi, \xi \rangle = \varphi\Big(\sum_{i,j=1}^{n} \overline{\xi_i} a_{ij} \xi_j \Big) = \varphi(\xi^* A \xi) \geq 0$$

since $\xi^* A \xi \geq 0$. ■

14.6.7. COROLLARY. *Every continuous linear map φ of an operator space $\mathcal{A}$ into a commutative C*-algebra $\mathfrak{B}$ is completely bounded with $\|\varphi\|_{cb} = \|\varphi\|$. Every positive continuous linear map φ of an operator system S into a commutative C*-algebra $\mathfrak{B}$ is completely positive.*

PROOF. We may take $\mathfrak{B} = C(X)$. The map φ is completely bounded (positive) if and only if $\delta_x \varphi(a) = \varphi(a)(x)$ is completely bounded (positive) uniformly for $x \in X$, and $\|\varphi\|_{cb} = \sup_X \|\delta_x \varphi\|_{cb}$. The result follows from Proposition 14.6.6. ■

Maps with $C(X)$ as a domain are also tractable for positive maps. However, Remark 14.6.5 shows that this is not valid for operator systems in $C(X)$.

14.6.8. PROPOSITION. *If $\mathfrak{A}$ is a unital abelian C*-algebra and $\varphi : \mathfrak{A} \to \mathfrak{B}$ is positive, then it is completely positive.*

PROOF. We may assume that $\mathfrak{A} = C(X)$ for a compact Hausdorff space X. By Remark 14.6.5, the map φ is continuous. Let $F = \big[f_{ij}\big] \in M_n(\mathfrak{A})_+ = C(X, M_{n+})$. Since F is continuous, for $\varepsilon > 0$, there is a finite open cover $U_1, \dots, U_k$ so that $\|F(x) - F(y)\| < \varepsilon$ for $x, y \in U_s$ and $1 \leq s \leq k$. Let $g_1, \dots, g_k$ be a partition of unity for this open cover (see Proposition 1.10.7), and select points $x_s \in U_s$. Then $G(x) = \sum_{s=1}^{k} F(x_s) g_s(x) \geq 0$. A routine estimate shows that $\|F - G\| \leq \varepsilon$. Now

$$\varphi_n(G) = \sum_{s=1}^{k} \big[f_{ij}(x_s)\varphi(g_s)\big] = \sum_{s=1}^{k} F(x_s)\big(\varphi(g_s) I_n\big) \geq 0$$

because this is a sum of products of commuting positive operators. Thus $\varphi_n(F)$ is a norm limit of positive operators and hence is positive. Therefore φ is completely positive. ■

The following characterizes completely positive maps with domain M_n.

14.6.9. CHOI THEOREM. *Let $\varphi : M_n \to \mathfrak{B}$ be a linear map into a C*-algebra $\mathfrak{B}$. The following are equivalent:*

(1) *φ is completely positive.*

(2) *φ is n-positive.*

(3) *The* Choi matrix $C_\varphi = \big[\varphi(E_{ij})\big]$ *is positive in $M_n(\mathfrak{B})$.*

PROOF. Clearly (1) implies (2). Since $\big[E_{ij}\big] \geq 0$ in $M_n(M_n)$, (2) implies (3). Assume (3). Consider $\mathfrak{B}$ as faithfully represented in $\mathcal{B}(H)$. Then C_φ is contained in $M_n(\mathcal{B}(H))$. Since it is positive, we may factor it as $C_\varphi = A^*A$ for $A \in M_n(\mathfrak{B})$. Let the n columns of A be $C_1, \dots, C_n$ in $M_{n,1}(\mathfrak{B})$, and note that $C_i^*C_j = \varphi(E_{ij})$. Define $V = \begin{bmatrix} C_1 \\ \vdots \\ C_n \end{bmatrix} \in M_{n^2,1}(\mathfrak{B})$. Think of $M_{n^2}(\mathcal{B}(H))$ as $M_n \otimes M_n \otimes \mathcal{B}(H)$. Define a representation $\pi : M_n \to M_n \otimes M_n \otimes \mathcal{B}(H)$ by $\pi(E_{ij}) = E_{ij} \otimes I_n \otimes I_H$. Then

$$V^*\pi(E_{ij})V = \begin{bmatrix} C_1^* & \dots & C_i^* & \dots & C_n^* \end{bmatrix} \begin{bmatrix} 0 & \dots & 0 & \dots & 0 \\ \vdots & \vdots & \vdots & \vdots & \vdots \\ 0 & \dots & I_n \otimes I_H & \dots & 0 \\ \vdots & \vdots & \vdots & \vdots & \vdots \\ 0 & \dots & 0 & \dots & 0 \end{bmatrix} \begin{bmatrix} C_1 \\ \vdots \\ C_j \\ \vdots \\ C_n \end{bmatrix}$$

$$= C_i^*C_j = \varphi(E_{ij}).$$

By linearity, we have $\varphi(A) = V^*\pi(A)V$. By Example 14.6.3(2), φ is completely positive. ∎

We can also characterize completely positive maps into M_n.

14.6.10. PROPOSITION. *Let $\mathcal{S} \subset \mathfrak{A}$ be an operator system, and let $\varphi : \mathcal{S} \to M_n$ be a linear map. The following are equivalent:*

(1) *φ is completely positive.*

(2) *φ is n-positive.*

(3) *Let $e_1, \dots, e_n$ be an orthonormal basis for $\mathbb{C}^n$. Define $s_\varphi : M_n(\mathcal{S}) \to \mathbb{C}$ by*

$$s_\varphi\big[a_{ij}\big] = \langle \varphi_n\big[a_{ij}\big]\xi, \xi\rangle = \sum_{i,j=1}^{n} \langle \varphi(a_{ij})e_j, e_i\rangle = \sum_{i,j=1}^{n} \varphi(a_{ij})_{ij},$$

where $\xi = \begin{bmatrix} e_1 \\ \vdots \\ e_n \end{bmatrix}$. Then s_φ is a positive linear functional.

PROOF. It is immediate that (1) implies (2) implies (3). Assume that $s_\varphi \geq 0$. Observe that for $a \in \mathcal{S}$,

$$\varphi(a) = \big[s_\varphi(aE_{ij})\big] = (s_\varphi)_n\big(a \otimes \big[E_{ij}\big]\big).$$

Now let $A = \big[a_{st}\big] \in M_k(\mathcal{S})$. Then

$$\varphi_k(A) = (s_\varphi)_{kn}\big(A \otimes \big[E_{ij}\big]\big).$$

By Proposition 14.6.6, s_φ is completely positive, and hence $\varphi_k \geq 0$ for all $k \geq 1$. That is, φ is completely positive. ∎

14.7. Dilation and Extension of CP Maps

In this section, we prove two major results. The first is Stinespring's theorem, which is a dilation theorem for completely positive maps on C*-algebras. Then we prove Arveson's extension theorem, an analogue of the Hahn–Banach theorem for completely positive maps.

The image C*-algebra may be considered as represented in some $\mathcal{B}(H)$. We will show that Example 14.6.3(2) is completely general. The argument is very similar to the GNS construction.

14.7.1. Stinespring Dilation Theorem. *Let $\mathfrak{A}$ be a unital C*-algebra, and let $\varphi : \mathfrak{A} \to \mathcal{B}(H)$ be completely positive. Then there is a Hilbert space K, a $*$-representation $\pi : \mathfrak{A} \to \mathcal{B}(K)$ and a bounded linear map $V \in \mathcal{B}(H, K)$ such that $\varphi(a) = V^*\pi(a)V$ and $\|V\|^2 = \|\varphi\|$.*

Proof. Consider the algebraic tensor product $\mathfrak{A} \odot H$. Define a sesquilinear form on $\mathfrak{A} \odot H$ by

$$[\![a \otimes \xi, b \otimes \eta]\!] = \langle \varphi(b^*a)\xi, \eta\rangle \quad \text{for} \quad a, b \in \mathfrak{A} \text{ and } \xi, \eta \in H$$

and extending by sesquilinearity. This is positive semidefinite because

$$\Big[\!\!\Big[\sum_{i=1}^n a_i \otimes \xi_i, \sum_{i=1}^n a_i \otimes \xi_i \Big]\!\!\Big] = \sum_{i=1}^n \sum_{j=1}^n \langle \varphi(a_j^* a_i)\xi_i, \xi_j\rangle = \langle \varphi_n\big[a_j^*a_i\big]\xi, \xi\rangle \geq 0$$

where $\xi = (\xi_1, \dots, \xi_n)$. This form satisfies the Cauchy–Schwarz inequality. Thus as in the GNS construction, we can define

$$N = \big\{u \in \mathfrak{A} \odot H : [\![u, u]\!] = 0\big\} = \big\{u \in \mathfrak{A} \odot H : [\![u, v]\!] = 0 \text{ for all } v \in \mathfrak{A} \odot H\big\}.$$

This is a subspace, and we obtain an induced positive definite inner product on $(\mathfrak{A} \odot H)/N$. The completion of this space in the norm arising from the inner product is a Hilbert space K.

Next we define a map $\pi_0 : \mathfrak{A} \to \mathcal{L}(\mathfrak{A} \odot H)$ by $\pi_0(a)(b \otimes \xi) = ab \otimes \xi$ and extend by linearity. If $u \in \mathfrak{A} \odot H$, the functional $\rho_u(a) = [\![\pi_0(a)u, u]\!]$ is positive. Hence

$$[\![\pi_0(a)u, \pi_0(a)u]\!] = \rho_u(a^*a) \leq \|a^*a\|\rho_u(1) = \|a\|^2 [\![u, u]\!].$$

This implies that N is invariant for $\pi_0(\mathfrak{A})$ and that π_0 is contractive. So it induces the map

$$\pi(a)(u + N) = \pi_0(a)u + N$$

which is a contractive $*$-homomorphism. This extends by continuity to a $*$-representation, also denoted by π, of $\mathfrak{A}$ into $\mathcal{B}(K)$.

Define a linear map $V : H \to K$ by $V\xi = (1 \otimes \xi) + N$. Note that

$$\|V\xi\|^2 = [\![1 \otimes \xi, 1 \otimes \xi]\!] = \langle \varphi(1)\xi, \xi\rangle \leq \|\varphi\|\|\xi\|^2.$$

Thus $\|V\|^2 \le \|\varphi\|$ is bounded. Compute for $a \in \mathfrak{A}$ and $\xi, \eta \in H$

$$\langle V^*\pi(a)V\xi, \eta\rangle = [\![\pi_0(a)(1 \otimes \xi), 1 \otimes \eta]\!] = \langle \varphi(a)\xi, \eta\rangle.$$

Therefore $\varphi(a) = V^*\pi(a)V$. Hence $V^*V = \varphi(1)$; so $\|V\|^2 = \|\varphi(1)\| = \|\varphi\|$. ∎

14.7.2. Remark. If $\varphi : \mathfrak{A} \to \mathcal{B}(H)$ is unital, then $V^*V = \varphi(1) = I$. Thus V is an isometry. Hence, we can identify H with VH and consider $K \supset H$. Then $\varphi(a) = P_H\pi(a)|_H$.

14.7.3. Definition. A Stinespring representation (π, V) of $\varphi : \mathfrak{A} \to \mathcal{B}(H)$ is called *minimal* if $K = \overline{\pi(\mathfrak{A})VH}$.

14.7.4. Proposition. *The minimal Stinespring representation (π, V) is unique up to unitary equivalence, i.e. if (π_1, V_1) is another minimal Stinespring representation of $\mathfrak{A}$ with a $*$-representation $\pi_1 : \mathfrak{A} \to \mathcal{B}(K_1)$ and $V_1 \in \mathcal{B}(H, K_1)$, then there is a unitary operator $U \in \mathcal{B}(K, K_1)$ so that $\pi_1(a) = U\pi(a)U^*$ and $V_1 = UV$.*

Proof. Let (π, V) and (π_1, V_1) be two minimal Stinespring representations of φ. Define

$$U\sum_{i=1}^{n} \pi(a_i)V\xi_i = \sum_{i=1}^{n} \pi_1(a_i)V_1\xi_i \quad \text{for} \quad a_i \in \mathfrak{A} \text{ and } \xi \in H.$$

We compute the norm of these two vectors

$$\Big\|\sum_{i=1}^{n} \pi(a_i)V\xi_i\Big\|^2 = \sum_{i,j=1}^{n} \langle \pi(a_i)V\xi_j, \pi(a_j)V\xi_j\rangle$$
$$= \sum_{i,j=1}^{n} \langle V^*\pi(a_j^*a_i)V\xi_j, \xi_j\rangle = \sum_{i,j=1}^{n} \langle \varphi(a_j^*a_i)\xi_j, \xi_j\rangle.$$

The right-hand side does not depend on the representation, and thus we obtain the same result using (π_1, V_1). Therefore, U is an isometry from a dense subspace of K onto a dense subspace of K_1. In particular, it is well defined. Thus U extends by continuity to a unitary operator. In particular, $UV\xi = V_1\xi$, so $V_1 = UV$. Hence

$$\pi_1(a)\sum_{i=1}^{n} \pi_1(b_i)V_1\xi_i = \sum_{i=1}^{n} \pi_1(ab_i)V_1\xi_i = U\sum_{i=1}^{n} \pi(ab_i)V\xi_i$$
$$= (U\pi(a)U^*)\, U\sum_{i=1}^{n} \pi(b_i)V\xi_i$$
$$= (U\pi(a)U^*)\sum_{i=1}^{n} \pi_1(b_i)V_1\xi_i.$$

These vectors are dense in K_1, and hence $\pi_1(a) = U\pi(a)U^*$. ∎

14.7.5. Example. Let $\varphi : M_n \to M_m$ be completely positive. Let (π, V) be its minimal Stinespring representation, where $\pi : M_n \to \mathcal{B}(K)$ and $V \in \mathcal{B}(\mathbb{C}^m, K)$. Every representation of M_n is a multiple of the identity representation, so $\pi \simeq id^{(\alpha)}$. The minimality condition $K = \overline{\pi(M_n) \operatorname{Ran} V}$ yields the estimate $\dim K \leq \dim M_n(\operatorname{rank} V) \leq n^2 m$, so that $\alpha \leq mn$. Write $K = H_1 \oplus \cdots \oplus H_\alpha$ where $H_i = \mathbb{C}^n$ and $\pi(A) = A \oplus \cdots \oplus A$. We can write $V^* = \begin{bmatrix} V_1^* & \dots & V_n^* \end{bmatrix}$ where $V_i = P_{H_i} V$ considered in $\mathcal{B}(\mathbb{C}^m, H_i)$. Then

$$\varphi(A) = V^* \pi(A) V = \sum_{i=1}^{\alpha} V_i^* A V_i.$$

In quantum computing, this is known as the *Kraus decomposition*. This result was rediscovered in the finite dimensional case long after Stinespring's theorem.

In the following proof, we will need to take a limit of CP maps into $\mathcal{B}(H)$, which requires an appropriate topology.

14.7.6. Definition. Let $\mathcal{A}$ be a closed subspace of a C*-algebra. Consider the set of bounded linear maps $\mathcal{B}(\mathcal{A}, \mathcal{B}(H))$ into $\mathcal{B}(H)$. Endow this with the *point-*WOT *topology* generated by the subbase of open neighbourhoods determined by $\varphi \in \mathcal{B}(\mathcal{A}, \mathcal{B}(H))$, $a \in \mathcal{A}$ and $\xi, \eta \in H$:

$$U(\varphi, a, \xi, \eta) = \{\psi \in \mathcal{B}(\mathcal{A}, \mathcal{B}(H)) : |\langle (\psi - \varphi)(a)\xi, \eta \rangle| < 1\}.$$

This topology is induced from the embedding of $\mathcal{B}(\mathcal{A}, \mathcal{B}(H))$ into $\prod_{\substack{a \in \mathcal{A} \\ \xi, \eta \in H}} \mathbb{C}$ with the product topology by sending φ to $\Phi(a, \xi, \eta) = \langle \varphi(a)\xi, \eta \rangle$. Thus, a net $\{\varphi_\lambda\}_\Lambda$ converges point-WOT to φ if and only if $\text{WOT-}\lim_\Lambda \varphi_\lambda(a) = \varphi(a)$ for each $a \in \mathcal{A}$.

The important point is that when we restrict to a set bounded by M, the image lies in the product of closed discs $\overline{\mathbb{D}_{M\|a\|\|\xi\|\|\eta\|}}$, which is compact by Tychonoff's theorem. Hence, a bounded net will have a convergent subnet. As in the proof of the Banach–Alaoglu theorem, it is easy to see that such a limit point defines a bounded linear map. In particular, the image of the closed unit ball of $\mathcal{B}(\mathcal{A}, \mathcal{B}(H))$ is mapped onto a compact set. This map is easily seen to be injective, and thus the closed unit ball of $\mathcal{B}(\mathcal{A}, \mathcal{B}(H))$ is compact and Hausdorff in the point-WOT topology.

14.7.7. Arveson Extension Theorem. *Let $\mathcal{S} \subset \mathfrak{A}$ be an operator system contained in a C*-algebra $\mathfrak{A}$. Suppose that $\varphi : \mathcal{S} \to \mathcal{B}(H)$ is a CP map. Then there is a CP map $\widetilde{\varphi} : \mathfrak{A} \to \mathcal{B}(H)$ such that $\widetilde{\varphi}|_{\mathcal{S}} = \varphi$.*

PROOF. First suppose that H is finite dimensional, so that $\mathcal{B}(H) \simeq M_n$. Let s_φ be the associated (completely) positive linear functional on $M_n(\mathcal{S})$ from Proposition 14.6.10. For convenience, scale φ so that $\|s_\varphi\| = s_\varphi(1) = 1$. By the Hahn–Banach theorem, there is a linear functional s on $M_n(\mathfrak{A})$ with $\|s\| = 1$ and $s|_{M_n(\mathcal{S})} = s_\varphi$. Since $s(1) = 1 = \|s\|$, s is a state by Lemma 9.6.5. By Proposition 14.6.10 again, there is a completely positive map $\widetilde{\varphi} : \mathfrak{A} \to M_n$ such that $s = s_{\widetilde{\varphi}}$ given by

$$\widetilde{\varphi}(a) = \big[s(aE_{ij})\big] = s_n\big(a \otimes \big[E_{ij}\big]\big).$$

This formula restricted to $\mathcal{S}$ yields φ, and hence $\widetilde{\varphi}$ extends φ.

Let Λ be the directed set of finite dimensional subspaces of H, ordered by inclusion. For each $F \in \Lambda$, let $\varphi_F(a) = P_F\varphi(a)P_F$. This is a CP map of $\mathcal{S}$ into $\mathcal{B}(F)$ considered as a subset of $\mathcal{B}(H)$. Let $\widetilde{\varphi}_F : \mathfrak{A} \to P_F\mathcal{B}(H)P_F$ be a CP extension of φ_F provided in the previous paragraph. This net is bounded by $\|\varphi(1)\|$. By the discussion preceding the theorem, there is a cofinal subnet Γ on which $\{\widetilde{\varphi}_{F_\gamma}\}_\Gamma$ converges point-WOT to a linear map $\widetilde{\varphi} : \mathfrak{A} \to \mathcal{B}(H)$. We check that this is the desired extension.

If $a \in \mathcal{S}$, then $\varphi_F(a) = P_F\varphi(a)P_F$ converges WOT to $\varphi(a)$, and thus $\widetilde{\varphi}|_{\mathcal{S}} = \varphi$. Now suppose that $A = \big[a_{ij}\big] \in M_n(\mathfrak{A})_+$. Then $\widetilde{\varphi}_n(A) = \lim_\Gamma(\widetilde{\varphi}_{F_\gamma})_n(A)$. The set of positive operators is WOT closed, and thus $\widetilde{\varphi}_n(A) \geq 0$. Therefore $\widetilde{\varphi}$ is completely positive. ∎

14.7.8. EXAMPLE. The extension is far from unique in general. Let $\mathcal{S} = \operatorname{span}\{1, x\} \subset C[0,1]$ and define $\varphi(a+bx) = a + \frac{b}{2}$. A CP extension is a state $\widetilde{\varphi}$ on $C[0,1]$ that extends φ. Thus, there is a Borel probability measure $\varphi(f) = \int f\,d\mu$ for $f \in \mathcal{S}$. This is known as a *representing measure*. One choice is $\mu_1 = \delta_{1/2}$ corresponding to $\widetilde{\varphi}_1(f) = f(\frac{1}{2}) = \varepsilon_{1/2}(f)$. This is multiplicative on $C[0,1]$, so the Stinespring dilation is $(\varepsilon_{1/2}, I)$. Another is $\widetilde{\varphi}_2(f) = \frac{1}{2}(f(0)+f(1))$ corresponding to $\mu_2 = \frac{1}{2}(\delta_0+\delta_1)$. The Stinespring dilation is $(\varepsilon_0 \oplus \varepsilon_1, V_1)$ where $V_1\lambda = (\frac{\lambda}{\sqrt{2}}, \frac{\lambda}{\sqrt{2}})$. A third choice is $\pi(f) = M_f \in \mathcal{B}(L^2(0,1))$ and $V\lambda = \lambda\mathbf{1}$. Then

$$\widetilde{\varphi}(f) = V^*M_fV = \langle M_f\mathbf{1}, \mathbf{1}\rangle = \int_0^1 f(t)\,dt.$$

We obtain the following immediate consequence. We call this *a* Stinespring dilation, not *the* Stinespring dilation, to recognize the lack of uniqueness.

14.7.9. COROLLARY. *Let $\mathcal{S} \subset \mathfrak{A} = C^*(\mathcal{S})$ be an operator system, and let $\varphi : \mathcal{S} \to \mathcal{B}(H)$ be completely positive. Then there is a Hilbert space K, a $*$-representation $\pi : \mathfrak{A} \to \mathcal{B}(K)$ and a bounded linear map $V \in \mathcal{B}(H,K)$ so that $\varphi(a) = V^*\pi(a)V$ for $a \in \mathcal{S}$.*

14.7.10. DEFINITION. The *multiplicative domain* of a unital completely positive map $\varphi : \mathfrak{A} \to \mathfrak{B}$ between unital C*-algebras is the set

$$\{a \in \mathfrak{A} : \varphi(a)^*\varphi(a) = \varphi(a^*a) \text{ and } \varphi(a)\varphi(a)^* = \varphi(aa^*)\}.$$

14.7.11. PROPOSITION. *Let $\mathfrak{A}, \mathfrak{B}$ be unital C*-algebras and let $\varphi : \mathfrak{A} \to \mathfrak{B}$ be unital and completely positive. Then*

$$\varphi(a)^*\varphi(a) \le \varphi(a^*a) \quad \textit{for all} \quad a \in \mathfrak{A}. \text{ (Kadison–Schwarz inequality)}$$

The multiplicative domain $\mathfrak{C}$ of φ is a C-algebra and $\rho = \varphi|_{\mathfrak{C}}$ is a $*$-homomorphism. Moreover φ is a $\mathfrak{C}$-bimodule map:*

$$\varphi(c_1 a c_2) = \rho(c_1)\varphi(a)\rho(c_2) \quad \textit{for} \quad a \in \mathfrak{A} \textit{ and } c_1, c_2 \in \mathfrak{C}.$$

PROOF. We may consider $\mathfrak{B}$ as a unital subalgebra of $\mathcal{B}(H)$. Let (π, V) be a Stinespring dilation of φ on K. Since φ is unital, V is an isometry. Let $P = VV^*$. The Kadison–Schwarz inequality follows from the computation:

$$\begin{aligned}\varphi(a^*a) - \varphi(a)^*\varphi(a) &= V^*\pi(a)^*\pi(a)V - V^*\pi(a)^*VV^*\pi(a)V \\ &= V\pi(a)^*P^\perp\pi(a)V = (P^\perp\pi(a)V)^*(P^\perp\pi(a)V) \ge 0.\end{aligned}$$

Equality occurs precisely when $P^\perp\pi(a)V = 0$, meaning that $\pi(a)$ leaves $\operatorname{Ran} V = PK$ invariant. Thus, the set of elements in $\mathfrak{A}$ for which this holds is a norm-closed algebra $\mathcal{A}$. Similarly $\varphi(a)\varphi(a)^* = \varphi(aa^*)$ implies that $a^* \in \mathcal{A}$. Thus, the set satisfying both inequalities is

$$\mathfrak{C} = \mathcal{A} \cap \mathcal{A}^* = \{a \in \mathfrak{A} : \pi(a)P = P\pi(a)\},$$

which is a C*-algebra. Let $a \in \mathfrak{A}$ and $c \in \mathcal{A}$. Then computing as above,

$$\varphi(ac) - \varphi(a)\varphi(c) = V^*\pi(a)(I - VV^*)\pi(c)V = V\pi(a)(P^\perp\pi(c)V) = 0.$$

Similarly, if $c \in \mathcal{A}^*$, then $\varphi(ca) = \varphi(c)\varphi(a)$ for $a \in \mathfrak{A}$. So the bimodule property holds for $c \in \mathfrak{C}$. In particular, $\rho = \varphi|_{\mathfrak{C}}$ is multiplicative and self-adjoint, and thus is a $*$-homomorphism. ■

14.7.12. DEFINITION. An operator system $\mathcal{S}$ is *injective* in the category $\mathfrak{S}$ of operator systems with completely positive maps as morphisms if whenever $\mathcal{E} \subset \mathcal{F}$ are operator systems and $\varphi : \mathcal{E} \to \mathcal{S}$ is completely positive, then there is a completely positive map $\widetilde{\varphi} : \mathcal{F} \to \mathcal{S}$ such that $\widetilde{\varphi}|_{\mathcal{E}} = \varphi$.

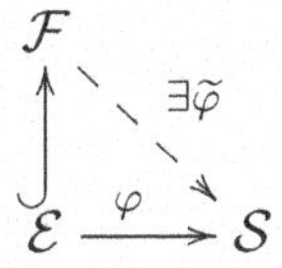

Observe that the extension from E to F is always norm preserving because

$$\|\widetilde{\varphi}\|_{cb} = \|\widetilde{\varphi}\| = \|\widetilde{\varphi}(I)\| = \|\varphi(I)\| = \|\varphi\| = \|\varphi\|_{cb}.$$

Arveson's extension theorem can be reformulated as

14.7.13. COROLLARY. *$\mathcal{B}(H)$ is injective.*

PROOF. We may assume that $\mathcal{E} \subset \mathcal{F} \subset \mathcal{B}(K)$. By Arveson's extension theorem, there is a completely positive map $\widetilde{\varphi} : \mathcal{B}(K) \to \mathcal{B}(H)$ extending φ. Restrict it to $\mathcal{F}$. ∎

In Definition 10.10.17, we provided a different definition of an injective von Neumann algebra. We will show that the two definitions are equivalent. Recall Tomiyama's theorem 10.6.2 that an expectation (norm 1 projection) of a C*-algebra $\mathfrak{A}$ onto a subalgebra $\mathfrak{B}$ is a positive $\mathfrak{B}$-bimodule map. We can now improve on that.

14.7.14. PROPOSITION. *Let $\mathfrak{B} \subset \mathfrak{A}$ be unital C*-algebras, and suppose that $\mathcal{E} : \mathfrak{A} \to \mathfrak{B}$ is an expectation. Then $\mathcal{E}$ is completely positive.*

PROOF. Let $A = \begin{bmatrix} a_{ij} \end{bmatrix} \in M_n(\mathfrak{A})_+$. Let σ be a representation of $\mathfrak{B}$ with cyclic vector $\xi \in H_\sigma$. The vectors of the form $\eta = (\sigma(b_1)\xi, \dots, \sigma(b_n)\xi)$ form a dense subset of $H_\sigma^{(n)}$. Let $B^* = \begin{bmatrix} b_1^* & \dots & b_n^* \end{bmatrix}$. Compute

$$\begin{aligned}\big\langle \sigma_n(\mathcal{E}_n(A))\eta, \eta \big\rangle &= \sum_{i,j=1}^{n} \langle \sigma\mathcal{E}(a_{ij})\sigma(b_j)\xi, \sigma(b_i)\xi \rangle \\ &= \Big\langle \sigma\mathcal{E}\Big(\sum_{i,j=1}^{n} b_i^* a_{ij} b_j \Big)\xi, \xi \Big\rangle = \langle \sigma\mathcal{E}(B^*AB)\xi, \xi \rangle \geq 0.\end{aligned}$$

It follows that $\sigma_n(\mathcal{E}_n(A)) \geq 0$ for all cyclic representations of $\mathfrak{B}$. So by the Gelfand–Naimark theorem, $\mathcal{E}_n(A) \geq 0$. Therefore $\mathcal{E}$ is completely positive. ∎

14.7.15. PROPOSITION. *Let $\mathcal{M}$ be a von Neumann algebra in $\mathcal{B}(H)$. The following are equivalent:*

(1) *$\mathcal{M}$ is an injective operator system.*

(2) *There is a completely positive idempotent map on $\mathcal{B}(H)$ with range $\mathcal{M}$.*

(3) *There is an expectation of $\mathcal{B}(H)$ onto $\mathcal{M}$.*

PROOF. If $\mathcal{M}$ is injective, consider the identity map $\varphi = id_{\mathcal{M}}$ as a CP map into $\mathcal{M}$. By injectivity, there is an extension $\widetilde{\varphi} : \mathcal{B}(H) \to \mathcal{M}$ extending φ. This is evidently idempotent, so (2) holds. When (2) holds say for a cp map E, then $\|E\| = \|E(I)\| = \|I\| = 1$. Thus E is an expectation onto $\mathcal{M}$. If (3) holds, say for E, then Proposition 14.7.14 shows that E is CP. Let $\mathcal{E} \subset \mathcal{F}$ be an inclusion of operator systems, and let $\varphi : \mathcal{E} \to \mathcal{M}$ be CP. By the Arveson extension theorem, there is a CP map $\widetilde{\varphi} : \mathcal{F} \to \mathcal{B}(H)$ extending φ. Then $E\widetilde{\varphi}$ is a CP extension of φ with range in $\mathcal{M}$ which extends φ. Therefore $\mathcal{M}$ is injective. ■

14.8. General Dilation Theory

We will reformulate dilation theory to the more general context of representations of (unital) operator algebras. The main result is Arveson's dilation theorem, which characterizes when there is a $\mathfrak{B}$-dilation.

14.8.1. DEFINITION. Let $\mathcal{A}$ be a unital (generally nonself-adjoint) subalgebra of a C*-algebra $\mathfrak{B}$ such that $\mathfrak{B} = C^*(\mathcal{A})$. If $\rho : \mathcal{A} \to \mathcal{B}(H)$ is a unital representation, a $\mathfrak{B}$-*dilation* is a $*$-representation $\pi : \mathfrak{B} \to \mathcal{B}(K)$, where $K \supset H$, such that $\rho(a) = P_H \pi(a)|_H$ for $a \in \mathcal{A}$.

Note that the map $P_H \pi(a)|_H$ is completely positive on $\mathfrak{B}$ and unital, and hence is completely contractive. Therefore, a necessary condition for a $\mathfrak{B}$-dilation is that ρ be completely contractive. By Proposition 14.1.2, H will be a semi-invariant subspace for $\pi(\mathcal{A})$.

14.8.2. EXAMPLE. Let the disc algebra $A(\mathbb{D})$ be considered as a subalgebra of $C(\mathbb{T})$. A contractive representation $\rho : A(\mathbb{D}) \to \mathcal{B}(H)$ is determined by $\rho(z) = T$, and $\|T\| \le \|z\| = 1$. Conversely von Neumann's inequality shows that any contraction T yields a contractive representation ρ_T with $\rho_T(z) = T$. The matrix von Neumann inequality shows that $\|\rho_T\|_{cb} = 1$. The Sz.-Nagy unitary dilation $U \in \mathcal{B}(K)$ of T determines a representation $\pi_U : C(\mathbb{T}) \to \mathcal{B}(K)$ by the continuous functional calculus. The fact that H is semi-invariant for U leads to the conclusion that $P_H p(U)|_H = p(T)$ for any analytic polynomial. By continuity, this extends to $A(\mathbb{D})$. Thus π_U is a $C(\mathbb{T})$-dilation of ρ_T.

14.8.3. LEMMA. *If $T \in \mathcal{B}(H)$, the* numerical radius $w(T) = \sup_{\|\xi\|=1} |\langle T\xi, \xi\rangle|$ *satisfies $\frac{1}{2}\|T\| \le w(T) \le \|T\|$. These inequalities are sharp.*

PROOF. Clearly $|\langle T\xi, \xi\rangle| \le \|T\|$, so that $w(T) \le \|T\|$. It is easy to check that if $A = A^*$, then $w(A) = \operatorname{spr}(A) = \|A\|$. If $T = A + iB$ for $A = A^*$ and $B = B^*$, then

$$|\langle T\xi, \xi\rangle| = |\langle A\xi, \xi\rangle + i\langle B\xi, \xi\rangle| \ge \max\{\langle A\xi, \xi\rangle, \langle B\xi, \xi\rangle|\}.$$

Hence

$$w(T) \ge \max\{w(A), w(B)\} = \max\{\|A\|, \|B\|\} \ge \frac{1}{2}\|T\|.$$

The 2×2 nilpotent $J = \left[\begin{smallmatrix} 0 & 1 \\ 0 & 0 \end{smallmatrix}\right]$ has $w(J) = \frac{1}{2}\|J\|$. ■

This next lemma is the key to relating a completely contractive map on $\mathcal{A}$ to a completely positive map on a related operator system.

14.8.4. LEMMA. *Let $\mathcal{A} \subset \mathfrak{B}$ be an operator space containing the unit 1 of a C*-algebra $\mathfrak{B}$. Let $\rho : \mathcal{A} \to \mathcal{B}(H)$ be a unital (completely) contractive map. Let $\mathcal{S} = \overline{\mathcal{A} + \mathcal{A}^*}$. Define a map on $\mathcal{A} + \mathcal{A}^*$ by $\widetilde{\rho}(a + b^*) = \rho(a) + \rho(b)^*$ for $a, b \in \mathcal{A}$. Then $\widetilde{\rho}$ is unital (completely) positive and extends by continuity to $\mathcal{S}$. Moreover, $\tilde{\rho}$ is the unique (completely) positive extension of ρ to $\mathcal{S}$.*

PROOF. Assume that ρ is contractive. For each unit vector $\xi \in H$, the functional on $\mathcal{A}$ given by $\psi(a) = \langle \rho(a)\xi, \xi\rangle$ has $\|\psi\| = \psi(1) = 1$. Extend it to a functional $\widetilde{\psi}$ on $\mathfrak{B}$ by the Hahn–Banach theorem. By Lemma 9.6.5, $\widetilde{\psi}$ is a state. In particular, it is self-adjoint and positive. Hence if $a, b \in \mathcal{A}$,

$$\begin{aligned}\langle \widetilde{\rho}(a + b^*)\xi, \xi\rangle &= \langle \rho(a)\xi, \xi\rangle + \langle \rho(b)^*\xi, \xi\rangle = \langle \rho(a)\xi, \xi\rangle + \overline{\langle \rho(b)\xi, \xi\rangle} \\ &= \psi(a) + \overline{\psi(b)} = \widetilde{\psi}(a + b^*).\end{aligned}$$

Applying Lemma 14.8.3 yields

$$\begin{aligned}\frac{1}{2}\|\widetilde{\rho}(a + b^*)\| \le w(\widetilde{\rho}(a + b^*)) &= \sup_{\|\xi\|=1} |\langle \rho(a) + \rho(b)^*\xi, \xi\rangle| \\ &\le \sup_{\widetilde{\psi}\in S(\mathfrak{A})} |\widetilde{\psi}(a + b^*)| \le \|a + b^*\|.\end{aligned}$$

This shows first of all that $\widetilde{\rho}$ is well defined. Also $\|\widetilde{\rho}\| \le 2$, so it extends by continuity to $\mathcal{S}$. Moreover if $s \in \mathcal{S}$, then $\langle \widetilde{\rho}(s)\xi, \xi\rangle = \widetilde{\psi}(s)$. To check positivity, take $s \in \mathcal{S}_+$ and note that $\langle \widetilde{\rho}(s)\xi, \xi\rangle = \widetilde{\psi}(s) \ge 0$. So $\widetilde{\rho}(s) \ge 0$.

Now suppose that ρ is CP. Then ρ_n is contractive on $M_n(\mathcal{A}) \subset M_n(\mathfrak{B})$. It is easily checked that $\overline{M_n(\mathcal{A}) + M_n(\mathcal{A})^*} = M_n(\mathcal{S})$ and $(\widetilde{\rho})_n = \widetilde{\rho_n}$, which is positive by the previous paragraph. Therefore $\widetilde{\rho}$ is unital and completely positive. Uniqueness is clear since positive maps are self-adjoint. ∎

If ρ is a complete isometry, we can also apply this result to $\rho^{-1} : \rho(\mathcal{A}) \to \mathcal{A}$.

14.8.5. COROLLARY. *Let $\mathcal{A} \subset \mathfrak{B}$ be an operator space containing the unit 1, and let $\rho : \mathcal{A} \to \mathcal{B}(H)$ be a unital completely isometric map. Then $\widetilde{\rho}$ is completely isometric on $\mathcal{S} = \overline{\mathcal{A} + \mathcal{A}^*}$.*

Now we are ready to prove Arveson's dilation theorem. It reduces the problem to checking whether a certain map is completely positive.

14.8.6. ARVESON DILATION THEOREM. *Let $1 \in \mathcal{A} \subset \mathfrak{B}$ be a unital subalgebra of a C*-algebra $\mathfrak{B}$. Suppose that $\rho : \mathcal{A} \to \mathcal{B}(H)$ is a unital contractive* *representation. Let $\mathcal{S} = \overline{\mathcal{A} + \mathcal{A}^*}$ and let $\widetilde{\rho}$ be the unique positive extension of ρ to $\mathcal{S}$. The following are equivalent:*

(1) *ρ has a $\mathfrak{B}$-dilation $\pi : \mathfrak{B} \to \mathcal{B}(K)$ for $K \supset H$.*

(2) *$\|\rho\|_{cb} = 1$, i.e. ρ is completely contractive.*

(3) *$\widetilde{\rho}$ is completely positive.*

Moreover H is a semi-invariant subspace for $\pi(\mathcal{A})$.

PROOF. We have already noted that (1) implies (2). Lemma 14.8.4 shows that (2) implies (3). If $\widetilde{\rho}$ is completely positive, use Arveson's extension theorem to extend this to a CP map $\varphi : \mathfrak{B} \to \mathcal{B}(H)$. Apply the Stinespring dilation theorem to obtain a (minimal) Stinespring representation (π, V) for ψ in $\mathcal{B}(K)$. Since $\psi(1) = \rho(1) = I$, we have V is an isometry. So we identify H with the subspace $VH \subset K$. Then

$$\rho(a) = V^*\pi(a)V = P_H\pi(a)|_H \quad \text{for all} \quad a \in \mathcal{A}.$$

By Proposition 14.1.2, H is a semi-invariant subspace for $\pi(\mathcal{A})$. ■

14.8.7. EXAMPLE. (Triangular algebras.) For $n < \infty$, let $\mathcal{T}_n$ denote the algebra of $n \times n$ upper triangular matrices acting on $\mathbb{C}^n$ with a fixed orthonormal basis. Let $\rho : \mathcal{T}_n \to \mathcal{B}(H)$ be a contractive representation. There is little lost if we assume that ρ is unital, since $\rho(T) = \rho(I_n)\rho(T)\rho(I_n)$ "lives" on $\rho(I_n)H$. We will assume that H is separable, but routine arguments will yield the general case.

Define $P_i = \rho(E_{ii})$ for $1 \le i \le n$, where E_{ij} are the standard matrix units. Since $P_i = P_i^2$ and $\|P_i\| \le \|E_{ii}\| = 1$, it follows that P_i are orthogonal projections. Also $E_{ii}E_{jj} = 0$ for $i \ne j$, so P_i are pairwise orthogonal and $\sum_{i=1}^n P_i = \rho(I_n) = I$. If some of the P_i are finite dimensional, replace ρ with $\rho \oplus id^{(\infty)}$. This does not affect the contractive property, but now each P_i will be infinite rank. Now we can identify H with a Hilbert space $H_0^{(n)}$ in such a way that $P_i \simeq E_{ii} \otimes I_{H_0}$ is the projection onto the ith summand.

Observe that $\rho(E_{ij}) = P_i\rho(E_{ij})P_j$ for all $1 \le i < j \le n$. In particular, there is an operator $T_i \in \mathcal{B}(H_0)$ such that $\rho(E_{i,i+1}) = E_{i,i+1} \otimes T_i$. Note that $\|T_i\| \le 1$ also. Then for $1 \le i < j \le n$,

$$\rho(E_{ij}) = \rho(E_{i,i+1})\rho(E_{i+1,i+2}) \cdots \rho(E_{j-1,j}) = E_{ij} \otimes T_i \cdots T_{j-1} =: E_{ij} \otimes T_{ij}.$$

Let U_i be a unitary dilation of T_i on $K_0 = K_+ \oplus H_0 \oplus K_-$ which is upper triangular and has T_i as the 2, 2 entry, and $K_\pm \simeq H_0^{(\infty)}$. Define $K = K_0^{(n)}$. Then we can decompose $K = K_+^{(n)} \oplus H_0^{(n)} \oplus K_-^{(n)}$. Let $Q_i = E_{ii} \otimes I_{K_0}$ be the orthogonal projection of K onto the ith summand. Observe that $U_{ij} := U_i \cdots U_{j-1}$ is upper triangular with T_{ij} as the 2, 2 entry.

Define a representation $\sigma : M_n \to \mathcal{B}(K)$ by setting $\sigma(E_{ij}) = E_{ij} \otimes U_{ij}$ and $\sigma(E_{ji}) = E_{ji} \otimes U_{ij}^*$ if $i \le j$. When restricted to $\mathcal{T}_n$, we see that the space $H_0^{(n)}$

is semi-invariant, and the compression of σ to $H_0^{(n)}$ is ρ. Thus σ is a dilation of ρ. Define $W = I \oplus U_1 \oplus U_1U_2 \oplus \cdots \oplus U_1U_2 \cdots U_{n-1}$. Then if $1 \le i < j \le n$,

$$\operatorname{Ad} W \circ \sigma(E_{ij}) = E_{ij} \otimes \big((U_1 \cdots U_{i-1})U_{ij}(U_{j-1}^* \cdots U_1^*)\big) = E_{ij} \otimes I.$$

Similarly $\operatorname{Ad} W \circ \sigma(E_{ij}) = E_{ij} \otimes I$ for $i = j$ and $i > j$. That is, $\operatorname{Ad} W \circ \sigma = id^{(\infty)}$ is a $*$-homomorphism. Hence $\sigma = \operatorname{Ad} W^* \circ id^{(\infty)} \simeq id^{(\infty)}$. This is an explicit M_n dilation of ρ. Therefore ρ is completely contractive. Note that we did not use the full strength of the contractive condition, but only that $\rho(E_{ij})$ are all contractive.

Now let $\mathcal{T}_\infty$ denote the algebra of upper triangular operators in $\mathcal{B}(H)$ with respect to an orthonormal basis $\xi_1, \xi_2, \ldots$. Let E_{ij} denote the matrix units with respect to this basis. Let $\rho : \mathcal{T}_\infty \to \mathcal{B}(K)$ be a weak-$*$ continuous contractive representation on a separable Hilbert space. As above, $P_i = \rho(E_{ii})$ are pairwise orthogonal projections. Since $\sum_{i \ge 1} E_{ii} = I$ is weak-$*$ convergent, we have $\sum_{i \ge 1} P_i = \rho(I)$, and we may assume that $\rho(I) = I$. Consider $\mathcal{T}_n$ to be the subalgebra $\mathcal{T}_\infty F_n$ of $\mathcal{T}_\infty$, where $F_n = \sum_{i=1}^n E_{ii}$. Then $\rho_n = \rho|_{\mathcal{T}_n}$ is contractive, and therefore $\|\rho_n\|_{cb} = 1$. It follows that ρ is completely contractive on $\overline{\bigcup_{n \ge 1} \mathcal{T}_n} = \mathcal{T}_\infty \cap \mathcal{K}(H)$. Now every $T \in M_k(\mathcal{T}_\infty)$ has $T = \lim_{n\to\infty} TF_n^{(k)}$ in the weak-$*$ topology, and $\|TF_n^{(k)}\| \le \|T\|$. By weak-$*$ continuity,

$$\|\rho_k(T)\| = \lim_{n\to\infty} \|\rho_k(TF_n^{(k)})\| \le \lim_{n\to\infty} \|TF_n^{(k)}\| = \|T\|.$$

Therefore $\|\rho\|_{cb} = 1$. By Arveson's dilation theorem, ρ has a $C^*(\mathcal{T}_\infty)$-dilation.

We claim that $C^*(\mathcal{T}_\infty) = \mathcal{B}(H)$. Care must be taken since $\overline{\mathcal{T}_\infty + \mathcal{T}_\infty^*} \ne \mathcal{B}(H)$ (see Exercise **13.13**). Let $A \in \mathcal{B}(H)$ be positive and invertible. Write $A^{1/2} = \big[\eta_1\ \eta_2\ \ldots\big]$ as a matrix where $\eta_i = A^{1/2}\xi_i$. Apply the Gram–Schmidt process to the sequence $\eta_1, \eta_2, \ldots$ to obtain an orthonormal set $\zeta_1, \zeta_2, \ldots$. Since A is invertible, the η_i's are independent and span H. Hence, we obtain an orthonormal basis such that $\eta_i \in \operatorname{span}\{\zeta_1, \ldots, \zeta_i\}$ for $i \ge 1$. Let U be the unitary such that $U\zeta_i = \xi_i$ for $i \ge 1$. Then $UA^{1/2}\xi_i = U\eta_i \in \operatorname{span}\{\xi_1, \ldots, \xi_i\}$ for $i \ge 1$. That is, $UA^{1/2} \in \mathcal{T}_\infty$. Therefore $A = (UA^{1/2})^*(UA^{1/2}) \in C^*(\mathcal{T}_\infty)$. Every operator in $\mathcal{B}(H)$ is a linear combination of four positive invertible operators, and hence $C^*(\mathcal{T}_\infty) = \mathcal{B}(H)$.

Let (π, V) be a minimal Stinespring dilation of ρ. By Theorem 9.8.2, the representation $\pi \simeq \pi_a \oplus \pi_s$ where $\pi_a = \operatorname{Ind} \pi|_{\mathcal{K}(H)}$ and π_s kills $\mathcal{K}(H)$. The projection Q onto the first summand belongs to $\pi(\mathcal{B}(H))'$. Since the only representations of $\mathcal{K}(H)$ are ampliations, $\pi_a(T) = T^{(\alpha)}$ for some cardinal α. However ρ is weak-$*$ continuous. For $T \in \mathcal{T}_\infty$, we noted that $T = w^*\text{–}\lim_{n\to\infty} TF_n$. Therefore

$$\begin{aligned}\rho(T) &= V^*(T^{(\alpha)} \oplus \pi_s(T))V = \underset{n\to\infty}{w^*\text{–}\lim}\, \rho(TF_n)\\ &= \underset{n\to\infty}{w^*\text{–}\lim}\, V^*((TF_n)^{(\alpha)} \oplus 0)V = V^*(T^{(\alpha)} \oplus 0)V.\end{aligned}$$

It follows that

$$I = \rho(I) = V^*V = V^*QV.$$

Hence V is an isometry with range in the first summand, and $\rho(T) = V^*T^{(\alpha)}V$. By minimality, $\pi = \pi_a$. By Proposition 14.1.2, the range of V is semi-invariant for $\mathcal{T}_\infty^{(\alpha)}$.

14.9. The minimal C*-algebra

14.9.1. EXAMPLE. An operator algebra can generate a variety of C*-algebras. Take $\mathcal{A} = A(\mathbb{D})$. We can consider this as a subalgebra of $C(\overline{\mathbb{D}})$. Since $\mathbb{T}$ is the Shilov boundary, we can also consider it as a subalgebra of $C(\mathbb{T})$. Also there are non-commutative C*-algebras such as the Toeplitz algebra $\mathcal{T}(C(\mathbb{T}))$, obtained by sending f to T_f, which is a completely isometric isomorphism on $A(\mathbb{D})$.

For a representation ρ of $A(\mathbb{D})$, we let $T = \rho(z)$. A unitary dilation U of T gives rise to a $C(\mathbb{T})$-dilation via the functional calculus. Now $C(\mathbb{T})$ is the quotient of $C(\overline{\mathbb{D}})$ by the ideal of functions vanishing on $\mathbb{T}$. It is also the quotient of $\mathcal{T}(C(\mathbb{T}))$ by the compact operators. The C*-algebra $C(\mathbb{T})$ is "smaller" than the other two because the quotient maps are completely isometric isomorphisms on the image of $A(\mathbb{D})$. This dilation can be considered as a $\mathfrak{B}$-dilation for any of these C*-algebras.

What we will show is that there is always a smallest choice, $C^*_{\min}(\mathcal{A})$, of the C*-algebra containing $\mathcal{A}$. Moreover the $C^*_{\min}(\mathcal{A})$-dilations have special features. Because of Lemma 14.8.4, Corollary 14.8.5 and Arveson's dilation theorem, we can talk about the operator system $\mathcal{S} = \overline{\mathcal{A} + \mathcal{A}^*}$ instead of $\mathcal{A}$. This is more general since not every operator system is generated by an algebra. Thus problems about dilations turn into problems about extensions of completely positive maps. It is not obvious, however, that one can recover the multiplicative nature of a homomorphism on $\mathcal{A}$ from a completely positive map on $\mathcal{S}$.

14.9.2. DEFINITION. A *dilation* of a u.c.p. map $\varphi : \mathcal{S} \to \mathcal{B}(H)$ is a unital completely positive map $\psi : \mathcal{S} \to \mathcal{B}(K)$ for $K \supset H$ such that $\varphi(a) = P_H\psi(a)|_H$ for $a \in \mathcal{S}$. Write $\varphi \prec \psi$. This extension is *trivial* if $\psi = \varphi \oplus \psi'$. If the only extensions of φ are trivial, we say that φ is *maximal* or *extremal*.

Likewise if $\mathcal{A}$ is an operator algebra, a dilation $\sigma : \mathcal{A} \to \mathcal{B}(K)$ of a unital representation $\rho : \mathcal{A} \to \mathcal{B}(H)$ is *trivial* if $\sigma = \rho \oplus \sigma'$. Say that σ is *maximal* or *extremal* if every dilation is trivial.

14.9.3. EXAMPLE. Take $\mathcal{A} = A(\mathbb{D})$ again. A (completely) contractive unital representation ρ is determined by the contraction $T = \rho(z)$. A dilation of ρ to σ is trivial if and only if $V = \sigma(z)$ is a trivial dilation of T. If T is not unitary, then the minimal unitary dilation is non-trivial. On the other hand, if U is already unitary, then any dilation of U, say

$$V = \begin{bmatrix} * & 0 & 0 \\ X & U & 0 \\ * & Y & * \end{bmatrix}$$

must be contractive. Hence $X = 0$ and $Y = 0$, so that $V = U \oplus \begin{bmatrix} * & 0 \\ * & * \end{bmatrix}$. This is a trivial dilation. Thus, the maximal dilations coincide with unitary or $C(\mathbb{T})$-dilations.

We introduce another key notion for u.c.p. maps on operator systems.

14.9.4. Definition. A unital completely positive map $\varphi : \mathcal{S} \to \mathcal{B}(H)$ has the *unique extension property* (u.e.p.) if there is a unique u.c.p. map $\pi : C^*(\mathcal{S}) \to \mathcal{B}(H)$ extending φ *and* π is a $*$-homomorphism. We may also say that π has the unique extension property if $\mathcal{S}$ is understood.

These notions also make sense for a unital algebra $\mathcal{A} \subset C^*(\mathcal{A})$ and a unital completely contractive map of $\mathcal{A}$ into $\mathcal{B}(H)$.

14.9.5. Proposition. *Let $\varphi : \mathcal{S} \to \mathcal{B}(H)$ be a unital completely positive map. Then φ is maximal if and only if it has the unique extension property.*

Proof. Suppose that φ is maximal. Let (π, V) be a minimal Stinespring dilation of φ on $K \supset H$. Since φ is maximal, $\pi|_{\mathcal{S}} = \varphi \oplus \pi'$. Since π is a $*$-representation of $C^*(\mathcal{S})$, H is reducing for $\pi(C^*(\mathcal{S}))$, whence $\pi = \widetilde{\varphi} \oplus \pi'$. Hence $\widetilde{\varphi}$ is multiplicative and $P_H \pi|_H$ is a $*$-homomorphism. In particular, *any* u.c.p. extension of φ to $C^*(\mathcal{S})$ must be a $*$-homomorphism. There is clearly a unique extension of φ to a $*$-representation (if it exists); so φ has the unique extension property.

Conversely suppose that there is a $*$-homomorphism $\pi : C^*(\mathcal{S}) \to \mathcal{B}(H)$ extending φ and it is the unique u.c.p. extension. Let $\psi : \mathcal{S} \to \mathcal{B}(K)$ be a dilation of φ on $K \supset H$; so $P_H \psi|_H = \varphi$. Let (σ, V) be a Stinespring dilation of ψ on $L \supset K$. Then $P_H \sigma|_H$ is a u.c.p. extension of φ to $C^*(\mathcal{S})$, and hence equals π. Since the compression of σ to H is multiplicative, Proposition 14.1.2 shows that H is semi-invariant for $\sigma(C^*(\mathcal{S}))$. But invariant subspaces for a C*-algebra are reducing, and thus H reduces σ, i.e. $\sigma \simeq \pi \oplus \sigma'$. Hence $\sigma|_{\mathcal{S}}$ is a trivial dilation of φ, whence ψ is a trivial dilation of φ. Therefore φ is maximal. ∎

Our next goal is to show that maximal dilations exist.

14.9.6. Definition. Let $\mathcal{S}$ be an operator system, and let $\varphi : \mathcal{S} \to \mathcal{B}(H)$ be a u.c.p. map. For $a \in \mathcal{S}$ and $\xi \in H$, we say that φ is *maximal at* (a, ξ) if $\|\varphi(a)\xi\| = \sup_{\varphi \prec \psi} \|\psi(a)\xi\|$.

14.9.7. LEMMA. *Let $\mathcal{S}$ be an operator system, let $\varphi : \mathcal{S} \to \mathcal{B}(H)$ be a u.c.p. map, and let $(a, \xi) \in \mathcal{S} \times H$. Then there is a dilation $\psi \succ \varphi$ on $K \supset H$ which is maximal at (a, ξ). Moreover we may take* $\dim K \ominus H \le 1$.

A u.c.p. map φ is maximal if and only if it is maximal at (a, ξ) for every $(a, \xi) \in \mathcal{S} \times H$.

PROOF. We may assume that

$$\|\varphi(a)\xi\| < L := \sup_{\varphi \prec \psi} \|\psi(a)\xi\| \le \|a\| \, \|\xi\|.$$

The second inequality follows because each ψ is unital and hence completely contractive. Given $\varphi \prec \psi$, we can restrict ψ to $K' = \operatorname{span}\{H, \psi(a)\xi\}$ without affecting $\|\psi(a)\xi\|$. Thus, we may consider only dilations ψ with range in a fixed Hilbert space $K = \operatorname{span}\{H, \eta\}$ for a unit vector $\eta \perp H$.

The set $C = \{\psi \in \mathrm{UCP}(\mathcal{S}, \mathcal{B}(K)) : \varphi \prec \psi\}$ is a bounded convex set which is closed under point-weak-$*$ limits. Hence it is point-weak-$*$ compact. The functional $l(\psi) = \langle \psi(a)\xi, \eta \rangle$ is point-weak-$*$ continuous and

$$L = \sup_{\psi \in C} \|\psi(a)\xi\|^2 = \|\varphi(a)\xi\|^2 + \sup_{\psi \in C} |l(\psi)|^2.$$

By compactness, there is some $\psi \in C$ with $\|\psi(a)\xi\| = L$. Every dilation σ of ψ is also a dilation of φ, so that

$$\|\psi(a)\xi\| \le \|\sigma(a)\xi\| \le L = \|\psi(a)\xi\|.$$

Therefore ψ is maximal at (a, ξ).

Suppose that φ is maximal at (a, ξ) for every $(a, \xi) \in \mathcal{S} \times H$. Let $\varphi \prec \psi$. Then for each (a, ξ), we have

$$\|\psi(a)\xi\| = \|\varphi(a)\xi\| = \|P_H \psi(a)\xi\|.$$

Thus $\psi(a)\xi = \varphi(a)\xi$ lies in H. So H reduces ψ (as ψ is self-adjoint), whence $\psi = \varphi \oplus \psi'$. Therefore φ is maximal. ■

We can now iterate this argument to obtain a maximal dilation.

14.9.8. DRITSCHEL–MCCULLOUGH THEOREM. *Let $\varphi : \mathcal{S} \to \mathcal{B}(H)$ be a unital completely positive map. Then φ has a maximal dilation ψ on some Hilbert space K, and the u.c.p. extension of ψ to $C^*(\mathcal{S})$ is a $*$-homomorphism. If $\mathcal{S}$ and H are separable, K can be separable.*

PROOF. Let $\{(a_\lambda, \xi_\lambda)\}_\Lambda$ be a dense subset of $\mathcal{S} \times H$. By the Axiom of Choice, we may suppose that Λ is well-ordered. We will use transfinite induction to build u.c.p. maps $\psi_\lambda : \mathcal{S} \to \mathcal{B}(K_\lambda)$ such that $\psi_\mu \prec \psi_\lambda$ if $\mu < \lambda$ and ψ_λ is maximal at (a_μ, ξ_μ) for all $\mu \le \lambda$. There are two cases. If λ is the successor of some μ, then use Lemma 14.9.7 to dilate $\psi_\mu : \mathcal{S} \to \mathcal{B}(K_\mu)$ to $\psi_\lambda : \mathcal{S} \to \mathcal{B}(K_\lambda)$, where

$K_\lambda = \operatorname{span}\{K_\mu, \mathbb{C}\}$, so that it is maximal at (a_λ, ξ_λ). It will already be maximal at (a_μ, ξ_μ) for $\mu < \lambda$.

If λ is a limit element, let $K'_\lambda = \bigcup_{\mu<\lambda} K_\mu$. This is an inner product space, and for each $a \in \mathcal{S}$ and $\xi, \eta \in K_\mu \subset K'_\lambda$ and $\mu \le \nu < \lambda$, $\langle \psi_\nu(a)\xi, \eta\rangle = \langle \psi_\mu(a)\xi, \eta\rangle$. Therefore, the net ψ_μ consists of contractive maps from $\mathcal{S}$ into K'_λ which converge point-weak-$*$ to a map ψ'_λ. We can complete K'_λ to a Hilbert space $\widetilde{K}'_\lambda$ and extend ψ'_λ to $\widetilde{\psi}'_\lambda$ on this Hilbert space by norm continuity. This will be a dilation of all ψ_μ for $\mu < \lambda$, and thus it will be maximal at (a_μ, ξ_μ) for all $\mu < \lambda$. Now use Lemma 14.9.7 to dilate $\widetilde{\psi}'_\lambda$ to ψ_λ on $K_\lambda \supset \widetilde{K}'_\lambda$ so that it is maximal at (a_λ, ξ_λ).

At the end, if Λ does not have a largest element, we can again build a dilation on the union of the K_λ's and complete it to obtain the desired dilation. This will be maximal at every (a_λ, ξ_λ), and so by continuity, it will be maximal on $\mathcal{S} \times H$. Call the dilation obtained above $\varphi_2 : \mathcal{S} \to \mathcal{B}(H_2)$ with $\varphi \prec \varphi_2$ which is maximal at every $(a, \xi) \in \mathcal{S} \times H$.

Choose a dense subset $\{(a_\lambda, \xi_\lambda)\}_\Lambda$ of $\mathcal{S} \times H_2$. Apply the same procedure again to obtain a dilation $\varphi_2 \prec \varphi_3$ on $H_3 \supset H_2$ which is maximal on $\mathcal{S} \times H_2$. Recursively we obtain dilations φ_n with $\varphi_n \prec \varphi_{n+1}$ acting on H_{n+1} so that φ_{n+1} is maximal on $\mathcal{S} \times H_n$. Again take the union $K' = \bigcup_{n\ge 2} H_n$ and define $\psi' : \mathcal{S} \to \mathcal{B}(K)$ as above. Then complete K' and ψ' to obtain a dilation ψ of φ on K which is maximal on $\bigcup_{n\ge 2} \mathcal{S} \times H_n = \mathcal{S} \times K'$. By continuity, ψ will be maximal on $\mathcal{S} \times K$. Then Lemma 14.9.7 shows that ψ is maximal. Proposition 14.9.5 now yields that ψ has the unique extension property. In particular, it extends to a $*$-homomorphism π on $C^*(\mathcal{S})$.

If $\mathcal{S}$ and H are separable, then $K_0 = \overline{\pi(C^*(\mathcal{S}))H}$ is separable and reduces ψ. Therefore $\psi_0(a) = \psi(a)|_{K_0}$ is a maximal dilation on a separable space. ∎

14.9.9. COROLLARY. *Let $\mathcal{A}$ be an algebra of operators and let $\rho : \mathcal{A} \to \mathcal{B}(H)$ be a unital completely contractive representation. Then ρ has a maximal dilation.*

PROOF. Use Lemma 14.8.4 to extend ρ to a u.c.p. map on $\mathcal{S} = \overline{\mathcal{A} + \mathcal{A}^*}$. By the Dritschel–McCullough theorem, this has a maximal dilation. But being maximal on $\mathcal{S}$ implies being maximal on $\mathcal{A}$. (In fact they are equivalent.) Thus we obtain a maximal dilation of ρ. ∎

14.9.10. DEFINITION. If $\mathcal{S}$ is an operator system, a *C*-cover* of $\mathcal{S}$ is a C*-algebra $\mathfrak{B} = C^*(j(\mathcal{S}))$ where $j : \mathcal{S} \to \mathfrak{B} = C^*(j(\mathcal{S}))$ is a unital complete isometry.

The *minimal C*-algebra* or *C*-envelope*, $C^*_{\min}(\mathcal{S})$, of an operator space $\mathcal{S}$ or operator algebra $\mathcal{A}$ is a pair $(\mathfrak{A}, \iota)$, where ι is a unital completely isometric imbedding of $\mathcal{S}$ or $\mathcal{A}$ into $\mathfrak{A} = C^*(\iota(\mathcal{S}))$ with the universal property:
(†) whenever $\mathfrak{B} = C^*(j(\mathcal{S}))$ is a C*-cover of $\mathcal{S}$, there exists a $*$-homomorphism

$\pi : \mathfrak{B} \to \mathfrak{A}$ such that $\pi j = \iota$, i.e.

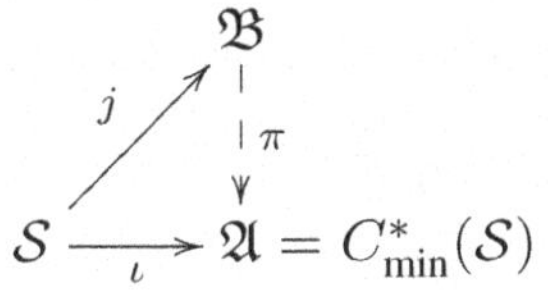

14.9.11. REMARK. $C^*_{\min}(\mathcal{S})$ is the unique minimal C*-algebra generated by $\mathcal{S}$ in a strong sense. If $\mathfrak{B}$ is another C*-algebra generated by a completely isometric unital copy of $\mathcal{S}$, the map π is necessarily unique and surjective; so it is a quotient map. The kernel is called the *Shilov ideal.*

If $\mathcal{A}$ is an operator algebra, Corollary 14.8.5 shows that any unital completely isometric copy of $\mathcal{A}$ generates a completely isometric copy of $\mathcal{S} = \overline{\mathcal{A} + \mathcal{A}^*}$ as an operator system. So there is no difference whether one uses $\mathcal{A}$ or $\mathcal{S}$.

The following theorem was established by Hamana using the theory of injective envelopes that he developed. See Section 14.11. The proof given here is due to Dritschel–McCullough and Arveson.

14.9.12. THEOREM. *Let $\mathcal{S} \subset \mathcal{B}(H)$ be an operator system. Then $C^*_{\min}(\mathcal{S})$ exists and is unique up to a $*$-isomorphism which fixes $\mathcal{S}$.*

PROOF. Let *id* $: \mathcal{S} \to \mathcal{B}(H)$ be the identity map, which is a completely isometric representation. By the Dritschel–McCullough theorem, there is a maximal dilation $\iota : \mathcal{S} \to \mathcal{B}(K)$ of *id*. Let $\mathfrak{A} = C^*(\iota(\mathcal{S}))$. We will show that $(\mathfrak{A}, \iota)$ is $C^*_{\min}(\mathcal{S})$. Let $j : \mathcal{S} \to \mathfrak{B} = C^*(j(\mathcal{S}))$ be a unital complete isometry. Define $\varphi : j(\mathcal{S}) \to \iota(\mathcal{S})$ by $\varphi = \iota j^{-1}$. This is a unital complete isometry. Since ι is a maximal dilation of *id*, Lemma 14.9.7 shows that ι is maximal at (a, ξ) for all $(a, \xi) \in \mathcal{S} \times K$. It follows that φ is also maximal at (a, ξ) for all $(a, \xi) \in \mathcal{S} \times K$. Thus by Lemma 14.9.7 again, φ is maximal. Hence, there is a (unique) $*$-homomorphism π of $\mathfrak{B}$ into $\mathfrak{A}$ extending φ. Thus $\pi j = \iota$ and $\pi(\mathfrak{B}) = C^*(\pi(j(\mathcal{S}))) = \mathfrak{A}$.

For uniqueness, suppose that $(\mathfrak{A}', \iota')$ is another minimal C*-algebra of $\mathfrak{S}$. Then there are $*$-homomorphisms $\pi : \mathfrak{A}' \to \mathfrak{A}$ and $\pi' : \mathfrak{A} \to \mathfrak{A}'$ such that $\pi\iota' = \iota$ and $\pi'\iota = \iota'$. Thus $\pi\pi'$ maps $\mathfrak{A}$ to itself and fixes $\mathcal{S}$, whence $\pi\pi' = id_{\mathfrak{A}}$. Similarly $\pi'\pi = id_{\mathfrak{A}'}$. Therefore, π and π' are isomorphisms that preserve the copy of $\mathcal{S}$. ∎

14.9.13. EXAMPLE. Let X be a compact Hausdorff space, and let $\mathcal{A} \subset C(X)$ be a *uniform algebra.* That is, $\mathcal{A}$ is a closed unital subalgebra which separates points of X. If we wish, we can take X to be the maximal ideal space $\mathcal{M}_{\mathcal{A}}$, which is the largest choice for X. There is a quotient map $\pi : C(X) \to C^*_{\min}(\mathcal{A})$ which is a complete isometry on $\mathcal{A}$. By Exercise **7.1**, there is a closed set $Y \subset X$ so that

$$\ker \pi = I(Y) = \{f \in C(X) : f|_Y = 0\}.$$

Hence $C^*_{\text{min}}(\mathcal{A}) \simeq C(Y)$. By definition, Y (completely) norms $\mathcal{A}$.

Suppose that $Z \subset X$ is another closed subset which norms $\mathcal{A}$, i.e. Z is a closed boundary for $\mathcal{A}$. Then the restriction map of $\mathcal{A}$ into $C(Z)$ is completely isometric. To see this, suppose that $x \in X$. By (the proof of) Proposition 7.8.6, there is a probability measure μ on Z such that

$$f(x) = \int_Z f\,d\mu \quad \text{for all} \quad f \in \mathcal{A}.$$

Hence if $F = [f_{ij}] \in M_n(\mathcal{A})$,

$$\|F(x)\| = \left\| \int_Z F(z)\,d\mu(z) \right\| \leq \int_Z \|F(z)\|\,d\mu(z) \leq \sup_{z \in Z} \|F(z)\|.$$

By the definition of the minimal C*-algebra, there is a quotient map $\sigma : C(Z) \to C(Y)$ which is the identity map on $\mathcal{A}$. By Exercise **7.23**, there is a map $h : Y \to Z$ so that $\sigma(f) = f \circ h$. Since $\mathcal{A}$ separates points, this means that $Y \subset Z$ and h is the inclusion of Y into Z. Thus Y is the smallest closed boundary for $\mathcal{A}$. This is a different proof that $\mathcal{A}$ has a Shilov boundary. (Compare with Theorem 7.8.3.)

14.9.14. PROPOSITION. *Let $\mathcal{S}$ be an operator system (or $\mathcal{A}$ an operator algebra). Suppose that $\varphi : \mathcal{S} \to \mathcal{S}$ (or $\varphi : \mathcal{A} \to \mathcal{A}$) is a surjective, unital complete isometry. Then there is a $*$-automorphism $\Phi \in \operatorname{Aut}(C^*_{\text{min}}(\mathcal{S}))$ which extends φ.*

PROOF. Corollary 14.8.5 shows that if $\varphi : \mathcal{A} \to \mathcal{A}$ is a unital complete isometry, then the unique extension to $\mathcal{S} = \overline{\mathcal{A} + \mathcal{A}^*}$ is also a unital complete isometry. Since φ is surjective, the extension has dense range, and is closed, so is also surjective. Hence, it suffices to deal with the operator system case. Let $\iota : \mathcal{S} \to C^*_{\text{min}}(\mathcal{S})$ be the canonical imbedding. Note that $\iota\varphi$ is another completely isometric unital imbedding of $\mathcal{S}$ into $C^*_{\text{min}}(\mathcal{S}) = C^*(\iota\varphi(\mathcal{S}))$. By the universal property, there is a $*$-homomorphism $\Phi : C^*_{\text{min}} \to C^*_{\text{min}}$ such that $\Phi\iota = \iota\varphi$. Since this procedure is reversible, Φ is an automorphism. ■

14.9.15. COROLLARY. *Let $\mathcal{A}$ be a unital operator algebra, and let $\varphi : \mathcal{A} \to \mathcal{A}$ be a unital complete isometry. Then φ is an automorphism.*

14.10. Boundary Representations

In this section, we will deduce the existence of "enough" irreducible representations with the unique extension property (boundary representations) to completely norm $\mathcal{S}$. These points are the analogue of the extreme points in the Shilov boundary. In convexity theory, these points form the Choquet boundary. See Section 16.2.

We need a result about the order structure of the set of CP maps on a C*-algebra.

14.10.1. THEOREM. *Let $\mathfrak{B}$ be a unital C*-algebra and let $\varphi : \mathfrak{B} \to \mathcal{B}(H)$ be a completely positive map. Let (π, V) be the minimal Stinespring dilation of φ on K. Then there is an order preserving bijection from $\mathcal{P} := \{0 \le T \le I : T \in \pi(\mathfrak{B})'\}$ onto the set $\{\psi \in \mathrm{CP}(\mathfrak{B}, H) : 0 \le \psi \le \varphi\}$ given by $T \mapsto \varphi_T(a) = V^*T\pi(a)V$.*

PROOF. If $T \in \mathcal{P}$, then

$$\varphi_T(a) = V^*T\pi(a)V = (T^{1/2}V)^*\pi(a)(T^{1/2}V)$$

is CP. Moreover $\varphi - \varphi_T = \varphi_{I-T}$ is also CP. So $0 \le \varphi_T \le \varphi$. If $T \ne 0$, then $T^{1/2}\pi(a)V\xi \ne 0$ for some $a \in \mathfrak{B}$ and $\xi \in H$ since $K = \overline{\pi(\mathfrak{B})VH}$. Thus

$$\langle \varphi_T(a)\xi, \xi\rangle = \langle V^*T\pi(a)V\xi, \xi\rangle = \|T^{1/2}\pi(a)V\xi\|^2 \ne 0.$$

If $S \lneq T$ in $\mathcal{P}$, then $\varphi_T - \varphi_S = \varphi_{T-S} \ne 0$ is CP, so this map preserves order and is injective.

Conversely, suppose that $0 \le \psi \le \varphi$ and let (σ, W) be the minimal Stinespring dilation for ψ on L. Note that $\psi_n \le \varphi_n$ for $n \ge 1$. Define a linear map A by

$$A\sum_{i=1}^{n} \pi(a_i)V\xi_i = \sum_{i=1}^{n} \sigma(a_i)W\xi_i \quad \text{for} \quad n \ge 1,\ a_i \in \mathfrak{B} \text{ and } \xi_i \in H.$$

Since (π, V) is a minimal dilation, this defines A on a dense subset of K with range in L. Compute using $\xi = (\xi_1, \dots, \xi_n) \in H^{(n)}$,

$$\begin{aligned}\Big\|\sum_{i=1}^{n} \sigma(a_i)W\xi_i\Big\|^2 &= \sum_{i,j=1}^{n} \langle W^*\sigma(a_j^*a_i)W\xi_i, \xi_j\rangle = \big\langle [\psi(a_j^*a_i)]\xi, \xi\big\rangle \\ &= \big\langle \psi_n\big[a_j^*a_i\big]\xi, \xi\big\rangle \le \big\langle \varphi_n\big[a_j^*a_i\big]\xi, \xi\big\rangle = \Big\|\sum_{i=1}^{n} \pi(a_i)V\xi_i\Big\|^2.\end{aligned}$$

This shows that $\|A\| \le 1$, and thus it extends by continuity to a contraction in $\mathcal{B}(K, L)$. By construction,

$$AV\xi = A\pi(1)V\xi = \sigma(1)W\xi = W\xi,$$

so that $AV = W$. For $a \in \mathfrak{B}$,

$$A\pi(a)\sum_{i=1}^{n} \pi(a_i)V\xi_i = \sigma(a)\sum_{i=1}^{n} \sigma(a_i)W\xi_i = \sigma(a)A\sum_{i=1}^{n} \pi(a_i)V\xi_i.$$

Thus $A\pi(a) = \sigma(a)A$. Define $T = A^*A$. Then

$$T\pi(a) = A^*A\pi(a) = A^*\sigma(a)A = \pi(a)A^*A = \pi(a)T \quad \text{for} \quad a \in \mathfrak{B}.$$

That is, $T \in \pi(\mathfrak{B})'$. Finally

$$\psi(a) = W^*\sigma(a)W = VA^*\sigma(a)AV = VA^*A\pi(a)V = VT\pi(a)V = \varphi_T(a).$$

Thus $\mathcal{P}$ is mapped onto the interval $[0, \varphi]$. ■

The set $\mathrm{CP}(\mathcal{S}, \mathcal{B}(H))$ is a cone. For a cone, the notion of extreme point is replaced by an extreme ray. A point on such a ray is called pure, as in the following definition.

14.10.2. DEFINITION. Let $\mathcal{S}$ be an operator system. Say that $\varphi \in \mathrm{CP}(\mathcal{S}, \mathcal{B}(H))$ is *pure* if $\psi \in \mathrm{CP}(\mathcal{S}, H)$ and $0 \le \psi \le \varphi$ implies that $\psi = t\varphi$ for some $0 \le t \le 1$.

14.10.3. COROLLARY. *Let $\mathfrak{B}$ be a C*-algebra. A CP map $\varphi \in \mathrm{CP}(\mathfrak{B}, \mathcal{B}(H))$ is pure if and only if its minimal Stinespring dilation (π, V) has an irreducible representation π.*

PROOF. By Theorem 14.10.1, the interval $[0, \varphi]$ is order isomorphic to $\mathcal{P} := \{0 \le T \le I : T \in \pi(\mathfrak{B})'\}$. This consists only of scalars if and only if $\pi(\mathfrak{B})' = \mathbb{C}I$, if and only if π is irreducible. ∎

14.10.4. LEMMA. *Let $\mathcal{S}$ be an operator system. A state $\varphi \in S(\mathcal{S})$ is pure if and only if it is an extreme point of $S(\mathcal{S})$.*

PROOF. If φ is not extreme, then $\varphi = \frac{1}{2}(\varphi_1 + \varphi_2)$ where $\varphi_i \ne \varphi$ are states. Then $0 \le \frac{1}{2}\varphi_1 \le \varphi$, but $\frac{1}{2}\varphi_1$ is not a scalar multiple of φ. So φ is not pure.

Conversely, if φ is extreme and $0 \lneq \psi \lneq \varphi$, let $t = \psi(1)$. Then $\varphi_1 = t^{-1}\psi$ and $\varphi_2 = (1-t)^{-1}(\varphi - \psi)$ are states on $\mathcal{S}$. Observe that $\varphi = t\varphi_1 + (1-t)\varphi_2$. Hence $\varphi_1 = \varphi_2 = \varphi$ and $\psi = t\varphi$. Therefore φ is pure. ∎

14.10.5. DEFINITION. Let $\mathcal{S}$ be an operator system or let $\mathcal{A}$ be a unital operator algebra with $\mathcal{S} = \overline{\mathcal{A} + \mathcal{A}^*}$. An *irreducible* representation π of $C^*(\mathcal{S})$ is a *boundary representation* if $\pi|_{\mathcal{S}}$ has the unique extension property.

14.10.6. LEMMA. *Let $\mathcal{S}$ be an operator system in $\mathfrak{B} = C^*(\mathcal{S})$. If a unital completely positive map φ of $\mathcal{S}$ into $\mathcal{B}(H)$ is pure and maximal, then the unique extension π to $\mathfrak{B}$ is a boundary representation of $\mathcal{S}$.*

PROOF. Maximality of φ corresponds to the unique extension property by Proposition 14.9.5. In particular, φ extends uniquely to a $*$-representation π of $\mathfrak{B}$. If π is not irreducible, there is a non-trivial projection $P \in \pi(\mathfrak{B})'$. Then $\widetilde{\psi}(a) = P\pi(a)$ is CP and $0 \le \widetilde{\psi} \le \pi$. Thus $\psi = \widetilde{\psi}|_{\mathcal{S}}$ satisfies $0 \le \psi \le \varphi$. But $\varphi(1) = I$ and $\psi(1) = P$, so ψ is not a scalar multiple of φ and thus φ is not pure. Hence if φ is pure, then π is irreducible with u.e.p., and thus is a boundary representation. ∎

The following refinement of Lemma 14.9.7 is the key to constructing boundary representations.

14.10.7. LEMMA. *Let $\mathcal{S}$ be an operator system, and let $\varphi : S \to \mathcal{B}(H))$ be a pure u.c.p. map. If $a_0 \in \mathcal{S}$ and $\xi_0 \in H$ and φ is not maximal at (a_0, ξ_0), there is a pure u.c.p. dilation $\psi : \mathcal{S} \to \mathcal{B}(H \oplus \mathbb{C})$ which is maximal at (a_0, ξ_0).*

PROOF. Set $L = \sup_{\varphi \prec \psi} \|\psi(a_0)\xi_0\|$ and $\eta = \big(L^2 - \|\varphi(a_0)\xi_0\|^2\big)^{1/2}$. In Lemma 14.9.7, an extension $\psi : \mathcal{S} \to \mathcal{B}(H \oplus \mathbb{C})$ was constructed with $\|\psi(a_0)\xi_0\| = L$. By conjugating by a unitary $I \oplus \lambda$, we can assume that $\psi(a_0)\xi_0 = \varphi(a_0)\xi_0 \oplus \eta$. Let

$$\Psi = \{\psi \in \mathrm{UCP}(\mathcal{S}, \mathcal{B}(H \oplus \mathbb{C})) : \varphi \prec \psi \text{ and } \psi(a_0)\xi_0 = \varphi(a_0)\xi_0 \oplus \eta\}.$$

Then Ψ is convex and compact in the point-weak-$*$ topology. We claim that Ψ is a face of $\Phi = \{\psi \in \mathrm{UCP}(\mathcal{S}, \mathcal{B}(H \oplus \mathbb{C})) : \varphi \prec \psi\}$. Indeed suppose that $\varphi_i \in \Phi$ and $\psi = \frac{1}{2}(\varphi_1 + \varphi_2) \in \Psi$. Then $\varphi_i(a_0)\xi_0 = \varphi(a_0)\xi_0 \oplus \eta_i$ and $|\eta_i| \le \eta$. However

$$\varphi(a_0)\xi_0 \oplus \eta = \psi(a_0)\xi_0 = \frac{1}{2}(\varphi_1(a_0)\xi_0 + \varphi_2(a_0)\xi_0) = \varphi(a_0)\xi_0 \oplus \frac{\eta_1 + \eta_2}{2}.$$

This forces $\eta_i = \eta$ and so $\varphi_i \in \Psi$. By the Krein–Milman theorem, Ψ has an extreme point ψ_0. To complete the proof, we will show that ψ_0 is pure.

Suppose that ψ_1 is a CP map into $H \oplus \mathbb{C}$ such that $0 \le \psi_1 \le \psi_0$. Define $\psi_2 = \psi_0 - \psi_1$. To avoid the possibility that $\psi_i(1)$ may not be invertible, take a small $\varepsilon > 0$ and use

$$\psi_i' = (1 - 2\varepsilon)\psi_i + \varepsilon\psi_0 \quad \text{for} \quad i = 1, 2.$$

Then $\psi_0 = \psi_1' + \psi_2'$ and $\psi_i'(1) =: Q_i \ge \varepsilon I$. Hence each Q_i is invertible, and $Q_1 + Q_2 = \psi_0(1) = I$. If we show that ψ_1' is a scalar multiple of ψ_0, then the same follows for ψ_1.

Notice that $P_H\psi_i'(\cdot)|_H \le \varphi$. Since φ is pure, there are positive scalars λ_i so that $P_H\psi_i'(\cdot)|_H = \lambda_i\varphi$. Clearly $\lambda_1 + \lambda_2 = 1$ and $\lambda_i \ge \varepsilon$. Thus writing Q_i as a matrix with respect to $H \oplus \mathbb{C}$, there is a vector $\xi_i \in H$ and $\alpha_i \in \mathbb{R}_+$ so that

$$Q_i = \begin{bmatrix} \lambda_i & \lambda_i^{1/2}\xi_i \\ \lambda_i^{1/2}\xi_i^* & \alpha_i \end{bmatrix} = \begin{bmatrix} \lambda_i^{1/2} & 0 \\ \xi_i^* & \beta_i \end{bmatrix} \begin{bmatrix} \lambda_i^{1/2} & \xi_i \\ 0 & \beta_i \end{bmatrix} =: R_i^* R_i$$

where $\beta_i = (\alpha_i - \|\xi_i\|^2)^{1/2} \ge 0$ and

$$R_i = \begin{bmatrix} \lambda_i^{1/2} & \xi_i \\ 0 & \beta_i \end{bmatrix}, \quad \text{and so} \quad R_i^{-1} = \begin{bmatrix} \lambda_i^{-1/2} & -\lambda_i^{-1/2}\beta_i^{-1}\xi_i \\ 0 & \beta_i^{-1} \end{bmatrix}.$$

The positivity and invertibility of Q_i guarantees that $\alpha_i > 0$. Since

$$R_1^* R_1 + R_2^* R_2 = Q_1 + Q_2 = I,$$

we obtain

$$\alpha_1 + \alpha_2 = 1 \quad \text{and} \quad \lambda_1^{1/2}\xi_1 + \lambda_2^{1/2}\xi_2 = 0.$$

Define two maps $\tau_i \in \mathrm{UCP}(\mathcal{S}, \mathcal{B}(H \oplus \mathbb{C}))$ by

$$\tau_i(s) = R_i^{-1*}\psi_i'(s)R_i^{-1} = \begin{bmatrix} \lambda_i^{-1/2} & 0 \\ * & * \end{bmatrix} \begin{bmatrix} \lambda_i\varphi(s) & * \\ * & * \end{bmatrix} \begin{bmatrix} \lambda_i^{-1/2} & * \\ 0 & * \end{bmatrix} =: \begin{bmatrix} \varphi(s) & \Gamma_i(s) \\ \Delta_i(s) & \omega_i(s) \end{bmatrix}.$$

Here $\Gamma_i \in \mathcal{B}(\mathcal{S}, H)$, $\Delta_i \in \mathcal{B}(\mathcal{S}, H^*)$ and ω_i is a state on $\mathcal{S}$. Note that $P_H \tau_i(\cdot)|_H = \varphi$, whence τ_i is a dilation of φ. Therefore $|\Delta_i(a_0)\xi_0| \le \eta$. Compute for $a \in \mathcal{S}$

$$\begin{aligned}\psi_i'(a) = R_i^* \tau_i(a) R_i &= \begin{bmatrix} \lambda_i^{1/2} & 0 \\ \xi_i^* & \beta_i \end{bmatrix} \begin{bmatrix} \varphi(a) & \Gamma_i(a) \\ \Delta_i(a) & \omega_i(a) \end{bmatrix} \begin{bmatrix} \lambda_i^{1/2} & \xi_i \\ 0 & \beta_i \end{bmatrix} \\ &= \begin{bmatrix} \lambda_i \varphi(a) & * \\ \lambda_i^{1/2}(\xi_i^* \varphi(a) + \beta_i \Delta_i(a)) & * \end{bmatrix}.\end{aligned}$$

Consideration of $\psi_i'(a_0)(\xi_0 \oplus 0)$ for $i = 1, 2$ yields

$$\begin{aligned}\eta = P_{\mathbb{C}} \psi_0(a_0)(\xi_0 \oplus 0) &= P_{\mathbb{C}} \psi_1'(a_0)(\xi_0 \oplus 0) + P_{\mathbb{C}} \psi_2'(a_0)(\xi_0 \oplus 0) \\ &= \lambda_1^{1/2}(\xi_1^* \varphi(a_0) + \beta_1 \Delta_1(a_0))\xi_0 + \lambda_2^{1/2}(\xi_2^* \varphi(a_0) + \beta_2 \Delta_2(a_0))\xi_0 \\ &= (\lambda_1^{1/2}\xi_1 + \lambda_2^{1/2}\xi_2)^* \varphi(a_0)\xi_0 + \lambda_1^{1/2}\beta_1 \Delta_1(a_0)\xi_0 + \lambda_2^{1/2}\beta_2 \Delta_2(a_0)\xi_0 \\ &= \lambda_1^{1/2}\beta_1 \Delta_1(a_0)\xi_0 + \lambda_2^{1/2}\beta_2 \Delta_2(a_0)\xi_0.\end{aligned}$$

By $|\Delta_i(a_0)\xi_0| \le \eta$, the formula for β_i and the Cauchy–Schwarz inequality,

$$\begin{aligned}\eta &\le (\lambda_1^{1/2}\beta_1 + \lambda_2^{1/2}\beta_2)\eta \le (\lambda_1^{1/2}\alpha_1^{1/2} + \lambda_2^{1/2}\alpha_2^{1/2})\eta \\ &\le (\lambda_1 + \lambda_2)^{1/2}(\alpha_1 + \alpha_2)^{1/2}\eta = \eta.\end{aligned}$$

Since this is an equality, the first inequality shows that $\Delta_i(a_0)\xi_0 = \eta$, whence $\tau_i \in \Psi$. The second inequality yields that $\beta_i^2 = \alpha_i$, and hence $\xi_i = 0$. The third equality implies that the unit vectors $(\lambda_1^{1/2}, \lambda_2^{1/2})$ and $(\alpha_1^{1/2}, \alpha_2^{1/2})$ are collinear, and hence equal. Thus $\beta_i = \alpha_i^{1/2} = \lambda_i^{1/2}$ and $R_i = \lambda_i^{1/2} I$ is a scalar matrix. Therefore

$$\psi_i'(\cdot) = R_i^* \tau_i(\cdot) R_i = \lambda_i \tau_i(\cdot).$$

Thus $\psi_0 = \lambda_1\tau_1 + \lambda_2\tau_2$ is a convex combination of τ_1 and τ_2. Since ψ_0 is an extreme point of Ψ, we obtain that $\tau_i = \psi_0$. So $\psi_1' = \lambda_1\psi_0$. Therefore ψ_0 is pure. ■

We can now establish the existence of enough boundary representations to completely norm $\mathcal{S}$.

14.10.8. Davidson–Kennedy Theorem. *Let $\mathcal{S}$ be an operator system, and let $\varphi : \mathcal{S} \to \mathcal{B}(H))$ be a pure u.c.p. map. Then φ dilates to a pure, maximal u.c.p. map ψ. Thus ψ extends to a boundary representation of $C^*(\mathcal{S})$.*

Moreover, for any $A = M_n(\mathcal{S})$, there is a boundary representation π such that $\|\pi_n(A)\| = \|A\|$.

Proof. When extending one dimension to achieve maximality at (a, ξ), we will use Lemma 14.10.7. At limit ordinals, we first take a union over all smaller ordinals and check purity. Hence, we obtain a maximal dilation in the same manner as in the proof of the Dritschel–McCullough theorem with the added condition that it is pure. Then Lemma 14.10.6 yields the boundary representation.

Let $A \in M_n(\mathcal{S})$, and normalize it so that $\|A\| = 1$. Define $P \in M_{2n}(\mathcal{S})_+$ by

$$P = \begin{bmatrix} I_n & A \\ A^* & I_n \end{bmatrix} \geq 0_{2n}.$$

We assume that $\mathcal{S} \subset \mathcal{B}(K)$. If $\xi, \eta \in K^{(n)}$ are unit vectors such that $\langle A\eta, \xi\rangle \approx 1$, then

$$\left\langle P \begin{pmatrix} \xi/\sqrt{2} \\ \eta/\sqrt{2} \end{pmatrix}, \begin{pmatrix} \xi/\sqrt{2} \\ \eta/\sqrt{2} \end{pmatrix} \right\rangle = \tfrac{1}{2}\|\xi\|^2 + \mathrm{Re}\langle A\eta, \xi\rangle + \tfrac{1}{2}\|\eta\|^2 \approx 2.$$

Therefore

$$2 \leq \|P\| \leq \left\| \begin{bmatrix} 1 & \|A\| \\ \|A\| & 1 \end{bmatrix} \right\| = 2.$$

Hence there is a state φ on $M_{2n}(C^*(\mathcal{S}))$ such that $\varphi(P) = 2$.

Consider $C = \{\varphi \in S(M_{2n}(\mathcal{S})) : \varphi(P) = 2\}$. This is a weak-$*$ compact convex set of states on $M_{2n}(\mathcal{S})$. Moreover, C is readily seen to be a face of the state space of $M_{2n}(\mathcal{S})$. By the Krein–Milman theorem, there is an extreme point φ_0 in C; so it is extreme in the full state space. Lemma 14.10.4 shows that φ_0 is pure. Dilate φ_0 to a boundary representation σ of $M_{2n}(C^*(\mathcal{S}))$ by the first paragraph of this proof. By Exercise **9.21**, there is an irreducible representation π of $C^*(\mathcal{S})$ such that $\sigma \simeq \pi_{2n}$.

We claim that π has the unique extension property. Let $\varphi = \pi|_{\mathcal{S}}$ and let $\widetilde{\varphi}$ be a CP extension to $C^*(\mathcal{S})$. Then $(\widetilde{\varphi})_{2n}$ is a CP extension of $\varphi_{2n} = \sigma|_{M_{2n}(\mathcal{S})}$, which implies that $(\widetilde{\varphi})_{2n} = \sigma = \pi_{2n}$, and thus $\widetilde{\varphi} = \pi$. Hence π is a boundary representation of $\mathcal{S}$ by Lemma 14.10.6.

Finally, we have

$$\begin{aligned} 2 = \|\sigma(P)\| = \|\pi_{2n}(P)\| &= \left\| \begin{bmatrix} I_n & \pi_n(A) \\ \pi_n(A)^* & I_n \end{bmatrix} \right\| \\ &\leq \left\| \begin{bmatrix} 1 & \|\pi_n(A)\| \\ \|\pi_n(A)\| & 1 \end{bmatrix} \right\| = 1 + \|\pi_n(A)\| \leq 2. \end{aligned}$$

Equality forces $\|\pi_n(A)\| = 1$. ■

14.10.9. COROLLARY. *Let $\mathcal{S}$ be an operator system, and let π be a direct sum of sufficiently many boundary representations to be completely isometric on $\mathcal{S}$. Then $\pi(C^*(\mathcal{S})) \simeq C^*_{\min}(\mathcal{S})$.*

PROOF. The direct sum π of maximal representations is maximal. Indeed, given any dilation ψ of $\pi|_{\mathcal{S}}$, each boundary representation $\sigma|_{\mathcal{S}}$ is dilated by ψ and by maximality, $\sigma|_{\mathcal{S}}$ is a summand of ψ. Thus π is also a summand of ψ, whence it is maximal. By the Davidson–Kennedy theorem, π is completely isometric on $\mathcal{S}$. Thus as in Theorem 14.9.12, $\pi(C^*(\mathcal{S}))$ is isomorphic to $C^*_{\text{min}}(\mathcal{S})$. ■

14.11. Injective Envelopes

In this section, we establish a theorem of Hamana that every operator system is contained in a minimal injective operator system. We recall Definition 14.7.12 that an operator system S is injective if whenever $E \subset F$ are two operator systems and $\varphi : E \to S$ is a completely positive map, there is a completely positive extension $\widetilde{\varphi} : F \to S$.

14.11.1. DEFINITION. Let S be an operator system. An *injective envelope* for S is a pair (T, ι) where T is an injective operator system, and $\iota : S \to T$ is a unital order injection such that if $S \subset T' \subset T$ is an injective operator system, then $T' = T$.

We can assume that $S \subset \mathcal{B}(H)$ with order unit I. Since $\mathcal{B}(H)$ is injective, S is always contained in an injective operator system. Unfortunately, it is not known whether the intersection of a nested family of injective operator systems is injective. So the proof has to be more round-about.

14.11.2. DEFINITION. Let $S \subset \mathcal{B}(H)$ be an operator system. An *S-projection* is a u.c.p. map $\varphi : \mathcal{B}(H) \to \mathcal{B}(H)$ such that $\varphi^2 = \varphi$ and $\varphi|_S = id_S$. If ψ is another S-projection, say that $\varphi \prec \psi$ if $\psi\varphi = \varphi = \varphi\psi$.

An *S-seminorm* on $\mathcal{B}(H)$ has the form $p_\varphi(A) = \|\varphi(A)\|$ for a u.c.p. map $\varphi : \mathcal{B}(H) \to \mathcal{B}(H)$ such that $\varphi|_S = id_S$.

14.11.3. HAMANA THEOREM.
Every operator system $S \subset \mathcal{B}(H)$ has an injective envelope.

Moreover, there is a minimal S-projection φ, and $(T = \text{Ran}\,\varphi, id_S)$ is an injective envelope for S. If (T_1, ι) is another injective envelope for S, there is a unique order isomorphism θ of T onto T_1 extending ι, and it is also a complete isometry. If $T_1 \subset \mathcal{B}(K)$, there is a minimal $\iota(S)$-projection ψ of $\mathcal{B}(K)$ onto T_1.

PROOF. We first show that minimal S-seminorms exist. Consider the set of all S-seminorms, which are ordered by $p_\varphi \le p_\psi$ if $p_\varphi(A) \le p_\psi(A)$ for all $A \in \mathcal{B}(H)$. If $\mathcal{C} = \{p_{\varphi_\lambda}\}_\Lambda$ is a decreasing chain of S-seminorms in this ordering, we show that there is an S-seminorm smaller than every element of $\mathcal{C}$. To this end, consider the net $\{\varphi_\lambda\}_\Lambda$, which all have norm 1. As in the proof of the Arveson extension theorem, there is a cofinal subnet $\{\varphi_\gamma\}_\Gamma$ with point-weak-$*$ limit ψ. This is a u.c.p. map such that $\psi|_S = id_S$. Moreover, for any $\lambda_0 \in \Lambda$, there is $\gamma_0 \in \Gamma$ so that for

$\gamma \geq \gamma_0$, $p_{\psi_\gamma} \leq p_{\varphi_{\lambda_0}}$. Therefore $p_\psi \leq p_{\varphi_{\lambda_0}}$, whence p_ψ is a lower bound for $\mathcal{C}$. By Zorn's lemma, minimal seminorms exist.

Let p_φ be a minimal S-seminorm. We show that φ is an S-projection. Since φ is contractive, $p_{\varphi^k} \leq p_\varphi$ for $k \geq 1$. For $n \geq 1$, let $\psi_n = \frac{1}{n}\sum_{k=1}^n \varphi^k$. Then $p_{\psi_n} \leq p_\varphi$ as well. By the minimality of p_φ, we get $p_{\varphi^k} = p_{\psi_n} = p_\varphi$. Therefore for $A \in \mathcal{B}(H)$,

$$\begin{aligned}\|\varphi(A) - \varphi^2(A)\| &= \|\varphi(A - \varphi(A))\| = \|\psi_n(A - \varphi(A))\| \\ &= \Big\|\frac{1}{n}\sum_{k-1}^{n}\big(\varphi^k(A) - \varphi^{k+1}(A)\big)\Big\| \\ &= \frac{1}{n}\|\varphi(A) - \varphi^{n+1}(A)\| \leq \frac{2\|A\|}{n}.\end{aligned}$$

Letting $n \to \infty$ yields $\varphi^2 = \varphi$.

Next we will show that φ is a minimal S-projection. Indeed, if $\psi \prec \varphi$, then $\psi\varphi = \psi = \varphi\psi$. In particular, $\|\psi(A)\| = \|\psi\varphi(A)\| \leq \|\varphi(A)\|$ for $A \in \mathcal{B}(H)$; so by minimality, $p_\psi = p_\varphi$. Therefore

$$\begin{aligned}\|\varphi(A) - \psi(A)\| &= \|\varphi(\varphi(A) - \psi(A))\| = \|\psi(\varphi(A) - \psi(A))\| \\ &= \|\psi(A) - \psi(A)\| = 0.\end{aligned}$$

That is, $\psi = \varphi$.

Let $T = \varphi(\mathcal{B}(H))$. It is easy to see that T is injective. For if $E \subset F$ are operator systems and $\psi : E \to T$ is a completely positive map, Arveson's extension theorem provides a completely positive extension $\widetilde{\psi} : F \to \mathcal{B}(H)$. Therefore $\varphi\widetilde{\psi} : F \to T$ and for $a \in E$, $\varphi\widetilde{\psi}(a) = id_T\,\psi(a) = \psi(a)$. Thus T is injective.

Suppose that T' is an injective operator space with $E \subset T' \subset T$. By injectivity, $id_{T'}$ extends to a u.c.p. map $\psi : \mathcal{B}(H) \to T'$; so $\psi^2 = \psi$. Since $T' \subset T$, we have $\varphi\psi = \psi$. On the other hand, $p_{\psi\varphi} \leq p_\varphi$ and hence $p_{\psi\varphi} = p_\varphi$. Thus

$$\begin{aligned}\|\varphi(A) - \psi\varphi(A)\| &= \|\varphi(\varphi(A) - \psi\varphi(A))\| = \|\psi\varphi(\varphi(A) - \psi\varphi(A))\| \\ &= \|\psi\varphi(A) - \psi\varphi(A)\| = 0.\end{aligned}$$

Hence $\varphi = \psi\varphi$ and $T \supset T' \supset \psi\varphi(\mathcal{B}(H)) = \varphi(\mathcal{B}(H)) = T$. Therefore T is minimal, and thus is an injective envelope for S.

Let (T_1, ι) be another injective envelope for S. By injectivity, there is a u.c.p. map $\theta : T \to T_1$ extending ι. Similarly, there is a u.c.p. map $\gamma : T_1 \to T$ extending $\iota^{-1} : \iota(S) \to S$. Hence $\gamma\theta : T \to T$ and $\gamma\theta|_S = id_S$. Let $\psi = \gamma\theta\varphi : \mathcal{B}(H) \to T$. Then $p_\psi \leq p_\varphi$, which is minimal; so $p_\psi = p_\varphi$. From the proof above, $\psi = \psi^2$ and $\operatorname{Ran}\psi = \operatorname{Ran}\gamma\theta = T$. Since $\varphi|_T = id_T$,

$$\psi = \psi^2 = \gamma\theta\varphi\psi = (\gamma\theta)\psi.$$

It follows that $\gamma\theta = id_T$.

Therefore θ is a complete order homomorphism and since γ and θ are completely contractive, θ is completely isometric. Hence, $\theta(T) \subset T_1$ is an injective operator system containing $\iota(S)$. As T_1 is an injective envelope, θ is surjective and $\gamma = \theta^{-1}$. Thus θ is a complete order isomorphism. If θ' were another u.c.p. extension of ι, $id_T = \gamma\theta' = \theta^{-1}\theta'$. Hence $\theta' = \theta$, and this map extending ι is unique.

Since $T_1 \subset \mathcal{B}(K)$ is injective, there is a u.c.p. projection φ' of $\mathcal{B}(K)$ onto T_1. As above, there is a u.c.p. map ψ' with $\psi'|_{\iota(S)} = id|_{\iota(S)}$ such that $p_{\psi'}$ is a minimal $\iota(S)$-seminorm. Then $\psi = \varphi'\psi'$ has $p_\psi \leq p_{\psi'}$ and it has range in T_1. By minimality, $p_\psi = p_{\psi'}$ is a minimal $\iota(S)$-seminorm, and ψ is an $\iota(S)$-projection. The range $T' = \operatorname{Ran}\psi$ is an injective operator space with $\iota(S) \subset T' \subset T_1$, and hence $T' = T_1$. Therefore, ψ is the desired $\iota(S)$-projection onto T_1. ∎

The latter parts of the proof show that the injective envelope has certain rigidity properties. This is formulated as follows:

14.11.4. DEFINITION. If S, T are operator systems and $\iota : S \to T$ is a unital complete isometry, we say that (T, ι) is an *extension* of S. An extension (T, ι) of S is *rigid* if a u.c.p. map $\varphi : T \to T$ such that $\varphi|_{\iota(S)} = id_{\iota(S)}$ must be the identity map on T. An extension (T, ι) of S is *essential* if a u.c.p. map $\varphi : T \to M$ such that $\varphi|_{\iota(S)}$ is completely isometric must be completely isometric on T.

14.11.5. COROLLARY. *Let S be an operator system, and let (T, ι) be an injective envelope of S. Then this extension is rigid and essential.*

PROOF. We may suppose that $T \subset \mathcal{B}(H)$ and that $\psi : \mathcal{B}(H) \to T$ is a u.c.p. projection onto T such that p_ψ is a minimal $\iota(S)$-seminorm. Let $\varphi : T \to T$ be a u.c.p. map such that $\varphi|_{\iota(S)} = id_{\iota(S)}$. Consider $\varphi\psi$. Since $p_{\varphi\psi} \leq p_\psi$, one sees that $p_{\varphi\psi}$ is a minimal $\iota(S)$-seminorm. Hence $\varphi\psi$ is a u.c.p. projection onto T (as in the last paragraph of the previous proof). In particular, $\varphi = id_T$, i.e. (T, ι) is a rigid extension of S.

Now suppose that $\varphi : T \to M$ is a u.c.p. map such that $\varphi\iota$ is completely isometric. Since T is injective, the map $\varphi|_{\iota(S)}^{-1} : \varphi\iota(S) \to \iota(S) \subset T$ extends to a u.c.p. map $\gamma : M \to T$. The map $\gamma\varphi : T \to T$ and $\gamma\varphi|_{\iota(S)} = id_{\iota(S)}$. By rigidity, $\gamma\varphi = id_T$. This map is completely isometric, and γ and φ are completely contractive. Therefore, φ must be completely isometric, i.e. (T, ι) is an essential extension of S. ∎

It turns out that these properties can be used to provide other characterizations of the injective envelope. See also Exercise **14.34**.

14.11.6. THEOREM. *Let S be an operator system and let (T, ι) be an extension of S. Then (T, ι) is a minimal injective extension of S, i.e. an injective envelope, if and only if (T, ι) is a maximal essential extension of S.*

PROOF. Assume that (T, ι) is an injective envelope for S, and suppose that (M, γ) is an extension of T such that $(M, \gamma\iota)$ is an essential extension of S. The injectivity of T provides a u.c.p. map $\varphi : M \to T$ such that $\varphi|_{\gamma(T)} = \gamma^{-1}$, i.e. $\varphi\gamma = id_T$. Since $\varphi|_{\gamma\iota(S)}$ is a complete isometry and M is essential, φ is a complete isometry. It follows that $\gamma(T) = M$. Therefore (T, ι) is a maximal essential extension of S.

Conversely suppose that (T, ι) is a maximal essential extension of S. Let $(T_1, \jmath)$ be an injective envelope of S. By injectivity, the u.c.p. map $\jmath\iota^{-1} : \iota(S) \to \jmath(S)$ extends to a u.c.p. map $\varphi : T \to T_1$. Since $\varphi|_{\iota(S)}$ is a complete isometry and T is essential, φ is a complete isometry. Thus (T_1, φ) is an extension of T, and by Corollary 14.11.5, $(T_1, \gamma\iota)$ is an essential extension of S. By maximality of (T, ι), $\varphi(T) = T_1$. Thus T is also an injective envelope of S. ■

The following theorem of Choi and Effros shows that an injective operator system can be endowed with a multiplication that makes it into a C*-algebra. This is called the *Choi–Effros product.*

14.11.7. THEOREM. *Let $S \subset \mathcal{B}(H)$ be an injective operator system, and let $E : \mathcal{B}(H) \to S$ be an S-projection. Define a product on S by $a \circ b = E(ab)$. Then $\mathfrak{A} := (S, \circ)$ is a unital C*-algebra and the identity map $\alpha : S \to \mathfrak{A}$ is a complete order isomorphism.*

If $\beta : S \to \mathfrak{B}$ is a unital complete order isomorphism onto another C-algebra $\mathfrak{B}$, then $\beta\alpha^{-1}$ is a $*$-isomorphism.*

PROOF. The binary operation $\circ$ maps $S \times S$ into S. Clearly $a \circ 1 = a = 1 \circ a$ for $a \in S$. Since E is linear, it is easy to check that this product is distributive. Also $\|a \circ b\| \le \|ab\| \le \|a\|\,\|b\|$. The adjoint operation is a conjugate linear involution, and

$$(a \circ b)^* = E(ab)^* = E((ab)^*) = E(b^*a^*) = b^* \circ a^*.$$

Associativity needs to be checked to make $(S, \circ)$ a Banach $*$-algebra.

Apply the Kadison–Schwarz inequality, Proposition 14.7.11, to E_2 and $B = \left[\begin{smallmatrix} 0 & 0 \\ a^* & x \end{smallmatrix}\right]$ for $a \in S$ and $x \in \mathcal{B}(H)$ to get $E_2(B)^*E_2(B) \le E_2(B^*B)$. And since E_2 is idempotent and completely positive,

$$\begin{aligned} E_2\big(E_2(B)^*E_2(B)\big) &= E_2\Big(\begin{bmatrix} 0 & a \\ 0 & E(x^*) \end{bmatrix} \begin{bmatrix} 0 & 0 \\ a^* & E(x) \end{bmatrix} \Big) \\ &= E_2\Big(\begin{bmatrix} aa^* & aE(x) \\ E(x^*)a^* & E(x^*)E(x) \end{bmatrix} \Big) \\ &= \begin{bmatrix} E(aa^*) & E(aE(x)) \\ E(E(x^*)a^*) & E(E(x^*)E(x)) \end{bmatrix} \\ &\le E_2(B^*B) = E_2\Big(\begin{bmatrix} aa^* & ax \\ x^*a^* & x^*x \end{bmatrix} \Big) = \begin{bmatrix} E(aa^*) & E(ax) \\ E(x^*a^*) & E(x^*x) \end{bmatrix}. \end{aligned}$$

Subtracting leaves

$$0 \leq \begin{bmatrix} 0 & E(ax) - E(aE(x)) \\ E(x^*a^*) - E(E(x^*)a^*) & E(x^*x) - E(E(x^*)E(x)) \end{bmatrix}.$$

Therefore, the off-diagonal terms vanish, so $E(ax) = E(aE(x))$ and $E(x^*a^*) = E(E(x^*)a^*)$.

Let $a, b, c \in S$. Then

$$a \circ (b \circ c) = E(aE(bc)) = E(abc) = E(E(ab)c) = (a \circ b) \circ c.$$

Hence $\circ$ is an associative product. So $\mathfrak{A}$ is a Banach $*$-algebra. The C*-norm condition also follows from the Kadison–Schwarz inequality. For $a \in S$,

$$a^*a = E(a)^*E(a) \leq E(a^*a) = a^* \circ a.$$

Hence $\|a\|^2 = \|a^*a\| \leq \|a^* \circ a\| \leq \|a^*\| \, \|a\| = \|a\|^2$. Therefore $\mathfrak{A}$ is a C*-algebra.

By Proposition 9.1.1, the positive elements of a unital C*-algebra are the self-adjoint elements $a = a^*$ such that $\|t - a\| \leq t$ for $t > \|a\|$. Hence, we see that $S^+ = \mathfrak{A}^+$. That is, the identity map $\alpha : S \to \mathfrak{A}$ is a unital, isometric order isomorphism. Now $M_n(\mathfrak{A})$ is a C*-algebra, and the product is defined by

$$\big[a_{ij}\big] \circ \big[b_{ij}\big] = \Big[\sum_{k=1}^{n} a_{ik} \circ b_{kj}\Big] = \Big[E\Big(\sum_{k=1}^{n} a_{ik}b_{kj}\Big)\Big] = E_n\big(\big[a_{ij}\big]\,\big[b_{ij}\big]\big).$$

Thus the product on $M_n(\mathfrak{A})$ corresponds to the Choi–Effros product on $M_n(S)$ determined by the projection E_n. Hence $\alpha_n : M_n(S) \to M_n(\mathfrak{A})$ is a unital, isometric order isomorphism for all $n \geq 1$. Therefore α is a complete order isomorphism.

Suppose that $\mathfrak{B}$ is another C*-algebra and $\beta : S \to \mathfrak{B}$ is a unital complete order isomorphism. Then $\gamma := \beta\alpha^{-1} : \mathfrak{A} \to \mathfrak{B}$ is a unital complete order isomorphism. Let $a = a^* \in \mathfrak{A}$. Since γ is a bijection, there is some $b \in \mathfrak{A}_{sa}$ so that $\gamma(a)^2 = \gamma(b)$. By the Kadison–Schwarz inequality, $\gamma(a^2) \geq \gamma(a)^2 = \gamma(b)$, and thus $a^2 \geq b$. Reversing the maps yields

$$b = \gamma^{-1}(\gamma(a)^2) \geq \big(\gamma^{-1}(\gamma(a))\big)^2 = a^2.$$

We deduce that $\gamma(a)^2 = \gamma(a^2)$. Therefore, $\mathfrak{A}_{sa}$ is contained in the multiplicative domain of γ, and by Proposition 14.7.11, γ is multiplicative on $\mathfrak{A}$. That is, γ is a $*$-isomorphism of $\mathfrak{A}$ onto $\mathfrak{B}$. ∎

Hamana used the injective envelope to provide the first proof of the existence of the C*-envelope, $C^*_{\text{min}}(S)$.

HAMANA'S PROOF OF THEOREM 14.9.12. Let S be an operator system, and let (T, ι) be an injective envelope. Let $\mathfrak{A} = (T, \circ)$ be the C*-algebra obtained by putting the Choi–Effros product on T, and let $\alpha : T \to \mathfrak{A}$ be the identity map. Since $\mathfrak{A}$ is completely order isomorphic to T, $(\mathfrak{A}, \alpha\iota)$ is also an injective envelope of S. Let $\mathfrak{A}_0 = C^*(\alpha\iota(S))$. This is a C*-cover of S.

To show that $\mathfrak{A}_0 \simeq C^*_{\min}(S)$, we verify the universal property. Let $(\mathfrak{B}, j)$ be a C*-cover of S. We may assume that $\mathfrak{B} \subset \mathcal{B}(H)$. Since $\mathfrak{A}$ is injective, there is a u.c.p. map $\varphi : \mathcal{B}(H) \to \mathfrak{A}$ such that $\varphi|_{j(S)} = \alpha\iota j^{-1}$. And since $\mathcal{B}(H)$ is injective, there is a u.c.p. map $\psi : \mathfrak{A} \to \mathcal{B}(H)$ such that $\psi|_{\alpha\iota(S)} = j(\alpha\iota)^{-1}$. Then $\varphi\psi : \mathfrak{A} \to \mathfrak{A}$ is u.c.p. and satisfies $\varphi\psi|_{\alpha\iota(S)} = id_{\alpha\iota(S)}$. Since $\mathfrak{A}$ is a rigid extension of S by Corollary 14.11.5, $\varphi\psi = id_{\mathfrak{A}}$.

Let $a = a^* \in S$ and let $A = \alpha\iota(a)$. By the Kadison–Schwarz inequality,

$$\varphi(j(a)^2)) \geq \varphi(j(a))^2 = \alpha\iota(a)^2 = A^2.$$

Conversely,

$$\psi(A^2) \geq \psi(A)^2 = j(a)^2.$$

Applying φ, we obtain $A^2 \geq \varphi(j(a)^2) \geq A^2$. Thus $\varphi(j(a)^2) = A^2 = \varphi(j(a))^2$. Therefore, $j(S_{sa})$ is contained in the multiplicative domain of φ. By Proposition 14.7.11, φ is multiplicative on $\mathfrak{B} = C^*(j(S))$. Thus, $\varphi|_{\mathfrak{B}}$ is a $*$-homomorphism into $\mathfrak{A}_0$ such that $\varphi j = \alpha\iota$. This shows that $\mathfrak{A}_0$ is isomorphic to $C^*_{\min}(S)$. ■

14.12. Completely Bounded Maps

In this section, we extend certain results about completely positive maps to completely bounded maps. We show that every c.b. map of an operator space into $\mathcal{B}(H)$ extends to a c.b. map on any C*-algebra containing it without increasing the norm. This yields Wittstock's extension of Stinespring's theorem to c.b. maps. The main application is Paulsen's theorem showing that every completely bounded representation of an operator algebra is similar to a completely contractive one. The proofs make use of an imbedding of any operator space as the corner of an operator system.

Let $\mathcal{A}$ be an operator space contained in a C*-algebra $\mathfrak{A}$, which we may assume to be unital. Form the *Paulsen system* in $M_2(\mathfrak{A})$:

$$\mathcal{S} = \left\{ \begin{bmatrix} \lambda & a \\ b^* & \mu \end{bmatrix} : \lambda, \mu \in \mathbb{C},\ a, b \in \mathcal{A} \right\}.$$

This is a closed, unital, self-adjoint subspace of $M_2(\mathfrak{A})$, and thus is an operator system. The main tool connecting cb maps to cp maps is the following lemma.

14.12.1. Lemma. *Let $\mathcal{A}$ be an operator space contained in a unital C*-algebra $\mathfrak{A}$, and let $\mathcal{S}$ be its Paulsen system. If $\varphi : \mathcal{A} \to \mathcal{B}(H)$ is a completely bounded map with $\|\varphi\|_{cb} \leq 1$, define $\Phi : \mathcal{S} \to \mathcal{B}(H \oplus H)$ by*

$$\Phi\left(\begin{bmatrix} \lambda & a \\ b^* & \mu \end{bmatrix}\right) = \begin{bmatrix} \lambda I & \varphi(a) \\ \varphi(b)^* & \mu I \end{bmatrix} \quad \textit{for} \quad \lambda, \mu \in \mathbb{C} \textit{ and } a, b \in \mathcal{A}.$$

Then Φ is unital and completely positive.

PROOF. A typical element of $M_n(\mathcal{S})$ is an $n \times n$ matrix with 2×2 entries. The *canonical shuffle* does a symmetric interchange of rows and simultaneously columns to produce a 2×2 matrix with $n \times n$ matrix entries, yielding the form:

$$S = \begin{bmatrix} \Lambda & A \\ B^* & M \end{bmatrix} \quad \text{where} \quad \Lambda, M \in M_n \text{ and } A, B \in M_n(\mathcal{A}).$$

Suppose that $S \geq 0$, so that in particular $\Lambda \geq 0$, $M \geq 0$ and $B = A$. Then for $\varepsilon > 0$ and setting $X_\varepsilon = (\Lambda + \varepsilon I_n)^{-1/2} A (M + \varepsilon I_n)^{-1/2}$, we get

$$\begin{aligned} 0 &\leq S + \varepsilon I_{2n} \\ &= \begin{bmatrix} (\Lambda+\varepsilon I_n)^{1/2} & 0 \\ 0 & (M+\varepsilon I_n)^{1/2} \end{bmatrix} \begin{bmatrix} I_n & X_\varepsilon \\ X_\varepsilon^* & I_n \end{bmatrix} \begin{bmatrix} (\Lambda+\varepsilon I_n)^{1/2} & 0 \\ 0 & (M+\varepsilon I_n)^{1/2} \end{bmatrix}. \end{aligned}$$

As we observed in Proposition 14.6.4, this implies that $\|X_\varepsilon\| \leq 1$. Hence

$$\|(\Lambda + \varepsilon I_n)^{-1/2} \varphi_n(A) (M + \varepsilon I_n)^{-1/2}\| = \|\varphi_n(X_\varepsilon)\| \leq 1.$$

Therefore

$$\begin{aligned} &\Phi_n(S + \varepsilon I_{2n}) \\ &= \begin{bmatrix} (\Lambda+\varepsilon I_n)^{1/2} & 0 \\ 0 & (M+\varepsilon I_n)^{1/2} \end{bmatrix} \begin{bmatrix} I_n & \varphi_n(X_\varepsilon) \\ \varphi_n(X_\varepsilon)^* & I_n \end{bmatrix} \begin{bmatrix} (\Lambda+\varepsilon I_n)^{1/2} & 0 \\ 0 & (M+\varepsilon I_n)^{1/2} \end{bmatrix}. \end{aligned}$$

The matrix in the centre is positive, and hence $\Phi_n(S + \varepsilon I_{2n}) \geq 0$. Letting $\varepsilon \to 0$ shows that Φ is completely positive. ∎

If $\varphi : \mathcal{A} \to \mathcal{B}(H)$, define $\varphi^* : \mathcal{A}^* \to \mathcal{B}(H)$ by $\varphi^*(b) := \varphi(b^*)^*$. When $\mathcal{A} = \mathcal{A}^*$, we can define $\operatorname{Re}\varphi := \frac{1}{2}(\varphi + \varphi^*)$ and $\operatorname{Im}\varphi = \frac{1}{2i}(\varphi - \varphi^*)$. Then $\operatorname{Re}\varphi = (\operatorname{Re}\varphi)^*$, $\operatorname{Im}\varphi = (\operatorname{Im}\varphi)^*$ and $\varphi = \operatorname{Re}\varphi + i \operatorname{Im}\varphi$.

14.12.2. COROLLARY. *Let $\mathcal{A}$ be an operator space contained in a unital C*-algebra $\mathfrak{A}$. If $\varphi : \mathcal{A} \to \mathcal{B}(H)$ is a completely bounded map with $\|\varphi\|_{cb} \leq 1$, then there are u.c.p. maps ψ_1 and ψ_2 from $\mathfrak{A}$ into $\mathcal{B}(H)$ and a c.b. map $\tilde{\varphi} : \mathfrak{A} \to \mathcal{B}(H)$ extending φ such that $\Psi : M_2(\mathfrak{A}) \to \mathcal{B}(H \oplus H)$ given by*

$$\Psi\left(\begin{bmatrix} a & b \\ c & d \end{bmatrix}\right) = \begin{bmatrix} \psi_1(a) & \tilde{\varphi}(b) \\ \tilde{\varphi}^*(c) & \psi_2(d) \end{bmatrix}$$

is a u.c.p. map and $\|\tilde{\varphi}\|_{cb} \leq 1$. Hence, every c.b. map from $\mathcal{A}$ to $\mathcal{B}(H)$ extends to a c.b. map on $\mathfrak{A}$ of the same c.b. norm.

PROOF. Let $\mathcal{S}$ be the Paulsen system for $\mathcal{A}$, and let $\Phi : \mathcal{S} \to \mathcal{B}(H \oplus H)$ be the u.c.p. map constructed in Lemma 14.12.1. Since $\mathcal{S} \subset M_2(\mathfrak{A})$, Arveson's extension theorem provides a u.c.p. extension $\Psi : M_2(\mathfrak{A}) \to \mathcal{B}(H \oplus H)$. Let E_{ij} be the matrix units in $M_2(\mathbb{C}) \subset M_2(\mathfrak{A})$. Note that $\Phi(E_{ii}) = E_{ii} \otimes I$ in $\mathcal{B}(H \oplus H)$. So Φ is multiplicative on the C*-algebra $\mathfrak{C} = \mathbb{C}E_{11} + \mathbb{C}E_{22}$, whence $\mathfrak{C}$ is contained in the multiplicative domain of Ψ. By Proposition 14.7.11, Ψ is a $\mathfrak{C}$ bimodule map. Therefore $\Psi(E_{ii}AE_{jj}) = (E_{ii} \otimes I)\Psi(A)(E_{jj} \otimes I)$, meaning that

$$\Psi\left(\begin{bmatrix} a & b \\ c & d \end{bmatrix}\right) = \begin{bmatrix} \psi_1(a) & \tilde{\varphi}(b) \\ \varphi_{21}(c) & \psi_2(d) \end{bmatrix}.$$

Moreover ψ_1 and ψ_2 are u.c.p. , and $\tilde{\varphi}$ extends φ. Since Ψ is self-adjoint, we have $\varphi_{21} = \tilde{\varphi}^*$. As Φ is unital, we have $\|\Psi\|_{cb} = \|\Psi(I)\| = 1$, and thus $\|\tilde{\varphi}\|_{cb} \leq 1$.

For an arbitrary c.b. map, normalize so that $\|\varphi\|_{cb} = 1$. ■

14.12.3. COROLLARY. *Let $\mathfrak{A}$ be a C*-algebra. Then*

$$\mathrm{CB}(\mathfrak{A}, \mathcal{B}(H)) = \mathrm{span}\{\mathrm{CP}(\mathfrak{A}, \mathcal{B}(H))\}.$$

Moreover, if $\varphi \in \mathrm{CB}(\mathfrak{A}, \mathcal{B}(H))$, there exists $\psi \in \mathrm{CP}(\mathfrak{A}, \mathcal{B}(H))$ with $\|\psi\|_{cb} = \|\varphi\|_{cb}$ such that $\psi \pm \mathrm{Re}\,\varphi$ and $\psi \pm \mathrm{Im}\,\varphi$ are all completely positive.

PROOF. Let $\varphi : \mathfrak{A} \to \mathcal{B}(H)$ be completely bounded, and scale it so that $\|\varphi\|_{cb} = 1$. Let ψ_1, ψ_2 be the u.c.p. maps from the previous corollary, and define the u.c.p. map $\psi = \frac{1}{2}(\psi_1 + \psi_2)$. If $\eta \in \{1, -1, i, -i\}$, then

$$0 \leq \begin{bmatrix} 1 \\ \eta \end{bmatrix}^* \begin{bmatrix} \psi_1 & \varphi \\ \varphi^* & \psi_2 \end{bmatrix} \begin{bmatrix} 1 \\ \eta \end{bmatrix} = (\psi_1 + \psi_2) + (\eta\varphi + \overline{\eta}\varphi^*) = 2\psi + (\eta\varphi + \overline{\eta}\varphi^*).$$

For $\eta = \pm 1$, we get $\rho_\pm := \psi \pm \mathrm{Re}\,\varphi$ are completely positive, and for $\eta = \pm i$, we get $\sigma_\pm := \psi \pm \mathrm{Im}\,\varphi$ are completely positive. Therefore

$$\varphi = \frac{1}{2}\rho_+ - \frac{1}{2}\rho_- + \frac{i}{2}\sigma_+ - \frac{i}{2}\sigma_-.$$

Hence $\mathrm{CB}(\mathfrak{A}, \mathcal{B}(H)) = \mathrm{span}\{\mathrm{CP}(\mathfrak{A}, \mathcal{B}(H))$. ■

The following variant of the Stinespring dilation theorem is very useful.

14.12.4. WITTSTOCK DILATION THEOREM. *Let $\mathfrak{A}$ be a C*-algebra, and let $\varphi : \mathfrak{A} \to \mathcal{B}(H)$ be a completely bounded map. Then there is a $*$-representation $\pi : \mathfrak{A} \to \mathcal{B}(K)$ and linear maps $V_1, V_2 \in \mathcal{B}(H, K)$ such that $\varphi(a) = V_1^*\pi(a)V_2$ and $\|V_1\|\,\|V_2\| = \|\varphi\|_{cb}$. If $\|\varphi\|_{cb} = 1$, we can choose V_i to be isometries.*

PROOF. Normalize so that $\|\varphi\|_{cb} = 1$. By Corollary 14.12.2, there is a u.c.p. map $\Psi : M_2(\mathfrak{A}) \to \mathcal{B}(H \oplus H)$ such that

$$\Psi\left(\begin{bmatrix} a & b \\ c & d \end{bmatrix}\right) = \begin{bmatrix} \psi_1(a) & \varphi(b) \\ \varphi^*(c) & \psi_2(d) \end{bmatrix}.$$

By Stinespring's theorem, there is a $*$-representation $\pi : M_2(\mathfrak{A}) \to \mathcal{B}(K)$ and an isometry $V : H \oplus H \to K$ so that $\Psi(A) = V^*\pi(A)V$.

Now $P_1 := \pi(E_{11})$ and $P_2 := \pi(E_{22})$ are complementary projections. Let $K_i = \pi(E_{ii})K$. Note that $U := \pi(E_{12})$ is a partial isometry mapping K_1 isometrically onto K_2. Also observe that $\pi_i : \mathfrak{A} \to \mathcal{B}(K_i)$ by $\pi_i(a) = \pi(aE_{ii})|_{K_i}$ are $*$-representations for $i = 1, 2$. Moreover

$$U^*\pi_2(a)U|_{K_1} = \pi(E_{21})\pi(aE_{22})\pi(E_{12})|_{K_1} = \pi(aE_{11})|_{K_1} = \pi_1(a).$$

It follows that

$$(P_1 \oplus U^*)\pi(A)(P_1 \oplus U) = \pi_1^{(2)}(A) = \begin{bmatrix} \pi_1(a) & \pi_1(b) \\ \pi_1(c) & \pi_1(d) \end{bmatrix}.$$

We will assume that this identification has been made.

As $\Psi(E_{ii}) = E_{ii} \otimes I = V^* P_i V$, we get $V(H \oplus 0) \subset K_1$ and $V(0 \oplus H) \subset K_2$. So we can write $V = \begin{bmatrix} V_1 & 0 \\ 0 & V_2 \end{bmatrix}$ where $V_i = V|_{\operatorname{Ran}\Psi(E_{ii})} \in \mathcal{B}(H, K_i)$. Therefore

$$\begin{bmatrix} 0 & \varphi(b) \\ 0 & 0 \end{bmatrix} = \begin{bmatrix} V_1^* & 0 \\ 0 & V_2^* \end{bmatrix} \begin{bmatrix} 0 & \pi_1(b) \\ 0 & 0 \end{bmatrix} \begin{bmatrix} V_1 & 0 \\ 0 & V_2 \end{bmatrix} = \begin{bmatrix} 0 & V_1^*\pi_1(b)V_2 \\ 0 & 0 \end{bmatrix}.$$

Hence $\varphi(b) = V_1^*\pi_1(b)V_2$. By construction, V_1, V_2 are isometries. ■

To see that there are "lots" of c.b. maps, we establish the following.

14.12.5. Proposition. *Let $\mathcal{A}$ be an operator space, and let $A \in M_n(\mathcal{A})$. Then there is a map $\varphi \in \mathrm{CB}(\mathcal{A}, M_{2n})$ with $\|\varphi\|_{cb} = 1$ such that $\|\varphi_n(A)\| = \|A\|$.*

Proof. We may suppose that $\mathcal{A}$ is contained in a C*-algebra $\mathfrak{A}$, and hence $A \in M_n(\mathfrak{A})$. There is a state f on $M_n(\mathfrak{A})$ such that $f(A^*A) = \|A\|^2$. Let π_f be the GNS representation of f with cyclic vector $\xi \in H_f$. We have

$$\|A\|^2 = \langle \pi_f(A^*A)\xi, \xi\rangle = \langle \pi(A)\xi, \pi(A)\xi\rangle = \|\pi(A)\xi\|^2.$$

Hence $\eta = A\xi$ has $\|\eta\| = \|A\|$.

Then there is a representation σ of $\mathfrak{A}$ on a Hilbert space K such that $\pi_f \simeq \sigma_n$ (see Exercise **9.21**). Thus, we can identify H_f with $K^{(n)}$, which matches ξ with $(\xi_1, \dots, \xi_n)$ and η with $(\eta_1, \dots, \eta_n)$. Let $M = \operatorname{span}\{\xi_1, \dots, \xi_n, \eta_1, \dots, \eta_n\}$. Then $\varphi(a) = P_M\pi(a)|_M$ is a u.c.p. map such that

$$\|A\| \geq \|\varphi_n(A)\| \geq \Big|\Big\langle \varphi_n(A)\xi, \frac{\eta}{\|\eta\|}\Big\rangle\Big| = \|A\|.$$

The restriction to $\mathcal{A}$ is a complete contraction in $\mathrm{CB}(\mathcal{A}, M_{2n})$. ■

An important application of these ideas is a generalization of Arveson's dilation theorem from the completely contractive case to the completely bounded case. In this situation, there cannot be a dilation to the C*-algebra, but we can show that such a representation is similar to one that does have a dilation.

14.12.6. Paulsen Theorem. *Let $1 \in \mathcal{A} \subset \mathfrak{B}$ be a unital subalgebra of a C*-algebra $\mathfrak{B}$. Suppose that $\rho : \mathcal{A} \to \mathcal{B}(H)$ is a unital representation. Then ρ is completely bounded if and only if there is an invertible operator $S \in \mathcal{B}(H)$*

such that $\operatorname{Ad} S \circ \rho$ *is completely contractive. In this case,* S *can be chosen so that* $\|S\|\,\|S^{-1}\| = \|\rho\|_{cb}$.

PROOF. If $\|\operatorname{Ad} S \circ \rho\|_{cb} = 1$, let $\sigma = \operatorname{Ad} S \circ \rho$. For $A \in M_n(\mathcal{A})$,

$$\|\rho_n(A)\| = \|S^{-1\,(n)}\sigma_n(A)S^{(n)}\| \le \|S^{-1}\|\|\sigma_n(A)\|\|S\| \le \|S\|\,\|S^{-1}\|\|A\|$$

Therefore $\|\rho\|_{cb} \le \|S\|\,\|S^{-1}\|$.

Now suppose that $\|\rho\|_{cb} < \infty$. By Corollary 14.12.2, we can extend ρ to a c.b. map $\tilde{\rho} : \mathfrak{B} \to \mathcal{B}(H)$ with $\|\tilde{\rho}\|_{cb} = \|\rho\|_{cb}$. Then by Wittstock's dilation theorem, there is a $*$-representation $\pi : \mathfrak{B} \to \mathcal{B}(K)$ and $V_1, V_2 \in \mathcal{B}(H, K)$ so that $\tilde{\rho}(b) = V_1^*\pi(b)V_2$ and $\|V_1\|\,\|V_2\| = \|\rho\|_{cb}$.

Define a new norm on H by

$$|\xi| = \inf\Big\{\Big\|\sum_{i=1}^{n} \pi(a_i)V_2\xi_i\Big\| : \xi = \sum_{i=1}^{n} \rho(a_i)\xi_i,\ a_i \in \mathcal{A},\ \xi_i \in H\Big\}.$$

Since $\xi = \rho(1)\xi$, we have $|\xi| \le \|V_2\xi\| \le \|V_2\|\,\|\xi\|$. Conversely, if

$$\xi = \sum_{i=1}^{n} \rho(a_i)\xi_i = V_1 \sum_{i=1}^{n} \pi(a_i)V_2\xi_i,$$

then

$$\|\xi\| \le \|V_1\|\Big\|\sum_{i=1}^{n} \pi(a_i)V_2\xi_i\Big\|.$$

Taking an infimum over all representations of ξ yields $\|\xi\| \le \|V_1\|\,|\xi|$. Hence $|\cdot|$ is an equivalent norm to $\|\cdot\|$.

We claim that $|\cdot|$ is a Hilbert space norm. To verify this, we use the parallelogram law (see Exercise **2.8**). Let $\xi, \eta \in H$ and write $\xi = \sum_{i=1}^{n} \rho(a_i)\xi_i$ and $\eta = \sum_{j=1}^{m} \rho(b_j)\eta_j$ for $a_i, b_j \in \mathcal{A}$ and $\xi_i, \eta_j \in H$. Then using the parallelogram law in K,

$$\begin{aligned}|\xi+\eta|^2 + |\xi-\eta|^2 &\le \Big\|\sum_{i=1}^{n} \pi(a_i)V_2\xi_i + \sum_{j=1}^{m} \pi(b_j)V_2\eta_j\Big\|^2 \\ &\qquad + \Big\|\sum_{i=1}^{n} \pi(a_i)V_2\xi_i - \sum_{j=1}^{m} \pi(b_j)V_2\eta_j\Big\|^2 \\ &= 2\Big\|\sum_{i=1}^{n} \pi(a_i)V_2\xi_i\Big\|^2 + 2\Big\|\sum_{j=1}^{m} \pi(b_j)V_2\eta_j\Big\|^2\end{aligned}$$

Taking an infimum over all representations of ξ and η yields

$$|\xi+\eta|^2 + |\xi-\eta|^2 \le 2|\xi|^2 + 2|\eta|^2.$$

Replacing ξ and η by $\xi+\eta$ and $\xi-\eta$ yields the reverse inequality. Therefore, the parallelogram law holds in $H' := (H, |\cdot|)$, and hence H' is a Hilbert space.

Define $S \in \mathcal{B}(H', H)$ by $S\xi = \xi$. Then $\|S\| \le \|V_1\|$ and $\|S^{-1}\| \le \|V_2\|$. Define $\sigma : \mathcal{A} \to \mathcal{B}(H')$ by $\sigma(a) = S^{-1}\rho(a)S$. For $a \in \mathcal{A}$ and $\xi \in H'$,

$$\begin{aligned}|\sigma(a)\xi| = |S^{-1}\rho(a)S\xi| &= \inf\Big\{\Big\|\sum_{i=1}^{n}\pi(a_i)V_2\xi_i\Big\| : \rho(a)\xi = \sum_{i=1}^{n}\rho(a_i)\xi_i\Big\}\\ &\le \inf\Big\{\Big\|\sum_{i=1}^{n}\pi(a)\pi(b_i)V_2\xi_i\Big\| : \xi = \sum_{i=1}^{n}\rho(b_i)\xi_i\Big\}\\ &\le \|\pi(a)\|\inf\Big\{\Big\|\sum_{i=1}^{n}\pi(b_i)V_2\xi_i\Big\| : \xi = \sum_{i=1}^{n}\rho(b_i)\xi_i\Big\}\\ &= \|\pi(a)\|\,|\xi| \le \|a\|\,|\xi|.\end{aligned}$$

Thus $\|\sigma\| = 1$.

Similarly, for $A = \begin{bmatrix} a_{ij}\end{bmatrix} \in M_n(\mathcal{A})$ and $\xi = (\xi_1, \dots, \xi_n) \in H'^{(n)}$, computing $|\rho_n(A)\xi|$ involves an infimum over all representations of this vector, which is smaller than taking the infimum over all representations of ξ, say as $\xi_j = \sum_{k=1}^{m}\rho(b_{jk})\eta_{jk}$. Thus

$$\begin{aligned}|\sigma_n(A)\xi|_{H'^{(n)}} = \big|S^{-1(n)}\rho_n(A)S^{(n)}\xi\big|_{H'^{(n)}} &= \left|\rho_n(A)\begin{bmatrix}\xi_1\\ \vdots\\ \xi_n\end{bmatrix}\right|_{H'^{(n)}}\\ &\le \inf\left\{\left\|\pi_n(A)\begin{bmatrix}\sum_{k=1}^{m}\pi(b_{1k})V_2\eta_{1k}\\ \vdots\\ \sum_{k=1}^{m}\pi(b_{nk})V_2\eta_{nk}\end{bmatrix}\right\|\right\}\\ &\le \|\pi_n(A)\|\inf\left\{\left\|\begin{bmatrix}\sum_{k=1}^{m}\pi(b_{1k})V_2\eta_{1k}\\ \vdots\\ \sum_{k=1}^{m}\pi(b_{nk})V_2\eta_{nk}\end{bmatrix}\right\|\right\}\\ &= \|\pi_n(A)\|\,|\xi|_{H'^{(n)}} \le \|A\|\,|\xi|_{H'^{(n)}}.\end{aligned}$$

Hence $\|\sigma\|_{cb} = 1$, i.e. σ is completely contractive. Therefore, $\rho = S\sigma S^{-1}$ is similar to a completely contractive representation and $\|S\|\,\|S^{-1}\| \le \|V_1\|\,\|V_2\| = \|\rho\|_{cb}$. The reverse inequality follows from the first paragraph. ∎

14.13. Some Similarity Problems

We will provide an overview of two famous problems related to similarity. The first, due to Kadison in 1955, remains open despite deep work by Haagerup. The second, the Sz.-Nagy–Halmos problem, was finally resolved negatively by Pisier.

14.13.1. KADISON PROBLEM. *Is every continuous (non-$*$) representation of a C*-algebra similar to a $*$-representation?*

Important partial progress has been made on this problem. See [239] for an overview. Christensen and Bunce independently showed that every representation of a nuclear C*-algebra is similar to a $*$-representation. Haagerup established the deep result that every cyclic representation of a C*-algebra is similar to a $*$-representation. We will record one result which is immediate from Paulsen's theorem (though it was established independently). It follows immediately because a contractive representation of a C*-algebra is a $*$-representation.

14.13.2. HAAGERUP THEOREM. *A representation ρ of a C*-algebra is similar to a $*$-representation if and only if $\|\rho\|_{cb} < \infty$.*

14.13.3. DEFINITION. Let $\mathfrak{A}$ be a C*-algebra and let $\pi : \mathfrak{A} \to \mathcal{B}(H)$ be a $*$-representation. A linear map $\delta : \mathfrak{A} \to \mathcal{B}(H)$ is a *π-derivation* if

$$\delta(ab) = \pi(a)\delta(b) + \delta(a)\pi(b) \quad \text{for} \quad a, b \in \mathfrak{A}.$$

The π-derivation δ is **-derivation* if $\delta(a^*) = \delta(a)^*$ for $a \in \mathfrak{A}$. It is *inner* if there is an element $X \in \mathcal{B}(H)$ such that $\delta(a) = \pi(a)X - X\pi(a)$ for $a \in \mathfrak{A}$.

Note that

$$\begin{aligned} \pi(a)\delta_X(b) + \delta_X(a)\pi(b) &= \pi(a)(\pi(b)X - X\pi(b)) + (\pi(a)X - X\pi(a))\pi(b) \\ &= \pi(ab)X - X\pi(ab) = \delta_X(ab). \end{aligned}$$

Hence $\delta_X(a) = \pi(a)X - X\pi(a)$ is a π-derivation.

The relationship between derivations and homomorphisms arises from the fact that if δ is a π-derivation, then

$$\sigma(a) = \begin{bmatrix} \pi(a) & \delta(a) \\ 0 & \pi(a) \end{bmatrix}$$

is a representation of $\mathfrak{A}$ on $H \oplus H$. Indeed,

$$\begin{aligned} \sigma(a)\sigma(b) &= \begin{bmatrix} \pi(a) & \delta(a) \\ 0 & \pi(a) \end{bmatrix} \begin{bmatrix} \pi(b) & \delta(b) \\ 0 & \pi(b) \end{bmatrix} \\ &= \begin{bmatrix} \pi(ab) & \pi(a)\delta(b) + \delta(a)\pi(b) \\ 0 & \pi(ab) \end{bmatrix} = \sigma(ab). \end{aligned}$$

14.13.4. PROPOSITION. *Let $\mathfrak{A}$ be a C*-algebra and let $\pi : \mathfrak{A} \to \mathcal{B}(H)$ be a $*$-representation. A derivation $\delta : \mathfrak{A} \to \mathcal{B}(H)$ is inner if and only if σ is similar to a $*$-representation.*

PROOF. If $\delta = \delta_X$, then

$$\begin{bmatrix} I & -X \\ 0 & I \end{bmatrix} \begin{bmatrix} \pi(a) & 0 \\ 0 & \pi(a) \end{bmatrix} \begin{bmatrix} I & X \\ 0 & I \end{bmatrix} = \begin{bmatrix} \pi(a) & \pi(a)X - X\pi(a) \\ 0 & \pi(a) \end{bmatrix} = \sigma(a).$$

Thus σ is similar to the $*$-representation $\pi \oplus \pi$.

Conversely suppose that $\sigma(a) = S\rho(a)S^{-1}$ for some $*$-homomorphism ρ. Let $T = SS^*$. Then

$$\sigma(a)T = S\rho(a)S^* = S\rho(a^*)^*S^* = SS^*(S\rho(a^*)S^{-1})^* = T\sigma(a^*)^*.$$

Writing $T = \begin{bmatrix} T_{11} & T_{12} \\ T_{21} & T_{22} \end{bmatrix}$, we have

$$\begin{bmatrix} \pi(a) & \delta(a) \\ 0 & \pi(a) \end{bmatrix} \begin{bmatrix} T_{11} & T_{12} \\ T_{21} & T_{22} \end{bmatrix} = \begin{bmatrix} T_{11} & T_{12} \\ T_{21} & T_{22} \end{bmatrix} \begin{bmatrix} \pi(a) & 0 \\ \delta(a^*)^* & \pi(a) \end{bmatrix},$$

that is,

$$\begin{bmatrix} \pi(a)T_{11} + \delta(a)T_{21} & \pi(a)T_{12} + \delta(a)T_{22} \\ \pi(a)T_{21} & \pi(a)T_{22} \end{bmatrix} = \begin{bmatrix} T_{11}\pi(a) + T_{12}\delta(a^*)^* & T_{12}\pi(a) \\ T_{21}\pi(a) + T_{22}\delta(a^*)^* & T_{22}\pi(a) \end{bmatrix}.$$

Thus $\pi(a)T_{22} = T_{22}\pi(a)$ and from the $1, 2$ entry, we obtain

$$\delta(a)T_{22} = T_{12}\pi(a) - \pi(a)T_{12}.$$

Now T is positive and invertible, so T_{22} is also invertible. Since T_{22} commutes with $\pi(a)$, so does T_{22}^{-1}, and we have

$$\delta(a) = T_{12}T_{22}^{-1}\pi(a) - \pi(a)T_{12}T_{22}^{-1}.$$

That is, $\delta = \delta_X$ is inner, where $X = T_{12}T_{22}^{-1}$. ■

An immediate consequence of Haagerup's theorem yields the following consequence due to Christensen.

14.13.5. COROLLARY. *Let $\mathfrak{A}$ be a C*-algebra and let $\pi : \mathfrak{A} \to \mathcal{B}(H)$ be a $*$-representation. A derivation $\delta : \mathfrak{A} \to \mathcal{B}(H)$ is inner if and only if it is completely bounded.*

A deep result of Kirchberg shows that if every derivation of a C*-algebra relative to any $*$-representation is inner, then every representation is similar to a $*$-representation.

Now we turn to another famous problem. This one has been resolved.

14.13.6. DEFINITION. An operator $T \in \mathcal{B}(H)$ is *polynomially bounded* if the representation $\rho_T(p) = p(T)$ extends to a bounded representation of $A(\mathbb{D})$, i.e. there is a constant C so that

$$\|p(T)\| \le C\|p\|_\infty = C \sup_{z \in \overline{\mathbb{D}}} |p(z)| \quad \text{for} \quad p \in \mathbb{C}[z].$$

The operator T is *completely polynomially bounded* if ρ_T extends to a completely bounded representation of $A(\mathbb{D})$, i.e. there is a constant C so that

$$\big\|\,[p_{ij}(T)]\,\big\| \le C\big\|\,[p_{ij}]\,\big\|_\infty = C \sup_{z\in\overline{\mathbb{D}}} \big\|\,[p_{ij}(z)]\,\big\| \quad \text{for} \quad [p_{ij}] \in M_n(\mathbb{C}[z]),\ n \ge 1.$$

This famous problem was open for many years:

14.13.7. Sz.-Nagy–Halmos Problem. *Is every polynomially bounded operator similar to a contraction?*

Observe that $\operatorname{Ad} S \circ \rho_T = \rho_{STS^{-1}}$. If $\|STS^{-1}\| \le 1$, then $\|\rho_{STS^{-1}}\|_{cb} = 1$ by the extended von Neumann inequality. Conversely if $\|\rho_{STS^{-1}}\| = 1$, then clearly $\|STS^{-1}\| \le 1$.

Dixmier's theorem 9.13.8 applies to bounded representations of the group $\mathbb{Z}$ and yields a result of Sz.-Nagy:

14.13.8. Theorem. *An invertible operator $T \in \mathcal{B}(H)$ is similar to a unitary operator if and only if* $\sup_{n\in\mathbb{Z}} \|T^n\| < \infty$.

Proof. Indeed if $\pi : \mathbb{Z} \to \mathcal{B}(H)$ given by $\pi(n) = T^n$ is bounded, then Dixmier's theorem shows that it is similar to a unitary representation. The converse is trivial. ∎

An operator is called *power bounded* if $\sup_{n\ge 1} \|T^n\| < \infty$. Sz.-Nagy asked if every power bounded operator is similar to a contraction. A counterexample was soon found. The revised version of the conjecture was raised by Halmos.

14.13.9. Example (Foguel). Let S be the unilateral shift on l_2 and let Q be the orthogonal projection onto $\operatorname{span}\{\xi_{3^k} : k \ge 0\}$. Define

$$T = \begin{bmatrix} S^* & Q \\ 0 & S \end{bmatrix} \quad \text{and} \quad Q = \sum_{k\ge 0} \xi_{3^k}\xi_{3^k}^*.$$

Then

$$T^n = \begin{bmatrix} S^{*n} & Q_n \\ 0 & S^n \end{bmatrix} \text{ and } Q_n = \sum_{i=0}^{n-1} S^{*i} Q S^{n-1-i} = \sum_{k\ge 0}\sum_{i=0}^{n-1} S^{*i}\xi_{3^k}(S^{*n-1-i}\xi_{3^k})^*.$$

We claim that Q_n is a partial isometry. Indeed, $Q_n\xi_m \ne 0$ only if there are integers $0 \le i \le n-1$ and $k \ge 0$ so that $m = 3^k + i + 1 - n$ and $i \le 3^k$. If there were two solutions to this, there would also be $0 \le j \le n-1$ and $l \ne k$ so that $m = 3^l + j + 1 - n$ and $j \le 3^l$. We may suppose that $k < l$, from which we deduce that $3^k \ge i - j = 3^l - 3^k \ge 2 \cdot 3^k$, a contradiction. Hence, $Q_n\xi_m = 0$ or $m = 3^k + i + 1 - n$ and $Q_n\xi_m = \xi_{3^k - i}$ for a unique choice of k, i. As an operator on l_2, $Q_n = Q_n^*$, and hence it is also the case that $Q_n^* \xi_m = 0$ or equals some other basis vector. Thus Q_n is a partial isometry permuting some elements of this basis. In particular, $\|Q_n\| = 1$ and therefore $\|T^n\| \le 2$, i.e. T is power bounded.

If $A \in \mathcal{B}(H)$, let $Z(A) = \{\xi \in H : \text{w-lim}_{n\to\infty} A^n \xi = 0\}$. If U is unitary, we claim that

$$Z(U) = \{\xi \in H : \lim_{n\to\infty} \langle U^n \xi, \xi\rangle = 0\}.$$

Clearly the RHS contains the LHS. The RHS implies that $\lim_{n\to\infty}\langle U^n\xi, \eta\rangle = 0$ for $\eta \in \text{span}\{U^k\xi : k \in \mathbb{Z}\} = M$. Now M is the smallest reducing subspace for U containing ξ, and clearly $\langle U^n\xi, \eta\rangle = 0$ for $\eta \in M^\perp$. Therefore we get equality. It also follows that $Z(U) = Z(U^*)$.

Now if A is a contraction, let U be its minimal unitary dilation on $K \supset H$. Then

$$\begin{aligned} Z(A) \supset Z(U) \cap H &= \{\xi \in H : \lim_{n\to\infty} \langle U^n\xi, \xi\rangle = 0\} \\ &= \{\xi \in H : \lim_{n\to\infty} \langle A^n\xi, \xi\rangle = 0\} \supset Z(A). \end{aligned}$$

Therefore $Z(A) = Z(U) \cap H = Z(U^*) \cap H = Z(A^*)$.

If $B = XAX^{-1}$ and $\|A\| \le 1$, then

$$Z(B) = \{\xi \in H : \text{w-lim}_{n\to\infty} XA^n(X^{-1}\xi) = 0\} = \{\xi : X^{-1}\xi \in Z(A)\} = XZ(A).$$

Similarly, $B^* = X^{*-1}A^*X^*$ and $Z(B^*) = X^{*-1}Z(A^*) = X^{*-1}Z(A)$. Hence

$$Z(B^*)^\perp = \{\eta \in H : 0 = \langle \eta, X^{*-1}\xi\rangle = \langle X^{-1}\eta, \xi\rangle \ \forall \xi \in Z(A)\} = XZ(A)^\perp.$$

In particular, $Z(B) \cap Z(B^*)^\perp = X(Z(A) \cap Z(A)^\perp) = \{0\}$.

Finally we check that this fails for T. Note that $\xi_0 \oplus 0 \in \ker T \subset Z(T)$. Observe that $Q_{2\cdot 3^k+1}\xi_0 = \xi_0$ by taking $i = 3^k$ in the first paragraph. Hence

$$\text{w-lim}_{k\to\infty} T^{2\cdot 3^k+1}(0 \oplus \xi_0) = \text{w-lim}_{k\to\infty} Q_{2\cdot 3^k+1}\xi_0 \oplus \xi_{2\cdot 3^k+1} = \xi_0 \oplus 0.$$

Therefore if $\eta \oplus \zeta \in Z(T^*)$, we have

$$\langle \xi_0 \oplus 0, \eta\oplus\zeta\rangle = \text{w-lim}_{k\to\infty}\langle T^{2\cdot 3^k+1}(0\oplus\xi_0), \eta\oplus\zeta\rangle = \text{w-lim}_{k\to\infty}\langle 0\oplus\xi_0, T^{*2\cdot 3^k+1}(\eta\oplus\zeta)\rangle = 0.$$

We see that $\xi_0 \oplus 0 \in Z(T) \cap Z(T^*)^\perp$. By the previous paragraph, T is not similar to a contraction.

Paulsen's theorem has the immediate consequence:

14.13.10. THEOREM. *An operator $T \in \mathcal{B}(H)$ is similar to a contraction if and only if T is completely polynomially bounded.*

PROOF. If T is similar to a contraction, the representation ρ_T is similar to a completely contractive representation of $A(\mathbb{D})$, and thus ρ_T is completely bounded and so T is completely polynomially bounded. Conversely, if T is completely polynomially bounded, then ρ_T is a completely bounded representation of $A(\mathbb{D})$. Hence by Paulsen's theorem, it is similar to a completely contractive representation. In particular, T is similar to a contraction. ■

The gap between polynomially bounded and completely polynomially bounded was finally established by Pisier. His counterexample requires a lot of work. See [227, 239] for an extensive treatment.

14.14. Abstract Operator Systems

Just as for C*-algebras, it is useful to have an abstract characterization for operator systems and operator spaces.

Recall that a concrete operator system $\mathcal{S} \subset \mathcal{B}(H)$ has an order structure on the self-adjoint spaces $M_n(\mathcal{S})_{sa}$ induced by the order structure on $\mathcal{B}(H^{(n)})_{sa}$; that is, $C_n = \{s \in M_n(\mathcal{S}) : s \geq 0\}$ is a convex cone such that $C_n \cap (-C_n) = \{0\}$ and $C_n - C_n = M_n(\mathcal{S})_{sa}$.

The identity is an *order unit*, meaning that if $s \in S_{sa}$, then there is a $t > 0$ such that $-tI \leq s \leq tI$. It is also *Archimedean*, meaning that if $s + \varepsilon I \geq 0$ for all $\varepsilon > 0$, then $s \geq 0$. The same is true for $I \otimes I_n$ in $M_n(\mathcal{S})$. We say that I is an *Archimedean matrix order unit*.

We can imbed $M_{m,n}$ into $\mathcal{B}(H^{(n)}, H^{(m)})$ as $m \times n$ matrices with scalar operator entries. Then observe that if $\alpha = [\alpha_{ij}] \in M_{m,n}$, $s = [s_{jk}] \in M_n(\mathcal{S})$ and $\beta = [\beta_{kl}] \in M_{n,p}$, then $\alpha s \in M_{m,n}(\mathcal{S})$ and $s\beta \in M_{n,p}(\mathcal{S})$. Moreover, $\alpha s \alpha^* \in M_m(\mathcal{S})$, and if $s \geq 0$, then $\alpha s \alpha^* \geq 0$. Thus $\alpha C_n \alpha^* \subset C_m$ for all $m, n \geq 1$ and $\alpha \in M_{m,n}$.

The abstract characterization formalizes all of these properties.

14.14.1. Definition. A *$*$-vector space* is a complex vector space V with a conjugate linear involution $v \to v^*$, i.e. $v^{**} = v$ and $(\alpha u + \beta v)^* = \bar{\alpha} u^* + \bar{\beta} v^*$ for $u, v \in V$ and $\alpha, \beta \in \mathbb{C}$. An element $v \in V$ is self-adjoint if $v = v^*$. Let $V_{sa} = \{v \in V : v = v^*\}$.

If $m, n \geq 1$, $M_{m,n}(V)$ is the vector space of $m \times n$ matrices with coefficients in V. If $\alpha = [\alpha_{ij}] \in M_{p,m}$, $v = [v_{jk}] \in M_{m,n}(V)$ and $\beta = [\beta_{kl}] \in M_{n,q}$, define $\alpha v \in M_{p,n}(V)$ and $v\beta \in M_{m,q}(V)$ by

$$(\alpha v)_{ik} = \sum_{j=1}^{m} \alpha_{ij} v_{jk} \quad \text{and} \quad (v\beta)_{jl} = \sum_{j=1}^{m} v_{jk} \beta_{kl}.$$

If V is a $*$-vector space, $M_n(V)$ becomes a $*$-vector space with $[v_{ij}]^* = [v_{ji}^*]$.

14.14.2. DEFINITION. An *abstract operator system* is a $*$-vector space S together with a family of cones $C_n \subset M_n(S)_{sa}$ and an *Archimedean matrix order unit* e such that

(1) $C_n \cap (-C_n) = \{0\}$ and $C_n - C_n = M_n(S)_{sa}$.

(2) $\gamma C_n \gamma^* \subset C_m$ for all $m, n \geq 1$ and $\gamma \in M_{m,n}$.

(3) If $s \in M_n(S)_{sa}$, there is a $t \in \mathbb{R}_+$ so that $-te \otimes I_n \leq s \leq te \otimes I_n$, where $s \leq s'$ if $s' - s \in C_n$. (*matrix order unit*)

(4) If $s \in M_n(S)_{sa}$ and $s + \varepsilon e \otimes I_n \geq 0$ for all $\varepsilon > 0$, then $s \geq 0$ (*Archimedean*).

We first show how to use the order unit to define a norm on $M_n(S)$.

14.14.3. PROPOSITION. *Let S be an abstract operator system with order unit e. For $m, n \geq 1$ and $s \in M_{m,n}(S)$, define*

$$\|s\|_{m,n} = \inf\left\{t > 0 : \begin{bmatrix} te \otimes I_m & s \\ s^* & te \otimes I_n \end{bmatrix} \in C_{m+n}\right\}.$$

(1) *This is a norm on $M_{m,n}(S)$.*

(2) *$\|e\| = 1$ and $\|s\|_{m,n} = \|s^*\|_{n,m}$ for $s \in M_{m,n}(S)$.*

(3) *The cone C_n is closed in $M_n(S)$.*

(4) *If $\alpha \in M_{p,m}$ and $\beta \in M_{n,q}$, then $\|\alpha s \beta\|_{p,q} \leq \|\alpha\|\|\beta\|\|s\|_{m,n}$.*

PROOF. It is clear that $\|s\|_{m,n} \geq 0$. Also $\|\lambda s\|_{m,n} = |\lambda|\|s\|_{m,n}$ for $\lambda \in \mathbb{C}$. To see this, let $\lambda = re^{i\theta}$ and let $t > \|s\|_{m,n}$. Then

$$\begin{bmatrix} 0_m & 0 \\ 0 & 0_n \end{bmatrix} \leq \begin{bmatrix} \sqrt{r} I_m & 0 \\ 0 & \sqrt{r} e^{i\theta} I_n \end{bmatrix} \begin{bmatrix} te \otimes I_m & s \\ s^* & te \otimes I_n \end{bmatrix} \begin{bmatrix} \sqrt{r} I_m & 0 \\ 0 & \sqrt{r} e^{-i\theta} I_n \end{bmatrix}$$
$$= \begin{bmatrix} rte \otimes I_m & \lambda s \\ (\lambda s)^* & rte \otimes I_n \end{bmatrix}.$$

Hence $\|\lambda s\|_{m,n} \leq |\lambda| t$. This is reversible when $\lambda \neq 0$, so $\|\lambda s\|_{m,n} = |\lambda|\|s\|_{m,n}$.

If $s_i \in M_{m,n}(S)$ and $t_i > \|s_i\|_{m,n}$, then

$$\begin{bmatrix} 0_m & 0 \\ 0 & 0_n \end{bmatrix} \leq \begin{bmatrix} t_1 e \otimes I_m & s_1 \\ s_1^* & t_1 e \otimes I_n \end{bmatrix} + \begin{bmatrix} t_2 e \otimes I_m & s_2 \\ s_2^* & t_2 e \otimes I_n \end{bmatrix}$$
$$= \begin{bmatrix} (t_1 + t_2) e \otimes I_m & s_1 + s_2 \\ (s_1 + s_2)^* & (t_1 + t_2) e \otimes I_n \end{bmatrix}.$$

Hence $\|s_1 + s_2\|_{m,n} \leq t_1 + t_2$, whence the triangle inequality follows.

Suppose that $\|s\|_{m,n} = 0$. Then $\|-s\|_{m,n} = 0$. Hence for $\varepsilon > 0$,

$$\begin{bmatrix} 0_m & 0 \\ 0 & 0_n \end{bmatrix} \leq \begin{bmatrix} \varepsilon e \otimes I_m & s \\ s^* & \varepsilon e \otimes I_n \end{bmatrix} \quad \text{and} \quad \begin{bmatrix} 0_m & 0 \\ 0 & 0_n \end{bmatrix} \leq \begin{bmatrix} \varepsilon e \otimes I_m & -s \\ -s^* & \varepsilon e \otimes I_n \end{bmatrix}.$$

By the Archimedean property,

$$\pm\begin{bmatrix}0_m & s\\ s^* & 0_n\end{bmatrix} \geq \begin{bmatrix}0_m & 0\\ 0 & 0_n\end{bmatrix}.$$

However $C_{m+n} \cap -C_{m+n} = \{0\}$, and thus $s = 0$. Hence $\|\cdot\|_{m,n}$ is a norm.

Observe that

$$\begin{bmatrix}te & e\\ e & te\end{bmatrix} = e \otimes \begin{bmatrix}t & 1\\ 1 & t\end{bmatrix}.$$

This is positive if and only if $t \geq 1$, and thus $\|e\| = 1$. Since

$$\begin{bmatrix}te\otimes I_m & s\\ s^* & te\otimes I_n\end{bmatrix} \simeq \begin{bmatrix}te\otimes I_n & s^*\\ s & te\otimes I_m\end{bmatrix},$$

we have $\|s\|_{m,n} = \|s^*\|_{n,m}$.

Suppose that $s_k \in C_n$ and $s_k \to s$ in $M_n(S)$. Let $t_k > \|s - s_k\|_n$ satisfy $t_k \to 0$. Then

$$\begin{aligned}0_n &\leq \begin{bmatrix}\frac{1}{\sqrt{2}}I_n & \frac{1}{\sqrt{2}}I_n\end{bmatrix}\begin{bmatrix}t_k e\otimes I_n & s - s_k\\ s - s_k & t_k e\otimes I_n\end{bmatrix}\begin{bmatrix}\frac{1}{\sqrt{2}}I_n\\ \frac{1}{\sqrt{2}}I_n\end{bmatrix}\\ &= t_k e\otimes I_n + s - s_k \leq t_k e\otimes I_n + s.\end{aligned}$$

By the Archimedean property, $s \geq 0$; so that C_n is closed.

For (4), we prove the inequality for $\|\alpha\| = \|\beta\| = 1$. The rest follows by scaling. If $t > \|s\|_{m,n}$, then

$$\begin{aligned}\begin{bmatrix}0_p & 0\\ 0 & 0_n\end{bmatrix} &\leq \begin{bmatrix}\alpha & 0\\ 0 & \beta^*\end{bmatrix}\begin{bmatrix}te\otimes I_m & s\\ s^* & te\otimes I_n\end{bmatrix}\begin{bmatrix}\alpha^* & 0\\ 0 & \beta\end{bmatrix}\\ &= \begin{bmatrix}te\otimes\alpha\alpha^* & \alpha s\beta\\ \beta^* s^*\alpha^* & te\otimes\beta^*\beta\end{bmatrix} \leq \begin{bmatrix}te\otimes I_p & \alpha s\beta\\ \beta^* s^*\alpha^* & te\otimes I_q\end{bmatrix}.\end{aligned}$$

Hence $\|\alpha s\beta\|_{p,n} \leq \inf\{t : t > \|s\|_{m,n}\} = \|s\|_{m,n}$. ∎

14.14.4. Definition. Let S and T be abstract operator systems with positive cones $\{C_n\}_{n\geq 1}$ and $\{D_n\}_{n\geq 1}$. A map $\varphi : S \to T$ is *completely positive* if $\varphi_n(C_n) \subset D_n$ for $n \geq 1$. If the map φ is a bijection such that both φ and φ^{-1} are completely positive, then φ is an *order isomorphism*.

The following analogue of Proposition 14.6.4 has essentially the same proof.

14.14.5. Proposition. *Let S be an abstract operator system, $\mathfrak{A}$ be a unital C*-algebra, and $\varphi : S \to \mathfrak{A}$ be completely positive. Then $\|\varphi\|_{cb} = \|\varphi\| = \|\varphi(e)\|$.*

PROOF. Since $\|e\| = 1$, we have $\|\varphi(e)\| \le \|\varphi\| \le \|\varphi\|_{cb}$. Let $P = \varphi(e)$. Let $s = \begin{bmatrix} s_{ij} \end{bmatrix} \in M_n(S)$ with $\|s\| < 1$. Then

$$0_{2n} \le \begin{bmatrix} e \otimes I_n & s \\ s^* & e \otimes I_n \end{bmatrix} \text{ whence } 0_{2n} \le \begin{bmatrix} P^{(n)} & \varphi(s) \\ \varphi(s^*) & P^{(n)} \end{bmatrix} \le \begin{bmatrix} \|P\| I_n & \varphi(s) \\ \varphi(s)^* & \|P\| I_n \end{bmatrix}.$$

This implies that $\|\varphi(s)\|^2 = \|\varphi(s)^*\varphi(s)\| \le \|P\|^2$. Thus $\|\varphi_n\| = \|P\| = \|\varphi(e)\|$ for all $n \ge 1$. ∎

The proofs of Propositions 14.6.6 and 14.6.10 work verbatim for abstract operator systems. Thus we obtain:

14.14.6. PROPOSITION. *Let S be an abstract operator system, and let $\varphi : S \to M_n$ be a linear map for $n \ge 1$. The following are equivalent:*

(1) *φ is completely positive.*

(2) *$\varphi_n(C_n) \subset (M_n \otimes M_n)_+$.*

(3) *Let $e_1, \dots, e_n$ be an orthonormal basis for $\mathbb{C}^n$. The linear functional $s_\varphi : M_n(S) \to \mathbb{C}$ given by*

$$s_\varphi\begin{bmatrix} a_{ij} \end{bmatrix} = \langle \varphi_n\begin{bmatrix} a_{ij} \end{bmatrix}\xi, \xi \rangle = \sum_{i,j=1}^{n} \langle \varphi(a_{ij})e_j, e_i \rangle, \quad \textit{where } \xi = \begin{bmatrix} e_1 \\ \vdots \\ e_n \end{bmatrix}$$

is positive.

For the following, we do not insist that a concrete operator system be complete. We could instead complete S in the norm constructed above.

14.14.7. CHOI–EFFROS THEOREM. *Let S be an abstract operator system. Then S is completely order isomorphic to a concrete operator system $\mathcal{S}$ via a map $\varphi : S \to \mathcal{S}$ such that $\varphi(e) = 1$. Moreover, φ is completely isometric.*

PROOF. Let $\mathcal{C} = \{\varphi \in \mathrm{CP}(S, M_n) : n \ge 1,\ \varphi(e) = I_n\}$. Define a Hilbert space $H = \sum_{\mathcal{C}}^{\oplus} H_\varphi$, where $\dim H_\varphi = n$ if $\varphi \in \mathrm{CP}(S, M_n)$, and set $\Phi = \sum_{\mathcal{C}}^{\oplus} \varphi : S \to \mathcal{B}(H)$. Set $\mathcal{S} = \Phi(S)$. Since each $\varphi \in \mathcal{C}$ is completely positive, so is Φ, and $\Phi(e) = I$. In particular, $\mathcal{S}$ is self-adjoint and unital, and thus is a concrete operator system. We need to check that $\Phi_n(s) \ge 0$ implies that $s \ge 0$ for $s \in M_n(S)$. This will imply that Φ is injective because $\Phi(s) = 0$ would imply that $s \ge 0$ and $-s \ge 0$, whence $s = 0$. Since Φ is self-adjoint, we can also suppose that $s = s^*$.

Suppose that $s = s^* \in M_n(S)$ and $s \notin C_n$. Since C_n is closed and convex, $\{s\}$ is compact, and S_{sa} is a real normed space, there is a real linear functional $f : M_n(S)_{sa} \to \mathbb{R}$ such that $f(s) < 0$ and $f(C_n) \subset [0, \infty)$. We complexify this by setting $f(a + ib) = f(a) + if(b)$ for $a, b \in M_n(S)_{sa}$ to obtain a positive linear functional on $M_n(S)$. By Proposition 14.14.6, this determines a completely positive map $\psi : S \to M_n$ such that $s_\psi = f$. In particular, with the notation from the proposition,

$$\langle \psi_n(s)\xi, \xi \rangle = f(s) < 0.$$

Thus $\psi_n(s)$ is not positive. We need to adjust ψ so that it is unital and preserves this property.

Let $\psi(e) = E$, let $\operatorname{Ran} E = M$ and let $m = \dim M$. If $a \in S_{sa}$, there is a $t > 0$ so that $-te \le a \le te$ and hence $-tE \le \varphi(a) \le tE$. Hence $\psi(S) \subset P_M M_n P_M$. There is a matrix $A = P_M A \in M_{n,m}$ so that $A^*EA = I_m$. Define $\varphi(a) = A^*\psi(a)A$. This is a completely positive map into M_m and $\varphi(e) = I_m$. Finally, $\varphi_n(s) = A^{(n)*}\psi(s)A^{(n)}$ is still not positive. It follows that $\Phi_n(s) \not\ge 0$.

This establishes a complete order isomorphism between S and $\mathcal{S}$. Since $\|\varphi\|_{cb} = \|\varphi(e)\| = 1$ and similarly $\|\varphi^{-1}\|_{cb} = 1$, φ must be completely isometric. ∎

Now we turn to abstract operator spaces. They do not have a positive cone, so we use matrix norms instead.

14.14.8. Definition. An *abstract operator space* is a Banach space V with a family of norms $\|\cdot\|_n$ on $M_n(V)$ for $n \ge 1$ such that

(1) $\left\| \begin{bmatrix} x_1 & 0_{n_1,n_2} \\ 0_{n_2,n_1} & x_2 \end{bmatrix} \right\|_{n_1+n_2} = \max\{\|x_1\|_{n_1}, \|x_2\|_{n_2}\}$ for $x_i \in M_{n_i}(V)$.

(2) $\|\alpha x \beta\|_n \le \|\alpha\|\|x\|_n\|\beta\|$ for $\alpha, \beta \in M_n$ and $x \in M_n(V)$.

We give a proof of the characterization of operator spaces by building an abstract operator system and applying the Choi–Effros theorem.

14.14.9. Ruan Theorem. *If V is an abstract operator space, there is a complete isometry $\varphi : V \to \mathcal{B}(H)$ for some Hilbert space.*

Proof. Let V^* denote the vector space with elements $\{v^* : v \in V\}$ and linear structure $\lambda v^* + \mu w^* := (\bar\lambda v + \bar\mu w)^*$. Provide matrix norms on $M_n(V^*)$ by

$$\big\| [v_{ji}^*] \big\|_n = \big\| [v_{ij}] \big\|_n.$$

Define the *Paulsen system* for V to be the space

$$S = \left\{ \begin{bmatrix} \lambda & v \\ w^* & \mu \end{bmatrix} : \lambda, \mu \in \mathbb{C},\ v \in V \text{ and } w^* \in V^* \right\}.$$

If we do a standard rearrangement of rows and columns called the *canonical shuffle*, an element of $M_n(S)$ has the form:

$$\begin{bmatrix} \alpha & v \\ w^* & \beta \end{bmatrix} \quad \text{for} \quad \alpha, \beta \in M_n,\ v = [v_{ij}] \in M_n(V) \text{ and } w^* = [w_{ji}^*] \in M_n(V^*).$$

The adjoint operation is

$$\begin{bmatrix} \alpha & v \\ w^* & \beta \end{bmatrix}^* = \begin{bmatrix} \alpha^* & w \\ v^* & \beta^* \end{bmatrix}.$$

We declare that $\begin{bmatrix} I_n & v \\ v^* & I_n \end{bmatrix} \geq 0$ if $\|v\|_n \leq 1$. This leads to

$$0 \leq \begin{bmatrix} \alpha^* & 0 \\ 0 & \beta^* \end{bmatrix} \begin{bmatrix} I_n & v \\ v^* & I_n \end{bmatrix} \begin{bmatrix} \alpha & 0 \\ 0 & \beta \end{bmatrix} = \begin{bmatrix} \alpha^*\alpha & \alpha^* v\beta \\ \beta^* v^* \alpha & \beta^*\beta \end{bmatrix}.$$

This leads us to define

$$C_n = \Big\{ \begin{bmatrix} \alpha & v \\ v^* & \beta \end{bmatrix} : \alpha,\beta \in (M_n)_+,\ v = [v_{ij}] \in M_n(V) \text{ and }$$
$$\|(\alpha + \varepsilon I_n)^{-1/2} v (\beta + \varepsilon I_n)^{-1/2}\|_n \leq 1 \text{ for all } \varepsilon > 0 \Big\}.$$

Let $e = \begin{bmatrix} 1 & 0 \\ 0 & 1 \end{bmatrix}$. We must verify that this is an abstract operator system.

We show that e is a matrix order unit. Let $X = \begin{bmatrix} \alpha & v \\ v^* & \beta \end{bmatrix} \in M_n(S)_{sa}$, so that α, β are self-adjoint. Let $t = \max\{\|\alpha\|, \|\beta\|\} + \|v\|$. Then

$$te \otimes I_n - X = \begin{bmatrix} tI_n - \alpha & -v \\ -v^* & tI_n - \beta \end{bmatrix} \geq \begin{bmatrix} \|v\| I_n & -v \\ -v^* & \|v\| I_n \end{bmatrix} \geq \begin{bmatrix} 0 & 0 \\ 0 & 0 \end{bmatrix}.$$

The same goes for $-X$ so $-te \otimes I_n \leq X \leq te \otimes I_n$.

For the Archimedean property, suppose that $-\varepsilon e \otimes I_n \leq X \leq \varepsilon e \otimes I_n$ for all $\varepsilon > 0$. Then in particular $-\varepsilon I_n \leq \alpha, \beta \leq \varepsilon I_n$ for all $\varepsilon > 0$, whence $\alpha = \beta = 0$. Then $\begin{bmatrix} \varepsilon I_n & -v \\ -v^* & \varepsilon I_n \end{bmatrix} \geq 0$ forces $\|v\| \leq \varepsilon$ for $\varepsilon > 0$, and hence $v = 0$.

Clearly $C_n \cap (-C_n) = \{0\}$. Since e is a matrix order unit, if $X \in M_n(S)_{sa}$, then as above there is a t so that $te \otimes I_n \pm X \in C_n$. Hence

$$X = \frac{1}{2}(te \otimes I_n + X) - \frac{1}{2}(te \otimes I_n - X) \in C_n - C_n.$$

Next we verify that $X \in C_n$ and $\gamma \in M_{m,n}$ implies that $\gamma X \gamma^* \in C_m$. Let $X = \begin{bmatrix} \alpha & v \\ v^* & \beta \end{bmatrix} \in C_n$. The product of a scalar matrix γ with X is obtained as multiplication by $\begin{bmatrix} \gamma & 0 \\ 0 & \gamma \end{bmatrix}$. So $\gamma X \gamma^* = \begin{bmatrix} \gamma\alpha\gamma^* & \gamma v \gamma^* \\ \gamma v^* \gamma^* & \gamma\beta\gamma^* \end{bmatrix}$. For arbitrary $\varepsilon > 0$, we will pick a certain $\delta > 0$ and consider

$$\begin{aligned} \|(\gamma\alpha\gamma^* + \varepsilon I_m)^{-\frac{1}{2}} (\gamma v \gamma^*)(\gamma\beta\gamma^* + \varepsilon I_m)^{-\frac{1}{2}}\| \leq \|&(\gamma\alpha\gamma^* + \varepsilon I_m)^{-\frac{1}{2}} \gamma (\alpha + \delta I_n)^{\frac{1}{2}}\| \\ &\|(\alpha + \delta I_n)^{-\frac{1}{2}} v (\beta + \delta I_n)^{-\frac{1}{2}}\| \\ &\|(\beta + \delta I_n)^{\frac{1}{2}} \gamma^* (\gamma\beta\gamma^* + \varepsilon I_m)^{-\frac{1}{2}}\|. \end{aligned}$$

The middle term is at most 1 by definition. The first term is at most 1 provided that

$$(\gamma\alpha\gamma^*+\varepsilon I_m)^{-\frac{1}{2}}\gamma(\alpha+\delta I_n)^{\frac{1}{2}}(\alpha+\delta I_n)^{\frac{1}{2}}\gamma^*(\gamma\alpha\gamma^*+\varepsilon I_m)^{-\frac{1}{2}}\leq I_m$$

which is equivalent to

$$\gamma(\alpha+\delta I_n)\gamma^*\leq\gamma\alpha\gamma^*+\varepsilon I_m.$$

This will certainly be true if $\delta\leq\varepsilon\|\gamma\|^{-2}$, and a similar calculation shows that the third term is bounded by 1. Thus $\gamma X\gamma^*\in C_m$.

Clearly each C_n is closed under positive real multiples. We show that it is closed under sums. Observe that if $X\in C_n$ and $Y\in C_m$, then $X\oplus Y\in C_{n+m}$ by a straightforward verification. So if $X,Y\in C_n$, then $X\oplus Y\in C_{2n}$, and hence C_n contains

$$\begin{bmatrix}I_n & I_n\end{bmatrix}\begin{bmatrix}X & 0\\ 0 & Y\end{bmatrix}\begin{bmatrix}I_n\\ I_n\end{bmatrix}=X+Y.$$

Therefore each C_n is a cone, and thus S is an abstract operator system.

By the Choi–Effros theorem, S is completely isometric to a concrete operator system, and hence V is completely isometric to a concrete operator space. ■

The abstract characterization of operator spaces allows us to define an operator space dual. The dual should result in pairings between matrices of all orders. It turns out that there is a natural way to accomplish this. This is known as the *standard operator space dual.*

14.14.10. THEOREM. *Let E be an operator space, and let E^* be its Banach space dual. Put an operator space structure on E^* by identifying $M_n(E^*)$ with $\mathrm{CB}(E,M_n)$. That is, if $\varphi=[\varphi_{ij}]\in M_n(E^*)$, then $\varphi(x)=[\varphi_{ij}(x)]$ for $x\in E$ defines a map from E into M_n and we define $\|[\varphi_{ij}]\|_n:=\|\varphi\|_{cb}$. Then $\|y\|_1=\|y\|$ for $y\in E^*$, and*

$$\|\varphi\|=\|\varphi\|_{cb}=\sup_{p\geq1}\sup_{x\in M_p(E)_1}\|\varphi^{(p)}(x)\|=\sup_{p\geq1}\sup_{x\in M_p(E)_1}\|[\varphi_{ij}(x_{kl})]_{pn\times pn}\|.$$

The induced weak-$*$ topology on $M_n(E^*)$ is the topology of pointwise convergence on E. There is an imbedding Γ of E^* into $\mathcal{B}(H)$ such that $\Gamma(E)$ is WOT*-closed, and Γ is completely isometric and a weak-$*$–weak-$*$ homeomorphism.*

The imbeddings of $M_n(E)$ into $M_n(E^{**})=\mathrm{CB}(E^*,M_n)$ are isometric.

PROOF. We verify the axioms of an abstract operator space. That the norm of a direct sum is the maximum of the two norms is immediate. Suppose that $\alpha=[\alpha_{ij}]\in M_n$ and $\varphi=[\varphi_{ij}]\in M_n(E^*)$. Then for $x\in M_p(E)_1$,

$$\|(\alpha\varphi)^{(p)}(x)\|=\|\alpha^{(p)}\varphi^{(p)}(x)\|\leq\|\alpha\|\|\varphi^{(p)}(x)\|.$$

Taking a supremum over all the unit balls of $M_p(E)$ yields $\|\alpha\varphi\|_n\leq\|\alpha\|\|\varphi\|_n$. Multiplication on the right is handled similarly. By Ruan's theorem, E^* is an operator space with these matrix norms. If $y\in E^*$, then $y\in\mathrm{CB}(E,\mathbb{C})=\mathcal{B}(E,\mathbb{C})=E^*$

isometrically by Proposition 14.6.6. The last equality is just an expansion of what is meant by the cb norm.

By definition of the weak-$*$ topology, this is just the topology of pointwise convergence of functions on E. To embed E^* into $\mathcal{B}(H)$, define $\Gamma : E^* \to \bigoplus_{x \in b_1(M_n(E)),\, n \geq 1} M_{n,x}$, where $M_{n,x} = M_n^{(\infty)}$ for $x \in M_n(E)$

$$\Gamma(\varphi) = \bigoplus_{x \in b_1(M_n(E)),\, n \geq 1} \varphi_n(x)^{(\infty)}.$$

This map is completely isometric since if $\varphi \in M_p(E^*) = CB(E, M_p)$, then

$$\|\Gamma_p(\varphi)\| = \sup_{x \in b_1(M_n(E)),\, n \geq 1} \|\varphi_n(x)\|_{np \times np} = \|\varphi\|_{cb}.$$

If $(\varphi_\lambda)_\Lambda$ is a bounded net in E^* converging weak-$*$ to φ, then $(\varphi_\lambda)_n(x) \to \varphi_n(x)$ for every $x \in M_n(E)$. Since this net is bounded, $\big(\Gamma(\varphi_\lambda)\big)_\Lambda$ converges weak-$*$ to $\Gamma(\varphi)$. So Γ is weak-$*$ continuous on $\overline{b_1(E^*)}$. By Corollary 3.6.4 of the Krein–Smulian theorem, Γ is weak-$*$ continuous. The unit ball $\overline{b_1(E^*)}$ is weak-$*$ compact, and thus $\Gamma(\overline{b_1(E^*)})$ is weak-$*$ compact, and Γ is a weak-$*$–weak-$*$ homeomorphism on $\overline{b_1(E^*)}$. Hence by the Krein–Smulian theorem, $\Gamma(E^*)$ is weak-$*$ closed. By Corollary 3.6.4 again, Γ^{-1} is also weak-$*$–weak-$*$ continuous on $\Gamma(E)$, whence Γ is a weak-$*$–weak-$*$ homeomorphism on E^*. Since the range of Γ has infinite multiplicity, every weak-$*$ continuous linear functional is WOT-continuous. Therefore $\Gamma(E)$ is WOT-closed.

Let $x \in M_p(E)$. Then x induces the map $\hat{x}(\varphi) = \varphi_p(x)$ on E^*. Thus if $\varphi = [\varphi_{ij}] \in M_n(E^*) = \mathrm{CB}(E, M_n)$, we have

$$\|(\hat{x})_n(\varphi)\| = \|\varphi_p(x)\| \leq \|\varphi\|_{cb}\|x\| = \|\varphi\|\,\|x\|.$$

Therefore $\|\hat{x}\|_{cb} \leq \|x\|$. Conversely, by Proposition 14.12.5, there is a map $\varphi \in \mathrm{CB}(E, M_{2p})$ such that $\|\varphi\|_{cb} = 1$ and $\|\varphi_p(x)\| = \|x\|$. Therefore, the natural map $x \to \hat{x}$ is isometric on each $M_p(E)$. ■

When an operator space is a dual Banach space, we can apply the previous result to E_*, which has a standard operator space structure induced from E^*.

14.14.11. COROLLARY. *If E is an operator space which is the dual of a Banach space E_*, then E_* can be made into an operator space by identifying $M_n(E_*)$ with the subspace w^*-$\mathrm{CB}(E, M_n)$ of τ_{E_*}-continuous CB maps in $M_n(E^*) = \mathrm{CB}(E, M_n)$. Then $M_n(E)$ is isometric to $CB(E_*, M_n)$, and E can be represented as a WOT-closed subspace $\Gamma(E)$ of some $\mathcal{B}(H)$ so that Γ is completely isometric and a weak-$*$–weak-$*$ homeomorphism.*

PROOF. From Theorem 14.14.10, we obtain the standard dual operator space structure on E^*. The natural imbeddings of $M_n(E_*)$ into $M_n(E^*)$ clearly are identified with the weak-$*$ (τ_{E_*}) continuous CB maps. Applying the theorem again to E_* yields the rest. ■

14.14.12. REMARKS.
(1) The norms on $M_n(E^*)$ have little in common with $M_n(E)^*$. For one thing, we always have

$$\|\varphi_1 \oplus \cdots \oplus \varphi_n\| = \max\{\|\varphi_i\| : 1 \le i \le n\}.$$

However, as an element of $M_n(E)^*$ one gets $\|\varphi_1 \oplus \cdots \oplus \varphi_n\| = \sum_{i=1}^n \|\varphi_i\|$.

(2) The operator space dual of an operator system need not be an operator system because it may not have an order unit.

(3) There are various ways to put an operator space structure on any Banach space X. The *minimal operator space structure* $\mathrm{MIN}(X)$ arises from the imbedding of X into $C(\overline{b_1(X^*)}, w^*)$, the space of continuous functions on the compact Hausdorff space $\overline{b_1(X^*)}$ endowed with its weak-$*$ topology. This is a C*-algebra, so the matrix norms on it induce matrix norms on $M_n(X)$.

There is also a *maximal operator space structure* $\mathrm{MAX}(X)$ obtained by taking a direct sum over all isometric imbeddings of X into $\mathcal{B}(H)$ for a Hilbert space of the dimension $\max\{\aleph_0, \delta_X\}$, where δ_X is the minimal cardinality of a dense subset of X. We refer the interested reader to [227].

14.14.13. EXAMPLE. Infinite matrices over an operator space. Let E be an operator space. If m, n are (possibly infinite) cardinals, we can form the vector space $M_{m,n}(E)$ consisting of all $m \times n$ matrices $X = [x_{ij}]$ with coefficients in E such that the finite by finite submatrices are uniformly bounded, i.e.

$$\|X\| = \big\|[x_{ij}]\big\| := \sup_{F \Subset m,\ G \Subset n} \big\|[x_{ij}]_{i \in F,\, j \in G}\big\| < \infty$$

and similarly we define norms $\|\cdot\|_k$ on $M_k(M_{m,n}(E))$. Let us write $X_{F \times G}$ to denote $[x_{ij}]_{i \in F,\, j \in G}$ and make a similar definition if $X \in M_k(M_{m,n}(E))$. It is easy to see that if $E = \mathbb{C}$, then $M_{m,n} = M_{m,n}(\mathbb{C})$ is completely isometrically isomorphic to $\mathcal{B}(H_n, H_m)$ where H_n is a Hilbert space of dimension n. We will show that $M_{m,n}(E)$ is an operator space and an M_m–M_n bimodule.

It is easy to verify the first condition about the norm of a direct sum. Suppose that $X \in M_k(M_{m,n}(E))$ and $Y \in M_l(M_{m,n}(E))$, then

$$\begin{aligned}\|X \oplus Y\| &= \sup_{F \Subset m,\ G \Subset n} \|X_{F \times G} \oplus Y_{F \times G}\| \\ &= \sup_{F \Subset m,\ G \Subset n} \max\big\{\|X_{F \times G}\|, \|Y_{F \times G}\|\big\} = \max\big\{\|X\|, \|Y\|\big\}.\end{aligned}$$

Now take $X \in M_k(M_{m,n}(E))$ and $\alpha, \beta \in M_k$. Then

$$\begin{aligned}\|\alpha X \beta\| &= \sup_{F \Subset m,\, G \Subset n} \|\alpha X_{F\times G} \beta\| \\ &\leq \sup_{F \Subset m,\, G \Subset n} \|\alpha\| \, \|X_{F\times G}\| \, \|\beta\| = \|\alpha\| \, \|X\| \, \|\beta\|.\end{aligned}$$

So the second condition is also satisfied.

Finally we must verify that $M_{m,n}(E)$ is complete. Let $\{X_k\}_{k\geq 1}$ be a Cauchy sequence in $M_{m,n}(E)$. Then, in particular, each matrix entry converges in E, defining an $m \times n$ matrix X with coefficients in E. The sequence is bounded, say by M, and each sequence $(X_k)_{F\times G}$ converges to $X_{F\times G}$ in $M_{|F|,|G|}(E)$. Therefore

$$\|X\| = \sup_{F \Subset m,\, G \Subset n} \|X_{F\times G}\| = \sup_{F \Subset m,\, G \Subset n} \lim_{k\to\infty} \|(X_k)_{F\times G}\| \leq M.$$

Thus $X \in M_{m,n}(E)$. Given $\varepsilon > 0$, use the Cauchy condition to choose k_0 so that $\sup_{k>k_0} \|X_k - X_{k_0}\| < \varepsilon$. A standard argument shows that $\|X - X_{k_0}\| \leq \varepsilon$, and thus $\lim_{k\to\infty} X_k = X$. Therefore $M_{m,n}(E)$ is an abstract operator space.

Now we show that $M_{m,n}(E)$ is an M_m–M_n bimodule. Let $\alpha = \big[\alpha_{ij}\big] \in M_m$ and $\beta = \big[\beta_{ij}\big] \in M_n$. Given $X = \big[x_{kl}\big] \in M_{m,n}(E)$, we first observe that the ij entry y_{ij} in $Y = \alpha X \beta$ is

$$y_{ij} = \begin{bmatrix}\alpha_{i1} & \alpha_{i2} & \dots\end{bmatrix} \begin{bmatrix} & & \\ & x_{kl} & \\ & & \end{bmatrix} \begin{bmatrix}\beta_{1j} \\ \beta_{2j} \\ \vdots\end{bmatrix} = \lim_{F \Subset m,\, G \Subset n} \big[\alpha_{is}\big]_{s\in F} X_{F\times G} \big[\beta_{tj}\big]^t_{t\in G}.$$

The RHS is bounded by $\|\alpha\| \, \|X\| \, \|\beta\|$. Moreover, since the ith row of α belongs to $l_2(m)$, there are finite subsets $F_k \Subset m$ so that $\|[\alpha_{is}]_{s\notin F_k}\|_2 < 2^{-k}$ and similarly there are finite subsets $G_k \Subset n$ so that $\|[\beta_{tj}]_{t\notin G_k}\|_2 < 2^{-k}$. A routine argument shows that $z_k = \big[\alpha_{is}\big]_{s\in F_k} X_{F_k\times G_k} \big[\beta_{tj}\big]^t_{t\in G_k}$ is Cauchy with limit $y_{ij} \in E$. The same holds for any finite rectangle in Y, namely, if $F \Subset m$ and $G \Subset n$, then

$$Y_{F\times G} = \lim_{F' \Subset m,\, G' \Subset n} \alpha_{F\times F'} X_{F'\times G'} \beta_{G'\times G}.$$

This limit exists, and $\|Y_{F\times G}\| \leq \|\alpha\| \, \|X\| \, \|\beta\|$. Therefore $\|Y\| \leq \|\alpha\| \, \|X\| \, \|\beta\|$, and thus belongs to $M_{m,n}(E)$. Checking the other bimodule properties is straightforward.

If E is a dual operator space with predual E_* and is represented as in Corollary 14.14.11, then $M_{m,n}(E)$ are all WOT-closed. Indeed, any WOT-limit has matrix coefficients in E and the finite subrectangles are uniformly bounded. As in Corollary 14.14.11, we can identify $M_n(E)$ with $\mathrm{CB}(E_*, M_n)$. This also works for infinite cardinals, where $M_n = \mathcal{B}(H_n)$ for a Hilbert space H_n of dimension n. To see this, note that an element of $M_n(E)$ is determined by its finite rectangles, which are identified with $\mathrm{CB}(E_*, M_{|F|,|G|})$. The norm obtained by taking the supremum over these finite rectangles is the c.b. norm of the map into M_n with coordinates in E_*, and thus is an element of $\mathrm{CB}(E_*, M_n)$. Conversely, every map in $\mathrm{CB}(E_*, M_n)$ determines a matrix with bounded c.b. norms over finite rectangles.

Exercises for Chapter 14

14.1 Show that $S \in \mathcal{B}(K)$ is unitarily equivalent to a coextension of $T \in \mathcal{B}(H)$ if and only if there is an isometry $V : H \to K$ such that $TV^* = V^*S$.

14.2 (a) Let $T \in \mathcal{B}(H)$ be a contraction. Show that $T\xi = \xi$ if and only if $T^*\xi = \xi$.
(b) Show that the subspace N of Proposition 14.2.2 equals
$$\{\xi \in H : \|T^n\xi\| = \|T^{*n}\xi\| = \|\xi\| \text{ for } n \geq 1\}.$$

14.3 Give an example of a contraction T for which the procedure to obtain the minimal unitary dilation in either proof 1 or 2 requires the last step of restricting to the minimal reducing subspace.
HINT: Look carefully at proof 3.

14.4 With the notation of Corollary 14.1.7, show that $M = \sum_{-\infty}^{\infty}{}^{\oplus} M_k = K$ if and only if $\text{SOT–}\lim_{n\to\infty} T^n = 0$. (Such operators are called $C_{0\bullet}$.)

14.5 Let T be a C_0 contraction.
(a) Prove that T^* is also C_0 and find m_{T^*}.
(b) Prove that $\text{SOT–}\lim_{n\to\infty} T^n = 0$ and $\text{SOT–}\lim_{n\to\infty} T^{*n} = 0$.

14.6 Find a C_0 contraction T with $\sigma(T) = \mathbb{T}$.

14.7 Let $T \in \mathcal{B}(H)$ be a contraction, and let $\lambda \in \mathbb{D}$.
(a) Show that $S = (T - \lambda I)(I - \overline{\lambda}T)^{-1}$ is a contraction.
HINT: Compute $\|(T - \lambda I)\xi\|^2 - \|(I - \overline{\lambda}T)\xi\|^2$.
(b) Show that if B is a finite Blaschke product, then $\|B(T)\| \leq 1$.
(c) Deduce von Neumann's inequality using Exercise **11.19**.

14.8 Let $T \in \mathcal{B}(H)$ be a contraction, and let $\lambda \in \mathbb{D}$.
(a) If V is an isometric dilation of T, show that $W = (V - \lambda I)(I - \overline{\lambda}V)^{-1}$ is an isometric dilation of $S = (T - \lambda I)(I - \overline{\lambda}T)^{-1}$.
(b) Show that W is minimal if and only if V is.

14.9 Provide an example to show that a minimal dilation in Ando's theorem is not unique.

14.10 Show that the contractive representations of $A(\mathbb{D}^2)$ are in a natural bijective correspondence with commuting pairs of contractions.

14.11 Suppose that $T_1, \dots, T_n$ are commuting contractions such that $T_iT_j^* = T_j^*T_i$ for $i \neq j$. Prove that they have a dilation to commuting isometries.
HINT: Start with a dilation of T_1 to $H^{(\infty)}$ and extend the rest to commuting contractions. Make sure that if some T_i is already an isometry, this property is preserved.

14.12 Fill in the details of the proof of Corollary 14.3.6.

14.13 Let $T_i \in \mathcal{B}(H_i)$ be two contractions, and let $V_i \in \mathcal{B}(K_i)$ be their minimal isometric dilations. Suppose that $X \in \mathcal{B}(H_2, H_1)$ satisfies $T_1X = XT_2$. Prove that there is an operator $Y \in \mathcal{B}(K_2, K_1)$ such that $V_1Y = YV_2$, $Y^*|_{H_1} = X^*$ and $\|Y\| = \|X\|$.
HINT: $T = T_1 \oplus T_2$ commutes with $\left[\begin{smallmatrix} 0 & X \\ 0 & 0 \end{smallmatrix}\right]$.

14.14 The Carathéodory problem is: given $a_0, \dots, a_{n-1} \in \mathbb{C}$, determine whether there is an H^∞ function with $\|h\|_\infty \le 1$ of the form $h(z) = \sum_{k=0}^{n-1} a_k z^k + z^n g(z)$ for $g \in H^\infty$. Show that this is possible if and only if $\|P_n T_p P_n\| \le 1$ where $p(z) = \sum_{k=0}^{n-1} a_k z^k$ and P_n is the projection onto $H^2 \ominus z^n H^2$.
HINT: Follow some of the ideas in the proof of Pick's theorem.

14.15 Let F be a $k \times k$ matrix with coefficients in H^∞. Let T_F be the corresponding Toeplitz operator of multiplication by $F(z)$ on $H^{2(k)}$ and let M_F be the multiplication operator on $L^{2(k)}$. Show that $\|M_F\| = \|T_F\| = \|F\|_\infty$.

14.16 Let $T_1, \dots, T_n \in M_2$ be commuting contractive 2×2 matrices. Show that they dilate to commuting isometries.
HINT: A non-scalar 2×2 matrix is either diagonalizable with distinct eigenvalues or is similar to a Jordan block. Which operators commute with it?

14.17 (a) Let $w \in \mathbb{C}^{nm}$ be a unit vector, written $w = \begin{bmatrix} w_1 \\ \vdots \\ w_n \end{bmatrix}$ for $w_i \in \mathbb{C}^m$, and suppose that $\varphi : M_n \to M_m$ has Choi matrix $\left[w_i w_j^*\right] = ww^*$. Show that there is an isometry $W : \mathbb{C}^n \to \mathbb{C}^{nm}$ so that $\varphi(A) = W^*(A \otimes I_m)W$.
(b) Use this to show that if $C_\varphi \ge 0$, then φ has a Kraus decomposition as in Example 14.7.5.

14.18 Show that a completely contractive homomorphism from a C*-algebra $\mathfrak{A}$ into a C*-algebra $\mathfrak{B}$ is a $*$-homomorphism.

14.19 The Schur product of two $n \times n$ matrices $A = \left[a_{ij}\right]$ and $B = \left[b_{ij}\right]$ is given by $A * B = \left[a_{ij} b_{ij}\right]$. Define $S_A(B) = A * B$.
(a) Show that $A * B = V^*(A \otimes B)V$ where $V : \mathbb{C}^n \to \mathbb{C}^{n^2}$ by $Ve_i = e_i \otimes e_i$.
(b) Prove that S_A is completely positive if and only if $A \ge 0$.
(c) Deduce that the Schur product of two positive matrices is positive.

14.20 (*Takesaki–Tomiyama*). Let $\varphi : M_n \to \mathcal{B}(H)$, and let C_φ be its Choi matrix. Show that the following are equivalent for $1 \le k \le n$:
(1) φ is k-positive.
(2) $(I_n \otimes P)C_\varphi(I_n \otimes P) \ge 0$ for all projections $P \in \mathcal{B}(H)$ with rank $P \le k$.
(3) $\varphi_k\left(\left[F_{ij}\right]_{k \times k}\right) \ge 0$ for all $F_{ij} \in M_n$ with $F_{ij}F_{pq} = \delta_{jp}F_{iq}$ and rank $F_{ij} \le 1$ for $1 \le i, j, p, q \le k$.

14.21 Let $n \in \mathbb{N}$ and let $\mathcal{T}_n = \{(T_1, \dots, T_n) \in \mathcal{B}(l_2)^n$ commuting contractions$\}$. Define a norm on $\mathbb{C}[z_1, \dots, z_n]$ by

$$\|p\|_s = \sup\{\|p(T_1, \dots, T_n)\| : (T_1, \dots, T_n) \in \mathcal{T}_n\}.$$

(a) Show that $\|p\|_\infty \leq \|p\|_s$, and that for $n \geq 3$, this inequality can be strict.
(b) Define corresponding norms on $M_n(\mathbb{C}[z_1, \dots, z_n])$.
(c) Let $\mathsf{T} = \sum_{T \in \mathcal{T}_n}^{\oplus} T = (\mathsf{T}_1, \dots, \mathsf{T}_n)$. Show that the norm-closed algebra generated by T is completely isometrically isomorphic to the completion $A(\mathbb{D}^n)_s$ of $\mathbb{C}[z_1, \dots, z_n]$ in $\|\cdot\|_s$.

14.22 Let $\mathcal{M} \subset \mathcal{B}(H)$ be a von Neumann algebra, and let $\varphi : \mathcal{M} \to \mathcal{B}(K)$ be a normal completely positive map. Let (π, V) be the minimal Stinespring dilation in $\mathcal{B}(L)$. Prove that π is normal.
HINT: If $\xi_i = \pi(A_i)V\eta_i \in L$ for $A_i \in \mathcal{M}$ and $\eta_i \in H$, show that

$$\langle \pi(A)\xi_1, \xi_2 \rangle = \langle \varphi(A_2^* A A_1)\eta_1, \eta_2 \rangle.$$

14.23 Let $E \subset \mathcal{B}(H)$ be an operator space containing I. Let $\varphi : E \to \mathcal{B}(K)$ be a completely contractive map with $\varphi(I) = I$. Show that there is a $*$-representation $\pi : \mathcal{B}(H) \to \mathcal{B}(L)$ and an isometry $V \in \mathcal{B}(K, L)$ so that $\varphi(a) = V^*\pi(a)V$ for $a \in E$.

14.24 Let $\mathfrak{A}$ be a unital C*-algebra, and $\varphi : \mathfrak{A} \to \mathcal{B}(H)$ a self-adjoint c.b. map.
(a) If ψ_i are c.p. maps such that $\varphi = \psi_1 - \psi_2$, then $\|\varphi\|_{cb} \leq \|\psi_1(1) + \psi_2(1)\|$.
HINT: Show that $\psi_1 + \psi_2 \pm \varphi$ are c.p.
(b) Show that there are such maps so that $\|\varphi\|_{cb} = \|\psi_1(1) + \psi_2(1)\|$.

14.25 (*Naimark*). Let $A_i \in \mathcal{B}(H)$ be positive operators so that SOT-$\sum_{i \geq 1} A_i = I$. Prove that there is a Hilbert space $K \supset H$ and projections $P_i \in \mathcal{B}(K)$ such that SOT-$\sum_{i \geq 1} P_i = I$ and $P_H P_i|_H = A_i$.
HINT: Consider $\varphi : c_0 \to \mathcal{B}(H)$ by $\varphi((a_i)) =$ SOT-$\sum_{i \geq 1} a_i A_i$. Apply Stinespring.

14.26 Let X be a compact subset of $\mathbb{C}$, and let $R(X)$ denote the closure of the rational functions with poles in $\mathbb{C} \setminus X$ in $C(X)$. Let $T \in \mathcal{B}(H)$ have $\sigma(T) \subset X$. Say that X is a *complete spectral set* for T if $\rho_T(f) = f(T)$ for f rational with poles off X extends to a completely contractive representation of $R(X)$. A normal ∂X-dilation is a normal dilation N of T with $\sigma(N) \subset \partial X$. Prove that T has a normal ∂X-dilation if and only if X is a complete spectral set for T.

14.27 (*Berger–Foiaş–Lebow*). Suppose that $R(X)$ is Dirichlet in $C(\partial X)$. Show that if X is a spectral set for $T \in \mathcal{B}(H)$, then T has a minimal normal ∂X-dilation which is unique up to unitary equivalence which fixes H.
HINT: Extend $\rho(f) = f(T)$ for $f \in R(X)$ to $C(\partial X)$.

14.28 Show that $J = \begin{bmatrix} 0 & 1 \\ 0 & 0 \end{bmatrix}$ has $w(J) = \frac{1}{2}\|J\|$.

14.29 (*Invariance principle*). Suppose that $\mathcal{S}_i \subset \mathcal{B}(H_i)$ are operator systems and $\theta : \mathcal{S}_1 \to \mathcal{S}_2$ is a unital order isomorphism.

(a) Show that a u.c.p. map $\varphi : \mathcal{S}_2 \to \mathcal{B}(H)$ has the unique extension property if and only if $\varphi\theta$ does.

(b) Show that θ extends to a $*$-isomorphism $\widetilde{\theta}$ of $C^*_{\min}(\mathcal{S}_1)$ onto $C^*_{\min}(\mathcal{S}_2)$.

(c) Show that ρ is a boundary representation of $\mathcal{S}_2$ if and only if $\rho\widetilde{\theta}$ is a boundary representation for $\mathcal{S}_1$.

14.30 An operator algebra $\mathcal{A}$ contained in a C*-algebra $\mathfrak{B}$ is *Dirichlet* in $\mathfrak{B}$ if $\overline{\mathcal{A} + \mathcal{A}^*} = \mathfrak{B}$. If $\mathcal{A}$ is Dirichlet in $\mathfrak{B}$, show that $C^*_{\min}(\mathcal{A}) = \mathfrak{B}$.
HINT: Use Lemma 14.8.4 and consider the quotient map of $\mathfrak{B}$ onto $C^*_{\min}(\mathcal{A})$.

14.31 (*Arveson*). Let $\mathcal{S} \subset \mathcal{B}(H)$ be an operator system such that $C^*(\mathcal{S})$ contains all compact operators $\mathcal{K}(H)$.

(a) Show that $C^*_{\min}(\mathcal{S}) = C^*(\mathcal{S})$ if and only if the quotient by $\mathcal{K}(H)$ is not completely isometric on $\mathcal{S}$.
HINT: Show that $\mathcal{K}(H)$ is the unique minimal ideal of $C^*(\mathcal{S})$.

(b) Deduce that in this case, the identity representation is a boundary representation. HINT: The boundary representations norm $C^*_{\min}(\mathcal{S})$.

14.32 Let $S \subset \mathcal{B}(H)$ be an operator system. Show that S is injective if and only if it is injective in the category of operator spaces with completely contractive maps, i.e. if $E \subset F$ are operator spaces and $\varphi : E \to S$ is a completely contractive map, then there is a completely contractive extension $\widetilde{\varphi} : F \to S$.
HINT: Both are equivalent to the existence of a u.c.p. projection of $\mathcal{B}(H)$ onto S.

14.33 (*Choi–Effros*). An operator system S is *monotone complete* if every increasing net in S_{sa} which is bounded above has a least upper bound in S. Prove that an injective operator system is monotone complete.
HINT: Use Lemma 9.3.5 and the projection onto S.

14.34 (*Hamana*). Let S, T be operator systems such that (T, ι) is an injective extension of S. Prove that the following are equivalent:

(1) (T, ι) is an injective envelope of S,
(2) (T, ι) is a rigid extension of S, and
(3) (T, ι) is an essential extension of S.

14.35 Let S be an operator system contained in an injective operator system E. Show that there is an injective envelope T of S with $S \subset T \subset E$.

14.36 Let $T \in \mathcal{B}(H)$.

(a) (*Rota*). If $\operatorname{spr}(T) < 1$, show that T is similar to a contraction.
HINT: find $r < 1$ and C so that $\|T^n\| \le Cr^n$ for $n \ge 0$. Estimate $\|\Phi_T\|_{cb}$.

(b) Show that $\operatorname{spr}(T) = \inf_{S \in \mathcal{B}(H)^{-1}} \|STS^{-1}\|$.

14.37 Let $K \in \mathcal{K}(H)$ be a compact operator. Show that if K is power bounded, then it is similar to a contraction.
HINT: Write K as a (non-orthogonal) direct sum of a matrix C with spectrum in $\mathbb{T}$ and an operator K_0 with $\operatorname{spr}(K_0) < 1$. Analyse the Jordan form of C.

14.38 (*Smith*). (*completely bounded maps into* M_n)

(a) Let $n < m$ and suppose that $\xi_i \in \mathbb{C}^n$ for $1 \le i \le m$. Show that there is a unitary $U \in M_m$ and vectors $\alpha_i \in \mathbb{C}^n$ for $1 \le i \le n$ so that $(I_n \otimes U)\begin{bmatrix}\xi_1\\ \vdots\\ \xi_m\end{bmatrix} = \begin{bmatrix}\alpha_1\\ \vdots\\ \alpha_n\\ 0\\ \vdots\\ 0\end{bmatrix}$.

HINT: Set $T(\lambda_1,\dots,\lambda_m) = \sum_{i=1}^m \lambda_i\xi_i$. Choose U so that the last $m-n$ rows of U lie in $\ker T$.

(b) Let $\mathcal{M}$ be an operator space and let $\varphi : \mathcal{M} \to M_n$ be a continuous linear map. Show that $\|\varphi\|_{cb} = \|\varphi_n\|$.

HINT: For $m > n$, choose $A = \begin{bmatrix}a_{ij}\end{bmatrix} \in M_m(\mathcal{M})$ with $\|A\| = 1$, unit vectors $\xi = \begin{bmatrix}\xi_1\\ \vdots\\ \xi_m\end{bmatrix}$ and $\eta = \begin{bmatrix}\eta_1\\ \vdots\\ \eta_m\end{bmatrix}$ in $\mathbb{C}^{mn}$, where ξ_i and η_i are in $\mathbb{C}^n$, such that $\langle\varphi_m(A)\xi,\eta\rangle \approx \|\varphi_m\|$. Use (a) to reduce this to an $n \times n$ matrix calculation.

14.39 Let E be an operator space and let P be a closed subspace. Endow the quotient $F = (E/P)_{opsp}$ with the operator space structure $M_n(F) := M_n(E)/M_n(P)$.

(a) Verify that F satisfies the operator space axioms.

(b) Let $q : E \to E/P$ be the quotient map. Show that $\varphi : F \to \mathcal{B}(H)$ is c.b. if and only if φq is c.b, and that $\|\varphi\|_{cb} = \|\varphi q\|_{cb}$.

(c) Show that $P^* \simeq E^*/P^\perp$ and $F^* \simeq P^\perp \subset E^*$ completely isometrically.

(d) A completely contractive map $\varphi : E \to G$ is a *complete quotient map* if the induced map from $(E/\ker\varphi)_{opsp} \to G$ is a completely isometric isomorphism. Show that this is equivalent to $\varphi^* : G^* \to E^*$ being a completely isometric imbedding.

14.40 (*Kavruk–Paulsen–Todorov–Tomforde*). Let $\mathcal{S}$ be an operator system. A subspace J is called an *operator system kernel* if there is a u.c.p. map $\varphi : \mathcal{S} \to \mathcal{B}(H)$ such that $\ker\varphi = J$. The quotient operator system is $(\mathcal{S}/J)_{opsys} := \overline{\varphi(\mathcal{S})}$, and $\varphi(e) = I$ is the order unit.

(a) Show that $J \subset \mathcal{S}$ is a kernel if and only if there are states $\{\varphi_\lambda\}_\Lambda$ on $\mathcal{S}$ such that $J = \bigcap_\Lambda \ker\varphi_\lambda$.

(b) Show that $\mathbb{C}E_{11}$ is not a kernel in M_2.

(c) Let $\mathcal{S}_4 = l^{(4)}_\infty$ be the four-dimensional l_∞ space. Let $y = y^* \in \mathcal{S}_4$ which is neither positive nor negative. Show that $\mathbb{C}y$ is an operator system kernel.

(d) Take $y_n = (-1, 0, n, 2n)$ and let $J_n = \mathbb{C}y_n$. Let $\mathcal{T}_n = (\mathcal{S}_4/J_n)_{opsys}$ and let $x_n = (0, 1, n+1, 0)$. Show that

$$\|x_n + J\|_{\mathcal{T}_n} = \inf\{t \ge 0 : t(e + J_n) - (x_n + J_n) \ge 0\} = 1$$

in the quotient *operator system*. However in the quotient *operator space* $(\mathcal{S}/J_n)_{opsp}$, $\|x_n + J_n\|_{opsp} = \frac{2}{3}(n+1)$.

(e) Build a u.c.p. map $\varphi : \mathcal{S}_4 \to \mathcal{B}(H)$ such that the operator space norm on $\mathcal{S}_4/\ker\varphi$ is not equivalent to its operator system norm.

Notes on Chapter 14

Sz.-Nagy established the unitary dilation of a contraction in [293] in 1953. The proof we give is due to Schäffer [277]. The von Neumann inequality was first established in [318] by different methods. The connection with unitary dilations is due to Sz.-Nagy. An H^∞-functional calculus for contractions with absolutely continuous unitary dilation is due to Schreiber [280]. Sz.-Nagy and Foiaş [295] developed it further, and showed that every c.n.u. contraction has such a calculus. In [296], they introduce C_0 contractions and compute their spectrum.

Ando [9] dilates two commuting contractions to commuting unitaries. Sarason [276] proved a special case of the commutant lifting theorem in order to provide an operator-theoretic proof of the Nevanlinna–Pick interpolation theorem. Sz.-Nagy and Foiaş [297] establish the commutant lifting theorem independently of Ando's theorem. While we deduced it from Ando's theorem, one can just as easily deduce Ando's theorem from commutant lifting. Pick [237] proved his interpolation theorem in 1916. Nevanlinna [217] proved it independently, and provided a complete parameterization of the solutions.

Parrott [222] provided his example of three commuting 4×4 contractions with no commuting unitary dilation in 1970, even though the von Neumann inequality is satisfied. Varopoulos [310] showed that the von Neumann inequality fails for $n \geq 3$ commuting contractions. His paper contains an addendum providing an explicit example (joint work with Kaijser) of three commuting 5×5 contractions which fail the von Neumann inequality. Crabb and Davie [64] provide another example using 8×8 matrices. Holbrook's 4×4 example is in [154]. The Choi–Davidson example is in [52]. Holbrook [153] shows that any family of commuting 2×2 matrices have commuting unitary dilations. Knese [180] has shown that the scalar von Neumann inequality is valid for any commuting family of 3×3 contractions.

Completely positive maps were defined by Stinespring [289] who proved his dilation theorem, and also that positive maps with a commutative C*-algebra as domain or range are completely positive. The central role of completely positive maps in operator algebras and abstract dilation theory is contained in Arveson [14] where he proved his extension and dilation theorems. Choi studied completely positive maps on matrix algebras in [50]. The dilation theory for triangular algebras comes from McAsey–Muhly [202] and Paulsen–Power–Ward [228], who extended the ideas to arbitrary nest algebras. The notion of the minimal C*-algebra of an operator algebra or operator system was formulated by Arveson [14, 15]. The first proof of its existence was due to Hamana [138], who constructed the injective envelope of an operator system, and found the C*-envelope as a C*-subalgebra of this envelope with the Choi–Effros product, which comes from [53]. Dritschel–McCullough found a new proof using ideas from dilation theory [101]. Arveson [19] used this to prove the existence of boundary representations in the separable case using disintegration of measures to decompose a representation into a direct integral of irreducible representations. Davidson–Kennedy [76] gave a dilation theoretic proof that was valid in complete generality.

The extension theorem for completely bounded maps was first established by Haagerup, but never published. Shortly after, this result was obtained by Wittstock [329], which also contains his dilation theorem. Paulsen [224] found the proof given here. Paulsen [225] showed that an operator is similar to a contraction if and only if it is completely polynomially bounded. This led him to Paulsen's theorem [226].

Kadison's problem dates from [164]. Bunce [42] and Christensen [56] solved the nuclear case. Haagerup [130] established that a homomorphism of a C*-algebra is similar to a $*$-representation whenever it has a cyclic vector, and also whenever the C*-algebra has no finite traces. The result on derivations was found earlier by Christensen [55].

Sz.-Nagy [294] showed that if K is compact and power bounded, i.e. $\|K^n\| \leq M$ for $n \geq 1$, then K is similar to a contraction. He asked if compactness was necessary. Quickly, Foguel [113] found a counterexample. Halmos, in his "ten problems" lecture [137], reformulated the problem in terms of polynomial boundedness. Lebow [190] showed that Foguel's example was not polynomially bounded. Paulsen's result [225] mentioned above gave a necessary and sufficient condition for similarity to a contraction, but the Sz.-Nagy–Halmos problem remained open. Pisier [238] constructed a very non-trivial counterexample, completing the problem. See [81, 227, 239] for treatments of his solution.

Axioms for abstract operator systems and the spatial representation theorem were given by Choi and Effros [53]. Ruan [265] gave axioms for abstract operator spaces and established the spatial representation theorem. The proof given here is taken from [227]. Blecher, Ruan and Sinclair [34] gave axioms for abstract unital operator algebras and established their spatial representation.

The dilation theory of a single contraction and its many ramifications is well covered in Sz.-Nagy–Foiaş [298]. The new edition has many notes and remarks about more modern developments. More general dilation theory is beautifully covered in Paulsen [227]. Another source with somewhat different focus is Pisier's book [239] on similarity problems. Some operator space theory can be found in [227], and there is more detail in Pisier [240] and Effros–Ruan [107]. The abstract theory of operator algebras can be found in Blecher–Le Merdy [33].

CHAPTER 15

Nonselfadjoint Operator Algebras

In this chapter, we take a look at three different classes of operator algebras. The first, semicrossed products, are the analogue of C*-crossed products, except the action comes from a semigroup and the algebra formed is nonself-adjoint. In this chapter, we only consider actions of $\mathbb{Z}_+$.

Then we look at free semigroup algebras, which are the WOT closed unital algebras generated by isometries $S_1, \dots, S_d$ with pairwise orthogonal ranges. So the C*-algebra that they generate is either the Cuntz algebra $\mathcal{O}_d$ or the Cuntz–Toeplitz algebra $\mathcal{E}_d$.

The third is a commutative operator algebra arising from the universal model for a commuting row contraction. While commutative, it is not a uniform algebra. The analysis of this algebra combines ideas from function theory, operator theory, free semigroup algebras and harmonic analysis.

In all three cases, we are only providing an introduction to the topics. The interested reader should consult the notes at the end of the chapter.

Semicrossed Products

15.1. Semicrossed Product Basics

A semicrossed product of an operator algebra by a semigroup action is a construction analogous to C*-crossed products. We begin with a general method for defining an operator algebra built from a family of representations of an algebra which may not have a norm itself.

15.1.1. DEFINITION. Let A be an algebra over $\mathbb{C}$. A non-empty family $\mathcal{F}$ of homomorphisms of A into $\mathcal{B}(H)$ for Hilbert spaces H is *admissible* if it is closed under arbitrary direct sums, unitary equivalence and restriction to reducing subspaces. If A is unital, we also require the homomorphisms to be unital. Then we can define a family of seminorms on $M_n(A)$ for $n \geq 1$ by

$$\big\| [a_{ij}] \big\|_{\mathcal{F}} = \sup_{\varphi \in \mathcal{F}} \big\| [\varphi(a_{ij})] \big\|.$$

K. R. Davidson, *Functional Analysis and Operator Algebras*, CMS/CAIMS Books in Mathematics 13,
https://doi.org/10.1007/978-3-031-63665-3_15

15.1.2. REMARK. The family $\mathcal{F}$ must be bounded at each $a \in A$. For otherwise, if $a \in A$ and $\varphi_n \in \mathcal{F}$ with $\sup_n \|\varphi_n(a)\| = \infty$, the homomorphism $\varphi = \sum_{n\geq 1}^{\oplus} \varphi_n$ is not defined at a.

15.1.3. PROPOSITION. *Let A be an algebra over $\mathbb{C}$, and let $\mathcal{F}$ be an admissible set of homomorphisms. The set $N = \{a \in \mathcal{A} : \|a\|_{\mathcal{F}} = 0\}$ is an ideal, and the completion $\mathcal{A}$ of $(A/N, \|\cdot\|)$ is an operator algebra. If A is unital, so is $\mathcal{A}$.*

PROOF. For any $\varphi \in \mathcal{F}$, $C^*(\varphi(A))$ decomposes as a direct sum of cyclic representations, and these representations live on Hilbert spaces of dimension bounded by the cardinality of a spanning subset of $\varphi(A)$, which is at most the cardinality of A. The restriction to reducing subspaces remains in $\mathcal{F}$, so every representation in $\mathcal{F}$ is unitarily equivalent to a direct sum of representations into a set of Hilbert spaces of bounded dimension. Let φ_0 denote the direct sum over this set of representations. This yields a single representation in $\mathcal{F}$ with the property that its norm dominates the norm of every other representation in $\mathcal{F}$. Therefore

$$\|[a_{ij}]\|_{\mathcal{F}} = \|[\varphi_0(a_{ij})]\|.$$

In particular, $N = \ker \varphi_0$ is an ideal. This identifies A/N with $\varphi_0(A)$, which by definition is a completely isometric identification. The completion is then identified with $\overline{\varphi_0(A)}$, which is a concrete operator algebra. If A has a unit 1, then $\varphi_0(1) = I$; so $\mathcal{A}$ is unital. ∎

A semigroup $P \ni 0$ contained in an *abelian* group G is a *positive cone* if $P \cap -P = \{0\}$, and is a *spanning cone* if $P - P = G$ as well.

15.1.4. DEFINITION. Let $\mathfrak{A}$ be a C*-algebra, and let $P \subset G$ be a spanning cone. A *C*-dynamical system* $(\mathfrak{A}, \alpha, P)$ is a homomorphism $\alpha : P \to \mathrm{End}(\mathfrak{A})$, where $\mathrm{End}(\mathfrak{A})$ is the set of $*$-endomorphisms of $\mathfrak{A}$. The system is automorphic or injective or surjective or unital if each α_s for $s \in P$ is automorphic or injective or surjective or unital.

A *covariant pair* (π, T) for $(\mathfrak{A}, \alpha, P)$ consists of

(1) $\pi : \mathfrak{A} \to \mathcal{B}(H)$ is a $*$-homomorphism, and

(2) $T : P \to \mathcal{B}(H)$ is a contractive homomorphism of P such that

(3) $\pi(a)T_s = T_s\pi\alpha_s(a)$ for all $s \in P$ and $a \in \mathfrak{A}$ (*covariance relation*).

This pair is isometric or unitary if T_s is an isometry or a unitary for all $s \in P$.

Given a C*-dynamical system $(\mathfrak{A}, \alpha, P)$, we form an algebra $A = P\mathfrak{A}$ consisting of all formal sums $\sum_{s\in P} e_s a_s$ for $a_s \in \mathfrak{A}$, with $a_s = 0$ except finitely often. Then $\{e_s : s \in P\}$ can be thought of as a basis with coefficients in $\mathfrak{A}$, and addition is defined in the obvious way. Define a multiplication by

$$(e_s a)(e_t b) = e_{s+t}\alpha_t(a)b \quad \text{for} \quad s, t \in P \text{ and } a, b \in \mathfrak{A}$$

and extend the definition to $A \times A$ using the distributive law. A covariant pair (π, T) yields a representation of A by

$$T\times\pi\Big(\sum_{s\in P} e_s a_s\Big) = \sum_{s\in P} T_s\pi(a_s).$$

This works because of the covariance relation:

$$\begin{aligned} T\times\pi(e_s a)\, T\times\pi(e_t b) &= T_s\pi(a)T_t\pi(b) \\ &= T_sT_t\pi\alpha_t(a)\pi(b) = T_{s+t}\pi(\alpha_t(a)b) \\ &= T\times\pi(e_{s+t}\alpha_t(a)b) = T\times\pi\big((e_s a)(e_t b)\big). \end{aligned}$$

There are several natural admissible families of representations. We will consider three, namely the set $\mathcal{F} = \{T\times\pi : (\pi, T) \text{ covariant}\}$, $\mathcal{F}_{is} = \{T\times\pi \in \mathcal{F} : T \text{ is isometric}\}$ and $\mathcal{F}_{un} = \{T\times\pi \in \mathcal{F} : T \text{ is unitary}\}$. It is routine to see that these families are closed under direct sums, unitary equivalence and the restriction to reducing subspaces. Thus they are admissible.

15.1.5. DEFINITION. Let $\mathfrak{A}$ be a C*-algebra, $P \subset G$ a spanning cone, and consider a C*-dynamical system $(\mathfrak{A}, \alpha, P)$. The *semicrossed product* $\mathfrak{A} \times_\alpha P$ is the completion of the algebra $A = P\mathfrak{A}$ with respect to the norm defined by all covariant representations, $\mathcal{F}$. The *isometric semicrossed product* $\mathfrak{A} \times_\alpha^{is} P$ has norm determined by $\mathcal{F}_{is}$; and the *unitary semicrossed product* $\mathfrak{A} \times_\alpha^{un} P$ has norm determined by $\mathcal{F}_{un}$.

15.1.6. EXAMPLE. Let X be a compact Hausdorff space, and let $\sigma : X \to X$ be a continuous map. The system (X, σ) is called a *topological dynamical system.* This induces an endomorphism α_1 of $C(X)$ by $\alpha_1(f) = f \circ \sigma$. Setting $\alpha_n = \alpha_1^n$ yields a C*-dynamical system $(C(X), \alpha, \mathbb{Z}_+)$. The semicrossed product algebra $C(X) \times_\alpha \mathbb{Z}_+$ contains $C(X)$ and an element $\mathfrak{s} = T(1)$ such that $f\mathfrak{s} = \mathfrak{s}(f \circ \sigma)$ for $f \in C(X)$.

If (Y, τ) is another topological dynamical system, we say that (X, σ) and (Y, τ) are *topologically conjugate* if there is a homeomorphism $h : X \to Y$ such that $h \circ \sigma = \tau \circ h$. Again we define endomorphisms of $C(Y)$ by $\beta_1(f) = f \circ \tau$ and $\beta_n = \beta_1^n$. The semicrossed product $C(Y) \times_\beta \mathbb{Z}_+$ contains $C(Y)$ and an element $\mathfrak{t}$ such that $f\mathfrak{t} = \mathfrak{t}(f \circ \tau)$ for $f \in C(Y)$. A topological conjugacy h induces a completely isometric isomorphism from $C(X) \times_\alpha \mathbb{Z}_+$ onto $C(Y) \times_\beta \mathbb{Z}_+$ by $\theta(f) = f \circ h^{-1}$ and $\theta(\mathfrak{s}) = \mathfrak{t}$.

15.1.7. EXAMPLE. A *Fock representation* of $\mathfrak{A} \times_\alpha P$ is obtained by taking a faithful representation $\pi : \mathfrak{A} \to \mathcal{B}(H)$ and forming a Hilbert space $K = H \otimes l_2(P)$. Let $\{\delta_s : s \in P\}$ be an orthonormal basis for $l_2(P)$. Define for $\xi \in H$ and $s, t \in P$:

$$\widetilde{\pi}(a)(\xi \otimes \delta_s) = \pi\alpha_s(a)\xi \otimes \delta_s \quad \text{and} \quad V_t(\xi \otimes \delta_s) = \xi \otimes \delta_{t+s}.$$

Since π is faithful, $\widetilde{\pi}$ is also faithful; and each V_t is an isometry. Compute

$$\widetilde{\pi}(a)V_t(\xi \otimes \delta_s) = \widetilde{\pi}(a)(\xi \otimes \delta_{t+s}) = \pi\alpha_{t+s}(a) \otimes \delta_{t+s}$$

and

$$V_t\widetilde{\pi}\alpha_t(a)(\xi \otimes \delta_s) = V_t\, \pi\alpha_s\alpha_t(a) \otimes \delta_s = \pi\alpha_{s+t}(a) \otimes \delta_{t+s}.$$

Since P is abelian, the covariance relation holds.

This shows that contractive and isometric representations always exist. Since this is faithful on $\mathfrak{A}$, both $\mathfrak{A} \times_\alpha P$ and $\mathfrak{A} \times_\alpha^{is} P$ contain isometric copies of $\mathfrak{A}$.

Next we show that the norm on $\mathfrak{A} \times_\alpha P$ in this representation does not depend on π, provided that π is faithful. Consider an element of $P\mathfrak{A}$, say a finite sum $X = \sum e_s a_s$. Then if $\xi \in H$ and $t \in P$,

$$\begin{aligned} V \times \widetilde{\pi}(X)(\xi \otimes \delta_t) &= \sum V_s \pi(a_s) V_t(\xi \otimes \delta_0) \\ &= \sum V_{s+t}(\pi\alpha_t(a_s)\,\xi \otimes \delta_0) = \sum \pi\alpha_t(a_s)\,\xi \otimes \delta_{s+t}. \end{aligned}$$

We can thus think of $V \times \widetilde{\pi}(X)$ as a $P \times P$ matrix with non-zero entries in the $(s+t,t)$ positions of $\pi\alpha_t(a_s)$ and 0 in any entry (s,t) where $s \not\geq t$, i.e. $s \notin t+P$. Because X has only finitely many terms, it is clear that $\|X\| \leq \sum \|a_s\| < \infty$. Thus the norm of this infinite matrix is the supremum over all finite submatrices. Define $X_F \in M_n(\mathfrak{A})$ for $F \Subset P$ with $|F| = n$ to have entry $\alpha_t(a_s)$ when $(s+t,t) \in F \times F$ and 0 in all other entries. Now π_n is a $*$-representation of $M_n(\mathfrak{A})$, and it is faithful if π is faithful, and thus is isometric. Therefore

$$\|V \times \widetilde{\pi}(X)\| = \sup_{F \Subset P} \|\pi_n(X_F)\| = \sup_{F \Subset P} \|X_F\|.$$

This shows that this norm is independent of the choice of π, provided that it is faithful. The same argument works for matrices, so the various faithful Fock representations are completely isometric to each other. We will call this the *Fock norm* $\|\cdot\|_{\text{Fock}}$. If π is not faithful, then the norm is generally smaller.

We will restrict ourselves to the special case of $P = \mathbb{Z}_+ \subset \mathbb{Z}$ and unital systems. In the next section, we will restrict our attention to injective systems. The idea is to provide a taste of the ideas in this area. When $P = \mathbb{Z}_+$, α is determined by $\alpha_1 \in \text{End}(\mathfrak{A})$ since $\alpha_n = \alpha_1^n$ for $n \geq 1$. A covariant pair is determined by (π, T_1) where $\pi(a)T_1 = T_1\pi\alpha_1(a)$. From this, we define $T_n = T_1^n$ and obtain a covariant pair (π, T).

15.1.8. Lemma. *Let $(\mathfrak{A}, \alpha, \mathbb{Z}_+)$ be a dynamical system. Then for each $z \in \mathbb{T}$, there is a completely isometric automorphism γ_z of $\mathfrak{A} \times_\alpha \mathbb{Z}_+$ such that*

$$\gamma_z(e_n a) = e_n z^n a \quad \textit{for} \quad n \geq 0.$$

The map $z \to \gamma_z(X)$ is continuous for each $X \in \mathfrak{A} \times_\alpha \mathbb{Z}_+$.

PROOF. Let (π, T) be a covariant representation such that $\pi \times T$ is a completely isometric representation of $\mathfrak{A} \times_\alpha \mathbb{Z}_+$. For each $z \in \mathbb{T}$, let $(zT)(n) = (zT)^n$. Then (π, zT) is also a covariant representation of $(\mathfrak{A}, \alpha, \mathbb{Z}_+)$. By the universal property, there is a completely contractive homomorphism σ_z of $\pi \times T$ $(\mathfrak{A} \times_\alpha \mathbb{Z}_+)$ into $\pi \times zT(\mathfrak{A} \times_\alpha \mathbb{Z}_+)$ that carries $T_n\pi(a)$ to $z^n T_n \pi(a)$ for $n \geq 0$. Clearly $\sigma_{\bar{z}}$ is the inverse of this map, and hence both are completely isometric automorphisms. By pulling this back to $\mathfrak{A} \times_\alpha \mathbb{Z}_+$, we obtain completely isometric automorphisms γ_z for $z \in \mathbb{T}$.

If $X \in \mathbb{Z}_+\mathfrak{A}$, say $X = \sum_{k=0}^n e_k a_k$, then

$$\|\gamma_z(X) - \gamma_w(X)\| = \Big\|\sum_{k=0}^n (z^k - w^k) e_k a_k\Big\| \leq \sum_{k=0}^n k|z-w|\|a_k\| \leq n|z-w|\|X\|_1.$$

Hence $z \longrightarrow \gamma_z(X)$ is continuous. Since the maps are contractive and $\mathbb{Z}_+\mathfrak{A}$ is dense in $X \in \mathfrak{A} \times_\alpha \mathbb{Z}_+$, the result extends to the whole algebra. ∎

This allows us to define a "Fourier series" for elements of $\mathfrak{A} \times_\alpha \mathbb{Z}_+$. Define

$$E_k(X) = \frac{1}{2\pi} \int_{-\pi}^{\pi} \gamma_{e^{it}}(X) e^{-ikt}\, dt \quad \text{for} \quad X \in \mathfrak{A} \times_\alpha \mathbb{Z}_+ \text{ and } k \in \mathbb{Z}.$$

Lemma 15.1.8 shows that the integrand is norm continuous, and thus this makes sense as a Riemann integral. Check for $X \in \mathbb{Z}_+\mathfrak{A}$, say $X = \sum_{j=0}^n e_j a_j$,

$$E_k(X) = \sum_{j=0}^n \frac{1}{2\pi} \int_{-\pi}^{\pi} e_j a_j e^{ijt} e^{-ikt}\, dt = e_k a_k.$$

This is clearly a completely contractive idempotent map of $\mathbb{Z}_+\mathfrak{A}$ onto $e_k\mathfrak{A}$. By continuity, it extends to $\mathfrak{A} \times_\alpha \mathbb{Z}_+$. For $k < 0$, $E_k = 0$. As for functions on the unit circle, we write $X \sim \sum_{k \geq 0} E_k(X)$ for $X \in \mathfrak{A} \times_\alpha \mathbb{Z}_+$.

As for Fourier series, we can define the Cesàro means $\Sigma_n(X)$ by

$$\Sigma_n(X) = \sum_{k=0}^n \big(1 - \tfrac{k}{n+1}\big) E_k(X).$$

We recall the basic facts about the Fejér kernel from Lemma 2.5.3:

$$k_n(t) = \sum_{k=-n}^n \big(1 - \tfrac{k}{n+1}\big) e^{ikt} = \frac{1}{n+1} \Big(\frac{\sin(n+1)t/2}{\sin t/2}\Big)^2.$$

Then k_n is even, continuous, 2π-periodic and $k_n \geq 0$. Moreover $\|k_n\|_1 = 1$ and k_n converges to 0 uniformly for $|t| \in [\delta, \pi]$.

15.1.9. THEOREM. *The maps Σ_n on $\mathfrak{A} \times_\alpha \mathbb{Z}_+$ are completely contractive maps. If $X \in \mathfrak{A} \times_\alpha \mathbb{Z}_+$, then $X = \lim_{n\to\infty} \Sigma_n(X)$. In particular, the map from X to $(E_0(X), E_1(X), E_2(X), \dots)$ is injective. Also E_0 is an expectation onto $\mathfrak{A}$.*

PROOF. As for functions on the circle, we obtain the formulae

$$\begin{aligned}\Sigma_n(X) &= \sum_{k=0}^{n} (1 - \tfrac{k}{n+1}) E_k(X) = \sum_{k=-n}^{n} (1 - \tfrac{k}{n+1}) E_k(X) \\ &= \frac{1}{2\pi} \int_{-\pi}^{\pi} \gamma_{e^{it}}(X) \sum_{k=-n}^{n} (1 - \tfrac{k}{n+1}) e^{-ikt}\, dt \\ &= \frac{1}{2\pi} \int_{-\pi}^{\pi} \gamma_{e^{it}}(X) k_n(t)\, dt.\end{aligned}$$

Since $\|k_n\|_1 = 1$ and $\gamma_{e^{it}}$ are completely isometric, it is easy to see that these maps are completely contractive. Now the proof of Fejér's theorem 2.5.4 applies to establish that $X = \lim_{n\to\infty} \Sigma_n(X)$. The reader should check this.

Since we can recover any $X \in \mathfrak{A} \times_\alpha \mathbb{Z}_+$ as a limit of terms $\Sigma_n(X)$ which only depend on the Fourier series, it follows that the map from X to its sequence of Fourier coefficients is injective. Since E_0 is a contractive projection onto $\mathfrak{A}$, it is an expectation. ■

We now show how to dilate contractive covariant representations to isometric ones.

15.1.10. THEOREM. *Let $\mathfrak{A}$ be an unital C*-algebra and let α_1 be an unital endomorphism of $\mathfrak{A}$. If (π, T) is a (contractive) covariant pair for $(\mathfrak{A}, \mathbb{Z}_+, \alpha)$, then there is an isometric covariant co-extension $(\widetilde{\pi}, S)$. The minimal co-extension is unique up to unitary equivalence fixing H, and the isometry S_1 is the minimal isometric dilation of T_1.*

PROOF. Set $\alpha_n = \alpha_1^n$ for $n \geq 1$. As usual, we let $D = (I - T_1^* T_1)^{1/2}$, $M = \overline{\operatorname{Ran} D}$ and $P = P_M$. Observe that for $a \in \mathfrak{A}$,

$$\begin{aligned}T_1^* T_1 \pi\alpha_1(a) &= T_1^* \pi(a) T_1 = \big(\pi(a^*) T_1\big)^* T_1 \\ &= \big(T_1 \pi\alpha_1(a^*)\big)^* T_1 = \pi\alpha_1(a) T_1^* T_1.\end{aligned}$$

So $D^2 = I - T_1^* T_1$ commutes with $\pi\alpha_1(\mathfrak{A})$, and thus so does D. In particular, M reduces $\pi\alpha_1(\mathfrak{A})$; so $P \in \pi\alpha_1(\mathfrak{A})'$. Since $\pi\alpha_n(\mathfrak{A}) \subset \pi\alpha_1(\mathfrak{A})$, P also commutes with $\pi\alpha_n\mathfrak{A}$ for $n \geq 1$.

The minimal isometric dilation S_1 of T_1 on $K = H \oplus M^{(\infty)}$ is given by

$$S_1(\xi, \eta_1, \eta_2, \dots, \eta_n, \dots) = (T\xi, PD\xi, \eta_1, \dots, \eta_{n-1}, \dots).$$

Define $\widetilde{\pi} = \pi \oplus \sum_{n\geq 1} P\pi\alpha_n$ by

$$\widetilde{\pi}(a)(\xi, \eta_1, \dots, \eta_n, \dots) = (\pi(a)\xi, P\pi\alpha_1(a)\eta_1, \dots, P\pi\alpha_n(a)\eta_n, \dots).$$

Then

$$\begin{aligned}\widetilde{\pi}(a)S_1(\xi,\eta_1,\eta_2,\dots,\eta_n,\dots) &= \widetilde{\pi}(a)(T\xi, PD\xi,\eta_1,\dots,\eta_{n-1},\dots)\\ &= (\pi(a)T\xi, \pi\alpha_1(a)PD\xi, P\pi\alpha_2(a)\eta_1,\dots,P\pi\alpha_n(a)\eta_{n-1},\dots)\\ &= (T\pi\alpha_1(a)\xi, PD\pi\alpha_1(a)\xi, P\pi\alpha_2(a)\eta_1,\dots,P\pi\alpha_n(a)\eta_{n-1},\dots)\\ &= S_1(\pi\alpha_1(a)\xi, P\pi\alpha_2(a)\eta_1,\dots,P\pi\alpha_{n+1}(a)\eta_n,\dots)\\ &= S_1\widetilde{\pi}\alpha_1(a)(\xi,\eta_1,\eta_2,\dots,\eta_n,\dots).\end{aligned}$$

Thus $\widetilde{\pi}(a)S_1 = S_1\widetilde{\pi}\alpha_1(a)$, and therefore $(\widetilde{\pi}, S)$, where $S_n = S_1^n$, is an isometric covariant pair which dilates (π, T).

If $(\widetilde{\pi}', S')$ is another minimal isometric co-extension on $K' = \bigvee_{n\geq 0} S'_n H$, define $W\sum S_n\pi(a_n)\xi_n = \sum S'_n\pi(a_n)\xi_n$ for finite sums, and check that this is isometric just as in the single operator situation. It follows that it is well defined and extends to a unitary such that for $\xi \in H$, $W\xi = WS_0\xi = S'_0\xi = \xi$. Observe that

$$\begin{aligned}WS_k\sum S_n\pi(a_n)\xi_n &= W\sum S_{n+k}\pi(a_n)\xi_n = \sum S'_{n+k}\pi(a_n)\xi_n\\ &= S'_k\sum S'_n\pi(a_n)\xi_n = S'_kW\sum S_n\pi(a_n)\xi_n.\end{aligned}$$

Hence $WS_k = S'_kW$. Similarly, for $b \in \mathfrak{A}$,

$$\begin{aligned}W\widetilde{\pi}(b)\sum S_n\pi(a_n)\xi_n &= W\sum S_n\pi\alpha_n(b)\pi(a_n)\xi_n = W\sum S_n\pi\big(\alpha_n(b)a_n\big)\xi_n\\ &= \sum S'_n\pi\big(\alpha_n(b)a_n\big)\xi_n = \widetilde{\pi}'(b)\sum S'_n\pi(a_n)\xi_n\\ &= \widetilde{\pi}'(b)W\sum S_n\pi(a_n)\xi_n.\end{aligned}$$

Therefore $W\widetilde{\pi} = \widetilde{\pi}'W$. This establishes the uniqueness of the minimal isometric co-extension. ■

15.1.11. COROLLARY. *Let $\mathfrak{A}$ be a C*-algebra and let $(\mathfrak{A},\alpha,\mathbb{Z}_+)$ be a C*-dynamical system. Then the natural maps from $\mathfrak{A}\times_\alpha\mathbb{Z}_+$ to $\mathfrak{A}\times_\alpha^{is}\mathbb{Z}_+$ to a Fock representation $V\times\widetilde{\pi}(\mathfrak{A}\times_\alpha^{is}\mathbb{Z}_+)$ (for any faithful representation π) are complete isometries.*

PROOF. The identity map on $\mathbb{Z}_+\otimes\mathfrak{A}$ extends to a completely contractive homomorphism $A\times_\alpha\mathbb{Z}_+ \longrightarrow A\times_\alpha^{is}\mathbb{Z}_+$. However, since every contractive covariant representation dilates to an isometric covariant representation, the norm of a contractive representation is dominated by the dilated isometric representation. Hence the two norms agree at all matrix levels. Since the Fock representation is isometric, the norm there is dominated by the universal norm on $\mathfrak{A}\times_\alpha^{is}\mathbb{Z}_+$.

For the converse, suppose that (π, V) is an isometric covariant pair such that V_1 is a *pure* isometry. Let $H_0 = \ker V_1^*$ and $H_n = V_nH_0 = V_1^nH_0$. By the Wold decomposition, $H = \sum_{n\geq 0}^{\oplus} H_n$. Compute

$$\pi(a)V_1V_1^* = V_1\pi\alpha(a)V_1^* = V_1\big(V_1\pi\alpha(a)\big)^* = V_1\big(\pi(a^*)V_1\big)^* = V_1V_1^*\pi(a).$$

Therefore $P_0 = I - V_1V_1^* = P_{H_0}$ commutes with $\pi(\mathfrak{A})$. Similarly $\pi(\mathfrak{A})$ commutes with $P_n := V_nV_n^* - V_{n+1}V_{n+1}^* = P_{H_n}$. Let $\pi_n = \pi|_{H_n}$ for $n \geq 0$. Then we get $\pi \simeq \sum_{n\geq 0}^{\oplus} \pi_n$. Moreover if $\xi \in H_0$, then $V_n\xi \in H_n$ and

$$\pi_n(a)V_n\xi = V_n\pi\alpha_n(a)\xi = V_n\pi_0\alpha_n(a)\xi \quad \text{for} \quad a \in \mathfrak{A}.$$

Thus $\pi_n \simeq \pi_0\alpha_n$. It is now clear that the representation $V\times\pi$ is unitarily equivalent to the Fock representation for π_0.

Not every isometric representation consists of pure isometries. Let (π, V) be an isometric covariant pair. Then for $0 < r < 1$, (π, rV) is a contractive covariant pair. Moreover the minimal isometric dilation S_r of rV is a pure isometry. By Theorem 15.1.10, there is an isometric covariant dilation (π_r, S_r) of (π, rV). For any element $X = \sum_{i=0}^{n} e_i a_i$ in $\mathbb{Z}_+\mathfrak{A}$, the previous paragraph shows that

$$\begin{aligned} \|V\times\pi(X)\| &= \Big\|\sum_{i=0}^{n} V_i\pi(a_i)\Big\| = \sup_{0<r<1} \Big\|\sum_{i=0}^{n} r^i V_i\pi(a_i)\Big\| \\ &= \sup_{0<r<1} \|S_r\times\pi_r(X)\| \leq \|X\|_{\text{Fock}}. \end{aligned}$$

The same argument works for matrices over $\mathbb{Z}_+\mathfrak{A}$. Thus the homomorphism of $\mathfrak{A}\times_\alpha^{is}\mathbb{Z}_+$ to a Fock representation is completely isometric. ■

15.2. Injective C*-dynamical Systems

In this section, we specialize to injective C*-dynamical systems over $\mathbb{Z}_+$. Our goal is to identify the C*-envelope in this case.

We will first dilate isometric covariant pairs to unitary pairs. However uniqueness may be lost because of the use of the Hahn–Banach theorem.

15.2.1. Lemma. *Let $\mathfrak{A}$ be a C*-algebra and let $\alpha_1 \in \mathrm{End}(\mathfrak{A})$ be unital and injective. If $\sigma : \mathfrak{A} \to \mathcal{B}(H)$ is a $*$-representation, then there is a Hilbert space K, a $*$-representation $\sigma' : \mathfrak{A} \to \mathcal{B}(K)$ and an isometry $V : H \to K$ such that $\sigma'\alpha V = V\sigma$.*

Proof. Every representation is a direct sum of cyclic representations, so we may suppose that σ has a unit cyclic vector ξ. Define a state $\varphi(a) = \langle\sigma(a)\xi, \xi\rangle$. Since α_1 is injective, we can define a state ψ on $\alpha(\mathfrak{A})$ by $\psi(\alpha(a)) = \varphi(a)$. Now extend ψ to a functional $\widetilde{\psi}$ of norm 1 on $\mathfrak{A}$ using the Hahn–Banach theorem. Since $1 = \|\widetilde{\psi}\| = \widetilde{\psi}(1)$, $\widetilde{\psi}$ is a state. Let $\sigma' = \pi_{\widetilde{\psi}}$ be the representation on $K = H_{\widetilde{\psi}}$ obtained from the GNS construction with cyclic vector ξ'. Define an operator V on $\sigma(\mathfrak{A})\xi$ by $V\sigma(a)\xi = \sigma'\alpha_1(a)\xi'$ for $a \in \mathfrak{A}$. This defines V on a dense subset of H

and

$$\begin{aligned}\|V\sigma(a)\xi\|^2 &= \langle \sigma'\alpha_1(a)\xi', \sigma'\alpha_1(a)\xi'\rangle = \langle \sigma'\alpha_1(a^*a)\xi', \xi'\rangle \\ &= \widetilde{\psi}(\alpha_1(a^*a)) = \varphi(a^*a) = \langle \sigma(a^*a)\xi, \xi\rangle = \|\sigma(a)\xi\|^2.\end{aligned}$$

Thus V is well defined and extends by continuity to an isometry of H into K. This calculation includes the fact that $\langle \sigma'\alpha_1(a^*a)\xi', \xi'\rangle = \langle \sigma(a^*a)\xi, \xi\rangle$. The positive elements span $\mathfrak{A}$, and thus $\langle \sigma'\alpha_1(a)\xi', \xi'\rangle = \langle \sigma(a)\xi, \xi\rangle$ for $a \in \mathfrak{A}$. The range projection VV^* of V commutes with $\sigma'\alpha(\mathfrak{A})$ because

$$\begin{aligned}VV^*\sigma'\alpha_1(a)\sigma'\alpha_1(b)\xi' &= V\sigma(ab)\xi = \sigma'\alpha_1(ab)\xi' \\ &= \sigma'\alpha_1(a)\sigma'\alpha_1(b)\xi' = \sigma'\alpha_1(a)VV^*\sigma'\alpha_1(b)\xi'.\end{aligned}$$

Hence $V^*\sigma'\alpha V$ is a $*$-representation such that $\langle V^*\sigma'\alpha(a)V\xi, \xi\rangle = \varphi(a)$. By Proposition 9.7.4, $V^*\sigma'\alpha V = \sigma$. ∎

15.2.2. THEOREM. *Let $\mathfrak{A}$ be a C*-algebra and let $\alpha_1 \in \mathrm{End}(\mathfrak{A})$ be unital and injective. If (π, T) is a (contractive) covariant pair for $(\mathfrak{A}, \mathbb{Z}_+, \alpha)$, then there is a co-isometric covariant extension $(\widetilde{\pi}, V^*)$.*

PROOF. As usual, $\alpha_n = \alpha_1^n$ for $n \geq 1$. Let $D_* = (I - T_1T_1^*)^{1/2}$, $N = \overline{\mathrm{Ran}\, D_*}$ and $P = P_N$. Observe that for $a \in \mathfrak{A}$,

$$\begin{aligned}\pi(a)T_1T_1^* &= T_1\pi\alpha(a)T_1^* = T_1\big(T_1\pi\alpha(a)\big)^* \\ &= T_1\big(\pi(a^*)T_1\big)^* = T_1T_1^*\pi(a).\end{aligned}$$

Thus D_* commutes with $\pi(\mathfrak{A})$. In particular, $N = \overline{\mathrm{Ran}(D_*)}$ reduces π.

Apply Lemma 15.2.1 to $\pi|_N$ to obtain a representation σ_0 on a Hilbert space N_0 and an isometry $V_0 : N \to N_0$ so that $\sigma_0\alpha V_0 = V_0\pi|_N$. Continue to apply Lemma 15.2.1 recursively to construct representations $\sigma_k : \mathfrak{A} \to \mathcal{B}(N_k)$ on Hilbert spaces N_k for $k \geq 1$ and isometries $V_k : N_{k-1} \to N_k$ such that $\sigma_k V_k = V_k\sigma_{k-1}$. Let $K = H \oplus \sum_{k\geq 0}^{\oplus} N_k$, and define a map

$$V^*(\xi, \eta_0, \eta_1, \eta_2, \dots, \eta_n, \dots) = (T\xi + PD_*V_0^*\eta_0, V_1^*\eta_1, V_2^*\eta_2, \dots, V_n^*\eta_n, \dots)$$

and a $*$-representation $\widetilde{\pi} = \pi \oplus \sum_{k\geq 0}^{\oplus} \sigma_k$ into $\mathcal{B}(K)$. Clearly V^* is a co-isometry that extends T. Compute

$$\begin{aligned}V^*\widetilde{\pi}\alpha(a)(\xi, \eta_0, \eta_1, \dots, \eta_n, \dots) &= V^*\big(\pi\alpha \oplus \textstyle\sum_{k\geq 0}^{\oplus} \sigma_k\alpha\big)(\xi, \eta_0, \eta_1, \dots, \eta_n, \dots) \\ &= V^*\big(\pi\alpha(a)\xi, \sigma_0\alpha(a)\eta_0, \sigma_1\alpha(a)\eta_1, \dots, \sigma_n\alpha(a)\eta_n, \dots\big) \\ &= \big(T\pi\alpha(a)\xi + D_*V_0^*\sigma_0\alpha(a)\eta_0, V_1^*\sigma_1\alpha(a)\eta_1, \dots, V_n^*\sigma_n\alpha(a)\eta_n, \dots\big) \\ &= \big(\pi(a)T\xi + D_*\pi(a)V_0^*\eta_0, V_1^*\sigma_1\alpha(a)\eta_1, \dots, V_n^*\sigma_n\alpha(a)\eta_n, \dots\big) \\ &= \big(\pi(a)T\xi + \pi(a)D_*V_0^*\eta_0, \sigma_0(a)V_1^*\eta_1, \dots, \sigma_{n-1}\alpha(a)V_n^*\eta_n, \dots\big) \\ &= \widetilde{\pi}(a)\big(T\xi + D_*V_0^*\eta_0, V_1^*\eta_1, \dots, \alpha(a)V_n^*\eta_n, \dots\big) \\ &= \widetilde{\pi}(a)V^*(\xi, \eta_0, \eta_1, \eta_2, \dots, \eta_n, \dots)\end{aligned}$$

Therefore $(\widetilde{\pi}, V^*)$ is a co-isometric covariant extension of (π, T). ∎

Note that V^* is not the minimal co-isometric extension of T because each V_k can be a proper isometry.

15.2.3. COROLLARY. *Let $\mathfrak{A}$ be a C*-algebra and let $\alpha_1 \in \mathrm{End}(\mathfrak{A})$ be unital and injective. If (π, T) is a (contractive) covariant pair for $(\mathfrak{A}, \mathbb{Z}_+, \alpha)$, then there is a unitary covariant dilation $(\widetilde{\pi}, U)$.*

PROOF. First apply Theorem 15.2.2 to (π, T) to obtain a co-isometric extension (π', V^*). Then apply Theorem 15.1.10 to obtain an isometric dilation $(\widetilde{\pi}, U)$. Moreover, the isometric extension U_1 of V_1^* is the minimal isometric dilation, which is unitary because V_1^* is a co-isometry. Thus $U_n = U^n$ are all unitary. ∎

15.2.4. COROLLARY. *Let $\mathfrak{A}$ be a C*-algebra and let $\alpha_1 \in \mathrm{End}(\mathfrak{A})$ be unital and injective. Then $\mathfrak{A} \times_\alpha \mathbb{Z}_+ = \mathfrak{A} \times_\alpha^{is} \mathbb{Z}_+ = \mathfrak{A} \times_\alpha^{un} \mathbb{Z}_+$ completely isometrically.*

PROOF. Clearly the identity map on $\mathbb{Z}_+\mathfrak{A}$ extends to completely contractive homomorphism $\mathfrak{A} \times_\alpha \mathbb{Z}_+ \longrightarrow \mathfrak{A} \times_\alpha^{un} \mathbb{Z}_+$. The fact that every contractive covariant pair dilates to a unitary covariant pair means that the supremum of norms over all unitary pairs is the same as the supremum over all contractive pairs. Thus these maps are completely isometric. ∎

Now we will consider C*-algebras containing $\mathfrak{A} \times_\alpha \mathbb{Z}_+$.

15.2.5. PROPOSITION. *Let $\mathfrak{A}$ be a C*-algebra and let $\alpha_1 \in \mathrm{End}(\mathfrak{A})$ be unital and injective. There is a C*-algebra $\mathfrak{A}_\infty$ containing $\mathfrak{A}$ and an automorphism $\widetilde{\alpha}_1 \in \mathrm{Aut}(\mathfrak{A}_\infty)$ such that $\widetilde{\alpha}_1|_{\mathfrak{A}} = \alpha_1$ and $\mathfrak{A}_\infty = \overline{\bigcup_{n\geq 1} \widetilde{\alpha}_1^{-n}(\mathfrak{A})}$.*

PROOF. Take $\mathfrak{A}_\infty$ to be the direct limit algebra $\mathfrak{A}_\infty = \varinjlim(\mathfrak{A}, \alpha_1)$ as in Theorem 9.10.2. That is, we take $\mathfrak{A}_n = \mathfrak{A}$ and $\rho_{n,n+1} = \alpha_1$ for all $n \geq 1$. Then

$$\begin{array}{ccccccccc}
\mathfrak{A}_0 & \xrightarrow{\alpha_1} & \mathfrak{A}_1 & \xrightarrow{\alpha_1} & \mathfrak{A}_2 & \xrightarrow{\alpha_1} & \dots & \longrightarrow & \mathfrak{A}_\infty \\
\downarrow{\scriptstyle \alpha_1} & & \downarrow{\scriptstyle \alpha_1} & & \downarrow{\scriptstyle \alpha_1} & & \downarrow & & \downarrow{\scriptstyle \widetilde{\alpha}_1} \\
\mathfrak{A}_0 & \xrightarrow{\alpha_1} & \mathfrak{A}_1 & \xrightarrow{\alpha_1} & \mathfrak{A}_2 & \xrightarrow{\alpha_1} & \dots & \longrightarrow & \mathfrak{A}_\infty
\end{array}$$

Since α_1 is injective, the embedding ρ_0 of $\mathfrak{A} = \mathfrak{A}_0$ into $\mathfrak{A}_\infty$ is injective. There are injective homomorphisms $\rho_n : \mathfrak{A}_n \to \mathfrak{A}_\infty$ such that $\rho_{n+1}\alpha_1 = \rho_n$. The vertical maps commute with the horizontal maps and determine a unique $*$-homomorphism $\widetilde{\alpha}_1 \in \mathrm{End}(\mathfrak{A}_\infty)$ such that $\widetilde{\alpha}_1\rho_n = \rho_n\alpha_1$. In particular, $\widetilde{\alpha}_1$ is injective and $\widetilde{\alpha}_1|_{\mathfrak{A}_0} = \alpha_1$. Finally

$$\operatorname{Ran}\widetilde{\alpha}_1 \supset \overline{\bigcup_{n\geq 0} \widetilde{\alpha}_1\rho_n(\mathfrak{A}_n)} = \overline{\bigcup_{n\geq 1} \rho_n\alpha_1(\mathfrak{A}_n)} = \overline{\bigcup_{n\geq 1} \rho_{n-1}(\mathfrak{A}_{n-1})} = \mathfrak{A}_\infty.$$

Therefore $\widetilde{\alpha}_1$ is surjective and thus is an automorphism. ∎

15.2.6. THEOREM. *Let $(\mathfrak{A}, \alpha, \mathbb{Z}_+)$ be an unital and injective dynamical system. Let $\mathfrak{A}_\infty$ be the C*-algebra obtained in Proposition 15.2.5, and let $(\mathfrak{A}_\infty, \widetilde{\alpha}, \mathbb{Z})$ be the $\mathbb{Z}$-dynamical system $\widetilde{\alpha}_n = \widetilde{\alpha}_1^n$ for $n \in \mathbb{Z}$. There is a completely isometric imbedding of $\mathfrak{A} \times_\alpha \mathbb{Z}_+$ into $\mathfrak{A}_\infty \rtimes_{\widetilde{\alpha}} \mathbb{Z}$. Moreover $\mathfrak{A}_\infty \rtimes_{\widetilde{\alpha}} \mathbb{Z}$ is a C*-cover of $\mathfrak{A} \times_\alpha \mathbb{Z}_+$.*

PROOF. Let π be a faithful representation of $\mathfrak{A}$ on H. By Exercise **9.26**, there is a representation σ of $\mathfrak{A}_\infty$ on a Hilbert space $K \supset H$ such that $\sigma(\mathfrak{A})$ reduces H and $\sigma(a)|_H = \pi(a)$ for $a \in \mathfrak{A}$. We may suppose that σ is also faithful. Let $\rho_0 : \mathfrak{A} \to \mathfrak{A}_\infty$ denote the inclusion.

By Corollary 15.1.11, the Fock representation $V \times \widetilde{\pi}$ is a completely isometric representation of $\mathfrak{A} \times_\alpha \mathbb{Z}_+$. By Theorem 9.14.9, $\mathfrak{A}_\infty \rtimes_{\widetilde{\alpha}} \mathbb{Z} = \mathfrak{A}_\infty \rtimes_{\widetilde{\alpha}r} \mathbb{Z}$; and thus by Proposition 9.14.7, the regular covariant representation $(1 \otimes \lambda, \widetilde{\sigma})$ yields a faithful, hence completely isometric, representation of $\mathfrak{A}_\infty \rtimes_{\widetilde{\alpha}} \mathbb{Z}$. Define a map from $\mathbb{Z}_+\mathfrak{A}$ into $\mathfrak{A}_\infty\mathbb{Z}$ by

$$\varphi \sum_{k=0}^{n} e_k a_k = \sum_{k=0}^{n} e_k \rho_0(a_k) = \sum_{k=0}^{n} \widetilde{\alpha}_k \rho_0(a_k) e_k.$$

It is routine to verify that this is an isomorphism based on the identity $\rho_0\alpha_k = \widetilde{\alpha}_k\rho_0$. Now $\widetilde{\sigma} \rtimes_{\widetilde{\alpha}} (1 \otimes \lambda)$ acts on $K \otimes l_2(\mathbb{Z})$. This has $H \otimes l_2(\mathbb{Z}_+)$ as a subspace which is invariant for $\widetilde{\sigma} \rtimes_{\widetilde{\alpha}} (1 \otimes \lambda)(\varphi(\mathbb{Z}_+\mathfrak{A}))$ and the restriction to this space is just $V \times \widetilde{\pi}(\mathbb{Z}_+\mathfrak{A})$. Therefore the norm on $\mathbb{Z}_+\mathfrak{A}$ in $\mathfrak{A}_\infty \rtimes_{\widetilde{\alpha}} \mathbb{Z}$ is at least as great as the norm in $\mathfrak{A} \times_\alpha \mathbb{Z}_+$.

On the other hand, $(1 \otimes \lambda|_{\mathbb{Z}_+}, \widetilde{\sigma}\rho_0)$ is easily seen to be a covariant representation of $(\mathbb{Z}_+, \mathfrak{A}, \alpha)$. Therefore the norm on $\mathbb{Z}_+\mathfrak{A}$ in $\mathfrak{A}_\infty \rtimes_{\widetilde{\alpha}} \mathbb{Z}$ is no larger than the norm in $\mathfrak{A} \times_\alpha \mathbb{Z}_+$. Hence they are equal. The same arguments work at the matrix level, and thus this imbedding is completely isometric.

Since $\mathfrak{A}_\infty = \overline{\bigcup_{n\geq 1} (1 \otimes \lambda_n)^* \rho_0(\mathfrak{A})(1 \otimes \lambda_n)}$, the smallest C*-algebra containing $\varphi(\mathbb{Z}_+\mathfrak{A})$ contains $\mathfrak{A}_\infty$ and is invariant under multiplication by $1 \otimes \lambda_n$ for $n \in \mathbb{Z}$, and thus $C^*(\varphi(\mathfrak{A} \times_\alpha \mathbb{Z}_+) = \mathfrak{A}_\infty \rtimes_{\widetilde{\alpha}} \mathbb{Z}$. Hence this is a C*-cover. ∎

We use the same notation in the next result.

15.2.7. THEOREM. *Let $(\mathfrak{A}, \alpha, \mathbb{Z}_+)$ be an unital and injective dynamical system. Then $C^*_{\min}(\mathfrak{A} \times_\alpha \mathbb{Z}_+) = \mathfrak{A}_\infty \rtimes_{\widetilde{\alpha}} \mathbb{Z}$.*

PROOF. By Lemma 15.1.8, for each $z \in \mathbb{T} = \widehat{\mathbb{Z}}$, there is a completely isometric automorphism γ_z of $A \times_\alpha \mathbb{Z}_+$ such that $\gamma_z(e_k a) = e_k z^k a$. By Proposition 14.9.14, this map extends to a $*$-automorphism Γ_z of $C^*_{\min}(\mathfrak{A} \times_\alpha \mathbb{Z}_+)$. Also by Remark 9.14.11, there is an automorphism σ_z of $\mathfrak{A}_\infty \rtimes_{\tilde{\alpha}} \mathbb{Z}$ such that $\sigma_z(\sum a_k u_k) = \sum z^k a_k u_k$; and these automorphisms leave $\mathfrak{A} \times_\alpha \mathbb{Z}_+$ invariant and extend γ_z. There is a canonical quotient map π of $\mathfrak{A}_\infty \rtimes_{\tilde{\alpha}} \mathbb{Z}$ onto $C^*_{\min}(\mathfrak{A} \times_\alpha \mathbb{Z}_+)$ which is the identity on $\mathfrak{A} \times_\alpha \mathbb{Z}_+$ Since π commutes the map γ_z on $\mathfrak{A} \times_\alpha \mathbb{Z}_+$, we must have $\Gamma_z \pi = \pi \sigma_z$. Therefore the Shilov ideal $J = \ker \pi$ is invariant under σ_z.

Again by Remark 9.14.11, the canonical (faithful) expectation $E : \mathfrak{A}_\infty \rtimes_{\tilde{\alpha}} \mathbb{Z} \to \mathfrak{A}_\infty$ is given by

$$E(x) = \frac{1}{2\pi} \int_{-\pi}^{\pi} \sigma_{e^{it}}(x)\, dt.$$

Likewise there is an expectation F on $C^*_{\min}(\mathfrak{A} \times_\alpha \mathbb{Z}_+)$ by

$$F(x) = \frac{1}{2\pi} \int_{-\pi}^{\pi} \Gamma_{e^{it}}(x)\, dt.$$

It follows that $F(\pi(x)) = \pi(E(x))$. In particular, $E(J) \subset J$. Thus if $J \neq 0$, since E is faithful, $J \cap \mathfrak{A}_\infty = J_0 \neq \{0\}$. Thus there is some integer n so that $J \cap \tilde{\alpha}_{-n}(\mathfrak{A}) \neq \{0\}$. Since J is an ideal, $\tilde{\alpha}_n(J) \cap \mathfrak{A} = u_n\big(J \cap \tilde{\alpha}_{-n}(\mathfrak{A})\big)u_n^* \neq \{0\}$. But π is isometric on $\mathfrak{A}$, so this is a contradiction. Therefore $J = \{0\}$ and hence $C^*_{\min}(\mathfrak{A} \times_\alpha \mathbb{Z}_+) = \mathfrak{A}_\infty \rtimes_{\tilde{\alpha}} \mathbb{Z}$. ∎

15.3. Recovering the Dynamics

One reason for the introduction of the semicrossed product was to provide an invariant for the dynamics. Even in the commutative case, complete invariants for a dynamical system have to be very complicated. In this section, we will show that for C*-systems over $\mathbb{Z}_+$ on an unital C*-algebra, the semicrossed product up to (completely) isometric isomorphism is such an invariant. This means that the dynamical system can be recovered, up to an appropriate equivalence, from the semicrossed product. In the case of a commutative C*-algebra, it is a complete invariant for dynamical systems up to topological conjugacy, which is the best one could hope for. In the noncommutative case, we need the notion of outer conjugacy.

15.3.1. DEFINITION. Let X be a compact Hausdorff space. A dynamical system on X is a homomorphism of a semigroup P, $h : P \to \mathrm{End}(X)$, into the continuous endomorphisms of X. Two dynamical systems (X, h, P) and (Y, k, P) are *topologically conjugate* if there is a homeomorphism $\eta : X \to Y$ such that $k_s = \eta h_s \eta^{-1}$ for $s \in P$.

Let $\mathfrak{A}$ and $\mathfrak{B}$ be unital C*-algebras. Two C*-dynamical systems $(\mathfrak{A}, \alpha, P)$ and $(\mathfrak{B}, \beta, P)$ are *conjugate* if there is a $*$-isomorphism $\gamma : \mathfrak{A} \to \mathfrak{B}$ such that

$\beta_s = \gamma\alpha_s\gamma^{-1}$ for $s \in P$. These systems are *outer conjugate* if there is a $*$-isomorphism $\gamma : \mathfrak{A} \to \mathfrak{B}$ and a unitary $w \in \mathfrak{B}$ such that

$$\beta_s(b) = \mathrm{ad}_w\, \gamma\alpha_s\gamma^{-1}(b) = w\big(\gamma\alpha_s\gamma^{-1}(b)\big)w^* \quad \text{for} \quad s \in P.$$

15.3.2. REMARK. A topological dynamical system (X, h, P) with an abelian semigroup P induces a C*-dynamical system $(C(X), \alpha, P)$ by $\alpha_s(f) = f \circ h_s$ for $s \in P$. Then

$$\alpha_t\alpha_s(f) = \alpha_t(f \circ h_s) = f \circ h_s \circ h_t = f \circ h_{s+t} = \alpha_{t+s}(f).$$

When P is non-abelian, define P^{op} to be P as a set with product $s * t := ts$. We then obtain a dynamical system $(C(X), \alpha, P^{op})$ by $\alpha_s(f) = f \circ h_s$ for $s \in P^{op}$. This is necessary since

$$\alpha_t\alpha_s(f) = \alpha_t(f \circ h_s) = f \circ h_s \circ h_t = f \circ h_{st} = f \circ h_{t*s} = \alpha_{t*s}(f).$$

Observe that α_s is injective precisely when h_s is surjective. Also α_s is surjective when h_s is injective because of Tietze's extension theorem.

Suppose that (X, h, P) and (X, k, P) are topologically conjugate via a homeomorphism $\eta : X \to Y$. Then η induces a $*$-isomorphism $\gamma : C(X) \to C(Y)$ by $\gamma(f) = f \circ \eta^{-1}$. Observe that

$$\gamma\alpha_s\gamma^{-1}(g) = g \circ \eta \circ h_s \circ \eta^{-1} = g \circ k_s = \beta_s(g).$$

Thus topologically conjugate systems yield conjugate C*-algebra systems, and this computation is reversible. Since $C(X)$ is abelian, conjugacy and outer conjugacy are the same.

When $\mathfrak{A}$ is noncommutative, outer conjugacy is a weaker relation than conjugacy. It turns out to be what we require here. It is easy to see that this relation is transitive (Exercise **15.4**). We need to show that it is symmetric.

15.3.3. PROPOSITION. *Suppose that two C*-dynamical systems $(\mathfrak{A}, \alpha, P)$ and $(\mathfrak{B}, \beta, P)$ are outer conjugate. This relation is symmetric, so that $(\mathfrak{B}, \beta, P)$ and $(\mathfrak{A}, \alpha, P)$ are outer conjugate. Moreover, $\mathfrak{A} \times_\alpha P$ and $\mathfrak{B} \times_\beta P$ are completely isometrically isomorphic.*

PROOF. Suppose that two C*-dynamical systems $(\mathfrak{A}, \alpha, P)$ and $(\mathfrak{B}, \beta, P)$ are outer conjugate via a $*$-isomorphism $\gamma : \mathfrak{A} \to \mathfrak{B}$ and a unitary $w \in \mathfrak{B}$. To reverse this, define $v = \gamma^{-1}(w)^*$ and compute

$$\begin{aligned}
\mathrm{ad}_v\, \gamma^{-1}\beta_s\gamma(a) &= \gamma^{-1}(w)^*\gamma^{-1}\beta_s\gamma(a)\gamma^{-1}(w) \\
&= \gamma^{-1}(w)^*\big(\gamma^{-1}\,\mathrm{ad}_w\, \gamma\alpha_s\gamma^{-1}\big)(\gamma(a))\gamma^{-1}(w) \\
&= \gamma^{-1}(w)^*\gamma^{-1}\big(w(\gamma\alpha_s\gamma^{-1})(\gamma(a))w^*\big)\gamma^{-1}(w) \\
&= \gamma^{-1}\big(\gamma\alpha_s(a)\big)\gamma^{-1}(w^*)\gamma^{-1}(w) = \alpha_s(a).
\end{aligned}$$

Thus $(\mathfrak{B}, \beta, P)$ and $(\mathfrak{A}, \alpha, P)$ are outer conjugate.

Let the elements of $P\mathfrak{B}$ be written as finite sums $\sum f_s b_s$ for $b_s \in \mathfrak{B}$. Define a covariant representation (γ, T) of $(P, \mathfrak{A}, \alpha)$ into $\mathfrak{B} \times_\beta P$ by $T_s = f_s w$. Compute

$$\begin{aligned}\gamma(a)T_s &= \gamma(a) f_s w = f_s \beta_s \gamma(a) w \\ &= f_s\big(w\big(\gamma\alpha_s\gamma^{-1}(\gamma(a)\big)w^*\big)w \\ &= (f_s w)\gamma\alpha_s(a) = T_s\gamma\alpha_s(a).\end{aligned}$$

Therefore there is a completely contractive homomorphism $T \times \gamma : \mathfrak{A} \times_\alpha P \to \mathfrak{B} \times_\beta P$. Similarly, (γ^{-1}, V), where $V_s = e_s\gamma^{-1}(w)^*$, is a covariant representation of $\mathfrak{B} \times_\beta P$ into $\mathfrak{A} \times_\alpha P$. Hence $V \times \gamma^{-1}$ is a completely contractive homomorphism. Observe that

$$\begin{aligned}(V \times \gamma^{-1})(T \times \gamma)(e_s a) &= V \times \gamma^{-1} f_s w \gamma(a) \\ &= e_s \gamma^{-1}(w)^* \gamma^{-1}(w\gamma(a)) = e_s a.\end{aligned}$$

Therefore $(V \times \gamma^{-1})(T \times \gamma)$ is the identity map on $\mathfrak{A} \times_\alpha P$. Similarly, $(T \times \gamma)(V \times \gamma^{-1})$ is the identity map on $\mathfrak{B} \times_\beta P$. So these maps are mutual inverses. As both are completely contractive, they must be completely isometric. ∎

The goal of this section is to establish the converse. The proof is quite algebraic and involves considerable calculation. We break the proof down into a series of lemmas to isolate the conceptual ideas. You may wish to read the statements of the lemmas and the proof of the main result first and return to check the details.

15.3.4. LEMMA. *Suppose that $\varphi : \mathfrak{A} \times_\alpha \mathbb{Z}_+ \to \mathfrak{B} \times_\beta \mathbb{Z}_+$ is an isometric isomorphism. Then $\gamma = \varphi|_{\mathfrak{A}}$ is a $*$-isomorphism of $\mathfrak{A}$ onto $\mathfrak{B}$, and $\varphi^{-1}|_{\mathfrak{B}} = \gamma^{-1}$.*

PROOF. We may suppose that $\mathfrak{B} \times_\beta \mathbb{Z}_+$ is represented completely isometrically on some Hilbert space H. A contractive homomorphism γ of a C*-algebra $\mathfrak{A}$ into $\mathcal{B}(H)$ must be a $*$-homomorphism. To see this, let $\xi \in H$ be a unit vector and consider $f(a) = \langle \gamma(a)\xi, \xi\rangle$. This is contractive and $f(1) = 1$. By Lemma 9.6.5, f is a state. In particular, if $a = a^*$, then $\langle \gamma(a)\xi, \xi\rangle \in \mathbb{R}$ for every unit vector $\xi \in H$. Therefore $\gamma(a) = \gamma(a)^*$. It follows that γ is a $*$-homomorphism.

In a Fock representation, it is easy to see that $(\mathfrak{B} \times_\beta \mathbb{Z}_+) \cap (\mathfrak{B} \times_\beta \mathbb{Z}_+)^* = \mathfrak{B}$. Such a representation is faithful by Corollary 15.1.11, so that $\mathfrak{B}$ is the largest C*-subalgebra of $\mathfrak{B} \times_\beta \mathbb{Z}_+$. Therefore $\gamma(\mathfrak{A}) \subset \mathfrak{B}$. Similarly $\varphi^{-1}|_{\mathfrak{B}}$ is a $*$-homomorphism and $\varphi^{-1}(\mathfrak{B}) \subset \mathfrak{A}$. It follows that γ is a $*$-isomorphism onto $\mathfrak{B}$ and $\varphi^{-1}|_{\mathfrak{B}} = \gamma^{-1}$. ∎

15.3.5. LEMMA. *Suppose that $\varphi : \mathfrak{A} \times_\alpha \mathbb{Z}_+ \to \mathfrak{B} \times_\beta \mathbb{Z}_+$ is an isometric isomorphism. If $E_0(\varphi(e_1)) = 0$, then $(\mathfrak{A}, \alpha, \mathbb{Z}_+)$ and $(\mathfrak{B}, \beta, \mathbb{Z}_+)$ are outer conjugate.*

PROOF. Let $V = \varphi(e_1)$ and let $E_1(V) = f_1 w$ for $w \in \mathfrak{B}$. Then $V = f_1 w + f_2 Y$ for some $Y \in \mathfrak{B} \times_\beta \mathbb{Z}_+$. Similarly, we can write $\varphi^{-1}(f_1) = a_0 + e_1 a_1 + e_2 X$. Then

$$\begin{aligned} f_1 &= \varphi(a_0 + e_1 a_1 + e_2 X) \\ &= \gamma(a_0) + (f_1 w + f_2 Y)\gamma(a_1) + (f_1 w + f_2 Y)^2 \varphi(X) \\ &= \gamma(a_0) + f_1(w\gamma(a_1)) + f_2 Z. \end{aligned}$$

Hence $\gamma(a_0) = 0$; whence $0 = a_0 = E_0\varphi^{-1}(f_1)$. Also $w\gamma(a_1) = 1$. Since $E_0\varphi^{-1}(f_1) = 0$, we can reverse the roles of $\mathfrak{A}$ and $\mathfrak{B}$. The corresponding calculation yields $a_1\gamma^{-1}(w) = 1$ as well; so $\gamma(a_1)w = 1$. Note that $\|w\| \le \|V\| = \|e_1\| = 1$ and likewise $\|a_1\| \le 1$. It follows that w is unitary and $\gamma(a_1) = w^*$.

For $a \in \mathfrak{A}$, compute

$$\begin{aligned} \varphi(ae_1) &= \gamma(a)(f_1 w + f_2 Y) = f_1 \beta\gamma(a) w + f_2 \beta^2\gamma(a) Y \\ &= \varphi(e_1 \alpha(a)) = f_1 w \gamma\alpha(a) + f_2 Y \gamma\alpha(a). \end{aligned}$$

By the uniqueness of the Fourier coefficients, we obtain

$$\beta\gamma(a) w = w\gamma\alpha(a) \quad \text{for} \quad a \in \mathfrak{A}.$$

Therefore, setting $a = \gamma^{-1}(b)$,

$$\beta(b) = w\big(\gamma\alpha\gamma^{-1}(b)\big)w^* \quad \text{for} \quad b \in \mathfrak{B},$$

which is outer conjugacy. ∎

15.3.6. LEMMA. *Suppose that $\varphi : \mathfrak{A} \times_\alpha \mathbb{Z}_+ \to \mathfrak{B} \times_\beta \mathbb{Z}_+$ is a completely isometric isomorphism. Then $\|E_0(\varphi(e_1)\| < 1$.*

PROOF. By Lemma 15.3.4, $\gamma = \varphi|_\mathfrak{A}$ is a $*$-isomorphism onto $\mathfrak{B}$. The system $(\mathfrak{B}, \beta, \mathbb{Z}_+)$ is conjugate to $(\mathfrak{A}, \gamma^{-1}\beta\gamma, \mathbb{Z}_+)$. This yields the completely isometric isomorphism $\psi : \mathfrak{B} \times_\beta \mathbb{Z}_+ \to \mathfrak{A} \times_{\gamma^{-1}\beta\gamma} \mathbb{Z}_+$ via the map

$$\psi(\sum f_i b_i) = \sum f_i \gamma^{-1}(b_i) \quad \text{for} \quad \sum f_i b_i \in \mathbb{Z}_+ \mathfrak{B}.$$

Note that $E_0\psi = \psi E_0$, so that $\|E_0(\psi\varphi(e_1))\| = \|E_0(\varphi(e_1))\|$. The map $\psi\varphi$ has $\psi\varphi|_\mathfrak{A} = id|_\mathfrak{A}$. So we may assume that $\mathfrak{B} = \mathfrak{A}$ and $\gamma = id_\mathfrak{A}$.

Let $v = \varphi(e_1) \sim \sum_{k \ge 0} f_k b_k$. Then $\|b_0\| = \|E_0(v)\| \le \|v\| = 1$. Suppose that $\|b_0\| = 1$. Choose a state g on $\mathfrak{A}$ such that $g(b_0^* b_0) = 1$, and let π_g be the GNS representation for g with cyclic vector ξ_g. Note that

$$1 = g(b_0^* b_0) = \langle \pi_g(b_0)^* \pi_g(b_0)\xi_g, \xi_g\rangle = \|\pi_g(b_0)\xi_g\|^2.$$

Thus $\eta = \pi_g(b_0)\xi_g$ is also norm 1.

If π_0 is any faithful representation of $\mathfrak{A}$, set $\pi = \pi_g \oplus \pi_0$, and consider the Fock representations $V \times \widetilde{\pi}$ of $\mathfrak{A} \times_\alpha \mathbb{Z}_+$ and $V \times \overline{\pi}$ of $\mathfrak{A} \times_\beta \mathbb{Z}_+$, where

$$\widetilde{\pi} = \pi \oplus \pi\alpha \oplus \pi\alpha^2 \oplus \dots \quad \text{and} \quad \overline{\pi} = \pi \oplus \pi\beta \oplus \pi\beta^2 \oplus \dots.$$

The unit vector $\xi_g \otimes \delta_0 \in H \otimes l_2(\mathbb{Z}_+)$ satisfies

$$V \times \overline{\pi}(v)(\xi_g \otimes \delta_0) = \sum_{k\geq 0} V_k\big(\pi(b_k)\xi_g \otimes \delta_0\big) = \eta \otimes \delta_0 + \sum_{k\geq 1} \pi(b_k)\xi_g \otimes \delta_k.$$

Since $\|v\| = 1 = \|\eta\|$, we have that $\pi(b_k)\xi_g = 0$ for $k \geq 1$.

Now $v = \lim_{n\to\infty} \Sigma_n(v)$. Therefore

$$e_1 = \varphi^{-1}(v) = \lim_{n\to\infty} \sum_{k=0}^{n} (1 - \tfrac{k}{n+1})\varphi^{-1}(f_k)b_k.$$

It follows that

$$\begin{aligned} V \times \widetilde{\pi}(e_1)(\xi_g \otimes \delta_0) &= \lim_{n\to\infty} \sum_{k=0}^{n} (1 - \tfrac{k}{n+1}) V \times \widetilde{\pi}(\varphi^{-1}(f_k))\pi(b_k)\xi_g \otimes \delta_0 \\ &= V \times \widetilde{\pi}(\varphi^{-1}(f_0))\eta \otimes \delta_0 = \eta \otimes \delta_0 \end{aligned}$$

because $\varphi^{-1}(f_0) = e_0$ is the identity element. This is absurd since $V \times \widetilde{\pi}(e_1) = V_1$ has range orthogonal to $H \otimes \mathbb{C}\delta_0$. Hence we must have $\|b_0\| < 1$. ∎

Recall that if $c \in \mathfrak{A}$ with $\|c\| \leq 1$, then $D_c = (1 - c^*c)^{1/2}$. This is invertible if $\|c\| < 1$. Moreover $D_{c^*}c = cD_c$ and $D_{c^*}^{-1}c = cD_c^{-1}$.

We need the following computational lemma. The definition of $\rho(v)$ should remind you of a conformal automorphism of $\mathbb{D}$. If $c \in \mathbb{D}$, then $f(z) = \frac{c-z}{1-\bar{c}z}$ takes 0 to c, maps $\mathbb{D}$ conformally onto itself, and $f(f(z)) = z$.

15.3.7. Lemma. *Let $(\mathfrak{A}, \alpha, \mathbb{Z}_+)$ be a C*-dynamical system, let $c \in \mathfrak{A}$ with $\|c\| < 1$ such that $ac = c\alpha(a)$ for all $a \in \mathfrak{A}$. Then there is a completely isometric automorphism ρ of $\mathfrak{A} \times_\alpha \mathbb{Z}_+$ such that $\rho|_{\mathfrak{A}} = \mathrm{id}_{\mathfrak{A}}$ and*

$$\rho(e_1) = D_{c^*}(1 - e_1c^*)^{-1}(c - e_1)D_c^{-1}$$

such that $\rho^2 = \mathrm{id}$ and $E_0(\rho(e_1)) = c$.

Proof. We can work in the Fock representation, so that $v = V \times \widetilde{\pi}(e_1)$ is an isometry. We suppress the use of $\widetilde{\pi}$ for clarity. Then

$$\begin{aligned} (1 - cv^*)^{-1}D_{c^*}^2(1 - vc^*)^{-1} &= (1 - cv^*)^{-1}(1 - cc^*)(1 - vc^*)^{-1} \\ &= (1 - cv^*)^{-1}\big((1 - cv^*) + (1 - vc^*) - (1 - cv^*)(1 - vc^*)\big)(1 - vc^*)^{-1} \\ &= (1 - cv^*)^{-1} + (1 - vc^*)^{-1} - 1. \end{aligned}$$

Let $w = D_{c^*}(1 - vc^*)^{-1}(c - v)D_c^{-1}$. Then

$$\begin{aligned} w^*w &= D_c^{-1}(c^* - v^*)\big((1 - cv^*)^{-1}D_{c^*}^2(1 - vc^*)^{-1}\big)(c - v)D_c^{-1} \\ &= D_c^{-1}v^*(vc^* - 1)\big((1 - cv^*)^{-1} + (1 - vc^*)^{-1} - 1\big)(cv^* - 1)vD_c^{-1} \\ &= D_c^{-1}v^*\big((1 - cv^*) + (1 - vc^*) - (1 - vc^*)(1 - cv^*)\big)vD_c^{-1} \\ &= D_c^{-1}v^*(1 - vc^*cv)vD_c^{-1} = D_c^{-1}(1 - c^*c)D_c^{-1} = 1. \end{aligned}$$

Therefore w is an isometry.

If $x_1, x_2 \in \mathfrak{A} \times_\alpha \mathbb{Z}_+$ satisfy $ax_i = x_i\alpha(a)$ for all $a \in \mathfrak{A}$, then

$$a(x_1x_2^*) = x_1\alpha(a)x_2^* = x_1(x_2\alpha(a^*))^* = x_1(a^*x_2)^* = x_1x_2^*a.$$

Thus cc^*, cv^* and vc^* lie in $\mathfrak{A}'$, as well as D_c and D_c^{-1}. Compute

$$\begin{aligned} aw &= aD_{c^*}(1 - vc^*)^{-1}(c - v)D_c^{-1} = D_{c^*}(1 - vc^*)^{-1}a(c - v)D_c^{-1} \\ &= D_{c^*}(1 - vc^*)^{-1}(c - v)\alpha(a)D_c^{-1} = D_{c^*}(1 - vc^*)^{-1}(c - v)D_c^{-1}\alpha(a) \\ &= w\alpha(a). \end{aligned}$$

It follows that there is a completely contractive representation ρ of $\mathfrak{A} \times_\alpha \mathbb{Z}_+$ such that $\rho|_{\mathfrak{A}} = id_{\mathfrak{A}}$ and $\rho(v) = w$.

Next compute

$$\begin{aligned} (1-vc^*)D_{c^*}^{-1}(1-wc^*) &= (1-vc^*)D_{c^*}^{-1}\big(1-D_{c^*}(1-vc^*)^{-1}(c-v)D_c^{-1}c^*\big) \\ &= (1-vc^*)D_{c^*}^{-1} - (c-v)c^*D_{c^*}^{-1} \\ &= (1-vc^*-cc^*+vc^*)D_{c^*}^{-1} = D_{c^*}. \end{aligned}$$

Hence $D_{c^*}(1-wc^*)^{-1} = (1 - vc^*)D_{c^*}^{-1}$. Similarly,

$$\begin{aligned} (1-vc^*)D_{c^*}^{-1}(c-w) &= (1-vc^*)D_{c^*}^{-1}c - (1-vc^*)D_{c^*}^{-1}D_{c^*}(1-vc^*)^{-1}(c-v)D_c^{-1} \\ &= (1-vc^*)cD_c^{-1} - (c-v)D_c^{-1} \\ &= (c - vc^*c - c + v)D_c^{-1} = vD_c. \end{aligned}$$

Putting this together, we obtain

$$\begin{aligned} \rho^2(v) = \rho(w) &= D_{c^*}(1 - wc^*)^{-1}(c - w)D_c^{-1} \\ &= (1-vc^*)D_{c^*}^{-1}(c - w)D_c^{-1} = vD_cD_c^{-1} = v. \end{aligned}$$

We deduce that $\rho^2|_{\mathfrak{A}} = id_{\mathfrak{A}}$ and $\rho^2(v) = v$, and hence $\rho^2 = id$. Since ρ is completely contractive and ρ^2 is completely isometric, ρ is a completely isometric automorphism. Finally, since E_0 is a homomorphism such that $E_0|_{\mathfrak{A}} = id_{\mathfrak{A}}$ and $E_0(v) = 0$, we get

$$E_0(w) = D_{c^*}(1 - 0c^*)^{-1}(c - 0)D_c^{-1} = cD_cD_c^{-1} = c. \qquad \blacksquare$$

Putting these lemmas together, the proof of the main result is now straightforward.

15.3.8. THEOREM. *Suppose that $\varphi : \mathfrak{A} \times_\alpha \mathbb{Z}_+ \to \mathfrak{B} \times_\beta \mathbb{Z}_+$ is an isometric isomorphism. Then $(\mathfrak{A}, \alpha, \mathbb{Z}_+)$ and $(\mathfrak{B}, \beta, \mathbb{Z}_+)$ are outer conjugate.*

PROOF. Let $b = E_0(\varphi(e_1))$. By Lemma 15.3.6, $\|b\| < 1$. Set $c = \gamma^{-1}(b)$. Compute for $a \in \mathfrak{A}$,

$$\begin{aligned} E_0(\varphi(ae_1)) &= E_0(\gamma(a)\varphi(e_1)) = \gamma(a)b \\ &= E_0(\varphi(e_1\alpha(a))) = E_0(\varphi(e_1)\gamma\alpha(a)) = b\gamma\alpha(a). \end{aligned}$$

Applying γ^{-1}, we obtain $ac = c\alpha(a)$ for $a \in \mathfrak{A}$.

Let ρ be the automorphism of $\mathfrak{A} \times_\alpha \mathbb{Z}_+$ determined by c as in Lemma 15.3.7. Define $\psi = \varphi\rho$. This is an isometric isomorphism of $\mathfrak{A} \times_\alpha \mathbb{Z}_+$ onto $\mathfrak{B} \times_\beta \mathbb{Z}_+$. Since E_0 is a homomorphism,

$$\begin{aligned} E_0\big(\psi(e_1)\big) &= E_0\big(\varphi\big(D_{c^*}(1 - e_1c^*)^{-1}(c - e_1)D_c^{-1}\big)\big) \\ &= E_0\big(D_{b^*}(1 - \varphi(e_1)b^*)^{-1}(b - \varphi(e_1))D_b^{-1}\big) \\ &= D_{b^*}D_{b^*}^{-2}(b - b)D_b^{-1} = 0. \end{aligned}$$

Therefore by Lemma 15.3.5, $(\mathfrak{A}, \alpha, \mathbb{Z}_+)$ and $(\mathfrak{B}, \beta, \mathbb{Z}_+)$ are outer conjugate. ∎

Free Semigroup Algebras

Free semigroup algebras are WOT-closed unital algebras generated by isometries $S_1, \dots, S_d$ with pairwise orthogonal ranges. We first show that these d-tuples are the universal models for row contractions. This yields a multivariable von Neumann inequality. The prototype free semigroup algebra is $\mathfrak{L}_d$, generated by the d canonical shifts on Fock space, $\ell_2(\mathbb{F}_d^+)$, coming from the left regular representation of the free semigroup $\mathbb{F}_d^+$. We examine this algebra in detail and obtain a functional calculus for pure row contractions. Moreover we establish a Beurling-type theorem for its invariant subspaces. We study general free semigroup algebras and obtain a structure theorem. Finally we return to $\mathfrak{L}_d$ to study its ideal lattice.

15.4. Row Contractions

The dilation theory for a row contraction closely parallels the Sz.-Nagy dilation theory of a contraction.

15.4.1. DEFINITION. A d-tuple $T = \big[T_1, \ \dots, \ T_d\big] \in \mathcal{B}(H^{(d)}, H)$ is a *row contraction* if $\|T\| \le 1$. It is a *row unitary* (*row isometry*) if T is a unitary (isometry).

Note that $\|T\|^2 = \|TT^*\| = \left\|\sum_{i=1}^d T_iT_i^*\right\|$. Thus T is a row contraction if $TT^* \le I$. To be a row isometry, each T_i must be an isometry, and since isometries take orthogonal subspaces to orthogonal subspaces, we must have $T_iH \perp T_jH$ for $i \le j$. Algebraically this means that $T_i^*T_j = \delta_{ij}I$ for $1 \le i,j \le d$. If each T_i is known to be an isometry, then $TT^* \le I$ implies that their ranges are orthogonal. To be a row unitary, we require in addition that $TT^* = \sum_{i=1}^d T_iT_i^* = I$. In this case, $C^*(T_1,\dots,T_d) \simeq \mathcal{O}_d$ is the Cuntz algebra. If T is a row isometry with $TT^* < I$, then $C^*(T_1,\dots,T_d) \simeq \mathcal{E}_n$, the Cuntz–Toeplitz algebra. See Section 9.11.

One way to obtain a proper row isometry arises from the left regular representation of $\mathbb{F}_d^+$, the free semigroup consisting of all words in $\{1,2,\dots,d\}$. The empty word $\varnothing$ is the identity element, and multiplication is just concatenation of words. That is, if $u = i_1 \dots i_k$ and $v = j_1 \dots j_l$ for $1 \le i_s, j_t \le d$, then $uv = i_1 \dots i_k j_1 \dots j_l$. Fock space is the space $F_d^2 := \ell_2(\mathbb{F}_d^+)$. Let $\{\xi_w : w \in \mathbb{F}_d^+\}$ be the standard orthonormal basis for F_d^2. Define $L_i \in \mathcal{B}(F_d^2)$ by

$$L_i\xi_w = \xi_{iw} \quad \text{for} \quad 1 \le i \le d \text{ and } w \in \mathbb{F}_d^+.$$

It is evident that each L_i is an isometry with $\operatorname{Ran} L_i = \operatorname{span}\{\xi_{iw} : w \in \mathbb{F}_d^+\}$, so their ranges are pairwise orthogonal. Moreover,

$$\sum_{i=1}^d L_iL_i^* = I - \xi_\varnothing\xi_\varnothing^* < I.$$

Thus $L = \begin{bmatrix} L_1, & \dots, & L_d \end{bmatrix}$ is a row isometry which is not a row unitary. We call it the *d-shift*. This plays the role of the unilateral shift in this noncommutative context.

We will write $L_u = L_{i_1} \dots L_{i_k}$ for $u = i_1 \dots i_k$, and similarly $T_u = T_{i_1} \dots T_{i_k}$ for any row contraction. If T is a row isometry, then each T_u is an isometry. In this case, because $T_j^*T_i = \delta_{ij}I$,

$$T_v^*T_u = T_{j_l}^* \dots T_{j_1}^*T_{i_1} \dots T_{i_k} = \begin{cases} T_{i_{l+1}} \dots T_{i_k} & \text{if } u = vu' \\ T_{j_l}^* \dots T_{j_{k+1}}^* & \text{if } v = uv' \\ 0 & \text{otherwise} \end{cases}.$$

There are many ways to get row unitaries. It is useful to recall Example 9.11.9. The details are repeated later in Example 15.7.3. For now, we just want to observe that the row unitary S obtained acts on $K = \mathbb{C}^k \oplus H^{(k(d-1))}$, where $H = F_d^2$. The restriction of S to one of the copies of F_d^2 is unitarily equivalent to L. Thus S is an extension of L to a row unitary.

The proof of the Wold decomposition is easily modified to obtain an analogous decomposition of row isometries. The role of the unilateral shift is played by L.

15.4.2. Frazho–Popescu–Wold Decomposition.
Let $T = \begin{bmatrix} T_1, & \dots, & T_d \end{bmatrix}$ in $\mathcal{B}(H^{(d)}, H)$ be a row isometry. Then there is a cardinal α and a representation π of $\mathcal{O}_d$ with $\pi(\mathfrak{s}_i) = S_i$ so that there is a simultaneous unitary equivalence $T_i \simeq L_i^{(\alpha)} \oplus S_i$ for $1 \le i \le d$.

PROOF. The proof follows the proof of the Wold decomposition. Let $W = \operatorname{Ran}(I - \sum_{i=1}^d T_i T_i^*)$. This is a *wandering space* for T in the sense that the subspaces $T_v W$ for $v \in \mathbb{F}_d^+$ are pairwise orthogonal. Indeed, if $\xi, \eta \in W$, then $\langle T_v \xi, \eta \rangle = 0$ because $\operatorname{Ran} T_v \subset W^\perp$ by definition. Therefore if $|v| \geq |w|$,

$$\langle T_v \xi, T_w \eta \rangle = \langle T_w^* T_v \xi, \eta \rangle = 0 \quad \text{if} \quad w \neq v.$$

Let $\alpha = \dim W$ and let $\{\xi_{j,\varnothing} : j \in A\}$ be an orthonormal basis for W, where $|A| = \alpha$. Define $\xi_{j,w} := T_w \xi_{j,\varnothing}$ for $j \in A$ and $w \in \mathbb{F}_d^+$. Then $\{\xi_{j,w} : j \in A\}$ is an orthonormal basis for $T_w W$. Each $\xi_{j,\varnothing}$ is a *wandering vector* for T because $\{\xi_{j,w} : w \in \mathbb{F}_d^+\}$ is orthonormal. Let $F_j = \operatorname{span}\{\xi_{j,w} : w \in \mathbb{F}_d^+\}$ for $j \in A$. Then F_j are pairwise orthogonal subspaces such that

$$F = \sum_{j \in A}^{\oplus} F_j = \sum_{w \in \mathbb{F}_d^+}^{\oplus} T_w W.$$

Observe that $T_i \xi_{j,w} = \xi_{j,iw}$. Thus F_j reduces T and $T_i|_{F_j} \simeq L_i$ in a canonical way. Therefore $T|_F \simeq L^{(\alpha)}$.

Let $K = F^\perp$. This also reduces T, and $T|_K$ is a row isometry. Let $S_i = T_i|_K$. Moreover $\operatorname{Ran} \sum_{i=1}^d T_i T_i^* = W^\perp \supset K$. Therefore $\sum_{i=1}^d S_i S_i^* = P_K$; whence $C^*(S_1, \ldots, S_d) \simeq \mathcal{O}_d$. Thus $T \simeq L^{(\alpha)} \oplus S$. ∎

15.4.3. DEFINITION. Say that a row isometry T is *pure* if $T \simeq L^{(\alpha)}$, meaning that there is no Cuntz part in the Wold decomposition.

15.4.4. REMARK. It is not hard to see that T is pure if and only if

$$\bigcap_{k \geq 1} \sum_{|w|=k} \operatorname{Ran} T_w = \{0\}.$$

Equivalently, SOT–$\lim_{n \to \infty} \sum_{|w|=k} T_w T_w^* = 0$.

15.4.5. COROLLARY. *Let $T = [T_1, \ldots, T_d] \in \mathcal{B}(H^{(d)}, H)$ be a row isometry. Then*

$$C^*(T_1, \ldots, T_d) \simeq \begin{cases} \mathcal{E}_d & \text{if } \alpha \geq 1 \\ \mathcal{O}_d & \text{if } \alpha = 0. \end{cases}$$

15.4.6. DEFINITION. The *noncommutative disc algebra* $\mathfrak{A}_d$ is the closed unital non-selfadjoint subalgebra of $\mathcal{O}_d$ generated by $\mathfrak{s}_1, \ldots, \mathfrak{s}_d$.

15.4.7. PROPOSITION. *Let $\mathcal{A}_d = \overline{\operatorname{Alg}\{I, L_1, \ldots, L_d\}} \subset \mathcal{E}_d$. Then there is a completely isometric map $\varphi : \mathfrak{A}_d \to \mathcal{A}_d$ such that $\varphi(\mathfrak{s}_i) = L_i$.*

PROOF. Consider the representation $\sigma_{1,1}$ of $\mathcal{O}_d$ given in Example 9.11.9. This is defined on $\mathbb{C}\eta \oplus F_d^{2\,(d-1)}$ by $S_1\eta = \eta$, $S_i\eta = \xi_{i,\varnothing}$ for $2 \le i \le d$ and $S_i|_{F_d^{2\,(d-1)}} = L_i^{(d-1)}$. The map φ is obtained by the compression to a copy of F_d^2, which is invariant, so this is a completely contractive homomorphism. Conversely, the quotient map $q : \mathcal{E}_d \to \mathcal{O}_d$ is a completely contractive homomorphism, and carries L_i onto $\mathfrak{s}_i$. It is clear that $q\varphi = id_{\mathfrak{A}_d}$ since this holds for the generators. Therefore φ is a complete isometry. ∎

We extend the definition of purity for a row isometry to row contractions.

15.4.8. DEFINITION. A row contraction $T = \big[T_1, \dots, T_d\big]$ is *pure* if

$$\text{SOT-}\lim_{n\to\infty} \sum_{|w|=k} T_w T_w^* = 0.$$

If we knew that every row contraction led to a completely contractive representation of $\mathfrak{A}_d$, we could invoke Arveson's dilation theorem in order to dilate a row contraction to a row isometry. However, that is not immediately evident; and the counterexamples for commuting tuples of contractions should suggest that it is not obvious. Instead we modify the proof of Sz.-Nagy's theorem.

15.4.9. FRAZHO–BUNCE–POPESCU DILATION THEOREM.
Let $T = \big[T_1, \dots, T_d\big] \in \mathcal{B}(H^{(d)}, H)$ be a row contraction. Then T has a co-extension to a row isometry. The minimal row isometric dilation is unique up to unitary equivalence fixing H. The minimal row dilation is pure if and only if T is a pure row contraction. In particular, this holds if $\|T\| < 1$.

PROOF. Note that $T^*T \le I_d$ in $\mathcal{B}(H^{(d)}) \simeq M_n(\mathcal{B}(H))$. So we can define $D_T = (I_d - T^*T)^{1/2}$; and let $D = \overline{\operatorname{Ran} D_T} = (\ker D_T)^\perp$. Set $K = H \oplus (D \otimes F_d^2)$. As in the case of a single operator, we have an isometry

$$\begin{bmatrix} T \\ P_D D_T \end{bmatrix} = \begin{bmatrix} T_1 & T_2 & \dots & T_d \\ D_1 & D_2 & \dots & D_d \end{bmatrix} \in \mathcal{B}(H^{(d)}, H \oplus D),$$

where D_i are the columns of $P_D D_T$. Therefore the columns $\begin{bmatrix} T_i \\ D_i \end{bmatrix} \in \mathcal{B}(H, H \oplus D)$ are isometries with pairwise orthogonal ranges. Let $J : D \to D \otimes F_d^2$ by $J\eta = \eta \otimes \xi_\varnothing$. Define $S_i \in \mathcal{B}(K)$ by

$$S_i = \begin{bmatrix} T_i & 0 \\ JD_i & I_D \otimes L_i \end{bmatrix} \quad \text{for} \quad 1 \le i \le d.$$

These are readily seen to be isometries with pairwise orthogonal ranges which co-extend T_i.

Uniqueness is established as in the Sz.-Nagy dilation. Minimality means that $K = \overline{\operatorname{span}\{S_w H : w \in \mathbb{F}_d^+\}}$. Suppose that $S' = \begin{bmatrix} S_1', \dots, S_d' \end{bmatrix}$ is another minimal dilation on $K' = \overline{\operatorname{span}\{S_w' H : w \in \mathbb{F}_d^+\}}$. Define

$$U \sum_{|w| \le n} S_w \xi_w = \sum_{|w| \le n} S_w' \xi_w \quad \text{for} \quad n \ge 1 \text{ and } \xi_w \in H.$$

Compute the norm

$$\begin{aligned}
\Big\| \sum_{|w| \le n} S_w \xi_w \Big\|^2 &= \sum_{|w| \le n} \sum_{|v| \le n} \langle S_w \xi_w, S_v \xi_v \rangle \\
&= \sum_{|v| \le |w| \le n} \langle S_v^* S_w \xi_w, \xi_v \rangle + \sum_{|w| < |v| \le n} \langle \xi_w, S_w^* S_v \xi_v \rangle \\
&= \sum_{\substack{w = vw' \\ |v| + |w'| \le n}} \langle S_{w'} \xi_w, \xi_v \rangle + \sum_{\substack{v = wv' \\ |w| + |v'| \le n \\ |v'| \ge 1}} \langle \xi_w, S_{v'} \xi_v \rangle \\
&= \sum_{\substack{w = vw' \\ |v| + |w'| \le n}} \langle T_{w'} \xi_w, \xi_v \rangle + \sum_{\substack{v = wv' \\ |w| + |v'| \le n \\ |v'| \ge 1}} \langle \xi_w, T_{v'} \xi_v \rangle
\end{aligned}$$

The final expression does not depend on the dilation, showing that U is isometric from a dense subset of K onto a dense subset of K'. Therefore it extends by continuity to a well-defined unitary operator. If $\xi \in H$, $U\xi = US_\varnothing \xi = S'_\varnothing \xi = \xi$; so in particular, $U|_H = I_H$. By construction, $US_i = S_i' U$; so the two dilations are unitarily equivalent.

Observe that there are operators D_w so that

$$S_w = \begin{bmatrix} T_w & 0 \\ D_w & I_D \otimes L_w \end{bmatrix} \quad \text{and} \quad S_w^* = \begin{bmatrix} T_w^* & D_w^* \\ 0 & I_D \otimes L_w^* \end{bmatrix}.$$

Because $I_D \otimes L$ is pure, we obtain

$$\begin{aligned}
\text{SOT--}\lim_{n \to \infty} \sum_{|w| = k} S_w S_w^* &= \begin{bmatrix} \text{SOT--}\lim\limits_{n \to \infty} \sum_{|w|=k} T_w T_w^* & * \\ * & \text{SOT--}\lim\limits_{n \to \infty} \sum_{|w|=k} I_D \otimes L_w L_w^* \end{bmatrix} \\
&= \begin{bmatrix} \text{SOT--}\lim\limits_{n \to \infty} \sum_{|w|=k} T_w T_w^* & * \\ * & 0 \end{bmatrix}
\end{aligned}$$

This limit is positive, and thus the $*$'s are 0. By Remark 15.4.4, the row isometry S is pure if and only if $\text{SOT--}\lim_{n \to \infty} \sum_{|w|=k} T_w T_w^* = 0$; i.e. if T is pure. If $\|T\| = r < 1$, then

$$\| \sum_{|w|=k} T_w T_w^* \| \le \|T\|^{2n} = r^{2n}.$$

Therefore this sequence converges to 0 in norm, and S is pure. ∎

Let $\mathbb{C}\langle Z_1,\dots,Z_d\rangle$ denote the space of polynomials in non-commuting variables $Z_1,\dots,Z_d$. This is the span of the monomials $Z_w = Z_{i_1}\dots Z_{i_k}$ for $w \in \mathbb{F}_d^+$.

15.4.10. COROLLARY. *Let $T = \begin{bmatrix} T_1, & \dots, & T_d \end{bmatrix} \in \mathcal{B}(H^{(d)}, H)$ be a row contraction. For any non-commuting matrix polynomial $P = [p_{ij}] \in M_n(\mathbb{C}\langle Z_1,\dots,Z_d\rangle)$,*

$$\|P(T)\| = \left\| \left[p_{ij}(T_1,\dots,T_d) \right] \right\| \le \left\| \left[p_{ij}(\mathfrak{s}_1,\dots,\mathfrak{s}_d) \right] \right\|_{M_n(\mathcal{O}_d)}.$$

This extends to a completely contractive representation of the noncommutative disc algebra, $\mathfrak{A}_d$. Moreover every completely contractive representation of $\mathfrak{A}_d$ arises in this way.

PROOF. The row contraction T dilates to a row isometry S. By the Frazho–Popescu–Wold decomposition, $S \simeq L^{(\alpha)} \oplus U$ where L is the row isometry generating $\mathcal{E}_d$ and U is a row unitary generating a copy of $\mathcal{O}_d$. By Proposition 15.4.7, the quotient map from $\mathcal{E}_d$ onto $\mathcal{O}_d$ is completely isometric from the algebra $\mathcal{A}_d$ onto $\mathfrak{A}_d$. Thus

$$\|P(T)\| \le \|P(L \oplus U)\| = \left\| \left[p_{ij}(\mathfrak{s}_1,\dots,\mathfrak{s}_d) \right] \right\|_{M_n(\mathcal{O}_d)}.$$

The norm on $M_n(\mathcal{O}_d)$ yields the norm on $M_n(\mathfrak{A}_d)$ by definition, so the map

$$\rho_T(p(\mathfrak{s}_1,\dots,\mathfrak{s}_d)) = p(T_1,\dots,T_d) \quad \text{for} \quad p \in \mathbb{C}\langle Z_1,\dots,Z_d\rangle$$

is completely contractive on a dense subset of $\mathfrak{A}_d$, and hence extends by continuity to a representation of $\mathfrak{A}_d$. Conversely, if ρ is a completely contractive representation of $\mathfrak{A}_d$, let $T_i = \rho(\mathfrak{s}_i)$. Then

$$T = \begin{bmatrix} T_1, & \dots, & T_d \end{bmatrix} = \rho_{1,n}\left(\begin{bmatrix} \mathfrak{s}_1, \dots, \mathfrak{s}_d \end{bmatrix} \right)$$

has $\|T\| \le \left\| \begin{bmatrix} \mathfrak{s}_1, \dots, \mathfrak{s}_d \end{bmatrix} \right\| = 1$, so it is a row contraction. Hence $\rho = \rho_T$. ∎

We now have a couple of ways to obtain the equivalent of a unitary dilation. Note, however, that there is no uniqueness.

15.4.11. COROLLARY. *Let $T = \begin{bmatrix} T_1, & \dots, & T_d \end{bmatrix} \in \mathcal{B}(H^{(d)}, H)$ be a row contraction. Then T has an $\mathcal{O}_d$-dilation to a row unitary $U = \begin{bmatrix} U_1, & \dots, & U_d \end{bmatrix}$.*

Also $\mathcal{O}_d = C^*_{\min}(\mathfrak{A}_d)$. Every irreducible representation of $\mathcal{O}_d$ is a boundary representation of $\mathfrak{A}_d$.

PROOF. One proof is to invoke Arveson's dilation theorem, now that we know that ρ_T is a completely contractive representation of $\mathfrak{A}_d$. This yields an $\mathcal{O}_d$-dilation π and thus a row unitary $U = \pi\begin{bmatrix} \mathfrak{s}_1, & \dots, & \mathfrak{s}_d \end{bmatrix}$.

However we already have an explicit description of the co-extension to a row isometry, $S \simeq L^{(\alpha)} \oplus U$. So we could instead dilate each copy of L to a row unitary in a variety of ways, such as Example 9.11.9. This provides uncountably many non-isomorphic ways of dilating L. Hence the dilation is generally not unique.

Since $\mathcal{O}_d = C^*(\mathfrak{A}_d)$ and it is simple by Corollary 9.11.7, it must be the minimal C*-algebra. A u.c.c. representation φ on $\mathfrak{A}_d$ is determined by the row contraction $\big[\varphi(\mathfrak{s}_1), \ \dots, \ \varphi(\mathfrak{s}_d)\big]$. If this is not a row unitary, then it has a non-trivial dilation to a row unitary. Any dilation of a row unitary to a larger space is necessarily just a direct sum; i.e. row unitaries correspond to maximal dilations. For φ to extend to a representation of $\mathcal{O}_d$, it is necessary and sufficient that this be a row unitary. Thus every representation of $\mathcal{O}_d$ has the unique extension property relative to $\mathfrak{A}_d$. Therefore it is a boundary representation precisely when it is irreducible. ■

All row contractions have a continuous functional calculus.

15.4.12. COROLLARY. *Let $T = \big[T_1, \dots, T_d\big]$ be a row contraction in $\mathcal{B}(H)^{(d)}$. Then there is a completely contractive representation Φ_T of $\mathfrak{A}_d$ into $\mathcal{B}(H)$ such that $\Phi_T(\mathfrak{s}_i) = T_i$ for $1 \le i \le d$.*

PROOF. By Corollary 15.4.11, T dilates to a row unitary U in $\mathcal{B}(K)^{(d)}$, and $H = K_2 \ominus K_1$ is the difference of two invariant subspaces for U. The compression map $\Pi : \mathcal{B}(K) \to \mathcal{B}(H)$ by $\Pi(B) = P_H B|_H$ is WOT-continuous and completely contractive. It is multiplicative on $\mathcal{A}(U)$ because H is semi-invariant. Now $\mathcal{A}(U)$ is completely isometrically isomorphic to $\mathfrak{A}_d$, say via $\Theta : \mathfrak{A}_d \to \mathcal{A}(U)$. Let $\Phi_T = \Pi\Theta$. This satisfies $\Phi_T(\mathfrak{s}_i) = \Pi(U_i) = T_i$ for $1 \le i \le d$. ■

15.5. Noncommutative Analytic Toeplitz Algebra

Recall that the operators L_i in $\mathcal{B}(F^2_d)$, with basis $\{\xi_w : w \in \mathbb{F}^+_d\}$, are given by the left regular representation of $\mathbb{F}^+_d$, $L_i\xi_w = \xi_{iw}$ for $1 \le i \le d$. We will also require the operators R_i arising from the right regular representation of $\mathbb{F}^+_d$, $R_i\xi_w = \rho(i)\xi_w = \xi_{wi}$ for $1 \le i \le d$.

15.5.1. DEFINITION. The *noncommutative analytic Toeplitz algebra*, $\mathfrak{L}_d$, is the WOT-closure of $\mathcal{A}_d$ in $\mathcal{B}(F^2_d)$, i.e. $\mathfrak{L}_d = \overline{\text{Alg}\{I, L_1, \dots, L_d\}}^{\text{WOT}}$. Also let $\mathfrak{R}_d = \overline{\text{Alg}\{I, R_1, \dots, R_d\}}^{\text{WOT}}$.

If $w = i_1 \dots i_n \in \mathbb{F}^+_d$, let $\widetilde{w} = i_n \dots i_1$ be the reversed word and let $|w| = n$ denote its length. Observe that multiplication on the right reverses order, that is, $R_i R_j \xi_w = \xi_{wji} = R_{ji}\xi_w$. We will write $R_w = \rho(\widetilde{w})$ so that $R_w \xi_v = \xi_{vw}$. Define a unitary operator $F\xi_w = \xi_{\widetilde{w}}$. Then it is easy to check that $F^2 = I$ and

$$FL_iF\xi_v = F\xi_{i\widetilde{v}} = \xi_{vi} = R_i\xi_v.$$

Thus $\mathfrak{R}_d = F\mathfrak{L}_dF$. Also it is clear that $L_vR_w = R_wL_v$, so that $\mathfrak{L}_d$ and $\mathfrak{R}_d$ commute.

15.5.2. DEFINITION. If $A \in \mathfrak{L}_n$ and $A\xi_\varnothing = \sum_{w\in\mathbb{F}_d^+} a_w\xi_w$, we define the *Fourier series* of A to be $A \sim \sum_{w\in\mathbb{F}_d^+} a_w L_w$.

Observe that if $A \in \mathfrak{L}_n$ and $v \in \mathbb{F}_d^+$, then

$$A\xi_v = AR_v\xi_\varnothing = R_v \sum_{w\in\mathbb{F}_d^+} a_w\xi_w = \sum_{w\in\mathbb{F}_d^+} a_w\xi_{wv} = \sum_{w\in\mathbb{F}_d^+} a_w L_w\xi_v.$$

So A is determined by the coefficients $\{a_w\}$. Moreover on finite linear combinations of the standard basis, it coincides with the action of $\sum_{w\in\mathbb{F}_d^+} a_w L_w$. The problem, as with classical Fourier series, is that this series does not converge in general. As usual, we can deal with this using either Cesàro or Abel means.

15.5.3. DEFINITION. If $A \in \mathfrak{L}_n$ and $A \sim \sum_{w\in\mathbb{F}_d^+} a_w L_w$, the *Cesàro means* are $\Sigma_n(A) = \sum_{|w|\le n} \big(1 - \frac{|w|}{n+1}\big) a_w L_w$.

15.5.4. THEOREM. *The maps $\Sigma_n : \mathfrak{L}_d \to \mathcal{A}_d$ are completely contractive and* WOT*-continuous. If $A \in \mathcal{A}_d$, then the Cesàro means $\Sigma_n(A)$ converge to A in norm. If $A \in \mathfrak{L}_d$, then the Cesàro means $\Sigma_n(A)$ converge* SOT *to A.*

PROOF. For $t \in \mathbb{R}$, let $D_t\xi_w = e^{it|w|}\xi_w$. Then D_t is a unitary operator, and $\gamma_t(A) = D_tAD_{-t}$ is a $*$-automorphism of $\mathcal{B}(F_d^2)$ such that $\gamma_t(L_w) = e^{it|w|}L_w$ for $w \in \mathbb{F}_d^+$. Define

$$\Phi_k(A) = \frac{1}{2\pi}\int_{-\pi}^{\pi} \gamma_t(A)e^{-ikt}\,dt.$$

Since γ_t are completely isometric and WOT-continuous, the map Φ_k is completely contractive and WOT-continuous for all $k \in \mathbb{Z}$. Moreover $\Phi_k = 0$ for $k < 0$. If $A = \sum_{w\in\mathbb{F}_d^+} a_w L_w$ is a finite sum, then $\Phi_k(A) = \sum_{|w|=k} a_w L_w$. Therefore by WOT-continuity, $\Phi_k(A) \in \operatorname{span}\{L_w : |w| = k\}$ for any $A \in \mathfrak{L}_n$. If $A \sim \sum_{w\in\mathbb{F}_d^+} a_w L_w$, then since the series for $A\xi_\varnothing$ converges unconditionally in F_d^2, we get

$$\begin{aligned}\Phi_k(A)\xi_\varnothing &= \frac{1}{2\pi}\int_{-\pi}^{\pi} \gamma_t(A)\xi_\varnothing e^{-ikt}\,dt = \frac{1}{2\pi}\int_{-\pi}^{\pi} \sum_{w\in\mathbb{F}_d^+} e^{i|w|t}a_w\xi_w e^{-ikt}\,dt \\ &= \sum_{w\in\mathbb{F}_d^+} \frac{1}{2\pi}\int_{-\pi}^{\pi} e^{i|w|t}e^{-ikt}\,dt\, a_w\xi_w = \sum_{|w|=k} a_w\xi_w.\end{aligned}$$

As $\Phi_k(A)$ is determined by $\Phi_k(A)\xi_\varnothing$, we have $\Phi_k(A) = \sum_{|w|=k} a_w L_w$.

Now we see that

$$\Sigma_n(A) = \sum_{k=-n}^{n} \big(1 - \tfrac{k}{n+1}\big)\Phi_k(A)$$
$$= \frac{1}{2\pi}\int_{-\pi}^{\pi} \gamma_t(A) \sum_{k=-n}^{n} \big(1 - \tfrac{k}{n+1}\big)e^{-ikt}\,dt = \frac{1}{2\pi}\int_{-\pi}^{\pi} \gamma_t(A)k_n(t)\,dt$$

where $k_n(t)$ is the Fejér kernel of Lemma 2.5.3. In particular, $k_n \geq 0$ and $\|k_n\|_1 = 1$. Thus we again conclude that Σ_n is completely contractive and WOT-continuous. It clearly maps $\mathfrak{L}_n$ into $\mathcal{A}_n$.

The proof of Fejér's theorem 2.5.4 works verbatim for $A \in \mathcal{A}_n$ to show that $\Sigma_n(A)$ converges in norm to A. Details are left to the reader. If $A \in \mathfrak{L}_n$ and $v \in \mathbb{F}_d^+$, then

$$\lim_{n\to\infty} \Sigma_n(A)\xi_v = \lim_{n\to\infty} \sum_{|w|\leq n} \big(1 - \tfrac{|w|}{n+1}\big)a_w\xi_{wv} = \sum_{w\in\mathbb{F}_d^+} a_w\xi_{wv} = A\xi_v.$$

Since the sequence $\Sigma_n(A)$ is bounded by $\|A\|$, this result extends to all vectors in H. That is, $\Sigma_n(A)$ converge SOT to A. ∎

15.5.5. Remark. If $w_1, \dots, w_{d^n}$ is a list of the words in $\mathbb{F}_d^+$ of length n, then because $\{L_w : |w| = n\}$ are isometries with pairwise orthogonal ranges, $\big[L_{w_1} \ \dots \ L_{w_{d^n}}\big]$ is an isometry onto $\sum^{\oplus}_{|w|=n} \mathfrak{L}_d[\xi_w] = \operatorname{span}\{\xi_v : |v| \geq n\}$. Also $\big\| \sum_{|w|=n} a_w L_w \big\| = \big(\sum_{|w|=n} |a_w|^2\big)^{1/2}$; and this sum is a scalar multiple of an isometry.

Several important results are easy consequences of this theorem.

15.5.6. Corollary. *$\mathfrak{L}_d' = \mathfrak{R}_d$ and $\mathfrak{R}_d' = \mathfrak{L}_d$. Thus $\mathfrak{L}_d'' = \mathfrak{L}_d$.*

Proof. Since $\mathfrak{R}_d = F\mathfrak{L}_dF$ and $F = F^* = F^{-1}$, it suffices to prove the second statement. Let A commute with $\mathfrak{R}_d$. Write $A\xi_\varnothing = \sum_{w\in\mathbb{F}_d^+} a_w\xi_w$. As for $A \in \mathfrak{L}_d$, we have

$$A\xi_v = AR_v\xi_\varnothing = R_vA\xi_\varnothing = \sum_{w\in\mathbb{F}_d^+} a_w\xi_{wv} = \sum_{w\in\mathbb{F}_d^+} a_wL_w\xi_v.$$

So A is determined by $A\xi_\varnothing$. Then it follows as in the proof of Theorem 15.5.4 that

$$\Phi_k(A) = \sum_{|w|=k} a_wL_w \quad\text{and}\quad \Sigma_n(A) = \sum_{|w|\leq n} \big(1 - \tfrac{|w|}{n+1}\big)a_wL_w.$$

Moreover the same proof shows that $\Sigma_n(A)$ converges SOT to A. Therefore A belongs to $\overline{\mathcal{A}_d}^{\text{SOT}} = \mathfrak{L}_d$. ∎

15.5.7. COROLLARY. *$\mathfrak{L}_d$ is inverse closed.*

PROOF. If $A \in \mathfrak{L}_d$ is invertible in $\mathcal{B}(F_d^2)$ and $R \in \mathfrak{R}_d$, then as in any ring,

$$A^{-1}R = A^{-1}RAA^{-1} = A^{-1}ARA^{-1} = RA^{-1}.$$

Thus $A^{-1} \in \mathfrak{R}_d' = \mathfrak{L}_d$. ∎

15.5.8. COROLLARY. *If $A \in \mathfrak{L}_d$, then $\|A\|_e = \|A\|$. In particular, $\mathfrak{L}_d$ contains no non-zero compact operators.*

PROOF. Let $M = \operatorname{span}\{\xi_{1^k} : k \geq 0\}$, and let

$$S = \{w = i1^k \in \mathbb{F}_d^+ : i \neq 1,\ k \geq 0\}.$$

Then $M^\perp = \sum_{w \in S}^{\oplus} R_w F_d^2$, which can be seen since these subspaces are orthogonal, and every standard basis vector ξ_w such that w does not consist entirely of 1's lies in exactly one of them. If $A \in \mathfrak{L}_d$, then $A|_{R_w F_d^2} \simeq A$ via the isometry R_w (which is a unitary when considered as a map into $R_w F_d^2$). Thus $A|_{M^\perp} \simeq A^{(\infty)}$. Therefore $\|A\|_e = \|A\|$. ∎

Every weak-$*$ continuous functional is given by a vector functional. See section 12.5 for definitions.

15.5.9. PROPOSITION. *$\mathfrak{L}_n$ has property $\mathbb{A}_1(1)$, meaning that if φ is a weak-$*$ continuous functional on $\mathfrak{L}_n$ with $\|\varphi\| < 1$, then there are vectors $\eta, \zeta \in F_d^2$ such that*

$$\varphi(A) = \langle A\eta, \zeta\rangle \quad \textit{and} \quad \|\eta\|\,\|\zeta\| < 1.$$

Thus the weak-$*$ and WOT *topologies on $\mathfrak{L}_d$ coincide.*

PROOF. Since $\mathfrak{L}_d$ is WOT-closed, it is weak-$*$ closed. It's canonical predual is $\mathcal{B}(F_d^2)_*/\mathfrak{L}_{d\perp}$, which is a quotient of $\mathcal{B}(F_d^2)_*$. Thus there is a functional $\widetilde{\varphi}$ in $\mathcal{B}(F_d^2)_*$ with $\|\widetilde{\varphi}\| < 1$ such that $\widetilde{\varphi}|_{\mathfrak{L}_d} = \varphi$. By Theorem 10.1.3, the predual of $\mathcal{B}(F_d^2)$ is the set $\mathfrak{S}_1(F_d^2)$ of trace class operators. Hence there is a trace class operator T with $\|T\|_1 < 1$ so that $\widetilde{\varphi}(A) = \operatorname{Tr}(AT)$ for $A \in \mathcal{B}(F_d^2)$. By Theorem 10.1.2, T has a polar decomposition $T = \sum_{n\geq 0} s_n \eta_n \zeta_n^*$ where $\{\eta_n\}$ and $\{\zeta_n\}$ are orthonormal sets, $s_n > 0$ and $\sum s_n = \|T\|_1 < 1$. For $A \in \mathfrak{L}_d$,

$$\varphi(A) = \sum_{n\geq 0} s_n \langle A\eta_n, \zeta_n\rangle = \sum_{n\geq 0} s_n \langle AR_{21^n}\eta_n, R_{21^n}\zeta_n\rangle$$

As in the proof of Corollary 15.5.8, the ranges $R_{21^n}F_d^2$ are pairwise orthogonal invariant subspaces for $\mathfrak{L}_d$. Define

$$\eta = \sum_{n\geq 0} \sqrt{s_n} R_{21^n}\eta_n \quad \text{and} \quad \zeta = \sum_{n\geq 0} \sqrt{s_n} R_{21^n}\zeta_n.$$

Then $\|\eta\|^2 = \|\zeta\|^2 = \|T\|_1 < 1$. Finally, since $AR_{21^n}\eta_n$ is orthogonal to $R_{21^m}\zeta_m$ for $m \neq n$,

$$\langle A\eta, \zeta\rangle = \sum_{n\geq 0} s_n \langle AR_{21^n}\eta_n, R_{21^n}\zeta_n\rangle = \varphi(A).$$

Since every weak-$*$ continuous linear functional on $\mathfrak{L}_d$ is WOT-continuous, the two topologies coincide on $\mathfrak{L}_d$. ∎

While the Fourier series is very useful, the lack of convergence is a problem. It is often useful to use a partial version of this series.

15.5.10. PROPOSITION. *Let $A \in \mathfrak{L}_d$ with $A \sim \sum_{w\in\mathbb{F}_d^+} a_w L_w$, and let $n \geq 1$. Then there are elements $A_w \in \mathfrak{L}_d$ for $|w| = n$ so that*

$$A = \sum_{|w|<n} a_w L_w + \sum_{|w|=n} L_w A_w.$$

PROOF. Let $B = A - \sum_{|w|<n} a_w L_w$. Then $\operatorname{Ran} B \subset \operatorname{span}\{\xi_w : |w| \geq n\}$. Note that

$$\operatorname{span}\{\xi_w : |w| \geq n\} = \sum_{|v|=n}^{\oplus} \mathfrak{L}_d[\xi_v] = \sum_{|v|=n} \operatorname{Ran} L_v$$

Thus $B = \sum_{|v|=n} L_v L_v^* B$ because the $\operatorname{Ran} L_v$ for $|v| = n$ are pairwise orthogonal. Define $A_w = L_w^* B$ for $|w| = n$. Then if $R \in \mathfrak{R}_d$,

$$A_w R = L_w^* RB = L_w^* R \sum_{|v|=n} L_v L_v^* B = \sum_{|v|=n} L_w^* L_v R L_v^* B = RL_w^* B = RA_w.$$

This holds since $L_w^* L_v = 0$ if $v \neq w$ and $|v| = |w|$. Hence $A_w \in \mathfrak{L}_d$. Therefore

$$B = \sum_{|v|=n} L_v L_v^* B = \sum_{|w|=n} L_w A_w.$$ ∎

An important class of WOT-continuous representations of $\mathfrak{L}_d$ arise as a functional calculus for certain row contractions. This is an immediate consequence of the Frahzo–Bunce–Popescu dilation theorem.

15.5.11. POPESCU FUNCTIONAL CALCULUS. *Let $T = [T_1, \dots, T_d]$ be a pure row contraction in $\mathcal{B}(H)^{(d)}$. Then there is a* WOT*-continuous, completely contractive representation Φ_T of $\mathfrak{L}_d$ such that $\Phi_T(L_i) = T_i$ for $1 \leq i \leq d$.*

PROOF. By Theorem 15.4.9, T dilates to a pure row isometry $S \simeq L^{(\alpha)}$ in $\mathcal{B}(K)^{(d)}$, say $S_i = UL_i^{(\alpha)}U^*$ for a unitary $U \in \mathcal{B}(F_d^{2(\alpha)}, K)$. The map Ψ : $\mathcal{B}(F_d^2) \to \mathcal{B}(K)$ by $\Phi(A) = UA^{(\alpha)}U^*$ is WOT-continuous $*$-isomorphism, and in particular is completely isometric . The compression map $\Pi : \mathcal{B}(K) \to \mathcal{B}(H)$

by $\Pi(B) = P_H B|_H$ is WOT-continuous and completely contractive. It is multiplicative on the space $\{B \in \mathcal{B}(K) : BH^\perp \subset H^\perp\}$, which includes $\mathcal{W}(S)$. Let $\Phi_T = \Pi\Psi|_{\mathfrak{L}_d}$. Then Ψ satisfies $\Psi(L_i) = S_i$ for $1 \le i \le d$, and hence it maps $\mathfrak{L}_d$ into $\mathcal{W}(S)$. Therefore Φ_T is a WOT-continuous and completely contractive homomorphism such that $\Phi_T(L_i) = T_i$ for $1 \le i \le d$. ∎

15.6. Invariant Subspaces of $\mathfrak{L}_d$

Now we can obtain a Beurling-style result about invariant subspaces for $\mathfrak{L}_d$. The parallel comes from comparing $\mathfrak{L}_d$ with the Toeplitz algebra $\mathcal{T}(H^\infty)$. The invariant subspaces for the unilateral shift, T_z, are the subspaces ωH^2 for inner functions ω. The isometries in $\mathcal{T}(H^\infty)$ are precisely T_ω for ω inner. Thus we see that the invariant subspaces for the shift are just the ranges of the isometries in $\mathcal{T}(H^\infty)$. However since our algebra $\mathfrak{L}_d$ is noncommutative, the closed ranges of elements of $\mathfrak{L}_d$ are not invariant. However the ranges of the commutant $\mathfrak{R}_d$ are invariant for $\mathfrak{L}_d$. One difference is that for the unilateral shift, every invariant subspace is cyclic.

15.6.1. THEOREM. *Let $\{0\} \ne M \in \operatorname{Lat} \mathfrak{L}_d$, and set $W = M \ominus \sum_{i=1}^d L_i M$. Then $M = \mathfrak{L}_d[W]$ is the direct sum of cyclic invariant subspaces. A cyclic invariant subspace is the range of an isometry in $\mathfrak{R}_d$, which is unique up to a scalar multiple. There is a row isometry $R = \begin{bmatrix} R_1 & R_2 & \dots \end{bmatrix}$ in $\mathcal{B}(F_d^{2(\infty)}, F_d^2)$ such that each R_i is an isometry in $\mathfrak{R}_d$ and $M = \operatorname{Ran} R$.*

PROOF. By the Wold decomposition for the row isometry $L|_M$, W is a wandering space for $\mathfrak{L}_d$. Suppose that $N := M \ominus \mathfrak{L}_d[W] \ne \{0\}$. Thus $N = \sum_{i=1}^n L_i N$. Among all non-zero vectors in N, choose $\eta \in N$ such that $\eta = \sum_{w \in \mathbb{F}_d^+} a_w \xi_w$ and a $w_0 \in \mathbb{F}_d^+$ with $a_{w_0} \ne 0$ so that $|w_0| = k_0$ is minimal among all possible choices of η and w_0. Then $N \perp \operatorname{span}\{\xi_w : |w| < k_0\}$. Therefore

$$N = \sum_{i=1}^n L_i N \perp \operatorname{span}\{\xi_w : |w| \le k_0\}.$$

But η is not orthogonal to this subspace, which is a contradiction. Therefore $M = \mathfrak{L}_d[W]$.

By the proof of the Frazho–Popescu–Wold decomposition 15.4.2, any orthonormal basis $\{\xi_{j,\varnothing} : j \in J\}$ for W allows us to define

$$M_j = \mathfrak{L}_d[\xi_{j,\varnothing}] = \overline{\operatorname{span}}\{\xi_{j,w} := L_w \xi_{j,\varnothing} : w \in \mathbb{F}_d^+\}.$$

The proof showed that they are pairwise orthogonal and $M = \sum_{j \in J}^{\oplus} M_j$. Define R_j on F_d^2 by $R_j \xi_w = \xi_{j,w}$. This carries an orthonormal basis onto another orthonormal set, and hence it extends by linearity and continuity to an isometry. For $v \in \mathbb{F}_d^+$,

$$R_j L_v \xi_w = R_j \xi_{vw} = \xi_{j,vw} = L_v \xi_{j,w} = L_v R_j \xi_w.$$

It follows that $R_j \in \mathfrak{L}_d' = \mathfrak{R}_d$. If S is another isometry in $\mathfrak{R}_d$ with range M_j, then

$$S \sum_{i=1}^{n} L_i F_d^2 = \sum_{i=1}^{n} L_i M_j = M \ominus \mathbb{C}\xi_{j,\varnothing}.$$

Therefore $S\xi_\varnothing = \lambda \xi_{j,\varnothing}$ for some $\lambda \in \mathbb{T}$. Since S is determined by $S\xi_\varnothing$, it follows that $S = \lambda R_j$. Finally,

$$M = \sum_{j \in J} M_j = \sum_{j \in J} \operatorname{Ran} R_j = \operatorname{Ran} \begin{bmatrix} R_1 & R_2 & \dots \end{bmatrix}. \qquad \blacksquare$$

We can easily extend this to pure row isometries, $S \simeq L^{(\alpha)}$.

15.6.2. COROLLARY. *Every invariant subspace $M \in \operatorname{Lat} \mathfrak{L}_d^{(\alpha)}$ has the form $M = \operatorname{Ran} R$, where $R = \begin{bmatrix} R_{ij} \end{bmatrix}_{\alpha \times \infty}$ is an isometry, and $R_{ij} \in \mathfrak{R}_d$ for all $A \in \mathfrak{L}_d$.*

PROOF. We prove this for $\alpha = \infty$. Let $M_0 = \operatorname{Ran} J_0$ where

$$J_0 = \begin{bmatrix} R_{21} & R_{21^2} & \dots R_{21^n} & \dots \end{bmatrix} \in \mathcal{B}(F_d^{2(\infty)}, F_d^2).$$

Thus $M_0 = \sum_{n\ge1}^{\oplus} \operatorname{Ran} R_{21^n}$. The subspaces $M_n = \operatorname{Ran} R_{21^n} \simeq F_d^2$ are invariant and pairwise orthogonal for $n \ge 1$. Hence $L|_{M_0} \simeq L^{(\infty)}$ and J_0 is an isometry that implements the equivalence. We can identify M with the invariant subspace $J_0 M \subset M_0$ for $\mathfrak{L}_d$. Therefore there are isometries $R_j \in \mathfrak{R}_d$ with pairwise orthogonal ranges so that $R = \begin{bmatrix} R_1 & R_2 & \dots \end{bmatrix}$ and $J_0 M = \operatorname{Ran} R$. Hence we see that $J_0^* R \in \mathcal{B}(F_d^{2(\infty)})$ is an isometry with range M.

We must verify that its coefficients $R_{21^n}^* R_j$ lie in $\mathfrak{R}_d$. Now $R_j \sim \sum_{w \in \mathbb{F}_d^+} a_w R_w$, and since $\operatorname{Ran} R_j \subset M_0$, $a_w = 0$ unless $w = v21^m$ for some $m \ge 1$. However $R_{21^n}^* R_{v21^m} = 0$ if $m \ne n$ and equals R_v if $m = n$. It then follows that if R_j is a finite sum of such terms, then $R_{21^n}^* R_j \in \mathfrak{R}_d$. Finally,

$$R_{21^n}^* R_j = \operatorname*{SOT-lim}_{k\to\infty} R_{21^n}^* \Sigma_k(R_j) \in \mathfrak{R}_d.$$

Therefore the matrix coefficients of $J_0^* R$ lie in $\mathfrak{R}_d$. $\blacksquare$

We now make some definitions by analogy with the factorization in $H^\infty(\mathbb{D})$. Compare with Remark 12.4.5.

15.6.3. DEFINITION. An element $S \in \mathfrak{L}_d$ is *inner* if S is an isometry. An element $B \in \mathfrak{L}_d$ is *outer* if $\overline{\operatorname{Ran} B} = F_d^2$.

15.6.4. COROLLARY. *Every $A \in \mathfrak{L}_d$ factors in $\mathfrak{L}_d$ as $A = SB$ where S is inner and B is outer.*

PROOF. Note that $M = \overline{\operatorname{Ran} A}$ lies in $\operatorname{Lat} \mathfrak{R}_d$ and $A\xi_\varnothing$ is a cyclic vector for M. Therefore there is an isometry $S \in \mathfrak{L}_n$ so that $M = \operatorname{Ran} S$. Define $B = S^*A$; so that $A = (SS^*)A = SB$. Then for $R \in \mathfrak{R}_d$ and $\xi \in F_d^2$,

$$\begin{aligned}(BR - RB)\xi &= (S^*R - RS^*)A\xi = (S^*R - RS^*)S(S^*A\xi) \\ &= \big((S^*S)R - R(S^*S)\big)S^*A\xi = 0.\end{aligned}$$

Therefore $B \in \mathfrak{R}_d' = \mathfrak{L}_d$. Finally $\overline{\operatorname{Ran} B} = S^*\overline{\operatorname{Ran} A} = F_d^2$. So B is outer. ■

We make a useful improvement on Proposition 15.5.9.

15.6.5. COROLLARY. *If φ is a weak-$*$ continuous linear functional on $\mathfrak{L}_d$, there are vectors $\eta, \zeta \in F_d^2$ such that η is cyclic and $\|\eta\|\,\|\zeta\| < 1$ such that $\varphi(A) = \langle A\eta, \zeta\rangle$.*

PROOF. Use Proposition 15.5.9 to obtain $\eta_0, \zeta_0 \in F_d^2$ such that $\|\eta_0\|\,\|\zeta_0\| < 1$ and $\varphi(A) = \langle A\eta_0, \zeta_0\rangle$. Let $M = \mathfrak{L}_d[\eta_0] = \overline{\mathfrak{L}_d\eta_0}$. This is invariant for $\mathfrak{L}_d$ and has a cyclic vector η_0. Therefore there is an isometry $R \in \mathfrak{R}_d$ so that $M = RF_d^2$. Define $\eta = R^*\eta_0$ and $\zeta = R^*\zeta_0$. Then for $A \in \mathfrak{L}_d$,

$$\langle A\eta, \zeta\rangle = \langle AR^*\eta_0, R^*\zeta_0\rangle = \langle ARR^*\eta_0, \zeta_0\rangle = \langle A\eta_0, \zeta_0\rangle = \varphi(A)$$

because RR^* is the projection onto M. Clearly $\|\eta\|\,\|\zeta\| \le \|\eta_0\|\,\|\zeta_0\| < 1$. Finally $RF_d^2 = \overline{\mathfrak{L}_d\eta_0} = \overline{\mathfrak{L}_d R\eta} = R\overline{\mathfrak{L}_d\eta}$. Since R is an isometry, $\overline{\mathfrak{L}_d\eta} = F_d^2$; so η is a cyclic vector. ■

We investigate joint eigenspaces of $\mathfrak{L}_d^*$ analogous to the kernel functions k_λ for H^2. These are the complements of the codimension 1 invariant subspaces for $\mathfrak{L}_d$.

If $\lambda = (\lambda_1, \dots, \lambda_d) \in \mathbb{C}^d$ and $w = i_1 \dots i_k \in \mathbb{F}_d^+$, define $w(\lambda) = \prod_{j=1}^k \lambda_{i_j}$. Let $\mathbb{B}_d = \{\lambda \in \mathbb{C}^d : \|\lambda\|_2 < 1\}$ denote the open unit ball.

15.6.6. THEOREM. *For $\lambda \in \mathbb{B}_d$, define*

$$\nu_\lambda = (1 - \|\lambda\|^2)^{1/2} \sum_{w \in \mathbb{F}_d^+} \overline{w(\lambda)}\xi_w = (1 - \|\lambda\|^2)^{1/2}\Big(I - \sum_{i=1}^d \overline{\lambda_i} L_i\Big)^{-1}\xi_\varnothing.$$

(1) *$L_i^*\nu_\lambda = \overline{\lambda_i}\nu_\lambda$ for $1 \le i \le d$.*

(2) *The codimension 1 invariant subspaces for $\mathfrak{L}_d$ are $(\mathbb{C}\nu_\lambda)^\perp$ for $\lambda \in \mathbb{B}_d$.*

PROOF. If $\lambda \in \mathbb{B}_d$, then a word $w = i_1 \dots i_n \in \mathbb{F}_d^+$ consisting of k_i copies of i for $1 \le i \le d$ will have $\binom{k}{k_1, \dots, k_d} = \frac{k!}{k_1! \dots k_d!}$ permutations that produce the same result $\prod_{i=1}^d \lambda_i^{k_i}$ when evaluated at the commuting tuple λ. Hence

$$\sum_{w\in\mathbb{F}_d^+} |\overline{w(\lambda)}|^2 = \sum_{n\geq 0}\sum_{|w|=n} |w(\lambda)|^2 = \sum_{n\geq 0}\sum_{k_1+\dots+k_d=n} \binom{k}{k_1,\dots,k_d}\prod_{i=1}^d |\lambda_i|^{2k_i}$$
$$= \sum_{n\geq 0}\Big(\sum_{i=1}^d |\lambda_i|^2\Big)^n = \frac{1}{1-\|\lambda\|^2}.$$

This series converges absolutely, and uniformly for $\|\lambda\| \leq r < 1$. Therefore $\|\nu_\lambda\| = 1$.

Since $\|\sum_{i=1}^d \overline{\lambda_i} L_i\|^2 = \sum_{i=1}^d |\lambda_i|^2 = \|\lambda\|^2 < 1$, $I - \sum_{i=1}^d \overline{\lambda_i} L_i$ is invertible. Reasoning as above,

$$\Big(I - \sum_{i=1}^d \overline{\lambda_i} L_i\Big)^{-1}\xi_\varnothing = \sum_{n\geq 0}\Big(\sum_{i=1}^d \overline{\lambda_i} L_i\Big)^n \xi_\varnothing = \sum_{w\in\mathbb{F}_d^+} \overline{w(\lambda)} L_w \xi_\varnothing$$
$$= \sum_{w\in\mathbb{F}_d^+} \overline{w(\lambda)}\xi_w = \frac{\nu_\lambda}{(1-\|\lambda\|^2)^{1/2}}.$$

Compute

$$L_i^*\nu_\lambda = (1-\|\lambda\|^2)^{1/2}\sum_{w\in\mathbb{F}_d^+} \overline{w(\lambda)} L_i^*\xi_w = (1-\|\lambda\|^2)^{1/2}\sum_{v\in\mathbb{F}_d^+} \overline{iv(\lambda)} L_i^*\xi_{iv}$$
$$= (1-\|\lambda\|^2)^{1/2}\sum_{v\in\mathbb{F}_d^+} \overline{\lambda_i}\,\overline{v(\lambda)}\xi_v = \overline{\lambda_i}\nu_\lambda.$$

Thus $\mathbb{C}\nu_\lambda$ is invariant for each L_i^*, and hence is invariant for $\mathfrak{L}_d^*$. Therefore $(\mathbb{C}\nu_\lambda)^\perp$ is invariant for $\mathfrak{L}_d$ and has codimension 1. Conversely, if $(\mathbb{C}\nu)^\perp$ is invariant for $\mathfrak{L}_d$ for a unit vector ν, then $\mathbb{C}\nu$ is invariant for $\mathfrak{L}_d^*$. Define λ_i so that $L_i^*\nu = \overline{\lambda_i}\nu$ for $1 \leq i \leq d$. It follows recursively that $L_w^*\nu = \overline{w(\lambda)}\nu$ for $w \in \mathbb{F}_d^+$. Write $\nu = \sum_{w\in\mathbb{F}_d^+} a_w\xi_w$. Then $\overline{w(\lambda)}a_\varnothing = a_w$ for all $\in \mathbb{F}_d^+$. Hence

$$\nu = a_\varnothing \sum_{w\in\mathbb{F}_d^+} \overline{w(\lambda)}\xi_w.$$

From the first paragraph, one sees that if $\|\lambda\| < 1$, then $\nu \in \mathbb{C}\nu_\lambda$; and if $\|\lambda\| \geq 1$, then $\|\nu\| = \infty$, a contradiction. ■

These eigenvalues for the adjoint yield multiplicative functionals on $\mathfrak{L}_d$.

15.6.7. THEOREM. *For* $\lambda \in \mathbb{B}_d$,

(1) *The map* $\varphi_\lambda(A) = \langle A\nu_\lambda, \nu_\lambda\rangle$ *is a* WOT*-continuous character of* $\mathfrak{L}_d$.

(2) *If* φ *is a character of* $\mathfrak{L}_d$ *such that* $\varphi(L_i) = \lambda_i$ *for* $1 \leq i \leq d$, *then* $\varphi = \varphi_\lambda$.

(3) *If* $A \in \mathfrak{L}_d$ *with* $A \sim \sum_{w \in \mathbb{F}_d^+} a_w L_w$, *then* $\varphi_\lambda(A) = \sum_{w \in \mathbb{F}_d^+} a_w w(\lambda)$. *This series converges uniformly on compact subsets of* $\mathbb{B}_d$.

(4) *The maps* $\{\varphi_\lambda : \lambda \in \mathbb{B}_d\}$ *are the only* WOT-*continuous characters of* $\mathfrak{L}_d$.

PROOF. The map $\varphi_\lambda(A) = \langle A\nu_\lambda, \nu_\lambda\rangle$ is clearly linear, contractive and WOT continuous. It satisfies $\varphi(L_w) = w(\lambda)$. Hence $\varphi_\lambda(p(L)) = p(\lambda)$ for all noncommutative polynomials in the semigroup algebra $\mathbb{C}[\mathbb{F}_d^+]$, and thus is multiplicative. Therefore it extends by WOT continuity to a character on $\mathfrak{L}_d$.

Suppose that φ is a character on $\mathfrak{L}_d$ such that $\varphi(L_i) = \lambda_i$. Let $r = \|\lambda\| < 1$. Given $0 < \varepsilon < 1$, choose N so that $r^N < \varepsilon$. Given $A \in \mathfrak{L}_d$ with $\|A\| = 1$, choose n so that

$$\sum_{k=0}^{N-1} \frac{k}{n} \Big\| \sum_{|w|=k} a_w L_w \Big\| \le \frac{1}{n} \sum_{k=0}^{N-1} k\|A\| < \varepsilon.$$

Then write $B = \sum_{k=0}^{N-1} \frac{k}{n} \sum_{|w|=k} a_w L_w$ and $A = \Sigma_n(A) + B + C$. Then $\|B\| < \varepsilon$, $\|C\| \le 2\|A\| + \varepsilon < 3$ and $\operatorname{Ran} C \subset \operatorname{span}\{\xi_v : |v| \ge N\}$. By Proposition 15.5.10, if $S_N = \begin{bmatrix} L_{w_1} & \dots & L_{w_{d^N}} \end{bmatrix}$ where $w_1, \dots, w_{d^N}$ is a list of the words in $\mathbb{F}_d^+$ of length N, then

$$C = S_N S_N^* C = \sum_{|w|=n} L_w C_w$$

and $\|S_N^* C\| \le \|C\|$ is a column operator with entries C_w in $\mathfrak{L}_d$. Applying φ, we obtain

$$\begin{aligned} |\varphi(C)| &= |\varphi^{(1,d^N)}(S_N)|\, |\varphi^{(d^N,1)}(S_N^* C)| \\ &\le \Big(\sum_{|w|=N} |w(\lambda)|^2 \Big)^{1/2} \|S_N^* C\| < r^N \|C\| < 3\varepsilon. \end{aligned}$$

Similarly $|\varphi_\lambda(C)| < 3\varepsilon$. Since φ and φ_λ agree on polynomials,

$$\begin{aligned} |\varphi(A) - \varphi_\lambda(A)| &\le \big|\varphi\big(\Sigma_n(A) + B\big) - \varphi_\lambda\big(\Sigma_n(A) + B\big)\big| + |\varphi(C) - \varphi_\lambda(C)| \\ &= |\varphi(C) - \varphi_\lambda(C)| < 6\varepsilon. \end{aligned}$$

But $\varepsilon > 0$ was arbitrary, so $\varphi = \varphi_\lambda$.

It also follows from the last calculation that

$$|\varphi_\lambda(A) - \varphi_\lambda(\Sigma_N(A))| < |\varphi_\lambda(B)| + |\varphi_\lambda(C)| < \varepsilon + 3\varepsilon = 4\varepsilon.$$

Therefore

$$\varphi_\lambda(A) = \lim_{n \to \infty} \varphi_\lambda(\Sigma_n(A)) = \lim_{n \to \infty} \sum_{|w| \le n} \big(1 - \tfrac{|w|}{n+1}\big) a_w w(\lambda).$$

We saw in the previous proof that $\sum_{w\in\mathbb{F}_d^+} \overline{w(\lambda)}\xi_w = (1-\|\lambda\|^2)^{-1/2}\nu_\lambda$. Therefore

$$\begin{aligned}\varphi_\lambda(A) &= \lim_{n\to\infty} \langle \Sigma_n(A)\xi_\varnothing, \sum_{w\in\mathbb{F}_d^+} \overline{w(\lambda)}\xi_w\rangle \\ &= \langle A\xi_\varnothing, \sum_{w\in\mathbb{F}_d^+} \overline{w(\lambda)}\xi_w\rangle = \sum_{w\in\mathbb{F}_d^+} a_w w(\lambda).\end{aligned}$$

The convergence of this series is uniform over $\{\lambda : \|\lambda\| \le r\}$ by Cauchy–Schwarz because the previous proof showed that $\sum_{w\in\mathbb{F}_d^+} |w(\lambda)|^2$ converges uniformly over this closed ball.

Suppose that φ is a WOT continuous character of $\mathfrak{L}_d$. Define $\lambda_i = \varphi(L_i)$ and $\lambda = (\lambda_1, \dots, \lambda_d)$. Multiplicative linear functionals have norm 1 by Proposition 7.1.2. Linear functionals are completely bounded with the same norm by Proposition 14.6.6. Therefore $\|\lambda\| = \|\varphi^{(1,d)}(L)\| \le \|L\| = 1$. If $\|\lambda\| < 1$, then $\varphi = \varphi_\lambda$. If $\|\lambda\| = 1$, let $V = \sum_{i=1}^d \overline{\lambda_i} L_i$. Then

$$\varphi(V) = \sum_{i=1}^d \overline{\lambda_i}\lambda_i = \|\lambda\|^2 = 1$$

and

$$V^*V = (\sum_{i=1}^d \overline{\lambda_i} L_i)^*(\sum_{i=1}^d \overline{\lambda_i} L_i) = \sum_{i=1}^d \lambda_i \overline{\lambda_i} I = I.$$

Now V is a proper isometry because $V^* \sum_{i=1}^d b_i\xi_i = \sum_{i=1}^d \lambda_i b_i\, \xi_\varnothing = 0$ for non-zero choices of b_i. In the proof of Theorem 15.6.1, it was shown that a proper isometry in $\mathfrak{L}_d$ is completely non-unitary. Therefore V^n converges WOT to 0, but $\varphi(V^n) = 1$. Hence φ is not WOT continuous. ∎

15.6.8. DEFINITION. If $A \in \mathfrak{L}_d$, we define a function on $\mathbb{B}_d$ by

$$\widehat{A}(\lambda) := \varphi_\lambda(A) = \langle A\nu_\lambda, \nu_\lambda\rangle.$$

15.6.9. PROPOSITION. *The map $A \to \widehat{A}$ is a completely contractive homomorphism. For $A \in \mathfrak{L}_d$, the function $\widehat{A}$ is bounded and analytic on $\mathbb{B}_d$. If $\{A_\alpha\}$ is a bounded net converging* WOT *to A, then $\widehat{A_\alpha}$ converges to $\widehat{A}$ uniformly on compact subsets of $\mathbb{B}_d$.*

PROOF. Each character φ_λ is completely contractive, and hence $A \to \widehat{A}$ is a completely contractive homomorphism.

If $w \in \mathbb{F}_d^+$, then $\widehat{L_w}(\lambda) = w(\lambda)$ and so $\widehat{p(L)}(\lambda) = p(\lambda)$ for all noncommutative polynomials. These are clearly analytic. For any $A \in \mathfrak{L}_d$, $|\varphi_\lambda(A)| \le \|A\|$ and hence $\|\widehat{A}\|_\infty \le \|A\|$. If $A \sim \sum_{w\in\mathbb{F}_d^+} a_w L_w$, then Theorem 15.6.7 showed that

$\varphi_\lambda(A) = \sum_{w\in\mathbb{F}_d^+} a_w w(\lambda)$ uniformly over compact subsets of $\mathbb{B}_d$. Therefore this limit is analytic on $\mathbb{B}_d$.

If A_α converges to A in the WOT topology, then in particular, $A_\alpha \nu_\lambda$ converges weakly to $A\nu_\lambda$. The map $\lambda \to \nu_\lambda$ is continuous, and for $\|\lambda\| \le r$, this is a compact set of vectors. Therefore, if this net $\{A_\alpha\}$ is bounded, convergence is uniform over this set. It follows that

$$\varphi_\lambda(A) - \varphi_\lambda(A_\alpha) = \big\langle (A - A_\alpha)\, \nu_\lambda, \nu_\lambda \big\rangle = 0$$

uniformly over compact subsets of $\mathbb{B}_d$. ■

15.7. Free Semigroup Algebras

In this section, we consider the larger class of operator algebras generated by row isometries.

15.7.1. Definition. A *free semigroup algebra* is the WOT-closed unital algebra $\mathfrak{S}$ generated by $S_1, \dots, S_d$ for a row isometry $S = \begin{bmatrix} S_1 & \dots & S_d \end{bmatrix}$. If $w = i_1 \dots i_k \in \mathbb{F}_d^+$, we write $S_w = S_{i_1} \dots S_{i_k}$.

Say that $\mathfrak{S}$ is *type L* or *analytic* if there is a completely isometric isomorphism φ of $\mathfrak{S}$ onto $\mathfrak{L}_d$ such that $\varphi(S_i) = L_i$ for $1 \le i \le d$.

15.7.2. Example. Let $x = i_{-1} i_{-2} i_{-3} \dots$ be an infinite word with each i_k in $\{1, \dots, d\}$, and set $x_m = i_{-1} i_{-2} \dots i_{-m}$ for $m \ge 0$. Let $\mathbb{F}_d^+ x^{-1} := \bigcup_{m\ge 0} \mathbb{F}_d^+ x_m^{-1}$ considered as a subset of the free group $\mathbb{F}_d$. This is an increasing union since $(\mathbb{F}_d^+ i_{-m}) x_m^{-1} = \mathbb{F}_d^+ x_{m-1}^{-1}$. Let $H = l_2(\mathbb{F}_d^+ x^{-1})$ with standard basis ξ_w for $w \in \mathbb{F}_d^+ x^{-1}$. Define $S_i \xi_w = \xi_{iw}$ where the composition is calculated in $\mathbb{F}_d$, so that words are written in reduced form, e.g. $i(i^{-1}v) = v$. The vector $\xi_{x_m^{-1}}$ is a wandering vector for S, and the cyclic subspace it generates is $H_m = l_2(\mathbb{F}_d^+ x_m^{-1})$. Thus the restrictions $S_i|_{H_m}$ are simultaneously unitarily equivalent to L_i, and therefore $\mathfrak{S}|_{H_m} \simeq \mathfrak{L}_d$. Moreover, we can think of $\mathfrak{S}$ as the (WOT-closed) inductive limit of these restrictions. The identifications of $S_i|_{H_m} \simeq L_i$ are compatible. Thus it is not hard to see that $\mathfrak{S}$ is completely isometrically isomorphic and weak-$*$ homeomorphic to $\mathfrak{L}_d$. Of course they are not spatially equivalent because S is a row unitary generating a copy of $\mathcal{O}_d$. This is an analytic free semigroup algebra.

15.7.3. Example. Recall Example 9.11.9. This is a class of representations of $\mathcal{O}_d$ determined by a word $u = i_k i_{k-1} \dots i_1 \in \mathbb{F}_d^+$ and $\lambda \in \mathbb{T}$ acting on the Hilbert space $K = \mathbb{C}^k \oplus H^{(k(d-1))}$, where $H = F_d^2$. Fix a basis $\eta_1, \dots, \eta_k$ for $M := \mathbb{C}^k$. For each $1 \le j \le k$, index $d-1$ copies of F_d^2 as $H_{j,s}$ for $1 \le s \le d$, $s \ne i_j$. Let

$H_{j,s}$ have orthonormal basis $\xi_{j,s,w}$ for $w \in \mathbb{F}_d^+$. Define isometries $S_1, \dots, S_d$ by

$$S_i \eta_j = \begin{cases} \eta_{j+1} & \text{if } 1 \le j < k \text{ and } i = i_j \\ \lambda \eta_1 & \text{if } j = k \text{ and } i = i_k \\ \xi_{j,i,\varnothing} & \text{if } i \ne i_j \end{cases}$$

$$S_i \xi_{j,s,w} = \xi_{j,s,iw} \quad \text{for } 1 \le j \le k,\ 1 \le s \le d,\ s \ne i_j,\ w \in \mathbb{F}_d^+.$$

Then $S = \begin{bmatrix} S_1 & \dots & S_d \end{bmatrix}$ is a row unitary. Let $\mathfrak{S}$ be the WOT-closed unital algebra generated by S.

We now assume that u is *primitive*, meaning that it is not a power $u = v^p$ for some $v \in \mathbb{F}_d^+$ and $p \ge 2$. Observe that $S_u \eta_1 = \lambda \eta_1$. For $2 \le j \le k$, the words u and the cyclical permutation $u_j = i_{j-1} i_{j-2} \dots i_1 i_k i_{k-1} \dots i_j$ are distinct. Hence S_u acts on η_j by mapping it to the next $\eta_{j'}$ as long as u and u_j agree on the last few terms, but at the point that they differ, and it is mapped into some $H_{j',s}$. Each $H_{j,s}$ is graded, and the action $S_i \xi_{j,s,w} = \xi_{j,s,iw}$ increases the word length. Therefore $(\overline{\lambda} S_u)^n \eta_1 = \eta_1$ while if ξ is any other vector in our basis, then $(\overline{\lambda} S_u)^n \xi$ converges weakly to 0. It follows that WOT–$\lim_{n\to\infty} (\overline{\lambda} S_u)^n = \eta_1 \eta_1^*$ belongs to $\mathfrak{S}$.

Similarly, $\eta_j \eta_j^* \in \mathfrak{S}$ for $1 \le j \le k$. Notice that

$$S_{i_{p-1}} S_{i_{p-2}} \dots S_{i_j} (\eta_j \eta_j^*) = \eta_p \eta_j^* \quad \text{for} \quad j < p \le k$$

and

$$S_{i_{p-1}} S_{i_{p-2}} \dots S_{i_1} (\overline{\lambda} S_{i_k}) S_{i_{k-1}} \dots S_{i_j} (\eta_j \eta_j^*) = \eta_p \eta_j^* \quad \text{for} \quad 1 \le p < j.$$

Hence $\mathfrak{S}$ contains the matrix units $\{\eta_p \eta_j^* : 1 \le j, p \le k\}$ spanning a copy of M_k. More precisely, if P is the projection onto $M = \text{span}\{\eta_j : 1 \le j \le k\}$, then M is coinvariant for each S_i and thus for $\mathfrak{S}$; and $P\mathfrak{S} = P\mathcal{B}(K)P = \mathcal{B}(M) \simeq M_k$.

Next we observe that any η_j is a cyclic vector for $\mathfrak{S}$, since it can be mapped to any η_p, and from there to any $\xi_{p,s,\varnothing}$ for $s \ne i_p$, and hence to any $\xi_{p,s,w}$ for $w \in \mathbb{F}_d^+$. It follows that $\xi \eta_j^* \in \mathfrak{S}$ for any $\xi \in K$. Therefore, $\mathfrak{S}P = \mathcal{B}(M, K)$.

When we restrict S_i to $M^\perp$, it is unitarily equivalent to $L_i^{(k(d-1))}$ on $F_d^{2\,(k(d-1))}$. Therefore the restriction $\mathfrak{S}|_{M^\perp}$ is unitarily equivalent to $\mathfrak{L}_d^{(k(d-1))}$ via a canonical map, and in particular, this restriction is completely isometrically isomorphic to $\mathfrak{L}_d$. Thus we have a complete description of $\mathfrak{S}$: with respect to $H = M \oplus M^\perp$,

$$\mathfrak{S} = \begin{bmatrix} \mathcal{B}(M) & 0 \\ \mathcal{B}(M, M^\perp) & \mathfrak{L}_d^{(k(d-1))} \end{bmatrix}.$$

To get a handle on the structure of these algebras in general, we make use of the following ideals.

15.7.4. DEFINITION. Let $\mathfrak{S}_0 = \overline{\langle S_1, \dots, S_d \rangle}^{\text{WOT}} = \overline{\text{span}\{S_w : |w| \ge 1\}}^{\text{WOT}}$. For $k \ge 1$, we set $\mathfrak{S}_0^k = \overline{\langle S_w : |w| = k \rangle}^{\text{WOT}} = \overline{\text{span}\{S_w : |w| \ge k\}}^{\text{WOT}}$. Finally let $\mathfrak{J} = \bigcap_{k \ge 1} \mathfrak{S}_0^k$.

Observe that $\mathfrak{S}_0 + \mathbb{C}I = \mathfrak{S}$, and hence $\dim \mathfrak{S}/\mathfrak{S}_0 \le 1$.

15.7.5. Lemma. *If $\mathfrak{S}_0 = \mathfrak{S}$, then $\mathfrak{S}$ is the von Neumann algebra $\mathfrak{W} = W^*(S_1, \dots, S_d)$.*

If $\mathfrak{S}_0 \ne \mathfrak{S}$, then for each $A \in \mathfrak{S}$ and $k \ge 1$, there are unique constants $a_w \in \mathbb{C}$ for $|w| < k$ and elements $A_w \in \mathfrak{S}$ for $|w| = k$ so that

$$A = \sum_{|w|<k} a_w S_w + \sum_{|w|=k} S_w A_w.$$

Proof. If $\mathfrak{S}_0 = \mathfrak{S}$, then there is a net $\{A_\lambda\}_\Lambda$, where $A_\lambda = \sum_{|w|\ge 1} a_{w,\lambda} S_w$ are finite sums, converging WOT to I. Now

$$S_i^* A_\lambda = \sum_{|w|\ge 1} a_{w,\lambda} S_i^* S_w \in \operatorname{span}\{S_w : |w| \ge 0\}$$

because $S_i^* S_w = S_v$ if $w = iv$ and 0 otherwise. Therefore $S_i^* = \text{WOT--}\lim_\Lambda S_i^* A_\lambda$ belongs to $\mathfrak{S}$. It follows that $\mathfrak{W} = W^*(S_1, \dots, S_d) \subset \mathfrak{S} \subset \mathfrak{W}$.

If $\mathfrak{S}_0 \ne \mathfrak{S}$, the quotient map $\psi : \mathfrak{S} \to \mathfrak{S}/\mathfrak{S}_0$ is a WOT-continuous character on $\mathfrak{S}$ such that $\psi(I) = 1$ and $\psi(S_w) = 0$ for $|w| \ge 1$. Clearly $\ker \psi = \mathfrak{S}_0$ and the unique scalar $a_0 := \psi(A)$ satisfies $A - a_0 I \in \mathfrak{S}_0$. Suppose that $A \in \mathfrak{S}_0$. Then as in the previous paragraph, $A = \text{WOT--}\lim_\Lambda A_\lambda$ where $A_\lambda = \sum_{|w|\ge 1} a_{w,\lambda} S_w$ are finite sums. It follows that $A = \big(\sum_{i=1}^d S_i S_i^*\big) A$ and $A_i := S_i^* A \in \mathfrak{S}$. Therefore $A = \sum_{i=1}^d S_i A_i$, and this is unique, since such an identity forces $A_i = S_i^* A$. For general A, we have $A = \psi(A) I + \sum_{i=1}^d S_i A_i$.

Assume that we have established the result for $k - 1$. Then we have a unique decomposition

$$A = \sum_{|w|<k-1} a_w S_w + \sum_{|w|=k-1} S_w A_w.$$

For each $|w| = k - 1$, we have a unique decomposition $A_w = a_w I + \sum_{i=1}^d S_i A_{wi}$. Hence

$$A = \sum_{|w|<k-1} a_w S_w + \sum_{|w|=k-1} S_w \big(a_w I + \sum_{i=1}^d S_i A_{wi}\big) = \sum_{|w|<k} a_w S_w + \sum_{|v|=k} S_v A_v. \qquad \blacksquare$$

15.7.6. Remark. Recall that a map $\varphi : \mathfrak{A} \to \mathfrak{B}$ is a *complete quotient map* if the induced map $\widetilde{\varphi} : \mathfrak{A}/\ker \varphi \to \mathfrak{B}$ is a completely isometric isomorphism. When a complete quotient map $\varphi : \mathfrak{A} \to \mathfrak{B}$ is also weak-$*$ continuous between dual spaces, every matrix in $M_k(\mathfrak{B})$ lifts to an element of the same norm in $M_k(\mathfrak{A})$. This is a straightforward application of weak-$*$ compactness of the unit ball.

15.7.7. Structure Theorem. *Let $\mathfrak{S}$ be a free semigroup algebra. Then either $\mathfrak{S}_0 = \mathfrak{S}$ and $\mathfrak{S} = \mathfrak{W}$ is a von Neumann algebra, or $\mathfrak{S}_0 \neq \mathfrak{S}$ and*

(1) *there is a canonical homomorphism $\varphi : \mathfrak{S} \to \mathfrak{L}_d$ such that $\varphi(S_i) = L_i$ which is completely contractive,* WOT*-continuous and surjective.*

(2) $\ker\varphi = \mathfrak{J} = \bigcap_{k\geq 1} \mathfrak{S}_0^k$ *and φ is a complete quotient map; and $\mathfrak{S}/\mathfrak{J}$ is completely isometrically isomorphic and weak-$*$ homeomorphic to $\mathfrak{L}_d$.*

(3) *$\mathfrak{J}$ is a left ideal of $\mathfrak{W}$ of the form $\mathfrak{J} = \mathfrak{W}P$ where $P = P^2 = P^*$ is a projection in $\mathfrak{S}$ and $P^\perp H \in \operatorname{Lat}\mathfrak{S}$.*

(4) *$\mathfrak{L} := \mathfrak{S}|_{P^\perp H}$ is completely isometrically isomorphic and weak-$*$ homeomorphic to $\mathfrak{L}_d$, and thus is an analytic free semigroup algebra.*

(5) *P is the largest projection in $\mathfrak{S}$ such that $P\mathfrak{S}P$ is self-adjoint.*

Thus on $H = PH \oplus P^\perp H$,

$$\mathfrak{S} = \begin{bmatrix} P\mathfrak{W}P & 0 \\ P^\perp\mathfrak{W}P & \mathfrak{L} \end{bmatrix}$$

Proof. By Lemma 15.7.5, if $\mathfrak{S}_0 = \mathfrak{S}$, then $\mathfrak{S} = \mathfrak{W}$ is a von Neumann algebra. So for the remainder, we assume that $\mathfrak{S}_0 \neq \mathfrak{S}$. We let $\psi : \mathfrak{S} \to \mathfrak{S}/\mathfrak{S}_0 \simeq \mathbb{C}$ be the character used in the previous proof.

(1) Since ψ is WOT-continuous, there are vectors η_j, ζ_j for $1 \leq j \leq k$ such that

$$\psi(A) = \sum_{j=1}^{k} \langle A\eta_j, \zeta_j\rangle.$$

Define $\eta = (\eta_1, \dots, \eta_k)$ and $\zeta = (\zeta_1, \dots, \zeta_k)$ in $H^{(k)}$. Then $\psi(A) = \langle A^{(k)}\eta, \zeta\rangle$. Since $\psi(I) = 1$, ζ is not orthogonal to $M = \overline{\mathfrak{S}^{(k)}\eta}$, but $\mathfrak{S}_0 = \ker\psi$, and hence ζ is orthogonal to $M_0 = \overline{\mathfrak{S}_0^{(k)}\eta}$. Select a unit vector $\xi \in M \ominus M_0$. Then since $\mathfrak{S}_0\xi \subset \mathfrak{S}_0 M \subset M_0$, we have

$$\langle \xi, S_w^{(k)}\xi\rangle = 0 \quad \text{for} \quad w \in \mathbb{F}_d^+,\ |w| \geq 1.$$

Then since $S_v^{(k)}$ is an isometry, $\langle S_v^{(k)}\xi, S_{vw}^{(k)}\xi\rangle = 0$ for $|w| \geq 1$. The isometries $S_1^{(k)}, \dots, S_d^{(k)}$ have pairwise orthogonal ranges, and thus $\langle S_v^{(k)}\xi, S_u^{(k)}\xi\rangle = 0$ for $u \neq v$. Hence ξ is a wandering vector. The map from $N = \operatorname{span}\{S_w^{(k)}\xi : w \in \mathbb{F}_d^+\}$ to F_d^2 by $US_w^{(k)}\xi = \xi_w$ implements a unitary equivalence from $\mathfrak{S}^{(k)}|_N$ onto $\mathfrak{L}_d$ sending $S_i^{(k)}|_N$ to L_i for $1 \leq i \leq d$. The map $\varphi : \mathfrak{S} \to \mathfrak{L}_d$ is obtained as the composition of the ampliation $A \to A^{(k)}$, the compression to N, and finally the unitary equivalence. It is clear that this map is completely contractive and WOT-continuous. It is also weak-$*$ continuous.

We show that φ is surjective. Let $\mathfrak{B} = \overline{\operatorname{Alg}\{I, S_1, \dots, S_n\}}$. Then $\varphi|_{\mathfrak{B}}$ is the canonical isomorphism of $\mathfrak{B}$ onto $\mathcal{A}_d$ taking S_i to L_i. It follows that $\varphi|_{\mathfrak{B}}$ is a complete isometry by Proposition 15.4.7. Let $A \in \mathfrak{L}_d$ with Fourier series

$A \sim \sum_{w\in\mathbb{F}_d^+} a_w L_w$. The Cesáro means are completely contractive and the sequence $\Sigma_n(A)$ converges SOT to A. We define

$$B_n = \sum_{|w|\le n} \Big(1 - \frac{|w|}{n+1}\Big) a_w S_w \quad \in \mathfrak{B} \text{ for } n \ge 1.$$

Since $\varphi|_{\mathfrak{B}}$ is an isometry, $\|B_n\| = \|\Sigma_n(A)\| \le \|A\|$. By WOT-compactness, there is a subsequence $\{B_{n_j}\}$ which converges WOT to some $B \in \mathfrak{S}$. Then by WOT-continuity,

$$\varphi(B) = \text{WOT–}\lim \varphi(B_{n_j}) = \text{WOT–}\lim \Sigma_{n_j}(A) = A.$$

Therefore φ is surjective. Moreover $\|B\| \le \sup \|B_n\| = \|A\|$, and since φ is contractive, $\|B\| = \|A\|$.

(2) If $A \in \mathfrak{S}_0^k \setminus \mathfrak{S}_0^{k+1}$, Lemma 15.7.5 shows that

$$A = \sum_{|w|=k} a_w S_w + \sum_{|v|=k+1} S_v A_v.$$

Moreover some $a_w \ne 0$ since otherwise A would lie in the ideal $\mathfrak{S}_0^{k+1}$. Hence

$$\varphi(A) = \sum_{|w|=k} a_w L_w + \sum_{|v|=k+1} L_v \varphi(A_v) \ne 0.$$

Therefore $\ker\varphi \subset \bigcap_{k\ge1} \mathfrak{S}_0^k = \mathfrak{J}$. Conversely, if $A \in \mathfrak{J}$, in the expansion above we have $a_w = 0$ for $|w| \le k$ and so $\varphi(A) \in \bigcap_{k\ge1} \mathfrak{L}_{d0}^k$ for all $k \ge 1$. Since $\bigcap_{k\ge1} \mathfrak{L}_{d0}^k = \{0\}$, we have $\ker\varphi = \mathfrak{J}$.

The surjective homomorphism φ induces an isomorphism $\widetilde{\varphi} : \mathfrak{S}/\mathfrak{J} \to \mathfrak{L}_d$. For $A \in \mathfrak{L}_d$, every $B \in \varphi^{-1}(A)$ satisfies $\|B\| \ge \|A\|$ because φ is contractive; and the previous paragraph shows that equality holds for the B which we constructed. Hence $\widetilde{\varphi}$ is isometric. The same argument works for matrices over $\mathfrak{S}$, and therefore $\widetilde{\varphi}$ is completely isometric, and so φ is a complete quotient map.

Since $\mathfrak{J}$ is weak-$*$ closed, there is a natural predual $\mathfrak{J}_\perp \subset \mathfrak{S}_*$ for $\mathfrak{S}/\mathfrak{J}$ and $\widetilde{\varphi}$ is also weak-$*$-continuous. Thus there is a bijective map $T : \mathfrak{L}_{d*} \to \mathfrak{J}_\perp$ such that $T^* = \widetilde{\varphi}$. Since $\widetilde{\varphi}$ is an isometry, so is T. The inverse $\widetilde{\varphi}^{-1} = T^{-1*}$ is also weak-$*$ continuous. Therefore $\widetilde{\varphi}$ is a weak-$*$–weak-$*$ homeomorphism.

(3) If $A \in \mathfrak{J}$, then for any $k \ge 0$, we have $A = \sum_{|w|=k+1} S_w A_w$; from which it follows that $S_i^* A = \sum_{|v|=k} S_v A_{iv} \in \mathfrak{S}_0^k$. Therefore $S_i^* A \in \mathfrak{J}$. Since $\mathfrak{J}$ is WOT-closed, $\mathfrak{J}$ is a left ideal for $\mathfrak{W}$. By Exercise **10.3**, there is a projection $P \in \mathfrak{J}$ such that $\mathfrak{J} = \mathfrak{W}P$. Now $\mathfrak{J} \cap \mathfrak{J}^* = \mathfrak{W}P \cap P\mathfrak{W} = P\mathfrak{W}P$ is a von Neumann algebra. The range of $\mathfrak{J}^*$ is PH. Since $\mathfrak{J}^*$ is an ideal in $\mathfrak{S}^*$, PH is invariant for $\mathfrak{S}^*$; and therefore $P^\perp H$ is invariant for $\mathfrak{S}$.

(4) Since $\mathfrak{J} = \mathfrak{S}P$, we have $\mathfrak{S}/\mathfrak{J} \simeq \mathfrak{S}P^\perp$, and this identification is completely isometric and a WOT-homeomorphism. By (2), $\mathfrak{S}P^\perp$ is completely isometrically isomorphic and weak-$*$ homeomorphic to $\mathfrak{L}_d$.

(5) $P\mathfrak{S}P = P\mathfrak{W}P = \mathfrak{J} \cap \mathfrak{J}^*$ is self-adjoint. If Q is a self-adjoint projection in $\mathfrak{S}$, then since $P^\perp H$ is invariant, $Q = Q_1 \oplus Q_2$ where $Q_1 \le P$ and $Q_2 \le P^\perp$. So $Q_2 \in P^\perp\mathfrak{S}$ which is isomorphic to $\mathfrak{L}_d$. But $\mathfrak{L}_d$ contains no non-scalar idempotents by Exercise **15.9**. Thus $Q_2 \in \{0, I\}$. However $P^\perp\mathfrak{S}$ is not self-adjoint since it contains ideals with no approximate identity. So if $Q\mathfrak{S}Q$ is self-adjoint, then $Q_2 = 0$ and hence $Q \le P$. ■

15.8. Ideals of $\mathfrak{L}_d$

15.8.1. Definition. Let $\mathit{Id}(\mathfrak{L}_d)$ and $\mathit{Id}_r(\mathfrak{L}_d)$ denote the set of WOT-closed 2-sided ideals and right ideals respectively. For $\mathfrak{J} \in \mathit{Id}_r(\mathfrak{L}_d)$, let $\mu(\mathfrak{J}) = \overline{\mathfrak{J}F^2_d}$.

Note that for $\mathfrak{J} \in \mathit{Id}_r(\mathfrak{L}_d)$,

$$\overline{\mathfrak{R}_d\mu(\mathfrak{J})} = \overline{\mathfrak{R}_d\mathfrak{J}F^2_d} = \overline{\mathfrak{J}\mathfrak{R}_dF^2_d} = \overline{\mathfrak{J}F^2_d} = \mu(\mathfrak{J}).$$

Thus $\mu(\mathfrak{J}) \in \operatorname{Lat}\mathfrak{R}_d$ for all $\mathfrak{J} \in \mathit{Id}_r(\mathfrak{L}_d)$. If $\mathfrak{J} \in \mathit{Id}(\mathfrak{L}_d)$, then we also have that $\overline{\mathfrak{L}_d\mathfrak{J}F^2_d} = \overline{\mathfrak{J}F^2_d} = \mu(\mathfrak{J})$, so $\mu(\mathfrak{J}) \in \operatorname{Lat}\mathfrak{L}_d \cap \operatorname{Lat}\mathfrak{R}_d$.

15.8.2. Theorem. *The map* $\mu : \mathit{Id}_r(\mathfrak{L}_d) \to \operatorname{Lat}\mathfrak{R}_d$ *is a complete lattice isomorphism. Moreover the restriction to* $\mathrm{Id}(\mathfrak{L}_d)$ *is a complete lattice isomorphism onto* $\operatorname{Lat}\mathfrak{L}_d \cap \operatorname{Lat}\mathfrak{R}_d$. *The inverse map is* $\iota(M) = \{A \in \mathfrak{L}_d : AF^2_d \subset M\}$.

Proof. The lattice operations in $\mathit{Id}_r(\mathfrak{L}_d)$ are

$$\bigvee_\Lambda \mathfrak{J}_\lambda = \overline{\sum_\Lambda \mathfrak{J}_\lambda}^{\mathrm{WOT}} \quad \text{and} \quad \bigwedge_\Lambda \mathfrak{J}_\lambda = \bigcap_\Lambda \mathfrak{J}_\lambda.$$

Thus

$$\mu\Big(\bigvee_\Lambda \mathfrak{J}_\lambda\Big) = \overline{\sum_\Lambda \mathfrak{J}_\lambda F^2_d} = \bigvee_\Lambda \mu(\mathfrak{J}_\lambda)$$

and

$$\begin{aligned}\iota\Big(\bigwedge_\Lambda M_\lambda\Big) &= \Big\{A \in \mathfrak{L}_d : AF^2_d \subset \bigcap_\Lambda M_\lambda\Big\} \\ &= \bigcap_\Lambda \{A \in \mathfrak{L}_d : AF^2_d \subset M_\lambda\} = \bigwedge_\Lambda \iota(M_\lambda).\end{aligned}$$

One readily checks that $\mu\iota\mu = \mu$ and $\iota\mu\iota = \iota$. Once we establish that μ and ι are inverses of one another, it will follow that they are complete lattice isomorphisms.

If $M \in \operatorname{Lat}\mathfrak{R}_d$, then if $B \in \iota(M)$ and $A \in \mathfrak{L}_d$, then $BAF^2_d \subset BF^2_d \subset M$. Thus $\iota(M)$ is a right ideal, and it is clearly WOT-closed. If $M \in \operatorname{Lat}\mathfrak{L}_d$ as well, then $ABF^2_d \subset AM \subset M$. So $\iota(M)$ is a 2-sided ideal.

If $M \in \operatorname{Lat}\mathfrak{R}_d$, then by the structure of invariant subspaces of $\mathfrak{R}_d$, there are isometries $S_i \in \mathfrak{L}_d$ so that $M = \overline{\sum_i^{\oplus} S_i F_d^2}$. Clearly $S_i \in \iota(M)$. Thus

$$M \supset \mu\iota(M) \supset \overline{\sum_i S_i F_d^2} = M.$$

Therefore $\mu\iota = id$ on $\operatorname{Lat}\mathfrak{R}_d$.

If $\mathfrak{J} \in Id_r(\mathfrak{L}_d)$, then it is immediate that $\mathfrak{J} \subset \iota\mu(\mathfrak{J})$. If they are not equal, then there is a WOT-continuous linear functional f on $\mathfrak{L}_d$ such that $\mathfrak{J} \subset \ker f$ but f is not zero on $\iota\mu(\mathfrak{J})$. By Proposition 15.5.9, there are vectors $\eta, \zeta \in F_d^2$ so that $f(A) = \langle A\eta, \zeta\rangle$. Now $N := \overline{\mathfrak{L}_d\eta} \in \operatorname{Lat}\mathfrak{L}_d$ and is cyclic, so there is an isometry $R \in \mathfrak{R}_d$ so that $RF_d^2 = N$. Thus

$$\overline{\mathfrak{J}\eta} = \overline{\mathfrak{J}\mathfrak{L}_d\eta} = \overline{\mathfrak{J}RF_d^2} = R\overline{\mathfrak{J}F_d^2} = R\mu(\mathfrak{J}).$$

Similarly

$$\begin{aligned}\overline{\iota\mu(\mathfrak{J})\eta} &= \overline{\iota\mu(\mathfrak{J})\mathfrak{L}_d\eta} = R\overline{\iota\mu(\mathfrak{J})F_d^2} \\ &= R\mu\iota\mu(\mathfrak{J}) = R\mu(\mathfrak{J}) = \overline{\mathfrak{J}\eta}.\end{aligned}$$

Since $f|_{\mathfrak{J}} = 0$, ζ is orthogonal to $\overline{\mathfrak{J}\eta} = \overline{\iota\mu(\mathfrak{J})\eta}$, and hence $\ker f \supset \iota\mu(\mathfrak{J})$. By the Hahn–Banach theorem, $\mathfrak{J} = \iota\mu(\mathfrak{J})$. Therefore μ and ι are inverses, and hence they are complete lattice isomorphisms. ■

In the next theorem, the map from $\mathfrak{L}_d$ to $P_M^{\perp}\mathfrak{L}_d$ is a complete quotient map.

15.8.3. THEOREM. *Let $\mathfrak{J} \in Id_r(\mathfrak{L}_d)$ and let $M = \mu(\mathfrak{J})$. Then $\mathfrak{L}_d/\mathfrak{J}$ is completely isometrically isomorphic to $P_M^{\perp}\mathfrak{L}_d$ via the right $\mathfrak{L}_d$-module map given by $\varphi(A + \mathfrak{J}) = P_M^{\perp}A$. If $k \geq 1$ and $B \in M_k(P_M^{\perp}\mathfrak{L}_d)$, there is an $A \in M_k(\mathfrak{L}_d)$ such that $P_M^{\perp(k)}A = B$ and $\|A\| = \|B\|$.*

PROOF. Suppose $A \in M_k(\mathfrak{L}_d)$ and $\operatorname{dist}(A, M_k(\mathfrak{J})) = 1$. Since $\mathfrak{J} \subset P_M\mathcal{B}(F_d^2)$,

$$\|P_M^{\perp}A\| = \operatorname{dist}\big(A, P_M^{(k)}\mathcal{B}(F_d^{2(k)})\big) \leq \operatorname{dist}(A, M_k(\mathfrak{J})).$$

Conversely, if $\varepsilon > 0$, there is a WOT-continuous functional f on $M_k(\mathfrak{L}_d)$ such that $f \perp M_k(\mathfrak{J})$ and $1 - \varepsilon < f(A) \leq \|f\| < 1$. $M_k(\mathfrak{L}_d)$ has $\mathbb{A}_1(1)$ like $\mathfrak{L}_d$ because $M_k(\mathfrak{L}_d)' = I_k \otimes \mathfrak{R}_d$ contains a countable set of isometries $I_k \otimes R_{21^n}$ with pairwise orthogonal ranges. The proof works verbatim. Hence there are vectors $\eta = (\eta_1, \dots, \eta_k)$ and $\zeta = (\zeta_1, \dots, \zeta_k)$ so that $f(T) = \langle T\eta, \zeta\rangle$ for $T \in M_k(\mathfrak{L}_d)$ and $\|\eta\|\,\|\zeta\| < 1$.

Let $N = \mathfrak{L}_d[\eta_1, \dots, \eta_k]$ in $\operatorname{Lat}\mathfrak{L}_d$; and let $N_0 = \overline{\mathfrak{J}N}$. By Theorem 15.6.1, there are isometries $R_j \in \mathfrak{R}_d$ so that $N = \sum_j^{\oplus} R_jF_d^2$. Thus

$$N_0 = \sum_j{}^{\oplus}\, \overline{\mathfrak{J}R_jF_d^2} = \sum_j{}^{\oplus}\, R_j\overline{\mathfrak{J}F_d^2} = \sum_j{}^{\oplus}\, R_jM.$$

Note that $\overline{M_k(\mathfrak{L}_d)\eta} = N^{(k)}$ and $\overline{M_k(\mathfrak{J})\eta} = N_0^{(k)}$. We can write $\eta = \sum_j R_j^{(k)} \nu_j$ for $\nu_j \in F_d^{2(k)}$. It follows that $\zeta \perp N_0^{(k)}$. We can replace ζ by $P_N^{(k)}\zeta$ without changing anything about f, and then we may write $\zeta = \sum_j R_j^{(k)} \theta_j$ for $\theta_j \in F_d^{2(k)}$. Since ζ is orthogonal to $N_0^{(k)} = \sum_j^{\oplus} R_j^{(k)} M^{(k)}$, we see that $\theta_j \in M^{\perp(k)}$.

Now since $R_j^{(k)} F_d^{2(k)}$ are pairwise orthogonal and invariant for $M_k(\mathfrak{L}_d)$, we have

$$\begin{aligned}
1 - \varepsilon < f(A) = \langle A\eta, \zeta\rangle &= \sum_j \langle AR_j^{(k)}\eta_j, R_j^{(k)}\theta_j\rangle = \sum_j \langle A\eta_j, P_M^{\perp(k)}\theta_j\rangle \\
&= \sum_j \langle P_M^{\perp(k)} A\eta_j, \theta_j\rangle \le \|P_M^{\perp(k)}A\| \sum_j \|\eta_j\|\,\|\theta_j\| \\
&\le \|P_M^{\perp(k)}A\| \Big(\sum_j \|\eta_j\|^2\Big)^{1/2} \Big(\sum_j \|\theta_j\|^2\Big)^{1/2} \\
&= \|P_M^{\perp(k)}A\|\,\|\eta\|\,\|\zeta\| < \|P_M^{\perp(k)}A\|.
\end{aligned}$$

Since $\varepsilon > 0$ was arbitrary, $\|A + M_k(\mathfrak{J})\| = 1 \le \|P_M^{\perp(k)}A\|$. The other direction was done in the first paragraph. Therefore $\|A + M_k(\mathfrak{J})\| = \|P_M^{\perp(k)}A\|$.

If $k \ge 1$ and $B \in M_k(P_M^{\perp}\mathfrak{L}_d)$, we have $\|B\| = \inf\{\|A\| : P_M^{\perp(k)}A = B\}$. So we can select a sequence $A_n \in M_k(\mathfrak{L}_d)$ with $\|A_n\| \le \|B\| + \frac{1}{n}$ and $P_M^{\perp(k)}A_n = B$. This is a bounded net, and thus there is a cofinal subnet with WOT limit A. Clearly $P_M^{\perp(k)}A = B$ and $\|B\| \le \|A\| \le \lim \|A_n\| = \|B\|$. ∎

15.8.4. THEOREM. *Let $\mathfrak{J} \in \mathrm{Id}_r(\mathfrak{L}_d)$ and let $M = \mu(\mathfrak{J})$, and represent $\mathfrak{L}_d/\mathfrak{J}$ as $P_M^{\perp}\mathfrak{L}_d$. Then every weak-$*$ continuous functional φ on $\mathfrak{L}_d/\mathfrak{J}$ with $\|\varphi\| < 1$ can be represented as $\varphi(A + \mathfrak{J}) = \langle A\eta, \zeta\rangle$ where $\zeta \in M^{\perp}$ and $\|\eta\|\,\|\zeta\| < 1$. Thus the* WOT *and weak-$*$ topologies coincide on $P_M^{\perp}\mathfrak{L}_d$, which is* WOT*-closed. Moreover, the map from $\mathfrak{L}_d/\mathfrak{J}$ onto $P_M^{\perp}\mathfrak{L}_d$ is a weak-$*$–weak-$*$ homeomorphism.*

PROOF. Define $\widetilde{\varphi}(A) = \varphi(A + \mathfrak{J})$ for $A \in \mathfrak{L}_d$. The quotient map and φ are both weak-$*$ continuous, so $\widetilde{\varphi}$ is weak-$*$ continuous. Also $\|\widetilde{\varphi}\| = \|\varphi\| < 1$. Thus by Corollary 15.6.5, there is a cyclic vector η for $\mathfrak{L}_d$ and a vector $\zeta \in F_d^2$ so that $\|\eta\|\,\|\zeta\| < 1$ and $\widetilde{\varphi}(A) = \langle A\eta, \zeta\rangle$ for $A \in \mathfrak{L}_d$. Since $\ker(\widetilde{\varphi}) \supset \mathfrak{J}$, ζ is orthogonal to $\overline{\mathfrak{J}\eta} = \overline{\mathfrak{J}\mathfrak{L}_d\eta} = \overline{\mathfrak{J}F_d^2} = M$. That is, $\zeta \in M^{\perp}$.

Hence every weak-$*$ continuous linear functional coincides with a WOT-continuous functional. Therefore the two topologies coincide on $P_M^{\perp}\mathfrak{L}_d$. So it suffices to prove that $P_M^{\perp}\mathfrak{L}_d$ is weak-$*$ closed.

If $A \in \mathfrak{L}_d$, there are $J_n \in \mathfrak{J}$ so that $\|A + J_n\| \to \|A + \mathfrak{J}\|$. Clearly $\{J_n\}$ is bounded; so we can drop to a subsequence with weak-$*$ limit J. Then $J \in \mathfrak{J}$ and

$$\|A + \mathfrak{J}\| \le \|A + J\| \le \liminf \|A + J_n\| = \|A + \mathfrak{J}\| = \|P_M^{\perp}A\|.$$

That is, $\|A+J\| = \|P_M^\perp A\|$.

Suppose that $\{B_\alpha\}$ is a bounded net in $P_M^\perp \mathfrak{L}_d$ converging weak-$*$ to B. Choose $A_\alpha \in \mathfrak{L}_d$ so that $B_\alpha = P_M^\perp A_\alpha$. By the previous paragraph, find $J_\alpha \in \mathfrak{J}$ so that $\|A_\alpha + J_\alpha\| = \|B_\alpha\|$. There is a cofinal subnet of $\{A_\alpha + J_\alpha\}$ which converges weak-$*$ to $A \in \mathfrak{L}_d$. Then $P_M^\perp A = B$. It follows that the unit ball of $P_M^\perp \mathfrak{L}_d$ is weak-$*$ closed. By the Krein–Smulian theorem, $P_M^\perp \mathfrak{L}_d$ is weak-$*$-closed. Since the weak-$*$ and WOT-topologies coincide, it is also WOT-closed.

The weak-$*$ continuous functionals on $\mathfrak{L}_d/\mathfrak{J}$ coincide with the weak-$*$ continuous functional on $P_M^\perp \mathfrak{L}_d$. Hence the map from $\mathfrak{L}_d/\mathfrak{J}$ onto $P_M^\perp \mathfrak{L}_d$ is a weak-$*$–weak-$*$ homeomorphism. ∎

15.8.5. COROLLARY. *Let $\mathfrak{J} \in Id(\mathfrak{L}_d)$ and let $M = \mu(\mathfrak{J})$. Then $\mathfrak{L}_d/\mathfrak{J}$ is completely isometrically isomorphic to $P_M^\perp \mathfrak{L}_d|_{M^\perp}$ via the* WOT*-continuous homomorphism $\psi(A+\mathfrak{J}) = P_M^\perp A|_{M^\perp}$. Moreover $P_M^\perp \mathfrak{L}_d|_{M^\perp}$ is* WOT*-closed in $\mathcal{B}(M^\perp)$ and has property $\mathbb{A}_1(1)$.*

PROOF. Since $M \in \operatorname{Lat} \mathfrak{L}_d$, compression to $M^\perp$ is a completely contractive and WOT-continuous homomorphism. Moreover, $P_M^\perp A = P_M^\perp A P_M^\perp$ on $\mathfrak{L}_d$, so this map ψ coincides with the map φ in Theorem 15.8.3. Hence it is completely isometric.

By Theorem 15.8.4, $P_M^\perp \mathfrak{L}_d = P_M^\perp \mathfrak{L}_d P_M^\perp$ is WOT-closed. But $P_M^\perp \mathfrak{L}_d = \begin{bmatrix} P_M^\perp \mathfrak{L}_d|_{M^\perp} & 0 \end{bmatrix}$ in $\mathcal{B}(F_d^2, M^\perp) = \begin{bmatrix} P_M^\perp \mathcal{B}(F_d^2)|_{M^\perp} & P_M^\perp \mathcal{B}(F_d^2)|_M \end{bmatrix}$. Thus it is clear that $P_M^\perp \mathfrak{L}_d|_{M^\perp}$ is also WOT-closed.

If φ is any weak-$*$ continuous functional on $P_M^\perp \mathfrak{L}_d|_{M^\perp}$ with $\|\varphi\| < 1$, it is naturally a weak-$*$ continuous functional on $P_M^\perp \mathfrak{L}_d$ as well. Again by Theorem 15.8.4, there is a cyclic vector η for $\mathfrak{L}_d$ and a vector $\zeta \in M^\perp$ so that $\|\eta\|\,\|\zeta\| < 1$ and

$$\varphi(A+\mathfrak{J}) = \langle A\eta, \zeta\rangle = \langle A\eta, P_M^\perp \zeta\rangle = \langle P_M^\perp A P_M^\perp \eta, \zeta\rangle.$$

Thus we can replace η by $P_M^\perp \eta$, so that $\eta, \zeta \in M^\perp$. Therefore $P_M^\perp \mathfrak{L}_d|_{M^\perp}$ has property $\mathbb{A}_1(1)$. ∎

As an example, we will compute the ideal

$$\mathfrak{C} = \bigcap_{\lambda \in \mathbb{B}_d} \ker \varphi_\lambda = \{A \in \mathfrak{L}_d : \widehat{A} = 0\}.$$

15.8.6. DEFINITION. Define a map on F_d^2 by sending $\xi = \sum_{w \in \mathbb{F}_d^+} a_w \xi_w$ to

$$\widehat{\xi}(\lambda) = \sum_{w \in \mathbb{F}_d^+} a_w w(\lambda) = \frac{\langle \xi, \nu_\lambda\rangle}{(1 - \|\lambda\|^2)^{1/2}} \quad \text{for} \quad \lambda \in \mathbb{B}_d.$$

The kernel of this map is *antisymmetric Fock space*, F_d^a. *Symmetric Fock space*, H_d^2, is the complement of F_d^a in F_d^2.

For $k = (k_1, \dots, k_d) \in \mathbb{Z}_+^d$, define $\lambda^k = \lambda_1^{k_1} \dots \lambda_d^{k_d}$ for $\lambda \in \mathbb{B}_d$. Let

$$\mathcal{P}_k = \{w \in \mathbb{F}_d^+ : w(\lambda) = \lambda^k\} \quad \text{so that} \quad |\mathcal{P}_k| = \binom{k}{k_1, \dots, k_d} = \frac{k!}{k_1! \dots k_d!}.$$

Define $\zeta_k = \frac{1}{|\mathcal{P}_k|} \sum_{w \in \mathcal{P}_k} \xi_w$. Then $\|\zeta_k\|^2 = \frac{1}{|\mathcal{P}_k|}$ and

$$\widehat{\zeta_k}(\lambda) = \frac{1}{|\mathcal{P}_k|} \sum_{w \in \mathcal{P}_k} \lambda^k = \lambda^k \quad \text{for} \quad \lambda \in \mathbb{B}_d.$$

15.8.7. Proposition. $H_d^2 = \overline{\operatorname{span}}\{\zeta_k : k \in \mathbb{Z}_+^d\} = \overline{\operatorname{span}}\{\nu_\lambda : \lambda \in \mathbb{B}_d\}$. *A vector* $\zeta = \sum_{w \in \mathbb{F}_d^+} a_w \xi_w \in H_d^2$ *if and only if* $a_w = a_{w'}$ *whenever* w' *is a permutation of* w.

Proof. If $\xi = \sum_{w \in \mathcal{P}_k} a_w \xi_w$, then

$$\widehat{\xi}(\lambda) = \sum_{w \in \mathcal{P}_k} a_w \lambda^k =: b\lambda^k \quad \text{and} \quad \widehat{\xi - b\zeta_k} = 0.$$

Thus $\xi - b\zeta_k \in F_d^a$. Hence the projection of $\operatorname{span}\{\xi_w : w \in \mathcal{P}_k\}$ onto H_d^2 has range $\mathbb{C}\zeta_k$. Therefore $H_d^2 = \operatorname{span}\{\zeta^k : k \in \mathbb{Z}_+^d\}$.

In particular, if w' is a permutation of w, then $\xi_w - \xi_{w'} \in F_d^a$. If a vector $\zeta = \sum_{w \in \mathbb{F}_d^+} a_w \xi_w$ is in H_d^2, then

$$0 = \langle \zeta, \xi_w - \xi_{w'} \rangle = a_w - a_{w'}.$$

Conversely, if this holds, then $\sum_{w \in \mathcal{P}_k} a_w \xi_w = a_w \, |\mathcal{P}_k| \zeta_k \in H_d^2$. Therefore setting $b_k = a_w$ for $w \in \mathcal{P}_k$,

$$\zeta = \sum_{k \in \mathbb{Z}_+} b_k \, |\mathcal{P}_k| \zeta_k \in H_d^2.$$

To simplify notation, write $k! := k_1! \cdots k_d!$ for $k \in \mathbb{Z}_+^d$; so $|\mathcal{P}_k| = \frac{|k|!}{k!}$. Now

$$\begin{aligned}\nu_\lambda &= (1 - \|\lambda\|^2)^{1/2} \sum_{w \in \mathbb{F}_d^+} \overline{w(\lambda)} \xi_w = (1 - \|\lambda\|^2)^{1/2} \sum_{k \in \mathbb{Z}_+^d} \overline{\lambda}^k \sum_{w \in \mathcal{P}_k} \xi_w \\ &= (1 - \|\lambda\|^2)^{1/2} \sum_{k \in \mathbb{Z}_+^d} \overline{\lambda}^k |\mathcal{P}_k| \zeta_k = (1 - \|\lambda\|^2)^{1/2} \sum_{k \in \mathbb{Z}_+^d} \frac{|k|!}{k!} \overline{\lambda}^k \zeta_k.\end{aligned}$$

Hence $\nu_\lambda \in H_d^2$. Conversely, fix $0 < r < \frac{1}{\sqrt{d}}$. For $s = (s_1, \dots, s_d) \in \mathbb{Z}_+^d$,

$$\begin{aligned}&\frac{1}{(2\pi)^d} \int_{[-\pi,\pi]^d} \nu_{re^{i\theta_1}, \dots, re^{i\theta_d}} e^{i(s_1\theta_1 + \dots + s_d\theta_d)} \, d\theta_1 \dots d\theta_d \\ &= (1 - \|\lambda\|^2)^{1/2} \sum_{k \in \mathbb{Z}_+^d} \frac{|k|!}{k!} \zeta_k \frac{1}{(2\pi)^d} \int \prod_{j=1}^d (re^{-i\theta_j})^{s_j} e^{i(s_1\theta_1 + \dots + s_d\theta_d)} \, d\theta_1 \dots d\theta_d \\ &= (1 - \|\lambda\|^2)^{1/2} r^{|s|} |\mathcal{P}_s| \zeta_s.\end{aligned}$$

Hence $H^2_d \subset \operatorname{span}\{\nu_\lambda : \lambda \in \mathbb{B}_d\}$. Whence they are equal. ■

Let $\langle S\rangle$ denote the ideal generated by a set S.

15.8.8. THEOREM. *Let $\mathfrak{C} = \bigcap_{\lambda\in\mathbb{B}_d} \ker\varphi_\lambda = \{A \in \mathfrak{L}_d : \widehat{A} = 0\}$. Then*

$$\mathfrak{C} = \overline{\langle AB - BA : A, B \in \mathfrak{L}_d\rangle}^{\text{WOT}} = \overline{\langle L_iL_j - L_jL_i : 1 \le i, j \le d\rangle}^{\text{WOT}}$$

is the WOT*-closure of the commutator ideal.*

The quotient $\mathfrak{L}_d/\mathfrak{C}$ is completely isometrically isomorphic and weak-∗–weak-∗ homeomorphic to $P_{H^2_d}\mathfrak{L}_d|_{H^2_d}$, which is a WOT*-closed abelian operator algebra on H^2_d with property $\mathbb{A}_1(1)$. It is generated by a commuting row contraction $\begin{bmatrix} S_1 & \dots & S_d\end{bmatrix}$ given by $S_i\zeta_k = \zeta_{\delta_i+k}$ for $k \in \mathbb{Z}^d_+$.*

PROOF. It is clear that $\mathfrak{C}$ is a WOT-closed ideal. Since each φ_λ is a character, $\widehat{AB-BA} = \widehat{A}\widehat{B} - \widehat{B}\widehat{A} = 0$. Hence the WOT-closed ideal these generate is contained in $\mathfrak{C}$. Similarly, this ideal contains $\mathfrak{J} = \overline{\langle L_iL_j - L_jL_i : 1 \le i, j \le d\rangle}^{\text{WOT}}$. Now $\mu(\mathfrak{J})$ contains $L_u(L_iL_j - L_jL_i)L_v\xi_\varnothing = \xi_{uijv} - \xi_{ujiv}$ for all $u, v \in \mathbb{F}^+_d$. Now every permutation of n terms is a product of simple interchanges of adjacent terms. Thus $\mu(\mathfrak{J})$ contains $\xi_w - \xi_{w'}$ if w' is a permutation of w. This means that $\mu(\mathfrak{J})$ contains $\xi_w - \xi_{w'}$ if $w, w' \in \mathcal{P}_k$. If $\zeta = \sum_{w\in\mathbb{F}^+_d} a_w\xi_w$ is orthogonal to $\mu(\mathfrak{J})$, then $a_{w'} = a_w$ if w' is a permutation of w. Therefore

$$\mu(\mathfrak{J})^\perp \subset H^2_d = \mu(\mathfrak{C})^\perp \subset \mu(\mathfrak{J})^\perp.$$

So $\mu(\mathfrak{J}) = \mu(\mathfrak{C})$ and thus $\mathfrak{J} = \mathfrak{C}$.

Since $\mathfrak{C}$ contains the commutator ideal, the quotient $\mathfrak{L}_d/\mathfrak{C}$ is abelian. By Corollary 15.8.5, this quotient is completely isometrically isomorphic and weak-∗–weak-∗ homeomorphic to $P_{H^2_d}\mathfrak{L}_d|_{H^2_d}$. This algebra $P_{H^2_d}\mathfrak{L}_d|_{H^2_d}$ is WOT-closed and has property $\mathbb{A}_1(1)$. The generators are $S_i = P_{H^2_d}L_i|_{H^2_d}$ for $1 \le i \le d$, which form a row contraction since it is a compression of L. The basis $\{\zeta_k : k \in \mathbb{Z}^d_+\}$ is orthogonal but $\|\zeta_k\| = |\mathcal{P}_k|^{-1/2} = \big(\frac{k!}{|k|!}\big)^{1/2}$; and

$$S_i\zeta_k \in \operatorname{span}\{\xi_v : v \in \mathcal{P}_{\delta_i+k}\} \cap H^2_d = \mathbb{C}\zeta_{\delta_i+k}.$$

To find the coefficient, compute

$$\Big\langle S_i\zeta_k, \frac{\zeta_{\delta_i+k}}{\|\zeta_{\delta_i+k}\|^2}\Big\rangle = \Big\langle \frac{1}{|\mathcal{P}_k|}\sum_{w\in\mathcal{P}_k}\xi_{iw}, |P_{\delta_i+k}|\frac{1}{|P_{\delta_i+k}|}\sum_{v\in P_{\delta_i+k}}\xi_v\Big\rangle$$
$$= \frac{|\mathcal{P}_k|}{|\mathcal{P}_k|} = 1.$$

Thus $S_i\zeta_k = \zeta_{\delta_i+k}$. ■

In the next section, we will identify H^2_d in another way which will bring analytic function theory into the picture.

Drury–Arveson Space

In this last part of the chapter, we examine the universal model for commuting row contractions. We begin with a quick primer on reproducing kernel Hilbert spaces (RKHS). Then we show that symmetric Fock space is a RKHS, and derive some of its properties. After that, we derive Drury's von Neumann inequality and the dilation theorem for commuting row contractions. Finally we establish the complete Pick property and deduce some consequences.

15.9. Reproducing Kernel Hilbert Spaces

In this short section, we develop a few basic properties about Hilbert spaces of functions on a set X for which point evaluation is continuous.

15.9.1. DEFINITION. If H is a Hilbert space of functions on a set X such that the point evaluations $\varepsilon_x(f) = f(x)$ are continuous, then H is a *reproducing kernel Hilbert space* (RKHS). There are vectors $k_x \in H$ such that $f(x) = \langle f, k_x \rangle$. The collection $\{k_x : x \in X\}$ is the reproducing kernel, and $k(x, y) = k_y(x) = \langle k_y, k_x \rangle$ is the kernel function. Say that k is a *non-degenerate kernel* if $k(x, x) > 0$ for all $x \in X$.

15.9.2. EXAMPLES.
(1) The Hardy space H^2 is an RKHS when considered as a space of analytic functions on $\mathbb{D}$ with the Szegö kernel $k_z(w) = \frac{1}{1-w\bar{z}}$ for $w, z \in \mathbb{D}$.

(2) $\mathbb{C}^n$ is an RKHS with $X = \{1. \dots, n\}$ and $k(i, j) = \delta_{ij}$. The kernel functions $k_1, \dots, k_n$ form an orthonormal basis.

(3) $L^2(0, 1)$ is not an RKHS because point evaluation is not continuous on $L^2(0, 1)$.

(4) In Example 2.1.5, we showed that the Hilbert space $L^2_a(\Omega)$ of all square integrable analytic functions on a bounded domain $\Omega \subset \mathbb{C}$ has the reproducing kernel property.

15.9.3. DEFINITION. A function $k : X \times X \to \mathbb{C}$ is *positive semidefinite* if for any $n \geq 1$ and distinct points $x_1, \dots, x_n$ in X, the $n \times n$ matrix $\big[k(x_i, x_j)\big]$ is positive semidefinite. Say that k is *strictly positive* if $\big[k(x_i, x_j)\big]$ is strictly positive for all finite subsets of X.

Note that if k is positive semidefinite, then $k(x, x) \geq 0$ for $x \in X$ and $k(y, x) = \overline{k(x, y)}$ for all $x, y \in X$.

15.9.4. PROPOSITION. *Let H be an RKHS on a set X with kernel function $k(x,y)$. Then k is positive semidefinite. Also $H = \overline{\operatorname{span}\{k_x : x \in X\}}$.*

PROOF. Note that $k(x,x) = \|k_x\|^2 \geq 0$. If $k(x,x) = 0$, then $k_x = 0$, and therefore $k(x,y) = 0 = k(y,x)$ for all $y \in X$. (We generally assume that $k_x \neq 0$, but this is not necessary.) For any $x_1,\dots,x_n \in X$ and $\xi = (\alpha_1,\dots,\alpha_n) \in \mathbb{C}^n$,

$$\begin{aligned}\big\langle \big[k(x_i,x_j)\big]\xi,\xi\big\rangle &= \sum_{i,j=1}^{n} \overline{\alpha_i} k(x_i,x_j)\alpha_j \\ &= \Big\langle \sum_{j=1}^{n} \alpha_j k_{x_j}, \sum_{i=1}^{n} \alpha_i k_{x_i} \Big\rangle = \Big\| \sum_{j=1}^{n} \alpha_j k_{x_j} \Big\|^2 \geq 0.\end{aligned}$$

Hence $\big[k(x_i,x_j)\big]$ is positive semidefinite.

If $f \in H$ is orthogonal to $\{k_x : x \in X\}$, then $f(x) = \langle f, k_x\rangle = 0$ for all $x \in H$; and hence $f = 0$. Therefore $H = \overline{\operatorname{span}\{k_x : x \in X\}}$. ∎

15.9.5. PROPOSITION. *Let H be an RKHS on a set X with kernel function $k(x,y)$. The following are equivalent:*

(1) *k is strictly positive.*

(2) *$\{k_x : x \in X\}$ are linearly independent.*

(3) *For any finite set $F = \{x_1,\dots,x_n\} \subset X$ with $|F| = n$, the restriction map $R_F : H \to \mathbb{C}^n$ given by $R_F(f) = (f(x_1),\dots,f(x_n))$ is surjective.*

In this case, given $F = \{x_1,\dots,x_n\} \subset X$ with $|F| = n$ and $\lambda = (\lambda_1,\dots,\lambda_n)$ in $\mathbb{C}^n$, there is a unique function $f \in H$ of minimal norm such that $R_F(f) = \lambda$; and moreover, $f \in \operatorname{span}\{k_{x_i} : 1 \leq i \leq n\}$.

PROOF. In the previous proof, we showed that for distinct points $x_1,\dots,x_n$ in X and $\xi = (\alpha_1,\dots,\alpha_n) \in \mathbb{C}^n$, $\big\langle \big[k(x_i,x_j)\big]\xi,\xi\big\rangle = \big\| \sum_{j=1}^{n} \alpha_j k_{x_j} \big\|^2$. If $\big[k(x_i,x_j)\big]$ is strictly positive, this quantity is strictly positive for $\xi \neq 0$, and thus $\{k_{x_1},\dots,k_{x_n}\}$ is linearly independent. Conversely, if they are linearly independent, the quantity is non-zero for $\xi \neq 0$ and so $\big[k(x_i,x_j)\big]$ is strictly positive.

Let P_F be the orthogonal projection of H onto $\operatorname{span}\{k_{x_i} : 1 \leq i \leq n\}$. Then $f \in \ker P_F$ if and only if $0 = \langle f, k_{x_i}\rangle = f(x_i)$ for $1 \leq i \leq n$ if and only if $f \in \ker R_F$. If (2) holds, $\operatorname{Ran} P_F$ has dimension n and thus $\ker R_F = \ker P_F$ is codimension n. Hence $\operatorname{Ran} R_F$ has dimension n, and thus it is surjective. Conversely, if (3) holds, the reversal of this argument shows that $\operatorname{rank} P_F = n$ and hence (2) holds.

Take any function $g \in H$ such that $R_F g = \lambda$, and let $f = P_F g$. Then f lies in $\operatorname{span}\{k_{x_i} : 1 \leq i \leq n\}$. Also $R_F f = R_F g - R_F P_F^{\perp} g = \lambda$ and $\|f\| \leq \|g\|$. If $R_F h = \lambda$, then $h - f \in \ker R_F = \ker P_F$, so that $\|h\|^2 = \|f\|^2 + \|h-f\|^2 \geq \|f\|^2$ with equality only if $h = f$. ∎

15.9.6. DEFINITION. If H is a RKHS on a set X, then a function h on X is a *multiplier* if $hf \in H$ for every $f \in H$. The set $\mathrm{Mult}(H)$ is the vector space of all multipliers on H.

15.9.7. PROPOSITION. *If H is a non-degenerate RKHS on a set X, then* $\mathrm{Mult}(H)$ *is a* WOT*-closed unital algebra of continuous linear operators on H. If h is a multiplier, then $M_h^* k_x = \overline{h(x)} k_x$ for $x \in X$. Conversely, if h is a function on X and there is a bounded operator T such that $Tk_x = \overline{h(x)} k_x$ for $x \in X$, then h is a multiplier and $M_h = T^*$. In particular, $\|M_h\| \geq \|h\|_\infty$.*

PROOF. Every multiplier determines a continuous linear map by $M_h f = hf$ as an application of the Closed graph theorem (see Exercise **2.11**). It is clear that $\mathrm{Mult}(H)$ is also an algebra. Since 1 is a multiplier, $M_1 = I \in \mathrm{Mult}(H)$. Moreover, $\mathrm{Mult}(H)$ is WOT-closed because if $\{M_{h_\alpha}\}$ is a net of multipliers with WOT limit T, then

$$Tf(x) = \langle Tf, k_x \rangle = \lim_\alpha \langle M_{h_\alpha} f, k_x \rangle = \lim_\alpha h_\alpha(x) f(x) =: h(x) f(x)$$

where $h(x) = \lim_\alpha h_\alpha(x)$. Indeed, this limit exists if $f(x) \neq 0$; and for every x, there is a function (such as k_x) such that $f(x) \neq 0$. Hence $T = M_h$ is a multiplier.

If h is a multiplier, then for $x, y \in X$,

$$\begin{aligned} \langle k_y, M_h^* k_x \rangle &= \langle M_h k_y, k_x \rangle = \langle h k_y, k_x \rangle \\ &= h(x) k_y(x) = h(x) \langle k_y, k_x \rangle = \langle k_y, \overline{h(x)} k_x \rangle. \end{aligned}$$

Since $\{k_y : y \in X\}$ span H, this shows that $M_h^* k_x = \overline{h(x)} k_x$ for $x \in X$. Conversely, if there is a bounded operator T such that $Tk_x = \overline{h(x)} k_x$ for $x \in X$, then

$$\begin{aligned} \langle T^* f, k_x \rangle &= \langle f, Tk_x \rangle = \langle f_y, \overline{h(x)} k_x \rangle \\ &= h(x) \langle f, k_x \rangle = h(x) f(x) = \langle hf, k_x \rangle. \end{aligned}$$

Thus $T^* f = hf$, so that h is a multiplier and $T^* = M_h$.

If h is a multiplier, $\sigma_p(M_h^*) \supset \{\overline{h(x)} : x \in X\}$ and hence

$$\|M_h\| \geq \mathrm{spr}(M_h) = \mathrm{spr}(M_h^*) \geq \|h\|_\infty. \qquad \blacksquare$$

15.10. H^2_d as a RKHS

The significance of symmetric Fock space H^2_d to dilation theory becomes a lot clearer when it is recognized in another way. We want to exhibit it as a space of analytic functions and show that it is a reproducing kernel Hilbert space. This makes the parallels with the Hardy space H^2 quite striking.

The orthogonal basis $\{\zeta_k : k \in \mathbb{Z}_+^d\}$ for H_d^2 has $\|\zeta_k\|^2 = \dfrac{k!}{|k|!}$. Therefore H_d^2 consists of vectors $\xi = \sum_{k\in\mathbb{Z}_+^d} a_k\zeta_k$ such that $\|\xi\|^2 = \sum_{k\in\mathbb{Z}_+^d} \frac{k!}{|k|!}|a_k|^2 < \infty$. We specialize Definition 15.8.6 to the subspace H_d^2 to obtain a map into a Hilbert space of power series. Since $\widehat{\zeta_k} = z^k := z_1^{k_1}\dots z_d^{k_d}$, one obtains $\widehat{\xi} = \sum_{k\in\mathbb{Z}_+^d} a_k z^k$. This is the space of power series on $\mathbb{B}_d$ of the form

$$f(z) = \sum_{k\in\mathbb{Z}_+^d} a_k z^k \quad \text{such that} \quad \|f\|^2 = \sum_{k\in\mathbb{Z}_+^d} \frac{k!}{|k|!}|a_k|^2 < \infty.$$

Moreover, we identified evaluation at $\lambda \in \mathbb{B}_d$ as $\widehat{\xi}(\lambda) = \big\langle \xi, \frac{\nu_\lambda}{(1-\|\lambda\|^2)^{1/2}} \big\rangle$. Thus

$$f(\lambda) = \langle f, k_\lambda\rangle \quad \text{where} \quad k_\lambda = \frac{\widehat{\nu_\lambda}}{(1-\|\lambda\|^2)^{1/2}} = \sum_{k\in\mathbb{Z}_+^d} \frac{|k|!}{k!}\overline{\lambda}^k z^k.$$

This makes H_d^2 into a reproducing kernel Hilbert space known as *Drury–Arveson space*. The following calculation shows that k_λ has a form very similar to the Szegö kernel for H^2.

15.10.1. PROPOSITION. *H_d^2 is a reproducing kernel Hilbert space with kernel*

$$k_\lambda(z) = \frac{1}{1-\langle z,\lambda\rangle} \quad \textit{for} \quad \lambda, z \in \mathbb{B}_d.$$

We have $\|k_\lambda\| = \frac{1}{(1-\|\lambda\|^2)^{1/2}}$. Every $f \in H_d^2$ has a power series $f = \sum_{k\in\mathbb{Z}_+^d} a_k z^k$ which converges on $\mathbb{B}_d$, and

$$|f(\lambda)| \le \frac{\|f\|}{(1-\|\lambda\|^2)^{1/2}}.$$

If $U \in \mathcal{U}_d$ is a $d\times d$ unitary matrix, then $f\circ U \in H_d^2$ and $\|f\circ U\| = \|f\|$. Hence there is a unitary composition operator $C_U \in \mathcal{B}(H_d^2)$ such that $C_U f = f\circ U$.

PROOF. First we calculate

$$k_\lambda(z) = \sum_{k\in\mathbb{Z}_+^d} \frac{|k|!}{k!}\overline{\lambda}^k z^k = \sum_{n\ge 0}\sum_{|k|=n} \binom{n}{k_1,\dots,k_d} (\overline{\lambda_1}z_1)^{k_1}\cdots(\overline{\lambda_d}z_d)^{k_d}$$

$$= \sum_{n\ge 0}\Big(\sum_{i=1}^d \overline{\lambda_i}z_i\Big)^n = (1-\langle z,\lambda\rangle)^{-1} = \frac{1}{1-\langle z,\lambda\rangle}.$$

We have $\|k_\lambda\| = \frac{1}{(1-\|\lambda\|^2)^{1/2}}\|\nu_\lambda\| = \frac{1}{(1-\|\lambda\|^2)^{1/2}}$. Hence

$$|f(\lambda)| = |\langle f, k_\lambda\rangle| \le \|f\|\,\|k_\lambda\| = \frac{\|f\|}{(1-\|\lambda\|^2)^{1/2}}.$$

In particular, this power series is defined on $\mathbb{B}_d$, and hence has a radius of convergence of at least 1.

Let $U \in \mathcal{U}_d$. Observe that for $\lambda, \mu \in \mathbb{B}_d$,

$$\langle k_{U\mu}, k_{U\lambda}\rangle = \frac{1}{1 - \langle U\mu, U\lambda\rangle} = \frac{1}{1 - \langle \mu, \lambda\rangle} = \langle k_\mu, k_\lambda\rangle.$$

Thus there is an isometric linear map C_U^* defined on $\operatorname{span}\{k_\lambda : \lambda \in \mathbb{B}_d\}$ such that $C_U^* k_\lambda = k_{U\lambda}$ for all $\lambda \in \mathbb{B}_d$. Now $\overline{\operatorname{span}\{k_\lambda : \lambda \in \mathbb{B}_d\}} = H^2_d$, and therefore C_U^* extends by continuity to a unitary operator in $\mathcal{B}(H^2_d)$. If $f \in H^2_d$,

$$\begin{aligned}\langle f \circ U, k_\lambda\rangle &= f \circ U(\lambda) = f(U\lambda) = \langle f, k_{U\lambda}\rangle \\ &= \langle f, C_U^* k_\lambda\rangle = \langle C_U f, k_\lambda\rangle.\end{aligned}$$

As $\{k_\lambda\}$ span H^2_d, we obtain $C_U f = f \circ U$ and so $\|f \circ U\| = \|f\|$. ∎

15.10.2. Corollary. *The set $\{k_\lambda : \lambda \in \mathbb{B}_d\}$ is linearly independent and the kernel $k(\mu, \lambda) = \frac{1}{1-\langle \mu,\lambda\rangle}$ is strictly positive.*

Proof. The polynomials in $\mathbb{C}[z_1, \dots, z_d]$ belong to H^2_d, and hence the restriction map R_F to any finite subset $F \subset \mathbb{B}_d$ is a surjective map onto $\mathbb{C}^{|F|}$. Therefore by Proposition 15.9.5, $\{k_\lambda : \lambda \in \mathbb{B}_d\}$ is linearly independent and $k(\mu, \lambda) = \dfrac{1}{1 - \langle \mu, \lambda\rangle}$ is strictly positive. ∎

Now we turn to the multipliers of H^2_d. Some of this information follows from the noncommutative case, but we give a direct argument first. The other approach will be noted a bit later.

15.10.3. Theorem. $\operatorname{Mult}(H^2_d)$ *contains the polynomials, and is contained in $H^\infty(\mathbb{B}_d)$. The d-tuple $M_z := \begin{bmatrix} M_{z_1} & \dots & M_{z_d}\end{bmatrix}$ is a commuting row contraction such that $M_z M_z^* = I - P_0$ where $P_0 = 11^*$. If $U \in \mathcal{U}_d$ is a $d \times d$ unitary and $h \in$* $\operatorname{Mult}(H^2_d)$, *then $h \circ U$ is a multiplier and $M_{h\circ U}$ is unitarily equivalent to M_h.*

Proof. If $f = \sum_{k\in\mathbb{Z}^d_+} a_k z^k \in H^2_d$, then $\|f\|^2 = \sum_{k\in\mathbb{Z}^d_+} \frac{k!}{|k|!} |a_k|^2$ and

$$\|z_i f\|^2 = \|\sum_{k\in\mathbb{Z}^d_+} a_k z^{\delta_i + k}\|^2 = \sum_{k\in\mathbb{Z}^d_+} \frac{k!}{|k|!}\frac{(k_i+1)}{(|k|+1)} |a_k|^2 \le \|f\|^2.$$

Thus M_{z_i} is a contractive multiplier. Moreover, it follows that $M_{z_i}^* z^k \in \mathbb{C} z^{k-\delta_i}$.

$$\langle M_{z_i}^* z^k, \frac{z^{k-\delta_i}}{\|z^{k-\delta_i}\|^2}\rangle = \frac{|k-\delta_i|!}{(k-\delta_i)!}\langle z^k, M_{z_i} z^{k-\delta_i}\rangle = \frac{|k-\delta_i|!}{(k-\delta_i)!}\frac{k!}{|k|!} = \frac{k_i}{|k|}.$$

Hence $M_{z_i}^* z^k = \frac{k_i}{|k|} z^{k-\delta_i}$. Moreover

$$M_z M_z^* z^k = \sum_{i=1}^d M_{z_i} M_{z_i}^* z^k = \begin{cases} 0 & \text{if } k = 0 \\ \sum_{i=1}^d \frac{k_i}{|k|} z^k = z^k & \text{if } |k| \geq 1. \end{cases}$$

Therefore $M_z M_z^* = I - P_0$ where $P_0 = z^0 z^{0*} = 11^*$. In particular, $\|M_z\| = 1$.

Since $1 \in H_d^2$, a multiplier h yields $h = M_h 1 \in H_d^2$. So h is analytic. Moreover h is bounded by $\|M_h\|$ by Proposition 15.9.7. Thus $\mathrm{Mult}(H_d^2)$ is contained in $H^\infty(\mathbb{B}_d)$. If $h \in \mathrm{Mult}(H_d^2)$ and $U \in \mathcal{U}_d$, for $f \in H_d^2$,

$$C_U M_h C_{U^*} f = C_U h(f \circ U^*) = (h \circ U) f.$$

Hence $C_U M_h C_{U^*} = M_{h \circ U}$; and thus $h \circ U$ is a multiplier. ■

One direction of the following has essentially the same proof as Proposition 15.6.9. But here the converse is also true.

15.10.4. Lemma. *Let $\{h_\alpha\}$ be a bounded net of multipliers in* $\mathrm{Mult}(H_d^2)$. *Then* WOT–lim $M_{h_\alpha} = M_h$ *if and only if $\{h_\alpha\}$ converges to h uniformly on compact subsets of $\mathbb{B}_d$.*

Proof. Recall that $\nu_\lambda = k_\lambda / \|k_\lambda\|$ and

$$\langle M_h \nu_\lambda, \nu_\lambda \rangle = \langle \nu_\lambda, M_h^* \nu_\lambda \rangle = \langle \nu_\lambda, \overline{h(\lambda)} \nu_\lambda \rangle = h(\lambda).$$

Hence if WOT–lim $M_{h_\alpha} = M_h$, then $\lim_\alpha h_\alpha(\lambda) \to h(\lambda)$. The set $\{\nu_\lambda : \|\lambda\| \leq r\}$ for $r < 1$ is a norm-compact set of vectors, and since this net is bounded, a routine estimate shows that convergence is uniform for $\|\lambda\| \leq r$. Thus $\{h_\alpha\}$ converges to h uniformly on compact subsets of $\mathbb{B}_d$.

Conversely, suppose that $\{h_\alpha\}$ is a bounded net which converges to h uniformly on compact subsets of $\mathbb{B}_d$. Then

$$\langle M_{h_\alpha} k_\mu, k_\lambda \rangle = \langle k_\mu, M_{h_\alpha}^* k_\lambda \rangle = \langle k_\mu, \overline{h_\alpha(\lambda)} k_\lambda \rangle = h_\alpha(\lambda) \langle k_\mu, k_\lambda \rangle.$$

Therefore

$$\lim_\alpha \langle M_{h_\alpha} k_\mu, k_\lambda \rangle = \lim_\alpha h_\alpha(\lambda) \langle k_\mu, k_\lambda \rangle = h(\lambda) \langle k_\mu, k_\lambda \rangle = \langle M_h k_\mu, k_\lambda \rangle.$$

The vectors $\{k_\lambda : \lambda \in \mathbb{B}_d\}$ form a spanning set for H_d^2. Since the net is bounded, it follows that WOT–lim $M_{h_\alpha} = M_h$. ■

An analytic function $h(z) = \sum_{k \in \mathbb{Z}_+^d} a_k z^k$ on $\mathbb{B}_d$ can be decomposed into its homogeneous components

$$h(z) = \sum_{n \geq 0} h_n(z) \quad \text{where} \quad h_n(z) = \sum_{\substack{k \in \mathbb{Z}_+^d \\ |k| = n}} a_k z^k$$

where convergence holds uniformly on compact subsets of $\mathbb{B}_d$. Then $h_n(tz) = t^n h(z)$ for $t \in \overline{\mathbb{D}}$. Hence $h(e^{it} z) = \sum_{n \geq 0} e^{int} h_n(z)$. If U_t is the unitary $U_t z = e^{it} z$

for $z \in \mathbb{B}_d$, then $M_{h(e^{it}z)} = C_{U_t} M_h$ and so $\|M_{h(e^{it}z)}\| = \|M_h\|$. Therefore we can read these terms off as for the Fourier series:

$$M_{h_n} = \frac{1}{2\pi} \int_{-\pi}^{\pi} M_{h(e^{it}z)} e^{-int}\, dt.$$

In particular, $\|M_{h_n}\| \le \|M_h\|$.

For $0 < r < 1$, set $h_r(\lambda) = h(r\lambda)$ for $\lambda \in \mathbb{B}_d$. Then

$$\begin{aligned} M_{h_r} &= \sum_{n \ge 0} r^n h_n(z) = \frac{1}{2\pi} \int_{-\pi}^{\pi} M_{h(e^{it}z)} \sum_{n=-\infty}^{\infty} r^{|n|} e^{-int}\, dt \\ &= \frac{1}{2\pi} \int_{-\pi}^{\pi} M_{h(e^{it}z)} P_r(t)\, dt \end{aligned}$$

where $P_r(t)$ is the Poisson kernel 11.1.5. Since $P_r \ge 0$ and $\|P_r\|_1 = 1$, we see that $\|M_{h_r}\| \le \|M_h\|$. Also h_r converges to h uniformly on compact subsets of $\mathbb{B}_d$. Hence by Lemma 15.10.4, WOT–$\lim_{r\to 1} M_{h_r} = M_h$.

Now we obtain a useful consequence of the lemma.

15.10.5. COROLLARY. *The* WOT*-closed algebra generated by* $\{M_{z_1}, \dots, M_{z_d}\}$ *is* $\mathrm{Mult}(H^2_d)$.

PROOF. To complete the proof, suppose that $h \in \mathrm{Mult}(H^2_d)$. It suffices to show that M_{h_r} belong to the norm-closed algebra generated by $\{M_{z_1}, \dots, M_{z_d}\}$. Now $M_{h_r} = \sum_{n \ge 0} r^n M_{h_n}$ is norm convergent since $\|M_{h_n}\| \le \|M_h\|$. Thus $M_h =$ WOT–$\lim_{r\to 1} M_{h_r}$ lies in the WOT-closed algebra generated by $\{M_{z_1}, \dots, M_{z_d}\}$. The reverse containment is obvious. ■

15.10.6. REMARK. We could have used Cesáro means instead of the Poisson kernel, but the Poisson kernel is very natural here. See Theorem 11.1.10 for Poisson's theorem.

Let the boundary of $\mathbb{B}_d$ be $\partial\mathbb{B}_d$, the unit sphere (of real dimension $2d - 1$).

15.10.7. THEOREM. *The commutators* $[M_{z_i}, M^*_{z_j}]$ *are compact, and there is a short exact sequence*

$$0 \longrightarrow \mathcal{K}(H^2_d) \longrightarrow \mathfrak{T}_d \longrightarrow C(\partial\mathbb{B}_d) \longrightarrow 0$$

where $\mathfrak{T}_d = C^*(M_{z_1}, \dots, M_{z_d})$ *and the quotient map takes* M_{z_i} *to* $z_i|_{\partial\mathbb{B}_d}$. $\mathfrak{T}_d$ *is the closed span of* $\{M_{z^k} M^*_{z^l} : k, l \in \mathbb{Z}^d_+\}$.

PROOF. The C*-algebra $\mathfrak{T}_d$ contains $P_0 = I - \sum_{i=1}^d M_{z_i} M^*_{z_i}$. Hence it also contains $M_{z^k} P_0 M^*_{z^l} = z^k (z^l)^*$. After normalization, we obtain a set of matrix units for the compact operators, $\mathcal{K} = \mathcal{K}(H^2_d)$. Whence $\mathcal{K} \subset \mathfrak{T}_d$. Next we compute for $i \neq j$, using $M^*_{z_i} z^k = \frac{k_i}{|k|} z^{k-\delta_i}$ from the previous proof,

$$[M_{z_i}, M^*_{z_j}]z^k = \frac{k_j}{|k|} z^{k+\delta_i-\delta_j} - \frac{k_j}{|k|+1} z^{k+\delta_i-\delta_j} = \frac{k_j}{|k|^2+|k|} z^{k+\delta_i-\delta_j}.$$

This commutator is a weighted shift with weights which tend to 0 as $|k|$ tends to $+\infty$, and thus it is compact. Similarly,

$$[M_{z_i}, M^*_{z_i}]z^k = \frac{k_i}{|k|} z^k - \frac{k_i+1}{|k|+1} z^k = \frac{k_i - |k|}{|k|^2+|k|} z^k.$$

This is a diagonal operator with weights tending to 0, and so it is also compact.

It follows that the elements of $\mathfrak{T}_d$ all commute modulo $\mathcal{K}$. Therefore $\mathfrak{T}_d/\mathcal{K}$ is abelian, and thus is isomorphic to some $C(X)$ where X is its maximal ideal space. It is generated by $\sigma_i = M_{z_i} + \mathcal{K}$ for $1 \le i \le d$. Note that $\sum_{i=1}^d \sigma_i \sigma_i^* = 1$; i.e. $\sigma = (\sigma_1, \dots, \sigma_d)$ is a row unitary. Hence any character χ of $\mathfrak{T}_d/\mathcal{K}$ satisfies $\sum_{i=1}^d |\chi(\sigma_i)|^2 = 1$. So $\chi^{(1,d)}(\sigma) \in \partial\mathbb{B}_d$. By identifying χ with this point, we have $\sigma_i = z_i$ are functions on $X \subset \partial\mathbb{B}_d$. This is a non-empty subset. For any unitary $U \in \mathcal{U}_d$, $\chi_U(A) = \chi(C_U A C_{U^*})$ is another character, and

$$\chi_U^{(1,d)}(\sigma) = U\chi^{(1,d)}(\sigma).$$

That is, the set X is invariant under $\mathcal{U}_d$, and hence $X = \partial\mathbb{B}_d$.

Let $\mathcal{S} = \overline{\operatorname{span}}\{M_{z^k} M^*_{z^l} : k, l \in \mathbb{Z}^d_+\}$. This contains the operators $M_{z^k} P_0 M^*_{z^l}$ for $k, l \in \mathbb{Z}^d_+$, and hence $\mathcal{S} \supset \mathcal{K}$. Since $\mathfrak{T}_d$ is commutative modulo $\mathcal{K}$, every product of terms of the form M_{z_i} and $M^*_{z_i}$ for $1 \le i \le d$ differs from a term $M_{z^k} M^*_{z^l}$ by a compact operator, and hence lies in $\mathcal{S}$. Therefore $\mathcal{S} = \mathfrak{T}_d$. ∎

We next translate the second part of Theorem 15.8.8 in our new language. It provided a completely contractive, WOT-continuous homomorphism φ of $\mathfrak{L}_d$ onto a WOT-closed abelian subalgebra of $\mathcal{B}(H^2_d)$ by compression, $\varphi(A) = P_{H^2_d} A|_{H^2_d}$. This also contains another proof that M_z is a row contraction.

15.10.8. Corollary. *For $A \in \mathfrak{L}_d$, we have $\varphi(A) = P_{H^2_d} A|_{H^2_d} = M_{\widehat{A}}$; and $\varphi(\mathfrak{L}_d) = \operatorname{Mult}(H^2_d)$. Thus $\operatorname{Mult}(H)$ is completely isometrically isomorphic and weak-$*$ homeomorphic to $\mathfrak{L}_d/\mathfrak{C}$. In particular $\varphi(L_i) = M_{z_i}$ commute and $M_z := [M_{z_1} \ \dots \ M_{z_d}]$ is a row contraction. The algebra $\operatorname{Mult}(H^2_d)$ is maximal abelian in $\mathcal{B}(H^2_d)$.*

Proof. Since $\mathbb{C}\nu_\lambda = \mathbb{C}k_\lambda$ is coinvariant for $\mathfrak{L}_d$, it is invariant for $\mathfrak{L}^*_d$. If $A \in \mathfrak{L}_d$, then $\varphi(A)^* k_\lambda = A^* k_\lambda = \overline{a(\lambda)} k_\lambda$ where

$$a(\lambda) = \langle \nu_\lambda, A^* \nu_\lambda \rangle = \langle A\nu_\lambda, \nu_\lambda \rangle = \widehat{A}(\lambda).$$

Hence by Proposition 15.9.7, $\varphi(A) = M_{\widehat{A}}$. In particular, $\varphi(L_i) = M_{z_i}$. Moreover, $M_z = \varphi^{(1,n)}(L)$ is a row contraction. Therefore $\varphi(\mathfrak{L}_d)$ is a WOT-closed algebra of multipliers of H^2_d containing multiplication by polynomials. By Lemma 15.10.4, this is all of $\mathrm{Mult}(H^2_d)$. We also showed in Theorem 15.8.8 that $\mathfrak{L}_d/\mathfrak{C}$ is completely isometrically isomorphic and weak-$*$ homeomorphic to $\varphi(\mathfrak{L}_d)$, which we have now identified with $\mathrm{Mult}(H^2_d)$.

We saw in Theorem 15.6.6 that the common eigenspaces for $L_1^*, \dots, L_d^*$ are $\{\mathbb{C}\nu_\lambda : \lambda \in \mathbb{B}_d\}$. Since H^2_d is an invariant subspace for $\mathfrak{L}_d^*$ and it contains these eigenspaces, we see that they are all of the common eigenspaces for $M_{z_1}, \dots, M_{z_d}$. Now suppose that T commutes with $\mathrm{Mult}(H^2_d)$. Then

$$M_{z_i}^*(T^*k_\lambda) = T^*(M_{z_i}^*k_\lambda) = \overline{\lambda_i}(T^*k_\lambda).$$

Thus T^*k_λ is a common eigenvector for $M_{z_1}, \dots, M_{z_d}$ and hence is a multiple of some k_μ; but since the multiples are $\overline{\lambda_i}$, $T^*k_\lambda \in \mathbb{C}k_\lambda$. By Proposition 15.9.7, T is a multiplier. Therefore $\mathrm{Mult}(H^2_d)$ is maximal abelian. ∎

It is not obvious whether $\mathrm{Mult}(H^2_d) = H^\infty(\mathbb{B}_d)$. This turns out to be false for $d \geq 2$ and can be established by showing that the multiplier norm is not bounded by a multiple of the sup norm.

15.10.9. PROPOSITION. $\mathrm{Mult}(H^2_d)$ *is a proper subalgebra of* $H^\infty(\mathbb{B}_d)$.

PROOF. Since $|2z_1z_2| \leq |z_1|^2 + |z_2|^2 \leq 1 = 2\frac{1}{\sqrt{2}}\frac{1}{\sqrt{2}}$, we obtain that

$$\|(2z_1z_2)^n\|_\infty = 1 \quad \text{for all} \quad n \geq 1.$$

It suffices to get a good estimate of $\|M_{(2z_1z_2)^n}\|$. Now

$$\|M_{(2z_1z_2)^n}\|^2 \geq \|M_{(2z_1z_2)^n}1\|^2 = \|(2z_1z_2)^n\|_2^2 = 2^{2n}\frac{n!n!}{(2n)!}.$$

(This is actually an equality.) By Stirling's formula,

$$\|M_{(2z_1z_2)^n}\| \approx \frac{2^n\sqrt{2\pi}\,n^{n+\frac{1}{2}}e^{-n}}{\sqrt{\sqrt{2\pi}(2n)^{2n+1/2}e^{-2n}}} = (\pi n)^{1/4}.$$

Hence the natural injection of $\mathrm{Mult}(H^2_d)$ into $H^\infty(\mathbb{B}_d)$ is not bounded below and therefore is not surjective. ∎

15.11. Commuting Row Contractions

In this section, we develop the dilation theory for a row contraction with commuting entries. Rather than construct the general dilation, we will first dilate a strict contraction to an ampliation of M_z. The first proof uses the noncommutative dilation theorem, and the second is more intrinsic.

15.11.1. THEOREM. *Let $T = [T_1, \dots, T_d]$ be a d-tuple of commuting operators in $\mathcal{B}(H)$ such that $\|T\|^2 = \|\sum_{i=1}^d T_iT_i^*\| < 1$. Then there is a cardinal α and an isometric imbedding $V : H \to H_d^{2(\alpha)}$ such that*

$$V^*M_{z_i}^{(\alpha)} = T_iV^* \quad \textit{for} \quad 1 \le i \le d.$$

This says that T coextends to an ampliation of M_z.

PROOF. *First proof.* By the Frazho–Bunce–Popescu dilation theorem 15.4.9, we can dilate T to a minimal row isometry. Moreover this dilation is pure, and hence unitarily equivalent to $L^{(\alpha)}$ acting on $F_d^{2(\alpha)}$. Let V be the isometric imbedding of H into $F_d^{2(\alpha)}$. Then VH is coinvariant for each $L_i^{(\alpha)}$ and

$$V^*L_i^{(\alpha)} = T_iV^* \quad \text{for} \quad 1 \le i \le d.$$

(See Exercise **14.1**.) We will show that VH is contained in $H_d^{2(\alpha)}$.

By Corollary 15.8.5 and Theorem 15.8.8, $H_d^{2\perp} = \overline{\mathfrak{C}F_d^2}$ where $\mathfrak{C}$ is the commutator ideal of $\mathfrak{L}_d$. So $H_d^{2\perp(\alpha)}$ is spanned by vectors of the form

$$A^{(\alpha)}(L_iL_j - L_jL_i)^{(\alpha)}B^{(\alpha)}\xi \quad \text{for} \quad \xi \in F_d^{2(\alpha)},\ 1 \le i, j \le d \text{ and } A, B \in \mathfrak{L}_d.$$

Now for $\eta \in H$, $\xi \in F_d^{2(\alpha)}$ and $A, B \in \mathfrak{L}_d$,

$$\begin{aligned}\langle V\eta, A^{(\alpha)}(L_iL_j - L_jL_i)^{(\alpha)}B^{(\alpha)}\xi\rangle &= \langle (L_iL_j - L_jL_i)^{*(\alpha)}A^{*(\alpha)}V\eta, B^{(\alpha)}\xi\rangle \\ &= \langle (L_iL_j - L_jL_i)^{*(\alpha)}VV^*A^{*(\alpha)}V\eta, B^{(\alpha)}\xi\rangle \\ &= \langle V(T_iT_j - T_jT_i)^*V^*A^{*(\alpha)}V\eta, B^{(\alpha)}\xi\rangle = 0\end{aligned}$$

because $VH = \operatorname{Ran} VV^*$ is coinvariant for $\mathfrak{L}_d^{(\alpha)}$ and so $A^{*(\alpha)}V = VV^*A^{*(\alpha)}V$. Therefore VH is orthogonal to $H_d^{2\perp(\alpha)}$; i.e. $VH \subset H_d^{2(\alpha)}$.

Hence we can consider V as an isometry of H into $H_d^{2(\alpha)}$ such that

$$V^*M_{z_i}^{(\alpha)} = T_iV^* \quad \text{for} \quad 1 \le i \le d.$$

Identifying H with VH, we have $H^\perp \in \operatorname{Lat}\operatorname{Mult}(H_d^2)^{(\alpha)}$; whence

$$M_{z_i}^{(\alpha)} \simeq \begin{bmatrix} T_i & 0 \\ * & * \end{bmatrix} \quad \text{for} \quad 1 \le i \le d.$$

Second proof. Let $\Delta = \left(I - \sum_{i=1}^d T_iT_i^*\right)^{1/2}$. Consider the completely positive map $\Theta(X) = \sum_{i=1}^d T_iXT_i^*$. This is completely contractive since

$$\|\Theta\| = \|\Theta(I)\| = \|TT^*\| < 1.$$

Then $\Delta^2 = I - \Theta(I)$ and $\|\Theta^n(I)\| \le \|T\|^{2n} \to 0$.

Define $V : H \to H^2_d \otimes H$ by

$$V\eta = \sum_{k\in\mathbb{Z}^d_+} \frac{z^k}{\|z^k\|^2} \otimes \Delta T^{*k}\eta.$$

Then V is an isometry because

$$\begin{aligned}\|V\eta\|^2 &= \sum_{k\in\mathbb{Z}^d_+} \frac{1}{\|z^k\|^2} \langle T^k \Delta^2 T^{*k}\eta, \eta\rangle = \sum_{n\geq 0} \Big\langle \sum_{|k|=n} \binom{n}{k_1,\dots,k_d} T^k\Delta^2 T^{*k}\eta, \eta \Big\rangle \\ &= \sum_{n\geq 0} \langle \Theta^n(\Delta^2)\eta, \eta\rangle = \sum_{n\geq 0} \langle (\Theta^n(I) - \Theta^{n+1}(I))\eta, \eta\rangle \\ &= \langle \eta, \eta\rangle = \|\eta\|^2.\end{aligned}$$

Compute

$$\begin{aligned}(M^*_{z_i} \otimes I)V\eta &= \sum_{k\in\mathbb{Z}^d_+} \frac{k_i}{|k|} \frac{z^{k-\delta_i}}{\|z^k\|^2} \otimes \Delta T^{*k}\eta = \sum_{l\in\mathbb{Z}^d_+} \frac{l_i+1}{|l|+1} \frac{z^l}{\|z^{l+\delta_i}\|^2} \otimes \Delta T^{*(l+\delta_i)}\eta \\ &= \sum_{l\in\mathbb{Z}^d_+} \frac{z^l}{\|z^l\|^2} \otimes \Delta T^{*l}\, T^*_i \eta = VT^*_i\eta.\end{aligned}$$

Thus $(M^*_{z_i} \otimes I)V = VT^*_i$, or equivalently $V^* M^{(\alpha)}_{z_i} = T_i V^*$ for $1 \leq i \leq d$. ∎

As in the situations that we have already studied, this dilation theorem leads to a von Neumann-type inequality. What makes it a bit different is that we have only proven the dilation theorem for strict contractions.

Let $A(H^2_d)$ denote the norm closure of the polynomials in $\mathrm{Mult}(H^2_d)$.

15.11.2. Drury's von Neumann Inequality. *Let $T = [T_1, \dots, T_d]$ be a row contraction of d-tuple commuting operators in $\mathcal{B}(H)$. Then if $p = [p_{ij}]$ is an $n \times n$ matrix with entries $p_{ij} \in \mathbb{C}[z_1, \dots, z_d]$, then*

$$\|p(T)\| = \big\| [p_{ij}(T_1, \dots, T_d)] \big\| \leq \|p(M_z)\|.$$

Thus there is a completely contractive homomorphism $\rho_T : A(H^2_d) \to \mathcal{B}(H)$ such that $\rho_T(z_i) = T_i$.

Proof. If $0 < r < 1$, then $\|rT\| \leq r < 1$. By Theorem 15.11.1, there is a cardinal α and an isometry $V_r : H \to H^{2(\alpha)}_d$ onto a coinvariant subspace such that $rT_i = V^*_r M^{(\alpha)}_{z_i} V_r$. It follows as before that this map extends to a homomorphism $\rho_{rT}(p) = V^*_r M^{(\alpha)}_p V_r$ for $p \in \mathbb{C}[z_1, \dots, z_d]$. Since this map is a compression of an ampliation, ρ_{rT} is completely contractive and extends to a homomorphism of $A(H^2_d)$. The inequality $\|p(rT)\| \leq \|p(M_z)\|$ is just a restatement of the completely contractive property. Now let r tend to 1 to get the inequality for ρ_T, and then extend to a map ρ_T on $A(H^2_d)$ again. ∎

15.11.3. COROLLARY. *Let $T = [T_1, \dots, T_d]$ be a contractive d-tuple of commuting operators in $\mathcal{B}(H)$. There is a completely positive map $\Phi : \mathfrak{T}_d \to C^*(T)$ such that for all $A, B \in A(H^2_d)$ and $X \in \mathfrak{T}_d$,*

$$\Phi(AXB^*) = \rho_T(A)\Phi(X)\rho_T(B)^*.$$

PROOF. By Drury's von Neumann inequality, there are unital completely contractive homomorphisms $\rho_{rT} : A(H^2_d) \to \mathcal{B}(H)$ for $0 < r \le 1$ such that $\rho_{rT}(M_{z^k}) = (rT)^k$ for $k \in \mathbb{Z}^d_+$. For any $A \in A(H^2_d)$, $r \to \rho_{rT}(A)$ is a norm continuous function. When $r < 1$, there are isometries $V_r : H \to H^{2(\alpha)}_d$ so that $\rho_{rT}(A) = V_r^* A^{(\alpha)} V_r$ and $\operatorname{Ran} V_r$ is coinvariant for $\text{Mult}(H^2_d)^{(\alpha)}$. Thus ρ_{rT} extends to the completely positive map on $\mathcal{B}(H^2_d)$ given by $\Phi_r(X) = V_r^* X^{(\alpha)} V_r$. Coinvariance means that $V_r^* A^{(\alpha)} = V_r^* A^{(\alpha)} V_r V_r^*$. Hence if $A, B \in A(H^2_d)$ and $X \in \mathfrak{T}_d$,

$$\begin{aligned}\Phi_r(AXB^*) &= V_r^* A^{(\alpha)} X^{(\alpha)} B^{*(\alpha)} V_r = V_r^* A^{(\alpha)} V_r V_r^* X^{(\alpha)} V_r V_r^* M^{*(\alpha)}_{z^l} V_r \\ &= \Phi_r(A)\Phi_r(X)\Phi_r(B^*) = \rho_{rT}(A)\Phi_r(X)\rho_{rT}(B)^*.\end{aligned}$$

Now $\rho_{rT}(p)$ converges to $\rho_T(p)$ for polynomials. Since these maps are completely contractive, this extends to $A(H^2_d)$. Therefore $\Theta_r(AB^*) = \rho_{rT}(A)\rho_{rT}(B)^*$ converges to the map $\Theta(AB^*) = \rho_T(A)\rho_T(B)^*$. This map can be extended to $\overline{\text{span}}\{AB^* : A, B \in A(H^2_d)\}$, which is all of $\mathfrak{T}_d$ by Theorem 15.10.7. Therefore Θ is a completely positive map of $\mathfrak{T}_d$ into $C^*(T)$. The last formula follows by taking limits as $r \to 1$. ∎

We now deduce the full dilation theorem. A special feature of this result is that like the Frazho–Bunce–Popescu dilation, only a co-extension is required.

15.11.4. DEFINITION. A d-tuple $S = [S_1, \dots, S_d]$ of commuting normal operators such that $\sum_{i=1}^d S_i S_i^* = I$ is called a *spherical unitary*.

15.11.5. THEOREM. *Let $T = [T_1, \dots, T_d]$ be a row contraction of d commuting operators in $\mathcal{B}(H)$. Then T has a co-extension to a commuting row contraction of the form $M_z^{(\alpha)} \oplus S$ where S is a spherical unitary.*

PROOF. Let $\Phi : \mathfrak{T}_d \to C^*(T)$ be the completely positive map of Corollary 15.11.3. By Stinespring's dilation theorem 14.7.1, there is a representation $\pi : \mathfrak{T}_d \to \mathcal{B}(K)$ and an isometry $V : H \to K$ such that $\Phi(X) = V^*\pi(X)V$ for $X \in \mathfrak{T}_d$ and $\Phi(AB^*) = \Phi(A)\Phi(B)^*$ for $A, B \in A(H^2_d)$. Therefore

$$\begin{aligned}V^*\pi(M_{z_i}M_{z_i}^*)V &= \Phi(M_{z_i}M_{z_i}^*) = \Phi(M_{z_i})\Phi(M_{z_i})^* \\ &= V^*\pi(M_{z_i})VV^*\pi(M_{z_i}^*)V.\end{aligned}$$

Thus

$$\begin{aligned} 0 &= V^*\pi(M_{z_i})(I - VV^*)\pi(M_{z_i}^*)V \\ &= \big(V^*\pi(M_{z_i})(I - VV^*)\big)\big(V^*\pi(M_{z_i})(I - VV^*)\big)^*. \end{aligned}$$

Hence $VV^*\pi(M_{z_i})(I - VV^*) = 0$, which shows that $(I - VV^*)K$ is invariant for $\pi(M_{z_i})$, and thus invariant for $\pi(A(H^2_d))$. So VH is coinvariant. Since $V^*\pi(M_{z_i})V = T_i$ for $1 \le i \le d$, this shows that $\pi(M_z) = (\pi(M_{z_1}), \dots, \pi(M_{z_d}))$ is a co-extension of T.

By Theorem 9.8.2, every representation of $\mathfrak{T}$ splits as $\pi = \pi_1 \oplus \pi_2$ where π_1 is induced by $\pi_1|_{\mathcal{K}}$ and π_2 annihilates $\mathcal{K}$. Every representation of $\mathcal{K}$ is a multiple of the identity representation, and so $\pi_1 = id^{(\alpha)}$; whence $\pi_1(M_z) = M_z^{(\alpha)}$. The representation π_2 factors through $C(\partial\mathbb{B}_d)$ as $\pi_2 = \tilde{\pi}_2 q$, its range in a commutative C*-algebra, and

$$\sum_{i=1}^{d} \pi_2(M_{z_i}M_{z_i})^* = \sum_{i=1}^{d} \tilde{\pi}_2(|z_i|^2) = \tilde{\pi}_2(1) = I.$$

Hence $\pi_2(M_z) = S$ is a spherical unitary. Therefore T coextends to $M_z^{(\alpha)} \oplus S$. ■

15.11.6. COROLLARY. *$C^*_{\min}(A(H^2_d)) = \mathfrak{T}_d$. The boundary representations of $A(H^2_d)$ are all irreducible representations of $\mathfrak{T}_d$, namely id and $\{\delta_z : z \in \partial\mathbb{B}_d\}$.*

PROOF. Since $A(H^2_d) \subset \mathfrak{T}_d$, the C*-envelope is a quotient of $\mathfrak{T}_d$. However $\mathcal{K}(H^2_d)$ is the unique minimal ideal of $\mathfrak{T}_d$ and the quotient map by $\mathcal{K}$ is not isometric on $A(H^2_d)$, so $C^*_{\min}(A(H^2_d)) = \mathfrak{T}_d$. The fact that every contractive representation of $A(H^2_d)$ has a $\mathfrak{T}_d$-dilation is established in Theorem 15.11.5.

If π is an irreducible representation of $\mathfrak{T}_d$, applying Theorem 9.8.2, either $\pi|_{\mathcal{K}(H^2_d)} \ne 0$ or $\pi|_{\mathcal{K}(H^2_d)} = 0$. In the first case, $\pi = id$ because the compacts have a unique irreducible representation. In the second case, π factors through $C(\partial\mathbb{B}_d)$ by Theorem 15.10.7. Since this is a commutative C*-algebra, the irreducible representations are point evaluations δ_z for $z \in \partial\mathbb{B}_d$. Every commuting row isometry has the form $M_z^{(\alpha)} \oplus S$ where S is a spherical unitary. The only dilations of these operators to row contractions are just direct sums with another row contraction, and thus every representation of $\mathfrak{T}_d$ is maximal and so has the unique extension property. Therefore every irreducible representation is a boundary representation. ■

15.12. The Complete Pick Property

In Section 14.4, we showed that $H^\infty(\mathbb{D})$ has the complete Pick property by exploiting properties of the Szegö kernel. This property can now be formulated in any reproducing kernel Hilbert space. In this section, we derive comparable results for multipliers on Drury–Arveson space.

This easy lemma sets the stage and explains why the Pick property is formulated as it is. The proof uses elements from Theorem 14.4.1.

15.12.1. LEMMA. *Let H be an RKHS on X with kernel $k(x,y)$. Then h in* $\mathrm{Mult}(H)$ *has $\|M_h\| \le 1$ if and only if for all finite $F = \{x_1,\dots,x_n\} \subset X$,*

$$\big[k(x_i,x_j)(1 - h(x_i)\overline{h(x_j)})\big] \ge 0.$$

PROOF. We have $M_h^* k_x = \overline{h(x)} k_x$. Let $N_F = \mathrm{span}\{k_{x_1},\dots,k_{x_n}\}$ and define a map $L : \mathbb{C}^n \to N_F$ by $L\xi_j = k_{x_j}$ for the standard orthonormal basis $\xi_1,\dots,\xi_n$. Then $\|M_h^* P_{N_F}\| \le 1$ if and only if $P_{N_F} - P_{N_F} M_h M_h^* P_{N_F} \ge 0$ if and only if

$$\begin{aligned} 0 &\le L^* P_{N_F}(I - M_h M_h^*) P_{N_F} L = \big[\langle (I - M_h M_h^*) k_{x_j}, k_{x_i}\rangle\big] \\ &= \big[k(x_i,x_j)(1 - h(x_i)\overline{h(x_j)})\big]. \end{aligned}$$

Since $\|M_h\| = \|M_h^*\| = \sup_{F \Subset X} \|M_h^*|_{N_F}\|$, the result follows. ■

15.12.2. DEFINITION. A reproducing kernel Hilbert space H on X has the *Pick property* if for all finite $F = \{x_1,\dots,x_n\} \subset X$,

$$\big[k(x_i,x_j)(1 - \lambda_i\overline{\lambda_j})\big] \ge 0 \quad \text{for} \quad \lambda_i \in \mathbb{C},\ 1 \le i \le n$$

implies there exists an $h \in \mathrm{Mult}(H)$ with $\|M_h\| \le 1$ such that $h(x_i) = \lambda_i$ for $1 \le i \le n$. Say that H has the *complete Pick property* if for all finite $F = \{x_1,\dots,x_n\} \subset X$,

$$\big[k(x_i,x_j)(I_k - A_i A_j^*)\big] \ge 0 \quad \text{for} \quad A_i \in M_k,\ 1 \le i \le n$$

implies there exists an $h \in M_k(\mathrm{Mult}(H))$ with $\|M_h\| \le 1$ such that $h(x_i) = A_i$ for $1 \le i \le n$.

15.12.3. THEOREM. *The space H^2_d has the complete Pick property.*

PROOF. Let $F = \{\lambda_1,\dots,\lambda_n\} \subset \mathbb{B}_d$ and set $N = \mathrm{span}\{k_{\lambda_1},\dots,k_{\lambda_n}\}$. Suppose that $A_1,\dots,A_n \in M_k$ such that

$$\big[k(\lambda_i,\lambda_j)(I_k - A_i A_j^*)\big] \ge 0.$$

Define a map B on $N^{(k)}$ by

$$B^*(k_{\lambda_i} \otimes \eta) = k_{\lambda_i} \otimes A_i^* \eta \quad \text{for} \quad 1 \le i \le n \text{ and } \eta \in \mathbb{C}^k$$

and extend by linearity. This makes sense since $\{k_{\lambda_1},\dots,k_{\lambda_n}\}$ are linearly independent. Define a linear isomorphism $L : \mathbb{C}^n \otimes \mathbb{C}^k \to N^{(k)}$ by $L\xi_i \otimes \eta = k_{\lambda_i} \otimes \eta$.

Then

$$\begin{aligned}\langle L^*(I_{kn}-BB^*)L(k_{\lambda_j}\otimes\eta),k_{\lambda_i}\otimes\zeta\rangle &= \langle (I_{kn}-BB^*)(k_{\lambda_j}\otimes A_j^*\eta),k_{\lambda_i}\otimes A_i^*\zeta\rangle \\ &= \langle k_{\lambda_j},k_{\lambda_i}\rangle\langle (I_n-A_iA_j^*)\eta,\zeta\rangle.\end{aligned}$$

That is,

$$L^*(I_{kn}-BB^*)L=\big[k(\lambda_i,\lambda_j)(I_k-A_iA_j^*)\big]\geq 0.$$

Hence $\|B\|\leq 1$. Note that there is a multiplier (even a polynomial multiplier) $p\in M_k(\mathrm{Mult}(H_d^2))$ so that $M_p^*|_N=B^*$. Thus $B\in P_N^{\perp(k)}M_k(\mathrm{Mult}(H_d^2))|_{N\otimes\mathbb{C}^k}$.

The ideal $\mathfrak{I}=\{h\in\mathrm{Mult}(H_d^2): h(\lambda_i)=0 \text{ for } 1\leq i\leq n\}$ is WOT-closed. By Corollary 15.10.8, $\mathrm{Mult}(H_d^2)\simeq\mathfrak{L}_d/\mathfrak{C}$ where $\mathfrak{C}$ is the WOT-closed commutator ideal of $\mathfrak{L}_d$. The ideal $\mathfrak{J}=\{A\in\mathfrak{L}_d:\widehat{A}(\lambda_i)=0 \text{ for } 1\leq i\leq n\}$ is the preimage of $\mathfrak{I}$ under this quotient map. The range $\overline{\mathfrak{J}F_d^2}=\{\nu_{\lambda_1},\dots,\nu_{\lambda_n}\}^\perp$. By Theorem 15.8.5 and Theorem 15.8.3, the quotient map $q:\mathfrak{L}_d\to\mathfrak{L}_d/\mathfrak{J}$ is completely isometric and weak-$*$ homeomorphic to $P_N\mathfrak{L}_d|_N$; and moreover, there is a norm preserving lift. That is, there is an $A\in M_k(\mathfrak{L}_d)$ so that $P_NA|_N=B$ and $\|A\|=\|B\|$. Let $M_h=A+M_k(\mathfrak{C})\in M_k(\mathrm{Mult}(H_d^2))$. Then $\|M_h\|=\|B\|$ and $M_h^*|_{N\otimes\mathbb{C}^k}=B^*$; whence $h(\lambda_i)=A_i$ for $1\leq i\leq d$. ∎

If H is a RKHS on a set X and $Y\subset X$, we obtain a RKHS on Y by setting $H_Y=\overline{\mathrm{span}}\{k_y: y\in Y\}$ and defining $k(y_1,y_2)$ by restriction of the kernel on $X\times X$. There is a natural map $\rho_Y:\mathrm{Mult}(H)\to\mathrm{Mult}(H_Y)$given by $\rho(M_h)=P_{H_Y}M_h|_{H_Y}$. Note that H_Y is invariant for M_h^* because each k_y is an eigenvalue. Thus H_Y is coinvariant for $\mathrm{Mult}(H)$ and ρ_Y is a completely contractive and WOT-continuous homomorphism.

The (complete) Pick property can be used to show that this restriction map is a (complete) quotient.

15.12.4. PROPOSITION. *Suppose that H is a RKHS on a set X with the (complete) Pick property, and let $Y\subset X$. Then $\rho_Y:\mathrm{Mult}(H)\to\mathrm{Mult}(H_Y)$ is a (complete) quotient map and every (matrix) multiplier on H_Y extends to a (matrix) multiplier in H of the same norm.*

PROOF. First observe that H_Y is also a (complete) Pick kernel. Indeed if $F=\{y_1,\dots,y_n\}\subset Y$ and $\lambda_i\in\mathbb{C}$ (or M_k) satisfy the Pick matrix condition, then since $F\subset X$, we use the (complete) Pick property for H to find a multiplier $h\in\mathrm{Mult}(H)$ with $\|M_h\|\leq 1$ which interpolates the data. The compression to H_Y yields a multiplier in $\mathrm{Mult}(H_Y)$ with the same properties.

Suppose that $h\in\mathrm{Mult}(H_Y)$ with $\|M_h\|=1$. For each finite subset $F\subset Y$, the Pick matrix for the data $h(y_i)$ for $y_i\in F$ is positive by Lemma 15.12.1. By the Pick property for H, there is a multiplier $g_F\in\mathrm{Mult}(H)$ with $\|M_{g_F}\|\leq 1$ and $g_F(y)=h(y)$ for $y\in F$. The bounded net $\{g_F\}_{F\Subset Y}$ has a cofinal WOT-convergent subnet with limit $g\in\mathrm{Mult}(H)$. Then $\|M_g\|\leq 1$ and $g(y)=h(y)$

for all $y \in Y$. The map ρ_Y is evidently completely contractive, and this argument shows that $\|M_h\| = \inf\{\|M_g\| : \rho_Y(g) = h\}$, meaning that ρ_Y is a quotient. If H has the complete Pick property, the same argument works for matrices of multipliers, and ρ_Y is a complete quotient map. ■

15.12.5. Definition. Let H be an RKHS on a set X. A subset V of X is a *variety* for Mult(H) if there is a subset $F \subset \text{Mult}(H)$ such that

$$V = V(F) = \{\lambda \in X : f(\lambda) = 0 \text{ for all } f \in F\}.$$

Similarly, we define a variety for a subset $F \subset H$. Likewise if $Y \subset X$, let

$$\mathfrak{J}(Y) = \{h \in \text{Mult}(H) : h|_Y = 0\}.$$

Note that $\mathfrak{J}(Y)$ is a WOT-closed ideal, and $V(\mathfrak{J}(Y))$ is the smallest variety containing Y. Let $H_V = \overline{\text{span}}\{k_x : x \in V\}$.

15.12.6. Remark. For H^2_d, where the functions are all analytic, we should compare this notion with the notion of an analytic variety. There are more analytic functions, and there are more varieties. For example, for H^2 and H^∞, the only discrete varieties in $\mathbb{D}$ are the Blaschke sequences. However by a theorem of Weierstrass, any discrete subset of $\mathbb{D}$ is the zero set of some analytic function on $\mathbb{D}$.

15.12.7. Proposition. *Let H be a non-degenerate RKHS on a set X with the Pick property, and let $V \subset X$. The following are equivalent:*

(1) *V is a variety for* Mult(H).

(2) *V is a variety for H.*

(3) $V = \{x \in X : k_x \in H_V\}$.

Proof. It is enough to show that if $h \in \text{Mult}(H)$, then

$$Z(h) = \{x \in X : h(x) = 0\}$$

is a variety for H. For each $x \in X \setminus Z(h)$, we have $k_x(x) = \|k_x\|^2 \neq 0$. Hence $x \notin Z(hk_x) \supset Z(h)$. Therefore $Z(h) = \bigcap_{x \in X \setminus Z(h)} Z(hk_x)$ is a variety for H. So (1) implies (2).

Suppose that $V = V(F)$ for $F \subset H$. Then if $f \in F$, $0 = f(y) = \langle f, k_y \rangle$ for all $y \in V$. Hence f is orthogonal to $H_V = \overline{\text{span}}\{k_y : y \in V\}$. In particular, if $k_x \in H_V$, then $f(x) = \langle f, k_x \rangle = 0$. Therefore $x \in V$. So (2) implies (3).

Suppose that (3) holds, and take $x \in X \setminus V$. Then $H_{V \cup \{x\}} = H_V \dotplus \mathbb{C}k_x$. There is an idempotent map $T \in \mathcal{B}(H_{V \cup \{x\}})$ with kernel H_V and range $\mathbb{C}k_x$. Since $Tk_x = k_x$ and $Tk_y = 0$, this is the adjoint of a multiplier $g \in \text{Mult}(H_{V \cup \{x\}})$ by Proposition 15.9.7. By the Pick property and Proposition 15.12.4, there is a multiplier $h \in \mathrm{Mult}(H)$ such that $h(x) = 1$ and $h|_V = 0$. Therefore $x \notin V(h) \supset V$. Taking an intersection over all $x \in X \setminus V$ yields (1). ■

15.12.8. COROLLARY. *Let V be a variety for* $\mathrm{Mult}(H^2_d)$ *and let* $\mathfrak{J} = \mathfrak{J}(V)$. *Then* $\rho_V : \mathrm{Mult}(H^2_d) \to \mathrm{Mult}(H_V)$ *is a complete quotient map with kernel* $\mathfrak{J}(V)$.

We can deduce another interesting relationship from the noncommutative case. Let *Id*$(\mathrm{Mult}(H^2_d))$ denote the set of WOT-closed ideals in $\mathrm{Mult}(H^2_d)$.

15.12.9. THEOREM. *There is a lattice isomorphism between* Id$(\mathrm{Mult}(H^2_d))$ *and* $\mathrm{Lat\,Mult}(H^2_d)$ *via maps* $\mu(\mathfrak{J}) = \overline{\mathfrak{J}H^2_d}$ *and* $\iota(M) = \{M_h : M_h H^2_d \subset M\}$, *its inverse. Moreover every invariant subspace is the range of a row partial isometry with coefficients in* $\mathrm{Mult}(H^2_d)$.

PROOF. By Corollary 15.10.8, $\mathrm{Mult}(H^2_d) \simeq \mathfrak{L}_d/\mathfrak{C}$. By Theorem 15.8.2, there is a pairing between WOT-closed right ideals of $\mathfrak{L}_d$ and $\mathrm{Lat}\,\mathfrak{R}_d$. The ideals containing $\mathfrak{C}$ are all 2-sided since $\mathfrak{L}_d/\mathfrak{C}$ is abelian, and they correspond to invariant subspaces of $\mathfrak{R}_d$ and $\mathfrak{L}_d$ containing the antisymmetric Fock space $F^a_d = \overline{\mathfrak{C}F^2_d}$. Thus we can identify an ideal $\mathcal{J} \supset \mathfrak{C}$ with the ideal $\mathfrak{J} = \mathcal{J}/\mathfrak{C}$ of $\mathrm{Mult}(H^2_d)$ and the invariant subspace $M = \overline{\mathcal{J}F^2_d}$ with the subspace $M \ominus F^a_d \subset H^2_d$. This a a pairing between all WOT-closed ideals of $\mathrm{Mult}(H^2_d)$ and all invariant subspaces. Finally $\mu(\mathfrak{J}) = \mu(\mathcal{J}) \ominus F^a_d$. That this map is a complete lattice isomorphism also follows from Theorem 15.8.2.

By Theorem 15.6.1, if $M \in \mathrm{Lat\,Mult}(H^2_d)$, there is a row isometry $R = [R_1\ R_2\ \dots]$ with range $M \oplus F^a_d$, and each $R_i \in \mathfrak{R}_d$. Hence

$$P_{H^2_d} R|_{H^{2(\alpha)}_d} = [P_{H^2_d} R_1|_{H^2_d}\ P_{H^2_d} R_2|_{H^2_d}\ \dots] = [S_1\ S_2\ \dots]$$

is a row partial isometry with range M. Now like $\mathfrak{L}_d$, $P_{H^2_d}\mathfrak{R}_d|_{H^2_d} = \mathrm{Mult}(H^2_d)$. ■

Exercises for Chapter 15

15.1 Let X be a compact Hausdorff space, and let $\sigma : X \to X$ be a continuous map. Form the semicrossed product $\mathcal{A} = C(X) \times_\alpha \mathbb{Z}_+$ as in Example 15.1.6 generated by $C(X)$ and an element $\mathfrak{s}$ such that $f\mathfrak{s} = \mathfrak{s}(f \circ \sigma)$ for $f \in C(X)$.

(a) Show that for each $x \in X$, there is a character $\theta_{x,0}$ of $\mathcal{A}$ such that $\theta(f) = f(x)$ and $\theta_{x,0}(\mathfrak{s}) = 0$.

(b) Show that if $\sigma(x) \neq x$, then this is the only character θ such that $\theta|_{C(X)} = \varepsilon_x$ is evaluation at x.

(c) Show that if $\sigma(x) = x$, then for each $z \in \overline{\mathbb{D}}$, there is a character $\theta_{x,z}$ such that $\theta_{x,z}(f) = f(x)$ and $\theta_{x,z}(\mathfrak{s}) = z$. HINT: Since $\mathfrak{s}$ is an isometry, the spectrum of $\mathfrak{s}$ is $\overline{\mathbb{D}}$. Use the universal property of $\mathcal{A}$ to show that $\theta_{x,z}$ exists.

(d) Show that $\widehat{A}(z) = \theta_{x,z}(A)$ is analytic on $\mathbb{D}$ when $\sigma(x) = x$.

15.2 Let $\sigma : \mathbb{T} \to \mathbb{T}$ by $\sigma(z) = z^p$ for $p \geq 2$. Let $\alpha_1(f) = f \circ \sigma$ be the corresponding endomorphism of $C(\mathbb{T})$. Show that the C*-algebra $\mathfrak{A}_\infty$ of Proposition 15.2.5 in this case is isomorphic to $C(X)$ where

$$X = \{x = (x_i) \in \mathbb{T}^\infty : \sigma(x_{i+1}) = x_i \text{ for } i \geq 1\};$$

and describe $\widetilde{\alpha}_1$.

15.3 Let $(\mathfrak{A}, \mathbb{Z}_+, \alpha)$ be a C*-dynamical system.

(a) Show that for any Banach algebra $\mathfrak{B}$, the Jacobson radical is invariant under automorphisms.

(b) Show that if $A \in \operatorname{rad}(\mathfrak{A} \times_\alpha \mathbb{Z}_+)$, then $E_0(A) = 0$ and $E_n(A) \in \operatorname{rad}(\mathfrak{A} \times_\alpha \mathbb{Z}_+)$ for $n \geq 1$.

(c) So if $\operatorname{rad}(\mathfrak{A} \times_\alpha \mathbb{Z}_+) \neq \{0\}$, there is some $e_n a \in \operatorname{rad}(\mathfrak{A} \times_\alpha \mathbb{Z}_+)$ for $0 \neq a \geq 0$.

15.4 Show that outer conjugacy is a transitive relation.

15.5 (*Donsig–Katavolos–Manoussos*).
Let X be locally compact, Hausdorff and $\sigma : X \to X$ be a proper continuous map; i.e. if $K \subset X$ is compact, then $\sigma^{-1}(K)$ is compact.

(a) Show that $\alpha_1(f) = f \circ \sigma$ is a $*$-endomorphism of $C_0(X)$. Let $\alpha_n = \alpha_1^n$ and $\mathcal{A} = C_0(X) \times_\alpha \mathbb{Z}_+$.

(b) Say that an open set $\varnothing \neq U \subset X$ is a *wandering set* if there is some N such that $\sigma^{-n}(U) \cap U = \varnothing$ for all $n \geq N$. If U is wandering, let $f \in C_0(X)$ with $\operatorname{supp}(f) \subset U$. Show that $e_N f \in \operatorname{rad}(\mathcal{A})$.

(c) Show that $X \setminus \sigma(X)$ is open, and is wandering or empty.

(d) A point $x \in X$ is *recurrent* if for every neighbourhood $U \ni x$, there is $n \geq 1$ with $\sigma^n(x) \in U$. Show that if the recurrent points are dense in X, then there are no wandering sets. (*In this case, $\mathcal{A}$ is semisimple* [96].)

(e) Assume that X is metrizable, say with metric ρ. Show that if there are no wandering sets, then the recurrent points are dense.
HINT: set $\delta(x) = \inf_{n \geq 1} \rho(x, \sigma^n(x))$. Show that δ is upper semicontinuous, i.e. $V_r = \{x : \delta(x) < r\}$ is open. If U is open and contains no recurrent points, then $U = \bigcup_{n \geq 1} U \setminus V_{1/n}$. Use the Baire category theorem.

15.6 Let $T = \begin{bmatrix} T_1 & \dots & T_d \end{bmatrix}$ be a row contraction. Show the following are equivalent:
(1) T is pure.
(2) $\bigcap_{k \geq 1} \sum_{|w|=k} \operatorname{Ran} T_w = \{0\}$.
(3) SOT–$\lim_{n \to \infty} \sum_{|w|=k} T_w T_w^* = 0$.

15.7 Modify the proof of Fejér's theorem 2.5.4 to show that $\Sigma_n(A)$ converges in norm to A for all $A \in \mathcal{A}_n$.

15.8 Show that the invariant subspace $\{\nu_\lambda\}^\perp$ for $\mathfrak{L}_d$ has a d-dimensional wandering space.

15.9 (*Davidson–Pitts*).
(a) Show that $\mathfrak{L}_d$ contains no non-scalar normal operators.
HINT: look at $N^*\xi_\varnothing$.
(b) Show that $\mathfrak{L}_d$ contains no non-scalar idempotents.
(c) Show that if $A \in \mathfrak{L}_d$, then $\sigma(A)$ is connected.

15.10 (a) Show that if $0 \neq A \in \mathfrak{L}_d$, then $\operatorname{spr}(A) > 0$ and $\ker A = \{0\}$.
HINT: look at a non-zero Fourier coefficient for a minimal word.
(b) Show that $\mathfrak{L}_d$ is semisimple.

15.11 Let $\mathfrak{S}$ be a free semigroup algebra over $\mathbb{F}_d^+$.
(a) Describe $\operatorname{rad}\mathfrak{S}$.
(b) If $\mathfrak{S}$ is semisimple, show that either it is a von Neumann algebra or type L.
(c) Show that if $\mathfrak{S}$ is not semisimple, $\mathfrak{S}/\operatorname{rad}\mathfrak{S}$ is isomorphic to $\mathcal{M} \oplus \mathfrak{L}_d$, where $\mathcal{M}$ is a von Neumann algebra.

15.12 Show that $\mathfrak{L}_d$ is not isomorphic to a von Neumann algebra.

15.13 Let $T = [T_1, \dots, T_d]$ be a row contraction with $T_i \in M_k$ with $\|T\| < 1$. Show that there is a unique representation Φ_T of $\mathfrak{L}_d$ such that $\Phi_T(L_i) = T_i$, and it is WOT-continuous.

15.14 Let $T = [T_1, \dots, T_d]$ be a row contraction. Show that T decomposes uniquely as $T = U \oplus T'$ where U is a row unitary and T' has no row unitary summand.
HINT: if $U = T|_M$ is a row unitary, then P_M commutes with T. Show that if $V = T|_N$ is another row unitary summand of T, then $T|_{M\vee N}$ is a row unitary.

15.15 (*Davidson–Katsoulis–Pitts*).
(a) Show that if $S = [S_1 \ \dots \ S_n]$ is a row isometry such that the wandering vectors span H, then the free semigroup algebra $\mathfrak{S}$ generated by S is type L.
HINT: if $T \in \bigcap_{k\geq 0}\mathfrak{S}_0^k$, consider $T|\mathcal{M}$ for every subspace generated by a wandering vector.
(b) Let $S = [S_1 \ \dots \ S_n]$ be an arbitrary row isometry. Form $H \otimes l^2(\mathbb{Z})$ from sequences $\xi = (\xi_j)_{j\in\mathbb{Z}}$ with $\xi_j \in H$ and $\|\xi\|^2 = \sum \|\xi_j\|^2 < \infty$. Let $T_i = S_i \otimes U$, where U is the bilateral shift, i.e. $T_i(\xi_j)_{j\in\mathbb{Z}} = (S_i\xi_{j-1})_{j\in\mathbb{Z}}$. Show that the free semigroup algebra $\mathfrak{T}$ generated by T is type L.

15.16 (*Davidson–Katsoulis–Pitts*).
Let $\mathfrak{S}$ be a free semigroup algebra for $d \geq 2$. Prove that the following are equivalent:
(i) $\mathfrak{S}$ is type L.

(ii) $\mathfrak{S}$ is algebraically isomorphic to a subalgebra of $\mathfrak{L}_d$.
(iii) $\mathfrak{S}$ contains no non-scalar idempotents.

15.17 (*Davidson–Kribs–Shpigel*).
Let $T = \begin{bmatrix} T_1 & \dots & T_d \end{bmatrix}$ be a row contraction such that $TT^* \neq I$. Let $S = \begin{bmatrix} S_1 & \dots & S_d \end{bmatrix}$ be the minimal row isometric dilation of T on a Hilbert space K containing H.
(a) Show that $W = \left(H + \sum_{i=1}^d S_i H\right) \ominus H$ is a wandering space, and that $\sum_{u \in \mathbb{F}_d^+} S_u W = H^\perp$.
(b) Let $X = \mathrm{Ran}(I - SS^*)$. Show that $X \cap H^\perp = \{0\}$.
HINT: Show that X is supported in $H \oplus M$ where $M = \overline{\mathrm{Ran}(D_T)}$.
(c) Suppose that $\dim H < \infty$. Show that

$$\dim X = \dim P_H X = \mathrm{rank}(I - TT^*).$$

15.18 Let $S_1, S_2 \in \mathcal{B}(L^2(0,1))$ be given by

$$S_1 f(x) = \sqrt{2} f(2x) \chi_{(0,.5)} \quad \text{and} \quad S_2 f(x) = \sqrt{2} f(2x-1) \chi_{(.5,1)}.$$

Let $\mathfrak{S}$ be the free semigroup algebra generated by $\{S_1, S_2\}$.
(a) Show that $\chi_{(0,.5)} - \chi_{(.5,1)}$ is a wandering vector.
(b) Find the complement of the wandering subspace from (a).
HINT: check out the Haar basis for $L^2(0,1)$.
(c) Describe S_1 and S_2 with respect to the decomposition of (b).
(d) Explain what the Structure theorem says about $\mathfrak{S}$.

15.19 (*Davidson–Dor-On–Li*). Suppose that S_1, S_2 are isometries with complementary ranges such that the free semigroup algebra $\mathfrak{S}$ that they generate is $\mathcal{B}(H)$.[1] Prove that for $d \geq 3$, there are isometries $T_1, \dots, T_d$ with orthogonal ranges such that they also generate $\mathfrak{T} = \mathcal{B}(H)$ as a free semigroup algebra. Follow this outline:
(a) Let $T_1 = S_1$, $T_i = S_2^{i-1} S_1$ for $2 \leq i \leq d-1$ and $T_d = S_2^{d-1}$. Let $\mathcal{A}$ be the (non-closed) algebra generated by $T_1, \dots, T_d$. Show that $\sum_{i=1}^d T_i T_i^* = I$.
(b) Show by induction on $|w|$ that for every $w \in \mathbb{F}_d^+$, that $S_w S_1 \in \mathcal{A}$.
(c) Deduce that $\mathrm{span}\{S_w : w \in \mathbb{F}_d^+\} S_1 \subset \mathcal{A}$, and hence that $\mathcal{B}(H) S_1 \subset \mathfrak{T}$.
(d) Show that $\mathfrak{T} = \mathcal{B}(H)$ by looking at rank one elements.

15.20 (*Davidson–Pitts*). Prove that every automorphism Θ of $\mathfrak{L}_d$, for $d \geq 2$, is continuous. Follow this outline of a *gliding bump argument*.
Let $B_i = \Theta^{-1}(L_i)$ for $1 \leq i \leq n$. If Θ is not continuous, select $A_k \in \mathfrak{L}_d$ with $\|A_k\| \leq 1/k!$ and $\|\Theta(A_k)\| > k$.
(a) Define $A = \sum_{k \geq 1} B_2^k B_1 A_k$. Show that this converges, and that A can be written as a *finite* sum that collects the sum over all $k > m$ as a single term of the form $B_2^{m+1} X_m$.
(b) Evaluate $L_1^* L_2^{*m} \Theta(A)$. Deduce a contradiction.

[1] Such a pair was constructed by Read [255].

15.21 Show that if H is a RKHS on X with orthonormal basis $\{\xi_n : n \geq 1\}$, then

$$k(x,y) = \sum_{n\geq 1} \xi_n(x)\overline{\xi_n(y)}.$$

15.22 (*Moore*). If k is positive semidefinite on $X \times X$, show that there is a unique RKHS H_k with kernel k.
HINT: define a sesquilinear form on the algebraic span of $\{k_x : x \in X\}$ by

$$\big\langle \sum a_i k_{x_i}, \sum b_j k_{x_j} \big\rangle = \sum_{i,j} a_i \overline{b_j} k(x_j, x_i).$$

15.23 (a) Show that $k(\mu,\lambda) = e^{\overline{\mu}\lambda}$ is a positive definite kernel on $\mathbb{C}$.
(b) Show that $\{e_n = z^n/\sqrt{n!} : n \geq 0\}$ forms an orthonormal basis for H_k.
(c) Show that the elements of H_k are entire functions.

15.24 Show that if $d' < d$ and $g \in H^2_{d'}$, then $g \in H^2_d$ and $\|g\|_{H^2_d} = \|g\|_{H^2_{d'}}$.

15.25 (*Hartz*). Suppose that H is an RKHS on $\mathbb{B}_d$ such that (i) the polynomials are dense in H, (ii) $\|f \circ U\| = \|f\|$ for $f \in H$ and $U \in \mathcal{U}_d$ and (iii) if $f = g(z_1)$ for $g \in H^2$, then $\|f\| = \|g\|_{H^2}$. Show that $H = H^2_d$ as follows:
(a) Show that $\{z^k : k \in \mathbb{Z}^d_+\}$ is an orthogonal basis.
(b) Show that $k_{\lambda e_1} = \frac{1}{1-\overline{\lambda} z_1} \in H$ and $\langle f, k_{\lambda e_1}\rangle_H = f(\lambda e_1)$ for $\lambda \in \mathbb{D}$, $f \in H$.
(c) For $w \in \mathbb{B}_d$, pick $U \in \mathcal{U}_d$ so that $Uw = \lambda e_1$. Use (ii) to compute k_w.

15.26 Show by direct computation that if $0 \neq \xi \in H^2_d$ and $M^*_{z_i}\xi = \overline{\lambda_i}\xi$ for $1 \leq i \leq d$, then $\|\lambda\| < 1$ and $\xi \in \mathbb{C}k_\lambda$.

15.27 (*Arveson*)
(a) Show *id* is a boundary representation for $\mathcal{S} = \operatorname{span}\{I, M_{z_i}, M^*_{z_i} : 1 \leq i \leq d\}$.
(b) Suppose that $T = \begin{bmatrix} T_1 & \dots & T_d \end{bmatrix}$ is a commuting row contraction on H and that K is a subspace of H such that $P_K T|_K$ is unitarily equivalent to the d-shift M_z. Show that K reduces T.
HINT: let ρ be the u.c.p. map from $C^*(T)$ to $B(H^2)$ by compression and unitary equivalence. Then $\rho\Theta_T$ maps $\mathfrak{T}_d$ to $\mathcal{B}(H^2)$ and agrees with *id* on $\mathcal{S}$.

15.28 (a) Show that every $f \in A(H^2_d)$ extends to a continuous function on $\overline{\mathbb{B}_d}$.
(b) Construct a function $g(z)$ which is analytic on $\mathbb{B}_d$ and continuous on $\overline{\mathbb{B}_d}$ which does not belong to $A(H^2_d)$.

15.29 (a) Compute the maximal ideal space of $A(H^2_d)$.
HINT: φ is determined by $\varphi_{1,d}([M_{z_1}, \dots, M_{z_d}])$.
(b) Show that if $f \in A(H^2_d)$ and $|f(z)| \geq \varepsilon > 0$ for $z \in \mathbb{B}_d$, then $1/f \in A(H^2_d)$.

15.30 Let V be a variety for $\operatorname{Mult}(H^2_d)$ and $\mathfrak{J} = \mathfrak{J}(V)$. Show that $H_V = H^2_d \ominus \mathfrak{J}(V)$.

Notes on Chapter 15

Arveson [13] took an automorphism α of $L^\infty(0,1)$ induced by an ergodic measure preserving transformation, and constructed a concrete nonself-adjoint operator algebra. He showed that two such automorphisms are conjugate if and only if the algebras are unitarily equivalent. With Josephson [20], he considered a homeomorphism of a locally compact Hausdorff space X and constructed a concrete operator algebra in a similar way. Subject to some strong conditions, they showed that these algebras are isomorphic if and only if the homeomorphisms are conjugate. McAsey, Muhly and Saito [202–204] studied *analytic crossed products*, which were concrete algebras like Arveson's. In particular, Muhly and Saito [208] introduced the idea that outer conjugate actions on noncommutative C*-algebras yield isomorphic algebras. Peters [235] was the first to define the semicrossed product (by $\mathbb{Z}_+$) as an abstract operator algebra with a universal property. He studied the actions of endomorphisms on C*-algebras, not just automorphisms, and improved on the results of Arveson–Josephson, removing any sort of measure theoretic condition but requiring no periodic points. Power [247] showed that $C_0(X) \times_\alpha \mathbb{Z}_+$ was a complete invariant for the action of a homeomorphism on a locally compact Hausdorff space. Hadwin and Hoover [132] were able to deal with a larger class of endomorphisms, still subject to a condition on fixed points. The isomorphism problem in the commutative case was completely settled by Davidson and Katsoulis [74]. The C*-envelope for injective systems was established by Kakariadis and Katsoulis [170]. Important progress on the isomorphism problem in the noncommutative case was made by Davidson and Kakariadis [73]. The problem was completely solved by Katsoulis and Ramsey [174], who also deal with actions of the free semigroup in the multivariable context. The paper by Davidson, Fuller and Kakariadis [72] develops semicrossed products by other semigroups.

The dilation theorem for row contractions was done by Frazho [115] for $n = 2$, Bunce [43] for $n < \infty$ and Popescu [241] for $n = \infty$. Davis [86] had established the Bunce result earlier as a consequence of the Naimark dilation theorem. The Wold decomposition for row isometries is due to Frazho [116] for $n = 2$ and in general by Popescu [241]. Popescu [243] introduced the noncommutative disc algebra and proved the von Neumann-type inequality. The noncommutative analytic Toeplitz algebra was introduced by Popescu [242]. Many of the results in section 15.5 are taken from Davidson–Pitts [84]. The Popescu functional calculus is from [245]. The invariant subspaces were found by Arias–Popescu [10] and Davidson–Pitts [84]. The ideal structure and automorphisms are examined in Davidson–Pitts [82]. The distance to ideals and the structure of the commutator ideal is in Davidson–Pitts [83]. Free semigroup algebras were defined in Davidson–Pitts [84], where in addition to a study of $\mathfrak{L}_d$, the atomic representations of $\mathcal{O}_n$ are classified including Examples 15.7.2 and 15.7.3. The structure theorem for free semigroup algebras is from Davidson–Katsoulis–Pitts [75]. This is extended in Davidson–Li–Pitts [78], where the notion of absolutely continuous functionals on $\mathfrak{A}_d$ are introduced. Kennedy [177] proves that every free semigroup algebra which is not a von Neumann algebra has wandering vectors. This leads to a more refined decomposition of row isometries [178], and the fact that the weak-$*$ closure of the algebra generated by a row isometry is always a free semigroup algebra.

Concrete examples of reproducing kernel Hilbert spaces have been studied for over a century, since the introduction of Hardy and Bergman spaces. The abstract theory was laid down by Aronszajn [11] in 1950. The books by Paulsen–Raghupathi [229] and Agler–McCarthy [3] cover this material. Drury [102] established his von Neumann inequality for row contractions in 1978. Müller and Vasilescu [209] found it again and established the dilation Theorem 15.11.5. Both defined the d-shift only as a d-tuple of weighted shifts, and these results were not well-known. Arveson [18] reproved the dilation theorem, and recognized H^2_d both as symmetric Fock space and as an RKHS. He also identified the C*-algebra of the d-shift. The Drury–Arveson space is also a Besov space [49], determined by an integral of certain radial derivatives, which has proven to be very important from the harmonic analysis viewpoint. This was used by Costea, Sawyer and Wick [63] to solve the corona problem positively in DA space. The complete NP property in general RKHS spaces was characterized by McCullough [205] and Quiggin [249]. The complete NP property for DA space was obtained in Davidson–Pitts [83] by non-commutative methods. Agler and McCarthy [2] had a new approach. They show that every irreducible RKHS with the complete NP property imbeds into the infinite DA space H^2_∞. Two nice surveys of DA space are Shalit [284] and Hartz [143]. Also see the monograph by Chen and Guo [48].

CHAPTER 16

Noncommutative Convexity

In this chapter, we explore a notion of convexity that is compatible with the matrix structures over operator systems. In classical convexity, we have the Krein–Milman theorem and its relatives which identify the extreme points as the distinguishing feature of compact convex sets. Further to that is the subject of Choquet theory.

The early parts of Choquet theory are explored briefly in the first three sections. We then define the notion of an nc convex set. There is a version of the Hahn–Banach theorem in this context due to Effros and Winkler. We define the space $\mathrm{A}(K)$ of continuous nc affine maps analogous to the affine functions $A(K)$. This leads us to a contravariant categorical duality between the category of operator systems with u.c.p. maps and the category of nc convex sets with nc affine maps. We show that the notion of nc extreme points corresponds to boundary representations, and establish a version of the Krein–Milman theorem.

We introduce the notion of nc functions on an nc convex set. The set of all nc functions, $\mathrm{B}(K)$, is a von Neumann algebra. We define $\mathrm{C}(K)$ to be the C*-subalgebra generated by $\mathrm{A}(K)$. It happens that $\mathrm{B}(K)$ is the universal von Neumann algebra of $\mathrm{C}(K)$, and that $\mathrm{C}(K)$ is the maximal C*-algebra of $\mathrm{A}(K)$. We establish the Takesaki–Bichteler representation of an arbitrary C*-algebra as a space of nc functions on its representation space. This yields a description of $\mathrm{C}(K)$ as the nc functions on K which are continuous in a certain topology.

Then we define convex nc functions and use them to define orders on the u.c.p. maps on an operator system. This leads to an nc analogue of the Choquet–Bishop–de Leeuw theorem.

Classical Choquet Theory

16.1. Affine and Convex Functions

We need to consider affine and convex functions on compact convex sets. Affine functions preserve convexity, and convex functions, which have convex epigraphs, are bigger on the boundary. So they are used to compare measures on K.

In the next three sections, suppose that E is a locally convex topological vector space in duality with another space E_*; and let K be a τ_{E_*}-compact, convex subset

K. R. Davidson, *Functional Analysis and Operator Algebras*, CMS/CAIMS Books in Mathematics 13,
https://doi.org/10.1007/978-3-031-63665-3_16

of E. We will refer to K as a compact convex set in E. A function on K is called continuous if it is τ_{E_*}-continuous.

16.1.1. DEFINITION. Let K be a compact, convex subset of an LCTV E. A function f on K is *affine* if

$$f(tx + (1-t)y) = tf(x) + (1-t)f(y) \quad \text{for} \quad x, y \in K \text{ and } 0 \le t \le 1,$$

and a real-valued function f is *convex* if

$$f(tx + (1-t)y) \le tf(x) + (1-t)f(y) \quad \text{for} \quad x, y \in K \text{ and } 0 \le t \le 1.$$

Say that f is a *strictly convex function* if this is a strict inequality for $x \neq y$ and $0 < t < 1$. A function is *concave* if $-f$ is convex.

Let $A(K)$ denote the space of all continuous affine functions with values in $\mathbb{C}$; write $A_{\mathbb{R}}(K)$ for the real-valued functions in $A(K)$; and let $A(K, E)$ be the space of restrictions to K of affine functions in $A(E) = \mathbb{C}\mathbf{1} + E_*$, where $\mathbf{1}(x) = 1$ is the constant function. Let $P(K)$ denote the set of all continuous convex functions with values in $\mathbb{R}$. A real function f on K is *lower semicontinuous* (l.s.c) if

$$\text{epi}(f) = \{(x, r) : x \in K \text{ and } r \ge f(x)\}$$

is closed; and *upper semicontinuous* if $-f$ is lower semicontinuous. Let $Q(K)$ denote the set of all bounded lower semicontinuous convex functions on K.

A real function a is affine if and only if $\pm a$ are both convex. A real function f is convex if and only if epi(f) is a convex subset of $K \times \mathbb{R}$.

Note that $A(K)$ is unital, self-adjoint and separates points in K since linear functionals in E_* are continuous and affine. By the Stone–Weierstrass theorem, $C^*(A(K)) = C(K)$. Thus $A(K)$ is an operator system in $C(K)$.

16.1.2. THEOREM. *Let K be a compact, convex subset of an LCTV E. Every $f \in Q(K)$ is the supremum of all continuous affine functions smaller than f:*

$$f(x) = \sup\{a(x) : a \in A_{\mathbb{R}}(K, E),\ a \le f\} \quad \textit{for all} \quad x \in K.$$

Moreover $P(K)$ is a closed convex cone. Every $f \in Q(K)$ is the pointwise limit of an increasing net in $P(K)$, and such a limit is lower semicontinuous and convex. Also $P(K) - P(K)$ is uniformly dense in $C_{\mathbb{R}}(K)$.

PROOF. Suppose $f \in Q(K)$. Then epi(f) is a closed convex set. If $x \in K$ and $s < f(x)$, then (x, s) is disjoint from epi(f) in $E \times \mathbb{R}$. By a version of the separation theorem (Corollary 3.3.9), there is a $(\varphi, t) \in E_* \times \mathbb{R}$ such that

$$\varphi(x) + ts = \beta < \alpha = \inf\{\varphi(y) + tr : (y, r) \in \text{epi}(f)\}.$$

This forces $t > 0$ and after scaling, we may suppose that $t = 1$. Then

$$(\beta\mathbf{1} - \varphi)(y) = \beta - \varphi(y) < f(y) \text{ for } y \in K \quad \text{and} \quad (\beta\mathbf{1} - \varphi)(x) = s$$

Therefore, since $\beta\mathbf{1} - \varphi \in A_{\mathbb{R}}(K, E)$,

$$f(x) = \sup\{a(x) : a \in A_{\mathbb{R}}(K, E) \text{ and } a < f\}.$$

Notice that $P(K)$ is closed under sums and multiplication by positive scalars, so it is a convex cone. It is also easy to check that $P(K)$ is closed under finite sups, i.e. if $f, g \in P(K)$, then $f \vee g = \max\{f, g\}$ is convex. The convexity condition is preserved even by pointwise limits, so in particular $P(K)$ is uniformly closed. If $f \in Q(K)$, then for any finite subset $F \Subset K$, we can select affine functions $a_x \in A(K, E)$ for $x \in F$ with $a_x \leq f$ and $a_x(x) > f(x) - |F|^{-1}$. Then $g_F := \bigvee_{x \in F} a_x$ lies in $P(K)$ and $g_F \leq f$ and $g_F(x) > f(x) - |F|^{-1}$ for $x \in F$. To make this an increasing net on Λ, the poset of finite subsets of K ordered by inclusion, define $h_F = \bigvee_{G \subseteq F} g_G$. Therefore the increasing net $\{h_F\}_\Lambda$ converges pointwise to f.

Conversely, if $\{h_\lambda\}_\Lambda$ is an increasing net in $P(K)$ with pointwise limit f, then

$$\operatorname{epi}(f) = \bigcap_{\lambda \in \Lambda} \operatorname{epi}(h_\lambda)$$

is closed and convex; whence f is l.s.c. and convex.

The set $P(K) - P(K)$ is a lattice. To see this, suppose that $f_i, g_i \in P(K)$. Then

$$(f_1 - g_1) \vee (f_2 - g_2) = \big((f_1 + g_2) \vee (f_2 + g_1)\big) - (g_1 + g_2)$$

Similarly $P(K) - P(K)$ is closed under taking minima. Since $P(K)$ contains $A(K, E) \supset E^*$, it separates points of K. By the lattice version of Stone's theorem, $P(K) - P(K)$ is uniformly dense in $C_{\mathbb{R}}(K)$. ■

We need a version of the monotone convergence theorem for nets. This is not generally valid, but we will have some extra hypotheses to make it work. We will think of finite regular Borel measures as linear functionals, and so we will write $\mu(f)$ to denote $\int_K f\, d\mu$.

16.1.3. PROPOSITION. *Let f be a bounded lower semicontinuous function on a compact set K. Suppose that $\{f_\lambda\}_\Lambda$ is an increasing net of continuous functions converging pointwise to f. Then for any finite regular Borel measure μ,*

$$\lim_\Lambda \mu(f_\lambda) = \sup_\Lambda \mu(f_\lambda) = \mu(f).$$

PROOF. By translating and scaling, we may suppose that $0 \leq f \leq 1$. Fix $\varepsilon > 0$. Since $f^{-1}(r, \infty)$ is open for $r \in \mathbb{R}$, f is Borel and thus measurable. The sets $U_{k,n} = \{x : f(x) > \frac{k}{2^n}\}$ are open and $g_n = 2^{-n} \sum_{k=1}^{2^n} \chi_{U_{k,n}}$ is an increasing sequence of Borel functions increasing uniformly to f. Also $f - g_n > 0$ for all n. Choose N so large that $\mu(g_N) > \mu(f) - \varepsilon$.

Since μ is regular, there are compact sets $C_k \subset U_{k,N}$ with measure sufficiently close to $\mu(U_{k,N})$ that $h' = 2^{-n} \sum_{k=1}^{2^n} \chi_{C_k}$ also satisfies $\mu(h') > \mu(f) - \varepsilon$. By

Urysohn's lemma, there are continuous functions h_k such that $\chi_{C_k} \le h_k \le \chi_{U_{k,n}}$. Define $h = 2^{-n} \sum_{k=1}^{2n} h_k \in C(K)$. Then $\mu(h) > \mu(f) - \varepsilon$ and $f - h \ge f - g_N > 0$.

For each $x \in K$, $\lim_\Lambda f_\lambda(x) = f(x) > h(x)$. Since $f_\lambda - h \in C(K)$, there is some $\lambda_x \in \Lambda$ and open set $O_x \ni x$ such that $f_{\lambda_x} > h$ on O_x. By compactness, extract a finite open cover $K \subset \bigcup_{i=1}^m O_{x_i}$. Select $\lambda_0 \in \Lambda$ such that $\lambda_0 \ge \lambda_{x_i}$ for $1 \le i \le m$. Then $f_{\lambda_0}(x) \ge f_{\lambda_{x_i}} > h$ on O_{x_i}; whence $f_{\lambda_0} > h$. Since $h < f_{\lambda_0} \le f$, we have $\mu(f_{\lambda_0}) > \mu(f) - \varepsilon$. Therefore

$$\mu(f) \ge \lim_\Lambda \mu(f_\lambda) = \sup_\Lambda \mu(f_\lambda) > \mu(f) - \varepsilon.$$

As $\varepsilon > 0$ was arbitrary, the result follows. ∎

16.1.4. DEFINITION. If $f \in C_{\mathbb{R}}(K)$, the *lower envelope* of f is

$$\check{f}(x) = \sup\{a(x) : a \in A_{\mathbb{R}}(K),\ a \le f\} \quad \text{for} \quad x \in K.$$

The *upper envelope* is

$$\hat{f}(x) = \inf\{a(x) : a \in A_{\mathbb{R}}(K),\ a \ge f\} \quad \text{for} \quad x \in K.$$

16.1.5. REMARK. Observe that $\operatorname{epi}(\check{f}) = \bigcap_{a \le f} \operatorname{epi}(a)$. This is the intersection of closed convex sets, and hence it is closed and convex. If f is bounded by L, then $-L \le \check{f} \le L$. Thus $\check{f}$ is a bounded lower semicontinuous convex function with $\check{f} \le f$. On the other hand, if $g \in Q(K)$ with $g \le f$, then if $a \in A_{\mathbb{R}}(K)$ and $a \le g$, then $a \le f$. By Theorem 16.1.2,

$$\begin{aligned} g(x) &= \sup\{b(x) : b \in A_{\mathbb{R}}(K),\ b \le g\} \\ &\le \sup\{a(x) : a \in A_{\mathbb{R}}(K),\ a \le f\} = \check{f}(x). \end{aligned}$$

Therefore $\check{f}$ is the largest element of $Q(K)$ smaller than f.

Similarly, $\hat{f}$ is a bounded u.s.c. concave function which is the smallest such function larger than f. It is easy to see that $\hat{f} = -(\widecheck{-f})$. Also $\check{f} \le f \le \hat{f}$.

We record a few simple properties of the lower envelope. Proofs are omitted.

16.1.6. PROPOSITION. *Let $f, g \in C_{\mathbb{R}}(K)$. Then*

(1) $\widecheck{tf} = t\check{f}$ *and* $\widehat{tf} = t\hat{f}$ *for* $t \ge 0$.

(2) *If* $f \le g$, *then* $\check{f} \le \check{g}$ *and* $\hat{f} \le \hat{g}$.

(3) $\check{f} + \check{g} \le \widecheck{f+g}$ *and* $\widehat{f+g} \le \hat{f} + \hat{g}$.

Now we turn to a more detailed look at affine functions.

16.1.7. THEOREM. *Let K be a compact, convex subset of an LCTV E. Every lower semicontinuous real affine function is the pointwise limit of an increasing net in $A(K, E)$; and every $a \in A_{\mathbb{R}}(K)$ is a uniform limit of an increasing sequence in $A(K, E)$. In particular, $A(K, E)$ is uniformly dense in $A(K)$.*

PROOF. Suppose a is a lower semicontinuous real affine function on K. By Theorem 16.1.2,

$$a(x) = \sup\{b(x) : b \in A_{\mathbb{R}}(K, E) \text{ and } b < a\}.$$

We wish to show that this set of affine functions is directed. Suppose that $b_1, b_2 \in A(K, E)$ with $b_i < a$. Let

$$m = \inf_{x \in K, i=1,2} b_i(x) \quad \text{and} \quad B_i = \{(x, s) : x \in K,\ m \le s \le b_i(x)\}.$$

Clearly B_i are compact convex sets in the convex set $\{(x, r) : x \in K,\ r < a(x)\}$. Hence $B = \text{conv}(B_1 \cup B_2)$ is a compact convex set which is disjoint from epi(a). By the separation theorem (Corollary 3.3.9), there is a $(\varphi, t) \in E_* \times \mathbb{R}$ such that

$$\sup\{\varphi(x) + ts : (x, s) \in B\} = \beta < \alpha = \inf\{f(y) + tr : (y, r) \in \text{epi}(a)\}.$$

Arguing as in the previous proof, we have $b_i \le \beta\mathbf{1} - \varphi =: b < a$. By induction, if $b_i \in A_{\mathbb{R}}(K, E)$ and $b_i < a$ for $1 \le i \le n$, there is a $b \in A_{\mathbb{R}}(K, E)$ such that $b_i \le b < a$ for $1 \le i \le n$.

Let Λ be the poset of finite subsets of K ordered by inclusion. For each F, we can find $b_F \in A_{\mathbb{R}}(K, E)$ with $b_F < a$, $b_F(x) \ge a(x) - |F|^{-1}$ for $x \in F$, and $b_G \le b_F$ if $G \subset F$. Such functions can be constructed recursively, dealing with all $|F| = k$ at the kth step. Then $\{b_F\}_\Lambda$ is an increasing net in $A_{\mathbb{R}}(K, E)$ converging pointwise to a.

Now suppose that a is continuous. Fix $k \ge 1$, and define open sets $U_F = \{x \in K : b_F(x) > a(x) - \frac{1}{k}\}$. Then $F \subset G$ implies that $U_F \subset U_G$. Pointwise convergence shows that $\bigcup_\Lambda U_F = K$. By compactness, there is a finite collection $F_1, \dots, F_n$ so that $U_{F_1} \cup \dots \cup U_{F_n} = K$. Setting $G_k = F_1 \cup \dots \cup F_n$, we have $U_{G_k} = K$. Thus $a - \frac{1}{k} \le b_{G_k} < a$. There is no harm in enlarging these sets so that $G_k \subset G_{k+1}$. Therefore we have constructed an increasing sequence in $A_{\mathbb{R}}(K, E)$ converging uniformly to a.

Every affine function is the sum of its real and imaginary parts, so $A(K, E)$ is uniformly dense in $A(K)$. ∎

Recall that a linear functional φ on an operator system S is *self-adjoint* if $\varphi(a^*) = \varphi(a)^*$ for all $a \in S$. For $x \in K$, there is a state δ_x acting on $A(K)$ by $\delta_x(a) = a(x)$. This is positive, norm 1 and unital, i.e. $\delta_x(\mathbf{1}) = 1$.

16.1.8. LEMMA. *The unit ball of $A(K)^*_{sa}$ is* conv$(K \cup -K)$, *and every element can be written as $t\delta_x - (1-t)\delta_y$ for $x, y \in K$ and $0 \le t \le 1$.*

PROOF. Let φ be a self-adjoint linear functional on $A(K)$ with $\|\varphi\| \leq 1$. By the Hahn–Banach and Riesz representation theorems, there is a real measure μ on K with $\|\mu\| = \|\varphi\|$ such that $\varphi(a) = \int_K a(x)\, d\mu(x)$ for $a \in A(K)$. By the Jordan decomposition of a signed measure, we have $\mu = \mu_+ - \mu_-$, where $\mu_\pm$ are positive measures and $\|\mu\| = \|\mu_+\| + \|\mu_-\|$. Let $s = \|\mu_+\|$ and $t = \|\mu_-\|$; and by Proposition 16.2.2, let $x, y \in K$ be the barycentres of μ_+/s and μ_-/t, respectively. By restriction to $A(K)$, we obtain $\varphi_+ = \mu_+|_{A(K)} = s\delta_x$ and $\varphi_- = \mu_-|_{A(K)} = t\delta_y$. So $\varphi = s\delta_x - t\delta_y$. If $\|\varphi\| = s + t < 1$, then

$$\begin{aligned}\varphi &= s\delta_x - t\delta_y + \frac{1-s-t}{2}(\delta_x - \delta_x) \\ &= \frac{1+s-t}{2}\delta_x + \frac{1-s+t}{2}\Big(\frac{2t}{1-s+t}\delta_y + \frac{1-s-t}{1-s+t}\delta_x\Big) \\ &= \frac{1+s-t}{2}\delta_x + \frac{1-s+t}{2}\delta_z\end{aligned}$$

where $z = \frac{2t}{1-s+t}y + \frac{1-s-t}{1-s+t}x \in K$. ■

Next we identify the bounded affine functions on K with the second dual of $A(K)$.

16.1.9. THEOREM. *Let K be a compact, convex subset of a LCTV E. Then $A(K)^{**}$ is naturally isometrically isomorphic to the space of bounded affine functions on K, $A_b(K)$.*

PROOF. Since $\{\delta_x : x \in K\}$ is a continuous affine image of K in $A(K)^*$, it is clear that any functional $f \in A(K)^{**}$ restricts to an affine function on K bounded by $\|f\|$.

Conversely, let f be a real bounded affine function on K. Define $f(-x) = -f(x)$ for $x \in K$, and try to extend f to an affine function on $\operatorname{conv}(K \cup -K)$. We need to check that this is well defined. If $z = \sum_i s_i x_i - \sum_j t_j y_j$ where $x_i, y_j \in K$, $s_i > 0$, $t_j > 0$ and $\sum_i s_i + \sum_j t_j = 1$ is a finite convex combination, we can simplify the expression by setting $s = \sum_i s_i$, $t = \sum_j t_j = 1 - s$ and

$$x = \sum_i \frac{s_i}{s} x_i \quad \text{and} \quad y = \sum_j \frac{t_j}{t} y_j$$

are points in K. Then $z = sx - (1-s)y$. By affine linearity,

$$\sum_i s_i f(x_i) - \sum_j t_j f(y_j) = sf(x) - (1-s)f(y).$$

Suppose that $z = sx - (1-s)y = tu - (1-t)v$ for $x, y, u, v \in K$ and $0 \leq s, t \leq 1$. Evaluation at 1 shows that $s = t$. Then $sx + (1-s)v = (1-s)y + su$ is an element of K. By affine linearity, $sf(x) + (1-s)f(v) = (1-s)f(y) + sf(u)$. Therefore $sf(x) - (1-s)f(y) = tf(u) - (1-t)f(v)$ provides a well-defined definition of $f(z)$. Moreover f is bounded on $\mathrm{conv}(K \cup -K)$ by $\|f|_K\|_\infty$. In particular, $f(0) = 0$, so f is linear. We can now extend f uniquely to a linear functional on $A(K)^*_{sa}$. The extension to $A(K)^*$ is the complexification of this real functional; and it has the same norm by the usual proof of the complex case of Hahn–Banach.

In general, we can write $f = g + ih$ where g, h are real. Both g and h extend uniquely to an element of $A(K)^{**}_{sa}$ and thus f extends. Since $A(K) \subset C(K)$, we have $A(K)^{**} \subset C(K)^{**}$. The latter is the universal von Neumann algebra of $C(K)$, and in particular it is abelian. Hence we can represent $A(K)^{**}$ isometrically as a space of normal operators. For any normal operator N, there is some $\theta \in \mathbb{R}$ so that $\|N\| = \|\operatorname{Re} e^{i\theta} N\|$. Hence

$$\|f\| = \|\operatorname{Re} e^{i\theta} f\| = \|\operatorname{Re} e^{i\theta} f|_K\| \le \|f|_K\|_\infty \le \|f\|.$$

That is, the norm on $A(K)^{**}$ coincides with the norm on $A_b(K)$. ■

16.2. The Choquet Theorem

In this section, we develop some initial parts of Choquet theory to explain boundary representations as a generalization of extreme points. The main result is Choquet's theorem, which says that in the metrizable case, evaluation at a point $x \in K$ on the space of affine functions always has a representing measure supported on the set of extreme points.

16.2.1. DEFINITION. Let $M_1^+(K)$ denote the space of regular Borel probability measures on K. As for uniform algebras, we say that a $\mu \in M_1^+(K)$ is a *representing measure* for $x \in K$ if

$$f(x) = \int f\,d\mu \quad \text{for all} \quad f \in A(K).$$

In this case, we say that x is the *barycentre* of μ. Let $M_{1,x}^+(K)$ denote the set of $\mu \in M_1^+(K)$ with barycentre x.

The point mass δ_x is always a representing measure for x.

16.2.2. PROPOSITION. *Let $\mu \in M_1^+(K)$. Then there is a point $x_\mu \in K$ which is the barycentre of μ.*

PROOF. Let $\varphi \in E_*$ and define

$$K_\varphi = \Big\{x \in K : \int_K \varphi(y)\,d\mu(y) = \varphi(x)\Big\}.$$

Since $\varphi(y)$ is continuous and linear, $\varphi(K)$ is a compact convex set. We claim that the collection $\{K_\varphi : \varphi \in E_*\}$ has the finite intersection property. So let

$F = \{\varphi_1, \dots, \varphi_n\} \subset E_*$ and define a continuous linear map $\Phi_F : K \to \mathbb{C}^n$ by

$$\Phi_F(x) = (\varphi_1(x), \dots, \varphi_n(x)).$$

Then $\Phi_F(K)$ is compact and convex.

The measure μ is the weak-$*$ limit of probability measures with finite support in K. Thus the integral $\int_K \Phi_F(y)\,d\mu(y)$ is a limit of convex combinations of points in $\Phi_F(K)$, and hence lies in $\Phi_F(K)$. In particular, there is a point $x \in K$ so that $\Phi_F(x) = \int_K \Phi_F(y)\,d\mu(y)$, and thus $x \in \bigcap_{i=1}^n K_{\varphi_i}$. By compactness, there is a point $x_\mu \in \bigcap_{\varphi \in E_*} K_\varphi$. This point is unique because E_* separates points in E. Hence $\mu(f) = f(x)$ for all $f \in A(K, E) = \mathbb{C}1 + E_*$. This extends to the uniform closure $A(K)$ by Theorem 16.1.2. Therefore μ is a representing measure for x_μ. ∎

The irreducible representations of $C(K)$ are the point evaluations ε_x for $x \in K$. We are interested in the set of points in K which have a unique representing measure, as these are the boundary representations of $A(K)$.

16.2.3. PROPOSITION. *A point $x \in K$ has a unique representing measure if and only if x is an extreme point of K.*

PROOF. If x is not an extreme point of K, there are distinct points $x_i \in K$ so that $x = \frac{1}{2}(x_1 + x_2)$. Then $\mu = \frac{1}{2}(\delta_{x_1} + \delta_{x_2})$ is another representing measure for x.

Conversely, suppose that $\mu \neq \delta_x$ is a representing measure for x. Let $y \neq x$ lie in the support of μ, and choose $\varphi \in E_*$ so that $\operatorname{Re} \varphi(y - x) = \alpha > 0$. Let

$$A = \{z \in K : \operatorname{Re} \varphi(z - x) > \alpha/2\}.$$

As this contains an open neighbourhood of y, $\mu(A) > 0$; and similarly $\mu(K \setminus A) > 0$ because

$$\operatorname{Re} \frac{1}{\mu(A)} \int_A \varphi(z)\,d\mu(z) > \operatorname{Re} \varphi(x) = \operatorname{Re} \int_K \varphi(z)\,d\mu(z).$$

Let x_1 be the barycentre of $\frac{\mu|_A}{\mu(A)}$ and let x_2 be the barycentre of $\frac{\mu|_{K \setminus A}}{\mu(K \setminus A)}$. Observe that $x_1 \neq x$ and $\mu(A)x_1 + \mu(K \setminus A)x_2 = x$. Therefore x is not extreme. ∎

16.2.4. COROLLARY. *The boundary representations of $A(K)$ are the point evaluations at the extreme points of K. Thus $C^*_{\min}(A(K)) = C(\overline{\operatorname{ext}(K)})$.*

16.2.5. DEFINITION. The set $\operatorname{ext}(K)$ is known as the *Choquet boundary* of $A(K)$. The closure $\overline{\operatorname{ext}(K)}$ is the *Shilov boundary.*

The following could be deduced from our results about boundary representations but is much more elementary in this setting. The set of extreme points is indeed a boundary in the sense of Definition 7.8.1.

16.2.6. Corollary. *Every function $F \in M_n(A(K))$ attains its norm at an extreme point.*

Proof. Let $F = \big[f_{ij}\big]$ be a matrix with coefficients in $A(K)$. Since K is compact and M_n is finite dimensional, there are $x_0 \in X$ and unit vectors $\xi, \eta \in \mathbb{C}^n$ so that

$$\|F\| = \sup \|F(x)\| = \|F(x_0)\| = \langle F(x_0)\xi, \eta\rangle.$$

The set

$$K_0 = \{x \in K : \langle F(x)\xi, \eta\rangle = \|F\|\}$$

is a non-empty closed face of K. Indeed, it is convex since F is affine, and closed by continuity. Suppose that $x \in K_0$ and $x = \frac{1}{2}(x_1 + x_2)$ for $x_i \in K$. Then

$$\|F\| = \langle F(x)\xi, \eta\rangle = \tfrac{1}{2}\langle F(x_1)\xi, \eta\rangle + \tfrac{1}{2}\langle F(x_2)\xi, \eta\rangle \le \|F\|.$$

Thus $\langle F(x_1)\xi, \eta\rangle = \langle F(x_2)\xi, \eta\rangle = \|F\|$, and so $x_i \in K_0$. Therefore K_0 is a face of K. By the Krein–Milman theorem, K_0 contains an extreme point y. Since K_0 is a face, y is an extreme point of K at which F attains its norm. ■

16.2.7. Example. The Choquet boundary is generally not closed. We give a simple example in $\mathbb{R}^3$. Let

$$C = \{(x, y, 0) : x^2 + y^2 = 1\} \quad \text{and} \quad K = \text{conv}(C \cup \{(0, 1, \pm 1)\}.$$

It is easy to check that $\text{ext}\, K = (C \setminus \{(0, 1, 0)\}) \cup \{(0, 1, \pm 1)\}$.

16.2.8. Proposition. *If K is metrizable, the set of extreme points is a G_δ.*

Proof. Let d be a metric on K so that the metric topology is the topology of K. For $n \ge 1$, let

$$K_n = \big\{x \in K : x = \tfrac{1}{2}(y + z),\ y, z \in K \text{ and } d(y, z) \ge \tfrac{1}{n}\big\}.$$

It is readily checked that K_n is closed and disjoint from ext(K). If x is not extreme, then $x = \frac{1}{2}(y + z)$ for some $y, z \in K$, and hence lies in K_n for n large enough. Since $\bigcup_{n \ge 1} K_n$ is an F_σ in K, the complement ext(K) is a G_δ. ■

In the non-separable case, the set of extreme points need not be Borel. This creates some technical difficulties which we discuss in the next section.

16.2.9. Lemma. *If K is metrizable, there is a strictly convex continuous function h on K.*

Proof. The dual of $C(K)$ is $M(K)$ and the space of regular complex Borel measures on K. The map sending $x \in K$ to the point mass $\delta_x \in (M_1^+(K), w^*)$ is a homeomorphism onto its range; so we identify K as a subset of $M_1^+(K)$. Since K is metrizable and compact, it is separable; and hence $C(K)$ and $A(K)$ are separable.

Choose a dense subset $\{h_n : n \geq 1\}$ of the unit ball of $A(K)$. Let H be a Hilbert space with orthonormal basis $\{e_n : n \geq 1\}$, and define

$$T : M(K) \to H \quad \text{by} \quad T\mu = \sum_{n \geq 1} 2^{-n} \mu(h_n) e_n.$$

Note that T is injective since $T\mu = T\nu$ implies that $\mu(h_n) = \nu(h_n)$ for $n \geq 1$, and thus $\mu(h) = \nu(h)$ for all $h \in A(K)$. If $\{\mu_\lambda\}_\Lambda$ is a net in $M(K)_1$ which converges weak-$*$ to μ, then $\mu_\lambda(h_n) \to \mu(h_n)$ for each $n \geq 1$. Thus for any $N > 1$,

$$\|T\mu_\lambda - T\mu\|_2^2 \leq \sum_{n=1}^{N} 4^{-n} |\mu_\lambda(h_n) - \mu(h_n)|^2 + \sum_{n>N} 4^{-n} 4$$

Since the first N terms are arbitrarily small and the second equals $\frac{4}{3}4^{-N}$, it follows that T is continuous from $(M(K)_1, w^*)$ to $(H, \|\cdot\|)$. In particular, since T is injective, TK is a compact convex set which is affinely homeomorphic to K.

Since $\xi \to \|\xi\|^2$ is continuous and convex, it follows that $h(x) = \|Tx\|^2$ is continuous and convex on K. We claim that h is strictly convex. Indeed, by the parallelogram law (see Exercise **2.8**)

$$h\Big(\frac{x+y}{2}\Big) < \|T(\frac{x+y}{2})\|^2 + \|T(\frac{x-y}{2})\|^2 = \frac{\|Tx\|^2 + \|Ty\|^2}{2} = \frac{h(x)+h(y)}{2}$$

when $x \neq y$. ∎

This next lemma uses the map T just defined.

16.2.10. Lemma. *Let $K_n = \{x \in K : x = \frac{y+z}{2},\ y, z \in K,\ \|Ty - Tz\| \geq \frac{1}{n}\}$. If $\mu \in M_1^+(K_n)$ has barycentre x, there is a measure $\nu \in M_1^+(K)$ with barycentre x such that $\int \|Tx\|^2 \, d\nu(x) \geq \int \|Tx\|^2 \, d\mu(x) + \frac{1}{4n^2}$.*

Proof. We prove it first for a measure μ with finite support, say $\mu = \sum_{i=1}^n t_i \delta_{x_i}$ where $t_i > 0$ and $\sum_i t_i = 1$. Select $y_i, z_i \in K$ with $x_i = \frac{y_i+z_i}{2}$ and $\|Ty_i - Tz_i\| \geq \frac{1}{n}$. Define $\nu = \sum_{i=1}^n \frac{t_i}{2}(\delta_{y_i} + \delta_{z_i})$. Then it is clear that $x_\nu = x_\mu$. By the calculation in the previous proof,

$$\begin{aligned}
\int \|Tx\|^2 \, d\nu(x) &= \sum_{i=1}^{n} t_i \frac{\|Ty_i\|^2 + \|Tz_i\|^2}{2} \\
&= \sum_{i=1}^{n} t_i \big(\|T(\frac{y_i+z_i}{2})\|^2 + \|T(\frac{y_i-z_i}{2})\|^2\big) \\
&\geq \sum_{i=1}^{n} t_i \|T(\frac{y_i+z_i}{2})\|^2 + \frac{1}{4n^2} = \int \|Tx\|^2 \, d\mu(x) + \frac{1}{4n^2}.
\end{aligned}$$

For general μ, we observe that measures with finite support are weak-$*$ dense in $M_{1,x}^+(K_n)$. The weak-$*$ topology on $M_1^+(K)$ is metrizable, so there is a sequence

$\mu_k \in M_{1,x}^+(K_n)$ of finitely supported measures converging to μ. For each, select a measure $\nu_k \in M_{1,x}^+(K)$ such that $\displaystyle\int \|Tx\|^2\,d\nu_n(x) \geq \int \|Tx\|^2\,d\mu_k(x) + \tfrac{1}{4n^2}$. Extract a weak-$*$ convergent subsequence with limit $\nu \in M_{1,x}^+(K)$. Then clearly

$$\int \|Tx\|^2\,d\nu(x) \geq \int \|Tx\|^2\,d\mu(x) + \tfrac{1}{4n^2}. \qquad ■$$

16.2.11. CHOQUET THEOREM. *Let K be a metrizable compact convex set, and let $x \in K$. There is a regular Borel probability measure μ with barycentre x supported on* $\mathrm{ext}(K)$.

PROOF. Define a continuous function $I(\mu) = \int \|Tx\|^2\,d\mu$ on $M_{1,x}^+(K)$. Since $M_{1,x}^+(K)$ is compact, I attains its maximum at some measure μ. Since $d(x,y) = \|Tx - Ty\|$ is a complete metric on K, Proposition 16.2.8 shows that

$$K \setminus \mathrm{ext}(K) = \bigcup_{n\geq 1} K_n = \bigcup_{n \geq 1} \{x \in K : x = \tfrac{y+z}{2},\ y, z \in K,\ \|Ty - Tz\| \geq \tfrac{1}{n}\}.$$

If $\mu(K \setminus \mathrm{ext}(K)) > 0$, there is an integer n so that $\mu(K_n) > 0$. Let

$$\mu_1(A) = \frac{\mu(A \setminus K_n)}{\mu(K \setminus K_n)} \quad \text{and} \quad \mu_2 = \frac{\mu(A \cap K_n)}{\mu(K_n)}.$$

Then $\mu_2 \in M_1^+(K_n)$. By the previous lemma, there is a measure $\nu_2 \in M_1^+(K)$ with $x_{\nu_2} = x_{\mu_2}$ and $\displaystyle\int \|Tx\|^2\,d\nu(x) \geq \int \|Tx\|^2\,d\mu(x) + \tfrac{1}{4n^2}$. Define

$$\nu = \mu(K \setminus K_n)\mu_1 + \mu(K_n)\nu_2.$$

Then

$$x_\nu = \mu(K \setminus K_n)x_{\mu_1} + \mu(K_n)x_{\nu_2} = \mu(K \setminus K_n)x_{\mu_1} + \mu(K_n)x_{\mu_2} = x_\mu = x.$$

However

$$\begin{aligned}
\int \|Tx\|^2\,d\nu &= \mu(K \setminus K_n)\int \|Tx\|^2\,d\mu_1 + \mu(K_n)\int \|Tx\|^2\,d\nu_2 \\
&\geq \mu(K \setminus K_n)\int \|Tx\|^2\,d\mu_1 + \mu(K_n)\Big(\int \|Tx\|^2\,d\mu_2 + \frac{1}{4n^2}\Big) \\
&= \int \|Tx\|^2\,d\mu + \frac{\mu(K_n)}{4n^2}.
\end{aligned}$$

This contradicts the maximality of I at μ. Thus μ is supported on $\mathrm{ext}(K)$. ■

16.3. Orders on Measures

We will use convex functions to put a partial order on measures on K. This leads to the notion of maximal measures and the theorem of Bishop–de Leeuw, which extends Choquet's theorem to the general case.

16.3.1. Definition. If $\mu, \nu \in M^+(K)$ are positive measures, say $\mu \prec \nu$ if $\mu(f) \le \nu(f)$ for all $f \in P(K)$. Say that μ is *maximal* if it is a maximal element in this partial order.

16.3.2. Remarks. Clearly $\prec$ is reflexive and transitive—so it is a partial order. If $\mu \prec \nu \prec \mu$, then $\mu(f - g) = \nu(f - g)$ for all $f, g \in P(K)$. By Theorem 16.1.2, μ and ν agree on $C(K)$; whence $\mu = \nu$. So this partial order is antisymmetric.

Suppose that $\mu \prec \nu$. If $a \in A(K)$, then $\pm a \in P(K)$, and hence $\mu(a) = \nu(a)$. In particular, $\|\mu\| = \mu(1) = \nu(1) = \|\nu\|$. When we restrict this partial order to $M_1^+(K)$, we see that μ and ν have the same barycentre.

By Theorem 16.1.2, every $g \in Q(K)$ is the increasing pointwise limit of a net $\{f_\lambda\}_\Lambda$ of elements in $P(K)$. By Proposition 16.1.3, $\mu(g) = \lim_\Lambda \mu(f_\lambda)$. Therefore if $\mu \prec \nu$, then $\mu(g) \le \nu(g)$. In particular, $\mu(\check{f}) \le \nu(\check{f})$ for all $f \in C(K)$.

If $f \in C(K)$ is concave, then $-f \in P(K)$; whence $\nu(f) \le \mu(f)$. As in the previous paragraph, we obtain $\nu(\hat{f}) \le \mu(\hat{f})$ for all $f \in C(K)$.

A routine Zorn's lemma argument shows that maximal measures exist.

16.3.3. Proposition. *For $\mu \in M^+(K)$, there is a maximal measure $\nu \succ \mu$.*

Proof. Consider the set $\{\nu \in M^+(K) : \mu \prec \nu\}$. By Zorn's lemma, to establish the existence of maximal elements, it suffices to show that if $\mathcal{C} = \{\nu_\alpha\}_A$ is a chain, i.e. A is totally ordered and $\nu_\alpha \prec \nu_\beta$ for $\alpha < \beta \in A$, then $\mathcal{C}$ has an upper bound. By the remarks above, this is a bounded set of measures. By compactness, there is a cofinal convergent subnet Γ with weak-$*$ limit ν. For any $\alpha \in A$ and $f \in P(K)$, $\nu_\alpha(f) \le \lim_\Gamma \nu_\beta(f) = \nu(f)$. Thus $\nu_\alpha \prec \nu$; i.e. ν is an upper bound for $\mathcal{C}$. Therefore there is a maximal element greater than μ. ■

Next we characterize maximal measures in a useful way.

16.3.4. Mokobodzki Theorem. *Let $\mu \in M^+(K)$. The following are equivalent:*

(1) *μ is maximal.*

(2) *$\mu(f) = \mu(\check{f})$ for all $f \in C(K)$.*

(3) *$\mu(f) = \mu(\hat{f})$ for all $f \in C(K)$.*

PROOF. Suppose that $\mu \prec \nu$. By Remarks 16.3.2, for any $f \in C(K)$,

$$\mu(\check{f}) \leq \nu(\check{f}) \leq \nu(f).$$

Conversely, if $\mu(\check{f}) \leq \nu(f)$ for all $f \in P(K)$, then $\check{f} = f$ so that $\mu(f) \leq \nu(f)$ and so $\mu \prec \nu$. Therefore $\mu \prec \nu$ if and only if $\mu(\check{f}) \leq \nu(f)$ for all $f \in C(K)$. Replacing f by $-f$, we see that $\mu \prec \nu$ if and only of $\nu(f) \leq \mu(\hat{f})$ for all $f \in C(K)$.

Next we show that given $g \in C(K)$, there is a measure ν with $\mu \prec \nu$ such that $\nu(g) = \mu(\hat{g})$. By Proposition 16.1.6(3), the function $p(f) = \mu(\hat{f})$ is sublinear on $C_{\mathbb{R}}(K)$. Define ν_0 on $\mathbb{R}g$ by $\nu_0(tg) = t\mu(\hat{g})$ for $t \in \mathbb{R}$. Note that for $t \geq 0$,

$$\nu_0(tg) = \mu(\widehat{tg}) = p(tg).$$

Now $0 = \hat{0} \leq \hat{g} + \widehat{-g}$; so that $-\mu(\hat{g}) \leq \mu(\widehat{-g})$. Thus

$$\nu_0(-tg) = -t\mu(\hat{g}) \leq t\mu(\widehat{-g}) = \mu(\widehat{-tg}) = p(-tg).$$

That is, $\nu_0 \leq p$ on $\mathbb{R}g$. Use the Hahn–Banach theorem to extend ν_0 to a functional ν on $C_{\mathbb{R}}(K)$ such that $\nu(f) \leq p(f) = \mu(\hat{f})$. Now if $f \in P(K)$, $-f$ is concave so $-f = \widehat{-f}$. Hence $\nu(-f) \leq \mu(\widehat{-f}) = \mu(-f)$; i.e. $\mu(f) \leq \nu(f)$. Therefore $\mu \prec \nu$. In particular, $\nu \in M^+(K)$.

Now suppose that μ is a maximal measure, and let $f \in C(K)$. By the preceding paragraph, there is some $\nu \succ \mu$ such that $\nu(f) = \mu(\hat{f})$. By maximality, $\mu = \nu$ and hence $\mu(f) = \mu(\hat{f})$. So (3) holds. Since $-(\widehat{-f}) = \check{f}$, it is apparent that (2) and (3) are equivalent. Finally suppose that (3) holds, and $\mu \prec \nu$. By the first paragraph, for $f \in P(K)$, $\nu(f) \leq \mu(\hat{f}) = \mu(f)$. Therefore $\nu \prec \mu$. This order is antisymmetric by Remarks 16.3.2, and thus $\mu = \nu$; i.e. μ is maximal. ■

16.3.5. BAIRE VERSUS BOREL. The σ-algebra of Borel sets of a locally compact space X is generated by the collection of all closed sets. The σ-algebra of Baire sets is generated by the collection of all compact G_δ sets. If X is metrizable, then these sets coincide, but in general, the collection of Baire sets is smaller.

The σ-algebra of Baire sets is the smallest σ-algebra in which all functions in $C_0(X)$ are measurable. Thus it plays a useful role in the discussion of linear functionals on $C_0(X)$. These functionals are sometimes called Radon measures. The Riesz–Markov theorem states that every positive linear functional on $C_0(X)$ is given by integration against a unique regular Borel measure. However, we can restrict this measure to the Baire sets, and obtain a Baire measure which represents the functional. Every Baire measure is regular, and the Baire measure determined by the functional is unique. A Baire measure has a unique extension to a regular Borel measure, but there may be non-regular extensions as well.

We refer the reader to Halmos [135, sections 51–56] for a nice treatment of these issues.

Difficulties arise in extending Choquet's theorem to the non-metrizable case. There are examples for which ext K is not Borel, and examples when it is Borel but there are points in K with no representing measure supported on $\operatorname{ext} K$. The Baire sets come into the picture because we will show that maximal measures vanish on compact G_δ sets disjoint from $\operatorname{ext} K$. Since Baire measures are regular, this implies that $\mu(A) = 0$ for any Baire set disjoint from $\operatorname{ext} K$.

16.3.6. LEMMA. *Let K be a compact convex set. Let $\{f_n\}_{n\geq 1}$ be a sequence in $Q(K)$ which is bounded above and $f = \limsup_n f_n$ satisfies $f(x) \leq 0$ for $x \in \operatorname{ext} K$. Then $f \leq 0$.*

PROOF. Fix $x_0 \in K$. By Theorem 16.1.2, find $a_n \in A(K)$ such that $a_n \leq f_n$ and $a_n(x_0) > f_n(x_0) - \frac{1}{n}$. Then $\limsup_n a_n(x_0) = f(x_0)$ and $\limsup_n a_n \leq f$.

First suppose that K is metrizable. By Choquet's theorem, there is a representing measure μ for x_0 supported on $\operatorname{ext} K$. Thus $f \leq 0$ a.e.(μ). According to Fatou's lemma,

$$f(x_0) = \limsup a_n(x_0) = \limsup \mu(a_n) \leq \mu(f) \leq 0.$$

If K is not metrizable, define a continuous affine map $F : K \to \mathbb{R}^{\mathbb{N}}$ by $F(x) = \big(a_n(x)\big)_{n\geq 1}$. Then $F(K)$ is a compact convex subset of a metrizable space. Let π_n be the coordinate projections on $\mathbb{R}^{\mathbb{N}}$; and let $g = \limsup_n \pi_n|_{F(K)}$. Note that $g(F(x_0)) = f(x_0)$.

If y is an extreme point of $F(K)$, then $F^{-1}(y)$ is a closed face of K. By the Krein–Milman theorem, there is an extreme point z of $F^{-1}(y)$; and thus $z \in \operatorname{ext} K$. Hence

$$g(y) = \limsup \pi_n(y) = \limsup a_n(z) \leq f(z) \leq 0 \quad \text{for} \quad y \in \operatorname{ext} F(K).$$

The maps π_n are continuous and affine (whence convex), so by the previous paragraph, $g \leq 0$ on $F(K)$. Therefore $f(x_0) = g(F(x_0)) \leq 0$. ■

The following central result extends Choquet's theorem to the non-separable setting. This is critical in many applications of the theory.

16.3.7. CHOQUET–BISHOP–DE LEEUW THEOREM.
Let K be a compact convex set, and let $x \in K$. Let μ be a maximal measure representing x. Then $\mu(A) = 0$ for any Baire set disjoint from $\operatorname{ext} K$.

PROOF. By Proposition 16.3.3, there is a maximal measure $\mu \succ \delta_x$. By Remarks 16.3.2, μ has barycentre x and $\mu \in M_1^+(K)$; i.e. μ is a representing measure for x.

Let C be a compact G_δ disjoint from $\operatorname{ext} K$. Let U_n be a decreasing family of open sets such that $\bigcap_{n\geq 1} U_n = C$. By Urysohn's lemma, there are continuous functions $f_n \in C(K)$ such that $\chi_C \leq f_n \leq \chi_{U_n}$. Observe that $\lim_{n\to\infty} f_n = \chi_C$. The sequence $\big(\breve{f_n}\big)_{n\geq 1}$ in $Q(K)$ is bounded above by 1, and $f = \limsup_n \breve{f_n}$ satisfies $f(x) \leq \lim_n f_n = 0$ for $x \in \operatorname{ext} K$. By Lemma 16.3.6, $f \leq 0$. Therefore by Mokobodzki's theorem and Fatou's lemma,

$$\mu(C) = \lim_{n\to\infty} \mu(f_n) = \lim_{n\to\infty} \mu(\breve{f_n}) \le \mu(f) = 0.$$

By regularity of Baire measures, if A is a Baire set disjoint from $\operatorname{ext} K$,

$$\mu(A) = \sup\{\mu(C) : C \subset A,\ C \text{ compact } G_\delta\} = 0. \qquad \blacksquare$$

Noncommutative Convex Sets

16.4. NC Convexity

An nc convex set is a disjoint union of convex subsets of matrices over some operator space. We allow matrix convex combinations that can combine elements from different levels. In this section, we give the definition and provide some examples and basic observations.

Let E be an operator space. For cardinal numbers m, n, we let $M_{m,n}(E)$ be the operator space of $m \times n$ matrices over E. When the cardinals are infinite, we use the discussion in Example 14.14.13. As usual $M_n(E) := M_{n,n}(E)$. For each cardinal n, fix a Hilbert space H_n of dimension n. We will write $M_{m,n} = M_{m,n}(\mathbb{C}) = \mathcal{B}(H_n, H_m)$; and use 1_n for the identity element of M_n. In particular, $M_{m,n}(E)$ is an M_m–M_n bimodule.

Fix an infinite cardinal κ at least as large as the cardinality of some dense subset of E. Let $\mathcal{M}(E) = \coprod_n M_n(E)$ be the disjoint union over cardinals $n \le \kappa$. We suppress κ in the notation, since it plays no role except to make this a set. So we will write "for all n" when we mean for all $n \le \kappa$, since it will also follow for larger cardinals if needed. If E is separable, we can take $\kappa = \aleph_0$. Let $\mathcal{M} = \mathcal{M}(\mathbb{C})$.

If E has a predual E_*, we can identify $M_n(E)$ with $\mathrm{CB}(E_*, M_n)$ where E_* has the standard operator space structure inherited from E^* as in Corollary 14.14.11. As noted in Example 14.14.13, this identification works for infinite matrices over E as well. In this case, we will use the weak-$*$ topology on $M_n(E)$, which is the point-weak-$*$ topology on $\mathrm{CB}(E_*, M_n)$.

16.4.1. Definition. Let E be an operator space. A subset $K = \coprod_n K_n$ of $\mathcal{M}(E)$ is *nc convex* if

(0) K is graded; i.e. $K_n \subset M_n(E)$ for all n.

(1) If $\{\alpha_i \in M_{n,n_i} : i \in I\}$ is a family of isometries such that $\sum_I \alpha_i\alpha_i^* = 1_n$ and $\{x_i \in K_{n_i} : i \in I\}$ is a bounded set, then $x = \sum_I \alpha_i x_i \alpha_i^* \in K_n$.

(2) If $\beta \in M_{m,n}$ is an isometry and $x \in K_m$, then $\beta^* x \beta$ belongs to K_n.

Say that K is closed or compact if each K_n is closed or compact.

16.4.2. CONVENTION. If $K \subset \mathcal{M}(E)$ is compact, we will assume that E is a dual operator space and the topology on K_n is induced from the point-weak-$*$ topology on $\mathrm{CB}(E_*, M_n)$. By Corollary 14.14.11, we may assume that E is a subspace of some $\mathcal{B}(H)$ in such a way that E is weak-$*$ closed and the τ_{E_*} topology is induced from the weak-$*$ topology on $\mathcal{B}(H)$.

16.4.3. REMARKS.
(1) The condition (1) just says that a direct sum of a bounded set of elements of K is an element of K. That is, if $\{x_i \in K_{n_i} : i \in I\}$ is a bounded set and $m = \sum_i n_i$, then $x := \bigoplus_i x_i$ belongs to K_m. In particular, if $x \in K_n$ and $y \in K_m$, then $x \oplus y \in K_{n+m}$. The case of a single element shows that K_n is invariant under unitary conjugation. When I is an infinite set, the sums are SOT convergent.

(2) Condition (2) says that K is closed under compressions.

(3) Given a bounded set $\{x_i \in K_{n_i}\}_I$ and $\beta_i \in M_{n_i,n}$ such that $\sum_i \beta_i^* \beta_i = 1_n$, we can form an *nc convex combination* by $x = \sum_i \beta_i^* x_i \beta_i$. This belongs to K_n because $y := \bigoplus_i x_i$ belongs to K_m for $m = \sum_i n_i$ by (1), and

$$\beta = \begin{bmatrix} \beta_1 \\ \beta_2 \\ \vdots \end{bmatrix} \in M_{m,n} \quad \beta^*\beta = \sum_i \beta_i^* \beta_i = 1_n \quad \text{and} \quad \beta^* y \beta = \sum_i \beta_i^* x_i \beta_i = x.$$

That is, β is an isometry, so that $x \in K_n$ by (2). Thus an nc convex set is a set which is closed under nc convex combinations.

(4) A *matrix convex set* $K = \coprod_{n \ge 1} K_n$, where $K_n \subset M_n(E)$ for $n < \infty$, is a set which satisfies (0), (2) and

(1′) If $x \in K_n$ and $y \in K_m$, then $x \oplus y \in K_{m+n}$.

There is no need for infinite direct sums. However the previous comment about nc convex combinations is only valid for finite combinations in the matrix case.

16.4.4. EXAMPLES.
(1) Fix $c, d \in \mathbb{R}$ with $c < d$. For $n \in \mathbb{N}$, let $K_n = [c1_n, d1_n]$, where

$$[c1_n, d1_n] = \{A \in (M_n)_{sa} : c1_n \le A \le d1_n\}.$$

Then $K = \coprod_n K_n$ is a weak-$*$ compact nc convex set.

Suppose that $L = \coprod_n L_n \subset \mathcal{M}$ is a compact nc convex set with $L_1 = K_1 = [c, d]$. Every 1-dimensional compression of $T \in L_n$, say $\langle T\xi, \xi\rangle$ for $\|\xi\| = 1$, lies in $[c, d]$. Now if $A = A^*$, then the numerical range

$$W(A) = \{\langle A\xi, \xi\rangle : \|\xi\| = 1\} \subset \overline{\mathrm{conv}(\sigma(A))}$$

follows mostly from the spectral theorem. It is dense in $\sigma(A)$, and hence contains the interior of the convex hull; but will miss an endpoint if it's not an eigenvalue. In particular, if $T = A + iB \in L_n$ with $A = A^*$ and $B = B^*$, and $W(T) \subset [c, d]$, then $W(B) = \{0\}$ and hence $B = 0$ (see Exercise **4.23**). Also $\sigma(T) \subset [c, d]$, so that $T \in [c1_n, d1_n]$. Conversely every $T \in [c1_n, d1_n]$ with finite spectrum is diagonalizable, and hence a direct sum of elements of L_1. By the spectral theorem, every self-adjoint operator is the norm limit of self-adjoint operators in $[c1_n, d1_n]$ of finite spectrum. Therefore $L = K$.

(2) Let E be a dual operator space. For each n, let $B_n(E)$ denote the closed unit ball of $M_n(E)$ and let $B(E) = \coprod_n B_n(E)$. Each $B_n(E)$ is weak-$*$ compact and convex in $M_n(E)$. It is evident that B is closed under direct sums and compressions. Thus $B(E)$ is a compact nc convex set.

Even when $E = \mathbb{C}$, $B = B(\mathbb{C}) = \coprod_n B_n$ is not determined by its first level. Let $K_n = \{A : W(A) \subset \overline{\mathbb{D}}\}$. Then $K_1 = \overline{\mathbb{D}} = B_1$, but K_2 contains $A = \left[\begin{smallmatrix} 0 & 2 \\ 0 & 0 \end{smallmatrix}\right]$, which is not in B_2.

However if $L = \coprod_n L_n \subset \mathcal{M}$ is a compact nc convex set with $L_2 = B_2$, then $L = B$. Indeed, if $A \in L_n$ has $\|A\| > 1$, there are unit vectors ξ, η such that $\langle A\xi, \eta\rangle > 1$. Thus the compression of A to $M = \operatorname{span}\{\xi, \eta\}$ lies in $L_2 \setminus B_2$. Therefore $L_n \subset B_n$. Conversely, if $A \in B_n$, then A has a unitary dilation U. Since U is normal, it can be approximated in norm by finite spectrum unitaries. These are diagonalizable and hence are direct sums of elements of $B_1 = L_1$. Hence every unitary belongs to L, and it is closed under compressions, so it contains B. Thus $L = B$.

(3) Free spectrahedra. Let $A = (A_1, \dots, A_d) \in \mathcal{B}(H)^{(d)}$. Define a *monic linear pencil* $L : M_n^{(d)} \to \mathcal{B}(H \otimes H_n)$ by

$$L(X_1, \dots, X_d) = I \otimes 1_n - \sum_{i=1}^{d} A_i \otimes X_i.$$

Let $D_{A,n} = \{X \in M_n^{(d)} : \operatorname{Re} L(X) \geq 0\}$, and define $D_A = \coprod_n D_{A,n}$.

We will show that this is a closed nc convex set. First it is clear that each D_A is weak-$*$ closed, since the positive operators are weak-$*$ closed. Moreover, $L(\bigoplus_i X_i) = \bigoplus_i L(X_i)$, and hence D_A is closed under direct sums. Lastly, if $\beta \in M_{m,n}$ is an isometry and $X \in D_{A,m}$, then

$$\operatorname{Re} L(\beta^* X \beta) = \operatorname{Re}(I \otimes \beta^*) L(X)(I \otimes \beta) \geq 0.$$

Hence the compression $\beta^* X \beta \in D_{A,n}$.

(4) The nc state space of an operator system. Let S be an operator system. Let $K_n = \mathrm{UCP}(S, M_n) \subset \mathrm{CB}(S, M_n) = M_n(S^*)$. Then the *nc state space* of S is $K = \coprod_n K_n$. It is clear that K_n is compact in the point-weak-$*$ topology. The direct sum of u.c.p. maps is a u.c.p. map, so K is closed under direct sums; and the compression of a u.c.p. map is a u.c.p. map, so K is also closed under compressions. Thus K is a compact nc convex set.

16.4.5. PROPOSITION. *Let E be a dual operator space. If K and L are closed nc convex sets in $\mathcal{M}(E)$ such that $K_n = L_n$ for $n < \infty$, then $K = L$.*

PROOF. By symmetry, it suffices to show that $K_n \subset L_n$ for all n. Let $x \in K_n$. For each subspace $F \subset H_n$ of finite dimension, say $\dim F = n_F$, choose an isometry $\alpha_F \in M_{n,n_F}$ such that $\operatorname{Ran}\alpha_F = F$. Then $\lim_\Lambda \alpha_F\alpha_F^* = \lim_\Lambda P_F = 1_n$, where Λ is the net of finite dimensional subspaces of H_n ordered by inclusion. Define $x_F = \alpha_F^* x \alpha_F \in K_{n_F}$ and observe that $x = \lim_\Lambda \alpha_F x_F \alpha_F^*$. Let $\beta_F \in M_{n,m_F}$ be isometries chosen so that $\beta_F\beta_F^* = 1_n - P_F$. Hence $\lim_\Lambda \beta_F\beta_F^* = 0$. Fix any $y \in K_1 = L_1$ and note that $y \otimes 1_{m_F} \in L_{m_F}$. Then

$$\alpha_F x_F \alpha_F^* + \beta_F(y \otimes 1_{m_F})\beta_F^* \in L_n.$$

Therefore L_n contains $\lim_\Lambda \alpha_F x_F \alpha_F^* + \beta_F(y \otimes 1_{m_F})\beta_F^* = x$. ∎

Motivated by the proof above, we provide a definition of a closed subset of K that allows us to say that $(\alpha_F x_F \alpha_F^*)_\Lambda$ converges to a point in K.

16.4.6. DEFINITION. Say that a subset X of an nc convex set K is *closed* if whenever (α_λ) is a net of isometries in M_{m,n_λ} such that $w^*\text{–}\lim_\Lambda \alpha_\lambda\alpha_\lambda^* = 1_m$ and $x_\lambda \in X \cap K_{n_\lambda}$ such that $w^*\text{–}\lim_\Lambda \alpha_\lambda x_\lambda \alpha_\lambda^* = y$ exists, then $y \in X$.

16.4.7. REMARK. A straightforward modification of the proof of Proposition 16.4.5 shows that any such limit, y, must belong to K. The Proposition shows that every $y \in K_m$ for infinite m is a limit of a net in the finite part of K. Thus $\bigcup_{n<\infty} K_n$ is dense in K, even though every level is closed. An easy modification of the proof of this Proposition shows that an nc convex set is closed in this new sense if and only if each level is closed.

16.5. The NC Hahn–Banach Theorem

In this section, we establish a separation theorem for nc convex sets due to Effros and Winkler.

In order to deal with both the norm closed and weak-$*$ closed situations simultaneously, we assume that an operator space E is in duality with another operator space E_*, which could be its dual or its predual.

16.5.1. LEMMA. *Let E be an operator space in duality with E_*, and put the τ_{E_*}-topology on $\coprod_n M_n(E)$. Let $K = \coprod_{n\ge1} K_n$ be a τ_{E_*}-closed matrix convex set in $\coprod_{n\ge1} M_n(E)$ with $0 \in K_1$; and let $y \in M_n(E) \setminus K_n$. Then there is a faithful state f on M_n and a τ_{E_*}-continuous linear functional φ on $M_n(E)$ so that $\operatorname{Re}\varphi(y) > 1$ and*

$$\operatorname{Re}\varphi(A^*xA) \le f(A^*A) \quad \textit{for} \quad x \in K_m \textit{ and } A \in M_{mn},\ m \ge 1.$$

PROOF. By the Hahn–Banach theorem, there is a τ_{E_*}-continuous linear functional $\varphi \in M_n(E)_* = M_n(E_*)$ so that

$$\sup_{x \in K_n} \operatorname{Re}\varphi(x) \le \alpha < \operatorname{Re}\varphi(y).$$

Since $0 \in K_n$, we may suppose, after scaling, that $\alpha = 1$. Let $A \in M_{mn}$ with $\|A\| = 1$, and define $D_A = (1_n - A^*A)^{1/2}$. As the matrix $\left[\begin{smallmatrix} A \\ D_A \end{smallmatrix}\right]$ is an isometry, for each $x \in K_m$,

$$A^*xA = A^*xA + D_A 0_n D_A \in K_n.$$

Thus $\operatorname{Re}\varphi(A^*xA) \le 1 = \|A^*A\|$. There is a state f on M_n with $f(A^*A) = 1$, and hence

$$C_{x,A} = \{f \in S(M_n) : \operatorname{Re}\varphi(A^*xA) \le f(A^*A)\}$$

is a non-empty, compact convex set. Scaling A does not change the set $C_{x,A}$.

We claim that $\{C_{x,A} : x \in K_m,\ A \in M_{mn},\ m \ge 1\}$ has the finite intersection property. If not, there is a finite collection C_{x_i,A_i} for $1 \le i \le k$ and $x_i \in K_{m_i}$ so that $\bigcap_{i=1}^k C_{x_i,A_i} = \varnothing$. Equivalently, the set D of points of the form

$$\delta(f) = \big(f(A_1^*A_1) - \operatorname{Re}\varphi(A_1^*x_1A_1),\ \dots,\ f(A_k^*A_k) - \operatorname{Re}\varphi(A_k^*x_kA_k)\big)$$

for $f \in S(M_n)$ is compact, convex, and disjoint from $\mathbb{R}_+^k$. By the Hahn–Banach theorem, there is a vector $\eta = (t_1, \dots, t_k) \in \mathbb{R}^k$ so that

$$\sup_{f \in S(M_n)} \langle \delta(f), \eta \rangle < \inf_{\xi \in \mathbb{R}_+^k} \langle \xi, \eta \rangle = 0.$$

This forces $t_i \ge 0$ and $\eta \ne 0$. But then setting $x = x_1 \oplus \dots \oplus x_k$ and $A = \begin{bmatrix} \sqrt{t_1}A_1 \\ \vdots \\ \sqrt{t_k}A_k \end{bmatrix}$, we get for all $f \in S(M_n)$ that

$$0 > \sum_{i=1}^k t_i\big(f(A_i^*A_i) - \operatorname{Re}\varphi(A_i^*x_iA_i)\big) = f(A^*A) - \operatorname{Re}\varphi(A^*xA).$$

This shows that $C_{x,A}$ is empty, a contradiction. By compactness, there is a state $f \in \bigcap_{(x,A)} C_{x,A}$; i.e.

$$\operatorname{Re}\varphi(A^*xA) \le f(A^*A) \quad \text{for} \quad x \in K_m \text{ and } A \in M_{mn},\ m \ge 1.$$

If f is not faithful, choose $\varepsilon > 0$ so small that $1 < (1-\varepsilon)\varphi(y)$. Replace φ by $(1-\varepsilon)\varphi$ and f by $(1-\varepsilon)f + \varepsilon\tau_n$ where τ_n is the normalized trace on M_n. It is easily checked that this works and now the state is faithful. ∎

We can now prove the separation theorem. To keep things simpler, we assume that $0 \in K_1$. This condition will be removed later.

16.5.2. EFFROS–WINKLER SEPARATION THEOREM. *Let E be an operator space in duality with E_*, and put the τ_{E_*}-topology on $\coprod_{n\ge1} M_n(E)$. Let $K = \coprod_{n\ge1} K_n$ be a τ_{E_*}-closed matrix convex set in $\coprod_{n\ge1} M_n(E)$ with $0 \in K_1$*

and let $y \in M_n(E) \setminus K_n$. Then there is a τ_{E_*}-continuous completely bounded map $\Psi : E \to M_n$ such that

$$\operatorname{Re}\Psi_m(x) \leq 1_n \otimes 1_m \quad \textit{for} \quad x \in K_m,\ m \geq 1 \quad \textit{and} \quad \operatorname{Re}\Psi_n(y) \not\leq 1_n \otimes 1_n.$$

PROOF. By Lemma 16.5.1, there is a τ_{E_*}-continuous linear functional φ in $M_n(E)_* = M_n(E_*)$ and a faithful state f on M_n so that $\operatorname{Re}\varphi(y) > 1$ and

$$\operatorname{Re}\varphi(A^*xA) \leq f(A^*A) \quad \text{for} \quad x \in K_m \text{ and } A \in M_{mn},\ m \geq 1.$$

Define a τ_{E_*}-continuous map $\Phi = \begin{bmatrix}\varphi_{ij}\end{bmatrix} \in \mathrm{CB}(E, M_n)$ by $\varphi_{ij}(x) = n\varphi(E_{ij}x)$ for $1 \leq i, j \leq n$. Then if $x = \begin{bmatrix}x_{ij}\end{bmatrix} \in M_n(E)$,

$$\varphi(x) = \varphi\Big(\sum_{i,j} E_{ij}x_{ij}\Big) = \frac{1}{n}\sum_{i,j}\varphi_{ij}(x_{ij}).$$

Let $\xi_1, \dots, \xi_n$ be the standard basis for $\mathbb{C}^n$ and let $\xi = \frac{1}{\sqrt{n}}\sum_{i=1}^n \xi_i \otimes \xi_i \in \mathbb{C}^{n^2}$. Then for $x \in M_n(E)$,

$$\langle \Phi_n(x)\xi, \xi\rangle = \frac{1}{n}\sum_{i,j=1}^{n}\langle \Phi(x_{ij})\xi_j, \xi_i\rangle = \frac{1}{n}\sum_{i,j}\varphi_{ij}(x_{ij}) = \varphi(x).$$

If $x = \begin{bmatrix}x_{ij}\end{bmatrix} \in M_m(E)$ and $A, B \in M_{mn}$, then

$$\Phi_n(B^*xA) = \begin{bmatrix}b_{ij}1_n\end{bmatrix}^*\begin{bmatrix}\Phi(x_{ij})\end{bmatrix}\begin{bmatrix}a_{ij}1_n\end{bmatrix} = (1_n \otimes B^*)\Phi_m(x)(1_n \otimes A).$$

Therefore,

$$\varphi(B^*xA) = \langle \Phi_n(B^*xA)\xi, \xi\rangle = \langle \Phi_m(x)(1_n \otimes A)\xi, (1_n \otimes B)\xi\rangle.$$

The representation of M_n on C^{n^2} given by $\pi(A) = 1_n \otimes A$ has ξ as a separating and cyclic vector because if $A = \begin{bmatrix}\alpha_1 & \dots & \alpha_n\end{bmatrix} \in M_n$, then

$$A \mapsto \pi(A)\xi = \frac{1}{\sqrt{n}}(\alpha_1 \oplus \dots \oplus \alpha_n) = \frac{1}{\sqrt{n}}\begin{bmatrix}\alpha_1 \\ \vdots \\ \alpha_n\end{bmatrix}$$

is injective and has n^2-dimensional range. The commutant $\pi(M_n)' = M_n \otimes \mathbb{C}1_n$ also has ξ as a separating and cyclic vector. The state f on M_n is given by $f(A) = \mathrm{Tr}(AT)$ where $T \geq 0$ and $\operatorname{Tr} T = 1$. By polar decomposition, $T \simeq \sum_{i=1}^n s_i\eta_i\eta_i^*$ for an orthonormal basis $\eta_1, \dots, \eta_n$. Since f is faithful, $s_i > 0$ for all i. Define a unit vector $\eta = \sqrt{s_1}\eta_1 \oplus \dots \sqrt{s_n}\eta_n$ in $\mathbb{C}^{n^2}$. Compute

$$f(A) = \sum_{i=1}^{n} s_i\langle A\eta_i, \eta_i\rangle = \langle \pi(A)\eta, \eta\rangle.$$

The defining property of f is that $\operatorname{Re}\varphi(A^*xA) \le f(A^*A)$ for all $x \in K_m$ and $A \in M_{mn}$. That is,

$$\begin{aligned}\operatorname{Re}\varphi(A^*xA) &= \operatorname{Re}\langle \Phi_m(x)(1_n \otimes A)\xi, (1_n \otimes A)\xi\rangle \\ &\le f(A^*A) = \langle \pi(A^*A)\eta, \eta\rangle = \|\pi(A)\eta\|^2.\end{aligned}$$

Observe that η is also a separating and cyclic vector for $\pi(M_n)$ and $\pi(M_n)'$. Thus there are elements $D, D' \in M_n$ such that

$$(1_n \otimes D)\eta = \xi = (D' \otimes 1_n)\eta.$$

Define a new τ_{E_*}-continuous map in $\mathrm{CB}(E, M_n)$ by

$$\Psi(x) = D'^*\Phi(x)D' \quad \text{for} \quad x \in E.$$

If $x \in M_m(E)$ and $A \in M_{mn}$,

$$\begin{aligned}&\big\langle \Psi_m(x)(1_n \otimes A)\xi, (1_n \otimes A)\xi\big\rangle \\ &\quad = \big\langle (D' \otimes 1_m)^*\Phi_m(x)(D' \otimes 1_m)(1_n \otimes A)\xi, (1_n \otimes A)\xi\big\rangle \\ &\quad = \big\langle \Phi_m(x)(1_n \otimes A)(D' \otimes 1_n)\xi, (1_n \otimes A)(D' \otimes 1_n)\xi\big\rangle \\ &\quad = \big\langle \Phi_m(x)(1_n \otimes A)(1_n \otimes D)\xi, (1_n \otimes A)(1_n \otimes D)\xi\big\rangle = \varphi(D^*A^*xAD).\end{aligned}$$

Therefore if $x \in K_m$ and $A \in M_{mn}$,

$$\begin{aligned}\operatorname{Re}\big\langle \Psi_m(x)(1_n \otimes A)\xi, (1_n \otimes A)\xi\big\rangle &= \operatorname{Re}\varphi(D^*A^*xAD) \le f(D^*A^*AD) \\ &= \|(1_n \otimes A)(1_n \otimes D)\eta\|^2 = \|(1_n \otimes A)\xi\|^2.\end{aligned}$$

If $A = \begin{bmatrix}\alpha_1 & \dots & \alpha_n\end{bmatrix}$ where $\alpha_i \in \mathbb{C}^m$, then $(1_n \otimes A)\xi = \frac{1}{\sqrt{n}}(\alpha_1 \oplus \cdots \oplus \alpha_n)$. This is clearly an arbitrary vector in $\mathbb{C}^{mn}$ as A runs over M_{mn}. It follows that

$$\operatorname{Re}\Psi_m(x) \le 1_n \otimes 1_m \quad \text{for all} \quad x \in K_m,\ m \ge 1.$$

Finally, there is a $C \in M_n$ such that $(1_n \otimes C)\xi = \eta$. Then $(1_n \otimes CD)\eta = \eta$. But η is a separating vector for $\pi(M_n)$, and thus $CD = I$. Hence

$$\begin{aligned}\operatorname{Re}\langle \Psi_n(y)\eta, \eta\rangle &= \operatorname{Re}\langle \Psi_n(y)(1_n \otimes C)\xi, (1_n \otimes C)\xi\rangle \\ &= \operatorname{Re}\varphi(D^*C^*yCD) = \operatorname{Re}\varphi(y) > 1.\end{aligned}$$

Therefore $\operatorname{Re}\Psi_n(y) \not\le 1_n \otimes 1_n$. ∎

Some routine considerations now yield an *nc separation theorem*. This should be thought of as a replacement for the Hahn–Banach separation theorem in the context of nc convexity.

16.5.3. COROLLARY. *Let K be a closed nc convex set over a dual operator space E. Suppose there is n and $y \in M_n(E)$ such that $y \notin K_n$. Then there is a finite $k \in \mathbb{N}$, a weak-$*$ continuous completely bounded linear map $\Psi : E \to M_k$ and self-adjoint $\gamma \in M_k$ such that*

$$\operatorname{Re}\Psi_n(y) \not\le \gamma \otimes 1_n \quad \textit{and} \quad \operatorname{Re}\Psi_p(x) \le \gamma \otimes 1_p \textit{ for all } p \textit{ and } x \in K_p.$$

If $0 \in K_1$*, we may take* $\gamma = 1_k$*. Furthermore, if* E *is an operator system,* $K = K^*$ *and* $y = y^*$*, then* Ψ *can be chosen self-adjoint.*

PROOF. First assume that $0 \in K_1$. If $n < \infty$, we apply the Effros–Winkler theorem. If n is infinite, arguing as in the proof of Proposition 16.4.5, taking Λ to be the net of finite dimensional subspaces of H_n ordered by inclusion, there are isometries $\alpha_F \in M_{n,n_F}$ with range F and a net of points $y_F = \alpha_F^* y \alpha_F \in M_{n_F}$ such that $y = \lim_\Lambda y_F$. Since K is closed and $y \notin K$, there is some F so that $y_F \notin K_{n_F}$. Apply the Effros–Winkler theorem to y_F to obtain a weak-$*$ continuous completely bounded linear map $\varphi : E \to M_k$, where $k = n_F$, so that

$$\operatorname{Re}\Psi_m(x) \leq 1_k \otimes 1_m \text{ for all } x \in K_m,\ m < \infty \quad \text{and} \quad \operatorname{Re}\Psi_n(y_F) \not\leq 1_k \otimes 1_k.$$

As in Proposition 16.4.5, each point $x \in K_m$, for infinite m, is a weak-$*$ limit of points in finite K_n's. Since Ψ is weak-$*$ continuous, $\operatorname{Re}\Psi_m(x) \leq 1_k \otimes 1_m$. On the other hand, $\alpha_F^* \operatorname{Re}\Psi(y)\alpha_F = \operatorname{Re}\Psi_n(y_F) \not\leq 1_k \otimes 1_k$, which implies that $\operatorname{Re}\Psi(y) \not\leq 1_k \otimes 1_n$.

If $0 \notin K_1$, pick $x_0 \in K_1$ and translate K by setting $L_n = K_n - x_0 \otimes 1_n$. Then $y - x_0 \otimes 1_n \notin L_n$. Find the map Ψ from the first paragraph, and set $\gamma = 1_k + \operatorname{Re}\Psi(x_0)$. Then if $x \in K_m$,

$$\begin{aligned}\operatorname{Re}\Psi(x) &= \operatorname{Re}\Psi(x - x_0 \otimes 1_m) + \operatorname{Re}\Psi(x_0 \otimes 1_m)\\ &\leq 1_k \otimes 1_m + \operatorname{Re}\Psi(x_0) \otimes 1_m = \gamma \otimes 1_m.\end{aligned}$$

Likewise, $\operatorname{Re}\Psi(y) \not\leq \gamma \otimes 1_n$.

Finally if E is an operator system and K and y are self-adjoint, we can replace Ψ by $\frac{1}{2}(\Psi + \Psi^*)$, where $\Psi^*(x) := \Psi(x^*)^*$. ■

16.6. Categorical Duality

In this section, we will build a categorical correspondence between nc convex sets and operator systems. These become categories using nc affine maps and u.c.p. maps, respectively. The correspondence takes K to $A(K)$, and an operator system to its state space as in Example 16.4.4 (4). The correspondence between maps is natural, but it reverses arrows (contravariant).

16.6.1. DEFINITION. A function $\theta : K \to L$ between two nc convex sets is *nc affine* if

(0) θ is graded; i.e. $\theta(K_n) \subset L_n$ for all n.

(1) If $\{\alpha_i \in M_{n,n_i} : i \in I\}$ is a family of isometries with $\sum_I \alpha_i \alpha_i^* = 1_n$ and $\{x_i \in K_{n_i} : i \in I\}$ is a bounded set, then $\theta(x) = \sum_I \alpha_i \theta(x_i)\alpha_i^* \in L_n$.

(2) If $x \in K_m$, $\beta \in M_{m,n}$ is an isometry, then $\theta(\beta^* x \beta) = \beta^* \theta(x)\beta$ in L_n.

Say that θ is *continuous* if $\theta|_{K_n}$ is continuous for all n. Say that θ is *bounded* if

$$\|\theta\| := \sup_{x \in K} \|\theta(x)\| < \infty.$$

If K is compact, we let $\mathrm{A}(K, L)$ be the set of all continuous, nc affine functions from K into L. In particular, $\mathrm{A}(K) = \mathrm{A}(K, \mathcal{M})$. A map $\theta : K \to L$ is an *affine homeomorphism* if $\theta \in \mathrm{A}(K, L)$ and has an inverse $\theta^{-1} \in \mathrm{A}(L, K)$.

16.6.2. Remarks.

(1) If $\theta : K \to L$ is bounded, then θ respects nc convex combinations: given a bounded set $\{x_i \in K_{n_i}\}_I$ and $\beta_i \in M_{n_i,n}$ such that $\sum_i \beta_i^* \beta_i = 1_n$, then

$$\theta\Big(\sum_i \beta_i^* x_i \beta_i\Big) = \sum_i \beta_i^* \theta(x_i) \beta_i.$$

This follows as in Remark 16.4.3. In particular, $\theta|_{K_n}$ is affine: if $0 < t < 1$ and $x, y \in K_n$, then $\theta(tx + (1-t)y) = t\theta(x) + (1-t)\theta(y)$.

(2) If K is compact, then

$$\|\theta\| = \|\theta|_{K_{\aleph_0}}\| = \sup_{n<\infty} \|\theta|_{K_n}\|.$$

By compactness, $\|\theta|_{K_n}\| < \infty$ for every n. If we fix some $x_0 \in K_1$, we can imbed K_n into K_m for $n < m$ by selecting an isometry $\alpha \in M_{m,n}$ and setting $f(x) = \alpha x \alpha^* + x_0 \otimes (1_m - \alpha\alpha^*)$. Then

$$\|\theta(f(x))\| = \max\big\{\|\alpha\theta(x)\alpha^*\|, \|\theta(x_0)\|\big\} = \max\big\{\|\theta(x)\|, \|\theta(x_0)\|\big\}.$$

Thus $\|\theta|_{K_n}\|$ is an increasing function of n. So $\sup_{n<\infty} \|\theta|_{K_n}\| \le \|\theta|_{K_{\aleph_0}}\| \le \|\theta\|$. Now arguing as in Proposition 16.4.5, we obtain $\|\theta\| = \sup_{n<\infty} \|\theta|_{K_n}\|$.

16.6.3. Proposition. *If K is a compact nc convex set, then an nc affine function $a : K \to \mathcal{M}$ is continuous if it is continuous on K_1; and it is positive if it is positive on K_1. Moreover, $\|a\| \le 2\|a|_{K_1}\|$. This inequality is sharp. The restriction map $\rho : a \mapsto a|_{K_1}$ is a unital order isomorphism of* $\mathrm{A}(K)$ *onto* $\mathrm{A}(K_1)$.

Proof. Note that a is continuous on K_n if $x \to \langle a(x)\xi, \eta\rangle$ is continuous for all $\xi, \eta \in H_n$, and is positive if $\langle a(x)\xi, \xi\rangle \ge 0$ for all $\xi \in H_n$. We may suppose that $\|\xi\| = 1$, and consider ξ as an isometry from $\mathbb{C}$ into H_n. Then

$$\langle a(x)\xi, \xi\rangle = \xi^* a(x)\xi = a(\xi^* x \xi).$$

Since $\xi^* x \xi \in K_1$, the continuity of $a|_{K_1}$ ensures the continuity of $x \to \langle a(x)\xi, \xi\rangle$ on K_n, and the positivity of $a|_{K_1}$ ensures the positivity of a. The polarization identity shows that

$$\langle a(x)\xi, \eta\rangle = \frac{1}{4}\sum_{k=0}^{3} i^k \big\langle a(x)\xi + i^k\eta, \xi + i^k\eta\big\rangle$$

is continuous. Moreover, this identity shows that

$$\begin{aligned}\|a(x)\| &= \sup_{\|\xi\|=\|\eta\|=1} |\langle a(x)\xi, \eta\rangle| \\ &\le \frac{1}{4}\|a|_{K_1}\| \sup_{\|\xi\|=\|\eta\|=1} \sum_{0\le k\le 3} \|\xi + i^k\eta\|^2 \\ &= \|a|_{K_1}\| \sup_{\|\xi\|=\|\eta\|=1} \|\xi\|^2 + \|\eta\|^2 = 2\|a|_{K_1}\|.\end{aligned}$$

In particular, a is determined by $a|_{K_1}$.

This inequality is sharp. Let $K_n = \{T \in M_n : W(T) \subset \overline{\mathbb{D}}\}$. Then the identity map *id* is affine and $\|\, id|_{K_1}\| = 1$ while $A = \left[\begin{smallmatrix} 0 & 2 \\ 0 & 0 \end{smallmatrix}\right] \in K_2$ and $\|A\| = 2$.

The restriction map ρ is bounded below, and hence the range is a closed subspace of $A(K_1)$. The range contains $A(K_1, E)$. Thus by Theorem 16.1.7, the range is the entire $A(K_1)$. It is evidently unital, positive and bijective. The first paragraph shows that ρ^{-1} is also positive. Therefore ρ is an order isomorphism. ■

Next we show that $\mathrm{A}(K)$ is an operator system with state space K.

16.6.4. THEOREM. *Let K be a compact nc convex set. Then $\mathrm{A}(K)$ is an operator system. For every $x \in K_n$, $\delta_x : \mathrm{A}(K) \to M_n$ by $\delta_x(f) = f(x)$ is a u.c.p. map. Conversely, every u.c.p. map $\varphi : \mathrm{A}(K) \to M_n$ is an evaluation δ_x for some $x \in K_n$. The map $\theta(x) = \delta_x$ is an nc affine homeomorphism of K onto the nc state space of $\mathrm{A}(K)$.*

PROOF. The space $\mathrm{A}(K)$ is self-adjoint with adjoint operation $a^*(x) := a(x)^*$. The positive cone $C_n = M_n(\mathrm{A}(K))_+$ of $M_n(\mathrm{A}(K))$ consists of all nc affine maps which are positive on positive operators. The constant function $\mathbf{1}(x) = 1_n$ for $x \in K_n$ is the order unit. We will verify the axioms for an abstract operator system from Definition 14.14.2. Every $a \in M_n(\mathrm{A}(K))$ is bounded by Remark 16.6.2(2). Clearly $C_n \cap (-C_n) = \{0\}$. Since $a \in M_n(\mathrm{A}(K))_{sa}$ is bounded, say by L, we have $-L\mathbf{1}_n \le a \le L\mathbf{1}_n$. Thus $\mathbf{1}$ is an order unit; and $a = (a + L\mathbf{1}_n) - L\mathbf{1}_n \in C_n - C_n$. If $\gamma \in M_{m,n}$ and $a \in C_n$, then

$$\gamma a \gamma^*(x) = a(\gamma x \gamma^*) \ge 0.$$

Thus $\gamma C_n \gamma^* \subset C_m$. Finally the Archimedean property is easy to check. Hence $\mathrm{A}(K)$ is an abstract operator system. An explicit unital order imbedding into some $\mathcal{B}(H)$ is given by

$$a \mapsto \prod_{x\in K} a(x) \in \prod_{x\in K} M_{n(x)}.$$

Let $x \in K_n$. By definition of C_k, we have $\delta_x^{(k)}(C_k) \subset M_k(M_n)_+$; which says that δ_x is completely positive. We have $\delta_x(\mathbf{1}) = 1_n$, and hence δ_x is u.c.p.

Let L be the nc state space of $\mathrm{A}(K)$ as defined in Example 16.4.4(4). Define $\theta : K \to L$ by $\theta(x) = \delta_x$. It is clear that θ preserves arbitrary direct sums. It also preserves compressions by an isometry $\beta \in M_{m,n}$ because if $x \in K_n$ and $a \in \mathrm{A}(K)$,

$$\theta(\beta^* x\beta)(a) = a(\beta^* x\beta) = \beta^* a(x)\beta = \beta^* \delta_x \beta(a).$$

Therefore θ is an nc affine map. It is continuous from the point-weak-$*$ topology on K to the point-weak-$*$ topology on L. Now $K_1 \subset E$, where E is a dual operator space. Thus the linear functionals in E_* are continuous, nc affine, and separate points of K. Hence θ is injective.

Lastly we show that θ is surjective. Observe that $\theta(K)$ is a compact nc convex subset of L. If there were some $y_0 \in L_n \setminus \theta(K_n)$, we could apply the nc separation Theorem 16.5.3 to obtain a finite $k \in \mathbb{N}$, a weak-$*$ continuous completely bounded linear map $\Psi : E \to M_k$ and self-adjoint $\gamma \in M_k$ such that

$$\operatorname{Re}\Psi_n(y_0) \not\leq \gamma \otimes 1_n \quad \text{and} \quad \operatorname{Re}\Psi_p(\delta_x) \leq \gamma \otimes 1_p \text{ for all } p \text{ and } x \in K_p.$$

The map

$$a(x) = \Psi_p(\delta_x) \quad \text{for} \quad x \in K_p$$

is nc affine and point-weak-$*$ continuous, so determines an element of $M_k(\mathrm{A}(K))$. By construction, $\operatorname{Re} a(x) \leq \gamma \otimes 1_p$ for $x \in K_p$; so that $\operatorname{Re} a \leq \gamma \otimes \mathbf{1}$. But then since y_0 is u.c.p.,

$$\operatorname{Re}\Psi_n(y_0) = \operatorname{Re} y_0^{(k)}(a) \leq y_0^{(k)}(\gamma \otimes \mathbf{1}) = \gamma \otimes 1_n.$$

This contradiction establishes our claim. ∎

In the reverse direction, every operator system arises as a space of nc affine functions up to a complete order isomorphism.

16.6.5. THEOREM. *Let S be a (complete) operator system with nc state space $K \subset \mathcal{M}(S^*)$. For $s \in S$, define $\hat{s}(x) = x(s)$ for $x \in K$. Then $\hat{s}$ is a continuous nc affine function, and $\theta : S \to \mathrm{A}(K)$ given by $\theta(s) := \hat{s}$ is a complete order isomorphism of S onto $\mathrm{A}(K)$.*

PROOF. We may suppose that $S \subset \mathcal{B}(H)$ and that

$$M_n(S)_+ = M_n(S) \cap \mathcal{B}(H^{(n)})_+.$$

Evidently $\hat{s}$ preserves direct sums. If $x \in K_n$ and $\beta \in M_{m,n}$ is an isometry, then

$$\hat{s}(\beta^* x\beta) = \beta^* x\beta(s) = \beta^* x(s)\beta = \beta^* \hat{s}(x)\beta.$$

Hence $\hat{s}$ is nc affine. It is continuous on K in the τ_S topology by definition. So $\hat{s} \in \mathrm{A}(K)$. By definition of the state space, if $s \in M_n(S)_+$, then $\theta_n(s) \geq 0$.

For any $s \in M_n(S)$ and $\varepsilon > 0$, there are unit vectors $\xi, \eta \in H^{(n)}$ so that $\langle s\xi, \eta\rangle > \|s\| - \varepsilon$. We can write $\xi = (\xi_1, \xi_2, \dots)$ and $\eta = (\eta_1, \eta_2, \dots)$ where $\xi_i, \eta_i \in H$. The compression γ to $\operatorname{span}\{\xi_i, \eta_i : i \leq n\}$ is a u.c.p. map into M_{2n}, and $\|\gamma_n(s)\| > \|s\| - \varepsilon$. It follows that θ is completely isometric. In particular, θ is injective and has closed range. Every τ_S-continuous linear functional on S^* is given by evaluation at some $s \in S$. Note that $A(K_1, S^*) = \mathbb{C}1 + S = S$

because $\hat{1} = 1$. It follows that $\theta(S)|_{K_1} = A(K_1)$ by Theorem 16.1.7. Therefore by Proposition 16.6.3, $\theta(S) = \mathrm{A}(K)$. Hence θ is a positive, unital, completely isometric linear map of S onto $\mathrm{A}(K)$.

The identity map *id* is a state on S. If $s \in M_n(S)_{sa}$ is not positive, then $id_n(s)$ is not positive. Thus θ_n^{-1} is also a positive map for all n. Therefore θ is an unital order isomorphism of S onto $\mathrm{A}(K)$. ■

This leads us to a duality between two categories.

16.6.6. DEFINITION. Let NCConv denote the *category of compact nc convex sets* with morphisms consisting of continuous nc affine maps. Let OpSys denote the *category of (complete) operator systems* with morphisms consisting of unital completely positive maps.

16.6.7. THEOREM. *Define a map A : NCConv $\to$ OpSys by $A : K \mapsto \mathrm{A}(K)$ and if $\theta : K \to L$ is a continuous nc affine map, define a unital completely positive map $A(\theta) : \mathrm{A}(L) \to \mathrm{A}(K)$ by*

$$A(\theta)(b)(x) = b(\theta(x)) \quad \text{for} \quad b \in \mathrm{A}(L) \text{ and } x \in K.$$

Then A is a contravariant functor with inverse given by $A^{-1}(S) = K$, the nc state space of S, and if $\varphi : S \to T$ is a u.c.p. map between operator systems, where $A^{-1}(T) = L$, then $A^{-1}(\varphi) : L \to K$ by

$$A^{-1}(\varphi)(y)(s) = y(\varphi(s)) \quad \text{for} \quad y \in L \text{ and } s \in S.$$

PROOF. Theorems 16.6.4 and 16.6.5 show that A establishes a bijection between compact nc convex sets and operator systems. It is straightforward to check that A and A^{-1} are functors which reverse arrows, and thus are contravariant. ■

We obtain an immediate useful consequence.

16.6.8. COROLLARY. *Two operator systems are unitally completely order isomorphic if and only if their nc state spaces are nc affinely homeomorphic.*

16.7. The NC Krein–Milman Theorem

In this section, we explore the notion of an nc extreme point. This has a fairly natural definition, but there are technicalities. We shall see that the nc extreme points correspond to the boundary representations of the operator system. As in the classical case, there is a Krein–Milman type result—which will follow from an application of the separation theorem.

16.7.1. DEFINITION. Let K be an nc convex set. A point $x \in K_n$ is *nc extreme* if whenever x is written as a finite nc convex combination $x = \sum \alpha_i^* x_i \alpha_i$ for $x_i \in K_{n_i}$ and non-zero $\alpha_i \in M_{n_i,n}$ satisfying $\sum \alpha_i^* \alpha_i = 1_n$, then each α_i is a positive scalar multiple of an isometry $\beta_i \in M_{n_i,n}$ satisfying $\beta_i^* x_i \beta_i = x$ and each x_i decomposes with respect to the range of α_i as a direct sum $x_i = y_i \oplus z_i$ for $y_i, z_i \in K$. The set of all nc extreme points is denoted $\partial K = \coprod_n (\partial K)_n$.

Note that y_i is unitarily equivalent to x via β_i.

Since we now can think of a compact nc convex set as the state space of an operator system, we can adopt some terminology from the dilation theory.

16.7.2. DEFINITION. Let K be an nc convex set. A point $x \in K_m$ is *dilated* to a point $y \in K_n$ if there is an isometry $\alpha \in M_{n,m}$ such that $x = \alpha^* y \alpha$; and we say that x is a *compression* of y. If y decomposes with respect to the range of α as $y = x' \oplus z$ where x' is unitarily equivalent to x via α, then the dilation is *trivial*. Say that x is *maximal* if it has no non-trivial dilations.

Categorical duality between nc convex sets and operator systems allows us to restate the Dritschel–McCullough theorem 14.9.8. We provide another proof in Theorem 16.16.1.

16.7.3. THEOREM. *Let K be a compact nc convex set. Then every point in K has a maximal dilation.*

16.7.4. DEFINITION. Let K be an nc convex set. A point $x \in K_n$ is *nc pure* if whenever x is written as a finite nc convex combination $x = \sum \alpha_i^* x_i \alpha_i$ for $\{x_i \in K_{n_i}\}$ and non-zero $\{\alpha_i \in M_{n_i,n}\}$ satisfying $\sum \alpha_i^* \alpha_i = 1_n$, then each α_i is a positive scalar multiple of an isometry $\beta_i \in M_{n_i,n}$ satisfying $\beta_i^* x_i \beta_i = x$.

16.7.5. PROPOSITION. *Let K be an nc convex set. A point $x \in K$ is nc extreme if and only if it is both nc pure and maximal.*

PROOF. Suppose x can be written as a finite nc convex combination $x = \sum \alpha_i^* x_i \alpha_i$ for $\{x_i \in K_{n_i}\}$ and non-zero $\{\alpha_i \in M_{n_i,n}\}$ satisfying $\sum \alpha_i^* \alpha_i = 1_n$. The condition that each α_i is a positive scalar multiple of an isometry $\beta_i \in M_{n_i,n}$ satisfying $\beta_i^* x_i \beta_i = x$ is equivalent to x being nc pure. The condition that each x_i decomposes with respect to the range of α_i as a direct sum $x_i = y_i \oplus z_i$ for $y_i, z_i \in K$ with y_i unitarily equivalent to x, combined with the preceding condition, is equivalent to the maximality of x. ■

By Lemma 14.10.6, a pure and maximal u.c.p. map on an operator system S has the unique extension property, and it extends to a boundary representation of

$C^*(S)$; and the converse holds. The Davidson–Kennedy Theorem 14.10.8 shows that every nc pure point dilates to an nc extreme point; and the set of nc extreme points is sufficient to completely norm S. Another proof, using nc convexity, is provided in Theorem 16.16.3.

16.7.6. COROLLARY. *Let S be an operator system and let K be its nc state space. The nc extreme points of K coincide with the boundary representations of S. There are sufficiently many nc extreme points to completely norm S. Let $x_e = \bigoplus_{x\in\partial K} x \in K_m$. Then $\theta : \mathrm{A}(K) \to \mathcal{B}(H_m)$ given by $\theta(a) = a(x_e)$ is a complete order monomorphism.*

We can now establish the nc analogue of the Krein–Milman theorem.

16.7.7. THEOREM. *A compact nc convex set is the closed nc convex hull of its nc extreme points; i.e. $K = \overline{\mathrm{ncconv}(\partial K)}$.*

PROOF. Let K be a compact nc convex set. By categorical duality, we can identify K with the nc state space of the operator system $\mathrm{A}(K)$ of continuous affine nc functions on K. In particular, K is a compact nc convex set over the dual operator space $\mathrm{A}(K)^*$.

Let L denote the closed nc convex hull of the nc extreme points of K. Clearly $L \subseteq K$. Suppose that there is an n and $y \in K_n \setminus L_n$. Then by the nc separation theorem (Corollary 16.5.3), there is a finite k, a weak-$*$ continuous completely bounded linear map $\Psi : \mathrm{A}(K)^* \to M_k$ and a self-adjoint $\gamma \in M_k$ such that

$$\operatorname{Re}\Psi_n(y) \not\leq \gamma \otimes 1_n \quad \text{and} \quad \operatorname{Re}\Psi_p(x) \leq \gamma \otimes 1_p$$

for every p and $x \in L_p$.

Since Ψ is weak-$*$ continuous, there is an $a \in M_k(\mathrm{A}(K))$ such that $\Psi_p(x) = a(x)$ for all $x \in K_p$. Let $b = (a + a^*)/2$. Then from above, $b(y) \not\leq \gamma \otimes 1_n$ but $b(x) \leq \gamma \otimes 1_p$ for all p and $x \in L_p$.

Let $x_e = \bigoplus_{x\in\partial K} x \in K_m$. By Corollary 16.7.6, the map $\theta : \mathrm{A}(K) \to \mathcal{B}(H_m)$ given by $\theta(a) = a(x_e)$ is a complete order monomorphism. From the previous paragraph, $\theta(b) \leq \gamma \otimes 1_m = \theta(\gamma \otimes 1_{\mathrm{A}(K)})$. It follows that $b \leq \gamma \otimes 1_{\mathrm{A}(K)}$. In particular $b(y) \leq \gamma \otimes 1_n$, giving a contradiction. Therefore $L = K$. ∎

16.7.8. EXAMPLE. Let $\mathfrak{s}_1, \dots, \mathfrak{s}_d$ be the standard generators of the Cuntz algebra $\mathcal{O}_d$. Let $S = \mathrm{span}\{1, \mathfrak{s}_1, \dots, \mathfrak{s}_d, \mathfrak{s}_1^*, \dots, \mathfrak{s}_d^*\}$. Let K be the nc state space of S. If $\varphi \in K_n$, then

$$\big\| \big[\varphi(\mathfrak{s}_1), \dots, \varphi(\mathfrak{s}_d)\big] \big\| \leq \big\| \big[\mathfrak{s}_1, \dots, \mathfrak{s}_d\big] \big\| = 1.$$

Hence $T = \big[\varphi(\mathfrak{s}_1), \dots, \varphi(\mathfrak{s}_d)\big]$ is a row contraction. By Corollary 15.4.11, every row contraction dilates to a row unitary. This is an $\mathcal{O}_d$-dilation, and it corresponds to a representation of $\mathcal{O}_d$. Since $\mathcal{O}_d$ is simple and infinite dimensional, these representations are all infinite dimensional. These are the maximal points in K. To be pure requires that the representation be irreducible, and so the boundary representations of S are all irreducible representations of $\mathcal{O}_d$. Thus ∂K is the set of irreducible row unitary d-tuples, and they all live in $K_{\aleph_0}$. There are no nc extreme points at finite levels of the state space.

16.7.9. EXAMPLE. Recall that $C^*_{\min}(\overline{A(H^2_d)}) = \mathfrak{T}_d$ by Corollary 15.11.6. Let K be the nc state space of $S_d = \overline{A(H^2_d) + A(H^2_d)^*} \subset \mathfrak{T}_d$. The nc extreme points correspond the boundary representations $id|_{S_d} \in K_{\aleph_0}$ and $\{\delta_z|_{S_d} : z \in \partial\mathbb{B}_d\} \subset K_1$.

16.7.10. EXAMPLE. Let $\mathfrak{A}$ be a C*-algebra with nc state space K. By Stinespring's theorem, for every u.c.p. map $x \in K_n$, there is a $*$-representation $y \in H_m$ and isometry $\alpha \in M_{m,n}$ such that $x = \alpha^* y \alpha$; which in our terminology says that y dilates x. Moreover if x is not a representation, then this is a non-trivial dilation as a summand of a representation is a representation; and hence x is not maximal. On the other hand, if x is a representation, then it must be a summand of any dilation. Thus the maximal elements of K are the $*$-representations.

Given $x \in K_n$, let $x = \alpha^* y \alpha$ be its minimal Stinespring dilation. If y is irreducible, Theorem 14.10.1 shows that every u.c.p. map φ with $0 \le \varphi \le x$ equals tx for some $0 \le t \le 1$. Hence if $x = \sum \alpha_i^* x_i \alpha_i$ for $x_i \in K_{n_i}$ and non-zero $\alpha_i \in M_{n_i,n}$ satisfying $\sum \alpha_i^* \alpha_i = 1_n$, then there are $0 < t_i$ such that $\alpha_i^* x_i \alpha_i = t_i x$. In particular, $\alpha_i^* \alpha_i = t_i$; so that $\beta = t_i^{-1/2} \alpha_i$ is an isometry and $\beta_i^* x \beta_i = x$. Therefore x is nc pure.

On the other hand, if y is not irreducible, say $y = y_1 \oplus y_2$ with respect to $H_m = pH_m \oplus p^\perp H_m$, then

$$x = \alpha^* y \alpha = (p\alpha)^* y_1 (p\alpha) + (p^\perp \alpha)^* y_2 (p^\perp \alpha).$$

Now $(p\alpha)^*(p\alpha) + (p^\perp\alpha)^*(p^\perp\alpha) = \alpha^*\alpha = 1_n$, so this expresses x as a non-trivial nc convex combination. Therefore Theorem 14.10.1 shows that $\varphi_1 = (p\alpha)^* y_1 (p\alpha)$ satisfies $0 \le \varphi_1 \le x$ is the u.c.p. map determined by $p \in y(\mathfrak{A})'$. This is not a scalar multiple of the identity, and as the correspondence is an order isomorphism, φ_1 is not a multiple of x. Hence x is not nc pure. We conclude that x is nc pure if and only if y is irreducible.

Finally, x is nc extreme if and only if it is nc pure and maximal, which is equivalent to x being an irreducible representation. Therefore K is the closed nc convex hull of the set of irreducible representations.

Noncommutative Functions

16.8. NC Functions

In this section, we develop the notion of nc functions and relate this to several operator algebras and spaces related to $\mathrm{A}(K)$. These functions are defined on a matrix convex set, with certain properties relating the functions on each level that are weaker than nc affine functions. We show that the set $\mathrm{B}(K)$ of all bounded nc functions on a compact nc convex set form a von Neumann algebra. The C*-algebra $\mathrm{C}(K)$ is the subalgebra generated by $\mathrm{A}(K)$. It turns out that $\mathrm{B}(K)$ is the universal von Neumann algebra of $\mathrm{C}(K)$.

16.8.1. Definition. Let K be a compact nc convex set. A function $f : K \to \mathcal{M}$ is an *nc function* if

(0) f is graded: $f(K_n) \subseteq M_n$ for all n,

(1) f respects direct sums: if $x_i \in K_{n_i}$ for $i \in I$ and there are isometries $\alpha_i \in M_{n_i,n}$ satisfying $\sum_{i\in I} \alpha_i\alpha_i^* = 1_n$, then

$$f\big(\sum \alpha_i x_i \alpha_i^*\big) = \sum \alpha_i f(x_i) \alpha_i^*.$$

(2) f is unitarily equivariant: if $x \in K_n$ and $\beta \in M_n$ is unitary, then $f(\beta x \beta^*) = \beta f(x) \beta^*$.

Say that f is *bounded* if $\|f\|_\infty = \sup_{x\in K} \|f(x)\| < \infty$. Let $\mathrm{B}(K)$ denote the space of all bounded nc functions on K.

16.8.2. Examples.
(1) Every nc affine function is an nc function.

(2) Let $S = \mathrm{span}\{1, \mathfrak{s}_1, \mathfrak{s}_2, \mathfrak{s}_1^*, \mathfrak{s}_2^*\} \subset \mathcal{O}_2$ be the Cuntz system, and let K be its nc state space. Every u.c.p. map $x \in K_n$ sends $[\mathfrak{s}_1,\ \mathfrak{s}_2]$ to $T = [x(\mathfrak{s}_1),\ x(\mathfrak{s}_2)]$, which is a row contraction; and this row contraction determines x. By Exercise **15.14**, there is a unique decomposition $T = U \oplus T'$ with respect to $H_n = M \oplus M^\perp$, where U is a row unitary and T' has no row unitary summand. Define an nc function f by $f(x) = P_M$. This function is graded and unitarily equivariant. To check direct sums, note that the direct sum of row contractions with no row unitary summand can have no unitary summand. The function f is a projection in $\mathrm{B}(K)$.

A significant feature of this function is that $f|_{K_n} = 0$ for all $n < \infty$. So f is not determined by its finite part. This is distinctly different from nc affine functions which are determined by their restriction to K_1 by Proposition 16.6.3.

16.8.3. Proposition. *Let K be a compact nc convex set. Then $\mathrm{B}(K)$ is a von Neumann algebra.*

PROOF. The space $\mathrm{B}(K)$ contains all nc affine functions, and in particular contains the constant function $\mathbf{1}$. The adjoint is given by $f^*(x) = f(x)^*$. This clearly is graded and preserves direct sums. If $x \in K_n$ and $\beta \in M_n$ is unitary, then

$$f^*(\beta x \beta^*) = f(\beta x \beta^*)^* = (\beta f(x)\beta^*)^* = \beta f(x)^* \beta^* = \beta f^*(x)\beta^*,$$

So f^* is an nc function. The product is defined pointwise: if $f, g \in \mathrm{B}(K)$, then $(fg)(x) = f(x)g(x)$. This is readily seen to be an nc function. It is easy to check that the norm limit of nc functions is an nc function, so $\mathrm{B}(K)$ is a Banach $*$-algebra.

There is a natural $*$-representation of $\mathrm{B}(K)$, $\pi_u : \mathrm{B}(K) \to \prod_n \prod_{x \in K_n} M_n$, given by

$$\pi_u(f) = \bigoplus_{x \in K} f(x) = f(x_u),$$

where $x_u = \bigoplus_{x \in K} x$. This is isometric by definition, preserves the adjoint, and is linear and multiplicative. Thus π_u is a faithful $*$-representation, and $\mathrm{B}(K)$ is a C*-algebra. To check that $\pi_u(\mathrm{B}(K))$ is WOT-closed, suppose $(f_\lambda)_\Lambda$ is a net such that $T = \text{WOT--}\lim_\Lambda \pi(f_\lambda)$. Then $T = \bigoplus_{x \in K} T(x)$, and hence determines a graded function. Since each f_λ preserves direct sums and is unitarily equivariant, the same is true for the WOT-limit. Therefore $\pi_u(\mathrm{B}(K))$ is a von Neumann algebra.

By Theorem 10.2.13, $\pi_u(\mathrm{B}(K))$ has a unique predual. So $\mathrm{B}(K)$ also has a unique predual. By Sakai's theorem 10.6.6, $\mathrm{B}(K)$ is weak-$*$–weak-$*$ homeomorphic to $\pi_u(\mathrm{B}(K))$. Thus $\mathrm{B}(K)$ is an abstract von Neumann algebra with faithful normal representation π_u. ∎

16.8.4. DEFINITION. If K is a compact nc convex set, let $\mathrm{C}(K)$ be the C*-subalgebra of $\mathrm{B}(K)$ generated by $\mathrm{A}(K)$.

16.8.5. DEFINITION. If $\mathfrak{A}$ is a C*-algebra, let κ be an infinite cardinal large enough that every cyclic representation of $\mathfrak{A}$ acts on a Hilbert space of dimension no greater than κ. Define $\mathrm{Rep}(\mathfrak{A}) = \coprod_{n \le \kappa} \mathrm{Rep}(\mathfrak{A}, H_n)$ where $\mathrm{Rep}(\mathfrak{A}, H_n)$ is the space of all non-degenerate $*$-representations of $\mathfrak{A}$ into $\mathcal{B}(H_n)$. The topology on $\mathrm{Rep}(\mathfrak{A})$ is the point-WOT topology. Each $a \in \mathfrak{A}$ determines a function $\hat{a}(\pi) = \pi(a)$ from $\mathrm{Rep}(\mathfrak{A}, H_n)$ to $\mathcal{B}(H_n)$.

16.8.6. LEMMA. *The topologies of pointwise convergence in the* WOT, *weak-$*$, strong, strong-$*$, ultrastrong and ultrastrong-$*$ topologies coincide on* $\mathrm{Rep}(\mathfrak{A})$.

For $a \in \mathfrak{A}$, $\hat{a}$ is continuous on $\mathrm{Rep}(\mathfrak{A})$.

PROOF. It is enough to show that if a net $(\pi_\lambda)_\Lambda$ in $\mathrm{Rep}(\mathfrak{A}, H_n)$ converges point-WOT to π, then it converges point-ultrastrong-$*$. However for any $a \in \mathfrak{A}$, the net $\pi_\lambda(a)$ is bounded, so the strong and ultrastrong topologies coincide on it; and the adjoint map is WOT-continuous, so we only need to check strong convergence. For

$a \in \mathfrak{A}$ and $\xi \in H_n$,

$$\|\pi_\lambda(a)\xi - \pi(a)\xi\|^2 = \langle \pi_\lambda(a^*a)\xi, \xi\rangle - \langle \pi_\lambda(a)\xi, \pi(a)\xi\rangle - \langle \pi_\lambda(a^*)\pi(a)\xi, \xi\rangle + \langle \pi(a^*a)\xi, \xi\rangle.$$

By point-WOT convergence, the right-hand side goes to 0. Thus $(\pi_\lambda)_\Lambda$ converges point-ultrastrong-$*$ to π.

The continuity of $\hat{a}$ for $a \in \mathfrak{A}$ is by definition. ■

16.8.7. REMARK. Assume that $\mathfrak{A} \neq \mathbb{C}$ is an unital separable C*-algebra. The nc state space L of $\mathfrak{A}$ is a compact nc convex set. The point-WOT and point-weak-$*$ topologies agree on L since it is bounded. The topology on Rep($\mathfrak{A}$) is induced from this topology, but Rep($\mathfrak{A}$) is in general not closed in L. Thus Rep($\mathfrak{A}$) is in general not compact.

To see this, let $\pi \in \mathrm{Rep}(\mathfrak{A}, H_{\aleph_0})$ be an infinite dimensional faithful representation, and let M be an infinite dimensional subspace which is not reducing for $\pi(\mathfrak{A})$. Choose an isometry $\alpha \in \mathcal{B}(H_n)$ with range M. Then $\rho(a) = \alpha^*\pi(a)\alpha$ is a u.c.p. map of $\mathfrak{A}$ into $\mathcal{B}(H)$ which is not a representation. Select an orthonormal basis $\{\xi_k : k \geq 1\}$ for $H_{\aleph_0}$ and let $N_k = \mathrm{span}\{\xi_1, \dots, \xi_k\}$. Select unitaries u_k such that $u_k|_{N_k} = \alpha|_{N_k}$. Then $\pi_k(a) = u_k^*\pi(a)u_k$ converges point-WOT to ρ.

This sequence does not converge point-SOT to ρ. The problem is that ρ is not multiplicative. If $\pi_k(a)$ converges SOT to $\rho(a)$ and $\pi_k(a^*)$ converges SOT to $\rho(a^*)$, then since multiplication is SOT continuous, we would have $\rho(a^*a) = \rho(a^*)\rho(a)$. This must fail for some $a \in \mathfrak{A}$.

16.8.8. DEFINITION. Let K be a compact nc convex set. For each $x \in K_n$, the map $\delta_x(f) = f(x)$ is a $*$-representation of $\mathrm{C}(K)$ in $\mathrm{Rep}(\mathrm{C}(K), H_n)$. This has a natural extension to a representation $\widetilde{\delta_x}$ of $\mathrm{B}(K)$ by $\widetilde{\delta_x}(f) = f(x)$.

The representations $\widetilde{\delta_x}$ of $\mathrm{B}(K)$ are normal. This is because π_u is a normal representation, and $\widetilde{\delta_x}$ is obtained by the restriction of π_u to a reducing subspace.

16.8.9. PROPOSITION. *The restriction map* $\rho : \mathrm{Rep}(\mathrm{C}(K)) \to K$ *given by* $\rho(\pi) = \pi|_{\mathrm{A}(K)}$ *is a continuous, bijective nc function. However the inverse map is generally not continuous.*

PROOF. The topology on $\mathrm{Rep}(\mathrm{C}(K))$ and K are the point-weak-$*$ topologies on $\mathrm{C}(K)$ and $\mathrm{A}(K)$, respectively. Hence ρ is continuous. If $\pi \in \mathrm{Rep}(\mathrm{C}(K), H_n)$, then $\pi|_{\mathrm{A}(K)}$ is a u.c.p. map into $\mathcal{B}(H_n)$, and thus is an element of K_n. Thus ρ is graded. It clearly preserves direct sums and is unitarily equivariant; so it is an nc function.

Each $x \in K$ gives rise to the representation δ_x, and $\rho(\delta_x) = x$. So ρ is surjective. Since $\mathrm{C}(K)$ is the C*-algebra generated by $\mathrm{A}(K)$, there is at most one way to extend any u.c.p. map on $\mathrm{A}(K)$ to a $*$-homomorphism. Thus ρ is a bijection.

The nc convex set K is compact, but $\mathrm{Rep}(\mathrm{C}(K))$ is usually not. Hence ρ^{-1} cannot be continuous in general. ∎

16.8.10. THEOREM. $\mathrm{B}(K)$ *is the universal von Neumann algebra* $\mathrm{C}(K)^{\dagger\dagger}$ *of* $\mathrm{C}(K)$. *The weak-$*$ closure of* $\mathrm{A}(K)$ *in* $\mathrm{B}(K)$ *is isometrically isomorphic to* $\mathrm{A}(K)^{\dagger\dagger}$, *and it coincides with the space of all bounded nc affine functions on* K.

PROOF. We work in $\pi_u(\mathrm{B}(K))$. Suppose that P is a projection in $\pi_u(\mathrm{C}(K))'$. Then PH_u reduces π_u, so that $\pi_u = \pi_1 \oplus \pi_2$ with respect to $H_u = PH_u \oplus P^\perp H_u$. By the previous proposition, there are points x_1 and x_2 in K so that $\pi_i = \delta_{x_i}$ and $x_u = x_1 \oplus x_2$. Therefore if $f \in \mathrm{B}(K)$,

$$\pi_u(f) = f(x_1 \oplus x_2) = f(x_1) \oplus f(x_2) = \pi_1(f) \oplus \pi_2(f).$$

In particular, $\pi_u(f)$ commutes with P. Since the von Neumann algebra $\pi_u(\mathrm{C}(K))'$ is spanned by its projections, $\pi_u(\mathrm{B}(K)) \subset \pi_u(\mathrm{C}(K))''$, which is the WOT-closure of $\pi_u(\mathrm{C}(K))$ by the double commutant theorem. Since $\mathrm{C}(K) \subset \mathrm{B}(K)$ and $\pi_u(\mathrm{B}(K))$ is a von Neumann algebra, $\pi_u(\mathrm{B}(K)) = \pi_u(\mathrm{C}(K))''$.

Every representation of $\mathrm{C}(K)$ is given by some δ_x for $x \in K$ by Proposition 16.8.9. Each δ_x has a (unique) normal extension to the representation $\widetilde{\delta}_x$ of $\mathrm{B}(K)$. This is the property which defines the universal enveloping von Neumann algebra of $\mathrm{C}(K)$. By Theorems 10.5.5 and 10.5.2, $\mathrm{B}(K)$ is isomorphic to the second dual of $\mathrm{C}(K)$.

Since $\mathrm{A}(K)$ is a closed subspace of $\mathrm{C}(K)$, the weak-$*$ closure of $\mathrm{A}(K)$ in $\mathrm{B}(K)$ is isometrically isomorphic to $\mathrm{A}(K)^{\dagger\dagger}$ by Proposition 2.6.16. Clearly $\overline{\mathrm{A}(K)}^{\text{weak-}*}$ is contained in the set of bounded nc affine functions. On the other hand, $\mathrm{A}(K)$ is linearly isomorphic to $A(K_1)$ by Proposition 16.6.3. By Theorem 16.1.9, $A(K_1)^{\dagger\dagger}$ coincides with the space of all bounded affine functions on K_1. Therefore $\mathrm{A}(K)^{\dagger\dagger}$ coincides with the space of all bounded nc affine functions on K. ∎

The proof above and Theorem 16.1.9 imply the following extension of Proposition 16.6.3.

16.8.11. COROLLARY. *The restriction map* $\rho : a \to a|_{K_1}$ *is an unital order isomorphism of* $\mathrm{A}(K)^{\dagger\dagger}$ *onto the space of all bounded affine functions on* K_1.

16.9. The Maximal C*-algebra

We have shown that every operator system sits inside a unique minimal C*-algebra. There is also a maximal C*-algebra containing any operator system. The existence is much easier than for the minimal C*-algebra. However it has a number of interesting features which require some more work.

16.9.1. DEFINITION. Let S be an operator system. The *maximal C*-algebra* of S, $C^*_{\max}(S)$, is a pair $(\mathfrak{A}, \kappa)$ where $\kappa : S \to \mathfrak{A} = C^*(\kappa(S))$ is a unital completely isometric imbedding of S into the C*-algebra $\mathfrak{A}$ with the universal property:

(‡) whenever $j : S \to \mathfrak{B} = C^*(j(S))$ is a unital complete isometry, there exists a $*$-homomorphism $\pi : \mathfrak{A} \to \mathfrak{B}$ such that $\pi\kappa = j$; i.e.

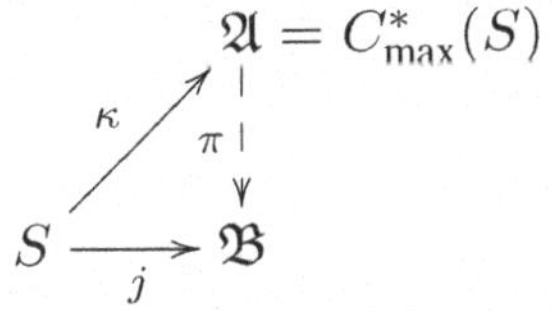

16.9.2. THEOREM. *Let S be an operator system. Then $C^*_{\max}(S)$ exists, and it is characterized by the property that whenever $\varphi : S \to \mathfrak{B}$ is a u.c.p. map into a C*-algebra $\mathfrak{B}$, there is a (unique) $*$-homomorphism $\pi_\varphi : C^*_{\max}(S) \to \mathfrak{B}$ extending φ. Also $C^*_{\max}(S)$ is unique up to a $*$-isomorphism that fixes S.*

PROOF. Let K be the nc state space of S. Define $x_u = \bigoplus_{x \in K} x : S \to \mathcal{B}(H_u)$. Set $\mathfrak{A} = C^*(x_u(S))$. For each $x \in K$, the compression to the summand H_x on which x acts yields a representation δ_x of $\mathfrak{A}$. Thus each $x \in K$ extends to a $*$-representation of $\mathfrak{A}$.

Consider a u.c.p. map $y : S \to \mathcal{B}(H')$. Then H' decomposes as a direct sum of cyclic subspaces for $C^*(y(S))$. Therefore $y \simeq \sum_\Lambda x_\lambda$ for a set of points $x_\lambda \in K$, possibly with repetitions. It follows that $\delta_y \simeq \sum_\Lambda^\oplus \delta_{x_\lambda}$ is a $*$-homomorphism of $\mathfrak{A}$ onto $C^*(y(S))$ which extends y. In particular, (‡) holds when y is an unital complete isometry.

If $(\mathfrak{A}, \kappa)$ and $(\mathfrak{B}, \lambda)$ are two C*-algebras with this universal property, then there are $*$-homomorphisms $\pi : \mathfrak{A} \to \mathfrak{B}$ such that $\pi\kappa = \lambda$ and $\sigma : \mathfrak{B} \to \mathfrak{A}$ such that $\sigma\lambda = \kappa$. It follows that $\sigma\pi : \mathfrak{A} \to \mathfrak{A}$ satisfies $\sigma\pi\kappa = \kappa\sigma\pi$. Since $\mathfrak{A} = C^*(\kappa(S))$, we have $\sigma\pi = id_{\mathfrak{A}}$. Similarly $\pi\sigma = id_{\mathfrak{B}}$. Therefore π is a $*$-isomorphism which intertwines κ and λ, establishing uniqueness. ■

Since we have seen that $\mathrm{C}(K)$ has this universal property relative to $\mathrm{A}(K)$, we obtain an immediate consequence.

16.9.3. COROLLARY. *Let K be a compact nc convex set. Then*

$$\mathrm{C}(K) \simeq C^*_{\max}(\mathrm{A}(K)).$$

16.10. Continuity and $\mathrm{C}(K)$

The goal of this section is to establish another characterization of $\mathrm{C}(K)$. For $f \in \mathrm{B}(K)$, we can define $\hat{f}$ on $\operatorname{Rep}(\mathrm{C}(K))$ because each $\pi \in \operatorname{Rep}(\mathrm{C}(K))$ extends to a normal representation on $\mathrm{B}(K)$. We want to be able to relate the continuity on $\operatorname{Rep}(\mathrm{C}(K))$ with continuity on K with an appropriate topology, but it turns out that the usual topology, obtained from the point-weak-$*$ topology on K as a subset of $\mathcal{M}(E)$, isn't the right one.

16.10.1. DEFINITION. For an nc convex set $K \subset E$, the point–ultrastrong-$*$ topology, τ_{us*}, is the weakest topology on K so that every $f \in \mathrm{A}(K)$ is continuous from K_n to $(M_n, \text{ultrastrong-}*)$.

Compare the following lemma with Proposition 16.8.9.

16.10.2. LEMMA. *Every $f \in \mathrm{C}(K)$ is continuous on (K, τ_{us*}). The restriction map $\rho : \operatorname{Rep}(\mathrm{C}(K), \textit{point-ultrastrong}^*) \to (K, \tau_{us*})$ is a homeomorphism.*

PROOF. As in Proposition 16.8.9, ρ is a continuous bijection.

Every $f \in \mathrm{A}(K)$ is continuous on (K, τ_{us*}) by definition. However multiplication and adjoints are continuous in the ultrastrong-$*$ topology. Thus polynomials of elements of $\mathrm{A}(K)$ are also continuous from K_n to $(M_n, \text{ultrastrong-}*)$. It follows that every element of $\mathrm{C}(K)$ is point-ultrastrong-$*$ continuous on (K, τ_{us*}). This means that ρ^{-1} is continuous. Thus the restriction map is a homeomorphism. ■

16.10.3. REMARK. In Remark 16.8.7, we saw that $\operatorname{Rep}(\mathrm{C}(K))$ is not closed in its nc state space, and thus is not compact. Lemma 16.8.6 shows that the point-ultrastrong-$*$ and point-WOT topologies coincide on $\operatorname{Rep}(\mathrm{C}(K))$. However this is not the case for (K, τ_{us*}) and K with its usual point-weak-$*$ topology, since the latter is compact.

The following result shows that the elements of $\mathrm{C}(K)$ are the nc functions which are continuous in an appropriate sense. We prove most of it based on our earlier work. Then we establish two technical lemmas to allow us to complete the proof.

16.10.4. THEOREM. *Let $f \in \mathrm{B}(K)$. The following are equivalent:*

(1) $f \in \mathrm{C}(K)$.

(2a) *$\hat{f}$ is continuous on $(\operatorname{Rep}(\mathrm{C}(K)), \textit{point-ultrastrong-}*)$.*

(2b) *$\hat{f}$ is continuous on $(\operatorname{Rep}(\mathrm{C}(K)), \textit{point-}\mathrm{SOT})$.*

(2c) *$\hat{f}$ is continuous on $(\operatorname{Rep}(\mathrm{C}(K)), \textit{point-}\mathrm{WOT})$.*

(3) *f is continuous on (K, τ_{us*}).*

(4) *f is continuous on $(L_1, \tau_{C(K)})$, where L_1 is the state space of $C(K)$.*

PROOF. Let L be the nc state space of $C(K)$. Then L_1 is the (classical) state space of $C(K)$. Define a map θ on $B(K)$ by $\theta(f)(s) = f(s)$ for $s \in L_1$. Every $s \in C(K)^* = B(K)_*$ is a weak-$*$ continuous linear functional on $B(K) = C(K)^{\dagger\dagger}$. L_1 is a compact convex set in the $\tau_{C(K)}$ topology, and $\theta(f)$ is a bounded affine function on L_1. Every unit vector $\xi \in H_u$ determines a state $s(f) = \langle \pi_u(f)\xi, \xi\rangle$ in L_1. It follows that $\|\theta(f)\| = \|f\|$ when $f = f^*$, and $\frac{1}{2}\|f\| \le \|\theta(f)\| \le \|f\|$ in general, as in the proof of Proposition 16.6.3.

By categorical duality, $C(K)$ is completely order isomorphic to $A(L)$. The map $\theta : C(K) \to A(L_1)$ is a linear isomorphism by Proposition 16.6.3. Since θ is injective, $\theta(f)$ is continuous on L_1 if and only if $f \in C(K)$. So (1) and (4) are equivalent.

Lemma 16.8.6 shows that (2a), (2b) and (2c) are equivalent. Also (1) implies that $\hat{f}$ is continuous on $\mathrm{Rep}(C(K))$ in all of these topologies. Lemma 16.10.2 shows that (2a) and (3) are equivalent.

It remains to establish that if $\hat{f}$ is continuous as a function on $\mathrm{Rep}(C(K))$ in the point-SOT topology, then f must be continuous on L_1. ■

We require two technical lemmas. The first is elementary linear algebra.

16.10.5. LEMMA. *Let H be a Hilbert space and let $\xi_1, \dots, \xi_n$ be a finite set of vectors in H. Given $\varepsilon > 0$, there is a $\delta > 0$ with the property that if $\eta_1, \dots, \eta_n$ is another set of vectors in a Hilbert space $H' \simeq H$ such that*

$$\big|\langle \xi_i, \xi_j\rangle - \langle \eta_i, \eta_j\rangle\big| < \delta \quad \text{for} \quad 1 \le i, j \le n,$$

then there is a unitary $U \in \mathcal{B}(H', H)$ so that $\|U\eta_i - \xi_i\| < \varepsilon$ for $1 \le i \le n$.

PROOF. Since we have no interest in the precise constants, we provide a proof in this spirit.

First we may assume that $\xi_1, \dots, \xi_n$ are linearly independent simply by making a very small perturbation ($\ll \varepsilon$) of those vectors which are linear combinations of the earlier ones. Let $X = \begin{bmatrix}\xi_1 & \dots & \xi_n\end{bmatrix}$ and $Y = \begin{bmatrix}\eta_1 & \dots & \eta_n\end{bmatrix}$. The hypothesis means that the matrix entries of X^*X and Y^*Y are close, and thus $\|X^*X - Y^*Y\|$ is small. Since $\xi_1, \dots, \xi_n$ are linearly independent, X^*X is an invertible $n \times n$ matrix. Thus for small enough δ, Y^*Y is also invertible, and so $\eta_1, \dots, \eta_n$ are also linearly independent. Finally $\|(X^*X)^{1/2} - (Y^*Y)^{1/2}\|$ must be small.

Write $X = V(X^*X)^{1/2}$ and $Y = W(Y^*Y)^{1/2}$ in polar decomposition. Let U be any unitary that extends VW^* (considered as a unitary from $\mathrm{span}\{\eta_1, \dots, \eta_n\}$ onto $\mathrm{span}\{\xi_1, \dots, \xi_n\}$) to a unitary from H' to H. Then

$$\|UY - X\| = \|(Y^*Y)^{1/2} - (X^*X)^{1/2}\|$$

is small. The columns of $UY - X$ are just $U\eta_i - \xi_i$, so we are done. ∎

Say that a pair (π, ξ) for $\pi \in \mathrm{Rep}(\mathfrak{A}, H_n)$ and $\xi \in H_n$ represents a state φ of $\mathfrak{A}$ if $\varphi(a) = \langle \pi(a)\xi, \xi\rangle$ for all $a \in \mathfrak{A}$. Note that this forces $\|\xi\| = 1$.

16.10.6. Lemma. *Let $\mathfrak{A}$ be an unital C*-algebra, and let L_1 be its state space. Given $\varphi \in L_1$, there is a representative pair (π, ξ) for φ so that for any open neighbourhood U of π in* (Rep($\mathfrak{A}$), *point-*SOT) *and open neighbourhood V of ξ in H_κ, there is an open neighbourhood W of φ in L_1 such that for each $\psi \in W$, there is a representation $\sigma \in U$ and unit vector $\eta \in V$ so that (σ, η) represents ψ.*

Proof. Let κ be an infinite cardinal so that every cyclic representation of $\mathfrak{A}$ acts on a Hilbert space of dimension no greater than κ. Fix an irreducible representation π_0 in $\mathrm{Rep}(\mathfrak{A}, H_0)$. Let $(\pi_\varphi, \xi_\varphi)$ be the GNS representation of φ. Set $\pi = \pi_\varphi \oplus (\pi_0 \otimes 1_\kappa)$ in $\mathrm{Rep}(\mathfrak{A}, H_\kappa)$ and $\xi = \xi_\varphi \oplus 0 \in H_\kappa = H_\varphi \oplus K_0$, where $K_0 = H_0 \otimes H_\kappa$. Clearly (π, ξ) represents φ.

A basic open neighbourhood of π is given by finite sets $a_i \in \mathfrak{A}$ and $\xi_i \in H_\kappa$ as

$$U = \big\{\sigma \in \mathrm{Rep}(\mathfrak{A}, H_\kappa) : \|\sigma(a_i)\xi_i - \pi(a_i)\xi_i\| < \varepsilon \text{ for } 1 \le i \le n\big\}.$$

We still obtain a basis of neighbourhoods if we assume that ξ_i either lies in H_φ or in $K_0 = H_\varphi^\perp$, and that $\|a_i\| \le 1$ and $\|\xi_i\| \le 1$. So we will instead assume that we have $a_i, b_k \in \mathfrak{A}_1$ and $\xi_i \in H_\varphi$ and $\zeta_k \in K_0$ for $1 \le i \le m$ and $1 \le k \le n$, and

$$U = \big\{\sigma \in \mathrm{Rep}(\mathfrak{A}, H_\kappa) : \|\sigma(a_i)\xi_i - \pi(a_i)\xi_i\| < \varepsilon \text{ and } \|\sigma(b_k)\zeta_k - \pi(b_k)\zeta_k\| < \varepsilon\big\}.$$

For simplicity, take $V = b_\varepsilon(\xi)$ using the same ε. Define vectors $\xi_{i+m} = \pi(a_i)\xi_i$ for $1 \le i \le m$. For $\xi_1, \dots, \xi_{2m}$, apply Lemma 16.10.5 using $\varepsilon' = \varepsilon/2$ to get some $\delta > 0$. Since π_φ is a cyclic representation, we can choose $c_i \in \mathfrak{A}$ so that $\|\pi(c_i)\xi - \xi_i\| < \delta/3$ for $1 \le i \le m$.

For $\psi \in L_1$, define $\sigma = \pi_\psi \oplus (\pi_0 \otimes 1_\kappa)$ in $\mathrm{Rep}(\mathfrak{A}, H'_\kappa)$, where $H'_\kappa = H_\psi \oplus K_0 \simeq H_\kappa$. Let $\eta = \xi_\psi \oplus 0$, and note that (σ, η) represents ψ. Define vectors $\eta_i = \sigma(c_i)\eta$ and $\eta_{i+m} = \sigma(a_i c_i)\eta$ for $1 \le i \le m$. For simpler notation, write $c_{i+m} = a_i c_i$ for $1 \le i \le m$; so that $\eta_i = \sigma(c_i)\eta$ for $1 \le i \le 2m$. Calculate for $1 \le i, j \le 2m$,

$$\begin{aligned}(16.10.7)\qquad \big|\langle \xi_i, \xi_j\rangle - \langle \eta_i, \eta_j\rangle\big| &\le \big|\langle \xi_i, \xi_j\rangle - \langle \pi(c_i)\xi, \pi(c_j)\xi\rangle\big| \\ &\quad + \big|\langle \pi(c_i)\xi, \pi(c_j)\xi\rangle - \langle \sigma(c_i)\eta, \sigma(c_j)\eta\rangle\big| \\ &\le \|\xi_i - \pi(c_i)\xi\| + \|\xi_j - \pi(c_j)\xi\| + \big|\varphi(c_j^* c_i) - \psi(c_j^* c_i)\big| \\ &< \tfrac{2}{3}\delta + \big|\varphi(c_j^* c_i) - \psi(c_j^* c_i)\big|.\end{aligned}$$

We now set

$$W = \big\{\psi \in L_1 : \big|\varphi(c_j^* c_i)) - \psi(c_j^* c_i))\big| < \tfrac{\delta}{3} \text{ for } 0 \le i, j \le 2m\big\}$$

to be a weak-$*$ open neighbourhood of φ. For $\psi \in W$, our estimate (16.10.7) is now bounded by δ. Define a unitary U in $\mathcal{B}(H'_\kappa, H_\kappa)$ by first setting $U|_Z = I|_Z$ where

$$Z = \mathrm{span}\{\zeta_k,\ (\pi_0 \otimes 1_\kappa)(b_k)\zeta_k : 1 \le k \le n\} \subset K_0,$$

and then applying Lemma 16.10.5 for $H_\kappa \ominus Z$ and $H'_\kappa \ominus Z$ so that

$$\|U\eta_i - \xi_i\| < \tfrac{\varepsilon}{2} \text{ for } 0 \le i \le 2m.$$

Now let $\sigma' = \operatorname{Ad} U \circ \sigma$ and $\eta' = U\eta$. Clearly (σ', η') represents ψ. We have

$$\|\sigma'(a_i)\xi_i - \pi(a_i)\xi_i\| \le \|a_i\|\|\xi_i - U\eta_i\| + \|U\eta_{i+m} - \xi_{i+m}\| < \tfrac{\varepsilon}{2} + \tfrac{\varepsilon}{2} = \varepsilon$$

for $1 \le i \le m$. Also

$$\sigma'(b_k)\zeta_k = (\pi_0 \otimes 1_\kappa)(b_k)\zeta_k = \pi(b_k)\zeta_k \quad \text{for} \quad 1 \le k \le n.$$

Thus σ' lies in U. In particular, $\|\eta' - \xi\| < \varepsilon$, so $\eta' \in V$. That is, $(\sigma', \eta') \in U \times V$ as required. ■

Proof of Theorem 16.10.4 continued. Suppose that $f \in \mathrm{B}(K)$ and $\hat{f}$ is continuous on $(\operatorname{Rep}(\mathrm{C}(K)), \text{point-SOT})$. Let $(\varphi_\lambda)_\Lambda$ be a net in $(L_1, \tau_{\mathrm{C}(K)})$ converging to φ. We will construct a cofinal subnet such that $\big(f(\varphi_{\lambda'})\big)_{\Lambda'}$ converges to $f(\varphi)$. The same argument then shows that every cofinal subnet has a cofinal subnet which converges to $f(\varphi)$, which implies that the original net converges to $f(\varphi)$; whence f is continuous on $(L_1, \tau_{\mathrm{C}(K)})$.

As in the previous lemma, we define $\pi = \pi_\varphi \oplus (\pi_0 \otimes 1_\kappa)$ in $\operatorname{Rep}(\mathfrak{A}, H_\kappa)$ and $\xi = \xi_\varphi \oplus 0 \in H_\kappa$; so that (π, ξ) represents φ. Let Λ' be the net consisting of finite subsets $F = \{(a_i, \xi_i) : a_i \in b_1(\mathfrak{A}),\ \xi_i \in b_1(H_\kappa) : 1 \le i \le n\}$ ordered by inclusion. To each F, we associate the open set $U_F \times V_F$ in $\operatorname{Rep}(\mathfrak{A}, H_\kappa) \times H_\kappa$ given by

$$U_F = \big\{\sigma \in \operatorname{Rep}(\mathfrak{A}, H_\kappa) : \|\sigma(a_i)\xi_i - \pi(a_i)\xi_i\| < \varepsilon \text{ for } 1 \le i \le n\big\}$$

and $V_F = b_{1/n}(\xi)$. By Lemma 16.10.6, there is a neighbourhood W_F of φ in $(L_1, \tau_{\mathrm{C}(K)})$ so that every $\psi \in W_F$ has a representative in $U_F \times V_F$. We construct our subnet ψ_F recursively, defining it for $|F| = n$ at the nth stage. Assuming that $\psi_G = \varphi_{\lambda_G}$ has been defined for all $G \subsetneq F$, select $\lambda > \lambda_G$ for all of these G with $\varphi_\lambda \in W$; and set $\lambda_F = \lambda$. Let $(\pi_F, \xi_F) \in U_F \times V_F$ represent ψ_F. By construction, (π_F, ξ_F) converges in $(\operatorname{Rep}(\mathrm{C}(K)), \text{point-SOT})$ to (π, ξ). Since $\hat{f}$ is continuous,

$$\lim_F \psi_F(f) = \lim_F \langle \pi_F(f)\xi_F, \xi_F\rangle = \langle \pi(f)\xi, \xi\rangle = \varphi(f).$$

Therefore (2b) implies (4), completing the proof. ■

16.11. The Takesaki–Bichteler Theorem

Much of the material in sections 16.8 and 16.10 was developed with a different purpose in mind. This section will explain that development. The goal is to provide a Gelfand style representation of an arbitrary (noncommutative) C*-algebra as a space of nc functions.

This is somewhat different than what we have done, since $\mathrm{C}(K)$ is always the maximal C*-algebra of some operator system, and that is not the case for an arbitrary C*-algebra.

Let $\mathfrak{A}$ be a C*-algebra. We fix an infinite cardinal κ sufficiently large that every cyclic representation lives on a Hilbert space of dimension no greater than κ. Let $\mathrm{Rep}(\mathfrak{A}, H_\kappa)$ denote the space of possibly degenerate representations of $\mathfrak{A}$ into H_κ. If $\pi \in \mathrm{Rep}(\mathfrak{A}, H_\kappa)$, let P_π denote the projection onto the range $\overline{\pi(\mathfrak{A})H_\kappa}$. If $\{\pi_\lambda\}_\Lambda$ is a family of representations such that $\{P_{\pi_\lambda}\}_\Lambda$ are pairwise orthogonal, then $\sum_\Lambda^\oplus \pi_\lambda$ denotes the direct sum, considered as a representation into H_κ with range projection SOT–$\sum_\Lambda P_{\pi_\lambda}$.

Two representations π and ρ are *equivalent* ($\pi \sim \rho$) if there is a partial isometry $U \in \mathcal{B}(H_\kappa)$ with $U^*U \geq P_\pi$, $UU^* \geq P_\rho$ and $\rho = U\pi U^*$. If $U^*U \geq P_\pi$, let $\pi^U := U\pi U^*$. Note that if $|\Lambda| \leq \kappa$, then for any set $\{\pi_\lambda\}_\Lambda$ of representations, there are equivalent representations $\rho_\lambda \sim \pi_\lambda$ such that the range projections are pairwise orthogonal. Hence $\sum_\Lambda^\oplus \rho_\lambda$ lives in $\mathrm{Rep}(\mathfrak{A}, H_\kappa)$ and is equivalent to the representation $\sum_\Lambda^\oplus \pi_\lambda \in \mathcal{B}(H_\kappa^{(\Lambda)})$. Alternatively, one can use a unitary equivalence between H_κ and $H_\kappa^{(\Lambda)}$ to make sense of these direct sums.

16.11.1. Definition. An *operator field* is a function $F : \mathrm{Rep}(\mathfrak{A}, H_\kappa) \to \mathcal{B}(H_\kappa)$ such that

(1) $\|F\| := \sup\{\|F(\pi)\| : \pi \in \mathrm{Rep}(\mathfrak{A}, H_\kappa)\} < \infty$.

(2) $F(\pi) = P_\pi F(\pi) P_\pi$.

(3) $F\big(\sum_\Lambda^\oplus \pi_\lambda\big) = \sum_\Lambda^\oplus F(\pi_\lambda)$ when $\{P_{\pi_\lambda}\}_\Lambda$ are pairwise orthogonal.

(4) $F(\pi^U) = U^*F(\pi)U$ when U is a partial isometry with $U^*U \geq P_\pi$.

If $a \in \mathfrak{A}$, we let $\hat{a}$ be the operator field given by $\hat{a}(\pi) = \pi(a)$. Let $\mathcal{B}$ denote the set of all operator fields.

While this is Takesaki's formulation, we can make it look a bit more like the previous situation by setting $\mathrm{Rep}(\mathfrak{A}) = \coprod_{n \leq \kappa} \mathrm{Rep}(\mathfrak{A}, H_n)$. A function $F : \mathrm{Rep}(\mathfrak{A}) \to \mathcal{M}$ is an operator field if it is graded: $F(\pi) \in \mathcal{B}(H_n)$ for $\pi \in \mathrm{Rep}(\mathfrak{A}, H_n)$, in addition to the properties above, where now the partial isometry can lie in $\mathcal{B}(H_n, H_m)$. The set $\mathrm{Rep}(\mathfrak{A})$ is not an nc convex set, but it is graded, closed under direct sums and equivalence of representations. It is not closed under compressions, but the definitions of operator field and nc function do not include anything about compressions either. So compressions are not required in the proofs except for statements about $\mathrm{A}(K)$ and its second dual, which doesn't arise here.

The proof of Proposition 16.8.3 applies equally well to $\mathcal{B}$, showing that the set of all operator fields is a C*-algebra which can be faithfully represented as a von Neumann algebra. It is easy to see that $\widehat{\mathfrak{A}} := \{\hat{a} : a \in \mathfrak{A}\}$ is a subalgebra of $\mathcal{B}$ $*$-isomorphic to $\mathfrak{A}$. The proof of Theorem 16.8.10 also applies to $\mathfrak{A}$ and $\mathcal{B}$, showing that $\mathcal{B}$ is the universal von Neumann algebra of $\mathfrak{A}$. Finally the proof of Theorem 16.10.4 applies to $\mathfrak{A}$ in place of $\mathrm{C}(K)$ and $\mathcal{B}$ in place of $\mathrm{B}(K)$ except for item (3). The important point is that $\mathfrak{A}$ is identified as the subalgebra of $\mathcal{B}$ which is continuous on $\mathrm{Rep}(\mathfrak{A})$ in any of the topologies point-WOT, point-SOT or point-ultrastrong-$*$. We record the following theorem without more discussion. The interested reader should trace through the details.

16.11.2. TAKESAKI–BICHTELER THEOREM.
The set $\mathcal{B}$ of all operator fields is a C-algebra which has a natural faithful representation as a von Neumann algebra. The C*-algebra $\mathfrak{A}$ is isomorphic to the subalgebra $\widehat{\mathfrak{A}}$ of $\mathcal{B}$ consisting of the operator fields $\{\hat{a} : a \in \mathfrak{A}\}$. $\mathcal{B}$ is the universal von Neumann algebra of $\mathfrak{A}$, isomorphic to $\mathfrak{A}^{\dagger\dagger}$. Moreover $\widehat{\mathfrak{A}}$ is the subalgebra of $\mathcal{B}$ consisting of operator fields which are point-*WOT *continuous on* $\mathrm{Rep}(\mathfrak{A})$.

16.12. NC states on $\mathrm{C}(K)$

Recall that a state on $C(C)$ for a compact convex set C is a Borel probability measure. Each state on $A(C)$ has at least one representing measure, and the Choquet–Bishop–de Leeuw theorem shows that it has a representing measure supported on the extreme points of C (in a technical sense for the non-metrizable case).

In our noncommutative setting, we have replaced states with nc states, or u.c.p. maps. In this section, we pursue this analogy. One important consequence is a partial converse to the nc Krein–Milman theorem akin to Milman's theorem 3.5.8.

16.12.1. DEFINITION. Let K be a compact nc convex set, and let L be the nc state space of $\mathrm{C}(K)$. If $\mu \in L_n$, then $x = \mu|_{\mathrm{A}(K)} \in K_n$ is the *barycentre* of μ, and μ is a *representing map for x*.

16.12.2. THEOREM. *Let K be a compact nc convex set. A point $x \in K$ has a unique representing map if and only if it is maximal.*

If x is maximal, then δ_x factors through $C^*_{\min}(\mathrm{A}(K))$. Conversely, if δ_x is the only representing map for x which factors through $C^*_{\min}(\mathrm{A}(K))$, then x is maximal.

PROOF. Suppose $x \in K_m$ has a unique representing map. Let $y \in K_n$ be a maximal dilation of x. Then there is an isometry $\alpha \in M_{n,m}$ such that $x = \alpha^* y \alpha$. Define a u.c.p. map $\mu : \mathrm{C}(K) \to M_m$ by $\mu = \alpha^* \delta_y \alpha$. Then μ has barycenter x. Since x has a unique representing map, one has $\mu = \delta_x$. Therefore, $\delta_y \simeq \delta_x \oplus \delta_z$ for some $z \in K_p$, where the decomposition is taken with respect to the range of α. In particular, $y \simeq x \oplus z$. Since summands of a maximal point in K are maximal, x is maximal.

Conversely, suppose that $x \in K_m$ is maximal. Let $\mu : \mathrm{C}(K) \to M_m$ be an unital completely positive map with barycenter x. Let $(y, \alpha) \in K_n \times M_{n,m}$ be a representation of μ. Then $x = \alpha^* y \alpha$, so y is a dilation of x. The fact that x is maximal implies that $y \simeq x \oplus z$ for some $z \in K_p$, where the decomposition is taken with respect to the range of α. Hence $\delta_y = \delta_x \oplus \delta_z$, so $\mu = \delta_x$.

Suppose $x \in K_n$ is maximal. Let $\iota : \mathrm{A}(K) \to C^*_{\text{min}}(\mathrm{A}(K))$ denote the canonical embedding and define $\varphi : \iota(\mathrm{A}(K)) \to M_n$ by $\varphi = x \circ \iota^{-1}$. By Arveson's extension theorem we can extend φ to a u.c.p. map $\psi : C^*_{\text{min}}(\mathrm{A}(K)) \to M_n$. Let $q : \mathrm{C}(K) \to C^*_{\text{min}}(\mathrm{A}(K))$ denote the canonical quotient map. Then $(\varphi \circ q)|_{\mathrm{A}(K)} = x$. Since x has a unique representing map, it follows that $\varphi \circ q = \delta_x$. In particular, $\ker \delta_x \supseteq \ker q$.

Conversely, suppose that the only representing map for x that factors through $C^*_{\text{min}}(\mathrm{A}(K))$ is δ_x. Let $y \in K_p$ be a maximal dilation of x and let $\alpha \in M_{p,n}$ be an isometry such that $x = \alpha^* y \alpha$. Define a unital completely positive map $\mu : \mathrm{C}(K) \to M_n$ by $\mu = \alpha^* \delta_y \alpha$. Then μ has barycenter x. From the previous paragraph, δ_y factors through $C^*_{\text{min}}(\mathrm{A}(K))$. Hence μ also factors through $C^*_{\text{min}}(\mathrm{A}(K))$. Therefore, by assumption $\mu = \delta_x$. Arguing as in the first paragraph shows x is maximal. ■

We need an elementary lemma about states on C*-algebras.

16.12.3. LEMMA. *Let $\mathfrak{A}$ be a C*-algebra and let $(\pi_\lambda)_\Lambda$ be a family of representations of $\mathfrak{A}$. Then every state φ with $\ker \varphi \supset \bigcap_\Lambda \ker \pi_\lambda$ is a weak-$*$ limit of finite convex combinations of states of the form $\varphi_{\lambda,\xi}(a) = \langle \pi_\lambda \xi, \xi \rangle$ where ξ is a unit vector in H_λ. If φ is a pure state, then φ is a weak-$*$ limit of $\Omega := \{\varphi_{\lambda,\xi} : \lambda \in \Lambda, \xi \in H_\lambda\}$.*

PROOF. Let $\pi = \bigoplus_\Lambda \pi_\lambda$. We may replace $\mathfrak{A}$ by $\pi(\mathfrak{A})$, and thereby assume that $\bigcap_\Lambda \ker \pi_\lambda = \{0\}$. Then if $a \in \mathfrak{A}_{sa}$, we have $a \geq 0$ if and only if $\pi_\lambda(a) \geq 0$ for $\lambda \in \Lambda$, and this holds if and only if $\varphi_{\lambda,\xi}(a) \geq 0$ for all unit vectors in H_λ, $\lambda \in \Lambda$. Thus if $\operatorname{conv}(\sigma(a)) = [r, s]$, then $\inf \varphi_{\lambda,\xi}(a) = r$ and $\sup \varphi_{\lambda,\xi}(a) = s$. Now $S(\mathfrak{A})$ lies in the real space of self-adjoint functionals on $\mathfrak{A}$, and every real weak-$*$ continuous linear functional on $S(\mathfrak{A})$ is given by evaluation at some $a \in \mathfrak{A}_{sa}$. Therefore $\varphi(a) \in [r, s]$ cannot be separated from $\operatorname{conv}(\Omega)$ by weak-$*$ continuous functionals. Thus by the Hahn–Banach theorem, $\varphi \in \overline{\operatorname{conv}(\Omega)}^{\text{weak-}*}$.

If φ is pure, then it is an extreme point of the state space of $\mathfrak{A}$ by Proposition 9.6.9. Therefore by Milman's theorem, $\varphi \in \overline{\Omega}^{\text{weak-}*}$. ■

16.12.4. COROLLARY. *Let L be the nc state space of a C*-algebra $\mathfrak{A}$, and let $(\pi_\lambda)_\Lambda$ be a family of representations of $\mathfrak{A}$. Let $y \in L_m$ be an irreducible representation such that $\ker y \supset \bigcap_\Lambda \ker \pi_\lambda$. Then y is a point-weak-$*$ limit of compressions of $\{\pi_\lambda : \lambda \in \Lambda\}$.*

PROOF. Let $n < \infty$ and let $\alpha \in M_{m,n}$ be an isometry. Set $z = \alpha^* y \alpha$. This is a u.c.p. map into M_n. Thus by Proposition 14.6.10, this is equivalent to the positivity of the functional s_z on $M_n(\mathfrak{A})$. This is a positive functional on $M_n(\mathfrak{A})$ which is obtained from a vector state for the irreducible representation $y \otimes id_n$; whence s_z is pure. By Lemma 16.12.3, s_z is a weak-$*$ limit of states $\{s_{\varphi_{\lambda,\xi}} : \lambda \in \Lambda,\ \xi \in H_\lambda\}$. This in turn implies that z is a point-weak-$*$ limit of n-dimensional compressions of $\{\pi_\lambda : \lambda \in \Lambda\}$. If m is finite, we are done. If m is infinite, as in Remark 16.4.7, y is a point-weak-$*$ limit of its finite compressions, and therefore is a point-weak-$*$ limit of compressions of $\{\pi_\lambda : \lambda \in \Lambda\}$. ■

We can now provide the analogue of Milman's theorem 3.5.8.

16.12.5. Theorem. *Let K be a compact nc convex set, and let $X \subset K$ be a closed set which is closed under compressions. If $\overline{\mathrm{ncconv}(X)} = K$, then $\partial K \subset X$.*

Proof. Let L be the nc state space of $\mathrm{C}(K)$, and set $Z = \overline{\mathrm{ncconv}\{\delta_x : x \in X\}}$. The barycentre map from Z into K is continuous and nc affine. The range is compact, and hence closed, and is nc convex. Since the range contains X, this map is surjective by hypothesis. Let $y \in (\partial K)_m$; so y is irreducible. There is some $\mu \in Z$ with barycentre y. By Theorem 16.12.2, y has a unique representing map δ_y, whence $\delta_y \in Z$. Since $\delta_y \in Z$, $\ker y \supset \bigcap_{x\in X} \ker \delta_x$.

By Corollary 16.12.4, y is a point-weak-$*$ limit of compressions of X. Since X is closed under compressions, y is in the closure of X. ■

16.12.6. Remark. It is necessary to include compressions. Consider Example 16.4.4(1), where $K_1 = [c, d] \subset \mathbb{R}$. Then $C^*_{\min}(\mathrm{A}(K_1)) = C(K_1)$. As this is abelian, $\partial K = \{c, d\} \subset K_1$. However $X = \{c1_2, d1_2\}$ is closed and $\overline{\mathrm{ncconv}(X)} = K$.

It is also necessary to allow limits as in Definition 16.4.6. This is because we could take $X = \bigcup_{n<\infty} K_n$. For certain examples such as the Cuntz system, Example 16.8.2(2), the nc extreme points all lie in $K_{\aleph_0}$.

Stinespring's theorem provides an important structure theorem for u.c.p. maps on a C*-algebra. Suppose that $\mu \in L_n$ is a u.c.p. map from $\mathrm{C}(K)$ into $\mathcal{B}(H_n)$. Then there is a $*$-representation $\pi : \mathrm{C}(K) \to \mathcal{B}(H_m)$ and an isometry $\alpha : H_n \to H_m$ such that $\mu = \alpha^* \pi \alpha$. By Proposition 16.8.9, there is an $x \in K_m$ so that $\pi = \delta_x$. Thus $\mu = \alpha^* \delta_x \alpha$.

16.12.7. Definition. If $\mu \in L_n$ is an nc state of $\mathrm{C}(K)$, and $x \in K_m$ and $\alpha : H_n \to H_m$ is an isometry such that $\mu = \alpha^* \delta_x \alpha$, say that (x, α) is a *representation* of μ. It is *minimal* if $\{f(x)\alpha H_n : f \in \mathrm{C}(K)\}$ is a dense subspace of H_m.

Stinespring's theorem shows that the minimal representation of μ is unique up to unitary equivalence. This means that if (x, α) and (y, β) are two minimal representations of μ, then $y \in K_m$ and there is a unitary $\gamma \in \mathcal{B}(H_m)$ so that $y = \gamma x \gamma^*$ and $\beta = \gamma\alpha$. From this, we can deduce that there is a unique normal extension to $\mathrm{B}(K)$ given by $\widetilde{\mu}(f) = \alpha^* f(x)\alpha$ which is independent of the choice of representation of μ. (See Exercise **16.13**.)

16.12.8. THEOREM. *Let K be a compact nc convex set, and let L be the nc state space of $\mathrm{C}(K)$. A u.c.p. map $\mu \in L$ factors through $C^*_{\min}(\mathrm{A}(K))$ if and only if μ lies in the closed nc convex hull of $\{\delta_x : x \in \partial K\}$.*

PROOF. Let $X = \{\delta_x : x \in \partial K\}$; and let $Z = \overline{\text{ncconv}(X)}$. Also let L_e be the set of u.c.p. maps in L which factor through $C^*_{\min}(\mathrm{A}(K))$. Then L_e is nc affinely homeomorphic to the nc state space of $C^*_{\min}(\mathrm{A}(K))$. Since the representation δ_x, for $x \in \partial K$, factors through $C^*_{\min}(\mathrm{A}(K))$, we have $Z \subset L_e$.

The representation $\sigma := \bigoplus_{x\in\partial K}\delta_x$ factors through $C^*_{\min}(\mathrm{A}(K))$. So we can view δ_x, for $x \in \partial K$, as a representation of $C^*_{\min}(\mathrm{A}(K))$. Corollary 16.7.6 shows that the restriction $\sigma|_{\mathrm{A}(K)}$ is a unital complete order embedding. Hence by the universal property of $C^*_{\min}(\mathrm{A}(K))$, σ is faithful on $C^*_{\min}(\mathrm{A}(K))$.

Let $y \in K$ be a point such that δ_y is irreducible and factors through $C^*_{\min}(\mathrm{A}(K))$. Then $\ker\delta_y \supseteq \bigcap_{x\in\partial K}\ker\delta_x = \ker\sigma$. By Corollary 16.12.4, y belongs to Z. By Example 16.7.10, ∂L_e is the set of all irreducible representations of $C^*_{\min}(\mathrm{A}(K))$. In particular, δ_y is an extreme point of L_e. It follows from Theorem 16.12.5 that $\delta_y \in \overline{X}$. Now we see that $\overline{X}$ contains ∂L_e, so by the nc Krein–Milman theorem, $Z = L_e$. ∎

Noncommutative Choquet Theory

16.13. Convex NC Functions

In this section we will introduce a notion of convexity for nc functions. It will play a role analogous to convex functions in the classical case.

We will need to consider matrices of bounded nc functions. For a compact nc convex set K and $f = (f_{ij}) \in M_p(\mathrm{B}(K))$, consider f as an nc function $f : K \to M_p(\mathcal{M})$ given by $f(x) = (f_{ij}(x))$ for $x \in K$. Note that f is graded, respects direct sums and is unitarily equivariant in the sense that if $x \in K_n$ and $\gamma \in M_n$ is unitary, then

$$f(\gamma^* x\gamma) = \big[f_{ij}(\gamma^* x\gamma)\big] = \big[\gamma^* f_{ij}(x)\gamma\big] = (1_p \otimes \gamma^*)f(x)(1_p \otimes \gamma).$$

We will refer to f as a nc function on K. We will say that f is *self-adjoint* if $f(x) \in M_p(M_k)_{sa}$ for all k and all $x \in K_k$.

16.13.1. DEFINITION. Let K be a compact nc convex set and let $f \in M_p(\mathrm{B}(K))$ be a self-adjoint bounded nc function. The *epigraph* of f, epi(f), is defined by

$$\mathrm{epi}_n(f) = \{(x, \alpha) \in K_n \times M_p(M_n) : x \in K_n \text{ and } \alpha \geq f(x)\}$$

as a subset of $\coprod_n K_n \times M_p(M_n)$. Then f is a *convex nc function* if epi(f) is an nc convex set, and f is *lower semicontinuous* if epi(f) is closed. Say that f is a *concave nc function* if $-f$ is convex.

Here are a few equivalent forms of convexity for nc functions. The equivalence of the first three is elementary. The fourth condition shows that being a (classical) convex function on each level is sufficient to imply convexity for nc functions.

16.13.2. PROPOSITION. *Let K be a compact nc convex set and let f be a self-adjoint bounded nc function in $M_p(\mathrm{B}(K))$. The following are equivalent:*

(1) *f is convex.*

(2) $$f\big(\textstyle\sum_i \alpha_i^* x_i \alpha_i\big) \leq \sum_i \alpha_i^* f(x_i) \alpha_i$$
for $x_i \in K_{n_i}$ and $\alpha_i \in M_{n_i,n}$ such that $\sum_i \alpha_i^* \alpha_i = 1_n$.

(3) $$f(\alpha^* x \alpha) \leq (1_p \otimes \alpha^*) f(x) (1_p \otimes \alpha)$$
for every m, n, every $x \in K_m$ and every isometry $\alpha \in M_{m,n}$.

(4) *$f|_{K_n}$ is a scalar convex function for every n; i.e.*

$$f(tx + (1-t)y) \leq t f(x) + (1-t) f(y)$$

for all n, all $x, y \in K_n$ and all $t \in [0,1]$.

PROOF. Suppose f is convex. Then if $x_i \in K_{n_i}$ and $\alpha_i \in M_{n_i,n}$ such that $\sum_i \alpha_i^* \alpha_i = 1_n$, then

$$\Big(\sum_i \alpha_i^* x_i \alpha_i, \sum_i \alpha_i^* f(x_i) \alpha_i\Big) \in \mathrm{epi}(f).$$

This means that $f\big(\sum_i \alpha_i^* x_i \alpha_i\big) \leq \sum_i \alpha_i^* f(x_i) \alpha_i$. So (1) implies (2), which implies (3) as a special case. But (3) says that epi(f) is closed under compressions. Since f is an nc function, epi(f) is also closed under direct sums, and thus is nc convex. So (3) implies (1).

Clearly (4) is a (very) special case of (2). Finally, suppose that (4) holds. We will verify (3). Suppose that $\alpha \in M_{m,n}$ is an isometry and $x \in K_m$. Define a unitary $\beta \in M_m$ by $\beta = \alpha\alpha^* - (1 - \alpha\alpha^*)$. Decompose $H_m = \mathrm{Ran}(\alpha\alpha^*) \oplus \mathrm{Ran}(1_m - \alpha\alpha^*)$ and identify $\mathrm{Ran}(\alpha\alpha^*)$ and $\mathrm{Ran}(1_m - \alpha\alpha^*)$ with H_n and H_{m-n} respectively. Then we can write

$$x = \begin{bmatrix} y & * \\ * & z \end{bmatrix}, \quad \beta = \begin{bmatrix} 1_n & 0 \\ 0 & -1_{m-n} \end{bmatrix}, \quad \text{and} \quad \alpha = \begin{bmatrix} \gamma \\ 0 \end{bmatrix}$$

for some $y \in K_n$ and $z \in K_{m-n}$, and $\gamma \in \mathcal{B}(H_n)$ is a unitary. Observe that by unitary equivariance and (4),

$$\begin{aligned} f(\alpha^* x \alpha) &= f(\gamma^* y \gamma) = (1_p \otimes \gamma)^* f(y)(1_p \otimes \gamma) \\ &= (1_p \otimes \alpha)^* \big(f(y) \oplus f(z)\big)(1_p \otimes \alpha) \\ &= (1_p \otimes \alpha)^* f(y \oplus z)(1_p \otimes \alpha) \\ &= (1_p \otimes \alpha)^* f\big(\tfrac{1}{2}(x + \beta^* x \beta)\big)(1_p \otimes \alpha) \\ &\leq (1_p \otimes \alpha)^* \tfrac{1}{2}\big(f(x) + f(\beta^* x \beta)\big)(1_p \otimes \alpha) \\ &= (1_p \otimes \alpha)^* \tfrac{1}{2}\big(f(x) + (1_p \otimes \beta^*) f(x)(1_p \otimes \beta)\big)(1_p \otimes \alpha) \\ &= (1_p \otimes \alpha)^* f(x)(1_p \otimes \alpha). \end{aligned}$$

Therefore (3) holds. ■

16.13.3. EXAMPLE. Let K be a compact nc convex set. It is easy to check that every continuous self-adjoint nc affine function $a \in M_p(\mathrm{A}(K))$ is convex and lower semicontinuous.

16.13.4. EXAMPLE. Fix a compact interval $I \subseteq \mathbb{R}$. A continuous real-valued function $f \in \mathrm{C}(I)$ is *operator convex* if

$$f(tA + (1-t)B) \leq tf(A) + (1-t)f(B)$$

for all n, all self-adjoint $A, B \in (M_n)_{sa}$ with spectrum in I and all $t \in [0, 1]$. Define an nc convex set K as in Examples 16.4.4(1) by setting

$$K_n(I) = \{A \in (M_n)_{sa} : \sigma(A) \subseteq I\}.$$

A continuous real-valued function $f \in \mathrm{C}(I)$ determines a continuous self-adjoint nc function in $\mathrm{C}(K)$ by the continuous functional calculus, while on the other hand, a continuous self-adjoint nc function in $\mathrm{C}(K)$ restricts to a continuous real-valued function in $\mathrm{C}(I)$. It follows immediately from Proposition 16.13.2 that a self-adjoint nc function $f \in \mathrm{C}(K)$ is convex if and only if it restricts to an operator convex function in $\mathrm{C}(I)$. This fact, that scalar operator convexity on each $K_n(I)$ implies nc convexity on K is the Hansen–Pedersen Jensen inequality [139].

Consider the function $f(x) = (1-x)^{-1}$ on $(-1, 1)$. By Proposition 9.1.7(4), f is operator monotone. We will show that it is also operator convex. Suppose that $A, B \in \mathcal{B}(H)_{sa}$ with $\|A\| < 1$ and $\|B\| < 1$ and $\xi \in H$. We need to show that $g(t) = \langle f((1-t)A + tB)\xi, \xi\rangle$ is a scalar convex function. For this, it is sufficient to show that $\frac{d^2}{dt^2} g(t) \geq 0$. After a routine change of variables, it suffices to show that if $\|X\| < 1$ and $X, Y \in \mathcal{B}(H)_{sa}$, then $\frac{d^2}{dt^2}(I - X - tY)^{-1}|_{t=0} \geq 0$. Now $\frac{d}{dt}(I - X - tY)^{-1} = (I - X - tY)^{-1} Y (I - X - tY)^{-1}$ follows from differentiating the identity $(I - X - tY)^{-1}(I - X - tY) = I$. Therefore

$$\begin{aligned}\frac{d^2}{dt^2}(I-X-tY)^{-1}|_{t=0} &= 2(I-X-tY)^{-1}Y(I-X-tY)^{-1}Y(I-X-tY)^{-1}|_{t=0} \\ &= 2(I-X)^{-1}Y(I-X)^{-1}Y(I-X)^{-1} \geq 0.\end{aligned}$$

So f is operator convex. Since $1 + x$ is nc affine,

$$h(x) = (1-x)^{-1} - 1 - x = x^2(1-x)^{-1}$$

is also operator convex on $(-1, 1)$. After a change of variables, one sees that the functions $h_t(x) = x^2(1 - tx)^{-1} = \sum_{n\geq 0} t^n x^{n+2}$ are operator convex on $(-1, 1)$ for $|t| \leq 1$.

16.13.5. Example. Suppose that $a \in M_n(\mathrm{A}(K))_{sa}$ is a self-adjoint nc affine function with $\|a\| < 1$. Then $\sigma(a) \subset (-1, 1)$ and we can form $f(a)$ for any operator convex nc function f on $(-1, 1)$. The functions $h_t(x) = x^2(1 - tx)^{-1}$ are nc functions which are operator convex by Example 16.13.4. It follows from Proposition 16.13.2 that $h_t(a)$ are convex nc functions on K.

For a compact nc convex set K, we do not know if convex nc functions in $\mathrm{C}(K)$ always span a dense subset of $\mathrm{C}(K)$. However, the next result will be sufficient our purposes.

16.13.6. Proposition. *Let K be a compact nc convex set. Suppose that $\mu, \nu : \mathrm{C}(K) \to \mathcal{M}$ are u.c.p. maps such that $\mu(f) = \nu(f)$ for every $n < \infty$ and every convex nc function $f \in M_n(\mathrm{C}(K))$. Then $\mu = \nu$.*

Proof. For $a = a^* \in M_n(\mathrm{A}(K))$ with $\|a\| < 1$, Example 16.13.5 shows that the nc function $h_t(a) \in \mathrm{C}(K)$ is convex for all $|t| \leq 1$. Hence

$$0 = (\mu - \nu)(h_t(a)) = \sum_{n\geq 0}(\mu - \nu)(a^{n+2})t^n \quad \text{for all} \quad t \in [-1, 1].$$

It follows that the analytic function $k(z) = \sum_{n\geq 0}(\mu - \nu)(a^{n+2})z^n$ is identically zero. Therefore $\mu(a^n) = \nu(a^n)$ for all $n \geq 2$. This also holds for $n = 0$ and $n = 1$ by hypothesis.

Suppose that $f = a_1 \cdots a_n$ for self-adjoint $a_1, \ldots, a_n \in \mathrm{A}(K)$. Define a self-adjoint element $b = (b_{ij}) \in M_{n+1}(\mathrm{A}(K))$ by $b_{i,i+1} = b_{i+1,i} = a_i$ for $1 \leq i \leq n$ and setting all other entries to zero. It is easy to check that the $(1, n+1)$ entry of b^n is f. From above, $\mu(b^n) = \nu(b^n)$. Therefore $\mu(f) = \nu(f)$. It follows that μ and ν agree on the C*-algebra generated by $\mathrm{A}(K)$, namely $\mathrm{C}(K)$, and hence $\mu = \nu$. ■

A major difficulty in the noncommutative setting is the fact that the self-adjoint elements of $\mathcal{B}(H)$ do not form a lattice. We can deal with this difficulty by working with multivalued functions.

16.13.7. DEFINITION. Let K be an nc convex set and let $F : K \to M_n(\mathcal{M})_{sa}$ for $n < \infty$ be a multivalued (or set valued) self-adjoint function. We say that F is a *multivalued nc function* if it is non-degenerate, graded, preserves direct sums, is unitarily equivariant and upwards directed, meaning:

(1) $F(x) \neq \emptyset$ for $x \in K$.

(2) $F(K_m) \subseteq M_n(M_m)$ for all m.

(3) $F(x \oplus y) = F(x) \oplus F(y)$ for $x, y \in K$.

(4) $F(\beta x \beta^*) = (1_n \otimes \beta)F(x)(1_n \otimes \beta^*)$ for $x \in K_m$ and unitary $\beta \in M_m$.

(5) $F(x) = F(x) + M_n(M_m)_+$ for every m and $x \in K_m$ (*upward directed*).

We say that F is *bounded* if there is a constant $C > 0$ such that for every $\beta \in F(x)$ there is $\alpha \in F(x)$ with $\alpha \leq \beta$ such that $\|\alpha\| \leq C$. If F is bounded, then we let $\|F\|$ denote the infimum of all C as above.

The *graph* of F is the subset $\text{graph}(F) \subseteq \coprod_m K_m \times M_n(M_m)$ defined by

$$\text{graph}_m(F) = \{(x, \alpha) \in K_m \times M_n(M_m) : x \in K_m \text{ and } \alpha \in F(x)\}.$$

We say that F is *convex* if $\text{graph}(F)$ is an nc convex set, and that F is *lower semi-continuous* if graph F is closed. If $G : K \to M_n(\mathcal{M})_{sa}$ is another multivalued nc function, then $F \leq G$ if $\text{graph}(F) \supseteq \text{graph}(G)$.

16.13.8. EXAMPLE. Let K be a compact nc convex set and let $f \in M_n(\mathrm{B}(K))$ be self-adjoint. There is a bounded multivalued nc function $F : K \to M_n(\mathcal{M})_{sa}$ naturally associated to f defined by $F(x) = [f(x), +\infty)$ for $x \in K$. Observe that $\text{graph}(F) = \text{epi}(f)$.

If K is a compact nc convex set and $\mu : \mathrm{C}(K) \to M_k$ is a u.c.p. map, then μ can be extended to a normal u.c.p. map on $\mathrm{B}(K)$ by Exercise **16.13**. We will extend μ further and make sense of the expression $\mu(F)$ when $F : K \to M_n(\mathcal{M})_{sa}$ is a bounded multivalued nc function.

16.13.9. DEFINITION. Let K be a compact nc convex set and let $\mu : \mathrm{C}(K) \to M_k$ be a unital completely positive map. Let $(x, \alpha) \in K_m \times M_{m,k}$ be a minimal representation for μ. For a bounded multivalued nc function $F : K \to M_n(\mathcal{M})_{sa}$, set

$$\mu(F) = (1_n \otimes \alpha^*)F(x)(1_n \otimes \alpha).$$

The fact that this extension of μ is well defined follows from unitary equivariance of multivalued nc functions and the uniqueness of the minimal Stinespring dilation.

Let $f \in M_n(\mathrm{B}(K))$ be self-adjoint and let $F : K \to M_n(\mathcal{M})_{sa}$ denote the corresponding bounded multivalued nc function as defined in Example 16.13.8. If $G : K \to M_n(\mathcal{M})_{sa}$ is a multivalued nc function, then we will write $f = G$, $f \leq G$ and $f \geq G$ if $F = G$, $F \leq G$ or $F \geq G$ respectively.

For cardinals m and n, we will view $f \in M_m(M_n(\mathrm{B}(K)))$ as a function from K to $M_m(M_n(\mathcal{M}))$ in the obvious way. For another function $g \in M_m(M_n(\mathrm{B}(K)))$, we will write $f \le g$ if f and g are self-adjoint and $f(x) \le g(x)$ for all $x \in K$. For a multivalued nc function $F : K \to M_n(\mathcal{M})_{sa}$, define a multivalued function $1_m \otimes F : K \to M_m(M_n(\mathcal{M}))$ by

$$(1_m \otimes F)(x) = \{1_m \otimes \alpha : \alpha \in F(x)\}, \quad x \in K.$$

Note that $1_m \otimes F$ is not an nc function since $1_m \otimes \alpha \le \beta$ does not imply that $\beta = 1_m \otimes \beta'$.

Now we introduce the notion of the convex envelope of an nc function that will play an important role in the noncommutative setting analogous to its role in classical Choquet theory. The non-trivial fact that this definition of the convex envelope is equivalent to an appropriate approximation from below by continuous nc affine functions will be the main result in this section.

16.13.10. Definition. Let K be a compact nc convex set. The *convex envelope* of a bounded multivalued function $F : K \to M_n(\mathcal{M})_{sa}$ is the multivalued nc function $\overline{F} : K \to M_n(\mathcal{M})_{sa}$ given by

$$\operatorname{graph} \overline{F} = \overline{\operatorname{ncconv}}(\operatorname{graph}(F));$$

i.e. the graph of $\overline{F}$ is the closed nc convex hull of the graph of F.

Here are a few easy consequences of this definition.

16.13.11. Proposition. *Let K be an nc convex set and let $F : K \to M_n(\mathcal{M})_{sa}$ be a bounded multivalued nc function with convex envelope $\overline{F}$. Then*

(1) *$\overline{F}$ is a lower semicontinuous convex multivalued nc function.*

(2) *$\overline{F} \le F$.*

(3) *If F is nc convex and lower semicontinuous, then $\overline{F} = F$.*

(4) *If F is bounded by C, then so is $\overline{F}$.*

(5) *If G is a lower semicontinuous convex nc function such that $G \le F$, then $G \le \overline{F}$.*

Proof. For (1), the graph of $\overline{F}$ is defined to be nc convex and closed. For (2), $\operatorname{graph} \overline{F} \supset \operatorname{graph}(F)$, so $\overline{F} \le F$. If $\operatorname{graph}(F)$ is already closed and nc convex, then $\overline{F} = F$; so (3) holds. For (5), it follows from the definition that $\overline{F}$ is the largest lower semicontinuous convex nc function smaller than F.

Suppose that F is bounded by C. Then $-C1_n \otimes 1_k \le F(x)$ for all $x \in K_k$, and this also holds for $\overline{F}$. Suppose that (x, β) belongs to the (algebraic) nc convex hull of $\operatorname{graph}(F)$; say $(x, \beta) = \sum \alpha_i^*(x_i, \beta_i)\alpha_i$ where $\sum \alpha_i^* \alpha_i = 1_k$. Then since F is bounded by C, there exist $\gamma_i \in F(x_i)$ with $\gamma_i \le \beta_i$ and $\|\gamma_i\| \le C$. Set

$\gamma = \sum \alpha_i^* \gamma_i \alpha_i \le \beta$. Then $(x, \gamma) \in \overline{F}(x)$ and $\|\gamma\| \le C$. In general, if (x, β) is a limit of a net of such points $\{(x_\lambda, \beta_\lambda)\}_\Lambda$. Choose points $(x_\lambda, \gamma_\lambda) \in \overline{F}$ with $\gamma_\lambda \le \beta_\lambda$ and $\|\gamma_\lambda\| \le C$. Extract a convergent cofinal subnet of $\{(x_\lambda, \gamma_\lambda)\}_\Lambda$ with limit (x, γ). Then $(x, \gamma) \in \overline{F}$, $\gamma \le \beta$ and $\|\gamma\| \le C$. Hence $\overline{F}$ is bounded by C. Therefore (4) holds. ■

We need the following easy lemma.

16.13.12. Lemma. *Set $\Delta(\varepsilon, B) = B + \varepsilon^{-1} B^2$ for $\varepsilon, B \in \mathbb{R}_+$. Whenever $A = \left[\begin{smallmatrix} a & b \\ b^* & c \end{smallmatrix}\right] \in \mathcal{B}(H \oplus H)_{sa}$ and $a_0 \in \mathcal{B}(H)_+$ satisfy $a \ge a_0 + \varepsilon$ and $\|A\| \le B$, then*

$$\begin{bmatrix} a_0 & 0 \\ 0 & -\Delta(\varepsilon, B) I \end{bmatrix} \le \begin{bmatrix} a & b \\ b^* & c \end{bmatrix}.$$

Proof. Compute

$$\begin{bmatrix} a & b \\ b^* & c \end{bmatrix} - \begin{bmatrix} a_0 & 0 \\ 0 & -(B + \varepsilon^{-1} B^2) I \end{bmatrix} \ge \begin{bmatrix} \varepsilon & b \\ b^* & \varepsilon^{-1} B^2 I \end{bmatrix}$$
$$= \begin{bmatrix} \varepsilon^{1/2} & 0 \\ \varepsilon^{-1/2} b^* & 0 \end{bmatrix} \begin{bmatrix} \varepsilon^{1/2} & \varepsilon^{-1/2} b \\ 0 & 0 \end{bmatrix} + \begin{bmatrix} 0 & 0 \\ 0 & \varepsilon^{-1}(B^2 I - b^* b) \end{bmatrix} \ge 0. \quad ■$$

The next result is a noncommutative analogue of the classical fact, Theorem 16.1.2, that the convex envelope of a function is obtained as the supremum of the continuous affine functions dominated by the function.

16.13.13. Theorem. *Let K be a compact nc convex set for $n < \infty$; let $F : K \to M_n(\mathcal{M})_{sa}$ be a bounded multivalued nc function. Then for $x \in K_p$,*

$$\overline{F}(x) = \bigcap_{m \in \mathbb{N}} \bigcap_{a \le 1_m \otimes F} \{\alpha \in (M_n(M_p))_{sa} : 1_m \otimes \alpha \ge a(x)\},$$

where the intersection is taken over all $m < \infty$ and all self-adjoint nc affine functions $a \in M_m(M_n(\mathrm{A}(K)))_{sa}$ satisfying $a \le 1_m \otimes F$.

Proof. Define

$$\widetilde{F}(x) := \bigcap_{m \in \mathbb{N}} \bigcap_{a \le 1_m \otimes F} \{\alpha \in (M_n(M_p))_{sa} : 1_m \otimes \alpha \ge a(x)\} \quad \text{for} \quad x \in K_p.$$

It is easy to see that

$$\mathrm{graph}(\widetilde{F}) = \bigcap_{m \in \mathbb{N}} \bigcap_{a \le 1_m \otimes F} \{(x, \alpha) \in K \times M_n(\mathcal{M})_{sa} : (x, 1_m \otimes \alpha) \in \mathrm{epi}(a)\},$$

where the intersection is taken over all $m < \infty$ and all self-adjoint nc affine functions a in $M_m(M_n(\mathrm{A}(K)))_{sa}$ satisfying $a \le 1_m \otimes F$. This is an intersection of closed nc convex sets, so $\widetilde{F}$ is a lower semicontinuous convex function with $\widetilde{F} \le F$.

Thus by definition of the convex envelope, we have that $\widetilde{F} \leq \overline{F}$. We will prove that $\overline{F} \leq \widetilde{F}$, and hence graph $\widetilde{F} =$ graph $\overline{F}$.

Since F is bounded, $\overline{F}$ is bounded by Proposition 16.13.11. By replacing F by $F(x) + (\|F\| + 1)$, we may assume that $\overline{F}(x) \subseteq [1_n \otimes 1_k, +\infty)$ for every k and every $x \in K_k$; i.e. for $(x, \alpha) \in \text{graph}_k(\overline{F})$, $\alpha \geq 1_n \otimes 1_k$.

By Proposition 16.4.5, an nc convex set is determined by its finite levels. Fix $x_0 \in K_k$ for $k < \infty$ and self-adjoint $\alpha_0 \in M_n(M_k)$ such that $(x_0, \alpha_0) \notin \text{graph}_k(\overline{F})$. To show that $(x_0, \alpha_0) \notin \text{graph}_k(\widetilde{F})$, we must find $m < \infty$ and an nc affine function $a \in (M_m(M_n(\mathrm{A}(K))))_{sa}$ such that $a \leq 1_m \otimes F$, in the sense that $[a(x), +\infty) \supseteq 1_m \otimes F(x)$ for all $x \in K$, but $a(x_0) \not\leq 1_m \otimes \alpha_0$.

Let E be the dual operator system containing K. Since graph$(\overline{F})$ is closed and nc convex, it follows from the nc separation theorem, Corollary 16.5.3, that there is a normal completely bounded self-adjoint map $\theta : E \oplus M_n \to M_k$ and self-adjoint $\gamma \in M_k$ such that $\theta_l((x, \alpha)) \leq \gamma \otimes 1_l$ for every l and every $(x, \alpha) \in \text{graph}_l(\overline{F})$, but $\theta_k((x_0, \alpha_0)) \not\leq \gamma \otimes 1_k$. (Recall that $\theta_l = \theta \otimes id_l$.)

Define normal completely bounded maps $\varphi : E \to M_k$ and $\psi : M_n \to M_k$ by

$$\varphi(x) = \theta(x, 0_n) \quad \text{for } x \in E \quad \text{and} \quad \psi(\alpha) = -\theta(0, \alpha) \quad \text{for } \alpha \in M_n.$$

Then for every l and every $(x, \alpha) \in \text{graph}_l(\overline{F})$,

$$\varphi_l(x) - \psi_l(\alpha) = \theta_l((x, \alpha)) \leq \gamma \otimes 1_l.$$

Rearranging gives

$$\varphi_l(x) - \gamma \otimes 1_l \leq \psi_l(\alpha).$$

We claim that ψ is completely positive. Note that for l and $(x, \alpha) \in \text{graph}_l(\overline{F})$, the normalization ensures that $\alpha \geq 1_n \otimes 1_l$. Since $\overline{F}$ is upwards directed, $(x, t\alpha) \in \text{graph}_l(\overline{F})$ for every $t \geq 1$. Hence by the previous paragraph,

$$\varphi_l(x) - \gamma \otimes 1_l \leq \psi_l(t\alpha) = t\psi_l(\alpha).$$

Dividing both sides by t and taking $t \to \infty$ yields $\psi_l(\alpha) \geq 0$.

For any positive $\beta \in M_n(M_l)$ and $\varepsilon > 0$, there is some $t > 0$ such that $t(\beta + \varepsilon 1_n \otimes 1_l) \geq \alpha$. Since $\overline{F}$ is upwards directed,

$$(x, t(\beta + \varepsilon 1_n \otimes 1_l)) \in \text{graph}_l(\overline{F}).$$

Hence by the previous paragraph,

$$t\psi_l(\beta + \varepsilon 1_n \otimes 1_l) = \psi_l(t(\beta + \varepsilon 1_n \otimes 1_l)) \geq 0.$$

Dividing by t implies $\psi_l(\beta + \varepsilon 1_n \otimes 1_l) \geq 0$, and taking $\varepsilon \to 0$ yields $\psi_l(\beta) \geq 0$. Therefore ψ is completely positive.

Next we modify θ so that $\psi(\alpha)$ is invertible for all $\alpha \in M_n^+$ with $\alpha \neq 0$. To this end, choose a faithful state $\tau : M_n \to \mathbb{C}$ and $\varepsilon > 0$ such that

$$\theta_k(x_0, \alpha_0) - \varepsilon \tau_k(\alpha_0) 1_k \otimes 1_k \not\leq \gamma \otimes 1_k.$$

Define a completely bounded self-adjoint map $\theta' : E \oplus M_n \to M_k$ by

$$\theta'(x, \alpha) = \theta(x, \alpha) - \varepsilon \tau(\alpha) 1_k.$$

We observe that $\theta'_l((x,\alpha)) \le \gamma \otimes 1_l$ for every l and every $(x,\alpha) \in \operatorname{graph}_l(\overline{F})$, but $\theta'_k((x_0,\alpha_0)) \not\le \gamma \otimes 1_k$. Furthermore, the map $\psi' : M_n \to M_k$ defined by $\psi'(\alpha) = -\theta'(0,\alpha)$ for $\alpha \in M_n$ satisfies

$$\psi'(\alpha) = -\theta(0,\alpha) + \varepsilon\tau(\alpha)1_k = \psi(\alpha) + \varepsilon\tau(\alpha)1_k.$$

In particular, the positivity of ψ and the faithfulness of τ implies that $\psi'(\alpha)$ is invertible for all $\alpha \in M_n^+$ with $\alpha \ne 0$. By replacing θ by θ', we can assume that ψ has this property.

The only representations of M_n are ampliations. Since ψ is completely positive and k is finite, Stinespring's theorem provides a finite integer m and an operator $\beta : H_k \to H_n^m$ such that

$$\psi(\alpha) = \beta^*(1_m \otimes \alpha)\beta \quad \text{for} \quad \alpha \in M_n.$$

Then

$$\psi_l(\alpha) = (\beta^* \otimes 1_l)(1_m \otimes \alpha)(\beta \otimes 1_l) \quad \text{for} \quad \alpha \in M_n(M_l).$$

The polar decomposition yields $\beta = \nu|\beta|$, where $\nu : H_k \to H_n^m$ is a partial isometry with initial space $(\ker\beta)^\perp$. Since $\psi(1_n)$ is invertible, it follows that $|\beta|$ is invertible. Hence $\nu^*\nu = 1_k$; so ν is an isometry. Let $q = \nu\nu^*$.

Since $|\beta|$ is invertible, there is an element $\beta' = |\beta|^{-1} \in M_k$. Then $\nu = \beta\beta'$, whence $\nu\beta'\beta^* = \nu\nu^* = q$. Thus for $\alpha \in M_n(M_l)$,

$$(\nu\beta' \otimes 1_l)\psi_l(\alpha)(\beta'\nu^* \otimes 1_l) = (q \otimes 1_l)(1_m \otimes \alpha)(q \otimes 1_l).$$

Decompose $1_m \otimes \alpha \in M_m(M_n(M_l))$ as a block matrix with respect to the projection $q \otimes 1_l$ to get

$$1_m \otimes \alpha = \begin{bmatrix} (1_m \otimes \alpha)_{11} & (1_m \otimes \alpha)_{12} \\ (1_m \otimes \alpha)_{21} & (1_m \otimes \alpha)_{22} \end{bmatrix}.$$

For $\varepsilon > 0$, define an nc affine function

$$a_{\varepsilon,11}(x) = (\nu\beta' \otimes 1_l)(\varphi_l(x) - \gamma \otimes 1_l - \varepsilon 1_k \otimes 1_l)(\beta'\nu^* \otimes 1_l).$$

Notice that on $(q \otimes 1_l)H$,

$$\begin{aligned}(1_m \otimes \alpha)_{11} - a_{\varepsilon,11}(x) &= (\nu\beta' \otimes 1_l)(\psi_l(\alpha) - \varphi_l(x) + \gamma \otimes 1_l + \varepsilon 1_k \otimes 1_l)(\beta'\nu^* \otimes 1_l) \\ &\ge (\nu\beta' \otimes 1_l)(\varepsilon 1_k \otimes 1_l)(\beta'\nu^* \otimes 1_l) \ge \varepsilon\|\beta\|^{-2} I.\end{aligned}$$

Now we define $a_{\varepsilon,22}(x) = -\Delta(\varepsilon\|\beta\|^{-2}, \|\overline{F}\|)I = -(\|\overline{F}\| + \varepsilon^{-1}\|\beta\|^2\|\overline{F}\|^2)I$ on $(q^\perp \otimes 1_l)H$, and set

$$a_\varepsilon = \begin{bmatrix} a_{\varepsilon,11} & 0 \\ 0 & a_{\varepsilon,22} \end{bmatrix}.$$

We claim that $a_\varepsilon \le 1_m \otimes \overline{F}$ in the sense that $[a_\varepsilon(x), +\infty) \supseteq 1_m \otimes \overline{F}(x)$ for all $x \in K$. The boundedness of $\overline{F}$ implies that for every l and every $(x,\alpha') \in \operatorname{graph}_l(\overline{F})$, there is $\alpha \in \overline{F}(x)$ such that $\alpha \le \alpha'$ and $\|\alpha\| \le \|\overline{F}\|$. Therefore, in order to show that $a_\varepsilon \le 1_m \otimes \overline{F}$, it suffices to show that $a_\varepsilon(x) \le 1_m \otimes \alpha$ for every l and every $(x,\alpha) \in \operatorname{graph}_l(\overline{F})$ with $\|\alpha\| \le \|\overline{F}\|$. This now follows from Lemma 16.13.12.

Finally, we claim there is an $\varepsilon > 0$ such that $a_\varepsilon(x_0) \not\leq 1_m \otimes \alpha_0$. To see this, suppose for the sake of contradiction that $a_\varepsilon(x_0) \leq 1_m \otimes \alpha_0$ for all $\varepsilon > 0$. Then in particular, looking at the top left corner of the block matrix of $1_m \otimes \alpha_0 - a_\varepsilon(x)$ with respect to the projection $q \otimes 1_k$ yields

$$\begin{aligned} 0 &\leq (1_m \otimes \alpha_0)_{11} - a_{\varepsilon,11}(x_0) \\ &= (\nu\beta' \otimes 1_k)(\psi_l(\alpha_0) - \varphi_k(x_0) + \gamma \otimes 1_k + \varepsilon 1_k \otimes 1_k)(\beta'\nu^* \otimes 1_k) \\ &= (\nu\beta' \otimes 1_k)(-\theta((x_0, \alpha_0)) + \gamma \otimes 1_k + \varepsilon 1_k \otimes 1_k)(\beta'\nu^* \otimes 1_k) \end{aligned}$$

Then multiplying on the left by $\beta^* \otimes 1_k$ and on the right by $\beta \otimes 1_k$ and taking $\varepsilon \to 0$ implies $\theta_k((x_0, \alpha_0)) \leq \gamma \otimes 1_k$, contradicting our original separation of (x_0, α_0) from the graph of $\overline{F}$. We conclude that for some $\varepsilon > 0$, the nc affine function a_ε achieves the desired separation. ■

16.14. UCP Maps of Convex NC Functions

In this section, we derive a useful formula for evaluating a u.c.p. map at a lower semicontinuous convex multivalued nc function. We will need a useful fact regarding the convex envelope and multiplicity.

16.14.1. COROLLARY. *Let K be an nc convex set, and let $F : K \to M_n(\mathcal{M})$ be a self-adjoint bounded multivalued nc function with convex envelope $\overline{F}$. Then*

$$\overline{1_l \otimes F} \leq 1_l \otimes \overline{F}.$$

PROOF. Note that

$$\text{graph}(\overline{1_l \otimes F}) = \overline{\text{ncconv}}\big(\text{graph}(1_l \otimes F)\big) = \overline{\text{ncconv}}\Big\{\coprod_{x \in K} (x, 1_l \otimes F(x))\Big\}.$$

The nc convex combinations include all points obtained using points x_i and contractions of the form $1_l \otimes \alpha_i$. Therefore

$$\text{graph}(\overline{1_l \otimes F}) \supset \coprod_{x \in K} (x, 1_l \otimes \overline{F}(x)) = \text{graph}(1_l \otimes \overline{F}).$$ ■

The next result shows that, as in the classical setting, the noncommutative convex envelope encodes information about the set of representing maps of a point. This is an analogue of Mokobodzki's theorem.

16.14.2. THEOREM. *Let K be a compact nc convex set, and let $f : K \to M_n(\mathcal{M})$ be a self-adjoint bounded nc function with convex envelope $\overline{f}$. Then for $x \in K_m$,*

$$\overline{f}(x) = \bigcup_{\mu} [\mu(f), +\infty),$$

where the union is taken over all unital completely positive maps $\mu : \mathrm{C}(K) \to M_m$ *with barycenter* x.

PROOF. Define $F : K \to \mathcal{M}$ by $F(x) = \cup_\mu [\mu(f), +\infty)$ for $x \in K_m$, where the union is taken over all u.c.p. maps $\mu : \mathrm{C}(K) \to M_m$ with barycenter x. Then F is a self-adjoint bounded multivalued nc function since it is clearly graded, preserves direct sums, is unitarily equivariant and upward directed.

We claim that F is lower semicontinuous. Let (x_i, α_i) be a net in $\mathrm{graph}_m(F)$ converging to $(x, \alpha) \in K_m \times M_n(M_m)$. Then there are unital completely positive maps $\mu_i : \mathrm{C}(K) \to M_m$ such that μ_i has barycenter x_i and $\mu_i(f) \leq \alpha_i$. Let $\mu : \mathrm{C}(K) \to M_m$ be a cluster point of the net (μ_i). Then μ has barycenter x and $\mu(f) \leq \alpha$, so $(x, \alpha) \in \mathrm{graph}_m(F)$.

Next we show that F is convex. Suppose that $(x_i, \alpha_i) \in \mathrm{graph}(F)$, where $x_i \in K_{n_i}$ and μ_i is a u.c.p. map $\mu_i : \mathrm{C}(K) \to M_m$ with barycenter x_i such that $\mu_i(f) \leq \alpha_i$. If $\beta_i \in M_{n,n_i}$ so that $\sum \beta_i^* \beta_i = 1_n$, let

$$x := \sum_i \beta_i^* x_i \beta_i \in K_n \quad \text{and} \quad \alpha := \sum_i (1_m \otimes \beta_i^*) \alpha_i (1_m \otimes \beta_i).$$

We need to verify that $(x, \alpha) \in \mathrm{graph}(F)$. Observe that

$$\mu := \sum (1_m \otimes \beta_i^*) \mu_i (1_m \otimes \beta_i)$$

is a unital completely positive map $\mu : \mathrm{C}(K) \to M_m$ with barycenter x. In addition,

$$\begin{aligned} \mu(f) &= \sum (1_m \otimes \beta_i^*) \mu_i(f) (1_m \otimes \beta_i) \\ &\leq \sum (1_m \otimes \beta_i^*) \alpha_i (1_m \otimes \beta_i) = \alpha. \end{aligned}$$

Therefore $(x, \alpha) \in \mathrm{graph}(F)$.

It now suffices to show that $\overline{f} = F$. We will accomplish using Theorem 16.13.13 by showing that if $a \in M_m(M_n(\mathrm{A}(K)))$, then $a \leq 1_m \otimes f$ if and only if $a \leq 1_m \otimes F$ for every $x \in K_m$.

If $a \leq 1_m \otimes F$ and $x \in K_m$, then $a(x) \leq 1_m \otimes \mu(f)$ for every unital completely positive map $\mu : \mathrm{C}(K) \to M_m$ with barycenter x. In particular, taking $\mu = \delta_x$ implies $a(x) \leq 1_m \otimes f(x)$. Hence $a \leq 1_m \otimes f$.

On the other hand, if $a \leq 1_m \otimes f$, then for $x \in K_m$ and every unital completely positive map $\mu : \mathrm{C}(K) \to M_m$ with barycenter x,

$$a(x) = \mu(a) \leq \mu(1_m \otimes f) = 1_m \otimes \mu(f).$$

Hence $a \leq 1_m \otimes F$. ■

We can use this result to obtain a formula for $\mu(\overline{F})$.

16.14.3. REMARK. In the following, we will use multivalued convex nc functions formed from a convex nc function g and a constant C:

$$H = g \cap -C := \mathrm{epi}(g) \cap \mathrm{epi}(-C) = \{(x, \alpha) : \alpha \geq g(x),\ \alpha \geq -C\}.$$

This function should not be confused with $h(x) = g(x) \vee (-C)$, which makes sense since $-C$ is a scalar function. This function h is usually not convex as an nc function, and epi(h) in general smaller than H. For example, the 2x2 self-adjoint matrix $A = \begin{bmatrix} 2 & \sqrt{2} \\ \sqrt{2} & 1 \end{bmatrix}$ satisfies

$$A \geq \begin{bmatrix} 1 & 0 \\ 0 & -1 \end{bmatrix} \quad \text{and} \quad A \geq \begin{bmatrix} 0 & 0 \\ 0 & 0 \end{bmatrix} \quad \text{but} \quad A \not\geq \begin{bmatrix} 1 & 0 \\ 0 & 0 \end{bmatrix}.$$

16.14.4. COROLLARY. *Let K be a compact nc convex set and let $F : K \to M_n(\mathcal{M})$ be a self-adjoint multivalued bounded nc function. Then for every u.c.p. map $\mu : \mathrm{C}(K) \to M_p$,*

$$\mu(\overline{F}) = \bigcap_{g \leq 1_m \otimes F} \{\alpha \in M_n(M_p) : 1_m \otimes \alpha \geq \mu(g)\},$$

where the intersection is over all m and all convex nc functions $g \in M_m(M_n(\mathrm{C}(K)))$ satisfying $g \leq 1_m \otimes F$.

PROOF. We first observe that for $x \in K_p$,

$$\overline{F}(x) = \bigcap_{g \leq 1_m \otimes F} \{\alpha \in M_n(M_p) : 1_m \otimes \alpha \geq g(x)\},$$

where the intersection is over all m and all convex nc functions $g \in M_m(M_n(\mathrm{C}(K)))$ satisfying $g \leq 1_m \otimes F$. By Theorem 16.13.13, $\overline{F}(x)$ is the intersection over such sets with respect to continuous affine nc functions. Since every affine nc function is a convex nc function, the intersection over all m and all convex nc functions $g \in M_m(M_n(\mathrm{C}(K)))$ satisfying $g \leq 1_m \otimes F$ is smaller. On the other hand, by Proposition 16.13.11, $\overline{1_m \otimes F}$ is the largest lower semicontinuous convex multivalued nc function dominated by $1_m \otimes F$, meaning that for all convex nc functions $g \leq 1_m \otimes F$, we have $g \leq \overline{1_m \otimes F}$. Hence by Corollary 16.14.1, $g \leq 1_m \otimes \overline{F}$. Therefore, the intersection is precisely graph$(\overline{F})$.

Now F is bounded, say by C. By Proposition 16.13.11(4), $\overline{F}$ is also bounded by C. In particular, $F \geq -C$. Thus for each convex nc function $g \in M_m(M_n(\mathrm{C}(K)))$ satisfying $g \leq 1_m \otimes F$, we also have $H = g \cap (-C) \leq 1_m \otimes F$. In particular, $\|H\| \leq C$ as well. Hence

$$\overline{F} = \bigcap_{g \leq 1_m \otimes F} \{\alpha \in M_n(M_q) : 1_m \otimes \alpha \geq H(x)\},$$

where $H = g \cap (-C)$ and the intersection is taken over all m and all convex nc functions $g \in M_m(M_n(\mathrm{C}(K)))$ satisfying $g \leq 1_m \otimes F$.

Fix a minimal Stinespring representation $(x, \nu) \in K_q \times M_{q,p}$ for μ so that $\mu(f) = \nu^* f(x) \nu$ for $f \in \mathrm{C}(K)$. By Exercise **16.19**, $\mu(\overline{F}) = (1_n \otimes \nu^*)\overline{F}(x)(1_n \otimes \nu)$. Since $\overline{F}(x)$ is upward directed and $\mu(\overline{F})$ is a corner of $\overline{F}(x)$, it is also upward directed. If $1_m \otimes \alpha \geq g(x)$, then setting $\beta = (1_n \otimes \nu^*)\alpha(1_n \otimes \nu)$, it follows that

$$1_m \otimes \beta \geq (1_m \otimes 1_n \otimes \nu^*)g(x)(1_m \otimes 1_n \otimes \nu) = \mu(g).$$

Therefore $\mu(\overline{F})$ is contained in the intersection on the RHS of the desired formula.

Conversely, suppose that $\beta \in M_n(M_p)$ satisfies $1_m \otimes \beta \geq \mu(g)$ for all m and all convex nc functions $g \in M_m(M_n(\mathrm{C}(K)))$ with $g \leq 1_m \otimes F$. Fix such a g and let $H = g \cap -C$ as above. Note that $\beta \geq -C$.

For $\varepsilon > 0$, define $\alpha_\varepsilon \in M_n(M_q)$ by

$$\begin{aligned}\alpha_\varepsilon &= (1_n \otimes \nu)(\beta + \varepsilon 1_p)(1_n \otimes \nu^*) + (1_n \otimes (1_q - \nu\nu^*))(\varepsilon^{-1}\|g\|^2 + \|g\|) \\ &\simeq \begin{bmatrix} \beta + \varepsilon & 0 \\ 0 & \varepsilon^{-1}\|g\|^2 + \|g\| \end{bmatrix},\end{aligned}$$

where the decomposition is taken with respect to the range of $1_n \otimes \nu$. If we also decompose $g(x)$ with respect to the range of $1_m \otimes 1_n \otimes \nu$, it has the form

$$g(x) \simeq \begin{bmatrix} g_{11} & g_{12} \\ g_{21} & g_{22} \end{bmatrix}$$

and by hypothesis $g_{11} \simeq \mu(g) \leq 1_m \otimes \beta$. Thus

$$\begin{aligned}1_m \otimes \alpha_\varepsilon - g(x) &\simeq \begin{bmatrix} 1_m \otimes \beta - g_{11} + \varepsilon(1_m \otimes 1_p) & -g_{12} \\ -g_{21} & \varepsilon^{-1}\|g\|^2 + (\|g\| - g_{22}) \end{bmatrix} \\ &\geq \begin{bmatrix} \varepsilon & -g_{12} \\ -g_{21} & \varepsilon^{-1}\|g\|^2 \end{bmatrix} \geq \begin{bmatrix} 0 & 0 \\ 0 & 0 \end{bmatrix}.\end{aligned}$$

Hence $1_m \otimes \alpha_\varepsilon \geq g(x)$ and $1_m \otimes \alpha_\varepsilon \geq -C$. Therefore, $1_m \otimes \alpha_\varepsilon \in H(x)$. Since $\|H\| \leq C$, there is some $\gamma_\varepsilon \in H(x)$ with $\|\gamma_\varepsilon\| \leq C$ such that $\gamma_\varepsilon \leq 1_m \otimes \alpha_\varepsilon$.

Define $\alpha'_\varepsilon \in M_n(M_q)_{sa}$ by

$$\begin{aligned}\alpha'_\varepsilon &= (1_n \otimes \nu)(\beta + \varepsilon 1_p)(1_n \otimes \nu^*) + (1_n \otimes (1_q - \nu\nu^*))(\varepsilon^{-1}C^2 + C) \\ &\simeq \begin{bmatrix} \beta + \varepsilon & 0 \\ 0 & \varepsilon^{-1}C^2 + C \end{bmatrix}.\end{aligned}$$

Then arguing as above, $1_m \otimes \alpha'_\varepsilon \geq \gamma_\varepsilon$. Since $\gamma_\varepsilon \in H(x)$ and $H(x)$ is upward directed, $1_m \otimes \alpha'_\varepsilon \in H(x)$.

Since the construction of α'_ε does not depend on H, it follows from our second expression for $\overline{F}(x)$ that $\alpha'_\varepsilon \in \overline{F}(x)$. Since $\|\overline{F}\| = C$, there is a $\gamma'_\varepsilon \in \overline{F}(x)$ with $\|\gamma'_\varepsilon\| \leq C$ such that $\gamma_\varepsilon \leq 1_m \otimes \alpha'_\varepsilon$. The net $(\gamma'_\varepsilon)_{\varepsilon > 0}$ is bounded and $\overline{F}$ is lower semicontinuous, so there is a cofinal subnet with limit $\gamma' \in \overline{F}(x)$. By construction

$$(1_n \otimes \nu^*)\gamma'(1_n \otimes \nu) \leq \lim_{\varepsilon \to 0}(1_n \otimes \nu^*)\alpha'_\varepsilon(1_n \otimes \nu) = \beta.$$

Therefore, since $(1_n \otimes \nu^*)\gamma'(1_n \otimes \nu) \in (1_n \otimes \nu^*)\overline{F}(x)(1_n \otimes \nu) = \mu(\overline{F})$ and $\mu(\overline{F})$ is upward directed, it follows that $\beta \in \mu(\overline{F})$. Therefore the RHS intersection equals $\mu(\overline{F})$. ∎

16.15. Orders on CP Maps

In this section, we introduce two different orders on u.c.p. maps on K, and eventually prove that they are equivalent. The nc Choquet order will be a direct analogy to the classical order, while the other is based on dilation theory. The equivalence of these orders is an additional tool available in the nc setting.

16.15.1. DEFINITION. Let K be a compact nc convex set and let L be the nc state space of $\mathrm{C}(K)$. Given two nc states $\mu, \nu \in L_m$, say that μ is dominated by ν in the *nc Choquet order* and write $\mu \prec_c \nu$ if $\mu(f) \leq \nu(f)$ for every n and every convex nc function $f \in M_n(\mathrm{C}(K))$.

We record some basic facts about the nc Choquet order.

16.15.2. PROPOSITION. *Let K be a compact nc convex set and let $\mu, \nu \in L_m$ be nc states on $\mathrm{C}(K)$ such that $\mu \prec_c \nu$.*

(1) *μ and ν have the same barycentre.*

(2) *$\mu(\overline{f}) \leq \nu(\overline{f})$ for every n and every nc function $f \in M_n(\mathrm{C}(K))_{sa}$.*

(3) *For each cardinal m, the nc Choquet order is a partial order on L_m.*

PROOF. For $a \in \mathrm{A}(K)$, both a and $-a$ are convex, so $\mu(a) \leq \nu(a)$ and $\mu(-a) \leq \nu(-a)$. Thus $\mu|_{\mathrm{A}(K)} = \nu|_{\mathrm{A}(K)} =: x \in K$; i.e. they have the same barycentre x.

If $\mu \prec_c \nu$ and $f \in \mathrm{C}(K)$, Corollary 16.14.4 shows that

$$\mu(\overline{f}) = \bigcap_{g \leq 1_m \otimes f} \{\alpha \in M_n(M_p) : 1_m \otimes \alpha \geq \mu(g)\}$$
$$\supseteq \bigcap_{g \leq 1_m \otimes f} \{\alpha \in M_n(M_p) : 1_m \otimes \alpha \geq \nu(g)\} = \nu(\overline{f}).$$

where the intersection is over all m and all convex nc functions g in $M_m(M_n(\mathrm{C}(K)))$ satisfying $g \leq 1_m \otimes f$. Since $\mu(g) \leq \nu(g)$ for convex $g \in \mathrm{C}(K)$, we get $\mu(\overline{f}) \leq \nu(\overline{f})$.

It is easy to see that the nc Choquet order is reflexive and transitive. Antisymmetry follows from Proposition 16.13.6. ∎

The following results are straightforward and are left as an exercise.

16.15.3. PROPOSITION. *Let K be a compact nc convex set and let L be the nc state space of $\mathrm{C}(K)$.*

(1) *If $\mu_i, \nu_i \in L_m$ for $1 \leq i \leq k$ with $\mu_i \prec_c \nu_i$ and $t_i \geq 0$ with $\sum_{i=1}^k t_i = 1$, then $\sum_{i=1}^k t_i\mu_i \prec_c \sum_{i=1}^k t_i\nu_i$.*

(2) *Suppose that* $(\mu_\lambda)_\Lambda$ *and* $(\nu_\lambda)_\Lambda$ *are nets in* L *with limits* μ *and* ν, *respectively. If* $\mu_\lambda \prec_c \nu_\lambda$ *for* $\lambda \in \Lambda$, *then* $\mu \prec_c \nu$.

The following is a straightforward analogue of Proposition 16.3.3.

16.15.4. PROPOSITION. *Let* K *be a compact nc convex set and let* $\mu : \mathrm{C}(K) \to M_n$ *be a u.c.p. map. Then there is a u.c.p. map* $\nu : \mathrm{C}(K) \to M_n$ *with* $\mu \prec_c \nu$ *which is maximal in the nc Choquet order.*

PROOF. Consider $N = \{\nu : \mathrm{C}(K) \to M_n : \mu \prec_c \nu\}$ with the $\prec_c$ partial order. We will show that N is inductive, and hence has maximal elements by Zorn's lemma. Suppose that $\mathcal{C} = \{\nu_\lambda\}_\Lambda$ is a chain in N. Since the state space L_n is compact, there is a cofinal net in $\mathcal{C}$ converging to $\nu \in L_n$. If f is any convex nc function, then $\lambda_1 \leq \lambda_2$ implies that $\nu_{\lambda_1}(f) \leq \nu_{\lambda_2}(f)$, and hence $\nu_\lambda(f) \leq \nu(f)$ for all $\lambda \in \Lambda$. Therefore ν is an upper bound for $\mathcal{C}$ in N. ∎

There is another natural order for unital completely positive maps that is derived from dilation theory. Recall Definition 16.12.7. This order has no parallel in the classical setting.

16.15.5. DEFINITION. Let K be a compact nc convex set and let $\mu, \nu : \mathrm{C}(K) \to M_m$ be u.c.p. maps. Say that μ is dominated by ν in the *dilation order* and write $\mu \prec_d \nu$ if there are representations (x, α) in $K_n \times M_{n,m}$ for μ and (y, β) in $K_p \times M_{p,m}$ for ν along with an isometry $\gamma \in M_{p,n}$ such that $\beta = \gamma\alpha$ and $x = \gamma^* y \gamma$. Say that μ is *dilation maximal* if it is maximal in the dilation order.

We record some basic facts about the dilation order.

16.15.6. PROPOSITION. *Let* K *be a compact nc convex set and let* $\mu, \nu : \mathrm{C}(K) \to M_m$ *be u.c.p. maps such that* $\mu \prec_d \nu$. *Let* (x, α) *in* $K_n \times M_{n,m}$ *and* (y, β) *in* $K_p \times M_{p,m}$ *be representations for* μ *and* ν, *and let* $\gamma \in M_{p,n}$ *be an isometry such that* $\beta = \gamma\alpha$ *and* $x = \gamma^* y \gamma$.

(1) *We can assume that* (x, α) *is a minimal dilation.*

(2) y *dilates* x.

(3) μ *and* ν *have the same barycentre.*

(4) $\delta_x \prec_d \mu$.

(5) $\mu \prec_d \nu$ *if and only if there is a representation* $(x, \alpha) \in K_n \times M_{n,m}$ *for* μ *and a u.c.p. map* $\tau : \mathrm{C}(K) \to M_n$ *with barycentre* x *satisfying* $\nu = \alpha^* \tau \alpha$. *Moreover such a* τ *exists for any representation* (x, α).

PROOF. (1) The uniqueness of the minimal Stinespring dilation means that any representation (x, α) can be decomposed at $x = x_0 \oplus x'$ and $\alpha = \left[\begin{smallmatrix}\alpha_0\\0\end{smallmatrix}\right]$ with respect to $H_m = H_{m_0} \oplus H_{m'}$, where (x_0, α_0) is a minimal dilation. If we replace γ by $\gamma_0 = \gamma|_{H_{m_0}}$, then (y, β) also dilates (x_0, α_0). It is not necessarily true that (y, β) will be a minimal representation of ν.

(2) The condition $x = \gamma^* y \gamma$ says that y dilates x.

(3) For $a \in \mathrm{A}(K)$,

$$\mu(a) = \alpha^* a(x)\alpha = \alpha^* a(\gamma^* y\gamma)\alpha = \alpha^*\gamma^* a(y)\gamma\alpha = \beta^* a(y)\beta = \nu(a).$$

Thus μ and ν have the same barycentre.

(4) For $x \in K_m$, $(x, 1_m)$ is a minimal representation of δ_x. Let $\mu : K_m \to M_m$ be a unital completely positive map with barycenter x, and let $(y, \alpha) \in K_n \times M_{n,m}$ be a minimal representation of μ. Then $x = \alpha^* y\alpha$ and $\mu = \alpha^*\delta_y\alpha$. So $\delta_x \prec_d \mu$.

(5) Suppose $\mu \prec_d \nu$ as above. By (1), we may assume that (x_0, α_0) is minimal, (y_0, β_0) represents ν and $\beta_0 = \gamma_0\alpha_0$. Then $\tau_0 = \gamma_0^*\delta_{y_0}\gamma_0$ is u.c.p. with barycentre $\gamma_0^* y_0\gamma_0 = x_0$, and $\nu = \alpha_0^*\tau_0\alpha_0$. In general, $(x, \alpha) \simeq (x_0, \alpha_0) \oplus (x', 0)$. So we can set $\tau = \tau_0 \oplus 0$.

Conversely, suppose there is a representation $(x, \alpha) \in K_n \times M_{n,m}$ for μ and a unital completely positive map $\tau : \mathrm{C}(K) \to M_n$ with barycenter x satisfying $\nu = \alpha^*\tau\alpha$. Choose a representation $(y, \gamma) \in K_p \times M_{p,m}$ for τ. Let $\beta = \gamma\alpha$. Then (y, β) is a representation for ν and $\gamma^* y\gamma = \tau|_{\mathrm{A}(K)} = x$. ■

Next we extend Theorem 16.14.2 to the evaluation of $\mu(\overline{f})$.

16.15.7. THEOREM. *Let K be a compact nc convex set. Let $f \in M_n(\mathrm{B}(K))$ be a self-adjoint bounded nc function with convex envelope $\overline{f}$. Then for a u.c.p. map $\mu : \mathrm{C}(K) \to M_k$,*

$$\mu(\overline{f}) = \bigcup_{\mu \prec_d \nu} [\nu(f), +\infty),$$

where the union is taken over all u.c.p. maps $\nu : \mathrm{C}(K) \to M_k$ such that $\mu \prec_d \nu$.

Moreover, μ is maximal in the dilation order if and only if $[\mu(f), +\infty) = \mu(\overline{f})$ for all $f \in \mathrm{C}(K)$.

PROOF. Let $(x, \alpha) \in K_m \times M_{m,k}$ be a minimal representation of μ. By Theorem 16.14.2,

$$\mu(\overline{f}) = \alpha^*\overline{f}(x)\alpha = \bigcup_{\tau} \alpha^*[\tau(f), +\infty)\alpha = \bigcup_{\tau} [\alpha^*\tau(f)\alpha, +\infty),$$

where the union is taken over all u.c.p. maps $\tau : \mathrm{C}(K) \to M_m$ with barycenter x. The result now follows from Proposition 16.15.6(5).

If μ is maximal in the dilation order, then the previous paragraph implies $[\mu(f), +\infty) = \mu(\overline{f})$ for all $f \in \mathrm{C}(K)$. Conversely, suppose that this holds. Let

$\nu : \mathrm{C}(K) \to M_k$ be a unital completely positive map with $\mu \prec_d \nu$. For $f \in \mathrm{C}(K)$,

$$[\mu(f), +\infty) = \mu(\overline{f}) \supseteq [\mu(f), +\infty) \cup [\nu(f), +\infty)$$

and

$$[\mu(-f), +\infty) = \mu(\overline{-f}) \supseteq [\mu(-f), +\infty) \cup [\nu(-f), +\infty).$$

Hence $\mu(f) \le \nu(f)$ and $\mu(-f) \le \nu(-f)$, implying $\mu(f) = \nu(f)$. Therefore $\mu = \nu$; whence μ is maximal. ■

We can now establish the main result of this section, the equivalence of these two orders.

16.15.8. THEOREM. *Let K be a compact nc convex set and let $\mu, \nu : \mathrm{C}(K) \to M_n$ be unital completely positive maps. Then $\mu \prec_c \nu$ if and only if $\mu \prec_d \nu$.*

PROOF. If $\mu \prec_d \nu$, then Theorem 16.15.7 implies that for every self-adjoint continuous convex nc function $f \in M_n(\mathrm{C}(K))$,

$$\mu(f) = \mu(\overline{f}) = \bigcup_{\mu \prec_d \lambda} [\lambda(f), +\infty) \supseteq \bigcup_{\nu \prec_d \lambda} [\lambda(f), +\infty) = \nu(\overline{f}) = \nu(f).$$

Therefore $\mu(f) \le \nu(f)$. That is, $\mu \prec_c \nu$.

Conversely, if $\mu \prec_c \nu$, then Proposition 16.15.2(2) implies $\mu(\overline{f}) \le \nu(\overline{f})$ for every n and every self-adjoint nc function $f \in M_n(\mathrm{C}(K))$. Let $\mathcal{F}$ be a dense family of functions in the unit ball of $\mathrm{C}(K)_{sa}$ such that $\mathcal{F} = -\mathcal{F}$. Define $h \in M_m(\mathrm{C}(K))$ by $h = \oplus_{f \in \mathcal{F}} f$. Then from above, $\mu(\overline{h}) \le \nu(\overline{h}) \le \nu(h)$. Hence by Theorem 16.15.7, there is a unital completely positive map $\lambda : \mathrm{C}(K) \to M_p$ such that $\mu \prec_d \lambda$ and $\lambda(h) \le \nu(h)$. But then $\lambda(f) \le \nu(f)$ and $\lambda(-f) \le \nu(-f)$ for all $f \in \mathcal{F}$, implying $\lambda(f) = \nu(f)$ for all $f \in \mathcal{F}$. Since $\mathcal{F}$ is dense in $\mathrm{C}(K)$, it follows that $\lambda = \nu$. Therefore $\mu \prec_d \nu$. ■

16.16. Maximal Dilations and Extreme Points

In this section, we use the equivalence of our two orders to provide new, quite different, proofs of the Dritschel–McCullough theorem 14.9.8 and the Davidson–Kennedy theorem 14.10.8.

It follows from Proposition 16.15.4 and Theorem 16.15.8 that there are u.c.p. maps which are maximal in the dilation order.

Here is the reformulation of the Dritschel–McCullough theorem.

16.16.1. THEOREM. *Let K be a compact nc convex set. Then every point in K has a maximal dilation.*

PROOF. Fix $x \in K_m$. Set $y_0 = x$ and $n_0 = m$. We will construct a sequence $y_i \in K_{n_i}$ for $i \geq 1$ such that y_i is a dilation of y_{i-1}, and thus y_i will also dilate x.

Suppose that we have y_j for $0 \leq j < i$, but y_{i-1} is not maximal. By Proposition 16.15.4, there is a $\mu_{i-1} : \mathrm{C}(K) \to M_{n_{i-1}}$ with $\delta_{y_{i-1}} \prec_c \mu_{i-1}$ which is maximal in the nc Choquet order. By Theorem 16.15.8, it is maximal in the dilation order. Note that y_{i-1} is the barycentre of μ_{i-1}. Let (y_i, α_i) be a minimal representation of μ_{i-1}; and let $y_i \in L_{n_i}$. Then y_i dilates y_{i-1}, and hence also dilates $x = y_0$. If at any stage, y_i is maximal, then we are done. Otherwise, we will construct a maximal dilation of x from the sequence (y_i, α_i).

The map α_i is an isometry in $\mathcal{B}(H_{n_{i-1}}, H_{n_i})$, and so we may identify $H_{n_{i-1}}$ with $\operatorname{Ran} \alpha_i$ as a subspace of $H_{n_i} = H_{n_{i-1}} \oplus \operatorname{Ran} \alpha_i^\perp$. Then $y_i \simeq \begin{bmatrix} y_{i-1} & * \\ * & x_i \end{bmatrix}$. For $i \geq 2$, we further obtain

$$y_i \simeq \begin{bmatrix} y_{i-2} & * & u \\ * & x_{i-1} & * \\ v & * & x_i \end{bmatrix} \quad \text{on } H_{n_i} = H_{n_{i-2}} \oplus (H_{n_{i-1}} \ominus H_{n_{i-2}}) \oplus \operatorname{Ran} \alpha_i^\perp,$$

and the upper left hand 2×2 corner is y_{i-1}.

We will show that $u = v = 0$. Indeed, we have $\mu_{i-2} = \alpha_{i-1}^* \delta_{y_{i-1}} \alpha_{i-1}$ and $\mu_{i-1} = \alpha_i^* \delta_{y_i} \alpha_i$. Set

$$\nu = \alpha_{i-1}^* \alpha_i^* \delta_{y_i} \alpha_i \alpha_{i-1} = \alpha_{i-1}^* \mu_{i-1} \alpha_{i-1}.$$

By Proposition 16.15.6(5), $\mu_{i-2} \prec_d \nu$. However μ_{i-2} is dilation maximal, and thus $\mu_{i-2} = \nu$. In particular for $a \in \mathrm{A}(K)$, looking at aa^*,

$$\begin{aligned} \mu_{i-2}(aa^*) &= \alpha_{i-1}^* a(y_{i-1}) a^*(y_{i-1}) \alpha_{i-1} \\ &= \nu(aa^*) = \alpha_{i-1}^* \alpha_i^* a(y_i) a^*(y_i) \alpha_i \alpha_{i-1} \\ &= \mu_{i-2}(aa^*) + a(u) a(u)^*. \end{aligned}$$

As a runs over all $\mathrm{A}(K)$, we see that u must be the 0 map. Similarly using $a^* a$, we see that $v = 0$.

Repeated application of this idea shows that each y_i for $y \geq 1$ has a tridiagonal form

$$y_i = \begin{bmatrix} x & r_1 & & & & \\ s_1 & x_2 & r_2 & & & \\ & s_2 & x_3 & r_3 & & \\ & & \ddots & \ddots & \ddots & \\ & & & s_{i-2} & x_{i-1} & r_{i-1} \\ & & & & s_{i-1} & x_i \end{bmatrix}.$$

Define

$$y = \begin{bmatrix} x & r_1 & & & & & \\ s_1 & x_2 & r_2 & & & & \\ & s_2 & x_3 & r_3 & & & \\ & & \ddots & \ddots & \ddots & & \\ & & & s_{i-2} & x_{i-1} & r_{i-1} & \\ & & & & s_{i-1} & x_i & r_i \\ & & & & & \ddots & \ddots & \ddots \end{bmatrix}.$$

We will show that this is a maximal dilation of x. Otherwise it has a non-trivial dilation $z = \left[\begin{smallmatrix} y & u \\ v & z_0 \end{smallmatrix}\right]$. If the column $u \neq 0$, there is some coefficient $u_i \neq 0$. However this would then be a dilation of y_i which with respect to the decomposition

$$H_z = H_{n_i} \oplus (H_{n_{i+1}} \ominus H_{n_i}) \oplus (H_z \ominus H_{n_{i+1}})$$

has a non-zero $1, 3$ entry. This isn't possible, and hence $u = 0$ and similarly $v = 0$. That is, y is maximal. ∎

This result has a rather striking consequence. It turns out that y_1 is already maximal, and the construction stops after one step.

16.16.2. Proposition. *Let K be a compact nc convex set, let $\mu : \mathrm{C}(K) \to M_m$ be a u.c.p. map, and let (y, α) be a minimal representation of μ. Then μ is maximal in the dilation order if and only if y is a maximal point.*

Proof. Suppose μ is maximal in the dilation order and (y, α) is minimal. By Theorem 16.16.1, there is maximal point $z \in K_p$ that dilates y. The maximality of z implies that it has a unique representing map, namely δ_z. Let $\beta \in M_{p,n}$ be an isometry such that $\beta^* z \beta = y$. Define a u.c.p. map $\nu : \mathrm{C}(K) \to M_m$ by $\nu = \alpha^* \beta^* \delta_z \beta \alpha$. Then by Proposition 16.15.6(5), $\mu \prec_d \nu$. Since μ is maximal, $\mu = \nu$. Thus the pair $(z, \beta\alpha)$ is a representation of μ. By the minimality of the representation (y, α) and the uniqueness of minimal representations, we must have $z \simeq y \oplus z'$ for some $z' \in K$. Since z is a maximal point, it follows that y is a maximal point.

Conversely, suppose y is a maximal point. Let $\nu : \mathrm{C}(K) \to M_m$ be an unital completely positive map such that $\mu \prec_d \nu$. Then by Proposition 16.15.6(5), there is a completely positive map $\tau : \mathrm{C}(K) \to M_n$ with barycenter y such that $\nu = \alpha^* \tau \alpha$. The map τ has barycenter y. Since y is maximal, it has a unique representing map, and hence $\tau = \delta_y$. Therefore $\nu = \alpha^* \delta_y \alpha = \mu$; and we conclude that μ is maximal in the dilation order. ∎

We now provide a new proof of the Davidson–Kennedy theorem. By Proposition 16.7.5, this is equivalent to saying that nc pure points in K dilate to nc extreme points.

16.16.3. THEOREM. *Let K be a compact nc convex set. Then every nc pure point in K has an nc extreme dilation.*

PROOF. Fix an nc pure point $x \in K_n$. Let L denote the nc state space of $\mathrm{C}(K)$, and set $F = \{\mu \in L_n : \mu \text{ has barycentre } x\}$. Note that F is a closed face of L_n. For if $\nu_i \in L_n$ and $\mu = \frac{1}{2}(\nu_1 + \nu_2)$, then restricting to $\mathrm{A}(K)$ yields $x = \frac{1}{2}(x_1 + x_2)$, where x_i is the barycentre of ν_i. This is an nc convex combination. By purity of x, one gets $x_1 = x_2 = x$; whence $\nu_i \in F$, showing that F is a face. Also F is hereditary in the dilation order, meaning that if $\mu \in F$ and $\mu \prec_d \nu$, then $\nu \in F$ because μ and ν have the same barycentre by Proposition 16.15.6(3).

Consider the collection of all (non-empty) closed hereditary faces contained in F. By compactness, the intersection of a nested family of these faces is a non-empty closed hereditary face. Thus a decreasing chain has a lower bound. By Zorn's lemma, there is a minimal closed hereditary face F_0. We claim that F_0 is a singleton. Otherwise, there are distinct maps $\mu, \nu \in F_0$. By Proposition 16.13.6, there is a convex nc function $f \in M_n(\mathrm{C}(K))$ for $n < \infty$ such that $\mu(f) \neq \nu(f)$. The set $\{\mu(f) : \mu \in F_0\}$ is a weak-$*$ compact subset of $M_{n\,sa}$. Choose a maximal element of this set in the usual order on self-adjoint operators, say A. Then $F_1 = \{\mu \in F_0 : \mu(f) = A\}$ is a closed face of F_0. If $\mu \in F_1$ and $\mu \prec_d \nu$, then $\nu \in F_0$ by the hereditary property and $A = \mu(f) \leq \nu(f) \leq A$, whence $\nu \in F_1$. That is, F_1 is a strictly smaller hereditary face, contrary to the minimality of F_0.

Therefore F_0 is a single point, say μ_0. As F_0 is hereditary, μ_0 is dilation maximal. Hence μ_0 is an extreme point of F. We claim that μ_0 is a pure c.p. map. Suppose that $0 < \psi_1 < \mu_0$ and set $\psi_2 = \mu_0 - \psi_1$. Restricting to $\mathrm{A}(K)$, $\psi_1|_{\mathrm{A}(K)} + \psi_2|_{\mathrm{A}(K)} = x$. Since x is pure, $\psi_i|_{\mathrm{A}(K)} = t_i x$, where $\psi_i(1) = t_i 1_n$ and $t_1 + t_2 = 1$. Let $\mu_i = t_i^{-1}\psi_i$. Then $\mu_i \in F$ and $\mu_0 = t_1\mu_1 + t_2\mu_2$. Since μ_0 is extreme, $\mu_1 = \mu_2 = \mu_0$, and so $\psi_1 = t_1\mu_0$. Therefore μ_0 is pure.

Let (y, α) be a minimal representation of μ_0. By Proposition 16.16.2, y is maximal. By Corollary 14.10.3, δ_y is irreducible, whence pure. By Proposition 16.7.5, y is extreme. ∎

We obtain the following refinement of Proposition 16.16.2 in the pure case.

16.16.4. COROLLARY. *Let K be a compact nc convex set, let $\mu : \mathrm{C}(K) \to M_m$ be a maximal nc pure u.c.p. map, and let (y, α) be a minimal representation of μ. Then $y \in \partial K$.*

PROOF. Use Theorem 16.16.3 to dilate μ to z so that z is nc extreme. Uniqueness of the minimal dilation shows that y is a summand of z, and thus is also pure and maximal, whence nc extreme. ∎

16.17. The Pedersen–Baire Envelope

In this section, we describe a canonical C*-algebra containing a given C*-algebra $\mathcal{A}$ which plays the role of the Baire measurable functions in the commutative case. In the case of separable $C(X)$ for metrizable X, this will just be the space of bounded Borel functions on X, which is the enveloping von Neumann algebra. This algebra will allow us to prove a version of the Choquet–Bishop–de Leeuw theorem for compact nc convex sets in the next section.

16.17.1. DEFINITION. The *monotone sequential completion* $\mathcal{B}^\infty(\mathfrak{A})$ of a C*-algebra $\mathfrak{A} \subset \mathcal{B}(H)$ is the smallest subspace of $\mathcal{B}(H)$ containing $\mathfrak{A}_{sa}$ which is closed under taking SOT limits of bounded monotone increasing or monotone decreasing sequences of self-adjoint elements. The *Pedersen–Baire envelope* $\mathrm{B}^\infty(\mathfrak{A})$ of $\mathfrak{A}$ is the monotone sequential completion of $\mathfrak{A}$ in its universal von Neumann algebra $\mathfrak{A}^{\dagger\dagger}$.

16.17.2. REMARK. The monotone sequential completion of a set $X \subset \mathcal{B}(H)_{sa}$ is the smallest subset $Y = \mathcal{B}^\infty(X) \subset \mathcal{B}(H)_{sa}$ containing X which is closed under taking SOT limits of monotone increasing or monotone decreasing bounded sequences. If X is a real subspace, then Y is a real subspace.

To see this, let $Y_0 = \{y \in Y : y + x \in Y \text{ for } x \in X\}$. This contains X, and if y_n is a monotone sequence in Y_0 with limit y, then $y + x$ is the limit of $y_n + x$; whence $y \in Y_0$. So Y_0 is monotone sequentially complete, and hence $Y_0 = Y$. Now let $Y_1 = \{y \in Y : y + z \in Y \text{ for } z \in Y\}$. We have shown that $X \subset Y_1$. If y_n is a monotone sequence in Y_1 with limit y, then $y + z$ is the limit of $y_n + z$. So Y_1 is monotone sequentially complete, and hence $Y_1 = Y$. Thus Y is closed under sums. An easier but similar argument shows that it is closed under real scalar multiples. Therefore Y is a real subspace.

16.17.3. EXAMPLE. Let X be a locally compact Hausdorff space, and consider $\mathrm{B}^\infty(X) := \mathrm{B}^\infty(C_0(X))$. If C is a compact G_δ set, then Exercise **1.19** yields a function $f \in C_0(X)$ with $0 \le f \le 1$ and $f^{-1}(1) = C$. Thus f^n converges monotonely and pointwise to χ_C, and hence $\chi_C \in \mathrm{B}^\infty(X)$. The σ-algebra of sets generated by the collection of compact G_δ sets is the collection of Baire sets. This is the smallest σ-algebra which makes every $f \in C_0(X)$ measureable. It follows that the set of real Baire measurable functions is monotone sequentially closed in the space of all bounded functions on X. Thus $\mathrm{B}^\infty(X)$ consists of Baire functions. It can be shown in a straightforward manner that $\mathrm{B}^\infty(X)$ consists of all bounded real Baire measurable functions which are supported on a σ-compact subset of X.

If X is metrizable and LCH, every compact set is a G_δ and the Baire and Borel functions coincide. The space X is σ-compact and every real Borel function belongs to $\mathrm{B}^\infty(X)$. In particular, $\mathrm{B}^\infty(\mathbb{R})$ is the set of all real bounded Borel functions on $\mathbb{R}$.

We establish some of the main properties of monotone sequential completions.

16.17.4. THEOREM. *Let $\mathfrak{A} \subset \mathcal{B}(H)$ be a concrete C*-algebra, and let $\mathcal{B}^\infty(\mathfrak{A})$ be its monotone sequential completion.*

(1) *Every countable subset of $\mathcal{B}^\infty(\mathfrak{A})$ lies in the monotone sequential completion of a separable subalgebra of $\mathfrak{A}$.*

(2) *$\mathcal{B}^\infty(\mathfrak{A}) + i\mathcal{B}^\infty(\mathfrak{A})$ is a C*-algebra.*

(3) *If $A \in \mathfrak{A}_{sa}$ and f is a bounded real-valued Borel function on $\mathbb{R}$, then $f(A) \in \mathcal{B}^\infty(\mathfrak{A})$.*

PROOF. (1) Let $Y = \{y \in \mathcal{B}^\infty(\mathfrak{A}_{sa}) : y \in \mathcal{B}^\infty(X_y),\ X_y \subset \mathfrak{A}_{sa} \text{ countable}\}$. Clearly $Y \supset \mathfrak{A}_{sa}$, and if $y_n \in Y$ is a monotone sequence with SOT limit y, then $y \in \mathcal{B}^\infty\big(\bigcup_{n\ge1} X_{y_n}\big)$. So Y is monotone sequentially complete, and hence $Y = \mathcal{B}^\infty(\mathfrak{A}_{sa})$. Then if $\{y_n : n \ge 1\}$ is any countable subset of $\mathcal{B}^\infty(\mathfrak{A}_{sa})$, then $\mathcal{B}^\infty\big(\bigcup_{n\ge1} X_{y_n}\big)$ contains them all.

(2) By (1), we can assume that $\mathfrak{A}$ is separable. Observe that in this case, $\mathcal{B}^\infty(\mathfrak{A}_{sa})$ is unital because $\mathfrak{A}$ contains an increasing sequential approximate identity; and it has SOT limit e, the unit for $\mathfrak{A}$.

Next we show that $\mathcal{B}^\infty(\mathfrak{A}_{sa})$ is norm closed. Suppose that $x_n \in \mathcal{B}^\infty(\mathfrak{A}_{sa})$ have norm limit x. We may suppose that $\|x_n - x_{n+1}\| < 2^{-n}$ for $n \ge 1$. The sequence $y_n = x_n - 2^{1-n}e$ is monotone increasing because

$$y_{n+1} - y_n = (x_{n+1} - x_n) + 2^{-n}e \ge 0.$$

Thus $x = \lim_n y_n \in \mathcal{B}^\infty(\mathfrak{A}_{sa})$.

Define $Y = \{x \in \mathcal{B}^\infty(\mathfrak{A}_{sa}) : x^k \in \mathcal{B}^\infty(\mathfrak{A}_{sa}) \text{ for } k \ge 1\}$. This set contains $\mathfrak{A}_{sa}$. Let y_n be a monotone sequence in Y with SOT limit y. This is a bounded sequence, so after scaling, we may suppose that $\|y_n\| \le 1$. Since $y_n^k \in Y$, it follows that for $0 \le t < 1$, the norm convergent series

$$(e - ty_n)^{-1} = e + \sum_{k\ge1} t^k y_n^k \in \mathcal{B}^\infty(\mathfrak{A}_{sa}).$$

The function $f_t(x) = (1 - tx)^{-1}$ is operator monotone by Proposition 9.1.7(4). Thus $(e - ty_n)^{-1}$ is a monotone sequence with SOT limit $(e - ty)^{-1}$ in $\mathcal{B}^\infty(\mathfrak{A}_{sa})$. As e and y are in $\mathcal{B}^\infty(\mathfrak{A}_{sa})$, so is $t^{-2}\big((1 - ty)^{-1} - 1 - y\big) = \sum_{n\ge0} t^n y^{n+2}$. This has norm limit y^2 as $t \to 0$; whence $y^2 \in \mathcal{B}^\infty(\mathfrak{A}_{sa})$. Recursively we pull off the initial terms, divide by t and deduce that $y^k \in \mathcal{B}^\infty(\mathfrak{A}_{sa})$ for $k \ge 2$. This shows that Y is monotone sequentially complete, and hence $Y = \mathcal{B}^\infty(\mathfrak{A}_{sa})$. That is, $\mathcal{B}^\infty(\mathfrak{A}_{sa})$ is closed under taking positive powers.

For products, we now take $x, y \in \mathcal{B}^\infty(\mathfrak{A}_{sa})$. Then

$$xy + yx = (x + y)^2 - x^2 - y^2 \in \mathcal{B}^\infty(\mathfrak{A}_{sa}).$$

Therefore

$$2xyx = (xy + yx)x + x(xy + yx) - (yx^2 + x^2y) \in \mathcal{B}^\infty(\mathfrak{A}_{sa}).$$

We also observe that

$$i(xy - yx) = (x + ie)^*y(x + ie) - xyx - y.$$

For fixed $x \in \mathfrak{A}_{sa}$, the set of $y \in \mathcal{B}^\infty(\mathfrak{A}_{sa})$ such that $(x+ie)^*y(x+ie) \in \mathcal{B}^\infty(\mathfrak{A}_{sa})$ contains $\mathfrak{A}_{sa}$ and is closed under monotone sequential limits. So this holds for all $y \in \mathcal{B}^\infty(\mathfrak{A}_{sa})$. Hence by the last identity, $i(xy - yx) \in \mathcal{B}^\infty(\mathfrak{A}_{sa})$ for $x \in \mathfrak{A}_{sa}$ and $y \in \mathcal{B}^\infty(\mathfrak{A}_{sa})$. Reversing the role of x and y in the identity shows that $\mathcal{B}^\infty(\mathfrak{A}_{sa})$ contains $(x + ie)^*y(x + ie)$ for $x \in \mathcal{B}^\infty(\mathfrak{A}_{sa})$ and $y \in \mathfrak{A}_{sa}$. But again the set of $y \in \mathcal{B}^\infty(\mathfrak{A}_{sa})$ such that $(x + ie)^*y(x + ie) \in \mathcal{B}^\infty(\mathfrak{A}_{sa})$ for all $x \in \mathcal{B}^\infty(\mathfrak{A}_{sa})$ is monotone sequentially closed. Therefore $(x + ie)^*y(x + ie) \in \mathcal{B}^\infty(\mathfrak{A}_{sa})$ for all $x, y \in \mathcal{B}^\infty(\mathfrak{A}_{sa})$. Therefore $i(xy - yx) \in \mathcal{B}^\infty(\mathfrak{A}_{sa})$ for all $x, y \in \mathcal{B}^\infty(\mathfrak{A}_{sa})$. By linearity, $\mathcal{B}^\infty(\mathfrak{A}) + i\mathcal{B}^\infty(\mathfrak{A})$ is closed under products, and thus is a C*-algebra.

(3) Now $\mathrm{B}^\infty(\mathfrak{A})$ contains $\mathrm{B}^\infty(C^*(A)) \simeq \mathrm{B}^\infty(\sigma(A))$. As noted in Example 16.17.3, this consists of all bounded Borel functions of A. There is a canonical normal homomorphism π of $\mathfrak{A}^{\dagger\dagger}$ onto $\mathfrak{A}'' \subset \mathcal{B}(H)$. In particular, π respects monotone sequential limits. So the restriction of π to $\mathrm{B}^\infty(\mathfrak{A})$ into $\mathcal{B}^\infty(\mathfrak{A})$ is a sequentially normal homomorphism. This takes $\mathrm{B}^\infty(\sigma(A))$ onto $C^*(A)''$ via the L^∞ functional calculus for A. ■

16.17.5. Proposition. *Every u.c.p. map $\mu : \mathfrak{A} \to \mathcal{B}(H)$ has a unique sequentially normal extension to a u.c.p. map $\bar{\mu} : \mathrm{B}^\infty(\mathfrak{A}) \to \mathcal{B}(H)$.*

Proof. This is established much like item (3) above. By Stinespring's theorem, there is a $*$-representation $\pi : \mathfrak{A} \to \mathcal{B}(K)$ and an isometry $\alpha : H \to K$ such that $\mu(a) = \alpha^*\pi(a)\alpha$. Let $\widetilde{\pi}$ be the unique normal homomorphism of $\mathfrak{A}^{\dagger\dagger}$ onto $\pi(\mathfrak{A})''$ extending π given by Theorem 10.5.5. Then $\widetilde{\mu} := \alpha^*\widetilde{\pi}\alpha$ is the unique normal extension of μ to a u.c.p. map of $\mathfrak{A}^{\dagger\dagger}$ into $\mathcal{B}(H)$ extending μ. The restriction $\bar{\mu}$ of $\widetilde{\mu}$ to $\mathrm{B}^\infty(\mathfrak{A})$ is the unique sequentially normal extension to $\mathrm{B}^\infty(\mathfrak{A})$. ■

16.18. The NC Choquet–Bishop–de Leeuw Theorem

In this section, we will establish the nc analogue of the Choquet–Bishop–de Leeuw theorem. As in the classical theorem, we need a technical definition of being supported on ∂K.

The *Pedersen–Baire envelope* of $\mathrm{C}(K)$ in $\mathrm{B}(K)$ is denoted by $\mathrm{B}^\infty(K)$.

16.18.1. DEFINITION. An nc state μ on $\mathrm{C}(K)$ is *supported on* ∂K if $\mu(f) = 0$ for all $f \in \mathrm{B}^\infty(K)$ such that $f|_{\partial K} = 0$.

The following easy observation will simplify the task to verifying this condition for scalar states.

16.18.2. PROPOSITION. *Suppose that every scalar state on* $\mathrm{C}(K)$ *which is maximal in the dilation order is supported on* ∂K. *Then every nc state on* $\mathrm{C}(K)$ *which is maximal in the dilation order is supported on* ∂K.

PROOF. Assume that $\mu \in L_p$ is an nc state on $\mathrm{C}(K)$ which is maximal in the dilation order. For every isometry $\alpha \in M_{p,1}$, the scalar state $\alpha^*\mu\alpha$ is maximal in the dilation order by Proposition 16.16.2. Let $f \in \mathrm{B}^\infty(K)$ such that $f|_{\partial K} = 0$. By hypothesis,

$$0 = (\alpha^*\mu\alpha)(f) = \alpha^*\mu(f)\alpha.$$

Now $\alpha(1) = \xi$ is an arbitrary unit vector in H_p, so this can be restated as saying that $\langle \mu(f)\xi, \xi\rangle = 0$ for all $\xi \in H_p$ with $\|\xi\| = 1$. This says that the numerical radius of $\mu(f)$ is 0; whence $\mu(f) = 0$. ■

We give an analysis of nc functions restricted to the space of scalar states on $\mathrm{C}(K)$. This provides a method of detecting dilation maximal states.

16.18.3. DEFINITION. For $f \in \mathrm{C}(K)_{sa}$, define scalar functions on L_1 by

$$\hat{f}(\mu) = \mu(f) \quad \text{and} \quad \check{f}(\mu) = \inf \mu(\bar{f}) \quad \text{for} \quad \mu \in L_1.$$

16.18.4. PROPOSITION. *Let* $f \in \mathrm{C}(K)_{sa}$. *Then*

(1) $\hat{f}$ *is continuous and affine.*

(2) $\check{f}$ *is convex and lower semicontinuous.*

(3) $\check{f} \le \hat{f}$.

PROOF. The function $\hat{f}$ is just the natural injection of f into $\mathrm{C}(K)^{\dagger\dagger}$ restricted to L_1. Hence it is weak-$*$ continuous and affine on L_1.

(2) Let $\mu, \nu \in L_1$ and $0 \le t \le 1$. Since $\bar{f}$ is nc convex, it is in particular convex on L_1, so that $\bar{f}(t\mu + (1-t)\nu) \supset t\bar{f}(\mu) + (1-t)\bar{f}(\nu)$. Thus

$$\begin{aligned}\check{f}(t\mu + (1-t)\nu) &= \inf \bar{f}(t\mu + (1-t)\nu)\\ &\le \inf t\bar{f}(\mu) + (1-t)\bar{f}(\nu) = t\check{f}(\mu) + (1-t)\check{f}(\nu).\end{aligned}$$

Therefore $\check{f}$ is convex.

Suppose that $(\mu_\lambda)_\Lambda$ is a net in L_1 converging weak-$*$ to μ. Since $\bar{f}$ is lower semicontinuous, the infimum $\breve{f}(\mu_\lambda)$ is attained. Drop to a cofinal subnet Λ' so that $\lim_{\Lambda'} \breve{f}(\mu_{\lambda'}) = \liminf_\Lambda \breve{f}(\mu_\lambda)$. Then $\big((\mu_{\lambda'}, \breve{f}(\mu_{\lambda'}))\big)_{\Lambda'} \subset \operatorname{epi}(\bar{f})$. As $\operatorname{epi}(\bar{f})$ is closed, $(\mu, \liminf_\Lambda \breve{f}(\mu_\lambda)) \in \operatorname{epi}(\bar{f})$. Therefore $\breve{f}(\mu) \le \liminf_\Lambda \breve{f}(\mu_\lambda)$; i.e. $\breve{f}$ is lower semicontinuous.

(3) Follows since $\bar{f} = \overline{\operatorname{ncconv}(\operatorname{epi}(f))} \supset \operatorname{epi}(f)$. ■

This leads to a checkable condition for dilation maximality of scalar states.

16.18.5. PROPOSITION. *Let $\mu \in L_1$. Then μ is maximal in the dilation order if and only if $\breve{f}(\mu) = \hat{f}(\mu)$ for all $f \in \mathrm{C}(K)_{sa}$.*

PROOF. This is now an immediate corollary of Theorem 16.15.7. If μ is maximal in the dilation order, then $\mu(\bar{f}) = [\mu(f), \infty)$ for all $f \in \mathrm{C}(K)_{sa}$. Thus $\breve{f}(\mu) = \mu(f) = \hat{f}(\mu)$. Conversely, if $\breve{f}(\mu) = \hat{f}(\mu)$, then $\mu(\bar{f}) = [\mu(f), \infty)$. So by Theorem 16.15.7, μ is maximal in the dilation order. ■

We need some results in the separable case that will be used to leverage the general result.

16.18.6. LEMMA. *Let K be a compact nc convex set such that $\mathrm{A}(K)$ is separable. Then the set*

$$Z = \{\nu \in L_1 : \nu \textit{ is pure and dilation maximal}\}$$

is a G_δ. If $\mu \in L_1$ is dilation maximal, there is a regular Borel probability measure ρ on Z with barycentre μ; i.e.

$$\mu(f) = \int_Z \nu(f)\, d\rho(\nu).$$

Moreoverany maximal measure on L_1 with barycentre μ is supported on Z.

PROOF. Since $\mathrm{A}(K)$ is separable, so is $\mathrm{C}(K)$; and L_1 is metrizable. Let $\{f_k\}_{k\ge 1}$ be a dense subset of $\mathrm{C}(K)_{sa}$. Define

$$X_{k,m} = \{\nu \in L_1 : \hat{f}_k(\nu) - \breve{f}_k(\nu) \ge \tfrac{1}{m}\} \quad \text{for} \quad k, m \ge 1.$$

As $\hat{f}$ is continuous and affine, and $\check{f}$ is convex and lower semicontinuous, $\hat{f} - \check{f}$ is concave and upper semicontinuous. Therefore $X_{k,m}$ is a closed set; and hence $Y = L_1 \setminus \bigcup_{k,m\geq 1} X_{k,m}$ is a G_δ. By Proposition 16.18.5, Y is the set of all dilation maximal elements of L_1. A point $\mu \in L_1$ is pure if and only if it is an extreme point. By Proposition 16.2.8, ∂L_1 is a G_δ. Therefore $Z = Y \cap \partial L_1$ is a G_δ.

By Choquet's theorem 16.2.11, there is a regular Borel probability measure ρ on ∂L_1 with barycentre μ; i.e.

$$\mu(f) = \int_{L_1} \nu(f)\, d\rho(\nu) \quad \text{for} \quad f \in C(K_1).$$

In particular, this holds for $\hat{f}$ when $f \in \mathrm{C}(K)$. We must show that ρ is supported on Y, which is equivalent to showing that $\rho(X_{k,m}) = 0$ for $k, m \geq 1$.

Suppose that $\rho(X_{k,m}) > 0$. Define measures

$$\sigma = \rho(X_{k,m})^{-1} \rho|_{X_{k,m}} \quad \text{and} \quad \tau = \rho(L_1 \setminus X_{k,m})^{-1} \rho|_{L_1 \setminus X_{k,m}}.$$

Let κ and λ be the barycentres of σ and τ in L_1. Then

$$\mu = \rho(X_{k,m})\kappa + \rho(L_1 \setminus X_{k,m})\lambda.$$

To compare $\widehat{f_k}(\kappa)$ and $\check{f_k}(\kappa)$, we can write σ as a limit of a sequence of finitely supported measures $\sigma_i = \sum t_{ij}\delta_{\nu_{ij}}$ with $\nu_{ij} \in X_{k,m}$. Let $\kappa_i = \sum t_{ij}\nu_{ij}$ be the barycentres of σ_i. The barycentre map is weak-$*$ continuous on $M_1^+(L_1)$. Therefore

$$\begin{aligned}
\widehat{f_k}(\kappa) &= \lim_i \widehat{f_k}(\kappa_i) = \lim_i \sum_j t_{ij} \widehat{f_k}(\nu_{ij}) \\
&\geq \frac{1}{m} + \liminf_i \sum_j t_{ij} \check{f_k}(\nu_{ij}) \\
&\geq \frac{1}{m} + \liminf_i \check{f_k}(\kappa_i) \geq \frac{1}{m} + \check{f_k}(\kappa).
\end{aligned}$$

The last line follows first because $\check{f_k}$ is convex and then because it is lower semicontinuous. However it now follows that

$$\begin{aligned}
\widehat{f_k}(\mu) &= \rho(X_{k,m})\widehat{f_k}(\kappa) + \rho(L_1 \setminus X_{k,m})\widehat{f_k}(\lambda) \\
&\geq \frac{\rho(X_{k,m})}{m} + \rho(X_{k,m})\check{f_k}(\kappa) + \rho(L_1 \setminus X_{k,m})\check{f_k}(\lambda) \\
&\geq \frac{\rho(X_{k,m})}{m} + \check{f_k}(\mu),
\end{aligned}$$

where convexity of $\check{f_k}$ is used again in the last inequality. But then $\hat{f_k}(\mu) \neq \check{f_k}(\mu)$, contrary to Proposition 16.18.5 and the maximality of μ in the dilation order. Therefore $\rho(X_{k,m}) = 0$ for $k, m \geq 1$, and so ρ is supported on $Y \cap \partial L_1 = Z$.

Indeed, the last paragraph shows that any representing measure for μ is supported on Y. Hence any maximal measure representing μ is supported on Z. ∎

The set L_1 of scalar states on $\mathrm{C}(K)$ lies in the predual of $\mathrm{B}(K) = \mathrm{C}(K)^{\dagger\dagger}$; so these states are normal states on $\mathrm{B}(K)$. If $f \in \mathrm{B}(K)$, the function $\hat{f}(\mu) = \mu(f)$ is bounded and affine on L_1. Say that f satisfies the *barycentre formula* if whenever ρ is a regular Borel probability measure on L_1 with barycentre μ, then

$$\mu(f) = \int_{L_1} \nu(f)\, d\rho(\nu).$$

In general, this is not the case. It is valid however for $f \in \mathrm{B}^\infty(K)$.

16.18.7. COROLLARY. *Let K be a compact nc convex set. Then the barycentre formula holds for $f \in \mathrm{B}^\infty(K)$.*

PROOF. Let Y denote the set of all functions in $\mathrm{B}^\infty(K)_{sa}$ for which the barycentre formula holds. By definition, the barycentre formula holds for $\mathrm{C}(K)$. Suppose that $f_n \in Y$ converge monotonely to f. In particular, this is a bounded sequence. Since states $\nu \in L_1$ are normal on $\mathrm{B}(K)$, it follows that $\hat{f}_n$ converges monotonely and pointwise to $\hat{f}$. Therefore the barycentre formula holds for f by the monotone convergence theorem. Hence Y is monotone sequentially closed, and thus coincides with $\mathrm{B}^\infty(K)$. ■

We are ready to establish the main result.

16.18.8. NC CHOQUET–BISHOP–DE LEEUW THEOREM. *Let K be a compact nc convex set. For each $x \in K_n$, there is a u.c.p. map $\mu : \mathrm{C}(K) \to M_n$ with barycentre x which is supported on ∂K.*

PROOF. Given $x \in K_n$, there is a u.c.p. map $\mu : \mathrm{C}(K) \to M_n$ representing x which is dilation maximal by Proposition 16.16.2. By Proposition 16.18.2, to show that μ is supported on ∂K, it suffices to prove this for dilation maximal scalar states. So we may take $n = 1$.

Fix $f \in \mathrm{B}^\infty(K)$ such that $f|_{\partial K} = 0$. By Theorem 16.17.4(2), f lies in the monotone sequential completion of a separable subalgebra of $\mathrm{C}(K)$, which is generated by a separable operator system $S \subset \mathrm{A}(K)$. Let K_0 be the state space of S. By the Hahn–Banach theorem, the restriction map $r : K \to K_0$ is surjective. We identify S with $\mathrm{A}(K_0)$ and observe that $g \in \mathrm{B}(K_0)$ can be identified with $g \circ r \in \mathrm{B}(K)$, and conversely, every function of the form $g \circ r \in \mathrm{B}(K)$ arises in this way. So we can identify $\mathrm{B}(K_0)$ with this subalgebra of $\mathrm{B}(K)$, identify $\mathrm{C}(K_0)$ with the subalgebra of $\mathrm{C}(K)$ generated by S, and identify $\mathrm{B}^\infty(K_0)$ with a subalgebra of $\mathrm{B}^\infty(K)$. Then $f \in \mathrm{B}^\infty(K_0)$; i.e. f is a Baire nc function on the nc state space of the separable operator subsystem $S = \mathrm{A}(K_0)$.

Next we show that $\mu_0 = \mu|_{\mathrm{C}(K_0)}$ is a dilation maximal state. For any g in $\mathrm{C}(K_0)_{sa}$, identified with $g \circ r \in \mathrm{C}(K)$, it is easy to see that $\overline{g \circ r} = \bar{g} \circ r$. Since μ

is dilation maximal,

$$\mu_0(\bar{g}) = \mu(\overline{g \circ r}) = \mu(g \circ r) = \mu_0(g).$$

By Proposition 16.18.5, μ_0 is dilation maximal.

Let $L_{0,1}$ denote the scalar state space of $\mathrm{C}(K_0)$. By Lemma 16.18.6,

$$Z = \{\nu_0 \in L_{0,1} : \nu_0 \text{ is pure and dilation maximal}\}$$

is a G_δ. Since μ_0 is dilation maximal, there is a regular Borel probability measure ρ on Z with barycentre μ_0; i.e.

$$\mu_0(g) = \int_Z \nu_0(g)\, d\rho(\nu_0) \quad \text{for} \quad g \in \mathrm{C}(K_0).$$

We claim that each $\nu_0 \in Z$ extends to a dilation maximal pure state on $\mathrm{C}(K)$. Now ν_0 has a minimal Stinespring dilation $\nu_0 = \alpha_0^* \delta_{x_0} \alpha_0$; i.e. (x_0, α_0) represents ν_0, where $x_0 \in (\partial K_0)_n$ by Proposition 16.16.2 and $\alpha_0 \in M_{n,1}$ is an isometry. Let $F = \{x \in K_n : r(x) = x_0\}$. This is a closed face of K_n. By the classical Krein–Milman theorem, F has an extreme point x, which means that x is pure. By the Davidson–Kennedy theorem, there is a maximal dilation $y \in (\partial K)_m$ and isometry $\beta \in M_{m,n}$ so that (y, β) represents x. Thus δ_y is a boundary representation of $\mathrm{C}(K)$; i.e. it is irreducible and dilation maximal. Define $\nu = \alpha_0^* \beta^* \delta_y \beta \alpha_0$. Thus $(y, \beta\alpha_0)$ is a representation of ν. Since δ_y is irreducible, this is a minimal representation and ν is pure. Since y is maximal, ν is dilation maximal by Proposition 16.16.2. Now $y_0 = y|_{K_0}$ is a dilation of x_0; but $x_0 \in \partial K_0$, so y_0 is a trivial dilation of x_0. Therefore $\nu|_{\mathrm{C}(K_0)} = \alpha_0^* \delta_{x_0} \alpha_0 = \nu_0$; whence ν_0 extends to ν.

The function $f \in \mathrm{B}^\infty(K)$ vanishes on ∂K. Since it is an nc function, it vanishes on all dilation maximal pure states ν. By the previous paragraph, if $\nu_0 \in Z$, there is a dilation maximal pure state ν on $\mathrm{C}(K)$ which extends it; whence $\nu_0(f) = \nu(f) = 0$. By Corollary 16.18.7, the barycentre formula holds for f, namely

$$\mu(f) = \mu_0(f) = \int_Z \nu(f)\, d\nu = 0.$$

It follows that μ is supported on ∂K. ∎

Exercises for Chapter 16

16.1 Show that if x is the barycentre of $\mu \in M_1^+(K)$, then $\delta_x \prec \mu$,

16.2 Let $f \in C(K)$.
(a) Show that $\hat{f} = \sup\{\mu(f) : \delta_x \prec \mu\}$ and $\check{f} = \inf\{\mu(f) : \delta_x \prec \mu\}$.
(b) If $x \in \operatorname{ext} K$, show that $\check{f}(x) = \hat{f}(x) = f(x)$.

16.3 Let $K = \{x = (x_n) \in l_2 : |x_n| \le 2^{-n}\}$. Define $f(x) = \sum_{n \ge 1} x_n$. Show that $f \in A(K)$, but is not in $A(K, l_2)$.

16.4 (*Rainwater*). Suppose that $(a_n)_{n\ge 1}$ is a bounded sequence in $A(K)$ such that $\lim_n a_n(x) = 0$ for $x \in \operatorname{ext} K$. Show that $\lim_n a_n = 0$ and 0 lies in $\overline{\operatorname{conv}}(\{a_n\}_{n\ge 1})$.

16.5 Show that every maximal measure is supported on $\overline{\operatorname{ext} K}$.

16.6 Let K be a metrizable compact convex set and let h be a strictly convex function on K.

(a) (*Hervé*). Show that $B_h := \{x \in K : \hat{h}(x) = h(x)\}$ is equal to ∂K.

(b) (*Mokobodzki*). Show that a measure on K is maximal if and only if it is supported on ∂K.

16.7 (*Bishop–de Leeuw*). Let Σ be the σ-algebra of subsets of $\operatorname{ext}(K)$ of the form $B \cap \operatorname{ext} K$ for B a Baire subset of K. Prove that there is a probability measure ν on $(\operatorname{ext} K, \Sigma)$ such that $\nu(a) = a(x)$ for $a \in A(K)$.
HINT: Let $\mu \succ \delta_x$ be maximal, and define $\nu(B \cap \operatorname{ext} K) := \mu(B)$.

16.8 (*Skau*). The Bishop–de Leeuw order on measures is given by $\mu \prec_{BL} \nu$ if $\mu(a^2) \le \nu(a^2)$ for all $a \in A(K)$. Fix $\frac{1}{2} < r < 1$. Let $A_k = e^{2k\pi i/3}$ and $B_k = -rA_k$ for $k = 0, 1, 2$. Let $K = \operatorname{conv}\{A_k, B_k : 0 \le k \le 2\} \subset \mathbb{C}$. Define $\mu = \frac{1}{3}\sum_{k=0}^{2} \delta_{A_k}$ and $\nu = \frac{1}{3}\sum_{k=0}^{2} \delta_{B_k}$. Show that both μ and ν are maximal in the Choquet order, but that $\nu \prec_{BL} \mu$ in the Bishop–de Leeuw order.

16.9 Let $A = (A_1, \dots, A_d)$ be a d-tuple in $\mathcal{B}(H)$, and let $\mathfrak{A} = C^*(A)$. Define

$$W_n(A) = \big\{(\varphi(A_1), \dots, \varphi(A_d)) : \varphi \in \mathrm{UCP}(\mathfrak{A}, \mathcal{B}(H_n))\big\}$$

for $n \le \kappa = \max\{\dim H, \aleph_0\}$.

(a) Show that $W = \coprod_{n\le\kappa} W_n$ is a compact nc convex set.

(b) (*Davidson–Dor-On–Shalit–Solel*). If $B = (B_1, \dots, B_d) \in \mathcal{B}(H')$ is another d-tuple, show that there is a u.c.p. map $\varphi : C^*(A) \to C^*(B)$ with $\varphi(A_i) = B_i$ for $1 \le i \le d$ if and only if $W(B) \subset W(A)$.

16.10 Using the notation from the proof of the Effros–Winkler theorem, define a functional on $M_n(E)$ by $\psi(x) = D^*xD$. This determines a map $\Psi \in CB(E, M_n)$. Show that this is the same map Ψ obtained in the proof.

16.11 (*Effros–Winkler*). If $K \subset \mathcal{M}(E)$ is an nc convex set, define it *nc polar* to be

$$(K^\circ)_n = \{\varphi \in M_n(E_*) : \mathrm{Re}(\varphi \otimes \mathrm{id}_p)(x) \le 1_{np} \text{ for } x \in K_p,\ p \le \kappa\}$$

Prove that if K is compact and $0 \in K_1$, then $K^{\circ\circ} = K$.

16.12 Let $A = (A_1, \dots, A_d)$ be a d-tuple in $\mathcal{B}(H)$. Show that

$$(W(A) \cup \{0\})^\circ = W(A)^\circ = D_A.$$

16.13 Show that every nc state on $\mathrm{C}(K)$ extends uniquely to a normal u.c.p. map on $\mathrm{B}(K)$.

16.14 Let K, L be nc convex sets and let $\theta : K \to L$ be an nc affine homeomorphism. Show that θ takes nc pure points to nc pure points, maximal points to maximal points, and nc extreme points to nc extreme points.

16.15 Let $S = \mathbb{C} \oplus \mathbb{C}$.
(a) Describe all u.c.p. maps from S into $\mathcal{B}(H)$.
(b) Show that $C^*_{\max}(S) \simeq C[0,1]$.

16.16 (*Kirchberg–Wassermann*). Let $S = \mathbb{C} \oplus \mathbb{C} \oplus \mathbb{C}$.
(a) Describe all u.c.p. maps from S into $\mathcal{B}(H)$.
(b) Deduce that for any $T \in \mathcal{B}(H)$, there is a $*$-homomorphism of $C^*_{\max}(S)$ onto $C^*(T)$.
(c) Show that the universal C*-algebra of a contraction is a quotient of $C^*_{\max}(S)$.

16.17 (*Kirchberg–Wassermann*). Let $S = M_2$.
(a) Let $T \in \mathcal{B}(H)$ with $\|T\| \le 1$. Show that $\varphi\big(\left[\begin{smallmatrix} a & b \\ c & d \end{smallmatrix}\right]\big) = \frac{1}{2}((a+d)I + bT + cT^*)$ is a u.c.p. map of S onto $\operatorname{span}\{I, T, T^*\}$.
(b) Deduce that the universal C*-algebra of a contraction is a quotient of $C^*_{\max}(S)$.

16.18 (*Kavruk–Paulsen–Todorov–Tomforde*). Let $\mathcal{S}$ be an operator system.
(a) Let J be an operator system kernel of $\mathcal{S}$ (Exercise **14.40**); and let $\mathcal{J}$ denote the ideal of $C^*_{\max}(\mathcal{S})$ generated by J. Show that $\mathcal{J} \cap \mathcal{S} = J$ and $C^*_{\max}(\mathcal{S})/\mathcal{J} \simeq C^*_{\max}\big((\mathcal{S}/J)_{opsys}\big)$.
HINT: Let $\varphi : \mathcal{S} \to \mathcal{B}(H)$ be a u.c.p. map with $\ker\varphi = J$, and let ψ denote the canonical embedding of $(\mathcal{S}/J)_{opsys}$ into $C^*_{\max}\big((\mathcal{S}/J)_{opsys}\big)$. Let $\pi : C^*_{\max}(\mathcal{S}) \to C^*_{\max}\big((\mathcal{S}/J)_{opsys}\big)$ be the unique $*$-homomorphism extending $\psi\varphi$. Let $\mathcal{I} = \ker\pi$. Show that π factors through $C^*_{\max}(\mathcal{S})/\mathcal{J}$, and that the induced map of $(\mathcal{S}/J)_{opsys}$ into $C^*_{\max}(\mathcal{S})/\mathcal{J}$ is completely isometric. Use the universal property of $C^*_{\max}((\mathcal{S}/J)_{opsys})$ to get a map back into $C^*_{\max}(\mathcal{S})/\mathcal{J}$.
(b) Show that a subspace $J \subset \mathcal{S}$ is an operator system kernel if and only if the ideal $\mathcal{J}$ of $C^*_{\max}(\mathcal{S})$ generated by J satisfies $\mathcal{J} \cap \mathcal{S} = J$.

16.19 Let K be a compact nc convex set and let $F : K \to M_n(\mathcal{M})$ be a self-adjoint multivalued bounded nc function with convex envelope $\overline{F}$. Prove that

$$\overline{F}(x) = \textstyle\bigcap_{g \le 1_m \otimes F} \{\alpha \in M_n(M_p) : 1_m \otimes \alpha \ge g(x)\}$$

for $x \in K_p$, where the intersection is taken over all m and all convex nc functions $g \in M_m(M_n(\mathrm{C}(K)))$ satisfying $g \le 1_m \otimes F$.

16.20 (*nc Jensen inequality*). Let K be a compact nc convex set and let $f \in \mathrm{B}(K)$ be a self-adjoint lower semicontinuous convex nc function. Show that for any u.c.p. map $\mu : \mathrm{C}(K) \to M_n$ with barycenter $x \in K_n$, $f(x) \le \mu(f)$.

16.21 Show that if μ is a u.c.p. map on $\mathrm{C}(K)$ with barycentre x, then $\delta_x \prec_c \mu$.
HINT: let (y, α) be a representation of μ; so $x = \alpha^* y \alpha$.

16.22 Verify Proposition 16.15.3.

16.23 Let $\mathfrak{A} \subset \mathcal{B}(H)$ be a C*-algebra.
(a) Show that if $\mathfrak{A}$ is not separable, $\mathcal{B}^\infty(\mathfrak{A})$ may not have a unit.
(b) Show that $\mathcal{B}^\infty(\mathfrak{A}) + \mathbb{C}I$ is monotone sequentially complete.

16.24 Let $\mathfrak{A} \subset \mathcal{B}(H)$ be a C*-algebra.
(a) If $a \in \mathfrak{A}_1^+$, show that $(\frac{1}{n} + a)^{-1}a$ increases to the range projection of a.
(b) Let $\{p_n : n \geq 1\}$ be a countable set of projections in $\mathcal{B}^\infty(\mathfrak{A})$. Show that $p := \bigvee_{n\geq 1} p_n \in \mathcal{B}^\infty(\mathfrak{A})$.
HINT: consider $a = \sum_{n\geq 1} 2^{-n}p_n$.
(c) Show that $p - \bigvee_{n\geq 1}(p - p_n)$ is the projection onto $\bigwedge_{n\geq 1} p_n$. So the set of projections $\mathrm{Pr}(\mathcal{B}^\infty(\mathfrak{A}))$ in $\mathcal{B}^\infty(\mathfrak{A})$ forms a σ-complete lattice.

16.25 Show that if X is a compact Hausdorff space such that $C(X)$ is monotone sequentially complete, then X is zero dimensional (i.e. if $x \neq y \in X$, there is a clopen set U such that $x \in U$ and $y \in U^c$).
HINT: use Urysohn's lemma to construct $f, g \in C(X)_1^+$ with $f(x) = 1 = g(y)$ and $fg = 0$. Let p be the range projection of f.

Notes on Chapter 16

The modern theory of representations of affine functions on convex sets begins with Choquet's theorem [54]. The proof of Choquet's theorem presented here is due to H. Rosenthal [264]. An order on measures was defined by Bishop and de Leeuw [30], who found an appropriate version of Choquet's theorem in the non-metrizable case. Their order used sums of squares, a somewhat smaller cone than the convex functions. Mokobodzki's theorem is from [207]. See the monographs Alfsen [6] and Phelps [236] for an introduction. Also see the more recent major treatise [196] by Lukeš, Malý, Netuka and Spurný.

Matrix ordered spaces were introduced by Choi and Effros [53]. This was taken up by Wittstock [329, 330], including a version of the Hahn–Banach theorem. The separation theorem for nc convex sets was established by Effros and Winkler [108]. Since nc convex sets are determined by their matrix convex part, the bulk of the categorical duality appears in Webster and Winkler [319]. They also prove a Krein–Milman type theorem, but the notion of extreme point includes many points that are not extreme in the definition used here. Indeed, since they only use finite levels, there may be no extreme points in our sense. The notion of nc convex sets is taken from Davidson–Kennedy [77], and the complete version of categorical duality and the nc version of Krein–Milman is there.

Nc functions occur somewhat disguised in Takesaki's paper [301] representing a noncommutative C*-algebra $\mathfrak{A}$ as a space of noncommutative functions on $\mathrm{Rep}(\mathfrak{A}, H)$. This was extended to the non-separable case by Bichteler [29]. These functions preserved direct sums and were equivariant under unitary conjugation.

Nevertheless, the results of sections 16.8, 16.10 and 16.11 are essentially from these two papers. A somewhat different notion was developed by J. Taylor [303] for the purpose of a functional calculus for non-commuting d-tuples. His domains were closed under direct sums, and his functions respect this and are equivariant under conjugation by similarities that remain in the domain. This determines a set of nc functions with analyticity. See [171] for a modern view of Taylor's work and more recent developments. Neither of these have a notion of matrix convex combinations, which came later with work of Choi–Effros and Wittstock.

The notion of the maximal C*-algebra of an operator system is due to Kirchberg and Wassermann [179] and to Blecher [32]. The connection between $\mathrm{C}(K)$ and $C^*_{\max}(\mathrm{A}(K))$ is from Davidson–Kennedy [77], which also contains the material of section 16.12.

Convex nc functions were introduced in Davidson–Kennedy [77] along with the main results. Operator monotone functions were defined and classified by Löwner [192]. For functions on $(-1,1)$, they have the form $f(x) = f(0)+\int_{-1}^{1} \frac{x}{1-tx}\, d\mu(t)$ for a positive Borel measure on $[-1,1]$. His student Kraus [183] defined operator convex functions, and found a useful characterization. Bendat and Sherman [25] characterized them as the functions $f(x) = f(0) + f'(0)x + \int_{-1}^{1} \frac{x^2}{1-tx}\, d\mu(t)$ for a positive Borel measure on $[-1,1]$. Hansen and Pedersen [139] studied the Jensen inequality for operator convex functions (see Exercise **16.20** for a generalization), and showed that operator convex functions on $[a,b]$ are nc convex. See Simon's book [286] for a complete treatment of operator monotone and convex functions.

The monotone sequential closure of a C*-algebra was introduced by Kadison [165], and further studied by Pedersen [230]. Pedersen [231] showed that the set of *universally measureable* elements $\mathcal{U}$ of a C*-algebra contains the monotone sequential closure, sometimes properly. Brown [40] showed that $\mathcal{U} + i\mathcal{U}$ is a C*-algebra which is the analogue of the Borel functions, now called the Brown–Borel envelope. He notes that Pedersen told him that the barycentre formula holds for this class. For more information, see [234] or [272].

The remaining results of this chapter are all taken from Davidson–Kennedy [77]. This paper also contains an integral representation when $\mathrm{A}(K)$ is separable for any u.c.p. map which is supported on ∂K. This requires some knowledge of measures taking values in the set of completely positive maps.

BIBLIOGRAPHY

[1] V.M. Adamjan, D.Z. Arov, and M.G. Kreĭn, *Infinite Hankel matrices and generalized problems of Carathéodory-Fejér and F. Riesz*, Funkcional. Anal. i Priložen. **2** (1968), 1–19 (Russian).

[2] J. Agler and J.E. McCarthy, *Complete Nevanlinna-Pick kernels*, J. Funct. Anal. **175** (2000), 111–124.

[3] J. Agler and J.E. McCarthy, *Pick interpolation and Hilbert function spaces*, Graduate Studies in Mathematics, vol. 44, American Mathematical Society, Providence, RI, 2002.

[4] L. Alaoglu, *Weak topologies of normed linear spaces*, Ann. of Math. (2) **41** (1940), 252–267.

[5] F. Albiac and N.J. Kalton, *Topics in Banach space theory*, 2nd ed., Graduate Texts in Mathematics, vol. 233, Springer, [Cham], 2016. With a foreword by Gilles Godefory.

[6] E.M. Alfsen, *Compact convex sets and boundary integrals*, Ergebnisse der Mathematik und ihrer Grenzgebiete, Band 57, Springer-Verlag, New York-Heidelberg, 1971.

[7] E.M. Alfsen and E.G. Effros, *Structure in real Banach spaces. I, II*, Ann. of Math. (2) **96** (1972), 98–128; ibid. (2) 96 (1972), 129–173.

[8] N.T. Andersen, *Similarity of continuous nests*, Bull. London Math. Soc. **15** (1983), 131–132.

[9] T. Andô, *On a pair of commutative contractions*, Acta Sci. Math. (Szeged) **24** (1963), 88–90.

[10] A. Arias and G. Popescu, *Factorization and reflexivity on Fock spaces*, Integral Equations Operator Theory **23** (1995), 268–286.

[11] N. Aronszajn, *Theory of reproducing kernels*, Trans. Amer. Math. Soc. **68** (1950), 337–404.

[12] N. Aronszajn and K.T. Smith, *Invariant subspaces of completely continuous operators*, Ann. of Math. (2) **60** (1954), 345–350.

[13] W.B. Arveson, *Operator algebras and measure preserving automorphisms*, Acta Math. **118** (1967), 95–109.

[14] W.B. Arveson, *Subalgebras of C^*-algebras*, Acta Math. **123** (1969), 141–224.

[15] W.B. Arveson, *Subalgebras of C^*-algebras. II*, Acta Math. **128** (1972), 271–308.

[16] W.B. Arveson, *Interpolation problems in nest algebras*, J. Functional Analysis **20** (1975), 208–233.

[17] W.B. Arveson, *Ten lectures on operator algebras*, CBMS Regional Conference Series in Mathematics, vol. **55**, Published for the Conference Board of the Mathematical Sciences, Washington, DC; by the American Mathematical Society, Providence, RI, 1984.

[18] W.B. Arveson, *Subalgebras of C^*-algebras. III. Multivariable operator theory*, Acta Math. **181** (1998), 159–228.

[19] W.B. Arveson, *The noncommutative Choquet boundary*, J. Amer. Math. Soc. **21** (2008), 1065–1084.

[20] W.B. Arveson and K.B. Josephson, *Operator algebras and measure preserving automorphisms. II*, J. Functional Analysis **4** (1969), 100–134.

[21] F.V. Atkinson, *The normal solubility of linear equations in normed spaces*, Mat. Sbornik N.S. **28(70)** (1951), 3–14 (Russian).

[22] S. Banach, *Sur les fonctionnelles linéaires, I, II*, Studia Math. **1** (1929), 211–216, 223–239 (French).

[23] S. Banach, *Théorie des opérations linéaires*, Chelsea Publishing Co., New York, 1955 (French).

[24] S. Banach and H. Steinhaus, *Sur le principe de la condensation de singularités*, Fundam. Math. **9** (1927), 50–61 (French).

[25] J. Bendat and S. Sherman, *Monotone and convex operator functions*, Trans. Amer. Math. Soc. **79** (1955), 58–71.

K. R. Davidson, *Functional Analysis and Operator Algebras*, CMS/CAIMS Books in Mathematics 13,
https://doi.org/10.1007/978-3-031-63665-3_0

[26] H. Bercovici, *Operator theory and arithmetic in H^∞*, Mathematical Surveys and Monographs, vol. **26**, American Mathematical Society, Providence, RI, 1988.

[27] J.F. Berglund, H.D. Junghenn, and P. Milnes, *Analysis on semigroups*, Canadian Mathematical Society Series of Monographs and Advanced Texts, John Wiley & Sons, Inc., New York, 1989. Function spaces, compactifications, representations; A Wiley-Interscience Publication.

[28] A. Beurling, *On two problems concerning linear transformations in Hilbert space*, Acta Math. **81** (1948), 239–255.

[29] K. Bichteler, *A generalization to the non-separable case of Takesaki's duality theorem for C^*-algebras*, Invent. Math. **9** (1969/70), 89–98.

[30] E. Bishop and K. de Leeuw, *The representations of linear functionals by measures on sets of extreme points*, Ann. Inst. Fourier (Grenoble) **9** (1959), 305–331.

[31] W. Blaschke, *Eine Erweiterung des Satzes von Vitali über Folgen analytischer Funktionen*, S.-B. Sächs. Akad. Wiss. Leipzig Math.-Natur. Kl. **67** (1915), 194–200 (German).

[32] D.P. Blecher, *Modules over operator algebras, and the maximal C^*-dilation*, J. Funct. Anal. **169** (1999), 251–288.

[33] D.P. Blecher and C. Le Merdy, *Operator algebras and their modules—an operator space approach*, London Mathematical Society Monographs. New Series, vol. 30, The Clarendon Press, Oxford University Press, Oxford, 2004. Oxford Science Publications.

[34] D.P. Blecher, Z.-J. Ruan, and A.M. Sinclair, *A characterization of operator algebras*, J. Funct. Anal. **89** (1990), 188–201.

[35] H.F. Bohnenblust and A. Sobczyk, *Extensions of functionals on complex linear spaces*, Bull. Amer. Math. Soc. **44** (1938), 91–93.

[36] W.M. Boothby, *On two classical theorems of algebraic topology*, Amer. Math. Monthly **78** (1971), 237–249.

[37] F.F. Bonsall and J. Duncan, *Complete normed algebras*, Ergebnisse der Mathematik und ihrer Grenzgebiete, Band **80**, Springer-Verlag, New York-Heidelberg, 1973.

[38] A. Böttcher and B. Silbermann, *Analysis of Toeplitz operators*, 2nd ed., Springer Monographs in Mathematics, Springer-Verlag, Berlin, 2006, Prepared jointly with A. Karlovich.

[39] A. Brown and P.R. Halmos, *Algebraic properties of Toeplitz operators*, J. Reine Angew. Math. **213** (1963/64), 89–102.

[40] L.G. Brown, *Large C^*-algebras of universally measurable operators*, Q. J. Math. **65** (2014), 851–855.

[41] N.P. Brown and N. Ozawa, *C^*-algebras and finite-dimensional approximations*, Graduate Studies in Mathematics, vol. 88, American Mathematical Society, Providence, RI, 2008.

[42] J.W. Bunce, *The similarity problem for representations of C^*-algebras*, Proc. Amer. Math. Soc. **81** (1981), 409–414.

[43] J.W. Bunce, *Models for n-tuples of noncommuting operators*, J. Funct. Anal. **57** (1984), 21–30.

[44] R.B. Burckel, *Weakly almost periodic functions on semigroups*, Gordon and Breach Science Publishers, New York-London-Paris, 1970.

[45] L. Carleson, *Representations of continuous functions*, Math. Z. **66** (1957), 447–451.

[46] I. Chalendar and J.R. Partington, *Modern approaches to the invariant-subspace problem*, Cambridge Tracts in Mathematics, vol. 188, Cambridge University Press, Cambridge, 2011.

[47] S.Y.A. Chang, *A characterization of Douglas subalgebras*, Acta Math. **137** (1976), no. 2, 82–89.

[48] X. Chen and K. Guo, *Analytic Hilbert modules*, Chapman & Hall/CRC Research Notes in Mathematics, vol. **433**, Chapman & Hall/CRC, Boca Raton, FL, 2003.

[49] Z. Chen, *Characterizations of Arveson's Hardy space*, Complex Var. Theory Appl. **48** (2003), 453–465.

[50] M.D. Choi, *Completely positive linear maps on complex matrices*, Linear Algebra Appl. **10** (1975), 285–290.

[51] M.D. Choi and K.R. Davidson, *Perturbations of matrix algebras*, Michigan Math. J. **33** (1986), 273–287.

[52] M.D. Choi and K.R. Davidson, *A 3×3 dilation counterexample*, Bull. Lond. Math. Soc. **45** (2013), 511–519.

[53] M.D. Choi and E.G. Effros, *Injectivity and operator spaces*, J. Functional Analysis **24** (1977), 156–209.

[54] G. Choquet, *Existence des représentations intégrales au moyen des points extrémaux dans les cônes convexes*, C. R. Acad. Sci. Paris **243** (1956), 699–702 (French).

[55] E. Christensen, *Extensions of derivations. II*, Math. Scand. **50** (1982), 111–122.

[56] E. Christensen, *On nonselfadjoint representations of C^*-algebras*, Amer. J. Math. **103** (1981), 817–833.

[57] L.A. Coburn, *Weyl's theorem for nonnormal operators*, Michigan Math. J. **13** (1966), 285–288.

[58] L.A. Coburn, *The C^*-algebra generated by an isometry*, Bull. Amer. Math. Soc. **73** (1967), 722–726.

[59] P.J. Cohen, *Factorization in group algebras*, Duke Math. J. **26** (1959), 199–205.

[60] P.J. Cohen, *The independence of the continuum hypothesis*, Proc. Nat. Acad. Sci. U.S.A. **50** (1963), 1143–1148.

[61] J. Conway, *A course in functional analysis*, 2nd ed., Graduate Texts in Mathematics, vol. 96, Springer-Verlag, New York, 1990.

[62] J. Conway, *A course in operator theory*, Graduate Studies in Mathematics, vol. 21, American Mathematical Society, Providence, RI, 2000.

[63] Ş. Costea, E.T. Sawyer, and B.D. Wick, *The corona theorem for the Drury-Arveson Hardy space and other holomorphic Besov-Sobolev spaces on the unit ball in $\mathbb{C}^n$*, Anal. PDE **4** (2011), 499–550.

[64] M.J. Crabb and A.M. Davie, *von Neumann's inequality for Hilbert space operators*, Bull. London Math. Soc. **7** (1975), 49–50.

[65] J. Cuntz, *Simple C^*-algebras generated by isometries*, Comm. Math. Phys. **57** (1977), 173–185.

[66] K.R. Davidson, *Similarity and compact perturbations of nest algebras*, J. Reine Angew. Math. **348** (1984), 72–87.

[67] K.R. Davidson, *Approximate unitary equivalence of continuous nests*, Proc. Amer. Math. Soc. **97** (1986), 655–660.

[68] K.R. Davidson, *Perturbations of reflexive operator algebras*, J. Operator Theory **15** (1986), 289–305.

[69] K.R. Davidson, *Nest algebras*, Pitman Research Notes in Mathematics Series, vol. 191, Longman Scientific & Technical, Harlow; copublished in the United States with John Wiley & Sons, Inc., New York, 1988. Triangular forms for operator algebras on Hilbert space.

[70] K.R. Davidson, *C^*-algebras by example*, Fields Institute Monographs, vol. **6**, American Mathematical Society, Providence, RI, 1996.

[71] K. R. Davidson, *Large perturbations of nest algebras*, Integral Equations Operator Theory **96** (2024), Paper No. 31, 9.

[72] K.R. Davidson, A.H. Fuller, and E.T.A. Kakariadis, *Semicrossed products of operator algebras by semigroups*, Mem. Amer. Math. Soc. **247** (2017).

[73] K.R. Davidson and E.T.A. Kakariadis, *Conjugate dynamical systems on* C^**-algebras*, Int. Math. Res. Not. IMRN (2014), 1289–1311.

[74] K.R. Davidson and E.G. Katsoulis, *Isomorphisms between topological conjugacy algebras*, J. Reine Angew. Math. **621** (2008), 29–51.

[75] K.R. Davidson, E. Katsoulis, and D.R. Pitts, *The structure of free semigroup algebras*, J. Reine Angew. Math. **533** (2001), 99–125.

[76] K.R. Davidson and M. Kennedy, *The Choquet boundary of an operator system*, Duke Math. J. **164** (2015), 2989–3004.

[77] K.R. Davidson and M. Kennedy, *Noncommutative Choquet theory*, Mem. Amer. Math. Soc., to appear, available at arXiv:1905.08436.

[78] K.R. Davidson, J. Li, and D.R. Pitts, *Absolutely continuous representations and a Kaplansky density theorem for free semigroup algebras*, J. Funct. Anal. **224** (2005), 160–191.

[79] K.R. Davidson and J.L. Orr, *The invertibles are connected in infinite multiplicity nest algebras*, Bull. London Math. Soc. **27** (1995), 155–161.

[80] K.R. Davidson, J.L. Orr, and D.R. Pitts, *Connectedness of the invertibles in certain nest algebras*, Canad. Math. Bull. **38** (1995), 412–420.

[81] K.R. Davidson and V.I. Paulsen, *Polynomially bounded operators*, J. Reine Angew. Math. **487** (1997), 153–170.

[82] K.R. Davidson and D.R. Pitts, *The algebraic structure of non-commutative analytic Toeplitz algebras*, Math. Ann. **311** (1998), 275–303.

[83] K.R. Davidson and D.R. Pitts, *Nevanlinna-Pick interpolation for non-commutative analytic Toeplitz algebras*, Integral Equations Operator Theory **31** (1998), 321–337.

[84] K.R. Davidson and D.R. Pitts, *Invariant subspaces and hyper-reflexivity for free semigroup algebras*, Proc. London Math. Soc. (3) **78** (1999), 401–430.

[85] K.R. Davidson and B.H. Wagner, *Automorphisms of quasitriangular algebras*, J. Funct. Anal. **59** (1984), 612–627.

[86] C. Davis, *Some dilation and representation theorems*, Proceedings of the Second International Symposium in West Africa on Functional Analysis and its applications, Kumasi, 1979.

[87] J.A. Deddens, *Every isometry is reflexive*, Proc. Amer. Math. Soc. **28** (1971), 509–512.

[88] A. Devinatz, *Toeplitz operators on H^2 spaces*, Trans. Amer. Math. Soc. **112** (1964), 304–317.

[89] J. Dieudonné, *Sur le théorème de Hahn-Banach*, Revue Sci. (Rev. Rose Illus.) **79** (1941), 642–643 (French).

[90] J. Diestel, *Sequences and series in Banach spaces*, Graduate Texts in Mathematics, vol. 92, Springer-Verlag, New York, 1984.

[91] J. Dixmier, *Les anneaux d'opérateurs de classe finie*, Ann. Sci. École Norm. Sup. (3) **66** (1949), 209–261 (French).

[92] J. Dixmier, *Les fonctionnelles linéaires sur l'ensemble des opérateurs bornés d'un espace de Hilbert*, Ann. of Math. (2) **51** (1950), 387–408 (French).

[93] J. Dixmier, *Formes linéaires sur un anneau d'opérateurs*, Bull. Soc. Math. France **81** (1953), 9–39 (French).

[94] J. Dixmier, *von Neumann algebras*, North-Holland Mathematical Library, vol. 27, North-Holland Publishing Co., Amsterdam-New York, 1981. With a preface by E. C. Lance; Translated from the second French edition by F. Jellett.

[95] J. Dixmier, *C^*-algebras*, North-Holland Mathematical Library, Vol. 15, North-Holland Publishing Co., Amsterdam-New York-Oxford, 1977. Translated from the French by Francis Jellett

[96] A.P. Donsig, A. Katavolos, and A. Manoussos, *The Jacobson radical for analytic crossed products*, J. Funct. Anal. **187** (2001), 129–145.

[97] R.G. Douglas, *Toeplitz and Wiener-Hopf operators in $H^\infty + C$*, Bull. Amer. Math. Soc. **74** (1968), 895–899.

[98] R.G. Douglas, *On the spectrum of Toeplitz and Wiener-Hopf operators*, Abstract Spaces and Approximation (Proc. Conf., Oberwolfach, 1968), Birkhäuser, Basel, 1969, pp. 53–66.

[99] R.G. Douglas, *Banach algebra techniques in operator theory*, 2nd ed., Graduate Texts in Mathematics, vol. 179, Springer-Verlag, New York, 1998.

[100] R.G. Douglas and W. Rudin, *Approximation by inner functions*, Pacific J. Math. **31** (1969), 313–320.

[101] M.A. Dritschel and S.A. McCullough, *Boundary representations for families of representations of operator algebras and spaces*, J. Operator Theory **53** (2005), 159–167.

[102] S.W. Drury, *A generalization of von Neumann's inequality to the complex ball*, Proc. Amer. Math. Soc. **68** (1978), 300–304.

[103] N. Dunford, *Spectral theory*, Bull. Amer. Math. Soc. **49** (1943), 637–651.

[104] N. Dunford and J.T. Schwartz, *Linear operators. Part I*, Wiley Classics Library, John Wiley & Sons, Inc., New York, 1988. General theory; With the assistance of William G. Bade and Robert G. Bartle; Reprint of the 1958 original; A Wiley-Interscience Publication.

[105] N. Dunford and J.T. Schwartz, *Linear operators. Part II*, Interscience Publishers John Wiley & Sons, New York-London, 1963. Spectral theory. Self adjoint operators in Hilbert space; With the assistance of W.G. Bade and R.G. Bartle.

[106] P.L. Duren, *Theory of H^p spaces*, Pure and Applied Mathematics, Vol. 38, Academic Press, New York-London, 1970.

[107] E.G. Effros and Z.-J. Ruan, *Theory of operator spaces*, AMS Chelsea Publishing, Providence, RI, 2022. Corrected reprint of the 2000 original.

[108] E.G. Effros and S. Winkler, *Matrix convexity: operator analogues of the bipolar and Hahn-Banach theorems*, J. Funct. Anal. **144** (1997), 117–152.

[109] W.F. Eberlein, *Weak compactness in Banach spaces. I*, Proc. Nat. Acad. Sci. U.S.A. **33** (1947), 51–53.

[110] P. Enflo, *A counterexample to the approximation problem in Banach spaces*, Acta Math. **130** (1973), 309–317.

[111] P. Enflo, *On the invariant subspace problem for Banach spaces*, Acta Math. **158** (1987), 213–313.

[112] J.A. Erdos, *Operators of finite rank in nest algebras*, J. London Math. Soc. **43** (1968), 391–397.

[113] S.R. Foguel, *A counterexample to a problem of Sz.-Nagy*, Proc. Amer. Math. Soc. **15** (1964), 788–790.

[114] G.B. Folland, *Real analysis*, 2nd ed., Pure and Applied Mathematics (New York), John Wiley & Sons, Inc., New York, 1999. Modern techniques and their applications; A Wiley-Interscience Publication.

[115] A.E. Frazho, *Models for noncommuting operators*, J. Functional Analysis **48** (1982), 1–11.

[116] A.E. Frazho, *Complements to models for noncommuting operators*, J. Funct. Anal. **59** (1984), 445–461.

[117] M. Fukamiya, *On a theorem of Gelfand and Neumark and the B^*-algebra*, Kumamoto J. Sci. Ser. A **1** (1952), 17–22.

[118] J.B. Garnett, *Bounded analytic functions*, 1st ed., Graduate Texts in Mathematics, vol. 236, Springer, New York, 2007.

[119] I. Gelfand, *Normierte Ringe*, Rec. Math. [Mat. Sbornik] N. S. **9 (51)** (1941), 3–24 (German, with Russian summary).

[120] I. Gelfand and M. Naimark, *On the imbedding of normed rings into the ring of operators in Hilbert space*, Rec. Math. [Mat. Sbornik] N.S. **12(54)** (1943), 197–213 (English, with Russian summary).

[121] I. Gelfand and M. Naimark, *Normed rings with involutions and their representations*, Izvestiya Akad. Nauk SSSR. Ser. Mat. **12** (1948), 445–480 (Russian).

[122] I. Gelfand, D. Raikov, and G. Shilov, *Commutative normed rings*, Chelsea Publishing Co., New York, 1964. Translated from the Russian, with a supplementary chapter.

[123] A.M. Gleason and H. Whitney, *The extension of linear functionals defined on H^∞*, Pacific J. Math. **12** (1962), 163–182.

[124] T. Giordano, D. Kerr, N.C. Phillips, and A. Toms, *Crossed products of C^*-algebras, topological dynamics, and classification* (F. Perera, ed.), Advanced Courses in Mathematics. CRM Barcelona, Birkhäuser/Springer, Cham, 2018. Lecture notes based on the course held at the Centre de Recerca Matemàtica (CRM) Barcelona, June 14–23, 2011; Edited by F. Perera.

[125] J.G. Glimm, *On a certain class of operator algebras*, Trans. Amer. Math. Soc. **95** (1960), 318–340.

[126] J.G. Glimm and R.V. Kadison, *Unitary operators in C^*-algebras*, Pacific J. Math. **10** (1960), 547–556.

[127] H.H. Goldstine, *Weakly complete Banach spaces*, Duke Math. J. **4** (1938), 125–131.

[128] K. Gödel, *The consistency of the axiom of choice and of the generalized continuum hypothesi*, Proc. Natl. Acad. Sci. USA **24** (1938), 556–557.

[129] A. Grothendieck, *Critères de compacité dans les espaces fonctionnels généraux*, Amer. J. Math. **74** (1952), 168–186 (French).

[130] U. Haagerup, *Solution of the similarity problem for cyclic representations of C^*-algebras*, Ann. of Math. (2) **118** (1983), 215–240.

[131] A. Haar, *Der Massbegriff in der Theorie der kontinuierlichen Gruppen*, Ann. of Math. (2) **34** (1933), 147–169 (German).

[132] D.W. Hadwin and T.B. Hoover, *Operator algebras and the conjugacy of transformations*, J. Funct. Anal. **77** (1988), 112–122.

[133] F. Hahn, *A fixed-point theorem*, Math. Systems Theory **1** (1967), 55–57.

[134] H. Hahn, *Über lineare Gleichungssysteme in linearen Räumen*, J. Reine Angew. Math. **157** (1927), 214–229 (German).

[135] P.R. Halmos, *Measure Theory*, D. Van Nostrand Co., Inc., New York, N. Y., 1950.

[136] P.R. Halmos, *Naive set theory*, Undergraduate Texts in Mathematics, Springer-Verlag, New York-Heidelberg, 1974. Reprint of the 1960 edition.

[137] P.R. Halmos, *Ten problems in Hilbert space*, Bull. Amer. Math. Soc. **76** (1970), 887–933.

[138] M. Hamana, *Injective envelopes of operator systems*, Publ. Res. Inst. Math. Sci. **15** (1979), 773–785.

[139] F. Hansen and G.K. Pedersen, *Jensen's operator inequality*, Bull. London Math. Soc. **35** (2003), 553–564.

[140] P. Harmand, D. Werner, and W. Werner, *M-ideals in Banach spaces and Banach algebras*, Lecture Notes in Mathematics, vol. 1547, Springer-Verlag, Berlin, 1993.

[141] P. Hartman, *On completely continuous Hankel matrices*, Proc. Amer. Math. Soc. **9** (1958), 862–866.

[142] P. Hartman and A. Wintner, *The spectra of Toeplitz's matrices*, Amer. J. Math. **76** (1954), 867–882.

[143] M. Hartz, *An invitation to the Drury-Arveson space*, Lectures on analytic function spaces and their applications, Fields Inst. Monogr., vol. 39, Springer, Cham, [2023] 2023, pp. 347–413.

[144] H. Helson, *Lectures on invariant subspaces*, Academic Press, New York-London, 1964.

[145] H. Helson and D. Lowdenslager, *Prediction theory and Fourier series in several variables*, Acta Math. **99** (1958), 165–202.

[146] H. Helson and D. Lowdenslager, *Invariant subspaces*, Proc. Internat. Sympos. Linear Spaces (Jerusalem, 1960), Jerusalem Academic Press, Jerusalem; Pergamon, Oxford, 1961, pp. 251–262.

[147] E. Hewitt and K.A. Ross, *Abstract harmonic analysis. Vol. I*, 2nd ed., Grundlehren der Mathematischen Wissenschaften, vol. 115, Springer-Verlag, Berlin-New York, 1979. Structure of topological groups, integration theory, group representations.

[148] P.J. Hilton and S. Wylie, *Homology theory: An introduction to algebraic topology*, Cambridge University Press, New York, 1960.

[149] M.W. Hirsch, *Differential topology*, Graduate Texts in Mathematics, vol. 33, Springer-Verlag, New York, 1994. Corrected reprint of the 1976 original.

[150] K. Hoffman, *Banach spaces of analytic functions*, Dover Publications, Inc., New York, 1988, Reprint of the 1962 original.

[151] K. Hoffman, *Bounded analytic functions and Gleason parts*, Ann. of Math. (2) **86** (1967), 74–111.

[152] K. Hoffman and I. M. Singer, *Maximal algebras of continuous functions*, Acta Math. **103** (1960), 217–241.

[153] J.A. Holbrook, *Inequalities of von Neumann type for small matrices*, Function spaces (Edwardsville, IL, 1990), Lecture Notes in Pure and Appl. Math., vol. 136, Dekker, New York, 1992, pp. 189–193.

[154] J.A. Holbrook, *Schur norms and the multivariate von Neumann inequality*, Recent advances in operator theory and related topics (Szeged, 1999), Oper. Theory Adv. Appl., vol. 127, Birkhäuser, Basel, 2001, pp. 375–386.

[155] A. Hulanicki, *Means and Følner condition on locally compact groups*, Studia Math. **27** (1966), 87–104.

[156] A.J. Izzo, *A simple proof of the existence of Haar measure on amenable groups*, Math. Scand. **120** (2017), 317–319.

[157] N. Jacobson, *Structure theory of simple rings without finiteness assumptions*, Trans. Amer. Math. Soc. **57** (1945), 228–245.

[158] N. Jacobson, *The radical and semi-simplicity for arbitrary rings*, Amer. J. Math. **67** (1945), 300–320.

[159] N. Jacobson, *Structure of rings*, Revised edition, American Mathematical Society Colloquium Publications, vol. 37, American Mathematical Society, Providence, R.I., 1964.

[160] T. Jech, *Set theory*, Springer Monographs in Mathematics, Springer-Verlag, Berlin, 2003. The third millennium edition, revised and expanded.

[161] Y.Q. Ji and Y.H. Zhang, *Connectedness of the invertibles in certain nest algebras of order type* ω, J. Funct. Anal. **263** (2012), 2584–2592.

[162] B.E. Johnson, *The uniqueness of the (complete) norm topology*, Bull. Amer. Math. Soc. **73** (1967), 537–539.

[163] V. Jones and V.S. Sunder, *Introduction to subfactors*, London Mathematical Society Lecture Note Series, vol. 234, Cambridge University Press, Cambridge, 1997.

[164] R.V. Kadison, *On the orthogonalization of operator representations*, Amer. J. Math. **77** (1955), 600–620.

[165] R.V. Kadison, *Unitary invariants for representations of operator algebras*, Ann. of Math. (2) **66** (1957), 304–379.

[166] R.V. Kadison, *Operator algebras—the first forty years*, Operator algebras and applications, Part 1 (Kingston, Ont., 1980), Proc. Sympos. Pure Math., vol. 38, Amer. Math. Soc., Providence, R.I., 1982, pp. 1–18.

[167] R.V. Kadison and J.R. Ringrose, *Fundamentals of the theory of operator algebras. Vol. I*, Graduate Studies in Mathematics, vol. 15, American Mathematical Society, Providence, RI, 1997. Elementary theory, Reprint of the 1983 original.

[168] S. Kakutani, *Two fixed-point theorems concerning bicompact convex sets*, Proc. Imp. Acad. Tokyo **14** (1938), 242–245.

[169] S. Kakutani, *Weak topology and regularity of Banach spaces*, Proc. Imp. Acad. Tokyo **15** (1939), 169–173.

[170] E.T.A. Kakariadis and E.G. Katsoulis, *Semicrossed products of operator algebras and their* C^**-envelopes*, J. Funct. Anal. **262** (2012), 3108–3124.

[171] D.S. Kaliuzhnyi-Verbovetskyi and V. Vinnikov, *Foundations of free noncommutative function theory*, Mathematical Surveys and Monographs, vol. 199, American Mathematical Society, Providence, RI, 2014.

[172] E. Kaniuth, *A course in commutative Banach algebras*, Graduate Texts in Mathematics, vol. 246, Springer, New York, 2009.

[173] I. Kaplansky, *A theorem on rings of operators*, Pacific J. Math. **1** (1951), 227–232.

[174] E.G. Katsoulis and C. Ramsey, *The isomorphism problem for tensor algebras of multivariable dynamical systems*, Forum Math. Sigma 10 (2022), Paper No. e81, 12.

[175] Y. Katznelson, *An introduction to harmonic analysis*, 3rd ed., Cambridge Mathematical Library, Cambridge University Press, Cambridge, 2004.

[176] J. Kelley, *General topology*, Graduate Texts in Mathematics, No. 27, Springer-Verlag, New York-Berlin, 1975. Reprint of the 1955 edition [Van Nostrand, Toronto, Ont.]

[177] M. Kennedy, *Wandering vectors and the reflexivity of free semigroup algebras*, J. Reine Angew. Math. **653** (2011), 47–73.

[178] M. Kennedy, *The structure of an isometric tuple*, Proc. Lond. Math. Soc. (3) **106** (2013), 1157–1177.

[179] E. Kirchberg and S. Wassermann, *C^*-algebras generated by operator systems*, J. Funct. Anal. **155** (1998), 324–351.

[180] G. Knese, *The von Neumann inequality for 3×3 matrices*, Bull. Lond. Math. Soc. **48** (2016), 53–57.

[181] A. Kolmogoroff, *Zur Normierbarkeit eines allgemeinen topologischen linearen Raumes*, Stud. Math. **5** (1934), 29–33 (German).

[182] P. Koosis, *Introduction to H_p spaces*, 2nd ed., Cambridge Tracts in Mathematics, vol. 115, Cambridge University Press, Cambridge, 1998. With two appendices by V. P. Havin [Viktor Petrovich Khavin].

[183] F. Kraus, *Über konvexe Matrixfunktionen*, Math. Z. **41** (1936), 18–42 (German).

[184] M. Krein and D. Milman, *On extreme points of regular convex sets*, Studia Math. **9** (1940), 133–138.

[185] M. Krein and V. Šmulian, *On regularly convex sets in the space conjugate to a Banach space*, Ann. of Math. (2) **41** (1940), 556–583.

[186] S. Kremp, *An elementary proof of the Eberlein-Šmulian theorem and the double limit criterion*, Arch. Math. (Basel) **47** (1986), 66–69.

[187] D.R. Larson, *Nest algebras and similarity transformations*, Ann. of Math. (2) **121** (1985), 409–427.

[188] E.C. Lance, *Cohomology and perturbations of nest algebras*, Proc. London Math. Soc. (3) **43** (1981), 334–356.

[189] D.R. Larson and D.R. Pitts, *Idempotents in nest algebras*, J. Funct. Anal. **97** (1991), 162–193.

[190] A. Lebow, *A power-bounded operator that is not polynomially bounded*, Michigan Math. J. **15** (1968), 397–399.

[191] Huaxin Lin, *An introduction to the classification of amenable C^*-algebras*, World Scientific Publishing Co., Inc., River Edge, NJ, 2001.

[192] K. Löwner, *Über monotone Matrixfunktionen*, Math. Z. **38** (1934), 177–216 (German).

[193] V.I. Lomonosov, *Invariant subspaces of the family of operators that commute with a completely continuous operator*, Funkcional. Anal. i Priložen. **7** (1973), 55–56 (Russian).

[194] V.I. Lomonosov and P. Rosenthal, *The simplest proof of Burnside's theorem on matrix algebras*, Linear Algebra Appl. **383** (2004), 45–47.

[195] L.H. Loomis, *An introduction to abstract harmonic analysis*, D. Van Nostrand Co., Inc., Toronto-New York-London, 1953.

[196] J. Lukeš, J. Malý, I. Netuka, and J. Spurný, *Integral representation theory*, De Gruyter Studies in Mathematics, vol. 35, Walter de Gruyter & Co., Berlin, 2010. Applications to convexity, Banach spaces and potential theory.

[197] J. Manheim, *The genesis of point set topology*, Pergamon Press, Oxford-Paris-Frankfurt; The Macmillan Company, New York, 1964.

[198] A. Markov, *Quelques théoremes sur les ensembles abeliens*, C. R. (Dokl.) Acad. Sci. URSS **10** (1936), 311–313 (French).

[199] D.E. Marshall, *Subalgebras of L^∞ containing H^∞*, Acta Math. **137** (1976), no.2, 91–98, DOI 10.1007/BF02392414. MR0428045

[200] D.E. Marshall, *Blaschke products generate H^∞*, Bull. Amer. Math. Soc. **82** (1976), 494–496.

[201] S. Masur, *Über konvexe Mengen in linearen normierten Räumen*, Studia Math. **4** (1933), 70–84 (German).

[202] M.J. McAsey and P.S. Muhly, *Representations of nonselfadjoint crossed products*, Proc. London Math. Soc. (3) **47** (1983), 128–144.

[203] M.J. McAsey, P.S. Muhly, and K.-S. Saito, *Nonselfadjoint crossed products (invariant subspaces and maximality)*, Trans. Amer. Math. Soc. **248** (1979), 381–409.

[204] M.J. McAsey, P.S. Muhly, and K.-S. Saito, *Nonselfadjoint crossed products. II*, J. Math. Soc. Japan **33** (1981), 485–495.

[205] S. McCullough, *Carathéodory interpolation kernels*, Integral Equations Operator Theory **15** (1992), 43–71.

[206] J. Milnor, *Analytic proofs of the "hairy ball theorem" and the Brouwer fixed-point theorem*, Amer. Math. Monthly **85** (1978), 521–524.

[207] G. Mokobodzki, *Balayage défini par un cône convexe de fonctions numériques sur un espace compact*, C. R. Acad. Sci. Paris **254** (1962), 803–805 (French).

[208] P.S. Muhly and K.-S. Saito, *Analytic crossed products and outer conjugacy classes of automorphisms of von Neumann algebras*, Math. Scand. **58** (1986), 55–68.

[209] V. Müller and F.-H. Vasilescu, *Standard models for some commuting multioperators*, Proc. Amer. Math. Soc. **117** (1993), 979–989.

[210] G.J. Murphy, *C^*-algebras and operator theory*, Academic Press, Inc., Boston, MA, 1990.

[211] F.J. Murray and J. von Neumann, *On rings of operators*, Ann. of Math. (2) **37** (1936), 116–229.

[212] F.J. Murray and J. von Neumann, *On rings of operators. II*, Trans. Amer. Math. Soc. **41** (1937), 208–248.

[213] F.J. Murray and J. von Neumann, *On rings of operators. IV*, Ann. of Math. (2) **44** (1943), 716–808.

[214] M. Nagumo, *Einige analytische Untersuchungen in linearen, metrischen Ringen*, Japanese J. Math. **13** (1936), 61–80 (German).

[215] I. Namioka, *Kakutani-type fixed point theorems: a survey*, J. Fixed Point Theory Appl. **9** (2011), 1–23.

[216] Z. Nehari, *On bounded bilinear forms*, Ann. of Math. (2) **65** (1957), 153–162.

[217] R. Nevanlinna, *Über beschränkte Funktionen, die in gegebenen Punkten vorgeschriebene Werte annehmen*, Ann. Acad. Sc. Fennicae (A) **13** (1919), 71p. (Finnish).

[218] A. Nica and R. Speicher, *Lectures on the combinatorics of free probability*, London Mathematical Society Lecture Note Series, vol. 335, Cambridge University Press, Cambridge, 2006.

[219] N.K. Nikolskii, *Treatise on the shift operator*, Grundlehren der mathematischen Wissenschaften [Fundamental Principles of Mathematical Sciences], vol. 273, Springer-Verlag, Berlin, 1986. Spectral function theory; With an appendix by S.V. Hruščev [S.V. Khrushchëv] and V.V. Peller; Translated from the Russian by Jaak Peetre.

[220] N.K. Nikolskii, *Operators, functions, and systems: an easy reading. Vol. 1*, Mathematical Surveys and Monographs, vol. 92, American Mathematical Society, Providence, RI, 2002. Hardy, Hankel, and Toeplitz; Translated from the French by Andreas Hartmann.

[221] J.L. Orr, *Triangular algebras and ideals of nest algebras*, Mem. Amer. Math. Soc. **117** (1995), no. 562.

[222] S. Parrott, *Unitary dilations for commuting contractions*, Pacific J. Math. **34** (1970), 481–490.

[223] S. Parrott, *On a quotient norm and the Sz.-Nagy - Foiaş lifting theorem*, J. Functional Analysis **30** (1978), 311–328.

[224] V.I. Paulsen, *Completely bounded maps on C^*-algebras and invariant operator ranges*, Proc. Amer. Math. Soc. **86** (1982), 91–96.

[225] V.I. Paulsen, *Every completely polynomially bounded operator is similar to a contraction*, J. Funct. Anal. **55** (1984), 1–17.

[226] V.I. Paulsen, *Completely bounded homomorphisms of operator algebras*, Proc. Amer. Math. Soc. **92** (1984), 225–228.

[227] V.I. Paulsen, *Completely bounded maps and operator algebras*, Cambridge Studies in Advanced Mathematics, vol. 78, Cambridge University Press, Cambridge, 2002.

[228] V.I. Paulsen, S.C. Power, and J.D. Ward, *Semi-discreteness and dilation theory for nest algebras*, J. Funct. Anal. **80** (1988), 76–87.

[229] V.I. Paulsen and M. Raghupathi, *An introduction to the theory of reproducing kernel Hilbert spaces*, Cambridge Studies in Advanced Mathematics, vol. 152, Cambridge University Press, Cambridge, 2016.

[230] G.K. Pedersen, *Measure theory for C^* algebras. III*, Math. Scand. **25** (1969), 71–93.

[231] G.K. Pedersen, *Applications of weak* semicontinuity in C^*-algebra theory*, Duke Math. J. **39** (1972), 431–450.

[232] G.K. Pedersen, *Analysis now*, Graduate Texts in Mathematics, vol. 118, Springer-Verlag, New York, 1989.

[233] G.K. Pedersen, *A commutator inequality*, Operator algebras, mathematical physics, and low-dimensional topology (Istanbul, 1991), Res. Notes Math., vol. 5, A K Peters, Wellesley, MA, 1993, pp. 233–235.

[234] G.K. Pedersen, *C^*-algebras and their automorphism groups*, Pure and Applied Mathematics (Amsterdam), Academic Press, London, 2018. Second edition; Edited and with a preface by S. Eilers and D. Olesen.

[235] J. Peters, *Semicrossed products of C^*-algebras*, J. Funct. Anal. **59** (1984), 498–534.

[236] R.R. Phelps, *Lectures on Choquet's theorem*, 2nd ed., Lecture Notes in Mathematics, vol. 1757, Springer-Verlag, Berlin, 2001.

[237] G. Pick, *Über die Beschränkungen analytischer Funktionen, welche durch vorgegebene Funktionswerte bewirkt werden*, Math. Ann. **77** (1915), 7–23 (German).

[238] G. Pisier, *A polynomially bounded operator on Hilbert space which is not similar to a contraction*, J. Amer. Math. Soc. **10** (1997), 351–369.

[239] G. Pisier, *Similarity problems and completely bounded maps*, Second expanded edition, Lecture Notes in Mathematics, vol. 1618, Springer-Verlag, Berlin, 2001. Includes the solution to "The Halmos problem".

[240] G. Pisier, *Introduction to operator space theory*, London Mathematical Society Lecture Note Series, vol. 294, Cambridge University Press, Cambridge, 2003.

[241] G. Popescu, *Isometric dilations for infinite sequences of noncommuting operators*, Trans. Amer. Math. Soc. **316** (1989), 523–536.

[242] G. Popescu, *Multi-analytic operators and some factorization theorems*, Indiana Univ. Math. J. **38** (1989), 693–710.

[243] G. Popescu, *von Neumann inequality for $(B(\mathcal{H})^n)_1$*, Math. Scand. **68** (1991), 292–304.

[244] G. Popescu, *On intertwining dilations for sequences of noncommuting operators*, J. Math. Anal. Appl. **167** (1992), 382–402.

[245] G. Popescu, *Functional calculus for noncommuting operators*, Michigan Math. J. **42** (1995), 345–356.

[246] S.C. Power, *The distance to upper triangular operators*, Math. Proc. Cambridge Philos. Soc. **88** (1980), 327–329.

[247] S.C. Power, *Classification of analytic crossed product algebras*, Bull. London Math. Soc. **24** (1992), 368–372.

[248] R.T. Powers, *Simplicity of the C^*-algebra associated with the free group on two generators*, Duke Math. J. **42** (1975), 151–156.

[249] P. Quiggin, *For which reproducing kernel Hilbert spaces is Pick's theorem true?*, Integral Equations Operator Theory **16** (1993), 244–266.

[250] H. Radjavi and P. Rosenthal, *Invariant subspaces*, 2nd ed., Dover Publications, Inc., Mineola, NY, 2003.

[251] H. Radjavi and P. Rosenthal, *Simultaneous triangularization*, Universitext, Springer-Verlag, New York, 2000.

[252] C.J. Read, *A solution to the invariant subspace problem*, Bull. London Math. Soc. **16** (1984), 337–401.

[253] C.J. Read, *A solution to the invariant subspace problem on the space l_1*, Bull. London Math. Soc. **17** (1985), 305–317.

[254] C.J. Read, *Quasinilpotent operators and the invariant subspace problem*, J. London Math. Soc. (2) **56** (1997), 595–606.

[255] C.J. Read, *A large weak operator closure for the algebra generated by two isometries*, J. Operator Theory **54** (2005), 305–316.

[256] M.I. Reed and B. Simon, *Methods of modern mathematical physics. I. Functional analysis*, Academic Press, New York-London, 1972.

[257] C.E. Rickart, *The uniqueness of norm problem in Banach algebras*, Ann. of Math. (2) **51** (1950), 615–628.

[258] F. Riesz, *Über die Randwerte einer analytischen Funktion*, Math. Z. **18** (1923), 87–95 (German).

[259] F. Riesz and M. Riesz, *Über die Randwerte einer analytischen Funktion*, Quatrième congrès des math. scand. 1916 (1916), 27–44 (German).

[260] J.R. Ringrose, *Super-diagonal forms for compact linear operators*, Proc. London Math. Soc. (3) **12** (1962), 367–384.

[261] J.R. Ringrose, *On some algebras of operators*, Proc. London Math. Soc. (3) **15** (1965), 61–83.

[262] J.R. Ringrose, *On some algebras of operators. II*, Proc. London Math. Soc. (3) **16** (1966), 385–402.

[263] M. Rørdam, F. Larsen, and N. Laustsen, *An introduction to K-theory for C^*-algebras*, London Mathematical Society Student Texts, vol. 49, Cambridge University Press, Cambridge, 2000.

[264] H. Rosenthal, *On the Choquet representation theorem*, Functional analysis (Austin, TX, 1986), Lecture Notes in Math., vol. 1332, Springer, Berlin, 1988, pp. 1–32.

[265] Z.-J. Ruan, *Subspaces of C^*-algebras*, J. Funct. Anal. **76** (1988), 217–230.

[266] W. Rudin, *Boundary values of continuous analytic functions*, Proc. Amer. Math. Soc. **7** (1956), 808–811.

[267] W. Rudin, *The closed ideals in an algebra of analytic functions*, Canadian J. Math. **9** (1957), 426–434.

[268] W. Rudin, *Fourier analysis on groups*, Wiley Classics Library, John Wiley & Sons, Inc., New York, 1990. Reprint of the 1962 original; A Wiley-Interscience Publication.

[269] W. Rudin, *Functional analysis*, 2nd ed., International Series in Pure and Applied Mathematics, McGraw-Hill, Inc., New York, 1991.

[270] V. Runde, *Lectures on amenability*, Lecture Notes in Mathematics, vol. 1774, Springer-Verlag, Berlin, 2002.

[271] C. Ryll-Nardzewski, *On fixed points of semigroups of endomorphisms of linear spaces*, Proc. Fifth Berkeley Sympos. Math. Statist. and Probability (Berkeley, Calif., 1965/66), Univ. California Press, Berkeley, Calif., 1967, pp. pp 55–61.

[272] K. Saitô and J.D.M. Wright, *Monotone complete C^*-algebras and generic dynamics*, Springer Monographs in Mathematics, Springer, London, 2015.

[273] S. Sakai, *A characterization of W^*-algebras*, Pacific J. Math. **6** (1956), 763–773.

[274] S. Sakai, *C^*-algebras and W^*-algebras*, Classics in Mathematics, Springer-Verlag, Berlin, 1998. Reprint of the 1971 edition.

[275] D. Sarason, *Invariant subspaces and unstarred operator algebras*, Pacific J. Math. **17** (1966), 511–517.

[276] D. Sarason, *Generalized interpolation in H^∞*, Trans. Amer. Math. Soc. **127** (1967), 179–203.

[277] J.J. Schäffer, *On unitary dilations of contractions*, Proc. Amer. Math. Soc. **6** (1955), 322.

[278] I. J. Schark, *Maximal ideals in an algebra of bounded analytic functions*, J. Math. Mech. **10** (1961), 735–746. "I. J. Schark" is a pseudonym for: Irving Kaplansky, John Wermer, Shizuo Kakutani, R. Creighton Buck, Halsey Royden, Andrew Gleason, Richard Arens and Kenneth Hoffman.

[279] J. Schauder, *Über lineare, vollstetige Funktionaloperationen*, Studia Math. **2** (1930), 183–196 (German).

[280] M. Schreiber, *A functional calculus for general operators in Hilbert space*, Trans. Amer. Math. Soc. **87** (1958), 108–118.

[281] L. Schwartz, *Théorie des distributions*, Publications de l'Institut de Mathématique de l'Université de Strasbourg, vol. IX-X, Hermann, Paris, 1966 (French). Nouvelle édition, entiérement corrigée, refondue et augmentée.

[282] I.E. Segal, *The group algebra of a locally compact group*, Trans. Amer. Math. Soc. **61** (1947), 69–105.

[283] I.E. Segal, *Irreducible representations of operator algebras*, Bull. Amer. Math. Soc. **53** (1947), 73–88.

[284] O.M. Shalit, *Operator theory and function theory in Drury-Arveson space and its quotients*, Operator theory (D. Alpay, ed.), Springer, Basel, 2015, pp. 1125–1180.

[285] S. Sherman, *The second adjoint of a C^*-algebra*, Proc. Inter. Congress Math. **1** (1950), 470.

[286] B. Simon, *Convexity*, Cambridge Tracts in Mathematics, vol. 187, Cambridge University Press, Cambridge, 2011. An analytic viewpoint.

[287] V. Smirnov, *Sur les valeurs limites des fonctions, régulières à l'intérieur d'un cercle*, Journ. Soc. Phys.-Math. Leningrad **2** (1929), 22–37 (French).

[288] V. Šmulian, *Über lineare topologische Räume*, Rec. Math. [Mat. Sbornik] N.S. **7/49** (1940), 425–448 (German, with Russian summary).

[289] W.F. Stinespring, *Positive functions on C^*-algebras*, Proc. Amer. Math. Soc. **6** (1955), 211–216.

[290] M. Stone, *Linear transformations in Hilbert space, I, II*, Proc. Natl. Acad. Sci. USA **15** (1929), 198–200, 423–425.

[291] Ş. Strătilă and L. Zsidó, *Lectures on von Neumann algebras*, Editura Academiei, Bucharest; Abacus Press, Tunbridge Wells, 1979. Revision of the 1975 original; Translated from the Romanian by Silviu Teleman.

[292] V. Sunder, *An invitation to von Neumann algebras*, Universitext, Springer-Verlag, New York, 1987.

[293] B. Sz.-Nagy, *Sur les contractions de l'espace de Hilbert*, Acta Sci. Math. (Szeged) **15** (1953), 87–92 (French).

[294] B. Sz.-Nagy, *Completely continuous operators with uniformly bounded iterates*, Magyar Tud. Akad. Mat. Kutató Int. Közl. **4** (1959), 89–93 (English, with Russian and Hungarian summaries).

[295] B. Sz.-Nagy and C. Foiaş, *Sur les contractions de l'espace de Hilbert. VI. Calcul fonctionnel*, Acta Sci. Math. (Szeged) **23** (1962), 130–167 (French).

[296] B. Sz.-Nagy and C. Foiaş, *Sur les contractions de l'espace de Hilbert. VII. Triangulations canoniques. Fonction minimum*, Acta Sci. Math. (Szeged) **25** (1964), 12–37 (French).

[297] B. Sz.-Nagy and C. Foiaş, *Dilatation des commutants d'opérateurs*, C. R. Acad. Sci. Paris Sér. A-B **266** (1968), A493–A495 (French).

[298] B. Sz.-Nagy, C. Foias, H. Bercovici, and L. Kérchy, *Harmonic analysis of operators on Hilbert space*, Revised and enlarged edition, Universitext, Springer, New York, 2010.

[299] Z. Takeda, *Conjugate spaces of operator algebras*, Proc. Japan Acad. **30** (1954), 90–95.

[300] M. Takesaki, *Theory of operator algebras. I*, Encyclopaedia of Mathematical Sciences, vol. 124, Springer-Verlag, Berlin, 2002. Reprint of the first (1979) edition; Operator Algebras and Non-commutative Geometry, 5.

[301] M. Takesaki, *A duality in the representation theory of C^*-algebras*, Ann. of Math. (2) **85** (1967), 370–382.

[302] A.E. Taylor, *The resolvent of a closed transformation*, Bull. Amer. Math. Soc. **44** (1938), 70–74.

[303] J.L. Taylor, *The analytic-functional calculus for several commuting operators*, Acta Math. **125** (1970), 1–38.

[304] H. Tietze, *Über Funktionen, die auf einer abgeschlossenen Menge stetig sind*, J. Reine Angew. Math. **145** (1915), 9–14 (German).

[305] J. Tomiyama, *On the projection of norm one in W^*-algebras*, Proc. Japan Acad. **33** (1957), 608–612.

[306] T. Turumaru, *Crossed product of operator algebra*, Tohoku Math. J. (2) **10** (1958), 355–365.

[307] A. Tychonoff, *Über die topologische Erweiterung von Räumen*, Math. Ann. **102** (1930), 544–561 (German).

[308] A. Tychonoff, *Ein Fixpunktsatz*, Math. Ann. **111** (1935), 767–776 (German).
[309] P. Urysohn, *Über die Mächtigkeit der zusammenhängenden Mengen*, Math. Ann. **94** (1925), 262–295 (German).
[310] N.Th. Varopoulos, *On an inequality of von Neumann and an application of the metric theory of tensor products to operators theory*, J. Functional Analysis **16** (1974), 83–100.
[311] D. Voiculescu, *A non-commutative Weyl-von Neumann theorem*, Rev. Roumaine Math. Pures Appl. **21** (1976), 97–113.
[312] J. von Neumann, *Zur allgemeinen Theorie des Masses*, Fund. Ann. **13** (1929), 73–116 (German).
[313] J. von Neumann, *Allgemeine Eigenwerttheorie Hermitescher Funktionaloperatoren*, Math. Ann. **102** (1929), 49–131 (German).
[314] J. von Neumann, *Zur Algebra der Funktionaloperationen und Theorie der normalen Operatoren*, Math. Ann. **102** (1930), 370–427 (German).
[315] J. von Neumann, *On complete topological spaces*, Trans. Amer. Math. Soc. **37** (1935), 1–20.
[316] J. von Neumann, *On a certain topology for rings of operators*, Ann. of Math. (2) **37** (1936), 111–115.
[317] J. von Neumann, *On rings of operators. III*, Ann. of Math. (2) **41** (1940), 94–161.
[318] J. von Neumann, *Eine Spektraltheorie für allgemeine Operatoren eines unitären Raumes*, Math. Nachr. **4** (1951), 258–281 (German).
[319] C. Webster and S. Winkler, *The Krein-Milman theorem in operator convexity*, Trans. Amer. Math. Soc. **351** (1999), 307–322.
[320] A. Weil, *L'intégration dans les groupes topologiques et ses applications*, Vol. No. 869, Hermann & Cie, Paris, 1940 (French). [This book has been republished by the author at Princeton, N. J., 1941.]
[321] J. Wermer, *On invariant subspaces of normal operators*, Proc. Amer. Math. Soc. **3** (1952), 270–277.
[322] J. Wermer, *On algebras of continuous functions*, Proc. Amer. Math. Soc. **4** (1953), 866–869.
[323] R. Whitley, *The Kreĭn-Šmulian theorem*, Proc. Amer. Math. Soc. **97** (1986), 376–377.
[324] H. Widom, *Inversion of Toeplitz matrices. II*, Illinois J. Math. **4** (1960), 88–99.
[325] S. Willard, *General topology*, Dover Publications, Inc., Mineola, NY, 2004. Reprint of the 1970 original [Addison-Wesley, Reading, MA; MR0264581].
[326] D.P. Williams, *Crossed products of C^*-algebras*, Mathematical Surveys and Monographs, vol. 134, American Mathematical Society, Providence, RI, 2007.
[327] S. Winkler, *The non-commutative Legendre-Fenchel transform*, Math. Scand. **85** (1999), 30–48.
[328] A. Wintner, *Zur Theorie der beschränkten Bilinearformen*, Math. Zeit. **30** (1929), 228–282 (German).
[329] G. Wittstock, *Ein operatorwertiger Hahn-Banach Satz*, J. Functional Analysis **40** (1981), 127–150 (German, with English summary).
[330] G. Wittstock, *On matrix order and convexity*, Functional analysis: surveys and recent results, III (Paderborn, 1983), North-Holland Math. Stud., vol. 90, North-Holland, Amsterdam, 1984, pp. 175–188.
[331] H. Wold, *A study in the analysis of stationary time series*, Almqvist & Wiksell, Stockholm, 1954. 2d ed.; With an appendix by Peter Whittle.
[332] T.H. Wolff, *Counterexamples to two variants of the Helson-Szegő theorem*, J. Anal. Math. **88** (2002), 41–62. Dedicated to the memory of Tom Wolff.
[333] F.J. Yeadon, *A new proof of the existence of a trace in a finite von Neumann algebra*, Bull. Amer. Math. Soc. **77** (1971), 257–260.
[334] G. Zeller-Meier, *Produits croisés d'une C^*-algèbre par un groupe d'automorphismes*, J. Math. Pures Appl. (9) **47** (1968), 101–239 (French).

INDEX OF NAMES

K. R. Davidson, *Functional Analysis and Operator Algebras*, CMS/CAIMS Books in Mathematics 13,
https://doi.org/10.1007/978-3-031-63665-3

INDEX

K. R. Davidson, *Functional Analysis and Operator Algebras*, CMS/CAIMS Books in Mathematics 13,
https://doi.org/10.1007/978-3-031-63665-3

T